# THE
# PLACE-NAMES
# OF
# STAFFORDSHIRE

David Horovitz

Published
by
David Horovitz
Brewood
2005

.

ISBN 0-9550309-0-0

Typeset and published by
David Horovitz
Kiddemore Cottage
Brewood, Stafford
ST19 9BH

Printed on acid-free paper
by
Woolnough Bookbinding Limited
Irthlingborough

# Preface

A little over one hundred years ago the Oxford University Press published a slim volume by W. H. Duignan, a Walsall solicitor and antiquary, entitled 'Notes on Staffordshire Place-Names'.[1] This was the first – and to date only – scholarly survey of the more important place-names of the whole of the county. Duignan's contribution to place-name research was to be recognised by no less an authority than Eilert Ekwall, arguably the foremost scholar of English place-names at the time, who credited Duignan and the great Victorian philologist W. W. Skeat with founding English place-name study on modern lines.[2] Although Duignan's book was undoubtedly a pioneering work for its date,[3] and is still of value today, advances in place-name studies and the availability of additional early spellings have thrown new light on many derivations and led to re-examination of Duignan's conclusions.

Prompted by the plaintive observation that 'Staffordshire historians could be described as topographically disadvantaged: there is no one convenient source of information about the place-names in the county...',[4] the present work is intended to fill that lacuna and provide a concise and accessible single-volume synthesis of early place-name spellings and derivations – or observations on possible derivations – based on modern research, together where appropriate with a note of earlier derivations, to serve as an interim study pending publication of further definitive volumes of the Staffordshire survey by The English Place-Name Society, which issued a single volume (on Cuttlestone Hundred) in 1984. Many name-forms from that volume, the fruit of many years of dedicated research by the late J. P. Oakden, have been incorporated into this work, though a number of Dr Oakden's derivations have been revised.[5] It is unlikely that a complete Staffordshire survey by the Society, to include field-names, will be available in the foreseeable future.

For historical records generally, on which place-name research is entirely reliant, Staffordshire is fortunate to possess in particular several dozen volumes relating to all aspects of the history of the county published by the Staffordshire Record Society as the Staffordshire Historical Collections, formerly the Transactions of the William Salt Archaeological Society. Only those who have made extensive use of this incomparable material are likely to recognise the prodigious labours of those who, particularly in the early years – publication of annual volumes began in 1880 – have provided such a treasury of raw material for later historians. Nor should the labours of such earlier

---

[1] Incorporating valuable observations added *post scriptum* by W. H. Stevenson, the historian and palæographer.
[2] Ekwall 1936: 12. In addition to various papers on philological matters relating to the early history of South Staffordshire and surrounding areas, Duignan also published a volume on Worcestershire place-names in 1905, and on Warwickshire place-names in 1912. A bibliography of his works appears in N & Q 11th Series XI 1915 373, 461; XII 1915 6, 39.
[3] Emphasised by, for example, a glance at some of the early twentieth-century volumes of the Transactions of the North Staffordshire Field Club, in particular that for 1908, where essays on Staffordshire place-names incorporate gloriously imaginative derivations from a startling array of languages, a particularly late manifestation of a theory developed by the antiquary Aylett Sammes in *Britannia Antiqua Illustrata* (1676) that the ancient Britons were descended from the Phoenicians, and that there were similarities between the Phoenician, Hebrew and Welsh languages: see Sweet 2004: 127. The theory was derided by Sammes' contemporaries.
[4] Paffard 1996: 1.
[5] Extracts from PN Wo and PN Wa relating to places formerly in Staffordshire have also been included.

Staffordshire researchers as Walter Chetwynd, Simon Degge, Sampson Erdeswick, John Huntbach, Thomas Loxdale, Robert Plot, Samuel Pipe Wolferstan, Stebbing Shaw and Richard Wilkes,[6] or in more recent times the great Shropshire medievalist the Reverend R. W. Eyton, be underestimated. Their remarkable achievements, in many cases from a period when communications and means of publication were still relatively primitive, with records and archives invariably scattered, uncatalogued and in many, if not most, cases untranscribed, form the bedrock of Staffordshire history, and our debt to them is incalculable. In this iconoclastic age it must not be forgotten that the range of learning of some of these pioneers was quite extraordinary, and we can but marvel how they amassed their immense erudition in the midst of careers devoted to practical affairs. Similarly, the zeal of those who ensured the preservation of Staffordshire's documentary history, the foremost of whom was the Staffordian bibliophile William Salt – whose surname derives, appropriately, from a Staffordshire village-name of Anglo-Saxon origin – places all Staffordshire researchers in their debt.

More recently, the millennium publication of Tim Cockin's formidably comprehensive 'Staffordshire Encyclopaedia', which marshalls and collates an extraordinary range of Staffordshire material, serves as a guide to many otherwise obscure sources (and Staffordshire places), and will prove an unrivalled sourcebook for historians. Furthermore, by recording early – and indeed more recent – thoughts on Staffordshire place-name etymology, the Encyclopaedia serves as an excellent summary of earlier thinking on the county's place-names, conveniently assembled under the relevant place entry, which has further emphasised the desirability for a new study on the place-names of the county.

It will be appreciated that a work of this size covering the whole of Staffordshire cannot hope to cover every name found on a modern large-scale map, and the list of entries is far from exhaustive. The basis for selecting names has been dictated mainly by the availability of reliable early spellings (and I must here express my thanks to Dr Nigel Tringham for making available unpublished spellings gathered during research for the various volumes of the Victoria History of the County of Stafford, and to the late Victor Watts, former Honorary Director of The English Place-Name Survey, for allowing access to the typescript draft of Dr Oakden's unpublished work on Totmonslow Hundred, from which valuable early spellings have been extracted),[7] and by an attempt to include the principal place-names, hill-names and river-names of Staffordshire. All 334 or so Staffordshire place-names in Domesday Book (the number is difficult to quantify for various reasons)[8] are treated, as well as most of those on Saxton's map of Staffordshire of 1577, Smith's manuscript Staffordshire map of 1599, and Browne's map of 1682 published in Plot's 'The Natural History of Staffordshire' of 1686.[9] An arbitrary selection,

---

[6] See especially SHC 4th Series 11 1982.
[7] Early documentary evidence for many names in the North Staffordshire moorlands is often scanty, illustrated by the mid-19th century Tithe Awards which affected only 12 of the 31 parishes in North Totmonslow: Elliott 1984: 95.
[8] Darby 1977: 336 gives 342 DB settlements. The first (1954) edition of Darby & Terrett 1971 mistakenly treats *Valencia* as the name of a lost Domesday vill in Staffordshire (p.168). In fact the word means 'value' or 'asset': VCH IV 43 fn. 47.
[9] Kip's map of Staffordshire 1607-10 and Speed's map of Staffordshire 1610, which contain virtually identical place-name forms, incorporate some extraordinarily corrupt spellings, evidently the result of misreading handwritten forms, making likely identification of some places possible only by their position on the map, for example (from Speed): *Langrose* (Rangemore), *Catnell* (Tatenhill), *Chomley wood* (Callingwood), *Hasker* (Haselour).

# The Place-Names of Staffordshire

To both Sylvias

not intended in any way to be comprehensive, of lesser, obsolete, or unlocated names (some of which, it must be emphasized, may relate to places outside Staffordshire), or those otherwise of particular interest, have also been included, in the hope that it may be of value to historians or researchers.

For the purposes of this work, the county of Stafford has been taken to include all areas which are or have at any time been within Staffordshire, but a handful of names of places just outside the county boundary have been included where of particular interest. The work is based primarily on spellings taken from unverified secondary sources, and should be treated accordingly. Likewise, the identification of a spelling with a particular place may not always be beyond doubt.

A number of place-name spellings are taken from Duignan's works, where sources are rarely cited, but those spellings that have been traced and checked for accuracy confirm that he was a meticulous transcriber, and his spellings extracted from lost charters and other sources now untraceable are likely to be reliable. The great majority of forms have been extracted from the Staffordshire Historical Collections, which seem generally to be accurate, but occasionally the authors, not being place-name scholars, have modernised spellings to help the reader, and there are deficiencies in the indexing of some volumes. Although Duignan's research (based to a very large extent, it must be noted, on the early volumes of the Salt Transactions) tended to concentrate on the southern part of the county, with which he was clearly more familiar, that is balanced in this work by the unpublished spellings made available from more recent Victoria County History ('VCH') working papers, which relate in the main to the northern part, and from Oakden's unpublished papers on the north-east of the county. It should be added that the policy at the VCH is to modernise the spelling of common place-name elements when found in a particular place-name,[10] and forms taken from VCH volumes cited in this work should be read in the light of that practice.

At a late stage in the preparation of this work the enormous range of catalogue entries on the Public Records Office *A2A* ('Access to Archives') and *Procat* websites became accessible, but those catalogues sometimes conceal pitfalls created quite innocently by archivists using modernised names to catalogue deeds, with those spellings becoming associated with early dates listed in an entry. The catalogues, daily increasing, are an extraordinary tool for researchers, and will become even more indispensible as the databases expand, but were never intended in themselves to support the work of place-name scholars, and pose obvious hazards when used for that purpose. Any references to Record Office archives in this work should be read in the light of that important caveat.

After a long gestation, 'The Cambridge Dictionary of English Place-Names', which is likely to become the standard general reference work for many years to come,[11] and which incorporates material from an early draft of this study, appeared in print shortly before this book went to press, and the opportunity has been taken to incorporate some suggested derivations from that volume.

A work of this kind would be impossible without assistance from others, and in addition to my obvious indebtedness to earlier researchers of English place-names in general and Staffordshire place-names in particular, I have accumulated numerous obligations to

---

[10] Personal communication 5 July 2001 from Dr Nigel Tringham, county editor of VCH.

[11] Despite a number of unfortunate shortcomings: see JEPNS 27 2004 133-42.

many other specialists in various disciplines. Special thanks for help during the prolonged research for this study are due to Paul Bibire, Jean Birrell, Dr Faith Cleverdon, the Reverend Michael Fisher, Dr Carole Hough, Peter Kitson, John Levitt, Dr Oliver Padel, the late Michael Paffard, Steve Potter, Dr Tania Styles, David Swinscoe, Edgar Tooth, and Chris Welch. The vigorous criticism and learned observations of Professor Richard Coates, Dr David Parsons, and Dr Paul Cullen have saved me from many errors and led to a number of place-name derivations that would not otherwise have occurred to me. I am also grateful to Dr Philip Morgan for making available to me his study of the Croxden Chronicle prior to publication, to Bob Meeson for providing details of his research on lost Roman roads in Staffordshire, to Andrew Kirkham for the care he has taken in the production of the maps in this volume, and to Sue and Matthew Martin for unravelling the complexities of electronic typesetting. But my particular debt is to Dr Margaret Gelling, not only for her inspirational writings – in all senses of that expression – over many years, which led indirectly to the compilation of this study, but for so generously sparing time from her own important researches to offer constructive counsel during various stages of this work. It is important to emphasise that none of those mentioned above is in any way responsible for the way I have used the information and advice so freely made available to me.

It is to be hoped that many suggestions or possibilities put forward in this work for the derivation of place-names in Staffordshire will, sooner rather than later, come to be seen as equally quaint and fanciful as some of Duignan's conclusions a century ago.[12] That is as it should be. Few names are likely to be insoluble, and it can be predicted with confidence that future research, whether philological, topographical or archaeological, or indeed the unearthing of hitherto undiscovered early spellings, will provide the crucial key to the origin of many names which presently remain unexplained or uncertain. Those with more detailed local knowledge will doubtless identify within this work suggested derivations which are topographically inappropriate, and recognise local features from which more likely derivations can be adduced. My intention has been to provide a corpus of early spellings and to set down some thoughts on possible derivations and etymological difficulties, which other researchers can build upon or demolish as appropriate. Throughout, the admonition of the great lexicographer Eric Partridge that 'A work is not necessarily the more scholarly for being written in philological shorthand'[13] has been a guiding influence.

In a work of this nature opportunities for errors of every kind are legion, and for those that have escaped detection, and for all other deficiencies, the responsibility must be mine alone.


David Horovitz
Brewood
June 2005

---

[12] See for example Duignan's entries relating to Pillaton and Watling Street.
[13] 'Origins: A short Etymological Dictionary of Modern English', London: Routledge & Kegan Paul 1958 10.

# Contents

# Staffordshire – the geology and geography

The county of Stafford, ranking 18th in size against the other historical (pre-1974) counties, takes the shape of an irregular diamond, some 38 miles wide at its central area and some 62 miles long, bounded by (clockwise from the north) Derbyshire, Leicestershire, Warwickshire, Worcestershire, Shropshire and Cheshire. Apart from its northernmost part, all of Staffordshire is geologically included in the Carboniferous, the Permian and Triassic systems. Stafford itself occupies a position in the centre of the county, lying in the wide central Midland Gap, an agricultural belt of Keuper Marl, which is bisected by a fault in the Triassic Sandstone which lies to the north and south. The fault to the south runs through Cannock Chase, a high wooded heathland plateau to the east of the county. The central part of Staffordshire is fertile, especially around the richer alluvial floodplain at the junctions of the rivers Tame and Trent and their tributaries, with remnants of the extensive woodland of Needwood Forest on the east. The northernmost part of the county forms the southern end of the Peak District, and can be categorised as Pennine uplands, the land above Leek rising from 800' to 1500', with Oliver Hill, the highest and almost northernmost point in Staffordshire, close to the border with Derbyshire, reaching 1,684 feet. The country hereabouts is high bleak Millstone Grit moorland with exposed outcrops such as The Roaches and Hen Cloud. Nearby, Flash, at 1,518' above sea level, claims to be the highest village in England. Folding and erosion of the softer shales has caused a series of edges, such as Ipstones Edge. The area to the north-east is Carboniferous Limestone which forms a rugged and broken landscape cut by the rivers Manifold and Dove, the latter serving as the border with Derbyshire. Moving south the landscape becomes less rugged and the river Dove turns eastwards to meet the Trent to the north-east of Burton upon Trent, with deposits of gypsum and alabaster to the west of Tutbury. In the central part of the county the valleys of the river Trent and its tributaries form a lowland corridor from east to west. The south-east of the county lies on the coalfields plateau, the southernmost part being broken by Silurian and Igneous rock ridges in the Rowley Hills. To the west the plateau is lower and cut by the river Stour and its tributaries. In the south-west is a gently undulating lowland zone lying on sandstone and forming part of the Midland watershed, with watercourses flowing eventually to the Severn to the south-west or the Trent to the north-east. The ancient forests of Morfe, Kinver and Brewood, all lying on Red Sandstone, covered much of this quarter, joining with Cannock Forest on Cannock Chase. The watershed continues to mark the western boundary with Shropshire, with the north-west of the county, which takes in the Potteries coalfield, formed by the long ridge on which lies Mow Cop, dividing Staffordshire from the Cheshire Plain. Parts of this area drain into the river Weaver, and then into the Mersey. The only sizeable natural area of water is Aqualate Mere near Newport on the Shropshire border. Rudyard Lake and Blithfield Reservoir were both created as reservoirs, the former c.1793 and the latter in 1953. As Camden succinctly (and diplomatically) observed of the topography of Staffordshire in 1586:

'The north part is mountainous and unpleasant; the middle rendered more agreeable by the river Trent, covered with woods and diversified with corn-fields and meadows: as is the south, which has also pit-coal and iron-mines; whether to its advantage or disadvantage can best be determined by its inhabitants'.

1

Long known as a county of little wealth – at the time of the Domesday survey it was one of the poorest of those surveyed[14] – it was the minerals of Staffordshire, particularly coal, ironstone, limestone and clay, combined with a ready supply of fuel from the areas of woodland, that allowed it to play an important role in the industrial revolution from the eighteenth century onwards, with a particular reputation for pottery manufacture, metalworking and engineering.

## Staffordshire – its early history and historical boundaries

In the pre-Roman Iron-Age what is now Staffordshire is likely to have formed the eastern part of the territory of the people known as the Cornovii, the precise meaning of which is uncertain, but which may be associated in some way with British *corno* 'horn', though the significance of the term remains unexplained.[15] Likewise the boundaries of Cornovian territory (if indeed there were fixed or identifiable boundaries) are unknown, but their heartland seems to have been near The Wrekin, for the Romans created a Civitas or tribal capital for the Cornovii at Viroconium (Wroxeter). In the absence of clear evidence, historians generally accept that the Cornovii occupied Shropshire, with their territory extending for an uncertain distance into what became Cheshire,[16] Staffordshire, Herefordshire and Worcestershire.[17] Much if not all of Staffordshire is assumed to have been under Cornovian influence until the Roman occupation, though it is possible that other Iron-Age peoples whose names are now lost inhabited some parts of the county.[18] It has been argued that the territory of the Cornovii is to be identified as what later became Powys, but the evidence is far from conclusive.[19]

The sub-Roman history of Staffordshire is veiled in obscurity, but the area which was to become Staffordshire lay at the core of the great kingdom of Mercia, which rose to

---

[14] See e.g. Eyton 1881: 21-4, but note Dyer 2002: 1-2 for evidence that by the fourteenth century 'Staffordshire was not a backward county, but one with a greater capacity for enterprise and innovation than some other parts of the country'.

[15] Possibly with reference to the profile of The Wrekin, which resembles a pair of down-curving horns similar to those of early cattle (which would not however explain other tribes of the same name in Caithness and Cornwall, both on peninsulas which might be considered 'horns'), or perhaps because the tribe worshipped a horned deity of the Cernunnos (stag-god) type: see Rivet & Smith 1979: 325. The Cornovii who gave their name to Cornwall may have been migrants from the Midlands.

[16] Ptolemy attributed the Cornovii to Chester, which is a strong indication that the legionary fortress lay in their territory: Rivet & Smith 1979: 325.

[17] See generally Webster 1991.

[18] It is uncertain whether Letocetum (Wall) fell within the territory of the Cornovii or the adjoining Coritani to the east. The agriculturally poor soils of Cannock Chase and the south Derbyshire hills may have formed the borderland between the two peoples: see Todd 1991: 14-15. The existence of a Civitas Letocetensium, distinct from the Civitas Cornovii, and based at Letocetum in the later Roman period, has been postulated: see Wacher 1966: 108-10; also Rivet 1964: 150. Desborough 1991: 4 mentions 'the Coritani in the Trent valley, the Cornovii centred on Wroxeter and the Dobunnii in southern Staffordshire'.

[19] The arguments are summarised in Gelling 1992: 27. The Tribal Hidage, generally held to be an eleventh-century copy of a much earlier (perhaps eighth century) tax or tribute listing of peoples and hidages, names an administrative district *Westerne/Westerna*, said by Stenton 1971: 296 to be 'probably…in Cheshire and north Staffordshire', but perhaps another name for the *Magonsæte*, who occupied an area to the west of Worcester (Gelling 1977: 192), or between the *Wreocensætan* and the *Magonsæte* (Hooke 1983: 11), or the British west of Wroxeter, e.g. the kings of Anglesey or Gwynedd (Higham 1993: 70, 72).

supremacy under the powerful and long-lived rulers Ethelbald (716-757) and Offa (757-796).[20] For reasons which remain unclear, Tamworth was the principal royal and administrative centre of the Mercian kings from at least the seventh century until the mid ninth century, and nearby Lichfield was chosen as the site of the first Mercian cathedral.[21] The county of Stafford as an administrative unit is likely to have been created, together with its neighbouring counties, in the tenth or early eleventh century, carved out of the territory of the bishopric of Lichfield and Coventry which was formed in the last quarter of the seventh century.[22] The boundary between Pirehill Hundred and Cuttlestone Hundred[23] (which may well pre-date the formation of the shire, perhaps by a considerable period)[24] runs along the river Sow, and through what must have been the ford or causeway after which the county town may well have been named, and it has been suggested, on the basis of an analysis of monumental sculpture, that a line running roughly south-east from above Stafford to the north of Lichfield may mark the division between North Mercia and South Mercia.[25] Possibly the Hundred boundary follows in part that ancient boundary, if indeed it is right to think of fixed boundaries for such early territories.

None of the Mercian shires is known to have been called a shire (from Old English *scir* 'a piece cut or sheared-off') before 1000 AD,[26] and the generally held view has been that Staffordshire is unlikely to have been a shire before the building at Stafford of a *burh* or fortification by Æthelflæd, the Lady of the Mercians, in 913, when the Anglo-Saxon Chronicle recording the event would doubtless have mentioned the shire. However, some sort of burghal system has been identified in Mercia in the late eighth century, and it is possible that the West Midland shires originated as territories of major Mercian fortified settlements (perhaps including Stafford) created as early as the mid-eighth century, which were greatly enhanced in the tenth century or early in the eleventh when they were given additional non-military functions.[27] Æthelflæd's choice of Stafford for a *burh*, part of a strategy launched by her brother, Edward the Elder, to create a network of forts or fortified towns regularly spaced throughout the West Midlands,[28] might not be considered at first sight an obvious position for a military post, for the town lies on low ground surrounded (apart from a neck of land on the north) by watercourses or wet ground, which even today flood regularly, overlooked by a prominent isolated 476-foot knoll (a fluvial glacial spoil dump) a mile to the south-west of the town which gives

---

[20] A gloss to the opening entry in the Tribal Hidage refers to 'the first land' of the Mercians, implying the existence of a district name *Ærest Myrcnaland*, 'original Mercia', the core of which is perhaps to be identified with the later county of Stafford (see Gelling 1992: 79-80), though that would place Tamworth at the eastern edge, which seems improbable.

[21] Zaluckyj 2001: 217-22. It has been suggested that the site chosen for the cathedral had formerly been a British centre, perhaps acquired by King Wulfhere: Yorke 2001: 20, citing Bassett 2000.

[22] Finberg 1972: 225-7; Gelling 1992: 97-8; also TSAS 57 157-60.

[23] SHC 1916 163.

[24] John 1964: 154-6 and Stenton 1971: 128-30, but also Gelling 1992: 141-2 and Hart 1992: 287, with Hart postulating that the Hundreds were modelled on the wapentakes of the Danelaw.

[25] Hart 1977: 43-61; Sidebotham 1994; Zaluckyj 2001: 16. Cramp 1977: 192 places the sandstone cross-shaft in Wolverhampton churchyard in a southern English group of sculpture.

[26] Taylor 1957: 21, 24; Gelling 1992: 14.

[27] Basset 1996: 156, citing Haslam 1987: 76-93. It is worth noting that the bounds recorded in an Anglo-Saxon charter of Madeley, dating from 975 (S.801, see Hooke 1984: 106-8), include *witena leage* 'the wood or clearing of the witan (counsellors)', at a point where the boundaries of Cheshire, Shropshire and Staffordshire meet, suggesting that at that date some sort of territorial divisions existed at that point: see Pantos 2003: 45.

[28] See Brooks 2000a: 39-40, 45.

sweeping views and formed a natural site for the Norman castle after the destruction or deterioration of the earlier fortification within the town itself. However, the wet and marshy ground surrounding Stafford provided a naturally defended site[29] – indeed marshy ground will have offered far better defence than water – which formed the apex of a triangular arrangement with the border *burhs* at *Cwatbrycg*, probably at or near Quatford, and Tamworth (which lie some 50 miles apart), and clearly was intended to defend the Midland Gap, the route of the Upper Trent between the southern end of the Pennines and the plateau of Cannock Chase, but also the principal and strategically important east-west highway in the region, Roman Watling Street, which lies to the south of Stafford. *Cwatbrycg* (some 45 miles from Stafford) defended the important Severn crossing associated with pillaging Danish armies on at least two occasions,[30] and Tamworth (some 40 miles from Stafford) was a particularly important Mercian royal site on a broad spur at the junction of the rivers Anker and Tame, formerly half-circled by protective marshland, and which lay on the direct route from Leicester, where a Danish army was located. All three *burhs* were constructed at about the same time;[31] indeed the same engineers and sappers may have been responsible for overseeing the layout and construction of all three defences.

It is said that the boundaries of the West Midland shires tend to disregard the traditional territories of the peoples who collectively formed the composite kingdom of Mercia[32] – indeed the Staffordshire-Warwickshire boundary actually bisects Tamworth, the principal Mercian centre during the earlier Anglo-Saxon period[33] – and that Staffordshire may have come into being (with the other West Midland shires) perhaps during the later part of the ninth century or the early part of the tenth century, when Mercia was controlled by Ealdorman Æthelred and his wife Æthelflæd, or more likely at some date after 920, when Edward the Elder was able to exercise control over local factions and impose a version of the West-Saxon system on those parts of Mercia under his domination.[34] However, the

---

[29] Possibly chosen by Æthelflæd because a short barrier across the northern neck of land would create a strong defensible site: see Higham & Hill 2001: 151. Æthelflæd's fortification, which Hill 1984: 143 suggests was between 35 and 40 acres in area, may have been limited to a palisaded bank and ditch along that line, or may have reinforced an earlier Anglo-Saxon fortification.

[30] In 895-6, when the Danes wintered at *Cwatbrycg*, and in 910, before they were defeated by a combined force of West Saxons and Mercians in the Tettenhall/Wednesfield area: Earle & Plummer 1892-99: I 89, 96.

[31] The *burh* at *Cwatbrycg* was created in 912, those at Stafford and Tamworth (the latter, from archaeological evidence, superimposed on a pre-Viking rampart: see Gould 1967-8: 18-23) in 913. The Danes had wintered at a 'work' or fortification at *Cwatbrycg* (probably at or near Quatford: see Groom 1992: 19) in 895-6, and it is not impossible that Æthelflæd's *burh* was superimposed on the Danish fortification there, perhaps as part of a double-burh with a fortification on each bank. The *burh* at Hereford attributed to Æthelflæd was also built over earlier defences, perhaps created by Mercians in the face of Welsh attacks: Rahtz 1968: 242-6; Shoesmith 1982: 74-7.

[32] The western boundary of Staffordshire, if indeed they had a fixed territory, is likely to have been based on the far older boundary of the Anglo-Saxon people known as the *Wreocensætan*, who lived in the area near The Wrekin, from which they took their name,. A charter of 963 (12th century, S.723) grants lands at Church Aston, which is said to be 'in provincia Wrocensetna' (Gelling 1992: 83). Church Aston is one mile south-west of Newport, eight and a half miles from the Wrekin, and and two and a half miles from the Staffordshire-Shropshire border. Hart 1977: 54 suggests that the north-western boundary of South Mercia is now represented by the western border of Staffordshire.

[33] Keynes 2001: 59; Hill 2001: 151.

[34] Bassett 1996: 155. Stubbs 1880: I 129 held that the Midland shires were not created before the recapture of the Danelaw by Edward the Elder, and Hart 1992: 286 considers that the shiring and Hundreding occurred simultaneously during the reign of Edward the Elder.

layout of Staffordshire, with Worcestershire, Herefordshire and Gloucestershire, has been shown to conform well with that of the territories which predated them, based around the settlements of Stafford, Worcester, Hereford and Gloucester.[35] Hart's attempt to map the territorial boundaries of Mercia in the late eighth century[36] suggests that the northern, western and southern boundaries of the county of Stafford may have followed the western boundary of the territory of the North Mercians and the South Mercians, but any such boundary may well have been much older than the eighth century. There is other evidence that the Midland territories dependant on towns may have been re-organised as shires in or about 1007, possibly associated with a ship levy made in 1008.[37] The shiring of the Midland counties continues to engage academic debate, but it is probably safe to say that the Mercian shires had developed as an informal arrangement whereby districts were associated with *burhs* for military purposes perhaps as early as the late eighth century, and probably by the early tenth century, although not formalised into the shire system proper until the early eleventh century.[38]

Also uncertain is the period during which the Germanic system of Hundreds was introduced into England, and into Staffordshire. The Hundreds (five in Staffordshire) were administrative areas with fiscal, judicial and military functions which possibly pre-dated the shires in Mercia, as they certainly did in Wessex,[39] and perhaps existed long before, or were created as part of the response to, the Danish invasions of the ninth century.[40] Like other aspects of royal administration, the name 'hundred' may have been borrowed from Carolingian Francia, where the *centena* was a similar administrative area.[41] No example has been traced of the word 'hundred' (or any word, in English or

---

[35] But Bassett 1996: 155 has argued that only the former provincial boundaries in Shropshire and Warwickshire are 'comprehensively ignored' in the pre-Conquest layout.

[36] Hart 1977: 47-54.

[37] See SHC 1916 157; Taylor 1957: 17-51; Finberg 1972: 228-30; Swanton 1997: 138; Lapidge *et al* 1999: 76.

[38] Whybra 1990: 1-15; 106; SHC 1916 166; Stenton 1971: 337. Campbell 1995: 53 claims that 'In the mid-ninth century Wessex certainly and Mercia probably had an ealdorman for each shire', but Keynes 2001a: 323 concludes 'The significant question is whether the kingdom [of Mercia] was divided, in the ninth century, into territorial divisions approximating to shires, each placed by the king under an ealdorman, and whether there is any sign of formally constituted sub-divisions set up for administrative, judicial, financial, military and social purposes ... the likelihood is that the process of extending such arrangements from Wessex into Mercia was not begun until the late ninth or early tenth century.'

[39] See Whitelock 1955: 393-4; Jewell 1972: 47-51. Hart 1975: 16-17 considers that it was Edward the Elder who grouped the existing Hundreds of the West Midlands to form new administrative areas to be known later as shires or counties.

[40] Staffordshire was assessed at 513 hides and 13 acres (at 120 acres to the hide) in Domesday Book: VCH IV 2. There is some slight evidence that the Hundreding of Staffordshire may have occurred at some time before 994, the date of Wulfrun's gift of several estates to the clergy of Wolverhampton, for half the parish lies in Seisdon, half in Offlow. VCH IV 5 suggests that the hidation of South Staffordshire had been assigned probably by the middle of the tenth century, and certainly by the end. Stafford, unlike other shire towns in the region, is not the caput of any Hundred, which might suggest that the Hundreding pre-dates Æthelflæd's *burh* of 913. Thorn 1991: 23 fn3 observes that Staffordshire may have evolved from 'a poor and wasted rump of land left after the rest of Mercia, apart from the Danelaw proper, had been laid out into Shires in multiples of 600 hides', but suggests that another and more likely possibility is that 'the 2404 (probably a slip for 2,400) hides allocated to Warwick by the Burghal Hideage included those of Tamworth and Stafford, and perhaps even land that later formed the western third of Leicestershire; Staffordshire may thus have been a later creation out of Warwickshire'.

[41] Yorke1995: 125.

Latin, meaning a numerical hundred) being used in England before the tenth century to denote a unit of government or of jurisdiction.[42] The number and size of Hundreds varies greatly from shire to shire, but notionally (and sometimes actually, as in Staffordshire) comprises 100 hides, or sometimes 100 long hundreds of 120 hides.[43]

The entries in Domesday Book help to identify the boundary of the county of Stafford as it existed at the end of the Anglo-Saxon period, though not with absolute certainty, for the borders of south-west Staffordshire in particular remain the subject of continuing academic research and debate. Early boundaries tended to follow indisputable physical features such as rivers, watersheds, ridges and ancient roads. The boundary of the ancient county of Stafford utilised various features including (clockwise, from the north) the rivers Dove, Trent and Stour, and (to the south-east of Kinver) the line of a lost Roman road,[44] on the west (in various parts) the watershed dividing streams and rivers which flow ultimately to the east or the west,[45] and Back Brook, Lonco Brook, a short length of Roman road between Whitleyford Bridge and Ellerton,[46] Waggs Brook, Coal Brook, the river Tern (with a deviation, perhaps post-dating the original boundary, to encompass the curious beak-like peninsula to the west of the Tern containing Knighton, which would otherwise be an indisputable part of Shropshire), the high ground between Mow Cop and The Cloud, and the river Dane. Over time, some watercourses have changed their line, sometimes with the boundary adjusted to follow the new course, in other cases leaving the old boundary as a relic of former landscapes. Severed oxbows explain how, for example, odd islands of land technically forming part of Derbyshire lay at one time along the west and south side of the river Dove.[47] The boundary on the west of the county appears to run deliberately through or very close to permanent features such as watersheds[48] and prehistoric monuments,[49] perhaps to be seen as evidence that when the

---

[42] Campbell 1986: 161, citing F. Liebermann, *Die Gesetze der Angelsachsen*, ii, pt. ii (Halle, 1912), p.516. According to Stubbs 1880: I 110 the Hundred first appears in the laws of Edgar, king of Mercia from 957, king of the English 959-75.

[43] Alecto 1988: 156. Whitelock 1952: 138 observed that 'the division into hundreds has a recent and artificial look in the Midlands, for the areas there are often neatly assessed at just one hundred hides...'. If the Hundred predates the shire, it might be assumed that the shire would be created by bundling together a number of pre-existing Hundreds. The natural boundaries adopted by the shire and the relatively artificial boundaries of the Hundred for Staffordshire might tend to suggest that the Hundreds were created as sub-divisions of the shire at the time the shire was formed, or a little later, and a meeting-place in each Hundred, probably already long-established, adopted as the Hundred meeting-place. Mander & Tildesley 1960: 4 conclude that in 994 the Hundreding of Staffordshire had occurred but the shire had not been created.

[44] Running south-east from Greensforge: Margary number 192. It is evident that at the time the boundary was fixed, some trace of the road remained as a landscape feature.

[45] A survey of the likely boundary of the kingdom of the Hwicce suggests that it followed in part natural watersheds between major river systems, and may have pre-dated the Anglo-Saxon period: Ford 1979: 147-8.

[46] Margary number 19.

[47] As Gregory King observed in 1679: 'The Dove does not divide Staffsh. & Darbysh. in all places, but about Rocester and Rolston (juxta Egginton in com. Derb.) it is divided by other branches, possibly the old Course of the River, which query concerning, there having been Suits at Law about it': SHC 1919 274.

[48] See TSAS 4th Series VI 1916-7 123-6.

[49] For example The Devil's Ring and Finger, a neolithic galleried tomb to the west of Mucklestone, and The Bridestones, to the north of Biddulph Common, a prehistoric gallery grave, the latter lying just within Staffordshire.

county was shired, the western border was superimposed on an existing, perhaps already ancient, territorial division.

The Domesday shire was certainly considerably larger than the present county, extending westwards to the river Severn and beyond to incorporate a sizeable tract of land on the west bank centred on the river-crossing at Upper Arley, one of the areas to the west of the Severn that formed part of the early See of Lichfield and Coventry. The present Staffordshire-Worcestershire boundary which forms an acute angle to the south-east of Kinver appears to follow this same early boundary between the bishopric of Lichfield and Coventry and the bishopric of Worcester.[50] The Staffordshire holding on the west of the Severn opposite Arley can be taken as evidence of an early river crossing at that point. Landholdings on the far side of a river at the site of a ford, ferry or bridge are not unusual.[51] Doubtless the reasons were related to security and the avoidance of disputes about maintenance and tolls. Early mapmakers were evidently unclear as to the precise extent of Staffordshire west of the Severn opposite Upper Arley, which varies considerably in early (and, in some cases, more recent) maps.

One particular long-lived curiosity was the existence of detached islands of one county set in another. Broom and Clent formed detached islands of Staffordshire within Worcestershire, and were supposedly acquired by the sheriff of Staffordshire, and later mised or transferred into his own county for administrative convenience. Tardebigge was in Staffordshire from c.1100, but transferred to Warwickshire in 1266,[52] and to Worcestershire in 1844. It may be noted that both Clent and Tardebigge are grouped at the end of the King's Worcestershire holdings in Domesday Book, possibly because they once shared the same holder, and Clent indeed gave its name to a Worcestershire Hundred. Rowley Regis was at the time of Domesday ecclesiastically in Worcestershire, as was Dudley, and it seems unlikely that either was in the original county of Stafford.[53] Until 1966 Dudley and some of the adjoining area formed a detached island of Worcestershire in Staffordshire, and there may have been other similar cases: some parts of the north of the county some distance from the border may have been in Cheshire in the 12th and 13th century, for in 1185 property at Leek was listed with the Earl of Cheshire's farms, and other references a century or so later suggest an earlier association of Leek and adjoining areas with Cheshire.[54]

---

[50] Perhaps itself based on earlier boundaries between the territories of the Anglo-Saxon sub-kingdoms of the Magonsæte and Hwicce: see Zaluckyj 2001: 87.
[51] Examples are the area of Quatford (in 1086 part of Eardington) on the west side of the Severn centred on the old ford (and the bridge recorded in 1086: see Mason 1961-4: 43); the area to the north of Seighford which encloses the site of the old ford at Great Bridgeford; the deviation of the county boundary from the river Dove to take in a river crossing to the east of Crakemarsh Hall (O.S. 1836); the curious boundary deviation to the east of Rocester where the border follows a secondary stream at the river Dove crossing; the deviation which appears on the first edition 1" O.S. map to the east of the Dove at Church Mayfield which includes the river crossing; the county boundary to the north of Edingale and Clifton Campville which deviates from the river Mease to include river crossings; and the bulge in the boundary on the south side of Chebsey to take in the site of the former ford.
[52] See for example SHC 1944 88; Palliser 1976: 29.
[53] SHC 1919 168-9; also King 1996: 73-91; SHC 1916 166.
[54] VCH Ch II 178.

A large area bordering the Severn, which included Alveley, Kingsnordley, Claverley, Worfield, and (possibly) Quatt, Romsley, Rudge and Shipley,[55] together with Cheswardine and Chipnall on the north-west border of the county, formed part of Staffordshire, but became part of Shropshire in the twelfth century, while Tyrley (in Hales) was taken from Shropshire at the same period, probably between 1099 and 1135.[56] The adjustments were doubtless made because the lords of the various manors preferred not to hold estates divided by a county boundary, and because the river Tern was seen as a more permanent division than the previous boundary. At about the same date Edingale was 'mised' (transferred) into Staffordshire from Derbyshire.

The county boundary seems to have remained to a large extent unchanged from the end of the twelfth century until the first of the modern local government boundary revisions was completed in 1844. Detached parts of counties were taken by those surrounding them, with Clent and Broom incorporated into Worcestershire. Curiously, however, the parish of Dudley was left untouched, although Dudley Castle and surrounding land remained in Staffordshire. From 1891 some small areas of Harborne which had become suburbs of Birmingham were incorporated into Warwickshire, with part of Harborne absorbed into Birmingham, and in 1894-5 two western parishes were lost, Sheriffhales to Shropshire, and Upper Arley (including the portion on the west side of the river Severn) to Worcestershire.[57] Four districts were taken from Derbyshire: Croxall, Edingale, Stapenhill, and Winshill, the last two now part of Burton-upon-Trent, although Oakley, part of Croxall, was in Staffordshire by 1086. At the same time, the whole of Tamworth, divided between Staffordshire and Warwickshire by the eleventh century, became part of Staffordshire. A large area adjoining the river Severn, which formed the south-west boundary of the county, was transferred to Shropshire, including part of Bobbington, which has since been recovered. Handsworth with its township of Perry Barr became part of Birmingham in 1911, and a few hundred square yards were transferred from Barton

---

[55] The last four of these places are entered under Warwickshire in Domesday Book, and although Eyton 1881: 2-4 considered 'unquestionably' that they properly belonged to Staffordshire, and it was subsequently suggested that the area was 'mised' or transferred into Shropshire either because the Earl of Shropshire owned that quarter of the Hundred 1068-1103, or because after the Earl lost it the Crown retained it and adminstered it from Shrewsbury with the rest of the Earl's late estates, with the boundary redrawn under Henry I, 100-1135, and certainly before 1157 (SHC 1916 163), that view is not free from doubt: see TSAS LVII 1961-4 157-160; VCH IV 1; Croom 1989: 306-315. An unresolved question is whether the parish of Worfield, assessed at no less than 30 hides with a value of £18 in DB (VCH IV 48), extended to the river Severn to encompass what is now the Low Town area of Bridgnorth on the east bank of the Severn: VCH Sa I 286 assumed that the Staffordshire-Shropshire boundary ran along the Severn from Newton, three miles north-east of Bridgnorth, and that the latter place did not exist, but see TSAS LVII 1961-4 39; SHC 1916 191. It is possible that in the late Anglo-Saxon period the river Severn formed the south-western boundary of Staffordshire from Newton to south of Arley, with part of Quatford forming a small projection of Shropshire on the east side of the river, doubtless explained by the existence of an early ford or bridge here. Bridgnorth, like Tamworth, is certainly omitted from Domesday Book, perhaps because both places held the status of royal burhs, but Bridgnorth may not have existed as a settlement in 1086: see Croom 1989: 294-368; Croom 1992: 19-20.
[56] Almost certainly because Almington was held at Domesday under the Earl of Shrewsbury by William Pantulf, Baron of Wem, who also held Tyrley under the same Earl. Pantulf mised Tyrley into the Staffordshire Hundred of Pirehill so that both places lay in the same Hundredal jurisdiction: Eyton 1881: 7. See also SHC 1945-6 24-5.
[57] The latter in 1897 according to VCH I 46. See also VCH IV 61.

and Yoxall wards to Scropton in Derbyshire in the same year.[58] It has been calculated that all these changes involved a net loss of only 2400 acres. In 1934 a small part of Edingale which had been in Derbyshire was transferred to Staffordshire. Further minor boundary changes were made in 1935 and 1936; in 1965 an area to the south and east of Tamworth hitherto forming part of Warwickshire was transferred into Staffordshire, bringing the whole of Tamworth into the county for the first time; and the following year the long-standing anomaly of Dudley was dealt with by moving it into Staffordshire.

In 1974, in the greatest reorganisation since 1066, the whole area generally known as the Black Country (the County Boroughs of Walsall, West Bromwich, Wolverhampton and Dudley), became the new West Midlands Metropolitan County.

**The linguistic background**

Within the last three thousand years five distinct groups of immigrants or invaders, each speaking a different language, have settled in Staffordshire, and each, to a greater or lesser extent, has influenced the formation of place-names in Staffordshire.

It is not possible to discover what the inhabitants of the area that became known as Staffordshire spoke in the hundreds of thousands of years of the so-called Stone Age periods, or the Bronze Age, or the earlier part of the Iron Age, but odd tantalising glimpses of this remote past may be discernible in the Staffordshire river-name Sow, which is likely to be pre-Celtic and derived from an earlier Old European language of which only fragmentary evidence in the form of place-names now survives.

The Celts were an ancient warrior people who originated in central Europe in about 1200 BC and settled in other parts of central and western Europe, including almost every part of England and Wales, in pre-Roman times. They spoke a Celtic language, known to philologists as British, technically a Brittonic or P-Celtic language, the ancestor of modern Welsh, Cornish and Breton, but which may have enshrined elements of a still earlier language. The Roman conquest of Britain began in 43AD, and by the end of the first century Latin was the language of officialdom throughout the conquered areas, including Staffordshire, and remained so for the next 300 years, although the native Celtic population will have continued to speak British. When the Romans withdrew early in the fifth century, the inhabitants they left behind were still essentially Celtic, and spoke much the same language as their ancestors at the time of the Roman invasion. That language was used by the indigenous population of Staffordshire well into the Anglo-Saxon period, in some places perhaps until the seventh century or even later, but it is customary to call British after approximately the middle of the sixth century by the names of the languages into which it developed, in this case Primitive Welsh.

Following – or perhaps even before – the withdrawal of the Romans early in the fifth century opportunistic incursions into England were made by pagan Anglo-Saxon groups from Northern Europe. Those groups or tribes included the original Angles or English who gave us the Old English language, from which the vast majority of English place-names can be traced. The influx of settlers that followed the earliest incursions increased in later centuries as permanent footholds became established, and variously merged and subdivided until they dominated most of the island, except for Wales, Scotland and

---

[58] TNSFC XLVII 1912-3 49. Under Enclosure Awards made in 1811, Scropton in Derbyshire gained a few hundred acres in Barton and Yoxall Wards. See also TNSFC XLVIII 1913-4 5 fn.

Cornwall, which remained Celtic strongholds. The settlers and invaders moved inland from several coastal areas, but in the main there were two separate routes of penetration into central England during the sixth and the early part of the seventh century. It was the Angles who advanced westwards into the Midlands from the eastern counties, reaching Staffordshire probably via the Trent and its tributaries in the mid sixth century and becoming firmly established during the first half of the seventh century,[59] while the Saxon people known as the Hwicce moved northwards and established settlements along the Severn and Avon focused around Worcester, reaching up as far as Wyre Forest and the forests of Kinver and Morfe in north Worcestershire and south Staffordshire.[60] From about the close of the sixth century it is likely that areas with Celtic-speaking populations gradually reduced to linguistically isolated enclaves which in the West Midlands may have all but vanished by the eighth century. The reasons why and how a relatively small number of Anglo-Saxon immigrants managed to impose their language on a much larger native population (which the Romans demonstrably failed to do after more than three centuries of occupation) is still the subject of academic speculation, but the presence of Welsh-speaking people in Staffordshire is supported by a number of place-names incorporating Old English words for 'Welshman', and further evidenced by river-names and place-names incorporating Celtic elements.

The groups of Anglo-Saxons (more specifically West Angles) involved in the westwards expansion into what was to become Staffordshire seem to have included tribes called the Bilsaeten, Hwicce, Pencersæten, Rhydware, Tomsaeten and Mierce, whose names are likely to be associated with Bilston, Wychnor, the area around the river Penk near Penkridge, Ridware, Tamworth and Mercia. The Mercians were named from the Old English word *mearc* 'boundary', but the nature and location of that boundary remains unresolved, though most historians opt for a boundary between the English of the West Midlands and the native British to the west.

Wherever their boundary, the Mercians expanded from a people concentrated around the basin of the upper Trent and its tributaries and gradually conquered or absorbed most of their neighbouring tribes, and became 'not so much a state as a group of peoples held together by an illustrious dynasty'.[61] It was in the early seventh century under Penda (c.632-55), a Mercian leader of royal blood, that the power of Mercia greatly increased. Penda seems to have conquered or absorbed the Middle Angles of the east, the Hwicce of the south-west Midlands and the Celts of what are now Shropshire and Herefordshire, including the people known as the Wreocensætan,[62] who lived in the area near The Wrekin, and so brought into being the united kingdom of Mercia, extending from the Humber and The Wash to Chester and Hereford, with its chief town at Tamworth. Mercia continued as an independent political unit for over 200 years, reaching its zenith during the reign of Offa (756-796), the most powerful English king of his day. From about 650 to 870 the Mercians were by far the most dominant of the Anglo-Saxon kingdoms, with greater Mercia eventually extending from the Humber to the Thames, but with its centre

---

[59] Jackson 1953: 209; 222. Higham 1993: 70 and 72 'very tentatively' identifies the *Wreocensetun* with the Cornovii of Roman Britain and believes they were a British people based at sub-Roman Wroxeter with their territory incorporating Cheshire, much of Shropshire and Western Staffordshire.
[60] From their links with Worcestershire evidenced in Domesday Book it is probable that the royal manors of Kinver and Kingswinford were settled from the south: see VCH Wo I 287.
[61] Stenton 1971: 202.
[62] Jackson 1953: 210.

of power remaining at its historic heartland in Staffordshire:[63] its earliest and principal cathedral was at Lichfield,[64] and its foremost royal residence at Tamworth, with a royal monastery, probably established by the late seventh century, at Repton in Derbyshire, a mere three miles east of the Staffordshire border. In 850 England was divided into the four large and wealthy kingdoms of Mercia, Northumbria, Wessex and East Anglia. By the end of the century only Wessex and fragments of western Mercia and northern Northumbria had escaped Danish depredation and domination.

The Danes, popularly known as the Norsemen ('men from the north') or Vikings, and frequently termed simply 'the heathen' or 'the pagans' in the early Chronicles, began their raids into England at the end of the eighth century. In 865 a great army of Danes invaded England, and in 874 launched raids from a base established at Repton and overran Mercia with great destruction, effectively ending Mercian supremacy. Danish settlers forming or following the army of conquest moved into unoccupied land in the north and east of England, and wealthy and powerful Viking kingdoms were established in East Anglia and at York. The language of the Danes, now called Old Norse, was used to adopt or translate existing place-names or to create new ones, but their influence was strongest in the Danelaw,[65] a word first recorded in 1008,[66] for an area which covered the north and east of England and was the subject of a treaty with king Ceolwulf c.877. The boundary of the Danelaw at that date is unknown from contemporary evidence, but a treaty of c.886[67] between king Alfred and the Danish king Guthrum mentions boundaries, the last of which is Watling Street.[68] It was long assumed from this that Watling Street in its entirety formed the boundary, but it is now believed that the treaty gives no indication how far the boundary follows the road because that was of no concern to the West Saxons (represented by Alfred), but only to the Mercians.[69] The line between English and Danish Mercia is likely to have deviated northwards from Watling Street at Mancetter, ignoring the rivers Anker and Tame, which meet near Tamworth, and followed a line which eventually became the boundary separating Staffordshire, Cheshire and Lancashire from Derbyshire and Yorkshire.[70] That would have left Tamworth itself, the early

---

[63] It has long been suggested that an early Mercian royal centre may have existed at *Wulfcester*, generally associated with the earthworks known as Bury Bank near Stone. VCH XIV 5 suggests that such identification is based more on long tradition than historical or archaeological evidence, but see Studd 1993: 55.

[64] The first Mercian bishops, many of whom came from Northumbria, appear to have had no cathedral -- their diocese was very large, and they may have been peripatetic -- but when Chad was consecrated in 669 he built a small church at Lichfield, perhaps at Stowe Pool to the east of the present cathedral, and made this the centre of his see: see Gould 1993: 101-4.

[65] The Danelaw was not a uniform entity, but is now used as a convenient shorthand expression for the complex arrangement of disconnected regional communities sharing a common respect for Danish law and a common pre-Danish ancestry: see Lapidge *et al* 1999: 137.

[66] Davis 1991: 29.

[67] Guthrum's treaty is generally held to date from 886x890, but see Dumville 1992: 1-23; Davis 1991: 47-54.

[68] 'First concerning our boundaries: up the Thames, and then up the Lea, and along the Lea to its source, then in a straight line to Bedford, then up the Ouse to the Watling Street': Whitelock 1955: 380.

[69] Davis 1991: 47-54.

[70] Sawyer 1971: 151; Davies 1982: 803-10. It may be remarked that a group of place-names ending in *-by*, including Ashby, Bretby and Smisby, all indicative of Danish occupation and marking Derbyshire's most intensive Danish settlement, lie relatively close to, but on the east side of, the river Trent, which forms the boundary between Staffordshire and Derbyshire in this area: Wainwright 1975: 297.

Mercian centre, several miles inside Mercia, and the whole of Staffordshire outside the Danelaw.[71] Evidence of Danish influence in Staffordshire is sparse,[72] but the river Trent evidently acted as a highway for Scandinavian penetration. Excavations at Catholme,[73] close to Wychnor and Ryknild Street, suggests that a settlement there may have been destroyed in the Danish invasions of the 890s, pointing towards Scandinavian penetration along the Roman roads (including Ryknild Street and Watling Street) as well as major rivers.

The treaty of c.886, which incorporates evidence that Guthrum's English subjects were treated as the social equal of the Danes,[74] did not prevent constant raiding into the others' territory by both the English and the Danes, culminating in a great battle between a Danish army, retreating from west of the Severn laden with booty, and a combined force of Mercians and West Saxons, in the vicinity of Tettenhall or Wednesfield c.910 when the Danes were decisively defeated and several of their kings killed.[75] That re-established English control in the area, with the Danes concentrating their attention on the European mainland, and allowed Alfred's son-in-law Æthelred, Alfred's daughter (and Æthelred's wife) Æthelflæd, 'Lady of the Mercians', and Alfred's son Edward the Elder, to create or strengthen a series of fortified centres from which to defend the English heartland, not only against threats from the Danelaw, but also against Irish Vikings moving into territory around North Wales and The Wirral. Those bases were the crucial element in a strategy which eventually led to the recovery of the Danelaw.

To what extent the Danish campaigns affected Staffordshire is difficult to establish, particularly since Scandinavian place-names do not occur in every area of the Danelaw, but from their Repton base, chosen for their main winter fortresses in 873-4, the Danes ravaged Mercia, and are likely to have occupied Tamworth,[76] Burton,[77] probably Lichfield,[78] and possibly Hanbury,[79] and since several of the *burhs* established by Æthelflæd and Edward the Elder were on sites known to have been occupied by the Danes, the *burh* established at Stafford in 913 might be seen as evidence, albeit slight, that Stafford suffered Danish occupation, if not destruction. Certainly it would have been militarily prudent for the Danes to have had a presence in areas close to their Repton base, even if only to act as forward posts to give advance warning of any enemy activity, and the Danes may themselves have utilised or refurbished an existing fortification at

---

[71] Holman 2001. The statement in Greenslade & Stuart 1984: 26 that 'Central and Northern Staffordshire...fell under Danish control' is based on the assumption that Watling Street formed the boundary of the Danelaw in Staffordshire.
[72] Darby & Terret 1971: 178; also SHC 1916 150; VCH IV 3, 6. Overall, the evidence provided by Domesday, the fact that the geld assessments were based on the hide, and the absence of Scandinavian tenurial terminology in early Staffordshire records supports the view that Danish influence was insignificant.
[73] Faull 1984: 101-14; Losco-Bradley & Kinsley 2002.
[74] A Mercian annalist records that in 909 St Oswald's body was taken from Bardsey in Lincolnshire, far within Danish territory, and carried to Gloucester, implying a peaceful interlude.
[75] Earle & Plummer 1892-99: I 94; Whitelock 1955: 193.
[76] The place disappears from the records from 874 to 913.
[77] VCH III 199.
[78] Ibid. 7.
[79] By tradition St. Werburgh's remains had been translated from Hanbury to Chester following Danish occupation of Hanbury or Repton (VCH Ch I 252; 268), but the tradition has not been traced to an earlier date than Ranulph Higden's 'Polychronicon' completed in 1452, and Chester is said to have been deserted in 893 (VCH Ch I 249), although that statement may be an exaggeration. .

Stafford created as early as the eighth century,[80] although no evidence of any such fortification has yet been identified. The relatively modest number of names of Scandinavian origin or influence in the county supports the generally accepted view that the county lay outside the area of Danish influence.

The Norman conquest of 1066 is the most recent and best-recorded conquest of England. The Normans, so-named as descendants of Norsemen who had conquered part of France, had assimilated French culture and language which they introduced into England. French became the official language of the ruling classes for the next two centuries, with formal documents written in Latin or Norman French. Perhaps surprisingly, given the nature of the total military conquest, just as the British language survived centuries of Roman occupation, so vernacular Anglo-Saxon survived and evolved and eventually replaced Latin and Norman French for official use, though some post-Conquest records, such as Feet of Fines, Subsidy Rolls and Assize Rolls, are in Latin or 'dog-Latin', which required the scribes or clerks to Latinise all names inserted in them, and it will be noticed that many Staffordshire names extracted from such records are Latinised or show evidence of Latin inflections.

## Sound changes, dialect and place-names

Words and personal-names used in the formation of place-names will have been pronounced in the local dialect, and whilst officials might have adopted formalised spellings, the 'vernacular' spellings used locally will generally reflect those dialect differences.[81] Certainly as early as the Anglo-Saxon period there were, as today, clear dialectal differences in the language spoken (and written) in different parts of the country, and the Old English language of that period has been divided into four distinct dialects: Kentish, West Saxon, Mercian and Northumbrian, the last two sometimes together known as Anglian. Phonological spellings often require the key provided by a knowledge of Old (and Middle) English dialect before they can be explained, but any dialectal analysis is outside the scope of this study, though it may briefly be noted by way of example that in the Midlands, Old English *a* was rounded in Middle English to *o*, so that Old English *ac* becomes modern *oak*. Old English *east* is a common first element in place-names, but in the Midlands often becomes *Ast-*, and the word for 'spring' (and occasionally 'stream', and very rarely 'a well'), a very common element in place-names, was *wella* in Anglian and Kentish areas, but in Mercian territory the word is *wælle*, which became *walle* in Middle English, and is very difficult to distinguish from *wall*, though the context will often provide a guide: Coldwall is more likely to be 'cold spring' than 'cold wall'.

## An analysis of Staffordshire place-names

### Introduction

The area of what was to become Staffordshire lay in the heartland of what is known as 'original' Mercia, probably a grouping of early unrecorded peoples created primarily by military means and based on the royal and ecclesiastical foci of Tamworth and Lichfield

---

[80] Basset 1996: 156, citing Halam 1987: 76-93.
[81] For West Midland dialect and sound changes see generally Clark 1919; EPNE i xxiv-xxxvi, Kristensson 1987; Levitt 1968; Miller 1891-1900; Nicholls 1934; Northall 1894; Orton 1969; Wilson 1974. On Poole 1880, however, see Levitt 1987: 195-206.

respectively. From those beginnings developed a kingdom which became by far the most successful of the various Anglo-Saxon kingdoms. It is therefore especially surprising to find that there are virtually no extant Mercian written records or other evidence. That dearth is attributable to a number of reasons.

As Bede had no Mercian contributors or informants, his Ecclesiastical History of the English People concentrates on non-Mercian areas. Various fragmentary annals purporting to date from the sixth or seventh century identified amongst the compilations of later chroniclers appear to be East Anglian in origin.[82] Although Mercia, with Lichfield as its religious hub, was an important literary and cultural centre,[83] no pre-Conquest manuscript, teacher or scholar can be definitely linked to Lichfield. Even a 9th century homily on the life of St Chad in the Mercian dialect may have been produced elsewhere, and the illuminated 8th century gospel book known as St Chad's Gospels originated outside Mercia, and reached Lichfield only in the 10th century.[84] No early biographies exist of the great Mercian kings Æthelbald or Offa, nor evidence of their laws.[85] Of at least 159 Latin diplomas recorded from Mercian rulers, almost all relate to estates outside Mercia proper, and coins issued by Mercian kings were seemingly minted elsewhere. Virtually no Mercian buildings survive above ground,[86] and there are no charters dated before 940. Even the Tribal Hidage, an enigmatic early listing of peoples and districts, may not have been Mercian in origin, for the first hidage assessment is of Mercia itself, and it has been argued that if it is indeed a taxation list, it is improbable that a Mercian king would have imposed tributes on his own people.[87]

The fact that Mercia was by far the last of the major kingdoms to weld itself into a single unit of the so-called Anglo-Saxon Heptarchy, and that there was no single Midland dynasty, together with the depredations of the Danes towards the end of the 9th century, probably explains the absence of annals or chronicles for the Midlands. Even the Anglo-Saxon Chronicle, with its later emphasis on matters pertaining to the West Saxons and King Alfred in particular, found little to record about what was to become Staffordshire until later in the Anglo-Saxon period, and then only fleetingly.[88] Indeed, were it not for the so-called Mercian Register, a brief record of the military exploits of Æthelflæd, Lady of the Mercians, in the early tenth century, which was incorporated en bloc in one version of the Chronicle,[89] and Irish annals of varying reliability known as The Three

---

[82] Davies 1977: 17-29.

[83] Wainwright observed that '... Lichfield seems to have been a cultural centre the importance of which is not sufficiently recognised': Finberg 1975: 71. It must not be forgotten that of King Alfred's seven named literary advisers, four came from Mercia: Stenton 1971: 270-1.

[84] VCH XIV 6-7. The view that the Gospels could not have originated in Wales because at the period in question there were no scriptoria has been challenged by Professor Wendy Davies, on the basis that in the Dark Ages, Wales had 36 monasteries in the south-east alone, and the author of a ninth-century poem describes himself as the librarian of the fathers of Tenby: Byron Rogers, 'Gospel Proof', Daily Telegraph, 31 December 2000 9.

[85] Chaney 1970: 177-8.

[86] One notable exception is All Saints' church, Brixworth, Northampton, perhaps dating from c.680.

[87] See Bassett 1989: 159-60, 225-30, but also the counter argument in Featherstone 2001: 29. See also Dumville 1989: 129; Higham 1992: 1-15 .

[88] The earliest reference in the Anglo-Saxon Chronicle to any place within Staffordshire is Lichfield, where Ceolred, king of Mercia, is said to have been buried in 716, with subsequent references to Tettenhall and Wednesfield c.910 in connection with the defeat of the Danes by a joint army of Mercians and West Saxons, and to Stafford in 913 when it was fortified by Æthelflæd. Staffordshire is first recorded by name in an entry for 1016: Earle & Plummer 1892-99: I 147.

[89] The 'C' version: Earle & Plummer 1892-99: I 99.

Fragments,[90] we would know next to nothing of her life and impact on the area. The Register serves to highlight the fact that many critical incidents in Mercian history must have gone unrecorded, or that the documentary evidence has been lost. In terms of documentary evidence, as opposed to archaeological or linguistic studies, the early Anglo-Saxon period in Staffordshire is in many ways a true Dark Age.

Nevertheless, Staffordshire toponymists are not denied some valuable, albeit limited, pre-Conquest material, in the form of place-names recorded in a number of late Anglo-Saxon writs and wills, most notably the will, dating from 1004, of the Mercian magnate Wulfric Spot, and of the relatively modest number of boundary clauses[91] from Anglo-Saxon estate charters (many surviving only as later copies), which have been the subject of considerable research and analysis and which provide some of the earliest evidence for the pre-Conquest history of Staffordshire, particularly in the southern and eastern parts of the county.

As would be expected, Domesday Book provides incomparable information on Staffordshire place-names at the end of the Anglo-Saxon period, with the great majority of ancient Staffordshire names first recorded in the great register, but some of the names from the survey pose particular difficulties of both identification and etymology.

For the purposes of this analysis, particular emphasis is placed on the more important names, that is names which are recorded in Domesday Book, or which have given their names to ancient parishes, or which are well-recorded as settlements from an early date, although other, lesser, names, including field-names, are included where appropriate.

## Celtic influence

In Staffordshire there are many place-names with probable (or possible) British or Primitive Welsh elements, or which refer to British people, including Barr (Great and Perry), Brewood, Chatterley, Cheadle, *Comberbach, Cumberfield*, Comberford, Crakelow, Creighton, Eccleshall, Gnosall, Hints, Ingestre, Kiddemore Green, Kinver, Leamonsley, Lichfield, Lizard, Minnbank, Morfe, Penkhull, Penkridge, Penn, Ridware, Saverley Green, Seisdon, Talke, Trysull, Walford, Walsall and Walton (four places of that name, and another now in Worcestershire), and perhaps Ocker Hill, as well as a number of hill-names which may be of British origin, including Mow Cop, Bar Hill and Barr Beacon. Celtic river- or stream-names are common, and include Anker, Churnet, *Cocker* (two), Dane, Dork, Dove, Hamps, *Kyre*, Penk, Tame, Tern and Trent. Tamworth and Tamhorn, both of which take their names from the river Tame, may well have existed as settlements in Celtic-speaking times, since the use of a river-name was one of the most common ways of denoting settlements in the Romano-British period.[92] Special significance has been attached by researchers to isolated Celtic names in areas where pre-

---

[90] See VCH Ch I 249; Cavill *et al* 2000.

[91] 17 for Staffordshire, compared with 96 for Worcestershire and 58 for Gloucestershire: Hooke 1983 25. For Shropshire, as well as documents connected with Wenlock Priory, 3 pre-Conquest charters (all of the 10th century), two wills and a writ of Edward the Confessor survive: PN Sa I xiv. Hart 1975: 165 suggests that a much larger collection of Anglo-Saxon charters is likely to have been preserved at Burton Abbey up to and after the Conquest, since most of the estates listed in Domesday Book as being in the possession of Burton are unrepresented in surviving charters, which must be seen as 'chance survivals from a much more comprehensive collection, which would have been particularly rich for Staffordshire and Derbyshire ...'.

[92] Gelling 1988: 51.

English names are rare, especially where such names are attached to relatively insignificant streams in an area where otherwise only some of the major rivers keep their ancient names, and there may be a correlation with the Romano-British archaeology of such areas.[93] *Cocker*, *Cun*[94], *Cund*,[95] *Kyre*, *Lemon*, *Severn*, and perhaps *Dork*, may be lost names of comparatively minor streams – Henry Bradley emphasised that '[t]here is evidence enough to show that the ancient Britons were in the habit of giving individual names to quite insignificant streams'[96] – which may provide tentative evidence for late Roman and post-Roman activity: the two cases of *Cocker* may be associated with the numerous Roman sites around Greensforge and the Roman road between Rocester and Stoke on Trent,[97] *Kyre* with the various Roman sites (including a villa) at Pennocrucium, and *Severn* with a possible Roman road running from Blythe Bridge to earthworks, possibly Roman, at Hollywood, south-east of Stone.

It may be noted that three ancient royal Forests (a legal term introduced by the Normans for areas, not necessarily wooded, subject to Forest law and in which the king had exclusive rights to hunt) in the southern half of the county – Brewood, Kinver and Morfe [98] – all have names of probable British origin, as does Cannock Chase, another vast tract of high ancient woodland. As elsewhere, Celtic place-names identified in Staffordshire typically incorporate topographical elements, rather than words for a habitation or settlement.

During the latter part of the sixth century and the earlier part of the seventh century Staffordshire was in transition from a predominantly Welsh populated and speaking area to a predominantly English peopled and English speaking area, and it was not until about the end of the ninth century that Welsh speech is likely to have died out.[99] The place-names listed above show other signs of a Welsh/English transition, particularly the large proportion of hybrid names, where an English element qualifies or explains a Welsh element, such as Brewood, Chatterley, Cheadle, Eccleshall, Gnosall, Lichfield, and Penkhull. These can be seen as the adaption by the English of existing Welsh names: they are English names with a Welsh basis, from a time when the English were beginning to dominate a Welsh area but the Welsh language was still recognisable. The other manifestation of the Welsh/English social overlap can be detected in the names Comberford, Comberbach, Cumberfield (and possibly Cumberstone), and in Walton, Walford, and (possibly) Walsall, collectively categorised by John Dodgson as 'Indian Reservation' names.[100] Comberford, Comberbach and Cumberfield are 'the ford of the Cumbre', 'the stream valley of the Cumbre', and 'the *feld* or open land of the Cumbre', noteworthy for the use of the Old English term *Cumbre*, the anglicised and non-derogatory version of the Welsh national name, *Cymry*. Walton, Walford and Walsall are respectively 'the farmstead or village of the Walas', 'the ford of the Walas', and (possibly) 'the *halh* of Walh or the Walas', *Walas* being the Old English word for 'the foreigners; the inferior race; the serfs', which was a derogatory term applied to the Welsh. With the *Comber-* names the Welsh are referred to in polite terms, whereas the *Walas* names are

[93] Gelling 1979: 115.
[94] Consall.
[95] Probably found in the name Scounslow.
[96] Bradley 1928: 110.
[97] Cocking Farm lies 1½ miles from this road (Margary number 181).
[98] VCH IV 21.
[99] Gelling 1992: 70.
[100] PN Ch VII (ii) 297.

less respectful of Welsh sentiment. These names might signify places associated with, or possibly even reserved for, the remnants of a displaced Welsh population, coined at a time when a concentration of Welsh people was a noteworthy survival in Staffordshire, meriting a special name. The *Comber-* names may be taken as evidence for enclaves of Welsh near Cannock and Tamworth (and perhaps near Seighford and Swynnerton) speaking their own language, with a degree of respectful recognition from the dominant English population.[101] But the possibility that every one of these names marked a historical, rather than contemporary, association with elements of the native Welsh population during the earliest period of Anglo-Saxon colonisation cannot be ignored, and it may be dangerous to read too much into these names, either individually or collectively. Finally, it may be noted that no English personal names with a Welsh element added have been noted as place-names in Staffordshire.

In his survey of Cheshire place-names John Dodgson noted:

'From the date-schemes proposed by Kenneth Jackson in Language & History in Early Britain[102] it will be found that Cheadle and Chathull contain a Welsh element in a form which has not undergone an eighth century Welsh change, whereas Tame and Lyme show English m, from a Welsh sound which was proceeding from m to v over a long period, sixth to tenth centuries, being heard by the English as m down to the tenth century, and as v as early as the late sixth century, with an overlap and alternation of English m, v, and sound-substitution in the seventh century...These names could be late sixth century adoptions into English'.[103]

For Cheadle (Cheshire) and Chathull (Cheshire), the Staffordshire names Cheadle and Chatcull, which contain correspondingly identical Welsh first elements, can be substituted to produce the same conclusions.

The name Hints is usually held to be the English plural of Welsh *hynt* 'road', from British *\*sento-*, presumably the Roman road, Watling Street, on which the place lies. If that is the case (the derivation is not free from doubt), the name must have been used by Welsh speakers through the period when *S-* changed to *H-*, usually considered to be the middle or second half of the 6th century.[104] In Wales Lane, Barton-under-Needwood, is Wales End, an early fifteenth-century timbered house, the name of which may denote Welsh-speakers, but the name has not been traced before the sixteenth century, is perhaps not much older in date, and may have a far more prosaic derivation.

## Roman influence

Only two places with Roman names have been identified in Staffordshire, *Letocetum* and *Pennocrucium*, which have given rise to the modern names Lichfield and Penkridge, although in neither case does the modern place stand on the site associated with the ancient name. The Roman names derive from yet older names, and incorporate the earliest recorded example of two of the commonest British place-name elements.

---

[101] These comments and conclusions are based largely on PN Ch V (ii) 296-8, and see particularly Cameron 1979-80.
[102] Jackson 1953: 555.
[103] PN Ch V (ii) 295-6.
[104] Jackson 1953: 521. Jackson himself considered such derivation for Hints 'quite uncertain', and the name possibly pre-Celtic: ibid.' 519.

*Letocetum* (also recorded as *Etoceto* and *Lectoceto*)[105] has been identified as the Roman settlement at Wall, which lies on Watling Street, 2 miles south-west of Lichfield, but the ghost of the name survives as the first part of the name Lichfield itself. The earliest spellings for the name Letocetum are found in a fourth-century copy, surviving in 8th century texts, of the Antonine Itinerary, a Roman road-book listing the distance between Roman cities, towns, and posting stations. The name derives from British *\*Letocaiton* meaning 'grey or brown wood', from *\*leto-* (Welsh *llwyd*) 'grey' and *\*caito-* (Welsh *coed*) 'wood'.[106] *Coed* is one of the most common British words in English place-names, but its British ancestor occurs only once – in Letocetum – in names recorded in Roman Britain.[107] When the name was first recorded in the Antonine Itinerary it had perhaps been taken from a pre-Roman wood or large estate (possibly with a so-far undiscovered pre-Roman settlement)[108] with the British name *\*Letged* (which gives Old Welsh *Luitcoyt*)[109] which in Romano-British times may have extended some distance north of Watling Street,[110] or was perhaps coined by troops from a continental Celtic area who formed part of the Roman army: it is known that such troops frequently gave Celtic names to uninhabited places.[111] The 'grey wood' might be identified as the extensive high woodland of Cannock Chase,[112] with the lower ground to the south and south-east later forming the *feld* of *Luitcoyt*,[113] or an extensive area of woodland which existed in the area of Wall in Romano-British times, with Lichfield in open land to the north.[114] Old English *feld* is generally interpreted as 'open land', but at the time of the English

---

[105] Rivet & Smith 1979: 387-8; see also Bradley 1928:117-19.

[106] Rivet & Smith 1979: 386-7. The usual rendering of the first element is 'grey', but the word 'crosses the English perceptual boundary between grey brown': see Hjelmslev 1968. The name was in late British *\*Ledoged*, developing via Primitive Welsh *\*Led'ged*, later *\*Luitged*, into OE *Liccid*. Letocetum may have been a semi-technical term for surviving areas of wildwood, the translation of which survives in the name Harwood (from OE *har* 'grey') found in several counties: see N & Q NS 44, No. 4, 453-8; PN Ch IV 227-8. The derivation of Letocetum was first solved by the philologist Henry Bradley (Bradley 1886: 296; 1889: 545f.), but his identification of *Lwytgoed* as Lichfield was discounted by Duignan 1902: 92-3, who concluded that the name was from Old English *lache*, *leche* meaning 'a morass, a bog', so giving 'the boggy field'.

[107] Gelling & Cole 2000: 223.

[108] Gelling 1988: 57.

[109] Jackson 1953: 332.

[110] A possible clue to the extent of the land or territorium of Letocetum may be the bishop of Chester's manor of Lichfield as recorded in Domesday Book. The manor may be evidence of an estate many centuries older. 23 settlements are listed, 3 of which cannot be identified, and 3 of which (Harborne, Smethwick and Tipton) can probably be disregarded given their distance from the area. The remainder are spread over a sizeable area bounded roughly by Cannock Chase on the west, on the east by the river Tame, and extending just north of the river Trent and just beyond Watling Street to the south. Letocetum lies close to the southern edge of this area: see TSSAHS X 1968-9 50-1. The ancient parish of St Michael's includes Wall within its southern boundary and incorporates (with St Mary's parish, which appears to have been taken out out of St Michael's parish) much of the early settlement of Lichfield: TSSAHS XXII 1980-1 33, 47.

[111] Rivet & Smith 1979: 22-3.

[112] Cf. the pre-English names Kinver and Morfe, and part pre-English name Brewood, for vast areas of forest.

[113] In that respect it is of interest that some places named *hean leah* ('high wood'), such as Henley in Oxfordshire and Hanley Castle and Hanley Swan in Worcestershire, are not themselves on high ground. It has been suggested that such places may have been so-named from woods on higher ground overlooking the settlement, which would be a striking visual marker for travellers: Gelling & Cole 2000: 239. Another possibility is that *heah* is to be interpreted in some contexts as 'chief, most important': cf. Wolverhampton.

[114] Gelling 1992: 60.

incursions may have held the special meaning 'common pasture', and been applied when the English began to cultivate that pasture.[115] In due course the name *Luitcoyt* became associated with the area, perhaps including several settlements, in which stands the place now known as Lichfield.

In the mid-13th century the idea developed that the name Lichfield was from Old English *lic* (ME *lich*) 'body, corpse', with the meaning 'field of corpses' attributable to a battle. The historian Matthew Paris (d.1259) suggested that the name commemorated the slaughter of 999 Christians, martyred under the emperor Diocletian between 284-305 AD, and associated the event with the fictitious St. Alban. The story may not have been widely known in Lichfield, for Leland does not mention the incident in his account of his visit to the city c.1540, but in 1549 the new corporation commemorated the massacre in the design of its seal, and attempts were later made to use local place-names to support the story: in the 1570s it was claimed that Boley and Spearhill near Lichfield referred to bows and spears used in the massacre, and as late as 1816, following the discovery of human bones, Elmhurst was identified as the scene of the massacre. A variant of the story claimed that the corpses referred to by the element *Lich-* were those of the army of three Christian kings defeated at Lichfield by Diocletian, and various places were identified as the burial place of the Christian dead, including Elmhurst, St. Michael's churchyard, Borrowcop Hill, and the site of the cathedral. Another theory, put forward in the later 17th century, proposed that the name is from Old English *\*lece* 'stream, boggy ground near a stream',[116] but that derivation is etymologically impossible. The name Lichfield is now held to mean 'the *feld* or open land near or belonging to Luitcoyt', or 'the common pasture in (or beside or near) the grey or brown wood', the colour presumably indicating the predominant species or perhaps denoting lichen-covered trees, and to have originated at some time during the seventh century.[117] Spellings for Lichfield dating from c.715 with the preposition *on* (or *an*) suggest (with parallels elsewhere) an area of some size called 'in Lichfield', and that 'Lichfield' came to mean the principal place in that area, in which case *Lyccidfelth* (which does not *per se* indicate a place of habitation) is likely to have been a late 7th century development of the existing name of a large area, gradually applied more narrowly to the cathedral and its immediate area.[118]

A Welsh elegy, the Marwnad Cynddylan, or 'Lament for Cynddylan',[119] possibly compiled as early as c.655, but perhaps more than a century later,[120] refers to a great raid on *Caer Lwytgoed* in which neither the bishop nor 'the book-clutching monks' were

[115] Gelling 1981: 14.
[116] Plot 1686: 398; SHC 1950-1 147.
[117] Jackson 1953: 332-5.
[118] VCH XIV 38. A very full account of the history of the name Lichfield can be found in VCH XIV 37-9 and TSSAHS XXVIII 1986-7 1-13; see also N & Q 242 453-8; Jackson 1953: 327, 332; Watts 1979: 123, 124 fn. 5. On Caer Lwytcoed and Letocetum see TSSAHS XXXIII 1986-7 7-10. Lichfield was created a county by royal charter in 1553 (Shaw 1798: I 309), and re-united with Staffordshire by statute in 1888 (VCH XIV 83).
[119] See VCH XIV 37, and Rowland 1990. The *Historia Brittonum*, a history of Britain to the 680s, compiled by an unknown author in north Wales c.829, and popularly attributed to Nennius (Lapidge *et al* 1999: 239-40) records *cair loit coit* (or *Cair Luitcoyt*) as one of the 28 (or 33, depending on the version) cities of Britain, which is presumed to be the same place as *Caer Llwtcoed* (VCH XIV 37). All the cities listed in the *Historia* are given the title *cair*. It is possible that the author of the *Historia* knew of the name from the 'Lament for Cynddylan'.
[120] The reference to an archbishop, for example, points towards knowledge of the archdiocese of Lichfield, created in 787: N. J. Higham 2002: 178.

spared,[121] and 1,500 cattle and 80 horses were seized as booty and taken to Powys.[122] The identification of *Caer Lwytgoed* has given rise to considerable academic speculation. *Caer* normally refers to a fortified place, and it has been observed that Lichfield could not then have been described as a *caer*, since its defences did not exist until about the mid-12th century.[123] The name may well refer to Letocetum (from which traces of a Christian presence have been recovered),[124] with its extensive stone defences, well recorded historically and archaeologically, but no evidence has been found (which does not of course mean that none exists) of any occupation later than the fifth century.[125] It is conceivable, but unlikely, that the prehistoric earthworks of Castle Old Fort near Stonnal (3 miles from Letocetum, and visible from the Roman site), or the enigmatic earthworks nearby at Loaches Bank, or Castle Ring on the highest point of Cannock Chase (5 miles from Letocetum), are to be identified with *Caer Lwytgoed*.[126]

However, the association of *Caer Lwytgoed* with the early site at Lichfield need not be dismissed entirely, for the identification as Letocetum based on a narrow meaning of *Caer*[127] fails to recognise that Old Welsh *cair* and *civitas* are evidently synonymous in the ninth-century list of the 28 *civitates* of Britain listed in a synchronising history to the 680s, probably composed c.830, known as the Historia Brittonum, the names of which all begin with *cair*,[128] and Welsh *caer* is often used in the Welsh forms of English place-names of Celtic origin, where it has the significance of Latin *urbs* 'city' in the earliest texts, but came in its Cornish and Breton cognates – and probably also in Primitive Welsh – to mean after the sixth century simply 'homestead' or 'village'.[129]

---

[121] An incident that might owe more to Bede's description of the Battle of Chester than any mysterious massacre in the Lichfield area: N. J. Higham 2002: 178.

[122] Usually considered to be a Welsh action against Mercia, but see Brooks 1989: 169: 169; Rowland 1990: 134; TSSAHS XXXIII 1991-2 9-10; Gelling 1992: 73.

[123] Brooks 1989: 169; Gould 1991-2: 7-8. The bishopric moved from Lichfield to Chester after a council held at Windsor in 1072 decided that bishops' sees should be in walled towns rather than in villages: Preest 2002: 208 fn.5. *Caer Luit Coyt* is identified as Lichfield on the 1974 O.S. map of Dark Age Britain.

[124] TSSAHS XIV 1972-3 30-31; TSSAHS XXXIV 1992-3 1-4. It has been suggested that there was a British see (and possibly a monastery) at sub-Roman Wall, which was succeeded by St. Michael's church in Lichfield: Basset 1992: 13-40; N. J. Higham 2002: 110. That would explain the statement of 'Eddius' Stephanus, the contemporary biographer of St Wilfrid, that Lichfield was '*a suitable place*' to establish the see of the Mercians in 669: Colgrave 1927: Chapter 15. But then so would the existence of a pagan shrine. It is not inconceivable that Stowe at Lichfield took its name not as popularly supposed from its legendary associations with St. Chad, but because it was already a holy Christian site when the place was chosen for the first Mercian cathedral: cf. Eccleshall and its Celtic Christian history.

[125] VCH XIV 4.

[126] A parallel would be Viroconium, founded on the banks of the Severn under the shadow of the Wrekin and its great hill-fort. In the case of Castle Ring, some nine acres in extent and the largest and most elaborate of the Iron-Age hillforts in Staffordshire, with views extending to the foothills of the Pennines and the South Shropshire hills, a Cistercian abbey was established a mile or so to the south at Radmore c.1143, perhaps on the site of a hermitage dating from the reign of Matilda (1102-67), before being moved to Stoneleigh in Warwickshire eleven years or so later: VCH III 225; TSSAHS VII 1965-6 30. It would be fanciful to believe that Radmore may have been chosen for long remembered local links with early Christianity.

[127] See for example Brooks 1989: 169.

[128] Padel 1985: 50, citing Jackson 1938: 44-55.

[129] EPNE i 76; Padel 1985: 50-4; also Coates & Breeze 2000: 150; 348.

In this case, therefore, *Caer Llwytgoed* could properly be interpreted as 'the town or village associated with Lwytgoed, 'the grey wood". Although indistinct traces of Saxon settlement have been recorded to the south of the cathedral,[130] historians conclude that Lichfield is unlikely to have been anything other than a modest ecclesiastical site until well after the Conquest: as late as the twelfth century the Anglo-Norman chronicler William of Malmesbury described Lichfield as 'a small village in Staffordshire far from the habitation of men'.[131] It might be surprising, however, if such an important ecclesistical centre with the shrine of St Chad had not attracted some limited commercial activity in the form of hostelries and suchlike to cater for the needs of pilgrims before the Conquest, but whatever the size of Lichfield it is not impossible that even a very modest community may have been described as a *caer*, as evidenced by other relatively insignificant places listed in the Historia Brittonum, such as Cair Guorthegern, Cair Peris, Caer Draithou, Cair Urhahc, and Cair Celemion, identified by John Morris as 'Vortigern's Fort', 'Llanberis (?)', 'Draitou Fort', 'Urnarc Fort', and 'Celemion Fort' respectively.[132] At present therefore it can only be concluded that *Caer Lwytgoed* probably refers to Letocetum (Wall), or possibly Lichfield, or perhaps even an earthwork in the area such as Castle Old Fort or Castle Ring, or some other location in the area that yet remains unidentified.[133]

The name Pennocrucium, recorded as *Pennocrucio* in the Antonine Itinerary for the Roman settlement straddling Watling Street ½ a mile west of Gailey,[134] long intrigued philologists. Imprudently rejecting a derivation from British *pen-crug* 'the head of the mound' proposed by Rhys, Duignan accepted the connection between Pennocrucium and Pencric (later Penkridge), but attributed the name to British *pen-cric*, with *pen* meaning 'head, end', and (as an adjective) 'chief', and the second element a supposed *crioch* or *criche*, cognate with German *crioch*, *criche*, Irish *crioc*, *crioch* 'a boundary, end, limit, frontier', so giving 'head or end of the border, a frontier'.[135] But more recent research has confirmed the derivation proposed by Rhys, and there is no doubt that the name is British *\*Pennocroucion*, with the first element from *\*penno-* (Welsh *pen*) 'head, end, headland, chief' (one of the commonest British words in English place-names, but found only once – in Pennocrucium – in names recorded in Roman Britain).[136] The second element is a derivative of British *\*crug* (Welsh *crug*, Old Cornish *cruc*) 'hill, mound, tumulus', so giving the meaning 'headland tumulus', 'chief mound', 'head of the mound' or similar, referring possibly to a prehistoric burial mound. It has been assumed that the tumulus

---

[130] TSSAHS XII 1980-1 3; 10.

[131] Hamilton 1870: 307; Preest 2002: 207 translates the passage: 'Lichfield is a small town in the county of Staffordshire, far away from crowded cities. It is surrounded by woods and has a small river flowing past'. See also TSSAHS XXII 1980-1 24. Stephen of Ripon (the so-called 'Eddius' Stephanus), writing probably between 710 and 720, refers to Lichfield as a *locus*: Colgrave 1927: chapter 15. Lichfield remained remote until a causeway across the wet valley to the south of the cathedral was constructed in 1296, creating a more direct route between London and Chester than Watling Street: Taylor 1979: 135-6.

[132] See e.g. Reno 2000: 30-6, 181-3.

[133] See also The Loaches. The situation may be compared with the lacunae in our knowledge of the situation in central Shropshire after the abandonment of Viroconium in the fifth century and before the creation of the town of Shrewsbury in the Anglo-Saxon period.

[134] Rivet & Smith 1979: 326.

[135] Duignan 1902: 116-7.

[136] Gelling & Cole 2000: 212. Hackwood 1915: 22 notes that Penncricket Lane, which forms the boundary between Oldbury and Halesowen and was the ancient county boundary when Oldbury was detached from Shropshire, may have derived from Penn Cruc.

which stands on Rowley Hill, ¾ of a mile north of the Roman site, is the mound referred to,[137] but there may be an inference that there were others in the area,[138] perhaps on the more prominent but heavily quarried summit of Beacon Hill nearby.[139] The name, which would have been in British *Pennocroucion*, was Latinised by the Romans by the alteration of the ending to -*io* or -*ium* and adopted for the civil settlement which lay astride Watling Street at Water Eaton (which is closer to Rowley Hill than Beacon Hill). As with Letocetum, the name may have been a Latinised version of an existing Celtic name, or created by Roman troops from continental Celtic areas, who are known to have created names for uninhabited areas.[140] Originally it was perhaps applied to a much larger area, and has been linked to a Mercian group or tribe recorded as the *Pencersæten* ('people of the Penk') mentioned in a charter of 849 relating to Cofton Hacket in Worcestershire,[141] suggesting that the southern boundary of that tribe lay in that area (the boundary was with the *Arosætna*, named from the river Arrow: Cofton Hackett is at or very close to the headwaters of the river), though it may be noted that this is the only known reference to the *Pencersæten* (they are not mentioned in the Tribal Hidage, an 11th-century manuscript of earlier material listing 34 Anglo-Saxon kingdoms south of the Humber, though other known kingdoms are also excluded), and the possibility of a scribal error cannot be entirely discounted: it has been observed that 'even the exact form of their name is uncertain, for Pencer may well be due to a misreading of Wencer [due to a similarity between the manuscript forms of Old English *w* and *p*]…The context in which they are mentioned makes it probable that they were afterwards incorporated in Worcestershire'.[142] Other explanations for the tribal name, such as a derivation from Old Welsh *pen caer* 'fort headland',[143] cannot be discounted entirely. By the 10th century the name of the Roman settlement of Pennocrucium had been transferred to the river, and to Penkridge, some two miles north-east of Pennocrucium,[144] via Old English *Pencric*,

---

[137] Gelling 1984: 138; Gelling & Cole 2000: 159. The mound was noted by antiquarians in the eighteenth century: see Shaw 1798: I 31.

[138] In the possible meaning '(place by) the chief mound', preferred by Padel: see Lapidge *et al* 1999: 367.

[139] In that respect, it may be noted that OE *becun* had a range of meanings including 'sign, signal, cross, memorial stone', and perhaps 'tumulus' (see VEPN I 68), and *Congreve fielde*, which may have included Beacon Hill, was also known as *le lowe fielde*, 'the field with the tumulus', in 1589 (SHC 1929 28; SHC 1930 16), *Lowe Field* 1682 (SRO D1057/F/1/3); and in the 17th or 18th century *Lowfield Leasow (alias Lowhill)* (SRO D260/M/T/5/46); and *lasselowe, lesselowe*, possibly 'the lesser low or tumulus', is recorded in Stretton in 1439: Oakden 1984: 179. A Roman road running slightly west of north has been traced across Beacon Hill: Horovitz 1992: 32. See also Shaw 1798: I 31.

[140] Rivet & Smith 1979: 436.

[141] S.1272; see Hooke 1983: 12-13; Hooke 1990: 135-42.

[142] PN Wa xvii, fn.4. It may also be pertinent to observe, conversely, that Wychnor is said to incorporate the name of the Anglo-Saxon people known as the Hwicce, although Wychnor is a considerable distance to the north of the generally accepted territory of the Hwicce: see Ford 1979: 146-8. If the supposed territory of the Pencersæten has been properly identified, it would appear to include that of the Husmerae, a tribe recorded in a charter of 736 (S.89), which occupied an area which perhaps corresponded to that now covered by the rural deanery of Kidderminster.

[143] Cf. Pencaer (Pencâr), Pembrokeshire, recorded as *Penker* in 1315: PN Pembrokeshire I 253.

[144] Cf. Lichfield and Wall. Like Lichfield, Penkridge has a long history and was an important ecclesiastical centre: a charter of Edgar dated there in 958 (Birch 1885-93: III 246), mentions 'in loco famoso qui dicitur Pencric' 'in that famous place called Pencric' (though we are tantalisingly deprived of the reasons for its fame), and Penkridge church, probably founded by the tenth century (traditionally by Edgar) was one of the nine royal free chapels named in a mandate of Henry II in 1318: Styles 1936: 56-95. The expression 'famous place' perhaps implies something other than a palace, which was usually referred to as a *villa regalis* or *villa regis*: see Zaluckyj 2001: 222.

which was taken by the Anglo-Saxons from speakers of Brittonic with the second *c* intact as a velar,[145] and has since changed into the modern form based on a wrongly-assumed river-name with an imaginary 'ridge' which exists neither at the place nor in the original name. Penkridge village indeed lies in a shallow valley. For completeness, it might be mentioned that the element *pen* 'head, end, headland' has been interpreted literally, to be associated with a Celtic cult: there is said to be evidence of pagan Celtic head worship at Wall,[146] and at Magh Slecht in County Cavan, Ireland, the Celtic god Cenn Cruaich (literally 'head of the mound') is said to have been worshipped. The theory has no support amongst place-name experts.[147]

Mention might be made of a handful of minor names which appear to contain the element *penk*, namely Penecford (Timmor), Penkshull (Claverley), and Penkholme (Pipe Ridware). All lie on or close to a stream or river. The significance of the names and the influence, if any, of the name of the river Penk is unclear, but the place at Timmor (and possibly the other names) may be from Old English *pennuc*, Middle English *penok*, *pinnok* 'a small animal pen', found in field-names in western England from the thirteenth century.[148]

## The earliest English names

A combination of British and English elements in a place-name is generally taken as evidence of British and English co-habitation (and thus dating from the earliest Anglo-Saxon period), and in tautological compounds that the meaning of the British element had become lost by the time the English element was added, although there is no obvious reason why an immigrant people should not add a descriptive or other element in their own language to a pre-existing name which they well understood. A careful analysis of the earliest recorded English names has allowed philologists to determine the likely period during which particular name-forming elements were used. For example, names incorporating Old English *tun*, very common throughout Staffordshire, are believed to date in the main from the two centuries between c.750 and c.950.[149]

Since the 1960s the long held theory[150] that place-names containing *-inga-* and *-ingas* provide evidence for the earliest phases of Anglo-Saxon colonisation has effectively been demolished, largely based on unsuccessful attempts to validate earlier research that purported to show a correlation between these names and the location of Anglo-Saxon burial sites.[151] The theory was based on the suffix *-ingas*, plural of *-ing*, which was added from an early date to a personal name to denote the dependants or people of the person so named, so creating a group-name, which sometimes evolved into a place-name when the

---

[145] Jackson 1953: 260.
[146] TSSAHS XXI 1979-80 3-11.
[147] The name Penkridge is not unique: see for example Penkridge Hall, 3 miles north-east of Church Stretton, Shropshire (built c.1590 by Rowland Whitbrooke, whose wife came from Penkridge in Staffordshire: VCH Sa X 27), and Penkridge Hall (otherwise Penkridge Lake Farm), recorded in the 13th century, 5 miles east of Runcorn (PN Ch II 155). The reference to *Penchrych-under-Lyme* in 1293 (SHC VI (i) 256) is presumably a transcriptional error for a spelling of Penkhull. Cf. Penchrise, south of Hawick: *Pencriz* 1380 (Watson 1926: 354).
[148] EPNE ii 62.
[149] Cox 1975-6.
[150] Formulated by J. M. Kemble in 1849: Cameron 1996: 66.
[151] Gelling 1988: 107-112.

23

group became associated with a particular locality. A study published in 1976[152] analysed 224 place-names recorded between c.670 and c.730, and showed that -*ingas*-type names were far from common, and that there was a preponderence of 'topographical' and 'habitation' names. One name in Staffordshire, the unlocated Elinges, may incorporate -*ingas*, and two names incorporate -*inga*-, namely Edingale and Essington. Edingale lies on the east side of the county on the river Mease, a tributary of the river Trent, some 1½ miles from the Derbyshire border, and within 2½ miles of the Roman Ryknild Street.[153] In the centre of the village is a low mound which might be Anglo-Saxon in origin.[154] Essington lies in the central part of the southern half of the county, either on or very close to the line of a lost Roman road from Pennocrucium to Metchley. No specifically Anglo-Saxon burials are recorded in the immediate area, but a great tumulus is well-recorded on Bushbury Hill, 2 miles to the west,[155] and a dozen or so tumuli of unknown date have been pinpointed in the Wolverhampton area from place-name evidence and accounts by early historians,[156] at least some of which may have dated from the Anglo-Saxon period.[157]

Pattingham, which incorporates the monothematic personal name P(e)atta,[158] is also an *ingaham* name (a late version of an -*inga*- and *ham* name, a folk- or group-name rather than a place-name proper, dating probably from a period later than the immigration phase of Anglo-Saxon colonisation represented by early pagan cemeteries, and possibly contemporary with the colonising phase which followed soon after),[159] with the Old English noun-forming suffix -*ing* in a palatalized and assibilated form (known as Brummagem-type) with the early pronunciation -*indge*, *inch*, which may be as old as the earliest English speakers in Britain.[160] The soft pronunciation of -*ing*- is said to have been used until recent times.[161] It has been suggested that Brummagem-type names, of which some fifty or so have been traced,[162] are of a different origin to -*ingham* names where the *g* was hard, and are not formed by the addition of *ham* to the genitive plural of a folk-name, but from an original place-name formed by the addition of *ing* to a man's name. Pattingham, for example, would mean 'place or settlement associated with P(e)atta', and the addition of *ham* would give a compound which meant 'village at P(e)attingi'. Since

---

[152] Cox 1975-6: 12-66.
[153] Plot 1686: 402 noted part of a raised road which he thought may have been Roman about a mile north-east of Edingale aligned towards Lullington in Derbyshire. See also StEnc 166, 197
[154] Perhaps the one mentioned in Plot 1686: 402
[155] See e.g. Plot 1686: 403; NSJFS 1965 59.
[156] See for example Shaw 1801: II 150; *172; VCH I 193.
[157] In which case contradicting the statement in Cox 1972-3 37 that Essington is in an area devoid of pagan burial sites.
[158] For a discussion of place-names incorporating P(e)atta see Wallenberg 1931: 41 and 1934: 310. The name is found in Patton, in Stanton Long parish, Shropshire (PN Sa I 133), which gave its name to a Hundred: see VCH Sa I 285; X 190-5; PN Sa III 277; Thacker 1985: 5. Patshull, which adjoins Pattingham, is held to derive from the OE personal name Pættel, an *l*- derivative of P(e)atta: PN W 158.
[159] Cameron 1976: 9.
[160] See PN Ch V (ii) 279-88, which includes a discussion of the relationship between -*ingham* place-names and Roman roads.
[161] SHC 1939 193. Duignan 1902: 114 states 'I believe this name is pronounced 'Pattinjam'', and indeed Forster 1981: 180 notes that this is the modern pronunciation. The author has lived in the area for over 50 years, and has never heard the name so pronounced, or found anyone else who has.
[162] See PN Ch V (ii) 286, fn. 92.

*ham*, probably not used to form place-names later than 800 AD,[163] was added to the locative of the earlier name, the case ending -*i* would cause the development of -*ing*- to -*inch*-.[164] The place lies 2 miles east of a lost Roman road from Burlaughton via Burnhill Green to Chesterton,[165] is on the continuation of the line of a lost Roman road traced from Giffard's Cross to Wrottesley Lodge Farm,[166] is 2 miles west of the presumed course of a lost Roman road which has been traced from Pennocrucium through Codsall,[167] and is less than 3 miles north of a Roman road from Greensforge to The Walls at Chesterton[168]and a nearby supposed Roman road running from Greensforge to Central Wales,[169] suggesting that the name may date from a phase of settlement which took place while some elements of Roman infrastructure survived. Within a two mile radius Burnhill Green, Kingslow and Stanlow (and possibly Copley) have names indicative of ancient burial mounds.[170] It may be noted that the adjoining estate of Patshull incorporates the Old English personal name Pættel: the individual involved may have been related to the P(e)atta associated with Pattingham,[171] for the Anglo-Saxons are known to have given their children alliterative names. There are no other place-names incorporating similar personal names in Staffordshire.

Studies have shown that the element *ham* was used at a particularly early period for the most desirable sites in topographical terms,[172] and is almost invariably associated with Roman roads[173] or major rivers which will have been utilised as the main routes by which the Anglo-Saxon settlers colonised the countryside, and with Anglo-Saxon pagan burial sites. There is only one certain and three possible place-names containing the element in

---

[163] Gelling 1984: 65. However, an alternative view is that place-names with personal names and *ham* may refer to whole districts rather than individual settlements, and that such names often appeared in the 10th century when large estates were split into smaller independent units: Fellows-Jensen 1990: 13-21.

[164] Nicolaisen *et al* 1970: 19-20.

[165] See for example Shaw 1798: I 34-5; VCH Sa I 273. It may be noted that Shaw suggests, without explanation, that Pattingham 'seems to have taken its name from the Roman road near which it stands': 1801: II 279. The route has been the subject of more detailed investigation in recent years by Tong Archaeological Group, which has yet to publish its findings.

[166] Horovitz 1992: 34-5.

[167] Ibid.

[168] Margary number 192. See Cox 1973: 15-73.

[169] See TSAS LVI 1957-60 237-41.

[170] Cox 1972-3: 37 states that like Essington, Pattingham is in an area devoid of pagan burial sites. The age of these mounds is not known, but the possibility that they are of Anglo-Saxon origin, or may have been used by the Anglo-Saxons for secondary pagan burials, cannot be overlooked. In 1841 the curate of nearby Beckbury wrote in the burial register about 'several mounds' in the district, although there is some doubt about the nature of such features: see TSAS LVII 1961-4 192-3; VCH Sa XX 240..

[171] The Staffordshire-Shropshire boundary appears to deviate from a natural line to incorporate Patshull and Burnhill Green into Staffordshire. The only other (former) Staffordshire name beginning in *Pat*- is Patmarsh, two miles north-east of Worfield. The age and history of the name are unknown, but if ancient it may share the same personal name as Pattingham.

[172] See for example Gelling 1988: 112.

[173] But note that the theory that the proximity of *ham* names to Roman roads has a causal significance is rejected in Kenyon 1986: 11-27, which by a statistical analysis of 30 such names in Lancashire and Cheshire shows no such relationship beyond that to be expected if such settlements had been distributed randomly.

Staffordshire: the certain name is Pattingham, and the possible place-names are Audnam, Burnaham/Bornam, and Trentham.[174]

Audnam, with a name of uncertain age, lies 1 mile east of the Roman road[175] which runs between Greensforge and Droitwich, which might point to an early origin. Trentham is probably '*ham* or village on the river Trent', but possibly from Old English *ham(m)*, *hom(m)* 'meadow, especially a flat low-lying meadow on a stream, a water meadow, land in a river bend', since the place lies on low ground on the river Trent, although not near a bend in the river. The elements *ham* and *ham(m)* (the latter also found in use before c.730)[176] are difficult to distinguish unless early spellings with -*mm* or -*o*- are available, but it may be noted that all the names incorporating *ham(m)* listed by Ekwall[177] are in the south of England.[178] Trentham lies several miles south of the Roman road from Rocester through Stoke on Trent.[179] The grave deposits in an isolated Anglo-Saxon burial at Barlaston, two miles south-east of Trentham, were similar to those from Anglo-Saxon burials in the Peak District, and suggest that the immediate area was settled, doubtless via the river Trent, late in the pagan period, probably in the first half of the seventh century.[180]

The compound *ham-stede* 'homestead, site of a *ham*', the precise meaning of which remains unclear, occurs only once in Staffordshire, in Hamstead three miles north-west of Birmingham. The place lies within two miles of Ryknild Street, and some three miles from the line of a supposed Roman road from Pennocrucium to Metchley.[181]

**Names associated with pagan religions**

Names associated with pagan religion have long held a particular fascination for historians and place-name scholars. Sir Frank Stenton observed of such names that

'[t]he worship of which they preserve traces was not hidden away in forest sanctuaries, and most of the sites to which they relate were either on the verge of cultivable ground or on conspicuous features of the countryside. Some of them occupy outstanding hills; many are within a short distance of Roman or other ancient roads...'.[182]

Six place-names perhaps associated with varying degrees of certainty with pagan religious beliefs have been identified in Staffordshire, namely Wednesfield, Wednesbury, *Wodesneswall*, Weeford, *Belstowe*, Gads Hill, and (possibly) Freeford.[183]

---

[174] No pre-nineteenth century spellings have been traced for the name Bentham, which has therefore been assumed to be modern.

[175] Margary number 192.

[176] Cox 1973: 61.

[177] 1936 214.

[178] But see Gelling & Cole 2000: 46-55, which notes (p.47) examples in Shropshire.

[179] Margary number 181; 70a.

[180] See Ozanne 1962-3: 41-7.

[181] Margary number 190.

[182] Stenton 1970: 287.

[183] Hart 1975: 96 reads the boundary mark *hedenan mós* in a charter of 975 relating to Madeley as *hǣðenan mos* 'the heathen swamp'. A place called Wodens near Combridge is recorded in Redfern 1886: 88, but has not been traced.

Wednesbury, 5½ miles south-east of Wolverhampton, takes its name from Old English Woden, a heathen Germanic god, corresponding to Old Norse Oðinn. Duignan[184] held the second element to be Old English *beorg*, a word which has been shown to have the very specific meaning 'hill with a continuously rounded profile, tumulus',[185] and noted the 'conspicuous somewhat conical hill', confirmed by bold hachuring on the first edition 1" Ordnance Survey map of 1836. The cult of Woden, both in England and the continent, was indeed especially connected with hills.[186] However, Duignan's conclusion was based on his confident identification of Wednesfield as the site of two battles recorded in the Anglo-Saxon Chronicle under the dates 592 and 715. In the first, the West Saxon King Ceawlin was defeated, and in the second Ine, King of Wessex, fought Ceolred, King of the Mercians. The Chronicle spellings for the 592 battle are *Woddesbeorge* (three times),[187] and *Wodnesbeorge* (once),[188] with John ('Florence') of Worcester naming *Wodnesbeorh* (with the helpful note *id est Mons Wodeni* 'that is Woden's Mount'),[189] Æthelweard (d.1118) naming the place *Wodnesbyrg*,[190] William of Malmesbury (before 1140) *Wodendic*,[191] and in the twelfth century Henry of Huntingdon *Wednesburie*.[192] For the 715 battle the Chronicle gives *Wodnesbeorge* (twice), and *Woddesbeorge* (three times). 'Florence' of Worcester gives *Wodnesbeorh*,[193] Æthelweard *Vothnesbeorhge*,[194] and Henry of Huntingdon *Wonebirih*.[195] In fact both battles are now held to have taken place at Adam's Grave, a burial mound at Alton Priors in Wiltshire (an identification reinforced by William of Malmesbury who records that the place was *apud Wodnes dic*, 'at Wansdyke'),[196] which was still a landmark in the ninth century.[197] A derivation from *beorg* would have an exact parallel in Woodnesborough, Kent, where the element, formerly thought to refer to a tumulus recorded near the church which lies atop a sizeable hill with a distinctively rounded profile, is now believed to refer to the hill itself.[198]

---

[184] 1902: vii-ix, 168.

[185] See Ekwall 1960: 503; Gelling & Cole 2000: 145.

[186] Stenton 1971: 40. The existence of King's Hill (q.v.) at Wednesbury, recorded from the early 14th century (perhaps to be identified with *Brynghull* (q.v.)), may be noted. A legendary hill called (improbably) *Wodenfreseford*, perhaps near Drakelow or Stapenhill, has been noted in an undated and untraced reference, probably from Shaw 1798: I. The name *Wadenesfale*, recorded in 1337 (SHC VI NS (ii) 158), is likely to be mistranscription for Waterfall, near the place where the river Hamps disappears underground, but poses the intriguing possibility that the place may have been known as 'Woden's pit or trap', i.e. the place where the waters of the river disappeared into Woden's domain.

[187] Ede 1962: 6.

[188] Ibid.

[189] Ede 1962: 7.

[190] Campbell 1962: 14.

[191] Ede 1962: 7.

[192] Ibid.

[193] Ibid.

[194] Campbell 1962: 21.

[195] Ede 1962: 7.

[196] Earle & Plummer 1892-99: II 118. Ceolred is known to have been buried at Lichfield in 716 (Earle & Plummer 1892-99: I 42; Lapidge *et al* 1999: 509), and it is possible that his death was not unconnected with the conflict, in which case a battle-site at Wednesbury, a mere dozen or so miles from Lichfield, is not inconceivable, although Ine's territory was the country south of the Thames. Plot 1686: 410 associates Bunbury in Staffordshire with the battle between Ceolred and Ine.

[197] Whitelock 1955: 158. Duignan 1902: viii notes 'There is only one Wednesbury in England (the *Wodensbeorg* in Alton [Hampshire] being out of the question) ...'.

[198] Gelling & Cole 2000: 148-9.

Although Duignan's derivation deduced from the spellings he cites is correct, the spellings on which he relied relate in the main to the two battles, and the derivation is not appropriate for the name Wednesbury in Staffordshire.

From the early spellings that do relate to the place in Staffordshire, the meaning of Wednesbury is beyond doubt 'Woden's earthworks or fortification', from Old English *burh*, perhaps implying that the feature was believed to have been established or protected by the pagan god or dedicated to him. There is no firm evidence of any earthworks on the hill on which Wednesbury lies, but on the west and south-west of the hill, which stands between the headstreams of the river Tame, the ground falls away abruptly with a terrace-like feature on the eastern side of Church Street, suggesting that the church of St Bartholomew on the summit may stand within an ancient earthwork[199] – the Old English element *burh* was frequently applied to Iron Age hillforts – and perhaps on the site of a pagan shrine. By tradition a castle on the hill was constructed by Æthelflæd, Lady of the Mercians,[200] and Shaw refers to a fortification here:

'some remains of which are still visible in a large graff[201] round the Church-yard hill, which, from its height and extensive prospect, was a very suitable situation for such a defence, and tradition likewise says there was a castle here in the time of the Saxons'.[202]

Wednesfield, 2 miles north-east of Wolverhampton and three miles or so from Wednesbury, means 'Woden's *feld* or open land'.[203] The earliest spelling preserves the god's name, but a *feld* associated with a pagan god is perhaps not quite as easy to explain as a hill, fortification, ford or other specific feature. The Anglo-Saxon Chronicle records a great battle here or at Tettenhall c.910 when the Saxons vanquished the Danes. Wednesfield is three miles or so from Wednesbury, both of which lie close to a line drawn between the Roman sites of Pennocrucium and Metchley, near Edgbaston, which was almost certainly connected by a Roman road: there is a road running south-east from Pennocrucium in the direction of Metchley[204] which is untraced after it reaches the outskirts of Wolverhampton.[205]

---

[199] Hackwood 1902: 7-8; TSSAHS 1982-3 XXIV 59. Sections of 'ramparts ?' are shown on a plan in TSSAHS 1990-1 XXXII 99, but the basis for the identification is unclear. It may be noted that Ric'o de Erbury is recorded in Wednesbury in 1327 (SHC VII 229). Erbury may be from OE *eorþburg* 'earthern fortification', perhaps providing further evidence of some earthwork here, though there can be no certainty that even if the derivation of the name is correct, it refers to an earthwork at Wednesbury.

[200] Recorded in Gough 1806: 495; see also Shaw 1798: I 37; Erdeswick 1844: 398 fn; see especially Ede 1962: 9-14. The association with Æthelflæd is due to the mistaken identification of this place with *Weard byrig*, fortified by Æthelflæd in 915, as recorded in the Mercian Register (*ibid.* 12-130, an identification first made by Camden: see Ede 1962: 13.

[201] 'a trench serving as a fortification': OED

[202] Shaw 1801: II 83. By local tradition the 'castle' was fortified by Æthelflæd. Apart from a reference to *Burgreves* in Wednesbury in 1686 (BCA MS 3145/114b), which may derive from OE *burh* 'fortification', and OE *græfe* 'coppice', or 'pit, trench', the location of which is uncertain, no evidence has been traced of any documentary reference to a *bury* or *buri*, and there is no Bury or Berry road-name in the town, but a reference to Campfield Lane occurs c.1776: SRO 1317/1/13/1/1/1-8.

[203] Ekwall 1960: 503.

[204] Margary number 190.

[205] The name Wednesfield is also recorded (but only in single spellings) in two lost field-names in Essex, *Wedynsfeld* in Theydon (1446), and *Wodnesfield* in Widdington (1303): PN Ess xxi; 579.

The name *Wodesneswalle*, 'Woden's spring or (less likely) stream', has been found recorded only once, in Wednesbury, in 1315,[206] but may be further evidence of the popularity of the cult of Woden in the area.

Weeford is a small village on the south side of Watling Street, 4 miles south-east of Lichfield. The first element of the name is generally held to be from Old English *wig*, *weoh* 'an idol', and perhaps 'holy place, shrine'. *Weoh* originated as *wih*, broken to *wioh* to give Kentish *wioh*, later West Saxon *weoh*, and Anglian *wih* and the alternative later spelling *wig*. The early spellings for Weeford are consistent with the use of *weoh* in the Lichfield area between c.500 and c.750 AD. However, Lichfield is in the heartland of Mercia, where a West Saxon form would not normally be expected.[207] It has been suggested that the compound *weo-ford* (with the -*h* of the first element dropped) dates from a time after *wih* had broken, but earlier than the date of smoothing before -*h* in Anglian speech.[208] Whilst that hypothesis is acknowledged to be not entirely satisfactory, no better explanation has yet been proposed.[209] The name (also probably found in Wyfordby, Leicestershire) is therefore likely to be 'ford by the heathen temple', the ford lying on Black Brook (which becomes Bourne Brook to the east), possibly where it is now crossed by Dog Lane.

The name Weeford has been seen as evidence of a remote area in which paganism survived longest rather than a name indicating an early phase of Anglo-Saxon settlement,[210] but since the place lies on Watling Street, which remained a major road throughout the Saxon period, and is close to the important and early Christian centre of Lichfield (one of a number of place-names with pagan associations which lie close to a See),[211] the supposed longevity of such a prominently-located name with pagan associations is perhaps surprising, It has been observed that most places incorporating in their name Old English *wig*, *weh* (and indeed many other places incorporating in their names evidence of pagan religion)[212] lie close to Roman roads, and the 'heathen temples' may have been road-side shrines or temples, perhaps in personal rather than group ownership.[213] Indeed, it is possible that Weeford is to be seen as a name coined by the Anglo-Saxons to record the remains of a typical roadside Roman shrine, possibly re-dedicated by the Anglo-Saxons, and that names of this type need not be associated with Anglo-Saxon pagan religion. In that respect, it may be noted that Weoley, which may

---

[206] SHC 1911 322. The location of the place has not been traced. It is unclear whether the modern name Waddens, found in various road names and a stream name (Waddens Brook) to the south of Wednesfield is ancient, and the derivation is uncertain, but the possibility that it incorporates a corruption of *Woden* cannot be ignored. *Wodneslega*, recorded in SHC V (i) 63, would appear to be Wensley in Derbyshire: PN Db 411.

[207] Cf. Weoley Castle in Warwickshire.

[208] Gelling 1975: 100, 101-2.

[209] 'An etymology 'ford by the *weg-hoh* (perhaps just possible on topographical grounds) is ruled out by the earliest spellings. Those from 1086 and 1200 might certainly be expected to show the -*g* of *weg* if that word entered the name': Cameron 1987: 102.

[210] Gelling 1975: 104. See also Stenton 1971: 40.

[211] '...the proximity of Woodnesborough and Wye to Canterbury is noteworthy, as is the proximity of Harrow to London, the site of another early see': Cameron 1987: 105.

[212] See e.g. Stenton 1970: 287.

[213] In contrast with sites incorporating OE *hearg*, which was probably used of pagan shrines or temples that served as communal places of worship for a particular group of people: Old English Newsletter 21 No.1 1987 155.

contain Old English *weoh*, also lies on a Roman road.[214] A little over a mile to the north-west of Weeford is the site of Offlow mound, which gave its name to Offlow Hundred. While there is no evidence of any connection with the great Mercian King, Offa, who died long after the pagan period from which mound-burials date, it is not inconceivable that it was used as a burial place for the Mercian royal family during the period when it was still pagan.[215]

Freeford is a very small settlement 1½ miles south-east of Lichfield, on an ancient road between Lichfield and Tamworth where it crosses Darnford Brook on the edge of the former Whittington Heath.[216] The forms indicate a straightforward derivation 'ford where no toll was payable' (from the Old English adjective *freo* 'free'), but as was noted over a century ago,[217] the small stream here would almost certainly have provided a ford too insignificant to have been anything other than free, and indeed no evidence has been traced to show that fords were ever subject to tolls,[218] so possibly here meaning 'open, accessible, unimpeded, unrestricted, free from undergrowth or waterweed or vegetation', distinguishing this ford from nearby Darnford ('hidden or overgrown') further downstream.[219] A derivation from the pagan goddess Frig or Friga (Freo),[220] though improbable, cannot be ruled out entirely. The deity, associated with the Norse goddess Freya (sister of Frey, and daughter of Oðin (Woden)) was evidently well-known, for Ælfric's tenth-century treatise *De falsis deis* refers to 'the shameless goddess Frigg',[221] and the goddesses' name may be found in the place-names Freefolk, Froyle and Frobury in Hampshire, Fryup in the North Riding of Yorkshire, and Frydaythorpe in the East Riding of Yorkshire, although none of these names is found in any pre-Conquest document, and it would be unsafe to draw any firm conclusions from such doubtful evidence.[222] It may be significant that Weeford, another ford-name with possible pagan associations, is two miles to the south-east; both Weeford and Freeford were held by the bishop of Lichfield in 1086 (they are adjacent entries in Domesday Book); and both are ancient prebends of Lichfield Cathedral, perhaps dating from before the Conquest.[223] It has also been suggested that places close together with pagan names may indicate

---

[214] Gelling 1975: 102. It may be noted that Gelling herself questions the presumption of an early date for this name.

[215] Gelling 1988: 155-6; VCH XIV 5.

[216] There is some slight evidence that Freeford may in early times have been of considerable extent, incorporating perhaps part of Whittington, Packington and Tamhorn, and possibly reaching the river Tame: Shaw 1801: II App. 14.

[217] Duignan 1902: 63.

[218] According to Shaw, Erdeswick states that Freeford was so-called 'because the London way passeth over a ford of the Black-brook, it being well and deep for the most part in other places' (Shaw 1801: II 23), but the reference has not been traced.

[219] As suggested in VCH XIV 253.

[220] Possibly to be found in Fretherne, Gloucestershire; Fryup, Yorkshire; Freefolk, Hampshire, although OE *frið, freoðu* 'sanctuary' may be the root of Fretherne. Frey, or Frea, a god of fertility and plenty, was worshipped in Sweden and superseded an original female cult of Freyja. It is likely that her cult continued in England, but no direct evidence has been found: see Chaney 1970: 26; 50-1; 56; Fitzhugh & Ward 2000: 58. See also Cameron 1987: 99-100.

[221] Niles 1991: 127.

[222] Branston 1974: 42,128, 135-6; Stenton 1970: 292 fn4. It may be added that in OE Frey had the meaning 'Lord', and Freya 'Lady'. The Anglo-Saxon poem The Dream of the Rood identifies Frey with 'Our Lord', i.e. Christ, and in Beowulf frea is used 17 times in the sense 'lord': Branston 1974: 139.

[223] TSSAHS 2 1960-1 38-52; VCH III 140-1.

sanctuaries set up by rival cults in opposition.[224] Finally the name might come from Old English *fræge*, *frege* 'known, famous', so 'the popular, the better known, the most frequented', to distinguish the ford from 'the hidden ford' nearby at Darnford.

One other name of special interest is *Belstowe/Belstowa* (the name appears in both forms) found in the boundary clause of an Anglo-Saxon charter of Ashwood of 994.[225] The name is evidently from Old English *bel* 'funeral pyre', with Old English *stow* 'place, a holy place, a place of assembly', here perhaps 'the assembly place at the funeral pyre',[226] possibly to be associated with the religious practices of pre-Christian Anglo-Saxons. The location of the place has not been identified with certainty, but is likely to have been in the north-western part of Kinver on the west side of Smestow Brook, near Lodge Plantation,[227] in which case it lay a mile or so west of the Roman forts at Greensforge, and close to the Roman road running in a westerly direction from Greensforge to Chesterton near its junction with a Roman road branching north-east through Hinksford.[228] It is possible that the place was in some way associated with the Roman sites: the proximity of place-names associated with pagan religious beliefs and Roman roads has already been noted.

Finally, there is one disused name in Staffordshire which, though not directly incorporating a pagan name, may provide indirect evidence of pagan religious practices or beliefs. Gads Hill was the name of a hill some two miles south-east of Biddulph.[229] No pre-eighteenth century spellings have been traced, but there is no reason to suspect that the name is corrupt. Sir Frank Stenton studied a small body of similar names, and concluded that '[t]here seems to be no direct evidence for the deliberate renaming of heathen sites under Christian influence, but there are place-names which suggest that some process of the kind has actually occurred. There are at least five known examples of the place-name Godshill, Gadshill or Godsell. All these names represent an Old English Godeshyll, which formally may mean the hill belonging to a man named God – a short form of a compound name such as the familiar Godwine or Godric. But the frequent Godeshyll makes this explanation unsatisfactory, and it is on the whole more probable that each of these names stands for a successful attempt to destroy unseemly associations of a hill-site once devoted to heathen worship.'. [230]

Of the four certain or likely Staffordshire place-names associated with pagan religious beliefs, Weeford, Wednesfield, Wednesbury, and *Wodeneswalle*, all lie well to the west of any known pagan-period cemeteries. The conventional wisdom is that place-names

---

[224] Meaney 1997: 228; also Gelling 1988: 257.

[225] S.1380. See Hooke 1983: 70; also Mander & Tildesley 1960: 14 fn.3.

[226] Hooke 1983: 46 offers the alternative translation 'place where a bell was cast'. Cf. *Bæl-stede* 'place of the funeral pyre' in *Beowulf* 3097. See also Bell Field Farm.

[227] At SO 846886: see Hooke 1983: 69-72. Aerial surveys have identified two rectangular enclosures 200 yards or so to the west and south-west of the Roman site at Greensforge (at SO 856888 and SO 857883), and a similar enclosure 1 mile north-west (at SO 840905). Excavations at the latter revealed Roman-type ditches, but charcoal within the enclosure was radiocarbon dated to 800 AD, indicating that at least one of the enclosures was being used in the Anglo-Saxon period: see Anon. 1985.

[228] Margary number 192. Baugh 1808 shows the line of a rectangular camp at Camp Farm, but that is almost certainly a mis-placing of the well-known camp at Greensforge.

[229] NGR SJ 907558. It may be noted that Ladymoor, a name with religious associations, lies nearby to the south-east.

[230] Stenton 1970: 286 fn.2.

incorporating evidence of the names of pagan gods either date from the earliest period of Anglo-Saxon migration, when pagan gods were still worshipped, or were adopted a generation or so later to reflect local pockets of pagan worship which had managed to survive the introduction of Christianity.[231] Places of pagan worship which managed to survive for decades because they were not accessible to [early missionaries] would be more likely to give rise to place-names because they would by that time be felt to be remarkable, as they might not have been when paganism was the only form of religion among English speakers'.[232]

It is generally held therefore that such names must date from the first half of the seventh century or earlier,[233] perhaps coined by the first Mercian immigrants moving into the area from the Trent and Tame valleys. The names (and the name Weoley in the extreme north of Worcestershire)[234] may indeed be surviving evidence of paganism from the time of the powerful pagan king Penda,[235] or possibly from the reign of Peada's brother Wulfhere, who ruled from 658 to c.674 and, though a pagan, was not unsympathetic to the Christian mission in Mercia.

It remains puzzling why these pagan place-names, where pagan practices and worship cannot have been carried out for longer than a few decades, continued to be tolerated in an area so close to a major Christian centre, and were not replaced as part of a cultural cleansing with the introduction of Christianity. All four Staffordshire places lie within a dozen or so miles of Lichfield, from where the Christian conversion of a huge area radiated from the mid seventh century, and they would have been readily accessible to early missionaries.[236] There is other evidence for changes of names with pagan associations (for example *Wodnes beorge* in Wiltshire, recorded in 825, now identified as a neolithic long-barrow known as Adam's Grave),[237] and it is indeed remarkable that the Church (including the clerical scribes) and a Christian population were willing to preserve and perpetuate the provocative names of pagan deities in close vicinity to what was the central hub of Christianity for the vast region of Mercia: changing a minor place-name is hardly to be compared with the difficulties associated with the change of the name of a week-day incorporating the name of a pagan god. It may be noted in passing

---

[231] Gelling 191988: 159.

[232] Cameron 1987: 104.

[233] Stenton 1970: 287 observes that '... it is probable, if not certain, that most of the heathen names which can now be identified arose before rather than after the year 600'.

[234] PN Wo xiii.

[235] Who reigned c.632-55 AD, and whose name may be commemorated in Pendeford, which lies less than four miles from Wednesfield. For a discussion on the possible association of Penda with Wednesbury see Ede 1962: 9-10.

[236] Those places are even closer to the Roman sites at Letocetum, and it has been noted that in Wessex and elsewhere the pagan Anglo-Saxons generally occupied areas which probably had a substantial rural population in the Roman-British period: see Bonney 1979: 43. Gelling 1988: 159 suggests that 'the proximity of unregenerate worshippers of Woden' may have been a factor in the choice of Lichfield as the site of the Mercian see.

[237] S.272, S.1403. Yorke 1995: 168 suggests that the name change may be evidence of a deliberate [Anglo-Saxon?] policy of substituting other place-names for pagan place-names, but the date of the substitution may be relatively recent: no early forms of Adam's Grave are recorded in PN W 318. Furthermore, *Wodnes beorge*, together with nearby *Wodnesdene* and *Woddesgeat* (the former recorded in S.449, a charter of 939), lie on or near Wansdyke, found as *wodnes dic* in 903 (13th century, S.368; PN W 17), and might be derived from the early name of Wansdyke, which may have been named from its supposed construction by the early king or god, without any direct pagan associations; see also Gelling 1988: 158-61.

that no priest or church is recorded in Domesday Book at Weeford, Wednesfield or Wednesbury,[238] but the inferences which might be drawn from that are unclear.

It is appropriate to consider whether alternative explanations might be put forward for some place-names involving the names of pagan deities. One possibility, which would admittedly run contrary to the great body of scholarly research which has built up over at least the last century, is that some place-names incorporating the names of pagan gods do not necessarily date from the pre-Christian era, and indeed may be considerably later. In that respect it should be emphasised that whilst place-name scholars have long held that a place-name incorporating the name of a pagan god is to be seen as firm evidence for the worship of that god at that place, no physical or documentary evidence of any kind has been found to support that conclusion. The Anglicization of Western Mercia, where Anglo-Saxon cemeteries are rarely found, is held by some historians to have come about by the acceptance of Christianity by the incoming Anglo-Saxons from the native British population, with the Anglo-Saxons otherwise retaining their cultural practices, principally language, but possibly also their ancient rites and superstitions, although others agree with Bede that the Britons made no effort to convert the Anglo-Saxons.[239] It is not inconceivable that enclaves of non-Christians co-existed in parts of the West Midlands late into the Anglo-Saxon period.[240] In fact there is surprisingly abundant evidence[241] from references to pagan practices in laws from the time of Alfred, Edward the Elder, Athelstan, Edmund, Aethelred the Unready, and Cnut – i.e. after the Viking invasions – to explain why Pope Formosus found it necessary to write to the bishops of the English in the 890s, a mere decade or so before the battle of Tettenhall/Wednesfield, to warn that he had learned 'that the abominable rites of the pagans have sprouted again in your parts'.[242] The warning was doubtless due to activities associated with the cult of heathenism in Anglo-Saxon England which had almost certainly never died out completely, and had been strengthened by the influx of Viking settlers, with a particular resurgance associated with the Great Army of the Danes which invaded Eastern England in the 860s and 870s, to be followed by settlers from Scandinavia in the later ninth and early tenth century, as evidenced by some dismissive references by Ælfric to the cults of Odin, Thor and Frigg (alias Friya or Freya).[243] Cult survivals are evidenced in England by a continuation or re-emergence of those superstitious practices which could be appended to a formal adoption of Christian worship, such as incantations and auguries, the veneration of particular stones, trees and wells, magic potions, ancient customs concerning the sun and the moon, and witchcraft.

---

[238] The absence of any Domesday Bookreference to a priest or church at Wednesbury, in 1086 a royal multiple estate, which included Bloxwich and Shelfield, and possibly Walsall (TSSAHS XIV 1982-3 1), might suggest that that the church is a post-Conquest foundation, though many churches recorded in pre-Conquest sources or with surviving Anglo-Saxon fabric are not recorded in Domesday Book. The first reference to a church at Wednesbury is c.1210: Ede 1962: 54; see also WL 40.

[239] N. J. Higham 2002: 65.

[240] The *pagani* recorded in *Wreocensetun*, 'the territory of the Wrekin-dwellers', in a Mercian charter of 855, less than 60 years earlier than the date of the battle of Tettenhall/Wednesfield (S.206 -- see Gelling 1992: 83; also Whitelock 1955: I 90; 485-6; Finberg 1961: 48 no. 77) are likely to have been Danish raiders, perhaps even the Danes recorded as the 'Black Gentiles' who devastated Anglesey c.853 (Wainwright 1942, in Cavill, Harding & Jesch 2000: 25), rather than established occupiers.

[241] See in particular Chaney 1970, on which much of the remainder of this paragraph is based, and North 1997.

[242] Whitelock 1955: 820.

[243] Niles 1991: 127.

In pre-Christian times the pre-eminent pagan god was evidently Woden, the subject of the 'official' or 'higher' religion of the Danes in Denmark, with the kings playing a key role in the cult,[244] although most of the evidence of pagan practices in Viking age Denmark comes from twelfth- and thirteenth-century Icelandic and Norse sagas.[245] There is no direct evidence of any worship in England of Woden, Thunor and the other deities of the old religion – indeed, Woden rarely appears in Old English literature, which was written entirely in the Christian epoch – but his name is well evidenced in by place-names both lost and current. Woden was above all associated in Anglo-Saxon England with victory in battle, and his principal rôle was the god of battles,[246] more specifically a god of the dead killed in battle, a haunter of battlefields, and described as *wælceasega* 'chooser of the slain' in the Anglo-Saxon poem Exodus.[247] Indeed, one particular form of sacrifice to the god for which evidence is available from the continent is the dedication to him of the war dead.[248] The sixth-century Byzantine historian Procopius records the human sacrifice of Goths by the Frankish King Theudebert in 539 A.D. as 'the first-fruits of the war. For these barbarians, though they have become Christians, preserve the greater part of their ancient religion'. Again, prior to battle near Uppsala, probably in the 980s, the Swedish King Erik the Victorious dedicated the army of his enemy Styrbjörn to Othin.[249] One source records that half those slain in battle belonged to Woden and half to the god Freya, the female deity who is sometimes said to have been Woden's wife, and according to the Icelandic writer Snorri Sturluson, who died in 1241, 'rode to battle and takes one half of the corpses and Odinn (Woden) the other half'.[250] That the pagans sacrificed to Woden for the acquisition of victory and courage was known among the English, as confirmed by the tenth-century chronicler Æthelweard.[251] But it should also be noted that Æthelweard, writing probably around the end of the tenth century, believed that Woden had been a real king from whom Æthelweard himself claimed descent,[252] and that Woden had received divine honours after his death.[253] There is no reason to imagine that during the Anglo-Saxon period that view was not widely shared.[254] Indeed, Anglo-Saxon royal genealogies invariably, and presumably proudly, traced lines of descent back to Woden, for even in Christian times the royal houses of Kent, Essex, Wessex, Deira, Bernicia and East Anglia all traced their origins to Woden.[255] Eschewing reservations about promoting a pagan deity, Bede noted that Woden was the ancestor of Æthelberht of Kent, and even the staunchly Christian king Alfred the Great took pride in a genealogy which carried his descent back to Woden.[256]

No semantic sophistry can help to explain the apparent longevity of the place-names Wednesfield and Wednesbury. That the name Wednesfield could be interpreted as 'the

---

[244] Hald 1963: 109. Traces of some 30 or so names associated with heathen religion have been recorded in Denmark, at least 600 in Norway, and well over 100 in Sweden: Hald 1963: 99.
[245] Abels 1998: 150.
[246] Cheney 1970: 35.
[247] Branston 1974: 107.
[248] Cheney 1970: 35, 39, 110; Branston 1974: 100.
[249] Cheney 1970: 39.
[250] Ibid. 35.
[251] Campbell 1962: 7.
[252] Cheney 1970: 39.
[253] Ibid. xxi.
[254] See Yorke 2001: 17-18, who notes: 'Whether Woden was actually seen as a divine ancestor has been questioned', citing John 1992.
[255] Branston 1974:93.
[256] Hodgson 1939: 686.

field or plain upon which Woden gave victory to the English' (with reference to the battle of Tettenhall/Wednesfield), or 'the field or plain upon which Woden was vanquished', coined after the Danish defeat in that battle as a contemptuous memorial to the site where a great slaughter occurred of those who worshipped the god, with the deaths of the pagans seen as evidence of the derisory protection afforded by their god and of the supremacy of the Christian beliefs of the victors, can safely be dismissed. But whilst it might seem unthinkable that a Christian society would incorporate the name of a pagan god into a place-name, it is equally difficult to explain how pagan place-names supposedly coined in the earliest period of Anglo-Saxon occupation, or adopted later to record long-lived pagan practices, remained unchanged thoughout the centuries of Christianity,[257] and indeed why some large areas of Anglo-Saxon England, notably East Anglia, are devoid of pagan place-names.

One possible explanation is that the name Woden in particular amongst the pagan gods should be treated as a name understood at two levels in tenth-century England, and possibly much earlier. Certainly the cult of Odin was dominent amongst the West Germanic peoples who lived nearest the Danes in the Migration Age, at least amongst the aristocracy.[258] Perhaps, like Æthelweard, who was clearly well educated, the Anglo-Saxons recognised Woden as a legendary but real king associated with victory in battle, who was proudly claimed as a lineal ancestor of Anglo-Saxon kings, and who was later worshipped by the pagans. The particular royal associations with Woden might help to explain the relationship which, according to tradition, was made between Woden and the harvest: even in Christian times Wednesday was considered a propitious day for sowing and planting crops, even though for other activities it was generally considered to be an unlucky day.[259] On that basis, his name may have been incorporated by the Anglo-Saxons into particular place-names as that of a real king associated with victories, rather than in his capacity as a pagan god, at the sites of great battles, with Wednesfield representing a particularly late manifestation of that practice. Evidence for that possibility may be found in the battles in 592 at 'Woden's mound or tumulus', and in 715 the battle between Ine and Ceolred at the same place.[260] It might be considered a remarkable coincidence if those battles, together with the great battle of Tettenhall/Wednesfield c.910, were all fought at places already incorporating the name Woden. To add weight to that possibility, it is worth noting that in a paper discussing Danish place-names incorporating the name Odin,[261] Kr. Hald has suggested that the name Wansdyke, applied to the famous earthwork in south-east England, which incorporates the name Woden, may post-date the heathen Anglo-Saxon period, and was perhaps so-named 'simply because it was a venerable and impressive relic of antiquity which might fittingly be attributed to the chief heathen god'.[262] Whatever the case for

---

[257] Stenton 1970: 286 fn.2 recognised that names associated with pagan religious beliefs are likely to have been changed, citing Gadshill (q.v.) as an example.

[258] Hald 1963: 108.

[259] Cheney 1970: 35.

[260] Swanton 1996: 42. Sir Cyril and Lady Fox held that Wandsdyke and the surrounding area in which place-names incorporating Woden's name are found was dedicated to the god Woden when it was constructed, whereas J. N. L. Myers felt that 'the name Wodnesdic is most naturally to be explained as a name given by the pagan Anglo-Saxons to an important monument whose origin and purpose they did not know and could not guess, and which was therefore attributed by them to supernatural agency': Cheney 1970: 36 fn.120.

[261] Hald 1963: 107.

[262] A parallel may be Torsburgen, the name of Scandinavia's largest defensive earthwork on Gotland, probably built during the migration period: Hald 1963: 107.

associating pagan practices with the names of other pagan gods, a prosaic explanation for place-names incorporating the name Woden is that the name of the god of war was applied to places where battles had been fought, places which at the time might well have been un-named: battles were presumably fought in open ground, rather than in named towns, villages or farmsteads. If the custom of dedicating the dead of battle to Woden were remembered well into the Christian period, the derivation of names such as Wednesfield might be self-evident. The theory might be compared to modern superstitions relating to salt, mirrors and ladders, where proponents of superstitious practices are probably unable to rationalise those beliefs, and the modern traveller's faith in trinkets bearing a representation of St Christopher.

It might be added, in respect of the second element of the name Wednesfield, that whilst Old English *feld* is normally held to mean 'an open space, an area of unenclosed ground', the meaning 'the ground on which a battle is fought' is recorded from c.1300,[263] and it is not inconceivable that this alternative specific meaning was also understood and in popular usage some four or more centuries earlier.

Parallels for names incorporating real kings are numerous, for example place-names recording Penda, king of the Mercians, which are concentrated in the West Midlands.[264] On the other hand, the chronicler Simeon of Durham records that in 937 King Athelstan defeated a combined army 'apud Weondune, quod alio nomine Aetbrunnanwerc vel Brunnanbyrig appellatur', which tells us that the site of the battle of Brunanburh was also known as *æt Weondune*, a name which, according to Stenton, seems to contain an inflected form of the adjective which is substantivised in the noun *weoh* 'idol, image, shrine',[265] usually associated with pagan religious practices – a battle site with a pagan, rather than kingly, connection, although that connection may be associated with, the not predate, the battle.

Whilst the proximity of Wednesbury to Wednesfield does not preclude the possibility that the battle of Tettenhall/Wednesfield extended in part to Wednesbury, and the place derives the first element of the place-name from such conflict, one slight piece of evidence that cannot properly be ignored is the name *Brynghul*, in or near Wednesbury, recorded in 1327,[266] and as *Brynghull* in 1332[267] and 1377.[268] Is it possible that the name incorporates Old English *bringe* 'that which is brought, an offering, a sacrifice', so giving 'the hill where offerings or sacrifices (to a pagan god?) were made'?

## The name Stafford

The town of Stafford lies on a tongue of Keuper marl overlain by gravelly deposits or fluroglacial drift laid down at the end of the last ice-age by the melting ice sheet amidst alluvium in the marshland of the river Sow, with the medieval town walls tracing the limit of the gravel,[269] and except for a route to the north, was effectively surrounded by

---

[263] OED.
[264] See Jones 1998.
[265] Stenton 1970: 290-1
[266] SHC V (ii) 229.
[267] SHC X 85.
[268] SHC 4th Series VI 10.
[269] VCH VI 185-6. The Saxon defences are likely to have enclosed a smaller area than the later defences: VCH VI 187; 199.

the river on the west and south, and by streams and marshes and the King's Pool (which may have been created out of one area of marsh) on the east.[270] Many areas around the town, and indeed well beyond the town to the north-west, are still subject to periodic flooding, and even areas within the town remain undrained marsh.[271] There have long been river or water crossings on the east, west and south side of the town.[272]

The early history of Stafford has yet to be unravelled.[273] The earliest documentary spellings for the place are *Stæf-forda*, *Stæfford*, *Staffordaburh* 913,[274] *Stadford*, *Statford* 1086,[275] *Statford* 1130,[276] *Stephordi* 1102,[277] *Stafford* 1162,[278] but earlier spellings can be gleaned from silver pennies, which normally give the name of the moneyer and place of

---

[270] The first reference to the King's Pool is in 1157 (VCH VI 210), but there is little doubt that even if the pool iteslf was not in existence much earlier, a large area of land on the east side of Stafford will have been very wet and marshy. Pollen studies of sediment show that the first forest clearance was in the late Neolithic period, with a greater human impact in the Bronze Age, and particularly at the beginning of the Iron Age, where a period of forest clearance was accompanied by agriculture, and evidence for a high level of agricultural activity from about the end of the Roman period to at least the 11th century: Bartley & Morgan 1990: 193. For reasons which remain unclear the king had the King's Pool dam removed in 1257, but in the same year permitted the townspeople to rebuild it at their expense to reinforce the town's defences (VCH VI 210). In 1644 the King's Pool meadows (*Kings poole Medes* 1610 Speed) and other land around Stafford was deliberately flooded (or allowed to flood) to improve the defences to the town (VCH VI 199). See also SHC 1914 113.

[271] Pennant 1782: 186 described Stafford as 'seated on a plain, bounded by rising grounds at a very small distance'.

[272] Gelling 1992: 153-5.

[273] Stafford existed as some form of settlement or industrial area from at least Roman times, evidenced by pottery kilns and at least two granaries of Roman date identified in excavations within the town, and the place was a centre for pottery-making in the ninth century, for Stafford ware has been found in Chester, Shrewsbury, Gloucester, Hereford and Worcester: see Gelling 1992: 153-5; Zaluckyj 2001: 214-7. The first documented references to salt production in Staffordshire date from the seventeenth century: VCH II 246-7. The absence of archaeological or early documentary evidence for salt-making associated with brine springs, the nearest of which appears to have been at Rickerscote, 1½ miles south-east of Stafford (the brine baths in Stafford, built in 1897 and demolished in 1977, were supplied with brine by a pipe from Stafford Common, 1 mile to the north-west: VCH VI 221), is curious, for it might be expected that the precious resource would have been utilised from a very early period, although the brine from this area is likely to have been much weaker than brine obtained from the traditional salt-producing areas of Cheshire and Worcestershire (see Hopkinson 1994: 5; 8). Furthermore, '[b]eing soluble, salt leaves no trace in the archaeological record [of the late Iron Age and early Roman period], except in very dry conditions not usually encountered in this country. Where produced by the 'open pan' method, salt produces few waste or by-products which may easily be recognised in the archaeological record. Silt (from settling of sediments contaminating brine), water vapour (from boiled brine) and fuel waste (from hearths) are all produced, but are hardly unequivocal identifiers of salt production. Additionally, few of the structures or implements used in salt production are solely characterised by that activity': Woodiwiss 1992: 183. It may be noted that of the wide range of pottery forms produced during the late Anglo-Saxon period, only one type appears to have been exported, namely a tall cooking pot (see WMA 26 1983 51), but it is unlikely that such pots were also used for the transportation of salt, and indeed improbable that if a salt-making industry had been established, even on a modest scale, it would have died out without trace.

[274] ASC 'C' and 'D', Earle & Plummer 1892-99: I 96-7.

[275] DB.

[276] SHC I 1.

[277] VCH VI 200.

[278] SHC I 35.

minting, usually in abbreviated form. The earliest abbreviations for Stafford taken from early coins are:[279]

| | |
|---|---|
| Æthelstan (924-939) | STFOR, STF, STEF, STED, ST |
| Edgar (959-975) | STAEð |
| Æthelred II (978-1016) | STAEFð, STÆTH, STAF, STÆ, STÆð, STÆFORA, STAE, STAEÐ, STAEF |
| Cnut (1016-1035) | STEF, STAE, STÆ, STÆðO, STÆF, STÆFFD, STA |
| Harold I (1035-1040) | STA, STAF, SAÆF, STF, STAEF |
| Edward the Confessor (1043-1066) | STAFORDE, STAFFO, STAE, STIE, STA |
| William I (1066-1087) | STAI, STEFFOR, STIEFF, STEFF, SIAEI, STAEFF, STAEF, STIEFR |
| William I or William II (1066-1100) | STEFFOR, STIFF, STAFFEO, STF |
| William II (1087-1100) | STAFRE, STF, STFRDI, STA, STAFFO, ST[?]D, STAFRED, STEIFR |
| Stephen (1135-1154) | STAFO |

Numismatic evidence is usually considered particularly reliable for early place-name spellings since coins will have been minted at the place, as opposed to evidence from documents which might have been compiled in a different part of the country by authors unfamiliar with the place about which they were writing.[280] The *d* in early spellings is not irregular philologically, nor its disappearance before *f*. There are parallels for -*d*- and -*t(h)f*- developing into -*f(f)*-, for example Blyford, Suffolk (*Blideford* 1086), Cloford, Somerset (*Clatford* c.1150, *Cloford* 1327), Parford, Devon (*Pathford* 739, *Patford* 1086, 1249); Trafford, Lancashire (*Statford* 1206, *Straforde* 1212).[281] The spellings for Stafford conform to this pattern. Occasional early spelling with *Ste*- could reflect West Midlands dialect *e* for *æ*, or, since they are generally found only in coins, may be attributable to epigraphical practice or accident: *Æ* on coins can appear very like *E*. Post-Conquest *e*-spellings for *æ* are common, as are *a*-spellings from the 10th century.

Addressing first the element 'ford' in the town name, it would be logical, if the word is to be given its conventional and usual meaning, to consider the crossing of the Sow on the south side of the town at Green Bridge (so-named by the 1590s, and probably to be identified as the 'great bridge' mentioned c.1200):[282] the axis of the town has long been the main street between the North and South Gates, and a crossing of some kind must have existed hereabouts from a very early date, certainly by the Roman period, and very probably in prehistoric times. However, the Old English word *ford* is often found where

---

[279] Compiled from Cherry 1890: 140-7; Mander 1944-5: 13-8; Gunstone 1971; Blunt 1973: 13-22; Early Medieval Corpus of Coin Finds, Fitwilliam Museum, Cambridge: http://www-cm.fitzwilliam.cam.ac/coins/emc_search_reply.php. No coins associated with Stafford earlier than the reign of Æthelstan are known: personal communication 4 April 2002 from Dr Anna Gannon, Assistant Curator of Early Medieval Coins, British Museum.

[280] From identifiable moneyers' names, it is clear that STAF was also used on coins for the more prolific Stamford (Lincolnshire) mint, and it seems possible that some of the forms given above are properly to be identified with that town. It may also be noted that there was a private mint at Stamford, Northamptonshire, to which some STA coins might properly be assigned.

[281] I owe these examples to Dr. Margaret Gelling.

[282] VCH VI 197 fn.64.

it is best translated as 'causeway',[283] and that may be the case here. In addition to the neck of land to the north of the town, there is some evidence to suggest that a causeway, which may have served as a dam at the southern end of King's Pool, carrying what is now Lammascote Road across the wet and marshy area on the east of the town beyond the former East Gate, long pre-dates the Anglo-Saxon period, and may have been constructed in the Iron Age,[284] in which case it will have formed a notable feature of the landscape, and is perhaps to be identified with the topographical formation from which the element *ford* is derived.[285]

As long ago as c.1600 the antiquarian Simon Degge deduced that: '...the true etymology is Stadeford, that is the strand, shore or bank of a ford, and we find it in Doomesday Booke writ Stadford',[286] and most modern studies adopt the prosaic derivation 'ford by a *stæþ* or landing-place'.[287] However, that derivation poses certain topographical difficulties. Old English *stæþ*, an uncommon place-name element,[288] has the meaning 'a shore, river-bank; land bordering on water'. The *ford* element, one of the most common topographical terms found in English place-names, and perhaps one of the earliest place-name elements, recorded for at least nine places by AD 730,[289] may be a later suffix (it makes its first appearance in abbreviated form on coins of the reign of Æthelstan, 924-939), although the use of the element as generic could be evidence that Stafford was one of the earliest settlements in the region to acquire its English name.[290] It has been suggested that the name might indicate that the place marked the limits of navigation on the river Sow, or imply that access to the settlement was primarily along the river Sow,[291] or indicate that the difficult road access on all sides of the town except the north 'may have led to regular use of the River Sow by local traffic, and this may have necessitated a

---

[283] See for example Gelling & Cole 2000: 72.

[284] TSHCS 1968-70 4-6. The conclusion is based on two red deer antler picks found in peat at depths of some 10' and 15' close together along the northern edge of Lammascote Road at approximately SJ 927232, traces of a possible Iron-Age hut in the lowest levels during the excavation of St. Bertelin's chapel in Stafford town centre, and evidence from pollen anlaysis at King's Pool of a period of forest clearance accompanied by agriculture in the early Iron Age: Bartley & Morgan 1990: 193. A parallel might be the enigmatic earthworks of unknown date at The Berth in Shropshire, which incorporate a causeway, and have produced pre-Conquest finds. Intriguingly, that site is considered to be the possible site of the legendary *Pengwern*, perhaps a successor to Viroconium before Shrewsbury became the county town, just as Stafford itself became a county town. Three worked timbers with pointed ends, identified as piles and radiometrically dated to the Iron Age, were discovered in 2002 in excavations on the site of a new filling station at ASDA in Queensway: *News From the Past 2002*, Stafford Borough Council, 2003 3.

[285] Wedgwood believed the track to the Staithe Ford came from Weston, where there was a ford across the Trent, and Uttoxeter, with a crossing of the Dove: SHC 1919 161.

[286] SHC 1982 74-5. William Harrison, the Tudor topographer, noted the variations in the early spellings of place-names, but chose to mention one only, Stafford, which he observed was in earlier times Stadtford: Harrison 1877-1909.

[287] For example Ekwall 1960: 435; Gelling 1984: 67; 70; Mills 1991: 305. 'There is no doubt about the identity of the word: it is *stæþ* ...': Gelling 1981: 8.

[288] Gelling 1981: 8. The element is found as a simplex name in Stathe near Athelney, Somerset, and in the dative plural in Statham near Lymm, Cheshire, and as the second element in Bickerstaffe, Lancashire; Birstwith, Yorkshire; Brimstage, Cheshire, and Croxteth, Lancashire, although in the last example the word is probably the cognate ON *stǫð*, since the first element is an ON personal name Tóki: Gelling 1981: 8.

[289] Cox 1975-6: 59.

[290] Gelling 1992: 153. For the early history of the town see Zaluckyj 2001: 214-6.

[291] Zaluckyj 2001: 214.

carefully maintained landing-stage'.[292] However, the river Sow is modest in both width and depth, and except in times of flood has probably always been so. It is difficult to believe that vessels of any size will have reached the town by water – certainly there appears to be little documentary or other evidence of medieval water-borne commerce[293] – and whilst small shallow-draught vessels may have been widely used, they would have had little difficulty discharging or taking on cargo almost at any point between Stafford and the Trent, some 5 miles to the east, of which the Sow is a tributary.

OE *hȳth* is the usual term used in place-names for inland ports,[294] and the use of *stæþ* in the sense of 'landing-place' is not otherwise evidenced until the 14th century, probably due to the influence of Old Norse *stoþ* which had that meaning, although it has been proposed that the meaning 'landing place' was an earlier meaning than 'bank' in Old English, and that it was revived from obsolescence by the influence of the Old Norse word.[295] Whatever the merits of that suggestion, it does seem improbable that a ford (in the usual sense of a shallow spot where a river could be waded) would be found at a place with a sufficient depth of water and sufficiently high banks to serve as a landing place or quay. Identifying *stæþ* with 'landing place' is in any event problematic, for some places which are held to contain the element are in topographically inappropriate situations, e.g. Bickerstaffe, Lancashire, which is not on a river, but which stands on a sandstone ridge,[296] and Brimstage in The Wirral, which is on a very small watercourse.[297] Such names may, it has been suggested, refer to 'settlements in marshes where communications were by log boat for part of the year'.[298] The early topography of Stafford would certainly support such a meaning.

Alternatively, if the Lammascote causeway is to be identified as the *ford* element, it might also hold another explanation for *stæþ*, for if the King's Pool existed (even if only as particularly wet marsh) in the centuries before the Conquest, the sense 'land bordering on water' would fit the causeway, the full interpretation of the name Stafford being 'the causeway bordered by water or very wet ground', or 'the causeway formed by a partially artificial bank'. It is unclear whether the ground from the south side of the Lammascote causeway to the river Sow was at any time permanently under water, but the ground is certainly low-lying, and at the very least would be very wet and marshy. Militarily, a settlement in very marshy ground is much more of an obstacle to attackers than open water, and less vulnerable than hilltop fortifications to capitulation under seige by reason of thirst and hunger.

The original name *Stæþ* or *Stæþford* was perhaps connected with Old English *stapol* (ME *staddle*) 'foundation, base, support' (compound Old English words beginning *stæþ* have a meaning associated with stability and firmness),[299] which may simply have denoted a paved ford or causeway or a settlement on firm ground by a river crossing. The meaning 'the firm road or causeway (through wet land)' might perhaps be associated with the

---

[292] Gelling 1992: 153.
[293] Plot 1686: 43 states that the river Severn was the only navigable river in Staffordshire.
[294] Gelling 1984: 76-8.
[295] Gelling 1981: 8; 1984: 80-1.
[296] It has been suggested that 'the interpretation of the second element in Bickerstaffe would more sensibly be that of 'place' rather than 'landing place' or river bank" [implying a derivation from ONorse *staðr*]: N. J. Higham 2002: 27.
[297] Gelling 1984: 80-1; PN Ch IV 235.
[298] Gelling 1981: 8.
[299] Cf. Staddle Bridge, West Yorkshire: VEPN II 56.

previously mentioned suspected Roman road running south towards Pennocrucium (Water Eaton) from Leek via Blythe Bridge which is likely to have crossed the river Sow at or near Stafford.[300] Indeed, if such a road, which would have been an important strategic north-south route, was still in use in the tenth century, it would help to explain why Stafford was chosen as the site of an Ælflædian *burh* in 913.[301] But if it was in use at that date, the reason for its subsequent disuse both north and south of Stafford remains to be explained.

Whatever the original meaning of the name, it seems that it evolved to reflect the more intelligible 'place of the ford (or causeway through wet ground) marked by stakes or posts', from Old English *stæf.* There is other place-name evidence of the use of planks or posts to improve or mark a river-crossing: Stapleford (found in at least eight counties) denotes the existence of posts, and Stakenford Bridge (Worcestershire), indicates a ford with stakes.[302]

Tradition says that before it acquired its present name, Stafford was called *Betheney* or *Bethenei* or *Bethnei*, by tradition after St. Bertelin or Berthelin (Beorhthelm), a legendary Saxon hermit and disciple of St. Guthlac who supposedly had a hermitage here,[303] but such a name could not in philological terms have had any association with St. Bertelin, whose cult seems to have been real even if the legend is baseless.[304] The name Betheney (like Broadeye in Stafford) may well, if genuinely ancient, have as its second element Old English *eg* 'island' (which would serve as further evidence of the wet areas surrounding the place), with an Old English personal name such as Betti as the first element,[305] but the derivation may be associated with the biblical Bethany, a village two miles or so east of Jerusalem on the south-east slopes of the Mount of Olives, frequently mentioned in connection with memorable incidents in the life of Jesus, and now known as el-Azariyeh or Lazariyeh.

### Angles, Saxons and others

Evidence of differing ethnic groups in Staffordshire may be detected in the place-names Engleton ('*tun* of the Angles'), which Stenton suggests may refer to East Anglian settlers in Mercian territory,[306] Normacot ('Northman's or Norwegian's cottage'), Seisdon ('Saxons' hill'), and possibly Wychnor (perhaps 'river bank of the Hwicce').[307] The

---

[300] Jermy & Breeze 2000: 109-10 suggest that Welsh *fford* may be found in English place-names such as Longford (q.v.), and have the meaning 'Roman road'. But such names probably incorporate OE *ford*, from which the Welsh word is derived, applied to particular sections of Roman roads which had been built on a causeway.

[301] ASC 'C' and 'D'.

[302] PN Wo 150. See also Stableford.

[303] Plot 1686: 409; SHC VIII (ii) 146; SHC 1914 112; VCH III 136; VCH VI 186; Oswald 1955: 7-9.

[304] Thacker 1985: 1-25.

[305] Ilam church has a shrine to St. Bertelin. The association of St. Bertelin with Stafford may be attributable to Æthelflæd, who built a burh at Runcorn (where there is another Bertelin dedication), in 915, two years after that at Stafford: VCH Ch I 253. Another church dedicated to the saint is at Thurstaston in The Wirral. The creation by Æthelflæd of Mercian cults (such as St. Milburg and St. Werburg) in newly founded *burhs* such as Gloucester is well recorded, and has been held to be a significant part of her strategy: Thacker 1984: 199-211.

[306] Ekwall 1960: 167.

[307] For a survey of racial and tribal names see Ekwall (a) xli.

kingdom of the Hwicce, first recorded in the later 7th century, does not appear in documents after Offa's reign, when it was subsumed into Mercia. Its territory extended into north Worcestershire, and its people may have had a Celtic ancestry.[308]

An analysis of the names included in this work has not produced any evidence to alter the generally accepted view that the Anglo-Saxon penetration of Staffordshire was led by the Angles penetrating from the east along the river Trent and its tributaries, and to a lesser extent the Saxons moving in from the south-west of the county.

## Early Christianity

The conversion of the pagan Anglo-Saxons to Christianity took place in Staffordshire from at least the time of Chad in 669.[309] Chad was given Lichfield as an episcopal seat by Wilfrid, who received it and other places in Mercia from Wulfhere,[310] the Mercian king who by tradition had a palace at Bury Bank,[311] the great earthworks near Darlaston, Stone, recorded in early records as *Wulfercester*, 'the camp or city of Wulfhere'.[312] Lichfield thus became the first seat of the peripatetic bishops of the Mercians, but having been raised to an archbishopric by King Cenwulf in 788, it reverted to a bishopric in 803, and its fortunes declined as Mercia gave way to Wessex in political importance. Evidence for early Christianity in the county is provided by the name Eccleshall, which (from the Eccles element, representing the Welsh word for 'Christian community') indicates the existence of an early Celtic Christian community, perhaps surviving from the Romano-British period. Eccleshall, one of the largest parishes in Staffordshire, was the centre of a number of estates held by the bishops of Lichfield from an early date – perhaps as early as the third quarter of the seventh century – and the name suggests that the bishops were attracted to the place by the existence of an established Christian community, or chose to re-assert Christianity from a place named from a former Christian community. No firm evidence has been traced for the continuity of Christianity at Eccleshall from the fifth century to the seventh century,[313] but such continuity is not improbable.

The other place-name element sometimes associated with early Chritianity is *stow*, meaning in many cases simply 'a place', but often with the specialised meaning 'a holy place'. The element occurs in three places in Staffordshire: Stowe in Lichfield; Stowe by Chartley, and Stow Heath, on the east side of Wolverhampton.

Stowe in Lichfield has long associations with early Christianity, being the reputed site of a hermitage established by St Chad in the seventh century. The ecclesiastical history of

---

[308] Lapidge *et al* 1999: 246.

[309] Chad's appointment may have followed a long period of missionary work undertaken from a Celtic community in the Lichfield area: TSSAHS XXII 1980-1 3. That might explain the enigmatic statement that Wulfhere gave Lichfield to Bishop Wilfrid in the 660s as 'a place made ready [paratum] as an episcopal see for himself or for any other': Colgrave 1927: chapter 15.

[310] Colgrave 1927: chapters 14, 15 and 17.

[311] See Shaw 1798: I 39, who also suggests that the Iron-Age hillfort at Kinver was constructed by Wulfhere. No archaeological evidence of Anglo-Saxon occupation has been traced at Bury Bank (Studd 1993: 55), or at Kinver.

[312] Lichfield lies 22 miles from Bury Bank.

[313] Eccleshall church contains a small Anglo-Saxon relief built into a wall, probably a fragment from a cross-shaft, showing a horseman tentatively identified as St Chad (Studd 1993: 59), although the fact that the figure holds a spear makes this unlikely: see Zaluckyj 2001: 74.

Stowe by Chartley is unclear – the church of St John is essentially Norman in date – but there is no evidence that the name has the meaning 'a holy place'. Stow Heath is the largest of the three manors of Wolverhampton,[314] and lies on or close to the 'plain of Wednesfield', the reputed site of a great battle between an English army of Mercians and West Saxons which defeated the Danish Army in about 910 AD.[315] The name Wednesfield incorporates the name of the pagan god Woden, and it is not inconceivable that a Christian centre was deliberately established at Stow Heath to exercise influence over a local centre of paganism, if indeed the name is properly to be seen as evidence of pagan worship. There is some slight evidence that Wulfrun, who founded, or perhaps more accurately refounded, the monastery at Wolverhampton (a place which a century or so later incorporated her own name) originally held land in or near Stow Heath,[316] and it is not inconceivable that Stow Heath may mark the site of the original monastery.

One other name which may imply Christian connections is Gads Hill near Biddulph, previously mentioned. It has been suggested that such hill-names may have been associated during the Anglo-Saxon period with pagan practices or beliefs, and been given names reflecting such activity, but were later renamed to reflect the abrogation of pagan customs.[317]

## The Hundred names and other meeting-place names

Even though all are attached to insignificant places – indeed in at least one case can no longer be precisely located – there are five Staffordshire place-names that demand particular consideration. The names are those of the five Staffordshire Hundreds of Cuttlestone, Pirehill, Totmonslow, Offlow and Seisdon. The Hundreds (in some Northern counties known as Wapentakes) were local administrative units consisting of an area of land[318] served by a Hundred-court at the Hundred meeting place.[319] Whilst the county boundary utilises natural features, it is noticeable that the Hundred boundaries, when they do not share the county boundary, make few attempts to follow landscape features. It has been noted that the typical Hundred meeting place was located away from settlements in

---

[314] Two of the manors (The Deanery and The Prebends) belonged to the church; Stow Heath was held by the king: Mander & Tildesley 1960: 25.

[315] Earle & Plummer 1892-99: I 94.

[316] From a 13th century reference to *fossatum Wulfrini* 'the entrenched place of Wulfrini' (Mander & Tildesley 1960: 28), perhaps the *vallum monasterium* of a monastic site. Wulfrini is almost certainly to be read as Wulfrun, from the not unusual misreading of minims. No name Wulfrini has been otherwise traced, and is not included in Searle 1897.

[317] Stenton 1970: 286 fn.2.

[318] Usually taken to be one hundred hides, but Yorke 1995: 125 has observed that the word Hundred is likely to have been borrowed, together with much royal ligislation, from Carolingian Francia, where the *centena* was a comparable local administrative unit, and so need not be interpreted as a literal measurement.

[319] Some of the duties of the Hundred courts, which met every four weeks, are set out in the mid-tenth century Hundred Ordinance: thieves were to be pursued by all the chief men of the Hundred, those who failed to appear on appointed trial days, and those who opposed the decisions of the Hundred court, were to be fined, and if they persisted, outlawed: see Lapidge *et al* 1999: 243. The five Staffordshire Hundreds are further evidence for the relative poverty of the county: Norfolk had 34, Shropshire 15, Worcestershire 12, and pre-Conquest Cheshire 12 (Studd 1993: 58). It has been suggested that Wolverhampton and its members formed a separate Hundred and was independent of the Hundred courts: Mander & Tildesley 1960: 5.

'neutral' territory, often near a main road or river crossing.[320] Of the Staffordshire meeting places, Offlow and Seisdon are more or less centrally placed within their respective Hundreds,[321] but Totmonslow is within a mile or so of the boundary with Pirehill Hundred.[322] In size, the adjoining Cuttlestone and Seisdon Hundred in the south-west of the county are about the same, and also the two smallest Hundreds, with Pirehill in the north-west the next largest. Totmonslow and Offlow, which adjoin and form the eastern part of the county, are about the same area. The size of the Hundreds in the north of the county has been ascribed to the inclusion of large tracts of unproductive upland.[323]

The age of these meeting places is not known, though the English names indicate that they are unlikely to be earlier than the seventh century, and there is even disagreement about the period at which the Hundreds – which the meeting places could well predate – were formed.[324] By the time of Domesday, only one of the Hundred assembly places (Seisdon) was of sufficient importance to be listed as a vill in its own right, which might suggest that in 1086 the other places were insignificant, although it is also true that several sizeable places known to have existed in 1086 are ignored by Domesday Book. What is very noticeable, however, is that with the possible exception of Pirehill, all the Staffordshire Hundred meeting places lie on or very close to Roman roads,[325] and in three cases close to junctions in those roads. That can be seen as evidence that most, if not all, of the Staffordshire Hundred assembly sites post-date the creation of the roads.

Typically, Anglo-Saxon meeting places were located in relatively out-of-the-way places away from other settlements and named after nearby prominent natural or topographical features. Indeed, those same features may have been sufficiently distinctive and well known in their own right to make a place a suitable choice for assemblies to be attended by participants who might need to travel some distance. Two of the five Staffordshire meeting-places (Offlow and Totmonslow) are, not untypically, named after mounds which may have been prehistoric tumuli, but archaeological evidence from the majority of meeting-place mounds so far excavated suggests that they were often created by the Anglo-Saxons to serve as assembly points, and do not support the view that it was the usual practice of the Anglo-Saxons to utilise tumuli for this purpose, although this did sometimes happen.[326] Two more meeting-places (Pirehill and Seisdon) are named after hills, and the fifth (Cuttlestone) takes its name from a stone. All the places are either on or close to high (or relatively higher) ground, and it may be that fire and smoke signalled

---

[320] Gelling 1988: 210. Meaney 1995: 29-42 places the earliest Hundred meeting places in three categories: primary, such as fords where people would expect to meet others when travelling; secondary, such as stones and other landmarks which may also have served as places of pagan sacrifice; and tertiary, such as pillars which might indicate places where heads were displayed before the coming of Christianity.
[321] Assuming that the early Hundred of Seisdon extended to the Severn.
[322] Thorn 1991: 27 fn14 suggests that Totmonslow and Pirehill may have formed a single 'long' Hundred of 120 hides, given the small size of Totmonslow (18 hides 25 acres) and the fact that its meeting place was not centrally placed.
[323] VCH IV 2.
[324] 'The Hundreds themselves often bear names of an earlier form, and met at places whose names suggest that they may have been places of assembly from early times… The later Hundred seems to have taken over the meeting-place as well as the function of the earlier popular assembly': Whitelock 1952: 138.
[325] The close relationship between Hundred meeting places and Roman roads has been noted elsewhere, for example Cambridgeshire (Meaney 1997: 228), and Leicestershire (Cox 1971-2: 14-21).
[326] Gelling & Cole 2000: 178-9.

the precise location to travellers on meeting days. None of the meeting places appears to have been visible from any of the others. All five meeting places are close to settlements which have place-names probably or possibly incorporating Celtic elements, i.e. Totmonslow (Cheadle), Offlow (Lichfield), Pirehill (Walton), Cuttlestone (Penkridge), and Seisdon (Seisdon).

The ancient mound of Offlow, now ploughed to destruction, lay 500 yards north of Watling Street less than a mile from its intersection with Ryknild Street, on the west side of a rounded hill which rises to 367', and within yards of a parish boundary on the south. Totmonslow lies on a high point of the Roman road from Rocester to Stoke on Trent,[327] two miles east of Blythe Bridge, where it is crossed by a lost Roman road from Leek. Cuttlestone Bridge, which now tends to be treated (though without firm evidence) as the Cuttlestone meeting place, is 2 miles north of Watling Street, but it is possible that the actual meeting-place stone lay closer to Watling Street, where no fewer than eight Roman roads intersect, with another untraced road probably running north-east. Seisdon lies 1 mile north of a lost Roman road from Greensforge which runs below the ridge of Abbot's Castle Hill,[328] and within a mile of the junction with the supposed line of a possible Roman road[329] running south from Giffard's Cross near Brewood which would meet the ridge of Abbot's Castle Hill where the county boundary bends one mile south-east of the western end of the ridge at Hillend.

No Roman road has been traced with certainty near Pirehill, a prominent headland with far-reaching views to the east, north and west,[330] but it lies just over two miles from a postulated Roman road running between Blythe Bridge and Stafford ,[331] and a mile or so

---

[327] Margary number 181.

[328] Margary number 192. Another road from Greensforge to mid-Wales which ran below Abbot's Castle Hill has also been traced: TSAS LVI 1957-60 237-40.

[329] Horovitz 1992: 34-5.

[330] In his study of English Hundred names, Anderson 1934: I 147 states that 'Pire Hill is the highest point for some distance; there is none higher between it and the river [Trent], and it seems to have a good view down the Trent valley'. However, Peasley Bank less than a mile south-west is 541' high (considerably higher than Pire Hill), is about the same distance from the Trent, and enjoys commanding views as far as Mow Cop and the high ground of Needwood, as well as to the south beyond Stafford. Browne's 1686 map of Staffordshire (in Plot 1686) shows two hills, the northen one named *Pyre Hill*. There is no name attached to the southern one. Bowen's map of 1749 is equivocal, since only one hill, captioned *Pire Hill,* is shown south-west of Aston. Yates' map of 1775 shows the northernmost of two hills as *Pyre-hill Hill*, with *Pyre Hill* on the west side of the southernmost hill, suggesting that the northernmost hill was 'the hill at (or near) Pirehill'. Yates' map of 1798 shows only the one name, *Pyre Hill*, at what is now known as Peasley Bank. The hill now known as Pirehill is shown but un-named. Teesdale's map of 1832 shows that hill as Pyre Hill. The 1836 1" O.S. map shows two Pirehills, one a short distance north-west of Peasley Bank, the other at the hill which now bears the name, and *Hundred-acres* (q.v.) on the west side of Peasley Bank, which suggests an association with the Hundred meeting place. A junction of three parishes lies on the south-east side of the hill. It seems possible, indeed likely, that the Hundred meeting place lay somewhere near the summit of Peasley Bank, perhaps at this boundary junction or close to what is now Pirehill Grange Farm, at SJ 896302, with the Stafford to Stone road less than a mile to the east. See also observations in Burne 1913: 47-8. The name Peasley Bank indeed might seem more appropriate for a road than a hill: cf. nearby Yarlett Bank for the steep incline on the Stafford-Stone road.

[331] The Independent, 11 June 1991, reported the discovery of a Roman road from Blythe Bridge via Fulford and Hilderstone to an undated earthwork at Hollywood (NGR SJ 932034), and from there to Common Road on the north side of Stafford; see also SOTMAS I 10-17; VCH VII 98; map in Phillips 1993: 50. The existence of such road has to date not been verified by archaeological

from a cropmark of a rectangular enclosure with rounded corners on the west bank of the river Trent, south of Aston-by-Stone, which has the distinctive appearance of a temporary Roman marching camp, and from which site a Roman coin is recorded.[332]

In the mid-14th century there was seemingly a place called Cuttlestone,[333] but its location remains unknown. Stebbing Shaw states that:

'there is no appearance of any town or village from whence [Cuttlestone] Hundred takes its name, but only a bridge so called over the river Penk', adding 'Tradition says, that at or near Cuddleston bridge was once a town or settlement of considerable magnitude and note'.[334]

As in Shaw's time, the name, which means 'Cuþwulf's stone', is preserved only in Cuttlestone Bridge,[335] one mile south-west of Penkridge which carries 'an ancient thoroughfare called King Street[336] across the river Penk'.[337] The exact location of the Hundred meeting place, which was presumably a prominent stone, possibly a glacial boulder or prehistoric megalith[338] near the bridge (which itself is not associated with the name until the 13th century), is not known, but there is no evidence that the river-crossing itself was the meeting-place (where an ancient name incorporating *ford* or *bridge* might be expected), although Hundred meeting places are not infrequently at the junction of roads and rivers, and five roads meet at the river crossing here. However, all other Staffordshire Hundred meeting places are on hills: Offlow 355', Pire Hill 463' (or Peasley Bank 541'), Seisdon over 300', and Totmonslow 588'. The possibility cannot be discounted that the meeting place was a prominent stone on a nearby hill, perhaps what is now known as Beacon Hill (389'), 1 mile south-west of the bridge, which has, as the name implies, far-reaching views, and across which passes a lost Roman road running slightly to the east of north from The Ivy House on Watling Street and through Whiston Mill.[339] In that respect, it is worthy of note that Old English *becun* had a range of

---

excavation. VCH I 192 records that 'In the meadows near Hilderstone Brook is another earthwork with a double fosse, the outer one representing a quadrilateral figure of 200 yards'.

[332] At NGR SJ 915309; see VCH I 192; Staffordshire Newsletter, 20 August 1993; Darlington 1994 11, 23; SMR 04606. Evidence of a Roman presence in the area is provided by a cache of 8 Roman coins found in the bank of Jolpool Brook near Yewtree Farm at Burston at SJ 938302: SMR 01810.

[333] VCH IV 61. Bowen's road book of 1720, which seems otherwise accurate, shows *Cudleston* on the south side of Watling Street to the north of Aldridge, for which no explanation can be offered. An early Shropshire Hundred recorded in Domesday Book was Culvestan (which included Hope Bowdler, Munslow, Church Stretton, and part of Cardington), and the name probably has the same derivation as Cuttlestone. The Hundred was combined with Patton Hundred to form Munslow Hundred in about 1100: PN Sa III 277-8.

[334] 1801: II 291. No archaeological evidence has been traced to support the claim.

[335] NGR SJ 9113, *Pontem de Cuthuluestan* early 13th century SRO D260/M/T/5/139, *(pons de) Cuthulueston'* 1225-59 Deed, *Cothelstonbrugge* 1307 SHC VII 179.

[336] Which ran westwards from the bridge: Yates 1798; O.S. 1833, the latter giving *King's Street* for the road running east to west to the north of Marston, and showing *Kingstreet Grange* 1 mile north-east of Sheriffhales. It is possible that the road is to be identified as an Anglo-Saxon Hundredway or Mootway, sometimes found mentioned in later records: see Bigmore 1979: 47-9. See also TSAS 1st Series I 1879 359.

[337] Duignan 1902: 48. King Street is shown on modern maps running east from Great Chatwell.

[338] Palliser 1976: 51.

[339] Horovitz 1992: 32. It is suggested in SHC 1916 139, 151, without explanation, that the name Cuttlestone Bridge may have been applied originally to the bridge carrying Watling Street over the river Penk near the Roman settlement at *Pennocrucium* (Water Eaton), 2 miles south of here. No

meanings including 'sign, signal, cross, memorial stone', and perhaps 'tumulus'.[340] Rather than a beacon site, the hill-name might conceivably record the existence of a lost stone or monument, possibly Cuþwulf's stone itself. It may be noted that Cuttlestone Bridge lies some two miles north of the Roman settlement of Pennocrucium (Water Eaton), the name of which, it has been suggested, 'could signify a special assembly point for the tribe, or a place of some local importance'.[341] That raises the possibility that the area (if not the actual meeting place) was the site of assemblies from pre-Roman times,[342] and was associated in the post-Roman period with the Anglo-Saxon people known as the Pencersætan whose territory may have been adminstered from the Penkridge area.[343]

The Hundred name Totmonslow is from 'Tatmann's *hlaw* or tumulus'. The Old English personal name Tatmann, recorded in charters of 947[344] and 963,[345] is found later as Tateman,[346] and may possibly be associated with Old English *tot-mann* 'look-out man, watchman', an occupation perhaps connected with the 665' hill that lies close to Totmonslow on the north-west. The actual mound or tumulus to which the name relates has not been identified: it is unlikely that it is from the nearby hill itself, since in Staffordshire the term *hlaw* is invariably associated with tumulus-sized mounds, although it is possible that such a mound once existed on the hill. A tumulus is recorded on Oakhill, a hill[347] of over 600' half a mile south-east of Totmonslow, and this is perhaps a more likely candidate. Shaw implies that there was formerly a tumulus at Totmonslow, 'though not now extant', but that may be mere supposition from the place-name.[348] It has been suggested that Totmonslow Hundred, the northernmost in Staffordshire, may have formed the southern part of the territory of the Anglo-Saxon peoples known as the Pecsæte. [349] The look-out association with this name is probably also to be found in the Hundred name Pirehill. Of further interest is the name Mobberley, attached to a small settlement less than a mile north-east of Totmonslow. If ancient (and it is possible that the name has been transferred in relatively recent times from Mobberley in Cheshire), the derivation may be from Old English *gemot-beorg leah* 'wood or clearing at the mound where moots or assemblies were held', perhaps to be associated with the large natural flat-topped circular mound at Mobberley[350] called Castle Croft.[351] The significance of the

---

evidence has been traced to support that suggestion, although the bridge does mark the meeting place of three parishes. It may also be noted that a parish boundary running north from this bridge follows the river Penk as far as a former ford below Beacon Hill; see also Shaw 1798: I 31.
[340] VEPN I 68.
[341] Webster 1991: 78.
[342] Aerial photographs show the corner of an earlier double ditched rectangular structure at an oblique angle to Watling Street and partly underlying (and so pre-dating) the north-west corner of the civil settlement that sits astride Watling Street: see Current Archaeology 145 (November 1995) 23. The structure appears to have been in existence before Watling Street itself, since it lies at an angle to that road.
[343] It is possible that Cuttlestone Hundred represents the ghostly remnants of the northern part of the territory of the Pencersten, and that the area was administered from Penkridge, which would explain why the charter of Edgar dated there in 958 (S.667), mentions *in loco famoso qui dicitur Pencric* 'in that famous place called Pencric'.
[344] S.714.
[345] S.525.
[346] 1190-1200, 1195 Anderson 1934: 147.
[347] Possibly *Tattemaneshull'* recorded in 1252 (Fees).
[348] 1798: I 37.
[349] See Higham 1993: 176; Studd 1993: 58; Brooks 2000b: 214-237.
[350] At SK 006413.

47

proximity of such a possible meeting place and the nearby Hundred meeting place is unclear, but it is not inconceivable that the Totmonslow meeting place was the mound at Mobberly.

Offlow Hundred takes its name from 'Offa's *hlaw* or tumulus', a mound 2½ miles south of Lichfield, in Swinfen on the northern boundary of Shenstone parish, 500 yards north of Watling Street. The mound, now obliterated by ploughing,[352] is well recorded by early historians,[353] but is not known to have been excavated. It was probably of Anglo-Saxon date, since its proximity to Watling Street may not be coincidental, and mounds to which Old English personal names have been applied may be more likely to date from the Anglo-Saxon period, but the possibility that it was a prehistoric tumulus cannot be ruled out. There is no evidence of any connection with the great Mercian King, Offa,[354] who died in 796, long after the pagan period from which mound-burials date, but it is not inconceivable that it was the burial mound of the Mercian royal family at the time when it was still pagan.[355] The place lies a little over a mile from the site of the supposed pagan shrine at Weeford, one mile from Freeford, the name of which might signify a pagan site, six miles from Tamworth, Offa's seat of power, and a mile or so from the Roman site of Letocetum (Wall). The northern boundary of Offlow Hundred may represent the territorial limits of the Anglo-Saxon tribal group known as the Tomsætan, whose name was taken from the river Tame.

Seisdon Hundred[356] covered the south-west corner of the county, which in 1086 may have extended as far as the river Severn, excluding Quatford.[357] The meeting place lies close to the present Hundred (and county) boundary,[358] and is less than a mile from the long ridge (with its ancient and enigmatic linear earthwork, perhaps Anglo-Saxon in date) on Abbot's Castle Hill,[359] which marks those boundaries. The meeting-place is probably remembered by fields to the east of Seisdon called *Musters*,[360] recorded as *le moustowe* (from Old English *(ge)mot* 'assembly' and *stow* 'a (sacred) place') in 1298,[361] meaning 'meeting place': one of Seisdon's open fields is recorded as *Mustowe field* in 1518,[362] *Mustowefyld* in 1549,[363] and *Musters* in 1842.[364] Associated with this area are

---

[351] SHC 1926 159-65. The earthworks are said to have been known also as *Huntley Castle, Mobberley Baen* and *Castle Cop* (*ibid.*).

[352] JNSFC LXVIII 1933-4 154-5.

[353] See for example Shaw 1798: 1 37.

[354] Who is said to have died on 29 July 794 and been buried in a chapel on the river Ouse, near Bedford: John ('Florence') of Worcester, DNB; but note also the doubts expressed by Shaw (1798: I 42) as to the site of Offa's burial, and Zaluckyj 2001: 161 (who gives Offa's date of death as 26 July). Offa was not an uncommon name: no fewer than 15 individuals so-named are recorded in Searle 1897: 364.

[355] Gelling 1988: 155-6; VCH XIV 5.

[356] Early references to the Hundred include *Saisdon(e) hvnd'*, *Seiesdon hvnd'* 1086 (DB), *hdr' de Saiesdona* 1130, *Seidon'hundredum* 1182, *Seisdon'hundredum* 1185 P, *Hundredum de Seisdon'* 1199 P, 1226-8 Fees, *hd of Seisduna* 1227 Ass, *Seysdon'* 1255 RH, *hd of Seylesdon* 1272 Ass.

[357] See Croom 1989: 306-15.

[358] But would have occupied a central position if the county had extended to the Severn. Indeed, the position of Seisdon might be seen as evidence for the county so extending.

[359] Plot 1686: 387; VCH I 372.

[360] WSL Misc 339 5.

[361] VCH XX 185.

[362] Ibid. 185, 190.

references to *Ploustowmere*, recorded c.1300,[365] and *Plewestowe green*, mentioned in 1549,[366] from Old English *plege-stow* 'a place used for sports and games'.

Other evidence of places of assembly may be detected in a cluster of meeting place-names in the Mayfield area. Mayfield is in north-east Staffordshire at a crossing of the river Dove, which forms the boundary with Derbyshire. The name Mayfield itself has been held to incorporate Old English *mǣddre* 'madder',[367] but botanists say that the plant is unlikely here, and the early spellings make a derivation from Old English *mæthel* 'meeting, council'[368] with Old English *feld* 'open land'[369] quite certain. The meeting place or places cannot now be identified, but the presence nearby of Harlow (for which early spellings have not been traced, but unparalleled in Staffordshire and perhaps from Old English *here* '(Viking) army, host, multitude', but also used for 'the whole people', with Old English *hlaw* 'mound, tumulus', so perhaps 'the mound associated with the Vikings', or 'the mound where the people met'),[370] and Motcarn Sprink (the age of the name being unknown, but possibly from Old English *mot* 'a meeting, an assembly', with Welsh *carn* 'a heap of stones, a cairn', so 'spring at the cairn where assemblies took place')[371] tend to suggest that at some period in history the area played host to assemblies of some importance. From the reference to Mayfield in Domesday Book, and the absence in early records of any mention of meetings here, it can be concluded that the gatherings took place before the late Anglo-Saxon period, quite possibly before the creation of the county of Stafford, in which case they may have been associated with the people known as the *Pecsǣtan*, 'the people of the Peak', mentioned in the mysterious document known as the Tribal Hidage, an 11th-century manuscript of earlier material listing 34 Anglo-Saxon kingdoms south of the Humber, and in a charter of 963 relating to Ballidon, Derbyshire.[372] The first edition 1" Ordnance Survey map of 1836 shows that the

---

[363] SRO D740/3/6. See especially Pantos 2003: 48. Gelling 1988: 210 in contrast has suggested that the meeting place was the 300' hill to the west of the village, called c.1300 *Penn hill*, from British *penn* 'head, end, headland'. By the late 16th century it was called *Round Hill* or *Whitney hill.*

[364] Pantos 2003: 48 fn.8.

[365] VCH XX 185.

[366] Ibid. 191.

[367] See for example Ekwall 1936: 110, 1960 318; Gelling & Cole 2000: 275.

[368] Which Clark suggests is one of the rarest place-name elements: Jackson 1995: 224. Aliki Pantos of St. Hugh's College, Oxford, (to whom I am indebted for sharing parts of her doctoral research on Anglo-Saxon assembly places) has traced 11 other place-names incorporating the element *mæthel,* 6 of which are also the names of parishes, and concludes that these sites were important meeting places named at an early date, and so more likely to give their names to parishes than tenth- or eleventh-century Hundred meeting places which only came into existence once the settlement pattern was well established. That conclusion cannot necessarily be applied to the Staffordshire Hundred meeting places, which might long pre-date any well established settlement pattern.

[369] As proposed by Duignan 1902: 100-1.

[370] See Nicolaisen *et al* 1970; Ekwall 1960: 220. The mound or tumulus may be that which lies ½ mile south-west of Mayfield at SK 143005, perhaps that called *The Low* in 1677: SRO D1134/1/1. A tumulus called Mayfield Low, which cannot now be identified, was excavated in 1849 (VCH I 31), and tumuli called *The Rowleys* (*Rowlows* c.1765: SRO D626/A/10/1-13) are recorded in *Mayfield* in 1916 (SHC 1916 208). Supposedly pre-Conquest earthworks known as The Cliffs and Hollow Lane have also been recorded in Mayfield (ibid. 207).

[371] *Mot* in Old English and Old Scandinavian also meant 'stream junction', and since the place lies in the junction of two streams, there must be a strong possibility, if not likelihood, that the name in fact is so derived.

[372] Brooks, Gelling & Johnson 2000. Ballidon is some 16 miles north-east of Mayfield. Whether the *Pecsǣtan* were British or Anglo-Saxon is unclear. It may be significant that the only sizeable part of the northern Danelaw in which Edward the Confessor held a demesne was in West Derbyshire

Staffordshire boundary then extended beyond the Dove into Derbyshire opposite Church Mayfield. If the deviation is ancient, it may mark the former course of the river which remained as the county boundary,[373] or simply reflect the not unusual arrangement whereby one authority had jurisdiction over both banks at a river crossing. Another possibility is that it may show that the river Dove served as a territorial boundary before Staffordshire was created, and two neighbouring peoples met on some 'neutral' territory on the east side of the Dove – it has been observed that 'it was a well-established custom for [Anglo-Saxon] kings to negotiate with one another on the boundary between their territories'[374] – but pending further research (which might reveal other lost place-names to reinforce the present indications that this area was an important area of assembly in Anglo-Saxon times),[375] these matters must remain conjectural. Matlock in Derbyshire, 18 miles north-east of Mayfield, also incorporates the element *mæthel*, and stands on an important crossing of the river Derwent. Both Mayfield and Matlock might be thought of as strategically significant places, and a major river crossing would be a place where travellers would naturally tend to congregate. The noticeable absence of ancient markets at meeting places, where trading would otherwise have been expected, might be explained if the meetings were limited by custom to appropriate representatives to conduct formal legal, judicial and administrative business, and were not generally attended by non-participants in any numbers, perhaps to prevent intimidation or confrontation, or because distracting commercial activity or entertainment was discouraged or forbidden.[376]

The only other place-names which might provide evidence of early meeting-places in Staffordshire are *Spellowe Field*, on the east side of Alrewas, and (more doubtfully) Mottley Pits on the north-west side of Stone.

**Danish influence**

Evidence of certain, likely or possible Scandinavian influence in Staffordshire place-names is rather more widespread than the traditional picture of Staffordshire outside the direct influence of the Danelaw might lead one to expect, although only a single -*by* name (Thevesby) has been traced, and that only in a single reference, and there is no

---

between the Noe at Hope and the river Dove at Alstonefield [i.e. opposite Mayfield]: Stenton 1970: 163; VCH Db I 297, 312.

[373] Plot 1686: 43 says: '... and the Dove sometimes will change its channel; which I suppose ha's been the cause that a part of Staffordshire in the parish of Mathfield lyes on Derbyshire side of the River, and a little below (near the bridge you pass over to Snelston) a part of Derbyshire on the Staffordshire side.'

[374] Stenton 1971: 332.

[375] The earliest spelling so far traced for Marten Hill, an 853' hill 1 mile north-west of Mayfield, is *Marting Hill* (1775 Yates). The name is perhaps from OE *gemǣrtun* 'boundary *tun*': three parish boundaries meet there. But the distinctive 1775 spelling suggests that the name may incorporate OE *þing* 'meeting, court of justice', a word normally limited to the Danelaw, so perhaps 'the meeting-place or tribunal at the boundaries'.

[376] It should be recorded that a cluster of minor names incorporating Rice, almost certainly from OE *hris* (or ON *hrís*) 'brushwood', can be found in the area around Mayfield, e.g. *Mafeild Ryse* recorded in 1538 (Survey), Bank Rice, Fords Rice, and two examples of *Rice* marked on the 1836 first edition 1" O.S. map, perhaps to be associated with *Rice house als. Rice* recorded in 1745 (SRO D1134/21/2). *Bank Rices* is recorded in 1810 (SRO D260/M/T/4/46B), and Woodside Farm in Mayfield was known as *Bagnolds Rice* before 1831 (SRO D514/M/38). Names incorporating Rice have not been noted in any other part of the county.

trace of -*thveit* names commonly found in the Danelaw.[377] That evidence is concentrated, as would be expected, mainly, but far from exclusively, in the north-east part of the county closest to Derbyshire, and generally near the principal rivers and in upland or even rugged country.[378] Those names include Basford, Beasley, Beeston Torr, *Beite Brige*, Car/Carr (several), Catton, Croxall, Croxden, Croxton, Dagdale, Densy Lodge, Dodslow, Drointon, Forsbrook, Foston, Gamesley Brook, *Garardesthorpe*, Gayton, Gill Bank, Gunstone, *Hacondale*, Hillswood, Holme, Houndhill, Hulme (a very common name, generally of minor places), Keele, *Ketelshul*, Kettlemoor, Kingstone, Knipe Wood, Knypersley, Ladderedge, Mickledale, Nabb Farm, Nabb Brook, Nabb End, The Neb, *Neuthorp*, Ossoms Hill, Ranslow, Roe Lane Farm, Rolleston, Rocester, The Rowe, Snail's End, Snelsdale, Swainsmoor, Swinscoe, *Thoraldeswod*, Thorpe Constantine, Thursfield, Tucklesholme, Turner's Knipe Thursfield, Winnothdale and Yoxall. It may be noted that Gunstone, the southernmost name with Scandinavian associations, well away from the main areas of Scandinavian influence, lies on a lost Roman road[379] running south-west from Pennocrucium (Water Eaton).

It has been observed that many if not most places with names of Scandinavian influence would not be habitation places of first choice, and that Scandinavian settlers may have been forced to take on land which remained unsettled by the Anglo-Saxons. One example cited[380] has been Leek, said to have a name perhaps of Danish origin. However, Leek, which lies on a prominent well-watered hill in a protective loop of the river Churnet, holds a strategic position above a pass known to geographers as the Rudyard-Churnet Gap,[381] and has the remains of at least four pre-Conquest stone crosses, one perhaps as early as the ninth century,[382] indicating that the place was a centre of some significance in the Anglo-Saxon period. Recent research supports the view that in the Anglo-Saxon period a large estate was centred on Leek,[383] and if the place-name has been influenced

---

[377] The name *Netherby House* is shown on the 6" O.S. map of 1890, and *Netherby* on the 1" O.S. map of 1962, between Sedgley and Gospel End (SO 9193), but is evidently of relatively recent coinage, probably influenced by nearby Netherton.

[378] Domesday Book shows that almost half the landholders in Edward the Confessor's time had Norse names, but they are to be seen as individuals rather than as part of any organised group: VCH IV 6. A rudimentary survey of Scandinavian place-names and personal names in Staffordshire is to be found in SHC 1916 150-4, and an analysis of Scandinavian personal names in Staffordshire in Tooth 2000a: 1-16, although the latter is based on the supposition that Watling Street formed the boundary of the Danelaw in Staffordshire.

[379] Horovitz 1992: 34-5. A settlement of Angles is commemorated in the name Engleton, which lies on the same road near its junction with Watling Street.

[380] Gelling 1992: 137.

[381] Beaver & Turton 1979: 13.

[382] SHC 4th Series 19 8. On the north side of one of the cross-shafts, near the porch of the church, is a fragmentary inscription in Anglo-Saxon runes, of which only a few letters can be made out: see Elliott 1964: 213-4. Dr David Parsons suggests that of the tentatively identified letters -*isath* and *bibae*-, bae is more likely to be the start of the word *baecun* 'beacon, monument', which is very common in such inscriptions, though if that is right the preceding *bi* ( the preposition 'by') does not conform to the common formula ('X raised this beacon in memory of Y'): personal communication to author, 10th May 2000. This is the only example of runic lettering in Staffordshire. Erdeswick 1844: 342 mentions a monument in Dudley churchyard 'with Saxon characters, as I take them ...', but nothing more appears to be known of this inscription, which was evidently not in runic script, nor is it clear whether Erdeswick saw it himself or based an opinion on the drawing commissioned from Mr Wyrley to send to William Camden: ibid.

[383] See in particular SHC 4th Series 19 1995 5-12.

by the Danes, they almost certainly adapted a cognate Anglo-Saxon name-forming element.

Further evidence of Danish influence is found in the element *kirk*, a Scandinavianised form of Old English *cirice*, found in Kirksteads, *Kyrkesleye*, *Kyrkelond*, and *Kirkmedwe*, and associated with a number of other places, including Leek (i.e. *le Kirkebrok*) and Church Mayfield, and possibly from the lost *Ratherseates*, in Alstonefield. The likelihood that the name Keele is of Danish origin may be slightly strengthened by the consistent *K*-, and the proximity of Kettlemoor, which might incorporate a Scandinavian personal name. A number of names incorporate a form of *booth* (including Birchenbooth, Boosley Grange, Boothen, Boothlow, and Hardings Booth), an element which is probably of Scandinavian origin.[384] Denstone and Normacot, whilst of English origin, imply occupation by Scandinavians.

Also noteworthy is the very sizeable number of names – well into three figures – mostly of minor or unlocated places, which incorporate the element *holm* or *hulme*, which may be seen as evidence of Scandinavian expansion: *holm* may have been used by the Anglo-Saxons, but it is evident from plotting the names on a map that they are concentrated, as might be expected of elements of Scandinavian origin, on the east of the county, and spread with decreasing frequency along the major rivers, including the Sow at Stafford. However, that evidence must be tempered by the fact that the major rivers in Staffordshire are on the east side of the county, and by the fact that the majority of names have been extracted from Stebbing Shaw's great The History and Antiquities of Staffordshire, which was never continued beyond parts of the central, eastern and southern parts of the county. Whilst Staffordshire lay outside the Danelaw, the Danes will have penetrated at one time or another into many, if not most, parts of the county,[385] and may for a time have occupied Stafford itself. There is possible evidence of more permanent occupation on the eastern edge of the county: Anderstaff Lane in Burton upon Trent is found in early records as Anlastoft,[386] clearly of Scandinvian derivation, and characteristic Danish *gata* ('street') names – Aldergate, Ellergate, Gumpegate and Gungate – are found in the northern part of Tamworth, and further evidence of Danish links is the church of St Editha in the same town: Editha, sister of king Athelstan, was married at Tamworth to the Danish king Sihtric in 926. The only piece of clearly identifiable Scandinavian sculpture in the county is a wheel-head type cross at Rolleston.[387]

---

[384] VEPN I 134.

[385] Whilst Tamworth and Repton are known to have been taken by the Danes in 940, there is, surprisingly, no evidence, either documentary, archaeological or legendary, of any Danish action or occupation at nearby Lichfield, but it is difficult to imagine that such an attractive target would have been ignored by the Danes: see TSSHS XXII 1990-1 31. The tradition that the remains of St Werburgh were removed from Hanbury in the face of the Danish occupation of Repton in 875 seems to be based on the life of the saint which appears in Ranulph Higden's 'Polychronicon', completed in 1452, but in 875 Chester was lying waste, and the tradition appears to have no earlier history. Field names in Oakden 1984 provide evidence for a handful of Scandinavian personal names in Cuttlestone Hundred.

[386] VCH IX 6.

[387] Studd 1993: 57; but note that there is some question over the origin of the cross -- StEnc 480 suggests that for many years it was part of the floor of Tatenhill church porch, and was removed to its present position in 1897. An analysis of the sculptural monuments of the region has suggested a degree of Scandinavian influence in the northern half of Staffordshire, with the boundary running to the north of Lichfield and Stafford: see Sidebotham 1994.

## Domesday Book

The great majority of English towns and villages are first recorded by name in Domesday Book, a major source of evidence for place-name scholars. However, the great survey was not intended as a list of settlements,[388] and is not altogether a reliable source of linguistic information about Old English names, mainly because it was written in Continental Latin (rather than pre-Conquest Anglo-Latin), and the name spellings are attempts to render vernacular words in the orthography and grammatical norms of medieval Latin. The spellings of place-names generally reflect neither Norman French influence nor the actual pronunciation of either scribes or informants,[389] although there is some evidence of Anglo-Norman or French scribal orthography: one example is the occurrence of prosthetic *e* before initial *S* plus a consonant, found in *Estretone* (Stretton near Penkridge).[390] Moreover, in the entries for Staffordshire there are a number of notable omissions from Domesday. Places not given specific entries include, remarkably, the Mercian royal site of Tamworth, as well as Burton upon Trent, Colwich, Ilam, Longdon, Newcastle under Lyme, Rowley Regis, Stone and Stow, although some of those places are named incidentally. A number of areas of woodland seem to be omitted (for example the forests of Kinver and Morfe, both recorded as early as 736),[391] doubtless because by 1086 they already formed part of royal hunting reserves, adopted from the Frankish pattern and better known after Domesday as the Royal Forests. Again, while most places mentioned in Domesday are easily recognisable, possibly because earlier surveys were available against which to check spellings, a number are clearly aberrant, possibly due to the recording of names by roving commissioners who were unfamiliar with the area (and local dialect), or because the clerks responsible for transcribing the information for the final record misread the local returns. Examples of some less obvious places-names include *Cetquille* (Chatcull), *Ceruernest* (Charnes), *Estendone* (Huntington), and *Lvfamesles* (Painsley).[392] Local charters of the twelfth and thirteenth centuries which offer consistent spellings are usually considered better evidence than an irregular form from Domesday Book, which might defy all laws of philology and represent a uniquely and inexplicably corrupt spelling.

Most place-names that appear in Domesday have now been identified with reasonable certainty. Two that have not been located are *Burouestone* and *Litelbech*, which are listed with Weeford as members of Lichfield. There is some very slight reason to believe that *Burouestone* might be identified with the obsolete Broughton (or Brocton) at Longdon, or indeed be Longdon itself, and *Litelbech* might be the same as (or close to) the lost *Bech* in Lichfield. The evidence is discussed in the alphabetical entries for each place.

---

[388] As indicated in the Anglo-Saxon Chronicle, the survey was intended to record the possible taxation yield of estates when they fell into the hands of the king by confiscation or death: Sawyer 1979: 3; Harvey 1979: 105-9.

[389] Clark 1993: 317-31.

[390] Sawyer 1956: 492-3.

[391] In a Grant by Æthelbald to Earl Cyneberht of land for a monastery at Husmere, Worcestershire: S.89; see Whitelock 1955: 453-4.

[392] The Domesday Book spellings which appear in the present work are based on the facsimile of the Domesday folios for Staffordshire published in Alecto 1988, supplemented by observations in Thorn 1991, and differ in some respects from those given, for example, in the translated transcript found in VCH IV 37-60 (from which the forms for Staffordshire names erroneously entered in the returns for other counties have been taken). In particular, the common practice of reading *u* for *v* where the context has seemed to require has not been followed, mainly because there is a generally a clear distinctinction between the two letters in the Domesday Book palæography.

*Mersetone* recorded in Domesday Book has hitherto been identified as Amerton, on the basis that the (interpolated) name appears with that of nearby Gayton,[393] and presumably in the light of the absence of any recognisable entry for Amerton. That identification may be correct but early spellings for Amerton bear no similarity to *Mersetone*, which could well be Marston near Stafford, some three miles from Gayton.[394]

## Norman influence

French influence on English place-names is generally small, but in Staffordshire French or French-influenced names include Armitage, Barnville (possibly), Bowgage, Carmounthead, Dieulacres, Enville, Frankwell, Grosvenor, Pickards, the Roaches, Reule (possibly), the mysterious manor of *la Desire*; transferred names such as *Doyle*, Foker, Fowlchurch, Jamage and Lysways; 'double-barrelled' names, where the name of a French family has been added as an indication of feudal overlordship to an existing (usually English) name, such as Clifton Campville, Drayton Bassett, Mavesyn Ridware and Weston Coyney; and derisory or laudatory names such as Beaudesert and possibly Belmont, Beyvill, Murdeford; and the mysterious Beuleg and Domvilles.

## Frankwell and *Le Desire*

From the early thirteenth century references begin to appear in Staffordshire records to a place near Ellenhall, a few miles west of the county town, called Frankville. Since it is mentioned by the Staffordshire historian Walter Chetwynd in 1679 without comment,[395] its name and location were then presumably well known, or at least well remembered, although the place would then seem to have been depopulated and levelled for more than a century: two documents of 1564 and one of 1599 refer to Frankville, some fields in it, and demolished tenements, etc., of the 'lost village'.[396] Loxdale, the eighteenth-century Staffordshire antiquary who knew the area well, speaks of 'a lane running from the Lawn [which] goes between Stubwood[397] and the old depopulated village of Frankwell to the head of Broad Heath'.[398] At the beginning of the twentieth century it was noted '... in a field near Ellenhall [is] a site known as Frankville, in which foundations have been dug up; one spot in the ploughland was known locally as the "Market place", and a kind of cobbled pavement has been exposed by the plough',[399] and in the 1920s a commentator mentions Frankwell Farm and a field named Frankwell Orchard where building rubble had been ploughed up.[400] The precise location of the vill of Frankwell, which may have existed by the early thirteenth century,[401] is uncertain, but seems to have been at or close

---

[393] VCH IV 46 fn.87.

[394] As accepted by Eyton 1881: 83-4.

[395] SHC 1914 92.

[396] SRO D798/1/6.

[397] The place has not been located.

[398] JNSFC LXIII 1928-9 165. Broad Heath is about a mile north of Ranton.

[399] JNSFC XXXVI 1901-2 118.

[400] Ibid. It is possible that *Ann's Well (Wood)*, (*Anneys Wood* 1836 O.S.), a spring-fed pond 1 mile south of Ellenhall (SJ 8424) may incorporate the ghost of the place-name, but more likely that it commemorates the hospital of St Anne established in the later 13th century which lay within the precincts of Ranton priory (see VCH III 136; 252), or perhaps 'the virtuous Madam, Ann Cope' (or Lady Anne Harcourt), who lived at Ranton in the 17th century: Plot 1686: 112; Erdeswick 1844: 134. *Frankwell Moor* in Hextall is recorded in 1509: SRO D798/1/6.

[401] Robert de Frankevill is recorded as a witness in a charter of land at Reule, south of Ranton: SHC 1928 280. The charter is undated, but the grantor died in 1206.

to SJ 844253.[402] The Ellenhall parish boundary bulges hereabouts, possibly to incorporate Frankwell. But this Staffordshire Frankwell holds more interest for historians than as a mere deserted medieval village. The name is one of a handful of French origin in the county, and means 'the free or French town' (Old French *franc*, *ville*), with the substitution of *well* for *ville* occurring from c.1600.[403] Whilst the meaning 'French town' is a literal interpretation of the name, the adjective French might be expected rather than a noun (cf. the well-recorded 'French borough', recorded as *Le Frencheborough* in 1384, in Nottingham).[404] The name might therefore have had some specialised technical meaning, since it seems unlikely to have been coined as a new name. It is one of only three places so-called in Britain. The others are in Shropshire and the Isle of Wight. Names incorporating an original -*ville* (to be distinguished from names where -*ville* has later replaced -*field*, e.g. Morville and Enville) are extremely rare, with Margaret Gelling citing Frankville in the Isle of Wight and Frankwell in Shrewsbury as probably the only examples.[405]

Frankwell in Shropshire is on the western outskirts of Shrewsbury, on the west bank of the river Severn. Its history is unclear, with the earliest reference to the place dating from the early thirteenth century.[406] Whilst Domesday records a sizeable French population before the Conquest, with almost a third of 150 burgages held by Frenchmen in 1086,[407] those burgages were near the castle, not outside the town, but the inhabitants may have been the predecessors of those who eventually established themselves outside the town walls in Frankwell.[408]

*Francheville*, on the Isle of Wight, sometimes also called Newtown, is known to have been a planted town established by the Bishop of Winchester in 1256 as a trading settlement enjoying tax concessions, hence Frank in the sense 'free of impositions'.[409] It is improbable that this explanation could account for the Staffordshire Frankwell, which lay in a rural area some distance from any urban centre.

What then might explain the origins of a medieval vill in an isolated position in rural Staffordshire with a name meaning 'the free or French town'? There are no obvious answers. The settlement cannot have been created by the prior of Ronton, since the name is earlier than the priory. One possible clue might be detected in Domesday Book, where the only specific references to Frenchmen (*franci[genae]*) in the folios for the whole of Staffordshire are two entries recording four Frenchmen and four thegns holding eight berewicks (Gerard's Bromley and Podmore, Tunstall, Swinchurch, Ellenhall, Walton in Eccleshall, Adbaston, Wootton and Knighton) of the manor of Sugnall, five miles north-west of Frankwell, and two Frenchmen who shared with a thegn four berewicks (Seighford, Doxey, Bridgeford and Coton Clanford) in the manor of Eccleshall, three

---

[402] TSSAHS XII 1970-1
[403] It may be noted that in the early 18th century Frankwell in Shrewsbury was also known as Frankville: SA 49/212.
[404] PN Nt 14; PN Sa IV 65-6.
[405] Darlington 2001: Appendix; PN Sa IV 65-6.
[406] See Gelling 1992: 166; PN Sa IV 65-6.
[407] Jackson 1995: 133.
[408] PN Sa IV 65-6.
[409] See MedA 3 1959 202ff.

miles north of Frankwell, all as subtenants of the bishop of Chester.[410] It has been suggested that the Frenchmen may have been Norman *milites*, professionals who had been settled on episcopal estates as part of the bishop's quota.[411]

It is known that the vill of Frankville was given to the prior of Ranton (otherwise Ronton) by John de Frankville.[412] In the same area were two other unlocated estates or settlements. One was called Le Desire – perhaps near Badenhall or Eyeswall or near a lost place called Newbolt at Hilcote[413] – first recorded in the thirteenth century and not recorded later than the fourteenth century.[414] The other, near Ranton, was known as Doyle, recorded in the mid-fifteenth century,[415] and named after the Doyley or D'Oyly family, lords of Ranton, who held the place from the late 14th century.[416] The surname is probably from one of the five Ouillys in Calvados: Ouilly-le-Basset, Ouilly-le-Vicomte, Ouilly-du-Houlley, Ouilly-la-Ribaude and Ouilly-le-Tesson. Doyle was subsequently known as Lewkenore's Manor, named after the Lewkenor or Lewknor family from Lewknor in Oxfordshire.

There is evidence that Frenchmen were sometimes attracted to a locality by special privileges introduced by the local lord, like the colony of French burgesses established at Hereford shortly after the Conquest and granted the 'laws and customs' enjoyed by the lord's French home town, a 'concession imitated by the lords of many towns on the Welsh border and elsewhere'.[417] There is no evidence of such process here, but by 1086 Frenchmen had an interest in no fewer than 12 vills encircling and within a five-mile radius of Eccleshall, and we know of at least three other places (Frankwell, Le Desire and Doyle) within that area with later French associations.[418]

Since both Doyle and Lewkenore's manor were family names, it might be considered that Frankville could have taken its name from a family of that name, perhaps from one of the several Frankvilles in France, or from Frankville in Shrewsbury. However, Margaret Gelling notes that '[Frankville in Staffordshire] is clearly not a manorial name conferred by a migrant from the [Shrewsbury] Frankwell',[419] and has pointed out that 1206 would be a very early date for a surname derived derived from a minor settlement in England, and if the de Frankevilles had brought their name from France the family might be expected to be more eminent.[420] No explanation can presently be put forward for a notable French presence, and perhaps a 'free town', in a commercially undeveloped, indeed relatively remote, part of Staffordshire in the centuries immediately after the Conquest.

---

[410] VCH IV 43.

[411] Desborough 1991: 17.

[412] SHC 1914 83; SHC IV 276.

[413] SHC VII (ii) 36.

[414] Chetwynd refers in 1679 to this manor 'which ... lay near to Eccleshall, but where I have not been able to discover', and to Sir Robert de Hastang who styled himself *dominus* [lord] *de la Desiree* in 1301: SHC 1914 60.

[415] SHC 1914 95; see also SHC III NS 213.

[416] SHC 1914 95-6.

[417] Stenton 1970: 200.

[418] Reule, which lies 3 miles south of Frankwell, may be a name of French origin. Further slight evidence of French influence may be the appellative adopted by or given to Robert le Frank of Wootton, near Eccleshall, who is recorded in 1351: SHC 1913 144.

[419] PN Sa IV 66.

[420] PN Sa IV 66 states that '[t]his is clearly not a manorial name conferred by a migrant from the Sa Frankwell', and suggests that the -*k*- in the Shrewsbury and Staffordshire names as opposed to the -*ch*- of the Isle of Wight name may be due to an earlier date of coinage.

## Beacons and look-out places

There are surprisingly few early references to beacons, given the long period during which they must have formed a key part of a sophisticated communication and warning system both regionally and nationally.[421] Beacons appear to have existed in more recent centuries (and perhaps long before) at *Beacon Stoop* on the Weaver Hills, and near Abbot's Bromley, Congreve, Milwich, Rolleston, Sedgley and Stafford,[422] and the name *Bekenfield* in Kingswinford, recorded in 1485,[423] and *Bekenhurste* and *Beckenhurst*, recorded in Upper Penn in the 13th century and 1319,[424] suggest the existence of beacons in those areas, although Old English *(ge)becon* also meant 'sign, signal, cross, memorial stone', and perhaps 'tumulus'.[425] The name Bignall may also incorporate the element. In addition, Old English *wearde*, *wearda* was sometimes used for a beacon, and the name may be found in Ward Hill near Dilhorne, Wardlow near Cauldon, and War Hill near Maer.[426] Noddy Field (two places, one near Cannock Wood and the other in Compton, Kinver), and Noddy Park in Aldridge and Rushall are on high ground and, if the names are ancient (which is far from certain), may mark the sites of early beacons. Lees Hill may possibly incorporate Old English *leg* 'fire', perhaps denoting a beacon. Beacon Bank Farm[427] 1 mile north-east of Abbots Bromley, and Beacon House[428] on the north-west side of Gillow Heath appear on modern maps, but older spellings have not been traced.

In early times advance warning of the approach of hostile or uninvited forces will have been a matter of considerable importance, and it seems reasonable to suppose that a network of look-out points will have been created. It is possibly no more than coincidence that several names beginning Tat-, Tet-, To(o)t-, Tut- lie on high ground – in some cases with particularly commanding views – but (notwithstanding other established or suggested etymologies) it may be that if such names do not indicate a look-out point *per se*, they derive from the occupational name of an individual who undertook look-out duties. Those names include Tatenhill, Tettenhall, Tittensor, Toot Hill, and Totmonslow.

## *Broc* and *burna*

The most common naming elements for streams in Staffordshire are Mercian Old English *wælle*, generally 'a spring', but occasionally 'a stream', Old English *broc* 'brook', and sometimes Old English *burna* 'a burn, a stream'. There is no evidence to support the theory that *burna* is earlier than *broc*, but the two elements appear to have been attached to differing types of watercourse, with *burna* tending to be applied to streams with gravel beds, clear water and submerged water plants, and *broc* used of muddy steams with sediment-laden water.[429] It has been noted that *burna* is not uncommonly found in the area formerly occupied by the Anglo-Saxon sub-kingdom of the Hwicce,[430] which

---

[421] No pre-Conquest place-name is known to incorporate *(ge)becon*: Rumble & Mills 1997: 157.

[422] Shaw 1798: I 99 mentions a mound of earth on an un-named hill near Yoxall used formerly as a beacon.

[423] Shaw 1801: II 229.

[424] SRO D593/B/1/17/1/3/5 and SRO D593/B/1/17/1/5/1.

[425] VEPN I 68.

[426] *Warda*, possibly in or near Ipstones, is recorded *temp.* Edward I: SRO D1229/1/4/41.

[427] NGR SK 0925.

[428] NGR SJ 8758.

[429] Gelling & Cole 2000: 67.

[430] Gelling & Cole 2000: 10, which cites 3 examples in Staffordshire: Bourne Vale, Harborne and Wombourne.

covered what is now Worcestershire, most of Gloucestershire, and the south-western half of Warwickshire, but is rarely found elewhere in the Midland counties outside this area: there is only one example in each of Huntingdonshire, Northamptonshire, Nottinghamshire, and Oxfordshire, and only two instances in each of Cheshire, Herefordshire, Leicestershire and Shropshire. The examples in the area covered by this study (Boney Hay, Bourne Vale, Burnaham/Bornam, Guthersburn, Harbourne, Lilleborn, Luthburn, *Patyngehamborn*, *Stanburneford*, Swarbourn, Wombourne and Womburnshawe) are certainly concentrated in the southern half of the county, closer to the Hwicce. It has been suggested that they reflect a Hwiccan place-naming vocabulary which differed from that of the Mercians.[431] A study of Anglo-Saxon charters has shown that the term was used by the 'old south-easterners', but not by the Midland Anglians, Thames Valley Saxons or the new immigrants in the south-west.[432] The evidence tends to show that *burna* was in early use across most of the country, that it remained in use in particular areas such as the south-east and north-east longer than elsewhere, and that it was superseded by *broc* before the later Anglo-Saxon period.[433]

## 'Corpse' names[434]

A number of minor names, many now obsolete, associated with dead bodies, occur in Staffordshire, including Dead Woman's Grave, 2 miles west of Codsall,[435] Dead Lad's Grave at the junction of Birches Barn Road and Trysull Road, 3 miles south-west of Wolverhampton,[436] Dead Knave 1 mile north-west of Sedgley,[437] Deadman's Dale or Dimsdale south of Bagot's Bromley, under what is now Blithfield Reservoir,[438] *Deadman's Grave* recorded in 1689 near Cellarhead,[439] *Dedemannesmore* and *Dedmanneslowe* recorded in 1401 in Castle Church,[440] *Dead Man's Lane* in Newcastle under Lyme in 1490,[441] also recorded as *Dead Man's Lane alias Marsh Street* in 1778,[442] *the dead woman's Buryall* at Friar's Park Corner, Walsall, recorded in 1606,[443] *dedmanslane* in Walsall recorded in 1545,[444] *Deadman'slane* in Handsworth, recorded in 1851,[445] *Deadman's Denne*, an unlocated place, possibly near Betley,[446] and Alice Hursts

---

[431] Gelling & Cole 2000: 9-14.

[432] Kitson 1995: 91

[433] VEPN II 90-93.

[434] See also Field 1972: 60-61.

[435] GR SJ 8404. *Dead Woman's Grave* 1676 Codsall ParReg. A name now applied to a glacial erratic at the junction of County Lane and Husphins Lane. By tradition named after a small house called Dead Woman's Grave, said to have taken its name from a woman who hanged herself there in a skein of wool, and was buried according to custom at the crossroads here: TSAS IV 1883 40.

[436] GR SO 8896. *The dead lads Grave* (1834 O.S.),

[437] GR SO 9094. *Dead Knave Field* 1736 SRO D3155/WH94, *Dead Knave Farm* 1769 WSL 30/35/12/42, *Dead Knave* 1798 Yates, *The dead Knave* 1834 O.S.

[438] 1836 O.S.

[439] Ward 1843: lx; also recorded in 1803: SHC 1933 150.

[440] Oakden 1984: 80.

[441] Pape 1928: 177.

[442] SRO D593/B/1/13/42.

[443] Sims 1882: 50. The place, recorded as *Dead Woman's Burial Gate* in 1709 (Willett 1882: 188) is said to be associated with the discovery of human bones: Willmore 1887: 245.

[444] Sims 1882: 31.

[445] White 1851: 698.

[446] *Deadmansden* 1590 Betley ParReg, *Deadman's denne* 1592 ibid.' The place is mentioned several times in the Betley parish registers, and was presumably a cave or hovel or suchlike where a body was found.

Grave, where in 1690 a suicide was buried in '... a certain parcel of ground parting Rolleston and Burton Road commonly called Alice Hursts Grave, who also hanged herself'.[447] Many of the names carry local traditions which have developed to explain them, but in most cases it is almost impossible to determine whether such traditions have any factual basis, although a violent death is recorded in a document which mentions Deadman's Grave in Cheddleton in 1689.[448] It may also be noted that some names have been corrupted and associated with dead bodies, such as Dimsdale, which appears on the first edition 1" Ordnance Survey map of 1836 as *Deadman's Dale or Dimsdale*. The name is almost certainly Dimsdale, with a macabre association having developed via popular etymology. Likewise Deadman's Green near Checkley, which has been so-named only in recent centuries, and has its origins in the Old English personal name *Dæda.[449]

## Customs

The place-name Morrey (and perhaps *Morghull*) incoporates traces of an expression associated with the ancient custom whereby money was given to a bride by her groom as her personal wedding settlement,[450] and the place-name Dowry perhaps records property given as a wedding gift.

## Dimmingsdale and similar names

Within Staffordshire is a handful of place-names which appear to incorporate a root *dimmin(g)*.[451] Those places are Dimmings Dale 3 miles east of Cheadle (first recorded in 1786), Dimmingsdale on the west side of Willenhall (1272), Dimmingsdale south-west of Wolverhampton (1753), Dimmins Dale on Cannock Chase (1840), Dimsdale north of Newcastle under Lyme (1086), and Dimsdale near Blithfield (1836). For the purposes of this analysis those names are assumed to have a common root. Ekwall[452] suggests that the first element of Dimsdale near Newcastle might be ME *dimple* 'dip in the ground', with the *p* lost early and *ml* becoming *lm*, but that explanation seems somewhat contrived and unsatisfactory, and is not supported by the early spellings.

The word Dimmin(g)s[453] seems to be found in place-names combined with no other element than Dale (from Old English *dæl* 'valley'), and in Staffordshire appears to be associated with mineral workings and/or dams, where a valley might be expected to be associated. Early ironworking is recorded at Dimmings Dale near Cheadle,[454] a medieval dam, reservoir and pond have been recorded at Dimsdale near Newcastle,[455] there has been a reservoir at Dimmingsdale south-west of Wolverhampton since at least 1832,[456] Dimmins Dale on Cannock Chase lies at the head of Fairoak Pools, almost certainly artificial, and Dimsdale near Blithfield adjoins a reservoir, albeit created in 1953, but

---

[447] Edwards 1949: 171.

[448] Reference lost: *mea culpa*.

[449] Redfern 1865: 348 notes that the place is recorded as Tetterton in Checkley ParReg.

[450] Fell 1984.

[451] Demmings in Cheadle, Cheshire; Dimin Dale in Taddington, Dimmon's Dale in Blackwell, Dimminsdale in Calke, and Dimons Dale in Cromford, *Dimminggesdale* Ed I *HardCh* (all in Derbyshire) may have the same root: see PN Db 69, 213, 358, 627

[452] 1936: 145.

[453] Cf. Dimmings, in Bengeo, Hertfordshire, *Dymmynges* 1590 PN Hrt 217.

[454] Old Furnace (*Old Ffurness* 1685 Alton ParReg; *Old Furnace* 1694 SHC 1947 63; 1760 SRO D240/D/110) lies at the west end of Dimmings Dale, built in the 1590s: see Welch 2000: 53.

[455] SMR 01196.

[456] Teesdale.

perhaps on the site of an earlier artificial pool. Early iron-mining is known to have taken place in the area around Dimsdale near Newcastle, for example at Holditch in the 2nd century AD,[457] in Madeley in 1293,[458] and in Knutton in 1315,[459] 1316[460] and 1321.[461] Support for an asociation with mineral working may be found in the place-name Dymsdale in Alwington, Devon, first recorded in 1371, which is believed to be from the topographical surname Dymmyngesdale, from a miner or miners from Derbyshire or Staffordshire working in the royal stannaries.[462]

A derivation from ME *dimmin* 'dim, gloomy' would not account for the *-s* endings in the place-names. The Oxford English Dictionary records *diminue* 'to break up small' (e.g. crush ore?), and *dimane*, 'to flow different ways; spread abroad' (e.g. to create sluices for ore washing?), which may be associated with the root of this name. One possibility is that the root may be an Old English word associated with Middle English *demming* 'a dam' (cf. Old English *\*demman* 'to dam, to obstruct the course of water'), perhaps in some cases associated with mineral working, so giving 'the dale with the dam or sluice', although the ubiquitous Dim- rather than Dem- spellings are not readily explicable.

**Fauna, flora, crops and soil**

Domestic animals evidenced in place-names include the bee (Beobridge), bull (Bonehill), calf (Caldon, Cauldon, Calf Heath, Calwich),[463] colt (Colton), goat (Tixall), lamb (Endon), ox (Oxenford, Oxley), ram (Ranton), sheep (Seabridge), swine (Kingswinford, Oldswinford, Swindon, Swinfen, Swinscoe, Swynnerton), and wether (Weatherworth). Wild creatures include badger (Brockhurst), beetle/chafer (Charnes), deer (Dosthill), fox (Foxt), frog (Froghall), hare (Harehills, Herbage), hart (Hartwell), louse (Luzlow), mouse (Mousehall, Musden), toad (Padwick, Podmore), wild boar (*Wildboarsegreave*), wild cat[464] (possibly in Catton and *Cattail Pool* (Head Pool)), and wolf (Woolley, Whittimere, Graiseley); and birds include bittern (Bemersley, Bitternsdale), bunting (*Hammersley*), cock (Hannell), crane[465] (*Cornbridge*, Crane Brook, Cranmere, Cranmoor, Cronk Hill,

---

[457] VCH II 108.

[458] SHC XII NS 203.

[459] SHC 1913 267.

[460] SHC XII NS 67.

[461] SHC IX 83. The Taddington, Blackwell and Cromford areas of Derbyshire are all in the limestone area of the Peak District, and certainly had lead mining from early times. *Dimingdale* (lost) is recorded in the West Riding of Yorkshire, with the suggested meaning 'dark, dim place' (PN Y West Riding 108), and *Dimmings Dale*, a field-name in Berkshire, is recorded from 1647 (PN Bk 111; this latter name may however be transferred: personal communication from Dr M. Gelling). *Dymmerdayk* in Cornwall is recorded in 1325 (SHC IX (i) 109), perhaps from ME *demman* 'dam, stop up', with ME *dic* 'dyke, ditch', possibly 'the watercourse associated with the dam'. Cf. Ricardo *Dymer*, recorded c.1250: SHC XI 313. Spellings for the Sussex Hundred of Dumpford include *Dymford, Demesford* in ME documents: Mawer 1929: 40.

[462] PN Db I xliv; Cameron 1996: 190.

[463] But note observation about names incorporating *calf* in Archaeology *infra*.

[464] Wild cats were found in Staffordshire well into the medieval period: in 1267 a licence was granted to hunt foxes, badgers and wild cats in the King's Forests of Shropshire and Staffordshire: Eyton 1854-60: II 243.

[465] Duignan 1902: 46 states that 'the dictionaries give [OE *cran*] as 'a crane'; but in the Midlands it meant, as it means now, a heron. It may have been otherwise in the fens, but I do not believe that cranes ever inhabited the Midland Counties'. It may be noted, however, that after the dove and the eagle, the crane is the most commonly illustrated bird in medieval manuscripts, and a significant number of crane bones have been identified from the late twentieth century Stafford Castle excavations.

Cronkledge), crow (Crowborough, Crateford, Cracow Moss), duck (Endon), eagle (Yarnfield), finch (Finchfield), goose (Gander Well, Goosemoor, Goosemoor Green, Gorsebrook, *Gorseholm*, Gospel End, Wildgoose), hawk (*Haukesclyf*, Haukeshill, Haukesmor, Hawkbach, Hawksley), jackdaw (Cowlow), kite (*Gleadley* and *Gledenhurst*), raven (Raven's Clough), sparrow (Sugnall), and water rail (Crakemarsh). Ousley (Brook and Cross) may incorporate the Old English word for blackbird. There is some slight evidence that the expression Frog-hole or Frog-hall may have been applied in the post-Conquest period to man-made excavations.

The types of trees for which evidence is found in Staffordshire place-names include alder (Aldershaw, Aldersley, Alrewas), apple (?Apedale), ash (Ashley), aspen (Aspley), birch (Birchills, Birchfield), hazel (Haselour, Hazelwood, etc), lime (Lynn), oak (Acton, Oaken), pear (Perry Hall, Perton), plum (?Blymhill), rowan (Wicken Walls, Wickeytree), and thorn (Castern). Shrubs, plants and crops include barley (Barton), bog myrtle (Gailey, Wyrley), briar (Brereton), broom (Bramhall, Broomhall), cress (Cresswell, Bilbrook), fern and bracken (Fawley), flax (Ellenhall, Flaxley Hill), hocks or mallows (Hoccum, Oakham), hops or similar plants (Himley), marigold (Goldenhill, Goldie Brook Bridge, Goldsitch), mistletoe (? Moisty Lane), oats (Pillaton), wild celery (Marchington), wild garlic (Ramshorn, Ravenscliffe), and wheat (Wheaton Aston). A cluster of names in the Pershall area (including *Pesecroft*, Peafield (Coppice), and Persbutt, as well as Pershall itself and Peasley Bank) attest to the cultivation of peas. A herb garden is found in the name Burlaughton, and Rue Barn (Farm.) may record the growing of the herb of that name. The nature of the soil is indicated in Clayton, Sandon, Gratwich, Gratton and Greets Green (clay, sand, and gravel/grit). Perhaps surprisingly, the name Orchard appears to be confined to high ground in the far north of the county.

### The hide

A number of place-names incorporating the Old English element *hid*, 'a hide of land', have been traced in Staffordshire, including The Hyde near Brewood, The Hyde at Kinver, and other places now lost at Freeford, Penkridge, Butterhill near Bradley, Weston-under-Lizard and Sheriffhales, as well as Halfhead, which means half-hide. The term 'hide' originally meant not a fixed area, but as much ground as would support one family, depending on the quality and nature of the land,[466] and later came to mean in Staffordshire an area of approximately 120 acres. The bounds of Wrottesley, set out in a charter of 1088 A.D.,[467] show that the area, expressed to enclose *duas hidas* 'two hides', actually enclosed some 1,600 acres, indicating that at that date it was a term of asessment and valuation, rather than acreage. It has been suggested that the holder of a hide owed lighter and less servile duties than villeins.[468]

### Industries and occupations

Place-names reflecting industries and occupations include Quarnford and Milton (milling), Colwich, Coley Hill and Colton (charcoal burning), Hammerwich (smithing), Salt (salt-working), Butterhill and Butterton (dairying), Felthouse Wood, Fields Farm, and two places called Felthouse (felt-making), Glascote, Glass House, Glasshouse Farm, the lost *Glashoushay*, and Glass Lane (glass-making), Chatkull, *Culnehill*, High and

---

[466] Henry of Huntington records: 'Hida autum Anglice vocatur terra unius aratri culturæ sufficiens per annum': Earle 1888: 459.
[467] SHC VI NS (ii) 7.
[468] Faith 1997: 20-23.

Little Onn, Keeling Ford and Oulton House Farm (kiln-working), Biddulph, Dilhorne, *Delph House* and Stonydelph (quarrying or mining), Huntington and Humpage Green (hunting), and Fisherwick (fishing).

## Kings

Pendeford, Pendlestone, *Wulfercester* (Bury Bank) and *Wulfursyde* probably incorporate the names of Penda and his son Wulfhere, early kings of Mercia.[469] It is noticeable that a number of names beginning *Wolf-* and *Wulf-* seem to be concentrated around Stone. Some at least may incorporate the name of Wulfhere, who by tradition is associated with Bury Bank hillfort on the north-west side of Stone.[470] King's Bromley, Kingsley, Kingsnordley, Kingstone, Kingswood, King's Wood and Kingswinford probably indicate early royal associations.

## The element *leah*

In recent years considerable attention has been paid to the use of topographical place-name elements such as *leah*, which is held to have meant in its earliest use 'a wood', developing later to mean 'a clearing'.[471] In particular, it has been held that *leah* is the generic which is frequent in wooded areas, while *tun* is characteristic of open areas:

'[the] two elements ... very much the commonest words in English place-names, have been shown to be characteristic of place-name formation in the period c.750 to c.950, and it is a reasonable supposition (though unfortunately no evidence survives for this) that some of the settlement names in which they occur are Mercian replacements for British names'.[472]

The frequency with which the element *leah* is recorded in place-names in a particular area has been taken as evidence of the extent of woodland within that area.[473] However, the well-known map of an area in the West Midlands including parts of south-east Staffordshire[474] used to illustrate the proposition may also be used to highlight inconsistencies in the theory. The area of what is now Cannock Chase, which can safely be assumed to have been heavily wooded in the early Anglo-Saxon period, is almost devoid of each element. The exercise is subject to a number of difficulties which make it necessary to treat the results of any such plotting with caution. Firstly, it assumes that all the names in a particular area were coined at about the same date. There is rarely any firm basis for reaching that conclusion, and smaller areas of woodland may over a period of several centuries or more have been obliterated and created, although it must be acknowledged that the longevity of landscape features is a regular and noticeable conclusion in many landscape studies. Secondly, we have no means of knowing for certain just what size of woodland would have been described as a *leah* in the earliest

---

[469] It is said that apart from its occurrence in place-names, the only example of the name Penda is that of the king of the Mercians: PN Wo 223; see also Searle 1897: 386; Zaluckyj 2001: 2-3.

[470] The tradition is recorded in additions made by Walter of Whittlesey to the 12th century Chronicle of Hugh Candidus, a monk at Peterborough who wrote a history of his house from the time of Wulfhere, its founder: see Mellows 1949: 146-148. See also SHC V (i) 9-10.

[471] See Johansson 1975. Wager 1998: 154-5 considers that the term is indicative of secondary woodland.

[472] PN Sa III x. See also Gelling 1988: 127; Gelling & Cole 2000: 237-9.

[473] See especially Gelling 1988: 126-29; Gelling & Cole 2000: 237-42.

[474] Gelling 1988: 127; Gelling & Cole 2000: 44.

period, or what size clearing in what size of woodland may have been so described at a later period. In an area which was heavily wooded, it might be expected that place-names denoting woodland would be less common, for place-names in themselves were coined, it must be supposed, to express the distinctive nature of the immediate area. The use of a word for woodland would suggest that the place was distinctive because of the very absence of woodland in the area as a whole. In an area with many woodland names, it might be argued that the area was relatively free of woodland, but dotted with small woods which were distinctive by the nature of the cleared area in which they lay. Indeed, that seems now to be recognised by place-name scholars, and the element is becoming increasingly flexible in its interpretation: Margaret Gelling for example observes that

'leah should be translated 'clearing' when it occurs in a cluster of names, but 'wood' or 'meadow' when it is isolated',[475]

and remarks of two *leah* names in Shropshire:

'it is possible that (despite the conventional translation 'clearing' given in the text ...) the reference in the two names is to small, jealously guarded stands of trees rather than to woodland clearings. The sense 'isolated wood' is always to be reckoned with when leah names occur in ones or twos in an area where tun is the more common of the two elements'.[476]

It must be remembered that the nature of woodland itself is uncertain. From the pre-Roman period woodland was treated as a valuable commodity, and carefully tended. Woodland is often assumed to be part of the 'original ancient wildwood' (which by historical times may have been all but obliterated), but the term woodland might just as easily have applied to cultivated areas of coppice. Whilst Anglo-Saxon topographical terms have in many cases been shown beyond doubt to be very specific and precise, we can never be certain how many trees were needed to form a grove, or the distinction between a grove and a copse, or how a large copse differed from a small wood. No doubt the terminology varied according to the locality: in an area with many trees, it may have taken a larger number to create a wood than in an area sparsely scattered with trees. In general terms, it might well be that a heavily wooded region would show little evidence for woodland names, for they would not serve to distinguish the specific from the general. Again, only by surviving in early records will authentic *leah* names be evidenced. Very many must have vanished without trace; others may be of relatively recent origin. Oliver Rackham has cautioned:

'Let us not make too much of the 'clearing' place-names. A clearing can arise as easily by the retreat of agriculture – by the surrounding fields becoming woodland – as by new fields being made. Place-names tell us nothing of when the clearings were made, or how, or by whom. Only archaeology can tell us how many leys and hursts were made by the Anglo-Saxons themselves'.[477]

Finally, it is clear that a number of *-ley* names in Staffordshire, for example Moxley, Muckley Corner, Rowley, Stanley and Swainsley, have probably developed from *-hlaw*

---

[475] Gelling 1988: 128. But the existence of an 'isolated' name today does not preclude the former existence of similar names which have not survived.

[476] PN Sa III xiv.

[477] Rackham 1990: 56.

'tumulus', and in some cases, for example Anslow, -*ley* has developed into -*low*, so that the derivation of many apparent -*ley* names may in reality be uncertain. Mapping exercises for relatively common elements of this type, particularly where early spellings have not been traced, may produce results which are difficult to interpret or perhaps misleading, and the absence of such elements cannot be taken as evidence that the feature would not have been found in that area.

## Monastic retreats

Sinai Park (and possibly Coena's Well, Sena Park and *Seyneshaulwe*) marks the site of a monastic building used for breaks ('seyneys') by monks.

## Salt

Salt has always been an important and valuable commodity. Evidence of salt-springs and the working and transporting of salt can be found in the names Salt, Salt Brook, Salter's Well Farm, Saltbrook Cottages, Salterford, Salter's Bridge, *Salter's Croft*, Saltershall Farm, *Saltersholme*, *Salters Lane*, *Salters Meadow Field*, Salters Park Farm, *Saltford*, Salthouse (Farm), *Saltmoor*, *Saltmoor Meadow*, *Saltmore*, Saltwell and Saltwells.[478]

## Social order

Social order or status is reflected by reference to freeholders (Franklin), peasant landowners (Bond End, Boon Hill, and possibly Bowers), churls (*Cherlecot*, Chorlton), free tenants (Drointon), knights or young retainers (Knighton), beggars (Latherford), hermits (Armitage), monks (Monkford, Monks Wood), and king, bishop and abbot are found in Kingsley, Kingswinford, Bishton, Bushbury, and Abbots Bromley. An interesting group of names in and around Yoxall indicating social rank includes *Reeve End*, *Swaynefield*,[479] and Bond End, from Old English *gerefa* 'a bailiff, steward, official of rank', Early Middle English *swein* 'boy, servant, retainer,' and Old English *bonde* (Old Norse *bondi)* 'householder; free man'.

## Superstitions

Superstitious beliefs are evidenced by the element *bug* 'goblin or boggart' (Buglaw), *scucca* 'spectre or evil spirit' (Shugborough), *hob* 'hobgoblin, sprite, elf' (Hobs Hole, Hob Hill, Hobriding), *grim*, another name for Woden (Grimditch, Grymsyll),[480] *pouke* (from Old English *puca*) 'demon, sprite, hobgoblin' (Pouke Hill), *thyrs* (from Old English *þyrs*) 'giant, demon' (Thor's Cave, Thorswood (House)). Drakelow, the lost *Drakeford*, Wormhill, Wormhough and Woundon, from Old English *draca* and *wrym*, record a belief in dragons and serpents. Grindley and *grendelsmere* may incorporate the name of Grendel, the monster immortalised in the Anglo-Saxon poem Beowulf, who lived in a

---

[478]Salt springs are also recorded near Adderley Hall, and at Branston (see Shaw 1798: I 98), Brierley Hill, Cradley Heath, Enson, Draycott, Ingestre, Kingstone, Rickerscote, Sandon, Sugnall and Tixall (cf. Salt). See also Wychdon.

[479] Recorded in 1665: NA DD/4P/24/99. The name is commemorated in Swainsfield Road. See also Stuart 1990: 7.

[480] PN Ch V I:1 xxviii suggests that there may be an alternative meaning for *grym*, *grim*, for the OE and ONorse words mean 'mask' (cf. modern dialect *grim* 'skull face' recorded in Yorkshire (EDD)), and place-names incorporating the element may refer to places where ancient burials were discovered.

lake or mere. The precise physiological distinction between a demon, goblin, hobgoblin, sprite, elf and boggart does not appear to have been the subject of any exhaustive academic research.

## Treasure

Names suggesting the discovery of treasure include *Goltherdesbeuch*, Goldthorn, Goldthorn Acre, *Gooldburynes*, Hordle Spring, and Hurdlow.[481]

## Archaeology

It is perhaps in the field of archaeology that place-names provide the richest source of information.

There are a number of places incorporating the word Berry or Bury or Borough in Staffordshire (including Berry Hill, Berry Ring, The Bury, *Burwey*, and perhaps *Borwey Foordes*), almost always referring to an ancient earthwork, often prehistoric, and field-names incorporating these words in areas where no archaeological feature is recorded may justify further research. In Staffordshire the element *hlaw*, which elsewhere has the general meaning 'hill, mound, tumulus', would appear to have a narrower meaning, and refer either to a burial mound, or an artificial (or possibly natural) mound resembling a burial mound. In places to which the element has been applied but no tumulus has been noted, local groundwork or an examination of aerial photographs may well provide evidence of the lost or unrecognised feature. It is possible that *arbour* has a similar meaning, from Old English *eorð-burh*, though in many cases it will be the modern 'arbour', which appeared in Middel English from French.[482]

The word *ceaster* (from Latin *castra*) was adopted by the Anglo-Saxons to mark the site of Roman or Romano-British towns, and is found in the two Staffordshire Chestertons, one on the west side of Newcastle under Lyme, the other near Worfield. The first place has a Roman fort, the second lies near the Iron-Age earthwork known as The Walls, which from the adjoining Roman road was almost certainly occupied by the Romans. The element is also found in Chesterfield, with reference to the Roman site at Wall, and in Rocester, on the site of a major Roman settlement. Stratford, Stretton and Streethay contain Old English *strēt* 'a paved street, a Roman road'.

Place-names incorporating the word Grym or Grim, for example *Grimditch* and *Grymsyll*, are frequently associated with ancient earthworks, especially linear dykes, probably from the name Grim used as a giant's name in Northern Europe,[483] rather than from Old English Grim, meaning 'the masked one', a nickname for Woden, who by tradition went about in disguise.[484]

---

[481] The field-name *Gooldburynes* in Brewood is recorded in 1453 (Oakden 1984: 47), the name meaning 'burial-place where gold was found', and *Goldhorde style* in the Shipley/Rudge Heath area is recorded in 1619 (SA 330/25). *Roman Pasture*, a field-name recorded in Lower Tean near Checkley 1746x1792 (SRO D1203/A/19/1-4) may also mark Roman remains or finds.
[482] Gelling 1988: 147.
[483] See Bronnenkant 1985: 72.
[484] PN W 15-6; Gelling 1988: 148-9.

Ancient names with *castle* often mark what was, or was thought to be, a castle in the conventional sense (e.g. Castle Church, referring to Stafford castle, and Castle Croft near Chesterfield, named from substantial stone walls of Roman date), but often a prehistoric or later earthwork, from Old English and Old French *castel*, but such names may sometimes be from Old English *ceastel* 'heap of stones', often with archaeological interest. Earthwork-type names include Abbot's Castle Hill, Burf Castle, Castle Ring, Castle Old Fort, and Knaves Castle, all of which are or were prehistoric earthworks. Occasionally the word *castle* is applied to castle-like rock outcrops or other natural features, or used in an ironical way for a humble dwelling, perhaps the case with Tinker's Castle.

Names incorporating Old English *w(e)all* 'a wall, a rampart' suggest the existence at an early period of some man-made feature sufficiently distinctive to be specially named. Examples of place-names likely to incorporate this element are Wall Acre near Butterton, Wall Hill near Claverley, Wall Heath near Himley, Wall Hill near Rushton Spencer, The Walls (near Audley and at Chesterton), and Walton Grange.

It is worthy of note that at every place in Staffordshire where Old English *calf* has been identified as a likely element in the place-name (i.e. Calf Heath, Calton, *Calver Croft*, Cauldon, Calwich, Hill Chorlton) there are well-recorded burial-mounds. The word 'calf' meant 'the young of any bovine animal, especially of the domestic cow', but the meaning 'a small island lying close to a larger one' is recorded from an early date in the well-known case of the Calf of Man, an island lying close to the Isle of Man,[485] and is also found applied to the names of tributaries of a larger river.[486] Some at least of those Staffordshire places incorporating the word calf are not perhaps the most likely calf-rearing localities, and it seems conceivable that the Old English word may have a secondary meaning applied to tumuli, perhaps in particular an arrangement of one or more smaller tumuli associated with one or more larger tumuli, but further research into similar names elsewhere would be necessary to support or disprove the theory.

---

[485] OED. No other example of the use of 'calf' in this way has been traced; the Calf of Man is said to be from ONorse *kalfr*.
[486] E.g. the river Riccal, in North Yorkshire: CDEPN 498.

# Abbreviations

Abbreviations printed in roman type refer to printed sources and those in italics to manuscript sources. County names are abbreviated in accordance with the convention now adopted in place-name studies.

\* = see Bibliography

| | |
|---|---|
| ABG | *Aris's Birmingham Gazette* |
| Act | Act of Parliament |
| AD | *Catalogue of Ancient Deeds* (PRO), London 1890 and in progress, referred to by volume |
| *AD* | *Ancient Deeds* in PRO (C 146, E 40, 210, 326) |
| *AddCh* | Additional Charters in BrMus |
| ArchIng | *Archaeological Evaluation of a Prehistoric Pit Alignment and Associated Deposits at Ingestre, Staffordshire,* 2000, Shrewsbury: Archaeophysica Limited. |
| AJ | *Archaeological Journal* |
| AngleseyCh | *Calendar of Staffordshire Charters in the possession of the Marquis of Anglesey,* ed. J.H. Jeayes and deposited in SRO (W)D 1734 |
| *AngleseyCh* | Unpublished Charters and Deeds in the Anglesey Collection, transcribed by I.H. Jeayes and deposited in SRO (W)D 1734 |
| Ant | *Antiquity* |
| AntJ | *Antiquaries Journal* |
| *Antrobus* | Deeds deposited by Col. Antrobus of Eaton Hall, Congleton in SRO(W) D 1490, 1535, 1761, 1921 |
| *AOMB* | Augmentation Office Miscellaneous Books (E 315) in PRO |
| app. | Appendix |
| *Aqualate* | Deeds from Aqualate in SRO(W) D 1788 |
| ASC | *The Anglo-Saxon Chronicle,* ed. B. Thorpe (RS), London 1861; *Two of the Saxon Chronicles Parallel,* ed. C. Plummer, Oxford 1892-9 |
| AScand | Anglo-Scandinavian |
| Ass | *Staffs Assize Rolls (1216-1405),* SHC IV, VI, VII, IX, XI, XII, XIII, XIV, XV, ed. Gen. Wrottesley 1883-94; *Justices in Eyre for Gloucestershire, Warwickshire and Staffordshire,* 1221, 1222, ed. D.M. Stenton, Seld. Soc. 59, 1940 |
| *Ass* | Unpublished Assize Rolls in PRO |
| ASWills | *Anglo-Saxon Wills,* ed. D. Whitelock, Cambridge 1930 |
| ASWrits | *Anglo-Saxon Writs,* ed. F.E. Harmer, Manchester 1952 |
| Bacon | J. Bacon, *Liber Regis,* London 1876 |
| Bagot | G. Wrottesley, *A History of the Bagot Family with copies of the deeds at Blithfield,* SHC NS XI, 1908 |
| Banco | *Banco Rolls 1216-1485,* ed. by Gen. Wrottesley, SHC III, IV, IX, XI, XII, XIII, XV, XVI, XVII, NS III, IV, VI, 1882-1903; *Index of Placita de Banco 1327-8* (PRO Lists and Indexes 32), London 1909 |
| BAR | *British Archaeological Reports* |

| | |
|---|---|
| Bateman | T. Bateman, *Ten Years Digging in Celtic and Saxon Grave Hills*, London 1861. |
| Baugh | Baugh's Map of Shropshire 1808 (SA) |
| BCA | Birmingham City Archives |
| BCS | *Cartularium Saxonicum*, ed. W.G. de Gray Birch, 3 vols, London 1885-93 |
| Bede | *Historia Ecclesiastica* in *Venerabilis Baedae opera Historica*, ed. C. Plummer, Oxford 1896 |
| Beresford | W. Beresford and S.B. Beresford, *Beresford of Beresford : A History of the Manor of Beresford*, Leek 1908 |
| Bk | Buckinghamshire |
| BL | British Library |
| Blaeu | Johann Blaeu, *Map of Staffordshire*, 1648 (SRO) |
| BLARS | Bedfordshire & Luton Archives & Record Service |
| Blithfield | D.S. Murray, *Notes on the Early History of the Parish of Blithfield*, SHC 1919; *Deeds from the Blithfield Papers* in R.F. Parker, *Some Account of Colton*, Birmingham 1897 |
| Blome | R. Blome, *Map of Staffordshire*, 1671, 1673 (SRO) |
| Blymhill | *The History of the Parish of Blymhill*, ed. by G.T.O. Bridgeman, SHC I, II (ii), and suppl. xii (ii), 1880-91 |
| Blythe | *Bishop Geoffrey Blythe's Visitations, c. 1515-25*, ed. P. Heath, SHC 4th Series, no. 7, 1973 |
| BodCh | *Calendar of Charters and Rolls in the Bodleian Library*, Oxford, 1878 |
| Bosworth-Toller | Toller 1898* |
| BM | *Index to Charters and Rolls in the Br.Mus.*, 1882-1900, 2 vols, 191-2 |
| Bowen | E. Bowen, *Map of Staffordshire*, 1749 & 1755 (SRO) |
| Bret | Breton |
| Bridgeman | C.G.O. Bridgeman, *Staffordshire Pre-Conquest Charters*, SHC 1916 |
| Brk | Berkshire |
| BrMus | British Museum; documents preserved in the BrMus |
| Brocklehurst | P.L. Brocklehurst, *Swythamley and its Neighbourhood*, London 1874 |
| Browne | J. Browne, *Map of Staffordshire*, 1682 in Plot 1686 |
| BSE | Suffolk Record Office, Bury St.Edmunds |
| Burton | *Descriptive Catalogue of the Charters and muniments of Burton Abbey belonging to the Marquis of Anglesey*, ed. by I.H. Jeayes and M. Deanesly, SHC 1937 |
| BurtonAbSurv | *The Burton Abbey Surveys* (12th century), by C.G.O. Bridgeman, SHC 1916 (pp. 209-300) |
| BurtonAn | *Annales de Burton*, in *Annales Monastici*, vol. 1 (RS), 1864 |
| c. | circa |
| CA | Constables' Accounts |
| CalComp | *Committee for the Compounding of Delinquents 1643-1660*, 5 vols, 1889-93. |
| CartAnt | Exchequer Transcripts of Charters (PRO) |
| Cary | J. Cary, *Map of Staffordshire* 1787 to 1832 (SRO) |

| | |
|---|---|
| CC | Calendar for Compounding |
| CDEPN | Watts 2004* |
| CensusRet | *Census Returns of Great Britain & Enumeration Abstracts,* 1841 |
| CEC | Barraclough 1988* |
| cf. | compare |
| Ch | *Cheshire* |
| CHS | *Cheshire Historical Society* |
| Ch | *Calendar of Charter Rolls* (PRO) 6 vols, 1903-27 |
| *Ch* | Unpublished Charter Rolls in PRO |
| *ChancM* | *A Gnosall Lawsuit of 1395 from Chancery Miscellanea* (PRO), SHC 1927 |
| *ChancP* | Unpublished Chancery Proceedings in PRO |
| Chell | *Chell Deeds from AddCh (BM),* in SHC 1911 |
| Cherry | J.L. Cherry, *Stafford in Olden Times,* Stafford 1890 |
| *ChwAccts* | Churchwardens' Accounts in various parishes |
| CIR | *Rotuli Litterarum Clausarum,* (RC), 2 vols, London 1833-44 |
| CKS | Centre for Kentish Studies |
| Coins | *Catalogue of English Coins, Anglo-Saxon Coins,* 2 vols, 1887, 1894; *Anglo-Norman Series,* 2 vols, 1916 |
| Comm | Exchequer Special Commissions in PRO (E178) |
| CoramR | *Coram Rege Rolls* (1307-1474), SHC X, XIV, XVI, XVII, NS III, IV ed. Gen. Wrottesley, 1882-1903; *Placita coram domino Rege 25 Ed I* (BRS 19), 1898 |
| Court | *Calendar of County Court, City Court & Eyre Rolls of Chester 1259-97,* Chetham Soc. NS. 84 |
| Croxall | R. Usher, *An Historical Sketch of the Parish of Croxall,* London 1881 |
| CroxdenChr | Croxden Chonicle |
| *Ct* | Unpublished Court Rolls in PRO, SRO(W), and in private hands |
| Ct | Published Court Rolls (Alrewas SHC 10, Standon *v.* Salt, Gnosall *v.* Hone, Tatenhill *v.* Hardy, Horton in TNSFC 1926, Audley ibid 1926, Norton-in-the-Moors ibid 1926-29, Tunstall ibid 1924-32 |
| *CtAugm* | Records of the Court of Augmentations in PRO (E 321) |
| Cur | *Curia Regis Rolls* (PRO), in progress |
| D | Devon |
| Db | Derbyshire |
| DB | Domesday Book; the St portion in VCH vol. iv, pp. 1-60; H.M. Fraser, *The Staffordshire Domesday,* Stone 1936 |
| DbA | *Journal of the Derbyshire Archaeological and Natural History Society* |
| DbCh | Jeayes 1906* |
| DCL | Madge 1938* |
| *Deed* | Unpublished deeds in SRO, Br.Mus., PRO and in private hands |
| DEPN | Ekwall 1960* |
| DES | Reaney & Wilson 1997* |
| Dieul | *The Cartulary of Dieulacres Abbey,* ed. Gen. Wrottesley, SHC NS IX, 1906 (a 17th century MS) |
| DNB | *Dictionary of National Biography* |
| DRO | Derbyshire Record Office |

| | |
|---|---|
| Du | Dutch |
| Dugd | Dugdale 1817-30* |
| Duig | Duignan 1902* |
| DuLaMB | *Duchy of Lancaster Miscellaneous Books* in Hardy's History of Tatenhill Vol. 2, pp. 216-20 |
| EEA 14 | Franklin 1997* |
| EEA 16 | Franklin 1998a* |
| EEA 17 | Franklin 1998b* |
| ECP | *Early Chancery Proceedings* (PRO) Lists and Indexes, 12, 16, 20, 29, 38, 48, 50); *Early Chancery Proceedings 1377-1509*, ed. by Gen. Wrottesley, SHC NS VII, 1904 |
| EDD | J. Wright, *The English Dialect Dictionary*, Oxford 1898-1905 |
| e.g. | exempli gratia |
| *Egerton* | Egerton MSS in BrMus |
| EHD | Whitelock 1955* |
| EHNMR | English Heritage National Monuments Record |
| Eliz ChancP or Eliz | *Elizabethan Chancery Proceedings* in PRO (Series II) Lists and Indexes 7, 24, 30, London 1896; *Calendar of Proceedings in Chancery in the reign of Queen Elizabeth* (RC), London 1827-32 |
| *EnclA* | Enclosure Awards in SRO, PRO and in private hands |
| EPNE | Smith 1956* |
| EPNS | Publications of the English Place-Name Society |
| Erdeswick | S. Erdeswick, *A Survey of Staffordshire*, c. 1600 (1717), in Harwood 1844* |
| ERN | E. Ekwall, *English River-Names*, Oxford 1928 |
| ERYARS | East Riding of Yorkshire Archives & Records Services |
| ES | Wolverhampton *Express & Star* |
| ESRO | East Sussex Record Office |
| Ess | Essex |
| etc. | *et cetera*; and the like |
| *et freq.* | *et frequenter*; and frequently (thereafter) |
| FA | *Feudal Aids* (PRO), 6 vols, London 1899-1920: *Testa de Nevill, Monina Villarum*, ed. Col. Wedgwood, SHC 1911 |
| Faden | W. Faden, *Map of Staffordshire* (2nd ed. of Yates *op. cit.*), 1799. (SRO) |
| Fine | *Calendar of Fine Rolls* (PRO), in progress: *The Fine Rolls*, t. Ed II, 1307-27, SHC 1888, ed. Gen. Wrottesley |
| FineR | *Excerpta e rotulis finium* (RC), London 1836 |
| fn. | footnote |
| f.n. | field-name |
| Fr | French |
| FrD | *Documents preserved in France*, Rolls Series 1899 |
| freq. | frequent(ly) |
| GDR | *Gaol Delivery Rolls 1294-1307*, ed. Gen. Wrottesley, SHC VII, 1886 |
| Gervase | Gervase of Canterbury, *Mappa Mundi*, Chron. and Mem. 28 (1867-9) |
| Gibson | Gibson 1695* |
| GKNB | *Gregory King's Notebook 1679-80*, SHC 1919, ed. by G.P. Mander |

| | |
|---|---|
| GLO | Gloucestershire Record Office |
| GM | *Gentleman's Magazine* |
| Greenwood | C. & J. Greenwood's map of Staffordshire 1820 (SRO) |
| Hales | *Court Rolls of the Manor of Hales,* Worcestershire Historical Society 1910-12 |
| HAME | Woolgar 1993* |
| HantsRO | Hampshire Record Office |
| Hardwicke MS | Hardwicke* |
| *Harl* | Harleian MSS in BrMus |
| HaRO | Hampshire Record Office |
| Harrison | W. Harrison, *Description of Britain* in Holinshed's Chronicles (vol. 1), 1577 & 1586 |
| Hatherton | *Calendar of Lord Hatherton's Charters,* ed. H.L.E. Garbett, SHC 1928 & 1931 |
| HB | *Historia Brittonum cum additamentis Nenni*, ed. Mommsen, MGH, Auct. ant. xiii (1899) |
| HL | House of Lords Library |
| HLS | Staffordshire Deeds, Harvard Law School Library |
| HMC | Historical Manuscripts Commission |
| HOK | Harrison 1986* |
| HRO | Herefordshire Record Office |
| Hrt | Hertfordshire |
| Huntbach | Huntbach MSS, WSL |
| ICG | *Inventory of Church Goods* (SHC NS VI (i); Hutchinson 1893; Sleigh 1883 199-301) |
| IE | Indo-European |
| Inq aqd and Indexes (17, 22) | *Calendarium Inquisitionum ad quod damnum* (RC), London 1803, *Inquisitions ad quod damnum* (PRO), Lists and Indexes (17, 22), London 1904, 1906 |
| IOW | Isle of Wight |
| Ipm | *Calendar of Inquisitions post mortem* (PRO), in progress; *Inquisitions post mortem* 1223-1366, SHC 1911, 1913; unpublished Inquisitions post mortem in PRO |
| Jansson | J. Jansson, *Map of Staffordshire* 1646 (SRO) |
| JEPNS | *The English Place-Name Society Journal*, in progress |
| JHMS | *Journal of the Historical Metallurgy Society* |
| JNSFC | *Journal of the North Staffordshire Field Club* |
| John of Worc | *The Chronicle of John of Worcester 1118-1140*, ed. J.R.H. Weaver (Anecdota Oxoniensis), 1908 |
| JRL | Manchester University, John Rylands Library |
| JRS | *Journal of Roman Studies* |
| K | Kent |
| Keer | P. Keer, *Map of Staffordshire*, 1880 (SRO) |
| Kip | Kip's map of Staffordshire 1607 (SRO) |
| Kitchen | T. Kitchen, *An accurate map of Warwickshire*, 1770 |
| Kelly | Kelly, *Directory of Staffordshire*, 1880 |
| La | Lancashire |
| Leland | Toumlin Smith 1906-10* |
| Li | Lincolnshire |
| LLRRO | Leicestershire, Leicester & Rutland Record Office |

| | |
|---|---|
| Longleat | Longleat House Archives |
| Loxdale | Collections of the Rev. Thomas Loxdale, SRO(W) 237M |
| Loxdale | *Deeds from the Loxdale MSS in the William Salt Library*, SHC 1911, pp. 438-443 |
| LP | *Letters and Papers Foreign and Domestic, Henry VIII* (PRO) London 1864-1933 (in progress) |
| LRMB | Land Revenue Miscellaneous Books in PRO |
| LyttCh | *Charters of the Lyttleton Family*, ed. J.H. Jeayes, London 1893 |
| M | Middle |
| MA | Monastic Accounts |
| ME | *Middle English* |
| MED | *Middle English Dictionary*, ed. H. Kurath, University of Michigan, Ann Arbor, in progress; Stratman 1891* |
| MedA | Medieval Archaeology |
| Mercator | Mercator's Atlas of Europe, 1564 |
| MidA | Carter 1882-7* |
| Middleton | University of Nottingham Middleton Collection |
| MinA | Ministers' Accounts in Lists and Indexes (PRO) 5, 8, 34 |
| ModE | Modern English |
| MLR | *Modern Language Review* |
| Moll | H. Moll's map of Staffordshire 1701 (SRO) |
| Morden | Morden's map of Staffordshire 1695 & 1701 (SRO) |
| MR | Muster Roll |
| MRA | *Magnum Registrum Album (1317-28)*, ed. by H. E. Savage, SHC 1924. |
| MS, MSS | Manuscript(s) |
| N&Q | *Notes and Queries*, in progress |
| N | note |
| n.d. | No date/not dated |
| Nf | Norfolk |
| NGR | Ordnance Survey National Grid reference |
| Nt | Nottinghamshire |
| n.p. | No publisher named |
| NS | New Series |
| NSFC | North Staffordshire Field Club |
| NSJFS | *North Staffordshire Journal of Field Studies* |
| Nightingale | J. Nightingale, *The Beauties of English and Wales, Staffordshire*, vol. xiii, London 1813 |
| Nomina | *Nomina : A Journal of Name Studies relating to Great Britain and Ireland published by English Name Studies*, in progress |
| O | Old |
| O | Oxfordshire |
| OblR | *Rotuli de Oblatis* (RC), London 1835. |
| OBrit | Old British. |
| ODan | Old Danish |
| OE | Old English |
| OEBede | *The Old English Version of Bede's Ecclesiastical History*, ed. Th. Miller, Early English Text Society 95ff (1890 etc.) |
| OED | *Oxford English Dictionary* |
| OEN | *Oxford English Newsletter*, Richard Rawlinson Center for Anglo-Saxon Studies and Manuscript Research, Western |

| | |
|---|---|
| | Michigan University |
| OFr | Old French |
| OHG | Old High German |
| *Okeover* | Okeover Deeds in DRO |
| Okeover Deeds | *The Okeover Family of Okeover*, ed. Gen. Wrottesley, SHC NS 7, 1904 |
| ON | Old Norse |
| Ord | Ordericus Vitalis, *Historia Ecclesiastica*, ed. A. le Prévost (Société de l'Histoire de France), Paris 1838-55 |
| Orig | *Originalia Rolls* (RC), 1805-10; Pipe Roll Society NS vol. xxi, 1943 |
| ORO | Oxfordshire Record Office |
| OS | Ordnance Survey |
| OS | Original Series |
| OSS | Transactions of The Old Stafford Society |
| Oxf | *Facsimiles of Early Charters in Oxford.* Oxford, 1929. |
| (p) | Place-name used as a personal name or surname |
| P | *The Staffordshire Pipe Rolls 1155-1216*, ed. R.W. Eyton, SHC I & II, 1880-81; The Pipe Roll Soc., in progress; *The Pipe Rolls* (RC), 3 vols, 1833-44; *The Great Roll of the Pipe for 26 Henry 3*, ed. Cannon 1918 |
| PA | Parliamentary Accounts |
| ParReg | Parish Register(s) |
| ParSurv | Parliamentary Survey |
| Pat | *Calendar of Patent Rolls* (PRO) in progress |
| PCC | *Index of Wills proved in the prerogative Court of Canterbury* (BRS), in progress |
| PDNPAS | Peak District National Park Archaeology Service |
| Peniarth | Peniarth MS 390 (14th century) in the National Library of Wales |
| *PlR* | *Plea Rolls* (1189-1327), ed.Gen. Wrottesley, SHC III, IV, VI, VII, X (1882-9) |
| PMA | Post Medieval Archaeology |
| PNEF | Owen 1994* |
| PN | Place Names |
| PN Sa | Place Names of Shropshire |
| PO | Post Office Directory |
| Poll | Copy of A Poll Taken for the County of Stafford the 9th, 10th, 11th, 13th and 14th Days of July, in the Year of Our Lord 1747, before George Hunt, Esq., High Sheriff for the said County |
| Pr | Primitive |
| PR(H) | *Shropshire Parish Registers*, Shropshire Parish Register Society, Hereford Diocese |
| PRO | Public Record Office (PROCAT on-line catalogue) |
| ProcJP | B.H. Putnam, *Proceedings before the Justices of the Peace in the 15th & 16th centuries*, London 1928 |
| PrWelsh | Primitive Welsh |
| QJF | *Quarterly Journal of Forestry*, Royal Forestry Society of England, Wales & Northern Ireland |
| QSR | *Quarter Session Rolls for Staffordshire* (1581-1609), ed. S.A.H. Burne, G.P. Mander and D.H.G. Salt, SHC 1927, 1929-30, |

| | |
|---|---|
| | 1932, 1935, 1940, 1948-9 |
| QSREnr | *Quarter Sessions Rolls Enrolled 1580-1621*, SHC 1934 |
| q.v. | *Quod vide* |
| RB | Wilson 1907* |
| RBE | *Red Book of the Exchequer*, 3 vols, 1896 |
| Reg | *Regesta Regum Anglo-Normannorum*. Oxford, 1913, 1956 |
| RegDiss | B. Donaldson, *The Registrations of Dissenting Chapels & Meeting Houses in Staffordshire 1689-1852*, SHC 4th Series 3 1960 |
| Rental | Unpublished rentals in BrMus, PRO, SRO, and in private hands |
| RH | *Rotuli Hundredorum, 2 vols.*, 1812-1818; *Hundred Rolls 1255 & 1275*, ED. Gen. Wrottesley, SHC V(I), 1884; *The Offlow Hundred Roll of 39 Hen III*, printed by Stebbing Shaw 1798-1801 |
| RHP | Johnson & Cronne 1956* |
| RontonC | *The Ronton Priory Cartulary*, ed. Gen. Wrottesley, (Text and Introduction), SHC IV 1883 |
| RRAN | Cronne & Davis 1968* |
| RydewareCh | *Rydeware Family Cartulary*, ed. I. H. Jeayes and Gen. Wrottesley (Text & Introduction), SHC XVI, 1895 |
| SA | Shropshire Archives, formerly Shropshire Records & Research Centre |
| SaDeeds | *Old Shropshire Deeds*. Shropshire Archaeological Society |
| SAHS | *Transactions of the Shropshire Archaeological & Historical Society* |
| Saints | *Die Heiligen Englandes*, ed. F. Liebermann, Hanover 1889 |
| Saxton | Christopher Saxton, *Map of Staffordshire 1577*; also his *Atlas of England and Wales*, London 1576. |
| SBT | Shakespeare Birthplace Trust Records Office |
| SCH | *Staffordshire Catholic History* |
| Senior | The Survey of Wetton; MS of William Senior's Plans in Chatsworth House, 1617, belonging to the Rt Hon. Lord Cavendish (copy in SRO) |
| SHC | *Staffordshire Historical Collections* (formerly *Transactions of the William Salt Archaeological Society)* |
| SHT | The Staffordshire Hearth Tax, 1666, ed. G. E. Grogan, SHC 1921 (Pirehill), 1923 (Seisdon and Offlow), 1925 (Totmonslow), 1927 (Cuttlestone), and for Lichfield ed. P. Laithwaite, SHC 1936. |
| Sketchley | *Sketchley & Adam's True Guide; or an Universal Directory for the towns of Birmingham, Wolverhampton, Walsall, Dudley [etc.]*, 4th edition 1763 |
| Sleigh | J. Sleigh, *A History of the Ancient Parish of Leek in Staffordshire*, London & Leek 1883 |
| Smith | Smith's MS map of Staffordshire 1599 (SRO); Smith's map of Staffordshire 1801 (SRO) |
| SMR | Sites & Monuments Record (Staffordshire County Council Planning Department unless other county shown) with index ref. |
| s.n. | sub. nom. = under the name of |
| SOT | Stoke on Trent City Archives |
| SOTMAS | City of Stoke on Trent Museum Archaeological Society |

| | |
|---|---|
| SP | *Staffordshire Pedigrees* (Harl. Soc. Pubs 63), 1912 |
| SPDom | State Papers Domestic in PRO (SP 14 16) |
| Speed | Speed's map of Staffordshire 1610 (SRO) |
| SPI | J. S. Roper, *Sedgley Probate Inventories 1614-1787*, Dudley n.d., n.p. |
| SR | *Subsidy Roll 1327, 1332-33,* ed. Gen. Wrottesley SHC vols viii & x 1886-9; *Subsidy Roll of 1640* ed. S.A.H. Burne, SHC 1941 |
| Sr | Surrey |
| SRO | Staffordshire Record Office |
| SRS | *Shropshire Records Series* (Keele: Centre for Local History) |
| SS | Seldon Society |
| St | Saint |
| StASt | *Staffordshire Archaeological Studies*, Stoke on Trent City Museum |
| *StEnc* | Cockin 2000* |
| StaffAcc | The Earl of Stafford, Manors, SRO D641/1 |
| StarCh | *Star Chamber Proceedings T.Hy 7 & Hy 8 1516-1549*, ed. W.K. Boyd, SHC 1910-1912, and X ns (i); *Lists of Proceedings in the Court of Star Chamber 1485-1558* (PRO Lists and Indexes 13) |
| StCart | *The Staffordshire Cartulary,* ed. R.W. Eyton, SHC II (i), 1881 and Gen. Wrottesley, SHC III, 1882, and SHC 1911 (pp. 416-37), ed. Col. Wedgwood; *Two Early Staffordshire Charters,* ed. C Swynnerton, SHC 1926 |
| StComm | *The Committee at Stafford 1643-45,* ed. G.D.H. Pennington & I.A. Roots, SHC 4th Series, Vol. I, 1957 |
| Stockdale | J. Stockdale, *Two Maps of Staffordshire* in J. Aikin, *A Description of the Country round Manchester*, Manchester 1794 |
| StSt | *Staffordshire Studies* |
| Stratmann | Stratmann 1891* |
| StThomas | *Deeds of St Thomas Priory Stafford* (Text and Introduction), ed. F.P. Parker, SHC VIII, 1887 |
| StV | *The Heraldic Visitations of Staffordshire 1614, 1663, and 1664* (SHC VII) 1884 |
| *Survey* | Unpublished Surveys in PRO, SRO and in private hands |
| Sx | Sussex |
| TA | Tithe Award |
| Taliesin | J. G. Evans, *Facsimile and Text of the Book of Taliesin*, Llandovery, 1910 |
| *TallAss* | Tallage Assessments in PRO (E 179/237/13) |
| Taylor | Taylor's map of Wolverhampton 1751 (SRO) |
| TBAS | *Transactions of the Birmingham Archaeological Society* |
| TBWAS | *Transactions of the Birmingham & Warwickshire Archaeological Society* |
| Teesdale | J. Teesdale's map of Staffordshire 1832 (SRO) |
| *temp.* | Temporore, in the time (reign) of, followed by the monarch's name |
| *Terrier* | Unpublished Terriers in various hands |
| TNFSC | *Transactions of the North Staffordshire Field Club*, from 1961 *North Staffordshire Journal of Field Studies* |
| TOE | Roberts, Kay & Grundy 1995* |
| TPS | *Transactions of the Philosophical Society* |

| | |
|---|---|
| TRS | *Philosophical Transactions of the Royal Society* |
| TSAS | *Transactions of Shropshire Archaeological & Historical Society* |
| TSAHS | *Transactions of Staffordshire Archaeological & Historical Society* |
| TSHCS | *Transactions of Stafford Historical & Civic Society* |
| TSSAHS | *Transactions of South Staffordshire Archaeological Society (formerly Lichfield and South Staffordshire Archaeological & Historical Society)* |
| Tunnicliff | Tunnicliff's map of Staffordshire 1786 (SRO) |
| *(Tw)* | F.R. Twemlow, *History of the Manor of Mere, Co. St, which includes the Township of Mere or Meretown, Acqualate, Forton, Sutton & Wharton,* copy in typescript 1916-20, SRO (W)182/261 |
| TWHS | *Transactions of Worcester Historical Society* |
| ValEccl | *Valor Ecclesiasticus.* Record Commission 1810 ff. |
| VCH | *Victoria County History* (of St unless otherwise indicated) |
| VE | *Valor Ecclesiasticus* (RC), 6 vols, 1810-34 |
| VEPN | Parsons & Styles 1997* |
| Visitation | *The Visitation of 1583,* ed. by H.S. Grazebrook, SHC III (ii), 1881. |
| W | Wiltshire |
| Wa | Warwickshire |
| WALS | Wolverhampton Archives & Local Studies Library |
| WC | *Wolverhampton Chronicle* |
| WaCRO | Warwickshire County Record Office |
| *Wills* | Unpublished Wills at Somerset House, the Lichfield Registry & in private hands |
| Wills | *A Register of Stafford and other local Wills,* ed. G.P. Mander, *SHC 1926.* |
| WJ | *Wolverhampton Journal* |
| WL | J. B. Hughes (ed.), *The Register of Walter Langton, Bishop of Coventry and Lichfield 1296-1321,* Vol. I, The Boydell Press for The Canterbury and York Society, 2001. |
| WMA | *West Midlands Archaeology* |
| WMANS | *West Midlands Archaeological News Sheet* |
| Wo | *Worcestershire* |
| Wodehouse | *Deeds at the Wodehouse, Wombourne,* ed. G.P. Mander, SHC 1928. |
| Worc | *Itineraria Symonis Simeonis et Willelmi de Worcestre,* ed. J.Nasmyth, Cambridge 1778 |
| WRO | Worcestershire Record Office |
| WSL | William Salt Library, Stafford |
| Wulst | Annals of the Hospital of St. Wulstan, with Cartulary, ed. F.T. Marsh 1890 |
| WYAS | West Yorkshire Archive Service |
| Yates | William Yates' maps of Staffordshire 1775 (based on a survey made 1769-75) and 1798 (SRO) |
| Yo | Yorkshire |

# Gazetteer - arrangement of entries

The modern name of each place is given in the gazetteer in bold capitals, and underlined if the place is recorded in Domesday Book. Places not found on a modern map, whether identifiable or nct, are shown in ordinary capitals. A place identified as Ancient Parish is an ancient parish as defined in Youngs 1991, i.e. parishes which existed before 1597. Counties are the historical counties existing before the 1974 local government reorganisation. An arbitrary selection of unlocated places has been included for the possible interest of researchers, with the important caveat that some may lie outside Staffordshire. Each identified name is followed by an approximate location (with approximate distances, given, without apology, in traditional English miles), including where possible a four-figure Ordnance Survey National Grid Reference. In the case of larger conurbations the Grid Reference has been based on the centre on the present place, rather than any historic core. The word 'lost' indicates that the name does not appear on the latest maps, and 'unlocated' that the exact location has not been identified. 'Obsolete' indicates that the name is not found on modern maps. When available, early spellings are given in italics, in chronological order, with their date and an abbreviated reference to the source. References are given by the Harvard system, with the author's name, the year of publication (or edition), and page number. A question mark before a name indicates uncertainty whether the name refers to the place; a question mark after the name shows that the spelling is uncertain; and a question mark before a date indicates that the date is uncertain. A second date in brackets is the date of the document or publication in which the first date appears. The letter $S$ followed by a number (in brackets) after a spelling indicates, in accordance with current convention, that the name is taken from an Anglo-Saxon charter of that number as listed in Sawyer 1968, though the spelling itself may well be taken from another source, e.g. Finberg 1972 or Hart 1975. (p) after a name indicates that it is recorded as part of a personal name. An asterisk before a word indicates that it is not recorded in early sources but its existence can be deduced with some certainty from philological research. An element that appears in italics indicates that the meaning will be found under the List of Elements towards the end of this volume, or in EPNE. Where appropriate, references to Roman roads cite the standard reference number from Margary 1973.

The text reflects the practice of English place-name studies at the time this work was started of citing words from which place-names may be derived in Old English forms in respect of names more likely to have been coined in the Middle English or Early Modern English periods, but contrary to the tradition long adopted by the English Place-Name Society, names have not been assembled into their relevant parish. The reason is that such a system only partly satisfies the desirability of relating adjoining or nearby names, for the names of adjoining places separated by a parish boundary will not be grouped together, and in any event place-names within individual parishes (some of which are particularly sizeable in Staffordshire) are conventionally listed alphabetically. In this volume names are simply listed in alphabetical order. Although Forster 1981 contains a number of pronunciations for Staffordshire names,[487] many of which are inaccurate, it has not proved possible to identify the modern local pronunciation of every minor place-name listed in this work, though many names which have unusual pronunciation have been recorded using the standard phonetic alphabet.

---

[487] For example Bushbury, Keele, Kinver, Leek, Pattingham.

# Gazetteer

**ABBEY GREEN**  1 mile north-west of Leek (SJ 9757). *Abbey Green* 1611 SRO DW1702/1/23, *Abbie Green* 1634 Leek ParReg, *Abbey Green Farm* 1677 SRO DW1702, *Abbey Green* 1696 Leek ParReg, 1842 O.S. 'The grassy open place at the abbey', from Dieulacres Abbey which lies nearby to the east.

**ABBEY HULTON**  near Burslem, 3 miles north-east of Stoke on Trent (SJ 9148). *Heltone* 1086 DB, *Hiltona* 1166 SHC 1923 297, *Hultone* 1242 SHC XI 314, *Hylton* 1281 SHC VI (i) 120, *Hulton* 1358 HLS, *Abbehilton* 1587 SHC XV 180, *Abbie-Hulton* 1591 SHC XVI 114, Hilton Abbey 1601 SHC 1935 341, *Abby Hulton* 1678 Norton-in-the-Moors ParReg, *Hilton Abbey* 1749 Bowen. From OE *hyll* and *tun*. 'Hill *tun*': the place is on high land. *Abbey* is an addition from the Cistercian abbey founded in 1219, with a charter of 1223 (VCH III 235), probably a much later forgery (Tomkinson 1994: 73-102). *Thabbay Mill*, recorded at Hulton in 1539 (MA), is evidently 'Th' Abbey Mill'.

**ABBOTS BROMLEY – see BROMLEY, ABBOTS.**

**ABBOT'S CASTLE HILL**  a 2-mile long escarpment on the Staffordshire-Shropshire border 2 miles west of Trysull (SO 8294). *Aquardescastell* 1295 SHC 1911 224, *Aguardescastel* 1295 SHC VII 29, *Aquardescastel* 1295 VCH XX 185, *Akewardes castel* 1298 TSAS LXXI 1996 27, *apewardes castel* 1301 Rees 1975: 248, *Apeward* and *Apeis Castle* 15th century VCH XX 185, *Abbots or rather Ape-wood Castle* 1686 Plot 397, *Abbots Castle Hill* 1752 Rocque, *Abbots Castle* 1775 Yates. The element *castel(l)* is found in Welsh, Cornish, OE, OFr, and ME, from Latin *castellum* 'a fort', but is usually a post-Conquest element introduced by the Normans. Occasionally it is from OE *ceastel* 'a heap of stones', but that is not likely here. *Aquard-* and *Aguard-* may, given the *castel* element, be from the OFr personal names *Aquart*, *Achart*, which are probably from OGerman *Akihart*, *Agihard* (the name is not unrecorded in Staffordshire: Walter Achard appears in connection with Bramshall c.1275 (SRO DW1733/A/2/64), and William Achart in connection with Stowe-by-Chartley in 1318 (SRO D938/493)). The change from *Aq-* or *Ag-* to *Ap-* is curious, and may perhaps be explained as misreadings. Whatever the derivation of the name, there is no connection with any abbot. Hardwick 1822 suggests that the hill was at that date also called *the Frife*, perhaps for Frith (q.v.). An intermittent ditch, which may be of Anglo-Saxon or earlier date, extends along the crest of the escarpment (*la Rugge* 'the ridge' in 1294: SHC 1911 224), and Plot 1686: 397 records 'a very ancient and considerable fortification'; see also VCH I 372. The county boundary formerly ran along the crest of the escarpment, but since 1895 only the northern part of the ridge marks the boundary. The 'castle' (almost certainly some kind of earthwork) from which the escarpment was named appears from its sequence in early perambulations and from early maps (e.g. Yates 1775) to have been at the north-west end of the escarpment, at what is now Hillend, where the county boundary turns around the steep headland and forms a pronounced curlicue. William atte Castell mentioned in a Pattingham court roll in 1327 and 1373, and William Castell mentioned in 1412 (Brighton 1942: 159; 24; SHC 1939 191) may have taken their surnames from this place: Pattingham is 2½ miles to the north-east, and no castle (other than the iron-age fortification at The Walls (q.v.)) is recorded in Pattingham or elsewhere in the area. That part of the escarpment cut by the road from Seisdon to Upper Aston is called Tinker's Castle (q.v.). A Roman road (? *stonystrete* 1298 TSAS LXXI 1993 27; ? *Tom or Thom Street* 15th century VCH XX 65, perhaps for Town Street: see Foxall 1980 7; cf. Tomhill Cottage, Tom Lane) from Greensforge lies below the ridge. At the south-eastern end of the ridge is a '...small square intrenchment with a single ditch, situated on a round promontory...' (VCH I 192), recorded as *Bekwyneburynesse* in a perambulation of 1295 (SHC 1911 224). This name (which Duignan associates with Beckbury, Shropshire: see TSAS IX 1897 388, where Duignan also notes correctly that '[the term ness] is usually applied to physical features such as a tongue or nose of land running out into the Sea, a cape or promontor', but adds inexplicably 'I am not aware of anything of that kind in the locality') would appear to contain a personal name (Duignan suggests Berchtwine: ibid. 388), with OE *burh*, dative singular *byrig*, 'earthwork, fortification', and Mercian OE *ness* 'a promontory, a headland, a projecting piece of high land', so giving 'Bekwin's fortification on the promontory'. Shaw 1801: II 278 mentions two small camps near the Hoar Stone (War Stone q.v.) at the south-eastern end of Abbot's Castle Hill, of which *Bekwyneburynesse* is likely to be one. The location of the other earthwork is uncertain, although aerial photographs have located an irregular quadrilateral cropmark at approximately SO 8392.

ABBOTTS FOREST  (obsolete, unlocated, in Leekfrith). *Abbotts forrest* 1538 Elliott 1984: 48, *Abbotts forest* 1539 SHC IX NS 301. So-named because the forest, which included High Forest (q.v.), was held by Dieulacres abbey.

**ABLEWELL**  in Walsall (SO 0198). *Abelwellsych* 1309 VCH 17 147, 221, TSSAHS XVII 1975-6 72, *Able Well* 1398-9 VCH XVII 143, *Abulwall Streets* 1511 BCA MS917/1258, *Aumblers (Street)* 1680x1690 TSSAHS XVII 1975-6 70, *Ablewell (Street)* 1756 SHC 1910 255. Duignan 1880: 26; 1902: xi gives a spelling *Avalwalle* from the 13th century, and proposes a derivation from Norman French *aval* 'below' with ME *walle* 'earthwork, entrenchment', linking the place to a street called The Ditch, from an earthwork still visible in the mid 19th century. However, the little evidence available suggests that the second element is from Mercian OE *wælle* 'a spring', and (sometimes) 'a stream', and (rarely) 'a well'. The first element is uncertain, but could be an unidentified personal name such as Eadbeald in abbreviated form, or the surname Abel(l): see for example SHC 1910 212-7. *Sych* is from OE *sic* 'water-course'. The name *Ablecott* is recorded in the Wednesbury area in 1684: BCA MS3145/91/2,

ABLOW  (obsolete) between Graiseley and Wolverhampton, above Graiseley Brook on the south-west side of Wolverhampton (SO 9098). *Abbelowefeld* 1361 BCA MS3145/117/1, *Abbelowefeld* 1420 D593/B/1/26/38/2, *Ablowfild* 1481 SRO D593/B/1/26/6/17, *Ablowfyld* 1498 SRO D593/B/1/26/6/29/12, *Abloo* 1516 SRO D593/B/1/26//6/29/11, *Ablowe (Field)* 1671 SRO D4092/C/1/57, *Ablow field* 1699 WA II 37. Perhaps 'Ab(b)a's *hlaw* or tumulus'. The mound was evidently still in existence c.1800, when it was said to be planted with a bush called *Iseley Cross* (*sic*): Shaw 1801: II *172; SHC 1982 72, and is said to lie beneath St Paul's church: Hackwood 1908: 7. The place, which gave its name to one of the open fields of Wolverhampton, is remembered in the name Ablow Street.

**ABNALLS**  1 mile north-west of Lichfield (SK 1010). *Abbenhale* 1274 SHC 1924 264, 1293 SHC VI (i) 255, 1347 SHC 1913 122, *? Habenhale* 1332 SHC X (i) 110, *Abenhale* 1349 SHC XVIII 117, *Abbenall* c.1537 SHC 1931 223, *Abnall* 1543 SHC XI 284, *Abbenall* 1601 SHC XVI 208. Probably 'Ab(b)a's *halh*'. Cf. Abinghall, Gloucestershire.

ABOVE PARK – see **CHEADLE**.

ACARD (unlocated, perhaps near Tutbury: SHC 1912 222) *Acard* (undated) SHC 1912 222.

**ACKBURY HEATH** 1 mile south of Brewood (SJ 8706). *Herkebarowe* 1199-1209 St Cart, *Erkebarghe* 1305 SHC VII 130, *Erkbarue, Erkbarwe, Erkbarow(e)* 1306, 1332 Ct, Erkenaroheth 1424 *Ct, Arkeborrowe heath* 1587 *Ct, Ackburyes* 1724 *Survey, Hackbury Heath* 1834 O.S. The first element is uncertain, but is unlikely to be the same as that in High Ercall, Childs Ercall, and The Ercall (an outlying outcrop of The Wrekin massif), all in Shropshire, as suggested by Oakden 1984: 36. A derivation from OE *earc* 'ark, chest' (see Gelling & Cole 2000: 118), perhaps used in a topographical sense for a flattened roof-like ridge with slight summit reminiscent of the lid of an early wooden ark or chest (which might fit the topography here, with nearby Giffard's Cross standing on the crest of a ridge formed by the Upper and Lower Avenue of Chillington), is unlikely, since OE *earc* would give ME *ark*, which is not found in the early spellings. Early ME *erk* would appear to be from OE *\*eorc*, but any such word is unrecorded. For completeness it may be added that there is an OE charm 'For unfruitful land' (BL MS Cotton Caligula A vii, Fol 176a-178a) which contains a verse passage referring to a pagan deity *'Erce, Erce, Erce, mother of earth'*, but nothing is known of this Erce from any other source, and the name in any event was probably pronounced Erch. The first element must therefore be considered unresolved. The early forms show that the second element of Ackbury is OE *bearu* 'wood, grove', and cannot be OE *beorg* meaning 'mountain, hill', and especially 'a barrow or tumulus'. The place is a farmstead at the fork of a lost Roman road (Margary number 191; see also Horovitz 1992: 34-5) running south from Pennocrucium (Water Eaton), which might provide some clue to the meaning of the first element.

**ACKLETON** 2 miles north-east of Worfield (SO 7798). *Aclinton* 1176 CDEPN, *Akelington* 1238x1250 Eyton 1854-60: III 112, *Akinton* 1256 ibid. 112, *Akelinton* 1272 ibid. 112, *Acliton* 1291 Eyton 1854-60: II 76, *Adelacton* 1292 CDEPN, *Akulton* 1505 TSAS 3rd Series III 120, *Aculton* 1532 SHC 4th Series 8 117, *Acleton* 1552 TSAS 3rd Series IX 1909 120, 1562 Worfield ParReg, *Ackleton* 1752 Rocque. Perhaps from OE *ac-hyll* 'oak hill', with OE *tun*, but the earliest spellings suggests an *–ingtun* derivation giving 'the *tun* associated with the oak hill', or 'the *tun* associated with [an unidentified personal name: CDEPN suggests Eadlac]'. In Shropshire since the 12th century.

**ACRE HEAD** (obsolete) 1 mile east of Meerbrook (SK 0060), *Acre Head* 1842 O.S.; **ACRE, UPPER** 2 miles north-west of Butterton (SK 0458), *Upper Acre, Lower Acre* 1842 O.S. From OE *æcer* 'field, ploughed land', in which case the first place is 'the top or head of the place with the ploughed land', and the second place 'the northern or higher place with the ploughed land'. It is unclear whether *Aker*, recorded in 1696 (Leek ParReg), refers to either of these places.

**ACTON** (unlocated, in Congreve) *? Acton (Bridge)* 1598 Ct, *Acton* 1689 SRO D1057/G/1/3. Seemingly 'oak *tun*'.

**ACTON HILL** 1 mile south-east of Eccleshall (SJ 8328). *Haketon* 1170 SHC I 61, 1190 SHC II 14, *Hakedene* 1254 SHC 1924 168, *Hakedone* 1314 SRO D938/39, *Hokedon* 1343 SHC XI 153, *Haketon* 1471 SHC IV NS 174, *Hakedonhill* 1588 SHC XII NS 96, *Hakedon, commonly called Acton, Hill...* 1679 SHC 1914 57, *Acton Hill* 1833 O.S. The earliest form suggests a derivation from OE *haca* 'a hook', with inconsistent terminals which could be OE *tun*, OE *denu* 'a dene, a valley', or OE *dun*, giving 'the hook-shaped or twisted *tun* or hill or valley'. The hachuring on the first edition O.S. map indicates that the place lies at the end of a long ridge with a hook-like spur at the south-east end, so perhaps 'the hill with the hooked end'.

**ACTON TRUSSELL** in Baswich parish, 3 miles south-east of Stafford (SJ 9318). *Actone* 1086 DB, *Actona Willelmi* 1166 SHC 1923 296, *Trussel'* 1204 OblR, *Aketon'* (p) 1206 Cur, *Acton'* 1242-3 Fees, *Acton Trussel(l)* 1481 Coram R, 1507 Ipm; **ACTON** 1 mile north-east of Whitmore (SJ 8241), *Acton* 13th century SRO D938/235, 1589 SHC 1929 326, 1836 O.S. 'Oak *tun*'. Willelmi (William) was presumably an early owner, and Trussell is from a local family who owned land here from at least 1342: VCH V 13. A full discussion of the name Acton which, it has been suggested, may have had the specific meaning 'subordinate settlement where the handling of oak timber was a specialised function', can be found in PN Sa I 1-4. The *Actun* mentioned in the will of Wulfric Spot of 1002x1004 (11th century, S.1536; see Whitelock 1930: 48, 157) could be either of these places (Acton near Whitmore lies some 5 miles north-west of Darlaston, the name of which follows that of Acton in the will, and some 7 miles from Cotwalton, which is also mentioned), or any of the Shropshire Actons.

**ADBASTON** Ancient Parish 4½ miles west of Eccleshall (SJ 7627; SMR 02095 places a deserted Anglo-Saxon settlement at SJ 76102780). *Edboldestone* 1086 DB, *Ædbaldestone* 1175 P, *Edbaldeston* 1200 SHC III 68, *Albodestun* 1224 SHC IV 30, *Atbaldestone, Ambaldeston* 1278 SHC 1911 170, *Atbaldeston* 1278 Ipm, *Adbaldestone, Albaldiston, Alboldestun* 12th and 13th century Duig. 'Eadbald's *tun'*. Eadbald was a common OE personal name.

**ADDERLEY** 1½ miles east of Dilhorne (SJ 9943), *Aderdeleye* c.1316 SRO D1229/1/4/10, *Adderley* 1332 SHC X (i) 117, *Adderley Greene* 1597 SHC 1932 333, *Adderley* 1833 O.S.; ADDERLEY GREEN 2 miles north of Madeley (SJ 7747), *Adderley Green* 165 ibid, 4 Betley ParReg, *Adderley Gr* 1775 Yates; **ADDERLEY GREEN** 1½ miles north-east of Longton (SJ 9244), *Addredeleg', Audredeleye* 1242-3 Fees, *Adderleigh* 1293 SHC 1931 263, *Adirdeeley*? 13th century SHC XI 306, *Adderley* 1327, 1332, 1356, 1465 ibid, *? Adurlegh* 1411 SHC XVI 76, *Adderley Green* 1582 Trentham ParReg, *Netherley Greene otherwise Adderley Greene* 1613 SHC 1934 (ii) 32, *Adderlegrine* 1655 ibid, *Atherley Green* 1693 ibid, *Upper Adderley Green(e)* 1663 SRO D3575/1, *Adderly Greene* 1644 SHC 4th Series I 100, *Atherley Green* 1836 O.S. From the OE personal name Aldred, so 'Aldred's *leah'*. The first two names may be transferred from Adderley near Longton (see SHC 1910 263-6), or possibly Adderley in Shropshire between Audlem and Market Drayton. *Adderleys Wood* appears on the 1836 O.S. map ½ mile north of Drointon (SK 0227), but earlier spellings have not been traced. See also Shaw 1798: I 81. The Green element suggests a squatter settlement.

**ADDER'S GREEN** 1 mile south of Flash (SK 0265). *Edders Grenehed* 1560s SHC 1938 99, *Eddersgreen* 1664 DRO D2375M/25/18, *Edars Green* 1688 Alstonefield ParReg. Perhaps from the recorded personal name Edda, rather than from OE *nĕddre* 'snake, adder', for which *E-* could be explained by the normal development of *æ* to *e* in West Midland dialect. *Grenehed* suggests a grassy area at the top or head of a hill.

ADELLAKESHUL (unlocated, in Swinscoe) *Adellakeshul* 15th century Okeover E192.

**ADMASTON** 4 miles north of Rugeley (SK 0523). *Ædmundeston* 1176 P, 1178 P, *Edmodeston* 1177 P, *Edmodistona* 1179 SHC I 92, *Ædmodeston* 1180 P, *Admodestun* 1181 SHC I 95, *Admundeston* 1203 SHC III 88, *Edmundeston* 1203 ibid.' 101, *Edmodeston* 1249 SHC VI NS (ii) 37, *Admondeston* 1276 SHC 1911 165, *? Acmundeston* 1277 ibid.' 169, *Edmundiston* late 13th century SRO (57/7915), *Admundeston* 1304 ibid, *Admodeston* 1306 SHC VII 151, *Oldmiston* 1341 SRO DW1781/1/43, *Edmundeston* 1635 WCRO CR1908/30/1-2. 'The *tun* of Eadmund', perhaps the person of the same name who held Blithfield in 1066 (DB), but the spellings

81

also suggest a variant 'the *tun* of Eadmod'. There is an Admaston 1½ miles north-west of Wellington, Shropshire, which is probably 'the *tun* of Eadmund': see Bocock 1923: 20; CDEPN.

ADSALL – see **YEATSALL**.

AGARDSLEY (obsolete) in Hanbury parish, 6 miles south-east of Uttoxeter (SK 1425). *Edgareslege* 1086 DB, 1280 Ass, *(ermitagium de) Adgaresleg'* 1192x1247 SHC 4th Series IV 79, *Adgareste, Addegaresleye* 13th century Duig, *Adgaresle* 1324 SHC X 50, *Agersley* 1686 Plot. 'Eadgar's *leah*'. The place is now known as Newborough (q.v.), although the name survives in Agardsley Park (SK 1327) in the north of the parish, recorded as *Aggardsbury Park* c.1659: DCL 393. The name *Eadgares lege*, found in a charter of land at Rolleston in 1008 (14th century, S.920), is associated with Agardsley by Hooke 1983: 96, but placed by Hart 1975: 217 between Rolleston and Marston-on-Dove. See also SHC 1912 222.

**AGGER HILL FARM** 1½ miles north-east of Madeley (SJ 7946). *Agger Hill Farm* 1834 O.S.

ALBURLEY (unlocated, possibly in or near Bagot's Park, perhaps marked by Abberley's Plantation (SK 1126)) *Alburghele* early 13th century SRO D986/43, *Alburleg* early 13th century SRO D986/67, *Alburley* c.1345 SRO D986/81, *Alburley (byrches)* 1402 SHC NS XI 208. Possibly 'the *leah* of Aluburg or Alhburg', both female names. Cf. Alberbury, Shropshire (Ekwall 1960: 4).

ALDEBURGE (unlocated, in Brocton near Milford) *Aldeburge* 1570 Oakden 1984: 35. From OE *ald, burh* 'old fortification'.

ALDEPORT (unlocated, possibly in the Wednesbury/Stonnal area) *Aldeport* 1286 SHC 4th Series XVIII 141, 176. From OE *ald, port* 'old (market) town', or possibly 'old gate'.

**ALDER CARR** north of Loxley Bank (SK 0631). *Alder Car* 1836 O.S.; ALDER CAR north-east of Ellastone (SK 1244), *Alder Car* 1836 O.S. From ON *kjarr* 'marsh, wet moor, boggy copse', a word so often found linked to that of the alder that the one word alone may mean 'wet place with alders': see Rackham 1990: 108; Field 1993: 40, 62. Alder is from OE *alor* 'alder'.

**ALDER LEE** at Meerbrook, 3 miles north of Leek (SJ 6199), *Alder Lee* 1775 Yates, *Alderley* 1842 O.S.; **ALDERLEY** on the south side of Meerbrook (SJ 9860), *Alderley* 1842 O.S. From OE *alor leah* 'alder *leah*'.

**ALDERS BROOK** a tributary of the river Dove. From OE *alor broc* 'alder brook'.

**ALDERSHAWE** 1 mile south-west of Lichfield (SK 1007). *Alreschawe* 1176x1182 EEA 16 66, *Alreshagh* 1255 SHC IV 246, *Alreshawe* 1266 ibid. 162, *Alresawe* 1272 ibid. 252, *Alreschawe* 1272 SHC IV 189, *Allershawe* 1349 SHC 1913 137, 1363 SHC XIII 29. From OE *scaga* 'small wood, copse', which in place-names usually becomes *shaw*, with OE *alor* 'alder', a frequent first element with words for a stream or wood, giving 'the alder copse'. It is unclear whether *Ollershay*, recorded in 1621 (SA 11/68), is to be associated with this place. See also Olrenschawe.

**ALDERSLEY** 1 mile north-east of Tettenhall (SJ 0109). *Allerleye* 1302 SHC VII 101, *Allerley* medieval VCH XX 13, 1332 SHC X (i) 131, 1412 SHC XVI 81, *Alderley* 1532 SHC 4th Series 8 147, *Autherley* 1588, *Alderley* 1592 SHC 1930 305, *Atherley* 1613 map of Tettenhall Hay PRO, *Alderley* 1649 map of William Fowler, *Aldersley* 1686 Plot,

*Aldersley* c.1800. From OE *alor* 'alder', with OE *leah*, giving 'alder *leah*'. Aldersley has been the spelling since the 19th century, with Autherley used for the junction of the Staffordshire & Worcestershire Canal and the Shropshire Union Canal here.

ALDREDESLEGA (unlocated) *Aldredeslega* 1129 SHC I 3. Seemingly from the OE personal name Aldred with OE *leah*, hence ' Aldred"s *leah*'.

**ALDRIDGE** Ancient Parish 3 miles north-east of Walsall (SK 0500). *Alrewic* 1086 DB, *Alrewich* 1166 SHC 1923 295, 1271 SHC V (i) 149, *Alrewyz* 1236 Fees, *Alrewyz* 1286 SHC V (i) 175, *Alrewych* 1302 SHC I 203, 1356 SHC 1913 168, *Allerwych* 1334 SHC XV 57, *Alrewyche* 1398 SHC XV 83, *Alriche otherwise Aldriche* 1580 SHC XV 130, *Aldrich* 1686 Plot. 'Alder *wic*', from OE *alor* 'alder', and OE *wic*, with palatalisation. This place was an independent estate in DB, and 'dwelling' seems a more likely derivation that 'dependant farm'. The combination of a word for a tree with OE *wic* is unusual, but alders are mentioned here frequently in medieval records. The name has evidently been interpreted as 'alder ridge' in more recent times.

ALFLEDEWEY, ALFLEDEFORD (obsolete, probably near Stafford Brook on Cannock Chase (SK 0218)) *Alfledewey, Alfledeford* 1290 Ch; SHC 1924 285. From the OE feminine personal name Ælffled, so 'Ælffled's way or road', and 'Ælffled's ford'. See also Oakden 1984: 64.

**ALLIMORE GREEN** 1 mile south-west of Haughton (SJ 8519). *Aliasmore* 1401 Staff Acc, *Alley-more-greene* 1661 Gnosall ParReg, *Allamore Greene* 1686 Bradley ParReg, *Allmoore-Green* 1747 Poll, *Allymoor Green* 1775 Yates, *Allymoor Green* 1832 Teesdale, *Allemoor Green (Meadow)* 1839 *TA*. Perhaps from OFr *alee*, ME *aley* 'path, passage-way', so 'the marshy ground with the path or passage-way', but *Allyne more*, recorded in 1593 (Oakden 1984: 163, where the name is cited as a Gnosall field-name) may conceivably refer to this place, in which case the name is likely to be from the surname Alleyn. The 1401 form is almost certainly an error. The place is noticeably marshy: see NSJFS 8 1968 114. The Green element suggests a squatter settlement.

**ALLSCOTT** 1 mile north-west of Worfield (SO 7396). *Halvescote* 1256 Eyton 1854-60: III 112, *Alvescote* 1272 ibid. 112, *Aluescot'* 1327 SR, *Alveskote* 1504 TSAS 3rd Series III 117, *Alvescote* 1525 SR, *Alstott* 1532 SHC 4th Series 8 118, *Alscott* 1602 SA 2028/1/5/8. From the OE personal name Ælfheah or similar, with OE *cot* 'cottage'. In Shropshire since the 12th century.

**ALMINGTON** 2 miles east of Market Drayton, in Drayton-in-Hales parish (SJ 7034). *Almentone* 1086 DB, *Alcminton* 1242 Fees, *Aleminton* 1243 SHC 1911 395, *Alkemonton* 1316 ibid. 410, *Ammington* 1686 Plot. The first element may be a short personal name with *-ingtun*, giving '*tun* associated with Alhmund', or perhaps more likely 'Alhmund's *tun*', with the not untypical West Midlands form of a personal name without genitival inflection.

ALOUMLEYE (unlocated, perhaps in the Kinver area) *Aloumleye* 1296x1307 SHC 4th series XVIII 188.

**ALREWAS** Ancient Parish (pronounced [? :lrəs]) 5 miles north-east of Lichfield (SK 1715). *Alrewasse* 941 (14th century, S.479), *Alrewas* 1086 DB, *Alrewæs* 11th or12th century Sawyer 1979a: xxxvii, *Alrewas* 1166 SHC 1923 295, *Allerwych, Alerewas* 1262 SHC V (i) 136, *Allirwyche* 1307 SHC 1911 68, *Allerwas* 1320 SHC 1911 344, *Alderwasshe* 1485 SHC VI NS (i) 159, *Alderwaies* 1577 Saxton, *Alderwais* 1607 Kip, *Aldesways* 1691 SHC 1925 42. From OE *alor* 'alder', with the second element *-wæsse*, meaning 'the plain liable to sudden flooding and draining, with alders': cf. Hopwas, Pur Brook, Bolas, Broadwas, Buildwas, Rotherwas, Sugwas, Wasperton. A full explanation of the term *wæsse* meaning 'alluvial land liable to sudden floods', perhaps first recognised by Duignan 'fen, swamp, or land liable to flood' (SAS XL 1897 388), and Rev. C. H. Drinkwater 'a watery place washed by the sea or other water, land formed by flooding' (TSAS 3rd Series I 406) will be found in Gelling 1984: 59-60; see also Gelling & Cole 2000: 64. Amongst various entries made in the parish registers over a period of some 50 years by the vicar, John Falkner, is the following: 'This 21 day of December Ano 1581 was the water of Trent dryed up, and sodenly fallen so ebbe that I, I[ohn] F[alkner] went over into the halle meddow in a Lowe peare of showes, about iiij of the cloke in the afternowne and so it was never in the remembraunce of any man then living in that time in the droughtest yeare that any man had knowen and the same water in the morning before was banke full which was very strannge': see also Shaw 1798: II 137; GM 1794 II 803. According to 'Nature' December 1930, 1581 was the driest year on record. See also Pirewasse. Cf. Allerwash, Northumberland.

**ALSAGER BANK**  2 miles south-east of Audley (SJ 8048). *Alsager Bank* 1833 O.S., *Alger Bank* 1850 Slater, *Alsager's Bank* 1851 White.  Alsager is probably derived from a surname derived from the place of that name in Cheshire. Early spellings for the Staffordshire place have not been traced, and its history is unclear.

ALSTANESAX (unlocated, in Horninglow) *Alstanesax* mid-13th century SHC 1937 65. Possibly 'Ælfstan's oaks'.

**ALSTONE**  in Bradley parish, 5½ miles south-west of Stafford (SJ 8518). *Aluerdestone* 1086 DB, *Aluredeston in Hylle* 1195 SHC II 47, *Aluredestona* 1197 P, *Aluredeston* '1199 Ass, *Alvedston* c.1200 SHC XIII 273, *Alvredeston* c.1235 SHC XII 273, *Alverestone* 1268 SHC 1911 141, *? Dalvestones, Alvestones* 1278 SHC V (i) 88, *Alstaneston* 1304 ibid.' 62, *Alvereston* 1324 ibid.' 101, *Alderstone* 1406 SHC XVII 76, *? Alderstone* 1420 SHC XVII 76, *Alson* 1586 SHC XVII 233, *Allston* 1666 Gnosall ParReg, *Alston* 1686 Plot. Alured is a form of the OE personal name Ælfred, hence 'Ælfred's *tun*'. *Hylle* in the 1195 form has not been identified: see SHC II 107. See also Aluredeshale.

**ALSTONEFIELD**  Ancient Parish 6 miles north-west of Ashbourne (SK 1355). *Ænestanefelt* 1086 DB, *Alfstanesfeld* 1179 P, *Alstilfeld* 1227 SHC IV 52, *Stanefeld* 1234 SHC 1910 295, *Allesfeld* 1290 SHC NS VI (i) 118, *Alstonesfelt* 1297 SHC 1911 253, *Alston Field* 1604 SHC 1940 125, *Alstonfield* 1686 Plot. 'Ælfstan's *feld* or open land': the place lies on a limestone plateau. The DB form results from a vocalisation of the consonant *f* represented by the letter *v*, written (in the usual way in DB) as *u* and mistranscribed as *n*: see Dodgson 1987: 121-137. *Alfstonewall* (undated) is recorded in Shaw 1798: I 171.

ALSTONEFIELD FOREST (obsolete) See Alstonefield. The forest existed by the early 12th century and covered an area including Fawfieldhead, Heathylee, Hollinsclough and Quarnford. By the 14th century it was known as the Forest of Mauban or Malbank Frith (*Mauban* 1302 SHC 1911 59, *Malbancfrith* 1329 SHC 1913 22, *Forest of Malbon or Malbanke* 1571 SHC 1931 125, *Malbon forest* 1608 VCH VII 6), from the Malbanc family who were lords of the manor of Alstonefield until 1176 (VCH VII 5) or 1214 (SHC 1912 22; SHC 1935 72). Nantwich in Cheshire was held by the Malbanc family, and was known as Wich-Malbanc or similar from c.1130: PN Ch III 30-31.

**ALTON** Ancient Parish 7½ miles north-west of Uttoxeter (SK 0742). *Elvetone* 1086 DB, *Aufeton* c.1247 SHC 1911 419, *Auneton* 1236 SHC 1911 404, *Alveton* 1283 ibid. 40,

*Alneton* (frequently) 13th and 14th century Duig, *Alveton* 1798 Yates, *Alveton or Alton* 1836 O.S.'*Ælfa's *tun*. *Ælfa is an unrecorded personal name. It would be short for a compound in Ælf-. Until the mid-19th century the place was known as Alveton: Lewis 1849: I 52.

**ALUMBROC** (unlocated, near Dunwood) *Alumbroc* 1275 SHC XI 334. Possibly a British stream-name (cf. river Alne, Northumberland; river Alham, Somerset (*Alum* 842, 14th century: S.292), which contains an unidentified, probably British, stream-name of unknown meaning: Ekwall 1928: 3-8), with OE *broc* 'brook, stream'.

**ALUREDESHALE, HALUREDESHALE** (unlocated, possibly near Marchington: see Rees 1997: 68) *Aluredeshale* c.1235 Rees 1997: 68, *Haluredeshale* c.1245 ibid. 68. 'Ælfred's *halh*'. See also Alstone.

**ALVELEY** Ancient Parish 6 miles south-east of Bridgnorth (SO 7684). *Alvidelege* 1086 DB, *Aluielea* 1160 P, *Ailuiel'* 1162 P, *Alveþelea* 1177 P, *Aluedelega* 1191 P, *Aluitheleg* 1195 P, *Alveleia* (p) 1199 Cur, *Avitheleg'* 1199 Cur, *Auvillers* 1266 Pat, *Aveley* 1577 Saxton, 1675 Ogilby. Probably from OE 'the *leah* of Ælfgȳþ', a feminine name. The 1266 form may be an echo of French names ending in *-vill(i)ers* (see PN Sa I 17-8). In Shropshire since the 12th century. Cf. Adderley, Shropshire (PN Sa I 7-10); Aveley, Essex (Ekwall 1960: 19).

**ALVERTON HALL FARM** ½ mile north-west of Denstone (SK 0941). *Alverton Hall Farm* 1891 SRO D240/K/15.

**AMBLECOTE** the Staffordshire portion of Oldswinford (Ancient Parish), near Stourbridge (SO 8985). *Elmelecote* 1086 DB, *Emelcote* 1236, *Amelcot* 1242, *Amelecote* 1255, *Hamelcote* 1317, *Amulcote* 1333, *Amelcote* 1338 (all PN Wo 309), *Amultone* 1377 SHC 4th Series VI 9, *Cote Hamele* 14th century Hundred Rolls Duig, *Hamblecote* 1540 FF, *Amblecote* 1622 WillsP. The first *l* in the DB spelling is almost certainly an error, in which case the first element is probably from OE *æmel* 'caterpillar', or, less likely, from the OE personal name *Æmela, a weak form of the name Æmele (the name is found as a witness to a charter of King Offa in 722 (13th century, S.108)). The second element is OE *cot* 'cottage, shelter, hut', hence 'caterpillar cot' or 'Æmela's cot'. Since 1866 in Worcestershire (Youngs 1991: 471).

**AMERTON** in Stowe parish, 5½ miles north-east of Stafford (SJ 9927). *Aunbriton, Ambrihiton* 1230 P, *Ambricton* 1251 Ch, *Embricton* c.1300, *Hambrighton* 1306 SHC VII 155, *Hambryton* 1307 ibid. 179, *Ambrighton* 1309 SHC 1911 73, *Ambrython (Brook)* 1349 SHC VIII 140, *Amarton nye Charteley* 1529 SRO 486[8015]. 'Eanbriht's or Eanbeorht's *tun*': see Ekwall 1960: 9; Stenton 1970: 98. VCH IV 46 fn. identifies DB *Mersetone* with Amerton, on the basis that the name (interpolated in the DB entry) appears with that of nearby Gayton, and presumably in the light of the passing resemblance to part of the modern name Amerton and the absence of any readily identifiable reference to Amerton. However, early spellings for Amerton bear no similarity to *Mersetone*, which may well be Marston near Stafford, as accepted by Eyton 1881: 83-4. Marston in Church Eaton is *Mersetone* in DB.

**AMETESAWE** (unlocated, at or near Wall Bridge) *Ametesawe* (probably for *Armetesawe*) 1275 SHC XI 334, perhaps 'the grove of the hermit', from ME *ermite* (from Latin *eremita*) with OE *scaga*, but the surname Armett is well recorded from an early date in this area, and may explain the derivation. Cf. Armathwaite, Cumberland.

**AMINGTON** on the north-eastern side of Tamworth (SK 2303; SMR 01177 places a deserted post-Conquest settlement at SK 23240566). *Ermendone* 1086 DB, *Aminton(a)* 1150 Mon, 1198 FF, *Arminton* 1221 Ass, *Hamontona* 1224 Bracton, *Amington* 1232 Ass, *Amigton* 1262 Ass, *Amyton* 1285 Ass, *Magna Aminton(a)* 1299 Ass, *Amynton juxta Tamworth* 1315 Ass, *Muckelamynton, Littelamynton* 1356 AddCh., *Amenton Magna, Amenton Parva* 1532 SHC 4th Series 8 20-21. Perhaps from the OE personal name *Earma or *Eamma (pet forms of Earnmund or similar), genitive singular *Earman, *Eamman, with the OE element *-ingtun*, so giving 'the *tun* associated with *Earma or *Eamma'. The 1356 spellings show that there were at that time a Great and Little Amington. Cf. Arngrove, Berkshire and Devon .

**ANC'S HILL(S)** in Forton parish (obsolete). *Anx(e)well hills* 1635 Oakden 1984 147, *Anc's Hill* 1686 Plot. A curious name, attached to two sandy 50' knolls connected by a shallow saddle. The second element *well* has disappeared, but was unlikely to be ancient, since OE *wælle* 'a spring', and (sometimes) 'a stream' would normally give *wall*. The first element cannot be identified, but a derivation from OE *ancor* 'hermit' is possible: the name may have attached originally to the rock-cut so-called *Roman Well* here: see Erdeswick 1844: 171 fn; VCH I 190; 1946 O.S.; StEnc 481.

**ANDERSLEY** (obsolete, an 'outlying district of Coseley': Roper 1952) *Andresley* 1539 Survey. Possibly 'Andrew's *leah*'. See also Underhill 1941: 142.

**ANDRESEY** (SK 2522) An island in the river Trent at Burton upon Trent ('that insulated meadow between the two branches of the river opposite to the church': Shaw 1798: I 1), the site of a church supposedly founded by St Modwen in the 7th century: VCH III 199. *Andreseya* 1229 VCH III 212, *Andreseya, Andresseye* 13th century SHC V (i) 9, *Andrew's Isle alias Mudwin's chappell* 1549 (1798) Shaw I 1. The name was evidently from St Andrew, to whom the church was dedicated, and OE *ea* 'a river, a stream': VCH III 199. Shaw 1798: I 1 states that the place 'was sometimes from her name called Mudwennestow', i.e. St Modwen's *stow* or holy place. See also VCH IX 6.

ANDREWSHALL – see CHILTERN HALL.

ANEWARDESHALE (unlocated, possibly near Rudyard or Wolfdale) *Anewardeshale, Auewardeshaleg, Eafwardeshileg* 13th century SHC IX NS 313-4. Evidently from an OE personal name ending *-weard*, possibly the very common Ælfweard, with OE *hale*, from OE *halh*.

**ANGLESEA COPPICE** on the east side of Chartley Moss (SK 0228) *Anglesea Coppice* 1836 O.S. An interesting name for which earlier spellings have not been traced. Possibly of recent origin for an island of land in a wet area, but Anglesey in Cambridgeshire is perhaps 'the island of the Angle(s)', from OE *eg* 'island': Ekwall 1960: 10. Englesea-brook (in Cheshire), ½ mile north-west of Balterley, appears as *Ingleshaw Brook* in 1733 (SHC 1944 12), *Inglesey Brook* in 1833 (O.S.).

**ANKER, RIVER** *Oncer* c.1000 Saints, *Ancre* 1332 Pat, c.1540 Leyland, *Auncre* 1295 Ipm, *Onker* 1421 SRO D187/1/5. Although Duignan 1902: 4 suggests a derivation from OE *ancra*, ME *ancre, ankre, anker*, 'anchorite, hermit, nun', which he associates with the two hermitages, both for anchoresses, and a nunnery (at Polesworth) along its course, the name is believed to be of pre-English origin of unknown etymology, but perhaps from an otherwise unevidenced Celtic *ankro* 'crooked, winding': see Ekwall 1928: 14-15; CDEPN. The Anker is a particularly winding river. Cf. Ankerwyke, near Staines.

**ANKERS LANE FARM** 1½ miles north-east of Leek (SJ 0057) *Anker's Lane Farm* 1880 Kelly, *Anchorlane* 1890 O.S. Probably from OE *ancor* 'an anchorite, a hermit or recluse', possibly associated with Dieulacres Abbey.

**ANKERTON** 2 miles north of Eccleshall (SJ 8331; SMR 02099 places a deserted post-Conquest settlement at SJ 83603157) *Emkerdon* 1240x1279 SRO D938/231, *Emkerton* 1272 SHC XI 321, *Emkerdon* 1281 SHC 1924 300, 1292 SHC 1926 171, 1293 SHC VI (i) 219, *Enikerdon* 13th century SRO D938233, *Emkerton, Emcorden* 1415 SHC XVII 55, *Emcurdon* 1421 ibid. 74, *Emcordon* 1423 ibid. 89, *Enkerton* 1581 ibid. 226, *Ankerton* 1606 Eccleshall ParReg, 1614-5 SHC 4th Series 16, *Ankerton farm* 1679 SHC 1919 230, *Ankerton* 1775 Yates, 1833 O.S. A curious name. The modern spelling suggests a derivation from OE *ancor* 'anchorite, hermit', with OE *tun*, giving 'the *tun* of the hermit', but the early forms show that the first element cannot be *ancor*. A personal name such as Emerca is possible, although that name is only recorded in the OE poem Widsith, so perhaps from an OE or ME word cognate with Modern German *imker* 'bee-keeper'. The second element is likely to be OE *dun* 'hill': the place lies on a 398' hill. *Ewkarton*, recorded in 1534 (SHC 1931 162), and in 1562 (SHC 1931 162), may be transcription errors for this place, with *m* misread as *w*. See also Bentham.

ANNOTS DALE (obsolete) in Huntley, near Cheadle (?SK 0091). *Annots Dale* 1732x1872 SRO D1203/C/5/1-16. Presumably from the ME feminine personal name Annot: see PN Ch V (I:1) xxv.

**ANN ROACH** 1 mile south-west of Flash (SK 0165). *Ann Roach* 1842 O.S., *Anroach* 1870 Rental. Possibly from OE *an* 'one', with ME *roche* 'rock', hence 'the lone rock or outcrop', but earlier spellings would be needed for a certain derivation. Cf. The Roaches. A parallel for *roche* producing Modern *roach* is Robert atte Roche, recorded at Roach Farm in Devon in 1330: PN D 578.

**ANSLOW** 2 miles north-west of Burton upon Trent (SK 2125). *ansidelege* 1008 (14th century, S.920), *Ansythlege, Eansyðelege* 1012 (13th century, S.930), *Ansedl(ega)* c.1180 Fr, *Andesley* c.1185 SHC 4th Series IV 111, 1412 SHC XVI 78, *Ansedeleg* c.1245 SHC 1937 39, *Aunsedeleye* 1297 SHC VII 45, *Ansedelee* 1300 BM, *Aunsedeleye* 1303 SHC 1911 59, *Aunesdele* 1309 SHC 1911 296, *Anzdeley* 1332 SHC X 108, *Ansedeleye* 1349 HLS, *Anesdeleye under Nedwode* 1366 SHC XIII 53, *Andusteye, Andusleye* 1373 HLS, *Anzedeleye* 1374 SHC XIII 111, *Anseley* 1546 SHC 1937 187, *Anneysley* 1563 HLS. Probably from the OE feminine personal name Eanswiþ (see Stenton 1970: 321), possibly a saint (found in a charter of 716x757 (13th century, S.100)), with OE *leah*, to be preferred to the suggestion in Ekwall 1960: 10 *ansetlleah* '*leah* with a hermitage'. The *low* element is a late corruption: there is no connection with OE *hlaw* 'hill, tumulus'.

**ANSONS BANK** 1½ miles south-east of Brocton, on Cannock Chase (SJ 9817). From the Anson family of nearby Shugborough Hall.

**APEDALE** 3 miles north-west of Newcastle-under-Lyme (SJ 8149). *Apedal* c.1230 Eyton 1854-60: XI 26, 1277 Misc, *Apedale* 1283 Ipm, 1371 SHC VIII NS 57. Perhaps OE *æppel-dael* 'apple valley', with the *l* lost by dissimilation, or the first element may be from the ON personal name Api. It is unclear whether *Appelton*, recorded in 1185 and 1186 (SHC I 121, 126), refers to this place, but it is certainly associated with nearby Audley. *Land called Apedale Moor near Pershall* is recorded in 1837 (SRO D641/3/R/5/4), but further references have not been traced. Cf. Ape Dale in Rushbury, Shropshire (PN Sa III 192); Apethorpe, Northamptonshire. Apes Dale (*Apesdale* 1832

O.S.) is 2 miles north-east of Bromsgrove, Worcestershire (SO 9972) but early spellings have not been found.

**APESFORD** 1 mile south-east of Bradnop (SK0153). *Apesford* 1223 VCH iii 235, *Arpisford(e)* c.1270 SHC 1911 429, 442, *Arpisford* 1270x1286 SRO DW1761/A/4/161, *Arpsford* c.1280 ibid. 442, *Harpesford* c.1286 ibid. 443, *Arplisford* 1304 ibid.' 433, *Harpeford* 1307 SHC VII 180, *Arpesforde* 1317 ibid. 433, *Arpisford* late 13th century SRO DW1761/A/4/161, *Arpesford* 1343 SRO D3272/5/13/7, *Arpisford* 1564 WSL 163/63, *Apysford* c.1573 SHC 1931 203, *Orpesford* 1637 Leek ParReg, *Alpesford* 1656 Okeover T699, *Orpsford or Apes Ford* 1836 O.S. Probably from the OE personal name *Earp, so '*Earp's ford'. The foundation charter of Hulton Abbey of 1223 naming *Arpesford*, printed in Ward 1843: app. ii, is almost certainly a much later forgery: Tomkinson 1994: 73-102.

**APES TOR** a hill on the river Manifold east of Warslow (SK 0958). *Apes Tor* 1842 O.S., *Apestor* 1884 O.S. The name incorporates OE *torr* 'a rock, a rocky peak', a word generally found only in the South West and the West Midlands. If ancient, the first element may be from the ON personal name Api.

**APETON** in Bradley parish, 5½ miles south-west of Stafford (SJ 8518). *Abetone* 1086 DB, *Abbetona* c.1200 SHC XII 273, *Ab(b)eton* 1203, 1242 Fees, 1272 Ass, *Abbeton*, *Albenton* 1253 SHC IV 128, *Apeton* 1285 SHC 1911 189, *Abbeton, Abbetone* 1302 SHC VIII 141, *Albeton* 1307 SHC VII 179, *Aboton* 1311 SRO (126/7900), *Apedon* 1320 Coram R, *Abbenton* 1253 Ass, *Apton* 1532 SHC 4th Series 8 97. From the OE personal name Ab(b)a, with OE *tun*. *Apetunesway*, perhaps near Maer, is recorded c.1235: SHC VIII 175.

**AQUALATE** 1 mile west of Newport (SJ 7719). *Aguilade* 1227 SHC IV 41, *Aquilad(e)*, *Aquilode, Aculote, Aculott, Akilade(e), Akylod* 1240-70 *Aqualate Deeds*, *Aquilado* c.1255 VCH IV 107, *Akilot* 1275, *Akilote* 1282 Ipm, *Aquilade, Aquilone* 13th century Duig, *Aquilot* 1327 SR, *Aquylott* 1535 VCH IV 107, *Acquilate, Acquylate* 1547 Pat, *Acquylott or Ackelatt* 1593 WCRO CR1291/233, *Aqualat* 1686 Plot. A name of particular interest to philologists, who have put forward various theories, many based on an erroneous connection with Latin *aqua* 'water'. Duignan 1902: 5 suggested that the name is from the family name de Aquila or L'Aigle (recorded in 1129: SHC I 12-3), citing transferred Continental names such as Cause in Shropshire and Montgomery in Montgomeryshire. The derivation proposed by Mawer (SHC 1923 303-4), and endorsed by Ekwall 1960: 11 and Oakden 1984: 146, is from OE *ac* 'oak', with OE *gelad*, meaning 'way, passage, path, road, course, ferry', (a word found twice in Beowulf), perhaps in the sense of 'difficult passage over water or wet ground, by the oak trees' (part of the ridge on which Aqualate stands is called Oakley Bank: TNSFC 1937 64). A very full discussion of the element can be found in Gelling 1984: 73-6, where doubt is cast on this derivation, since the spellings show a development to *-lot(e)*, which is not found in other *gelad* names, but those doubts are overcome in Gelling & Cole 2000: 81. The second element may have been influenced by OE *lad*, ME *lode* 'way, journey, course', dialect 'road', or 'a watercourse; an aquaduct, channel; an open drain in fenny districts'. Both the *gelad* and *lad* might suit the topography of this place: 'There has been a road through Aqualate from the earliest times, inevitably because it occupied a ridge of sound ground, with swamps on either side of it. The present main road from Newport to Stafford has been made since 1800. Before that time the route was less direct and passed by Meretown, Pinker's Hill, and through the middle of the present deer park of Aqualate. The line of the old road is still shown by the lines of trees, mostly oak and beech, which bounded it': SHC 1923

303-4; see also NSJFS 6 1994 27-43. It is very likely that such ridge explains the derivation of this name. Walkley Lane below Broom Hill runs across a bog and suffers from subsidence; the Staffordshire-Shropshire boundary to the west runs through very marshy land prone to flooding; and the Pennocrucium to Chester Roman road to the east of Aqualate (Margary number 19) crosses some very boggy ground: see Nomina 17 1994 10. Indeed, the 'difficult passage' may refer to the Roman road itself, since the element has been found to apply to other Roman roads: cf. Cricklade, Wiltshire ibid. Cf. Hampton Loade. *Aquilone*, recorded in 1227 (SHC IV 225), may have been in or near Brocton near Milford.

**AQUALATE MERE** 2 miles north-east of Newport (SJ 7720). *Water of Mere* 1227 Ass. From OE *mere* 'lake'. The mere took its name from Aqualate (q.v.). There is a reference in 1593 to *the river Ee of the [Aqualate] mere*: WRO CR1291/233. The name is from OE *ea* 'river', and may refer here to the river Mease (q.v.).

**AQUAMOOR** to the east of Marston (SJ 8213). *Hackeymoor Pits* 1833 O.S. Possibly from the dialect adjective *haggy*, with several meanings including 'soft bog in a moor or morass; islet of grass in the midst of a bog; rough, broken or boggy land' (EDD). Whatever the case, the name is not from Latin *aqua*.

ARBLASTER (obsolete, near Broughton Hall in Longdon (SK 0913)) *Arblaster* 1379 SHC XVI 168. The place took its name from the Arblaster family (see SHC 1925 1-24) who lived here for many centuries: Adam le Arblaster is recorded in this area in 1348: WCRO CR1908/3/1-26. The family name is from 'a soldier armed with an arbalest, a cross-bowman, a maker of cross-bows' (DES 4). A pedigree of the Arblaster family is given in Shaw 1798: I 225; see also SHC 1925 1-24. For Robert le Arbelastir recorded in 1227 see SHC IV 53. Shaw 1798: I 223 mentions '...*Leswes (afterwards called Arblaster) hall...*': see Lysways.

ARBLASTER HAYES (obsolete) *Arblaster Hay* c.1598 Erdeswick 1844: 243. The place appears from a survey made in 1763 (SRO D603/N/5/2) to have been what is now Brereton Hayes (Wood), 2 miles south of Rugeley (SK 0414). *Arblaster Hays* 1568 SRO D603/E/3/4, *Arblaster Hayes* 1595 (SRO D1720/13), *Arblaster Hays* 1615 SHC VI NS (i) 5, *Arblaster Hay* 1801 Shaw II 223. Held by the Arblaster family until sold to the Pagets in 1615: SRO D/603/A/3/12-26. See also Shaw 1798: I 223 fn.5; SHC 1925 3; StSt 12 2000 68.

**ARBOUR, THE** 1 mile north-west of Mucklestone (SJ 7137). *Windy Arbour* 1833 O.S., *Arbour Farm* 1946 O.S. If ancient, possibly from OE *eorþ-burg* 'earthern fortification', or OE *here-beorg* 'shelter or protection for a number of men; army quarters' (EPNE ii 244): to the north of the farm of this name is said to be a rectangular earthwork some 900' by 700': see TNSFC 1908 114-5; TNSFC LXXII 1937-8 117-8; TNSFC LXXIII 1948-9 112; but see also TSAS XLIX 1937-8 88. *Arbar leso* ('Arbour leasow') is recorded in Gorsebrook, near Wolverhampton, in 1537: SHC 1912 93. Shaw mentions *Arbour-close*, a tumulus 2 or 3 miles north-west of Okeover (1798: I 33), and a tumulus on *Arbour Hill* near Throwley Hall is mentioned in VCH I 171, suggesting that the word arbour may have been applied to archaeological features such as tumuli. *Arbor*, on the north-west side of Grindon (? SK 0854), appears on Yates' map of 1775. *Arboryes, lately called The Near Arberry, Little Arberry*, etc., is recorded at Pillaton in 1712 (SRO D260/M/T/5/77), to be associated with *Arborow Flatt meadow other wise Stoneyford Meadow* recorded in the 18th century (SRO D260/M/T//5/54). *Rough Arbora Wood* in Aldridge is recorded in 1676 (SRO D260/M/T/2/12; cf. Loaches Bank). See also Beechenhill; Windy Arbour.

**ARCAL, HIGH** – see **ERCALL, HIGH**.

ARCHBERRY (obsolete) ½ mile north-east of Whiston near Cheadle (SK 0447). *Archberry* 1815 *EnclA*, 1837 O.S, *Archbury* 1840 TA. A name of uncertain origin. Possibly from OE *byrig*, dative singular of OE *burh* 'fortified place'. The first element is unidentified: *arch* is a French word first recorded in 1297, and it is unlikely that it is to be found here. Possibly OE *ersc* 'a ploughed field, a stubble field'.

ARCHFORD BRIDGE (unlocated, in Alstonefield: VCH VII 10-11 suggests that the bridge, recorded in 1403, was on the river Manifold on the west side of Hulme End (SK 1059)) *Archford Bridge* 1608 SHC 1934 (ii) 143, SHC 1948-9 81, *Archard Bridge* 1678 Alstonefield ParReg, *Archers Bridg* 1682 ibid, *Archford Bridge* 1686 ibid. 'The bridge at Archford (q.v.)'. The bridge presumably replaced the ford.

**ARCHFORD MOOR** 2 miles north-west of Alstonefield (SK 1158). *Archer Moore* 1686 Plot 115, *Archel Moor* 1840 O.S. Perhaps 'the moor (or ford) of the archer or the man named Archer'.

ARCHILL BROOK (obsolete) 'Archill-brook, issuing from rising ground of that name, near the Holly-hall...serves as the [Staffordshire-Worcestershire] county boundary': Scott 1832: 138-140. Perhaps associated with High Ercall (q.v.), near Sedgley.

**ARELEY KINGS** 1 mile south of Stourport-on-Severn (SO 8070). *(H)erneleia* c.1138, *Ernele* 1156 (1266), *Ernleie* c.1200 and c.1250, *Arneley* 1275 (all PN Wo 29-30), *Suth Erlee* 1276 SHC VI (i) 80, *Alrelege* 1283, *Arleye* 1291, 1428, *Ardley Regis*, *Kyngges Arley* 1405, *Areley* 1453, 1535, 1549 (all PN Wo 29-30), *Lower Arley alias Arley Kings* 1749 SRO D666/19/11. Duignan 1894: 3 suggested a derivation from OE *ærn* 'a house, a habitation, a building', in place-names usually in the sense 'a building used for a specific purpose', which might be considered appropriate for a site at an early river crossing, but the element is normally found in compounds to define a specific use, for example *mæðel-ærn* 'meeting-place', *win-ærn* 'tavern'. The derivation is therefore more likely to be from OE *earn* 'eagle', probably the white-tailed eagle (see Gelling 1987: 173-81), and OE *leah*, giving 'the *leah* of the eagle', although a derivation from the OE personal name Earna, a pet-form of OE names beginning Earn- cannot be ruled out. The place is 7 miles down the Severn from Upper Arley (q.v.), too far away for the same *leah* to be referred to in both names. In the Middle Ages it was in the royal manor of Martley, hence *Kings*. It was also known as Nether Areley in contrast to Upper or Over Arley. Since 1895 in Worcestershire.

**ARKALL FARM** 1½ miles north-east of Tamworth (SK 2206). *Arcull(suche)* 1524 (1798) Shaw I 314, *Arkhall* 1834 O.S. It is unclear whether Rad'o Arkle, recorded in Elford in 1332 (SHC X (i) 105) can be associated with this place. The place-name is possibly 'the hill (or the hall) shaped like a coffer or ark'. The earliest spelling is from 'Arcall stream', from OE *sic*.

**ARLEY, UPPER** Ancient Parish 3 miles north-west of Bewdley (SO 7680). *Ernlege* 963 (14th century, S.720), *Earnleie* 994 (17th century, S.1380), *Ernlege* 1086 DB, *Ernlege* 1100 P, *Erneslea* 1166 SHC 1923 298, *Erneleche* pre-1172 SHC 1924 152, *Erleia* 1188 P, 1197 P, 1200 Cur, *Arnlege*, *Arnleye* 1276 Pat, 1316 FA, 1327 SR, *Ar(e)leye* 1330 SR, 1401, 1408, FA, 1465 Pat, *Arneley vel Arley* 1432 Pat. For the likely derivation see Areley Kings. PN Wo 30 mentions a river name Earn associated with Earnshill, Somerset, and notes that there is a stream that joins the Severn at Arley, but concedes that compounds of *leah* with a river-name are unlikely. The place was also known as Over Arley and Arley de Port, from the Port family (from Port-en-Bessin) who are recorded here from the 12th century: VCH Wo III 5. The appellation Upper or Over was not to distinguish this place from a Lower Arley that has disappeared, as suggested in VCH IV 45, but from Areley Kings in Worcestershire: PN Wo 29. Finberg 1972: 111-2 prefers to identify the first form with this place rather than Arley in Warwickshire, as suggested by Ekwall 1960: 12. Since 1895 this place has been in Worcestershire.

ARMISTON (unlocated, said to be in the parish of Stowe: SHC NS IV 147) *Armiston* 1467 SHC NS IV 147. Possibly from ME *ermite* (from Latin *eremita*) 'hermit'. The second element is uncertain: 'the hermit's *tun*' seems improbable, as also 'the hermit's stone', so possibly a mistranscription of Armistow, 'the holy place associated with the hermit', from OE *stow* (which had a variety of meaning, including 'hermitage'), in which case the place is perhaps more likely to have been at Stowe, Lichfield, rather than Stow by Chartley .

**ARMITAGE** Ancient Parish 5½ miles north-west of Lichfield (SK 0716), *Hermitage* 13th century Duig, *Ermytage* 1306 SHC VII 164, 1432 SHC XVII 138, *the hermitage* 1479 SRO DW1781/6/3/4, *Armytage* 1520 BM, *Armetage* 1532 SHC 4th Series 8 77, *The armitage* 1577 Saxton; **ARMITAGE** 1 mile west of Rocester (SK 0939), *the Armitage* 1664 SRO D786/25/3i; **ARMITAGE (HILL)** between Barton and Tatenhill (SK 1919), *Harmytage* 1503 Ct, *Upper Armitage, Nether Armitage* 1661 DRO D3155/WH13. From the root OFr *(h)ermite*, ME *(h)ermite* 'hermit', giving 'the hermitage'. There was a hermitage at the first place in the 13th century: Shaw 1798: I *209 mentions a local tradition that it lay to the north of the church. The name is not uncommon in Staffordshire: *The Hermitage* is recorded c.1530 between Ellenhall and Ranton (SHC 1912 46), doubtless associated with Ranton abbey, and *the Armitage* in Newcastle under Lyme is mentioned in 1619 (SRO D1229/1/3/43). See also Admaston; The Hermitage.

**ARMSHEAD** 1½ miles north-west of Cellarhead (SJ 9348). *Armshead* 1708 Stoke on Trent ParReg, *Armes head* 1709 ibid, *Gt. Arms Head* 1816 SRO DW1909/E/9/1, *Armishead* (p) c.1836 SRO D1798/579/6/1-10. If ancient, the first element may be an abbreviated form of an OE personal name such as Eadmund or Earnmund, with OE *heafod* 'headland, summit': the place lies on high land on Wetley Moor. No spellings earlier than the 18th century have been traced, however, and the name may be relatively modern, with the second element perhaps influenced by nearby Cellarhead (or vice versa).

ARMYSCOTE (unlocated, perhaps near Walsall). *Arumscote* 1310 Willmore 1887: 235, *Armyscote* 1330 (1801) Shaw 1801: II 71.

**ARNOTT'S GRAVE** 1 mile north-west of Cannock (SJ 9711). *Arnott's Grave* 1887 O.S.

**ASH BROOK** a tributary of the river Blithe. From OE *æsc-broc* 'ash-tree brook'. A watercourse called *Aschbroke* is recorded between Ranton and Ellenhall in 1531: SHC 1912 46.

**ASHCOMBE** 1 mile south of Cheddleton (SJ 9751). *Ashcomb(e)* 1813 Nightingale; 1844 Garner, *Ashcombe Hall, Park & Wood* 1837 O.S. The estate was created 1806-10, and the name would appear to have originated at that date: previously the house in Ashcombe Park was known as Botham Hall: Erdeswick 1844: 496. See also Bothom.

**ASHCROFT FARM** on the north-west side of Shenstone (SK 1005). *Ashcrofts* 1794 Sanders 1794: 110, *Ash Crofts* 1834 O.S. From the Ashcroft family who lived here.

**ASHFIELD BROOK** a tributary of the river Trent. From OE *æsc-feld* 'the open ground with the ash-trees', with *broc* 'brook, stream'.

**ASH HALL, ASH BANK** 1 mile north of Hulme (SJ 9247). *Asse ?*1283 SHC 1911 182, 1303 SRO D847/1, *Asshe* 1306 ibid.' 1911 67, 1346 SRO (463/8016), *Aysshe* 1327 SHC VII (i) 198, *? Ash otherwise Hales* 1599 SHC XVI 196, *Ash* 1663 SRO D664/A/1/A, *The Ash* 1934 SHC XII NS 33. From OE *æsc* 'ash-tree'. See also SHC XII NS 33 fn.

**ASHEHAY** (lost, a wood in Weeford). *Ashehay* 1289 Erdeswick 1844: 565, Shaw 1801: II 3, *? Hastay* 1295 BodCh. 'The hay or enclosure with the ash-trees'.

**ASHENHURST** (obsolete) 1 mile south-west of Bradnop (SK 0154). *Ashenhurst* 1275 VCH VII 172, 1318 (1883) Sleigh 1883: 108, 1586 SHC 1927 158, 1634 (1883) Sleigh 1883: 17, 1686 Plot 387, 1836 O.S., *Asshunhurst* 1401 SHC XV 95, *Ashun Hurst* 1786 Tunnicliff. From OE *æscen* 'ashen', in place-names usually 'growing with ash-trees', with OE *hyrst* 'hillock, knoll, copse, wooded eminence', hence 'the small hill with the ash-trees'.

**ASHENWOOD** (obsolete) in Woodhouses (SK 0809). *Ashenwood* 1834 White 104. From OE *æscen* 'ash-trees', so 'the wood with the ash-trees'.

**ASHES, THE** 2 miles west of Horton (SJ 9256). *Ashes* 1623 JNSFC LXIV 1929-30 131, 1626 Burslem ParReg, *Ashes* 1656 Leek ParReg, 1775 Yates, 1815 *EnclA*, 1842 O.S. 'The ash-trees'.

**ASHFORD MILL** (obsolete) on the east side of Claverley (SO 8093). *Asshefordmyll* 1493 SA 4229/2/24, *Ashford Mill* 1892 O.S. 'The ford to the east'. Probably to be associated with *Asteford*, recorded in 1412 (SA 4229/2/48).

**ASH HALL** (unlocated, possibly near Clough Hall). *Ash Hall* 1855 SRO D3272/1/9/9. See also Hashall.

**ASHOLME** (unlocated, perhaps near Grindon or Beresford) *Assheholme* 1480 SHC NS VI (i) 121, *Asholme* 1611 SHC NS IV 8. From OE *æsc, holmr* 'the island or dry land in the marsh with the ash-trees'. *Asholm*, recorded in 1277 (SHC 1911 168), appears to have lain in the south-west of the county.

**ASHLEY** Ancient Parish 5 miles north-east of Market Drayton (SJ 7636). *Eselie* 1086 DB, *Eslega* 1203 SHC III 94, *? Eslea* 1203 ibid. 110, *Esseleg* 1211 Eyton 1854-60: II 8, 1231 SHC IV 81, c.1235 SHC 1911 424, *Essele* 1243 SHC 1911 117, *Astheleye* 1259 ibid.' 133, *Assel* 1253 ibid.' 135, *Asseleye* 1273 ibid.' 149, *Asscheleye* 1290 ibid.' 197, 1300 ibid.' 263, *Assyngelegh* 1293 SHC VI (i) 250, *Asshelegh* 1302 SHC VII 98, *Asshele* 1318 ibid. 91. From OE *æscen leah* 'the *leah* growing with ash-trees'. A derivation from the OE personal name Æsca cannot be ruled out, but is unlikely.

**ASHMORE BROOK** 2 miles north-west of Lichfield, a farmstead (SK 1011) and tributary of the river Trent. *Estmorebroc* 1242 Fees SHC 1911 402, *Estmeresbrok* 1254 CoramR, *Essemeresbrok* c.1270 SHC 1924 264, *Asschmorebroke, Ashmeresbroke* 13th and 14th centuries Duig, *Ashmores* 1558 SHC XVII 229, *Ashmers Park* 1686 Plot 235, *Ashen brook* 1695 Morden. Originally a stream name which gave its name to the nearby settlement. It would seem that the first element orginated from OE *east* 'east', which was replaced by *æsc* 'ash-tree'. It is uncertain whether the second element is OE *mor* 'marshland' or OE *mere* 'a pool', but the latter element was also sometimes applied to wet ground.

**ASHMORE HEATH** 2 miles south-west of Rushton Spencer (SJ 9160). *Heighassemore* 1369 *Antrobus*. From OE *heah, æsc, mor* 'high ash moor', to which *heath* was later added. See also Tallash.

**ASHMORE PARK** 1 mile north-west of Wednesfield (SJ 9602). *Assever* 1262 SHC 4th Series XVIII 36, *Assenovere* 1286 ibid. 143, *Asshemore* 1427 SHC XVII 122, *Ashemores* 1604 SHC 1940 177, *Ashmores* 1665 WA I 327, *Ashmers* 1686 Plot, 1747 Bowen, *Ashmoor Hall* 1834 O.S., *Ashmore Park Farm* 1895 O.S. The first element is OE *æsc* 'ash-tree', with OE *mor* 'marshland', though the earliest spellings suggest an original derivation from OE *ofer* 'hill, steep slope', and a derivative *\*yfre*: see Gelling & Cole 2000: 199-202.

**ASHNOUGH** 1 mile south-east of Talke (SJ 8352). *Assenesaue* c.1223 SHC 1911 444, *Assenesawe* c.1281 ibid.' 445. A puzzling name. The *-es* of the spellings suggests that the first element may be a personal name such as Æsne, with an uncertain second element, possibly ON *saurr* 'sour ground, mud, dirt, marshy ground'. The place lies in a raised valley between two prominent hills.

ASHWICH FOORD, ASHEWOOD FOORD (obsolete) on Smestow Brook at Greensforge (SO 8688). *Aschwycheford* 1296 SHC 4th Series XVIII 188, *Ashwich Foord or Ashewood Foord* 1608 Foley Deeds E12/F/V1/KAC/58. From Ashwood (q.v.). See also Ashford, Over.

**ASHWOOD, ASHWOOD LODGE** 1½ miles west of Kingswinford (SO 8887). *(Epswich, for) Eswich* 994 (17th century, S.1380), *Haswic* 1086 DB, *Estwood* 1240 SHC 1911 8, *Estwode* 1262 SHC 4th Series XVIII 67, *Aswode* 1286 SHC V (i) 158, *Aswude* 1232 Cl, *Asshewode* 1292 Ch, *Asshewode* 1306 SHC 1911 289. From OE *æsc-wic* 'wic with the ash-trees', later *æsc-wudu* 'ash-tree wood'. Ashwood was one of the Hays of Kinver Forest. At Greensforge, on Ashwood Heath, a number of Roman forts have been identified: VCH I 190, 344; TNSFC 1964 27; Ant XL 1966. Plot 1686: 406 records that Ashwood was 'commonly known as Wolverhampton churchyard', perhaps explained by the fact that the manor of *Eswich*, which has been associated with this place, formed part of Wulfrun's endowment to Wolverhampton Church in 994: 996 (S.1380), see also SHC 1916 111; SHC SHC 1927 185-206; also VCH IV 45. For a legend connecting 'Wolverhampton churchyard' with the devil see Shaw 1801: II 233. For the association of *Haswic* with this place see SHC 1916 111-2. VCH IV 45 fn.65 suggests that *Eswich/Haswic*, if not Ashwood itself, perhaps lay between modern Ashwood and the southern border of Trysull parish, on the basis that the boundary clause of *Eswich* recorded in Wulfrun's endowment mentions *Tresel*. However, that reference is to *Tresel river* (Smestow Brook), not Trysull village. See also Ashwich Foord, Ashewood Foord; Ashford, Over.

**ASH WOOD, ASHWOODHEAD FARM** 3 miles south of Eccleshall (SJ 8224). *Ayswode* 1327 SHC VII (i) 240, *Hasshwode* (p) 1381 Tax, *the Ashwood-head* 1638 Gnosall ParReg, *Ashwood Head* 1833 O.S. From OE *æsc-wudu* 'wood consisting mainly of ash-trees'. The wood from which Ashwoodhead Farm takes its name lies to the east.

**ASKEW BRIDGE** 2 miles east of Himley (SO 9091). *Askew Bridge* 1808 Baugh, 1814 Himley ParReg, 1834 O.S. Possibly to be associated with *Hasco*, recorded here in 1626 (TSAHS 1996-7 XXXVIII 61), and *Hascod* and *Horsecroft*: StEnc 272. The derivation of the name is uncertain, but John Askewe is recorded in 1576 in Sedgley ParReg.

**ASPLEY** 3 miles north-west of Eccleshall (SJ 8133), *Haspeleia* 1086 DB, *Aspleg* 1227 Duig, *Espelega* 1310 SHC IX 20, *Aspley* 1566 SHC XIII 258, *Asplaye* 1600 Eccleshall

ParReg; **ASPLEY** in Brewood parish, 1 mile east of Coven (SJ 9207), *Aspelega* 1227 Ass, *Esp(e)leg'* 1253 Ipm, Hy 3 Wodehouse, *Espele* c.1275 SHC 1928 108, *Aspeley(e)* 1309, 1511 Wodehouse, 1557 Eliz Chanc P, *Aspley* 1605 Lane, 1834 O.S. From OE *æspe-leah* '*leah* with aspen trees'. Aspley in Brewood is not mentioned in DB, *contra* the entry in Oakden 1984: 36, which confuses the place with Aspley near Eccleshall.

ASSHFORD, OVER ASHFORD (unlocated, in Kingswinford). *Assheford(slade)* 1342 SHC 1913 91, *Over Ashford* 1360 SHC IX (ii) 59. From OE *æsc, ford* 'the ford with the ash-trees'. Perhaps to be associated with *Asford* in Kingswinford, recorded in 1273: SHC 1911 157. See also Ashwich Foord, Ashewood Foord.

**ASTLEY** 1 mile north-east of Alveley (SO 7785). *Estleg* 1211 Eyton 1854-60: III 152, *Estleye* 1305 ibid.' 153, *Astley* 1601 SA 2028/1/2/87, *Asly* 1697 SA 2922/11/5/6. From OE *east-leah* 'eastern *leah*'.

**ASTON** 2 miles south-east of Stone (SJ 9131), *Estone* 1086 DB, *Estona* 1166 SHC 1923 297, *Eston'* 1203 Pleas, *Eston* c.1220 SHC VI (i) 18, c.1226 SHC III (i) 231, *Little Aston* 1266 Ass, *Aston near Stanes* 1276 SHC VI (i) 84, *Aston juxta stone* 1403 SHC XI 308; **LITTLE ASTON** 4½ miles west of Walsall (SK 0900), *Eastun* 957 (12th century, S.574), *Little Aston upon Colefeld, Little Aston upon le Colefield* 13th century, *Aston on le Colefeld* 14th century Duig, *Aston in Colfeld* 1422 SHC XVII 84, *Aston on the Colefeld, Aston super le Colfeld* 1435 SHC XVI 151, *Little Aston in Coffilde* 1608 SHC 1948-9 32; **WHEATON ASTON** 4 miles north-west of Brewood (SJ 8512), *Estone* 1086 DB, *Estona* 1166 SHC 1923 296, *Wetenaston* 1248 Ipm, *Eston* 1281 SHC VI (i) 151, *Aston by Brewode* 1293 SHC 1911 49, *Weetnaston* 1413 SHC XVII 47; **ASTON** 3 miles south-west of Madeley (SJ 7541), *Aston* 1298 SHC 1911 57, *Aston in Hales under Lyme* 1472 SHC NS IV 184, *Aston Meyreway* 1609 SHC III NS 46; **ASTON** 2 miles west of Stafford (SJ 8923), *Estone* 1086 DB, *Aston* c.1243 SHC 1914 89, 1666 SHC 1921 107, 1836 O.S.; **ASTON** 1 mile east of Claverley (SO 8093), in Shropshire since the 12th century, *Aston* 1221 Eyton 1854-60: III 93, c.1266 SHC V (i) 142, 1271 SHC 4th Series XVIII 68, 1419 SHC XVI 65. From OE *east-tun* 'eastern *tun*'. Aston-by-Stone was perhaps east of Pirehill; Little Aston east of Aldridge; Aston near Madeley perhaps of Whitmore (SHC XII 237 suggests Maer); Aston near Stafford perhaps of Coton Clanford; and Wheaton Aston perhaps of Brockhurst. Colfield was a vast heath, partly in Cannock Forest and partly in Sutton Chase. The name, recorded from at least 1269 (PN Wa 12, 49; see also TSSAHS XX 1978-9 50), which formed part of the place-name Sutton Coldfield, is said to be from OE *colfeld* 'the cold open land', (from OE *col* 'cool') rather than 'the open land where charcoal was made' (see TSSAHS XXXII 1990-1 87-95; and for imaginative etymology Shaw 1798: I 10 fn.), although in Staffordshire *cald* is the usual term for 'cold', and the regular spellings for Colfield (with *o* and without *d*) do not prevent a derivation from charcoal-burning, recorded from at least the early 14th century: TSSAHS XX 1978-9 50. The first part of Wheaton Aston is from the OE adjective *hwǣten* 'wheaten, growing with wheat' (the significance of which is unclear, since wheat will have been widely grown), to distinguish it from other Astons, though Aston Eyre near Morville in Shropshire is said to have been sometimes called Wheaton Aston (see TSAS VI 1883 33; TSAS 2nd Series IV 1892 xv), for example *Wheaton Aston* 1545 Eyton 1854-60: I 41; 1602 TSAS LI 1941-3 7. *Meyreway* refers to the way or road to Maer Hills, which lie south-east of Aston near Madeley. Plot 1686: 398 mentions *Aston*, '*A place under Kinfare edg*' [Kinver Edge], which has not been traced.

**ASTONSITCH** 1 mile south-east of Bradnop (SK 0254). *Astonsich* 1668 Okeover E5018. Possibly from OE *east-tun* 'eastern *tun*', with OE *sic* 'watercourse'.

**ASTONTHYNK** (unlocated, near Tamworth) *Astonthynk* 1509 SHC 1923 321.

**ASTRELL** (unlocated, in Worfield parish) *Esterhul* 1238x1250 Eyton 1854-60: III 112, *Astenhull* 1292 ibid. 113, *Asterhull* 1327 SHC X 103, 1349 SHC XII 63, *Astrell* 1709 SA 2028/1/5/8. From OE *easterre hyll* 'the eastern hill'.

**AUDLEY** Ancient Parish 4½ miles north-west of Newcastle-under-Lyme (SJ 7950). *Aldidelege* 1086 DB, *Aldedeslega* 1130 SHC I 3, *Aldithelega* 1182 P, *Aldedalega* 1185 SHC I 121, *Aldithlege, Aldithleia* 12th century Duig, *? Aldithelee* c.1235 Rees 1997: 68, *Auditheleg'* 1242 Fees, *Auddedelegh, Audedeleye* 1271 SHC V (i) 152, *Audeyeleg* 1272 SHC XI NS 242, *Aldithlegh* 1679 SHC 1926 215. 'The *leah* of Aldgȳþ', an OE feminine name. Stratton Audley in Oxfordshire takes the second part of its name from a family who probably came from this place.

**AUDLEY'S CROSS** 1 mile south of Mucklestone (SJ 7135). A much-repaired medieval stone cross erected as a memorial to Lord Audley, who led the Lancastrian forces to defeat and was killed at the battle of Blore Heath here on 23 September 1459: see JNSFC 20 1980 9-17. A cross is mentioned by Plot 1686: 449, and may have been erected soon after the battle: the adjoining field was known as *Barn Cross* (JNSFC LXVI 1931-2 186) or *Cross field* (SHC 1945-6 93 fn.) in 1553, and may be associated with the pasture called *le brocken Crosse* mentioned in 1594 (SHC XVI 59). See also StEnc 24.

**AUDMORE** 1 mile north-east of Gnosall (SJ 8321). *Aldermore* 1597 HRO F78/II/327, *Aldmore* 1645 ParReg, *Oldmore* 1677 Gnosall ParReg, *Aulmer* c.1680 GKNB, *Auldmore* 1880 Kelly. The earliest spelling suggests 'the moor with the alders', with later spellings pointing towards an abbreviated form rationalised as 'the old marshland', with the modern form reflecting the typical Staffordshire dialect pronunciation of 'old'. The precise meaning of the term 'old' here is unclear: the word is ambiguous, and could mean 'disused' or 'in use for a long time', but here (if ancient, which seems doubtful) possibly indicating a former, drained, marsh.

**AUDNAM** 2 miles south-east of Kingswinford (SO 8986). *? Aldenham* 1312 SHC X (i) 11, *Audenham (brook)* 1727 *Will, Audenham Brook, Audenham Bank* 1774 canal map, *Audenham* 1775 Yates, *Audenham Bank* 1834 O.S., *Audnam* 1895 O.S. Early certain spellings have not been traced (Audnam glassworks was built here on land bought in 1662 (VCH II 226), but it is unclear whether the name pre-dates the glassworks), but if ancient perhaps *\*æt þēm aldan ham* '(at) the old village, estate, manor, homestead', or the final element may be OE *ham(m), hom(m)* 'meadow, especially a low-lying meadow near a stream, a hemmed-in place', since the place is on the north side of the river Stour, which forms the boundary with Worcestershire here. However, it has been noted that place-names with *ham* are generally early in date and often close to major Roman roads, and this place lies 1 mile east of the Roman road which runs between Greensforge and Droitwich (Margary number 192), which might point to an early origin. Audnam Brook is said to have run through Pensnett Chase to Audnam: StEnc 24. Cf. Aldenham, Hertfordshire.

**AUST(E)** (unlocated, possibly in the Freeford or Pelsall area) *Auste* 1284 SHC 1911 315, 1312 ibid. 40, 1331 SRO D948/3/40-44, *Aust* 1343 SHC XII 25. Possibly from OE *\*estr, eowestre* 'sheepfold'. *Auster*, recorded in 1650 (Sedgley ParReg), has not been located.

**AUSTIN FRIARS** on the south side of Stafford (SJ 9222). *the Freers Augustines* c.1540 Leland. From the Augustinian friary established here in 1343 or 1344: see VCH III 273-4; VCH V 92.

AUSTINS WOOD (obsolete) north of Bucknall (SJ 9049). *Austins Wood* 1836 O.S. The name is unlikely to be from 'Augustinians' (the place lies ½ mile north-east of Abbey Hulton), since Hulton was a Cistercian foundation. John Austyn is recorded in this area in 1316 (SHC 1911 86), so almost certainly from a personal name.

AUSTRELLS, THE (unlocated; described in Duignan 1902: 8 as fields standing on high ground at Aldridge) *Estrehull* 1277 SHC 1911 167, *Asterhul* 1286 SHC V (i) 173, *Asterhull* 1312 SHC 1911 315, 1313 SHC IX (i) 36, *Aschul* 1315 SHC 1911 326, *Asturhull* 1328 SHC XVII 271, *Ostrill* (undated) SHC 1931 82, *Asterhall* (undated) 1801 Shaw II 102. Possibly 'hill of the hearth', from ME *aster* 'hearth', from a simple forge or iron-bloomery, of which several are known to have existed in the area, although the word is not normally found in place-names. However, a more likely derivation is from OE *easterra* 'more easterly', given the existence nearby of Aston ('east-*tun*') Forge, recorded from at least 1329, hence 'the more easterly hill', or from OE *\*estr, eowestre* 'sheepfold'. Cf. Asterleigh, Oxfordshire; Asterley, Shropshire; Austhorpe, Austby, Owston, Lancashire; Austwick, Owstwick, Yorkshire.

**AVERILL SIDE** 2½ miles west of Hulme End (SK 0659). *Averellsyde, Overelsyde* 1527 SHC NS IX 63-5, *Overelside* 1570 ibid. 97, *Arillsyde, Avrilsyde* 1600 Alstonefield ParReg, *Averilside* 1602 ibid, *Aprillside* 1633 Rental, *Averyside* 1653 PCC, *Avril Side* 1775 Yates, *Averil(l) Side* 1839 EnclA, *Aver Hill* 1840 O.S, *April Side* 1851 White. 'April hill-side', from ME *averil* (from OFr *Avril*), an association that has been made since at least the 16th century. The precise meaning behind the name is uncertain, but *Winter Side*, seemingly near Hollinsclough, is recorded in 1683 (Alstonefield ParReg), and may provide a clue to solving this name. It may also be noted that April Hill in Ewloe Town, East Flintshire, is said to be from the wild or single daffodil, called locally April: PNEF 144. Some of the spellings suggest the possibility of 'over hillside', but the early forms show that this is not the derivation.

AVIS HEIRON (unlocated) south of Draycott in the Clay (? SK 1527). *Avis Heiron* 1686 Plot, *Hieron* 1695 Morden, *Avis Hieron* 1747 Bowen. From a *messuage called Avis Hyron* mentioned in 1609-10 (SRO D786/22/5). A curious name for a place evidently of sufficient importance to appear on early county maps. *Heiron* and *Hieras* are perhaps from the alternative disyllabic forms *hyren* or *hyryn*, from OE *hyrne* 'an angle, a corner', frequently used in a topographical sense of 'a recess in hills, a curving valley, a spit of land in a river bend', and often found in ME field-names (cf. John in *le Huyron* of Branston recorded in 1416: SHC 1937 194; see also Foxall 1980: 22; PN Ch V (II) 348-50). *Avis* is Latin for 'bird', but it is improbable that it is used in that sense here (although ME *heiron* means 'heron'). A derivation from the personal name Avis, from OFr Avice (see DES 20), seems the most likely explanation, giving 'the corner of land held by Avice'. Cf. unidentified land called *Avice bruche, Avys Bruche, Avysbrych, Auetzbrucce* recorded in SHC 1928 76, 77, 82, 84, 86, 93, 106.

**AXE EDGE END** 1 mile north-east of Flash (SK 0268). Axe Edge and Axe Edge Moor lie in Derbyshire, but this place straddles the border with Staffordshire, *Axeedge* 1533 Bateman, *? Axen* 1564 SHC 1938 99, *Axen* 1795 Aitken, *Ax Edge Common* 1775 Yates, *Axe Edge* 1842 O.S.; **AXE EDGE GREEN FARM** on the west side of Flash (SK 0167), *Axe Edge Green* 1842 O.S. Perhaps from ON *askr* or a Norse-influenced form of Mercian OE *esc* 'ash tree(s)', giving 'the steep ridge with the ash-trees': such trees often marked boundaries. The river Dove rises from a natural spring on Axe Edge at a height of 1,684': VCH I 46.

**AXSTONES SPRING** ½ mile south of Heaton (SJ 9562). *Axstones Spring* 1842 O.S. A curious name of uncertain age. The name may be associated in some way with a cylindrical stone shaft which is recorded here: see StEnc 24. The word *spring* could mean either a water source or a young wood or copse, but here probably in the former sense: a spring of water is shown here on the 1898 O.S. map.

**AYLESLADE FORD** (obsolete) to the south of Stafford Brook Farm on Cannock Chase (SK 0114). *Eyleslate Forde* 1554 Survey, *Ayleslade Ford* 1826 SRO DW1781/11/1 map. Probably from the OE personal name *Ægel*, with OE *slæd* 'valley, dell', so 'the ford in Ægel's valley'.

**AYLEWARDSTY WAY** (obsolete) *Aylwardesty* 1192x1247 SHC 4th Series IV 79, *Aylewardsty way* 1558 (1798) Shaw I 60. A path through Needwood Forest from Tatenhill Gate to Ravensnest Gate, via Byrkley Lodge: VCH II 350 plan, 352 (where the spelling given is *Aylwardley*). The root is presumably a personal name, with OE *leah*.

**BABYLON** (unlocated, in Gnosall parish) *Babylon* 1757 SRO DW1736/iv/11. Evidently a biblical name, perhaps for particularly fruitful land.

**BACSTONESLEY** (unlocated, possibly near Elkstone) *Bacstonesley* 1332 SHC X (i) 116. Perhaps from OE *bæc-stan* 'a baking stone, a flat stone for baking', when found in place-names usually denoting a place where such stones were obtained. Cf. Baxstonehurst.

**BADDELEY, BADDELEY GREEN** and **BADDELEY EDGE** 3 miles north-west of Hanley (SJ 9151). *Baddilige* 1227 Ch, *Badeleye* 1270, *Badilegh* 1271 Ass, *Baddylegh* 1347 SHC XII 69, *Badgley* 1572 SHC XIII 287, *Badeles, Badeley Edge* 1572 BCA MS917/1515, *Badeley Edge otherwise Baxley Edge* 1601 SHC XVI 212, *Boddeley, Baddeley greene* 1613-4 SHC 1934 31, *Baddeley Green* 1771 SRO HL SRM14. Perhaps 'the *leah* of B(e)adda', but the recurring medial *-i-* is noteworthy, and may be an abbreviated form of *-ing*. The Green element suggests a squatter settlement.

**BADEN HALL** 2 miles north-east of Eccleshall (SJ 8431). *Badehale* 1086 DB, *Bladenhala* 1172 SHC I 68, *Badenhale* 12th century SRO D(W)1082/L/10/10, 1228 SHC 1914 42, *Badenal* c.1250 SHC XI 319, *Badinghale* 1284 SHC 1914 42, *Badenhale* 1296 ibid.' 42, 1314 SHC XII NS 278, *Badenshall* 1568 SRO D5721/1/24//1-24, *Badenhall* 1679 SHC 1914 55. 'Bada's *halh*'. Chetwynd suggests that during the reign of Henry VIII *Alde-Badenhall* and Chebsey were disposed of as Newbold (q.v.): SHC 1914 68. The reference to *Alde-Badenhall* implies the existence of a 'New' Badenhall.

**BADGER'S CROFT** 3 miles north-east of Upper Hulme (SK 0463). *Bochardescroft* 1307 SHC XI NS 257, 1422 SRO D2375M/1/1, *Badgers Croft or Butchers Croft* 1571-3 SRO D2375M/105/36, *Bacherscroft* 1601 Alstonefield ParReg, *Bogerscroft* 1602 ibid, *Butchers Croft* 1623 SRO D2375M/54/3/10, *Badgers Croft* 1842 O.S. From the personal name Bochard, the French form of the Old German Burchard (cf. Botcherby, Cumberland), and OE *croft* 'a piece of enclosed pasture land, a small piece of arable land adjoining a house'.

**BAGGERIDGE** 4½ miles south-west of Wolverhampton (SO 8993). *(chase of) Baggerugge* 1286 SHC V (i) 160, *Bagerugh* 1295 SHC 1928 23, *Baggerugge(feld)* 1317 ibid. 28, *Bagerwge (wood)* 1424 ibid. 68. From OE *bagga*, thought to have been the name of a wild animal, an element often found with terms denoting woods and sometimes hills, probably identical with OE *bagga* 'a bag', so 'a bag-like or fat animal', perhaps a bag-like animal such a badger (a word not recorded independently by OED before 1523; cf. Middle Dutch *bagghe* 'a small pig'). Alternatively, used in a topographical sense 'hill resembling a bag' (which might fit the topography here), with OE *hrycg* 'ridge'. The place lay within Kinver Forest. For *chase* see Cannock Chase. *Baggaridge* in Wightwick (Tettenhall parish) is recorded in 1728 (SRO D571/A/PO/161), *Bagridge* 1730 (SRO D1364/2/27), but *the Chase of Baggeridge which is in the Forest of Cannock*, recorded in 1286 (SHC 4th Series XVIII 119, to be distinguished from the 1286 form given above), cannot be the same place, since Wightwick was not in Cannock Forest. *Sedgley Hay*, recorded in 1255, may be Baggeridge Hay or Wood, recorded in the early 13th century: VCH II 344. Cf. Begeridge, Dorset; Beggeridge, Somerset.

**BAGNALL** 5 miles north-east of Stoke on Trent (SJ 9250), *Baggenhall* 12th century Duig, *Badegenhall* 1203 Ass, *Baginhal* 1269 SHC 1910 127, *Baginholt, Bagynholt* 1271 Ass, *Badegenhall* 1273 ibid, *Bagenholte* 1278 SHC 1911 170, *Bagenhold, Baginhold* c.1280-6 Loxdale, *Bagenholt* 1281 Ass, *Bakenholt* 1293 ibid, *Bogenholt* (p) 1297 Ipm, 1298 SHC XI NS 248, *Bagenald* 1329 Banco, 1417 (p) Fine, *Bagenhalt* 1332 SHC X 112, *Bagenhald* 1332 SR, *Baknold* 1435 FF, *Bagnald(e)* 1470 MinA, *Bagnald* 1512 FF, *Bagenold* 1532 SHC 4th Series 8 54, *Bagnall* 1547 HortonCt; **BAGNALL** on the west side of Alrewas (SK 1614), *Bagenhale* early 13th century SRO DW1851/8/33, *Bagganhal, Baginhal* 13th century Ct, *Baginhale* c.1300 c.1300 TSSAHS XX 1978-9 loose map, *Bagnold Field* 1703 SRO DW1851/8/12. See also SHC 1910 102; TSSAHS XX 1978-9 loose map. The first element comes from an OE personal name Badeca or Bacga, the second is OE *halh*, but in Bagnall near Stoke on Trent the spellings show variation between OE *holt* 'a wood', *hall* and *halh*. *Bagnallditch* appears on modern maps 1½ miles south of Church Eaton (SJ 8415), but the history of the name is not known.

**BAGOTS BARN** 1 mile south-west of Shenstone (SK 0908). *Bagots Barn* 1834 O.S. 'Bagots are certain lands lying to the upper part of Footherley…formerly the property of Edward Fryth…': Sanders 1794: 159.

**BAGOT'S BROMLEY** – see **BROMLEY, BAGOT'S.**

BAGOTS LITTLEMORE (unlocated, in Uttoxeter) *Bagots Littlemore* 1763 SRO D(W)1721/3/12/12. Seemingly land held from at least 1412 by the Bagot family: see SRO D(W)1721/3/31/10; D(W)1721/3/8/9.

**BAILEY'S WOOD** on the north side of Biddulph (SJ 8859). *Bailey Wood* 1791 SHC 4th Series XIII 132.

**BALD STONE** 2 miles north of Upper Hulme (SK 0164). *Bald Stone* 1842 O.S. From a large boulder of this name, from *balled* 'rounded or smooth'. See also TNSFC XLVIII 162.

**BALDWINS GATE** 1½ miles south-west of Whitmore (SJ 7939). *Baldwin's gate* 1676 SHC 1914 16, *Balding Gate* 1775 Yates, *Balding gate* 1834 White 646, *Baldwins Gate* 1833 O.S. The place may have taken its name from William Baldwin, the parker of Madeley Park, who is recorded in 1293 (NSJFS 3 1963 39) – Baldwin's Gate is at the south end of the former park (ibid.), and marks the site of a gate to the park. He or a relative may have been associated with the name *Baldewyneforlong*, recorded in this area c.1275 (SHC 1913 242), *furlong* probably meaning here 'a division of the common field'. *Baldwynes-pitte* is recorded in this area in 1450-1: NSJFS 3 1963 42.

**BALLAMONT GRANGE** 1 mile south of Winkhill (SK 0649). *Bellyband Grange* 1769 *map*, *Ballington Grange* 1836 O.S. A bellyband was the girth-band of a pack-horse, but the significance of the word in the 1769 spelling is inexplicable, unless a jocular corruption of another (possibly the present) name. See also Ballington.

**BALLANCE HILL** on the south-west side of Uttoxeter (SK 0832). *Ballonds Hill* c.1584 SRO DW1733/A/2/52/A-B[7], *Balance Hill (croft)* 1674 ibid. [53], *Balance Hill* 1836 O.S. The name is presumably to be associated with a street called *Baulond*, recorded in 1439 (SRO DW1733/A/2/48), and as *Balondes* in 1466 (SRO DW1733/A/2/49). Perhaps from a personal name Bolland or similar (see Bolland's Hall). Cf. Balance, a field-name in Dunchurch, Warwickshire: Field 1993: 16.

**BALL BANK** 1 mile south of Hollinsclough (SK 0665). *Balebank* 1444 DRO D2375M/1/1, *Ballebank* 1461 ibid, *Balebancke* 1601 Alstonefield ParReg, *Barr banke* 1651 *Rental*, *Bawbank* 1655, 1659 ParReg, *Ballbank* 1690 Alstonefield ParReg, *Ball Bank* 1842 Miller 1891-1900. The word 'ball' was often applied to a rounded hill or boundary mound: Field 1972: 11. The hill here is 1391' high.

**BALL EDGE** ½ mile west of Endon (SJ 9252). Possibly to be associated with *Ball Heath*, recorded c.1664: Okeover T754. Perhaps 'the ridge or steep slope associated with the rounded hill': the word 'ball' was often applied to a rounded hill or boundary mound: Field 1972: 11. *Ball Torre* (from OE *torr* 'a rock, a rocky peak', hence 'the rounded rock or outcrop', recorded c.1664 (Okeover T754) may be associated with this area.

**BALLFIELDS** 1 mile south of Bradnop (SK 0053). *Ballfields* 1635 Leek ParReg, *Ball Fields* 1638 ibid, *Ball Field* 1775 Yates, 1837 O.S. 'The fields with the rounded hillock'. Yates' map of 1775 shows the old road deviating to the west of a conspicuously rounded hill here.

**BALL GREEN** on north-east side of Tunstall (SJ 8852). *Ballgreene* 1623 JNSFC LXII 71, 1625 BCA MS917/1670, *Ball Green* 1836 O.S. Possibly 'the green with the ball-shaped mound or hillock', or 'the green where ball games were played', but the context of the 1625 spelling mentions Richard Ball, so perhaps from a personal name.

**BALL HAYE GREEN** ½ mile north of Leek (SJ 9857). *Balle Heys* c.1539 LRMB, *Ball(e) Haye* c.1540 *AOMB*, *Ball-haye* 1565 (1883) Sleigh 1883: 86, *Ball Heys Greene* 1615 FF, *? Ballyes Hay* c.1615 SRO D3272/1/17/4/6-8, *Bawhey* 1667 Leek ParReg, *Ball Hay* 1697 Leek ParReg, 1798 Yates. Not recorded before the 16th century, and of uncertain origin. Sleigh 1862 suggests a derivation from French *la belle haye* 'the fair enclosure' (perhaps with nearby Dieulacres in mind), but although French *haie* is well-recorded, the spellings show no evidence of *belle*. It is possible that there is a connection with ME *bal(le)* 'a ball', used in place-names to mean 'a rounded hill, a hillock', sometimes meaning 'a mound of earth set up as a boundary mark'. Balle is an ODan personal name (cf. Balby, Yorkshire), and Adam Balle is recorded in Leek in 1332: SHC X (i) 115.

**BALLINGTON** 1 mile south of Leek (SJ 9954). *Baliden* c.1220 StSt 5 3 (p), 1241 SHC IX NS 315, *? Baglington* 1278 SHC 1911 35, *Balydone* 1332 SHC X (i) 115, *Ballidon* 1618 SHC VI NS (i) 45, *Burlington Wood* 1836 O.S., *Ballington Wood* c.1862 map. Perhaps from OE *balg* 'bulging, rounded' with OE *denu* 'valley, so 'the bulging or rounded valley': cf. Balladen, Lancashire; Ballidon, Derbyshire. The name of Cowhay Farm (*? Cowhey* 1608 SHC 1948-9 103, *Cowhay Farm* 1608 VCH VII 89, *Cow Hay* 1775 Yates, *Cowhay* 1836 O.S.) was changed to Ballington Grange c.1960 (local information). The 1836 form for Ballington is evidently an error. The first edition of the 1" O.S. map of 1836 shows Ballington Grange, 1 mile south of Winkhill (SK 0549), at what is now Ballamont Grange.

**BALLS HILL** on the north-west side of Walsall (SO 0198). *Balls Hill* 1834 O.S. Perhaps from the Boweles family: see Bowelles.

**BALTERLEY** in Barthomley parish, 6½ miles north-west of Newcastle-under-Lyme (SJ 7650). *Baltrypeleag(e)* 1004 (11th century, S.906; 11th century, S.1536), *Baltredelege* 1086 DB, *Baldrithelega* 1279 SHC VI (i) 144, *Baltedereleye* 1282-3 SHC XI NS 246, *Baltredelegh* 1289 Ct, *Balridele, Barderdeleye* 1327 SHC 1913 8, *Balturdeley* 1357-8 JNSFC 1924-5 51. '*Baldþrȳþ's *leah*'. The female name *Baldþrȳþ is otherwise unrecorded. See also Stenton 1970: 321.

**BANCROFT** 1 mile north-west of King's Bromley (SK 1117), *Banecroft* 1323 SHC IX (i) 94, 1335 SHC 1939 119, 1345 SHC XII 48, *Bancroft* 1327 SHC VII (i) 230, 1367 SHC VIII NS 40, 1414 Hardy 1908: 106, 1592 SHC 1930 216, 1695 Morden, 1749 Bowen, 1834 O.S.; BANCROFT (unlocated, in Chell), *Banecroft* 1272x1290 SHC 1911 445, *Banchrofte* (p) 1459, *Bancroft(e)* 1467 Ward 1843: app v-vi. Perhaps from OE *bean* 'bean', and OE *croft* 'a piece of enclosed pasture land, a small piece of arable land adjoining a house', so 'the croft where beans were grown'. Some of the spellings shown for the first place may refer to the second place.

**BANES BROKE** (unlocated, in the Swynnerton area) *Banes Brook* 1368 SRO D641/5/T/1/1.

**BANGLEY FARM** 1 mile east of Penkridge (SJ 9414). *Bangley Park* 1601 Comm, 1963 O.S. The history of this name has not been traced, but possibly from ME *bank(e)-ley*, with *bank(e)* here meaning 'hill'. See VCH V 106, 128.

**BANGLEY (GREAT & LOWER)** 4 miles south-west of Tamworth (SK 1500 & SK 1601). *Bangeley* 1397 TSSAHS XXX 1988-9 45, *Bangle* 1538 SHC 1912 128, *Bangeley parke* 1601 (1798) Shaw I 76, *Bangly* 1798 ibid. 9, *Great Bangley, Lower Bangley* 1834 O.S. Possibly from ME *bank(e)-ley*, with *bank(e)* here meaning 'hill': the places respectively are near the summit and near the foot of a 437' hill. For Bangley Park see TSSAHS XXX 1988-9 45.

**BANK END FARM** 1 mile north-east of Norton-in-the-Moors (SJ 9053). *Banke House* 1576 HLS 588, *ye Bancke end* 1613 Norton-in-the-Moors ParReg, *Bank End* 1836 O.S. '(The place at) the end of the bank'.

BANK HOUSE (obsolete) on the north side of Caverswall (SJ 9543). *Bank House* 1640 Caverswall ParReg, 1836 O.S. Presumably 'the house on the bank'.

**BANK TOP** in Tunstall (SJ 8751), *Banke House* 1576 HLS, *Bank House* 1836 O.S.; **BANK TOP** 1 mile east of Kingsley (SK 0246); **BANKTOP** 1½ miles south-west of Consall (SJ 9746), *Bank Top* 1836 O.S. From ME *bank(e)* 'a bank, the slope of a hill or ridge'. *the Bank Toppe* is recorded in 1660 (Kingsley ParReg), and probably refers to Banktop on the east side of Kingsley Holt (SK 0246).

BANNERSTONES (unlocated, near Leek). *Bannerstones* c.1665 Leek ParReg, 1684 ibid. The printed Leek ParReg I 167 notes that the location of this place is 'not known'.

BANNSALL (unlocated, in Balterley or Betley) *Bannsall* 1611 BCA MS3810/125.

**BANNUTT TREE (THE)** in Compton, near Kinver (SO 8284); **BANNUT TREE FARM** 1 mile south-west of Upper Arley (SO 7579), *Banuttred house* 1695 Morden, *Bannut-tree* 1756 (1801) Shaw II 254. Bannut (ME *banne-note* 'a nut, probably a walnut') is a word of unknown origin used in the western counties for 'the walnut, fruit of

Juglans regia; also applied to the grown tree itself and in Warwickshire and Shropshire to soft-shelled walnuts of a larger kind' (EDD; see also PN Sa III 173; VEPN I 47-8). In place-names the word is almost always combined with *tree*. Halliwell suggests that in Worcestershire the word, dating from at least 1609 (VEPN I 48), was only used for the growing tree, with the timber known as walnut. The second place lies on the west side of the river Severn, and has been in Worcestershire since 1895.

**BANTOCK PARK** on the north-west side of Wolverhampton (SO 8997). *Merridale Farm*, built in 1788, and sold in 1854 to Thomas Bantock: ES 5 April 2003 6.

BARDELEY (unlocated) *Bardeley* 1611 SHC NS III 45.

BAREWEHULL (unlocated, near Drointon) *Barewehull* 1249 SRO 510[7922], *Barwhul* pre-1284 SRO 543[7902]. From OE *bearu* 'wood, grove', with OE *hyll* 'hill', so 'the hill with the grove'.

**BAR HILL** 1 mile south-west of Madeley (SJ 7643). *Barr Hill* 1698 Betley ParReg, 1698 SRO DW1082/B/7/1-22, *Barhill* 1749 Bowen, *Bar Hill* 1775 Yates, 1833 O.S. Almost certainly a pre-English name from Welsh *barr* 'top, summit' (cf. Great Barr), though it may be noted that the name is not included in Coates & Breeze 2000.

**BARLASTON** Ancient Parish 3½ miles north-west of Stone (SJ 8938). *Bernvlvestone* 1086 DB, *Berliston* 1140x147 (13th-14th century) EEA 14 40, c.1142 SHC III 322, *Barlaston* 1166 SHC 1923 297, *Berleston* 1212 SHC 1911 387, 1287 ibid. 43, *Berleston Bolnton* c.1235 (1798) Shaw I xxv, *Borlaston* 1293 ibid.' 49, *Barlaston* 1300 ibid. 259, *Borweston* 1303 ibid. 49, *Berlaston* 1316 ibid.' 411, *Borlstone* 1466 SHC NS IV 138. 'Beornwulf's *tun'*. *Bolnton* in the Shaw reference is unexplained, but may be another place altogether. *Beorelfestun*e 1004 (11th century, S.906; 11th century, S.1536) may refer to this place (see Whitelock 1930: 154; Hart 1975: 98), but is more likely to be Barlestone, Leicestershire: Sawyer 1979a: xxiv.

**BARLEY FIELDS** in Callingwood (SK 1924). *Barley-fields Farm* 1649 (1798) Shaw I 30, *Barley Fields* 1836 O.S. Self-explanatory.

BARNDEGNOST, BERNEDEKNOST (unlocated) *Bernedeknost* 1199 SHC III (i) 49, *Barndegnost* 1204 ibid.' 123, *Bernetnoste, Bernecnoste, Bernecost* c.1205 Rees 1997: 45-7. Bosworth-Toller gives OE *gnast* to mean 'spark', while Stratmann offers 'burning ashes', so perhaps 'the ashes at the burned place', from ME *bernde, brende* 'burnt'. *Barndegorst* near Caverswall, or possibly near Hardwick, is recorded in the late 13th century (SRO 3764/21[27574]), and may be associated with this place.

**BARN FARM** 1 mile south of Handsacre (SK 0914). Possibly associated with *Barne Leasowes*, recorded in 1676: WCRO CR1291/169.

**BARNHURST** 3 miles north-west of Wolverhampton (SJ 8902). *Barinhurst* 1250 (1801) Shaw II 204, *Barnchurch* 1305 SHC VII 168, *Barnehurst* 1327 SHC VII (i) 255, *Barehulst* 1337 SHC VI NS (ii) 158, *Barndhurst* 1377 HRO B47/524, *Barnthurst, Barnehurst, Barnhurst, Barnthurst* 14th century Duig, *Bernthurst* 1414 SHC XVII 24, *Barndhurst, Barnhurst* 1419 SRO D593/A/2/16/6, *Barndhurst* 1515 SHC 1928 90, *Barneshurst* 1532 SHC 4th Series 8 148, *Barneshurst More* 1649 TSSAHS XXI 1979-80 16, *Barnehurst* 1686 Plot. The earliest spellings suggest a derivation from OE *bere-ærn* 'barn', with later spellings pointing towards ME *barnde, brende* 'burnt', with OE *hyrst* 'copse, small wood on a hill' (cf. Burntwood). The 1305 spelling is evidently an aberration. Shaw 1801: II 201 and Jones 1894: 84 mistakenly identify this place as Bovenhill (q.v.): see VCH III 319 fn.91.

**BARNS LEE** near the north end of Rudyard Lake (SJ 9360). *Barnards Lee* 1566 SRO DW1761/A/4/179, *Barnslee* 1659 ParReg, 1725 *Will, Barns Lee* 1775 Yates. 'Bernard's woodland glade'; the personal name derives from the OFr name Bernart. It is possible that the same person gave his name to Barnswood (q.v.), 1 mile to the east.

**BARNSLEY** 1½ miles south of Worfield (SO 7592). *Barndelegh* 1301 Rees 1975 249, *Barnesley* (p) 1516 (1801) Shaw II 208, *Barnley* 1525 SR, *Berndeley, Berneley* (p) 1539 SHC VI (i) NS 70, *Barnsley* 1602 SA 2028/1/5/8, *Barnesley* 1731 SA 5586/1/509, *Barnsley* 1752 Rocque. Confusingly, some spellings suggest a derivation from a personal name Beorn or similar, with others pointing towards OE *brend* 'burnt', with OE *leah*. In Shropshire since the 12th century. Shaw 1801: II 209 suggests that the surname Barnsley found in this area is transferred from *Barnsley-hall* in Worcestershire.

**BARNSWOOD** on the east side of Rudyard Lake (SJ 9460). *Bernardiswode* c.1330 (1883) Sleigh, *Bernarduswoodes* 1389 ibid, *Barnswood* 1606 VCH VII 218, 1644 Leek ParReg, *Barnswood Farm* c.1680 SP, 1842 O.S. 'Bernard's wood'. Possibly associated with *Bernardscroft* and *Bernardsmore*, recorded in the 13th century (SHC IX NS 313-4), and *Bernardescroft* recorded in 1315 (SHC 1911 326). See also Barns Lee.

BARNVILLE A name frequently found in early records, possibly to be associated with one or more of Barnfields Farm at Oulton Heath, 1½ miles north-east of Stone (SJ 9036); a lost *Barnfields* (1836 O.S.) at Cauldon (SK 0849); Barnsfields (*Barnfield* c.1889 O.S.) on the south-west side of Leek (SJ 9755); or Barnfields (*Barnvile* 1327 SHC VII (i) 213, *Barnfield* 1775 Yates, *Barnfields* 1836 O.S.), 2 miles south-east of Stafford: alhough the history of the last name is not known, an ancient barn incorporating early stonework is recorded (see VCH V 4), and it may have been the site, or associated with the site, of a medieval leper house – the hospital of St Lazarus of the Holy Sepulchre (perhaps connected with *Spyttellfield*, recorded in 1422: SHC XII (i) 311) – at Radford, recorded from the mid-13th century: StEnc 31, 297. *Berneville* 1269 SHC IV 175, *Banvile* 1272 Ass, *Baumvyle* c.1280 St. Thomas Ch, *Barnevill* 1293 Ass, 1302 Ass, *Barnevile* 1327 SHC VII (i) 197, *Barnevyle* 1332 SR, *Bernewell* 1374 SHC XIII 107, *Barnevyle* 1375 ibid.' 126, *Bernewale* 1383 ibid.' 177, *Barnevyle* 1405 SHC XVI 41. The first element is likely to be from OE *bere-ærn* 'barn', but the second element is uncertain: sometimes OE *feld* 'open land' has been influenced by French *ville*: cf. Enville. However, in this case - *ville* is early and surprisingly consistent in the forms (if indeed the forms relate to the same place), and it is possible that it is a genuine *ville* name, though with an English first element is unlikely. The surname Barnvyll, Barnevyle, Bernevyle is recorded in association with Shenstone in the 14th century: see MidA III 99; SHC XVII 289-91. See also Bernefeld.

**BARR, GREAT** in Aldridge parish, 3 miles south-east of Walsall (SP 0599), *æt Bearre* 957 (12th century, S.574), *Barra* 1086 DB, *Little Barre* 1208 FF, *Barr* 1209 Pleas, *Bare, la Bare* 1242 SHC IV 96, *Great Barre* 1322 Ipm, *Magna Barra* 1428 SHC XII 313; **PERRY BARR** 3 miles north of Birmingham (SO 0791), *Pirio* 1086 DB, *Piri* 1176 SHC I 78, *Peri, Perii* 1199 SHC III (i) 55, *Pirihe* 1236 Fees, *Pirie* 1242 Fees, *Pyrie* frequently 13th century Duig, *Pyrryber, Pyrrybar* 1529 SHC 27-8, *Pereford (sic)* 1532 SHC 4th Series 8 74, *Pirrie Barr* 1561 SHC 1931 137, *Pury-barr* 1686 Plot. From Welsh *barr*, from OCeltic *\*barro-,* Welsh *bar* 'top, summit'. The prefix *Great* in Great Barr is a ME addition, referring not to the conspicuous 744' hill of Barr Beacon (*Barbeacon* 1686 Plot), but to distinguish this manor from the adjoining manor of Perry Barr or Barr Parva (*Parva Barra* 1242 Fees). Beacon, from OE *beacon* 'sign, signal, cross, memorial stone', later 'beacon fire', is a common name for a commanding hill, but the name Beacon Hill is a suprisingly rare hill-name before the sixteenth century. There is a reference in a boundary clause in a charter of Little Aston and Barr of 957 AD to an *ealdan ad* 'an old beacon', which may to refer to Barr Beacon: Forsberg 1970: 20-82; Hooke 1983: 45; but see also Gould 1987: 82-9; Forsberg 1987: 82-89. Round Hill, an ancient earthwork, stands on the hill. For early observations on the name Barr see Shaw 1798: I 11 fn. Perry is from OE *pirige* 'pear tree'. Since 1928 Perry Barr has formed part of Birmingham: StEnc 454. It may be noted that *Barre-moor* is recorded in Colton in 1492 (Parker 1897: 119), and *Barre Moor* in 1538 (ibid.' 370), perhaps to be associated with the 367' hill on the east side of Colton.

BARRE (unlocated) *Barre* 1086 DB. Probably with the same derivation as Bar Hill and Great Barr (q.v.). This may be the same place as *Parva Barr* ('Little Barr') mentioned in 1327 (SR), where it is linked to Perry Barr: VCH IV 55 fn.

BARRIDGE MOOR (unlocated, near Horton) *Barridge Moor* 1810 SRO DW1909/N/2. Possibly from Welsh *barr* 'top, summit', with OE *hrycg* 'ridge', so 'the ridge at the hill called Barr', but in the absence of early spellings the derivation must remain speculative.

**BARROW COP HILL** – see **BORROWCOP HILL**.

**BARROW HILL**  1 mile north of Rocester (SK 1040), *Barrow Hill* 1662 SRO D1380/1/1, *Barrowhill* 1834 O.S.; **BARROW HILL** a 518' hill 2 miles south of Lower Gornal on the Staffordshire-Worcestershire border (SO 9189), *Barrow-hill* 1686 Plot 175, 414, 1695 Gibson, *Barrow Hill* 1775 Yates, 1799 (1801) Shaw II 6, 1834 O.S. Probably from OE *beorg* 'hill, mound, tumulus' (for the first place, on which a rectangular earthwork is recorded in VCH I 192, see Redfern 1865: 71-2; for the second place VCH I 376 records a 99' diameter circular tumulus thirty feet east of Pensnett churchyard; see also JNSFC 5 1965 43; White 1834: 265), or possibly OE *bearu* 'grove, small wood': *Burwe* and *Berwe*, recorded in 1286 (SHC 4th Series XVIII 114, 119) may have been associated with Barrow Hill near Pensnett, in which case that name is from *bearu*. *Two Barrow Hills*, recorded in 1713 (Okeover T740) is evidently to be associated with Rocester.

**BARROW MOOR**  1½ miles south-west of Hollinsclough (SK 0564). *Barrow Moor(e)* 1670 Dep, *Barrow Moor* 1735 Alstonefield ParReg, *Barrow Moor Side* 1839 *EnclA*. Probably from OE *bearu* 'grove, small wood', rather than OE *beorg* 'hill, mound, tumulus', with OE *mor* 'moor, upland waste, fen'.

**BARTHOMLEY** Ancient Parish 1½ miles north of Balterley (SJ 7652). *Bertemeleu* 1086 DB, *Bertamelegh temp.* Henry III, *Bertumleg'* early 13th century *et freq*, as *Bertumlega, Bertumlehga, Bertumley(e), Bertumlegh(e), Beretumle, Bertumleigh* to 1518 (all from PN Ch III 5-6), *Birchinley* 13th century Dieulacres, *Bertoveleye* 1282 Ct, *Bartumilegh* 1287 Ct, *Berethumlegh* 1288 Ct, *Birchinlegh* 1289 Ct, *Birthimleg'* 1289 Ct, *Bertomlegh* 1290 Ipm, *Bertmil'* 1291 Tax, *Bertymleg'* 1297 PN Ch III 5-6, *Bartomelegh* 1325 ibid., *Bartomlegh* 1337 ibid., *Bertumeley* 1419 ibid., *Burtumlegh* 1459 ibid., *Barthomley* 1549 Pat. Place-name scholars disagree about the origins of this difficult name. Perhaps 'glade at (a place called) *Brighthamm*', from OE *leah* with a compound place-name *\*beorht-hamm* or *\*beorht-homm*, in which *hamm*/*homm* sometimes alternates with *hemm* (PN Ch V (II) xx), or 'woodland clearing of the dwellers at a place called Brightmead or Brightwell or the like', or from a personal name having a first element with a final *t*, perhaps Beorht, or OE *beorht* 'bright' or similar, with OE *hǣme, leah* (PN Ch V (I:1) xxxiii; Mills 1998: 27), or from OE *beretunum* 'at the granges' (PN Sa I 194-197, CDEPN 39, but see PN Ch V (I:1) xxxiii). The place was part of Balterley township until 1866 when it was taken into Cheshire.

BARTHON, BASTON (unlocated) Possibly near Dilhorne or Tean. *Barthon* 1244 SHC IV 112, *Baston* 1272 ibid. 211. From the two inconsistent spellings (assuming they relate to the same place) no derivations can be suggested for this name.

**BARTON** ½ mile west of Bradley, 5 miles south-west of Stafford (SJ 8618). *Bernertone* 1086 DB, *? Berderton* 1199 SHC III 54, *Berthertan* 1203 ibid.' 91, *Becterton* 1243 VHC IV 77, *Bertherton* 1314 ibid, *Berton, Benton* 1387 to 1460 ibid, *Betherton* 1391 SHC XV 45, *? Bertherton* 1403 ibid. 105, *Barton* 1532 SHC 4th Series 8 97, 1616 SHC VI NS (i) 21, *Upper & Lower Barton* 1834 O.S. The forms are too varied for any certainty, but the name may contain an OE personal name, perhaps a compound of Beorht-, with OE *tun*. The name *Bederetan*, which appears in a fragment of charter possibly dating from 982x1006 (SRO D603/A/Add/3; see also SHC 1937 6-7), has not been identified, but may be associated with this place or Bedintun (q.v.). *Bartons More* in *Kynston* (? Kingstone) is recorded in 1560: SHC XIII 209.

**BARTON UNDER NEEDWOOD** in Tatenhill parish, 5 miles south-west of Burton-upon-Trent (SK 1818). *Barton* 942 (14th century, S.479), *Bertone* 1086 DB, *Barton sub Nedwode temp.* Henry III BM, *Barton (Park)* 1296 SHC 1911 251, *Barton Undere Nedewode* 1499 NA157DD/70/22. 'Demesne farm or outlying grange in Needwood Forest (q.v.)', from OE *bere-tun*, in the Midlands variant *bær-tun*, originally 'barley enclosure, barley farm', later 'outlying grange, demesne farm' (especially one retained for the lord's use and not let to tenants). The first spelling is from the 14th century, with the *a* representing an earlier *æ*. *Under Needwood* is a medieval addition to distinguish it from other Bartons, and because it lay below the Forest of Needwood.

**BASFORD** 1½ miles east of Cheddleton (SJ 9951). *Bechesword* 1086 DB, *Barkeford* 1199 (1265) Ch, *Barchisfort* c.1250 SHC 1911 426, *Barclesford* c.1255 ibid. 440, *Barclisford* 1261 ibid. 427, *Blacford, Barkeford* 1265 Rees 1997: 136-7, *Barkesford* 1273 SHC VI (i) 52, *Barkisford* 1281 Ipm, *Barclesford* c.1282 SHC 1911 442, *Barsford otherwise Basford* 1590 SHC XVI 106, *Basford* 1686 Plot 43. A puzzling name. The spellings with a medial -*l*- suggest the possibility of a derivation from the OE personal name Beorcol (cf. Baswich), but if so one would expect ME spellings with Berc-, so possibly from the ON personal name Börkr or Barkr (cf. Barkestone, Leicestershire; Barkisland, Yorkshire; Barkston, Lincolnshire), with ME genitive singular -*es* (cf. PN Ch III 48-9), although that would not explain the spellings with -*l*-. There is a Basford 1 mile north-east of Newcastle-under-Lyme (SJ 8646), early spellings for which have not been traced, and other places of the same name in Cheshire and Nottinghamshire, though the latter has a quite different derivation.

**BASFORD BROOK** a tributary of the river Churnet. *Barkesfordebroc* (ancient deed without date: see Oakden 1984: 5). See Basford.

BASSETFRYTHE (unlocated, possibly in the area around Bradnop or Hollinsclough, although the 1414 spelling is associated with *Asshenfeld* (? Alstonefield)) *Bassetfryth* 1401 SHC XVI 82, *Bassetfrythe* 1414 SHC XVII 13. The first element is presumably from the Basset family who held estates in various parts of Staffordshire. For the second element see Frith.

BASSET'S CROSS (unlocated) *Bassets Crosse* 1447 SHC III NS 178, *Basset's Cross* c.1540 Leland. Leland mentions the cross in describing 'the right way to Coventrie from Lichfeld' (Toulmin Smith 1906-10: II 103). The place has been assumed to be Drayton Bassett, but Leland adds that 'there is no building' there, and the context of the 1447 reference shows that it was close to *Hillewode* (Hillwood, 1 mile north-west of Little

Sutton (SK 1200)) and *Theffesoke* ('Thieves Oak', a place where thieves were hanged). The latter name is of interest since the stream forming the Staffordshire-Warwickshire boundary a mile or so to the east of Basset's Pole (q.v.) is Gallows Brook. It seems likely that Basset's Cross was a mile or so to the west of Bassett's Pole.

BASSET'S MILL (unlocated, in Pattingham: VCH XX 178) *Basset's Mill* 1403 VCH XX 178. Associated with Ralph Basset who held land in Pattingham in 1257 (ibid.).

BASSETTHAYES (unlocated, in Swinfen) *Bassetthayes* 1480 SRO D948/3/68. *Bassetesfeld*, recorded in 1306 (SHC 1911 65), 'the *feld* or open ground of the man named Basset', has not been located, but may be associated with this place.

BASSETTS HEATH (obsolete) The extensive heath at Bassetts Pole on the west side of Drayton Bassett: Duignan 1902: 11. *Bassetts heath* 1756 Duignan 1902: 11, *Bassettes heath* 1798 Shaw I 9, *Bassetts Heath* c.1800 SHC 1941 11.

**BASSETTS POLE** 4 miles south-west of Tamworth (SO 1499). Formerly within a vast heath, on the boundary of Staffordshire and Warwickshire, and the meeting point of the manors of Sutton Coldfield, Middleton, Canwell, Drayton Bassett, and the ancient boundaries of Cannock Forest and Sutton Chase: Duignan 1902: 11. From at least the 17th century a tall pole, evidently to guide wayfarers, existed at the place (at 485' the highest point in the area), which took its name from the Bassetts of Drayton Bassett. The pole appears in Ogilby ('111 miles and 4 furlongs from London', on the east side of the road), and on Bowen's map of Staffordshire 1749. Shaw 1801: II 9 quotes the Staffordshire historian Pipe Wolferstan, writing in 1756: 'Basset's Pole, which is a noted mark and guide for travellers, as it stands just on the spot where the road from Tamworth to Birmingham cuts the great road from Lichfield to London at right angles. It used to stand like a May-pole, twenty five or thirty feet high; but is worn to a stump ...'. See also Basset's Cross; StEnc 35.

**BASWICH, BERKSWICH** Ancient Parish 2 miles south-east of Stafford (SJ 9422). *Bercheswic, Berchesuuic* 1086 DB, *Berchlewich* 1151 MRA, *Bercleswich, Bercleswych* 1174 StThomas, *Berccleswitgh, Berceleswich* 1161x82 SRO D939/1, *Bercleswiz* 1241 MRA, *Berkswich* 1259 AngleseyCh, *Berkewic* 1259 MRA, *Bertelewyk* 1293 SHC 1939 74, *Bersewych* 1475 Banco, *Baswiche als Berkeswiche* 1617 FF, *Baswidge* 1690 VCH V 10. Perhaps 'Beorcol's *wic*', or from OE \**bircel*, a rare diminutive of OE *birce* 'birch tree' (cf. the recurring name Bircheles in Cheshire and Derbyshire, possibly from the same word: VEPN I 103), with OE *wic*. *Baswich* is the name used by the Ordnance Survey, *Berkswich* the form used for the ecclesiastical parish. From the 16th century the manor of Baswich was usually referred to as the manors of Sowe and Brocton: VCH V 5. See also Sowe.

**BATCHACRE** 1½ miles south-west of Adbaston (SJ 7525). *Badingesacre* 1272 SHC 1924 243, *Batingacre* 1275 SHC IV 283, *Batyngacre* 1313 Ipm, *Battysacre* 1540 SHC V NS 118, *Bachaker* c.1567 SHC IX NS 57, *Bachecar* 1680 SHC 1919 247, *Beacher* c.1538 Erdeswick 1844: 126, *Batchacre* 1833 O.S. Although the later forms suggest a derivation from OE *bece* 'steep-sided valley with a stream', with OE *æcer* 'plot of arable or cultivated land' (and the Lonco Brook runs through a steep-sided valley here), the earlier spellings indicate that the first element is an OE personal name, giving '\*Bad(d)ing's plot of arable or cultivated land'. A grange of Ranton Abbey lay here: StEnc 36.

BATCHLEY (obsolete) 1 mile west of Redditch (SO 0267). *Bacheley* 1464 Pat. From OE *bece* 'steep-sided river valley', with OE *leah*. In Tardebigge parish, forming part of

Staffordshire from c.1100 until 1266, in Warwickshire until 1844, and since that date in Worcestershire.

**BATES FARM** on the east side of Maer (SJ 7838). *The Holme* 1833 O.S. The present name is from a family name. For the earlier name see Hulme.

**BATH FARM** 2 miles south-west of Brewood (SJ 8507). *Bath* 1775 Yates, *Bath Farm* 1808 Baugh, *The Bath* 1834 O.S. Said to be from a cold bath which existed here from at least 1727: see Horovitz 1992: 327. It is uncertain whether Roger de Bathe, recorded in 1358 as a witness to a deed relating to nearby Weston under Lizard (SHC II NS 65), was connected with the place.

BATH HILL (unlocated, in Wolstanton) *? Bathull* 1362-3 JNSFC LIX 1924-5 58, *? Bathon* 1427 SHC XVII 117, *Bath House* 1620 SRO D997/V/2/1-4, 1656 Wolstanton ParReg, *Bath Hill* c.1787 SRO D997/1/1. Presumably from some pool or bathing place.

BATHIS HILL (unlocated, near The Lowe in Worfield) *Bathis Hill* 1858 SA 1190/87/21. Perhaps to be associated with *Bathehill*, recorded in 1585 (SHC 1929 145).

**BATH POOL** 1 mile east of Talke (SJ 8352). *? la Bathe* 1343 SHC 1913 102. Possibly from the use of the pool for bathing, or with reference to an artificial water container: see Gelling 1984: 13.

BATMAN'S HILL (obsolete, on north-east side of Coseley (SO 9494)) *Batesman Hill* 1652 Sedgley ParReg, *Batman's Hill* 1655 Assessment, *Batemans hill* 1667 Sedgley ParReg, *Batmans (Colliery)* 1834 O.S. Probably from the personal name Batman (recorded in Dudley in 1681: Roper 1980: 147), meaning 'the servant of Bartholomew': DES 31.

**BATTLEFIELD** an area adjacent to the Stourbridge road east of Wombourne (SO 8893). *Battlefield* 1841 VCH XX 197, 1895 O.S. Probably a 19th century name from supposed tumuli here called *Soldier's Hill* (c.1750 (1798) Shaw I 38) which were thought by Plot 1686: 397 to be Roman burial mounds, and by the 18th century antiquary Richard Wilkes to be connected with the battle of Tettenhall c.910 (cf. Tettenhall, Wednesfield): Erdeswick 1844: 368. See also StEnc 682; JNSFC 3 1965 59. The place is commemorated in Battlefield Lane.

**BATTLESTEAD HILL** on the east side of Tatenhill (SK 2022). Early spellings have not been traced, but *Battlestead Lane* is recorded in the area in 1820 (SRO D615/M/7/7). OE *stede* had the meaning 'place, site', so possibly 'the site of a battle', but no evidence has been found to suggest that the name is ancient.

BATTRIDGE NEST (obsolete, in Chesterton near Newcastle) *Partridge Farm* c.1715 Heathcote MSS xviii, *Betteridge Nest* 1767 SRO D1229/1/3/91, *Battridge Nest* c.1775 Heathcote MSS xviii. It has been suggested that a 40' high furnace in Springwood Road known as Partridge Nest or Throstles Nest gave its name to this place (see StEnc 532), but the furnace is said to have been constructed c.1765, in which case the furnace name seems to have been a whimsical version of the farm name.

BAXSTONEHURST (unlocated, in Whitmore) *Baxstonehurst* (undated) SHC 1913 240. OE *bæc-stan* 'a baking stone, a flat stone for baking', with OE *hyrst* 'hillock, copse, wooded eminence'. See also Bacstonesley.

BAYSHALL (unlocated, in Handsworth) *Bayshall* 1549 (1798) Shaw I 108.

**BEACON HILL** 1½ miles north-east of Stafford (SJ 9425), *Beacon feild* 1646 SHC 4th Series I 290, *Beacon Farm* 1798 Yates, *Beacon Hill, Beacon Farm* 1836 O.S.; **BEACON HILL** 1½ miles south-west of Penkridge (SJ 8913), *the Beacon Hill* 1612 Penkridge ParReg, *Beacon (flatt)* 1652 Survey, *Beacon Hill* 1706 Penkridge ParReg, 1720 SRO D1057/A/1/5/1-7, *Beacon hill* 1725 Ct, *Beacon-hill* 1798: Shaw I 31; **BEACON HILL** ½ mile south-east of Rolleston (SK 2426), *Beacon Hill* 1836 O.S.; **BEACON HILL** 1 mile north of Sedgley (SO 9194), *Beacon* 1798 Yates, *Beacon Hill (Quarry)* 1834 O.S.; **BEACON BANK (FARM)** 1 mile north-east of Abbots Bromley (SK 0925); **BEACON BANK** 1 mile south of Milwich (SK 9730), *Bacan Bank (sic)* 1775 Yates. From OE *(ge)be(a)con* 'a sign, a signal, a beacon', so 'the hill (or bank) with the beacon'. All places are on hills or high ground with commanding views. Perhaps surprisingly, no spellings pre-dating the 17th century have been found for any of these places, but VEPN I 68 records only one example (in Essex) before 1600. Stukeley 1776:II observes of Beacon Hill near Stafford: '...a large parcel of rocks, laid on a level eminence , and covered with grass, having a steep ascent on every side, like a Camp...'. The name *Bekenfeld*, possibly in Kingswinford, is recorded in 1485: Shaw 1801: II 229.

BEACON STOOP (obsolete. At 1217' the highest point of the Weaver Hills (SK 0946)) *Beacon Hill* 1775 Yates, *Beacon Stoop* 1836 O.S. From OE *(ge)be(a)con* 'a sign, a signal, a beacon'. *Stoop* is from dialect *stulpe, stolpe*, from ON *stólpi* 'a post, a pillar, a stake, a stump', recorded in English from the 15th century (EPNE ii 157): Yates' map of 1798 shows a beacon here, which perhaps once took the form of an iron brazier mounted on a post. The element *stólpi* was also sometimes applied to boundary-posts or similar (see PNEF 127). See also Stoop.

**BEAM HILL** 2 miles north-west of Burton-upon-Trent (SK 2326). *Beinhull* 1247 BL Stowe Ch 173, *Bemhul* 1286 SHC 4th Series XX 17, *Bemhull* c.1293 SHC V (i) 68, *Bernehull* 1385 SRO DW1734/2/1/103[iii],m.21d, *Beanhill (Field)* 1710 SRO D603/H/5/49, *Beamhill (Field)* 1737 SRO D4219/8/49. The spellings show that the derivation is probably from OE *bean* 'bean', so giving 'hill where beans were grown'. OE *beam* had various meanings, including 'tree, beam, piece of timber, post, footbridge, cross (of the crucifixion), long, straight, gallows, vein of ore, balance for weighing' (see VEPN I 63). If that is the root here, the meaning is uncertain, but perhaps 'hill where timber was obtained', or 'hill with a post'. The precise purpose of any such post, if such is the meaning, cannot be known, but a beacon (and Beacon Hill a short distance to the north-east may be noted) or gallows or a gibbet might be expected to have given their specific name to the hill. Hart 1985: 240 associates Beam Hill (which is not named on the first edition 1" O.S. map of 1836) with *hunger hylle* in a charter relating to Wetmore of 1012 (14th century, S.930). *Hunger* is normally held in Staffordshire to be a West Midlands development of OE *hangra* 'slope, wood on a steep hill-side', from OE *hangian* to suspend'. See also Beamhurst.

**BEAMHURST** 2 miles north-west of Uttoxeter (SK 0636). *Bemhurst* c.1250 SHC XII NS 44, *Bemhurst* 1327 SHC VII 206, *Beymhurst* 1477 SHC VI NS (i) 111, *Beamehurst* 1544 SHC 1910 76, *Beamehurst* 1563 SHC 1938 36, *Bemhurst* 1580 SHC XV 129, 1601 SHC 1935 392, *Beamhurst* 1599 Smith, 1646 SHC 4th Series I 283. From OE *beam*, literally 'tree, beam', but also 'piece of timber, post, footbridge, cross (of the crucifixion), long, straight, gallows, vein of ore, balance for weighing' (see VEPN I 63). Here perhaps with the meaning 'footbridge, bridge formed by a single beam': the road from Uttoxeter to Upper Tean crosses the river Tean at Beamhurst Bridge, with two streams flowing from the north to join the river Tean passing through Beamhurst. The second element is OE *hyrst* 'hillock, wooded eminence', so perhaps 'the wooded hill near the foot-bridge':

Oldwood (*Old Wood* 1836 O.S.) lies on higher ground to the north-east, and may be associated with the hurst. *Beamhurstleys*, recorded in the 16th century (Erdeswick 1844: 18) could be Bemersley (q.v.). Cf. Bamford, Derbyshire; Benfleet, Essex. See also Beam Hill.

**BEARDA** 1 mile north of Heaton (SJ 9664). Evidently associated with *Berdeholm* 1340 VCH VII 186, *Berdehulme, Bordhulm mill in Heyton* 1539 MA, *Berdehulme (Mill)* 1546 SHC 1912 350, *Berdhulme* 1601 BL AddCh 46709, *Berdehulme* 1605 (1883) Sleigh 20, *Beardhall or Berdhulme* 1677 SRO DW1702/1/2, *Beardall* 1697 Leek ParReg, *Bard Hall* 1707 ibid.' The place lies on the side of the Dane valley very close to the Staffordshire-Cheshire border. Ekwall 1936: 163-4 discusses OE *\*byrde*, a derivative from OE *bord* 'border, rim'; see also VEPN I 127. This may be the root of this name. ON *barp* has the meaning 'edge, rim, margin', and is found in Norwegian, Danish and (especially) Icelandic place-names in the sense 'verge, edge of a hill', and ME *berde* is given the meaning 'rim of a vessel'. It has been suggested that Beard in Derbyshire (*Berd(e)* 1236, *Brede* 1251 PN Db 151) is perhaps from OE *beard* 'beard', in the sense 'hillside, edge', or from OE *brerd* 'brim, margin, hill-side', with the loss of the first *r* by disssimilation: Bearda Farm (perhaps, with a mill here, the property listed as belonging to Dieulacres at the Dissolution: VCH VII 189-90) stands on a hillside. The second element is evidently ON *holmr*, usually meaning 'a small island, a piece of ground near water'. Heyton in the 159 spelling is Heaton. See also Hulme.

BEARDMOORS (unlocated, in Farley) *Berdesmor* 1290 SHC 1911 199, 1361 (1883) Sleigh 51, *Berdemore, Berdemor* 1296 SHC VII 41-2, *Berdmor* 1310 ibid.' 303, 1311 SHC 1911 303, *Berdmore* 1481 (1883) Sleigh 182, *land in Farley called Beardmores* 1798 SRO D554/41. The spellings suggest a derivation from an OE personal name \*Beord, with OE *mor*, so '\*Beord's moor'. The place is the origin of the common Staffordshire surname Beardsmore: see Tooth 2000b:136-7.

**BEARNETT (FARM, HOUSE, DRIVE, LANE)** in Lower Penn, 3 miles south-west of Wolverhampton (SO 8894). Perhaps from the Burnett family, who held land here from at least 1434: see for example VCH XX 203; SHC 1928 5, 56, 59; SHC NS III 17, fn.1. See also Putley.

**BEAR'S BROOK** a tributary of the river Trent. No early spellings have been traced.

BEARSHAY (unlocated, 'between Lichfield and Alrewas turnpike road and the Coventry Canal': SRO D615/M/1/6) *Bears Hay* 1775 Yates, *Bearshay* c.1800 SRO D615/M/1/6. The first element is probably a corrupted personal name not now identifiable. The second is OE Mercian OE *(ge)heg* 'fence, enclosure'.

**BEARSTONE MILL** 1½ miles north of Mucklestone, on the Staffordshire-Shropshire border (SJ 7239). *Bearson mill* 1644 SHC 4th Series I 114.

**BEARWOOD** in Smethwick, 3 miles west of Birmingham (SO 0286). *Barewood* 1736 BCA MS917/1774, *? Berwood (Meadow & Moor)* 1773 BCA MS3369/Acc1961-9/82, *Berwoods Hill Farm* 1796 BCA MS3602/284. Possibly from the Bear Inn (*Bear of Smythwick* 1798 Yates, *Bear* 1808 Baugh, *Bear Inn* 1834 O.S.) which may have existed in 1718 (VCH XVII 96), but the area was heavily wooded until the late 18th century (ibid.), and a plot of arable on the south side of Three Shires Oak Road was known as *the Bear Wood* in the 1830s (VCH XVII 96), so perhaps from OE *bēr* 'pasture, especially woodland-pasture for swine', with OE *wudu* 'wood', so 'swine pasture wood', i.e. one where rights of pannage were exercised, or possibly from OE *bearu* 'grove, wood', giving a tautological 'wood called Bearu' when the meaning of *bearu* was forgotten, with the

108

inn-name a natural successor. It is unclear whether *Berewode Strete*, recorded in Eardington *temp*. Edward I (MidA I 140), is to be associated with this place.

**BEASLEY** on the east side of Chesterton near Newcastle (SJ 8349). *Little Beasley* 1801 SOT SD4845, *Peaseley* 1836 O.S. The origins of this name are uncertain, but if ancient (and it may be a field-name), perhaps 'Besy's woodland clearing', and possibly to be associated with the surnames Beezly (1725) and Beasley (1875) recorded in the area: Tooth 2000b: 80. The personal name Bøsi (recorded in Lincolnshire in DB) is ON: see PN Li 12; Ekwall 1960: 35 *sub nom* Beesby. Beasley Bank (*Beasley Bank* 1892 O.S.), a hill of 612', lies to the east of Mucklestone (SJ 7337), but early spellings have not been traced.

**BEATTY HALL** in Yarnfield (SJ 8634). One of a number of Government establishments built in the area in the early 1940s and named after naval heroes. See also Drake Hall, Duncan Hall, Frobisher Hall, Howard Hall, Nelson Hall, Raleigh Hall, Rodney Hall.

**BEAUDESERT** (pronounced bo-dussair [bəʊdəsɛə(r)]) 4 miles west of Lichfield (SK 0313). *Beaudesert* 1259 SHC 1924 101, 1291 SHC VI (i) 277, *Bellum desertum* 1332 SHC I 258, *Beudesherd (parke)* 1385 BCA MS3415/200, *Bewdethert* 1461 SHC 1939 109, *Bewdesert* 1546 ibid.' 110, *Bewdysert* 1576 BCA MS3069/Acc1930-02/371503, *Bewdezarte* 1578 SRO D1734, *Bewdeserte* 1589 ibid.' 112, *Bewdezert Parke* 1594 ibid. 80, *Beawdezarte* 1594 ibid.' 139. 'Beautiful wilderness', a Norman name, from OFr *beau desert. Desert* in English is often used for 'wild, mountain or forest land'. There is another Beaudesert in Warwickshire, east of Henley-in-Arden (see PN Wa 199), and *Beaudesport* is recorded in Shropshire in 1391: SHC XV 33.

BECH (unlocated, recorded in DB in Offlow Hundred) *Bech* 1086 DB. Perhaps the place of the same name south-east of Stowe in Lichfield recorded in the 12th or 13th century (VCH XIV 7), presumably to be associated with *Bechefyld temp*. Edward II, *Bechefield temp*. Edward III & Richard II (SHC VI (ii) 185), which is possibly the same place as *Beche iuxta Lichfield* recorded c.1250 (SRO D948/3/13) and *Bechefeld* (near Stow Street) recorded in 1336 (SHC 1939 94) and 1374 (SHC 1939 100). *Bech* is recorded in 1179 (SHC I 93), 1199 (SHC III (i) 43), *Bec* 1208 (SHC III 141), and *Bec* (1220 SHC IV 12). See also Litelbech. The name would appear to be from OE *bece* 'beech-tree', but that derivation poses difficulties, since beech is generally held to have been native in England south of a line from The Wash to south Wales (Clapham, *et al* 1962: 74; Rackham 1980: 10, 141, 211; Rodwell 1991: 23; Rackham 1990: 5), and some authorities question whether it is native at all (see QJF April 2002 97). See also Beech.

BECHETON (unlocated, in or near Audley) *Becheton manor* 1492 SHC 1912 257.

**BECKMINSTER** on the south-west side of Wolverhampton (SO 9097). *Beckmaster* 1647 Manorial Survey, 1839 Tithe map, *Beckminster* 1760 SRO D593/B/1/17/19. No suggested derivation can be offered for the curious early spelling of this name.

BECKNELLS FIELD (obsolete, 1 mile north-east of Kingswinford (SO 8989)) *? Bickenhill* 1671 Tipton ParReg, *Bickingsfield* 1749 Bowen, *Becknells Field* 1834 O.S. Early spellings have not been traced, but if ancient, possibly from ME *bigging* 'a building', later 'an outbuilding, an outhouse'. Cf. Bickenhall, Somerset; Bickenhill, Warwickshire.

BEDERETAN (unlocated) The place called *Bederetan*, which appears in a fragment of charter possibly dating from 982x1006 (SRO D603/A/Add/3; see also SHC 1937 6-7), has not been identified, but might be Bedintun (q.v.) or Barton near Bradley (q.v.).

BEDINTUN (obsolete) 1½ miles south-east of Penkridge (SJ 9413). *Bedintun* 993x996 (11th century, S.879), 1004 (11th century, S.906; 11th century, S.1536), *Beddintone* 1086 DB, *Bedintonæ* 11th century Sawyer 1979a: xxxv, *Bedintona* 1114x1118 SHC 1916 228. The earliest spellings suggest 'Beda's *tun'*. The half hide held here in DB was in the early 12th century in two equal parts called *Bedintun* and *Pilatehala* (Sawyer 1979a: xxix), where Bedintun is described as *vasta* 'waste', and by 1185 *Bedintun* appears to be included in Pillaton (q.v.): SHC 1916 290-1; VCH IV 44 fn. The bounds given in the charter 993x996 (S.879) suggest that the estate extended north-east, east and south-east of Pillaton: Hooke 1983: 90-94. The personal name Beda is found in Bednall (q.v.), three miles away: the same person may have given his name to both places. It is unclear whether *Bederetan* (q.v.) refers to this place.

**BEDNALL** in Baswich parish, 4 miles south-east of Stafford (SJ 9517). *Bedehala* 1086 DB, *Bedenhal'*, *Bedenhala* (p) 1194 StCart, *Bodenhale* 1242 Fees, *Bedenhale* 1257 (p) FineR, *Bedenhale*, *Bedenhulle* 1271 SHC V (i) 145, 149, *Bydenhale* 1323 CoramR, *Bedenhal* 1332 SHC X (i) 118, *Bedenall* 1369 RB 1907 127, *Bednoll* 1610 Speed. From the OE personal name Beda with OE *halh*, in this case with the sense 'valley' or 'hollow': the place lies in a large shallow basin. The same personal name is found in Bedintun (q.v.), three miles away, and the same person may have given his name to both places.

**BEECH** in the Hanchurch Hills, 5 miles south of Newcastle under Lyme (SJ 8538). *Beche* 1176 EEA 16 104, *la Bech* 1178 SHC I 93, *Beche* 1199 SHC II 83, *Bleche* (*sic*) 1235 SHC 1911 43, *Beche* 1240 (1798) Shaw I xxv, 1288 SHC 19 1 43, 1311 SHC X (i) 9, 1550-1 SHC 1939 36, 1566 SHC 1931 207, *le Beh* 1284 SHC I 173, *Le Bech* 1284 FA, *La Besch* 1306 SHC VII 162, *The Beach* 1673 Blome, *Beche Beach* 1747 Bowen. From OE *bece* 'stream, stream valley': there is a ravine-like valley here which has long been worked for building stone. Philologically a derivation from OE *bece* 'beech tree' is possible, but can probably be discounted not only on topographical grounds, but because beech is generally held to have been native in England south of a line from The Wash to south Wales (Clapham, *et al* 1962: 74; Rackham 1980: 10, 141, 211; Rodwell 1991: 23; Rackham 1990: 5), though some authorities question whether it is native at all (see QJF April 2002 97). *Beech Eyrs*, recorded in this area c.1684 (SRO D641/5/T/2/49), is probably to be associated with this place, though the second word is unexplained. See also Bech.

**BEECH DALE** 1 mile south of Trentham (SJ 8537). *Beech-dale* 1741 Synnerton ParReg, *Beech Dale* 1836 O.S. 'The dale or valley at Beech (q.v.)'.

**BEECHENHILL (FARM)** 1 mile north-west of Ilam (SK 1252). *Bychenne Hill*, *Bychen Hill*, *Bechenhill*, *Bechen Hill* 1542 SHC 1912 144-6, *Bytchyn Hyll Bothome* 1552 SHC 1912 184, *Bitchen hill* 1611 *Senior*, *Beechin Hill* 1732 Ilam ParReg, *Beechin Hill* 1880 Kelly. The earliest spellings support a derivation from OE *bicce* 'a bitch', perhaps used in names for a place linked in folklore to hauntings by bitches, or where hounds were bred and kept (EPNE i 34; VEPN I 96), with the not unusual change to *beech* when the word bitch became used as a term of opprobrium, probably in the 15th century, so 'hill of the bitches' becoming 'hill of the beeches': this place lies on the side of a hill of 1103' on which burial mounds and earthworks are recorded: see VCH I 190; StEnc 42. The appearance of *bicce* in a number of names with 'hill' has not been satisfactorily explained, but see VEPN I 96. Two burial mounds are recorded in 1845 (Bateman 1861:

152) as *Bitchinhill Harbour*, the Harbour element perhaps from OE *eorþburg* 'earthern fortification': a D-shaped earthwork and enclosure of Iron Age/Romano-British date lies south of the farm: PDNPAS. See also VCH I 190 and 376; WMA 1987 30 28. Cf. Beechingstoke, Wiltshire; Beechen Cliff in Bath, Somerset.

**BEELOW HILL** an 852' hill ½ mile east of Oakamoor (SK 0644). *Below Hill* 1775 Yates, *Beelow* 1787 SRO D240/D/239, *Bee Low Hill* 1798 Yates, *Beelow Hill* 1836 O.S. Early spellings have not been traced (and the name might result from 18th century antiquarianism), but if ancient possibly from OE *beo, hlaw* 'bumblebee tumulus': there is a tumulus near the summit. Or possibly from the ME *bilooghe*, originally an adverb which developed into *below* in the 1500s, and is frequently used by Shakespeare, perhaps applied here in some topographical sense. The name is not unique in this region: see PN Db 160, 184.

BEERLEY unlocated, perhaps near Callowhill, between Newton and Bagot's Bromley, or at Stansley: SRO D603/A/Add/117-8. *Berleg* 1252 SHC 1937 47, *Hay de Berley* 1257 StThomas, *Burleya* 1271 SHC 1937 71, *Berleye* 1300 SHC VII (i) 71, *Berley* 1304 SHC 1911 273, *Berleye* c.1315 D938/557, *Birleye* 1323 SHC 1937 123, *Berleye* 1325 ibid. 124, *Berlegh* 1348 SRO D(W)1721/3/2/22, *Beerley* 1541 SHC 1916 330. The various forms preclude any firm derivation. The place may be associated with *Berleyard*, recorded in 1374: SRO D(W)1721/3/32/20. See also Stansley (Wood).

**BEESTON TOR** a pronounced hill ½ mile south of Wetton, on the river Manifold near its junction with the river Hamps (SK 1054). *Grene Beston* c.1547 SHC 1910 63, *Beistone, Grene Beistone, Greene Beyston* 1552 SHC 1912 197-8-9, *Bystorn* 1671 Alstonefield ParReg, *Beeston Torr* 1686 Plot 89, *Beeston Ptar* 1750 Bococke, *Beeston Tor* 1836 O.S. A name of uncertain derivation. There are four examples of Beeston in Norfolk, one in Nottinghamshire, and one in Cheshire. The latter is probably from OE *byge* 'commerce', with OE *stan* 'stone', hence 'the stone where trading took place': Ekwall 1960: 35. Perhaps here from OE *beos* 'coarse grass', or possibly from the ON personal name Bøsi, with OE *tun*, but neither suggestion seems entirely satisfactory. It has been noted (StEnc 42) that the hill contains many small holes in the rock, so possibly from OE *beo* 'bumblebee'. The 1552 forms are probably an aberration. The name *Bisech* (recorded in 1199 SHC III 36) may possibly be associated with this place. *Ptar* appears to be Bococke's idiosycratic rendering of *tor*, a word found in moorland areas meaning 'a high outcrop': see Yelpursley Torr. It is unclear whether *Beeston*, recorded in 1281 (SHC 1911 107), is to be associated with this place.

**BEFFCOTE** in Gnosall parish, 5 miles south-west of Stafford (SJ 8019). *Beffcote* 1086 DB, *Beffecote* 1277 FF, *Deffecote (sic)* 1293 SHC VI (i) 289, *Berscote, Bescott, Besscot* 13th century Duig, *Beofcote* 1343, 1352 Banco, *Berscote* 1367 SHC VIII NS 35, *Beffcott* 1618 SHC VI NS (i) 53, *Beffcote* 1833 O.S. From the OE personal name Beffa, with OE *cot* 'cot, cottage, shelter, hut', so 'Beffa' cottage'. It is evident that medieval scribes confused the place with Bescot (q.v.), possibly due to *f* and *s* being much alike in early writing, an error often repeated: Bowen's 1749 map of Staffordshire shows the place as Bescot. But the many examples of misspellings of Beffcote itself, even in the parish registers, suggest that locally the names were interchangeable.

BEGGARS BUSH (obsolete) at New Oscott, 2 miles south-west of Sutton Coldfield (SO 0994), *beggers bush* 1686 Plot, *Beggar's Bush* 1821 PN Wa, 1834 O.S.; BEGGARS BUSH (obsolete) on the Stourbridge-Wolverhampton road, 1 mile south of Wombourne (SO 8792), near the north-west corner of Himley Park, *beggers bush* 1686 Plot 213, *Beggars Bush* 1834 O.S.; BEGGARS BUSH (unlocated, fields in Castle Church), *Near and Far Beggars Bush* 1788 Oakden 1984: 78; BEGGARS BUSH (unlocated, between Whiston and Penkridge), *Congreve Field voc' the Beggar's Bush* ('Congreve Field called the Beggar's Bush') 1732 perambulation (Oakden 1984: 99); BEGGARS BUSH (unlocated, north-west of Wigginton), *Beggars-bush* 1798 Shaw I 432; BEGGARS BUSH (obslete) at The Lea, Wolverhampton (SO 9097), *Beggars Bush* 1832 WA I 385; BEGGARS BUSHES (unlocated, in King's Bromley), *Beggars Bushes* 1775 map. It has been suggested that this common name may be a derogatory term for a poor dwelling or infertile land: see PN Nt 298; PN Sr 371; PN Wo 338; Field 1993: 108; VEPN I 67-8. Halliwell states that 'To go by beggar's bush' is to go on the road to ruin, confirmed by Brewer 1894, which adds that Beggars Bush is the name of a tree which once stood on the left hand of the London road from Huntingdon to Caxton, so-called because it was a noted rendezvous for beggars. *Beggar* is not recorded in OE: Duignan 1902: 12 cites *Beggares-thorn*, but the name is from a spurious charter of 975 (12th century, S.804), which contains ME names. It seems possible that the name has been linked to particular bushes. The name Beggars Bush at New Oscott is associated with a large hawthorn on the boundary of the parishes of Sutton Coldfield and Perry Barr, and of the counties of Stafford and Warwick. Beggars Bush near Wombourne may be associated with a great white thorn recorded in the area in 1317 (1801 Shaw II 213), perhaps *le Merethorne* (from OE *(ge)mǣre* 'boundary, border', hence 'the boundary thorn') recorded in 1320: SHC 1928 31. It is also possible that some places are so-named ironically: from medieval times a bush was the symbol for a beerhouse, and a 1622 comedy by Fletcher and Massinger was called The Beggar's Bush. Teesdale's map of 1834 and the tithe map of 1838 show *Tinker's Bush* on the north side of Brome. Cf. Beggar's Bush in Powys, 3 miles west of Presteigne.

**BEIGHTERTON** 1 mile south-west of Blymhill, in Weston-under-Lizard parish (SJ 8011; SMR 02379 places a deserted Anglo-Saxon settlement at SJ 80701150). *Bertone* 1086 DB, *Bethertona* c.1200 SHC VIII 194, *Becterton'* 1242 Fees, *Bertherton* 1534 W, *Beghterton* (p) 1302 Ass, *Beyghterton* 1392 (p) CovRoll, *Beytharton* 1565 SHC XVII 214, *Biterton* 1682 Browne, *Beighterton* 1833 O.S. Perhaps 'Beorhthere's *tun*', or possibly from OE *\*beg-þorn* 'berry thorn', perhaps the Midland hawthorn (*Crataegus laevigata*), known as the waythorn in Shropshire (Grigson 1958: 154; see JEPNS 1999-2000 32 21-2).

BELANESTON (unlocated) *Belaneston* 1227 SHC IV 53. Possibly Barlaston.

**BELLAMOUR** 2 miles north of Rugeley ( SK 0420). *Bellamore* 1680 SHC 1919 285, *Bellamore hall* 1747 Bowen, *Bellamour* 1778 SRO D538/A/5/52. Named after a house built here in about 1639 by Herbert Aston, of the Astons of Tixall, which was given the name *Bell amore* (Italian 'good love') because it was completed by the benevolence and assistance of his friends, and probably with reference to his wife: see Erdeswick 1844: 225; SHC 1914 155-6; Parker 1897: 143; StEnc 43. The name survives in Bellamour Lodge Farm.

**BELL FIELD FARM** on Watling Street, 1 mile south-west of Wheaton Aston (SJ 8610). Perhaps to be associated with *Belle*, recorded in 1356 (SHC XI 173). Shaw (1801: II 304) records *Belne* in this area in 1307; Elienor' de Belle is recorded in Engleton in 1332 (SHC X 121), Rad'o de Belne is recorded in Stretton in the same year (ibid. 122), and Rog'o de Belle of Stretton in 1327 (SHC VII 244). Ann Bell of Stretton is recorded in 1605 (Penkridge ParReg). Engleton and Stretton both lie within a mile of so of Bell Field Farm. Belvide Reservoir (q.v.) lies immediately south of this place, and The Bell public house (*Bell* 1775 Yates) nearby to the east is recorded (as *Bleu Bell*) from at least

1648 (PA). It is possible that the name is from OE *beolone, belene* 'henbane', perhaps applied to a stream (cf. Belbroughton, Worcestershire, *on (aqua que vocatur) Beolne, (aqua de) Beolne* 1300: Ekwall 1928: 32-3; PN Wo 274-5) in the sense 'the stream on whose banks henbane grows', perhaps the brook that runs on the south side of Watling Street through Horsebrook. Henbane appears to have been widely found in Staffordshire: see Shaw 1798: I 105. It has been suggested that OE *beolone* is a word of Celtic origin, derived from Belenos, the name of a deity, but see Ekwall 1928: 33; VEPN I 83. Other possible derivations for this place are from OE *bel* 'fire' (the place lies on a hill on Watling Street with far-reaching views east and west: cf. Beacon Hill; VEPN I 78), or OE *belle* used as a hill-name: there are long rising approaches along Watling Street from both west and east.

**BELMONT HALL** 1 mile west of Ipstones (SK 0049). *Bellemont* 1775 Yates, *Belmont* 1798 Yates, 1836 O.S. Evidently from French *bel mont* 'beautiful mount'. The name is said to date from the building of Belmont Hall c.1770: Brighton 1937: 44. Curiously, the 1836 1" O.S. map shows two places called Belmont south-west of Ipstones. *Beumund* recorded in 1176 (SHC 1924 239), *Belmunt* c.1177 (ibid. 83), *Bellomonte* 1203 (SHC III 117), 1280 (SHC VI (i) 108), *Bealmonnd*, recorded in 1344 (SRO DW1733/A/2/20), and *Bellemont*, recorded in 1378 (SHC VIII NS 74), have not been located.

**BELMOT GATE** 2 miles south-west of Tutbury (SK 1926). *Bellmott gate* 1559 (1798) Shaw I 60, *Belmot* 1686 Plot, *Belmoat* 1798 Yates. The age of the name is uncertain, but perhaps from ME *belle* 'bell' and ME *mote* 'hillock, mound, tumulus', so 'the bell-shaped mound or tumulus'. Possibly one of the gates to Stockley Park.

BELOCKES BRIDGE (unlocated, on the river Tame, possibly near Salter's Bridge (q.v.)) *Belockes bridge* 1609 (1798) Shaw I 138, Alrewas ParReg. DES 38 gives Bellock as a dialectal form of the common French place-name Beaulieu, from which there seems to have been also a personal name Beloc, so 'the bridge associated with a man named Beloc'.

**BELVIDE RESERVOIR** 2 miles north-west of Brewood (SJ 8610*).* Belvide Reservoir* 1895 O.S. Built as a feeder for the Shropshire Union Canal in 1834 on Belvide Fields (*Belvide Fields* 1815 SRO D1798/656/1; 1832 Teesdale), perhaps from dog-French *bel vide* 'beautiful view', but see also Bell Field Farm, which lies on the north side of the reservoir, and which doubtless influenced this name.

BELWOODE (unlocated) *Belwoode* 1526 SRO D1810/f290.

**BEMERSLEY GREEN** 3 miles north-east of Tunstall (SJ 8854). *Bemreslega* 1199 SHC III (i) 40, *Bemeresleg* 1252 Ch, *Bembersleye* 1326 JNSFC LIX 1924-5, *Bemerusley* 1362-3 ibid. 58, *Bembersley* 1426 SHC XVII 109, *Bemmasley* 1581 Biddulph ParReg, *Bemmersley* 1616 SHC IV NS 90, *Bemersley* 1619 SHC VII NS 204, *Bemersley Green* 1635 SHC 1910 251. Literally 'the *leah* of the trumpeter', from OE *bemere* (OE *beme, bȳme* 'trumpet'), but the word is now also believed to mean a type of bird, perhaps a bittern, also called in modern English 'boomer', named from its trumpet-like call, so in this case more likely to be 'the *leah* of the bittern': see Hough 1998: 60-76. The word *bittern*, not recorded in English before the 14th century, is apparently from OFr. Cf. Bemerhills and Bemerton, Wiltshire. The Green element suggests a squatter settlement.

BENNETSHAY (unlocated, in Keele) *Bennethay* 1410 Harrison 1986: 21. 'The hay or enclosure of the man called Bennet', from Mercian OE *(ge)heg*.

BENNYSTOILS (unlocated, in the Bentley/Rushall/Great Bloxwich area) *Bluningshall* 1272x1307 SHC 1910 199, *Benyngeshale (brouk)* 1300 SHC V (i) 177, *Bettyngeshale* 1306 SHC VII 171, *Beryngeshale* 1362 SRO DW1082/A/4/5, *Brennydhalle* 1396 SHC 1950-1 25, *Mag's Benyngsale in Great Bloxwich* 1430 SRO D593/B/1/26/6/5/8, *Benyngsall (alias Hathes leasow alias Wylbruche in Wednesfield)* 1542 SRO D593/B/1/26/1/8/14-15, *Benningsall* 1638 SRO 35/6/4, *Beningsalls* 1663 SRO D260/M/T/2/7, *Benningsalls in Great Bloxwich* 1668 SRO D260/M/T/2/63, *Benningsalls in Rushall* 1703 SRO D260/M/T/2/63, *Bennystoils* 1734 SRO D260/M/E/426/2. Presumably connected with Benynscrosse (q.v.). Henr' Benynes is recorded at Wrottesley in 1327 (SHC VII (i) 251), and Ric'o Benyn in 1333 ( SHC X 131). From the OE personal name *Benning (DES 39), with OE *halh*.

**BENT END FARM** 1 mile north-east of Heaton (SJ 9663). *Bent* 1842 O.S. Probably from OE *beonet* 'bent-grass'.

**BENT FARM** between Blymhill and Wheaton Aston (SJ 8312), *The Bent* 1832 Blymhill ParReg, *Bent Farm* 1833 O.S.; **THE BENT** 1 mile south-west of Longnor (SK 0762), *The Bent* 1642 ParReg, 1775 Yates. From OE *beonet* 'place with rough coarse grass': cf. Bentley.

**BENTELEE (WOOD)** 1½ miles north of Whitmore (SJ 8143). *? Bentlee* 1240 (1798) Shaw I xxx, *? Bentelee* c.1295 SHC 4th IV 243, *Bentelegh* 1348 SHC 1913 248, *Benteley* 1586 SHC 1927 172; **BENTILEE PARK** 2 miles south-east of Abbots Bromley (SK 1022), *Bentileye* 1286 SHC 4th Series XX 29, *Benetleyahurst, Benethleyehurst, Benethlega* ? 13th century SHC V (i) 71, *Bentileghehurste* c.1305 SHC 1937 27, *Benethelehurst* c.1245 ibid. 63, *Beneteleyehurst temp.* Edward I ibid. 113, *Bentylee Farm* 1741 DW1734/2/3/159, *Bentilee* 1836 O.S.; **BENTILEE** on the north side of Ubberley (SK 9146), *Bentteley* 1775 Yates. From OE *beonet* 'coarse wildgrass, bent-grass', sometimes alternating with the ME adjective *\*benti* 'growing with rushes or coarse grass' (see VEPN I 83, 81, which cites a similar alternation in West Yorkshire), with OE *leah*, hence '*leah* overgrown with rushes or coarse grass', with the second example showing influence from the OE proposition *beneoþan* 'beneath, under, below', perhaps for 'beneath Le Hurst'. *Hurst(e)* is from OE *hyrst* 'hillock, copse, wooded eminence'.

BENTHAM (obsolete, 1 mile north of Eccleshall (SJ 8331)) *Bentham* 1833 O.S. The age of the name is unknown, and the place is now known as Little Ankerton. The earlier name may be associated with the Bentham family: see Shaw 1801: II app. 5-8.

**BENTHEAD** ½ mile north-east of Bradnop (SK 0155), *Benthead* 1836 O.S.; **BENT HEAD** ½ mile west of Upper Hulme (SK 0061), *Bent Head* 1842 O.S. Probably from OE *beonet* 'bent-grass', with OE *heafod* 'head, headland, summit, upper end'. There was a dialect word *bent* 'the slope or hollow of a hill', which might be appropriate with *heafod*, but it is mainly recorded in the south of England (OED). *Benthead*, recorded in 1637 (Leek ParReg), may refer to either (or neither) of these places.

**BENTLEY** 2 miles west of Walsall (SO 9899), *Benætlea, Benetlegh* 12th century Duig; *? Benetclee, Benetcle* 1250 Fees, *Beneytleg'* 1255 Fees, *Benetley, Benetleye* 13th century Duig; **BENTLEY FARM** 1½ miles west of Hamstall Ridware (SK 0818), *Benetlegh* 1277 SHC 1911 169, *Bonedleg, Bonedle* c.*temp.* Richard I (1798) Shaw I 200, *Benteleyes* 1627 SBT DR18/1/1908, *Old Bentley* 1834 O.S.; **BENTLEY PAUNCEFOTE** 3 miles south-east of Bromsgrove (SO 9866), *Beneslei* 1086 DB, *Benetlega, Benetlege* 1185, *Bunetleg* 1280, *Benetley in Fekenham forest* 1281, *Benteley in Tardebigg* 1499, *Stretch Bentley* 1578 (all PN Wo 366), *Bentley-Pauncefoot* c.1616 Erdeswick 1844: 377;

**BENTLEY** (obsolete) between Hoccum and Swancote, 1 mile south-west of Worfield (SO 7493), *Bentley* 1752 Rocque, 1833 O.S. From OE *beonet* 'bent-grass', a word found only in place-names and later meaning a coarse, stiff reedy or rush-like grass found on the acid soils of high moorland, with OE *leah*, hence '*leah* overgrown with bent-grass'. Bentley near Walsall was one of the Hays of Cannock Forest. *Pauncefote* is from Richard Panzeuot who held land here in 1185, and *Stretch* is from the Streche family who held land here in 1275: PN Wo 366. Bentley Pauncefote is in Tardebigge parish, in Staffordshire from c.1100 until 1266, in Warwickshire until 1844, and since that date in Worcestershire. *Fenney Bentley*, recorded in 1315 (SHC 1911 87), has not been identified.

**BENTLEY BROOK**   a tributary of the river Trent, *Bentley Brook* 1769 Survey; **BENTLEY BROOK**   a tributary of the river Tame. From OE *beonet* 'bent-grass' with OE *leah* and OE *broc* 'brook'. Cf. Bentley.

**BENTS, THE**   on the south-west side of Withington (SK 0234). *bent* 1327 Tax, *The Bents* 1836 O.S. From OE *beonet* 'bent-grass'.

**BENTY GRANGE**   1 mile north-west of Waterhouses (SK 0650). *Benty Grange* 1840 O.S. Early spellings have not been traced, but probably from a ME adjective *\*benti* 'growing with rushes or coarse grass', from OE *beonet* 'bent grass', with the OE adjectival suffix *-ig*.

**BENYNSCROSSE** (unlocated, near Oldfallings) *Benynscrosse* 1431 SHC XVII 130. Presumably associated with Bennystoils (q.v.).

**BEOBRIDGE**   1 mile south of Claverley (SO 7991). *Beebrugiam, Bewbridge, Beebrugia* c.1180-6 Rees 1985 48, *Beebrugge* 1194, *Bebrig* 1200, 1203, *Bebrug* 1255, *Bewbrugge* 1272-81, *Bebruge* 1274, *Bebrugg* 1292 (all from Bocock 1923: 41), *Bebrugger* 1298 TSAS LXXI 1996 27, *brugg* 1301 Rees 1975: 248, *Bewbreche* 1525 SR, *Baybryge* 1532 SHC 4th Series 8 125, *Bewbridge* 1535-6 Bocock 1923: 41, *Beybriche* 1536 MA, *Bewbridge otherwise Bebridge* 1588 SHC XVII 236, *Beabridge* 1700 Bocock. Perhaps from OE *beo brycg*, here meaning 'the bridge at which bees were found' (cf. Beeford, Yorkshire). In Shropshire since the 12th century.

**BERDINGESTON** (unlocated) *Berdingeston'* 1235 Fees. Perhaps from an OE personal name *\*Beording* (a derivative of *\*Beord*: see Beardmoors), so here '*\*Beording's tun*'.

**BERESFORD**   2 miles south-east of Sheen, on the river Dove (SK 1259). *Beveresfort* 1275 SHC V (i) 117, *Beveresford* 1277 SHC 1911 168, 1304 SHC VII 120, 1306 ibid.' 177, 1307 ibid. 172, *Bereresford* 1304 ibid. 280, *Beresford* 1414 SHC XVII 13, *Berisford* 1561 ibid. 210. 'Beaver's ford', from OE *beofor* 'beaver': the animal is said to have become extinct by about the 12th century in southern Britain (JEPNS 34 2000-1 19). A mistaken etymology of the name led the Beresford family to adopt a crest of arms incorporating a bear: VHC VII 11.

**BERKHAMSYTCH (UPPER & LOWER)**   2 miles north-east of Ipstones (SK 0452). *Birkham Sitch* 1678 Okeover T755, *Birkemstich* 1851 White, *Upper Bircumsitch, Lower Bircumsitch* 1890 O.S., *Bircumsych* TNSFC 1908 133. A curious name. The last element is evidently OE *sic* 'stream'. The first word may be from ON *birki* 'a place overgrown with birch trees, a birch copse, a birch', but would not normally be combined with an OE word, here *ham* 'village', or perhaps more likely *hamm, homm* 'an enclosure; a meadow, especially a flat low-lying meadow on a stream', perhaps more specifically 'a place hemmed in by some feature of the topography, often by water or marsh', often found as -

ham. The first element may be from the northern form of OE *berc* (usually found in place-names as Bark-, Berk-: see EPNE i 28-9) 'birch-tree', which does not seem to be found with habitative terminals such as *ham* (EPNE i 28-9), but could be combined with *hamm, homm*, which would sit comfortably with OE *sic.*

**BERKSWICH** – see **BASWICH**.

BERNARDSCROFT, BARNARDSMORE (unlocated, possibly near Rudyard or Wolve Dale) *Bernardscroft, Bernardsmore* 13th century SHC IX NS 313-4, *Bernardescroftes* 1343 SRO D3272/1/14/1/14-19. See also Barns Lee, Barnswood.

BERNARD'S HALL (unlocated, in Morfe Forest: Eyton 1854-60: III 114) *Bernard's Hall* 1292 Eyton 1854-60: III 114.

BERNARDUM (unlocated) *Bernardum* 1203 SHC III 104.

BERNE (unlocated, possibly near Swynnerton) *Berne* 1304 SHC 1911 273, 1415 SHC XVII 55.

BERNEFELD (unlocated, in Tipton) *? Bernesfeld* 1306 SHC 1911 65, *Bernefeld* 1396 SHC XV 68.

**BERRY HILL** 1 mile north of Fenton (SJ 9046), *Berihul* 1212 SHC 1911 386, *Buryhill* 1567 SHC XIII 264, *Berryshill* 1573 SHC 1938 138, *Berriehill* 1586 SHC 1927 128 (but may refer to Berryhill Farm near Barlaston), *Bury Hill* 1798 Yates, 1799 Faden, *Berry Hill* 1836 O.S.; **BERRYHILL FARM** 2 miles east of Barlaston (SJ 9238), *Beryhulle* 1377 SHC 4th Series VI 12, 1549 SHC IV NS 114, *Bere Hill* 1532 SHC 4th Series 8 134, *Beyryhill* 1552 SHC XII 209, *Buryhill* 1567 SHC XIII 264, *Berriehill* 1586 SHC 1927 128, *Berrihill* 1697 SHC 1910 231; **BERRYHILL** 2 miles north-east of Stafford (SJ 9525), *Berry Hill* 1836 O.S.; **BERRYHILL** a 402' hill on the north side of Knowle Farm, 1½ miles south of Lichfield (SK 1107). *Berry Hill, Bery Hill* 1552 Harwood 1806: 389-90, *Berry-hill* 1585 (1798) Shaw I 336, *Broughton Hill* 1599 Smith, 1610 Speed, *Bury Hill* 1619 Erdeswick 1844: 310, *Berryhill (Field)* 1649 SRO D15/10/2/17, *Lyttylle Beryhylle, Lytle Bery Hylle* 1659 (1798) Shaw I 313, *Bury Hill* 1775 Yates, 1834 O.S.; **BERRY HILL** 1 mile north of Maer (SJ 7939), *Berry Hill* 1833 O.S.; **BERRY HILL** between Tixall and Milford (SJ 9722), *Berry Hill* 1950 O.S.; **BERRY HILL** ½ mile east of Oat Hill, Cannock Chase (SJ 9820), *Berry Hill* 1949 O.S.; BERRYHILL FARM (unlocated, in Kibblestone), *Berryhill farm* 1693 SRO 49/14/44; BERRYHILL (unlocated, in Tutbury), *Berryhill* 1782 SRO D15/11/14/87. *Buri* is a ME form of *byrig,* dative of OE *burh* 'fortified place' – many of these names mark the site of ancient earthworks. Some names may be from OE *beorg* 'hill', but the element appears to have been used for hills with a regular well rounded shape, and whilst the element is recorded as meaning 'mound', *hlaw* is the usual term for mounds and tumuli in Staffordshire. From OE *burh* 'fortified place'. Berryhill near Lichfield is likely to be *Broughto(n) hill* which is shown between Swinfen and Wall (with a drawing of a rounded hill) on Smith's MS map of Staffordshire 1599, and is also found on Kip's and Speed's maps of 1610: it seems likely that the drawing and name were taken from Saxton's map of 1577 where the name appears as *Borugcop hill*, which could well have been 'regularised' by the later map makers. *Bryhill* in Rugeley is recorded in the late 15th century: SRO DW1781/5/12/1. See also Bury Hill; Burouestone.

**BERRY RING** an iron-age hillfort 2 miles south-west of Stafford (SJ 8821). *Bury* 1278 SHC VI (i) 88, *Buryhil* 1470 Rental VCH I 335-6, *Buryhill* 1471 VCH IV 74, *? Les Buroughs* 1616 SHC VI NS (i) 21. *Buri* is a ME form of *byrig*, dative of OE *burh* 'fortified place'. The name *Billandbury* (SHC VIII (i) 122) may refer to this place, but its meaning is uncertain unless a corrupt form of 'Billington bury': see Billington.

**BERRY RING FARM** 2 miles south-west of Stafford (SJ 8821). *Bury Farm* 1836 O.S. From OE *burh* 'fortification, fortified place', with reference to the nearby Berry Ring (q.v.) hillfort.

**BERTH, THE** an iron-age hillfort ½ mile north-west of Maer (SJ 7939). *? Erthbiri* 1227 SHC IV 43, *? Burth, Burgh* 1281 SHC VI (i) 121, *the Broughe, Brugh* c.1565 SHC 1938 24-5, *the Borough* c.1598 Erdeswick, *The Brough, Bruff* 1686 Plot 408, *The Borough* 1747 Bowen, *Byrth* 1775 Yates, *the Byrth, Little Byrth* 1812 GM Pt II 602-6. From OE *burh* 'a fortified place': for details of the earthwork here see JNSFC LXVI 1931-2 91-100. OE final *-h* of *burh* can develop into modern *-f* (cf. the pronunciation of Modern *laugh, tough*), and this had led to *burh* becoming Burf rather than Borough in names such as Abdon Burf and Clee Burf (Shropshire), and Burfa Banks (Radnorshire), which are all hill-forts (Gelling 1988: 144), see also Burf Castle. In areas where the form Burf was common, the confusion of *-f* with *-th* has led to the names of some hill-forts developing into Berth, for example The Berth near Baschurch, Shropshire: see Gelling 1988: 144. The 1227 spelling is evidently from OE *eorð-burh* 'earthern fortification'. The 1281 spellings may refer to Brough Hall near Gnosall: SHC VI (i) 121.

BERTHERTON (unlocated, in or near Blithfield) *Bertherton* 1391 SRO D986/40.

BERTRAMSCOTE (unlocated, possibly Bescot (q.v.)) *Bertramscote* 1302 SHC VII 100. 'Bertram's cottage'. Bertram is from OFr Bertran(t), OGerman Bertram, Bertran(d): DES 30.

**BESCOT** 1½ miles west of Walsall (SP 0097). *Bresmvndescote* 1086 DB, *Bermonscot* 1240 (1801) Shaw II 89, *Bercumdescote* 1271 SHC IV 183, *Bermundescote* 1273 Ass, *Bermondscote, Bermonscote, Beremundescote, Bermundescote* 13th century Duig, *Bertmescote* 1306 (1798) Shaw I 71, *Berkumscote* 1345 ibid.' 94, *Berscote* 1349 Sims 1882: 13, *Berpmescote* 1350 SHC 1928 78, *Berkescote* 1356 SHC 1913 168, *Berkunscote* 1360 SHCVIII NS 6, *Berkmondescote* 14th century Duig, *Bascot* 1686 Plot 315, *Berkenscote or Bescote* c.1727 SOT D260/M/T/2/74. 'Beorhtmund's cottage', from OE *cot* 'cot, cottage, shelter, hut'. Shaw 1801: II 81 suggests that the place was anciently called *Berkenscot*. It would appear that the place was also known as Littleton in the 16th century (*Littleton* 1596 SHC 1932 183, *Lytteldon* 1597 ibid. 347). See also Bertramscote.

BESSY BANKS GRAVE (obsolete) on the north-western side of Lichfield (SK 1010). *Bessy Banks Grave* 1815 O.S. No information has been traced to explain this name.

BETHENEY, BETHENEI – see **STAFFORD**; **BROADEYE**.

**BETLEY** 6½ miles north-west of Newcastle-under-Lyme (SJ 7548). *Betelege* 1086 DB, *Bettelega* 1175 P, *Bettelegh* 13th century Duig, *Bettileg* 1272-3 SHC XI NS 245, *Betunlegh* 1289 SHC 1911 45, *Bettyleye* 1325 ibid. 365, *Betteley* 1532 SHC 4th Series 8 46, *Betle* 1605 SHC 1940 223. Perhaps from an OE masculine personal name *Betta, with OE *leah*. The *-an* genitive singular ending, thought to be represented by persistent *-e-*, only indicates a weak noun, not its gender. It was at one time said (e.g. Ekwall 1960: 40) that this place, and nearby Audley, Balterley and Barthomley (the last in Cheshire) all have the names of females, but this place need not, and it is not certain that Barthomley (q.v.) is from a personal name.

117

BETSFEILDE (unlocated, between Newport and Market Drayton) *Betsfeilde* 1598 SHC 1935 158. Perhaps from the OE personal name Bette or similar, with OE *feld* 'open ground', so 'the open ground associated with Bette'.

BEULEG (unlocated, said to be Bewley, near Pendeford (SHC 1928 111), which has not been traced, though *Clewley* is recorded (1946 O.S.) ½ mile north-east of Pendeford Mill (SJ 8904)) *Beuleg'* 1310 SHC 1928 111. An interesting name, perhaps from French Beaulieu 'beautiful place': see Bewdley.

BEVERLEY HALL (obsolete) A prebendal manor-house in Gnosall. *Beverley Hall* 1366 Lich Episc Reg, *Berlay hall* 1399 Pat, *Beverlehall* 1496 SHC 4th Series VII 169, *Beverly hall at Gnostall* c.1502 Bod. 28. Probably after Thomas de Beverley, a prebendary of Gnosall in 1223: SHC 1927 110; 1923 271. The hall was still standing in 1580, but not by 1677 (VCH IV 115). It may have been on the Stafford to Gnosall road in a field called Barley (Beverley ?) Orchard, where moated remains have been recorded: SHC 1927 110.

**BEWDLEY** 3 miles south-east of Upper Arley (SO 7875). This extra-parochial place was included in Staffordshire in the 15th century, but usually considered to be in Worcestershire, to which it was assigned by statute in 1543: Youngs 1991: 403. *Beuleu, Beauleu, Beaulieu* from 1275 SR to 1424 Ipm, *Bellum Locum* 1308 Pat, *Buleye* 1316 Ipm, *Beudle* 1335 Ipm, *Beudeley* 1349 Pat, *Beaudeley* 1381 Ipm, 1365 Pat, *Bewdeley* 1547 Pat. The name, from French *beau* 'beautiful, fine, splendid', and *lieu* 'place, locality, spot', so 'beautiful place', is a common type of laudatory French place-name. See also Beuleg.

BEXMORE FARM (obsolete) 1 mile east of Lichfield (SK 1310). *Berkesmoor* 1574 VCH XIV 277. Perhaps from the OE personal name Beorcol or similar, or OE *berc* (usually found in place-names as Bark-, Berk-: see EPNE i 28-9) 'birch-tree', with OE *mor* 'moorland', so 'Beorcol's moor' or 'birch moor'.

BEYVIL (unlocated; if in Staffordshire possibly near Chelle, or perhaps simply a family name, from Biéville in France: see CEC 191) *? Boevilla* 1171 CEC 191, *Beyvil, Beyvill* 1250 SHC XI 319, *Bevile* c.1250 ibid. 310, *Beville* 1254 SHC 1911 123, *Bevvul* 1256 ibid.' 126, *Beyvile* c.1230 SHC 1911 445, *Boyville* 1326 SHC 1911 372, *Bevil, Burvile, Buivil, Bevile, Beyvil* ? 14th century SHC XI 309. The spellings *Buville, Bevill, Beuvil, Beuvilla* are also recorded: SHC XI 320-1. Possibly *beau ville* 'beautiful village', a Norman-French name. See also SHC 1911 445.

BEYTE BRIGE (unlocated, in Alstonefield parish) *Beyte Brige* c.1576 SHC 1931 186. Possibly from ON *beit* 'bite, pasture', so 'the bridge at the pasture called Beit'.

**BHYLLS** 2 miles south-west of Wolverhampton (SJ 8797). *the bill, the Bills* 1647 Survey, *the Bill* 1775 Yates, *Bhylls* 1895 O.S. Possibly from OE *bill* 'a sword, a bill, a prominent hill', or OE *bile* 'a beak, a hill', used in a topographical sense of a promontory. The intrusive -*h*- appears to be a relatively recent affectation. The name is remembered in Bhylls Lane.

**BICKFORD** 2 miles west of Penkridge (SJ 8814). *Bigeford* 1086 DB, *Bykford, Bicford* 1251 Ass, *Bikeford, Bykeford* 1263 Ipm, *Bigesford* 1307 (p) GDR, *Whistone Bikforde* 1379-81 PollTax, *Bick(e)ford(e), Byckford(e)* 1547 Pat, *Bickford* 1614 Ct, 1834 O.S. Bick forms the first element to a large number of place-names, such as Bickenhall, Bickenhill, Bickham, Bickley, Bicton, Bickmarsh, etc. Possibly 'Bic(c)a's ford', or from OE *\*bica*, a word believed to denote a topographical feature having the appearance of a bill or a beak, and thought to be the root of names such as Purbeck, Dorset: see VEPN I 96. Cf. Bicton, Shropshire. The place lies near Whiston Brook at the eastern end of a pronounced ridge or promontory which may be the *bica.

**BIDDLE, RIVER** or **BIDDULPH BROOK**  a tributary of the river Dane. *Bidle water* 1577 Saxton. A a back-formation from the place-name Biddulph (q.v.).

**BIDDULPH** Ancient Parish (pronounced biddle [bɪdl]), 7 miles north of Stoke-on-Trent (SJ 8856). *Bidolf* 1086 DB, 1227 CH, *Bidulf* 1205-10 Ekwall 1928: 33, *Middle Bidolf*, *Holm-Bidolf* 1208 SHC III 173, *Middel Bidulf* 1272 SHC 1911 31, *Bydolf* 1283 Ekwall 1928: 33, *Bydoulf* 1284-5 SHC 1910 299, *Bydulf* 1291 Tax, *Bydulf* 1305 Ekwall 1928: 33, *Bedulf* ibid. 33, *Bydulfe* 1332 SHC X 97, *Bedell* 1425 SHC XVII 109, *Nether Biddulph* 1427 Ct, *Bedyll* 1532 SHC 4th Series 8 35, *Biddle* c.1540 Leland. From OE *dylf* 'a digging', which would be expected to give *dulf* in the West Midlands, from OE *delfan* 'to dig' (cf. modern *delve*), with OE *bi* 'by' (typically followed by no definite article), giving '(the place) by the mine or quarry', almost certainly with reference to ancient stone quarries, which still exist: see Ekwall 1928: 33-4. The 1208 *Holm* reference (from ON *holmr* 'a small island, a piece of land on a stream, dry ground in a marsh') is unclear. There were three manors in Biddulph: Over, Middle and Nether Biddulph: SHC IV 21. Chetwynd states that in 1769 Upper Biddulph was also known as Overton (q.v.), and Middle Biddulph as Middleston (q.v.): SHC XII 7. For a reference in 1563 to *'mines, delphes and quarries'*, see SHC 1931 183.

**BIDDULPH'S POOL**  3 miles east of Cannock (SK 0309). *Biddulph's Pool* 1834 O.S. Created in 1734 by John Biddulph, the pool was also known as *Lichfield heath pool* in the early 19th century: VCH XIV 195, 214.

**BIGNALL HILL, BIGNALL END,** on the east side of Audley (SJ 8150, SJ 8051). *Bikenhou* 1252 Ch, *Bicenhou* 1253 Ward 1843: app. iv, *? Bydonue* 1278 SHC XI NS 251, *Bygmowe* 1306 SHC VII 163, *Bygenowe* 1306 ibid.' 164, *Bygenon* 1307 SHC XI NS 263, *Bygenou* 1307 ibid.' 266, *Bygemor* 1307 ibid. 265, *Bykenou* 1327 SHC VII 206, *Bygenowe* 1377 SHC 4th Series VI 12, *Bugnall, Bignoo, Bygnohull* 1492 SHC 1912 256, *Bygnowe* 1492 ibid. 257, *Bignoll ende* 1576 SHC XIV 188, 1611 SHC III NS 60, *Bignall Hill* 1588 Audley ParReg, *Bygnowende* 1592 SHC XVIII 6, *Biggenhall* 1609 SHC III NS 34, *Biggnar-end* 1666 SHC 1921 113, *Bignal-hill, Bignall hill* 1686 Plot, *Bigney Hill* 1744 Burslem ParReg, *Bignole Hill* 1775 Yates, *Bignal End, Bignal Hill* 1833 O.S. A curious and difficult name. The adjective *big* can be ruled out: *micel* is the usual term for 'big, great' in place-names. Some of the forms (which may have transcription errors) seem to suggest a derivation from OWelsh *genou*, the first element in Gnosall (q.v.), with the OE preposition *bi* 'by' (cf. Biddulph, 6 miles to the north-east), so perhaps '(the place) by the mouth (of a valley or stream)', but that combination is very unlikely, and the earliest spellings point towards OE *bican hoh* 'hill-spur with a point', or 'Bica's hill-spur', a fitting description for the 774' Bignall Hill on which stands Wedgwood's Monument: *bica* is an element found mainly in the West Midlands and the south of England (see VEPN I 96). The spellings do not preclude completely a derivation from OE *becun* 'beacon', which would fit the topography. For completeness, the possibility that the second element is from ON *haugr* 'a hill, a mound, a tumulus' cannot be ignored (cf. Wraggoe Wappentake, Lincolnshire (Cameron 1998: 143-4); Gelling & Cole 2000: 174; see also Houndhill). The word *end* (which is very common in this area) meant not a terminal point, but simply 'a place', and was often applied to squatter dwellings on the outskirts of a settlement. The 1833 1" O.S. map shows Bignal End at what is now Bignall

End Farm, 1 mile north of what is now Bignall End on Raven's Lane, suggesting that the original name was of sufficient importance to cover a considerable area.

**BIKERSDALE WOOD** 1 mile south-west of Tatenhill (SK 1921). Early spellings have not been traced, and the name is not shown on the first edition 1" O.S. map of 1836, but possibly (if ancient) from OE *beocere* 'bee-keeper', so 'the dale or valley of the bee-keeper'.

**BILBROOK** 4 miles north-west of Wolverhampton (SJ 8803). *Bilrebroch* 1086 DB, *Bilrebroc, Billebroc* 1166 SHC 1923 296, 1227 Ass, *Billebroc* 1167 P, *Balebroc* 1227 SHC IV 46, *Bulebroc* 1228 SHC XVII 46, *Bilroba* c.1250 SHC VI NS (ii) 49, *Bilrebroc* 1271 SHC VI (i) 51, *Byllerbrok* 1275 SHC VI (i) 70, *Billingbrok* 1307 SHC VII 178, *Billurbroke* 1327 SHC VII (i) 251, *Billerbrok* 1376 SHC XIV 140, *Billenbroke* 1425 SHC XVII 106, *Bulderbroke* 1419 ibid.' 36, *Bellesbrok* 1434 SHC XII (ii) 22, *Bellerbrok* 1435 SHC XI 244, *Bylderbrok, Billerbrok* c.1457 SHC VI NS (ii) 101, 108, *Bylbroke, Bylderbroke* c.1482 SHC VI NS I 152, *Bilbrooke* 1635 SRO 3764/153. Probably 'brook of the billers or bilders', a dialect name for several water-plants, including watercress, cow parsley, etc. *Billers* (OE *billere*, ME *billure, byllerne*) may be of Celtic origin: cf. Welsh *berwr*. VEPN I 100. The brook here might have been expected to bear the same name (VCH I 48 gives it the name Billbrook, but cites no source), but is called Moat Brook (*ye mot-brooke* 1638 Codsall ParReg), possibly from moated sites at Moor Hall (obsolete) (q.v.) and Wood Hall (q.v.), Codsall, both on tributaries (cf. *the Mote House* in Codsall, recorded in 1783: WALS DX83/13). Billsmore Wood, Warwickshire (PN Wa 68), Billacombe, Devon (PN D 184), and Bilbrook near Minehead in Somerset (*Bilrebroc* 1227 Pat) are believed to have the same root as this place. *Over Billers Wood* 1 mile south-west of Madeley (SJ 7643) appears on the 1834 1" O.S. map. Its history and derivation are unknown. *Bilbroke* in the Clipston area of Derbyshire is recorded in 1613: DRO D3155/WH217.

**BILL HEATH** 2 miles south-west of Brewood (SJ 8705). *The byll heathe* 1562, 1569 Ct, *the Byll* 1567 Ct, *bill heathe* 1597 Codsall ParReg, *the Bill* 1842 TA. Perhaps from OE *bill* 'a sword, a bill, a prominent hill', or OE *bile* 'a beak, a hill', used in a topographical sense of a promontory (perhaps *Byle* recorded in 1414: SHC XVII 17). The place lies below a pronounced headland.

**BILLINGTON** in Bradley parish, 3 miles south-west of Stafford (SJ 8820). *Belintone* 1086 DB, *Billinton* 1204 SHC III (i) 101, *Biledon* 1208 ibid.'142, *Billenton* 1208 Cur, *Belinton* 1213 SHC III (i) 161, *Belintton* 1214 Cur, *Belinton* 1285 FA, *Bylington* 1299 SHC VII 64, *Belyngton* 1304 Ass, *Byllynton* 1307 SHC VII 178, *Belenton* 1532 SHC 4th Series 8 97, *Billington* 1602 SHC 1935 478. A difficult name. Possibly from OE *\*billing*, based upon OE *bill* 'a sword, an edge, a bill, a prominent hill', or OE *bile* 'beak', with an OE *-ing* suffix giving '*tun* at *\*Billing*, tun at the *\*billing*', with *\*billing* meaning 'hill-place' (VEPN I 100-101; see also Finberg 1975: 41-5; PN Ch V (II) 282). Some place-names (as here) indicate a parallel formation from OE *belle* 'bell-shaped hill', and some point towards a folk-name *Bil(l)ingas/Belingas* 'people living at the *Bil(l)/Bell*', alternating with the topographical terms (VEPN I 101), but other derivations for this name cannot be ruled out, such as 'Bill's or *\*Billa's tun*'. The place lies on a ridge, with the hill-fort of Berry Ring standing on a headland to the north-west, perhaps making a topographical or folk-name more likely.

**BILSON** 3 miles north-west of Lichfield (SK 0912). *Bilston* 1437 BCA MS3415/215, *Bilstone* (p) 1456 Harwood 1806: 277, *Bilston* (p) 1496 OSS 1936 48, *Bilson* 1589 SHC XVIII 4, *Bilston (Brooke)* 1608 SHC 1948-9 12, *Billson* 1623 SHC X NS I, *Bilson*

*(Brook)* 1695 Morden, *Bilston (Brooke)* 1734 SA 1987/3/3, *Bilston Brook, Bilston Brook Farm* 1834 O.S. Perhaps from the OE personal name Bil(l) or the OScandinavian personal name Bildr, with OE *tun*. Bilson Brook is said to have been previously known as Bourne Brook: VCH XIV 195.

**BILSTON** 3 miles south-east of Wolverhampton (SO 9596). *Bilsetnatun* 996 (17th century, S.1380) 17th century, *Billestune* 1086 DB, *Billistan* 1173 SHC I 68, *Billeston'* 1190 Pipe, *? Belsten* 1293 SHC 1911 229, *Bilestun, Billeston, Bilestone* 13th century Duig, *Billesdone* 1302 SHC VII 102, *Bylston* 1304 SHC 1911 277, *Bilstun* 1327 SHC VII (i) 250, *Bylleston* 1414 SHC XVII 49, *Bryerley Byllson* 1621 Worfield CA. 'The dwellers at Bil', from OE *sǣte* (genitive plural *sǣtna*) 'settlers'. Bil might be a contraction of a longer name, but no such name is known locally. It cannot represent a personal name, since *sǣte* is not added to personal names. Another possibility is that Bill may have been the name of a nearby hill, connected in some way with OE *bile* 'bill, beak', used topographically of a promontary or pointed hill, or OE *bill* 'sword', perhaps used in a similar topographical application: VEPN I 99-100. The boundary of the Bilsǣtan is recorded in 985 as *Bilsatena gemǣro* 985 (12th century, S.860). *Bryerley* is Brierley (q.v.) near Coseley. Cf. Billinge, Merseyside; Billingham, Cleveland; and Billingshurst, West Sussex. See also WA II 85-93. It may be noted that *Bilston Lane* is recorded between Colwich and Colton (presumably what is now Ballamour Way) in 1759 (Burne n.d.: 21), but the history of the name is not known.

**BINCLIFF (MINES)** 1 mile south-east of Wetton (SK 1153). *Byncliff* 1547 AugOffice, 1861 Bateman 1861. Perhaps associated with the lead mines here.

**BINE FARM** 2 miles south-west of Bobbington, on the south side of a pronounced hill of 469' (SO 7789). *one parcel of wast 'The Binde'* 1614 SA 2922/2/25, *the Bynde* 1616 SA 2922/2/26, *The Bind* 1632 SA 2922/1/3/30, *Bind* 1751 Rocque, *The Bind* 1766 Claverley ParReg, *The Binde* 1833 O.S. Possibly from *bind* in the sense 'to subject to a specific legal obligation' (OED), or perhaps with reference to a plant such as honeysuckle or woodbine (OED), through neither suggestion seems entirely satisfactory. The place is recorded *temp.* Edward I: SA 2922.

**BINNS FARM** 1½ miles north-west of Lichfield (SK 1011). *the Beenes* 1640 VCH XIV 233, *Bean's Farm* 1834 O.S. From land here called *the Bynds* c.1468 (VCH XIV 235). Perhaps from a personal name, or from the type of crop formerly grown here.

**BIRCHALL** 1 mile south of Leek (SJ 9854). *Byrchehull, Byrcheshull* 1271 SHC VI (i) 52, *Birchel* (p) 1285, *Byrchulle, Birchull* 1292 SHC VI (i) 219, *Byrchoure* 1293 SHC 1911 217, *Byrchull, Birchull* 1296 Ch, 1327 (p) SR, *Birchow, Bircholt, Birchou, Byrch(e)holt(e), Birch(e)holt(e)* 13th century Dieul, *Birchenehull* 1330 SHC XVI 4, *Byrcholt* 1538 (1883) Sleigh 1883: 17, 1539 SHC IX NS 300, *Byrchold* c.1540 AOMB, *Byrcholte* 1546 SHC 1912 351, *Byrchehold* 1556 FF, *Byrcholt* c.1560 SHC 1938 159, *Byrche Holte* 1560 SHC XIII 207, *Birchowe als Bircholte* 1565 FF, *Berchall als Berholt* 1623 FF, *Birchall* 1641 Leek ParReg, *Big & Little Birch Hill* 1775 Yates. The first element is clearly OE *birce* 'birch', but the forms of the second element show much variation, and it seems possible that the alternate forms of OE *hyll* 'hill', *holt* 'wood', and *hoh* 'the end of a ridge where the ground falls away sharply' would all be appropriate here. There is no likelihood that the word 'hall' (a rare element in place-names) is incorporated in the name. Cf. Birchall Green, Worcestershire.

BIRCHALL GRANGE (obsolete, 1 mile south of Leek (SJ 9854)) *Burchehowgraunge* 1345 SHC 1938 159, SHC IX NS 296, *Graunge de Burch(e)holte, Graunge de*

*Byrcholt(e)* c.1540 *AOMB, Birchehall Graunge, Birchehall Graunge* 1611 QSR. The place was the grange or outlying farm of Dieulacres Abbey at Birchall (q.v.).

BIRCH DALE (obsolete) 2 miles south-west of Stone (SJ 8830). *Birch Dale* 1775 Yates, *Birchdale* 1836 O.S. 'The dale or valley with the birches'.

BIRCHEN BOWER (obsolete) 2 miles south of Uttoxeter (SK 0729). *Byrches Bowers* 1735 D240/E(A)2/157, *Birchen Bower* 1798 Yates, 1836 O.S. From the OE adjective *\*bircen* 'growing with birch trees', with OE *bur* 'cottage', so 'the cottage at the birches'.

BIRCHENBOOTH (unlocated, in Quarnford, near the Staffordshire-Derbyshire boundary) *Birchinbooth* 1598 SHC XVI 170, *Birchenbough* 1682 Brown, 1749 Bowen, *Birchen Booth* 1798 Yates. From the OE adjective *\*bircen* 'growing with birch trees', and the north country dialect *booth* 'a cowhouse, a herdsman's hut', from ODan *boð* 'a bothy, a temporary shelter', giving 'the shelter at the birch trees'.

**BIRCHENFIELDS** 1 mile north-east of Dilhorne (SK 9844). *Bouthes Birchenfeld* 1609 SHC III NS 52. 'Open land with the birches near Booth': Booth Hall and Booths Farm (q.v.) lie to the east.

BIRCHEN FIELDS (obsolete) on the south-west side of Armitage (SK 0715). *Birchen Fields* 1834 O.S. 'Open land with the birches'.

**BIRCHES** 1 mile south-east of Codsall (SJ 8702). *Birches* 1716 Ct, 1775 Yates, *ye Birches* 1730 Codsall ParReg, *the Birches* 1762 (1801) Shaw II 289. See The Bratches. The estate gave its name to Birches Bridge (which crosses a railway) here.

**BIRCHES BARN** 1 mile south-west of Wolverhampton (SO 8997). *Barndeleye* 1327 WA I 282, *Birch-his-barne* 1654 Wolverhampton ParReg. The earlier spelling seems to be from 'the burnt *leah*'; the later spelling suggests 'the barn of the man named Birch', or a rationalisation of *Birches-barn* 'barn in the birches', or '(at a place called) Birches'. See also WA I 282.

**BIRCHES HEAD** on south-east of Burslem (SJ 8948). *Bircheshead* 1641 Burslem ParReg, *Byrcheshead* 1697 ibid, *Bircheshead* 1755 DRO D3155/7037. 'The head or top end of the hill with the birches'.

**BIRCHFIELD** 1 mile south of Perry (SO 0690). *Birchfield House* 1834 O.S. 'This place derives its name from an ancient family named Birch, who long resided here': P.O. 1870 538. Hackwood 1905a: 162 suggests that the place 'was so called from the family of Wyrley-Birch, who were once owners of the manor of Hamstead'.

**BIRCHILLS** 2 miles north-west of Walsall (SK 0000). *Burchles* 1271 SHC 4th Series XVIII 78, *Bircheles* 1325 Willmore 1887: 241, *Byrchelles* 1587 SRO D260/M/T/1/12, *Bircheleses, Byrchylles, Burchelles, Rough Byrchells covered with a wood, Byrchells* 16th century Duig, *Burch-hills* 1686 Plot. Almost certainly from OE *\*bircel* 'little birch trees' (cf. Birchill, Derbyshire; Birtles, Cheshire: see VEPN I 103). Birchills was within the bounds of Cannock Forest.

BIRCHINLEE (unlocated, in Rushton Spencer) *Byrchynlee yate* 1485 SRO DW1761/A/4/36, *Great Birchen Lee, Birchinlee Meadows* 1673 SRO DW1761/A/4/92. 'The *leah* with the birches'. *Yate* is from OE *geat* 'a gate'.

BIRCHOVER (unlocated, possibly near Butterton) *? Bircho* 1208 SHC III (i) 99, *Birchovre* 1293 SHC 1911 217, *Birchover* 1386 SHC XVI 25, *Bircheover* 1586 SHC

1927 139. From OE *birce* 'birch-tree', with OE *ofer* 'slope, ridge, hill', giving 'the ridge with the birch-trees'. Some of the above spellings may relate to Birchover in Derbyshire.

**BIRCHWOOD PARK** 2 miles south-west of Church Leigh (SK 0133). *? Birchewode* 1396 SHC XV 69, *Birchwood parke* 1600 SHC 1935 365, *Birchwood Park* 1610 Speed, *Birchwood (Park)* 1686 Plot 107. Self-explanatory.

**BIRDSLEY FARM** 2 miles north of Tamworth (SK 2008). *Burgley Farm* 1834 O.S. The place lies on a 262' hill.

**BIRKS BARN** on south side of Belvide reservoir, Brewood (SJ 8609). *Birks Barn* 1834 O.S. From the Birk family who farmed here. The original farmhouse was originally some 400 yards to the east, and was relocated when the site was flooded in the creation of Belvide reservoir, constructed in 1835: Horovitz 1992: 222.

BISHOP'S HILL (unlocated, in Longdon, possibly near Arblaster) *Bishop's Hull* c.1569 SHC IX NS 72. The place was a manor held by the Bishops of Lichfield. According to Shaw 1798: I 291 the prebend of Bishopshull (*Byshopshull* 1528 (1798) Shaw I 285) took its name from premises in Lichfield, but see also Bispill Plantation. VCH XIV 135 suggests that Bishopshull took its name from land at Lichfield later known as Bispels. See also Bisphills. A house called Bishop's Hill (*Bishop's Hill* 1887 O.S.) lies ½ mile south-east of Newborough (SK1525), but the history of the name has not been traced.

**BISHOPS OFFLEY** – see **OFFLEY, BISHOPS**.

**BISHOP'S WOOD** 2 miles west of Brewood on the Staffordshire-Shropshire border (SJ 8309). *Stryfwode* 1314 VCH IV 26, *Bishop's wood* 1538 VCH IV 26, 1549 (1801) Shaw II 303, *Bysshoppes woodd* 1597 SHC 1910 258, *Bishops Wood* 1624 Brewood ParReg, 1661 SRO D590/29/1, 1747 Bowen. DB records a wood in Brewood held by the bishop of Lichfield: VCH IV 41. The location is uncertain, but in 1139 and 1144 the Pope confirmed the bishop's possession of a wood of Brewood, and in 1314 an area of the same wood called *Stryfwode* (i.e. 'strife-wood', or 'the wood being the subject of dispute') was bought by the bishop of Lichfield from Sir Fulk Pembrugge, lord of Tong (VCH IV 26; see also VHC V 25; SHC VI (ii) 109). This was the wood, probably *Brewude Wood* recorded in 1247 (Eyton 1854-60: II 221) and the 'high wood' recorded in 1315 (WL 104), which gave its name to Bishop's Wood. The place lay in Brewood Forest. It is possible that *Bishopesbrewode*, recorded in 1302 (SHC VII 95) refers to Bishop's Wood: Brewood was also held by the bishop, but in the case of that place the descriptor would be unnecessary.

**BISHOP'S WOOD** an area of woodland, said to have contained 1300 acres (Erdeswick 1844: 132), 5 miles to the west of Eccleshall, lying against the Shropshire border (SJ 7531), which was held by the bishops of Lichfield. *Bishop's woods* 1694 SRO EXD&CB/C/5, *Bishop's wood* 1798 Shaw I 93, 1801 Shaw II 113. The larger area of woodland, to the west of Blorepipe, is called locally Big Bishop's, and the smaller area to the east of Burnt Wood is known as Little Bishop's: local information.

**BISHTON** 2 miles north-west of Rugeley (SK 0120). *Bispestone* 1086 DB, *Bissopeston'* 1198x1208 (13th century) EEA 17 74, *Bussepeston* 1199 SHC III (i) 53, *Bissopesdon* 1204 SHC III 114, *Bispbiston* 1273 SHC 1911 158, *Bispeston* 1278 SHC XI NS 245, *Bissopeston* 1283 SHC 1911 182, *Bissopestun* 13th century Duig, *Bysopeston'* 1316 WL 106, *Bisshopeston* 1324 SHC 1911 101, *Bisshopestone* 1337 SHC 1934 (ii) 60, *Byston* 1532 SHC 4th Series 8 64, *Bisheton* 1618 SHC VII NS 193, *Bishton* 1798 Yates. From OE *biscop* 'a bishop', hence 'bishop's *tun*'. It is not always clear whether the name is from the OE personal name Biscop (which as the title of a senior ecclesiastic is unlikely to have been common), or has ecclesiastical links. According to Duignan 1902: 16 the bishops of Lichfield had a residence at Bishton, but evidence supporting that statement has not been traced. There is another Bishton on the north side of Patshull Park (SJ 8001), just within Shropshire, recorded as *Byspeston* in 1437 (SHC NS III 135). Cf. Bushton, Wiltshire.

**BISPHILL PLANTATION** ½ mile north-west of Elford (SK 1711). Seemingly 'the bishop's hills': the prebend of Bishopshull in Lichfield cathedral is said to take its name from this place (StEnc 62), but see also Bishop's Hill; Bisphills.

**BISPHILLS** (unlocated, in Freeford: Shaw 1798: I 316) *Bispells* 1664 (1798) Shaw I 316, *The Bispills* 1723 SRO (LJRO/D88), *Bispells* 1798 Shaw I 316, *Bispell's Farm* 1845 SRO D661/8/1/1/4. Seemingly 'the bishop's hills': according to Shaw (1798: I 291) the prebend of Bishopshull in Lichfield cathedral was 'denominated from premises in the liberty of Lichfield city'. VCH XIV 135 suggests that Bishopshull took its name from from land at Lichfield later known as Bispels. *Bysphill Marsh* is recorded in 1558 (Shaw 1801: II *28), and may be associated with this place. See also Bishop's Hill; Bisphill Plantation.

**BITHAM** (obsolete, south-west of Penkridge, between Wolgarston and Otherton (SJ 9313)) *? Bucheme* 1261 SHC 1950-1 49, *Belehem* 1284 FA, *Betlehem, Bethem* 1332 SHC X 119, *Butheme* 1327 SHC VII (i) 242, SHC XII 47, 1340 SHC 1913 75, *Botheme* 1340 ibid.' 75, *? Bucheme, Buthene* 1345 SHC XII 41,47, *Buthem* 1398 SHC XV 82, *? Bethum* 1428 FA, *Bedam* 1583 SHC III (ii) 4, *Bythom(e)* 1598 *Ct*, 1637 Penkridge ParReg, *Betham* 1599 ibid, *Bithome* 1657 SRO D260/M/F/5/98, *Bithom* 1673 Bradley ParReg, *Bitham* 1679 SHC 1919 219, 1695 Morden, 1749 Bowen, *Blytham* 1779 Bradley ParReg. A curious name, with inconsistent early spellings precluding a firm derivation (if indeed all relate to this place), but possibly from OE *byðme* 'head of a valley, valley bottom', an element that occurs in Mercian areas, and which may be a Mercian variant of West Saxon *botm* (see JEPNS 20 1987-8 46), in some cases influenced by the name Bethlehem. *Bythom greene als Lynehill greene* is recorded in the bounds of Penkridge in 1598 (Oakden 1984: 102). Will'o de Bethlehem is recorded in Tutbury in 1332: SHC X (i) 107.

**BITTERNSDALE** 2 miles south-east of Draycott in the Moors (SJ 9936). *Bitterns Dale* 1775 Yates, 1836 O.S. Self-explanatory. The name is unlikely to be ancient: 'bittern' in OE was *hæferblæte, raredumle, raradumbla*. The word *bittern*, not recorded in English before the 14th century, is apparently from OFr.

**BITTERSCOTE** 1 mile south-west of Tamworth (SK 1903). *Bittrescote* 1300 SHC V (i) 178, *Butrescote* 1391 SHC XV37, *Bytterscote* 1437 LLRRO 23D66/14, *Birtirscote* 1459 SHC NS IV 114, *Bythyrscote* 1459 ibid. 105, *Betercott* 1532 SHC 4th Series 8 23, *Bistercote* 1556 LLRRO 23D66/17, *Bytteston* 1562 SHC 1938 123, *Bytterscote* 1585 SHC XVII 233, *Bitterscott* 1623 SHC X NS I 51, *Biterscote* 1834 O.S. Perhaps from the personal name Brihthere (Beorhthere), with the first -*r*- lost owing to dissimilation (cf. Bittering, Norfolk), with OE *cot* 'cottage, shelter, hut'; one of the forms suggests OE *tun*.

**BLACK BROOK** occurs frequently (e.g. *Blakebrucke* 1616 SHC VI NS (i) 18) for the upper reaches of several rivers, including Dane, Hamps, Manifold, Tame, and Smestow Brook. From OE *blæc* 'dark-coloured', perhaps from the black organic silt found in many headwaters.

**BLACK COUNTRY** a 19th century generalised expression first noted in 1834 '... in the densely populated black country ...', from C Young, 'Memoirs of C M Young', 1871 212 (OED) for the nebulous and indeterminate industrialised district to the west of Birmingham, the extent of which has been, and assuredly will continue to be, the subject of much debate, even though it is certain that when first coined the expression was not used with specific boundaries in mind, so that no definitive area can ever be drawn. The expression has evolved and been interpreted in various ways since first coined, but by usage now tends to include (in, or formerly in, Staffordshire) Bilston, Blackheath, Brierley Hill, Coseley, Darlaston, Pensnett, Upper and Lower Gornal, Netherton, Old Hill, Rowley Regis, Sedgley, Smethwick, Tipton, Tividale, Walsall, Wednesbury, Wednesfield, West Bromwich, Willenhall and Wolverhampton, but also including (in, or formerly in, Worcestershire) Cradley Heath, Dudley, Lye, Oldbury and sometimes Halesowen, Stourbridge and Wordsley (see Oakden 1984: 2, which also includes, surprisingly, Tettenhall). The expression, which seems to have been in popular use by the mid-19th century (see The Illustrated London News, 14 April 1849 reporting a speech of J. P. Dyott, Mayor of Walsall, on the official opening of the South Staffordshire Railway line from Walsall to Alrewas on 9 April), came from the smoke and pollution caused by the collieries, quarries, blast furnaces, foundries, kilns, potteries, glassworks, ironworks, forges and factories based on the South Staffordshire coalfield, with its coal, iron ore, fireclays and limestone. It may be noted that OED defines the expression as 'a name given to parts of Staffordshire and Warwickshire [*sic*] blackened by the coal and iron trades', so including parts of Warwickshire without mention of parts of Worcestershire. One of the earliest references to the Black Country (with capitals) may be in White 1860: 6. The Daily Telegraph of 12 December 1864 recorded: 'By night the Black Country blazes up lurid and red with fires which ... are never extinguished', and on 8 and 29 December 1866 Punch used the term Black Country in derogatory essays about the area. The expression doubtless became more widely known with the publication of Burritt 1868 (Burritt being the American consul in Birmingham in the 1860s), which opens with the words 'THE BLACK COUNTRY, black by day and red by night...', and describes Birmingham as 'the capital ... of the Black Country' (p.5), a statement unlikely to be accepted today, and demonstrating the debateable and changing boundaries of the area encompassed by the term. No evidence has been traced to support popular folklore that the expression was associated originally with the extent of a particular coal-seam, notwithstanding Beaver 1945 which suggests that it could be associated with the exposed coalfield, but excludes the centres of Walsall, Wolverhampton and West Bromwich. See also StEnc 63; ES 12 February 2004 28-9.

**BLACK FIELD** in Bagot's Park, north-east of Dunstall (SK 0827). *Blakefeld* 1402 SHC XI NS 207.

BLACKFORDS (obsolete) 1½ miles north-east of Cannock (SJ 9912). *Blacke fourdes* 1570 *Rental, Blackfords* 1834 O.S.

**BLACKHALVES** ½ mile south-west of Essington (SJ 9502). *Blacke Hove* 1601 SHC 1935 345, *Blackhaughe* 1603 SHC 1940 60, *Blakehalfe* 1608 SHC 1928 120, *blackhalve* 1633 Wolverhampton ParReg, *Blackhalve* 1834 O.S. The earliest spellings suggests a derivation from ME *blæc halh*. OE final -*h* of *halh* sometimes develops into modern -*f* (cf. the pronunciation of Modern *laugh*, *tough*), and this may have led to *halh* becoming *halve*.

BLACKHILL (unlocated, in Hopton) *Blackhull temp.* John SHC VIII 170.

**BLACK HILL**  on the north-east side of Sandon (SJ 9529). *Blackehill (Filde)* 1591 SHC 1934 (ii) 19.

**BLACK HOUGH**  (unlocated, in Haughton parish) *the Black Hough, Blakehalghe* Eliz ChancP, *Blackhough* 1697 WALS 277/14/31, *the Black Hough* 1836 O.S. From OE *blæc, halh* 'black *halh* or haugh', with the second element later replaced by *hoh* 'a spur of land, a promontory'.

**BLACKLADIES**  2 miles west of Brewood (SJ 8409). *nigris monialibus de Brewde* c.1200x1210 SHC 1939 185, *blakladys* 1362 Ipm, *Blackladies* 1632 SHC II (ii) 15, *Blacke ladies* 1666 SHC 1927 71, *Blackladyes* 1686 Plot. Named from the black habits of Benedictine nuns who had a priory here from c.1147: VCH III 220-2. The earliest deeds describe the nuns as 'of the church of St Mary of Brewood' (the dedication of the local parish church), and the description *Black* does not occur before the turn of the 12th century (SHC 1939 180), to distinguish the place from the Cistercian nuns of White Ladies (*Whitladies* 1384 SHC XIV 240), 2 miles to the west in Shropshire, who wore white habits. Both Blackladies and White Ladies lay within Brewood Forest, and early references to the nuns of Brewood may refer to either house. See also VCH III 220-2.

**BLACK LAKE**  1 mile north of West Bromwich (SO 9992). VCH XVII 7 suggests that this was the name of a plot of land here (as *Whyte lake*) by the end of the 14th century, and of a house or cottage by 1502, but Willet 1882: 209 records a reference to *the Blake Lake* in 1474. OE *blæc* 'black' can be confused with OE *blac* 'shining, white', which is apparently the derivation of this name.

**BLACK LAKE** (unlocated, in Forton parish) *Blake Layke* 1527 Survey, *Black Lake* 1686 Plot. Perhaps 'the black-looking lake', but see Black Lake above.

**BLACK LEES**  3 miles south-west of Cannock (SJ 9606). *le Blakele(ye)* 1290 Hatherton, 1342 Vernon, *Blakelye* 1342 VCH V 177, *le Blakelie* 1380, 1515 Vernon, *Great Blake Leys, the Black Lyese* 1526 Vernon, *Black Lees* 1608 VCH V 177, *Blakelees* 1631 SRO D260/M/T/2/12, *Blake Leys* 1775 Yates. From OE *blæc* 'black, dark', with OE *leah*.

**BLACKSHAW MOOR**  1 mile south of Upper Hulme (SK 0159). *Blakeschawe* 1250-9 SHC 1911 428, *Blakeshawemore* 1539 LRMB, *Blackshawmoor* 1648 Leek ParReg, *Blackseymore* 1656 ibid, *Blackshaw* 1775 Yates. 'Black copse', from OE *scaga* 'copse, small wood', with OE *mor* 'moor'.

**BLACKWATERS**  1 mile north-west of Croxton (SJ 7732). *Blackwater* 1696 Eccleshall ParReg. Self-explanatory.

**BLACKWELL**  2 miles north-east of Bromsgrove (SO 9971). *Blakewell* Henry III PN Wo 362. 'The black spring or stream', from Mercian OE *wælle* 'a spring', and (sometimes) 'a stream'. In Staffordshire from c.1100 until 1266, in Warwickshire until 1844, and since that date in Worcestershire.

**BLACKWOOD**  1 mile west of Horton (SJ 9257). *Blacwode* 1299 SHC NS XI 253, 1325 SHC 1911 366, 1332 SHC X (i) 101, 1361 JNSFC LIX 1924-5 57, *Blakewood* 1514 JNSFC LX 1925-6 37, *Blackwood* 1579 Biddulph ParReg, 1608 SHC III NS 27. OE *blac* 'pale, white, bleak', is difficult to distinguish from OE *blæc* 'black, dark', but the latter is more likely here.

**BLACKWOOD HILL**  2 miles south-west of Horton (SJ 9255). *Blackwood Hill* 1596 BCA MS3558/170, *Blackwood hill* 1662 Norton-in-the-Moors ParReg. See Blackwood.

**BLAKE BROOK** a tributary of the river Manifold. Probably the same derivation as Black Brook (q.v.).

**BLAKE HALL** (obsolete) 1 mile south-east of Dilhorne (SJ 9942). *Blakehagh* 1449 HLS 389, *Blakehalghe* 1538 SHC 1910 267, *Blackhaghe* 1583 ibid, *Blake Hall* 1836 O.S. The second element is from OE *halh*, so giving 'the black haugh'. The place may have been known formerly as *Heghe Halghe*. SHC 1910 265-6 (but see also Delph House). It seems likely that *Jakehall*, recorded in 1539 (MA), refers to this place.

**BLAKELANDS** on the Staffordshire-Shropshire border, 1 mile north-east of Bobbington (SO 8291). *Blakeland* 1433 VCH XX 68, 1452 SHC 1939 233, 1486 SA 5735/2/22/1/29, 1595 SHC 1932 127, 1663 SHC II (ii) 32, *Blakelands* 1749 Bowen. 'The black land'. *The Black Wall*, recorded in Bobbington in 1683 (SRO 3764/155[40683]) may be associated with this place, but see also Blakenhall. Shaw 1801: II 278 states that there were two places of this name in this area. The names mark the site of drained bogland.

**BLAKELEY** ½ mile east of Whiston (SK 0547), *Blake legh* 1292 Ipm, *Blakeleye* 1307 (p) GDR, *Blakelegh* 1327 Ipm, *Blake Lee* 1775 Yates, *Blakeley* 1837 O.S.; **BLAKELEY GREEN** 2 miles north-west of Wolverhampton (SJ 8900), *Blakelegh* 1334 SHC 1928 138, *blakeleye* 1338 Tooth 2000b: 79, *Blakeley Green* 1834 O.S.; **BLAKELEY LANE** 1 mile south-west of Consall (SJ 9747), *Blakeley Lane* 1607 QSR, *Blacklee Lane* 1606 ibid, *Blakelee* 1646 SHC 4th Series I 287, *Blake Lee* 1697 ParReg; **BLAKELEY (POOL FARM)** 1 mile north-west of Seisdon (SO 8295), *Blakeley (Common)* 1774 SRO cp443/788rot24. From OE *blæc*, *leah* 'the dark woodland glade'. *Blakeleye* near Wrottesley, recorded in 1317 (SHC IX 62), may be Blakeley Green. The Green element suggests a squatter settlement.

**BLAKELOW** 2 miles north-west of Stone (SJ 8635), *Blakelowe* 1292 SHC VI (i) 233, 1327 SHC VII 207, *Blakelow* 1263 Duig, *Blakelawe* ? 13th century SHC VI (i) 8, *Blakelowe* 1306 SHC 1911 67, 1475 SHC NS VI (i) 95, *Blaklowe* 1477 ibid.' 111, *Blakelow* 1567 (1801) Shaw II app. 12, *? Blakely* 1608 SHC 1948-9 177, *Blakelows* 1710 Swynnerton ParReg; **BLAKELOW** 2 miles south-west of Onecote (SK 0352), *Blakelow* 1836 O.S.; **BLAKELOW** 2 miles north-east of Hilderstone (SJ 9736), Blakelow 1891 O.S.; BLAKELOW (obsolete) 1 mile north-east of Bucknall (SJ 9148), *Blakeloue* 1263 SHC IV 157, *Blakelowe* 1332 SHC X 94, *Blakeley* 1775 Yates, *Blakelow* 1836 O.S.; BLAKELOW (unlocated, near Warslow Green), *Blakelow grene under Orueslow* c.1284 Loxdale, *Bleaklow Quarry* 1822 EnclA. Probably from OE *blæc* 'black, dark', with OE *hlaw* 'tumulus, burial mound', but perhaps in some cases from an unrecorded early form of *bleak* 'bare of vegetation, exposed, wind-swept', not recorded by OED before the 16th century. *Blakelowfeld* is recorded in Chapel Chorlton in 1355 (SRO D593/B/1/6/1D/5), *Hatton Blakelowe* is recorded at The Hattons, near Brewood, in 1588 (SRO D260/M/E/428/2), and *Blakelow, Cheadulton* is recorded in 1687 (Caverswall ParReg), but has not been located.

**BLAKEMERE** 1½ miles east of Upper Hulme (SK 0361). *Blakemere* 1348 VCH VII 212, *Blakemere House* 1638 ibid, *black-meer of Morridge* 1686 Plot 1686: 44, *Black Mare of Morridge* 1749 Bowen, *Blake Meer or Black Meer, Blake Meer House, Blake Meer Holes* 1842 O.S. 'The black mere or pool'. The small, dark, peaty pool may have been man-made for watering beasts: it lies near the highest point of Morridge alongside an ancient drover's road.

**BLAKEMERE POOL** in Weston Jones parish, 1 mile north of Norbury (SJ 7824). *Blakemere* 1327, 1332 (p) SR, 1590 FF, *Blakemereheth* n.d. AD vi, *Blake meere* 1668 Survey, *Black Meer* 1686 Plot, *Pool or water called Blakemear* c.1740 SRO D1717/A/1/29-40. 'The black pool'. See also Blakemore (House), to which some of the spellings may relate.

**BLAKEMORE (HOUSE)** ½ mile south-east of Norbury (SJ 7822). *? Blakemore* 1268 CoramR (p), *Blakemere* 1327 SHC VII (i) 240, *Blakemerehouse* 1413 VCH V 159, *Blake Morehouse* 1668 *Ct*, *Blakemore Ho* 1725 *Deed*, *Blakemere House* 1833 O.S. 'The dark mere or pool'. The *mere* element has later become confused with *moor*. See also Blakemere Pool, to which some of the spellings may relate. There is some slight evidence that a place called Blakemore may have existed in the Blymhill area (see SRO 155[7985]), where the surname is well recorded.

**BLAKENHALL** 1 mile south of Wolverhampton (SO 9297), *Blakenhall* 1895 O.S.; **BLAKENALL** 2 miles north of Walsall (SK 0099), *Blakenhullesfeld* 1375 SRO D4407/37[SF54], *Blakenhale* 1415 SHC XVII 54, *Blakenall Heath* 1834 O.S.; **BLAKENHALL** 1 mile west of Barton-under-Needwood (SK 1718), *? Blakenhale* c.1235 Rees 1997: 68, *Blakenale* 1322 MinA, *Blakenhale* 13th and 14th century Duig, *Blakenall* 1536 SHC XI 275, *Blakenoll* 1549 SHC XII 203, 1613 SHC IV NS 50, *Blakenhall* 1559 (1798) Shaw I 117. Probably from the OE adjective *blæc* (nominative), *blacan* (in oblique cases), with the meaning 'dark coloured', rather than the OE personal name Blaca, which is unlikely to be found three times with *halh*. The second element is almost certainly from the more common OE *halh* rather than the rare element *hall*. *La Blakenalle* at Morfe, recorded in 1300 (SHC 1911 266) has not been identified, but may be associated with *The Black Wall*, recorded in Bobbington in 1683: see Blakelands.

**BLAKE STREET** the name of an ancient road, probably Roman, forming part of the boundary between Staffordshire and Warwickshire, and the parishes of Shenstone and Sutton Coldfield. *Blakestrete* 1300 SHC V (i) 177. From OE *blæc* 'black, dark'. Duignan 1902: 17 states that another ancient road called Blake Street (*Blakestret* 1294 SHC VI (i) 295, *Black street* 1595 Duig), once part of the great London to Chester road, ran across Cannock Chase between Brownhills and Hednesford, forming a manorial boundary. The route is said to have been from Birmingham via Barr Beacon, crossing the Chester road near Brownhills, passing Knaves Castle and though what has since become Chasewater (formerly Norton Pool), then to the north of Heath Hayes and past the Cross Keys Inn at Old Hednesford, up Stafford Lane and on the Huntington Belt, passing Deakin's Grave, and crossing the course of Chad's Ditch and swinging well to the west of the Cank Thorn at Broadhurst Green to leave Cannock Chase at Brocton Gate (Anon. 1957: 67; map p.16). The route is said to have been called variously Salter's Street, Coventry Road and London Road, but was better known as Blake Street in 'the old coaching journals': ibid. 67. There is another Blake Street in Cheshire, 2 miles east of Nantwich.

BLAZING STAR (obsolete, possibly to be associated with Old Star (*Old Star* 1836 O.S.), 1 mile west of Ramshorn (SK 0645)) *Blazeing Star*, *Blazing-star* 1686 Plot, *Blazeing Star* 1747 Bowen, *Blazing Star* c.1768 SRO D240/D/214. Possibly from a smelting furnace, or a public house name, or perhaps commemorating Halley's Comet, which was seen in 1682, and popularised the expression 'blazing star', recorded by OED as early as 1460.

**BLITHBURY** 3 miles north-east of Rugeley (SK 0820). *Blithebery* 1129x1147 EEA 14 9, *Blidebire* 1200 SHC II 91, *Blitheburgh*, *Blithburie* 12th century Duig, *Blythbury* 1307

SHC XI NS 40, *Blythebury* 1414 SHC XI NS 40. From OE *burh*, giving 'the fortified place or manor on the River Blithe'.

**BLITHE, RIVER** a tributary of the river Trent. *bliðe, up æfter bliþe* 993 (11th century, S.878), *aqua de Blye* 1224 Stone Ch, *rivam de Blithe* 1279 SHC IV 280, *acqua de Bliye* 13th century St Thomas, *Riva de Blithe* 14th century Ronton Ch, *aqua de Blie* 1526 Trentham Ch. From OE *bliðe*, an adjective with various meanings, including 'gentle, quiet; cheerful, merry', chiefly as a river name, in which either of these extremes may be applicable, or the sense may be no more specific than 'pleasant': see VEPN I 115-6. The river is also known as the River Blythe, but Blithe is the spelling adopted by the O.S. There is also a tributary forming a loop with the river called The Little Blithe south of Abbots Bromley. The Blithe has given its name to several places on its banks, including Blithewood Moat, Blithfield, Blithbury, Blythe, Blythe Bridge, Blythe Marsh, and the unlocated *Blythmore*, the latter recorded in 1407 (SHC XI NS 33).

BLITHEWOOD MOAT (obsolete) to the west of Dairyhouse Farm, Leigh (SK 9936). *Blithewood Moat* 1891 O.S. Redfern 1886: 87 mentions *Blithard Moat or Mot*, a square ditched earthwork. 'The moated enclosure at the wood near the river Blithe'.

**BLITHFIELD** Ancient Parish 2 miles south-west of Abbots Bromley (SK 0424; SMR 00872 places a deserted post-Conquest settlement at SK 04552455). *Blidevelt* 1086 DB, *Blithefeld Ferrour* 1232 SHC XII 7, *Blithefeud* 1235 SHC 1911 387, *Blihefeud* 1236 Fees, *Blythefeld* c.1245 SHC XI NS 154, 1336 ibid. 218, *Blithefeld* 1367 ibid, 1361 ibid, *Blyffeld* 1467 SRO DW1733/A/2/91, *Blefeld* 1532 SHC 4th Series 8 106. 'The *feld* or open land on the River Blithe'. For a possible explanation of *Ferrour* (? Ferrers) see SHC 1919 3, 8. Blithfield reservoir was created here in 1953.

**BLITHFORD (FARM)** 2 miles south of Abbots Bromley (SK 0821), *Blitheford* 1322 Ipm; BLITHEFORD (obsolete) near Cookshill, *Blitheforde, Blitheford* 1255 SHC XI 306, 315, *Blythefourde* 1544 SHC 1934 19, *Blythe Fourde* c.1555 SHC 1910 77. 'The ford on the River Blithe'. Ward 1843: app. ii mentions *Oldblitheford* in 1223, suggesting a new and an old (or disused) ford at that date.

**BLOOMSBURY** on the Staffordshire-Shropshire border, 1½ miles north of Sheriffhales (SJ 7614). *Blomesbury* 1695 Morden, 1749 Bowen. The second element would appear to be OE *burh* 'fortification, fortified manor, manor'. The first element is uncertain (see Blymhill), but it may be noted that Plot 1686: 159 mentions iron-stone to be found at Sheriffhales, a mile or so from Bloomesbury, so possibly from ironworking (cf. OE *bloma* 'a mass of iron').

**BLORE** Ancient Parish 3 miles north-west of Ashbourne (SK 1349; SMR 00460 places a deserted settlement at SK 14004935), *Blora* 1086 DB, 1176x1182 (13th century) EEA 16 13, *Blore* 1203 SHC III 117, 1227 Ass, 1240 FF, *Bloie* 1204 SHC III (i) 119, *Blore on les Mores* 1408 FF, *Bloer* c.1540 Leland, *Blower* 1562 Pat, *Blowre* 1644 SHC 4th Series I 165, *Blore Ray* 1786 Bacon; **BLORE** 3 miles east of Market Drayton (SJ 7234), *Blore* c.1194 SHC II 266, *Bloie* 1203 SHC III 118, *Blora, Blore* 1209 Pleas, *Blore* 1237 Cur, 1239 SHC IV 90, 1293 Ass, QW, *Netherblore* 1339 SHC 1945-6 56, *Blorehales* 1598 SHC XVI 40. An interesting name incorporating an element which would seem to be found nowhere outside Staffordshire, and for which no satisfactory derivation can be put forward. Almost certainly not from ON *blar* 'blue' and perhaps 'cold, cheerless', since (despite appearances), it would not account for the *-r* in the forms. Perhaps from the onomatopoeic word *blore* (not recorded in England before 1400) meaning 'a violent gust or gale', or perhaps associated with dialect *blore, blare, blair, blaar, blear, blar*, and similar, one meaning of which is 'to roar' (EDD), which might be appropriate for windswept places. An alternative possible derivation is from unrecorded (and uncertain) OE *blor, ME blure 'blister, swelling' , associated with OE blawan 'to blow, to inflate', which is the root of the obsolete words blure, bloure, blowre, given by the OED to mean 'a blister, a swelling', perhaps applied topographically to gently rounded hills or hillocks on rumps of rising ground or larger hills: see VEPN I 117. Blore near Ashbourne lies above the 650' contour on the slopes of Calton Moor which reaches 1088', with the most striking feature in the immediate area Hazleton (or Hazelton) Hill, a conical hill to the north-west, on which stands Hazelton Clump (q.v.), forming a very distinctive 'blister or swelling'. Blore near Market Drayton lies at 477' on a lesser rounded hill which forms part of a massif of hills which reaches 714', and Blurton (q.v.) lies on the side of a 721' hill with ground reaching 833' two miles or so to the south-east. Though exposed, none of these places is the highest point in the area. The fact that all the places containing this element are close to the county boundary or (in the case of Blurton) another boundary is doubtless coincidence. The Ray element in Blore near Ashbourne (the ecclesiastical parish is properly Blore Ray), for which there is no early evidence, is perhaps from OE æt þēre ea 'at the river' (probably to distinguish it from Blore near Market Drayton), with a later mis-division of the elements, a form likely to date from about the 13th century (see Ekwall 1928: 337), although in this case the element was perhaps added by a learned cleric in more recent times. The river is either the Manifold or the Dove, both of which are overlooked by Blore: see also Ray Hall; Rea Cliff Farm. The word ray is sometimes derived from ON vrá 'corner', and that derivation may apply here, since Scandinavian influence is evident in the immediate area: cf. Swinscoe. Hales is from the parish in which the second place lies.

**BLORE DALE** (obsolete) ½ mile south of Mucklestone (SJ 7236). Blore dale 1679: SHC 1919 250. 'The dale near Blore (q.v.)'. See also Ekwall 1936: 130-1.

**BLOREHEATH** 2 miles north-east of Market Drayton (SJ 7135). Blorehet 1356 SRO 3764/12[40015], Bloreheth 1403 SHC XV 113, Blore heathe 1577 Saxton, Boower Heath 1752 Rocque, Bloore Heath 1776 DRO D3155/7345. From nearby Blore (q.v.). The site of the battle of Blore Heath on 23 September 1459. For an analysis of various medieval accounts of the battle see Rowney 1980 9-17. See also Blorepipe.

**BLORE PARK** in Bishop's Wood, north-west of Blorepipe (SJ 7531). Blore Park 1298 NSJFS 4 1964 63, the park of Blore 1351 SHC 1913 143, c.1540 Leland, Blore Park 1686 Plot 89, 1694 SRO EXD&CB/C/5, 1833 O.S. The name of the Bishop of Lichfield's park (sometimes called Eccleshall Park), in Blore Wood (the Chace of Blore Wood 1377 SHC 1939 74), some three miles from Blore (q.v.) and 1 mile from Blorepipe (q.v.). The north-east part of the park contains an un-named 530' hill, but it is probable that the place takes its name from Blore. The wood of Blore and Gongles, recorded in 1292 (SHC VI (i) 251) may refer to this place. The meaning of the Gongles element is uncertain, but may be from ME gang, gong 'privy', so possibly 'gong holes or privy holes', perhaps with reference to some natural feature in the area (see for example Blore Pipe). StEnc 236 mentions Gonder Hall, apparently in the Fairoak area, but the place has not been traced. StEnc 236 also mentions The Songles near Fairoak, and fields at Fairoak Grange near Hookgate called Songes. These places do not appear on maps, but the 1292 spelling may possibly be a misreading of Sondles. OED gives songles to mean 'a handful of gleaned corn' in counties near the Welsh border. For a map showing the probable boundaries of Blore Park in 1298 see Spufford 2000: 295.

**BLOREPIPE** 2 miles south-west of Croxton, on the river Sow (SJ 7730). *Bloore Pipe* 1617 Eccleshall ParReg, *Bloor-pipe* 1660 Blount 1660: 40, *Blore-Pipe* 1676 SHC 1914 16, *Blore Pipe* 1683 SRO 26/6/2/22, *Blore-pipe* 1686 Plot 1686: 311, *Blore Pipe* 1798 Yates, *Blore pikes* 1834 White 1851: 638, *Blore Pipe* 1833 O.S. See Blore for the first element: the place lies at the foot of a hill of 572', which is (surprisingly) un-named on maps. There is a possibility that Blore was the name of the hill, which gave its name to this place and Blore Park (q.v.). However, the pipe element is from OE *pipe* 'a pipe, a conduit', perhaps here meaning 'a natural underground watercourse', possibly the feature described by Plot 1686: 89: '... there is a Rivulet comeing from West of Broughton Chappel, and running by Fair-oak, that two meddows below the houses, falls into the ground within Blore Park, belonging to the right Reverend the Lord Bishop of Lichfield, which but two Meddows beyond, rises again under a flat stone before it comes to Blore Pipe: This 'tis true is but inconsiderable, it being but a small Rindle, and running underground but a little way, and not very deep; yet the greatest flood (as I was told) never causing it to run above ground, as Hamps and Manyfold doe ...'. It is likely that Blorepipe was known originally as Pipe, with Blore added, probably from nearby Blore Park, to distinguish it from Pipe near Lichfield. A heath here (probably what is now Great Heath: SHC 1945-6 3) is recorded as *Bloreheth* 1355 SHC 1913 160, not to be confused with Bloreheath near Market Drayton.

**BLOUNT'S GREEN** 1 mile south of Uttoxeter (SK 0832). *Blunts green* 1686 Plot, *Blunts Green* 1747 Bowen. From the Blount family who held the place, which was known as *Blountes Hall* in the 16th century (SHC 1910 131, 1910 78), *Blowntis Hall* 1523 (SHC XI 264), *Blunt haule* c.1540 Leland, *Blounts'-hall* c.1595 Erdeswick 1844: 514. The moated hall here (probably built by Walter Blount, d.1524: SHC 1917-8 273) was demolished in 1770: Erdeswick 1844: 515. Blount's Green may be associated with *Grenefield*, recorded in 1557 (SROD786/17/3), *Gryne Feyld* 1559 (SRO D786/17/4).

BLOW-O'RAM (unlocated, in Waterfall) *Blow-O'ram* 1851 White 787. A puzzling name for which no explanation can be offered.

**BLOXWICH** 2 miles north-west of Walsall (SJ 9902). *Blocheswic* 1086 DB, *Blockeswych* 1286 SHC V (i) 172, *Blokeswyke* 1292 SHC VI (i) 251, *Blockswich*, *Blokeswych* 13th century Duig, *Blakeswych*, *Great Blockeswych* 1300 SHC V (i) 178, *Bloctuswych* 1303 SHC VII 105, *Blockeswich*, *Blakeswych*, *Great Blockeswyz*, *Little Blockewyz* 1307 SHC VII 186, *Blokkeswich*, *Bloxwych* 14th century Duig, *Greott Bloxsugh*, *Parva Bloxsuche* 1532 SHC 4th Series 8 144, *Bleckysreyche* 1537 SHC 1912 101. Possibly from an OE personal name *Blocc (cf. *Bloccan leah* in Blockley, Gloucestershire), with OE *wic*. There were evidently two distinct places, Little Bloxwich and Great Bloxwich.

**BLUE HILLS** 1 mile north of Upper Hulme (SK 0162). *blew-Hills* 1686 Plot 98, *Blue Hill* 1747 Bowen, *Blue Hills* 1842 O.S. Perhaps from ME *blew* 'blue', but also 'dark coloured, variegated', perhaps here 'the dark or variegated hills', or from the streams coloured by coal deposits mined here from at least the early 15th century, as suggested in VCH VII 33. Or possibly from Northern dialect *blae* 'cheerless, cold, exposed' (VEPN I 109), perhaps influenced by ME *blou* 'blast of wind': the gritstone hills here are particularly exposed and windswept. See also PN Ch III 145.

**BLUNDIES** ½ mile north-east of Enville (SO 8287). ? *Blundel* 1271 SHC V (i) 141, *Blunders* 1827 O.S., *Blundies* 1834 O.S. Perhaps from 'Blunt's *halh* or hill'. See also Hoo.

**BLURTON** in Stoke on Trent, 2 miles south-west of Longton (SJ 8941). *Bloetona* 1176 EEA 16 104, *Blozton' (sic)* 1194 Pipe, *Blorton* 1195 f. P, 1248 FF, 1324 SHC X 45, *Blortun* 1250 SHC XI 309, *Bloerton* 1473 SHC IV NS 192, c.1540 Leland. For the first element see Blore. The second element is OE *tun*. Blurton lies near a 721' hill at Cocknage (and near a boundary: cf. Meir). *Bloremedewe* ('the meadow at Blore') and *Blorewallesich* ('the stream of the spring at Blore', from Mercian OE *wælle* 'spring' and *sic* 'watrercourse') are recorded in Blurton c.1342: SHC XI 312. Cf. Blore.

BLURTON WASTE (obsolete) 1 mile west of Blurton (SJ 8842). *vaso de Blurton* 1302 SHC XI 311, *Blurton Waste* 1836 O.S. 'The wasteland at Blurton'.

**BLYMHILL** Ancient Parish (pronounced Blimmul [blɪməl]). 6 miles south-east of Newport (SJ 8012). *Brvmhelle* 1086 DB, *Blumehil* 1166 SHC 1923 296, *Blumenhall* 1194 (p) CartAnt, *Blumenhull* c.1199 SHC II NS 294, 1218 SHC VI (i) 32, *Blumhilla* 1221 Bracton, *Blumenhul* 1223 SHC I 292, *Blimhill* 1225 ibid.' 292, *Blumhull* 1225 Cur, 1313 Banco, 1376 Pat, 1394 Fine, *Blimenhul* 1236 Fees, *Blimenhull* c.1240 SHC I 293, *Blomenhull* 1248 ibid.' 313, *Blemenhull* 1254 SHC II NS 17, *Blumenhull* 1259 ibid. 293, *Blemenhull* 1276 SHC 1911 31, *Blimenhull* 1284 SHC I 294, *Blyminhill* 1290 SHC XII 9, *Blomenhale* 1306 SHC VII 157, *Dilmenhale* 1308 SHC II 81, *Blomhull* 1362 SHC XIII 16, *Blemhull* 1375 Ipm, *Blumonhull* 12th, 13th and 14th century freq. Duig, *Bleumenhull* 1331 SHC I 334, *Blomehill* 1355 SHC XII 132, *Plymylle* 1423 SHC XVII 89, *Blumhulle* 1432 ibid. 138, *Blemell* 1532 SHC 4th Series 8 98, *Blym hill* 1682 Browne. A puzzling name. Ekwall 1936: 125 suggests a derivation from an original form *Plӯman-hyll* or *Plӯmena-hyll*, from OE *plӯme* 'wild plum-tree', which may well be correct: the change of *p-* to *b-* can be likened to the change from *t-* to *d-* (see for example Tunstall/Dunstall), but as observed in Gelling & Cole 2000: 194, no change from *p-* to *b-* is evidenced in other names containing plum.* The suggestion in Duignan 1902: 19-20 that the name may be associated with OE *bloma* 'bloom, metal ingot', meaning the site of a bloomery hearth (or perhaps a hill in the shape of a typical domed iron bloom) is doubtful, given the *Blimen-* and *Blemen-* spellings. The spellings seem to point towards an OE *\*blӯm(e)* 'bloomery' or similar, The oblique form *\*blӯman* would be expected to develop into ME *blumen-, blimen-, blomen-*. This derivation would require support from the presence of ironstone on or close to the hill, or traces of smelting. No such evidence has been traced, either documentary, archaeologically or from field-names (see Oakden 1984: 131-3), but it may be noted that Bloomsbury (wood), some three miles to the north-west of Blymhill, is shown on Morden's map of 1695 and Bowen's map of 1749 as *Blomesbury*, and appears as *Bloomsbury* in 1749 Blymhill ParReg, and Plot 1686: 159 mentions iron-stone to be found some three miles away at Sheriffhales (and, in another reference not to this area, to iron ore or blemish called, intriguingly, Boylom (1686: 169)). Before water power was used to power ironworks, bloomeries were sited on hills to utilise the wind as a draught for the furnace (cf. Gatherwynd Farm, Blymhill), and ironstone may have been conveyed to this area for processing. See also Bloomsbury.

---

*The plum, Prunus domestica, is not native to Britain, but is a post-Conquest introduction. The two native plum-like species (fruit having a single hard stone enclosing a seed) are the sloe (Prunus spinosa, the fruit of the blackthorn), and the bullace. The sloe was known in OE as sla(h), slag, so that could not have been a plyme. But the bullace (cf. Bolacetre 'bullace tree' in Wombourne, recorded in 1318 (1801) Shaw II 213) has a name said to be ME from OFrench beloce (although TOE records bulentse under 'unidentified plants'). It seems possible that OE plyme was what we now know as bullace, and that bullace (first recorded in OED in 1616) acquired its present name when the word plum began to be used for Prunus domestica. It is also of interest that the many dialect words for bullace include bullum and similar, which could provide a clue to the root of this name.*

**BLYMHILL LAWN** 1 mile south-east of Blymhill (SJ 8111). *Blymhill Lawn* 1812 *EnclA*, 1833 O.S. From Blymhill (q.v.) with ME *launde* 'an open space in woodland, a forest glade, woodland pasture', meaning in this area 'an open passage through woodland': cf. Coven Lawn and Langley Lawn.

**BLYTHE BRIDGE** 1 mile south of Caverswall (SJ 9541). *? Blye* c.1230 SHC VI (i) 11, *Blythbryge* 1475 SHC VI NS (i) 94, *Blythebridge* 1573 SHC XIII 295, *Blithe-bridge* 1686 Plot. The place lies on the river Blithe (q.v.). It is unclear whether *Blitheforde*, recorded c.1250 (SHC XI 306) relates to this place. If it does, it might help to date the first bridge here.

**BLYTHE MARSH** on the south-west side of Forsbrook (SJ 9641). *Blythe Marsh* 1782 SRO D1798/601. 'The marshaland near the river Blithe'.

**BLYTHE, RIVER – see BLITHE, RIVER.**

**BOARSGROVE** 2 miles north-east of Upper Hulme (SK 0462). *Borisgreve* 1566 *Deed*, *Boresgre(a)ve* 1582 FF, *Boorsgreene* 1605 Alstonefield ParReg, *Boarsgreene* 1624 ibid, *Boresgreave* 1655 DRO D2375M/58/3, 1700 Alstonefield ParReg, *Boarsgrove* c.1768 VCH VII 35. The first word is likely to be from the animal rather than OE *\*bor* 'elevation, hill', or the personal name Boar. The second element is from *græfe* 'grove, thicket', hence 'the thicket of the wild boar'. The word *græfe* (also found as *graf, grafe,* and *grafa*), is probably associated in some way with OE *grafan* 'to dig', *grafa, græf* 'pit, trench'. 'Grove' is probably the more likely interpretation unless the first element suggests 'pit' or similar, or there is such a feature in the vicinity. Some of the parish register entries may incorporate misreadings of *n* for *v*.

**BOBBINGTON** 6 miles east of Bridgnorth (SO 8090). *Bvbintone* 1086 DB, *Bobinton* Eyton 1854-60: I 109, *Bubington* 1236 Fees, *Bobynton* 1588 SHC XVII 236, *Bovington otherwise Bubbington otherwise Bublington* 1600 SHC III NS 17, *Bovington* 1603 SHC 1940 40, *Bovington or Bobbington* 1689 SRO 5623/2. 'The *tun* or estate associated with a man called Bubba'.

**BODNETS, THE** 2 miles west of Tamworth (SK 1703). *Bodnets* 1834 O.S. It is unclear whether *Bawndawns Hille*, recorded c.1524 ( 1798 Shaw I 314) is to be associated with this place.

BOKELEG(E) (unlocated, possibly in the Leigh area (SHC 4th Series VI 7); possibly associated with Leigh itself, in the sense 'Leigh granted in writing', i.e. bookland) *Bokeleg* 1272 SHC IV 187, *Bokeleye* 1377 SHC 4th Series VI 7. Perhaps to be associated with Bukkeleg, Bukkelyh (q.v.).

BOLD – see **BOOTH**.

**BOLEHALL, BOLEBRIDGE** on the north-east side of Tamworth (SK 2103). *Bolebrugge* 1166 P, *Bolebrige* 1198 FF, *Bollehall(e)* 1390 FF, *Bollumhulle* 1391 SHC XV 37, *Bol Hall* 1460 BM, *Bolle Bridge* 1497 SHC 1917-8 262, *Bow Bridge* 1508 ibid.' 282, *The Bow Burge* 1532 SHC 4th Series 8 19, *Bollebrigestrete* 1538 FF, *Bowbrydge* 1543 FF, *Bolehall* 1610 FF, *Bowl-bridg* 1656 Dugdale, *Bolehall streete* 1693 FF, *Bolebridge* c.1750 K, *Bolehall* 1872 P.O. Both names are to be associated with the ancient Bullring (*le Bulryng* 1314 Palmer 1845: 189) and Bullstake (*le Bolestake, Bulstake* ibid.), both connected with bull-baiting, *bol* and *bolle* being common ME forms of bull. Formerly in Warwickshire, the place became part of Staffordshire in 1965. See also Bonehill.

**BOLEHEVED** (unlocated, in Mavesyn Ridware) *Boleheved* n.d. (1798) Shaw I: 170. From OE *bula* 'bull, bullock', with OE *heafod* 'head, headland, summit', so perhaps 'the headland shaped like the head of a bull': cf. Swinchurch.

**BOLEWYK** (unlocated, in Bishton). *Bolewyk* 1295 SRO DW1781/1/19. From OE *bula* 'bull, bullock', with OE *wic* 'dwelling, hamlet, village', and more especially 'a farm, a dairy-farm'.

**BOLEY PARK** on the east side of Lichfield (SK 1309). *Bolele* 1215 TSSAHS XVIII 1986-7 9, *? Boleleye* 1271 SHC IV 184, *Bollele* later 13th century TSSAHS XVIII 1986-7 9, *Bolley field* 13th century VCH XIV 110, later *Boley field* ibid, *Bowley* c.1535 SHC VI (ii) 166, *Boley* 1798 Shaw I 316. From OE *bola leah* 'wood where logs were obtained': VCH XIV 38.

**BOLLAND'S HALL** on the north-west side of Butterton (SK 0656). Early forms have not been traced, and the age of the name is uncertain, but presumably from a personal name. See also Ballance Hill.

**BOLTONGATE (FARM)** on the north side of Weston Coyney (SJ 9344). *Bolton Gate* 1760 SRO D641/5/TS/4/20.

**BOLTSTONE** a bulky tapering standing stone in Compton, near Kinver, destroyed in 1840, which may have been a prehistoric monolith. It was known as *the Boltstone, Battlestone* (VCH XX 119), or *Baston* (Plot 1686: 397; Shaw 1798: I 13; see also VCH I 191), and seemingly gave its name to one of the open fields (presumably *Bothestonesfeld* VCH XX 137; 1342 SHC 1913 91) and perhaps to the later field-name *Bowstone Field, Boltstone Field* c.1831 (SRO D801/2/9): any connection with *Belstowa* (? 'holy place; place with the pyre') found in the boundary clause of Ashwood in an Anglo-Saxon charter of 985 AD (12th century, S.860) is doubtful, since the boundary mark seems to have been further north: see Hooke 1983: 70-72. Matthew Bolestone is recorded in 1348: SHC XIV 66-72. See also Foxearth.

**BOND END** on south side of Yoxall (SK 1418). *Bond-end* 1499 (1798) Shaw I 98, *The Bondend* 1532 SHC 4th Series 8 173, *Bond End* 1631 NA DD/P/6/3/40/1, 1798 Yates. From ME *bond*, derived from ON *bóndi*, ODan *bunde* 'a peasant landowner'. The element is remembered in Bondfield Lane on the west side of Yoxall. See also Reeve End; Burton Extra.

BONEFORD (unlocated, between Ingestre and Hopton) *Boneford* 1291 SRO/ 416/7912.

**BONEHILL** 2 miles south-west of Tamworth (SK 1902). *Bolenhull* 1230 P, 1271 For, 1286 SHC V (i) 175, *Bulenhall* 1230 P, *Bollehul* c.1280 SHC 1921 5, *Bollenhull* 1327 SHC VII (i) 234, 1352 Pat, *Bolunhull* 1377 SHC 4th Series VI 10, *Bollunhull* 1472 SBT DR3575 at 570, *Bonell* 1532 SHC 4th Series 8 23, *? Bonill* 1601 SHC 1935 366, *Bonehill* 1608 SHC III NS 19, 1834 O.S. Duignan 1902: 20-22 suggests a derivation from a word associated with ME *bole* 'a place where ore was smelted' (see VEPN I 123-4), citing Bole Hill in Derbyshire and elsewhere signifying a place where lead (or other metals) was smelted. However, it seems more likely that the name is from OE *bulena*, genitive plural of ME *bule* 'bull', giving 'hill of the bulls'. See also Bolehall.

**BONEY HAY** 1 mile north of Burntwood (SK 0510). *Le Brendewode called Le Burne hew* 1361 SRO DW1734/2/1/598, *Borne heymedwe* SRO DW1734/2/3/112D, *... land lately called Bornehay, Borne Meadow ...* 16th century SRO DW1734/2/1/744 258, *Bournhay Meadow* ibid 51, *Bourn Heay* LJRO D110, *Bourne Hay* 1679 SRO D1734, *Boney Hay* 1822 SRO D4045/7/2. The first element is OE *burna* 'spring, brook, stream'.

The second element is perhaps more likely to be from Mercian OE *(ge)heg* 'an enclosure (especially in Forest areas)', notwithstanding the earliest spellings (especially *heymedwe*) which point towards OE *heg* 'hay'.

**BONTHORN** 1 mile south-west of Barton under Needwood (SK 1817). *? Borntthoun, Bonthoun* 1301 SHC 4th Series XVIII 195-6, *Bonethorn* 1495 Hardy 1908: 141, *Bochenthorn* 1503 ibid.' 147, *Bonthorn* 1834 O.S. Perhaps 'Bana's or Bunt's thorn': the 1503 spelling which suggests a derivation from OE *\*bocen* 'of beech' is doubtless an aberation, and in any event 'thorn of beech' is impossible.

**BOODEN FARM** ½ mile south of Haughton (SJ 8619). *le Hall of Bolde* 1548 Survey, *Boden* 1539 MR, *Bold Hall als Bowldhall* 1555 FF, *the hall of Bowle* 16th century Ct Augm, *Bouldhall* 1577 Saxton, 1605 FF, *Bowdon* 1625 SRO DW1781/5/15/1, *Booden House* 1725 ParReg, 1836 O.S., *Boldon* 1775 Yates. The forms suggest a derivation from OE *bold* 'dwelling-place' (and possibly 'a superior hall, a castle, a mansion': see Campbell 1986: 113-4), with *hall* (cf. Bold), but the unlocated *Bowode* 1277 (SHC 1911 168), 1327 (SHC VII (i) 244), 1295 SRO D938/121, *Bouwood* 1290 (SHC 1911 198), *Bouwod* 1318 (ibid. 90), which is believed to have been in this area, could be early forms of this name, in which case the derivation may be from OE *boga* and *wudu* 'bow wood, wood where bows were obtained', or possibly 'the wood with bow-shaped boundaries'. *Bowode* in Stafford is recorded in 1436 (SHC XII 312).

**BOON HILL** 1 mile east of Audley (SJ 8050). *Boundhill* 1571 Audley ParReg, *Boondhill* 1733 SHC 1944 11, *Boon Hill* 1775 Yates, *Bound Hill* 1833 O.S., *Boond Hill* 1872 P.O. Possibly 'the hill of the unfree tenants', from ME *bond*, derived from ON *bondi* 'a peasant landowner'.

**BOOSLEY GRANGE** 2 miles south-west of Longnor (SK 0662). *Bothesleye* 1328 SHC 1913 21, *Boothesley Grange* 1571 (reciting earlier deed) VCH VII 27, *Booseley Grange* 1840 O.S., *Boozeley Grange* 1880 Kelly. The early forms indicate a derivation from ODan *boð*, ME *both* 'booth, temporary shelter', with later spellings suggesting an association with OE *\*bos* 'a cattle stall', found in the north country dialect word *boose*, with the same meaning, with OE *leah*, giving '*leah* with a cattle-stall'. The word Grange (ME *grange, graunge*) originally meant 'a granary, a barn', later 'a farm', and frequently denoted a monastic farm, but there is no record of any monastic landholding here, so the word may refer to buildings, perhaps used for a feudal lord, where crops were stored.

**BOOT HALL** on north-west side of Horton (SJ 9357). *Boot Hall* 1891 O.S. If ancient, perhaps from the Boot family, recorded in the early 18th century: VCH VII 65.

**BOOTH, UPPER & LOWER** north of Newton, on the west side of the river Blithe, 5 miles north of Rugeley (SK 0426 & SK 0427; SMR 01217 places a deserted post-Conquest settlement at SK 04322716). *la Bolde* 1170 SHC 1919 38, 1175 SHC 1914 165, *Bold* 1199 SHC III (i) 41, *laboude* 13th century SHC XI NS 148, *Bolda* c.1240 *et freq.* ibid, *la Bold* 1257 SHC VIII 145, *Le Bolde* 1290 SHC XI NS 112, *Boolde* 1309 SRO D59[7917], 1371 SRO D938/593, *Boolde* 1405 SRO D786/3/3. From OE *bold* 'a house, dwelling-place', one form of an element which also occurs as *bopl, botl.* The northern form is usually *bopl*, but *botl* occurs both in the North and in the East Midlands. *Bold* is the only form found in the West Midlands. The place became known as Booth in the 19th century. *Boldewode* 'the wood at Bolde' is recorded in this area in 1299 (SRO D938/553), and *le Bolde Mos* 'the moss or bog at Bolde' c.1315 (SRO D938/557). See also SHC 1919 34-41; also Bull Bridge and Booden Farm.

**BOOTH HALL, BOOTHS FARM** 1½ miles south-west of Kingsley (SK 0045). *Kyngesleghe Bothes* 1293 SHC VI (i) 236, *? Bothes* 1327 SHC VII (i) 215, *Bothehall* 1583 SHC III 9, 1609 SHC III NS 52, *Boothall, Boothes* 1598 SHC XVI 185, *Booths* 1610 Kingsley ParReg, *Boothall* 1621 SA 11/68, *Booth Hall* 1836 O.S. An interesting name. ME *both-hall* had the meaning 'market-hall, town-hall', with examples of the term recorded in Shrewsbury, Shropshire, and three Gloucestershire towns (VEPN I 132). That etymology is hardly appropriate for these rural places, which may be derived from ODanish *boð* 'booth, temporary shelter'.

**BOOTHEN** 2 miles north of Trentham (SJ 8744). *? Bothes* 1461 HLS, *Bowdon, Bowthon* c.1569 SHC 1926 104, *Bowthen* c.1560 ibid. 107, *Booden brig* 1683 Stoke on Trent CA, *Boothen (Bowden) Bridge* 1689 ibid, *Boothen* 1755 Stoke on Trent ParReg, *Boden* 1803 ibid, *Booden, Boothden, Boothen* 1806 ibid, *Boothen Ville* 1836 O.S. Possibly from the dative plural of ME *both*, from ODan *boð*, meaning '(at the) booths or temporary shelters'.

**BOOTHLOW, OVER BOOTHLOW** 1 mile south-east of Longnor (SK 0963). *Boothlo* 1670 Alstonefield ParReg, *Booth Low, Nether Booth Low* 1840 O.S. Perhaps from ODan *boð*, ME *both* 'booth, temporary shelter', with OE *hlaw* 'mound, tumulus'.

**BOOTH'S HALL FARM, BOOTH'S WOOD** 1 mile south-west of Ipstones (SK 0148, SJ 0048). *? Buwothus* 1335 D1229/1/4/25, *le Bothus* 1352 SRO D1229/1/4/27, *the Bethes* 1593 SRO D1229/1/4/16, *Booth, Booth Wood* 1836 O.S. Perhaps from ODan *boð* 'booth, temporary shelter', with ODan *hús* 'house', so 'booth-house'. See also Boothen.

**BORDESLEY** 2 miles north of Redditch (SO 0470). *Bordeslega* 1138 (1266) Ch, *Bordeslea* 1159 SHC I 29, *Bordesley, Borsly* 1535 VE, *Bursley* 1577 Saxton, *Boresley* 1650 *Survey*. From a personal name *Bord (cf. Borda, Derbyshire), or from a genitival composition involving OE *bord* 'board, plank; border', with OE *leah*. In Tardebigge parish, forming part of Staffordshire from c.1100 until 1266, in Warwickshire until 1844, and since that date in Worcestershire. Cf. Bordesley, Warwickshire; Balsall Heath, Birmungham.

BORESHANKY, BORESHANKS (unlocated, between Kiddemore Green and Broom Hall, Brewood (SJ 8709)) *Bereshankes* 1611 SHC 1934 (ii) 39, *Boreshanky* 1643 Brewood ParReg, *Boreshanck* 1683 *Dep, Boreshanks* 1685 ibid. A curious name, perhaps from OE *bar* 'boar', with OE *sc(e)anca* 'shank, shin, leg', sometimes found in place-names to describe something long and narrow (see PN Db 747; Halliwell gives a northern meaning of shank 'the projecting point of a hill, joining it with the plain'), hence perhaps 'the narrow projecting point with the boar'.

BORKESTALLES (unlocated, near Sedgley) *Borkestalles* c.1270 SHC 1941 77. See also Bukstalles.

**BORROWCOP HILL** 1 mile south-east of Lichfield (SK 1208). *Burwey (field)* 1303 SHC 1939 91, *Burwhay* 1372 SRO 3764/60[27543], *Burghay* 1376 SRO 3764/71[40945], *Burghewhaye* 1380 SRO 3764/73[40945], *Burweycop* 1444 VCH XIV 110, *Burrowcop (Field)* 1514 SRO DW1851/8/50, *Bowwe Cope (Field)* 1547 SRO 3764/82[27543], *Borrowcopfeld* 1552 Harwood 1806: 390, *Borughcop hill* 1577 Saxton, *Barrow-cop-fields* 1585 (1798) Shaw I 336, *Burwaycop* (before 17th century) VCH XIV 7, *Borrowcop* 1613 SRO D15/10/1/52, *Burway or Borrowcop* 1719 ibid. 110, *Boroughcop Hill* 1720 Bowen, *Barrow Cop Hill* 1834 O.S. From OE *burh* 'fortified place', with OE *weg* 'way, path, highway', hence 'the trackway of (or to) the fortification', with OE *cop* 'a hill-top'. SHC VI (ii) 186 records *Bradway alias Burghay*, and *Burghay* *or Burway*. There is some doubt whether such earthworks ever existed – Shaw 1798: I 231 states '... there are no remains of a camp on the top of it ...', and see also SHC 1950-1 147. But a circular earthwork is shown on the 1882 6" O.S. map as 'Supposed site of Saxon fort'; and an earthwork 310' in diameter is recorded in OSS 1949-50 18; see also TSSAHS XXII 1980-1 15; VCH XIV 7, 40. The element *burh* is also found in Oxbury (*Oxenbury* 1391 SHC VI (i) 187), a name found north of the hill by a ditch known as Castle Ditch (*Casteldyke*, part of the moat of the Close, c.1535: SHC VI (i) 165-6. See also TSSAHS XXII 1980-1 112; 114). The *bury* element in *Oxenbury* is likely to be used there in the sense 'borough', and Castle Ditch probably referred to the Cathedral Close (see TSSAHS XXII 1980-1 114): Lichfield had no castle, although there were 'perimeter earthworks' (ibid. 112-4). It is possible that Borrowcop is to be associated with DB *Burouestone* (q.v.). See also Borwey Foordes; Bradway. Stukeley's reference to *Mawcop* in 1724 (Itinerary II 21) possibly refers to this hill.

BORWAY FORD (unlocated, but probably on the river Trent north-west side of Alrewas: see TSSAHS XX 1978-9 loose map) *Burweyford* 1341 SHC 4th Series XX 81, *Borwey Foordes* 1573 SHC XIII 297, *Borway Ford* c.1745 SRO D615/D/148. The name is presumably to be associated with field-names *Oldeburwey* and *New Burwey*, recorded in 1327 (Alrewas Ct), *Burwey* recorded in 1331 (ibid.), *Borewey* recorded in 1334 (ibid.), *Burwaye House* recorded in 1617 (SHC 1934 25), and *Burway Meadow* recorded in 1658 (SRO DW1851/8/4; SK 174147). See also Burwey, Bradway. Probably from OE *burh-weg* 'the way to the fortification (or borough; see also Broadfields)', or possibly *burh-(i)eg* 'fort-island' (see JNSFC 21 18), but the significance of that form is not fully understood. PN Sa I 39 records three examples of *burh-(i)eg* regularly spaced along the river Thames (Laleham Burway, Surrey; Borough Marsh near Sonning, Berkshire; and Burroway near Bampton, Oxfordshire), and one (Burway) near Ludlow on the river Teme.

BOSCASTLE (unlocated, in Gnosall parish) *Boscastle* 1650 SRO DW1736/1/ix11. A curious name for which no explanation can be offered.

**BOSCOBEL** a hunting-lodge built c.1600 by the Giffards of nearby Chillington Hall just inside Shropshire on the Staffordshire-Shropshire border, 8 miles north-west of Wolverhampton (SJ 8308). *Boscobell* 1624 SPDom, *Boskevile* 1631 (1680) SHC 1919 273, *Boscobell* 1632 Brewood ParReg, *Bascaveale* 1644 (*et freq.*) Brewood ParReg, *Bassobel* 1647 CC, *Boscabell* 1660 Blount, *Boscobel* 1680 SHC 1919 241, *Boscofield* 1690 PRO C205/19/12, *Baskabell* 1707 ibid. The traditional derivation, given in Blount 1660: 12, is that 'John Giffard, Esq., who first built this house, invited Sir Basil Brook [1576-?1646, of Madeley Hall, Shropshire, and the grandfather of Frances Cotton who owned the Boscobel and Whiteladies estate c.1651] with other friends and neighbours, to a housewarming feast; at which time Sir Basil was desired by Mr Giffard to give the house a name, he aptly calls it BOSCOBEL (from the Italian Bosco-bello, which in that language signifies fair wood), because seated in the midst of many fair woods'. The traditional derivation may well be correct, but some of the spellings (including, curiously and perhaps significantly, that used locally in the parish registers) suggest an association with Boscherville-le-Perry, in the province of Eure west of Rouen in Normandy, or Baskeville (variously Bascevilla, Basqueville, Balkierville) in the Pays-de-Caux, which gave its name to the abbey of St George there (see Eyton 1854-60: I 231). It is possible that the Giffards, from Longueville-la-Giffard, in the valley of Scie in Normandy some 20 miles north of Boscherville (and also lords of Bolbec in Normandy), had connections with and named the lodge after Boscherville: various members of the Giffard family spent time in northern France around the latter part of the 16th century (see SHC V NS

144-61; 168). The family name Boscherville is frequently found in early records: at the beginning of the 13th century there were Boschervilles in Herefordshire, Northamptonshire, Shropshire, Warwickshire, Norfolk, Buckinghamshire, and Wiltshire: Eyton 1854-60: I 237-8. References in Staffordshire records include *Boesavilla* c.1130 (SHC II 201), *Boschervile* (c.1177 SHC III 228), *Boschervile* (1199 ibid. 34), *Baschervill* (1200 ibid. 66), *Boscherville* (c.1200 SHC XI 315), *Buschervill* (1224 SHC IV 32), *Boschomele* (1256 Selden Soc. 96 1980 232; see also SHC II 203, SHC XI 315 fn, SHC XVI 264, 265, 275, 296; SHC XII NS 106-7), but no connections with this place have been traced, other than a tenuous link between the Giffard and Baskerville families: Eyton 1864-60: I 237-8. The lodge at Boscobel sheltered Charles II after the battle of Worcester in September 1651. The Royal Oak, said to have been grown from an acorn of the tree in which the king hid, is nearby. The reference to *Bosco Bello Woods, near Madeley* (Shropshire) in 1648 (SHC X (ii) 30) is almost certainly an error, since Boscobel is nine miles from Madeley (and see also TSAS 3rd Series X 94-5), but if not may possibly explain the tradition recorded in Blount, above. Francis Baskerfeld is recorded in 1595 (SHC XVI 143), and Andreus Kelled de Baskerfield in 1597 (SHC 4th Series 9 70), and may be associated with *Baskerfelde House* (unlocated, between Quatt and Arley) which appears on a map of c.1560 of the Bridgnorth area (see Bellett 1856: 206), possibly to be associated with *Baskervile*, recorded c.1250: Rees 1997: 128. An intriguing reference to Andreus Kelled (cf. Avery Kellet of Knightley recorded in 1604: SA 1037/23/44) of *Baskerfield* in a Summons requiring recusants to appear at Stafford assizes in 1597 (SHC 4th Series IX 70) may refer to that place, but *Baskerfield* is said to have been in Staffordshire (Boscobel, which may well have housed recusants – the Catholic Giffards of Chillington, Blackladies and Church Eaton, who owned Boscobel, are also named in the Summons – is just within Shropshire). *Baskeyfields Farm* appears on modern maps at Chatterley (SJ 8451), but the age and history of the name have not been traced, although *Baskerville* is recorded in this area c.1337 (SHC XI 73), perhaps to be associated with *Boschervilla*, recorded in 1166x1176 (EEA 16 104). On the Baskervill or Boskervill family see also Dugdale 1730: I 50.

**BOSCOMOOR** (pronounced Boozmoor [buːzm? :]) ½ mile south of Penkridge (SJ 9213). *Bowes more* 1598 Ct, *Boothsmoor* 1644 Penkridge ParReg, *Boosemoore (fflatt)* 1646 Ct, *Boosmore* 1681 ibid, *Booth(s)moor* 1644 to 1763 ibid, *Boothmore otherwise Boscomoore* 1789 SRO D260/M/7/5/63, *Boosemore* 1823 Penkridge ParReg, *Boscomore* 1825 ibid, *Bosco Moor* 1832 Teesdale, *Boscomoor* 1834 O.S. Perhaps from a personal name Bows or Bowes, rather than from ODan *boð* 'temporary shelter, bothy', with OE *mor* 'moorland'. The change in spelling in the 18th century to *Bosco-* (Latin 'wood') suggests a learned affectation, probably influenced by Boscobel (q.v.), 6 miles to the south-west, but the earlier pronunciation has been retained.

**BOSSES, THE** an area of low-lying ground in Shenstone parish (SK 0902). *in bosco suo de Boshay* 12th century Duig, *bosco de Bossay* 1209 SHC 1923 277, *Boscehaye* 1262 SHC V (i) 137, *Boscehaya* 1262 SHC 4th Series XVIII 50, *Bossey* 1636 SRO DW1784/2, *The Bosses* 1798 Shaw I 41, *Great Bosses* 1834 White 377. Probably from OE *\*bors* 'a spiky or bristly plant', with *hay*, from Mercian OE *(ge)heg* 'fence, enclosure, clearing', now lost. Cf. Boasley, Devon; Boscombe, Wiltshire.

**BOSTHAYE** (unlocated, possibly near Stanton) *Bosthaye* c.1235 SHC 4th Series IV 200.

**BOSTY LANE** in the parishes of Rushall and Aldridge, part of an old drove road between North Wales and London. From the name *Boltstyle*, *Bolestile*, which appears frequently in medieval perambulations of Cannock Forest (e.g. *Bolestyle* 1286 SHC V (i) 166, *Boltstyle* 1285-6 SHC 1924 330, *Boltestile* Inq. 1309-10), possibly derived from OE *bolt* 'bolt', and OE *stigel* 'stile, crossing place' (EDD), giving 'the bolted stile'. The word *bolt* is also recorded with the meaning 'bundle of withies' or 'wood from which lathes were split', and it is possible that the word is to be taken in that sense here. Duignan 1880: 60 gives a derivation from Boarstone Lane, later Boston, becoming Bosty, but no evidence to support such derivation has been traced.

BOTHAM HALL (obsolete) 1 mile south of Cheddleton (SJ 9751). *Bothom* 1306 Banco, 1327 SHC 1912 253, 1530 SHC 1910 19, 1644 SHC 4th Series I 138, *Bottom Hall* 1775 Yates. From OE *\*boðm*, which seems originally to have referred to damp valley floors, but which came to mean 'a short, level stretch of valley floor with abruptly rising sides, prone to flooding': see Cole 1987-8; VEPN I 133. A house called Ashcombe (q.v.) was built on the site of Bothom Hall c.1806: see Erdeswick 1844: 496. *Bothom Feld* and *the Botham* are recorded in 1553 and 1698 respectively near Uttoxeter: SRO D786/11/1-3. A *Bothom* is recorded in Chell c.1275 (SHC 1911 442), and appears to be the same place as *Botham* 1586 (SHC 1929 145). See also Botham Hall.

BOTTERAMS (obsolete) on the south side of Smestow (SO 8591). *Botterams* 1834 O.S., *Bottrams* 1834 White, *Botterham (Lock)* 1895 O.S. Possibly from a personal name.

**BOTTESLOW** 1 mile north of Fenton (SJ 8946). *Buttelowe* 13th century SHC XVI 285, *? Bochilewe* 1307 SHC XI NS 259, *Boteslowe* 1377 SHC 4th Series VI 7, *Botteslowe* 1471 HLS, *Botteslowe* 1573 SHC 1938 138, 1590 SHC XVI 102, 1595 ibid.' 149, 1607 SHC III NS 4. The first element is perhaps an OE personal name *Bott or Botta, with OE *hlaw* 'burial mound', so 'Bott(a)'s tumulus'. See also Bukkeleg, Bukkelyh.

**BOTTOM** 2 miles north-east of Ipstones (SK 0452). *Bothum* 1309 SHC X (i) 6, 1566 SHC 1926 141, *le Bothome* 1344 SHC 1913 109, *Botham* 1562 SHC 1931 146, *Bothom* 1591 SHC XVI 110, *Bottom House* 1775 Yates, 1836 O.S. From OE *\*boðm* 'a short, level stretch of valley floor with abruptly rising sides, prone to flooding': see Cole 1987-8; VEPN I 133. Cf. Botham Hall.

**BOTT'S COPPICE** ½ mile south-west of Hanbury (SK 1627). From the Bott family, well recorded in this area from at least 1666: SA 513/2/18/7/22. *Bottscroft* is recorded hereabouts in 1666: SA 513/2/18/7/22.

**BOUGHEY HALL FARM** ½ mile west of Colton (SK 0420). *Boughey's Farm* 1773 SRO DW1792/1. From the Boghay or Boughey family who purchased the estate c.1579: SHC 1914 153 note, 155. William Buirghay, recorded in 1379 (SHC XVI 168), and John Bogeys of Rugeley, recorded in 1430 (SHC XVII 136), may have been associated with this place. See also Bowhill Farm; Bowseywood.

**BOUNDARY** 1 mile south-east of Dilhorne (SJ 9842). *Boundary* 1836 O.S. The place lies at the junction of three parishes. The name appears to be relatively modern.

BOURCHIERS WOOD (unlocated, in Betley) *Bourchiers Wood, Bourchers Wood, Bowchiers Wood* c.1543 SHC X NS I 167-8; SHC 1912 142. From a surname Bourchiers, Bourchers or similar (see for example SHC XII (i) 310; cf. Boozer, DES 54). OFr Bouchier meant 'pursemaker'.

BOURKE GRANGE – see **WINDGATES**.

BOURNE (unlocated, in Longdon parish, possibly near Burntwood or Abnalls) *Burne iuxta Lichesfeld* 1160x1182 EEA 16 7, *Le Bourn* c.1310 SRO D1734, *Bourne* 1318 VCH

III 223, 1341 VCH XIV 217, 1377 SRO D1734. Possibly associated with Boney Hay (q.v.). It may be noted that the name Bilson Brook (q.v.) was formerly known as Bourne Brook: VCH XIV 195. This place may perhaps have taken its name from such brook: see Bourne Brook, Bourn Brook.

**BOURNE BROOK** running though Little Aston, *bradanburnan* 957 (12th century, S.574), *le Burne* c.1213 MRA, *aqua de Burne* 1235 Ch, 1276 RH, 1315 IPM, *La Burne, La Bourne* 1286 SHC V (i) 161, 166, 1379 Ct, *la Burne* 1307 SHC VII 173, *the old water course or stream of Bowrne* 1601 (1801) Shaw II 16, *the waters of Burne* 1601 SHC XVI 207, *Bourne* 1749 Bowen; **BOURN BROOK** a tributary of the river Trent south of King's Bromley, *Burnebroc* 1198x1208 EEA 17 74, *Bourn Brook* 1834 O.S. From OE *burna* 'stream' (with OE *brad* 'broad, wide' in the first form), referring in the first case to a stream which formed the boundary between Cannock Forest and Sutton Chase: Duignan 1902: 23. The upper part of that brook was known as *Blakewatar* c.1540 (Leland ii 99), and *Black-brook* c.1595 (Erdeswick 1844: 431). The lower part of Bourn Brook is now known as Moreton Brook. See also Bilson.

**BOURNE POOL** 1 mile east of Aldridge (SP 0800). From Bourne Brook (q.v.): the pool was created in the 15th century for an iron mill, and reduced to its present size before 1902: TSSAHS XXXII 1990-1 90.

**BOURNE VALE** 1 mile south-east of Aldridge (SO 0699). *Bournevale* late 18th century StEnc 77, *Bourn Vale* 1834 O.S. From Bourne Brook (q.v.) which runs through this place. Vale is from ME *vale* 'a vale, a wide valley', from OFr *val*, a rare element in Staffordshire, but early spellings for the name have not been traced, and the name may be of relatively recent origin.

BOVENHILL (unlocated, in Tettenhall parish, perhaps near Barnhurst) *Bobenhull* 1369 SHC XIII 76, *Bovenhill* 1398 VCH III 317-8, XX 19, 1642 Ct. From OE *a-bufan* 'above', with OE *hyll* 'hill', hence '(the place) above the hill'. Shaw 1801: II 201 incorrectly identifies this place as Barnhurst (q.v.): a list of 1719 and White 1851 show Barnhurst and Bovenhill separately: VCH III 319. If near Barnhurst, the hill may be Pendeford Hill, which overlooks Barnhurst. *Buvenheorth*, recorded in 1228 (SHC IV 63), has not been located.

BOWELES (unlocated, probably at Caudy Fields near Walsall: see Willmore 1887: 68, 254) *Bueles* 1228 SHC IV 65, *? Buell* 1242 SHC IV 96, *Boeles* 1242-3 Fees, *Boeles, Boweles* 1271 SHC V (i) 152, *Bowelis* 1276 SHC 1911 166, *Boueles* 1286 ibid. 174, *Bouwelis* 1300 ibid.' 178, *Bouvelys* 1309 SSAHS XVII 1975-6 71. Perhaps associated with Bowelles Felde (q.v.). See also Balls Hill. The name is from the Boweles family: Alice of Rushall married Sir Hugh Boweles (d. before 1271): see Willmore 1887: 249-54. The family name is probably from Bouelles in Seine-Infériere: DES 57. *Bowelles* in Shropshire is recorded in the 16th century: SA 2028/1/2/55.

BOWELLES FELDE (unlocated) *Bowelles Felde* 1550 SHC XII 204. See also Boweles.

**BOWEREND** ½ mile west of Madeley (SJ 7644). *Bowre-Ende* 1536 SHC XI 275, *Bowre End* 1602 AD, *Bower-ende* 1610 SHC III NS 31. From OE *bur* 'cottage', or possibly OE *(ge)bur* 'a peasant who held land in return for rent and services'. The word *end* did not mean a terminal point, but simply 'place', and was often waste or common land at the end of an inhabited area later used by squatters.

**BOWER HOUSE** on the north-west side of Rugeley (SK 0218). *Bower Farm* 1887 O.S. Probably to be associated with the Bowier/Bowyer family recorded in this area since at least the 14th century, e.g. John Bowier 1390 (BCA MS3415/224), John Bowyer 1435 (BCA MS3415/226).

**BOWERS** 4 miles north of Eccleshall (SJ 8135). *Boures* 1278 SHC 1911 170, 1286 SHC VI (i) 177, 1320 SHC 1921 23, *Bowres* 1401 SHC XV 114, *Boures* 1422 SHC XVII 85, *Bowers* 1597 SHC 1935 28, *Standon Bowers* 1807 SRO D615/M/7/37. Possibly from OE *bur* 'cottage', but the element is very difficult to distinguish in place-names from OE *(ge)bur* 'a peasant who held land in return for rent and services', but since the place lies 1½ miles south-east of Chapel Chorlton, which incorporates an element indicating the status of the occupiers, *(ge)bur* cannot be ruled out completely, though a simplex name of this type would be very surprising, and the term was ina ny event archaic by 1086, having been replaced by *villan*. The place is sometimes called Standon Bowers, from nearby Standon (q.v.).

**BOWERS BENT** on the east side of Bowers (SJ 8235). *Bent* 1696 Standon ParReg, *Bowers Bent* 1729 ibid, *the Bent* 1818 Salt 1888: 143, 1833 O.S. The *bent* element is from dialect *bent* 'the slope or hollow of a hill'.

**BOWGAGE FARM** ½ mile north-west of Grindley (SK 0329). *Bowgage Farm* 1909 EHNMR SC00936. Said to have been recorded as *Bow Gage* and *Beau Gage*. TNSFC 1886 38; StEnc 38. The second element is perhaps from OFr *cage* (see Cage Hill), used in place-names with reference to fenced enclosures, sometimes (perhaps especially) in parks (see VEPN II 122), so here possibly *beau cage* 'the beautiful fenced-off place' – the place lies near Chartley Park.

**BOWHILL FARM, BOWHILL BANK** ½ mile north-east of Betley (SJ 7648). *Bowhill Farm* 1705 SRO D1461/4/1, *The Bowhill* 1752 SRO D210/M/22, *Bowey (?Lone)* 1833 O.S. Although SHC 1933 31 states 'there does not appear to be any place-name in Staffordshire from which [the surname Boughay, Boghay, Bughay] could have been taken', there is evidence to associate this place with the name (recorded as *Boghay* 1199 SHC 1933 (ii) 31-45, *Bouheye* ibid. 32, *Boghay* 1332 SHC X (i) 101, *Boughay* 1407 SHC XVI 59, *Bughay* 1411 ibid.' 75, *Boghay* 1432 SHC III NS 129, *Bowghey* 1460 SHC III NS 213, *Boghay* 1486 SHC VII 131): this may be the place in Balterley associated with the family in 1288 (SHC 1933 (ii) 32). It is unclear whether the place gave rise to the family name or vice-versa. The name may be from OE *boga* 'bow', possibly referring to the shape of the *(ge)heg* 'clearing, enclosure', or perhaps from a bend in a river, so 'the enclosure or clearing near the river bend'. This place lies to the north-west of a bend in a stream. See also Boughey Hall Farm; Bowsey.

BOWODE – see **BOODEN FARM.**

**BOWSEYWOOD** 1 mile south-east of Betley (SJ 7646). *Bowsiewood* 1583 Betley ParReg, *Bowseywood* 1592 ibid, *Bowsyewood* 1600 ibid, *Bowsywood* 1694 ibid, *Bowsey Wood* 1795 SRO D3272/1/22/7/1, 1833 O.S. Perhaps from OE *\*bosig* 'animal stall, cow-stall' or 'right of pasture', in the second sense applied to land on which an outgoing tenant was allowed to continue pasturing animals (see VEPN I 131). The wood was evidently of some size and lies on both sides of Checkley Brook.

**BOYLES HALL** ½ mile east of Audley (SJ 8050). *Boyleshall* 1492 SHC 1912 256, *Boylle hall* 1539 SHC NS V 270, *Boylshall* c.1569 SHC 1938 96, *Boyleshall* 1611 Audley ParReg, *Boyls Hall* 1733 SHC 1944 2, *Boils Hall (Colliery)* 1790 SRO D1788/58/10. The first element would appear to be the personal name Boyle (DES 58).

BOYLSTON (unlocated) *Boylston* 1599 SHC 1935 110.

**BRABSON** or **BRABASONS** (unlocated, in or near Uttoxeter) *Brabasons* 1514 (1798) Shaw I 87, *Brabsons* 1737 SRO D543/B/3/4/1-6, *Brabson* 1786 SRO D543/C/7/14/1-2. Presumably from occupiers of that name.

**BRADDOCKS HAY** on east side of Biddulph (SJ 8957). *Braddockshey* 1662 JNSFC LXIV 1929-30 97, *Braddocks Hay* 1842 O.S. Perhaps 'the hay or enclosure with the broad oak', from OE *brad ac* with Mercian OE *(ge)heg*, or from the name Braddock: John de Brodock is recorded here in 1427 (Kennedy 1980: 25), and the place-name *Brodork* is recorded in 1427 (ibid. 27).

**BRADELIE** in Pirehill (unlocated) *Bradelie* 1086 DB. The location is now uncertain, but possibly Bagot's Bromley or Bradley Green in Burslem: VCH IV 51 fn.30; SHC XI NS 11-12; SHC 1916 169. Perhaps to be associated with *Bradley Moor*, recorded in 1781: SRO D615/D/310/1, or *Brondeye Heath*, marked to the south-east of King's Bromley on Slater's map of Staffordshire 1850, though the latter may be an error for *Bromley Heath*.

**BRADEMORE** (unlocated, possibly in the Cheadle area) *(manor of) Brademore* 1275 SHC V (i) 118.

**BRADENBROOK** a tributary of the river Manifold. From the oblique case *bradan* of OE *brad* 'broad'. The early forms of nearby Bradnop (q.v.) show similar development: *Bradenhop(e)* 1219 to 1346 FF.

**BRADEN HEATH** (unlocated, between Blymhill and Sheriffhales: see Plot 1686: 161) *Bradenham* ? 13th century SHC XIV (ii) 14, *Braden heath* 1686 Plot 161. 'The broad heath', from OE *bradan*, the oblique case of *brad* 'broad, spacious', and OE *hǣð*.

**BRADES VILLAGE** 1 mile west of Sandwell (SO 9890). *Brades* 1654 Roper 1980: 97, *Brades (Hall)* 1834 O.S. Perhaps from the OE personal name Brægd.

**BRADLEY** Ancient Parish (pronounced Bradeley [breɪdliː]) 4 miles south-west of Stafford (SJ 8717), *Bradelie*, *Bradelia* 1086 DB, *Bradley* 1798 Yates; **BRADLEY** (pronounced Bradeley [breɪdliː]) 1 mile south-west of Bilston (SO 9595), *Bradeleg* 1086 DB, *Bradele* 1222 SHC IV (i) 223, *Bradelea* 1227 ibid. 52, *Bradeleye* 1290 Misc, 1308 SHC X 4, *Bradele* 13th century, *Bradeleye* 14th century Duig; **BRADELEY** 1 mile west of Norton-in-the-Moors (SJ 8851), *Bredleye* 1274 SHC 1911 161, *Bredley, Bredeleye, Bredelee, Berdele* 1293 SHC VI (i) 245; **BRADLEY GREEN** in Biddulph (SJ8857), *? Baddele* 1243 SHC IV 98, *Bradlegreene* 1582 Biddulph ParReg, *Bradley greene* 1662 ibid, *Bradley Green* 1793 SRO D4842/14/1/61, *Bradley Green* 1842 O.S.; **BRADELEY (FARM)** 1 mile north-east of Dilhorne (SJ 9844), *Bradeley* 1836 O.S. The common medial -e-, and lengthened modern pronunciation, indicate a derivation from inflected OE *bradan*, from OE *brad* 'broad', with OE *leah*, hence 'broad or wide *leah*'. Bradley Hall (*Bradley Hall* 1801 Shaw II 233), a 17th century timbered mansion built in 1596 in High Street, Kingswinford, and originally occupied by Dennis Bradley, a local yeoman, was dismantled and rebuilt in Stratford-upon-Avon as Bradley Lodge in 1924. Early forms of the name have not been traced, unless *Bradleigh*, recorded in the early 13th century (SHC 1921 4) refers to this place, and the name may be from the surname. The Green element suggests a squatter settlement.

**BRADLEY IN THE MOORS** 4 miles east of Cheadle (SK 0641). *Bretlei* 1086 DB, 1166 SHC I 225, *Bredelege, Bredleyge* c.1192 CEC 261, *Bredleye* 1274 Ipm, *Bredleye* 1327 SHC VII (i) 216, *Brodley* 1532 SHC 4th Series 8 111, *Bradley* 1695 Morden, 1706

SRO D240/A/2/13/1-3, 1836 O.S. From OE *bred* 'board, plank', or OE *brǽdu*, in the later recorded sense 'a space in a field', such as 'broad strip of cultivated land', with OE *leah*, giving '*leah* where boards were produced', or 'the *leah* with the strip of uncultivated land'. The element *bred* is sometimes found in the sense 'plank bridge', but this place lies on a slight ridge, not on a stream.

**BRADMORE** 2 miles south-west of Wolverhampton (SO 8997). *Breadmere, Bradmere* 1303 SRO D593/B/1/17/1/4/5, *Brademere(heth)* 1318 SRO D593/B/1/17/1/4/13, *Bradmore* 1647 Manorial Survey, 1659 StEnc 80, 1895 O.S. From OE *brad* with OE *mere* 'mere, pool', which has developed into 'the broad moor'.

BRADNAYS (unlocated, possibly near Cheadle). *Bradnays* 1663 SHC II (ii) 63. Probably the same derivation as Bradney (q.v.).

**BRADNEY** ½ mile east of Worfield (SO 7695). *Bradeney, Bradeneye* 1327 SR, *bradeney (p)* 1514 Worfield CA, *Bradley* 1752 Rocque, *Bradney* 1833 O.S. From OE *bradan-eg* 'the broad island or land beside a stream, or broad raised land in marshland': the place lies on the north side of Stratford Brook. In Shropshire since the 12th century.

**BRADNEY (WOOD)** on the south side of Penn (SO 8895). Probably from a personal name, perhaps to be associated with John Bradeney recorded in 1506 (SHC 1928 85), Richard Bradeney recorded in 1516 (ibid. 69). The place lies on high ground, which makes a derivation from OE *bradan-eg* 'the broad island or land beside a stream' improbable.

**BRADNOP** 2 miles south-east of Leek (SK 0155). *Bradehope* 1167 Eyton 1854-60: X 36, *Bradenhop* 1219 FF, 1233 BM, *Bradnap(p)e* 1227 Harl to 1477 Banco, *Bradonhop* ' 1256 Ch, *Bradenap* 1522 FF, *Bradnop Syde* 1532 SHC 4th Series 8 30, *Bradnoppe* 1564 *Antrobus*, *Bradnappe* 1616 SHC IV NS 91, *Bradnup* 1686 Plot. From OE *bradan* (weak oblique case of OE *brad*) 'broad, spacious' and *hop* 'small enclosed valley; a plot of enclosed land, especially in a marsh', probably here in the second sense, as the land is very marshy, so 'the broad plot of enclosed land in the marsh'. References to Great Bradnop (*Bradnipe-grete* c.1240 *Deed* (1883) Sleigh) and Lesser Bradnop (*Bradnipe-lesse* c.1240 *Deed* (1883) Sleigh) and to *Upper Bradnopp* in the 1260s (*Loxdale*) indicate that there was then another settlement, which was perhaps at the north end of the Bradnop valley: VCH VII 170.

**BRADSHAW** 1 mile south of Horton (SJ 9455), *Bradesawe* c.1300 SRO DW1761/A/4/163, *Bradeschawe* 1325 SHC 1911 366, *Bradsehawe* 1333 (1883) Sleigh 188, *Bradeschagh* 1353 SHC XII (i) 122, *Bradshawe* 1357 (1883) Sleigh 80, 1409 SHC XVI 72, 1556 SHC 1938 43, *Bradschawe* 1359 SRO DW1761/A/4/14, *Bradshaw* 1586 SHC 1927 130, *Bradshaw juxta Horton* 1664 SHC II (ii) 63; BRADSHAW (unlocated) in Alstonefield, *Bradshaw* 1429 DRO D2375M/1/1, *Bradshawe* 1571 DRO D2375M/189/14. From the OE adjective *brad* 'broad, spacious', and OE *scaga* 'a small wood, a copse, a strip of undergrowth or wood', so 'the broad small wood'.

**BRADSHAWS FARM** 1½ miles south-west of Codsall (SJ 8401). *Bradshaws Farm* 1920 O.S. The place would appear to be that shown as *Brandhill* on Baugh's map of Shropshire, 1808. If so, the name originated from ME *brend* or *brand* 'burnt', with OE *hyll* 'hill', so 'the hill where the fire occurred', and later took the name of an owner or occupier.

BRADSTON BOTHES (unlocated, possibly in Rushton James) *Bradston Bothes* 1430 SRO DW1761/A/4/25.

**BRADWAY** (unlocated, possibly at Borrowcop Hill near Lichfield) *Bradway* c.1195 SHC VI (ii) 186, *? Bradeweia* 1199 SHC III 58. The forms suggest 'the broad road', from OE *brad, weg*, but SHC VI (ii) 186 records *Bradway alias Burghay*, and *Burghay or Burway*. The *burg* element suggests a possible association with Borrowcop Hill (q.v.). See also Borwey Foordes.

**BRADWELL** 2 miles north of Newcastle-under-Lyme (SJ 8449). *Bradewulle* 1217x1227 CEC 393, *Bradewell* 1223 SHC IV 27, 1240 (1798) Shaw I xxx, 1237 SHC 1910 296, 1260 SHC 4th Series 13 15, 1282-3 SHC XI NS 252, *Bradwelle* 1302 SHC VII 102, *Bradewalle* 1325 SHC IX 107, *Bradwall, Bradwalle* 1332 SHC X 82, *Bradewalle* 1416 SHC XVII 58, *Brodewal* c.1540 Leland ii 172, *Broadwall* 1571 GM 1811, *Bradwall* 1600 SHC 1935 283, 1686 Plot 152. 'Broad stream', from OE *brad* 'broad, spacious', and Mercian OE *wælle* 'a spring', and (sometimes) 'a stream'.

**BRAKENHURST (FARM), BRAKEN HURST (WOOD)** ½ mile north and ½ mile south-west respectively of Newchurch (SK 1423 and SK1422). *bracanhyrst* 942 (14th century, S.479), *bracan hyrst* 1008 (14th century, S.920), *Brakenhurst* 1836 O.S. From OE *\*bræcen* (which may have an ON origin: see VEPN I 141), with OE *hyrst*, giving 'the wooded hill overgrown with bracken or fern'. The land rises to 429' at Newchurch. See also Hooke 1983: 96. *Brakenhurst*, recorded *temp.* Edward I (Okeover T298), appears to have been in the area around Stretton, near Burton upon Trent.

**BRAMPTON, THE** on the north-east side of Newcastle under Lyme (SJ 8584). *Brampton* 1616 SHC NS IV 83, *Brampton Cottage* 1836 O.S. From OE *brom-tun* '*tun* where broom grew'.

**BRAMSHALL** Ancient Parish 2 miles west of Uttoxeter (SK 0633). *Branselle* 1086 DB, *Brumeshel* 1195 P, *Brendeshulf* 1211 SHC III (i) 151, *? Brimshill* 1306 SHC VII 161, *Bromschulf* 1327 SR, *Bronsulf* 1242 Fees, *Bromsshulf* 1366 SHC VIII NS 28, *Bromshell* 1532 SHC 4th Series 8 67. From OE *brom* 'broom', with OE *scelf* 'shelf, ledge, shelving land', hence 'broom-covered shelf'. The church lies on the rim of a conspicuously flat hill-top, which is almost certainly the shelf of the place-name (see also Gelling 1981: 11). For Little Bramshall (*Little Bramshall* 1662 SRO DW1733/D/18) on the north-east side of Bramshall see SHC XI NS 3-4; VCH IV 51 fn.42.

**BRAMSTEAD, BRAMSTEAD HEATH** 4 miles south-east of Newport (SJ 7917). *Brams(t)on Heath* 1674, 1749 ParReg, *Bromstead Heath* 1793 Open Fds, *Bromston heath recte Bromstead* 1795 Reg Diss. The earliest spellings point towards OE *brom stan* 'the broomy place with the stone', with the alternative '–place where broom grew', from OE *brom-stede*. Bromstead represents Staffordshire dialect pronunciation.

**BRANCOTE** 2 miles east of Stafford (SJ 9622). *Bromcote* 14th century freq. Duig, *Brancott* 1836 O.S. From OE *brom* 'broom', and OE *cot* 'cot, cottage, shelter, hut', hence 'the cottage with the broom'. See also Broncott.

**BRANTLEY** (obsolete) on the north-west side of Bobbington (SO 9090). *Brantley* 1892 O.S. Probably from OE *brand* 'burnt', with OE *leah*.

**BRANDY-LEA** ½ mile north-west of Heaton (SJ 9463). *Brandleyhouse* 1543 NSFC Vol. LXVII 51-70, *Brandylee* 1686 SRO C/8/12. Probably from OE *brand-leah* 'burnt *leah*'.

**BRANKLEY (FARM, HOUSE, LODGE & COVERT** 2 miles north-east of Yoxall (SK 1520, SK 1521, SK 1621). *Brankeley* 1658 DCL 380.

**BRANSTON**  2 miles south-west of Burton-on Trent (SK 2221). *Brontiston* 942 (14th century, S.479), *Branteston* 956 (not listed in Sawyer 1979) SHC 1937 6; Shaw 1798: I 21-2, *Brantestone* 1086 DB, *Brantisto'* c.1114 (13th century) SHC 1916 215, *Brantheston* c.1132 SHC 1937 9, *Brantest'* 1159-9 SHC 1937 13, *Brontestona* 1175 SHC I 81, *Bruntiston* 1284 SHC 1911 191, *Brawnston* 1569 SHC XVII 217. The genitive *-s-* indicates a derivation from a OE personal name *Brant (rather than from OE *brant* 'burnt'), with OE *tun*. The place was renamed Branston from Branstone in 1958: Youngs 1991: 404. *Branston Holm* is recorded in Burton upon Trent in 1771: SHC 1931 91.

**BRATCH, THE**  1 mile north-west of Wombourne (SO 8693). *the Bratches* early 17th century VCH XX 210, *the Braches* 1612 SRO D740/8/7, *ye Brach* 1767 Trysull ParReg, *Bratch* 1775 Yates, 1825 ibid, *The Bratch* 1834 O.S. 'The new enclosure', from OE *brec* 'land broken up for cultivation', commonly found as field-names as Brache, Breche, Breach, Bridge, Britch, Birch, etc. – see Foxall 1980: 33. The Bratches, 2 miles south-east of Cannock, probably has the same derivation. Cf. Breach Mill; The Breatch.

BRAUNDGRENES  (unlocated, near Abbots Bromley). *Brandgrenes* 1317 SRO D(W)1721/3/12/2, *Braundgrenes* 1338 SRO D(W)1721/3/7/2. Probably from OE *brand* 'burnt', so 'the grassy open spaces where burning had taken place'.

**BRAZENHILL**  1 mile north of Haughton (SJ 8621). *Brussenhull* 1238 SHC IV 89, *Brusenhulle* 1299 SHC VII (i) 54, *Bruzenhul* 13th century SHC IV 273, *Bresonhulle* 1381 SHC XVII 197, *Brasenell, Bressenell* Eliz Chanc P, *Brasenhill* 1557 SHC 1926 14, *Bressenell* c.1562 SHC IX NS 29, *Brasenell* 1566 SHC IX 86, *Brasnyll* 1586 SHC XVII 233, *Brasnill* 1686 Plot, 1747 Bowen. Possibly from OE *borsten* the past participle of OE *berstan* to burst, to crack' (see VEPN II 115), with OE *hyll*, hence 'the hill with the breach or landslip', perhaps explained by the fact that the place lies on a small deposit of glacial sand and gravel, presumably once quarried. Cf. Bristnall; Bristnall Hall, Worcestershire; Burshill, West Yorkshire.

**BREACH HOUSE FARM** 1 mile south-east of Abbots Bromley (SK 0923); **BREECH LANE, BREECH COPPICE** 1½ miles north-west of Abbots Bromley (SK 0726), *Le Breche* late 13th century SRO D(W)1721/3/12/6; **BREACH, THE**  1 mile south of Cheadle (SK 0040), *the Breach* 1767x1830 SRO D953/20; **BREACH, THE** on the north-west side of Upper Tean (SK 0040), *The Breach* 1836 O.S.; **BREACH, THE** in Halesowen; **BREATCH, THE** near Belbroughton; **BREACH MILL** near Hagley. 'The new enclosure', from OE *brec* 'land broken up for cultivation', commonly found as field-names as Brache, Breche, Breach, Bridge, Britch, Birch, etc. – see Foxall 1980: 33. Cf. The Bratch.

BREADIHURST (unlocated) *Breadihurst* 1590 SHC 1930 (ii) 197.

**BREDON, RIVER**  a tributary of the river Hamps. Although not marked on maps, Oakden 1984: 6 records that the name is widely used locally. Early forms are not available, but if it is ancient the name might be related to the river Bride in Dorset, perhaps of Celtic origin with the meaning 'boiling, bubbling': see Ekwall 1928: 52.

**BREECH COPPICE**  in Bagot's Park (SK 0726). *le Breche* 1402 SHC NS XI 208, *Breech Close* 1724 Survey, *Breech Coppice* 1836 O.S. From OE *brec* 'land broken up for cultivation', commonly found as field-names as Brache, Breche, Breach, Bridge, Britch, Birch, etc. The name is also recorded in Breech Lane here.

**BRENDEHET**  (unlocated, near Blurton) *Brendehet* 1302 SHC XI 311. 'The burned heath'.

**BRENDWOOD** (unlocated, in Okeover) *Brentewode* 1439 SHC VII NS 51, *Brendwoodd, Bryndwoodd* 1571 ibid. 73, *Branwode* 15th century ibid. 171. 'The burned wood'.

**BRERETON** 1 mile south-east of Rugeley (SK 0516). *Brerdun* 1292 Banco, *Breredon* 1279 Cl, 1317 Ch, *Brerdon* 1329 SHC XI 8, *Brereton* 1412 AD 1, 1834 O.S., *Brewarton* 1547 FF, *Bruerton* 1577 Saxton, 1755 Bowen, *Brereton vulg Brewerton* 1675 Ogilby. From OE *brer* 'briar, bramble, thorn', with OE *dun* 'hill'.

**BRETELL, BRETELL LANE** 1 mile south-west of Brierley Hill (SO 9084). *Breydehille* 1300 SHC VII (i) 80, *Bredhull* 1307 SHC VII 175, *Bredhul* 1327 SHC VII (i) 247, *Bredhull* 1332 SHC X 87, *Brethill* (p) *temp.* Henry VIII SHC X NS I 114, *Bretyll* 1539 SHC NS VI (i) 72, *Bretill* 1603 SHC 1946 63, *Brettell* 1614 Duig, 1663 SHC II (ii) 29, *Britwell* 1686 Plot, *Brettell Lane* 1834 O.S. Duignan 1902: 24 suggests a derivation from the personal name Brihtelm (Brihthelm), but the forms suggest that the root may be OE *brǣdu*, in the later recorded sense 'a space in a field', such as 'broad strip of cultivated land', or possibly OE *bred* 'board, plank', so 'the place from which planks were obtained', with OE *hyll* 'hill'. Cf. Bredfield, Suffolk; Bredhurst, Kent. The name *Brechelemore* is recorded in the area in 1300 (SHC VII (i) 80), and may be associated with this place. Guttery 1950: 21 refers to *Bryt Hill* at Hawbush and *Brettell* on the western borders of Pensnett Chase. Some of the above spellings may relate to Bryt Hill.

**BREWOOD** Ancient Parish (pronounced Brood [bruːd]) 6 miles north-west of Wolverhampton (SJ 8808). *Brevde* 1086 DB, *Breoda* 1139 MRA, *Broude* c.1150 Brewood Ch, *Breode* 1151, 1166 MRA, *Browoda* 1152x1159 (13th century) EEA 17 65, *Brewuda* 1188 SHC I 137, *Brewuda* 1188 to 1196 P, *Bri(e)wude* 1196 Chanc R, 1202 P, *Brewod(e), Brewud(e) freq.* 1200 to 1834 O, *Brewde* 1201 P, *Breuwud'* 1207 ClR, *Browd(e)* c.1211x1216 Brewood Ch, 1330 Ch, 1236 Fees *et freq* to 1559 Pat, *Bruwode* 1245 Lib, 1315 to 1496 Ipm, 1415 Coram R, *Brehude* 1290 Ch, *Breuwode* 1306 GDR, 1322 Ch, *Brewod'* 1461 HAME 470, *Brude* 1462 CA, *Brywoode* 1562 SRRL 1514/310, *Breewood als Braywoode* 1571 Pat, *Bruwood* 1558 FF, *Bruyde* 1561 Will, *Brude* 1597 Pattingham CA, *Breawood* 1643 Will, *Breewood* 1659 Tong ParReg, *Breerwood* 1690 ibid. A hybrid name, the first element from Welsh (and Cornish) *bre* 'hill', deriving from British *\*briga*, late British *\*brega*. In Welsh, *bre* is found only in place-names and early poetry, and names containing the element are likely to be early (Padel 1985: 30). The precise meaning of *bre* is unresolved, but it has been suggested that it may have been applied to the most significant hill in an area (Padel 1985: 30). Brewood, however, is centred on the edge of a bank sloping down to stream meadows, rather than on a hill in the conventional sense. The second element is OE *wudu* 'wood' (a common element, but rare before c.700: JEPNS 8 43; see also Jackson 1953: 701-2), perhaps demonstrating that at the time of English colonisation there were still Welsh-speaking natives in the area, or that the colonists simply added *wudu* to a surviving hill-name Bre. The meaning of the name is therefore 'the wood at the hill called Bre'. Brewood, Charnwood (Leicestershire), Chetwode (Berkshire) and Crewood (Cheshire) are probably the only examples of a combination of *wudu* with an earlier place-name or pre-English element. Brewood gave its name to Brewood Forest (q.v.), in which it lay. *Breudewode* in Fulfen, recorded in the 16th century (SHC XI 276, 284), is almost certainly a misreading of *Brendewode* 'burnt wood'.

**BREWOOD FOREST** *foresta de Brewuda* 1187 SHC I 139. See Brewood. An area subject to Forest law dating from at least the 12th century which covered a large area of south Staffordshire to the west of the river Penk and extended into Shropshire, including

Blymhill, Wheaton Aston, Bilbrook, Perton and (in Shropshire) Hawkswell (see SHC VI NS (ii) 49), Albrighton, Donington, White Ladies and Tong. The southern boundary ran more or less in line with, and may indeed have followed, the boundary between Cuttlestone and Seisdon Hundreds: see Anderson 1934: 145. Brewood Forest was disafforested in 1204: VCH V 18 fn.2.

BRIDDESHUS or BRIDDESHALL (unlocated, 'a small spot in [Tatenhill], not noticed on any of the maps, and with difficulty now to be discovered': Shaw 1798: I 107) *Briddeshous* c.1262 (1798) Shaw I 107, *Briddeshale* 1277 SHC VI (i) 81, *Breyesdeshale* 1280 SHC 1911 177, *Briddeshus* 1290 Ch, *Briddleshall* 1301 SHC VII 79, *Briddeshall* 1301 ibid. 80, SHC XI 147, *Briddeshale* 1302 ibid. 102, *Briddeshalle* 1355 SHC 1913 158, *Bredshall* 1357 SHC XII NS 281, *Briddeshalle* 1380 Ipm XV 97, *Briddushalle* 1413 SHC XVII 46, *Byrdeshall* 1414 Ct, *Bryddesale(feldes)* 1422 Ct, *Birdhouse, Birdshill, Briddeshalle, Bredshall, Birdshouse* 1559 (1798) Shaw I 123, 107, *Birdsall* 1614 SHC IV NS 18, 1660 DRO D3155/7112. Perhaps from OE *bridd* 'a young bird', with OE *hus* 'house', possibly meaning here 'the building where young birds were reared' or 'house or hall with the fowl or the coop'. However, a derivation from a personal name *Bridd cannot be ruled out: William Brid appears in an early undated deed in this area (Shaw 1798: I 109), and Robert Bridde is recorded in 1440 (SHC 1938 307). The personal name is probably a nickname from OE *bridd*: see DES 45. Hardy 1907: 132 states that this manor lay between Dunstall, Fernhill Farm, Rangemore and Highlands Park, and that the name survived in *Birdshall Meadow*, which lay near Rangemore Park next to the Dunstall Brook.

**BRIDESTONES** 2½ miles east of Congleton, on the Staffordshire-Cheshire border (SJ 9062). *Bridestones* 1766 Gough 1806: II 506, *Bride Stones* 1775 Yates, 1823 GM II 217-22, 1844 Garner, *Bryde Stones* 1791 SHC 4th Series 13 133, *Bridestones* 1831 Lewis. The remains of a gallery grave (first recorded in Rowlands 1766: 319-20), possibly built before 2,500 BC, associated until the 18th century with an elongated mound of stones variously recorded as 60 and 120 yards in length, about 14 yards wide, and orientated east-west, of which 'several hundred' loads were removed for road-making in 1764: JNSFC LX 1925-6 188; JNSFC 5 1965 33. 18th century accounts show that there were two other chambers covered by the mound, about 55 yards west of the gallery: AJ 120 1963 247. A detailed account and history of the monument is in VCH Ch I 43-6. Various fanciful legends have become attached to the stones. The age of the name is unknown, but a standing stone called *the Stepmother Stone* and a mound called *Mystylowe* recorded in the early 17th century may be associated with the monument: VCH VII 223. There may have been another monument called Bride Stones near Cheddleton (see Ring Hey): Shaw 1801: II 2 quotes from the notes to Holliday's poem 'The British Oak': '... Bride Stones, as they have been called time out of memory, have been found in or near Chedleton ... [t]hese Bride Stones the author has not yet had an opportunity of seeing ...'. No such monument has been traced near Cheddleton. Other Bride Stones are said to have existed near Dilhorne: Shaw 1801: II 2. It is evident that Bride Stones was a colloquial name, certainly in Staffordshire, for prehistoric monuments. EDD suggests that in North Yorkshire, bride-stones was the name given to pillars of rocks found on the moors, at which marriage ceremonies were said to have been practised. A prehistoric burial chamber called The Bridestones is at Stansfield, Yorkshire. See also JNSFC 5 1965 33; PN Ch II 291.

BRIDEWODE (unlocated) *Bridewode* 1306 SHC VII 156.

**BRIDGE CROSS** 1 mile west of Burntwood (SK 0409). *Byrdes Crosse* 1578 VCH XIV 201 – perhaps from the name Byrd (DES 45), which by metathesis has become Brid- and so Bridge.

**BRIDGE END** over the river Churnet on the north-west side of Leek (SJ 9757). *Bridgend* 1695 Leek ParReg, *White's bridge or Bridge end* 1851 White. The place evidently had an alternative name (recorded as *? Whitbige* 1692 Leek ParReg, *Whites Bridge* 1704 ibid, *Whitesbridge* c.1800 SRO D3272/1/4/3/17-20, *White's Bridge* 1842 O.S.), presumably from a personal name.

**BRIDGEFORD** (unlocated, perhaps near Tutbury: SHC 1912 222) *Bridgeford* (undated) SHC 1912 222.

**BRIDGEFORD, GREAT** in Seighford parish, 3½ miles north-west of Stafford (SJ 8826). *Brigeford* 1086 DB, *Bruggeford* 13th century Duig, *Bregeford* 1532 SHC 4th Series 8 66, *Bridgford* 1586 SHC 1929 111. 'The ford with the bridge', from OE *brycg* 'bridge, causeway'. Little Bridgeford (*Little Bridgford* 1679 SHC XII NS 100) lies ½ mile north-west of Great Bridgeford. The division of Great and Little Bridgeford had occurred by 1333: SHC 1914 98. This is a rare Staffordshire example of a pre-Conquest place-name incorporating *brycg*. *Bruggefordia*, recorded in a spurious charter 1191x1194, has been associated with this place (CEC 261), but that is unlikely; the place may be East Bridgeford in Nottinghamshire.

**BRIDGEWOOD** (unlocated, in or near Biddulph) *Brugge Wode, Bruggewode* 1327 SHC VII 198, *Bruggewode* 1332 SHC X 94, *Brygewode* 1447 HLS, *Bridgwood* 1646 Norton-in-the-Moors ParReg. From OE *brycg* 'bridge', with OE *wudu* 'a wood, a forest', so 'the wood at the bridge or causeway'. See also Tooth 2000b: 48-9.

**BRIDLEY MOOR** (obsolete) on west side of Redditch (SO 0367). *Bridleymore* 1464 Pat, *Bridley Moor* 1832 O.S. Possibly from OE *bridd* 'bird', with OE *leah*, and OE *mor* 'moorland, marsh'. In Tardebigge parish, forming part of Staffordshire from c.1100 until 1266, in Warwickshire until 1844, and since that date in Worcestershire.

**BRIERLEY HILL** 3 miles north-east of Dudley (SO 9186), *Brerelay* 1248 SHC IV 106, *Brierley* 1273 SHC 1911 155, *Brerley* 1396 SHC XV 70, *Brereley* 1417 SHC XVII 63, *brireley hill* 1619 Guttery 1950: 3, *Breerley otherwise Bryerley* c.1621 SHC VII NS 235, *Brierley Hill* 1642 Guttery 1950: 4; BRIERLEY (obsolete) in Coseley, probably at or near what is now Bradley (SO 9595), *Brerleia* c.1175 SHC 1941 73, *Brereley* 1594 Sedgley ParReg, *Bryerley* 1621 Worfield CA, *Briarly* 1655 Sedgley ParReg, *Brayerley* 1696 BCA MS3549/276. 'The thorny *leah*', from OE *brer* 'brier, brambles, thorns'. Some of the forms given for Brierley may relate to Brierley Hill. Brierley near Coseley is remembered in Brierley Lane, to the south of Bradley.

**BRIERYHURST** 1 mile north of Kidsgrove (SJ 8455). *Bryeris Hurst, Bryerie Hurst* 1597/8 SHC 1935 31, *Breeryhurst* 1608 SHC 1948-9 113, *Brerehurst* 1644 SHC 4th Series I 175, *Bryery-Hurst* 1738 BCA MS3375/447261, *Brerehurst or Brieryhurst* 1834 White 551. From OE *brerig* 'growing with or overgrown with briars', with OE *hyrst* 'a hillock; a copse, a wood, a wooded eminence', here meaning 'the wooded hill overgrown with briars'.

BRIGGEND, BRIDGGEND – see **WOLSELEY BRIDGE**.

BRIMLANDES (unlocated, in Charnes) *Brimlandes* 1227 SHC IV 73. The single spelling does not allow a derivation to be put forward (*brim* meaning 'border or edge' is not recorded before the 16th century: OED), but the second element is OE *land*, a

148

common second element in English place-names, with meanings ranging from 'estate; landed property', to 'district; portion of a village or estate', and even to the nature of the soil where the first element is a descriptive word.

BRIN – see MARSH, THE.

**BRINDLEY FORD** 1 mile south of Biddulph (SJ 8854). *Brindley Ford* 1872 P.O. A place said in 1872 to have '... sprung into existence during the last few years; its inhabitants are chiefly employed in the extensive coal and iron works ...': P.O. 1872 532. The place appears to have been named after James Brindley (1716-72), the canal engineer, who lived at nearby Turnhurst, near Tunstall.

**BRINDLEY HEATH** 2 miles north-east of Cannock (SJ 9914), *Brinsy Coppice* 1698 Fiennes, *Brindley Heath* 1708 Constables Book, 1833 O.S.; BRINDLEY POOL (UPPER & LOWER) (obsolete) 1½ miles north-east of Hednesford (SK 0014), *Brindley Pool* 1743 SRO D1057/F/2/2, *Upper Brindley Pool, Lower Brindley Pool* 1834 O.S. From ME *brend* 'burnt', and OE *leah* 'clearing', giving 'the *leah* cleared by burning' (cf. Burntwood). It is unclear whether *Breyndelehalghe*, recorded c.1300 (SHC 4th Series XVIII 189) is to be associated with either place.

BRINEPITS (obsolete) at Shirleywich (q.v.). *Brinepits* 1686 Plot, *Brine Pitts* 1747 Bowen. See Shirleywich.

**BRINETON** 1 mile north of Blymhill (SJ 8013). *Brvnitone* 1086 DB, *Brintona* 1116 RBE (p), c.1182 SHC II 256, *Bruntan* late 12th century SRO 620[7910], *Brienton* 1204 Cur, *Brininton* 1211 Cur, *Brim(s)tona* 1221 Bracton, *Brunton* 1223 Blymhill, 1225 Coram, 1301 Ass, 1306 GDR, 1327 SR, *Brumton*'1225 Cur, 1279 Blymhill, *Brimengton* 1236 Fees, *Bruminton* 1272 Ipm, *Brimton, Brimtone* 1305 SHC 1911 63, *Brynton, Brinton* 1305 Cur, 1242 Fees, 1252 (p) Cl, 1305 FF, *Bruynton* 1310 Banco, 1348 Ass, 1373 FF, *Broynton* 1364 Banco, *Brugnton* 1436 Banco, *Brineton* 1775 Yates. The forms are inconsistent, but some appear to show traces of -*ing*-, so possibly 'the *tun* associated with ? Brun(a) or Brȳni', or (as proposed by Ekwall 1960: 65) 'the *tun* of Brȳni's people'. Some of the spellings may have been influenced by the spellings of nearby Blymhill (q.v.).

**BRINSFORD** 4 miles north of Wolverhampton (SJ 9105). *Brenesford, Brunesford* 996 (17th century, S.1380), *Brunesford* 1176 P, *Brunnesfort* c.1215 WA II, *Bruneford* 1227 Ass, *Brumesford* 1286 SHC V (i) 170, *Brunnesford* 1317 SHC 1911 111, *Bruynesford* 1381 Duig. 'Brun's ford'. Brun (modern Brown) was a common OE name.

**BRISTNALL FIELDS** 2 miles east of Smethwick (SO 9986). *Brussenhulle* 13th century *freq.* Duig, *Bristnall Fields, Bristnal End* 1834 O.S. From OE *(ge)brist* 'a burst, a crack in the earth or soil', ME *bursten, brusten*, past participle of 'to burst' (see VEPN II 115), with OE *hyll* 'hill', hence 'the hill with the breach or landslip'. Cf. Brazenhill. See also PN Wo 303.

**BROADEYE** a district on the north-west side of Stafford, within the town walls (SJ 9222). *? Brodhe* 1290 SHC 1911 199, *Brodeye* 1290 Ipm, *Brodey* c.1610 map, *Broade eye* 1610 Speed. From OE *brad-eg* 'broad or spacious island or land beside a stream or raised ground in marshland'. There was a tradition in Elizabethan times that a mound then known as Castle Hill to the south-east of Broadeye Bridge was the site of the fortification built in 913 by Æthelflæd, daughter of King Alfred and sister of Edward the Elder. Historians cannot agree whether the tradition is well-founded: see Eyton 1881: 20-1; VCH VI 189, 200; Darlington 1995. It is possible that 'the broad raised raised ground in marshland' was an early description of the site of what became the town of Stafford, and that the *eg* element is also to be traced in the name *Betheney* or *Bethenei* (q.v.).

**BROADFIELD** near Goldenhill, Stoke on Trent (SJ 8553). *Brodfield* 1586 SHC 1927 133. 'The wide open space'. The name is preserved in Broadfield Road.

**BROADFIELDS** 1½ miles south-east of Alrewas (SK 1813). *Borugh* 1325 Hardy 1908: 20, *Borough Field* 1775 Yates, *Broad Fields* 1798 Yates, 1834 O.S. Evidently from OE *burh* 'a fortification', presumably referring to some ancient earthwork here: see also Borway Ford. The present name appears to have been adopted towards the end of the 18th century.

**BROADFIELDS FARM** on north-west side of Edingale (SK 2112). *le Brodefylde*, *Brodfeld* 1541 Croxall. From OE *brad, feld* 'the wide open land'. Transferred from Derbyshire in 1894.

BROAD FORD (obsolete) a ford across the river Tame at Elford Mill (SK 1909). The ford is recorded in the 13th century, with the road crossing it *Sropstreteweye*, 'the street-way to Shropshire': VCH XIV 240.

**BROADGATE HALL** 1 mile north-east of Checkley (SK 0438). *Bradenhevet* 1255 SHC 1911 124, *Bradehened* 1281 Okeover E5104, *Bradhode* 1323 SHC IX (i) 93, *Bradhead Haull* 1559 SHC 1938 40, *Brodhead Haull* 1569 ibid. 29, *Brodheadhall* 1592 SHC 1930 (ii) 343, *Broadgate Hall* 1836 O.S. From OE *brad* and *heafod* 'the broad headland'. The name was presumably changed to protect the sensibilities of the occupiers.

**BROAD HEATH** 1 mile north of Ranton (SJ 8525). *(The Graunge upon) the Heyth* 1539 SHC V (i) NS 322, *Broad heath [grange]* 1679 SHC 1919 227, *Broadheath Grange* 1680 SRO D1798/617/15, *Broad Heath* 1720 SHC 1931 90. The site of a grange of Ranton abbey. See also Frankwell.

**BROADHURST GREEN** on Cannock Chase (SJ 9815). This may be the same place as *Braddockes Green* recorded in the 1590s (SRO D1720/13), which appears as *Box Green* in the 1820s (map). The Green element suggests a squatter settlement.

**BROADOAK** 3 miles north-west of Uttoxeter (SK 0535), *Brodhok* 1277 SHC VII NS 20, *Brodok* 1280 SHC VI (i) 147, *? Brodeoke* 1296 SHC VII 37, *? Brodock* 1569 SHC XVII 217, *Broade Oke, Broad Oke* 1599 SHC 1935 106; **BROADOAK** 1 mile south-east of Consall (SJ 9847), *? Brodehoc* 1253 SHC XI 310, *? Brodeoke* 1297 SHC VII 37, *Broodwoke, Broodoke* 1482 SHC VI NS (i) 144, 147, *Brodde Oke* 1585 SHC 1929 116, *the Broad Ocke* 1668 Kingsley ParReg. Self-explanatory. *Brodhok* on Cannock Chase is recorded in 1292: SHC VI (i) 273.

**BROCKHILL FARM** 2 miles south-west of Alvechurch (SO 0169). *Brokhyll* 15th century PN Wo 362, *Brockehull* 1535 ibid. Probably from OE *brocc, hyll* 'badger-hill'. In Tardebigge parish, forming part of Staffordshire from c.1100 until 1266, in Warwickshire until 1844, and since that date in Worcestershire.

BROCKHOLES, BROKHOLES (unlocated, in Abbots Bromley), *Brokholes* 1230 SHC 1937 61, *Brocolis* 1272 ibid.' 71, *Brokholis* 1283 ibid.' 83, *Brocolis* c.1292 ibid.' 95, *Brockholes* c.1300 SRO DW1721/3/6/2; BROCKHOLES (obsolete) 1 mile north-west of Ipstones (SK 0151), *Brockholes* 1836 O.S.; BROCKHOLES (unlocated) in Tixall, *Brockholes* 13th century SHC VIII (i) 192. From OE *brocc-hol* 'badger hole or burrow'.

**BROCKHURST** 1 mile south-east of Blymhill (SJ 8211). *Ruscote* 1086 DB, *Brokhurst* 1322 SHC IX (i) 87, 1349, 1365 Ipm, 1361 Pat (*et freq.*), *Brokehurst* 1421 SHC XVII 97, *Brokehurst* 1424 Banco, *Bruchhurste temp.* Henry VIII SHC VIII 111, *Brockhurst(e)* 1533, 1544 Staff Acc, *Brockhurst Farm* 1673 SHC II (ii) 123. From OE *brocc-hyrst* 'the copse frequented by badgers'. There are other Brockhursts in Cheshire, Derbyshire, Warwickshire and in Hampshire, near Gosport. This place has been identified with *Ruscote* in DB (SHC 1923 31-2; VCH IV 53) from OE *rysc-cot(e)* 'rush cottage', perhaps meaning 'the cottage by the rushes', or possibly 'the cottage thatched with rushes (rather than the usual straw)'.

**BROCKHURST (FARM)** 1 mile south-west of Hints (SK 1501). *Brockehurst* 1312 (1801) Shaw II 14, *Brockhouse (moore)* 1606 (1801) Shaw II 16, *Brockhurst* 1834 O.S. From OE *brocc-hyrst* 'the copse frequented by badgers'.

BROCKLEY MOOR (obsolete) 2 miles south-east of Market Drayton (SJ 6932). *Brockley Moor (Farm)* 1675 SRO 1/279/73, *Brockley Moor* 1696 SRO 828/6, *Brockley Moore* 1699 SRO D681/E/5/21. Probably from OE *broc* 'brook', hence 'the moor at brook *leah*'. OE *brocc* is 'badger', but where a stream or river runs through a place, as here, the brook derivation is far more likely.

**BROCKMOOR** 1 mile south-east of Kingswinford (SO 9087). *Brockmeer* 1749 Bowen, *Brochmeer* 1775 Yates, *Brockmoor* 1834 O.S. Probably from OE *broc* 'brook', rather than OE *brocc* 'badger', with OE *mere* 'mere, lake (and sometimes marsh)', since 'badger lake' is improbable. The 1834 O.S. map shows Brockmoor further east than the place marked on modern maps.

**BROCKTON** 1½ miles north of Eccleshall (SJ 8131), *Broctune* 1086 DB, *Brocton* 1295 SHC VII 27, 1297 SRO D (W)1734/J2268, 1327 SR, 1332 SHC X 98, 1679 SHC 1914 16; **BROCKTON GRANGE** 4½ miles south-east of Newport, in Sheriffhales (SJ 7913), *Brotone* 1086 DB, *Brocton* c.1215 Rees 1997: 28, *Brokton Graunge* 1422 SHC XVII 83, *Bocton* 1534 Eyton 1854-60: XI 91, *Broketon* 1537 ibid. Probably from OE *broc* 'brook', hence '*tun* or settlement on a brook'. OE *brocc* is 'badger', but where a stream or river runs through a place, as with these places (there is a stream to the east of Brockton Grange, and Brockton lies on Brockton Brook), the brook derivation is far more likely. It has been noted that Brockton Grange (in Shropshire since 1895) was also known as *Brocton Oliver* (Sheriffhales printed register iv), perhaps from an early owner or occupier, although no evidence of that alternative name for the place has been traced. PN Sa I 59-60 observes that there are six names derived from OE *broc tun* in Shropshire, with five now Brockton and one Bratton. As well as the two Staffordshire examples, and others in Somerset and Yorkshire North, which have become Bratton and Brotton, elsewhere *broc tun* has evolved into Broughton in at least 19 major settlement names. See also Brocton; Broughton.

**BROCKWOOD HILL** 2 miles north-west of Audley (SJ 7852). *Brockwood Hill* 1733 SHC 1944 19, 1833 O.S. From OE *brocc-wudu hyll* 'badger-wood hill'. Perhaps to be associated with *Brockwall Hill*, recorded in 1573 (Audley ParReg.), *Brackwall Hill* in 1609 (ibid.).

**BROCTON** in Baswich parish, 4 miles south-east of Stafford (SJ 9619), *Broctone* 1086 DB, *Broctune* 1176 Blymhill, *Broctuna* 1166 SHC 1923 296, *Brocton* 1199 Ass, 1221 FF, 1242-3 Fees, 1272 SHC VIII (i) 148, *Brohton* 1227 SHC IV 65, *Broghton* 1289 SHC 1911 195, *Borchton* 1294 SHC 1911 227, *Brocton o' the Cank* 1325 Banco, *Brocton juxta Bedenhale* 1345 Banco, *Brocton next Bastwyche* 1545 SHC XI 287. Probably from

OE *broc* 'brook', hence '*tun* or settlement on a brook'. OE *brocc* is 'badger', but where a stream or river runs through a place, as here (the place lies on Oldacre Brook), the brook derivation is far more likely. See also Brockton; Broughton.

BRODEWOOD (unlocated, possibly near Audley) *(Hay in) Brodewood* 1493 SHC 1912 257.

BROMBAR (unlocated, possibly between Pershall and Eccleshall) *Brombar* 1298 Spufford 2000: 295.

BROMEHURST (unlocated, near St Thomas' priory, between Stafford and Hopton) *Bromhurst* 1349 SRO D938/482, *Bromehurst* 1454 SHC III NS 212.

BROMISPITT (unlocated) *Bromispitt* 1602 SHC 1935 467.

**BROMLEY (FARM)** 1 mile north-west of Upper Arley (SO 7581), *Bromiley* 1295, 1317 PN Wo 30; **BROMLEY GREEN** 1½ miles north-west of Whitmore (SJ 8043), *Bromley, Bromleye* 1281 SHC VI (i) 121, *Bromeleye* 1324 SHC 1913 232, *Grene* 1333 SHC 1913 228, *campum quod vocatur Bromleye* ('the open land which was known as Bromleye') 14th century ibid. 227, *Bramley* 1558 SHC 1926 22, ? *Bromylowe* 1608 SHC 1948-9 134, **BROMLEY** 1 mile east of Kingswinford (SO 9088), *Bromle* 1300 SHC VII 80, *Bromley* 1749 Bowen, 1798 Yates; **BROMLEY** 1½ miles west of Worfield (SO 7395), *Bromley* 1525 SR, 1563 Worfield ParReg, 1752 Rocque; **BROMLEY LANE (FARM)** 1½ mile north-east of Armitage (SK 0916), *Bromley Lane* 1836 O.S.; **BROMLEY FARM, BROMLEY WOOD** 2 miles north-east of Hilderstone (SJ 9735): BROMLEY (unlocated, in Blymhill), *Bromley* 1332 SHC X (i) 123, *Bromley Hawk Bank, Bromley Hay Meadow* 1735 *Terrier*, BROMLEY (unlocated, on the border of Teddesley Hay), *Bromley* 1586 Oakden 1984: 122; see also SHC 4th Series XVIII 131. The spellings for Bromley near Upper Arley suggest a derivation from OE *bromig* 'broomy', with the other places derived from OE *brom* 'broom', with OE *leah* 'broomy clearing'. The first place has been in Worcestershire since 1895, and Bromley near Worfield has been in Shropshire since the 12th century. *Bromleymore*, recorded in 1385 (BCA MS3415/200), may have been near Beaudesert, and is perhaps to be associated with Bromley Lane (Farm) (q.v.). The Green element suggests a squatter settlement.

**BROMLEY, ABBOTS** Ancient Parish 6 miles south of Uttoxeter (SK 0824). *Bromleage* 942 (14th century, S.479), *Bromleg(e)* 996 (11th century, S.878), *Bromleage* 1004 (11th century, S.1536), *Brvnlege* 1086 DB, *Bromleia Abbatis* 1203 SHC III 107, *Bromleigh Abbatis* 1304 Ass, *Bromley the Abbottes* 1532 SHC 4th Series 8 100, *Pagetsbramley* 1577 Saxton, *Pagettesbromlye* c.1578 SHC 1926 80, *Heckley otherwise Heckeley Abbottes otherwise Pagettes Bromley* 1614 SHC NS IV 53, *Pagettes Bromley* 1616 SHC VI NS (i) 18, *Pagets Bromley formerly Bromley Abbots* 1749 Bowen. From OE *brom-leah* 'broom *leah*'. The place belonged to Burton Abbey, to which it was given in 1004 by Wulfric Spot. After the Dissolution it was granted to Sir William Paget, hence *Paget's*. Hart 1975: 105 corrects Sawyer's date for S.878 from 993 to 996, and suggests (p.91) that the reference in S.479 is to both Abbots Bromley and King's Bromley. *Heckley* in the 1614 form is Heatley (q.v.).

**BROMLEY, BAGOT'S** 1½ miles north-west of Abbots Bromley (SK 0626). *Brumley Bagod* 1290 SHC VI (i) 197, *Bromlega Bagot* ? 13th century SHC V (i) 71, *Bromlegh Bagot* 1314 SHC 1911 83, *Brompleye Bagot* 1325 ibid.' 103, *Bromlegha Bagod* 1338 SHC XI NS 188, *Bagottes Syde* 1532 SHC 4th Series 8 101, *Bromley Barcottes* 1586 SHC 1927 166. From OE *brom-leah* 'broom *leah*'. *Bagot's* is from the family who held the place. *Syde* is from OE *sid* 'side', ME 'slope of a hill'. It has been suggested that

152

Bagot's Bromley might be the place recorded in DB as *Bradelie* in Pirehill (SHC XI NS 11-12), but that name has also been associated with Bradley Green in Burslem (SHC 1916 169).

**BROMLEY, GERRARD'S** 1½ miles south-east of Ashley (SJ 7734). *Bramelie* 1086 DB, *Bromeley* 1339 SRO D(W)1082/L/10/3, *Bromley-in-halys* 15th century Duig, *Bromley* 1532 SHC 4th Series 8 103, *Bromley Gerards* 1695 Morden, *Gerrards Bromley* 1798 Yates, *Garrards Bromley* 1833 O.S. From OE *brom-leah* 'broom *leah*'. Gerrard is from Sir Gilbert Gerard who acquired the place in the time of Elizabeth I. *In halys* (hales) means 'in the meadows', or is a reference to its proximity to Hales, 4 miles to the west, to distinguish it from other Bromleys.

**BROMLEY HURST** 1 mile south-east of Abbots Bromley (SK 0822). *Herst* 1204 SHC III (i) 105, *Hurst* early 13th century SRO D(W)1721/3/14/2, *The Hurst* 1532 SHC 4th Series 8 101, *Bromley hurst* 1643 SHC 4th Series I 16. 'The hurst or wooded hill near Bromley'.

**BROMLEY, KING'S** Ancient Parish 5 miles north-west of Lichfield (SK 1216). *Bromlege, Bromle* 942 Duig, *Bromelei* 1086 DB, *Bramlea Reg[is]* 1166 SHC 1923 295, *Bromley Regis alias Bromley Corbet* 1358 SHC 1913 322, *Bromley the Kynge* 1532 SHC 4th Series 8 108. For Bromley see Abbotts Bromley. *King's* records the holding of the manor by the king at Domesday and afterwards. *Corbet* is from Roger Corbet who held the place in 1358. The chronicler John ('Florence') of Worcester (d.1118) records that earl Leofric died in 1057 at *Bromleage*, which has been identified with this place: SHC 1916 134, Sawyer 1979a: xxix.

BROMLEY, PAGET'S – see **BROMLEY, ABBOT'S**.

**BROMLEY WOOD** east of Abbots Bromley (SK 0924). *Bromley Wood* 1836 O.S. The first edition 1" O.S. map of 1836 shows two places of this name, one almost a mile east of Abbots Bromley, the other almost a mile to the south-east of the first place.

BROMLOW EDGE (unlocated, probably to the north of Winkhill near the river Hamps) *Bromlow Edge* c.1753 SRO D260/M/T/4/89. Perhaps from OE *brom* 'a thornbush or thorny shrub', with OE *hlaw* 'mound, tumulus', so 'the burial mound with the thorny bush', with OE *ecg* 'edge', used in place-names in the sense 'the crest of a sharply pointed ridge, a steep hill or hillside'. Remembered in Bromleyhedge Lane between Winkhill and Waterfall Cross.

**BROMWICH, WEST** Ancient Parish 6 miles west of Birmingham (SO 0091). *Bromwic* 1086 DB (listed under Northamptonshire), 1199 SHC III 49, *Bramwic* 1152x1159 EEA 14 75, *Bramvic* 1203 SHC III 123, *Bramwic* 1224 SHC IV 31, *Bramwys* 1275 SHC VI (i) 56, *Brumwyc* 1275 ibid. 58, *Bromwic, Bromwig, Bramwic, West Bromwich, West Bromwych, Bromwych* 12th and 13th century Duig, *Westbromwich* 1322 Ipm. From OE *brom* 'thorny bush or shrub, furze, gorse, bramble, brier', hence '*wic* on the heath' or '*wic* where broom grew', or even '*wic* where broom was worked or traded': the element *wic* had several meanings, including 'a building for a particular occupation'. The place was becoming known as West Bromwich by the early 14th century to distinguish it from Castle Bromwich (which had no castle, but was named after the de Castello family: SHC V (i) 78) and Little Bromwich, both in Aston (Warwickshire).

BRONCOTT (obsolete) on east side of Upper Hulme (SK 0160). *Bromkote* 1282-3 SHC XI NS 257, *Broncott* 1299 VCH VII 32, *Bramcote* 1302, 1580 FF, *Bromkot* 1307 Ipm,

*Bramcott* 1620 FF, *Broncote* 1591 FF, *Broncott* 1620 FF, *Bramcott* 1842 O.S. From OE *brom cot* 'the cottage where broom grew'. See also Brancote.

**BROOKEND** on the south-east side of Combridge (SK 0937). *Brookend* 1695 Morden, 1836 O.S. Self-explanatory.

**BROOK (FARM)** 1 mile north-west of Audley (SJ 7851). *Brooke* 1589 Audley ParReg. From OE *broc* 'a stream, a brook'.

**BROOK HOUSE (FARM)** 1 mile west of Marchington (SK 1030), *Brook House* 1674 SA 513/2/18/19/15; BROOK HOUSE (obsolete, south of Brindley Ford (SJ 8854)), *Brochouse* 1326 JNSFC LIX 1924-5 38, *Brockehouse* 1609 JNSFC LXIII 63, *Brook House* 1836 O.S. Probably 'brook house' rather than from OE *brocc* badger', since the latter element is less common and both places lie on a stream.

**BROOKHOUSE** 1 mile south of Bagnall (SJ 9147). *Brookhowse* 1586 SHC 1927 172, 1598 SHC 1935 79, *Brookhouse* 1704 Stoke on Trent ParReg. The 1836 O.S. map shows *Little Brookhouse*, *Big Brookhouse*, *Brookhouse Green* and *Brookhouse* spread along the stream running east to west to the south of Bagnall.

**BROOKHOUSES** 1 mile south-east of Dilhorne (SJ 9942). *Brookhowses* 1598 SHC XVI 185, *Brockhowses* 1610 SHC III NS 52, *Brookhouses* 1698 SRO D1229/2/2/7. Self-explanatory. The place lies on the river Tean.

**BROOM(E)** Ancient Parish 4 miles south of Stourbridge (SO 9078). *Brome* 1169 P, *Broome* 1343 PN Wo 278, *Brome* 1379 SHC XV 32. From OE *brom*, originally meaning 'a thorny bush or shrub', and applied to furze, gorse, bramble, etc. Since 1844 in Worcestershire. The O.S. shows Broom as the name of the village, Broome as the parish; see also Youngs 1991: 474.

BROOM EDGE (obsolete) 1 mile south-west of Hollinsclough (SK 0565). *Brooms Edge* 1687 Alstonefield ParReg, *Broom Edge* 1840 O.S. From OE *brom* 'a thornbush or shrub', with OE *ecg* 'edge', used in place-names in the sense 'the crest of a sharply pointed ridge, a steep hill or hillside'.

**BROOM HALL** 1 mile north-west of Brewood (SJ 8710), *Bromhale* 1149x1155 (17th century) EEA 17 62, *Bromhale* 1266 SHC 1913 317, *Brumhale* 1306 SHC VII 159, *Bromhale* 1332 SHC X 121, *Bromhaste* 1423 SHC XVII 90, *villa de Brome* c.1538 AOMB, *Brumhall* 1664 Brewood ParReg; **BROOMHALL GRANGE** on the Staffordshire-Shropshire border, east of Market Drayton (SJ 6834), *Bromhale* 1166 SHC I 227, c.1177 Rees 1997: 55, 1266 SHC 1945-6 317, 1271 Eyton 1854-60: IX 193, *Brumehull* c.1300 SHC 1913 251, *Bromhall* c.1300 SHC 1945-6 30, *Broomehall alias Broomewell grange* 1600 SHC 1945-6 123, *Broomer Grange* 1775 Yates, *Broomhall Grange* 1833 O.S. 'The *halh* where broom grows', a common name. The second element has become confused, as often happens, with OE *heall* 'hall', an element rarely found in Staffordshire. *Bromhull* in Newton is recorded in 1299 (SRO D938/553), *Bromeshall* (unlocated) in Checkley is recorded in 1562 (SHC 1938 127), and may be *Bromale* recorded in 1377 (SHC 4th Series VI 12), *Brumeshel'* in 1195 (Pipe Rolls); *Bromehill* is recorded in Seabridge/Clayton in 1162 (Pape 1928: 98); *Bromhal, Bromhullewey* in Chebsey are recorded in 1326 (SRO 3764/28[27576]); and *le Bromhull* is recorded in Lichfield *temp.* Edward II (SHC 1939 94), perhaps the same place as *Bromihyll* in Streethay recorded in 1470 (SHC 1939 122).

**BROUGH HALL** 1½ miles north-east of Gnosall (SJ 8322). *Buchale* 1086 DB, *Burgh* c.1182 SHC II 256, *Burgo, Burht* 1199 VCH IV 125, *? Bure* 1208 SHC III 173, *?*

*Ourbure*, *Ourburg* 1213 SHC III 161-2, *? Burre* c.1233 Rees 1997: 82, *Burgo extra Romtona* 1262 SHC 4th Series XVIII 67, *Burgo extra Ronton* 1271 SHC V (i) 142, *Burgo* c.1280 SHC 1928 107, *? Burth, Burgh* 1281 SHC VI (i) 121, *Burgh'* 1283 Ass, *? Burgh* 1285 SHC VIII (ii) 27, *? Burgo* 1290 SHC 1911 198, *? Bourgh* 1326 SHC X 75, *? Burgh* 1333 SHC X 125, *Bourgh* 1346 VCH IV 125, *Over Bourgh* 1393 ibid.' 126, *Burgh* 1417 VCH IV 126, *? Burby* ?1434 SRO DW1721/1/363-385, *Burg'* 1472 SHC 1912 223, *Burrowehall* 1537 MinA, *Brough Hall* 1550 VCH IV 126, *Broughall* 1578 SHC IX NS 208, *Borrough hall* 1616 SHC VI NS (i) 21. From OE *burh*, usually meaning 'fortification, fortified place, manor', here probably 'the fortification': JNSFC XXXVI 1901-2 118 states 'Brough Hall is a strongly entrenched position on high ground ... connected with Ranton by a remarkable earthwork or vallum about 25' wide, traces of which are also to be seen in the wood to the north of Ranton Abbey on the way to Ellenhall', such earthwork (presumably the 'vallum or raised road in the neightbourhood of Ranton Abbey' mentioned in VCH I 186) perhaps the *Wal* (from OE *w(e)all* 'a wall, a rampart of earth or stone') recorded in 1213 (SHC III 161), or *Burgo extra Ronton* recorded in 1271 (SHC V (i) 142), or *Netherburgh* recorded in 1471 (SHC NS IV 175). The second element is OE *halh*. It appears that the place was also known as Over Burh 'Upper *burh*', or perhaps 'the place above the earthworks', to be distinguished from Nether burh 'Lower *burh*', perhaps 'the place below the earthworks'. For the identification of Buchale (the manuscript form in DB may be read as *Bughale*, as in VCH IV 49, following Shaw 1798: viii, *xi, Eyton 1881: 31, 98-9, SHC 1916 167 and SHC 1919 158) with this place see SHC I 229; SHC 1919 158; VCH IV 49; also SHC II 119-20. The 1281 spellings may relate to Berth Hill (q.v.).

**BROUGHTON** 1½ miles south-east of Claverley (SO 8091). *Burgton* 1194 P, 1292 Ipm, *Burton* 1300 SHC 1911 266, *Bureton* 1212 Fees, *Burton* 1296 SHC 1911 266, *Borughton* 1327 SHC VII 211, *Burgh'ton* 1327 SR, *Borouhton* 1400 SA 52/9, *Burghton* 1516 (1801) Shaw II 208, *Broghton* 1525 SR, *Broughton* 1752 Rocque. From OE *burh-tun* '*tun* at the fortified place' or 'fortified *tun*' or '*tun* at the manor-house'. *Burh-* became *Bruh-* owing to metathesis. No fortification has been recorded here, but the place (first recorded in 1191: PN Sa I 40) lies on the course of a suspected Roman road from Greensforge to Central Wales: see TSAS LVI 1957-60 237. See also PN Sa I 38-41. In Shropshire since the 12th century.

**BROUGHTON** 6 miles north-west of Eccleshall (SJ 7633; SMR 02399 places a deserted post-Conquest settlement at SJ 76713380). *Hereborgestone* 1086 DB, *Borchton* 1258 SHC 1914 31, *Borton* 1275 ibid.' 29, *Burghton* 1281 Ass, *Burghton, Burgton* 1311 SHC 1914 30-1, *Burhton* 1314 SRO D939/39, *Borghton, Boroughton* 1327 SHC VII (i) 211, *Bructon* 1339 SRO D(W)1082/L/10/3, *Burghton* 1472 SHC NS IV 173, *Broughton* c.1560 SHC IX NS 130. Although Erdeswick 1844: 544 believed the DB spelling referred to Horsley, historians (e.g. Eyton 1881: Table VI, VCH IV 42) are satisfied that the name is connected with Broughton. Most Broughtons derive from *Brocton*, but in this case the DB form points to OE 'Hereburh's *tun*' (the personal name is feminine), although one would not expect the medial *-s-*, and the absence of the first part of the DB form in later spellings suggests that the name may in fact be from OE *here* 'army', with OE *burg* 'fortified place, manor', emphasising a special military connection (see JNSFC 21 1981 18), perhaps the site of a defended military post created during the political expansion of Mercia in the early 7th or 8th century (Gelling 1981: 5-7, 17-18; Studd 1993: 56), although no earthworks or other archaeological features appear to have been recorded here, and a derivation from OE *here-beorg* 'shelter or protection for a number of men; army quarters' (EPNE i 244) seems more likely. The second element is OE *tun*. *Hereburge byrig* mentioned in the foundation deed of Burton Abbey and the will of

Wulfric Spot c.1004 (11th century, S.1536) almost certainly refers to Harbury, Warwickshire: SHC 1916 38; Whitelock 1955: 543; Sawyer 1979a: xxxiii.

BROUGHTON (obsolete, in Longdon Green (SK 0813)) 'Brocton or Broughton Hall is situated about one furlong North-east of the turnpike road at Longdon green, and near to Liswis-hall': Shaw 1798: I 226; see also VCH V 6; SHC 1921 7. In about 1596 it was stated that the place (*Brocton*) lay next to Arblaster: Erdeswick 1844: 245. *Broughton temp.* King John (1798) Shaw I 226, *Brocton* c.1308 ibid. 148, *Broughton* 1316 ibid, *Brocton* 1403 ibid, *Broughton or Brocton* 1798 ibid.' Most modern Broughtons are from OE *broc tun* '*tun* at the brook', and the spellings support such a derivation here. Erdeswick 1844: 244 mentions a brook, and Brook House appears on modern maps to the east of Lysways Hall. Some Broughtons are from OE *burh-tun* '*tun* at the fortified place', with *Burh-* becoming *Bruh-* owing to metathesis (cf. Broughton near Eccleshall), and an ancient earthwork, Longdon Camp, is recorded near Longdon church (see VCH I 346; SHC 1916 207), with Borough Lane, a road running south-west from Longdon, shown as *Burrough Lane* (perhaps from OE *burh* 'fortification') on Yates' map of 1775. (*Burrowfield* and *Burrowfield Meadowe* are recorded in Longdon in 1661 (WCRO CR1908/23), and *Burrow Lane* appears as a place on the south side of Upper Longdon (SK 062141) on the first edition 1" O.S. map of 1834). However, the medieval spellings with *Broc-* show that a derivation of this place-name from *burh* can be ruled out, although the place may be associated with the unlocated DB *Burouestone* (q.v.), which might indeed relate to Longdon, which is recorded before 1086 but not in DB itself. A possible moat to the east of the Moat House, some 400 yards west-south-west of Longdon has been tentatively identified as the site of Broughton Hall: TSSAHS XXIV 1982-3 44.

BROWN BANK 2 miles east of Cheadle (SK 03422). Possibly to be associated with *Le Brounesyde*, recorded in 1330 (CroxdenChr).

BROWN BROOK a tributary of the river Dove. Self explanatory

BROWN EDGE 2 miles south-east of Biddulph (SJ 5390), *Browenage* 1298 SHC XI NS 248, *Browene Edge* 1572 SHC XIII 287, *Browne edge* 1572 BCA MS917/1515, *Brome Edge otherwise Browne Edge* 1601 SHC XVI 212, *Brown Edge* 1630 JNSFC LXII 1926-7 74, 1695 Morden, 1747 Bowen, 1836 O.S.; **BROWNEDGE** 2 miles south-east of Bradnop (SK 0252), *Brown Edge* 1836 O.S. Perhaps from the colour of the vegetation at these places, both of which are on high ground which slopes steeply, although ON *brún* 'a brow, the edge of a hill, a moor' may be found in some Brown Edge names, giving 'the edge of the brow called Brun': see VEPN II 49.

BROWNHILLS 5 miles north-east of Walsall (SK 0405), *Browne Hills* 1683 Armitage ParReg, *Brownhill* 1686 Plot, *Brown Hill* 1747 Bowen, *The Brown Hills* 1834 O.S.; BROWNHILLS (obsolete) on the south side of Burslem (SJ 8650), *Brownhills* 1590 Ward 1843: 152, *Brown Hills* 1765 SRO SD4842/14/1/1, *Brownhills* 1836 O.S. Perhaps from the colour of the soil or vegetation, possibly bracken. *Brownhills* in the Callow Hill area is recorded c.1310 (SRO D(W)1721/3/31/1), and *Brownehill* in Streethay in 1604 (SHC 1940 205).

BROWN LEES on south-west side of Biddulph (SJ 8756). *Brown Leise* 1596 SRO D1229/1/3/18, *Brownlees* 1637 Wolstanton ParReg, *Brown Lees* 1836 O.S. From OE *lǣs* 'meadow, pastureland'.

BROWNSETT ½ mile north of Roche Grange (SJ 9963). *Brownesford* 1608 SHC 1948-9 118, SRO D538/A/2/4, 1631 SHC II (ii) 15, *Brownsford* 1634 Leek ParReg, *Bromsford*

1636 Stoke on Trent ParReg, *Brownsort* 1666 Leek ParReg, *Brownsott* 1668 ibid, *Bronsot* 1669 ibid, *Braunsote* 1673 ibid, *Bromsott* 1680 ibid, *Brownsword* 1675 Leek ParReg, *Brownsort* 1680 LJRO B/C/11 Edmund Brough. The name is evidently from 'Brown's ford', with the *-f-* dropped from the difficult consonantal cluster *-nsf-*, and the later confusion of the second element with 'sword', reduced to 'sett'.

**BROWNSFIELDS** 1 mile north-east of Lichfield (SK 1210). *Brounfeldes* c.1535 SHC II (ii) 166, *? Brownefeldes otherwise Brownes feldes* 1562 SHC XIII 221, *Browns Fields* 1834 O.S. Evidently from the Brown family of Lichfield, who held land here from the 1440s: VCH XIV 278.

**BROWN'S GREEN** on the north side of Hamstead (SO 0491). *Brown's Green* 1695 Morden, 1749 Bowen. Probably from Richard Browne who held a croft here in 1538: StEnc 96. The Green element suggests a squatter settlement.

**BROWNSHORE** in Essington (SJ 9603). *Upper Broune Share* 1685 *Vernon, Broune shares* 1665 ibid, *Brownshaw* 1834 O.S. From OE *brun* 'brown', or the name Brown, with OE *sc(e)aru* 'a share of land, a share of the common land'.

**BROWNSPIT** 2 miles north-west of Sheen (SK 0863). *Broun(e)spit(e)* 1599, 1604 ParReg, *Brounds Pit* 1656 Leek ParReg, *Brounespitt* 1626, 1633 Rental, *Browne spitte* 1651 ibid, *Brown Spit* 1775 Yates. Probably 'the dark-coloured spit or tongue of land', rather than 'Brown's spit or tongue of land': the place lies on a blunt headland on the side of a stream valley.

**BRUND** 1 mile west of Sheen (SK 1061), *Brunde* 1598 SHC 1935 147, *Brund* 1840 O.S.; BRUND, UPPER & LOWER (obsolete) 1 mile south-west of Cheddleton (SJ 9651), *? Broned* 1695 Leek ParReg, *Brund* 1729 ParReg, *Lower Brund* 1837 O.S. From ME *brend* 'burnt', generally in the sense 'cleared by burning' (as of a wood, field, etc.), or sometimes 'consumed by fire' (as of a house or village); the element is frequently found in minor names and field-names. Cf. Barnhurst; Burntwood.

**BRUND HAYS** 1 mile north-west of Butterton (SK 0557). *Brand Hays* 1775 ParReg, *Brund Hees* 1778 ibid, *Burndheys* 1880 Kelly. From OE *brand* 'fire, flame; place where burning has occurred', with OE *\*hǣs* 'brushwood', so 'the place where brushwood has been cleared by burning'.

**BRYAN'S HAY** ½ mile south-west of Longsdon (SJ 9554). *Brianshay (House)* 1613 SRO DW1702/8/21, *Bryan's Hay* 1815 *EnclA, Brineshay (House)* 1837 O.S. Perhaps from the ME personal names Brian, with Mercian OE *(ge)heg* 'a fence, an enclosure'.

BRYNGHULL (unlocated, in or near Wednesbury) *Brynghul* 1327 SHC V (ii) 229, *Brynghull* 1332 SHC X 85. Possibly from OE *bringe* 'that which is brought, an offering, a sacrifice', in some uncertain sense, though the association of the name Wednesbury (q.v.) with the god Woden (and possible shrine to the god) should be noted. It is uncertain whether *Brichull*, recorded in 1307 (SHC 1911 287), is to be associated with this place.

BUBELENHAY (unlocated, possibly near Horninglow) *Bubelenhay* 1248 (1798) Shaw I 23.

**BUCKNALL** (pronounced locally as Buckner [bʌknə]) 1½ miles east of Hanley, in Stoke-on-Trent (SJ 9047). *Bvchenole* 1086 DB, *Buchenhale* late 12th century Rees 1997: 42, *Buccenhal* 1227 Ch, *Becchenhale* 1253 SHC 1911 123, *Buccanhal* 1274 SHC 1911 161, *Boukkunhall cum Bydoulf* 1284 SHC 1910 299, *Bukenhale, Bokenhowe* 13th century, *Bockenhale* 1327 SHC 1913 8, *Bokenhale* 1447 HLS. From the OE personal name Bucca, or from OE *bucca* (*buccan* genitive singular, *buccena* genitive plural) 'buck, he-goat' (see VEPN II 61-2. The second element is more difficult, possibly OE *halh*. The element *hall* is extremely rare in pre-Conquest place-names. There were two manors in Bucknall, Bucknall cum Bagnall and Bucknall Eaves *(? Eves* 1586 SHC 1927 172; *Buckenhall Eves, Bucknall Eves* 1591 SHC XVI 110): SHC VI (i) 71 fn; SHC XII NS 32). See also Bukkeleg, Bukkelyh.

**BUDDILEIGH** ½ mile north-west of Betley (SJ 7449). *? Bothilih, Bothilige* 1227 Ch, *Buderlea* 1654 Betley ParReg, *Booddeleighe* 1607 ibid, *Buderlea* 1654 ibid, *Buddyley* 1798 Yates. The forms are too disparate for any derivation to be offered.

**BUGHOLE** on north-west side of Darlaston Green (SO 9797). *Bug Hole* 1775 Yates. The name survives in Bughole Bridge, over the Walsall Canal. Probably from ME *bugge* 'boggart, goblin', so 'the goblin's hole or hollow'. See SHC 4th Series VI 174.

**BUGLAWE** (unlocated, possibly in the north-west of the county) *Buglawe* 1478 SHC VI NS (i) 115. Probably from ME *bugge* 'boggart, goblin', with OE *hlaw* 'mound, tumulus', so 'the goblin's tumulus'. It is difficult to believe that this place is not to be associated with Buglawton in Cheshire, 1½ miles north-east of The Cloud, and within a mile or so of the Staffordshire border, but the context in which the name appears suggests that Buglawe lay in Staffordshire. A field-name Buglaw is recorded in Cheshire: PN Ch I 191; VEPN II 63-4.

**BUGSDALE** (obsolete) 2 miles north-west of Whitgreave (SJ 8930). *Bugsdale* 1890 O.S. Perhaps from the ON personal byname Buggi, with OE *dæl* 'valley'.

**BUKKELEG, BUKKELYH** (unlocated, perhaps near Bucknall) *Bukkeleg* 1223 Hulton, *Bukkeley* 1223 VCH III 235, *(a lawn called) Bukkelyh* 1227 Ch. Perhaps to be associated with Bokeleg (q.v.), and *Bulkylegh*, recorded in 1358 (SHC XII (i) 149). The name may be from OE *bucca* 'he-goat, male deer', or a personal nickname Bucca from the same word: DES 70.

BUKSTALLES (unlocated, possibly in the Tatenhill area) *Bukstalles* 1424 Hardy 1908: 113. From a ME word meaning 'deer traps'. See also Borkestalles.

**BULL BRIDGE** on the north side of Penkridge, over the river Penk (SJ 9214). *(le) Bold(e) brugge* 1375 Vernon, *the Bouldbridge* 1587 Extent, *Bull Bridge* 1754 Plan, *Bole Bridge* 1749 Bowen. From OE *bold* 'dwelling place', with OE *brycg* 'bridge'. *le Bold(e)* (p) is mentioned from 1327 (SR), and was almost certainly *le Bolde*, possibly near Dunston, mentioned in 1342 and 1547-8: SHC 1950-1 25; 1589 SHC 1934 20. Cf. Booden Farm.

**BULLCLOUGH** 1½ miles west of Grindon (SK 0554). *Bull's Clough* 1635 D694/1-6/1, *Bull Clough* 1775 Yates. From OE *cloh* 'a ravine or narrow valley', so 'the narrow valley with the bull', or from a personal name such as *Bula.

**BULL GAP** on the west side of Swinscoe (SK 8048). *Bull Gap* 1635 Ellastone ParReg, 1728 SOT D240/D/269, 1836 O.S. The place lies in a deep stream-valley.

**BULLMOOR (LANE)** the lane running north-west from Chesterfield near Shenstone (SK 0905). *Bolmore* 1274-5 SHC 1923 275, 1316 SHC 4th Series XVIII 200, *Bulmore* 1549 SHC VI (ii) 190, *Churche Bomers, Bulmers otherwise Church Bulmers* 1564 SHC 1931 145-6, *Bulmores* 1616 ibid. 196, *the Bulmores* 1646 SHC VI (ii) 196, *Bullmoor Lane* 1834 O.S. Possibly from OE **bol* 'a smooth, rounded hill' (a rare element in place-names), but much more likely to be from OE *bula* 'bull', with OE *mor* 'moorland, marsh', so 'the marsh or moorland with the bull'. The *Churche* element is unexplained.

**BULWARDINE** 1½ miles south-east of Claverley (SO 9991). *Bulewardin* 1228 Eyton 1854-60: III 100, *Bolewardyn* 1326 ibid.' 93, *Bollewardyn* 1410 Peace Roll, *Bulwardine* 1833 O.S. From OE *bula* 'a bull', with OE *worþign* 'an enclosure', which is often found as -*wardine* in the West Midlands, for example Ellardine, Ingardine, Leintwardine, Ridgwardine, Shrawardine, Stanwardine, Wrockwardine (all in Shropshire), Bredwardine, Lugwardine, Pedwardine (Herefordshire). In Shropshire since the 12th century.

**BUNBURY HILL** north of Alton (SK 0742). *Bunbury* 1686 Plot 410, 1798 Shaw I 36, *Banbury (Wood)* 1836 O.S., *Bunbury* 1849 Lewis 1849: I 52. Perhaps 'Buna's *burh*', from OE *burh* 'fortified place', from the iron-age hillfort on the hill: see VCH I 334; StEnc 98. The place is by tradition associated with a battle in 716 A.D. between Ceolred, King of Mercia, and Ine, King of the West Saxons, at nearby Slain Hollow (q.v.). The tradition appears to date from John Brompton's '*Chronicon*', an untrustworthy compendium of material based on other authors, compiled c.1437, which refers to a battle at *Bonebury* (cited in Plot 1686: 410; Shaw 1798: I 36-7). Although early O.S. maps mark a battle here in 716 A.D., there is no evidence that such battle ever took place. Brompton's material is evidently a corrupt account of the only battle between Ine and Ceolred recorded in ASC, which occurred in 715 at a place now held to be Adam's Grave in Wiltshire: Earle & Plummer 1892-9: i 42, ii 38; PN W 318; Swanton 1996: 42; Ashley 1998: 309. Cf. Bunbury, Cheshire. See also Ina's Rock; Yornburi.

**BUNSTER HILL** 1 mile north-east of Ilam (SK 1451). *Bonster(re)* 1542 StarCh, *Bunsterre* 1542 SHC 1912 144, *Bonster* 1542 ibid. 144, *Bunster* 1543 SHC XI 285, c.1579 Okeover E418 & F18, 1836 O.S.; *Bounster* 1586 SHC XV 169. Perhaps from the OE personal name *Bunt (Searle 1897: 120; DES 73), with OE *stæger* 'stair', perhaps used in a topographical sense of 'stepped hill', or possibly *torr* 'a rock, a rocky peak', a word generally found only in the South West and the West Midlands. The name *Bunt is found only in place-names, including Bonsall, Derbyshire (*Bunteshale* DB, PN Db 345), as well as Birchgrove, Sussex (PN Sx 271), Bountisborough Hundred (Hampshire), and the weak form in Benton, Devon (PN D 30).

**BURBROOK, LITTLE** 1 mile north of Seisdon (SO 8396)) The age of this name is not known, but the place lies on Smestow Brook (known in its upper reaches as Black Brook). The name, if ancient, could be from OE *burh* 'manor, fortification'. No earthwork or other fortification is recorded in the area, but a quadrangular cropmark is recorded at SO 84029682 (SMR 04025), and another at SO 83689654 (SMR 04045).

**BURCOTE** 1 mile south-west of Worfield (SO 7495). *Burechote* 1086 Eyton 1854-60: I 109, *Burchota* 1176 P, *Burcote* 1283 Eyton 1854-60: III 113, *Burkot* 13th century Misc, *Wurcote (sic)* 1301 Rees 1975: 249, *Burcott, Burcot* 1562-3 Worfield ParReg, *borcote* c.1575 Worfield TSAS 3rd Series X 65. From OE *bur-cot* 'dwelling-place, cottage', or, perhaps more likely, since the place was a component part of a large estate, OE *burh-cot* 'cot belonging to a fortification or manor'. The place has been in Shropshire since the 12th century. Other Burcot(e)s are recorded in Wrockwardine, Shropshire; Homer, Herefordshire; Bierton and Wing, Berkshire; Bromsgrove, Worcestershire; and one in Somerset: see PN Sa III 75.

BURDON MILL (unlocated, possibly near The Wergs or Dam Mill) *Bordensmulne* 1381 SHC VI NS (ii) 153, *Burdon Mill* 1441 ibid. 207, 1463 ibid. 208. From a personal name Burden: Richard Burden of Tettenhall is recorded in 1286 (SHC 4th Series XVIII 156), perhaps the person of the same name recorded in 1315 (SHC IX (i) 50).

**BURF CASTLE** an ancient 1½ acre double-banked enclosure 1½ miles east of Quatford (SO 7690). *Burf Castle* 1833 O.S., 1840 TA. Almost certainly from OE *burh* 'fortified place', with the common dialectal confusion between *f* and *th* (see The Berth near Maer). The suggestion in Lias 1991: 85 that in some cases places with this element may take their name from Welsh *perth* 'thornbush, thicket', on the basis that early earthworks may have succeeded strongholds defended by thorn hedges alone, or the defences strengthened by thorn hedges, is highly unlikely. The *castle* element is post-Conquest and was often applied to earthworks. In Shropshire since the 12th century. The place has sometimes been recorded as *Burn Castle* according to Watkins-Pitchford 1937. See also VCH Sa I 379-80.

BURFORD (unlocated, possibly in Coppenhall) *Burford* 1601 SHC 1935 368. Seemingly from OE *burh* 'fortified place', so 'ford at the earthworks or fortification'.

BURGHANDESLEGH (unlocated, possibly in the Eccleshall area, but probably Broseley in Shropshire: see PN Sa I 63-4) *Burwardelega* 1227 SHC IV 56, *Burwardele* 1232 SHC IV 80, *Burgardeslee* c.1233 Rees 1997: 82, *Burghandeslegh* 1326 SHC IX (i) 113. A name of archaeological interest, since it would appear to be 'the *leah* of the *burgweard* or fort-guardian'. The 1326 form is evidently a mistranscription.

BURHALL (obsolete) near Ewdness. A field-name (cf. *Burhull* 1283 Eyton 1854-60: III 112) recorded in the 18th century (Shropshire SMR), possibly associated with cropmarks of a complex of irregular enclosures at SO 727983 in a field known as *Burhall* in the 18th century (ibid).

BURLAKE HEAD (unlocated, on the river Trent, perhaps near Alrewas) *Burlake* 1341 SHC 4th Series XX 81, *Burlake Head* 1510 Dent & Hill 1896: 115, *Burlake* 1613 Alrewas ParReg.

**BURLAUGHTON** on Watling Street, 1 mile south-east of Sheriffhales (SJ 7711). *Brerelectone* 1166 Red Book of the Exchequer, *Breclleton, Brellet', Brellect'* 1215 Rees 1997: 28, *Brereletona, Brerlatton* 1217 (1285) Ch, *Brelatthon* c.1236 Rees 1997: 123, *Brerlatton* 1265 ibid. 136, *Brerlahton'* 1271-2 Ass, *? Brelaughton* 1346 SHC 1921 184, *Burlaghton* 1532 SHC 4th Series 8 99, *Burlaughton* c.1565 SHC 1938 196, *Burlaton* 1577 Saxton, *Burlawton* 1608 SHC 1948-9 44, *Burlauton* 1686 Plot, *Burlington* 1783 ParReg, *Burlaughton commonly called Burlington* 1909 ParReg. Although Roman forts have been found here (see JRS lix 104; lxii 223-4), which might suggest a derivation from OE *burh* 'fortification', the earliest spellings show that the name is from OE *leac-tun*, literally 'leek enclosure', meaning 'herb garden', and later 'kitchen garden, vegetable garden', with OE *brer* 'brier'. Since 1895 in Shropshire. Burlaughton was the name applied to the area to the north of Watling Street, Burlington (q.v.) to the south. Both share the same derivation. The O.S. now shows both places as Burlington.

**BURLEYFIELDS** 1 mile west of Stafford (SJ 9023). *? Burley* 1275 SHC VIII 132, *Burleighton nigh Stafford Castle* 1705 Penkridge ParReg, *Burley field* 1722 SRO D641/2/D/1/2, *Burley Fields* 1775 Yates, *Burley Fields (Farm)* 1788 VCH V 83, *Birley Fields* 1836 O.S. From OE *burh* 'fortification', with OE *leah*: the place lies on the north side of Stafford Castle, which may have been built on the site of a prehistoric hill-fort.

BURLEYS, THE (unlocated, in Chesterton near Newcastle under Lyme) *Burlegh* 1227 SHC XI NS 240, *The Burleys* 1767 Heathcote MSS Box XIII. From OE *burh* 'fortification' (cf. Chesterton), with OE *leah*. *Burley Cottage* is ½ mile north-east of Podmore (SJ 7836), but early spellings have not been traced.

**BURLINGTON** 1 mile south-east of Sheriffhales (SJ 7710). *Burlington or Burlaughton* 1833 O.S. The derivation is the same as Burlaughton (q.v.), of which this name is a variant. Since 1895 in Shropshire.

BURNAHAM, BORNAM (unlocated, possibly in the Stonnal/Shenstone/Bosses area, perhaps to be associated with Bourne Brook) *Bornam* 1209 SHC XVII 247, *Burnaha*, *Burnaham* 1209 SHC 1923 277. The spellings are variant transcriptions from the same deed. Seemingly from OE *burna* 'spring, brook, stream', generally denoting a more substantial stream or even a river, with OE *ham* 'a village, a manor, a homestead', here probably 'the homestead at the stream'. The place appears to have been in the Stonnal/Shenstone/Bosses area, perhaps to be associated with Bourne Brook, and close to a number of Roman sites centred on Wall and the junction of the Roman roads of Watling Street and Ryknild Street. It has been noted that *burna* is found combined with *ham* but not with *tun* (except in the Northumberland name Brunton), whereas OE *broc*, the word later used for 'a brook, a stream', is very often combined with *tun*, but never certainly with *ham*: EPNE i 224 fn.3; VEPN II 36-9.

**BURNDHURST MILL** on the river Blithe, 1 mile south-east of Gratwich (SK 0431). *Bournhurst Mill* 1798 Yates, *Burnt Hurst Mill* 1836 O.S., *Burndhurst Mills* 1851 White. The 1798 spelling suggests a possible derivation from OE *burna* 'spring, brook, stream', with OE *hyrst* 'copse, wooded eminence', perhaps from a minor watercourse in a nearby wood, but the combination of *burna* and *hyrst* seems improbable, so perhaps 'the burnt small wood'.

BURNED HILL (unlocated, in Rugeley) *Burndhill* 1623 WCRO 1908/3/1-26, *Burnt-hill* 1653 WCRO 1908/17/1, *Brundhill* 1730 WCRO 1908/6, *Burned Hill* 1732 SOT DW1885/1/2. From ME *brend hyll* 'the burnt hill'.

**BURNHILL GREEN** 2½ miles north-west of Pattingham, on the Staffordshire-Shropshire border (SJ 7800). *Byrnhill* 1279 SHC XI NS 133, *Birnhulle* 1309 SHC 1928 24, *Byrunhull* 1327 SHC VII (i) 251, *Brynhull* 1333 SHC X (i) 131, *Burnhull* 1334 (1801) Shaw II 281, *Birnhull* 1413 SHC XVII 46, *Burnell Grene* 1566 SHC 1931 132, *Burnell Greene* 1599 SHC 1930 90. It is possible that the name is from OE *byrgen-hyll* 'burial mound hill; hill with a tumulus' (cf. Burnhill, Buckinghamshire, *Burnhill* 1276, where there is a hill marked by a tumulus; Bernwood, Berkshire, *Byrnewuda* 921; and Burn, West Yorkshire, *Byrne* c.1030, *Birne* 1279-81). Although there is no tumulus recorded at Burnhill Green, in 1841 the curate of nearby Beckbury wrote in the burial register about 'several mounds' in the district, some of which may have been artificial, and fields north of Beckbury are said to have been called Golden hill and Urn field: VCH Sa X 240, but see PN Sa III 105. Smith 1990: 220 fn.2 suggests that the element *byrgen* was probably coined during during the Christian Anglo-Saxon period out of traditional folk-memory and antiquarian interest to describe heathen burial sites, but *byrgen* is well-recorded, and there seems no reason to believe it was restricted to heathen burials. The county boundary follows a marked deviation to include this place which would otherwise lie in Shropshire. See also Goldthorn. The Green element suggests a squatter settlement.

**BURNTOAK HOLLINS** 1 mile west of Meerbrook (SJ 9760). *Brandockholyns* c.1539 LRMB, *Burnt Oak Hollins* 1662 Leek ParReg, 1775 Yates, 1842 O.S. From ME *brend*, *ac* 'burnt oaks' with OE *holegn* 'holly', so 'the hollies at the place of the burnt oaks'.

**BURNT TREE** on the north-east side of Dudley (SO 9590). *Burnt Tree* 1749 WRO 1/1/354/34, *Burnt-Tree* 1851 White. Self-explanatory.

**BURNTWOOD** 3 miles north of Brownhills (SK 0609), *Brendewode* 1298 SRO DW1734/J2268 fl, *Brendwode* 1370 SRO DW1734/2/1/598, *Brendwodde* 1546 SHC 1939 111, *Brundwood* 1571 SRO DW1734/2/1/609 m19, 1600 SHC 1935 205, *Brandewodd* 1570 SHC XIII 277, *Brendwood, Brendewoode* 16th century Duig, *Burndwood* 1680 SHC 1919 268; **BURNT WOOD** 5 miles east of Market Drayton (SJ 7434), *The Brand Wood* 1567 SHC 1945-6 152, *Burnt Wood* 1712 SRO 828/236, 1833 O.S. From ME *brend* 'burnt', hence 'the burnt wood'. The first place is in the parish of Hammerwich, which lay within Cannock Forest. In 1262 a Forest jury determined that 'a certain heath was burnt by the vill of Hammerwich, to the injury of the King's game': VCH XIV 198. But fires, accidental and deliberate, within the Forest must have been frequent, and the names perhaps arose from the practice of clearing areas for agricultural use by burning. The spelling *Brundwood* became normal in the later 16th century, and the modern name by the 17th century: VCH XIV 198. For Burnt Wood near Market Drayton – which may have taken its name from charcoal burners who provided fuel for iron (and wood for glassworkers) here, and is said (StEnc 99) to have formerly been called *Rowney Wood*, then *Rounhay*, then *Brand* – see SHC 1945-6 24; 156. *Burnt Hill* in Rugeley is shown on the first edition 1" O.S. map 1834. Cf. Brandwood and Brantwood, Lancashire; Brentwood, Essex.

BUROUESTONE (unlocated) *Burouestone* 1086 DB. A puzzling name. The spelling strongly suggests that it is to be identified with Burston (q.v.) in Pirehill Hundred, as proposed in Duignan 1902: 28-9 and implied in Ekwall 1960: 77, and whilst that may be correct, it is recorded in DB under Offlow Hundred (SHC 1907 227-30; Derby & Terrett 1971: 173; VCH IV 43), whereas Burston is in Pirehill. Since it is listed in DB together with *Weforde* (Weeford) andf the unidentified *Litelbech*, historians have tended to locate it in the area around Weeford, and it may indeed be associated with Borrowcop Hill (q.v.), the name of which may share the same root (see SHC 1916 170), or Berryhill, south of Lichfield, but the three places in DB might have been listed together because they were held by the same person, and they may well have been some distance apart. Clues to the location of *Litelbech* and *Burouestone* are few. All three places were members of the manor of Lichfield. A possible contender for Burouestone may be Broughton (otherwise Brocton) (q.v.), in Longdon: Longdon was a prebend of Lichfield cathedral, possibly before DB (see SHC II 89-90; VCH III 141). Although Longdon is not mentioned in DB (which might also make it a contender for *Burouestone*), its existence is confirmed in earlier charters (11th century, S.906; 11th century, S.1536), and it was granted to the community at Tamworth in the Will of Wulfric Spott 1002x1004 (Whitelock 1955: 543). Plot 1686: 406 mentions traces of an ancient square earthwork (later known as Longdon Camp), on the east side of Longdon church (see also VCH I 346; StEnc 366), perhaps to be associated with Borough Lane, a road running south-west from Longdon shown as *Burrough Lane* (from OE *burh* 'fortification') on Yates' map of 1775, perhaps associated with *Burrow Lane*, marked as a place on the O.S. map of 1834 to the south of Upper Longdon, and *Burwaye Lane* recorded in 1506 (OSS 1936 51), although the *burh* may be Castle Ring (q.v.): see also Burwey; Longdon. Intriguingly, Milo de Beche (cf. Litelbech) is recorded in Longdon in 1189-90: SHC II 83. Another curiosity is the reference to an unidentified *Burestan* (which may be Burston) in the Armitage Parish Registers in 1814 and 1815: cf. *Burgeston* 1208 SHC III (i) 145, *Bureston* 1212 ibid. 160. Curborough (q.v.), which was a prebend of Lichfield cathedral, is also a possible location for *Burouestone*. The name *Burouestone* itself (if not corrupt, which is very possible) is unlikely to incorporate OE *stan* 'stone' (which would give -

*stan*), and the terminal is almost certainly OE *tun*, probably preceded by a personal name (perhaps Burgwine or Burgwulf: see Ekwall 1960: 77 *sub. nom.* Burston) with a genitive -*s*-. See also Litelbech.

**BURSLEM**   3 miles north of Stoke on Trent (SJ 8745). *Barcardeslim* 1086 DB, *Burewardesleg' Lime*, *Borewardeslyme* 1242 Fees, *Burewardeslime* 1252 Ch, *Borewardeslyme* 1363 SHC 1909 40, *? Burghashelam* 1423 SHC XVII 94, *Burseleym* 1485 SHC VI NS(i) 160, *Byrdyslyme, Burdeslyme, Bureslyme* 1539 MA, *Burselem* 1576 HLS, *Burslem* 1599 Smith. The first element is a contracted form (with the possessive *s*) of the OE personal name (or occupation) Burgweard meaning 'fort-watchman, fort-guardian': the DB spelling is evidently a scribal error. The second element is from Lyme Forest (see Lyme), which itself means literally 'elm place', and is found in several nearby places, including Newcastle under Lyme. The meaning is therefore 'estate in the district called Lyme belonging to a fort-guardian', or 'estate in the district called Lyme belonging to Burgweard'. Cf. Broseley, Burwarton and Treverward, Shropshire; Burwardesley, Cheshire; Buscot, Berkshire. See also Borewardes Croft.

**BURSNIPS**   ½ mile east of Essington (SJ 9703). *Birches nape* 1290 Vernon, *Bursnape* 1384 SRO D1790/A/6/3, *Birchenapps* 1543 SRO D593/A/2/1/15, *Burs(e)napps* 1556 to 1631 Vernon, *The Bursnips* 1834 O.S. Perhaps from OE *birce* 'birch tree', and OE *\*snæp*, probably meaning 'a boggy piece of land', hence 'the boggy land with birch-trees'.

**BURSTON**   4 miles south-east of Stone (SJ 9330). *Burgestun* 1188x1204 SRO D938/397, *Burexton'* 1235 Fees, *Bautharston* 1234-40 TestNev, *Bureweston* 1242 Fees, 1255 Ass, *Burcheston* 1278 Ass, *Burgheston* 1293 SHC 1913 193, *Burghstone* 1310 SHC X 7, *Burgheston* 1314 SHC 1911 81, *Borgheston* 1321 SHC 1913 205, *Bourgweston* 1326 SHC 1911 371, *Burnstone, Burneston'* 1333 SHC X 92, *Borwgheston* 1381 SHC XVII 181, *Bureston, Burweston* freq. 14th century Duig, *Boroostone* 1477 SHC VI NS (i) 108, *Borrowston* 1550 SHC XII 202, *Borowston* 1557 SHC XII 226, *Burston alias Burweston* 1621 SA 11/68, *Burston* 1749 Bowen, 1836 O.S. A difficult name. The first element may be from OE *burh* 'a fortification' with OE *west-tun*, so 'the fortification at the west *tun*' (perhaps here meaning west of Sandon), or even 'the fortification at the east *tun'*, or the first element may be an unidentified OE personal name (Ekwall 1960: 77, having associated *Burouestone* (q.v.) with this place, suggests Burgwine or Burgwulf) with OE *tun*.

BURTHEY       (unlocated, probably in the Lichfield area) *Burthey* 1297 SRO DW1734/J2268.

**BURTON**   in Castle Church parish, 2 miles south of Stafford (SJ 9122). *Burtone* 1086 DB, *Burton juxta Stafford* 1295 FF, *Burghton* 1480 Banco, *Burton* 1749 Bowen. From OE *burh* 'fortified place', with OE *tun*, meaning 'the farmstead by the fortification, the fortified manor, *tun* by or belonging to a *burh'*. Cf. Burton upon Trent, Broughton.

BURTONE Domesday Book (f.248v) identifies two different lost places named *Burtone*. The first is stated to be on the site of Henry de Ferrers' castle at Tutbury, perhaps a prehistoric earthwork, presumably subsumed when the castle was constructed (cf. a charter of 1012 of land at *Burtune*, which has not been identified: see Hart 1975: 241-4), and the location of the second place remains unknown: see Thorn 1991: 26, 40; StEnc 104. Both places are likely to have names deriving from OE *burh* 'fortified place', with OE *tun*, meaning 'the farmstead by the fortification, the fortified manor, *tun* by or belonging to a *burh'*. In connection with the second, a reference to *le bury in Stapenhull* 'the fortification in Stapenhill' (then in Derbyshire) in 1437 (*Burton*: PN Db 662; SHC

1937 166; GM 1794), possibly to be associated with *le Bere*, recorded in the early 14th century (Shaw 1798 I 6), may be noted.

**BURTON EXTRA** on the south side of Burton upon Trent (q.v): see Shaw 1798: I 11. *Burton Extra* 1595 SRO D1734, 1836 O.S. The suffix is from Latin *extra* 'without, outside', to denote parts of a parish outside the bounds of a town. White 1834: 319 suggests that the place was also known as *Bond-End* (*Bondende* 1373 SRO D1734, *le Bondhyerd* 1454 Underhill 1976: 68, *le Bondende* c.1500 ibid.'154, *Bondende* 1554 SHC XII 194), probably from ME *bond*, derived from ON *bondi* 'a peasant landowner': Joh'e le Bonde and Will'o Boonde are recorded in Rolleston in 1327 (SHC VII (i) 227). However, both Bond End and Burton Extra appear on the 1877 O.S. map. See also Bond End.

BURTON HALL (unlocated, in Tutbury) Shaw 1801: II 56 records *Burton hall, Burton Holmes* and *Burton Bache*, named from the Burton family of Leicestershire.

**BURTON UPON TRENT** Ancient Parish on the eastern edge of Staffordshire on the river Trent (SK 2423). *Byrtun* 1004 (11th century, S.906; 11th century, S.1536); 1012 (14th century, S.929), *Bertone, Bvrtone* 1086 DB, *Birtune* 11th century Sawyer 1979a: xxxv, *Burton super Trente* 1234 Ep. Burton is a very common place-name, and in most cases derives from OE *burh-tun*, but occasionally (as here) from OE *byrh*, genitive singular of *burh* 'stronghold, fortification, fortified manor', hence '*tun* by or belonging to a *burh*'. The *burh* may well have been nearby Tutbury, which is actually recorded as *Burtone* in DB, but probably as a result of a misreading of Burg- as Burt-: see VCH IV 48 fn.7; but see also Burton Hall. Trent, the river on which Burton lies, was added from the early 14th century to distinguish it from many other Burtons: VCH IX 3. 84 names have been recorded incorporating OE *burh-tun*, and it has been suggested (see PN Sa I 40) that in Mercia, which has at least 43 examples, the name may refer to places forming a network of fortifications for the defence of the country until the Danish wars of the late 9th century (cf. Burton, Broughton (q.v.)), although further research, presently unpublished, does not support such theory. The place is also recorded as *Mudwennastow* (Shaw 1798: I 1), after the bones of St. Modwen which were reburied in the monastery here after it was rebuilt following its destruction by the Danes in the 870s. In fact St. Modwen was a fiction based on the life of St. Monenna, an early Irish royal abbess saint who never left Ireland: the 11th century biographer Conchubranus muddled the 5th century Darerca alias Moninna with the 7th century Monenna alias Modwenna: SHC 4th Series 11 2-3; Bartlett 2002. OE *stow* meant 'place' or sometimes (as here) 'holy place'.

BUR WALLS (unlocated, near Wolseley Bridge) *Bur walls* 1798 Shaw I 105. Possibly from OE *burh* 'fortified place', so 'the walls of the fortification', perhaps marking some early earthworks.

BURWEY (unlocated, probably near Longdon) *Boreweya* 1250 SHC 1911 119, *Boruwey* 1269 SHC 1910 103, *Borewey* 1269 ibid.' 108, *Burway* 1269 ibid. 130, *Burewey, Le Burywey* 13th century BCA MS 3415/137-8, *le Bury Weystrete* 1322 BCA MS3415/153, *Burywey* 1325 BCA MS3415/158, *Burwey* 1327 SHC 1912 250, *Boreweye* 1347 SHC 1913 121, *Borewey* 1347 SHC XII 48, *Buriwey* 1362 BCA MS3415/88, *Burway, Burwaye* 1376 SHC X NS (ii) 93, *Berewey* 1383 BCA MS3415/196, *le Berywey* 1383 BCA MS3415/197, *le Bureway* 1444 BCA MS3415/217. Perhaps to be associated with Borough Lane, a road running south-west from Longdon shown as *Burrough Lane* (from OE *burh* 'fortification') on Yates' map of 1775, presumably associated with *Burrow Lane*, marked as a place on the O.S. map of 1834 to the south of Upper Longdon, and *Burwaye Lane* recorded in 1506 (OSS 1936 51), although the *burh* may be Castle Ring (q.v.): see

Longdon. The name is from OE *burh* 'fortification, fortified place, manor-house', with OE *weg* 'a way, a path, a road', hence 'the road to the fortification or manor-house'. Borough Lane runs in the direction of Castle Ring (q.v.), but an earthwork is said to have existed at Longdon (q.v.), though the evidence is inconclusive. See also Borrowcop Hill; Borway Ford.

**BURY** (obsolete) an iron-age hillfort at Kinver (SO 8383), *Bury* 1293 VCH XX 122; BURY (unlocated, in Stapenhill, perhaps to be associated with the 336' hill on the east side of the river Trent) *le Bury* 1436 SHC 1937 166. From OE *burh* 'a fortified place'. See also Burtone.

**BURY, THE** – see **SPRINGHILL**.

**BURY BANK** an iron-age hillfort at Darlaston, 1 mile north-west of Stone (SJ 8835). *Wulfcestre ... le Buri* 13th century SHC VI (i) 10; SHC XII NS 10, *Wlferecestria* 13th century Mellows 1949: 146, *Wlferecestriam* 13th century ibid. 148, *montis qui vocatur Wulfecestre* n.d. SHC VI (i) 9, *Wulfcastre* n.d. SHC VI (i) 10, *Welfercester* n.d. (1798) Shaw I 36, *Byri hille* c.1540 Leland v 20, *Berry-bank* 1686 Plot 406, *Berry Bank* 1718 SRO D593/B/1/28/11, 1748 BCA MS3145/103/2a&b, *Wulfercester, now Bury Bank* 1798 Shaw I 232 fn.8, *Wulferecester* 1946 O.S. Bury Bank is from OE *burh* 'a fortified place', with OE *banc* 'a bank'. By ancient tradition the old name of this place (shown as *Wulferecester* in Gothic script on the 1920 1" O.S. map) is from Wulfhere, the son of Penda and king of Mercia 658-75 ('not veri far from Stone priori appereth the place wher King Woulphers castel or manor place was. This Byri hille stode on a rok by a broke side. Ther appere great dikes and squarid stones': Leland c.1540 v 20; see also Plot 1686: 406), with OE *cester* 'castle, fortress'. No archaeological evidence has been found to support the Wulfhere tradition, but two 12th-century charters refer to *Wulf(e)cestre*, and in one case the name by which the place was known in the 12th-century, *le Buri*, an OE term for a fortification: neither charter uses the name Darlaston, the estate in which the place lay, probably coined between c.750 and c.950, which, it has been suggested, may imply that the name *Wulf(e)cestre* predates the mid-8th century, with the earthworks perhaps refortified during the expansion of Mercia in the early 7th century: see Studd 1993: 55; SHC VI (i) 9-10. It is noteworthy that *cester*, generally applied to Roman towns and ancient fortifications, was preferred to the more usual OE *burh* 'fortification', suggesting that the name may have been coined as 'Wulfhere's city' in the post-Conquest period to support the legend associated with the origins of nearby Stone (q.v.). One survival from the legendary place-name *Wulferecester* is the Staffordshire surname Walchester. See also Jones 1998: 29-62 (which does not mention this place). See also Wolfesbrigg; Woolley.

**BURY BROOK** on the north side of Wolverhampton. Bury Brook is recorded (as *Byribroc* in 994 (17th century, S.1380; see Hooke 1983: 72-5)), from OE *burh* (dative *byrig*) 'fortification', to the north-east of Bushbury Hill (see StEnc 110), and as Berry Brook (or Waterhead Brook) formed the boundary between Bushbury and Wolverhampton. It may have taken its name from a *burh* on Bushbury Hill (see Bushbury) or in or near Wednesfield: *Burycroft* is recorded frequently in Wednesfield in the 14th and 15th centries – see for example SRO D593/B/1/26/6/92. The junction of Seisdon, Offlow and Cuttlestone Hundreds lies on Bury Brook: see Mander & Tildesley 1960: 29.

**BURY FARM** 1 mile south-west of Aldridge (SO 0499). *Berry Farm* 1775 Yates, *Bery F(arm)* 1798 Yates, *Bury* 1834 O.S., *Bury Farm* 1895 O.S. From OE *burh* 'fortification, fortified place', presumably with reference to ancient earthworks here.

BURY (FARM) (unlocated, at Branston) *Bury farm, Bury or Brampton hill* 1551 (1798) Shaw 118. From OE *burh* 'fortification, fortified place', presumably with reference to ancient earthworks here.

BURY HILL (obsolete) in Moreton near Newport: see JNSFC 1937 62; StEnc 110. From OE *burh* 'fortified place', possibly associated with the 'raised work here...which seems to be of [Roman] fashion': Plot 1686: 395. See also JNSFC 1902 119. A Bury Hill (*Burihul, Byryhyll* 1325 SHC 1928 129) lay on the north side of Wolverhampton (see Hooke & Slater 1986: 39-44), and another is recorded in Meretown (JNSFC LXXII 1937-8 62), possibly the same place as *Buryhill* recorded in Forton in 1487 (Oakden 1984: 151). See also Bushbury; Berry Hill; Bury (Farm).

**BURY RING** – see **BERRY RING**.

BUSHBURNE (unlocated) *Bushburne* c.1567 SHC IX NS 159.

**BUSHBURY** Ancient Parish 2 miles north of Wolverhampton (SJ 9202). *Byscopesbyri* 994 (17th century, S.1380), *Biscopesberie* 1086 DB, *Bisopisbury* c.1250 SRO D938/178, *Bissopesbiri, Biscopesbiri, Bishbiri, Bischbury, Bissopeburi* 12th and 13th century Duig, *Bissopbure* c.1272 SHC 1928 73, *Bissopburi, Bussopburi* 1275 SHC VI (i) 56, *Byspesbury* 1315 SHC 1911 331, *Byssburi, Bysshebury* 1352 SRO D938/79, *Bisshebury* 1374 BodCh, *Bysbere* 1532 SHC 4th Series 8 112, *Bushbury* 1749 Bowen. Evidently 'the bishop's manor'. However, OE *burh* from which the second element derives has several meanings, including 'fortification', 'fortified manor house', and 'manor': a passage in the laws of King Alfred which sets out the penalties for breaking into the premises of a nobleman or other distinguished person calls such an offence *burhbryce*, literally 'fortress breaking', which idicates that the residence of a king, bishop, ealdorman or nobleman might be called a *burh*. It is impossible to determine the precise meaning in this case (though PN Sa I 300 suggests that, in the West Midlands, nominative *burh* often has reference to an ancient feature, while dative *byrig*, as here, may more commonly denote a manor-house), but the place lies on the side of a particularly prominent 590' hill (*? Hulle*, recorded in 1249: SHC 1911 118), and a lost earthwork may explain the meaning – a large tumulus (now destroyed) is recorded on the south end of the hill (see Low Hill), and the summit has been heavily quarried, although 'the bishop's fortification' seems unlikely. No bishop is known to have had any connection with the place, but it has been suggested that Bushbury may have formed part of Wolverhampton (SHC 1916 197), or of Brewood and the bishop of Lichfield's holdings, on the basis that the extra-parochial portion of Wolverhampton to the north was surrounded by the bishop's lands (SHC 1919 183-4). Stenton 1970: 320-21 holds that one specialised meaning of OE *burh* was 'monastery', perhaps from the enclosure which surrounded monastic buildings, cf. Hanbury, also Prestbury, Gloucestershire, and Abbotsbury, Dorset, from OE *preost* 'piest' and OE *abbod* 'abbot' with OE *byrig*. It is possible that the first element is from the OE personal name Biscop: there is evidence that names such as Bishop and King (perhaps nicknames or from roles in pageants or similar) were not unusual (e.g. Bishopdale, North Yorkshire): see Redin 1919: 18 fn.1. The fact that the Hundred boundary runs through the parish indicates that the parish was created at a later date than the Hundred (Mander & Tildesley 1960: 17 fn.2). In the 19th century the name was still pronounced Bishbury by local people. See also Bury Brook.

**BUSHTON** 1½ miles south of Tutbury (SK 2026). *Busson(ibus)* 1201 SHC III (i) 73, *Bussums* 1253 SHC 1911 125, *le Bussons* 1286 SHC VI (i) 168, *Busshumes* 1305 SHC 1911 63, *Boussones* 1326 HLS 267, *Busschones* 1327 SHC VII 209, *Busshones, Busshonnes* 1332 SHC X (i) 89, 108, *le Busshonnez* 1384 HLS 413, *Bussons* 1396 SHC

XV 73, *Bushouses* 1414 (1798) Shaw I 46, *Bushouse* 1415 (1798) Shaw I 43, *Bussuns* c.1450 SHC 4th Series IV 220, *Bushtons* 1657 SRO DW1743/T/283. Almost certainly from the dative plural *\*byscum*, from OE *\*bysce* 'a copse of bushes', or *buscum* from OE *\*busc* 'a bush'. The suffix *-ton* is an arbitrary modern alteration of the second syllable.

BUSTLEHOLME (MILL) (obsolete) 1 mile north-east of Stone Cross, 3 miles south of Walsall (SO 0293). *Bustelhome alias Bustleholme* 1594 Willett 1882: 186; Dilworth 1976: 58; *Bustellhome temp*. Elizabeth I MidA III 72-3, *Nether Bustlehome* 1570 WALS 34/14/6, *Nether Bustley Holme, Over Bustley Holme* 16th century SOT D260/M/T/1/1b, *Bustleholme* 1622 BCA MS3301/Acc 1941-008, 1628 TSAHS 1996-7 XXXVIII 72, 1669 WHS New Series 9 (i) 114, *Bussleholme* 1669 ibid.' (ii) 6, *Buslam* 1820 Greenwood, *Bustleholm, Bustleholm Mill* 1834 O.S. The second element may be from ON *holmr* 'river island, meadow'. Bustleholm is marked on the 1834 O.S. map ½ mile south-west of Bustleholm Mill, and it is likely that the mill was named from that place. At Bustleholm Mill was an island formed by a loop in the river Tame. The first element may incorporate the word *bustle* (recorded from 1622: OED) in its conventional sense: a mill existed here from at least 1595 (VCH XVII 32), and iron-works, including a slitting-mill, operated hereabouts from at least 1633 (VCH II 114). Bustle, meaning 'activity with excitement, noise and commotion', would be particularly apt for such an activity. The word *holm* is not necessarily evidence of any early origin, and indeed the 1594 spellings suggests that it may be a relatively modern alternative, though OE *homm, hamm* 'an enclosure, a meadow, a water-meadow' is unlikely, and the consistent spellings with *holm(e)* are noteworthy. There is some slight evidence that the mill may have been known formerly as Grinder's Mill: cf. *Gryndersford* 1526 Dilworth 1976: 53, *Grinder's Ford* 1585 ibid. 53.

**BUSTOMLEY LANE, BUSTOMLEY FARM**   on the south-east side of Morrilow Heath (SJ 9835). *Bustomley Farm* 1836 O.S.

**BUTTERBANK BROOK**   a tributary of the river Sow. *ye broke* 1496 Ct. Probably from the name of a field adjoining the brook, *Butter Bank*, perhaps meaning 'the bank which gave rich grazing'. Cf. Butterton, Butterhill.

**BUTTERHILL**   3 miles south-west of Stafford, in Bradley parish, near the boundary with Coppenhall parish (SJ 8919). *Buterales* 1160 Stone Ch, *Buterhale* 12th century, *Buttrehelle* c.1251 SRO 75[7909], *Buterhale, Butrehale* 13th century Duig, *Butterall* 1558 VCH V 138, *Butterall* 1590 SHC 1930 105. The first element is from OE *butere* 'butter', and the name perhaps means 'the place with rich grass for plentiful butter'; the second element appears originally to have been OE *halh*, which from the proximity of the neighbouring hill has developed into modern 'hill'. Cf. Butterbank Brook, Butterton. It may be noted however that there is an intriguing reference in Plot 1686: 111 to a phenomenon which caused the teeth of cattle to turn a golden or brassy colour. This gilding, according to Plot, had also been widely observed in Westmorland, and was tentatively attributed to feeding on particular types of plant. No evidence has been traced to prove or disprove the phenomenon (by tradition Staffordshire gentry took delight in 'humbugging old Plot' with implausible tales), but if the report carries any truth it seems at least possible that such gilding might explain some names with the element *butter* (and perhaps *golden*).

**BUTTERLANDS**   1 mile north of Biddulph Moor (SJ 9159). *Butterlands* 1842 O.S. 'The place with rich grass for plentiful butter'.

**BUTTERMILK HILL**  a 491' hill 2½ miles south-west of Marchington (SK 1028). *Buttermilk hill* 1695 Morden, *Butermilk Hill* 1775 Yates – see Butterhill.

**BUTTERS GREEN**  1½ miles east of Audley (SJ 8150). *Butters Fields* 1733 SHC 1944 23. Named after the Butters family: SHC 1944 23. The Green element suggests a squatter settlement.

**BUTTERTON**  6 miles east of Leek (SK 0756), *Buterdon* 1201 SHC III 72, *Buterden'* 1222 Pleas, *Butterdon* 1223 FF, *Boterdon* 1236 FF, *Buterdon* 1277 SHC VII NS 20, *Boterdone* 1337 SHC VI NS (ii) 157, *Boturton* 1422 SHC XVII 91; **BUTTERTON** 3 miles south of Newcastle-under-Lyme (SJ 8242), (some of the following forms may relate to the other Butterton) *Botertun, Buterton, Buterdon* 12th century, *Boterdon* 1200 SHC III 69, *Buterdon* 1201, *Botterton-juxta-Lyme* 1208 SHC 1913 234, *Butterdon, Buterden, Buterdon* 1223 SHC III NS 10, *Botredon* 13th century ibid.' 10, *Botterton* 1335 SHC 1913 235, *Bottertone* 1342 ibid.' 238, *? Betton under Lyme* 1413 SRO D(W)1082/A/5, *Butter otherwise Butterton otherwise Butterdon* 1608 SHC III NS 10. Almost certainly from OE *butere* 'butter', supposedly meaning 'land which gave rich grazing for cows'. The terminal element is clearly from OE *dun* 'hill' in the Leek name. The Newcastle name is less clear, and may well be from OE *tun*, although the topography does not exclude OE *dun*. *Butterton Clowes* is recorded in Mayfield in 1774 (SHC 1931 91), but has not been located. Cf. Butterbank Brook; Butterhill (q.v.).

**BUXTON BROW**  1½ miles north-west of Upper Hulme (SJ 9864). *Buckstone Brow* 1640 VCH VII 194, *Buxton Brow* 1842 O.S. Presumably 'the brow or summit on the road to Buxton', but a derivation from dialect *buck-stone* 'a stone on which linen is beaten as part of the process of buck-washing', i.e. 'the place where buck-stones were obtained' (see PN Ch V (I:1) 121) cannot be discounted.

**BYANNA**  ½ mile north of Eccleshall (SJ 8321). *? Byyondence* 1351 SHC 1913 144, *Byrendeneo* 1386 WLCH 35/42/6, *Byrondenes* 1387 WLCH 35/1/7, *Berondnee* 1435 WLCH 35/1/4, *Boyondeney Dales* 1509 SHC XI 256, *Byyonney* c.1511 SHC X NS I 169, *Byanwey* c.1532 ibid. 135, *Berondenee* 1539 WLCH 35/1/5, *Byonney* 1571 SHC XIII 285, *Beonny otherwise Berondeney* 1603 SHC XVIII 33, *the Hall of Byony* 1606 Eccleshall ParReg, *Bionna* 1628 ibid, *Bianno where the Bishop lives* 1680 SHC 1919 230, *Byanna* 1686 Plot, 1747 Bowen, *Biana* 1721, 1745 SHC 1912 289-90, 1833 O.S. A name, probably originally a field-name, which became identitified with an ancient timbered house of '... the Bosvile family, whose ancient seat, Byam [*sic*], situated to the north-east of the palace [of the bishops of Lichfield], was afterwards converted into a farmhouse ...': GM 1823 II 217-22. The various spellings indicate a derivation from OE *begeondan* '(place) beyond, on the other side of', with the (unusual) survival of the initial *be-*, with the second element OE *ea* 'river': the place lies on the north side of the river Sow, which separates it from Eccleshall. The 1351 spelling may be a mistranscription of *Byyondenee*. The place-name 'beyond the river' is not uncommon: see VEPN I 72-3. Johanna de Bienheshal is recorded in 1199 (SHC III (i) 31), and may be associated with this place, and the surname Andyence is recorded in the area in the 16th century (SHC IX NS 221), and may be associated with this place-name. See also SHC 1914 57; Shaw 1801: II (unpublished plates 24). Cf. *Byendeyebrok* 1322-54 SHC 1937 121, *Byondebrok* 1324 ibid. 123.

**BYCARS**  in Burslem (SK 8750). *The Biker* 1658 Ward 1843: App. viii, *Bykers (Colliery)* 1836 O.S. The place is remembered in Bycars Lane. Probably from the OE preposition *bi* 'by, beside', with ON *kjarr* 'a marsh', so '(the place) near or by the marsh' (cf. Bicker, Lincolnshire; Byker, Northumberland). A derivation from ME *biker* 'a fight, a dispute' cannot be ruled out completely (see EPNE i 35; VEPN I 99), but seems improbable as a simplex name.

**BYRKLEY (LODGE)** 1 mile south-west of Needwood (SK 1723). *Birkeleye* 1337 Ct, *Birkeleyloge* 1524 Hardy 1908: 200, *Birkley Lodge* 1658 DCL 380, *Berkley Lodge* 1798 Shaw I 66, *Byrkley Lodge* 1836 O.S. 'Thomas de Berkley, baron of Berkley in Gloucestershire, was keeper of Tutbury ward, and resided at Berkley Lodge': Shaw 1798: I 66.

**CABBATIE** (unlocated, in Bramshall) *Cabbatie* 1636 SRO D5684/8.

**CADEBRIGE** (unlocated) *Cadebrige* c.1570 SHC 1931 222.

**CADEHAM** (unlocated) *Cadeham* 1190 Pipe.

**CADSEY** (unlocated, possibly in the Longridge/Preston area, to the west of Penkridge) *Cadsey* 1547 SHC 1950-1 41.

**CAGE HILL** on the south side of Chartley (SK 0027). Early spellings have not been traced, but almost certainly from OFr, ME *cage*, used in place-names with reference to fenced enclosures, sometimes (perhaps especially) in parks, usually a pen for deer or livestock and occasionally for enclosures for fish: see VEPN II 122-3. Cf. Cage Hill, Cheshire (PN Ch I 199). See also Bowgage Farm.

**CALCOT HILL** an 800' hill 1 mile south-east of Clent (SO 9478). *Caldecote* 1327 SR, *Kalotthyll* 1609 PN Wo 280. From OE *cald* and *cot* 'cold cottage'. In Staffordshire from the early 13th century until 1844, when it became part of Worcestershire.

**CALD(E)WELL** – see COLDWELL.

**CALDHOCK** (unlocated, in Penkhull) *Caldhock* c.1249 SHC 1911 145.

**CALDMORE** (pronounced Karmer [kɑːmə]) 1 mile south-west of Walsall (SO 0097). *Caldemor* 1306 SHC 1911 288, *Caldmore* 1513 SHC XI 258, *Coldmore* 1564 SHC 1938 41, *Calmore* 1596 SHC 1932 233, *Caldemore* 1605 SHC 1940 319, *Caldmore* c.1621 SHC VII NS 228, *Cauldmore* 1632 SHC II (ii) 42, *Callmore* 1644 SHC 4th Series I 191, *Calmoor* 1808 Baugh, *Colmore* 1834 O.S. From OE *cald* 'cold', and OE *mor* 'marsh'. The 1922 O.S. map shows *Cauldmore* (*Caldmore Meadow* 1796 SRO D1066/7) on the north side of Wheaton Aston (SJ 8513), perhaps to be associated with *Caudwell Leasow* recorded in 1735 (SRO DW1738/C/31. The name is remembered in Cauldmore Lane.

**CALEAWAY EDGE** (unlocated) *Caleaway Edge* 1663 Leek ParReg.

**CALF HEATH** 3 miles west of Cannock (SJ 9309). *Kalfre heie* 994 (17th century, S.1380), *Calnheth* 1286 SHC V (i) 165, *Caleshuve* c.1290 Hatherton deed, *Calonhethe* 1300 SHC V (i) 177, *Calwehet* 1311 SHC 1911 311, *Calghet* 1332 SHC XIV 21. Although the earliest spelling (if reliable) suggests 'the enclosure of the calves', from OE genitive plural *calfra* 'calves', with Mercian OE *(ge)heg* 'hedge, enclosure', the later spellings evidently incorporate OE *hæþ* 'heathland'. Formerly a vast heath and part of Gailey Hay, one of the Seven Hays (*the vii Hayes* c.1540 Leland v 22) of Cannock Forest. It is possible that the 'calves' were the tumuli which stood on the heath until destroyed by gravel working in the early 20th century (see SHC 1938 297; JNSFC 3 1965 35): it is not unknown for adjoining large and small hills and islands to have names like Cow and Calf (cf. Cowes). See also Hill Chorlton.

**CALLINGWOOD** 4 miles west of Burton-on-Trent (SK 1923). *Le Chaleng* 1247 Ch, *Borschaleng* 1251 Ch, *Borcheslang* 1252 Ch, *Bosco Calumpniato* 1273 SHC 1911 28, *Calyngewode* 1280 Ass, *Kalangewode* c.1290 SHC 1937 97, *Calyngewode*, *Chalengwode, Boscum Calumniatum* 13th century SHC XVI 276, *Boischallenge* 1306 SHC VII 165, *Chalaungwode* 1327 SHC VII 224, *Kalangewode* 1332 SHC X (i) 106, *Callenwood* 1532 SHC 4th Series 8 141, *Chalyngewood* 1577 Saxton. This name is a French-English hybrid, originally the ONFr *calenge* 'challenge', to which was added ONFr *bois* 'wood', the latter replaced by ME *wode* 'wood', hence 'disputed wood': the place lies near a parish boundary. The dispute was presumably post-Conquest, given the French root. By local tradition Robert de Ferrers I promised before the battle of Northallerton in 1138 to award the forest to the bravest of his troops, hence the 'challenge' (VCH II 349). The tradition is not recorded earlier than the 19th century, and is doubtless apocryphal, but the simplex form of the earliest spelling is noteworthy. Cf. *Calyngwoodleys*, recorded in 1462 in Castle Church: Oakden 1984: 80; Callans Wood, Worcestershire.

**CALLOW HILL** 1½ miles north of Blithfield (SK 0426), *Caluhull* 1257 SHC 1919 20, *Caluhull, Kalewhull* 13th century SHC VIII 145-7, *Cawlhull* c.1300 SRO D(W)1721/3/24/15, *Calughull* 1323 SHC 1937 123, *Calvhull* 1325 ibid.' 124, *Callow Hill* 1399 SRO D(W)1721/3/31/1, *Kalughulle* 14th century Duig, *Cailowehill* 1583 SHC 1927 178, *Calohill* 1610 SHC III NS 41, *Calley Hill* 1775 Yates, *Callow Hill* 1798 Yates; **CALLOWHILL** 1 mile south of Dilhorne (SJ 9742), *Callow Hill* 1836 O.S.; CALLOWHILL (unlocated) possibly near Thornes, 2 miles north-east of Aldridge (see SHC XVII 245), evidently 'the lost township of Caunhulle' mentioned in SHC 1923 274, but see also Willmore 1887: 68, which locates the place at Calder Fields near Longwood Bridge, *Calewenhulla* c.1190 SHC XVII 244-5, *? Calvehul* c.1200 SHC VI (i) 20, *Calewenhulle* c.1200 SHC 1923 274, *Calewenhulle* c.1205 SHC 1923 274, *Kallowhull* c.1238 SHC 1923 274, *Grange of Calewenhull* 1329 (1801) Shaw II 71; CALLOWHILL (unlocated, in Dunston), *Kanhull-meirs, Caluhulmeres* 13th century SHC VIII 161, 165; KALEWHULL (unlocated, in Tillington), *Kalewhull* late 13th century SRO 3764/114[36347]. A common name, from OE *calu* 'bald, bare, lacking vegetation', with OE *hyll* 'hill'. Callow Hill near Blithfield is doubtless to be associated with *Caluulford*, recorded c.1257: SRO D938/62.

**CALTON** 1½ miles west of Ilam (SK 1050). *? Canton* 942 (14th century, S.1606), *Caltone* 1191x1194 CEC 261, *Kalton* 1227 SHC IV 44, *Caldon* 1228 Pat, *Calton(e)* 1229 Harl, *Kouton', Couton'* 1236 Fees, *Caltone* 1311 SHC 1911 303, *Calveton* 1340 Ipm, *Calton'* c.1450 SHC 4th Series IV 253, *Calton* 1532 SHC 4th Series 8 10, *Cawton* 1577 Saxton, 1644 SHC 4th Series I 237, *Caulton* 1666 SHC 1925 194, 1682 Brown. The 1340 spelling strongly suggests a derivation from OE genitive plural *calfra* 'calves' (possibly also meaning 'tumuli' in Staffordshire: see Calf Heath), but the other spellings do not support that etymology, and point towards OE *cald* 'cold' with OE *dun* 'hill', giving 'the cold or bleak hill' – the place lies on a hill of 1034' – or OE *tun*. The spellings support the latter, the topography the former. Calton and Cauldon are in such close proximity that there is often confusion between the forms. A footnote in SHC 1925 109 observes: 'Hitherto Calton has generally been described as in Derbyshire, but ... [i]t seems to be an indeterminate place on the border of the two counties, and in four or five parishes'. The areas which form Calton are in fact in four Staffordshire parishes, but in a random arrangement, with alternate fields and farms sometimes in different parishes. *Calton Buds* (also known as *Buds* and now *The Budds*) was the name for the steep ground above the river Hamps at Calton: TNSFC 1948 44. The term is unexplained. There is a Calton in Ashbourne, Derbyshire.

CALVEHUL (unlocated, near Tittensor: SHC VI (i) 20) *Calvehul*?13th century SHC VI (i) 20. 'Calf hill'. See also Calton.

CALVERECROFT (unlocated, near Abbots Bromley), *Calverecroft* 1260 SHC 1937 61; CALVER CROFT (unlocated, in Caverswall, possibly associated with Calverhay (q.v.)), *Caln(er)croft* 1562 SRO D5100/24, *the Calver Croft* 1630 SRO D1275/3; CALVERCROFT (unlocated, in Apeton), *Calvercroft* 1355 SROD938/136. From the OE genitive plural *calfra* 'calves', with OE *croft* 'a piece of enclosed pasture land, a small piece of arable land adjoining a house', so 'the croft of the calves', but see also Calf Heath.

CALVERHAY (unlocated, in Caverswall, possibly associated with Calver Croft (q.v.)) *Calverhey (Meadow)* 1652 SRO D593/B/1/20/20, *Calverhay* 1813 D593/B/1/23/30. From the OE genitive plural *calfra* 'calves', with Mercian OE *(ge)heg* 'a hedge, an enclosure', so 'the enclosure of the calves', but see also Calf Heath.

**CALVING HILL** on the north-east part of Cannock (SJ 9810). *Calughul(l)* 1304 MRA, 1369 Ct, *Calughhul(le)feld* 1341 to 1370 Ct, *Caloughehyllfilde als Calfe hylles fylde* 1580 SHC 1939 79, *Calloughe-Hyll ... als ... Calf Hylls ...* 1580 SHC 1939 79, *Calving Hill* 1840 TA, *Calven Hill* 1843 VCH V 60. From OE *calu* (weak oblique case *cal(e)wan*) 'bare, bald, lacking vegetation', later corrupted to *calf* and *calving*.

**CALWICH** ½ mile east of Ellastone (SK 1243; SMR 02616 places a deserted post-Conquest settlement at SK 12904340). *Calwich* c.1130 VCH III 237, *Calewiz* c.1175 SHC VII NS 134, *Calowic* 1177 SHC XII NS 278, *Calewich* 1197 P, *Calowic temp.* Henry II (1314) Ch, *Caleviz* c.1200 DRO D258/27/1/6, *Culewich* 1203 SHC III 107, *Calewiz, Calewis* c.1235 SHC 4th Series IV 200, *Calwyche* 1521 SHC 4th Series VII 60, *Caldwich* 1539 Ellastone ParReg, *Caldewiche* 1543 SRO D626/A/1/1-2, *Caldwiche, Caldwich, Callowiche* 1566 SHC 1926 93-4, *Calewiche* 1583 SHC III 7, *Calwithe* 1599 SHC 1935 139, *Colwich-common* 1686 Plot 404. Possibly from OE *calf-wic* 'farm where calves were kept', but the *-o-* is puzzling, perhaps explained as a voiced 'glide' between *Cal-* and *-wic*. There was a priory here founded c.1125, possibly on the site of a hermitage – a *heremitorium de Calwich* is recorded at an early date: VCH III 237. *Calwidge* and *Colwich*, recorded in 1604 (SHC XVIII 43) would seem to be two separate but adjoining places: it is unclear whether they refer to this place or Colwich (q.v.).

**CAMP FARM** on the Shropshire border, 2½ miles south-west of Adbaston (SJ 7324), *Camp Farm* 1775 Yates, *Camp* 1808 Baugh, *The Camp* 1833 O.S.; **CAMP FARM** on the Warwickshire border, 2 miles south of Shenstone (SK 1101), *Camp Farm* 1895 O.S.; **CAMP FARM** 1 mile south-west of Swindon (SO 8489), *Camp* 1775 Yates, *Camp Hill* 1819 GM 1819 i 396, *Camp Farm* 1834 O.S.; THE CAMP (unlocated, in Adbaston: SHC 1924 191), *The Camp* 1924 SHC 1924 191. *Camp* may be a perpetuation of the term *campus*, used in the late Roman period for areas of uncultivated land on the fringes of a villa or town, which was reintroduced in later times as *camp*, meaning simply 'ancient earthwork, especially from the Roman period' (see Gelling 1988: 74-8), but it is unlikely that these are ancient names. The first place is on the Roman road from Pennocrucium (Water Eaton) to Chester (Margary number 19), on the west side of which (in Shropshire) Yates' map of 1775 identifies 'Lines of a very Ancient Incampment' by a gently undulating line which seems to run roughly parallel with the Roman road; the second is in an area in which many Roman finds have been made; and the third is near Greensforge where several Roman forts have been identified: see SHC 1927 185-206; VCH I 344-6. Baugh's map of Shropshire, 1808, shows the outline of a rectangular camp (at SO

851892) at Camp Farm itself, but Baugh probably misplaced the nearby camp at Greensforge (q.v.). Joh'e de Campo is recorded in Kinver in 1327 (SHC VII (i) 246), and may be associated with this last place, but *campo* is Latin for 'field, plain'. For The Camp in Adbaston see also Chesterfield.

**CAMP HILL** 1 mile north-west of Maer (SJ 7840). *Camphills* 1686 Plot, *the Camp Hills* 1768 (1801) Shaw II xviii. The hill stands above Berth Hill (q.v.), and evidently takes its name from the earthwork on that hill.

**CAMPFIELD (WOOD)** at Hollywood, 2 miles south-east of Stone (SJ 9333). *Campfield* 1908 VCH I 192. From a small rectangular earthwork here: VCH I 192; SMR 00215.

CAMTON (unlocated) possibly near Ilam or Musden. *Camton* 1569 SHC XVII 217.

**CANK THORN or CANNOCK THORN** on Cannock Chase, marking the meeting place of the boundaries of the manors of Teddesley, Baswich and Cannock, and of the parishes of Penkridge, Cannock and Rugeley (SJ 9815): '[A] small circular bank of earth about 10 yards south-west of the old London road': 1805 perambulation of Rugeley (Burne 1957: 13). *Naughmarethorn* 1290 SHC 1924 286, *Nantmarethorn* 1294 SHC VI (i) 296, *Naurmar Thorn* 13th century Burne 1957: 13, *Canck Thorne, Cannock Thorne* 1595 Duig, *Cannock Thorn* 1719 Baswich ParReg, Cank Thorn 1754 SRO D260/M/E/353a. The 13th century forms may be *mere-thorn* 'boundary thorn', from OE *gemǣre* 'boundary' (the tree stands at the junction of three manors), with an unrecognisable first element, but there is a possibility that they incorporate OE *\*niht-maere* 'nightmare': c.f. field-names *Negtmereslond* and *Nahtmarefurlong* recorded in Cheshire (PN Ch IV 70, 87). A nightmare was a female monster which supposedly settled on sleeping people and animals and caused a feeling of suffocation (OED). This isolated boundary thorn may have been thought to be haunted. The decayed bush (which could not be found during perambulations of Walton and Brocton in 1728 and 1732: Baswich ParReg) was replaced by a blackthorn in the early 18th century, and in 1949 was said to be a Cockspur thorn, a North American species: JNSFC LXXXIV 1949-50 119. That tree was uprooted in 1972, but cuttings taken for replanting: NSJFS 12 1972 141. A map of Teddesley Warren of 1754 is endorsed with a note of 1886 concerning the position of Cank Thorn, which is shown on the map some three quarters of a mile to the north of the position accepted by the Enclosure Commissioners in 1814: see SRO D260/M/E1-424 11; SRO D260/M/430/143; SRO D260/M/E/353a. See also StEnc 116.

CAN LANE (obsolete, 1 mile east of Sedgley (SO 9394)) *Can Lane* 1834 O.S. Supposedly named after the Cann family, originally from Devon, famed for its champion wrestlers in the 18th century: see StEnc 117. Now known as Hurst Road: ibid.

**CANNOCK** Ancient Parish 8 miles north-east of Wolverhampton (SJ 9810). *Chenet* 1086 DB, *Chnoc* 1130 P, *Can(n)oc, Can(n)ok(e), Kan(n)oc, Kan(n)ok(e)* c.1135-40 St Cart, 1153-1272 MRA, 1198, 1212, 1236 Fees, *Can(n)ock(e), Kannock(e)* 1151, 1285, MRA, *Cancia (sic)* 1155 St Cart, *Cnot* 1156-1196 P, *Canot* 1157-1215 P, *Chenot* 1162, 1173 P, *Knot* 1166 P, *Chnot* 1170 P, *Kanot* 1187 P, *Canet* 1203 Ass, *Gannok vel Kannok* 1221 ClR, *Gan(n)ok, Gan(n)oc* 1245 BM, 1262 Lib, *Canocbir', Kanocbir', Kanocbur'* 1259 Ch, *Kannokbury, Cannokbury* 1286 SHC V (i) 167, *Cankbur'* 1293 QW, *Cannokburi* 1327 SHC VII (i) 245, *Cannocbury* 1348 SHC 1939 75, *Kankbury* 1352 Banco, *Kannockbury* 1319, 1377 (1801) Shaw II unpublished sheets 316, *Cannockbury* 1377 BCA MS3033/Acc-1914-020, *le Cank(e)* 1403 Cl, 1415 Coram R, *Cannock als Canck* 1493 Ipm, *le Cank* 1415 SHC VIII 30. It was formerly thought that this name (spellings ending –*t* are likely to be misreadings for –*c*, since the two letters are virtually indistinguishable in medieval manuscripts) was of Celtic origin, from the form *Canuc* in a charter of 956 (12th century, S.608), interpreted as a supposed word *\*cunaco-*, 'hill'. It has now been shown not only that the charter in question refers to a place in Hampshire or Wiltshire, not Staffordshire, but that the word *\*cunaco-*, and the suggested PrWelsh *\*cuno-*, 'high', may not exist: see Oakden 1984: 56; but also JEPNS 16 1983-4 1-24, where *\*cunaco* is held to be pre-British and non-Celtic, a hill-name but not necessarily 'hill'. The name Cannock is now believed to derive from Welsh *cnwc* 'a hill, a lump, a hillock', or OE *cnocc* 'hill' or 'hillock', a native Germanic word (cognate with Danish dialect *knok* 'little hillock', and related to ON *knjukr* 'a high steep rouded hill'), modified by Norman pronunciation to *canoc* with the usual insertion of a vowel between two consonants, in the same way that King Cnut came to be known as Canute. The hill in question has been taken to be the 650' Shoal Hill (q.v.), 1 mile north-west in nearby Huntington (see Nicolaisen *et al* 1970: 66; Oakden 1984: 56; CDEPN 113), but that can hardly be called a hillock. It seems more likely that the feature after which the place was named was 'the slight hill of gravelly soil' (VCH V 52) on which the place lies. Early writers and historians record a 'great stone' in a field south of Cannock Church: Erdeswick 1844: 192 fn. states 'near the church is a stone of great weight and magnitude, which has been sunk under the surface of the ground, and the plough passes over it. Several large single stones there are objects of antiquity ...'; see also Dudley 1665; Plot 1686: 397; Shaw 1801: II 13; and note *Great Stone Field* recorded in 1594 SRO D260/M/T/6/132; *Great Stone Stile Field* recorded c.1843 VCH V 60). The reference in Plot suggests that the stone may have been something more than a rounded boulder: 'And for other British antiquities that are in any way probably such, I met with none, unless the great stone in a field South of Cannock Church...may be accounted such'. Shaw 1798: I 13 observes that '... the great Stone in a field South of Cannock church may also be numbered amongst the British antiquities ...'. However, there is little likelihood that the name is from Welsh *cnwc* 'lump' with reference to that particular feature, since a more appropriate Welsh word, such as *maen*, would be expected. The *bury* element in some of the spellings is from OE *burh*, which often meant 'fortification', perhaps referring to Castle Ring (q.v.), which lies in the north-east corner of the parish, but it is possible that a fortification of some kind once existed in Cannock itself: Shaw (1801: II unpublished sheets page 316) mentions a record in Dugdale of a payment in the time of Henry I to the Constable of the Castle. No other reference to a castle has been traced, though there was a prison at Cannock in 1286 (SHC 4th Series XVIII 110, 128-9, 143), and Duignan believed that the foundations of the stone structure within the earthworks of Castle Ring were the remains of a 12th century castle which was never completed (MidA III 141-2; Hackwood 1905b: 146). OE *burh* also came to mean 'a manor house' and similar, and that may well be the meaning here. For early observations on the name Cannock see Shaw 1798: I 10 fn. See also Cank Barn, Warwickshire (PN Wa 292); Cank in Inkberrow, Worcestershire; Consett, Durham; Conock and Knook, Wiltshire. *Cnokomalay* (possibly in or near Audley) is recorded in 1307: SHC XI NS 265.

**CANNOCK CHASE**   to the north of Cannock. *Le Cank Chase* c.1326 SHC I 252, *Cannock Chase* 1481 SRO DW1781/7/1/1-4, *the forest or chace of Cannok wood alias Cank Wood* c.1540 Leland. From OFr *chace* 'chasing, hunting, a hunting ground, wild park-land'**,** which was a term applied to an area used for hunting with its administration subject to chase courts and the civil law, rather than the Forest courts of the royal Forests. The Chase, granted to the bishop of Lichfield in 1290 after some 60 years of dispute between the bishop and the king (VCH II 338, 343, map 336), strictly consisted of the bailiwicks of Trumwyn and of Puys (or Rugeley), and descended with the manors of

Cannock and Rugeley: VCH V 58-9. It is now a vast area of heath and woodland. A reference to *Chaciam de Kannock* supposedly dating from 1175x1209 (SHC VIII (i) 134), is in fact from a forged charter of the 14th century purporting to date from 1176x1182: EEA 16 90. For early woodland names in Cannock Chase see Welch 2001: 17-73. Cf. Cannock Forest.

**CANNOCK FOREST** created or enlarged by William I, the Forest (an area subject to Forest law, and not necessarily wooded), was known as such by the 1140s (VCH II 338), and occupied a vast area in the centre of the county extending from Radford Bridge near Stafford in the north to Wolver`hampton and Walsall in the south, and from the river Penk in the west to the Tame on the east. The original nine nine Bailiwicks or Hays (from Mercian OE *(ge)heg* 'hedge, fence, enclosure') of Cannock Forest were Alrewas, Bentley, Cheslyn, Gailey, Hopwas, Ogley, Teddesley, Cannock and Rugeley. They were reduced to seven (*the vii Hayes* c.1540 Leland) when Rugeley and Cannock were granted to the Bishop of Lichfield and Coventry in 1189 (SHC X NS I 213; see also VCH II 338, 342-3). That area became Cannock Chase (q.v.). Cannock had the largest single wood recorded in DB, possibly 24,000 acres: Rackham 1980: 115-7. The name Cannock Forest is now applied to that part of Cannock Chase which was the subject of the Rugeley *EnclA* of 1885: StEnc 121.

**CANNOCK WOOD** 4½ miles north-east of Cannock, below Castle Ring (SK 0412). *Conikwode* 1564 Mercator, *Canckwood* 1623 PCC, *Cannock Wood* 1666 SHT, 1775 Yates. It is unclear whether these references are to the relatively modern village of Cannock Wood: Leland refers c.1540 to *the forest or chace of Cannok wood alias Cank Wood*. Toulmin Smith 1964: 102. *The Survey of the Cankewood*, made in 1554 (SRO (DW1734/2/3/43), covers the wooded parts of the manors of Cannock and Rugeley, and it seems likely that Cannock Wood was so defined in earlier times.

CANTERTON (unlocated) *Canterton* 1199 SHC III (i) 48.

CANTRELL (unlocated, possibly in the Alstonefield/Calton/Throwley area) *Cantrell* 1439 SHC III NS 146

**CANWELL** 5 miles south-west of Tamworth (SK 1400; SMR 02617 places a deserted post-medieval settlement at SK 14100040). *Canewell* c.1120 (1801) Shaw II 2, *Kanewell* c.1185 SHC XVII 251, *Canewelle* 12th century Duig, later *Canewall, Canwalle, Kanewall, Kanewell, Canwell, Canwelle, Canewelle* Duig, *Canewell* 1209-35 Ep, *Canewall* 1272 SHC IV (i) 214, 1304 SRO D938/58, *Kanewell* 1285-6 Rees 1997: 178, 1332 SHC X 104, 1391 SHC XV 37, *Kannewell* 1370 SHC VIII NS 240, *Canewella* 1410 Mon, *Cawnoll* 1606 (1798) Shaw I 16, *Cannall* 1640 SHC XV 212, 1686 Plot, 1694 (1798) Shaw I 16, 1749 Bowen. Perhaps from the OE personal name *Can(n)a, with Mercian OE *wælle* 'a spring', and (sometimes) 'a stream', hence '*Can(n)a's spring'. A derivation from OE *canne* 'a recepticle or vessel for holding liquids', meaning 'the spring provided with a cup', would have numerous parallels (e.g. Beardwell, Wiltshire; Bedlars Green, Essex and Hertfordshire; Biddles, Berkshire; Bidwell, in Bedfordshire, Dorset, Northamptonshire and Somerset, all with OE *byden* 'a vessel, a tub, a butt' as the first element), but here the absence of spellings with –nn- makes such a derivation unlikely (but see also Cumbwell Brook). There is little likelihood that the name has any connection with Welsh *can, cain* 'beautiful, clear'. A spring known as St Modwen's Well is at Canwell: see Burton upon Trent.

CAPAXWODE (unlocated) *Capaxwode* 1481 SHC VI NS (i) 130.

**CAPE HILL** a district in Smethwick (SO 0387), so-named from the Cape of Good Hope inn recorded in 1814 which stood at the junction of Grove Lane and Cape Hill: VCH XVII 94.

**CAR HOUSE** (obsolete) 1 mile south-west of Audley (SJ 7850), *Carr* 1585 Speake 1972: 25, *Carre House* 1599 SHC 1935 182, (1801) Shaw II Appendix 12, *Car* 1609 Audley ParReg, *the Carr* 1612 SHC 1944 83, *Carr* 1649 Audley ParReg, *Carr House* 1733 SHC 1944 4, *Car House* 1775 Yates; **CAR HOUSE** 2 miles south-east of Stone (SJ 9231), *Carre* 1549 SHC XII 202, *Carrhouse* 1594 SHC 1932 79, *Carre House* 1657 SRO 358/1/34, *Carr (Hill)* 1834 White 1834: 681, *Carr House* 1836 O.S. Possibly from OE *carr* 'rock', but see also Carr Wood, Carr Bank. Bowen's map of 1749 records *Car* between Maer and Shelton under Harley, but no other evidence has been traced to confirm the existence of such place.

**CARMOUNTHEAD** 1 mile south-west of Bagnall (SJ 9150). *Keversmunt* 1218 SHC XI NS 219, *Hay of Kenvermunt* 1223 SHC XII NS 30, *Kevremunt* 1227 SHC XI NS 240, *Kenermunt, Keversmunt* 1228 (17th century) SHC XII NS 218-9, *Kevermunt* 1228 SHC IV 67, *Keuermunt (closed hay of)* 1256 Ch. Rolls, *Couremunt, Kevermunt* 1288 SHC VI (i) 167, 175, *Kevermund* 1293 ibid.' 239, *Kenmund* 1299 SHC XI 311, *Kervermund* 1303 Stafford MS (Erdeswick 1844: 98), *Kermond* 1394 SHC XV 60, 1402 SHC 1912 242, *Carmonhead* 1606 SHC 1946 299, *Carman head* 1608 Norton-in-the-Moors ParReg. A name showing clear evidence of Norman influence: Chevremont is common in France, and Quevremont is also found. The initial K- is evidence that the influence is likely to be from the Norman-Picard variant of this name, since the Central French form would have been expected to have an initial Ch-: see PN L 75. Kirmond le Mire, Lincolnshire, is *Cheuremont* (1086), *Kuermunt* (c.1150), and *Keuermunt* (c.1152). The name means 'goat hill', with the second element here developing via ME *munt* (from OFr *mont*) 'mound, hill' into modern *mount*, with the *head* suffix a late addition, denoting the end or head of what was evidently a Hay (from Mercian OE *(ge)heg* 'an enclosure; an area, often part of a forest, fenced of for hunting') of a sizeable area in the region of what is now Wetley Moor, since several places on the margins of that high ground have the name Kevermunt as a qualifier: Weston Coyney, Norton-in-the-Moors and Hulme appear as *Weston sub Kevermont* (1242 Fees), *Norton under Kevremunt* (1227 SHC XI NS 240), and *Holm under Kevermund* (1293 SHC VI (i) 239) respectively. The name was evidently applied originally to a single hill. The form *Caermuned*, recorded *temp.* Henry III (Erdeswick 1844: 98), suggests the intriguing possibility of an original derivation from Welsh *caer mynydd* 'fortification on the hill', which was later Normanised into the similar-sounding Kevermunt, but it is more likely that Keveremont is the authentic name and the 'Welsh' form is an early indication of the pronunciation recorded in 1394 as Kermond. See also Ward 1843: 289 fn. (where the name is held to be Kenermunt), and Barns 1907-8.

**CARMOUNTSIDE** to the north of Abbey Hulton (SJ 9049). *the Carmountside* 1744 Burslem ParReg, *Car Mount* 1775 Yates, *Carmountside* 1836 O.S. For Carmount see Carmounthead. Side is from OE *side* 'the long side of a slope or hill, a hillside', and in ME 'the land extending alongside a river or lake', perhaps originally applied to the long hillside to the north of the stream which flows from east to west on the north side of Holehouse Farm. Ward 1843: 289 fn. records 'two farms taken out of the ancient park [of Kevermunt], and called Carmont, and Carmont side'.

**CARNCOE, ROUGH** (obsolete) ½ mile south-west of Willoughbridge Lodge (SJ 7338). *Carnecole* 1585 TSAS LI 1941-3 114, *Garnecoale* 1617 SHC 1917-8 390, *Rough Carncoe* 1833 O.S. A curious name, seemingly identical with a stream-name and .

meadow name in Abbey Foregate, Shrewsbury, recorded as *Carnecowe temp.* Henry VI, *Carnecoll* 1508, *Carnecolle* 1548, *water comenlye called Cauncole* 1594, *Colum Meadow or Carne Cole Meadow* 1634 (1703), all from PN Sa IV 88. The name appears to be a recurrent minor name of unsolved etymology, but see also Lonco Brook.

**CARROWAY HEAD** 5 miles south-west of Tamworth (SO 1599). *Caraway-head* 1686 Plot, *Caraway head* 1695 Morden, *Caraway head* 1747 Bowen, *Carway Head* 1775 Yates, *Caraway Head* 1777 Bowen, *Carr-way or Carraway-Head* 1801 Shaw II 9, *Carroway Head* 1834 O.S. Possibly from the Caraway plant (*Carum carvi*, the source of the seed for seed-cake), which was once grown in England, although the name is rarely found, and then in the south of England (see Field 1993: 100; also Shaw 1801: II i Adv. iv). Or, if the name is ancient, possibly from OWelsh *\*carrou*, plural of *carr* 'a rock' (cf. Carraw, Northumberland): Plot 1686: 157 states: '... in the hollow way between the hills on Weeford heath, as you pass between Swynfen and Cannel yate [Canwell Gate], there lie divers little heaps [of stones], and one great one at the top of the hill at Weeford Park corner, which according to the tradition of the Country, was placed there in memory of as Bishop of Lichfield, who rideing thither with a large attendance, was set upon by Robbers; and Himself and all his men being slain, that these heaps of stones were layd where each dead body was found: whence by the Country people and travellers they are call'd the Bishops stones. But this is merely a fable of them, the truth follows, as I received it from the learned and judicious Antiquary Sr. William Dugdale Kt. Garter King of Armes. About the later end of the raigne of King Henry the 8 or shortly after, John Vessy then Bishop of Exeter, a man of publick spirit, and borne close by, at Sutton Cofield in Warwickshire, resolving with himself to become a benefactor to that place and the parts adjacent, procured for that towne not only a Mercat and fairs, but ... finding the road above mention'd much annoyed with these rolling pebbles, which frequently accasion'd travellers horses to stumble and sometimes to fall, amongst others of his works of Charity, He hired poore people to gather them out of the way, and lay them thus on heaps; and this is the true reason they are call'd Bishops Stones [*Bishop's heap of Stones* 1720 Bowen] ...'. The *Head* element means 'top, end', so possibly giving 'the top or end of the way to Carrow'. The Staffordshire historian Samuel Pipe Wolferstan was clearly bemused by the name, since in 1756 he observes 'perhaps it should be Carriage-way Head' (Shaw 1801: II 9). For completeness, it may be added that Dent & Hill 1896: 150 associate this place (which they call *Garroway Head*, a form cited by Duignan in MidA II 172) with a verse from a ballad of c.1564, 'The Tanner of Tamworth', which refers to a 'payre of gallowes' here, suggesting that the name may be a corruption of Gallow-way Head: the stream which forms the Staffordshire-Warwickshire boundary to the east is called Gallows Brook. Duignan (MidA II 172) suggests a derivation from Welsh *garw* 'rough', and records that 'It was a rough bit of road when I was a boy [c.1840], but is now good'. It may be noted that Thomas Galawey of Willenhall is recorded in 1482 (SHC NS VI (i) 136), and William Callawey is recorded in Lichfield in 1540 (SHC 1912 137).

**CARR WOOD, CARR BANK** on the north side of Oakamoor (SK 0645). *Cares* 1593 StSt 12 2000 70, *Carr* 1594 SHC 1932 49, *Carr Bank* 1860 SRO D953/34, *Car Wood, Car Bank* 1836 O.S.; CARR (unlocated, in Barton under Needwood), *the Carr* 1720 DRO D3155/WH92-92a. From ON *kjarr* 'marsh, wet moor, boggy copse'. The name is so often found linked to that of the alder that the word may mean 'wet place with alders': see Rackham 1990: 108. Shaw 1798: I 149 records *the Carr* in Yoxall in 1735.

**CARRY COPPICE** two small woods, 1 mile west of Bramshall (SK 0432) and 1½ miles west of Bramshall (SK 0332), doubtless the remnants of a larger single wood. *Carr*

*Coppice* 1775 Yates, *Car Coppice* 1798 Yates, *Carriscopice* 1865 Redfern 1865: 87. From ON *kjarr* 'marsh, wet moor, boggy copse'. The name is so often found linked to that of the alder that the word may mean 'wet place with alders'. The place is also marked by Carry Lane.

**CARTER'S GREEN** West Bromwich (SO 9992). *Carter's Green* 1764 VCH XVII 7. Self-explanatory, from a personal name or occupation. Hackwood 1895: 7 suggests a connection with John Carter, High Constable of the Hundred in 1647. The Green element suggests a squatter settlement.

**CARTLEDGE** 1½ miles north of Bradnop (SK 0057). *Cartelage by Knivedon* c.1290 SHC 1911 432, *Cartelache* 1356 SRO D3272/5/13/10, 1598 Alstonefield ParReg. ME *carte* 'a cart' seems improbable as the first element, so possibly from OE *cert, cært* 'rough ground', cognate with ON *kartr* 'rocky ground', Norwegian *kart* 'rocky ground', with OE *\*lece* 'a stream, a bog, swampy boggy land', hence 'the swampy ground with rocks'. The evidence of an OE personal name \*Cearda (EPNE i 91) is doubtful. Cf. Cartmel, Lancashire.

**CARTLEDGE BROOK** a tributary of the river Churnet, not marked on maps – see Cartledge.

**CASEY, THE** a stretch of road which runs south-west from Winkhill towards Foxt (SK 0550). The age of the name is unknown. Brighton 1937: 11 describes the road as Roman, which is unlikely. The name is from *causee, causey* (see VEPN II 51-2), from ME *cauce* 'a mound, an embankment, a raised way across low wet ground'.

**CASHHEATH FARM** 1 mile south-east of Caverswall (SJ 9642), *Cash Heath* 1832 Teesdale, 1836 O.S. A name of uncertain origin, perhaps to be associated with the word in its monetary sense, recorded from the late 16th century (OED), but possibly from the family name Cash. See also Whitleygreaves; Whitley Heath.

**CASTERN** 1½ miles north-west of Ilam (SK 1252; SMR 02618 places a deserted post-Conquest settlement at SK 12305250). *Cӗtespyrne* 1004 (11th century, S.906; 11th century, S.1536), *Casterne, Casturne, Chatesturne* 1116-33 Burton, *Chatesturne* 1150x1159 SRO D603/A/Add/21, *Catesturne* 1185 ibid, *? Cathesthorne* 1199 SHC III 38, *Catsterne* 1203 Ass, *Catesturn* 1227 Ass, *Cattesterne* c.1240 SHC 4th Series IV 117, *Casterne* 1327 SR, *Castern* 1428 DbCh, *Kasterne* 1436 Fine, *Castorn(e)* 1493-1500 ECP *et freq*, *Coston* 1569 Pat, *Casterton* 1600 Alstonefield ParReg to 1851 White. The first element is possibly an OE personal name \*Catt, with OE *þyrne* 'thorn bush', so '\*Catt's thorn bush'. Upper and Lower Castern (now obsolete) are recorded as *Overcasto(u)rne, Overcasterton* 1538 Ipm, *Over Casterne* 1565 to 1617 FF, and *Nethercastorne* 1538 Ipm, *Over & Nether Casterne* 1568 FF.

**CASTERS BRIDGE** over Black Brook, a tributary of the river Dane, near Gradbach (SJ 9965). *Caister's Bridge* 1862 (1883) Sleigh, *Castor's Bridge or Smelter's Bridge* 1874 Brocklehurst. Possibly from *caster*, from casting, as in smelting. A forge lay close to the bridge: NSFC LXVII 51-70; Elliott 1984: 50.

**CASTLE CHURCH** Ancient Parish ½ mile south of Stafford (SJ 9022). *Castell* 1208 SHC III 172, *villa castri Stafford* 1293 Ass, *Castello* 1293 SHC VI (i) 238, *Castre* 1307 SHC VII 179, *Castel* 1332 FF, *le Castelparke* 1439 MinA, *Castell Parysshe* 1553 SHC XII 212, *Chastel chuerche* 1562-6 Harl, *Stafford Castle alias Castle parish* 1715 WRO 705:24/1109. Originally simply 'the castle'. The place is close to the hill on which stands

Stafford Castle. The manor may also have been known as Forebridge (q.v.): VCH V 86. It became a separate parish from Stafford c.1546. See also Monetville.

CASTLE CLIFF (obsolete) in Newcastle under Lyme. *(haiya de) Clive* 1234-40 TestNev, *(the King's wood called) Le Cliff* 1263 SHC 1911 134, *(haya nostra de) Clyf juxta novum castrum subtus Limam* 1271 SHC V (i) 155, *? Scherteclivie* 1297 SHC 1911 245, *Castelcliff* 1422 SHC VIII 219, SHC XII NS 74, SHC 1912 219. From the castle at Newcastle under Lyme.

CASTLE CLIFF ROCKS near Ludchurch (SK 9865). Early spellings have not been traced, but the outcrop is possibly the *knokled knarre? with knorned stonez* mentioned in the 14th century poem *Sir Gawain and the Green Knight* (see Elliott 1984: 47; see also Flash; Knotbury; Ludchurch), perhaps associated with Castle Cliff across the river Dane in Derbyshire (*Castell Clyff, Castell Cliff* 1503 ChFor, *the Castle Cliff* 1611 *LRMB* 200), which appear castle-like in profile. See also SHC 1910 304.

CASTLE CROFT near Chesterfield, 2½ miles south-west of Lichfield (SK 0905), *Castle-croft* 1695 Morden, *the castle* 1776 Stukeley; CASTLE CROFT at Huntley Hall, 1 mile south of Cheadle (SK 0041), *Castle Croft* 1686 Plot, 1747 Bowen, 1775 Yates; CASTLECROFT 4 miles south-west of Wolverhampton (SO 8697), *Castle Crofte* 1647 Survey, *Castle-croft* 1686 Plot 156, *Castle Croft* 1798 Yates, *Castlecroft* 1801 Shaw II *221, *Castle Croft* 1834 O.S. Ancient names with *castle* often mark what was, or was thought to be, 'a castle, a camp, an earthwork', from OE, OFr *castel*, but sometimes from OE *ceastel* 'heap of stones'. Of Castle Croft near Chesterfield, Horsley 1732: 420 recorded certain Roman walls called Castle Croft which encompassed about two acres, which were still visible in 1817 (Pitt 1817: 128-9). The remains had disappeared by 1872, but excavations revealed traces of a wall 150' long and 11' thick: TSSAHS V 1963-4 1; see also Erdeswick 1844: 301. A flat-topped circular mound at Mobberley, known as *Huntley Castle, Mobberley Baen* and *Castle Cop* (SHC 1926 159-65), is now known to be a natural sandy knoll (SMR 00039), but no record of any early remains at the place near Wolverhampton (unless *Tilbury Camp*, or 'some remains of an old fort or earthworks' opposite (*sic*) Wightwick Mill, recorded in Jones 1894: 7, 10; or a square cropmark at Pool Hall (SO 86209740: see StEnc 461) are to be associated with the 'castle', although the place lies on a steep bank on the east side of Smestow Brook. The age of the name *Tilbury Camp*, and its site, are uncertain, but the name is remembered in Tilbury Close on Castlecroft Hill. *Castle Croft* in Longdon, near Russell's Bank, is recorded c.1737 (SRO D260/M/T/5/125), a field-name *Castle Croft* is recorded adjoining the south side of Colton church (Parker 1897: 7), and *Castle Croft* is marked in Gothic lettering on the north-west side of Tamworth (SK 2104) on the first edition 1" O.S. map of 1834. See also Chesterfield; Mobberley.

CASTLE HAYES PARK (FARM) 1 mile south-west of Tutbury (SK 1927). *Le Castelhaye* 1297 SHC 1911 251, *Castelhay* 1324 ibid. 359, *Castellhaye, Castell Haye* 1540 SHC 1910 45, *Castlehey, Castelhey* c.1540 ibid. 49, *Castle Hayes* 1702 SRO D3629/9/1. One of the hays (bailiwicks or clearings) of Needwood Forest, from Mecian OE *(ge)heg*. The name is doubtless from Tutbury Castle. A gate to the park is recorded as *Nedewodeges* in 1540: SHC 1910 47.

CASTLE HILL in Audley (SJ 7950). The site of a flat-topped mound of Audley castle, dating from 1227: SHC 1944 xviii; StEnc 24.

**CASTLE HILL** ½ mile north of Ashley (SJ 7637). *Castle Hill* 1920 O.S. The age of this name is unknown, but if ancient, perhaps indicates ancient earthworks. *Castle Hill* in Beech is recorded in 1814: SRO D641/5/T/12. See also Castle Old Fort.

**CASTLE OLD FORT** an iron-age hillfort 3 miles north of Aldridge (SK 0603). *Castle-old-ford, Castle old-fort* 1686 Plot 396, *Roman Camp* 1798 Yates, *Castle Old Fort* 1834 O.S. From OE *castel* (from Latin *castellum*) 'a castle, a camp, a fortification, an earthwork', a word often applied to ancient earthworks. Shaw (1798: I 12 footnote) notes that the place is called *Castle-old-ford* in 'ancient writings'; see also VCH I 373; NSJFS 1964 34; WMA 2001 44 202-3; StEnc 124. The name is remembered in Castlehill Road (cf. *Castle Hill* 1798 Shaw I 53) and Castlebank Plantation.

**CASTLE RING** an iron-age hillfort 5 miles north-east of Cannock on the highest point (801') of Cannock Chase (SK 0412). *Castle-hill* 1686 Plot 39, 418, *Castle hill* 1798 Shaw I 221, *Castle Ring* 1907 O.S. Such earthworks are commonly known as castles, from OE *castel* 'a castle, a camp, a fortification, an earthwork', but in this case the stone foundations of a rectangular medieval structure in the north-west corner of the earthworks may account for this element (see also Cannock). *Ring* is from the roughly circular (but actually pentagonal) shape of the earthworks here, the northern ramparts of which lie on the boundary between Cuttlestone and Offlow. A coalmine near here is decribed as *Subter Castrum* ('below the fortification') c.1445: VCH II 72 fn.32.

CASTLOW CROSS (unlocated) on Cauldon Hills. *Castlow Cross* 1686 Plot. Possibly from OE *ceastel* 'a pile of stones', with OE *hlaw* 'tumulus', hence 'burial mound with a cairn'. Perhaps the same place as *Astlow Cross*, mentioned in Hackwood 1924: 74.

**CAT HOLME** 2 miles north-east of Wychnor (SK 1916). *Catholm* 1325 Hardy 1908: 20, *Catholme* 1415 ibid.' 72, 1495 ibid. 141, *Barton Catholme* 1561 DRO D3155/7512, *Catholme* 1827 SRO 5166/1/177, *Cats Holme* 1834 O.S. The first element is possibly the OE personal name *Catta or *Catt, or OE *catt(e)*, the animal, with ON *holmr* 'a piece of land in a marsh or almost surrounded by a stream'. The place lies in what was a loop of the river Trent, which marked the border with Derbyshire, but the flat alluvial meadow land (on which an Anglo-Saxon settlement has been excavated: see Gelling 1992: 29-30; Losco-Bradley & Kinsley 2002) is remarkably dry. Across the river lies Catton, which probably incoporates the same first element, and to which this place was presumably linked when it was named: a ford in recorded across the river here in the 15th century: Faull 1984: 101-14. The O.S. shows *Catholme* at SK 1915, and *Cat Holme* to the north at SK 1916. To the north *Burrow Holme* (in Derbyshire) and *Fat Holme* (q.v.) appear on the 1834 O.S. map, and Tucklesholme (q.v.) lies nearby. Individual fields named Thornholme and Reedholme are also recorded: Faull 1984: 101-14. See also Hulme, Upper Hulme.

**CATCHEMS END** ½ mile north-east of Brewood (SJ 8908), *Catchems End* 1834 O.S., *Catchems Inn* 1838 TA; **CATCH 'EMS CORNER** in Ettingshall (SO 9396), *Catchems Corner* 1775 Yates, 1813 Nightingale, *Catchem Corner* 1791 Penn ParReg, *Catchems Corners* 1808 Baugh, *Catch 'ems Corner* 1834 O.S., *Ketchem's Corner* 1845 doc. A not uncommon name of unexplained derivation, but probably whimsical in origin and not ancient (but note Willam Cattecham recorded in Wyken in 1524: SRS 3 34). It has been suggested that the place at Ettingshall is 'a humorously descriptive name for the toll-bar erected at Ettingshall to 'catch' the tolls of those who had used this 'back way' from Wolverhampton to evade the toll-gate on the main road': Hackwood 1898: 7; and that by local tradition it was the custom for those beating the bounds of Sedgley and Bilston to meet here for a friendly fight: printed Sedgley ParReg. *Catchems Inn* is recorded in

Alrewas in 1801 (SRO DW1851/3/1/1), and *Catchams End* appears near the northern boundary of Worfield, near Badger Mill, in 1839 (TA). The fact that *Catchems Inn* is recorded at two places suggests that the name may have originated, or been used in some cases, as a colloquial term for a beerhouse. Cf. Catchems End in Hatton, Warwickshire; Catchems End to north of Bewdley, Worcestershire.

**CAT'S EDGE** 2 miles west of Cheddleton (SJ 9452). *Catsall Edge* 1731 ParReg, *Catsaw Edge* 1728 ibid, *Cats Edge* 1836 O.S. The first word is from Catswall (q.v.), with OE *ecg* 'escarpment'.

**CATSHILL** ½ mile north-east of Walsall (SK 0505). *Catteshulle, Kateshulle* 1251 SHC XVI 287-8, *Cutteslowe* 13th century Duig, *Catteslowe* 1300 SHC V (i) 177, *Cattes Lowe alias Cattes Hill* 1576 VCH XVII 278, *Cattlowe als Catshill* 1617 Willmore 1887: 440, *Catts-hill* 1686 Plot 403, *Cat's Hill or Canute's Hill* 1851 White. From a now-vanished tumulus cut by the canal on the south of the old Chester Road, which formed the boundary of the manors of Walsall, Ogley Hay, and Little Wyrley: see JNSFC 5 1965 46. The name is probably from the OE personal name *Catt, or OE *catt(e)*, the animal, with OE *hlaw* 'tumulus, burial mound' alternating with OE *hyll* 'hill'. The 1851 alternative name is an example of popular etymology. See also Catteslowe.

**CAT'S HILL CROSS** 1 mile north of Eccleshall (SJ 8230). *Catsell Cross* 1655 Eccleshall ParReg, *Catchill-cross* 1692 ibid. Cats Hill lies a short distance to the north. Perhaps from the OE personal name *Catt or *Catta, or OE *catt(e)*, the animal, with OE *hyll* 'hill'. A cross may have existed, or the Cross element may refer to the crossroads at which this place stands.

**CATSTREE** 1 mile north-west of Worfield (SO 7496). *Cattystre* (p) 1500 Worfield CA, *Catstre* 1525 SR, *Catstree* 1752 Rocque, 1833 O.S. Perhaps '*Catt's tree'. In Shropshire since the 12th century.

CATSWALL (obsolete) 2 miles west of Cheddleton (SJ 9452). *Catteswall(e)* (p) 1340 AD 5, *Catswall* 1651 Deed, 1775 Yates, *Catswall* 1836 O.S. From 'Catta's or Catt's spring or stream', or 'the spring or stream of the wild-cat', with the Mercian OE *wælle*, which frequently becomes -*wall*. See also Cat's Edge.

CATTESLOWE (obsolete) A tumulus some 40 yards in diameter and 18 feet high which lay on the south-west side of Leek, between Waterloo Road and Spring Gardens (SJ 9756), destoyed in 1907: VCH VII 85. *Catteslowe* 1587 (1883) Sleigh 22, later 16th century VCH VII 85, *Cocklow* 1686 Plot 404, *Cock Lowe or Great Lowe* 1723 VCH VII 85. '*Catta's or *Catt's *hlaw* or tumulus', or 'the tumulus of the wildcat'. The change from *Catte*- to *Cock* is curious. See also Leek; Lowe Hill; Catshill.

**CATT HAYES (FARM)** 2 miles west of Horton (SJ 9157). *? Cateshouse* 1531-2 SRO DW1400/109, C. If the spelling is correctly identified as this place, perhaps '*Catt's house', but no other spellings have been traced, and it is unclear whether the name is ancient.

**CATTON** in Croxall parish, 2½ miles east of Alrewas (SK 2015). *Chetun* 1086 DB (listed under Derbyshire), *Cathton(a)* 1162-3 SHC 4th Series IV 23, *Katton'* 1198-1208 ibid. 43, *Catiton* 1208 Cur, *Catton* 1236 Fees, 1307 SHC VII 173. Probably from the ON personal name Káti, or (less likely) from the OE personal name *Catta, with OE *tun'*, or 'the *tun* of the wild cat', from OE *catt*. Croxall contains a Scandinavian personal name, which increases the likelihood that this name has a Scandinavian derivation. There are other places of this name in Norfolk, Northumberland and Yorkshire.

CAUDEWELL (unlocated, in Blurton, perhaps associated with Coldriding at SJ 9140), *Caldewall* 1275 SHC XI NS 243, SHC XI 311-2, *Caldwelle* 1300 SHC XI 307, *Caudewell* 1302 SHC XI 307, SRO D593/B/1/23/3/1/9. See Coldwell.

CAUDY FIELDS (obsolete, 1 mile east of Walsall (SO 0398)) *Caldwell* 1317 SHC IX (i) 64, *Caldwelle* 1332 SRO D1790/A/2/906, *Caudy Fields* 1834 O.S. See also TSSAHS XXIV 1982-3 49. From OE *cald* 'cold', with Mercian OE *wælle* 'a spring', and (sometimes) 'a stream', a common name in Staffordshire. It is unclear whether *Caldwallemore* near Cheslyn Hay, recorded in 1300 (SHC V (i) 177), and in 1340 as *Caldewallemore* (SRO D593/B/1/26/6/2/11), are to be associated with this place. See also Boweles.

CAULDFORD (unlocated, on the river Tame, on the boundary of Hopwas: Shaw 1798: I 433) *Caldefford, Caldford* 1798 Shaw I 433. 'The cold ford'.

**CAULDON** 6 miles north-east of Cheadle (SK 0749). *Cealdun, Celfdun* 1004 (11th century, S.906; 11th century, S.1536), *Caldone* 1086 DB, *Caluedon* 1196 FF, *Calfdon* c.1200 Bodl, *Caldon'* 1224 Cur, 1228 Pat *et freq.*, *Caldona* 1226 SHC V (i) 54, *Chaueledon, Caldon'* 1242 Fees, *Calveduna* c.1270 SHC 1913 318, *Caldone* 1311 SHC 1911 303, *Coldon* 1332 SR, *Caldon* 1532 SHC 4th Series 8 9, *Calden* 1562-6 *Harl, Caudon, Cawdon* 1598 *Dep, Cauldon* 1599 Smith, *Cawdon* 1601 SHC XVI 225, *Cauldon* 1686 Plot. 'Calf hill', from OE Mercian *cælf, dun. Cælf,* the i-mutated form of *calf,* in the West Midlands normally becomes *calf,* which might conceivably refer to tumuli in some Staffordshire place-names (see Calves Heath), perhaps in this case associated with Cauldon Lowe (q.v.). The O.S. used the spelling Cauldon for the village, but Caldon for the Caldon Canal here which opened c.1777. *Cauldon Grange* is recorded c.1650: DRO DD/SR/225/140. It is possible that some of the above forms may refer to nearby Calton (q.v.). It is unclear whether *Kalwedun,* recorded in 1236 (Fees), is to be associated with this place. Some spellings above may relate to Chorlton (q.v.), and vice versa.

**CAULDON LOWE** a 1190' hill 1 mile south of Cauldon (SK 0748). *? montem de Caldona* c.1220 SHC V (i) 52, *la Lowe* (p) 1322 *AddCh, Caldon Low* 1799 SRO D240/D/240, 1836 O.S. From OE *hlaw* 'tumulus, burial mound', from a burial mound to the south of Caldon. The name has become 'the low or tumulus at Cauldon (q.v.)'. The 7th Series 1" O.S. sheet shows Cauldon Lowe as the hill name, with Caldon Low to the north-east.

**CAVE** in Upper Elkstone (SK 0457). *? Kave* 1286 SHC 1911 40, *? Cave* 1287 SHC ibid.' 192, 1288 SHC VI (i) 185, *Cave* 1775 Yates, 1870 P.O. No cave is recorded here, but the place lies close to the river Hamps, and a derivation from a stream-name OE *caf* 'swift, quick' cannot be discounted (cf. Cave, Humberside).

**CAVERSWALL** Ancient Parish 5 miles south-east of Stoke-on-Trent (SJ 9542). *Cavreswelle* 1086 DB, *Cauereswell* 1167, *Kevereswell* 1177 SHC XII NS 278, *Chaverswella* 1186 I 126, *Chauereswella* 1189 ibid. 140, *Kavreswall(e)* 1221 P, *Kaveriswalle* c.1230 SHC VI (i) 11, *Caverswal* c.1238 (1798) Shaw I xxvi, *Caverreswall* 1242 Fees, *Caveriswelle* c.1266 SHC 1924 360, *Chawerreswell* 1269 SHC IV 177, *Caverswall* 1272 Ass, *Cariswell, Cariswel* 1274 SHC 1911 160, *Kavereswell* 1275 SHC VI (i) 69, *Kaverswalle* 13th century Dieul., *Careswell* 1327 SR *et freq, Carriswall* c.1355 NA DD/FJ/4/26/7, *Kareswall* 1388 Fine, *Cariswall'* 1461 HAME 485, *Caerswall* 1561 SHC 1931 230, *Careswall* 1601 SHC XVI 223, *Careswall als Caverswall* 1618 FF, *Carswall* 1643 SHC 1957 6. The first element is almost certainly a personal name connected with OE *caf* – perhaps *Cafhere, or a derivative with an *-r* suffix (cf. Caversfield, Oxfordshire; Caversham, Berkshire). The second element is Mercian OE *wælle* 'a spring', and (sometimes) 'a stream'. See also Caverswall in Loxley. A field-name *Caverswalle* is recorded in Lilleshall, Shropshire (PN Sa III 31), possibly associated with *Caverswall*, recorded in Sheriffhales in 1524 (SRO D593/J/22/20/2).

**CAVERSWALL** in Loxley 3 miles south-west of Uttoxeter (SK 0431). *? Caverswall* 1272x1327 SRO DW1733/A/2/16, 1286 SHC 1919 39, *? Caverswalle* c.1286 NA DD/FJ/1/180/1, *? Careswelle* 1296 SHC VII (i) 41, *? Caverswall* 1493 SRO DW1733/A/2/33, *Caverswall House* 1671 SRO DW1733/C/1/4, *Caraswell* 1679 SHC 1919 270, *Caverswall* 1836 O.S., *Caveswall* 1851 White. This place was evidently associated from an early date with the Caverswall family from the better-known place of the same name (see e.g. references *temp.* late Edward I: SRO DW1733/A/2/7), and it is possible that some of the spellings which appear for that place are properly to be associated with this place. It is curious and unusual for two places with the same name and no distinguishing place-name elements to have existed in such close proximity. Yates' map of 1775 shows two buildings some distance apart in this area, one marked as *Cavers*, the other *Wall*.

**CAWARDEN SPRINGS** 1½ miles east of Rugeley (SK 0618). *? Cawardyn* 1414 SHC XVII 17, *? Kawardyne* 1437 SHC III NS 135, *Cawardyne* 1439 SHC III 149, 1528 SHC 1939 78, *Cawardyn* 1453 ibid.' 206, *Cawerden* 1551 SHC 1912 189, *Cawardine* 1653 WCRO CR1908/17/5, *Carden-spring* 1798 Shaw I 187, *Cawarden-spring* 1798 ibid. 200, *Cawarden Springs* 1834 O.S. The name is from the Cawarden family of Cawarden in Cheshire who acquired the place by marriage at about the end of the 13th century: Shaw 1798: I 180. The *spring* element is not connected with water, but from *spring* 'a newly-planted wood; a coppice': it 'had its name from an ancient adjoining spring wood, formerly owned by the Cawardens ...' : Shaw 1798: I 187; 200. The pronunciation is recorded by Shaw (ibid. 205) as Ca'rden. See also Shaw 1798: I 196 fn.4.

**CAWBROOK** a tributary of the river Manifold. Early forms are unrecorded, but perhaps from *caw-*, a local version of OE *cald* 'cold'.

CAWNE HILL (unlocated, near Cannock Wood: SRO D603/E/3/25) *Cawne Hill* 1608 SRO D603/E/3/25.

**CAWNEY HILL** ½ mile south-east of Dudley (SO 9589). *Cawenhal* 1655 Roper 1980: 100, *Cawney Hill* 1834 O.S., *Cawney* (*Moor, Meadow* and *Grounds*) is recorded in Aston, north of Birmingham, in 1652: BCA MS3369/Acc 1961-9/13a.

CAWYTON unlocated, perhaps near Ranton. *Cawyton* 1539 SHC V NS 322.

**CELLARHEAD** 4 miles east of Stoke on Trent (SJ 9547). *Sellarhead* 1736 *EnclA*, *Cellar Head* 1775 Yates, *Sellar Head* 1794 GM II 1078, *Cellar Head* 1803 SHC 1933 150, 1836 O.S. Elsewhere the element *cellar* is associated with land endowed for the office of an ecclesiastical cellarer (see Cellers Grove (unlocated), Worcestershire: PN Wo 104), but no ecclesiastical connections with this place have been traced. Although twice-yearly fairs were formerly held here (White 1851; VCH VI 94), it seems likely that the name is not ancient. EDD gives *cellar-head* as 'the landing or shelf at the top of the stairs leading to the cellar', and Halliwell gives *cellar* to include 'a canopy', so perhaps an 18th century jocular name, or from one or more of the public houses here, or possibly 'the high ground with the shelter': the place lies at a high, exposed crossroads. However, the expression cellar-head seemingly in the sense 'public-house' is recorded in 1798 (see Shaw 1798: I 200 fn.1), and it seems very likely that this is the meaning here. Yates' map of 1775 shows only one building (which may have been the feature from which the place was named) on the south-east corner of the crossroads here. The second element may have been influenced by nearby Armshead (q.v.), or vice versa.

**CHADDESHOLM** (unlocated, perhaps near Wolseley or Colton) *Chaddesholm* 1176 SHC 1914 137, *Chaddeholm* 12th century SRO DW1781/1/2, *Chadholme field* 1546 SRO DW1781/5/2/1, *Chadholme* 1587 SRO DW1781/9/2/2. Seemingly from ON *holmr* 'an island, higher ground amidst marshland', so perhaps 'St. Chad's island or dry ground in a marsh'.

**CHADSMOOR** 1 mile north of Cannock (SJ 9811). *Chadsmoor* 1895 O.S. A 19th century housing development named after St. Chad, the patron saint of the church built here in 1891: VCH V 67.

**CHADSMOOR** 2 miles north-east of Cannock (SJ 9912). Duignan 1902: 34-5 records that the north-western boundary of Cannock manor was marked by a deep, broad ditch called in ancient deeds *Fossa beati Cedde* (1549 (1801) Shaw II 312), 'the fosse of the Blessed St. Chad' (see SHC VI (i) 296; also NSJFS 8 1968 44-6; StEnc 128), the patron saint of Lichfield Cathedral, the manors of Cannock and Rugeley being held by the bishops of Lichfield to the time of Henry VIII. The ditch also defined part of the boundary of Cannock Chase: JNSFC 8 1968 49; TSSAHS VII 1965-6 36. A gate called *St. Chad's Gate* stood on the ditch, and the adjoining moor was called *Chad's Moor*. See also SHC 1924 286.

**CHADWELL** 2 miles north-west of Blymhill (SJ 7814). *Little Chatwall* 1356 SRO DW1082/L/17/1, *Little Chetwell* 1547 SHC XI 292, *Lytel Chatwalle* 1570 SHC IV NS 185-6, *Lyttle Chattwall* 1592 SHC 1934 18. 'Ceatta's spring', from OE Mercian *wælle* 'a spring', or (sometimes) 'a stream'. The terminals *welle* and *wall* are frequently interchanged in ME forms. The name was seemingly pronounced Chattle in the 19th century: Hope 1893: 160. The 1833 O.S. map shows *St Chadds Well* at Little Chatwell (now Chadwell), showing that the former name became popularly but erroneously linked with St Chad, a saint with strong Staffordshire connections. There are still springs in Chadwell, the largest, which feeds the pool of Chadwell Mill, being St Chad's Well. It has been suggested (see StEnc 197) that *Elder-well* recorded in Plot 1686: 106 may be this well, but Plot locates Elder-well beween Blymhill and Brineton, which if correct rules out St Chad's Well. Chadwell has been in Shropshire since 1895. See also Chatwell, Great & Little, and St Chadds Well.

**CHAPEL ASH** on the south-west side of Wolverhampton (SO 9098). *Chapel Leasowe* 1515x1547 TSSAHS XXXVII 1995-6 130, *Chappell Ash* 1707 BCA MS3145/63/1a&b, *Chapel Ash* c.1743 (1801) Shaw II 163, 1810 Codsall ParReg, 1834 O.S. In 1550 there is reference to a chapel of the Hermitage here, with a burial ground and adjacent croft, probably the place recorded as *Chapel Leasowe*. TSSAHS XXXVII 1995-6 122; see also Roper 1966: 8-9. The place was formerly known as *Oxenford* (q.v.). Wilkes notes c.1750 the discovery of medicinal springs here which gave rise to the name *Wolverhampton Spa*. Shaw 1801: II 163.

**CHAPEL CHORLTON** – see **CHORLTON, CHAPEL**.

**CHAPEL HILL** ½ mile south of Bishton (SK 0219). *Chapel Hill* late 15th century SRO DW1781/5/12/1.

**CHAPEL HOUSE FARM** on the south-west side of Tutbury (SK 2027). From a chapel which once stood here: see Shaw 1798: I 56, 57.

**CHARLEMONT** 2 miles south-east of Wednesbury (SO 0193). *Charlemont* 1723 VCH XVII 21, 1806 SRO D742/A/33/1-9, *Charley Mount* 1758 ibid, BCA MS3375/456901, 1775 Yates, *Charleymount* 1834 O.S. Charlemont is said to have been called originally Crump Hall, built by John Lowe (d.1729), with the name changed to reflect its owner's Whig sympathies after Charlemont, a French fortress on the Meuse built in 1555, or Charlemont castle in Munster: see Hackwood 1895: 70-71. However, after the Dissolution, Sandwell Priory and much of the surrounding land (including the area incorporating Charlemont) passed eventually to the Legge family. William Legge (?1609-1670) was the son of Edward Legge, Vice-President of Munster, and spent his youth in Ireland. William served Charles I and Charles II, and his son was created Baron Dartmouth in 1682. This place may have been named by the family after the castle in Munster (for very tenuous evidence see SRO D742/N/2/40), or perhaps after the Earls of Charlemont. See also StEnc 128.

**CHARNES** 4 miles north-west of Eccleshall (SJ 7733). *Ceruernest* 1086 DB, *Chauernese* 1197 P, *Chauernes* 1242 Fees, *Chaunes* 1227 SHC IV 73, *Cauernessa* 1230 P, *Chavernes, Charneves, Charneles, Chaunes, Chavernesse* 13th century Duig, *Charnez* 1377 SHC 4th Series VI 15, *Chavernes* 1380 SHC XVII 202. Possibly from OE *ceafor* 'a beetle', with Mercian OE *ness*, a term related to modern *nose*, meaning 'headland, promontory', thus 'headland infested with beetles'. It is unlikely that the second element is OE *nest* 'nest'.

**CHARTLEY** 5 miles south-west of Uttoxeter (SK 0028; SMR 00747 places a deserted pre-Conquest settlement at SK 00862851). *Certelie* 1086 DB, *Certelea* 1192 P, *Cerdel* 1232 Cl, *Scerteley* 1236 Fees, *Certele, Cartesleg'* 1242 Fees, *Certeley* 12th to 14th century Duig, *Chartley (Castle & Park)* 1798 Yates. Probably from OE *\*cert* 'rough ground' (cognate with ON *\*kartr* 'rough, rocky soil'), in which case the element may be plural, but the word is little used outside south-east England, and an OE personal name such as *\*Cearda* cannot be ruled out completely. The second element is OE *leah*.

CHARTLEY HOLME (obsolete) an extra-parochial liberty corresponding roughly with Old Chartley Park, containing Chartley Castle and Chartley Hall: see Erdeswick 1844: 55 fn.a; Youngs 1991: 407; StEnc 131. *Charteley holme* 1586 SHC 1927 167, *Chartley Holme* 1592 SHC 1930 295, 1640 SHC XV 208. *Holme* is from ON *holmr* 'a small island, a piece of dry land in a marsh; a piece of land partly surrounded by streams or by a stream': there are large areas of marshland at Chartley.

**CHARTLEY MOSS** a 104-acre raft of floating peat to the east of Chartley (q.v.) (SK 0228). The name is from OE *mos* 'a moss, a bog, a marsh'.

**CHASEPOOL (LODGE)** 2 miles west of Kingswinford (SO 8589; SMR 01908 places a deserted post-Conquest settlement at SO 84408990). *Catspelle* 1086 DB, *Chacepel* 1271 SHC V (i) 155, *Cachepol* 1273 SHC 1911 155, *Chacepol* 13th century, *Chaspell otherwise Chasboll* 1580 SHC XVII 225, *Chaspell Chase* 1581 SRO D260/M/F/1/2/f100d-101d, *Chaspell* 1598 SHC VI NS (ii) 289, *Chasbold otherwise Chaspell* 1624 SHC NS X (i) 64, *Chasphill* 1686 Plot 397. Formerly assumed (e.g. by Erdeswick 1844: 551 and Eyton 1881: 68) to be at Gospel End in Sedgley, the DB place is now held to have been in Kinver Forest (see SHC 1944 89), with the name surviving in Chasepool Lodge, near Swindon (SHC XI 253; VCH IV 54 fn.). The derivation of the name is uncertain, but may be from OE *\*ceas* 'a heap'. The second element is not identified. It is unclear whether *the King's forest of Chaspeth*, recorded in 1543 (SA 2089/2/2/46), refers to Chasepool, which was probably a hay or enclosure in Kinver Forest: NSJFS 8 1968 42.

**CHASE TERRACE** 1 mile west of Burntwood (SK 0409). A mining village which had developed by 1870: VCH XIV 201. The name is from nearby Chasetown (q.v.).

**CHASETOWN** 2 miles north of Brownhills (SK 0408). Created in the mid-19th century to house colliers for the local mines, the place was first known as Cannock Chase, and later as Chasetown (a name possibly created by George Poole, vicar of Burntwood and his wife, or Elijah Wills, a local schoolmaster) by 1867, when it became an ecclesiastical parish: VCH XIV 199; Youngs 1991: 407.

**CHASEWATER** a lake 1 mile east of Norton Canes (SK 0307). Crane Brook on Norton Bog was dammed c.1798 to create Norton Pool, a reservoir to feed the Wyrley & Essington Canal: VCH XIV 195. The pool was renamed Chasewater (from Chasetown to the east) in 1956.

**CHATCULL** 4½ miles north-west of Eccleshall (SJ 7934; SMR 02407 places a deserted post-Conquest settlement at SJ 79503440). *Cetquille* 1086 DB, *Chatkull* 1199 SHC III 56, *Katkulne* 1199 ibid.169, *Chatculne* 1199 FF, *Chatkiln* 1200 SHC III 65, 1327 SR, *Chatkull* 12th century, *Chachull, Chatchull, Chatculne* 13th century Duig, *Shatkelve* 1227 SHC IV 43, *Chatkehull* 1273 SHC 1911 149, *Scatculne* 1327 SHC VII (i) 92, *Chekkulne* 1332 SHC XI 173, *Chatkyll* 1551 SHC 1914 23, *Chattell* 1666 SHC 1921 118, *Chat-Kilne* 1676 SHC 1914 22. The second element is from OE *cyl(e)n* 'kiln, a furnace for baking or burning materials', so perhaps '*Ceatta's kiln'. It is unclear what material might have been processed in the kiln. The first element is unlikely to be from PrWelsh *ced*, British *ceto* 'a wood', since other spellings with *Cet-* would be expected.

CHATFIELD (unlocated, possibly remembered in Chatfield Place in Longton (SJ 9142)) An interesting name, found as a surname in the northen part of the county (see Tooth 2000: 51), which is perhaps a hybrid with the first element from PrWelsh *ced*, British *ceto* 'a wood' (cf. Modern Welsh *coed*), with OE *feld* 'open land', probably used by the Anglo-Saxons for uncultivated areas used for common pasture, and incorporated into settlement names when arable encroachments forming part of new settlements were made.

**CHATTERLEY** 3½ miles north of Newcastle-under-Lyme (SJ 8451; SMR 04848 places a deserted post-Conquest settlement at SJ 84505125). *? Cattelega* 1187 SHC I 131, *Chaderleg* 1212 Fees, *Chaderlyhe* 1217x1227 CEC 393, *Chaterlyh* 1227 Ch, *Chaderleia* 1227 SHC XI NS 240, *Chadderley* 1252 Ch, *Chaddenelle* 1273 SHC VI (i) 59, *Chatterley* 1377 SHC 4th Series VI 14, *Chadderlegh, Chaddendelle* 13th century Duig, *Chetterton (sic)* 1532 SHC 4th Series 8 49, *Chatterley* 1592 SHC 1935 219, 1643 Erdeswick 1844: 26, *Chaterley* 1686 Plot, *Big Chatterley, Little Chatterley* 1836 O.S. The first element is almost certainly from PrWelsh *cadeir* 'chair' from Latin *cathedra*, used of a hill or elevated place, hence 'wood by a hill called Cader'. The development of an initial palatal, especially found in the West Midlands, is OE: EPNE i 75. The second element is OE *leah*, but there appears to have been a variant with OE *dell* 'pit, dale, valley'.

**CHATWELL, GREAT** 4 miles south-east of Newport (SJ 7914). *Chattewell* 1203 SHC III (i) 99, 1273 FF, *Chatewell* 1275 SHC 1911 29, *? Schadewall* 1294 SHC 1911 222, *Chattwelle* 1315 SHC 1911 318, *Chatwall* 1327 SR, *Chatwalle* 1331 SHC 1913 34, *Chetewalle* 1349 SRO D590/127, *More Chatwalle, Lytel Chatwalle* 1473 SHC IV NS 185-6, *Moche Chatwall* 1535 SHC 1912 82, *Chatwel* 1454 ECP, *More Chatwalle* 1570 SHC IV NS 185-6, *Challeton otherwise Great Chatwell* 1592 SHC XVI 119, *Chattwell* 1608 Gnosall ParReg, *Chatwall* 1666 SHC 1927 70. '*Ceatta's spring', from OE Mercian *wælle* 'a spring', and (sometimes) 'a stream'. The personal name is not recorded in early records, but is inferred from place-name evidence: see PN Sa III 120; VEPN II 149-50. The terminals *welle* and *wall* are frequently interchanged in ME forms. The name was seemingly pronounced Chattle in the 19th century: Hope 1893 160. *Moche* (i.e. Much) in the 1535 form means 'Great': cf. Much Wenlock. Little Chatwell is now known as Chadwell (q.v.). See also St Chadds Well.

**CHEADLE** Ancient Parish 8 miles east of Stoke on Trent (SK 0043). There were two manors in Cheadle, one held by the family of Bassett (of Sapcote) from at least 1176 (SHC 1923 36), the other by the Bishops of Lichfield and later Croxden Abbey (SHC 1923 35-41). The latter is almost certainly the *Cedla* of DB, which became Hounds Cheadle and later Cheadle Grange (q.v.). *Celle* 1086 DB, *Chedle* 1162x1173 CEC 170, *Chelle* 1176 SHC III (i) 225, *Chedelea* 1195 Pipe, *Chedele* c.1196 SHC II 68, *Chedle* 1227 Ass, 1253 FF, *Chedlhe Basset* 1236 Fees, *Chedele Basset* c.1238 (1798) Shaw I xxviii, *Chedlee* 1293 SHC VI (i) 229, *Chedille* 1377 SHC XIII 141, *Chetelle* 1423 SHC XVII 90, *Chedyll'* c.1447 SHC 4th IV 239, *Chedyll* 1532 SHC 4th Series 8 27, *Chedull* 1597 SHC 1934 (ii) 4. The first element is British *ceto* (Welsh *coed*) 'wood', an element in a form which has not undergone an 8th-century Welsh change (see PN Ch V (II) 296), with OE *leah* 'forest, wood, glade, clearing', and later 'pasture, meadow'. If the meaning here is (as seems quite possible) 'forest, wood', the result is the tautologous '*chet* wood' or 'forest called Ched', which does not necessarily indicate that the meaning of the first word had been forgotten when the element was added. For the association of DB *Celle* with this place see SHC 1923 35-7. Cheadle was divided into four 'quarters': Above Park (*Above Parke* 1596 SHC XVI 185; *Above Park, Little Above Park* 1836 O.S.), Cheadle (town), Huntley (q.v.) and Cheadle Grange. The first three were included in Basset Cheadle. The fourth (with some other places including Thornbury Hall and Woodhead) was effectively Cheadle Grange (q.v.): SHC 1926 155-6; JNSFC 48 142-161. See also Parkhall.

CHEADLE GRANGE There were two manors in Cheadle, one held by the Bassetts, the other by the Bishops of Lichfield and later Croxden Abbey. The latter became Hounds Cheadle (recorded from at least c.1236: SRO D593/A/2/23/23) and later Cheadle Grange: see SHC 1923 35-41. *Cedla* 1086 DB, *Dogg(e)-Chedle, Dogg-Chedile* c.1275 SHC V (i) 118-9, *Chirche Chedle* c.1282 SHC 1923 39, *Doggechedile* 1290 SHC VI (i) 203, *Doggechedle* 1323 SHC IX (i) 92, *Hundchedial* c.1536 SHC 1947 50, *Hundchedull, Hundis Chedull* 1539 MA, *Houndes Chedull* 1616 SHC VI NS (i) 9. For the first element see Cheadle. The element *Dogg(e)-*, with the variant *Hound's* (which survived until the 17th century: SHC 1947 50), remains explained, but Basset was an old word for a short-legged dog (not recorded in the OED before 1616), and it seems possible that Dog and Hound developed as jocular canine alternatives to the *Basset* suffix for the other manor of Cheadle held by the Bassetts, and perhaps to distinguish this place from Cheadle Bulkeley and Cheadle Moseley (otherwise Cheadle Hulme) in Cheshire. *Chirche Chedle* may have been a variation used after the place was granted to Croxden Abbey by the Bassetts: SHC 1923 39.

**CHEBSEY** Ancient Parish 2 miles east of Eccleshall (SJ 8628). *Cebbesio* 1086 DB, *Chebeseya* 1160x1176 (14th century) EEA 16 55, *Chebesey* 1220 SHC III (i) 210, *Chebbesee* 1222 FF, *Chebbeshey* 1236 Fees, *Chebbesey* c.1272 SHC III (i) 213, *? Shepefeie* 1291 (1798) Shaw I xvi, *Schebseye* 1325 SHC 1911 365, *Chebbeseye* 1351 SHC 1913 143, *Chebbesey* 1461 HAME 485. OE *Cebbi is a normal formation from the name Ceabba or Ceobba, hence '*Cebbi's island or raised dry area', from OE *eg* 'island, place near water': the church lies on a raised sub-circular churchyard on the south side of the village projecting into low lying wet ground though which passes the river Sow. At the time of DB it included the area later known as Cold Norton, which is not listed in DB. See also Badenhall; Newbold.

**CHECKHILL** 2 miles south of Swindon (SO 8587). The name of this place is believed to be associated with the Forest of Sechehulle: see SHC 1925 244-5; VCH II 335 fn.1. *Schecheel* 1190 VCH II 335 fn.1, *Forestæ de Schechell* 1191 SHC II 11, *Forestæ de Scethelle* 1192 ibid. 17, *Forestæ de Schethell* 1193 ibid. 24, *? Sethill', Sethull* 1194 Pipe, *Forestæ de Secchevill* 1197 SHC II 64, *Forestæ de Secchulle* 1198 ibid. 71, *Forestæ de Sechull* 1199 ibid. 78, *Forestæ de Secchenil* 1201 ibid. 101, *Sechehulle* 1201 SHC 1925 244, *Chekhull* 1286 SHC 1925 244, *Great Checkhill Common, Little Checkhill Common* 1801 SHC 1931 94, *Check hill (Mill), Check hill Plantation* 1834 O.S. The place lies on a promontory in a particularly boggy valley in an angle between two streams, but the root of the name is uncertain. There is a hill reaching 366' to the north-west, and a steeper mass to the south which reaches a height of 443'. Perhaps from OE *sceat* 'corner, angle, edge, point, promontory', with OE *hyll* 'hill'. All the early references describe the Forest, mentioned until the time of Edward I, as waste: VCH II 335 fn.1. See also VCH II 335 fn.1; SHC 1925 244-5. Cf. Checkendon, Oxfordshire; Checkley, Herefordshire. The early spellings attributed to Shottle in PN Db 573 almost certainly relate to this place.

**CHECKLEY** Ancient Parish 3 miles south-east of Cheadle (SK 0237). *Cedla* 1086 DB, *Chakele* 1114x1118 SHC 1916 227, *Checkele* c.1187 SHC II 261, *Checkelega, Checkelee* 1196 SHC 1923 41, *Chekkesleye, Checkele* 1227 Duig, *Chagewell (sic)* 1276 SHC VI (i) 78, *Chekkeleye* 1292 ibid.' 208, *Checcholey* 1366 SRO D938/199, *Chekeley* 1569 SHC 1938 41, *Checkley* 1580 SHC XV 129. If the DB form (which is identical to the DB entry for Cheadle) is discounted as a scribal error, the later forms suggest an OE personal name *Cæcca or similar, but may be from OE West Mercian *cecce (see VEPN II 154), perhaps 'a lump', as applied to a hill, with OE *leah*. For the identification of DB *Cedla* (perhaps a scribal error for *Cekla*) with this place see SHC 1923 35-6. The 1276 spelling is evidently an aberration. Cf. Checkley, Cheshire (PN Ch III 56).

**CHEDDLETON** Ancient Parish 3½ miles south-west of Leek (SJ 9752). *Celtetone* 1086 DB, *Chetilton* 1201 Cur, *Celtilton* c.1216 Rees 1997: 61, *Cheteltun* 1227 Ass, *Chetulton* 1325 SRO DW1781/1/41, *Chetelton* 14th century Duig. '*Tun* in a narrow valley', from OE *cetel* 'kettle', generally with a topographical meaning such as 'deep valley surrounded by hills', which is particularly appropriate for the topography here, where the hills fall steeply to the river Churnet.

**CHEEKS HILL** the northernmost point in Staffordshire, 2 miles north of Flash (SK 0269). *Cheeks Hill* 1842 O.S. Early spellings have not been traced, but Cheek is a surname recorded in this region: SRO D1229/6/6/6.

**CHELL, GREAT & LITTLE** 2 miles north of Burslem (SJ 8652). *Chelle* 1200-5 *et freq.* SHC 1911 443, 1217x1227 CEC 393, *Ceolegh* 1313 Duig, *Chelle Parva* 1334 SHC XVI 6, *Chell* 1576 HLS, *Chell otherwise Great Chell* 1581 SHC XV 137, *Great Chell, Little Chell* 1583 ibid. 146, *Chell Heathe apud Norton* 1597 Eccleshall ParReg, *Great Chell, Little Chell, Chell Green* 1836 O.S. The place lies at the southern end of a long ridge, with a narrow valley running parallel on the west, so possibly from OE *ceole* 'throat', used topographically in the sense of 'a channel, gorge, ravine', or (though much less likely) from OE *ceol* 'keel', from the resemblance of the ridge to an upturned boat: *c* before *e* in OE would be pronounced [tʃ]. It may be noted that modern 'keel' is from the OScandinavian word, rather than the OE word, which explains the *k*. Great and Little Chell probably reflect the division of the manor at some date before 1212: VCH VIII 89.

It is unclear whether *Chelhall*, recorded in 1308 (SHC XI NS 261), is to be associated with these places.

CHELSFORD (unlocated, possibly in Reule) *Cheteleford* c.1290 SHC XII (i) 277, *Chenesford (Br)* 1564 *Ct*, *Chentford (Br)* 1597 *Ct*, *Chednesford (bridge)* 1599 *Ct*, *Chednesford (meadow)* 1599 *Ct*, *Chenfford (meadow)* 1649 SHC IV (ii) 89, *Chelsford (Highway, Footway & Sandy)* 1838 *TA*. An intriguing name. Oakden 1984: 145 suggests 'the ford over the stream called Kennet', a well-known British river-name, from British *\*cunetiu*, the meaning of which is uncertain (see Ekwall 1928: 225-8; Ekwall 1960: 271-2; Jackson 1953: 331-2, 676), but this seems improbable. The forms are inconsistent, but are evidently to be associated with *Ched(d)nes(se)fordbrigge*, a field-name recorded in 1460 in nearby Bradley (Oakden 1984: 139), which Oakden suggests is from 'Cedda's headland', from OE *ness* 'nose, headland'. The various spellings point towards 'Ceadela's ford', with the not unusual confusion of *n* and *l*. Or, less likely, from Chednete, a form of the family name Cheney or Cheyne, cf. Chenies, Berkshire (a very tenuous association of the Cheney family with this area is recorded in SHC IV (ii) 80, but is doubtless coincidental). The ford/bridge may have been where the road from Church Eaton to Bradley crosses the Church Eaton Brook at Church Eaton Common. Cf. Chaddenwick alias Charnage, and Chadlanger, Wiltshire (PN W 178-9, 158-9).

CHERE, LE (unlocated, in Chapel Chorlton: SHC 1913 232) *? la Char* 1308 SHC XI NS 265, *le Chere* 1368 SRO D641/5/T/1/1, *Cheres* 14th century SHC 1913 232. An interesting name. Possibly from Latin *schura*, *shira* 'shed' or OE *\*scȳr* 'hut': see PN Sa IV 7.

CHERLECOT (unlocated, possibly in the Eccleshall area) *Cherlecot* c.1238 (1798) Shaw I xxvi. From OE *ceorl* 'free peasant', with OE *cot* 'cottage', so 'the cot of the *ceorls* or free peasants'. Possibly associated with Chorlton (q.v.). Cf. Charlcott, Shropshire. See also Bowers.

**CHERRYEYE BRIDGE** 1 mile north of Kingsley (SK 0148). A bridge over the Caldon Canal named after the nearby Cherryeye Mine (which closed in 1921) from which was extracted a bright red iron ore which by tradition is said to have made the miners' eyes red. If older (early spellings have not been traced), the name may be from 'the island of land or place by the river where wild cherries grew', from OE *eg* 'island, piece of land on or between streams'.

**CHESLYN HAY** 2 miles south-west of Cannock (SJ 9707). *Chistlin* 1236 Fees, *? Chystehaye* 1250 Fees (but which may be Chestall q.v.), *Chistelin* 1251 Ch, *Chisteling* 1252 Cl, *Hay of Chistelyn* 1293 Ass, *Chisteyn Haye, Chistlyn Haye, Chistline Haye, Chystline Haye* 1538 SHC 1912 117-8. The first element may be from OE *ci(e)st* 'a chest, coffin', perhaps applied to places where ancient cist burials have been found (although no such burials are recorded in this area), or from OE *cistel, cist* 'chestnut tree'. The second element is almost certainly OE *hlinc* 'a terrace, a bank, a ledge': see Gelling 1984: 163-5. Hay is from Mercian OE *(ge)heg* 'an enclosure, a clearing in the forest (cf. Haywood): Cheslyn Hay was one of the Hays of Cannock Forest, and was extra-parochial until 1857. The village here was formerly known as Wyrley Bank (*Wirley Banck* 1691 WCRO CR1291/194/1-3; *Wyrley Bank* 1788 Reg Diss): VCH V 100.

**CHESTALL** in Longdon parish, 3½ miles south of Rugeley (SK 0512). *Chirstalleia* 1129-40 VCH III 223, *Chirshalleiam* 1129x1147 EEA 14 19, *Chistalea* c.1140 ibid, *?Chystehaye* 1250 Fees (but which may be Cheslyn Hay, q.v.), *Chestall* 1577 Saxton, 1584 Comm, 1603 PCC, *Chestals* 16th century Duig, *Chesthall Hall* 1682 Dep. The meaning of this name is uncertain, but possibly a compound of OE *ceast* (*c* being pronounced [tʃ]), meaning 'strife, contention' (a word linked to places which lie on parish boundaries, as here), or OE *cistel*, *cist* 'chestnut tree', with OE *halh* and *leah*. The compound has been noted in Chesthill, a lost place in Moreton Say, Shropshire: see PN Sa I 77. The brook which formed the southern boundary, now called Maple Brook (*Maplebrook* 1595 SRO D603/E/2/45-66), had the same name, recorded as *Christale brook* in 1376 (StSt 10 1998 97), presumably the brook called *Chistals* or *Chistalea* in 1140 (Shaw 1798: I 229; VCH III 223).

**CHESTERFIELD** 2½ miles south-west of Lichfield (SK 0905), *Cestrefeld* 1167 SHC I 47, *Chasterfeld* 1227 SHC IV 70, *Cestrefeud, Chestrefewde* 1262 Duig, *Cestrefeud* 1273 SHC VI (i) 61, *Cestrefend, Chesterfeld* 1332 SHC X 103, *? Chesterford* 1419 SHC XVII 72, *Chesterffeld* 1532 SHC 4th Series 8 165; CHESTERFIELD (unlocated, in Adbaston), *Chesterfield* ?c.1300 SHC 1924 191. From Mercian OE *cester* 'ancient fortification, city', with OE *feld* 'open country'. The first place lies half a mile south of Wall (q.v.), Roman *Letocetum* or *Etocetum*. Above Chesterfield is Castle Croft (*Castlecroft* 1695 Gibson), which has produced Roman remains. The fact that SHC 1924 191 associates the second place with The Camp (unlocated) supports a derivation from an early fortification.

CHESTERHURST (unlocated, in Dilhorne) *Chesterhurst* 13th century SHC VIII 159, *Chestrehurst(well)* 14th century D260/M/T/5/134. From OE *cester* 'ancient fortification, city', with OE *hyrst* 'a hillock, a bank, a copse, a wood', but usually 'a wooded eminence'. Dilhorne lies to the north of the Roman road running between Rocester and Stoke on Trent (Margary number 181), and this lost name may record some ancient archaeological feature. *Well* is from Mercian OE *wælle* 'a spring', and (sometimes) 'a stream'.

**CHESTERTON** 3 miles north of Newcastle under Lyme (SJ 8349), *Cestretone* 1201 SHC II 101, *Cestreton* 1214 CDEPN, 1227 SHC XI NS 240, *? Cheston* 1272 SHC VIII (i) 151, *Cesterhunte, Cesterton* 1280 SHC VI (i) 101, *Chestreton* 1282-3 SHC XI NS 268, *Chasterton* 1293 SHC VI (i) 214, *Castreton* 1298 SHC XI NS 250, *Chesterton* 1335 SHC XI 65; **CHESTERTON** 7 miles west of Wolverhampton (SO 7897), *Cheterton* 1272 Eyton 1854-60: III 112, *Chesterton* 1327 SR, 1349 SRO D593/A/1/19/1, *Chester* 1499 Worfield CA, *Chesterton* 1563 Worfield ParReg, 1602 SA 2028/1/5/8, 1636 Claverley ParReg, 1752 Rocque, *The Walls of Chesterton* 1798 Yates, *Chesterton* 1808 Baugh. From Mercian OE *cester* 'ancient fortification, city', with OE *tun*. The first place is on the site of a Roman settlement (perhaps the 'old' castle, to be distinguished from the 'new' castle at Newcastle): Erdeswick 1844: 22 records c.1595 'ruins of a very ancient town or castle, there yet remaining some rubbish of lime and stone; whereby may be perceived that the walls have been of marvellous thickness'. It is possible that Adam de Camp, recorded in 1307 (SHC XI NS 266), took his name from the Roman remains. Adjoining the second place, which has been in Shropshire since the 12th century, is The Walls, an earthwork, probably of Iron Age date. The brook here is called *Stratford Brook* (1719 Gale 1780: 123, Eyton 1854-60: III 213; 1833 O.S., 1952 O.S.), and the road was known as *Stony Street* (TSAS 4th Series XI xiii), which suggests a Roman road or *stret*, doubtless a continuation of the road from Greensforge which has been traced as far as the north-west end of Abbot's Castle Hill, and perhaps to be associated with Stretford (q.v.).

**CHESWARDINE** Ancient Parish (pronounced Chezadine [tʃɛzədaɪn]) 4 miles south-east of Market Drayton (SJ 7129). *Ciseworde* 1086 DB, *Cheswardin* 1159-60 frequently, with *Cheswurthin, Cheswardin, Cheswerdyn, Cheswardyn, Cheswordyn* to 1428, *Chesworthin* 1210-12 Gelling 1990: 78, *Cheswardine* 1662 (PN Sa I 78). From OE *cese* 'cheese', with OE *worþign* 'an enclosure, a farm', alternating with OE *worþ*, with the same meaning, hence 'the cheese-making farm'. Since at least 1166 in Shropshire.

CHETEWIK (unlocated) *Chetewik* 1227 SHC IV 48. The place has not been identified, and may not be in Staffordshire.

**CHETWYND BRIDGE** – see **SALTER'S BRIDGE**.

CHEVELESDON (unlocated; perhaps Hill Chorlton) *Chevelesdon* 1240 SHC IV 237.

CHIBRICTON (unlocated, perhaps near Hints/Packington) *Chibricton* 1234-40 (1798) Shaw I xxvii, *Chibritton* 1243 SHC 1911 396.

CHIDLOWE (unlocated) *Chidlowe* 1484 SHC VI NS (i) 159.

CHILDERHEY (obsolete, in the north-east part of Burntwood), *Childerhay* 1298 SHC XIV 199, *Childephay* 1302 SHC 1911 59, *Chylderhey* c.1563 SHC IX NS 36, *Childerhey* 1565 SHC XVII 214, *Childerhaye* 1599 FF; CHILDERPLAY (obsolete, south of Stadmorslow (SJ 8755)), *Childerplawe* 1320 Pape 1928 76, 1323 Ass, 1334 SHC XI 55, *Childerplaye* 1360 FF, *Chylderplaye* 1410 Ass, *Childerplay* 1426 SHC XVII 109, *Childerley* 1544 SHC 1910 247, *Childerpler* 1576 Biddulph ParReg, *Childerspleay* 1668 ibid., *Childer Play* 1832 Teesdale. From OE *cildru hege* (or *(ge)heg*) 'enclosed area where children or young creatures play', and OE *cildru plega* '(place where) children or young creatures play', respectively. OE *cild* probably meant 'young living creature', from which the more specialised senses 'young human' and 'young animal' developed: JEPNS 36 2004 63-82. The element *(ge)heg* is more readily explained with reference to young animals than children. An area west of Childerhay in Burntwood around Spade Green was known as *Childerheyende* in 1608 (Farewell ParReg), *Chelderheyend* in 1609 (ibid.'), and *Childerend Pipe* in the later 16th century (VCH XIV 199). For a 12th/13th century reference to *Childescroft* near Stapenhill, Burton upon Trent, see SHC V (i) 43. The field-names *Near*, *Far* and *Upper Childer Play* are recorded in Cauldon in 1845 (TA). The name Childerplay near Stadmorslow is preserved in Childerplay Road. *Childrecroft meadow* and *Childrecroft Gorstes* are recorded near Penkridge in 1471 (SHC 1931 237, 240), presumably to be associated with *Childercroft Barn near Pileton* (Pillaton) recorded in 1782 (SHC 1970 150).

**CHILLINGTON** 1½ miles south-west of Brewood (SJ 8606; SMR 01898 places a deserted post-Conquest settlement at SJ 86550655). *Cillentone* 1086 DB (listed under Warwickshire), *Cildentona* c.1129 SHC I 2, *Chi(l)linton(a)* 1162 to 1181 Giffard, 1292 ibid, *Chylinton, Chylynton, Chilinton* 1175x1182 StCart, *Chilintonam* c.1180 SHC III (i) 203, *Chilintona* 1176x1182 EEA 16 24, *Chylton* c.1214 SHC III (i) 193, *Chyllington, Chillington, Chillyngton* 1236, 1278 MRA, 1278 to 1516 FF, *Chelenton* 1532 SHC 4th Series 8 86, *Chellnigton* 1577 Saxton. Perhaps '*Cilla's tun*', or possibly from OE *cille* 'spring': see Gelling & Cole 2000: 285 (in OE *c* was pronounced [tʃ] when it appeared before *i*), meaning '*tun* at the spring' (cf. Kellington,Yorkshire), but the 1129 spelling points towards a derivation from OE *cildena*, genitive plural of OE *cild*, 'child' (not infrequently found in place-names), so 'the children's *tun*', perhaps denoting use by the young heirs of a landowner (see EPNE i 93-4). *Cild* is found in the 11th century as a title of honour, sometimes for the sons of royal or noble families. However, more recent research has produced evidence that OE *cild* may mean 'young, living creature', from which the more specialised senses 'young human' and 'young animal' developed, and the numerous examples of the place-name Chilton (at least 14 have been identified) and similar (including this Chillington) may be interpreted to mean 'breeding farm'. At some date between 1086 and 1162 Chillington was mised into Shropshire, and subsequently into Staffordshire: SHC I 37. There are other places of this name in Stokenham, Devon, and in Somerset, but each has a different derivation. See also Gelling 1988: 176.

CHILTERN HALL (obsolete, a prebendal manor house in Gnosall) *Chiltern(e)hall* 1395 Chanc M, *Chilternha* 1496 SHC 4th Series VII 169. Perhaps from Walter de Chilterne who was appointed prebendary in 1347. Before this date it seems to have been known, for reasons which are uncertain, as *Andrewshall*: SHC 1927 110. The hall was still standing in the later 18th century: VCH IV 115.

**CHIPNALL** 1 mile north-east of Cheswardine (SJ 7231). *Ceppecanole* 1086 DB, *Cipenol* 1180 For, *de Chipnoll* 1250 Bowcock, *Chipknol* 1256 Rees 1997: 124, *Chippeknol* 1260 Eyton, *Chippechol* 1278 SHC 1911 170, *Chipnolesti* 1298 NSJFS 4 1964 63, *Chipknol* 1320 SHC IX (i) 76, *Schippenol* 1332 SHC X (i) 99, *Chipnolle* 1370 SHC VIII NS 233. Ekwall 1960: 105 suggests 'Cippa's knoll', but the personal name appears to be otherwise unrecorded, and a more likely derivation may be from OE *cipp(a) cnoll* 'the knoll or small hill where logs were obtained', from OE *cipp* 'a beam, a log': cf. Chippenham; Chipley, Suffolk and Somerset. Bowcock 1923: 70 suggests that the name may have been 'the chapman's knoll', where a market was held, but the forms do not support such derivation. The place has been in Shropshire since at least 1166. *Chipcnollesmedewe* in Drointon is recorded in the mid-13th century (SRO 499[7922]), and *Chipenol* mid-13th century (SRO 521[7922]), *Chippenol* and *Chipnol* late 13th century (SRO 537[7922] and SRO D938/544), *Chippechnoll'* 1272 (SRO 542[7922]), but it is unclear whether all forms relate to the same place, which has not been identified.

CHISBEMHULLE (unlocated, in Abbots Bromley: Stuart 1987: 50; SHC V (i) 46) *Chisbemhulle* c.1220 SHC 1937 35, *Chisbill Hill* 1600 SRO DW1734/J/1134. From OE *cist-beam hylle* 'chestnut-tree hill', a rare example of *cist-beam*.

**CHITLINGS BROOK** a tributary of the river Trent. Perhaps an onomatopoeic name from the sound made by the stream.

**CHITTA RIVER** a tributary of the Manifold which flows from Grindon – see Chitlings Brook.

CHOLPESDALE – see **DIEULACRES**.

**CHORLEY** 3 miles west of Lichfield (SK 0711). *Cherlec* 1231 CDEPN, *Churleye* 1306 SHC VII 160, *Scherleg'* c.1310 SRO D1734, *Cherleymor* pre-1311 SHC 1921 7, *Chorley* 14th century Duig., *Chorley* 1589 SHC 1929 326, *Chorley, alias Charley* 16th century Duig. Probably from OE *ceorl* 'free peasant', rather than the OE personal name Ceorl, with the second element *leah*. Cf. Chapel Chorlton.

**CHORLTON, CHAPEL** 6 miles north-west of Stone (SJ 8137). *Cerletone* 1086 DB, *Cherletona* 1166 SHC I 147, *Cherleton* 1267 Ass, 1273 SHC 1913 245, 1278 SHC VI (i) 85, *Charleton* 1532 SHC 4th Series 8 176, *Chorleton* 1539 SHC V (i) NS 299, *Chawlton* 1559 SHC XVII 210, *Chorleton* 1565 SHC 1938 61, 1575 SHC 1926 85, *Chalton* 1586 SHC 1927 132, *Chauton* 1666 SHC 1921 119, *Chorlton* 1686 Plot, *Chorlton Devisover* 1747 Bowen. The original form was probably *Ceorlatun*, a very common place-name compound from OE *ceorl* 'free peasant', with OE *tun*, giving 'the settlement of a group of ceorls', or (more likely) 'the peasants' enclosure', denoting 'a village on an estate which includes more than one unit of settlement. It is not the principal unit, being situated a mile or more away from the seat of the lordship, but is subject to the same lord ...': Finberg 1964: 159. The name indicates that other places had a different status (see for example Bowers, Cherlecot), and is evidence of early manorialism: see PN Sa III xii; 76. Evidence for the meaning 'enclosure' is found in the Anglo-Saxon Law of Ine (King of Wessex 688-726) which required *ceorles weorþig sceal beon betyned*: 'a peasant's homestead must be fenced (Law 40: EHD I 368). The French loan-word *chapel* is a later addition: the place was a chapelry of Eccleshall parish. *Devisover* in the 1777 form is unexplained. It is clear that the names of Chapel Chorlton and Hill Chorlton (q.v.) have different derivations, and that Cherleton and its variations is now Chapel Chorlton, and Chauldon is Hill Chorlton, which is supported by the references to *Cherleton otherwise Chorleton, Chawton otherwise Chelvedon* 1604 (SHC 1945-6 169), and *Chaldon otherwise Chawton otherwise Chalvedon* 1613-4 (SHC 1934 32; see also SRO D593/B/1/6). See also Stableford.

**CHORLTON, HILL** 1½ miles south-west of Whitmore (SJ 7939). *Cerueldone* 1086 DB, *Hylle* 1194 P, *Hulle* 1227 SHC IV 43, 1267 ForPleas, *Chaveldona* c.1270 SHC 1913 318, *Chalvedon* 1273 SHC 1913 245, *? Schelvedon* 1283 SHC 1911 183, *Chauelden temp.* Edward II SRO DW1082/L/12/4, *Chalveldon* 1323 SHC 1913 245, *Chauldon* 1323 (p) Ass, *Schaldon* 1326 SHC 1911 365, *Chalvedon* 1343 SHC 1913 245, *Charldon* 1386 SRO DW1082/A/3/3, *Callton* 1532 SHC 4th Series 8 177, *Chalden* 1539 SHC V (i) NS 297, *Chawton* c.1565 SHC 1938 61, *Chauton* 1666 SHC 1921 119, *Chawton on ye Hill* 1686 Plot. The name is probably derived from OE *calf* 'calf', genitive plural *calfra* (later *calver*), with OE *dun* 'hill', so 'the hill of the calves': cf. Chaldon Herring or East Chaldon, West Chaldon, Dorset; Chaldon, Surrey. The pronunciation of *calf*, etc., would normally be expected to be *karf*, but parallels for a *chalv* pronunciation include Chalvey, Berkshire, and Chalvington, Sussex, though both are in the south. Over time a superficial similarity to the name of nearby Chapel Chorlton (q.v.), 1 mile to the south-east, led to both places using the name Chorlton, with distinguishing descriptors. It is possible that the word *calf* was applied to tumuli (cf. Calf Heath), and at least one tumulus some 15' high survives to the south of the hamlet (TNSFC 1983 12 ). See also Stableford.

**CHORLTON MOSS** 1 mile south-west of Whitmore (SJ 7939). *Mosseffeld* c.1270 SHC 1913 242, *Chorlton Moss* 1833 O.S. From OE *mos* 'moss, lichen', but also (as here) 'a bog, a swamp', with OE *feld* 'open land'. For Chorlton see Chorlton, Chapel.

CHOTER BRIDGE (unlocated, possibly near Alrewas) *Choter bridge* 1612 (1798) Shaw I 138.

CHOTES – see **COTTON, NEAR & UPPER**.

**CHRISTIANSFIELD** on the south side of Stychbrook, 1 mile north of Lichfield (SK 1111). *Cristiannsfeld* 1365 TSSAHS XXVIII 1996-7 9, *Christian Field* 1637x1749 BCA MS3558/150/63B. Plot 1686: 398-9 provides the following explanation for the name: 'said to be the place where St Amphibalus taught the British Christians converted by the Martyrdom of St Alban, who flying from the bloody persecution of Maximian raised in Britain An. 286, followed him hither 84 miles ... from the place of their conversion; where the Romans that were sent after them (some say from Verulam, others from Etocetum now Wall as the tradition goes here) finding them in the exercise of their Religion, tooke them and carryed them to the place where Lichfield now is, and martyred 1000 of them there, leaving their bodies unburyed to be devowered by birds and beasts, whence the place yet retains the name of Lichfield or Cadaverum campus, the field of dead bodies to this very day ...'. The legend, for which there is no factual basis, is supposedly first recorded by John Rous of Warwick, a herald in the early 15th century (Shaw 1798: I 232; Beresford 1883: 7), but seems to date from marginalia by the chronicler Matthew Paris in his copy of William of St. Albans: 'Hoc apud Lichefeld evenit. Inde Lichfeld dicitur quasi campus cadaverum. Lich enim Anglice cadaver sive corpus dicitur' – 'This [massacre] happened at Lichfield. Whence it is called "Lichfield", as it were "field of corpses". For "Lich" in English means a corpse or the body of a dead person': TSSAHS XXVIII 1986-7 5. The discovery of quantities of human bones near here recorded in 1819 was held to be evidence in support of the derivation: see JNSFC 4 1964 28; TSSAHS XXVIII 1986-7 1-13. It seems possible that the name was applied in the light of the massacre legend after burials had been discovered in the area in the early medieval period. However, at some date before 1257 a Christana Venatrix ('huntress') owned land at Elmhurst and Stitchbrook, and may have given her name to the place. For a full account of the history of the legends see TSSAHS XXVIII 1986-7 1-13.

**CHUCKERY, THE** in south-east Walsall (SO 0197). *Chirche-greve*, *Chirchegrevefeld* 13th century Duig, *Church Grevefield* 16th century SOT D260/M/T/1/16, *Churchgreavefield* 1630 SOT D260/M/T/2/12. A development of 'church grove field' (from OE *græfe* 'grove, copse, thicket', and in some cases 'pit, trench'), one of the medieval common fields which existed by the late 13th century and remained open until at least 1735. See also VCH XVII 153, 180.

**CHURCHBRIDGE** 1½ miles south-east of Cannock (SJ 9808). *Chyrche Brugge* 1385 Vernon, *Chirche Bridge* 1538 Deed, *? Chirchebrygge* 1659 (1798) Shaw I 314, *Churchbridge* 1775 Yates. The land adjoining the bridge, which carries Watling Street, belonged to a Lichfield guild, which may have built or rebuilt a bridge here. The present bridge was described as new in 1830: VCH V 77.

CHURCH GAUSE GREEN (obsolete, 1½ miles south-east of Consall (SJ 9947)) *Church gause Green* 1836 O.S. In the absence of early spellings no suggestions can be offered for this curious name, but a wood named Church Gorse on the north-west side of Enville (SO 8287) may be noted.

CHURCHHOUSE (unlocated, probably at Chillington, although no church is recorded there) *La Chirchehusse* 1289 SHC VI NS (ii) 185, *Chirchehous* 1327 SHC VII (i) 236, *Chirchehouse* 1332 SHC X 121, *Chirchehous* 1468 SRO D260/M/T/4/1, *The Churchehouse* 1538 SRO D260/M/E/428/2. 'The church house'.

**CHURCH LEIGH** – see **LEIGH, CHURCH.**

CHURCHSTEADS (unlocated, in Mayfield) *Churchsteads* c.1637 D1132/1/19. 'The site of a church'. Perhaps to be associated with *Church-towne field in upper Mathfield* recorded in Plot 1686: 404, and *Church-town field in Upper Chalkfield* (?) recorded in Shaw 1798: I 33, the latter doubtless a mistranscription of the former. The name may perhaps to be associated with the site of an earlier church from which Church Mayfield may take its name. *Chalkfield* appears to be a misreading of *Mathfield.*

**CHURNET, RIVER** *Chirned* 1239 Su, *Chirnet* 1240 *et freq.*, Su, 1250 Dieul, 1293 Ass, 1318 AD 5, *Chirnete* 1284 SHC XI 333, *Chyrnet* 1318 AD 5, 1372 Croxden Chr, *Chyrned* 14th century AD 5, *Chernet(t)* 13th century AD 5, 1272 Ass, 1612 FF, *Churne* 1586 Harrison. A pre-English river name, possibly of British origin, of unknown etymology and meaning: see Ekwall 1928: 79-80. For details of early woodland names in the Churnet Valley see Welch 2001: 17-73.

**CHYKNELL** 1 mile west of Claverley (SO 7794). *Chekenhull* 1209 Eyton 1854-60: III 98, *Chikenehull* 1224, *Chykenhul* 1292, *Chikenhulle* 1316 (all from Bowcock 1923: 72), *Chickenhulle* 1326 Eyton 1854-60: III 93, *Chykenhull* 1327 SR, *Chikynhyll* 1538 SHC XI

277, *Chicnal* 1602 Claverley ParReg, *Chicknall* 1745 SA 867/280. From OE *cicen hyll* 'chicken or chickens' hill'. In Shropshire since the 12th century.

**CINDER HILL** ½ mile north-east of Sedgley (SO 9294), *Synderhill* 1657 Sedgley ParReg, *Sinder Hill* 1749 Bowen, *Cinder Hill* 1798 Yates, 1834 O.S.; **CINDER HILL** ½ mile north-west of Normacot (SJ 9243), *Sinderhill* 1563 SRO D5100/26, *Synderhill House* 1663 SRO D3575/1, *Syndrill House* 1700 Okeover T763. The name is normally assumed to be from cinders produced by iron-works, but in the first case it has been suggested that the name is from sundered or separated land on the border of Ettingshall manor which, though in Sedgley, was independent of it for manorial purposes: Hackwood 1898: 6. *Sinderhills* in Walsall is recorded in 1635 (SRO D260/M/T/7/8), in 1649 as *Sinder Hills near Powke Laughton in Walsall Foreign* (WLHC 276/250), and 1760 (SBT DR 42/706).

CIPPEMORE (unlocated, probably between Enville and Kinver: VCH II 343) *Cippemore* 1086 DB, ? c.1598 Erdeswick 1844: 343. On the questionabe association of the first element with OE *scip, sce(a)p* 'sheep', it has been suggested that this place may have been in Kinver Forest, or near Great Moor, Pattingham, or in Kinver (see The Compa), or the area of high ground known as The Sheepwalks south-west of Enville (possibly associated with *Scipricg* in Wulfrun's charter of 994 (17th century, S.1380) to the monastery of Wolverhampton): SHC 1916 170; VCH IV 54 fn; VCH XX 93, 107. The name (if not corrupt, as are many DB spellings) may be from OE *cipp* 'a beam, a log', with OE *mor* 'moorland, bog', meaning 'the moor or bog with the bog-oaks'. Seemingly Erdeswick knew of its location in the 16th century: Erdeswick 1844: 343.

CIRCHEBURY (unlocated, at Wallbridge) *Circhebury* 1257 StEnc 359.

**CIRCUIT BROOK** a tributary of the river Trent.

**CLAMGOOSE LANE** 1 mile north-west of Cheadle. *Clamgoose Lane* 1836 O.S. The absence of early spellings makes any suggested derivation dangerously speculative, but the name possibly incorporates OE *clam* 'mud, clay', or OE *\*clǣme* 'a clayey or muddy place'.

**CLANBROOK** on the north side of Trysull (SO 8484). Early spellings have not been traced, but almost certainly from OE *clan* 'clear, pure', with OE *broc* 'a brook'.

**CLANFORD** 1 mile south-west of Seighford (SJ 8724). *Claneford* 1290 SHC IV 269, *the farm or grange called Slamford Grange* 1541 SHC V NS 118, *Clanford alias Clamford alias Slamford Grange* 1589 SRO D3089/1/1, *Clanford* 1679 SHC 1914 13. From OE *clan* 'clear; free from undergrowth; pure', with *ford* (cf. Coton Clanford). A grange of Ranton Abbey lay here: VCH III 254. See also Coton Clanford; Down House Farm.

**CLAN PARK FARM** 1½ miles south-west of Trysull (SO 8392). *Clam Park* 1832 Teesdale, *Clenn Park* 1834 O.S. The late spellings preclude any derivation, but the name is unlikely to be ancient.

**CLAP GATE** in Wombourne (SO 8693). *Clappgate* (p) 1666 SHC 1923 113. From dialect *clap-gate* 'a gate which shuts on either of two posts joined with bars to a third post; a small hunting gate wide enough for a horse to pass': EDD.

**CLAREGATE** 2 miles north-west of Wolverhampton (SJ 8801). *Clare* 1260 SHC 4th Series 13 7, 1271 SHC V (i) 144, 1286 SHC 4th Series XVIII 156, 1381 SHC 1928 81, *Claregate* 1699 VCH XX 7. No suggestion can be offered for this name. A derivation from a British stream-name (cf. Clare, Suffolk; Clere, Hampshire) is unlikely, since the place lies on relatively higher ground with no significant watercourses, though *Clare* may have been some distance from the present Claregate: the 1889 6" O.S. map, for example, shows *Clare Gate* at Lothians Road in Tettenhall (SJ 891007). Richard de Clare is recorded in this area in 1286 (SHC 4th Series XVIII 156), Nich'o le Clare is recorded in Tettenhall parish (in which lies this place) in 1327 (SHC VII 252, VCH XX 7), and Thomas de Clare in nearby Wolverhampton in 1380 (SHC 4th Series IV 4).

**CLAREHEYS** (unlocated, said to be 'a mile from Enville': SHC 1912 73) *Clareheyes, Clareheys* 1534 SHC 1912 73, *Clare Heyes* c.1535 SHC NS X (i) 141.

**CLAUSTON** (obsolete, on the south-east side of Kingslow (SO 793982)) *Clauston, Crawston temp.* Edward III, *Clayston* 1903 field-name map.

**CLAVERLEY** Ancient Parish 5 miles east of Bridgnorth (SO 7993). *Claverlege* 1086 DB, *Clauerlai* 1163 P, *Clauerlea* 1166 SHC 1923 298, *Claverleg* 1221 SHC IV 17, *Claverlegh* 1222 ibid. 21, *Claverleye* 1301 Rees 1975, *Claverlei* 1498 SHC 1928 69. From OE *clǣfre-leah* 'clover *leah*'. Since the 12th century in Shopshire. *Claverley home* is recorded in 1745 (SA 867/280), incorporating OE *ham* in the sense 'manor', so 'Claverley manor': see PN Sa II 10.

**CLAYHANGER** 4 miles north of Walsall (SK 0404). *Cleyhungre* Hy III BM, *Cleyhungermore* 1300 SHC V (i) 177, *Cleohongre* 1392 Ipm, *Cleyhungre* 1407 SHC 1931 282, *Clehonger* 1588 Walsall ParReg, *Clehanger* 1606 ibid. 'Clayey wooded slope', from OE *clæg* 'clay', or *clǣig* 'clayey', with OE *hangra* 'hanging wood, i.e. wood on a slope'. There is a great deposit of red marl here, and a sloping bank. It is uncertain whether *Chyancland, Clianeland,* mentioned in a charter of 1226 (WA I 296; Mander & Tildesley 1960: 30) refers to this place.

**CLAY MILLS** on the north-east side of Stretton near Burton upon Trent (SK 2627). *Clay Mills* c.1780 DRO D5236/32/55.

**CLAYTON, CLAYTON GRIFFITH** 2 miles south of Newcastle-under-Lyme (SJ 8543). *Claitone* 1086 DB, *Claiton* 1240 (1798) Shaw I xxx, c.1249 SHC 1911 145, *Cleyton* 1254 Ipm, *Clayton Griffyn* 1293 SHC 1911 51, 1487 SHC XI 329, *Cleyton* 1306 SHC VII 159, *Parva Clayton* 1487 SHC XI 330, *Clayton Griffin* 1526 SHC ibid. 329. '*Tun* on clayey ground', from OE *clæg* 'clay', or *clǣig* 'clayey'. Clayton was divided by the mid 13th century, with Great Clayton, the southern portion, absorbed into Newcastle, and Clayton Griffith, or Little Clayton, lying between Great Clayton and Newcastle held by the Griffyn family: VCH VIII 77; see also SHC IV 170; SHC VI (i) 19.

**CLEAT HILL** a 370' hill on the south side of Longdon Green (SK 0812). *Cleithul* c.1175 SHC 1924 290, 1160x1181 EEA 16 57, 1177x1183 ibid. 291, *Clehithul* c.1270x1298 SHC 1921 8, *Claychul* (? for *Claythul*) c.1293 ibid. 10, *Cleathill* 1798 Yates, 1834 O.S. Probably from OE *clête*, a side form of *clate* 'burdock', so 'the hill where burdock grew': see Mawer 1929: 50, 60. See also SHC 1924 93, 290, 291; VCH III 21; VCH Wa VI 122. See also Ridgeacre.

**CLEDERE, GRANGE DE** (unlocated) *Grange de Cledere* 1291 (1798) Shaw I xxiii.

**CLENT** Ancient Parish 4 miles south-east of Stourbridge (SO 9279). *Clent* 1086 DB, 1169 P, 1242 Fees, 1273 SHC 1911 152. A difficult name. Perhaps from OE *\*clent* 'stone, rock', related to OSwed *klinter*, ON and Icelandic *klettr* 'a hard, flinty rock' (modern Swedish *klint* 'a precipice, a cliff'), used in English before 1300 as *clint*. The word *clent* is found in a ME text c.1400 in the phrase *a clent hille*, perhaps meaning 'a hill of rock': PN Wo 279. The place stands at the foot of the Clent Hills, with the summit of Walton Hill at 1,036', and since 1844 has been in Worcestershire. There is place-name evidence that the name once applied to a fairly large district: see Anderson 1934: 143-4. For a detailed discussion of this name see PN Wo 279. Cf. Clint, Yorkshire. See also Mawer 1929: 76; Cavill, Harding & Jesch 2000: 116.

**CLEWLEY (COPPICE)** ½ mile north-east of Pendeford Mill (SJ 9004). *Clewley* 1946 O.S. The name is remembered in Clewley Drive, Pendeford. See also Beuleg.

CLIFF (unlocated, in Abbots Bromley) *Clyf* 1330 SHC XI NS 187, *Clyve* 1345 SHC XII 43. From OE *clif* 'declivity, cliff, river bank'.

**CLIFF'S ROUGH** on the north side of Hanchurch (SJ 8441). *? the Clyves* 1329 SRO D593/B/1/23/7/1/1.

**CLIFF VALE** 1 mile south-west of Hanley (SJ 8646). *Clyf* 1204 SRO 154[7969], *Le Clif* 1253 SHC 1911 121, *Le Cliff* 1263 ibid. 134, *Clyf* 1271 SHC V (i) 155, *Castlecliff* 1423 VCH II 348. From OE *clif,* which had varied meanings, including 'a slope, not necessarily steep', 'the bank of a river', 'escarpment', and 'cliff'. The meaning here is probably 'slope'. Vale is ME from OFr *val(s)*, a post-Conquest word meaning 'a wide valley', evidently used here in that sense but adopted relatively recently. Cliff or Clive was one of the Hays of The New Forest (q.v.): VCH II 348-9; *haiya de Clive* c.1238 (1798) Shaw I xxviii. *Castle* is from the 'new castle' at nearby Newcastle-under-Lyme. VCH VIII 200-1 fn.16-17 incorrectly assigns the 1204 spelling to Penkhull.

**CLIFF WOOD** 1 mile west of Horton (SJ 9257). *boscum le cliffe* 1253 IpmR, *Cliff Wood* 1842 O.S. Self-explanatory.

**CLIFTON CAMPVILLE** Ancient Parish 5½ miles north-east of Tamworth (SK 2510). *Clyfton* 941 (14th century, S.479), *Cliftone, Clistone* 1086 DB, *Cliftun* c.1100 Duig, *Caunvilla, Clyftona* 1194 SHC III (i) 25, *Canvill, Kanvill* 1195 ibid. 27, *Kanvill* 1203 SHC ibid. 111, *Clifton'* 1236 Fees, 1242 ibid., *Clifton Caunvil* 1284 Ass, *Clifton-Chamville* 1293 SHC VI (i) 240, *Camvill* 1294 SHC VII 9, *Clifton Caunvill* 1298 WL 3, *Kaumpville, Clyfton upon Hundenho* 1306 SHC VII 169, *Clyston* 1532 SHC 4th Series 8 69, *Clyfton Cambvyle* 1604 SHC 1940 167. '*Tun* near a cliff or bank or on the edge of a river', from OE *clif* 'cliff, bank'. The place was held by Richard de Camvill in 1231. Camvill is a family name from Canville in l'Eure-Inférieur in Normandy: Duignan 1902: 41; DES 82. *Hundenho* may be Haunton.

**CLIVE, THE** 5½ miles west of Wolverhampton (SO 8297). *Clive* 1327 Duig, *Clyve* 1332 SHC X 131, *The Clyff* 1532 SHC 4th Series 8 186, *Clieve* 1686 Plot, *Cleive* 1695 Morden, 1749 Bowen. From OE *clif* 'cliff, bank'. The place lies on a pronounced ridge.

CLOCK MILL (obsolete) in Pelsall (SK 0102). *Clockmill* 1568 Walsall ParReg, 1576 Dilworth 1976: 79, *Clokmyll* 1613 ibid, *Clock mill* 1617 Willmore 1887: 440. The name perhaps derives from ME *clack*, meaning the clapper of a corn-mill, which by striking the hopper causes the corn to be shaken into the millstones, but the sound produced by the process may have led to the association with the ticking of a clock. The dandelion was sometimes called the clock, but the usage is late, and can be ruled out here. The mill evidently gave its name to Clock Mill Brook (... *Perle called Clock mill perle* ... 1617 Willmore 1887: 440; Dilworth 1976: 93) in Pelsall. *Perle* is from OE *pyrle* bubbling', a common stream-name.

CLONEWOOD (unlocated, in Leek) *Clonewood* 1562 SHC IX NS 113.

**CLOUD, THE** a prominent hill on the Cheshire/Staffordshire border, 2½ miles east of Congleton (SJ 9063). *Cloud* 1637 Leek ParReg, *The Cloud* 1747 Bowen. From OE *clud* 'mass of rock, outcrop, hill' (cf. Welsh *clud* 'load, bundle'), found in North Staffordshire (cf. Hen Cloud). Evidently associated with *Clowdewoods*, recorded in 1581 (SRO DW1761/A/4/49), *Clowde Wood*, *Clowdwood* recorded in 1590 (SHC 1930 72, 96). An outcrop known as Five Clouds (from its five buttresses) lies beneath The Roaches (q.v.).

**CLOUGH** a common name in the North Staffordshire moorlands and with variants in the northern counties, but not found south of Stone. From OE *cloh*, ME *clough* which Duignan 1902: 42 believed to mean 'a ravine or narrow valley with steep sides, usually forming the bed of a stream', but now believed to have been applied to less pronounced or secondary features: see Gelling 1984: 88. The old pronunciation was as in 'bough', but is now 'cluff'.

**CLOUGH HALL** 2 miles north of Newcastle under Lyme (SJ 8353). *the Clough Hall* 1697 SRO D260/M/T/4/103, *Clough Hall* c.1760 SRO D260/M/E/428/3.

**CLOUGH HEAD** ½ mile south-east of Ipstones (SK 0248). *the Cloughehead* 1576 Brighton 1937: 193, *The Cloughhead* 1670 SRO D1065/1/2, *Clough Head* 1836 O.S. From OE *cloh*, ME *clough* which Duignan 1902: 42 believed to mean 'a ravine or narrow valley with steep sides, usually forming the bed of a stream', but now believed to have been applied to less pronounced or secondary features: see Gelling 1984: 88. The word *head* was applied to the top or end of something, so 'the top or end of the valley'.

**CLOUGH HOUSE** 1 mile north-east of Heaton (SJ 9663). Perhaps to be associated with *Clou*, recorded in the Heaton area in 1327: SHC VII (i) 219. From OE *cloh*, ME *clough* which Duignan 1902: 42 believed to mean 'a ravine or narrow valley with steep sides, usually forming the bed of a stream', but now believed to have been applied to less pronounced or secondary features: see Gelling 1984: 88.

CLULOW (unlocated, possibly in Leek) *Clulow* 1568 (1883) Sleigh 33, 1604 SHC 1946 116, *Clewlo* 1576 SHC 1926 39, *Clulowe* 1594 SHC 1932 49, *Clewlow* 1666 SHC 1921 143, 1784 SHC 1947 81. Possibly from OE *\*cleo(w)* 'a clew, a ball' (see especially PN Sa I 82-7), with OE *hlaw* 'mound, tumulus', perhaps from some rounded tumulus or mound here. *Clewlow Sprink* (i.e. 'the coppice or newly-planted woodland at Clewlow') appears on the 1836 1" O.S. map ½ mile south of Whiston near Cheadle (SK 0346), but the history of the name has not been traced: see Tooth 2000b: 115. *Clewlows Bank* is shown to the south of Stanley Moor on the 1890 O.S. map. Some of the spellings may refer to those places, or to Cleulow Cross (*Clulow Cross* 1842 O.S.) 1 mile north of Wincle in Cheshire (SJ 9467), the site of an Anglo-Saxon cross-shaft on a mound: see PN Ch I 165-6; Gelling 1992: 189.

CLYVES, THE (unlocated, in Longton) *Le Clyvis* c.1280 SRO D593/B/1/23/4/1/13, *the Clyves* 1329 SRO D593/B/1/23/7/1/1. Fom OE *clif* 'cliff, bank'.

CNOKOMALAY (unlocated, perhaps in the north-west of the county) *Cnokomalay* 1308 SHC XI NS 265. The first part of this curious name (which may well be a corrupt transcription) might be Welsh *cnwc* 'a hill, a lump, a hillock', or OE *cnocc* 'a hill, a hillock': see Cannock.

**COAL BROOK** Drayton-in-Hales. A tributary of the river Tern. *colbrok* 1387 SHC 1945-6 61, *Colbrowke* 1554 ibid.' 132. Probably from OE *col* 'cool'. It is uncertain whether *Colbrook*, recorded in 1328 (SHC XI NS 262) refers to this stream.

**COALPIT HILL** on the north side of Talke (SJ 8253), *Coalpitt Hill* 1733 SHC 1944 67; COAL PITS (obsolete) on the north side of Orchard Farm, near the northernmost point in the county (SK 0269), *cole pitts* 1599 Plan PRO MPC 214, *Coal Pits* 1834 O.S.; COAL PITS (obsolete) 1½ miles south-west of Flash (SK 0164), *the Colepyttes* 1564 SHC 1938 99, *Coal Pits* 1842 O.S. Self-explanatory. At Coal Pits near Orchard Farm were mines which were worked from at least 1401 to c.1930: see StSt 8 1996 66-95. *Colepytford*, recorded in 1608 (SHC 1948-9 152), may be associated with those mines, or the ones south-west of Flash.

**COAL POOL** 2 miles north of Walsall (SO 0100). *Colepoole* 1580 Walsall ParReg, *Collpoole* 1615 ibid, *Coale Poole* 1620 ibid, *Colepool* 1627 ibid. A flooded coal-pit is a more likely derivation than 'cold pool'. See Colepool Brook.

**COATESTOWN** ½ mile south-west of Hollinsclough (SK 0666). *Coatestown* 1775 Yates. 'The farmstead of a man named Coates': Isaac Coates, a dealer and chapman recorded in the second half of the 18th century, may be associated with the place: VCH VII 38.

**COBB MOOR** (obsolete, on north side of Kidsgrove (SJ 8555)) *Cobbmore* 1656 Wolstanton ParReg, *Cobmore* 1660 ibid, *Cobmoore* 1713 JNSFC LXV 1930-1 46, *Cobb Moor* 1836 O.S. From OE *\*cobb(e)* 'roundish mass, lump, etc.', with OE *mor* 'upland waste, marshy land'.

COBINTONE (unlocated) *Cobintone* 1086 DB. The place has not been identified, but is listed in DB in Seisdon Hundred. The places listed after the entry for this place in DB are all in Cuttlestone Hundred, starting with Sheriffhales, held by Rainald, who also held neighbouring Weston-under-Lizard, Beighterton and Brockton Grange in Sheriffhales. Cobintone may have been part of Sheriffhales (perhaps Cuttesdon (q.v.)), with the Hundredal heading (or the entry) inserted in the wrong place. Certainly those who held Sheriffhales from the 12th to the 14th century under the FitzAlans were also lords of Kibblestone (SHC II 224), which points towards that place being Cobintone (as suggested by Erdeswick 1844: 30-31), but (quite apart from the philological difficulties) Kibblestone is said to have been formed in the 12th century from a conglomeration of Moddershall, Cotwalton and other manors (see SHC 1911 388-9; SHC 1916 169), and not to be identified with Cobintone. The sequence in which Cobintone is recorded in DB (Claverley-Kingsnordley-Alveley-*Cobintone*), if correct, suggests that the place may have lain to the south of Alveley, or even in that part of Staffordshire that lay to the west of the Severn opposite Arley. The most likely identification of Cobintone, however, is Cubbington in Warwickshire (which is recorded once as *Cobintone* and twice as *Cubintone* in DB: VCH Wa I 305, 316, 327; PN Wa 169), rather than a lost place in Staffordshire. The name itself, which appears to be 'Cubba's *tun*', Cubba representing a pet-form of an OE personal name such as Cuþbeorht (PN Wa 169), offers no clue to its location.

**COBLEY HILL** 3 miles east of Bromsgrove (SO 0171). *Cobesleie* 12th century Dugd. v 409, *Cobbele(ye)* 1271, 1299 (18th century), *Cobley Hill* 1535 (all PN Wo 362). From the personal name Cobbe (cf. Cobley, Hampshire), or from OE *\*cobb(e)* 'roundish mass, lump', with OE *leah* and *hyll* 'hill'. In Tardebigge parish, forming part of Staffordshire from c.1100 until 1266, in Warwickshire until 1844, and since that date in Worcestershire.

**COBRIDGE** between Hanley and Burslem (SJ 8748). *Cobbrage Yate* 1687 Stoke on Trent ParReg, *Cobridge gate* 1679 SHC 1919 258, 1733 Stoke on Trent ParReg,

*Cobberidge* 1705 ibid, *Coe Bridge* 1799 Faden, *Corbridge* 1836 O.S. Since the place is on a ridge, probably from OE *\*cobbe* 'roundish mass, lump', with OE *hrycg* 'back, ridge'. A mis-division of the original elements has led to confusion with *bridge*. The earliest spelling refers to Cobridge Gate, said to be named after a gate opening into the lane from Rushton Grange to Hulton Abbey: Ward 1843: 273.

COBSHURST (obsolete, 2 miles south-east of Longton (SJ 9240)) *Copt hurst, Copshurst, Copte hurst, Coppehurst* 1544 SHC 1910 75-6, *Cobhurst* 1729 Okeover T768, *Copshurst* 1738 ibid. T769, *Cobshurst* 1836 O.S. The first element is uncertain, but perhaps from the OE adjective *\*coppod* 'having a peak or top', also 'polled, cut down somewhat', and frequently found with tree-names (EPNE i 107), with OE *hyrst* 'a hillock, a bank, a wooded eminence', so perhaps 'the pollarded coppice on the small hill' for the first place, and 'the nook frequented by woodcock' for the second.

COCKET KNOB (obsolete) 2 miles east of Cheddleton (SK 0053), *Cocket-know* 1730 Alstonefield ParReg, *Cocket Knob* 1836 O.S., *Cocket Knowl* 1880 Kelly. Perhaps from *cocket* or *cocked* 'set erect, having a pronounced upward turn' (OED), or from OE *cocc* 'woodcock', with the OE noun suffix *-et* 'a place characterised by what is named', so giving 'place frequented by woodcock', or from OE *cocc-wudu* 'the wood frequented by cocks or wild birds' (cf. river Coquet, Northumberland), with OE *cnoll* 'knoll'.

COCKETS NOOK (obsolete), 1 mile north-west of Rugeley (SK 0138), *Cockets Nook* 1834 O.S.

**COCKING FARM** ½ mile north of Caverswall (SJ 9544). *Cocking, Upper Cocking* 1836 O.S., *Cocking* 1890 O.S. Probably to be associated with *Cocker*, recorded in 1691 (Okeover T762), in which case perhaps reflecting a British stream-name (see *Cocretone*): the first edition 1" O.S. map shows a stream rising at Cocking and running south to join a tributary of the river Blithe. The name may have been influenced by the word *cocking*, which appears to have been associated with places used for cock-fighting. It may be noted that Cocking Farm lies some three miles from Saverley Green (q.v.), a name which may be of British origin.

**COCKLEY** 1 mile north-east of Ellastone (SK 1343). *? Cokkylegh* (p) 1372 SHC XIII 92, *Cockley Farm* 1785 SRO D626/A/2/1-2. From OE *cocc* 'cock', or OE *cocc* 'a heap, a hillock, a clump of trees', with OE *leah*.

**COCKNAGE** 1½ miles north-east of Barlaston (SJ 9140). *Cochenache* 1176 EEA 16 104, *Kokenache* 1194 Pipe, *Cokenache* 1195 ff, 1200 SHC II 91, *Cokenach* 1198 P, *Chokeneche* c.1208 SHC 1911 417, *Kokenache* c.1230 SHC 1921 18, *Cocsache* c.1231 SHC 1911 425, *Cokenach', Kokenhache* 1240-1 Cur, *Cocenag* 1272-3 SHC XI NS 243, *Cockenegge* 1283 SHC 1911 184, *Cognage* 1575 SHC 1926 44, *Cockurge* 1598 SHC 1935 49, *Cockindge* 1600 ibid.' 213, *Cockenage* 1605 SHC 1940 297. The first element is from the OE personal name Cocca. The terminal is perhaps OE *ǣc*, the dative of *ac* 'oak tree', which gives the ending *-age* in place-names (cf. Radnage; Cressage; Stevenage), rather than Mercian OE *hec(c)* 'a hatch, a grating, a half-gate' (cf. Cockenhatch, Hertfordshire), as suggested by Ekwall 1960: 115.

COCKSHUT HAY (unlocated, said to be 2 km west of Rudyard: CEC 384) *Cockstuth, Cocsute, Cocksuth, Cocsuche* 1221x1226 CEC 384, *Cokshete* 1293 SHC VI (i) 220. Very commonly found as a minor name from the 13th century, especially in the south-west Midlands counties: EPNE i 104-5. The expression is said to mean 'a broad glade in a wood, through which woodcock might dart or shoot, so as to be caught by nets stretched across the opening', although Mawer 1929: 47-8 prefers 'corner of land frequented by woodcock'. In this area the Anglian OE form *cocc-scyte* would be expected, giving -*shete*, but the spellings show that West Saxon OE *cocc-sciete*, *coc-scȳte*, giving -*shute*, are closer roots. Hay is from Mercian OE *(ge)heg* 'a fence, an enclosed piece of land', often in Staffordshire meaning 'an administrative bailiwick within a Forest area'. *Cockshoot Hill* is recorded south-east of Swindon (VCH XX 213), and *Cockshatt flatte* near The Delph House (q.v.) in 1605 (SHC 1934 47).

**COCKSTER BROOK** a tributary of Chitlings Brook, which flows into the river Trent. No early forms are available to suggest a derivation.

COCRETONE, COCORTONE (unlocated, between Trysull and Seisdon) *Cocretone*, *Cocortone* 1086 DB, *Cocretone* 1288 SHC 1911 194, *? Cokton* (p) 1286 SHC 4th Series XVIII 133, *Corcorton* c.1598 Erdeswick. Shaw 1801: II 208 mispells the name as *Colverton*. Erdeswick 1844: 541 lists the 'Modern Name' as *Cocretone*, without *Q[uare]* attached to other names he was unable to identify, implying that in the late sixteenth century the location of the place was still known., and indeed land called *Cockerton* is recorded in 1623 in the area later known as The Beeches to the west of Trysull (SO 841937): VCH XX 185-7 (where the spelling *Corcortone* is an error); VCH IV 38 fn, 54 fn. The land may have been called *Crockington* as recently as 1928: WSL Misc 339 5. The name survives in Crockington (formerly Cockerton) Lane connecting Trysull and Seisdon. TSSAHS XII 1970-1 34 locates the lost village (which was still inhabited in the later 13th century: SHC 1911 194) at SO 843943. It may be noted that DB records William fitz Ansculf holding both Trysull and Seisdon, as well as 1½ hides of this place, with ½ hide of this place belonging to Kinswinford held by the king. Both holdings in this place are described as waste (VCH IV 38). It has been suggested that the name is possibly 'potters' settlement', from OE *croccere*, which could become *Cocor-* (see for example Morris 1976: 1,1; 12,16) but no archaeological evidence of pottery working has been recorded and no early pottery-associated surnames have been found. A more obvious derivation would be from a Celtic river-name Cocker (found in Cumbria, Durham, Lancashire, Lincolnshire, Nottinghamshire, and Somerset), probably from a British form *\*Kukra*, which would give *\*Cocra*, *\*Cocr* (Welsh *Cogr*) 'crooked': see Ekwall 1928: 83-4; Coates & Breeze 2000: 135; 361; CDEPN 147. Cf. Cockermouth, Cumbria; Cockerham and Cockersand, Lancashire; Cockerton, Durham. However, while the nearby Smestow Brook could properly be described as 'winding, crooked', the ancient name of this stream is well-evidenced as Tresel – see Trescott; Trysull. There is no other significant stream to which the name Cocker might apply, though the first edition 1" O.S. map shows a minor watercourse running from east of The Beeches into Smestow Brook. If not from a river-name, perhaps from OE *cokken* 'to fight', from which the ME surname Cockere is recorded (both noun and verb being recorded in 'The Proverbs of Alfred': see Clark 1995: 210), so giving here 'the settlement of the brawler', which has developed into Croc-. Henry of Cokton is recorded in 1286 (SHC 4th Series XVIII 133), and the surname Crochintones and similar is found in the 16th and 17th centuries in the Penn ParReg (but not in the Trysull ParReg). It should be noted that the names Crogeton, Crogynton, and variants recorded in Worfield and the surrounding area from at least 1483 (see e.g. TSAS 3rd Series VI 12; SRS 3 34 and 101), are said to be from Crudgington in Shropshire, and Crochintones may have the same derivation. See also SHC 1919 164. See also Cocking Farm.

COCSTAL, COCKESTALL, COKSTALL (unlocated, near Longton) *Cocestal* c.1250 SRO D593/B/1/23/4/1/6, *Kocstal* c.1260 SRO D593/B/1/23/4/1/15, *Kocstal*, *Cocstal* (undated) SHC XI 320-21, *Cockestall*, *Cokstall* (undated) ibid. 322. See also Cookshill.

**CODNYNNESHEVED** (unlocated, possibly in Horton) *Codnynnesheved* 1307 SHC XI NS 255. The last element is from OE *heafod* 'head', meaning topographically 'end, top. summit, head of a stream', with what appears to be the OE personal name Codda. Cf. *le Codyngeheye* in Cheshire: PN Ch V (ii) 281 fn.76. See also Codyngton.

**CODSALL** 4 miles north-west of Wolverhampton (SJ 8603). *Codeshale* 1086 DB, 1271 Ass, *Coddeshal* 1167 SHC 1923 296, 1248 Cl, *Codeshal* 1293 SHC 1911 230, *Cotteshale* 1293 SHC VI (i) 260, *Cottussale* 1421 SHC XVII 40, *Coddessale* 1484 SHC VI NS (i) 152, *Codssowle* 1547 TSAS 3rd Series VIII 1908 138, *Codshall otherwise Codsall otherwise Codsoll* 1606 SHC XVIII 65. Probably 'Cod's *halh*', although Gelling & Cole 2000: 127 suggests without further explanation that this derivation 'rests on slender foundations'. The same or a similar personal name (not found recorded other than in place-names: see PN Wo 116) is the first element of Cotswolds; Codnor, Derbyshire; Codford, Wiltshire; Cotheridge and Cutsdean, Worcestershire. Codsall Wood (*Codsall Woodde* 1592 Codsall ParReg) lies 2 miles to the north-west (SJ 8405), and Codsall Lanes (obsolete) lay on the west side of Codsall (SJ 8504), recorded as *le Lone* in 1334 (SHC XVI 3), *Codsall Lanes* in 1799 (Yates), *Lanes (near C.)* in 1805 (Codsall ParReg).

**CODUSBAG** (unlocated; possibly Cottesbach in Leicestershire: see SHC I 206) *Codusbag* 1274 SHC 1911 162.

**CODYNGTON** (unlocated) *Codyngton* 1325 SHC X 57, 1395 SHC XV 65. Perhaps from an *-ing* suffix added to the OE personal name Codda. Cf. *le Codyngeheye* in Cheshire: PN Ch V (II) 281 fn.76. See also Codnynnesheved.

**COENA'S WELL** a spring which runs into a pool south-west of Wall Grange Farm (SJ 9754). *Cena de Wal* c.13th century Su, *Signe Wall(s)* 1627 (1883) Sleigh, *St Ann's Well, Senus Well, Sinners Well* 1849 ibid, *Coena's well* 1870s ibid.' VCH VII 203 suggests a derivation from St Agnes' or St Ann's Well, but the spellings point towards 'the well of Sainte Cene', the name of several holy wells in Normandy, or possibly a derivation from *sene, senyie*, from Fr *sene, senne*, Latin *synodus* 'a meeting of clergy for deliberations, a synod': *seyney-houses* were buildings belonging to monastic houses where breaks ('seyneys') were taken by monks in need of rest and recuperation after the regular blood lettings they undertook for health reasons, or after illness: see Sinai Park. Perhaps to be associated with Ametesawe (q.v.).

**COKEFELD** (unlocated, in Rownall) *Kokfeld* 1273 SHC VI (i) 54, *Kokfeud* 1275 ibid. 56, *Cokefeld* 1284 SHC VI (i) 136, *Cokefeud, Cokefeld* 1286 ibid. 166-7. Perhaps from OE *cocc* 'wild bird' or OE *cocc* 'heap or hill', with OE *feld* 'open ground'.

**COKESALLE** or **COXALL** (unlocated, possibly near Tatenhill) *Cokeshalle* 1420 SHC XVII 75, *Cokesall otherwise Coxall* 1618 SHC VI (i) NS 57. The first element is more likely to be an unrecorded personal name *Cocc, rather than OE *cocc* 'wild bird' or OE *cocc* 'heap or hill', since the second element seems in this case to be *hall*, an element rare in Staffordshire, rather than the ubiquitous OE *halh*, although the latter cannot be ruled out.

**COKKESHOLM** (unlocated, in Rickerscote or Silkmore) *Cockeshelm, Cocholm, Cockulm, Cokkesholm* 1349 SRO 89-92[7904], *Cockesholm* 1350 SHC VIII 130. Almost certainly the same place as *Cowesholm*, recorded in 1356 (SHC VIII (ii) 129), and *Kocholme* n.d. (SHC VIII (i) 134). The first element is an unrecorded personal name *Cocc, or OE *cocc* 'wild bird' or OE *cocc* 'heap or hill'. The second element is *holme*, from ON *holmr* 'small island, a piece of dry land in a fen, a piece of land partly surrounded by streams or by a stream',

**COLCLOUGH** 1½ miles north of Tunstall (SJ 8553). *Colleclogh* 1362 SHC 1913 328, *Colcloghe* 1376 ibid.' 134, *Colclogh* 1413 SHC 1911 470, *Colclogh Lane* 1532 SHC 4th Series 8 50, *Coulclughe* 1585 SHC 1929 56, *Coleclugh* 1582 Betley ParReg, *Colklowghe* 1590 SHC 1930 54. 'The stream-valley where charcoal was made or coal dug', from OE *col* 'coal, charcoal'. See also Clough.

**COLDFIELD** – see **ASTON**.

**COLD NORTON** – see **NORTON, COLD**.

**COLDHAM** 2 miles west of Brewood (SJ 8408). *Coldhome* 1581, 1660, 1729 Ct, *Coldham* 1594 SRO D590/446/1-5, *Colldome* 1626 Brewood ParReg. Probably a 16th century name 'cold home': the place is in an elevated position. See also Hilton near Brewood.

**COLDMEECE** 3 miles north-east of Eccleshall (SJ 8632). *Mess* 1086 DB, *Mes* 1208 Curia, *Coldemes* 1273 Ass, *Coldemeys* 1292 SHC 1926 172. From OE *meos* 'moss, marsh, bog', or the nearby river Meece and Meece Brook (q.v.), with OE *cald* 'cold'. The place is near Cold Norton.

COLDMOOR (obsolete, on the north-west side of Wheaton Aston (SJ 8413)) *Coldmoor* 1836 O.S. The place is remembered in Cauldmore Lane.

**COLDRIDGE WOOD** in Upper Arley, 1 mile south-east of Romsley (SO 8182). *Colrugge* (p) 16th century PN Wo 30. From OE *col-hrycg* 'coal-ridge or charcoal-ridge'. The place lies on coal measures, but *col* also meant charcoal. In Worcestershire since 1895.

**COLDRIDING FARM** 1½ miles north-east of Barlaston (SJ 9140). *Cold Ridding* 1755 SRO D4092/C/2/18. From OE *\*ryding* 'clearing', so 'the cold clearing'.

**COLDSHAW** 2 miles north-west of Hollinsclough (SK 0467). *Coldshawe* 1429 DRO D2375M/1/1, *Cold Shaw* 1678 Alstonefield ParReg. From OE *cald scaga* 'the cold or exposed copse'. The second element is commonly found in the northern counties.

**COLDWALL** ½ mile east of Blore with Swinscoe (SK1449). *Coldewalle* 1245 Okeover T3, *Caldewall* 1275 SHC XI NS 243, *? Coldewalle* 1293 SHC 1911 215, *Coldwell* 1309 SHC 1911 72, *Caldewell* 1311 Okeover, *Could Wall* 1775 Yates, *Cold Wall* 1640 Ipm, 1836 O.S.; **COLD WELL** between Gentleshaw and Goosemoor Green (SK 0511), *a well called le calde Walle* 13th century BCA MS3415/130, *Coldewalle* 1492 OSS 1936 47, *le Cold Walle* 1505 ibid.' 51, *Cold Well* 1834 O.S.; COLDWELL (unlocated, in Streethay), *Coldervelle* 1271 SHC IV 184; COLDEWALLE (unlocated, in or near Wombourne), CALDWELL FIELD (unlocated, in Codsall), *Caldwallefeld* 1349 SRO D593/A/2/5/7, *Caldwell Field* 1522 SRO D593/A/2/5/16; COLEWALL (unlcated, in Haywood), *Colewele, Colewall* c.1225 SHC VIII 155; see also Coley; COLEWALL (unlocated, near Lount Farm, Colton), *(Lund sub) Colewall*, c.1215x1225 SRO D938/104, *Caldewalle* 1286 SHC IX 81; CALDWELLS (unlocated, in Orton or Wombourne), *Caldewalle* 1314 (1801) Shaw II 213, *Caldewall (Mulne)* 1362 Inq., *Caldwalls* in 1489 (1801) Shaw II 215, *Caldwalles* 1576 SRO D740/8/1, *Cawdells alias Caldwells* 1681 SRO D740/8/18; CALDEWALLE (unlocated, in Loynton), *Caldewalle* n.d. WSL D1564/6; CALDWELL (unlocated, in Bowers), *Caldwell* 1777 SRO D174/M/T/1. A common name in Staffordshire, from OE *cald* 'cold', with Mercian OE *wælle*, 'a spring', and (sometimes) 'a stream'. See also Caudewell; Caudy Fields; Coldwall.

202

**COLEPOOL BROOK** a tributary of the river Tame. *Colepool Brook* 1768 Survey. Probably from OE *col* 'coal, charcoal', with *pol* 'pool', so 'the pool created by coal extraction' to which *broc* has been added, though a derivation from OE *col* 'cool' cannot be ruled out. See also Coal Pool.

COLESHALE (unlocated, possible in the Chebsey area) *Coleshale* 1326 SRO 3764/28.

**COLEY** 1 mile north-west of Colwich (SK 0122; SMR 02418 places a deserted Anglo-Saxon settlement at SK 01202250). *? Colewele* 1198x1208 EEA 17 74, *? (the bishop's royal land in) Colweleye* 1316 WL 106, *Coley* 1317 SRO D(W)1721/3/12/2, *? Coly* (p) 1368 SHC VIII NS 214, *Coley temp.* Elizabeth I SHC 1914 131, *Haywood Coley* 1625 ibid. 133, *Coley(mill)* 1656 ibid. 152, *Coley* 1768 Erdeswick 1844: 73, 1775 Yates, *Upper Coley, Near Coley* 1836 O.S. In 1916 Wedgwood suggested that *Scoteslei* in DB 'can I think be identified with Coley ... [t]he transformation is in accordance with the laws of euphony' (SHC 1916 168), and that identification has been accepted since that date: see e.g. VCH IV 42. However, there is no philological reason to associate *Scoteslei* with Coley, and indeed there is no firm evidence that the name Coley is older than the 14th century (but note e.g. Nicholas Coli, recorded in 1300: SHC VII (i) 68). Chewynd records that Coley belonged to the bishop of Lichfield, and later to the Paget family and by attainder passed to Elizabeth I: SHC 1914 131. Possibly *Scoteslei* is to be associated with Colwich (q.v.), or perhaps Shugborough (q.v.), neither of which is recorded in DB. If the two earliest forms can be ignored, the name Coley is probably from OE *col-leah* 'the *leah* where charcoal was made', but the two earliest spellings (which may not relate to Coley) suggest a derivation from OE *col-wælle* 'the charcoal spring' (the first spelling is found as *Lund sub Colewele...iuta Burnebroc*, i.e. evidently to be associated with Lount Farm (q.v.), which lies close to Moreton Brook, still called Bourn Brook in its upper reaches, where charcoal burning and glass-making is recorded), or possibly OE *col-wælle* 'cool spring', perhaps (cf. the 1316 spelling) with OE *eg* 'land by a river or stream', or OE *ea* 'river', perhaps associated with *Caldewalle* recorded in the Colton area in 1286 (SHC IX 81). Coley lies on a 415' hill some 2 miles to the north-west of Lount Farm, a mile from the river Trent and a little further from Moreton Brook, and the spellings may refer to Coley or possibly a lost place closer to Lount Farm. The possibility that Coley is the otherwise unidentified DB holding listed as *Colt* (q.v.), part of Colton, cannot be discounted entirely, but it was not held by the church at DB. See also Colton.

**COLEY MILL** 3 miles south-west of Gnosall (SJ 7819). *Coley Hall* 1540 StEnc 152, 1651 SRO D4038/D/13, *Colyemill* 1656 ParReg, *Coley* 1775 Yates, *Coley Mill* 1798 Yates, 1833 O.S., *Coley Farm* 1963 O.S From OE *col-leah* and *myll* 'the mill at the *leah* where charcoal was made'.

**COLLYHOLE** 1 mile north-west of Ipstones (SK 0050). *Colleys Croft* 1749 SRO D538/A/5/52, *Collymoor* 1836 O.S. Seemingly from the surname Colly (DES 105).

**COLSHAW** 1 mile north-west of Hollinsclough (SK 0467). *Co(u)ldshaw* 1566 *Deed*, 1605 to 1680 Alstonefield ParReg, *Col(d)shaw(e)* 1626, 1651 *Rental*, *Coldshaw* 1842 O.S. From OE *cald, scaga* 'the cold or exposed copse'.

COLT (unlocated, listed as part of Colton in DB: VCH IV 46) *Colt* 1086 DB. Possibly Littlehay in Colton (VCH IV 46 fn.), but see also Coley.

**COLTON** Ancient Parish 2 miles north of Rugeley (SK 0520). *Coltone, Coltvne* 1086 DB, *Colton* 1176 P, *Colton'* 1198x1208 (13th century) EEA 17 74, *Coltuna* 1201 SHC II 105, *Coltun* 1227 Ass, 1240 SHC VIII 158, *Kuton* 1229 SHC XVII 75, *Knoleton* (*sic*) 1269 SHC IV 172, *Coltun* 1271 SRO D603/A/Add/202, *Cunton* 1280 SHC VI (i) 100,

*Couton* 1288 SHC 1911 43, *Coltone* 1293 SHC VI (i) 273, *Colton, called Marshalles and Gryffyns* 1392 SHC XI 199, *Coulton* 1656 Barton under Needwood ParReg. Possibly OE *colt-tun*, '*tun* where colts were reared', or (more likely, given the absence of spellings with *Colte-*), from OE *col* 'charcoal'. The presence of Coley (although no early spellings are available) and Colwich to the west, and references to the surname Coleman in 1322 (SHC IX (ii) 87) and Colemon in 1327 (SHC VII 198), and to *Coleman's-more* in Colton in 1374 (Parker 1897: 352) also suggests that charcoal burning is the probable origin. Colton was a double manor, each part held by a different overlord. One part was held by the Wasteneys and subsequently by the Gresleys, and the other part by the Griffins and Mareschalls, which explains the reference to *Marshalles* and *Gryffyns* in 1391: SHC XI 199. See also SHC II 247. Cf. Colwich.

**COLTON HILLS** on east side of Upper Penn (SO 9095). *Colton Hilles* 1593 SHC 1932 12, *Colton Hylls* 1598 SHC 1935 12, *Colton Hills* 1699 WA II 36. Seemingly a corruption of Coton (q.v.): some 97 acres of land here were subject to Coton tithes (ibid.).

**COLTS MOOR** 1½ miles north of Bradnop (SK 0057). *le Coltesmor* 1344 SRO D3272/5/13/8, *Colts Moor* 1345 VCH VII 171, *Coltsmore* 1586 SHC 1927 135, *Coletesmere* 1608 SHC 1948-9 41, *Coults Moore* 1644 SHC 4th Series I 192, *Coltsmoor* 1666 SHC 1925 242, 1836 O.S., *Coulch Moor* 1697 Leek ParReg. 'The marshland or moorland where the colts grazed', or perhaps from the Colt family, recorded c.1275 (SHC 1911 429, 430), 1319 (1883) Sleigh 126. A reference to Thomas called Kolte in 1281 (SRO DW1761/A/4/168) may also be noted. The present Coltsmoor Farm dates from the later 18th century. The site of the earlier Coltsmoor House is uncertain: VCH VII 171.

**COLTSTONE** ½ mile north-west of Ipstones. (SK 0150). *Coltstone* 1775 Yates, 1836 O.S, *Coldstone Heath* 1777 OpenFds, 1813 Deed, *Coltstone Heath & Common* 1780 *EnclA*. Possibly 'the stone where the colt was kept', or from the name Colt, hence 'Colt's stone'.

**COLWICH** Ancient Parish 3 miles north-west of Rugeley (SK 0121). *Calewich* 1166 Duig, *Colwich* c.1177 SHC XI 323, *Colwych* c.1180 SHC 1914 135, *Colewich*' 1198x1208 (13th century) EEA 17 74, SRO D938/103, *Kolewich* c.1230 SHC VIII 156, *Colewich* 1240 CL, c.1255 SHC VIII 156, *Colewyz, Colwich* 1247 Ass, *Kolewich* c.1230 SRO D938/105, *Colwych*' 1461 HAME 463, *Colwhich* 1518 SHC 1910 10, *Colwyche, Calwyche* 1532 SHC 4th Series 8 63, 131, *Collwyche* 1586 SHC 1927 167, *Collige* 1607 Kip, *College* 1761 Swynnerton ParReg, *Colledge* 1762 ibid. Possibly from the OE personal name Col(l)a, but more likely to be from OE *col* 'coal, charcoal', with OE *wic*. Cf. Coley; Colwich, Nottinghamshire. Colwich was evidently a variant spelling of Calwich (q.v.): Plot 1686: 404 refers to *Colwich-common* between Mayfield and Ellastone. Colwich Home Farm lies to the east of Calwich Abbey. See also Scoteslei.

COMBERBACH (unlocated, in Rugeley) *? Cumberbache* (p) *temp.* Henry VIII SHC X NS I 166, *Comberbach* 1570 *Survey*. From the OE loanword *Cumbre* (genitive plural *Cumbra*), from the PrWelsh ancestor of Welsh *Cymro* (plural *Cymry*), with OE *bece*, so giving 'valley or stream of the Britons or Welshmen', possibly associated with Cumberledge Hill (q.v.).

**COMBERFORD** 2 miles north-west of Tamworth (SK 1907). *Cumbreford* 1183 SHC II (i) 259, 1186 SHC I 129, *Cumberford* 1247 SHC 1911 118, 1313 ibid. 312, *Comberford* 1266 ibid. 136, 1280 ibid. 172, *Cumbreford* 1278 ibid. 32, 1286 SHC V (i) 169, *Cumberford* 1467 Hatherton, *Coumbforde, Coumberford hall* c.1562 SHC 1938 193,

*Cummerforde* 1608 SHC 1948-9 161. From OE *cumbre*, borrowed from the PrWelsh form of modern Welsh *Cymro* 'Welshman', probably meaning 'the ford of the Britons', perhaps a more courteous way of referring to Britons than *Walh*, since it is the OE version of *Cymro* (plural *Cymry*), the Welsh name for themselves. Cf. Cumberstone Wood; Combertbach; Cumberfield; Cumberledge.

**COMBES BROOK** a tributary of the river Churnet. *Cwms Brook* 1686 Plot 43, *Cooms Brook* 1810 *EnclA*. *Combes* is shown on the west bank 2 miles south of Bradnop on the 1836 O.S. map. The name is from OE *cumb* 'a cup, vessel', found in place-names in a transferred topographical sense 'a short spoon-shaped valley': the stream runs through a deep and narrow valley between steep hills here, draining the high moorland of Morridge into the river Churnet, but the valley could not be described as a *cumb*. It is uncertain whether *Cambes*, recorded in 1307 (SHC XI NS 258) refers to this place.

**COMBRIDGE** 2½ miles north of Uttoxeter (SK 0937). *Combrigge* 1191x1194 CEC 261, *Kanbrugge* 1246 Ch, *Combruggee* 1258 FF, *Combruge* 13th century frequently Duig, *Combruch* 1400 SHC XV 93, *Combrugge* 1467 SHC NS IV 152, *Combryge* 1532 SHC 4th Series 8 109, *Canbridge* c.1595 Erdeswick 1844: 504, *Combridge* 1615 SHC VI (i) NS 4. Possibly from OE *camb*, ON *kambr* 'a comb, crest' (an element rare in place-names), with OE *brycg* 'bridge', perhaps denoting a comb-like handrail or rails, or with OE *hrycg* 'a ridge', so 'the ridge with the undulating profile'. A derivation from OE *comb* 'a cup, vessel', found in place-names with the transferred topographical meaning 'short spoon-shaped valley', with OE *hrycg* 'ridge' is unlikely: the valley here appears from the map not to conform to a typical *cumb* form.

**COMMON PLOT** on the north side of Stone (SJ 8935). *Common Plot, Common Plots* 1798 Act. The land was held by the town with certain residents having rights of pasture.

**COMPA, THE** in Kinver (SO 8483). *? Coumbere* 1332 SHC X (i) 86, *? Combere* 1377 SHC 4th Series VI 9, *Coumber* (p) 1428 SHC XVII 119, *Comber* (p) 1539 SHC VI (i) NS 73, (p) 1584 SHC 1930 73, *? Compey* 1627 VCH XX 142, *Cumber* (p) 1666 SHC 1923 108, (p) 1670 SHC 1923 108. Perhaps from OE *cumb* 'a short spoon-shaped valley', but the name may be from ME *cumber* 'an encumbrance', probably used of ground encumbered with rocks, stumps, etc. (see PN Ch V (I:1) 152), or perhaps from OFr *combre* 'a heap of stones' (PN Ch V (I:1) 140). It has been noted, however, that second names or surnames ending in -er were common in medieval times, e.g. baker, bridger, laker. '[T]he family who lived at Coumbe became Coumber ... the family of atte Compe (OE *comp* 'field') became Comper ...': Mawer 1929: 68. Comber Road in Kinver may have the same derivation as The Compa, which it adjoins. Comber has been tentatively identified as DB *Cippemore* (q.v.): SHC 1916 170; see also VCH IV 54 fn.70. There is no philological (or indeed other) evidence to support any such connection.

**COMPTON** 2 miles south-west of Wolverhampton (SO 8898), *Contone* 1086 DB, *Conton* 1166 SHC 1923 298, *Cumpton* 1227 Ass, *Compthon* 1260 SHC 4th Series 13 6, *Comptone* 14th century Duig, *Cumpton* 1539 SHC VI (i) NS 64; **COMPTON** 1 mile north-west of Kinver (SO 8284), *Conton* 1166 SHC 1923 298, *Cumptun* 1227-8 SHC IV (i) 69, *Cumpton in Kenefare* 1293 SHC 1911 47, *Compton* 1577 Saxton; **COMPTON** on the south side of Leek (SJ 9856), *Cumton* 1256 Lib, *Compton* 1873 SRO D4855/3/1/1-12; COMPTON (unlocated) in Hanley (? SJ 8747), *Kumton* 1360 SHC XIII 11; **LONG COMPTON** 1 mile north of Haughton (SJ 8522), *? Compton'* 1327 SHC VII 244, *Long Compton* early 18th century SRO D590/364, 1775 Yates, 1836 O.S.; **LONG COMPTON (FARM)** 1 mile north-east of Swynnerton (SJ 8536), *Long-Compton* 1755 Swynnerton ParReg., *Long Compton* 1920 O.S.; COMPTON HALLOWES or

COMPTON WHORWOOD (obsolete), 2½ miles west of Kinver (SO 8083). *Horewode* 1267 SRO C143/2/34, *Horewood* 1268 SHC 1911 140-1, *Horewood* late 13th century VCH XX 132, *Hawloo* 1327 SHC VII (i) 247, *Haulowe* 1387 VCH XX 132, *Horewood* 15th century ibid, *Halowes in Compton* late 15th century ibid, *le Horewood otherwise le Halowes* 1527 ibid, *Hallowes* 1565 SHC XIII 248, *Compton Hallowes* later 16th century, *Compton Hallowes alias Compton Wherewoodes* 1631 SRO D660/19/2, *Whorwood*, *Compton Whorwood* 17th and 18th century VCH XX 132. '*Tun* in a *cumb* or short spoon-shaped valley'. There are 32 Comptons in DB, all with the spelling *Contone*. In the first place the name may refer to Compton Holloway (*Compton Hollowayes* 1586 SHC 1927 174), the steep road to Tettenhall Wood. The second place has narrow valley nearby (cf. Congreve). *Horwood* (*Whorwood* 1269 Ipm) appears on an 18th century estate plan at SO 813834 (WMANS 18 1975 9), and is from OE *horh*, *horu* 'filth', 'dirt', here probably meaning 'the muddy wood', and *Haulowe* is probably from the Haudlo family (from Hadlow, Kent) who held the place in the early 14th century (VCH XX 132). The moated site of the manor house is now occupied by (Compton) Park Farm. See also Camton.

CONDELEG(E) (unlocated) *Condeleg* 1251 Ch, *Condelege* 1252 (1798) Shaw I 38. See also Cuniteleg.

**CONEYGREAVE** 1 mile north-east of Draycott in the Moors (SJ 9940), *? Le Conyecrofte* 1279 SRO 98[7939], *Conyngreve* 1334 SHC 1913 47, *the conyngree* 1592 SHC 1930 355; **CONEY GREAVE** 1 mile south-west of Whitmore (SJ 8040), *Coney Grey* 1833 O.S.; **CONEYGREAVE HAFT** 1 mile north-west of Norbury (SJ 7923), *Coneygreve (Pool)* 1833 O.S. A common name, from ME *coninger* 'rabbit warren'. EDD defines *haft* as 'a little island or raised bank in a pond on which water-fowl build their nests', and Halliwell, giving the same meaning, suggests that the word is a Staffordshire term. It is found in Plot 1686: 232-3, who describes '...Hafts or Islands in the pooles...' at Shebden, and provides an illustration.

**CONEY LODGE FARM** 1 mile north of Norton Canes (SK 0410). *Coney Lodge* 1834 O.S. From ME *cunin* (from OFr *con(n)in*) 'rabbit': *Warren Field* lies to the west of this place on the 1834 O.S. map. The lodge doubtless originated as the home of the warrener.

**CONGREVE** 1 mile south-west of Penkridge (SJ 9013). *Comegrave* 1086 DB, *Cungrave* 1203 SHC III 122, *Conegrave* 1203 ibid. 122, *Cumgrave* 1236 FF, *Cunegrave* 1293 SHC VI (i) 289, *Cungreve* 1372 SHC XIII 92, *Connegrove* 1372 SHC 1931. OE *cumb-grǣfe* 'grove or small wood in a cumb or valley': there is a short spoon-shaped valley on the north-east side of Beacon Hill with its mouth between Congreve Manor and Congreve Farm (cf. Compton). The word *grǣfe* (also found as *graf*, *grafe*, and *grafa*), is probably associated in some way with OE *grafan* 'to dig', *grafa*, *grǣf* 'pit, trench'. 'Grove' is probably the more likely interpretation unless the first element suggests 'pit' or similar, or there is such a feature in the vicinity (see Gelling & Cole 2000: 226-30), and it may be noted that quarrying has historically taken place on nearby Beacon Hill. See also Somerford.

CONIGRE PARK (obsolete, on the north-east side of Dudley) *Cunhigre Parke in St Edmund* 1700 HRO E12/VI/NC/35. From ME *coningre* 'a rabbit-warren'.

**CONSALL** 2½ miles south of Cheddleton (SJ 9848). *Cvneshala* 1086 DB, *Cunshall*, *Cuneshale*, *Conleshale* 1227 SHC IV 49, 53, *Koneshull*, *Coneshill* 1265 Rees 1997: 136-7, *Culeshale* 1277 SHC VI (i) 84, *Conishale* 1281 Ipm, *Coneshill* 1285 Ch, *Coneshale* 1302 SHC 1911 59, 1309 SHC IX 6, 1456 SHC IV NS 95, *Conshale* 1306 GDR, *Consale*

1331 SHC XI 71, 1338 SHC XIV 51, 1339 SHC 1913 79, *Cunsale* 1348 to 1421 Banco, *Cunstall* 1386 Banco, *Conesale* 1456 ibid, *Cunsall* 1529 SHC 1912 30, *Co(u)nsall* 1561 AD 6, *Knut(e)shall* 1577 Saxton, *Counsall otherwise Counshall* 1583 SHC XV 147, *Consall* 1608 SHC III NS 26, *Knudshall* 1646 Jansson, *Cunsall* 1652 JRL RYCH/1804, 1674 SRO 49/14/44. A difficult name, for which no certain explanation can be offered. If derived from from ON *kunungr* 'king', the medial syllable would normally be expected in at least some of the spellings, but it is possible (though unlikely) that the name was shortened at an early date: King Street in Norwich, for example, derived from ON *kunung*, has spellings with Cunes-, Cuns, Conis- (PN Nf 114), and a local royal connection is implied in the name of the adjoining parish of Kingsley (q.v.). A common element in Welsh personal names was *cu* 'dog, hound', genitive *cunos* (cf. the british king Cunobelinos, Shakespeare's Cymbeline), and river-names incorporating the element (and the names of other animals) are found in Wales, e.g. Cynlais in south Powys, where the second element is Welsh *glais* 'stream'. The powerful stream which runs from Consall into the river Churnet is not named on maps, but may have been known as 'the hound (river)': see Bromwich 1972: 10; Breeze 2001: 76-7. *Cunsall Smythes* is recorded in 1662 (Kingsley ParReg), presumably Consallforge (*Consall Forge* 1836 O.S.), 1½ miles north-east of Consall (SJ 9949). See also Kingstone.

**COOK'S COPPICE** on the north-east side of Bagot's Bromley (SK 0726). Probably from a personal name – the Cook family are recorded in the area from an early date, e.g. Adam Cocus (Cook) in 1227 (SHC IV 52), Johannes Cook in 1402 (SHC XI NS 208).

**COOKSGATE** 1½ miles east of Betley (SJ 7648). *Cooks Gate* 1712 Audley ParReg, 1733 SHC 1944 34, 1798 Yates, 1833 O.S. From the surname Cooke: SHC 1944 34. The place may have been one of the gates to Heighley Park: TNSFC 1919 23.

**COOKSHILL** 1 mile north-west of Caverswall (SJ 9443). *Cookes Hill* 1608 SHC 1948-9 6, *Cook Hill* 1677 Caverswall ParReg, *Cookshill Green* 1836 O.S. Evidently from the personal name Cook (DES 108), with OE *hyll*. However, *Kocstall, Cocstal, Cokestall* recorded in the 12th-13th century (SHC XI 320-2), may be early spellings for this place, in which case the name may be from OE *coc* 'cook', with OE *stall* 'a place, the site where a building or other object stood', so possibly 'the place where the kitchen stood'. See also Cocknage..

**COOKSLAND** ½ mile north of Seighford (SJ 8825). *Cvchesland* 1086 DB, *Cokeslonia* c.1145 SHC II 219, *? Cuccessone* c.1150 VCH III 251, *Cokeslane* c.1180 SHC I 186, *Cokeslanie* c.1182 SHC II 256, *Cokeslone* c.1245 BodCh, *Cokeslond* 1377 SHC 4th Series VI 14, *Cokeslane* 1534 SRO D(W)1721, *Cokslande* 1537 JNSFC LXI 1926-7 30, *Cokis Lane* 1539 MA, *Cookes Lande* 1608 SHC III NS 15, *Cookesland* 1610 ibid. 46. Possibly from the OE personal name Cok or Cok, or from OE *\*coc(e)* 'lump of earth, a hillock', with the second element OE *land*, a word with several meanings, but here perhaps in the sense 'newly cultivated land', alternating with OE *lone* 'lane'.

**COOMBES, THE** a coppice ½ mile south of Ashley (SJ 7535). Early spellings have not been traced, but probably from the plural of OE *cumb* 'short spoon-shaped valley': hachuring on the the first edition 1" O.S. map shows what seems to be a group of short well-defined valleys here.

**COOMBESDALE** 1 mile north-east of Maer (SJ 8038). *? Lechombas* 13th century D593/B/1/6/1D/3, *Coomes* 1676 SHC 1914 16, *Cooms Hole* 1832 Teesdale, *Coombs Hole, Coombs Rough*s 1833 O.S, *Coombs* 1872 P.O. From the plural of OE *cumb* 'short spoon-shaped valley', with the recent addition of modern *dale*. The place lies in a steep hollow. *Combeshurst, Cumbesiches*, recorded in c.1300 (SHC 1913 239), appear to have been in the Whitmore/Limepits area.

COOPERS GREEN (obsolete) 1 mile west of Audley (SJ 7850). *Coopers Green* 1612 SHC 1944 83, 1733 ibid. 3. The Green element suggests a squatter settlement.

**COPLEY** 1 mile south-west of Pattingham (SO 8198). *Copley* 1314 VCH XX 173, 1490 SRO D593/A/1/18/18, 1532 SHC 4th Series 8 186, 1686 Plot, 1695 Morden, 1752 Rocque, 1833 O.S. From OE *cop(p)* 'a summit, a peak', in dialect 'a mound, a ridge of earth', with OE *leah* or (perhaps more likely) OE *hlaw* 'tumulus' (of which there are others in this area: see Kingslow; Stanlow), so 'the peak or summit with the burial mound'. The tendency to substitute *-ley* for *-low* is not uncommon in Staffordshire.

**COPLOW** a tumulus between Camp Hills and Berth Hill, near Maer: Plot 1686 409. *Coplow* 1686 Plot 1686: 409. From OE *cop(p)* 'a summit, a peak', in dialect 'a mound, a ridge of earth', with OE *hlaw* 'mound, tumulus', so 'the summit with the tumulus'.

**COP MERE** 2 miles west of Eccleshall (SJ 8029). *Cockesmere* 1250 SHC IV 116, *Cokemere* c.1272 SHC IV 189, 1278 SHC 1911 35, *Cokkemere* 1274 SHC 1921 22, *Corkemear* 1606 Eccleshall ParReg, *Cockemayer* 1608 ibid, *Cockmere* 1651 ibid, *Cop Meer* 1747 Bowen, *Cockmere* 1775 Yates, *Cop Mere* 1833 O.S. The early forms suggest that the first word is from OE *cocc* 'cock, a wild bird', with OE *mere* 'pool'. *Cokkesmere* near Stapenhill, Burton-upon-Trent, is recorded in 1462 (SHC 1937 173), but has not been identified.

COPNAL (obsolete, on the east side of Shobnall (?SK 2423). *Copenhale* 1286 SHC 4th Series XX 24, *Copnal* c.1758 SRO DW1734/2/3/136. Evidently the same derivation as Coppenhall (q.v.).

COPPEDLOWE CLOUGHS a deep ravine on the north side of Thorncliff (SK 0158). *Cowpedelowe Clowe* c.1280 SHC 1911 442, *le coppedlowesclogh* 1353 VCH VII 233. Perhaps from OE *\*coppod-hlaw* 'the flat-topped tumulus', from OE *\*coppian* 'to pollard', in this case meaning 'having had the top removed, polled, reduced'. See also Clough. *Le Coppedelowe* also occurs as a field-name in Derbyshire: Gelling 1988: 134.

**COPPENHALL** in Penkridge parish, 3 miles south-west of Stafford (SJ 9019). *Copehale* 1086 DB, *Coppenhale* c.1187 SHC II 261, *Kopenhale* 1235 SHC 1911 387, *Coppenhull, Copenhale* 1243 ibid.' 395, 399, *Coppenhale* 1323 ibid. 98, *Copenhull* 1383 Rees 1997: 172. 'Coppa's *halh*', with *halh* here perhaps meaning 'promontory into a marsh': CDEPN. Cf. Coppenhall, Cheshire. See also Copnal.

COPPY HALL (obsolete, 1 mile north of Aldridge (SK 0602)) *Coppye hall* 1596 SHC 1932 235, *Coppie Hall* 1597 ibid. 256, *Coppiehall* 1608 SHC 1948-9 84, *Copyhall* 1767 SRO D1317/1/15/7/35, *Coppy Hall* 1775 Yates. Perhaps 'hall at the coppice'. It is unclear whether *Coppice Hall*, recorded in 1663 (SHC II (ii) 58), refers to this place.

**COPSHURST** 1 mile south of Longton (SJ 9240). *Copt hurst, Coppehurst, Copte hurst* c.1555 SHC 1910 74-6, *Cobshurst* 1836 O.S. Perhaps from OE *\*coppod*, from OE *\*coppian* 'to pollard', and OE *hyrst* 'hill, wood, wooded hill, so giving 'the wood on the hill with the pollarded trees'.

CORBIN'S HALL (obsolete) in Kingswinford (SO 8888). *Corbyns hall* 1650 Guttery 1950: 10, *Corbins-hall* 1686 Plot 212, *Corbin's Hall* 1798 Yates. 'Which took its name from the owners thereof': Shaw 1801: II 228. The place was acquired by marriage by

Thomas Corbin in 1291: ibid.' A genealogy of the Corbin family is given in Shaw 1801: II 230.

CORBRIDGE (unlocated, near Syerscote (? SK 2106)). *Corbrigge* 1302 SHC VII 97, *Corbrugge* 1302 ibid. 108. This name is linked to Syerscote in the 1302 records, but the place has not been identified. Bridge Cottages lie 1 mile south-west of Syerscote, and may be associated with the name, the precise meaning of which is uncertain.

CORNBRIDGE (unlocated) It appears that the bridge lay on the river Trent to the west of King's Bromley: SHC V (i) 176; Shaw 1798: I 132. *Cornbrugge* 1290 SHC XI NS 176, 1297 SHC I 213, *Cornbrug* 1300 SHC V (i) 176, *Cornbrigge* 1455 SHC III NS 216. Probably from the OE bird-name *corn*, modern *crane*, or perhaps 'the bridge over which corn was carried'. It seems likely however that *Cornbrugge* 1266 (SHC IV 162) refers to Combridge, as may some of the above forms.

**CORNPARK** 1 mile south-east of Swinscoe (SK 1447). *Corn Park Farm* 1771 Okeover F425, *Corn Park* 1798 Yates.

**COSELEY** 3 miles south-east of Wolverhampton (SO 9493). *? Coleshai* 1204 SHC III (i) 102, *Colseleye* 1292 Inq, *Colseleie* 1317 SHC 1928 27, *Coleseley* 1325 (1801) Shaw II 214, *Colseleye* 1336 SHC 1928 33, *Colseleghe* 1357 Hackwood 1898: 4, *Colseley*, *Coulsley*, *Colsley*, later *Coseley* 1357 to 1664 Duig, *Colsley* 1680 BCA MS3549/262. Probably 'the wood from which charcoal was obtained', from OE *col* 'charcoal', with OE *leah*, rather than from the OE personal name Cole or Col.

COTEHILL (obsolete) in Werrington (SJ 9347). *Cote Hill* 1836 O.S., *Cotehill* 1889 O.S. From OE *cot*, *hyll* 'the cottage or shelter on the hill'.

**COTES** 4 miles north of Eccleshall (SJ 8335). *Cota* 1086 DB, *Cotis* 1260 SHC IV 142, *Cotes* 1280 SHC 1911 37, 1299 ibid.' 257, 1308 ibid.' 299, *Cotetes* 1324 ibid. 361, *Cottes* 1532 SHC 4th Series 8 104, *Cootes* 1585 SHC 1928 163, *Cotes* 1590 SOT SD4842/17/6, *Coats* 1702 Eccleshall ParReg, *Coates* 1834 O.S. From plural forms of OE *cot* 'cottage, hut, shelter'. *Cotelandis*, recorded c.1130 (SHC II 204) may refer to this place, or Coton to the east of Stone, or may be an unlocated place.

**COTON** 6 miles east of Stone (SJ 9732; SMR 01763 places a deserted post-Conquest at SJ 98453267), *Cote* 1086 DB, *? Cottin* 1598 Norton-in-the-Moors ParReg; **COTON** 1 mile north-east of Stafford (SJ 9324), *Cote* 1086 DB, *Cotes* 1209 Cur, *Kotes juxta Stafford* 1235 Fees, *Coton* 1285 FA, *Cotus* 1292 SHC II NS 148; **COTON** in Wiggington, 1½ miles north-west of Tamworth (SK 1805), *Coten* 1313 Ipm, *Cotoun* 1324 SHC X 48, *? Cawton* c.1570 SHC IX NS 238; **COTON** 1 mile south-west of Gnosall (SJ 8120), *Coten* 1327 SR, *Coton* c.1260 MRA, *Coton juxta Gnoushale* 1346, 1358 Banco, *Cotton*, *Coton* 1557 VCH IV 126; **COTON CLANFORD** 3 miles west of Stafford (SJ 8723), *Cote* 1086 DB, *Coton* 1291 Tax, 1356 SHC XII 130, *Cotun juxta Claneford*, *Cotus juxta Claneford* 1298 SHC IV 278, *Coton (Hayes)* 1414 Inq, *Coton juxta Clanford* 1421 BCA MS3525/Acc1935-043; **COTON IN THE CLAY** 1 mile north-east of Draycott in the Clay (SK 1729), *Cotton Bache* 1627 SRO D1522/1, *Coton* 1686 Plot, 798 Yates, *Coton in the Clay* 1836 O.S.; **COTON** 1½ miles north-east of Alveley (SO 7786), in Shropshire since the 12th century, *Coton* 1833 O.S.; COTON (obsolete, in Goldthorn Hill, Upper Penn, south of Wolverhampton (SO 9196)), *Cotes*, *Haye of Cotes temp.* King John Eyton 1881: 35, *Cotes* 1273 SHC 1911 154, 1291 Mander & Tildesley 1960: 8, *Cotone* 13th century SRO D593/B/1/17/1/3/10, *Coton* 1327 SHC VII (i) 249, 1332 SHC X (i) 126, *Cotone in Overpenne* 1375 SHC 1928 62, *Coton* 1457 ibid. 62, *Cotten* 1615 SRO D1717/A/2/1; **COTON (HALL)** 1 mile north-east of

Alveley (SO 7786), *Cotton Hall* 1752 Rocque, *Coton Hall* 1833 O.S. (in Shropshire since the 12th century). From OE *cot* 'cottage, shelter, hut' (*cotan* 'cottages'). *Clanford* is from the place of this name (q.v.) in Seighford; *Bache* is unexplained, unless from OE *bece* 'a stream or steep-sided valley'. *Cote*, recorded in DB as held by the church of Wolverhampton, is generally identified as Trescott (see SHC 1916 104; VCH IV 45), which was held by the church, but may be Coton south of Wolverhampton, which lies closer to Bushbury and Tettenhall, the entries for which precede and follow it in DB (see VCH IV 45; SHC 1916 104), and which was also held by the church: Slater & Hooke 1986: 24. *Cotes* and the *Haye of Cotes* recorded *temp.* King John (? *Cotetes* 1324 SHC 1911 361) which lay against the boundary of Penn (Eyton 1881: 35, 67), may be Coton south of Wolverhampton. See also Trescott. *Showcotone* near Mansty is recorded in 1588 (SHC 1928 164-5), but has not been located.

**COTON END**  in Gnosall (SJ 8020). *Cotenend* 1573 Ct, *Cootenende* 1619 ParReg, *Coaten End* 1773 WALS DX3/11. 'The end quarter of Coton (q.v.)'. The word *end* did not mean a terminal point, but simply 'place', and was often waste or common land at the end of an inhabited area later used by squatters.

**COTTON, NEAR & UPPER**  2 miles south-west of Cauldon (SK 0646 & SK 0547). *?* *Cotton* 1277 SHC 1911 30, *Cawton* 1610 Speed, *Cotton* 1686 Plot, *Upper Cotton, Nether Cotton* 1798 Yates. Probably from OE *cotan* 'cottages'. It has been suggested that this place is to be associated with *Chotes* or *Chotene*, recorded in 1176 (VCH III 226; Croxden Chr), the site of an early monastic house transferred to Croxden (q.v.) in 1179 (CEC 208), but there is no philological evidence to support the association. [The reference to this place in CDEPN is an error: the entry properly relates to Cottons in Derbyshire.]

**COTWALL END**  ½ mile south-west of Sedgley (SO 9192). *?* *Cottewwelle* c.1270 SHC 1941 78, *Cutwalle* 1327 SHC X 249, *Cattewall* 1332 SHC VII (i) 128, *Cotewallend* c.1400 SBT DR37/2/Box 122/36, *Cotwalle End* 1532 SHC 4th Series 8 115, *Cotwalend* 1578 Sedgley ParReg, *Cottwallend* 1591 SHC XVI 107, *Cotwall end* 1834 O.S. The spellings point towards OE *cot* 'cottage, hut, shelter', or perhaps ME *cutte* 'a cut, a water channel' (recorded from the 13th century in place-names: EPNE i 120), with OE *wælle*, giving 'the place of the cottage or shelter with the spring', or 'the spring with the water channel'. Hackwood 1898: 105 states that 'the ancient name of Upper or Over Gornal is 'Sheep Cotwall' as appears by the Court Rolls', in which case 'the [sheep]cot wall' or 'the spring or stream at the [sheep]cot', but no evidence has been traced to support such derivation.

**COTWALTON**  2 miles north-east of Stone (SJ 9234). *Cotewaltune* 1004 (11th century, S.906; 11th century, S.1536), *Codeuualle, Cotewoldestvne* 1086 DB, *Codewalton* 1176 SHC I 229, SHC XII NS 91, *Cotwaldeston* 1679 ibid. 86. The first element may be OE *cot* 'cottage, hut, shelter', or possibly the personal name Cotta, with OE *wald* 'wood' (a very rare element in Staffordshire), or (perhaps more likely), a stream-name with Mercian OE *wælle* 'a spring', and (sometimes) 'a stream', to which was added *tun*.

COUDRY, CAWDRY  (unlocated, in Bradnop). *Coudray* 1214-7 SHC 1913 314, CEC 371, *Cawdry* 1222 SHC 1911 423, *Le Coudray* 1227 SHC XI NS 240, *Coundrey* c.1240 SRO DW1761/A/4/9-10, *Coudray* 1241, c.1265 SHC 1911 438, 1271 SHC V (i) 145, *Coudry* c.1250 SHC ibid.' 426, *Koudrai* 1281 *Antrobus, la Coudrey* 1284 SHC VI (i) 140, *Cowdray* 1284 SHC XI 333, *Coudre* 1293 SHC VI (i) 224, *Coudrey* 1402 SHC XI NS 213, *Cowaderey* 1451 SHC III NS 194, *Cowdrey* 1776 Butterton *EnclA, Caudery* 1811 Census. Camden 1674: 129 notes 'Couldray, that is, Haslewood'; White 1834: 713

suggests that *Caudery* is recorded in association with Rudyard; and Sleigh 1883: 124 states that Cawdry adjoined Onecote, which seems to be confirmed by *Coudr(ey) Onecoat*, recorded in 1307 (SHC XI NS 257). Possibly there was more than one place of this name, which is from OFr *coudraie* 'a hazel copse': cf. Cowdray Park, Sussex. Clark observes 'There seems little reason to doubt that among literate English people of the mid-thirteenth century Old French co(u)dre and co(u)draie could have been familiar enough to have sprung to the mind of a dog-Latinist improvising terms for landmarks': Jackson 1995: 369.

**COUNSLOW** 1½ miles east of Cheadle (SK 0342). *Cundeslowe* c.1282 SHC 1923 39, 1284 SHC 1911 186, *Conndeslowe* 1318 SRO D1275/2, *Counesley* 1409 SHC XI NS 54, *Cownslowe* 1598 SHC XVI 185, *Cowneslowe* 1605 SHC 1940 270, *Comslowe* 1609 SHC III NS 52, *Counslow Plantation* 1836 O.S. The first element appears to be an unidentified personal name (perhaps Cundhere or similar), but see also Scounslow Green (to which some of the spellings may relate). The second element is OE *hlaw* 'mound, tumulus'. The place lies on a hill of 829'.

COVELE (unlocated, possibly Cowley, near Gnosall) *Covele* 1314 SHC XII NS 278.

**COVEN** (pronounced coh-vun [kəʊvən]) 2 miles south-east of Brewood (SJ 9006). *Cove* 1086 DB, *Couena* 1175, *Couene* 1176 P, *Koven* 1236 Fees, *Covene* 1242 Fees, *Cone* 1262-72 SHC 1939 30, *Covewode* 1283 SHC 1911 41, *Covene (wode)* 1311 SHC X 11, *Kovene* 1332 SR (p), *Covun* 1342 Wodehouse, *Cowyn* 1532 SHC 4th Series 8 85. The meaning of this name remains unresolved: the root is generally held (see e.g. Ekwall 1960: 126; Oakden 1984: 37) to be OE *cofa*, dative plural *cofum* (later *cofan*), 'an inner chamber, a cell', a meaning well-attested in OE literary sources in compounds such as *bed-cofa* 'bed-chamber', *hord-cofa* 'treasure-chamber', etc. It is not until much later that the meanings 'hollow in a rock, cave, cove' become evident. The meaning 'small bay, cove, inlet' is not recorded by OED before 1590, and 'a sheltered place amongst hills or woods' not before 1786, although a reference to *þe coue* in the medieval poem The Wars of Alexandrer (line 5422), seemingly with the meaning 'vale' (Elliott 1984: 104) may be noted. The usual meaning 'chamber' is normally taken to include 'shelter, hut', with 'cave', or 'recess in the steep side of a hill' in later place-names, and the word may be connected with Middle Low German *cove*, *coven*, Modern German *koven*, ON *kofi* 'cell, hut, shed', Swedish *kofva*, dialect *kove*, *kuvi*, Norwegian *kove* 'hut'. None of the topographical meanings fits the local landscape of gently undulating countryside near the meandering river Penk. The accepted derivation for Runcorn (*Rumcofan* ASC, *Rumcofa* 1086) is 'wide or roomy bay', although that derivation would long pre-date the supposed age of 'cove' in the OED. *Cofa* may conceivably be associated with streams or water, perhaps '(place at) the coves or inlets or creeks', from some now obscure features in the nearby river Penk, possibly a widening on the south side of the river on the north-west side of the village, where the riverside meadows are low-lying: Ekwall 1960: 114 suggests that the element which occurs in Cobham, Surrey refers to a river bend. Cove in Hampshire (*Cove* DB) is, interestingly, also in level countryside for which the conventional topographical meaning would be innappropriate. As here, the derivation remains uncertain (see Coates 1989: 60), and tends to support the possibility that the name may be connected with some impermanent non-topographical feature. There is a reference to *'the wood of Covenhue'* (cf. *Covewode, Covene wode* above) in Whittington near Kinver (a place known for its ironworks: see VCH II 108) in an undated charter SRO T p.1273/12 Bk of Dds of Evidences 13, and Yates' map 1775 shows *Covton Brook* in the same area, but no obvious feature to account for the element has been noted there. A *Cove Wood* is also recorded on a tributary of the Coundmoor Brook in Harnage

(Shropshire) in the 19th century: PN Sa II 132. Those associations with 'wood' (cf. Conholt, Wiltshire, *Covenholt* 1251) may provide some clue to the derivation of the name. The name *Karrecouein* is recorded in 1195 (Pipe), and may be associated with this place: see Kiddemore Green. *Covencestor*, recorded in 1268 (SHC 1911 141) has not been located. See also Cover river (Ekwall 1928: 100). *Pencovan*, possibly in Herefordshire, is recorded in 757x796: Finberg 1972: 140, and *stancofan*, recorded in the bounds of Oldswinford in 951x955 (S.579, 16th century) is tentatively interpreted by Hooke 1990: 162, 165 as 'stone cave', perhaps meaning the remains of a prehistoric burial chamber (on which see also Featherstone, which lies a mile or so south-east of Coven). Cf. Coaley, Gloucestershire; Cofton Hackett, Hereford & Worcester.

**COVEN LAWN** 1 mile south of Coven (SJ 9005). *Coven Lawn* 1834 O.S. See Coven. *Lawn* is from ME *launde* 'an open space in woodland, a forest glade, woodland pasture', meaning in this area 'a cleared passage in woodland': cf. Blymhill Lawn and Langley Lawn.

**COWALL** 1 mile south east of Biddulph (SJ 9055). *Couhale* 1325 SHC 1911 366, 1327 SHC VII (i) 206, *Couhale, Cowale, Kowale, Couwale* 1348x1369 Tunstall Ct, *Cowall* 1594 SHC 1934 17. From OE *cu* 'cow' with OE *halh*, dative *hale*.

**COWHAY** 1 mile west of Warslow (SK 0658), *Cowhey* c.1615 SRO D3272/1/17/4/6-8, *Cow Hay* 1697 Leek ParReg; COWHAYE (unlocated, in Barlaston), *Cowhaye* 1525 D(W)1721/1/1/64-65. From OE *cu* 'cow' and Mercian OE *(ge)heg* 'enclosure'. See also Ballington.

**COWLEY** 1 mile south of Gnosall (SJ 8219; SMR 02098 places a deserted post-Conquest settlement at SJ 82801930). *Covelav* 1086 DB, *Culeg'* 1215 MRA, *Kuleg(a)* 1199 Ass (P), *Couleg'* 1225 Cur, *Coulee* 1292 SHC VI (i) 226, *Cullega* 1293 ibid.' 289, *Coule* frequently 12th and 13th century Duig, *Couleye* 1324 SHC 1911 101, *Caweleye* 1327 SHC VII 239. The DB form is probably to be read as *Coue-*, so probably from OE *cu* 'cow', with OE *leah*, but Ekwall 1960: 126 suggests that some Cowleys may contain a descriptive first element, noting that Cowleys are near hills (this place lies in undulating countryside), and that there may have existed an OE word for a hill cognate with Norwegian *kuv* 'a rounded top', or even a word denoting something obtained from a wood, for example an OE *\*cufl* 'a block of wood, a log, a stump', similar to Swedish *kubb* 'a log'.

**COWLEY** ½ mile south-east of Hamstall Ridware (SK 1018). *Cuweleye* 1266 SHC IV 163, *Cowley* 1279 (1798) Shaw I 152, *Couleye* 1285 ibid, *Coulee* 1327 ibid, *Cowleas* 1585 SHC 4th Series 9 13, *Cowley* 1798 ibid, *Cowley (Hill)* 1834 O.S. See Cowley near Gnosall. This place has been identified as *Rideware* (DB): VCH IV 52 fn.50. From OE *cu* 'cow', with OE *leah*. An unlocated *Cuweleye* (Park) is recorded in 1266 (SHC IV 163), and land called *le Cowley* in Marchington in 1401 (SRO DW1733/A/3/8).

**COWLOW** in Holme End (SK 1059). *Calah* 1659 Alstonefield ParReg, *Cowloe* 1600 ibid, *Calowe* 1655 ibid, *Calah* 1658 ibid, *Cawlan* 1690 ibid, *Cawlaw* 1701 ibid, *Cawla* 1711 ibid, *Cawlow* 1715 ibid, *Cawlo* 1750 ibid, *Cawley, Cauley* 1767 ibid, *Cawlow* 1839 EnclA, *Cow low* 1840 O.S., *Cow Low* 1861 Bateman. Possibly from OE *ca* 'jackdaw', with OE *hlaw* 'mound, tumulus'.

**COWPERS GREEN** (unlocated, in or near Audley) *Cowpers Greene* 1577 Audley ParReg, *Coupers Greene* 1584 ibid. In Northern dialect a cowper is one who barters, deals, buys and sells: OED. The Green element suggests a squatter settlement.

**COXENGREEN** on south side of Butterton (SK 0756). *Coxe Greene* 1616 FF, *Coxen Green* 1832 Teesdale, 1838 O.S. 'Cox's Green'. The Green element suggests a squatter settlement.

**CRABTREE** (unlocated, in Wolstanton) *Crabbetre* 1379 Pape 1928: 200, *Crabtree* 1619 Biddulph ParReg, *the Crabtree* 1672 ibid.' 'The crab-apple tree'. An early example of the name: see EPNE i 110.

**CRACKLEY BANK** on Watling Street, ½ mile south of Sheriffhales (SJ 7610). *Crackeley banke* 1664 Sheriffhales ParReg, *Crackley banck* 1679 SHC 1919 243, *Crackley-bank* 1686 Plot 400, *Crackley Banke* 1692 Sheriffhales ParReg. The name is recorded c.1603: TSAHS LXXXVI 2001 75. Perhaps from OE *\*craca* 'a crow, a raven', with OE *leah*. The suggestion in Duignan 1902: 116 that the name is from a Germanic or Celtic word meaning 'boundary' is baseless, though the place lies on the Staffordshire-Shropshire (and perhaps a much earlier) border.

**CRACOW MOSS** 1 mile south-west of Betley (SJ 7447). *Cracalmosse* 1360 AD, 1429 ibid, *le Cracalmos* 1373 ibid, *Craca Moss* 1695 Betley ParReg, *Cracamoss* 1696 ibid, *Creca Moss* 1832 Teesdale, *Creka Moss* 1842 TA. Moss is from OE *mos* 'bog, marsh, swamp'. The first part of the name may be from OE *\*craca* 'a crow, a raven', with OE *halh* (note the adjoining name Ravenshall (q.v.)), but the spellings would also support a derivation from the surname Crachale (see DES 114). Corruption of the name has occurred, with the common loss of the medial *l*, influenced by Cracow (Krakow), Poland.

**CRADDOCKS MOSS** 1 mile east of Betley (SJ 7748). *Cradocks Mosse* 1733 SHC 1944 26. From the Cradock family: see SHC 1944 26ff. Moss here means 'bog, swamp, marsh', from OE *mos*.

**CRADLEY HEATH** (pronounced Crayd-lee [kreɪdli:]) 2 miles south-west of Rowley Regis near the Worcestershire border (SO 9486). From Cradley in Worcestershire, *Cradelei* 1086 DB, *Crandelega, Cradelega* 1179 SHC I 93, 96, *Cradelega* 1180-8 P *passim, Cradelea* c.1189 SHC II 2, *Cradele* 1204 SHC SHC III (i) 95, *Cradeleg* ' c.1238 (1798) Shaw I xxvi, *Cradele(ye)* 1272 Ct, 1273 SHC IX (ii) 27, 1275 *Ass*, 1485 Pat, *Cratley* 1749 Bowen. The first element is probably an OE personal name *\*Crad(d)a* or Cradel, or (less likely) from OE *cradol* 'cradle', a word which seems originally to have meant something plaited or woven, perhaps here used for a place which provided material for making hurdle fences, or (as suggested by PN Wo 294) because with high ground to the north and south it lay in a cradle of land. The second element is OE *leah*. The place, on the banks of the river Stour, was formerly an area of heathland.

**CRAKELOWE** ½ mile south-east of Sheen (SK 1160). *Crychelaw* (p) 1535 SHC 1912 79, *Crichelowe* (p) c.1556 SHC 1939 235, *Crycheloe* 1570 Alstonefield ParReg, *Crychloe* 1587 ibid, *Crychlow, Crichlowe* (p) 1598 SHC 1935 157, 173, *Crichlow* 1784 SHC 1947 85. The spellings indicates a derivation from OE *cryc* (from British *cruc*) 'a hill' with OE *hlaw* 'mound, tumulus', giving 'the hill called Cryc with the tumulus'. It is possible that some or all of the spellings do not relate to this place, but the surname Crichloe is strongly associated with Sheen: see for example SHC 1925 203-4. If they do, the change from Crich- to Crake- is remarkable.

**CRAKEMARSH** 2 miles north of Uttoxeter (SK 0936). *Crachemers* 1086 DB, *Crakemess* 1189x1199 NA DD/FJ/1/181/1, *Krakemers* 1235 SHC 1911 387, *Krakemerz* 1236 Fees, *Crakemers* 1242 Fees, *Crakemershe* 13th century Duig, *Crakemarche* 1532 SHC 4th Series 8 5, *Craykemarsshe* c.1569 SHC 1926 109. From OE *\*craca* 'a crow or a raven', with OE *mersc* 'a marsh', hence 'marsh of the crows or ravens'. The place is in an area which is particularly flat and poorly-drained. Duignan 1902: 46 suggests that *crake* as part of a place-name is not found elsewhere south of Yorkshire. See, however, Crateford, in Brewood.

**CRANE BROOK** 4 miles south of Lichfield. *Cronebrouk* 1300 SHC V (i) 177-8, SHC 1916 113, *Crone-brook* 1308 (1801) Shaw II 58. From OE *cran, cron*, in West Midlands perhaps meaning 'heron'. It is doubtful whether the crane was found in the Midland counties.

**CRANMERE FARM** 1 mile north of Worfield (SO 7597). *Cranmere (Farm)* c.1750 WRO 899:749/8782/83/v/1-75, *Cranmoor (Heath)* 1752 Rocque, *Cranmere* 1793 SA 2161/157-8. In Shropshire since the 12th century. From OE *cran, cron*, in the West Midlands possibly meaning 'heron', with OE *mor*, probably here in the sense 'marshland'.

**CRANMOOR** 1½ miles north-east of Pattingham (SJ 8400). *Cranemore* 1088 SHC II 183, *Cranemere* 1598 SHC 1934 (ii) 11, *Cranmores (Oxe, Vpper & Calues)* 1634 map SRO D3548/1, *Cranmoor* 1686 Plot 214, *Cranmoor (Wood)* 1834 O.S. From OE *cran, cron* 'crane', perhaps in the West Midlands 'heron', with OE *mor*, probably here in the sense 'marshland'. The 1843 Penn TA shows an area called *Cranmoor* (*Cronmore* 1324 (1801) Shaw II 210, *Cranmere* 1598 SHC 1934 (ii) 11, *Cranmoore* 1647 Survey) in Lower Penn, some 3 miles south-east of Cranmoor near Pattingham.

CRANNAGE (unlocated, in Worfield) *Crannage* 1790 SA 445/241-2.

**CRATEFORD** 1 mile north-east of Brewood (SJ 9009). *Crakeford* 1327 SR (p), 1332 ibid, 1655 PCC, 1682 Browne, 1755 Bowen, *Crackford* 1660 PCC, *Crateford* 1834 O.S. 'The ford of the crows or ravens', from OE *\*craca*: see Crakemarsh. Another Crateford lies on the west side of the river Severn, ½ miles north-east of Chelmarsh (SO 7288), in Shropshire.

CRAWFORD (obsolete) on the east bank of the river Tame, 2 miles south-east of Whittington (SK 1806). *Crawford* 1834 O.S. An interesting name, possibly (if ancient) from Welsh *cryw* 'a ford'. Perhaps to be associated with *Crawelake, Crolake* recorded in 1498 and 1506 (OSS 1936 49-51), but the trackway leading to the west bank of the river Tame from the direction of Whittington is recorded as *Caldefordwey* 'cold ford way' c.1300 (TSSAHS XX 1978-9 loose map), so the name may be a corruption of Caldeford.

**CRAWLEY** on the south side of King's Bromley (SK 1216). *Crawley* 1744 SRO D357/A/4/1-7. From OE *crawe leah* 'the *leah* where crows nested'.

**CRAWLEY BROOK** a tributary of the river Trent. See Crawley.

**CRAYTHORNE** ½ mile south-east of Rolleston (SK 2426). This name may be associated with *greatan porne* 'great thorn', a boundary mark in a charter relating to Rolleston of 1008 AD (14th century, S.920). The 1837 TA map gives the name *Craythorne Field*. Hooke 1983: 95.

CREAME (unlocated, in Worfield parish) *Creame* 1815 SA 2161/113-4.

**CREIGHTON** 2 miles north-west of Uttoxeter (SK 0836). *Crectone* 1166 RBE, *Creighton* 1222 Ass, *Cratton* 1241 Duig, *Cracton'* 1242 Fees, *Creghton* 1327 SHC XI 139, *Creygthon* 1337 SHC 1913 59, *Creghtton* 1532 SHC 4th Series 8 4, *Croyton, Croyton (parke)* 1645 SHC 4th Series I 250, *Chreton* 1666 SHC 1925 225. From OWelsh *creic*, Welsh *craig* 'a rock', with OE *tun*, so 'the *tun* at the rock called Creic'. The 1222 and 1327 spellings indicates that the root is unlikely to be OE *cræt* 'cart'.

**CRESSWELL FARM** on the north-west side of Brewood (SJ 8809). *Creswell's Barn* 1834 O.S. From the Creswell family who farmed here from at least the 17th century: see SHC 1919 240.

**CRESWELL** Ancient Parish. A lost medieval village, marked only by fragments of a 12th century stone chapel, 2 miles north-west of Stafford (SJ 8826), *Kersewell, Karsewell* 1203 SHC III 78-9, *Kerseuell'* c.1206 SHC 1928 280, *Keyeswell* late 13th century SRO 3764/114[36347], *Cresswall* c.1595 Erdeswick 1844: 139; **CRESSWELL** on the river Blithe, 1 mile south-west of Draycott in the Moors (SJ 9739), *Cressvale* 1086 DB, *Cresswellam* c.1160 SHC III (i) 224, *Cresswella* 1190 P, *Creswal* c.1238 (1798) Shaw I xxvi, *Cresswelle* 1242 SHC I 228, *Crestewall* 1284 ibid, *Kersewelle* 1288 SHC 1911 194, *Creswall* 1611 SHC IV NS 13; **CRESSWELL GREEN** 2 miles west of Lichfield (SK 0710), *Cressewalle* 1380 VCH XIV 199, *Cressewelle* 1381 SHC XVII 178, *Cresswell Green* 1834 O.S.; CRESSWELL (unlocated, near Tettenhall), *Cressewalle* 1321 SHC X 37, *Cressewall* 1332 ibid.' 131, *Creswall* 1539 SHC VI NS (i) 64. From OE *cærse* 'cress, watercress', with Mercian OE *wælle* 'a spring', and (sometimes) 'a stream', hence 'watercress spring'. Lewis 1849: I 724 records a 'copious spring' at Cresswell near Draycott in the Moors. Cresswell Ford (possibly *Karsewalle* recorded in 1293: SHC VI (i) 229), lies 2 miles to the north of Cresswell near Draycott (SJ 9743), and Cresswell's Piece 4 miles north-west (SJ 9545). The latter is likely to be from a personal name. *Cressewalle* in Whitmore is recorded in the early 14th century: SHC 1913 250. The Green element suggests a squatter settlement.

CRISTAGE (unlocated, below Ribden) *Cristage* 1686 Plot 89. An interesting name, probably from OE *Crist-āc* 'Christ's oak', with the second element in the dative form (though the absence of the genitive *-s* is curious), perhaps here meaning 'Christ's cross', an oak or cross at which the gospel was preached: cf. Cressage, Shropshire. The location of the place has not been traced; it is unclear whether there is any association with *Hoftons Cross* (q.v.).

CRISTEBRUGGE (unlocated, in Hatherton) *Cristebrugge, Crystbrygge*? 14th century SHC 1928 140-1. Possibly to be associated with *Kestbridge meadow but now known as Russells Meadow* recorded in 1813 (SRO D260/M/T/5/1).

CROFTIS (unlocated, perhaps to be associated with West Croft Plantation, 2 miles north-west of Wheaton Aston (SJ 8213)) *Croftis* 1279 SHC 1911 176.

CROKEMAREBOTH, CROKEMAREHOUS (unlocated, possibly in Audley) *Crokemareboth, Crokemarehous* 1308 SHC XI NS 264-5. Perhaps from ME *crook* 'to bend, to curve', with OE *(ge)mære* 'boundary', so 'the booth or house at the place where the boundary changes direction'.

**CROMER HILL** 1 mile south of Milwich (SJ 9631). *Crammer Hill* 1781 SRO D637/1/2. If ancient, perhaps from OE *cran, cron* 'a crane', or (more likely) 'a heron', with OE *mere* 'pool, wet ground', so 'the hill at the mere or pool with the heron', although no pool appears on the modern map. OE *mere*, and *mor* 'upland; moorland; marsh', often become interchanged in place-names, and may have done so here.

CROMSLEY (obsolete) 1 mile north-west of Bagot's Bromley (SK 0527). *Cromburley* early 13th century SRO DW1721/3/11/3, *Cromberley* 1299 SHC XIV (ii) 23, *Cromberley Welle* c.1321 SRO DW1721/3/21/1, *Cromberleigh (field)* 1332 SRO

D(W)1721/3/3/10, *Cromburleye* 1375 SRO DW1721/3/23/11, *Cromberleye* 1397 D(W)1721/3/29/16, *Cromsley* 1724 *Survey*. A curious name. Possibly OE *burh* 'fortification', and OE *leah*, with OE **crumbe* 'a bend, especially in a river or stream': the place lay near Tad Brook and the river Blithe.

CROMWELL'S GREEN – see **OLIVER'S GREEN**.

CRONKHALL (obsolete) 1 mile north of Tettenhall (SJ 8801), possibly at what is now Windermere Road: StEnc 165. *Cronkwall* c.1225, *Kronekwall* 1271 SHC V (i) 144, *Cronekwelle* 1271 SHC 4th Series XVIII 73, *Cronetwell* 1286 SHC 4th Series XVIII 151, *Crunkwelle* 1302 SHC VII 100, *Croukewall* 1332 SHC X 126, *Croukwalle* 1347 SHC 1913 120, *Crougwalle* 1357 SHC XII (i) 146, *Croucwall* 1402 SHC VI NS (ii) 199, *Crouggewall* 1413 SHC XVII 8, *Cronkhall Green* 1419 VCH XX 173, *Croukwall* 1463 SHC VI NS (ii) 209, *Crankewall* 1532 SHC 4th Series 8 147, *Crankall* 1591 SHC VI NS (ii) 286, *Cronckall* 1612 Codsall ParReg, *Cronkhall* 1834 O.S. The first element is likely to be OE *cranuc* 'a heron, or similar bird', with OE Mercian *wælle* 'a spring', and (sometimes) 'a stream', but an alternative derivation may be from OE **cronc* 'a winding path, bent, crooked, twisted', with OE *wall, weall* 'wall', giving 'the crooked wall'. For completeness it may also be noted that an unrecorded OE personal name Crannuc has been postulated: see PN Ch I 107. Cf. Cronk Hill, Shropshire; Conksbury, Derbyshire; Crankland, Devon.

**CRONK HILL, CRANK HILL** 2 miles east of Wednesbury (SO 0194). *Crank Hall, Crankhall Farm* 1834 O.S. Possibly from OE *cranuc, halh* 'the nook or corner frequented by cranes', although a derivation from OE **cronc* 'a winding path, bent, crooked, twisted', cannot be ruled out: the word is not recorded in England before 1552, but Duignan 1902: 46-7 records that the base of the hill is noticeably tortuous. Cf. Cronkhall. *Cronk Hill* appears 1 mile east of Upper Elkstone (SK 0659) on the first edition O.S. map of 1842. Earlier forms have not been traced. There is also a Cronk Hill in Atcham, Shropshire. The fact that the first element is not uncommonly found with 'hill' suggests that a derivation from **cronc* may be more likely.

CRONKLEDGE (unlocated, in Warslow) *Cronkeslach* 1633 Rental, *Cronklach* 1660 Alstonefield ParReg, *Chronklidge* 1685 ibid. From OE *cranuc,*lece* 'the boggy stream frequented by cranes'.

**CROSS OF THE HANDS** 1 mile south-east of Abbots Bromley (SK 0822). *Cross in the Hand* 1775 Yates. Perhaps from a sign-post with traditional finger pointers which may have stood on the island in the centre of this crossroads. It is unclear whether *the wod crosse*, recorded in 1539 (SHC V NS 321) is to be associated with this name. See also Cross o' th' Hands. Cf. Cross o' th' Hands, Hulland, Derbyshire (PN Db 576).

CROSS O' TH' HANDS (obsolete, near Ashmore Brook) *Cross-o'th'hand* 1664 (1798) Shaw I 316, *Cross o' th' Hand* 1720 Bowen, *Cross in the Hand Close* 1766 SRO D661/4/13/2, *Cross o' the Hand Piece* 1767 SRO D 615/D/235/2, *the Cross-o' th'Hand* 1798 Shaw I 316. From a direction post in the form of a cross with a hand mentioned in the later 15th century, recorded here in 1675, but gone by 1828: VCH XIV 203. The post, which stood on the city of Lichfield boundary, is preserved in the name Cross in Hand Lane. See also Cross of the Hands.

**CROWBOROUGH** 1 mile east of Biddulph (SJ 9057). *Crowebarewe* 1298 SHC XI NS 253, *Crowbarwe hows* 1299 SHC NS XI 253, *Crowebarwe* 1359 SHC XII 166, 1391 SHC XVI 29-30, *Crowbaro* 1514 JNSFC LX 1925-6 36, *Crowborow* 1634 Biddulph

ParReg, *Crowbery* 1644 SHC 4th Series I 258, *Crowburrough* 1733 SHC 1944 11. 'Crow grove; rookery', from OE *crawe* 'crow, raven', with OE *bearu* 'a wood'.

**CROWCROFTS** 1 mile north of Barlaston (SJ 8939). *Crowecroft'* c.1342 SHC XI 312, *Crowcrofts* 1723 Barlaston ParReg, 1836 O.S., *Crow Croft* 1741 SRO D593/B/1/1/1. Possibly from OE *crawe* 'a crow', with OE *croft* 'a small enclosed field, a small enclosure of arable or pasture land, a pasture enclosure near a house', so 'the small pasture enclosure with the crows', but the first element may be the obsolete dialect word *crew(e)*, also found as *creuh, crow, crough, crue* (cf. Kidsgrove), meaning 'animal pen, sty, hovel, hut', giving 'the small pasture enclosure with the animal pen'.

**CROWDICOTE** (in Derbyshire) on east bank of the river Dove, ½ mile east of Longnor (SK 1065). *Crudcote* 1223 FF, *Croudecote temp.* John SHC 4th Series IV 73, c.1450 ibid.' 225, *Croudecote* 1339 SRO 1110/1, *Crowde(n)cote* 1577 Saxton, *Crowdey Cote* 1775 Yates, *Crowdecote* 1840 O.S. Possibly from an unrecorded OE personal name *Cruda, hence 'Cruda's cottage': PN Db II 365. This place is said to have been a member of Wetton although situated on the Derbyshire side of the Dove: SHC 4th Series IV 73.

CROWESBRIDGE (obsolete, south of Hurst Hill, Coseley (SO 9393)) *Croksbridges* 1539 Survey, *Crowesbridge* 1637 Underhill 1941: 143. The place is said to have been near the junction of Upper Ettingshall Road and Coppice Road, and is also recorded as *Crows Britch, Croksbridge, Croksbritches*. Roper 1952. A curious name, the derivation of which is made difficult by the inconsistent forms, which suggest alternative names. The first element may be from the personal name Croc, with an uncertain second element: *bridge* is supported by the early forms, but there appears to have been no watercourse here, so possibly from OE *brēc* 'newly cultivated land'. *Crokebruge*, possibly in Marchington Woodlands, is recorded in 1482: Shaw 1798: I 86.

**CROWGUTTER** on the north-east side of Ipstones (SK 0250). *Crowgutter* 1650 BCA MS917/1556, 1668 Okeover T700, 1694 SHC 1947 64, 1775 Yates, *Crowgutter (Farm)* 1791 SRO D1134/18/1. See also Grimditch.

**CROWHOLT** 2 miles north-east of Cheddleton (SJ 9953). *Crowholt Farm* 1728 ParReg, *Crow Holt* 1837 O.S. From OE *crawe, holt* 'the wood frequented by crows'.

CROWS HEATH (unlocated, near Rudge Heath) *Crows Heath* 1806 SA 5586/13/5.

**CROWTREES** in Waterhouses (SK 0750). *Crowtrees* 1661 Ilam ParReg. '(the place) where crows nested'. Crowtrees in Elkstone is recorded c.1870 (*Rental*).

**CROXALL** Ancient Parish 6 miles north-east of Lichfield (SK 1913; SMR 00947 places a deserted Anglo-Saxon settlement at SK 19911355). *Crokeshalle'* 942 (14th century, S.1606), *Crocheshalle* 1086 DB (listed in Derbyshire), *Croxhale* c.1200 DbCh, 1208 FF, 1259 SHC X NS I 269, *Crokeshal'* 1209 Pleas, *Croxhall(e)* 1209 FF, 1296 Ipm, *Crokeshal(e)* 1239 FF, 1296 Ipm, *Crocsal(e)* 1276 RH, *Croxsale* 1291 *Tax*, *Croxsall* 1569 BCA MS3878/108, *Croxall* 1577 Saxton. Perhaps 'Croc's *halh*', although the early spellingfs with *-hall(e)* make a derivation from OE *h(e)all* hall', a rare element in Staffordshire, quite possible. OE Croc is ON Krókr, ODan Krok, OSw Kroker, originally a by-name from *krokr* 'a hook' (Ekwall 1960: 133; cf. Croxton, Lincolnshire), and provides evidence here of Scandinavian influence. Part of Croxall (Oakley) was in Staffordshire in 1086. The remainder was transferred from Derbyshire in 1895. In 1934 Croxall parish and part of Alrewas were joined to the parish of Edingale to form the civil parish of Edingale: Youngs 1991: 409. See also PN Db 631-2.

**CROXALL MILL** ½ mile south of Croxall (SK 1912). *Myl(fylde)*, *Myl(meydowe)* 1541. See Croxall.

**CROXDEN** Ancient Parish 5 miles north-west of Uttoxeter (SK 0639). *Crochesdene* 1086 DB, *Crokesdene* ?1187x1194 CEC 208, *Crokesdene* 1212 Fees, *Crokesden* 1232 SHC IV 89, *Crokesdone* 1247 ibid.' 109, *Crockesdene* 1290 SHC VI (i) 193, *Crokesdun*, *Crokesden* 1227 Duig. From the OE personal name Croc, which is ON Krókr, ODan Krok, OSw Kroker, originally a by-name from *krokr* 'a hook' (Ekwall 1960: 133; cf. Croxton, Lincolnshire), providing evidence of Scandinavian influence. The second element is OE *denu* 'valley'. Croxden abbey was known as *Hounds Chedull* abbey in 1543: TSSAHS XXXVII 1995-6 131 (see also Cheadle). There is some uncertainty about the history of Croxden parish – see Youngs 1991: 409. See also Croxall, Croxton.

**CROXTON** 4 miles north-west of Eccleshall (SJ 7831). *Crochestone* 1086 DB, *Croxton* 1327 SR, *Crofton* 1435 SHC XVII 151, *Croxson* 1583 ibid. 229. 'Croc's *tun*' (cf. Croxall, Croxden). The OE personal name Croc is ON Krókr, ODan Krok, OSw Kroker, originally a by-name from *krokr* 'a hook' (Ekwall 1960: 133; cf. Croxton, Lincolnshire), providing evidence of Scandinavian influence.

**CRUMPWOOD FARM** ½ mile east of Alton (SK 0842). *Crumpwood* 1742x1766 SRO D240/D/91, *Crump Wood* 1775 Yates, *Crumpwood* 1836 O.S. Possibly from OE *crump*, a by-form of OE *crumb* 'crooked', or a family name Crumpe. *Crumpwood Farm* in Haywood is recorded in 1852: SRO D240/E(A)2/184/1.

**CRUMWITHIES** 1 mile north-east of Ipstones (SK 0350). *Cromwitheyes* 1702 Okeover E5089, *Cromwillies* 1709 Okeover T739. Perhaps from OE *crumb* 'crooked', so 'the twisted withies or willows'.

CUCKOLDS HAVEN (obsolete) ½ mile east of Kingstone (SK 0729). *Cuckolds Haven* 1836 O.S., *Cuckolds Haven (toll gatehouse)* 1878 SRO D240/B/3/30. An intriguing name, doubtless explained by a colourful local incident now lost to history. Hart 1975: 207 identifies this place as *cumb welle léa*, recorded in a charter of 996 of Abbot's Bromley.

**CUCKOO LANE (FARM)** on the south-east side of Withington (SK 0335). *Cucknow (Field)* 1788 SRO D543/B/3/1. The single spelling makes any derivation problematic, but the name is not necessarily from the bird: ME *cuck* meant 'to void excrement'.

**CUCKOOSTONES** 1½ miles north-west of Warslow (SK 0760). *Cuckoo Stone* 1840 O.S, c.1870 Rental. The precise meaning of the bird name here is unclear.

CUDEL(S)FORD unlocated, in Mucklestone: see Rees 1997: 84. *Cudelford*, *Cudeslesford*, *Chudelesford* c.1205 Rees 1997: 84, *Cudesleford*, *Cudeslesford* c.1305 ibid. 84. An interesting name, perhaps incorporating the OE personal name Cuþwulf or similar.

**CULLAMOOR (BIG, MIDDLE & LOWER)** ½ mile south-east of Darlaston (SJ 9137, SJ 9138, SJ 9038), *Mayford Cullamores* 1701 SRO D/593/B/1/19/1, *Collow Moor* 1775 Yates, *Colmoor* 1798 Yates, *Coalamore Farm* 18th century SRO D3098/14/46, *Cull-moor* 1836 O.S.; CULAMOOR (obsolete) ½ mile south of Bucknall (SJ 9046), *Cole a Moore (House)*, *Collomoore* 1717 SRO D1798/590/16, *Culamoor* 1836 O.S., *Colamore* 1843 Ward 1843: 526. Difficult names. Perhaps from ME *colyer* 'charcoal maker, coal miner', hence 'moorland with the coal or charcoal worker' (the area overlies coal deposits – cf. *Colleyhale*, perhaps in the Cocknage area, recorded in 1495: SRO D593/B/1/23/7/1/9), but the lack of early spellings makes any derivation uncertain. Big

Cullamoor is on steep and broken land. *Mayford* is Meaford, evidently applied to distinguish the place from other Cullamoors. *Coleamore* and *Colaymoor* are recorded in 1734 and 1746 respectively in Stoke ParReg, but it is unclear whether they refer to either of these two places. *Colleyhale*, perhaps near Hanchurch, is recorded in 1495 (SRO D593/B/1/23/7/1/9), *The Cullimore* is recorded in 1625, at Oulton Heath, Stone (WSL 49/18/44), and as *Oulton Cullymans* in 1663 (ibid. 49/2/44), and *The Cullinores* in 1699 (SOT SD4842/42/42). *Colmoor*, on the north-east side of Hardings Wood, appears on Yates' map of 1775. Cullamore Lane (*Cullamore Lane* 1836 O.S.) is 1 mile south-west of Uttoxeter.

CULNEHILL (unlocated, near Munkford) *Culnehill* 1274-90 SHC 1911 442. Presumably from OE *cyl(e)n* 'a kiln, a furnace for baking or burning', with OE *hyll* 'hill'.

CULVERDSLOW (obsolete, in Fenton (SJ 8944), 'on the Longton side of the former boundary between Fenton Culvert and Longton south of Grove Road': VCH VIII 205 fn.5) *Culverdislow, Culverdslaw* 'Culverd's-low', from OE *hlaw* 'tumulus'. The name is found in a charter of the late 12th century: VCH VIII 212; SHC XII NS 12. The mound is also recorded as *Mole Cop* (VCH VIII 205), and *Cop Low* (TNSFC 1926 136). See also Fenton.

CUMBERFIELD (unlocated, in Seighford) *Cumber Field* 1617 SRO D798/1/8/4, *Comberfield* 1652 SRO D798/1/8/7, *Cumbowe Field* c.1600 D748/1/8/7, *Cumber Field* 1617 SRO D798/1/8/4. From OE *cumbre*, borrowed from the PrWelsh form of modern Welsh *Cymro* 'Welshman', with OE *feld* 'open land', so originally perhaps 'the open land with the Welshmen'. The place is remembered in The Cumbers, a road in Seighford. It may be noted, however, that PN Li 36 and CDEPN 175 cite an OE personal name Cumbra in the derivation of Cumberworth, Lincolnshire. Cf. Comberbach, Cheshire (PN Ch I 11; V (II) 297). See also Comberford; Cumberledge; Cumberstone Wood.

**CUMBERLEDGE HILL** a 671' hill on the west side of Cannock Wood, close to Castle Ring prehistoric hill-fort (SK 0412). The age of this name is unknown, but the surname Comberledge and similar is commonly found in the Walsall ParReg in the 16th and 17th centuries (see also SHC SHC 1923 140; SHC 1931 114), possibly from this place, which may be associated with Comberbach (q.v.). No place called Cumberledge is recorded in the adjoining counties of Cheshire, Derbyshire, Nottinghamshire, Warwickshire or Worcestershire, or has been traced elsewhere. The name suggests a derivation from OE *cumbra*, \**lece* 'the boggy stream of the Welsh': see Cumberledge (Park).

**CUMBERLEDGE (PARK)** 1½ miles north-west of Cheddleton (SJ 9653). *? Cumylache* 1539 SHC VI NS (i) 80, 1616 ibid.'18, *? Comilatch* 1600 SHC 1935 266, *?Cumberidge* (p) 1601 SHC 1935 393, *? Comerlage* 1601 ibid.' 309, *? Comerledge* 1601 ibid. 361, *?Cumberledge* (p) 1604 SHC 1940 173, *? Comylache* 1606 SHC XVIII 54, *Cumberledge* 1635 Leek ParReg, *Cummerletch* (p) 1658 Leek ParReg, *Comilech* (p) 1666 SHC 1925 220, *Cumberledge Park* 1836 O.S. If the earlier spellings are to be relied on, then possibly from OE *cumbra*, \**lece* 'the boggy stream of the Welsh'. See also Cumberledge Hill.

**CUMBERSTONE (WOOD)** 1½ miles north-east of Swynnerton (SJ 8637). *Cumberstone (Nook)* 1721x1816 SRO D593/B/1/22/30, *Comberstone (Nook)* 1825 SRO D593/B/1/22/36, *Cumbersome Hill* 'before 1830' VCH VI 100, *Cumberstone Hill* 1836 O.S. Early spellings have not been traced, but the first element may be from OE *cumbre*, borrowed from the PrWelsh form of modern Welsh *cymro* 'Welshman': the place is 3 miles north-west of Walton near Stone (q.v.), and 5 miles north-east of Walton near

Eccleshall (q.v.). The second element may be OE *stan* 'a stone, a rock' (*cumbre* does not have -*s* in the genitive, so it could not be OE *tun*), hence 'the Welshman's stone'. If so, the stone may be the *sceortan stane* (cf. Scortestona) mentioned in a charter of Darlaston 956 AD (12th century, S.601): Hart 1975: 178, but see also Hooke 1983: 88.

CUMBER STETCH (unlocated, a field-name in Forsbrook) *Cumber Stetch* 1841 TA. Earlier forms have not been traced, but is ancient possibly from OE *cumbre, stycce* 'the Welshman's meadow beside the stream', though it would be unsafe to rely on such a late spelling.

**CUMBWELL BROOK** a tributary of the river Smestow. *Cumbwelle* 1334 SHC 1928 138. From OE *cumb* 'a cup, vessel', sometimes found in place-names in a transferred topographical sense 'short spoon-shaped valley', with Mercian OE *wælle* 'a spring', and (rarely) 'a stream', hence 'the spring or stream in the hollow', or perhaps 'the spring with the drinking cup': cf. Canwell. Beardwell (Wiltshire), Bedlars Green (Essex and Hertfordshire), Biddles (Berkshire), and Bidwell (Bedfordshire, Dorset, Northamptonshire and Somerset), are all names with OE *byden* 'a vessel, a tub, a butt' as the first element.

CUNITELEG (unlocated, in Cuttlestone Hundred, possibly near Haughton, or in the Maer area) *Cuniteleg* 1234x1240 (1798) Shaw I xxv, *Cunteleg* 1284 FA, *Cunleley* 1615 SHC NS IV 65. Possibly the same place as *Cunyton* recorded in 1208 (SHC III 173). The name may have as its root OE *cunet* or *cunete*, from British *cunetio*, perhaps meaning 'hound', possibly applied to stream-names: see TSAHS LXXVI 2001 76-7. The terminal element is evidently *leah*. See also Condelege.

**CURBOROUGH** 2 miles north of Lichfield (SK 1212; SMR 02089 places a deserted post-Conquest settlement at SK 13551255). *Coreburhe* 1260 SHC X NS I 280, *Churbur'* 1216x1272 SRO D(W)1734/J/1631, *Curborud* 1280 Ipm, *Curberue* 1280 SHC 1911 173, *Curbur'* 1285 FA, *Curburgh* 1290 (1798) Shaw I 119, *Corrun* 1291 StSt I 1988, *Corbrun* 1291 Tax, *Curburgh* 1291(1798) Shaw I 119, *Curborowe* 1293 ibid, *Little Carboru* 1297 SHC 1911 241, *Coreburgh* 1305 WL 59, *Curburghthornes* 1341 SHC 4th Series XX 68, *Curburg* 1346 SHC XI 157, *Corburgh* 1363 SHC X NS (ii)115, *Corburgh* 1392 SHC XV 44, *Curborowe, Curborough, Currebourgh* 14th century Duig, *Curborough Turvile* 1415 VCH XIV 233, *Curburn* 1428 FA, *Curburgh* 1461 SRO 3005/3, *Caurboro'* 1535 (1798) Shaw I 144, *Carborowe* 1540 ibid. *156, *Courboroughe-Somervyle, Courboroughe-Darvyle* 1567 SHC XVII 216, *Curborowe, Corborogh, Corboroughe* 1582 SHC XV 141, *Courboroughe Somervill, Courboroughe Darvile otherwise Courboroughe Dervill* 1601 SHC XVI 208, *Curborrowe* 1657 BCA MS307/Acc1904-008/183812. A difficult name. Ekwall 1960: 136, 137 explains this name as OE *cweorn burna* 'mill stream', with *cweorn* 'certainly' meaning here 'watermill', and *burna* replaced by *burg* 'fortification'; see also VCH XIV 229, but the forms do not conclusively support a derivation from *cweorn* or *burna*. The identification of the first element must remain uncertain: Modern *cur* 'a dog' is not recorded in OE, but ME *curre* might possibly be found in some post-Domesday place-names, of which this may be one. If the second element was originally *burg* (the second 1280 spelling points towards OE *bearu* 'a wood, a grove'), it may be included in the list of possible sites of the unlocated DB *Burouestone* (q.v.): Shaw 1798: I 350 suggests that Curborough 'was antiently a member of the bishop's barony of Lichfield, as appears in Domesday Book...it was afterwards held of the manor of Longdon'. But there is no reference to Curborough in DB, and *Burouestone* is omitted from Shaw's list of DB names. Somervyle and Turvill (which later became Darvell or Darvile) were local families: John de Somerville held land in Little Curborough in the later 13th century, which by 1327 was known as Curborough Somerville (VCH XIV 278, 282), and Philip de Turvill was prebendary of Curborough 1309-37 (VCH XIV 233). VCH XIV 229 suggests that the two settlements of Great Curborough and Little Curborough are to be identified with the sites of what are now Curborough Hall Farm and Curborough House respectively. The former lies in a stream valley; the latter (from the hachuring on the first edition 1" O.S. map of 1834) on a long north-south ridge. Both places are named Curborough on that map. *Curborow* in Horton parish is mentioned in 1597 (Biddulph ParReg), but has not been traced.

CURSONS (unlocated, a manor in Alrewas) *Curcun* 1203 SHC III 109, *Cursons* 1462 SHC NS IV 125.

CUTTESDON (obsolete) The field-name *Cudsons* (1844 TA) suggests that the place is likely to have lain on the north side of Burlaughton at SJ 775113. *Cutteston juxta Hales* 1175 Eyton 1854-60: IX 163 fn.4, *Cuttested'* c.1226 Rees 1997: 107, *Cuttesd'* c.1227 ibid.' 76, *Cuttesdon* c.1228 ibid.' 92, *? Cuttesdown* 1262 SHC 4th Series XVIII 55, *Cuttesdon* 186 SRO D593/B/1/19/2/1/11, c.1300 SRO D593/B/1/19/2/1/9, *Cutteston* early 14th century Rees 1997: 153, *Couttesdon* c.1324 ibid. 156, *? Cubbeston* 1547 SHC IV (ii) 123, TSAS 3rd Series VIII 1908 138, *? Cubston* 1547 ibid. The second element is inconsistent, but the earliest spelling suggests OE *tun*, alternating with OE *dun* 'hill', but there is no marked hill here, so possibly '*Cutt's (or *Cuþ's) tun*', though that derivation must be very uncertain. In Shropshire since 1895. See Eyton 1864-60: V1 363, VII 286 fn.13, 388, VIII 286, XI 162, 163; TSAS 3rd Series I 282; SHC I 221.

**CUTTLESTONE** *(Hundredum de) Codwestan, Colvestan, Cudolvestan, Cudulvestan, Culvestan* 1086 DB, *Cuthulvestan, Cuthuluestan* 1203, 1227 Ass, *Cuthulfestan'* c.1255 RH, *Cuduluestan', Cudeluestan'* 1130, 1185 *et freq.* to 1202 P, *Kudolveston, Kudolvestan* 1199 Fees, 1199 Ass, *Codulvestan, Coduluestan* 1193 P, *Cuteluestan* 1187 P, *Cutolvestan* 1188 P. 'Cuþwulf's stone'. A more detailed discussion of the name will be found in the Introduction, pages 46-47. See also Stretwyle.

**CUTTLESTONE BRIDGE** an ancient stone bridge crossing the river Penk 1 mile south-west of Penkridge (SJ 9113). *Pontem de Cuthuluestan* early 13th century SRO D260/M/T/5/139, *(pons de) Cuthulueston'* 1225-59 Deed, *Cothelstonbrugge* 1307 SHC VII 179. See Cuttlestone.

**DAB GREEN** – see **PYE GREEN**.

DADNALL HILL (obsolete) 1 mile east of Pattingham (SO 8399). *Dadnall Hill* c.1650 VCH XX 173, 1801 Smith's map, *Dadnal Hall* 1808 Baugh, *Dodnall Hill* 1832 Teesdale. Possibly to be associated with *The Hill*, recorded in 1392: VCH XX 173. The derivation of the name is uncertain, but perhaps from an unidentified personal name with OE *halh*, or connected in some way with OE *\*dod* 'hill', represented by northern dialect *dod* 'rounded summit'.

DAFFODIL FARM (obsolete) 1½ miles east of Walsall (SO 0497). *Hurst's House* 1546 SHC XVII 175, *Hurst House or Wood End House* 1621 ibid.' 175, *Daffodilly House* 1816 ibid. 175, *Daffodilly* 1834 O.S, *Daffydowndilly House* 1843 VCH XVII 175, *Daffodil Farm* 1907 O.S. Held by the Hurst family from the 14th century: VCH XVII 175, TSSAHS 1992-3 XXIV 50.

**DAGDALE** 2 miles west of Uttoxeter (SK 0534). *Dagdale* 1474 SRO D5684, 1480 SRO D5684/2, 1484 SRO D5684/2, 1503 SRO D5684/5, 1613 SHC IV NS 49, 1775 Yates, 1836 O.S., *Daggars Dale* 1656 Leek ParReg. A curious name, perhaps to be asociated with *Dagdale Field*, recorded in Bramshall in 1851 (SBT DR636/32). The word *dag* was a dialect word used for light rain or heavy mist (EDD), so perhaps 'the misty valley', from OE *dæl*. Deggs Leasow (*Deggs Leasow* 1836 O.S.) is 1 mile to the north-east (SK0635), but there is no evidence to connect the two names.

DAGGER HALL  a house that formerly stood at the junction of what is now Dagger's Lane and Salter's Lane, Mayers Green (SO 0292), demolished 1894-5: Hackwood 1895: 67; VCH XVII 4, 21. *Dagger Hall* 1625 Willett 1882: 165, 1639 SRO D742/A/17/1-7, 1754 SRO D1250/1, 1820 Greenwood, *Dager Hall* 1728 Willett 1882: 227. Presumably from the surname Dagger: see DES 12.

**DAIRY HOUSE**  2 miles south-east of Market Drayton (SJ 7032), *Derihouse* 1644 SHC 1945-6 187, *Dayry House Farm* 1675 SRO 1/279/73, *Daryhouse* 1687 SRO D 681/E/5/21, *Dayry House* 1709 SRO D 828/22; **DAIRY HOUSE FARM** 1 mile west of Eccleshall (SJ 8128), *ye deariehouse* 1672 Eccleshall ParReg, *Dearyhouse* 1691 ibid, *Dairyhouse* 1749 Bowen, *Dairy House* 1775 Yates, *Dairy* 1833 O.S.; **DAIRY HOUSE FARM** 1½ miles north-west of Church Leigh (SK 9936), *Dairy House Farm* 1886 Redfern 1886: 87, *Dairy House* 1891 O.S. Self-explanatory. Dairy House near Market Drayton was formerly known as Tagg Moor (*Tagmore* c.1570 SHC 1945-6 148): SHC 1945-6 187. *Dayry House* in Haughton parish is recorded in 1644 (Erdeswick 1844: 27), and *Dairy House Farm* in Cheadle is recorded in 1747 (SRO D260/M/E/1).

**DAISY BANK**  2 miles east of Walsall (SO 0497), *Deasy Bank* 1709 HRO E12/V1/NC/2; **DAISY BANK**  2 miles north-west of Newborough (SK 1126), *Daisey Bank* 1836 O.S. 'The bank with the daisies'.

**DAISY LAKE**  1 mile south-west of Mucklestone (SJ 7035). *Daisy Lake* 1833 O.S. The pool here is mentioned in 1669: SHC 1945-6 7. 'The lake with the daisies'.

**DALE, THE**  1½ miles south-east of Cauldon (SK 0948). *the Dale* 1691 Ellastone ParReg, *Pantones in the dale* 1695 Morden, *Pantons in the Dale* 1749 Bowen, *The Dale* 1836 O.S. From OE *dæl*, or possibly ON *dalr*, 'valley'. The element *Pantones* is unclear: perhaps from the family of that name who held land in north Staffordshire: see Erdeswick 1844: 493-4; SHC VII NS 134, 136. It may be noted that the place-name Panton in Lincolnshire is probably from OE *panne* 'a pan', used in a transferred topographical sense for a depression or hollow, with OE *tun*, appropriate for a place in a valley (although Ekwall 1936: 145-6 gives a derivation from OE *\*pamp* 'a hill'), but the terminal *s* in the Staffordshire spellings points towards a family name.

**DALE BROOK**  a tributary of the river Trent. From OE *dæl* 'valley', with *broc*, hence 'brook flowing through a valley'.

**DALE FARM**  on the north-west side of Fulford (SJ 9438). Perhaps to be associated with *Dalehouse*, recorded c.1655 (SRO D1798/685/399).

**DALE HOUSE**  on the south-west side of Cheddleton (SJ 9651). *Dalehouse* 1681 SRO D3272/1/4/3/25-36, *Dale House* 1836 O.S. 'The house in the dale or valley'.

DALE TORR  (obsolete) 1 mile south of Calton (SK 1048). *Dale Torr* 1836 O.S. From OE *dæl* (ME *dale*) 'a dale, valley', with OE *torr* 'a rock, a rocky peak'. Perhaps to be associated with *Torr Piece* recorded on Calton Moor in 1802: SRO D3597/1-4.

**DALESGAP**  1 mile north of Rocester (SK 1140). *the Dale Gap* 1713 Okeover T740. From OE *dæl* 'valley', so 'the gap in the valley'.

**DALICOTT** 1 mile north-west of Claverley (SO 7794). *Dalicote* 1274 Eyton 1854-60: III 98, *Dedecott* 1532 SHC 4th Series 8 123, *Dallicott* 1636 Claverley ParReg, *Dallicot* 1752 Rocque, *Dalicote* 1833 O.S. The 1532 spelling is clearly an aberration, but a derivation incorporating OE *dæl* 'valley' can be ruled out, since the place lies on a 280' hill. The surname Dalley is recorded in this area in the later 15th century ('The Reliquary' 1882-3 14), and may be the root of this place-name. The second element is evidently OE *cot* 'cottage, shelter'. PN Wa 182 offers no explanation for the first element in the place-names Dallies and Dallimore in Warwickshire In Shropshire since the 12th century.

**DAMBRIDGE MILL** 1 mile south-west of Marchington (SK 1128). *Damberidge Mill* 1836 O.S.

**DAM MILL** 2 miles south-east of Codsall (SJ 8802). *Doun* 1300 WA I 267, *Dom mulne* 1341 SHC 1913 82, *Dommulne* 1412 SHC XVII 17-18, *Damme mylne* 1616 Codsall ParReg, *Damm Mills* 1652 WA I 268. From ME *damme* 'a dam, a pool formed by a bank across a stream', with OE *myl(e)n* 'mill'. The spellings (the first of which is evidently aberrant or mistranscribed) reflect the West Midlands dialect pronunciation of *o* for *a*. The Dean of Wolverhampton had a mill here which was working until at least 1616: VCH XX 85. See also WA I 266-8.

**DAMGATE** 1½ miles north-west of Ilam (SK 1253). *Dameyatte* (undated) CtRequests, *Dam Yate* 1775 Alstonefield ParReg, *Damgate* 1838 O.S. From ME *damme* 'a dam, a bank across a stream to form a pond, a pond so formed', with OE *geat* 'a hole, an opening, a gap'. The application of the first element is uncertain (the word sometimes carried the meaning 'causeway through wet ground': CDEPN 178, with reference to Damgate, Norfolk), but the place lies on a saddle of high ground, which is presumably the gap. The place may be associated with *the Mile dam* (? 'the Mill Dam') recorded in Alstonefield ParReg in 1658.

DAMPORT – see DEARNEFORD.

DAMSEN HILL (unlocated, in Marchington Woodlands) *Damsen Hill* 1698 SRO D599/5.

DANDILLIONS (obsolete) on the south side of Cheadle (SK 0042). *Dandillions Farm* c.1739 SRO D953/9, *Dandilions* 1836 O.S., *Dandillions* 1891 O.S. The surname Dandelion is recorded: DES 125.

**DANE, RIVER** a tributary of the river Weaver, forming the boundary between North Staffordshire and Cheshire. *Davene* 13th century Dieul, 1345 Coram R, *Daan* 1416 AD 4, *Dane* c.1540 Leland, 1577 Saxton, *Daven* 1596 SHC XVI 158. Ekwall 1928: 112 proposes a derivation from MWelsh *dafn* 'a drop, a trickle', suggesting 'the slow, trickling stream', perhaps used ironically, since the description is hardly appropriate here, but another suggestion is from PrWelsh *daven*, derived from British *Damina* 'the river of the ox-goddess': see PN Ch V (II) 292.

**DANEBRIDGE** 2 miles north-east of Heaton (SJ 9665). *Scliderford* c.1190 VCH VII 187, *Slideresford* c.1255 SHC 1911 428, *Dauenbrugge* 1357 ChFor, *Sliderfordbrugge in Haukesyerd* 1347 Eyre, *Slyderford brugge* 1384 Rental, *Sliderford bridge* 1545 VCH VII 187, *Sliderford Bridge* 1608 SHC 1948-9 82, *Danebridge* 1703 SRO D1109/1, *Dane Bridge* 1842 O.S. From the river Dane. The earliest name, which evidently alternated with the present name, is from OE *slidor* 'slippery place', perhaps the only instance of this word in a place-name. *Haukesyerd* is from The Hawkesyord (unlocated), from ME *haueksherd*, *hauekshord* 'a clearing in which hawks were flown': PN Ch I 166; see also

223

Hawksyard (Priory). Cf. *Slideford-siche*? 13th century SHC XII NS 12; n.d. Erdeswick 1844: 9; SHC XII NS 12: from OE *sic* 'watercourse, stream'. See also SHC 1948-9 82.

**DANE'S COURT** – see **TETTENHALL**.

**DANESFORD & UPPER DANESFORD** 1 mile north-east of Claverley (SO 8093). *Danesford* 1833 O.S. Early spellings have not been traced, but it is improbable that the name is ancient. Almost certainly from local antiquarianism in the 18th or 19th century associating the place with the supposed passage of the Danish army which crossed from the west side of the Severn and was defeated by an army of West Saxons and Mercians in the Tettenhall area c.910: Earle & Plummer 1892-9: i 95-7. There is no evidence of any kind to support such association.

**DANESFORD** on the river Severn 1 mile south-east of Bridgnorth (SO 7391). *Darneford* 1420 TSAS LVII 1961-4 39, *Daneford* 1612 SA 796/78, *Danesford* 1833 O.S. In Shropshire since the 12th century. The name originated as 'the hidden ford', from OE *derne* 'hidden, secret, obscure', and doubtless developed into its present form as a result of later antiquarianism associating the place (and name) with the Danish armies which wintered in the area in 896 and crossed the Severn hereabouts c.910: Earle & Plummer 1892-9: i 89; 95-7.

DANEWOOD (unlocated, in Swythamley) *Danewood* 1591 SHC 1930 174. From the river Dane (q.v.).

**DANFORD** on the north-west side of Kinver on the river Stour (SO 8483). The age of this name has not been traced. Perhaps from OE *derne* 'hidden, secret, obscure', but see also Danesford and Upper Danesford.

**DAPPLE HEATH** 1 mile north-east of Hixon (SJ 0426). *dapple heath* 169 SHC 1919 27, *Dapple heath* 1677 ibid. 30, *Dabble Heath* 1775 Yates, *Dapple Heath* 1812 SRO D3259/2/15. Perhaps 'the variegated heath'.

DARFUR BRIDGE, CAVE & CRAGS (obsolete) ½ mile north-west of Wetton (SK 0955). *Darffall* 1538 Survey, *Darfort Ptar* 1750 Pococke 1888-9, *Darfa* 1844 Garner, *Darfar Bridge* 1838 O.S. A curious name. It is here that the river Manifold disappears into the ground, to re-emerge near Ilam, and it would perhaps be surprising if the name did not relate to this phenomenon, so the second element may be from *\*(ge)fall* 'place where something falls'. The first element is more difficult, possibly from Welsh *dwfr* 'water, the waters', or from MWelsh *dafn* 'a drop, a trickle', so 'the place where the waters fall'. However, *dar* is sometimes found meaning 'deer', so perhaps 'deer fall', denoting a place where deer were chased over the crags, or simply 'deer fold' (cf. Deerfold, Herefordshire; Dorfold Hall, Cheshire), or 'deer ford'. *Ptar* appears to be an idiosycratic spelling for *tor*: see Yelpersley Torr. Cf. River Dane; River Dove. It may be noted that Darfoulde House (? from 'deer-fold') lies near the river Dove on the north-west side of Stretton (SK 2527), but the history of the name is not known.

**DARLASTON** Ancient Parish 2 miles north-west of Stone (SJ 8735), *Deorlafestune* 956 (11th century, S.602, incorrectly indexed as 601), *Deorlafestun* 1004 (11th century, S.906; 11th century, S.1536), *Derlavestone* 1086 DB, *Derlavestune* 11th century Sawyer 1979a: xxxv, *Dorlaveston'* 1198 Fees, *Derlaweston, Derlaveston* 1277 SHC VI (i) 91, *Dorlaston* 1341 HLS; **DARLASTON** 1½ miles north of Wednesbury (SO 9796), *derlauest[onia]* c.1154 SHC 1941 61, *Dalaueston* 1166 SHC 1923 295, *Derlaveston* 1262 SHC V (i) 137, *Derlaston* 1316 FA. 'Deorlaf's *tun*'. Speed's map of 1610 inexplicably shows *Darlaston* on the north-west side of Wolverhampton.

**DARLEY OAKS** ½ mile south of Newchurch (SK 1422). *Darley* 1834 O.S. If ancient, probably from OE *deor-leah* ' *leah* frequented by deer'.

**DARLING, THE** a tributary of the river Sow. Presumably an affectionate pet-name for a favourite stream.

**DARNFORD** 1½ miles east of Lichfield (SK 1308). *Dernford* 1243 SHC 1924 97, *Darneford* 1260-1 SHC X NS I 289, *Derneford* 1274 SHC 1923 275, *Darneford* c.1285 SHC 1939 92, *Dorneford Mill* 1478 SHC VI NS (i) 115, *Danford (Mill)* 1834 O.S. From OE *derne* 'hidden, secret, obscure', an element commonly found with words for ford, water, etc., perhaps meaning here 'the ford overgrown with vegetation and the like': cf. Darnford, Suffolk. The place is perhaps to be associated with *Dernemoth*, recorded in the Lichfield area ?c.1175: SHC 1921 28. *Danford Lake* in Bordesley is recorded in 1797 (BCA MS3375/446231), and as *Dearnford* in 1657 (BCA MS3069/Acc1904-005/181602). See also Denford.

**DAVENPORT HOUSE** on the south-west side of Worfield (SO 7595). *Davenport House* 1752 Rocque. From the Davenport family who lived here from at least 1635: James 1878: 51. The house was built in 1726 to the design of Francis Smith of Warwick. In Shropshire since the 12th century.

**DAVETRIPORT** (unlocated, perhaps outside Staffordshire, possibly Davenport, Cheshire) *Davetriport* 1199 SHC III (i) 36.

**DAW END** in Rushall, 2 miles north-east of Walsall (SK 0299). *Dawende* 1597 SHC 1932 346, *Daw-end* 1686 Plot, *Daw, Dawe End* 17th century Duig, *Daw End* 1834 O.S. Duignan 1902: 49 suggests a derivation from the personal name David, abbreviated in the medieval period to Dau, Daw, Dawe, but the root may be OE *\*dawe* 'a crow, a jackdaw'. The word *end* meant not a terminal point, but simply a place at the edge of a settlement, hence perhaps 'the outlying property with the jackdaws'.

**DAWFORD BROOK** runs north of Great Chatwell.

**DAWLEY** (unlocated, near Wombourne, possibly between Lloyd Hill and The Woodhouse) *Daweleye* 1313 SHC 1938 24, *Daueweleie* 1316 ibid.' 25, *Dauweleie* 1316 ibid.' 27, *Dauwleye* 1317 ibid. 27, *Dawley* 1322 (1801) Shaw II 213. Perhaps from the personal name Dealla with OE *leah* 'the woodland clearing associated with Dealla' (see PN Sa III 3).

**DAYHILLS (FARM)** 1 mile north-west of Milwich (SJ 9633). *Day-hills, Day-hills Common* 1775 Yates, *Day Hills* 1834 O.S., *Dayhills* 1852 SRO D917/11/7. A curious name for which no explanation can be offered.

**DEADMAN'S GREEN** ½ mile east of Checkley (SK 0337). *Dadelond* 1317 SHC 1911 339, 1327 SHC VII 220, *Dadelound* 1323 SHC XIV 15, *Dadlond* 1327 SHC 1912 250, *Dadelound* 1329 SHC XIV 15, *Dadland Grene* 1514 SHC XI 259, *Dadland Green* 1518 SRO DW1733/A/3/34, *Dadlands* 1520 SRO DW1733/A/3/11[12], *Dodland Green* 1602 SHC 1935 457, *Dadland greene* 1644 SHC 4th Series I 198, 1647 BCA MS3307/ACC1927-020/337138, *Deadman's Green* 1872 P.O. Possibly from an OE personal name \*Dæda, giving 'the open space at \*Dæda's estate'. The relatively modern association with a corpse is doubtless the result of local popular etymology.

DEAKINS GRAVE (obsolete) – see **PYE GREEN**.

**DEAN BROOK** a tributary of the river Weaver. From OE *denu* 'valley'.

**DEANERY FARM**  2 miles west of Tatenhill (SK 1721), *Allotment Farm* 1836 O.S.; **DEAN'S HALL FARM**  on the south side of Brewood (SJ 8808), *Deanes Hall* 1629 WCRO CR1291/550/1-10, *Deans Hall* 1653 SRO D590/433/29-86, *Dean's Hill* (for *Dean's Hall*) 1680 SHC 1919 240. Both places belonged to the Dean of Lichfield, the first as rector of Tatenhill, and the second as part of the prebend of Brewood.

**DEANS HILL**  between Stafford and Castle Church (SJ 9122). *Deans Hill* 1836 O.S. *Deanes Meadow* and *Deans Hill Meadows* are recorded here in 1629: SRO D4715/1-23.

**DEARNDALES**  1 mile south-west of Uttoxeter (SK 0732). *Derndale Magna, Derndale Parva* 1474 SRO DW1733/A/1/1, the *Little Derndale* 1536 SRO DW1733/A/2/34, *Dearndales* 1836 O.S. From OE *derne* 'hidden, secluded', with OE *dæl* 'valley' . There were evidently two places here, *Magna* ('Great') Derndale, and *Parva* ('Little') Derndale.

DEARNEFORD  (unlocated, in Hatherton) *Damport alias Dearneford* 1629 SRO D260/M/T/4/60. From OE *derne ford* 'hidden or secluded ford'. The alternative name is unexplained.

**DEARNSDALE**  1 mile south-west of Derrington (SJ 8821). *Dearn(e)sdale* 1551 to 1616 FF, *Dernsdale* 1605 ibid, *Dearmesdall* 1605 VCH IV 84, *Dearnesdale* 1616 SHC VI NS (i) 20, *Darnsdale* 1682 Browne. There are several river-names similar to this name, perhaps derived from OE *derne* 'hidden, secluded', although the evidence is not altogether conclusive. This place is on the west side of the hill on which stands Berry Ring hill-fort, so a derivation 'valley of the stream called Dierne', from OE *derne*, with OE *dæl* 'valley' (see particularly Ekwall 1928: 116-7) must be speculative, although there is a stream to the east. Another possibility is a derivation from a personal name such as *De(a)rne: see Derneslowe. *Dearnsdale Wood* is recorded in this area in 1792: SRO D798/1/10/4.

**DEEPDALE**  ½ mile south of Grindon (SK 0853). *Dup(p)edal(e)* 1227 Ass, *Depedal(e)* 1227 Ass, *Depedale* 1318 SHC IX (i) 71, *Deep Dale* 1775 Yates, *Depdale* 1796 ParReg, *Dip Dale* 1816 BCA MS3628/Ass-1937-059/4700100, *Deepdale* 1838 O.S. Self-explanatory. Curiously, Deepdale in Lancashire has the spelling *Dupedale* in 1228: Ekwall 1960: 141. *Depedale*, recorded in 1493 (SRO DW1733/A/2/33), appears to have been in the Lees Hill area.

**DEEPFIELDS**  2 miles south-west of Bilston (SO 9494). *Deepfield* 1834 O.S. The derivation of this name is uncertain. Hackwood 1906: 119 suggests that the name is from *Dip Fields*, associated with wells here, but that is unlikely.

**DEEP MOOR**  2 miles north-west of Shareshill (SJ 9208), *Deepemore* 1604 PCC; DEEPMORE COPPICE (obsolete) in Bentley near Willenhall (SO 9999), *Dupemore* pre-1290 SRO D1790/A/2/3, *Deepmore (Coppice)* 1834 O.S.; DEEPEMORE (HAY) (unlocated, at Walton near Eccleshall), *Deepemore (Hay)* 1593 SRO D590/1. 'The deep boggy marshland'.

DELPH HOUSE  (obsolete) 1 mile west of Cheadle (SJ 9942). *Delf* 1281 Ipm, *Huythehalg* c.1311 SHC 1911 436, *Huuehalgh* 1313 Ch, *Delff* 1316 SRO D1229/1/4/10, *The Heyghe Halghe* 1380 SHC 1910 265, *Delf* 1331 SHC 1913 27; 1384: SRO D1229/1/4/13, *the Heyge Halghe* 1538 SHC 1913 266, *the Heyghe Halghe* 1538 SHC 1910 267, *Delf Felde* 1539 MA, *Hyghehawghe* 1556 SRO D1229/1/4/1, *Delfehowse* c.1590 SHC 1930 54, *Delfhowse* 1598 SHC XVI 185, *the Delfe howse* c.1606 SHC 1934 (ii) 47, *Highhaugh al. Dephe House* 1615 SRO D1229/1/4/2, *Delphouse* 1677 SRO D1788/42/11, *Delphhouse* c.1697 SRO D1229/2/2/7, *Delf Ho.* 1798 Yates, *Delph House*

1836 O.S. From OE *(ge)delf* 'a digging, a trench, a pit, a quarry': in the 19th century the place was a colliery, but mining or quarrying has a much longer history here. An alternative name was evidently 'the high *halh* or *haugh*', possibly to be associated with *Heghegge*, recorded in 1380 (SHC 1910 265), but see also Blake Hall. See also Erdeswick 1844: 495.

DELVES HALL (unlocated, in Apedale) *Delves Hall* 1679 SHC XII NS 46. Named from the Delves family who came originally from Delves Hall near Uttoxeter (SHC XII NS 45), recorded from at least 1303: see Shaw 1798: I xxxv.

**DELVES, THE; DELVES GREEN** 2 miles south of Walsall (SO 0295), *Delves* 1575 Walsall ParReg, *Dalves* 1600 BCA D260/M/T/1/33, *Dealves Wednesbury* 1604 SHC 1940 137, presumably the *Walstwude* recorded in 1300 (SHC V (i) 177): see Walstead Green: DELPH, DELVES in Tunstall appears as *Deffes* in 1282-3 (SHC XI NS 253), *Delves* in 1307 (ibid. 265), and 1372 (SHC XIII 119), and the 'mines, delphes and quarries' there are recorded in 1563 (SHC 1931 183. A name commonly-found in Staffordshire, particularly in the mining areas, from OE *(ge)delf* 'a digging, a trench, a pit, a quarry', in the plural meaning 'the diggings, the workings'.

**DENFORD** 4 miles south-west of Leek (SJ 9553). *Derneford* 1341 VCH VII 203, *Dernford bridge* 1529 SRO DW1490/106 n25, *Dern Ford* 1794 Stockdale, *Dearneford* early 19th century DW1761 Box 1 pt, *Denford* 1836 O.S. From OE *derne* 'hidden, secret, obscure', an element commonly found with words for ford, water, etc., perhaps meaning here 'the ford overgrown with vegetation and the like'. Sleigh 1883: 143 mentions, without explanation, *'Dearneford, formerly called Darple'*.

DENNE, LA (unlocated) *la Denne* 1283 SHC 1911 183.

**DENNIS (PARK)** in Amblecote (SO 9085). *Deneys* 1271 SHC 4th Series XVIII 70, *Dennis* 1798 Shaw I xxi, 1801 Shaw II 237, 1834 O.S. From the Denis family of Amblecote who were living here in the 12th century: VCH XX 51.

**DENSTONE** 5 miles north of Uttoxeter (SK 1040). *Denestone* 1086 DB, *Denstun*, *Denstone* 1191x1194 CEC 261, *Deneston* 1199 SHC III (i) 53, c.1200 DRO D258/27/1/6, 1208 FF, c.1230 SHC VI (i) 11, *Daneston* 1201 SHC III (i) 70, *Devestona* c.1210 SHC 4th Series IV 105, *Benston (sic)* 1242 SHC IV 94, *Denstones* 1282-3 SHC XI NS 259, *Daneston' temp.* John-early Henry III ibid.' 114, *Deneston* early 13th century SRO D593/E/6/15, *Denston, Denstyn* 1339 SHC 1913 77, *Denstone* 1366 SRO DW1733/A/2/78, *Denston* 1377 SRO DW1733/A/2/46. 'Dene's or Dane's *tun*'. Dene was an OE (and probably ON) personal name, and this place-name may perhaps be seen as evidence of Danish influence. Cf. Densy Lodge.

**DENSY LODGE** 1 mile south of Sudbury (SK 1630). *Denseye* late Henry III TutCart, *Densey* 1632 SRO D15/33/4. This place evidently took its name from the first element of *deues broke* (*denes broc*) mentioned in a charter of land at Marchington dated 951 (14th century, S.557): see Sawyer 1979a: 18; Hooke 1983: 103. 'In the Needwood enclosures ... several plots east of this Lodge are called Densey Meadow ...': Sawyer 1979a: 18. The second element is OE *eg* 'island, land by water': the place lies close to the river Dove. The root is perhaps the OE (and probably ON: cf. Halfdene) personal name Dene, so 'Dene's brook', perhaps to be seen as evidence of Danish influence. Cf. Denstone.

**DEPNERS, THE** on the south side of Willoughbridge Park (SJ 7438). A curious name, for which no explanation can be offered, but perhaps associated with *Deercote* ('the shelter or cottage with the deer') 1833 O.S.

**DERCUSHALL** (unlocated, in or near Bagot's Bromley) *Dercushall* 1448 SRO DW1733/A/3/24. Perhaps to be associated with Darcel's Rough, 1 mile south-east of Kingstone (SK 0628).

**DERHULL** (unlocated, possibly near Rushall) *Derhull* 1222 SHC IV 21, c.14th century SRO D260/M/T/7/1. From OE *deor hyll* 'deer hill'.

**DERNESLOWE** (unlocated, in Whittington) *Derneslowe* 1227 SHC IV 42, 1272 ibid. 259. The second element is from OE *hlaw* 'tumulus, mound', with an unidentified personal name, which might also be found in other names, such as Dearnsdale.

**DERPLAWE** (unlocated, probably in Biddulph) *Deerplay* 1293 Kennedy 1980: 23, *Derplaus* 1293 SHC VI (i) 274, *Derplawe* 1313 SHC IX (i) 41, 1327 SHC VII (i) 198. From OE *deor* 'deer', with OE *plaga* 'play, sport', found in place-names for 'a place for games or where animals played', so 'the place where deer play'. Cf. Deerplay, Lancashire; Deer Play, Yorkshire; Durpley, Devon.

**DERRINGTON** in Seighford parish, 2 miles west of Stafford (SJ 8922). *Dodintone* 1086 DB, *Duddinton* 1203 SHC III 107, *Doddinton* 1228 Duig, *Dudington* 1236 Fees, *Dodington* 1242 SHC 1914 87, *Dudinton'* 1242-3 Fees, *Dodinton* 1282 SHC 1911 39, *Dotinton* 1290 SHC 1911 198, *Dodinton* 13th century SHC IV 277, *Dudynton* 1318 SHC IX (i) 69, *Dodinton* 1294 ibid. 226, *Dudynton* 1320 SHC IX 69, *Dodynton* 1532 SHC 4th Series 8 65, *Derington* 1601 SHC 1935 398. 'The *tun* associated with Dod(d)a'. The personal name Dod(d)a is well-evidenced in OE, and at least 20 places deriving from OE *Dod(d)ingtun* have been identified. It has been suggested that not all contain the connective *-ing-* used to link a place-name generic to a personal name or other first element, and that some may be associated with an OE *\*dod* 'hill', represented by northern dialect *dod* 'rounded summit', creating 'place near a dod' with the addition of *tun* to an earlier place-name \*Doding – see more particularly PN Sa I 106-7. In that respect it may be noted that Derrington lies within a mile of the great rounded hill on which stands Stafford Castle. Chetwynd mentions *Derrington* in Mucklestone in 1679 (SHC XII NS 262), which is to be identified with Dorrington (*Derington* c.1205 Rees 1997: 84), now in Shropshire. For the interchange of *d* and *r* see Carrington (earlier Caddington), Hertfordshire (PN Hrt 30-1); Dorrington (earlier Doddington) in Northumberland & Durham.

**DESIRE, LA** (unlocated, possibly near Badenhall or Eyeswall or near the lost Newbolt at Hilcote (q.v.).) *la Desirée* 1216-72 SHC 1914 60, *la Disure* 1276 ibid. 80, *la Desiree* 1301 Baron's Letter to the Pope, *la Desire(e)* 1349 SHC 1914 57, SHC VII (ii) 36, 74, 81, *Desere* 1349 Erdeswick 1844: 112, *le Desirre* 1365 SHC XIII 52, *le Desirre* 1366 SHC XIII 52. Chetwynd refers in 1679 to this manor 'which ... lay near to Eccleshall, but where I have not been able to discover', and to Sir Robert de Hastang who styled himself *dominus* [lord] *de la Desirée* in 1301: SHC 1914 60. A mysterious name, clearly of French origin, about which nothing further has been discovered. A more detailed discussion will be found in the Introduction, pages 59-61.

**DEUKERHULL** (unlocated) *Deukerhull* 1315 SHC IX (ii) 51. Possibly Deuxhill in Shropshire: see PN Sa I 107-8.

**DEVIL'S RING AND FINGER** two large stones (not in their original position), one circular and perforated, which formed part of a Neolithic galleried tomb (see JNSFC 5 1965 44) on the Staffordshire-Shropshire boundary, 1 mile west of Mucklestone (SJ 7037). *? mere stones* 17th century map WSL H.M.10. The 17th century name is from OE *(ge)mære* 'a boundary, a border'. The age of the present name is unknown, but is probably

228

relatively modern. The O.S. Archaeological Division record reads: 'Originally called Whirlstones (1. Meadow 7.3.27) (1) see Transactions of the North Staffs Field Club 1909 xliii 195; see Staffs Sentinel 23.10.26; see T Pape 'Antiquity' 1 229 photo'. TSAS XI 1927-8 132 gives the names *Devil's Ring & Finger or Whirl Stones*. The word *whirl*, meaning a fly-wheel or similar (which might be considered appropriate for a perforated stone) is not recorded by OED before the fifteenth century. *le Stones*, recorded in 1332 (SHC X 101), may refer to this monument, and Joh'e de *Stone* of Oakley, recorded in 1327 (SHC VII 215), may have taken his name from the feature, rather than the place of this name. In the same year Will'o de *Beadeston'* is recorded in nearby Muckleston (ibid.). The word *bead*, meaning a perforated object used as jewellery, is first recorded in 1400 (OED), but it is not inconceivable that the monument was once known as The Beadstone. However, since Will'o de Beadeston' is not mentioned in 1332, but Will'o de *Bereston'* (from Bearstone in Shropshire, 2 miles north of Mucklestone) is (SHC X 100), the first spelling (if the second is correct) is almost certainly a transcription error.

**DEVIL'S DRESSING ROOM, THE** 4 miles west of Tamworth (SK 1703). *The Devils Dressing room* 1834 O.S. A name of unknown age, but possibly a jocular or ironical 18th or 19th century adoption for the site of a former quarry which is said to have provided the stone for Tamworth church. Possibly so-named because the stone was particularly difficult to extract. The devil is commonly linked with pits, hollows and depressions of various kinds, for example Devil's Pulpit is a deep declivity below Offa's Dyke in Tidenham; The Devil's Chapel in Newland, Gloucestershire, refers to Roman ironworkings in the Forest of Dean; and Devil's Dyke is a common name for ancient ditches in various parts of the country. It would appear that such places were associated with the devil's subterranean kingdom. It is not impossible that there is a link with OE *delfan* 'to dig'. Cf. Devil's Drumble; Grimditch.

DEYSEMORE (unlocated, in Handsworth) *Deysemore* 1566 SHC 1938 77. Possibly 'daisy moor', from OE *dæges eage* 'day's eye', so-called in allusion to the flower opening in the morning (OED), with OE *mor*.

**DIAL LANE (FARM)** on the north side of Biddulph Moor (SJ 9061). *The Dyall* 1698 SRO DW1761/A/4/111-113[4]. Said to be from ME *dial* 'a dial, a sun-dial', sometimes denoting a sundial cut in turf. The element is commonly found in field-names from the 16th century: Field 1972: 62; Foxall 1980: 55.

**DICK SLEE'S CAVE** on Cannock Chase, 1 mile west of Brocton (SJ 9919). *Old Dick Slee's Cave* 1867 Trubshaw, *Dick Slee's Cave* 1946 O.S. A former hermit's dwelling. One tradition is that Richard Slee, a 19th century hermit lived here in a two-roomed turf hermitage, under which was a brick vault in which Slee intended that he be buried (JNSFC LVIII 1923-4 142), but see also Trubshaw 1867, and StEnc 178, the latter mentioning a hermit's rustic dwelling built c.1770, and citing without explanation the names *Abdullam's Cave* (later 19th century O.S.) and *Abdulalhs Cave*. Teesdale 1832 shows *Adullam's Cave*.

**DIEULACRES** (pronounced Jew-lack-ress [dʒuːlækrɛs]) 1 mile north of Leek (SJ 9857). *Deulecresse* 1214-16 BM, 1228 CH, *Deulacresse* 1214x1216 CEC 371, *Deulacres(se)* c.1222-30 StCart, *Diolacchrescha* 1274x1296 JNSFC 10 1970 92, *Deulacresse* 1334 SHC 1913 43, *Dewenleucres* 1532 SHC 4th Series 8 28, *Delencres* 1558 VCH III 234, *Dulencryst* 1561 SHC 1938 174, *Dewlincres* 1565 SHC XIII 52, *Dieulacresse* 1566 SRO SD4842/42/5, *Delyencrise* 1594 SHC XVI 134, *De la Cress* 1686 Plot, *Del-a-cross* 1775 Bowen, *Dieu la Croix or Delacres Abbey* 1775 Yates. An abbey founded in 1214, when the place was seemingly known as *Cholpesdale* c.1200

(VCH III 230; VCH VII 193). Dugdale, quoting from Henry of Huntingdon's 'Historia Anglorum', records that Ralph, Earl of Chester, returning from abroad, abandoned his wife and married Clemence, of French family (see Fowlchurch). Whilst in bed, a vision of his grandfather Ralph instructed him to 'go to Cholpesdale, in the territory of Leeke, and in that place where there was formerly a chapel to the Blessed Virgin, you shall found an abbey of the white monks, and Clemence then said to him in French 'Deux encres', and the Earl thereupon determined that the place should be called Dieulacres'. The precise meaning of Clemence's *'Deux encres'* is uncertain, with various possibilities: *Deux lecree* 'God creates it', *Deux l'egree* 'God sanctions it', but the most likely seems to be from the Anglo-French verb *encrestre* 'to increase', giving ' *[Que] Dieu L'encresse,* 'may God increase it': see SHC NS IX 293; NSJFS 2 1962 85. Ekwall 1960: 144 observes that the name is analogous to *Dieulouard* 'Dieu le garde' and *Dieu s'en Souvienne*, monasteries in France. The name Dieulacres is also recorded in 1279 as an assart in Clewer, Berkshire belonging to Salisbury Cathedral (Gelling 1978: 238), and an Irish Premonstratensian abbey named Dieulacresse is also recorded (VCH III 231). *Cholpesdale* may be associated with a small cave and inner chamber, which show evidence of a former attached structure, and may perhaps have been a hermitage with a grotto chapel, in a rock face between Abbey Farm and Abbey Green: see Fisher 1969: 14. The etymology of Cholpesdale (assuming it to be an authentic place-name) is not known.

**DIGBETH** in Walsall (SP 0198). *Digbathe* (p) 1573 Walsall ParReg, 1635 SRO D260/M/T/7/8, *Digbath* 1583 SHC 1910 168, *Diggbath* 1638 SRO D260/M/T/1/71. For the first element see Diglake. Bath(e) may be from a man-made pool, See also Willmore 1887: 99.

**DIGLAKE** 1 mile north-east of Audley (SJ 8151). *Viglake* [sic] 1632 Audley ParReg, *Diglake* 1646 SHC 4th Series I 237, 1733 SHC 1944 40, 1833 O.S., *the Dig-Lake* 1750 BCA MS917/1261. The derivation of this name, which is not uncommon in Staffordshire, is uncertain. *Digg Lake* is recorded on the boundary of Stoke-on-Trent in 1689: see Ward 1843: app lxiii. *Diglakes*, recorded in 1615 (Pape 1928: 127) and *the Diglake*, recorded in 1690 (JNSFC LXVI 1931-2 131), may refer to the latter place. Tipping Street in Stafford was formerly known as *Diglake*, though not before the end of the 16th century (VCH VI 187-8), and a field-name *Diggelake medowe* is recorded in Leek in c.1535 (LRMB), possibly to be associated with *Diglake* near Leek recorded in 1542 (1883): Sleigh 71. A field in Forton known as *Diglake* in 1839 (TA) is recorded as *Dyglake* in 1487 (*Rental*). A possible derivation is from OE *dic, lacu* 'watercourse in a ditch' (see PN Ch II 291; III 82; also PNEF 32, which suggests that *dic* is sometimes used of historic defence works), but the persistent –*g* makes such derivation unlikely: Field 1993: 51 suggests that *dig* may be from the dialect word for 'duck' (see also Harrison 1986: 34 for *Dygge* meaning 'duck' in 1561), giving 'duck stream'.

**DILHORNE** Ancient Parish (pronounced Dillon [dɪlən]) 2 miles west of Cheadle (SJ 9743). *Dvlverne* 1086 DB, *Dulverne* c.1187 SHC II 261, 1200 P, *Dilverne* 1236 Fees, *Dulverne* 1242 Fees, *Delverne* 1281 Misc, *Dilverne, Dylverne* 1286 Fees, *Dellren* c.1355 NA DD/FJ/4/26/7, *Delveron* 1387 Banco, *Dulron(e)* 1405 FF, *Dylryne* 1453 SHC NS III 204, *Dylren* 1503 Ipm, *Delos* 1532 SHC 4th Series 8 137, *Dyllern* 1534 StarCh, *Dillerne* 1607 QSR, *Dilhorne* 1665 BM, *Dilhorne als Dillerne* 1786 Bacon. Perhaps 'place by a mine or quarry', from OE *(ge)delf* 'a digging, a trench, a pit, a quarry', with OE *ærn*, an element sparingly used in place-names with the specialised meaning 'place of production, workshop', normally associated with a product such as stone, charcoal, etc. Ekwall 1928: 33-4 observes: 'Both [Biddulph and Dilhorne] are in the hilly parts of St[affordshire], and the element *dulf, Dulv-* is no doubt connected with OE *delfan* 'to dig', the meaning being 'a digging, a mine', as indeed suggested for Dilhorne by Duignan. Whether we have to start from OE *gedeolf*, and unrecorded form with fracture of *e* before *lf*, as in Mercian *seolf* 'self', or we have to assume a formation from *delfan* with the ablaut stage of the past participle, it is not easy to say. Dial[ect] *dolver* S[uffolk], N[orfolk] 'a piece of bog where peat is cut for fuel' may give some countenance to the latter suggestion. An *i*-stem OE *\*dylf*, analagous to OE *drync, spryng* from *drincan, springan*, would perhaps meet the case best'. See also Biddulph.

**DILHORNE BROOK** a tributary of the river Churnet.

**DIMBLE** – see **DRUMBLE.**

**DIMMINGS DALE** a deep gorge, 1½ miles long, 3 miles east of Cheadle (SK 0443), *Dimmingsdale* 1786 SRO D240/D/130, *Dimmings Dale* 1836 O.S.; DIMMINGSDALE on west side of Willenhall (SO 9698), *Dymmingsdale* c.1272 (1801) Shaw II 150, *Dymmesdale* c.1295 Mander & Tildesley 1960: 31, *Diminsdale* late 13th century SRO D593/B/1/26/6/39/5, *Dimmisdale* late 13th century SRO D593/B/1/26/6/39/5, *Dymmynggesdale* 1326 SRO D593/B/1/26/6/14/16, *? Dimsdale* 1428 SHC 1910 301; **DIMMINGSDALE** 4½ miles south-west of Wolverhampton (SO 8696), *? Dimpoole* 1670 Survey, *Dimings Dale* 1753 WALS DX-240-22, *Dimmingsdale* 1813 Trysull ParReg, *Dimings Dale* 1815 O.S., *Dimminsdale (Reservoir)* 1832 Teesdale, *Dimmingsdale* 1840 TA; **DIMMINS DALE** on Cannock Chase 2 miles south-west of Rugeley (SK 0116), *Demon's Dale* 1840 TA. A more detailed discussion of these names will be found in the Introduction, pages 59-60. See also Dimsdale.

**DIMSDALE** 1½ miles north of Newcastle under Lyme (SJ 8448), *Dvlmesdene* 1086 DB, *Dimsdal* 1212 SHC 1933 (ii) 11, *Dimesdal* 1212 SHC 1911 386, 1242 Fees, 1260 SHC 4th Series 13 15, *Dimmiesdale* 1228 SHC IV 72, *Demisdale* 1278 SHC XI NS 245, *Dimesdale* c.1278 SHC 1911 430, *Dymmesdale* 1282-3 SHC XI NS 252, 1315 SHC 1911 85, 1356 SHC 1913 166, *Dymnesdale* 1302 SHC VII 103, *Dymmesdale, Dymesdale* 1333 SHC X 80, 95, *Dimsdale* 1686 Plot 354, 1747 Bowen; **DIMSDALE** on north side of Blithfield Reservoir (SK 0525). *Deadman's Dale or Dimsdale* 1836 O.S. The DB form of Dimsdale near Newcastle under Lyme is almost certainly aberrate. A more detailed discussion of these names will be found in the Introduction, pages 59-60.

**DINGLE BROOK** rises in Rushton Spencer and formerly joined the river Churnet to the west of Leek. Dammed in 1799 to form Rudyard reservoir. See also Fule (River).

DIPDALE (obsolete) on the south-west side of Lower Gornal (SO 9291). *Depedale* 1580 Sedgley ParReg, *Dipdale* 1627 ibid, *Dibdale* 1652 ibid, *Dibdale in Neither g.* ['Nether (i.e. Lower) G[ornal'] 1659 ibid, *Dipdale Bank* 1669 SPI, *Dipdale Furnace, Dipdale Bank* 1834 O.S. The earliest spelling shows that the name is 'deep dale or valley'.

**DIPPONS** 3 miles west of Wolverhampton (SO 8699). *Dipping, Dippings* 1711 VCH XX 17, *Dippings* 1726 Sanders 1794, 1798 Yates, *Diptons farm* 1780 (1801) Shaw II 202, *the Dippens* 1781 SA 1067/7, *Dippins* 1808 Baugh, *Dipen* 1834 O.S. A name of uncertain etymology: it seems unlikely that the name is from OE *deoping* 'deep fen; hollow, depression in land': Deeping St James, Deeping St Nicholas, Market Deeping and West Deeping, all in Lincolnshire, are probably a very early example of a unique formation. Possibly from a place where sheep were dipped, but it may be noted that c.1700 *dipping* was a term used in the area for baptisms (TSSAHS XXXIV 1992-3 37), though there is little likelihood of any association with baptism here. The name was originally attached to an area to the west of Tettenhall Wood, but is now found near Wightwick.

**DIRTY GUTTER** 1 mile south-west of Onecote (SJ 0353). *Dirty Gutter* 1841 O.S. The word gutter 'watercourse' is found in England from the 13th century, from OFr *gutiere*, *goutiere*, from *goute* 'a drop'. The name here is probably relatively modern.

**DIXON'S GREEN** 1 mile south-east of Dudley (SO 9589). *Dixon's greene* 1655 Roper 1980: 101. From the personal name Dixon: the surname Dixsone/Dixon/Dickson is recorded in Dudley from at least 1618: see Roper 1980. The Green element suggests a squatter settlement.

**DODDEWELL** (unlocated) *Doddewell* 1305 SHC VII 135.

**DODDLESPOOL (HALL)** 1 mile north-west of Betley (SJ 7449). *Daddlespoole* 1593 Betley ParReg, *Dadlespool House* 1733 SHC 1944 35, *Doddlespool* 1799 Faden. Possibly from an OE personal name such as Dædhild or similar.

**DODEHAM** (unlocated) *Dodeham* 1199 SHC III 169.

**DODS LEIGH** 1 mile south-west of Church Leigh (SK 0134). *Dadesleia* 1114x1118 SHC 1916 227, *Dedeslega* 1166 SHC 1923 296, *Dadesleia temp.* Henry I Burton, *? Doddecota* 1266 SHC 1913 317, *Daddesleye* 1272 SHC IV 187, 1327 SHC VII 235, *Daddesley* 1323 SHC IX 94, 1332 SHC X 108, *Dadeles* 1572 SHC III 287, *Dadesleyfeld* 1590 SHC XVI 99. The first element may be an unrecorded OE personal name, with OE *leah*. There is a reference to an adjoining place called *Bilestanes Legam*, 'Bilstone's Ley' in the first half of the 12th century: SHC V (i) 34. See also Church Leigh.

**DODSLOW** – see HORNINGLOW CROSS.

**DOGGELEYHETHE** (unlocated, possibly near Biddulph) *Doggeleyhethe* 1401 SHC XV 96.

**DOGKENNEL** (obsolete) ½ mile north-east of Stewponey (SO 8685). *the Dogkennell* 1657 Sedgley ParReg, *Dogkennel* 1834 O.S. Self-explanatory.

**DOGLANDS, THE** 1 mile south-west of Fradswell (SK 9830). *Doglands* 1605 SHC XVIII 63, *Dog Lane* 1798 Yates, *Doglands* 1836 O.S.

**DOGLANE** ½ mile east of Calton (SK 1150). *Doggelone* 1390 SHC XV 18, *Doglane* 1419 SHC XVII 77, *Doglone* 1422 ibid. 77, *Doglon* 1584 SHC XVII 231, *(the) Dog Lane* 1597 FF, *Dog Lane* 1836 O.S. Presumably from a dog, perhaps memorably savage or vicious, with which the lane was once associated, but names of this type remain unexplained: see PNEF 133.

**DOLEFOOT GATE** (obsolete) in Needwood Forest, 1 mile north-west of Christchurch (SK 1424). *Dolesfore* Gate 1778 VCH II 350 map, *Dolefoot Gate* 1836 O.S. The place, which was one of the gates to Needwood Forest, is remembered in Dolefoot Lane.

**DOLEY** 1½ miles north-west of Adbaston (SJ 7429). *Doley* 1833 O.S. Early spellings have not been traced, but if ancient perhaps from OE *dole* 'a share of the common field', or (less likely) from Welsh *dôl* 'a meadow, pasture usually beside a stream', with OE *eg* 'island; a piece of land in a marsh and land on a stream or between streams': the place lies on Waggs Brook, which forms the Staffordshire-Shropshire boundary. There is no evidence to link this place with Doley Common or Doley Gate (q.v.), 7 miles to the south-east.

**DOLEY COMMON, DOLEY GATE** 1 mile north-west of Gnosall (SJ 8121). *Darley* 1435 SHC XVII, *Dorley* 1586 Ct, *Dorel(e)y Common* 1686 Plot 214, *Dauley Common*

1833 O.S. Possibly from OE *dor* 'door', in the topographical sense 'narrowing valley' with OE *leah*: Doley Gate to the south may use OE *geat* 'a hole, an opening, a gap' in the same sense, perhaps associated with the topographical feature associated with the first element of the name Gnosall (q.v.). A derivation from OE *deor-leah* 'the wood or clearing frequented by deer', supported by the earliest spelling, is also possible. Cf. Darley Abbey, Derbyshire.

**DOMVILLES** 2 miles north-west of Audley (SJ 7751). *? Damenevilla* c.1233 CEC 440, *? Damvill* 13th century SHC XI 333, *? Domville* (p) 1316 SHC X 68, *? Donnuill, Donuyll, Donnuyll, Donnvyle, Domuill, Domnuyle* 1371-83 SHC VIII NS 59-85, *? Dingles* 1775 Yates, *Dingles* 1799 Faden, *Dumbles* 1786 Tunnicliff, *Dumvells* 1833 O.S., 1890 O.S. A puzzling name, for which no suggestions can be offered, unless a transferred name from Donville in Calcados: see Dumville's Farm in Nether Alderley, Cheshire (PN Ch I 97). If all the spellings relate to this place, the inconsistencies are inexplicable, and the similarity between the earliest and present forms remarkable. It may be noted that this place lies some three miles from Jamage (q.v.).

DORK, THE – see **NEWFIELDS**.

DORUESLAU, DORESLEY (lost, in Little Sugnall: SJ 800333 according to TSSAHS XII 1970-1 35, but see also StEnc 182) *Dorueslau, Doresley* 1086 DB, *Deureslawa* 1187 SHC I 136, *Derueslawe* 1199 SHC III NS 74, *Derueslowe* 1199 SHC 1923 33, *Derneslowe* 1272 ibid. 34, *Derueslowe* 1281 ibid. 34, *Dorslow* 1284-5 ibid. 34, *Derselowe* 1327 SHC VII (i) 212, *Doreslowe* 1408 SHC XVI 64, *Doryslowe* 1433 SHC XVII 142, *Doreslowe* 1462 SRO 3764/96, *Doreslowe otherwise Dorneslowe otherwise Doresley* 1584 SHC XV 158, *Derneslowe otherwise Dorslow* 1590 SHC XVI 101, *Dorslow* c.1600 Erdeswick 1844: 123, *Dorslowe* 1618 SHC VI NS (i) 59, *Doreslaw farm* 1610 (1679) SHC 1914 49. The place (still evidently known as *Doreslaw* in 1679: see SHC 1914 50) has been identified with Sugnall and Little Sugnall (SHC 1923 32-5, VCH IV 42 fn; Derby & Terrett 1971: 173; TSSAHS XII 1970-1 35), and a deed of 1553 mentions a pasture in Little Sugnall called *Doreslow Buttes* (SHC 1923 35), with *Dorsley Ley Field* recorded in Little Sugnall in 1689 (SRO D1798/H.M.Chetwynd/36), probably to be associated with two pastures remembered as *Doesley* or *Dorsley* in 1900 (SHC 1923 33), although there is an unexplained reference in 1539 to *Doreslowe in Hilreston*, i.e. Hilderstone (MA). The first element may be an OE personal name, such as Deora, Deore, but the varied forms preclude a certain derivation. The second element is OE *hlaw* 'mound, tumulus'.

**DOSTHILL** 2 miles south of Tamworth (SK 2100). *Dercelai* 1086 DB, *Derteulla* 1166 RBE, *Dercetehull* (p) 1195, *Dercetehille* 1242 Fees, *Dersethull* 1247 FF, *Derstill* 1273 Ass, *Donestholle, Dunesholle* 1309 SHC IX 6, *Dersthull juxta Kynnesbury* 1315 Ass, *Dorsthull* 1316 FA, *Dorsthull, Derchetehulle* 1391 SHC XV 37, *Dostell* 1526 FF, *Dastell* 1549 Pat., *Dorsthyll al. Dastyll* 1550 FF. It is possible that some of the forms are corrupt, *c* having been transcribed, as frequently the case, as *t*, and vice versa. PN Wa 17 and 267 is hesitant in offering a derivation, but the 1391 form supports Ekwall 1960: 149 which proposes 'hill with a shelter for deer', from OE *deor-cete* 'deer shelter', perhaps a roofed shelter where deer could feed and shelter, with OE *hyll* 'hill'; see also TBWAS 86 1974 75. CDEPN proposes a derivation from PrWelsh *\*derw, \*ced* 'oak-tree forest', with reference to the Nottinghamshire, Leicestershire, Northamptonshire and Warwickshire wolds, against which the village lies, but the forms make such derivation improbable. Cf. Avon Dassett and Burton Dassett, Warwickshire. In Warwickshire until transferred to Staffordshire in 1965. *Kynnesbury* is Kingsbury, 1 mile south in Warwickshire.

**DOVE, RIVER** One of the main head-streams of the Trent, forming part of the boundary between Derbyshire and Staffordshire. *Dufan* 951 (13th century, S.557), *Dufan* 1008 (14th century, S.920), *Duve* late 12th century Okeover, *Dove* 1200-25, 1294 Derby, 1314 Abbr, c.1540 Leland, *Douve* 1255 BurtonAn, *Doue* 1298 Ipm, 1394 Pat, *Douue* 1281 Ass 148 m 33d, 1290 Ass 147 m 17, *Dowe* 1306 Ct, *Doff-water* 15th century Worc 357, *Dow* 1577 H. 'The dark river', from a Celtic adjective *\*dubo-* 'black, dark' (Welsh *du*, Irish *dubh*). River-names containing this element are common in Wales and Ireland. Not all Dove rivers have dark water, but they invariably have a dark bed or lie in a deep valley.

**DOVE BRIDGE** 1 mile north-east of Uttoxeter on the river Dove (SK 1134). *Dubrige* 1086 DB, *Doubrig* 13th century Duig, *Dobbrugge* 1332 SHC X 113, *Dufbrigge* 1360 SHC VIII NS 177, *Dovebridge* 1593 Rocester ParReg, *Dove Bridge* 1836 O.S. 'Bridge on the Dove', from OE *brycg*.

**DOVEDALE** (unlocated, in Ellastone) *Duvesdale* 1269 Pat, *Duuedale* 1296 Abbr, *Duvedale* 1293 SHC VI (i) 299, *Douuedale* 1326 SHC IX (i) 132, 1346 CoramR, 1364 SHC 1913 333, *Douuedale* 1332 *CPG*, *Douvdale* 1369 RB IV 43, *Dowesdale* 1418 Banco, *Duffdale* 1542 SHC XI 285, *Dowffe Dale* 1542 SHC 1912 144. 'The valley of the river Dove (q.v.)', with OE *dæl* 'valley'.

**DOVEFLATS** on the north-east side of Rocester (SK 1140). *Dove Flatts* 1668 SRO D1380/1/4, *Dove Flat* 1836 O.S. Probably from ME *flat* (from ON *flöt*) 'flat', so 'the flat land by the river Dove'.

**DOWN HOUSE FARM** 2 miles north-west of Penkridge (SJ 8916). *Donne* 1273 SHC VI (i) 59, *la Duoune* 1289 SHC 1911 45, *la Doune* 1293 SHC VI (i) 217, *la Donne* 1294 SHC VII 18, *la Doune* 1295 ibid.' 27, *la Dune* 1295 SHC 1911 54, *la Doune* 1300 ibid.' 269, *Dunmowe* 1313 SHC X 78, *Doune, la Doune* 1332 ibid. 125, *The Downe* 1588 VCH IV 84, *The Downs* 1614 ibid. From OE *dun* 'hill, an expanse of open hill-country', or (perhaps the meaning here) 'open country'. The original house may have been the moated site a short distance south-east of the present farmhouse. Cf. Gun. See also SHC VIII (ii) iv. A reference to *the Downe of Coton juxta Clanford* (Coton adjoining Clanford) in 1421 (BCA MS 3525/Acc1935-043), suggests that some of the spellings cited above may relate to that place. *Down Top*, south of Marston, is shown on Yates' map of 1798; see also Highdown (Cottages). *La Doune* in Salt is recorded in 1366: SRO D938/431.

**DOWRY** 1 mile south-west of Kingstone (SK 0528). *Dowrey* 1775 Yates, *Dowry Farm* 1782 SRO D24o/E/V/1/17, *Dowry* 1836 O.S. Presumably the property was the subject of a wedding gift.

**DOXEY** 1 mile north-west of Stafford (SJ 9023). *Dochesig* 1086 DB, *Dokeseia* 1168 P, *Doteshay* 1203 SHC III 111, *Dokeseia* c.1205 Rees 1997: 84, *de Akesey* 1275 SHC VI (i) 72, *Dokushey* 13th century SHC IV 275, *Dokeseye* 1355 SHC VIII 99, *Docsy* 1532 SHC 4th Series 8 65, *Doxie* 1666 SHC 1921 107. From the OE personal name *\*Docc*, with OE *eg* 'island, land partly surrounded by water, dry ground in marshland', here meaning either '*\*Docc's dry ground in marshland' (the place is on the south side of the river Sow, and gave its name to Doxey Marshes, an area of low-lying wetland to the north-east), or 'Docc's island': the place lies above the 300-foot contour, with the river Sow on one side and small streams on the other sides. The only other place where the personal name Docc has been recorded is Doxford, Northumberland.

**DOXEY WOOD** (obsolete) 2 miles south-west of Stafford, between Thorneyfields and Butterhill (SJ 9020). *Doxey Woods* 1591 Bradley ParReg, *Doxie Wood* 1680 ibid, *Doxeywood* 1704 ibid, *Doxey Wood* 1836 O.S. 'The wood associated with Doxey'.

DOYLE an unlocated manor in Ronton. *Doyle* 1454 SHC 1914 95. See also SHC III NS 213. Named after the Doyley or D'Oyly family, who held the place (which was later known as Lewkenore's Manor) from the late 14th century: SHC 1914 95-6. The surname is probably from one of the five Ouillys in Calvados: Ouilly-le-Basset, Ouilly-le-Vicomte, Ouilly-du-Houlley, Ouilly-la-Ribaude and Ouilly-le-Tesson. See also Lewkenore; Leukenore; Oils Heath; Stokedoily.

[DRAITON *Draiton*, recorded in DB under Staffordshire as held by Turstin has been identified as Drayton near Wroxton, Oxfordshire: see VCH IV 55, fn.82.]

DRAKEFORD (unlocated: an inquision of 1467 implies that the place was in the Tunstall/Colclough/Ridgeway/Bancroft area: Ward 1843: App. v. Perhaps remembered by Drakeford Grove, in Norton-in-the-Moors (SJ 8951)) *Drakeford* 1206 SHC III (i) 135; 1467 Ward 1843: App. v. The place is mentioned frequently in the medieval manor court rolls of Tunstall (Tooth 2000: 51), and is recorded as *Drakeforde* (1579) and *Drakesford* (1616) in the Norton-in-the-Moors ParReg. From OE *draca-ford* 'dragon ford', perhaps in this case 'the ford with the water serpent'. The surnames Drakefeilde and Drakeforde are found in the Rushall/Goscote area in 1666 (SHC 1923 138-9), and in the early 18th century in Seighford (ParReg), and the surname Drakeford is recorded in Talke in 1666 (SHC 1921 114).

**DRAKE HALL** 1½ miles north of Eccleshall (SJ 8331). Originally one of a number of Government establishments built in the area in the early 1940s and named after naval heroes, now an open prison. See also Beatty Hall, Duncan Hall, Frobisher Hall, Howard Hall, Nelson Hall, Raleigh Hall, Rodney Hall.

**DRAKELOW** 2½ miles south-west of Kinver (SO 8180). *Brakelowe* 1240 WoP, *Drakelow, Dracloe* 1582 Wills, 1649 Survey, *Drakeley* 1832 O.S. From OE *draca-hlaw* 'dragon tumulus', a name which records the ancient folk belief that buried treasure (and mounds believed to contain it) are guarded by a dragons. One of the Anglo-Saxon poems known as the Gnomic Verses includes a reference to 'the dragon in the tumulus'. Other places of the same name include Drakelow 2 miles south of Burton-upon-Trent, in Derbyshire, close to the Staffordshire border. The field-names *Great Drakeley* and *Drakeley Bank* are recorded in Seighford, perhaps near Cooksland, in 1762 (WALS DX240/40), *Drakelow Covert* appears on modern maps ½ mile south of Great Bridgeford (SJ 8826), and the field-name *Drakelaw* or *Drakilaw* is recorded in Sandon in 1844 (TA), but their history has not been traced: Drakelow Covert appears as *Drakeley Pits* on the 1836 1" O.S. map. OE *hlaw* often becomes -*ley* in Staffordshire, and these places may incorporate a genuine Drakelow name, though it may be noted that the surname Drakefield/Drakefoot (Drakeford?) is recorded in Seighford ParReg in the 1730s. Whilst names incorporating Drake- are held to refer to mythological dragons, a rhinocerus skeleton was discovered in fluvio-glacial deposits of the river Trent near Alrewas in 2001, with the skull having the appearance (from the single hole at the front which had contained the horn) of a one-eyed dragon. It is said that such skulls were once known as dragon's skulls (information from Andy Currant, Natural History Museum), and it is of interest that Drakelow near Burton-upon-Trent (which is little more than a mile from an important Anglian cemetery of the heathen period) lies on the deposits of a second river terrace; Drakelow Covert near Great Bridgeford is immediately adjacent to fluvio-glacial deposits along the river Sow; and Drakelow south of Kinver is adjacent to a strip of glacial deposits. *Drakenage*, recorded in 1271 (SHC 4th Series XVIII 94; see also PN Wa 17) is Drakenage Farm, a moated site south-east of Hemlingford Green in Kingsbury parish, Warwickshire, south of Tamworth, on the east side of the river Tame (SO 2295), meaning 'Dragon's edge': the place lies at the end of a low hill. Further research may show a wider association of such names with glacial deposits, and provide stronger evidence that such names might in some cases derive from the exposure by erosion or activities such as quarrying of the skulls of rhinocerus or similar animals which were taken as evidence of dragon burials. See also Drakenage.

**DRAYCOTT IN THE CLAY** in Hanbury parish, 5 miles south-east of Uttoxeter (SK 1528), *Draicote* 1086 DB, 1251 Ch, *Draycote* c.1286 NA DD/FJ/1/180/1, *Draicote* 1445-50 TutCart, *Dracote* late Henry III TutCart, *Draycote, Draycott under Nedwode* 1435 SRO DW1733/A/3/12; **DRAYCOTT IN THE MOORS** Ancient Parish 3 miles south-west of Cheadle (SJ 9840), *Draycoten* c.1240 SHC 1925 77, *Draycoht* c.1248 SHC 1911 420, *Draicot* 1251 CH, *Draycote* 1291 Tax, *Drycote in le More* 1420 Oakden 1984: 3; **DRAYCOTT CROSS** 2 miles south-west of Cheadle (SJ 9841), *Draycott Cross* 1836 O.S.; **DRAYCOTT** 1½ miles east of Claverley (SO 8192), *Draicote* c.1250 Eyton 1854-60: III 99, *Dayncot* 1256 ibid. 99, *Draycot* late 13th century ibid. 99, *Draycote* 1833 O.S. (in Shropshire since the 12th century); DRAYCOT (unlocated) in Sandon, *? Draycote* 1377 SHC 4th Series VI 13, *Dracote* c.1600 Erdeswick 1844: 44; **DRAYCOTT WASTE** (obsolete) on the north side of Bromley Wood, 2 miles north-east of Hilderstone (SJ 9736), *Draycott Waste* 1836 O.S. A common name – there are 10 places called Draicote in DB – which has not been satisfactorily explained. The first element *dray-* is generally held (see Ekwall 1936: 151; EPNE i 134-6) to be from OE *dræg* (corresponding to ON *drag*), a word found (curiously, and perhaps significantly) only in place-names, and derived from OE *dragan* 'to draw, to pull', also 'to go, progress, travel (usually on land)', and generally understood to be used in two senses in English place-names, namely 'a portage; a place where boats are dragged around an obstruction in a river' and 'a dray, something which can be dragged'. Ekwall 1936: 151 suggested also 'a steep slope where extra effort is required', but that has been questioned, since OE *cot* is not usually combined with topographical elements, and it would be surprising if, when so combined, it is always with *dræg*. It seems possible that at least one meaning of the element in some of place-names in which it is found is some kind of shelter or accomodation for travellers. The most common place-names with *dray-* are Drayton, followed by Draycott, but the element is also found (though much less frequently) with OE *dun* 'hill', *stan* 'stone', *mere* 'pool', and *ford*. Curiously, there is little evidence, archaeological, literary, pictorial or otherwise, to suggest that the Anglo-Saxons used horse-drawn transport. If manpower was normally used for haulage, places incorporating the element *dray* were perhaps so-named because they held horses or additional manpower to assist with haulage (and perhaps guiding) through difficult areas: in two of the Staffordshire names at least there is some indication in the descriptive locational elements of the nature of the difficult ground (*in-the-clay* and *in-the-moors*, which, perhaps significantly, are descriptors not – except for Bradley in the Moors and Norton-in-the-Moors – found attached to any other Staffordshire place-names), suggesting perhaps that this meaning of the names was known when those elements were first attached. Drayton Bassett, for example, was in a vast heath where navigation may have been difficult. The proximity of Draycots and Draytons to Roman roads has been noted (see *Nomina* 17 1994 11-12): Draycott in the Moors lies on the Roman road from Rocester to Stoke on Trent (Margary number 181); Draycott near Claverley lies between two lost Roman roads, running a mile or so north-east and south-west of the settlement; Drayton Hales lies on the north side of the lost Roman road from Pennocrucium to Chester (Margary number 19); Drayton near

Penkridge lies a mile or so to the east of a lost Roman road from Pennocrucium which ran north from Watling Street through Preston Vale Farm: Horovitz 1992: 32; and Drayton Basset lies 1 mile south of Watling Street. In that respect it is of interest that Shaw 1801: II 1 suggests that Dray may be from a Celtic word, which he does not cite, signifying a to\'n lying on a straight road. For completeness it may be added that although the word *dray*, meaning 'squirrel's nest', is unrecorded before the 17th century, its derivation is unknown and it may well have a much older origin with a differing meaning – Ekwall 1936: 151 proposed 'house of shelter at the head of a pass or long hill' for some Draycotts, including those in Staffordshire, although the 'house of shelter' was taken from the *cot* element. It is not impossible that some Dray- names incorporate OE *dræge* 'a drag-net'. Finally, it must be noted that Draycott in the Clay lies in a tiny dale created by Salt Brook, with a saline spring near Draycott Mill, 1 mile north-east of Draycott, which suggests a possible derivation from a word associated with OE *dryge* 'dry', perhaps a place where the water caused a thirst, or on particularly well-drained ground. The lost Draycot near Sandon is also in an area known for its salt: see Salt and Shirleywich. In that respect, a reference has been noted to 'a saltpan called in English Draiburne', recorded c.1150: PN Wo 390; EPNE i 135. The qualifying descriptions of the Staffordshire names serve to distinguish them from each other and similar names elsewhere. Cf. Drayton; Drayton Bassett; Glascote. The correct ecclesiastical form of Draycott in the Moors is Draycott le Moors: Youngs 1991: 409.

**DRAYTON**  2 miles north of Penkridge (SJ 1692). *Draitone* 1086 DB, *Drayton* 1194 SHC VIII 160, *Draiton* 1211 SHC III 151, *Draton* c.1220 ibid. 161, *Draytun* 1327 ibid. 162, 1682 Bradley ParReg. A common place-name – there are 37 places called Draitone in DB. From OE *dræg tun*: see Draycott; Drayton Bassett.

**DRAYTON BASSETT**  Ancient Parish 3 miles south-west of Tamworth (SK 1900). *Draitone, Draiton* 1086 DB, *Draiton* 12th century Duig, *Dreyton Park* 1285 SHC 1924 330, *Braiton Basset* (*sic*) 1610 Speed, *Drayton Basset* 1682 Browne. From OE *dræg tun* (see Drayton, Draycott) with the name Bassett from its early lords of the manor of that name to distinguish it from other Draytons. For Drayton Park see TSSAHS XXXX 1988-9 40-44. See also Bassett's Cross.

**DRAYTON HALES** – see HALES.

**DRESDEN**  on south side of Longton (SJ 9042). An ecclesiastical parish formed in 1853. The history of the name in uncertain, but is probably transferred from the German city in the light of its ceramic connections with this pottery-making area. The parish is also known as Redbank: Youngs 1991: 410.

**DROINTON**  5½ miles north-west of Abbots Bromley (SK 0226). *Dregetone* 1086 DB, *Drengeton'* 1198x1208 (13th century) EEA 17 74, *Drengeton, Drengenton, Dregenton* 1199 SHC III 56, *Dreuton* 1294 SHC 1911 109, *Drengetone* 1299 SRO 553[7902], *Drenkton* 1372 SRO 571[7922], *Dreynton* 1619 SHC VII NS 195, *Drointon* 1836 O.S. The first element is from OE *dreng*, 'a free tenant holding by tenure combining rent, service and military duty', a word borrowed from ON *drengr* 'a young man, a servant'. Hence 'the *tun* of the drengs'.

**DRUIDS HEATH**  1 mile north of Aldridge (SK 0501). *Druwode* 1326 (1801) Shaw II 98, 1343 ibid.' 98, *le Drewed field* 1592 SRO 3005/13-14, *Drued Field* 1631 SRO D260/M/T/2/12, *Drewed field* 1684 TSSAHS XX 1978-9 43, *Drude Meer* 1686 Plot 46, *Drewed Heath, Drewed Field* 1712 SRO 3005/91, *Drood, or Druid heath* 1798 Shaw I 11, *Druidsmeer* 1798 ibid. 40, *Druid Heath* 1798 Yates. Formerly heath in Cannock

Forest adjoining Sutton Chase. According to Duignan 1902: 53, a Norman family of Dru, deriving their name from Dreux, department of Eure-et-loir in Normandy, were medieval lords of Aldridge, and acquired the heath here. The diminutive of Drogo (the Latinised form of the name – Drogo of Aldridge is recorded c.1175x1208 SRO 3005/1) is Dru, giving Druwood, later Drewed and so Druid (cf. *Drewood Field*, recorded c.1740; see also TSSAHS XX 1978-9 47). Shaw 1801: II 93 mentions a Drew who was granted the nearby manor of Bentley by William I. A reference to *Drywode* in Aldridge in 1343 (SRO 3005/2) may be noted. The name has no connection of any kind with druids.

**DRUMBLE, DEVIL'S** 1 mile east of Brocton, on west side of Sherbrook Valley (SJ 9819), *Devil's Dimble* 1605, 1618 Tatenhill ParReg; **DRUMBLE, PARROTS** 1 mile north-west of Talk (SJ 8252), *Parrot's Drumble* 1833 O.S.; **DRUMBLE, CARTWRIGHTS** 1 mile north-east of Caverswall (SJ 9644); **DRUMBLES, FOXLEY** on the Staffordshire-Cheshire border, 2 miles north-west of Audley (SJ 7852), *Foxley Drumble* 1833 O.S. Drumble is a common element in minor place-names in Staffordshire, from OE *\*dumbel, \*dymbel*, of uncertain origin but perhaps connected with OE *dimple*, recorded in the topographical sense 'depression in the ground' c.1205 (Ekwall 1960: xxxiii), or associated with 'dim' ('gloom, obscurity') and 'dingle', not evidenced before the 14th century when it appears in place-names in Midlands dialect as *dimble, dumble*, 'a hollow, a ravine through which a watercourse runs, a deep shady dell, a wooded valley, a belt of trees along the bed of a small stream' (perhaps with Scandinavian links: cf. Norwegian *dembel* 'a pool' – see Dembleby, West Yorkshire), or (from Jackson 1879: 127) 'a rough wooded dip in the ground; a dingle'. It often becomes *drumble*. Parrot is from the Parrott family (see SHC 1944 xviii), as also Cartwright. Foxley is from the place of this name 2 miles north of Audley. For Devil's Drumble, see also The Devil's Dressing Room. EDD records *dumble* in Cheshire, Derbyshire, Leicestershire, Nottinghamshire, Shropshire and Warwickshire, in the sense 'a wooded valley, a belt of trees along the bed of a small stream, a ravine through which a watercourse runs': a full discussion of the word can be found in PN Nt 280-1. In Yorkshire *dumble-pit* appears to have been the name given to a pit or dell containing stagnant water: see PN Nt 279-80.

**DRYBROOK** a tributary of the river Pipe. *Driebrouk* 1286 For. From OE *drȳge* 'dry', presumably because the stream on occasions ran dry.

**DRY STONES** 1½ miles north-east of Upper Hulme (SK 0362). *Drystones (Road)* 1839 EnclA, *Drystones* 1842 O.S.

**DUDHILL** 2 miles south-west of Bobbington (SO 7887). In Shropshire since the 12th century. Early spellings have not been traced, but if ancient, possibly from northern dialect *dod* 'a rounded summit'.

**DUDLEY** Ancient Parish 5 miles south-west of Birmingham (SO 9490). *Dudelei* 1086 DB, 1199 Cur, *Duddelaege, Duddelege, Duddeleye* c.1140 Chronicle John of Worc, 1229 Ch, 1264 Ipm, 1275 SR, *Duddele* c.1185 SHC III (i) 218, *Dudele, Dudeleia* c.1200 SHC III (i) 215, *Doddele(ye)* 1289 Wigorn, 1327 SR, *Duddeley* c.1540 Leland v 20. 'Dudda's *leah*'. This is one of a group of names ending in 'ley' that are concentrated in the area immediately west of Birmingham, which may show that the area had a considerable area of woodland at the time of the Anglo-Saxon settlements: see Gelling 1988: 126-8. The place was an island of Worcestershire wholly within Staffordshire until 1966, when it became part of Staffordshire.

**DUDLEY PORT** 1 mile south-east of Tipton (SO 9691). *Dudley Port* 1802 Rowley Regis ParReg, 1834 O.S. So-called because the settlement developed from 1777 as a port by the Birmingham Canal Old Cut to serve Dudley, which was for long inaccessible by water, before the Dudley Canal tunnel was built in 1792: StEnc 190.

**DUDMASTON** 1 mile north-west of Quatt (SO 7488). *Dodemanestun temp.* Henry I Eyton 1854-60: III 186, *Dudemanneston* 1165 ibid. 187, *Dudemaneston* 1240 Bowcock, *Dudemoneston* 1253 ibid, *Dudemaston* 1270 Eyton 1854-60: III 188, *Dodemanston* 1279 Bowcock, *Dodemoneston* 1296 SHC 1911 266, *Dudmanston* 1300 SHC VII 66, *Dodemoneston* 1300 SHC 1911 266, *Dudingston* 1562 SHC 1938 166. 'Dudeman's *tun*'. Since the 12th century in Shropshire.

DUGDALE (obsolete, on the south side of Wimblebury, at what is now Heath Hayes (SK 0110)) *Dugdale* 1834 O.S.

**DUNBROOK** in Longnor (SK 0864). *Dunbrook* 1884 O.S. Possibly from OE *dunn* 'dark brown'.

**DUN COW'S GROVE** 1 mile west of Hollinsclough (SK 0466). *Duncosgreve* 1566 Deed, *Duncote Greave* 1600 VCH VII 38, *Duncoategreave* 1601 Alstonefield ParReg, *Duncotes Greave* 1614 ibid, *Dunkets greave* 1665 ibid, *Dunkerd Greve* 1714 ibid, *Dimcalf* 1803 ibid, *Dun Cows* 1842 O.S. Probably from the OE personal name Dunn, with OE *cot* 'a cottage, a shelter', and OE *græfe* 'grove, coppice' (and in some cases 'pit, trench'), so 'Dunn's cottage or shelter in the grove or at the pit', which has become much corrupted.

**DUNCAN HALL** in Yarnfield (SJ 8633). One of a number of Government establishments built in the area in the early 1940s and named after naval heroes. See also Beatty Hall, Drake Hall, Frobisher Hall, Howard Hall, Nelson Hall, Raleigh Hall, Rodney Hall.

DUNGEON (obsolete, in West Bromwich (SO 0093)), *Dungeon* 1775 Yates; DUNGEONS (unlocated, in or near Walsall, possibly at Rushall), *? Little Donges* 1589 SRO D260/M/T/1/19, *? Dungens* c.1593 (SRO D260/M/F/1/2/f33), *Dungeons* 1617 Willmore 1887: 440, 1708 SRO D260/M/T/1/120c. From ME *dungeon* 'an underground chamber', possibly with reference to deep pits, holes or caverns, natural or man-made, which are not uncommonly found in these areas.

**DUNGE WOOD** 1 mile west of Keele (SJ 7945), *le Dunge* 1476 NSJFS 3 1963 58, *Dunge Wood* 1833 O.S.; **DUNGE BOTTOM** 1 mile north-east of Alstonefield (SK 1357), *Dunge* 1840 O.S. A derivation from OE *dyncge* 'manured land', or ME *dunge* 'dung', perhaps meaning 'manured land', seems the most likely derivation, but the first place lies in a valley in which the Hazeley Brook rises in a number of boggy hollows. The name in that case may therefore be 'the foul marsh or boggy place'. A reference to '... the 6 seats for the "dunge" or wilderness and garden are almost finished...' in 1725 (SRRL 112/1/2717) suggests that the word may have been applied to overgown areas or perhaps grottoes. A derivation from OE *dung* 'dungeon, underground room', recorded only in poetic use, can be ruled out in the first place, but may have been applied to some natural feature such as a cave at the second place. Dunge Falls, Upper Hulme, are recorded in Dent & Hill 1896: 83; see also PN Sa III 110.

**DUNIMERE FARM** 1 mile south of Harlaston (SK 2109). *Donimere* 1798 Shaw I 402. Possibly associated with *Dounfield*, recorded in 1758: BCA MS3878/631.

DUNKFORD BRIDGE (obsolete) at Hilcote, where the road between Eccleshall and Stone crosses the river Sow (SJ 8429). *Dunkesford temp.* Henry II SHC 1914 68, *Dunkesford* (in 'Alde-Badenhall') *temp.* Henry VIII (1679) SHC 1914 68, later *Dunforde (Bridge)*: SHC 1934 29, *Dunkford Bridge* 1621 SRO DE615/EX/1. Possibly from the OE personal name Dunnic or similar. The bridge was presumably erected in or post *temp.* Henry VIII.

**DUNKIRK** 1 mile north-east of Audley (SJ 8152); *Dunkerk Estate* 1776 SRO D738/8, *Dunkirk* 1833 O.S.; DUNKIRK in Newcastle under Lyme (SJ 8446); *Dunkirk* late 18th century SRO D593/T/10/12. Possibly so-named to commemorate the surrender of Dunkirk by the Spanish to English forces on 14th June 1657.

**DUNLEA FARM** 1½ miles north-west of Onecote (SK 0257). *Dunleigh over Morridge* 1674 Leek ParReg, *the Dunley* 1683 ibid, *Dunleigh* 1680 ibid, *Dun Lee* 1775 Yates, 1842 O.S., *Dunlee* 1962 O.S. Seemingly from OE *dunn* 'dark, dusky, swarthy', with OE *leah*, giving 'the dark woodland glade', but VCH VII 211 identifies *Duncowleye*, recorded in 1405, in this area. If that name relates to this place, the derivation may be 'dun-cow *leah*' rather than 'dun cowley' (see Cowley), but the loss of a complete element in such a relatively short period would be surprising. See also Dunley.

DUNLEY (unlocated, possibly near Abbey Hulton, or the same place as Dunlea (q.v.)) *Dunley* 1227 SHC XII NS 218. Probably from OE *dunn* 'dark, dusky, swarthy', with OE *leah*.

**DUNSLEY** ½ miles east of Kinver (SO 8583). *Dunsley* 1274, 1316, 1324 SHC 1911 28-9, *Dunnesleye, Dunneslegh* 1324 ibid. 363, *Donnesleye* 1346 SHC 1913 113, *Dunnesley* 1412 SHC XVI 80, *Downysley, Dounysley* 1504 SHC 1928 89-90, *Dounsley* 1532 SHC 4th Series 8 15. From the OE personal name Dunn, so 'Dunn's-*leah*'.

DUNSMOOR (obsolete) in the area around Bradshaw, 1 mile north-west of Longsdon (SJ 9455). *Dunnismour* 1278 Ipm, *Dunsmore alias Bradshaw More* 1558 SRO DW1702/2/1-10. Possibly from the personal name Dunn: cf. Dunsmore, Warwickshire. It seems that the name of this place is not to be associated with Dunsmore Brook (q.v.).

**DUNSMORE BROOK** a tributary of the river Churnet which runs from Rushton James to the river Churnet, north of Harracles. *Dunsmore flu.* 1577 Saxton, *Dunsmore brook, Dun* c.1600 Erdeswick 1844: 493-4, 1610 Speed, *Dunsmore River* 1689 Lea. *Dunsmoore Meddowes near Rudyard (over against Herracles)* is recorded in 1686: Plot. See also Dunsmoor.

**DUNSTALL** 1½ miles west of Tamworth (SK 1803), anciently within the Hay of Hopwas in Cannock Forest. *Tunstall (Wood)* 1300 SHC V (i) 176, *Tunstal* 1323 SHC IX (i) 102, 1324 SHC 1911 101, *Dunstall* 1472 SBT DR3/575 at 570, *Dunstal* 1798 Shaw I 433; **DUNSTALL** 4½ miles south-west of Burton-on-Trent (SK 1820; SMR 02619 places a deserted post-Conquest settlement at SK 18602025), formerly in Needwood Forest, *? Stunstull* 1269 SHC 1910 129, *Tunstall* 1272, 13th century Duig; **DUNSTALL** 1 mile north-west of Abbots Bromley (SK 0726), formerly within Needwood Forest, *Tunstall* 1327, *Tunstal Maner* (Manor) 1355 Duig, *Tunstall* 1463 SRO D3155/7112; **DUNSTALL** 1 mile north-west of Wolverhampton (SJ 0000), *Tunstall* 1327, *Tunstall near Hampton* 1356, *Tunstall* 1450 and 1563 all Duig, 1686 Plot 8. From OE *tun-stall* 'site of a farm'. It has been observed that the name is frequently found on the borders of ancient wastes, as if they had been outlying farmyards without homesteads: see Field 1993: 215. A 'hermitage of the well of Dunstall' belonging to Trentham priory, probably near the priory itself, is recorded in 1162: VCH III 136. See also Tunstall.

**DUNSTON** 2 miles north-east of Penkridge (SJ 9217). *Dvnestone* 1086 DB, *Doneston* c.1220 SHC VIII 161, *Doneston'* 1242 Fees, *Donston(e)* 1272 FF, 1403 Ipm, 1587 *Survey, Dunstone* 1281 SHC VI (i) 154, *Donston, Dunston* 1327 SHC VII (i) 244, *Dunstane* 1367 Pat, *Dunson* 1552 FF. Dunn (from OE *dunn* 'dark, dusky, swarthy') was a common OE personal name, hence 'Dunn's *tun*'.

**DUNWOOD** 3 miles west of Leek (SJ 9455). *Dunewode* 1275 SHC XI 334, 1278 Ipm, *Dunewode* 1275 SHC XI 334, *Donewode* 1280 SHC VI (i) 100, *Dunwood* 1587 QSR, 1634 Leek ParReg, *Nether Longsdon otherwise Donwood* 1622 SHC X NS I 29. From OE *dun-wudu* 'wood on or at the hill'.

DUTTON The name is often found in early records associated with the Maer/Aston area, e.g. *Dutton* 1272 SHC IV (i) 193, 1304 SHC XII NS 278, 1313 SHC XI 334, 1316 SHC 1921 16, 1327 SHC VII 203, 1410 SHC XVI 71, 1594 SHC 1932 45, *Dotton* 1326 SHC 1911 368, *Dotton'* 1333 SHC X 100. 'The [family of] Duttons had been settled in Staffordshire since the reign of Henry III, when Vivian de Standon gave one fourth part of Mere and Aston to Thomas de Dutton in frank marriage with his daughter Philippa': SHC 1913 233; SHC IX NS 257; SHC XII NS 242-5; SHC 1938 26. See also Salt 1888: 53; PN Ch III 112; and CEC 260 for a reference to the Dutton family of Cheadle. *Dutton Water* (unlocated, possibly near Longdon) is recorded in the late 13th century: SHC 1921 7.

**DYDON** 1 mile east of Middle Mayfield (SK 1344). *? Doiton, Doyton* c.1568 SHC 1938 143-4, *Dydon Wood Farm, als Dydon Farm* 1705x1875 SRO D626/A/9/1-29, *Dibden Mathfield* 1798 Yates, *Dydon & Dydon Woods* 1836 O.S.

EADMORE or IDMORE HEATH (unlocated, probably near New House Farm near Little Sugnall) *Eadmore, Idmore Heath* 1725x1809 SRO D1192/39, SRO D1192/35.

**EARDLEYEND** 1 mile north of Audley (SJ 7952). *Erdele* 1332 SHC X (i) 101, *Yerdleyende* 1397 DRO D3155/WH42, *Erdeleyende* 1512 JNSFC LX 1925-6 34, *Yeardley Ende* 1531 X NS I 166, *Yerdley End* 1540 DRO D3155/WH44, *Eardley* 1686 Plot 214, *Eardley End* 1833 O.S. Perhaps from OE *gyrd-leah* 'wood where yards or poles were obtained'.

**EARLSWAY HOUSE** 1½ miles south-west of Rushton Spencer (SJ 9160). *Herlesweygrene* 1350 *Antrobus, Herleswey, Urlesweye* 1359 ibid, DW1761/A/4/15, *Earles Way* 1673 SRO DW1761/A/4/111-3[4], *Earles Way* 1673 SRO DW1761/A/4/111-113, *Earls Way house* 1775 Yates, 1862 (1883) Sleigh. From OE *eorl* 'earl', with OE *weg* 'way', giving 'the earl's way', the *via comitis* of the Earls of Chester (or pre-Conquest Earls of Mercia: SHC 4th Series 19 11), which ran from Chester to Leicester and Nottingham, via Leek and Derby, also found in Yelsway or Yarlsway (Lane), Caldon. See also VCH II 279; Palliser 1976: 79-81; Elliott 1984: 64-5.

**EASING FARM** 2 miles south of Upper Hulme (SK 0157). *Hesing* 1250 SHC 1911 428, *Esynge* 1274 SHC 1911 429, *Esyngge* 1280 SHC VI (i) 109, *Esinge, Essinge* c.1292 SHC NS IX 314, *Esyng* 1332 SHC X 115, 1534 ValEccl, *Heysynge* 1535 SHC 1912 77, *Esyns, Esynges* 1546 SHC 1912 351, *Essynge* 1552 SHC IX NS 301, *Eesinge, Essinge* 1640 Leek ParReg. The Isinge in Macclesfield, Cheshire, has the forms *Esyng* 1274, *Hessyng'* 1467, *Hesyng* 1508 (PN Ch V (II) 283), and it is suggested that the basis of that name may be OE *hǣs* 'brushwood, heath', giving '(the place called) the *hǣs* or brushwood or heath; that which has to do with brushwood or heath' (ibid.), but a more likely derivation may be from an early instance of late-ME *esing* 'the eaves of a house', hence 'a roof, shelter, dwelling', so giving 'the dwelling place' or 'the place at the edge of a wood

241

or hill': see PN Ch I 119; also Wilson 1974: 32: this place lies on the long hillside of Morridge. For an example of *esyng* used *temp.* Henry VIII in the sense 'eaves' see SHC X NS I 160.

**EASTWALL** on the south bank of the river Churnet, 1 mile west of Oakamoor (SK 0344). *Esteswelle* c.1290 Chester 1979: 3, *East Wall* 1836 O.S. If the early spelling is reliable, the medial -*s*- suggests that the first element of the name is an unidentified personal name, with Mercian OE *wælle* 'a spring', and (sometimes) 'a stream', which invariably develops into *wall* in the West Midlands. Modern maps show a small pool here, which may be asociated with the name, but the pool does not appear on the first edition 1" O.S. map of 1836.

**EATON BROOK** a tributary of the river Penk. From Water Eaton (q.v.).

**EATON, CHURCH** Ancient Parish 2 miles south-east of Gnosall (SJ 8417). *Eitone* 1086 DB, *Eitun'* 1176 MRA, *Eton* 1198 CurR, 1242 Fees, *Eiton* 1200 Ass, 1203 Fine, *Eyton, Eaiton'* 1236 Fees, *Chirch(e) Eyton* 1261 MRA, *Chirche-Eyton* 1293 SHC VI (i) 245, *Eytona* 1298 RontonC, *Chirche Eton* 1481 Pat, *Churche Eyton* 1532 SHC 4th Series 8 95, *Churcheton* 1570 Pat. From OE *eg-tun* '*tun* on an island or on dry ground in a marsh'. The place is on a small raised area in wet ground with the church on the highest point. It is possible that *tun* is a later addition to an original name which was simply *eg*, and although there was a church here before Domesday, the *Church* element is a medieval addition. Morden 1695 shows an *Eaton* on the north side of Crakemarsh. See also Wood Eaton.

**EATON, WATER** 2 miles south-west of Penkridge (SJ 9011, SMR 02444 places a deserted Anglo-Saxon settlement at SJ 90201100). *? Eatun* 940 (14th century, S.392), *Etone* 1086 DB, *Eton'* 1242 Fees, *Eton* 1262 For, 1286 SHC V (i) 162, *Eten' super Watlinstrete* 1315 SHC IV (ii) 106, *Watter Eyton* 1532 SHC 4th Series 8 89. From OE *ea* 'river', usually applied to a larger watercourse, with OE *tun*, hence '*tun* on a river'. The place lies on the river Penk on the north side of Watling Street.

**EATON, WOOD** 2 miles south-east of Gnosall (SJ 8417). *Wode-Eyton* 1293 SHC VI (i) 245. From OE *eg tun* '*tun* on an island or dry ground in a marsh', with OE *wudu* 'a wood'. The place shares the same area of raised land as Church Eaton (q.v.), which it adjoins on the north-west.

**EAVES, THE** 1 mile south of Cheadle (SK 0141). *Cheadle-Eaves* 1663 SHC II (ii) 63, *Chedle-eves* 1695 Morden, *Cheadley eves* 1749 Bowen, *The Eaves* 1836 O.S. From OE *efes* 'eaves; an edge or border, especially of a wood', and in place-names 'the brow of a hill, the edge of a precipice or bank', or as here, 'the place on the edge of the township (of Cheadle)'. See also Whiston Eaves.

EBON ASH (unlocated, in Horseley) *Ebon Ash* 1851 White.

**EBSTREE** 1 mile north of Trysull (SO 8595). *Epstree* 1759 Trysull ParReg, *Hebstry* 1794 ibid, *Ebstree* 1832 Teesdale, *Ebstrey* 1834 O.S, *Ebstree* 1840 TA, 1853 Trysull ParReg. A curious name, early spellings for which have not been traced. From spellings with *Ep-* in the parish registers possibly (though considerable doubt remains) from OE *heope* 'fruit of the wild rose', from which derives Modern dialect *hip, hep*, so giving 'the tree with the hips': in the 1840 TA the field name following *Upper Ebstree* is *Hepstile* 'the stile at the hips'. Or, if ancient, perhaps from the OE personal name *Hebbi: cf. Hepscott, Northumberland.

**ECCELWALL** (unlocated; possibly a scribal error for Eccleshall (q.v.), or perhaps Eccleswall, Herefordshire) *Eccelwall* 1293 SHC VI (i) 215. If a genuine name, the second element is Mercian OE *wælla* 'a spring', and (sometimes) 'a stream'. The first is uncertain, but perhaps from the OE personal name *Eccel, a derivative of Ecca (cf. Eccles, La; Ecclesall Bierlow, West Yorkshire: Ekwall 1960: 159), though the absence of a possessive indicator would be difficult to explain.

**ECCLESHALL** Ancient Parish 7 miles north-west of Stafford (SJ 8329). *? Eccleshale* 1002x1004 (11th century, S.1536), *Ecleshelle* 1086 DB, *Ecclesale* 1227 Ass, *Ecclyshale* 1255 FF, *Ecclesale* 1262 SHC V (i) 136, *Eccleshall* 13th century, *Ekeleshale* 1319 SHC X 28, *Egleshale* 1415 SHC XVII 55, *Ekylsall* 1532 SHC 4th Series 8 102, *Eckleshall* 1577: Saxton, *Eccelshal* 1610: Speed. The second element is OE *halh*, often meaning 'sheltered place', sometimes applied to a small valley or hollow, which would be suitable for this place, which lies in a shallow curve of gently rising ground south of the river Sow (see Gelling 1988: 97), or perhaps here 'land not included in the general administrative arrangement of a region', as proposed by CDEPN. The first element is OE *eles*, from Latin *eclesia*, late British *egles*, Welsh *eglwys* 'a body of Christians, a church', found in three west Midlands place-names, of which this is one, with OE *halh*. The others are the two places called Exhall in Warwickshire (see PN Wa 107-8, 2080, and Eccleshall south-west of Sheffield with a similar derivation. The *eles* element (early OE had no medial -g-, and -c- was substituted for it) is found combined in place-names with a small range of second elements (or alone, as in Eccles, north-west of Chapel en le Frith, and near Hope, both in Derbyshire), but is never itself used as a second element, suggesting that the Anglo-Saxons did not attach a distinguishing name to one of these particular communities or churches, but occasionally used the community or church as a defining term when naming something else. Eccleshall in Staffordshire is an exceptionally large parish at the centre of a composite estate revealed in DB: see Palliser 1976: 47. The bishops of Lichfield held property here from an early date – perhaps as early as the third quarter of the 7th century, during the reign of Wulfhere – perhaps attracted to a place with an existing Christian community, or by a name and tradition recording the earlier existence of such a group. The name suggests that Celtic Christian communities (and perhaps churches) continued into the Anglo-Saxon pagan period, and the possible existence of a pre-English administrative unit, although Ekwall 1960: xxiii observes that *eles* is a little too common to be sure that every name incorporating the element shows the adoption of a British place-name, and suggests that it is not improbable that the British word was adopted into English and used for some time before the word church (OE *cirice*) came into use, with Cameron 1987: 3 emphasising that there is no independant evidence that the British word was ever taken over into colloquial use in OE (see also EPNE i 145). The stress in Celtic was on the final syllable, but in English on the first, so that a change of stress eventually led to Eccles-: Cameron 1996: 32. *Eccleshall Holme* is recorded in 1602 (SHC 1935 IV 423), from ON *holmr* 'a small island, a piece of dry land in a marsh; a piece of land partly surrounded by streams or by a stream', presumably from an island in the river Sow near Eccleshall, or a piece of land enclosed by other watercourses. For a full discussion of names containing *eles* see Gelling 1988: 96-9. For Eccleston Church near Leek see Elkstone.

**ECHELLS** a common minor name, from OE *ecels* 'an addition, land added to an estate by reclamation'. Ekwall 1963: 33-5 cites Echeles near Wombourne (which has not been traced), Drayton Basset and Wolverhampton (see Nechells). See also Etching Hill.

**ECLINGES** (unlocated, possibly near Rowley Regis) *Ecling* 1280 SHC VI (i) 108, *Eclynges* 1281 ibid.' 117, *Eclynge* 1282 ibid. 121, *Eclingh* 1289 ibid. 187, *Eclingge* 1293 ibid. 242. An intriguing name. Perhaps from the OE personal name *Eccel, a derivative of Ecca (cf. Eccles, La; Ecclesall Bierlow, West Yorkshire: Ekwall 1960: 159), with -*ingas* 'people of', hence '(the place of) *Eccles' people'.

**ECTON, ECTON HILL** 3 miles south-west of Sheen (SK 0958). *? Ekedon'* c.1240 TutCart, *Ekeyton* 1293 QW, *Ecton, Brode Ecton* 1559 SHC IX NS 165, *Eekton* 1666 SHC 1925 189. From the OE personal name Ecca, hence 'Ecca's *tun*' (cf. Ecton, Northamptonshire).

**EDEFORD** (unlocated, in the Cannock area, perhaps Hednesford) *Edeford* 1531 SHC 1910 20.

**EDGE HILL** onb the north side of Stone (SJ 8935). *Edge Hill* 1891 O.S.

**EDGELAND** (obsolete) on the south side of Whiston Brook, south-west of Longnor Hall in Lapley (SJ 8614). *Hydesland, Hydeslond, Hidesland,Hideslond* 1263 Ipm, 1265 MRA, 1293 QW, *Hydeslaund* 1280 Ass, *Hydeslaund* 1280 SHC VI (i) 149, *Hysdelunde* 1292 Ch, *Hydeslond, Hudeslond* 1293 SHC VI (i) 247, *Hideslond* 1383 Rees 1997: 172, *Heddesland(es)* 1548 Pat, *Heddesland* 1574 SHC XVII 220, *Hyddeslaunde otherwise Hydgelande* 1576 SHC XIV 188, *Higdland* 1604 PCC, *Hidgland* 1616 StSt 13 2001 51, 1666 SHC 1927 63, 1680 SHC 1919 220, *Hildersland* 1765 SOT D15/10/3/7, *Edgeland* 1834 O.S. Probably from the OE personal name Hiddi (cf. Hedgeley, Northumberland), rather than from OE *hid* 'a hide of land', a word normally found alone (cf. The Hyde), or as a second element. The second element of this name is OE *land*, rarely found as a first element, but common as a second, the precise meaning of which varies from 'estate, landed property', to 'district', and (as perhaps here) 'portion of a village or estate'. The name Hideslonde is also recorded in Oswestry and Ercall Magna in Shropshire (Rees 1985).

**EDGES, THE** (unlocated, in Hanchurch) *? Super-Egge* 1227 SHC XII NS 219, *? super Egge* mid-13th century SRO D593/B/1/23/7/21, *the Edges* 1660x1721 SRO D593/B/1/9/3.

**EDIAL** (pronounced Eddyal [εdɪəl]) 3 miles south-west of Lichfield (SK 0708). *Edichalewode* 1299 SRO DW1734/1/4/3A, *Edysale* 1379, *Edihall* 1416, *Edihale* 1453 SRO DW1734/2/1/603 m41, *Edyalewod* 1474 ibid.' 598, *Edyall, Edihall* 16th century Duig, *Edgiall* 1686 Plot, *Edjall* 1834 O.S. The earliest spelling suggests a derivation from OE *edisc* 'enclosed pasture, a park', with OE *halh*. White 1834: 104 suggests that in the 1830s the place was known as *Edgehill*. The place appears originally to have been centred some distance to the east where Pipe Grange Farm now stands: VCH XIV 201. See also Walsall Wood.

**EDINGALE** 5 miles north of Tamworth (SK 2112). *Ednunghalle, Ednunghale* 1086 DB (listed in Derbyshire), *Edelinghal* 1100x1107 RHP 77, *Edenyghale* c.1170 SHC 1939 27, *Ederingehale* 1191 SHC II 11, *Eadinghall, Ederingehale* 12th century Duig, *Edeling(e)-hale* 1208 FF, *Edenighale* 1259-60 SHC X NS I 273, *Hedenighale* 1269 SHC 1910 122, *Edenyng-hale* 1272 ff. Ass, *Yedyhale* 1457 SHC III NS 222. Possibly an OE *-inga-* name, giving 'the *halh* associated with Edin's or Eadwine's people': cf. Edensor, Derbyshire, 2 miles north-east of Bakewell, which may incorporate the same personal name. *Halh* here may have the meaning 'land in a river bend', or 'promontory into a marsh': CDEPN. In 1086 Edingale consisted of two manors, one of which now forms part of the manor and parish of Croxall, Derbyshire, the other forming part of the parish of Alrewas in Staffordshire.

EDRITHESHURST (unlocated) *Edritheshurst* 1255 (1798) Shaw I xxvii. Probably from the OE personal name Eadþryþ (see Searle 1897: 189), with OE *hyrst* 'wooded hill'.

EDULUESMOR (unlocated, perhaps in the Pipe/Lichfield area) *Eduluesmor* 1292 *Deed.*

EE, RIVER – see **AQUALATE MERE**.

**EFFLINCH** 1 mile south of Barton under Needwood (SK 1917). *Heffallyngelake* 1415 Hardy 1908: 84, *Hethfallynglake* 1415 ibid.' 89, *Effallyngelake* 1415 ibid. 90, *Efflinch* 1812 *EnclA.* Perhaps 'the fallow-land at the heath', from Mercian OE *fælging* 'fallow land', with OE *hǣð* 'heath' and *lacu* 'stream'.

**EGG WELL** at Roost Hill, 2 miles south-east of Leek (SK 0053). Early forms have not been traced, but if ancient (and this may be the un-named well at Ashenhurst mentioned by Sleigh 1883: 106), perhaps from ME *egge* 'edge, ridge', or OE *ece* 'perpetual', so 'the spring that does not dry up'. Or possibly from the shape of the well (which may however have taken its shape from the name of the well), or from the oval stone bearing an inscription which is said to have been placed over the spring by William Stanley of Ashenhurst Hall between 1744 and 1752 (VCH VII 171), or perhaps a jocular association with Roost Hill. It may also be noted that the Egge (later Edge) family are recorded from the 14th century in Horton (SHC 1912 350; VCH VII 71) and Caverswall (SHC X (i) 118): see also TSSAHS XXXVIII 1996-7 56; PN Ch IV xiii.

EGGERTON (unlocated, but possibly Egginton in Derbyshire) *Eggerton* 1416 SHC XVII 57.

**EGGINTON BRIDGE** the old bridge carrying Icknield Street over the river Dove north of Burton upon Trent (SK 2623): see Plot 1686: 400. *pontem de Egintona super aquam de Dove* 1255 BurtAbSurv, *pontem Monachorum* 1330 Ass, *le Munkbrugg'* 1383 *Cor*, *Monkebrigge* 1394 Pat, *Lytulmonkbryge* 1406 *Ch*, *Monkysbryge* 1465 SHC 1939 128, *Monks-bridg* 1686 Plot 400, *Monks Bridge* 1775 Yates. The former name is from OE *munuc, brycg* 'the monk's bridge': Shaw 1798: I 26 states that it was so-named because it was erected by John de Stretton, prior of Burton. The present name is from nearby Egginton in Derbyshire. See also PN Db 459-60.

EID LOW (obsolete) in Wootton Park (SK 0944). *Eid Low* 1836 O.S. A curious name for which no derivation can be offered. The second element is OE *hlaw* 'mound, tumulus'. The name is remembered in Eid Low Plantation.

**ELAND BROOK** a tributary of the river Swarbourn. *Ealandbrook temp.* Henry VI (1798) Shaw I 66, *Ealands Brook* 1650 ParSurv. A back-formation from OE *ea-land* 'land by a river': see Eland Lodge.

**ELAND LODGE** 1 mile south-west of Draycott in the Clay (SK 1427). *Eland Lodge* 1773 SRO D861/T/2/44/1-29, *Eland Lodge & Farm* 1792 PRO (DL31/244), *Ealand Lodge* 1798 Shaw I 67. From OE *ea-land* 'land by a river': the place lies near Eland Brook (q.v.).

**ELFORD** Ancient Parish 4½ miles north-west of Tamworth (SK 1810). *(æt) Elleforda* 1002x1004 (11th century, S.906; 11th century, S.1536), *Elefurd* 1086 DB, *Elleford* 11th or 12th century Sawyer 1979a: xxxvii, *Eyleford* 1286 SHC VI (i) 164, *Elfold* 1286 SHC 1924 334, *Elford, Eleford* 1413 SHC XVII 6, *Elleford* 1413 ibid. 16, *Earleford* 1537 SHC 1931 223, *Elseford* 1562 SHC 1938 123, *Eirleforde* 1567 SHC XVII 216, *Eireleford otherwise Elford* 1601 SHC XVI 208, *Elford* 1833 O.S. Probably 'the ford of

Ælla', a common OE personal name. Before the erection of a bridge here in the early 19th century, the river Tame was also forded at Willyford, 1 mile to the west.

**ELFORD HEATH**  ½ mile west of Eccleshall (SJ 8229). *? Hulleforde, Hullforde* 1351 SHC 1913 144, *? Elmton* 1749 Bowen, *Helford Heath* 1833 O.S. If the spellings refer to this place, seemingly 'the ford by the hill'. It is unclear whether *Wollfordes Marshe*, recorded in 1603 (SHC 1934 (ii) 52), refers to this place.

**ELFORD LOW FARM**  1 mile south-east of Elford (SK 1909). *Elford-low* c.1750 (1798) Shaw I 381, *Elford low Farm* 1834 O.S. From a burial mound (OE *hlaw* 'mound, tumulus') to the north-west of the farm, recorded (and opened) by Plot: see Plot 1686: 405. See also Shaw 1798: I 381.

ELIOTES BRIDGE  (unlocated, on the river Tame or Trent, possibly near Salters Bridge) *Eliotes Bridge* 1612 Alrewas ParReg.

**ELKSTONE, UPPER & LOWER**  6 miles north-east of Leek (SK 0559). *Helkesdon'* c.1175 (p) Okeover, *Olkesdon* 1210-5 SHC XI 331, *Elkesdon* 1227 Ass *et freq.* to 1341 ibid, *Elkesdun'* 1251 Ch, *Elkysdon* (p) 1332 SR, *Ulkeston* 1424 Banco, *Elkyston'* 1491 Ct, *Elkeston* 1486 Ct *et freq.*, *Elkyston* 1515 DRO D2375M/171/1/2, *Elkyston* 1532 SHC 4th Series 8 41, *Eccleston Ch.* 1599 Smith, *Elkerston* 1553 ICG, *Eccleston Ch.* 1599 Smith, *Elkinston* 1604 ParSurv, *Eccleston* 1607 Kip, 1610 Speed, *Elxton* c.1641 DRO D258/30/26, *Eccleston (church)* 1673 Blome, *Elkstone (Upper & Lower)* 1689 Plot. '*Ealac's dun* or hill'. The name Ealac is a variant of *Eanlac. Upper Elkstone is recorded as *Over Elkesdon* 1272 Ass, and Lower Elkestone as *nether Elkysdon* 1286 Banco. Cf. Elkesley, Nottinghamshire; Elkstone, Gloucestershire; North & South Elkington, Lincolnshire. It may be noted that various early maps show Eccleston Church near Leek (e.g. *Eccleston Ch.* 1599 Smith, *Eccleston Ch.* 1607 Kip, *Eccleston Church* 1610 Speed, *Eccleston church* 1673 Blome). The name is a perpetuated mis-spelling of Elkstone.

**ELLASTONE**  Ancient Parish 5 miles south-west of Ashbourne (SK 1143). *Edelachestone, Elachestone* 1086 DB, *Athelast'* 1177 SHC XII NS 278, *Adeloheston, Adelakeston, Edelaghestone* c.1196 SHC II 68, 70 (& index), *Aselacston* 1227 SHC IV 61, *Adlacston* 1236 Fees, *Adthelaxton* 1236 SHC 1911 393, *Athelaxton* 1242 Fees, *Ethelaston* 1327 SHC 1913 12, *Ellaston or Gaston* 1749 Bowen. 'Eadlac's or (though much less likely) *Æþelac's tun*'. The 1749 variant presumably reflects local pronunciation. See also Verdon.

**ELLENHALL**  Ancient Parish 2 miles south-east of Eccleshall (SJ 8426). *Linehalle* 1086 DB, *? Linhale* 1203 SHC III 103, *Ellinhale* 12th century Duig, *Ælinhale* c.1200 DC, *Elynhale* c.1200 *Rees 1985*, *Elinhale* 1242 Fees, 1258 Ipm, *Helinhall* 1243 SHC 1911 396, *Elnal(park)* 1395 SHC 1927 105, *Elynhall* 1531 SHC 1912 46, *Elenall* 1532 SHC 4th Series 8 66, *Elnol, Elnehaul* c.1540 Leland. The regular -i- in the second syllable suggests that the DB form may be correct, with the original name OE *lin-halh* 'flax *halh*', later prefixed with OE *ea* 'river', giving 'flax *halh* by a river', with *halh* here perhaps with the meaning 'tongue of land between two streams': the place lies near a pronounced tongue of land formed by two tributaries of the river Sow.

**ELLERTON GRANGE**  3 miles south-west of Adbaston (SJ 7225). *Ellerton Grange* c.1569 SHC IX NS 94, 1749 Bowen, 1833 O.S. From nearby Ellerton in Shropshire, recorded as *Alarton* 1203, 1212 Bowcock, *Edlarton, Athelarton, Allarton* 1253-60, 1262, 1272 Eyton 1854-60: VIII 93, *Adhelardeston* 1272 SHC 1924 243. 'Æþelheard's *tun*'. The place was a grange of Ranton Abbey: VCH III 253-4.

**ELLISHALL BROOK** a tributary of the Ordley Brook, which flows into Northwood Brook, and via Tit Brook into the river Dove. See Ellis Hill, from which the brook takes its name.

**ELLIS HILL** 1½ miles north-west of Mayfield (SK 1347). *Ellis Hyll* 1538 *AOMB*, *Ellyeshyll* 1539 MinA, *Ellis Hill Farm* 1768 SRO D3437/1/1-30, *Ellis Mill (sic)* 1836 O.S. From the surname Ely or Elly: Richard Ely is recorded in 1538 (*AOMB*), and the name Elly in 1539 (MinA).

**ELLOWES HALL** 2 miles south of Sedgley (SO 9192). *Ellenvale* 1242 Duig, 1273 SHC IX (ii) 29, SHC 1911 156, *Ellenualle, Ellenvalle* 1273 SHC 1911 156, *Elwals* 1527 SHC XI 266, *Ellavales* 1563 Erdeswick 1844: 241, *Elwells* 1736 SRO D3155/WH94, *Ellowes Hall* 1834 O.S. Perhaps from OE *ellern, ellen, elle* 'elder tree' rather than the OE personal name Elle or Ella, although it may be noted that *Ellsbarn Plantation* appears on the 1834 1" O.S. map on the north side of Upper Penn, some 2 miles to the north-west, and the surname Ellewalle is recorded in the same area c.1378: SHC 1928 39. The word *vale*, rare in Staffordshire, is French for 'valley', so not found in English place-names before the Norman Conquest.

ELMDON (unlocated, near Penkridge) *Helemdon* 1215 SHC VIII (i) 137, *Elemedone* 1279 SHC VI (i) 93, *Elmedon* 1292 ibid. 251, 1327 SHC VII (i) 240, *Elmedone* 1308 SRO D938/35, *Elmedon* 1317 SHC 1928 151, 1386 SHC XIII 196, *Elmedon'* 1332 SHC X 120. From *elm-dun* 'hill with the elm or elms'. The name may have been transferred from Elmdon in Warwickshire: the Walter family of Pillatonhall took their name from Walter de Elmedon, of Elmdon in Warwickshire, who held Pillatonhall *temp.* Edward I: SHC XII 101 fn.

**ELMHURST** 2 miles north of Lichfield (SK 1112). *Elmhurst* 1208 SHC III (i) 173, 1259 SHC 1924 207, 1271 SHC V (i) 151, 1283 SHC 1939 90, *Helmhurst* 1269 SHC 1910 122, *Henhurst* 1280 SHC VI (i) 106, *Elinghurst* 13th century Duig, *Elmhurst* 1416 SHC XVII 59, frequently thereafter *Elmhurst, Elmehurst, Elmeshurst* Duig, *Elmehurste* 1532 SHC 4th Series 8 183. OE *elm-hyrst* 'elm-copse'.

**ELMHURST** 3½ miles north-west of Stafford (SJ 9029). *Elberson Farm* 1832 Teesdale, *Elmerson* 1836 O.S., *Elmstone* 1890 O.S. Probably to be identified with *Elmershull* 1252 SHC VIII (i) 200. From the late OE personal name Ælmer (representing earlier Ælfmer or Æþelmer), although the change in the ME terminal *hull* 'hill' is unusual, and also the further change to *hurst* 'wooded hill'. JNSFC XLII 1907-8 119 states: 'near the summit [of Pirehill] is a place formerly known as Elmstone'.

ELMSTONE – see ELMHURST, near Stafford.

**ELSTON** to the west of Bushbury (SJ 9102). *Alleston* 1527 (1801) Shaw II 182, *Elston Hall* 1694 Bushbury ParReg, *Elston or Ailston* 1798 (1801) Shaw II 182, *Elston Hall* 1834 O.S. The first element is evidently a personal name, so perhaps 'Ælf's *tun*'. See also Erdeswick 1844: 347.

EMMOTTES HAY (unlocated, in Cheadle) *Emmottes Hey* 1599 SHC 1935 107.

**ENDON** 4 miles south-west of Leek (SJ 9253). *Enedvn* 1086 DB, 1227 Ass, *Enedon* 1252 Ch, *Henedon* 1274 SHC 1911 160, *Hennedon* 1298 SHC XI NS 253, *Henedun* 13th century Duig, *? Yenn* 1586 SHC 1927 135, *Endon otherwise Yondon* c.1621 SHC VII NS 234, *Yen* 1665 Leek ParReg, *Endon or Enerden* 1747 Bowen. The first element is possibly OE *ened* 'a duck' (although more frequently found in place-names in combination with elements for 'water'), but more likely to be OE *\*ean* 'lamb' (which corresponds to Latin *agnus*, and is a word found only in place-names, unrecorded in any Germanic language except in derivations, e.g. *eanian* 'to lamb', *(ge)ean* 'with lamb'), or from the personal name Eana (cf. *eanan-dun* in a charter of 1003 (c.1300, S.1664) relating to Bengeworth, Worcestershire: see Hooke 1990: 344-6), with the second element *dun* 'hill': the place lies at the end of a pronounced hill, and the earliest settlement may have been in the area of Endon Bank: VCH VII 177. The medial *e* in the early spellings rules against a derivation from OE *ende* 'end'. It is said that in the early 19th century the name was pronounced Yan by local people (GM 1829 II 28-31). The spellings suggest that Yen or Yan was an alternative pronunciation from at least the late 16th century.

**ENDON BROOK** a tributary of the river Churnet. *Yendon River* 1577 Saxton, 1689 Lea. Named from Endon village.

**ENGLETON** 1 mile north-east of Brewood (SJ 8910; SMR 01906 places a deserted post-Conquest settlement at SJ 89891000). *Engelton* 1152x1155 (17th century) EEA 17 60, *Hengleton* 1204 SHC III (i) 90, *Engleton* (p) 1206-30 StCart, *Hengleton* 1289 SHC VI (i) 184, *Engleton* 1322 SRO D(W)1721/2-3, *Engulton* 1330 Coram, 1391 FF, *Engletone* 1381 SHC XVII 183, *Ingleton* 1478 Coram R, *Inggylton* 1480 SHC VI NS (i) 126, *Yengylton* 1491 Hatherton, *Engleton* 1749 Bowen. From OE *Engle* 'the Angles', later in OE meaning 'the English', with OE *tun*, hence 'the *tun* of the Angles', perhaps referring here to an isolated settlement or group of Angles in a mainly British area. The name Brewood (q.v.) is held to form part of the evidence for the existence until well into the Anglo-Saxon period of a British community here. A Roman villa lay by the river Penk west of Engleton Hall Farm: see SHC 1938 267-93.

ENGLONDFEILD (unlocated, perhaps in or near Bishop's Wood, west of Bishop's Offley) *Englondfeild* 1609 SHC III NS 27.

ENDDELEY (unlocated, in Morfe Forest) *Enddeley in foresta de Morff* 1414 Kimball 1959: 107.

**ENSON** 4 miles north-east of Stafford (SJ 9428). *Hentone* 1086 DB, 1272 FF, *Eneston* 1275 Ass, *Enestan* pre-1290 SRO (393/7999), *Enson, Henestone, Enestone* 13th century Duig, *Enston* 1433 SHC XVII 147, *Henstone* 1471 SHC IV NS 172, *Enstone* c.1598 Erdeswick 1844: 48. Possibly '*Ean's tun*', with the late disappearance of *-t-*.

**ENVILLE** Ancient Parish 5 miles west of Stourbridge (SO 8286). *Efnefeld* 1086 DB, *Cuniefeld* (*sic*) 1152 (14th century) EEA 14 96, *Evenesfeld* 1175-6 SHC I 78, *Euenfeld* 1183 P, *Efnefeld* 1194-1203 SHC III (i) 217, *Eveniford* 1206 SHC III (i) 138, *Inefeld* 1207 SS lxxxiv 3, *Evenefeud* 1240 FF, *Evenefeld* 1332 *et freq* SHC 1913 39, *Evil* 1610 Speed, *Envile* 1660 (1801) Shaw II 272, *Endfield alias Envield* 1770 (1801) ibid. 272. The first element is the OE adjective *efn* 'even, smooth, flat, level', hence 'flat *feld* or open land'. The *-ville* substitution is 'partly due to a native change of *f* to *v*, partly to popular etymology': Ekwall 1960: xxix (cf. Evenwood, Durham). The place-name is particularly appropriate for the flat valley-bottom overlooked by the church here.

**ERCALL, HIGH** 1 mile south-east of Sedgley (SO 9292), *High Arcall* 1701-25 Sedgley RentRolls, 1813 Himley ParReg, *High Ercall* 1834 O.S.; **HIGH ARCAL WOOD** (obsolete) on east side of Himley Park (SO 9091), *High Arcal Wood* 1834 O.S. Both names, some two miles apart, are probably relatively recent, adopted in a transferred sense from High Ercall in Shropshire (see PN Sa I 124-6; Gelling 1992: 72), but see also Archill Brook, with which these names may be associated. High Ercall Farm is said to have been formerly known as The Flaxhall (*? Flaxle* 1307 SHC VII (i) 173, *Flaxhale* 1273 SHC 1911 155; Hackwood 1898: 11).

**ERCHENEBRUG** (unlocated, near Clayhanger) *Erchenebrug* 1300 SHC V (i) 177.

**ERDYNTON** (unlocated, in Stafford) *Erdynton* 1436 SHC XII 312.

**ESKEW BRIDGE** 1 mile north-west of Brewood (SJ 8801). *Eskew Bridge* 1834 O.S. 'The bridge which crosses at an angle'. Skew Bridge, on the south-east side of Forton (SJ 7520), was a combined road bridge and aquaduct: VCH IV 103; see also Robinson 1988: 80.

**ESSEX BRIDGE** A narrow stone packhorse bridge of with 14 arches, 100 yards long and 4' wide, over the river Trent at Shugborough (SJ 9922). *Shutborrowe Bridge* 1647 Levy. The bridge was known as Shugborough Bridge or Haywood Bridge until the 19th century. There is no evidence to support the popular tradition, seemingly dating only from the 19th century (SHC 1970 90), that the bridge was so-named because it was built by the county as a compliment to the last Devereux, Earl of Essex, of Chartley. Chetwynd, writing c.1679, says that there was 'a Wooden Bridge, which being ruinous was in ye last Age rebuilt wth stone, & contains 43 Arches': SHC 1914 129. A levy in 1647 includes 'ffourescore pounds for building of 16 Arches of Shutborrowe bridge': AJ 120 1963 287. See also Erdeswick 1844: 72. Pennant 1782 remembered 42 arches; the 14 remaining still make it the longest packhorse bridge in England.

**ESSEX FALL** (obsolete) ½ mile south-east of Bridgnorth (SO 7392). *Essex Fall* 1739 TSAS IX 1886 196, 1752 Rocque. It is said that in the early 18th century there was a tradition that the place was named after Henry Bourchier, the young Earl of Essex, or one of his men who fell and was killed here in 1540: TSAS IX 1886 196; Watkins-Pitchford 1932: 22-3..

**ESSINGTON** 4½ miles north-west of Walsall (SJ 9603). *Esingetun* 996 (for 994) (17th century, S.1380), *Eseningetone* 1086 (DB, listed also in Warwickshire), *Esenington* 1227 Ass, 1240 FF, *Eselington'* 1236 Fees, *Esynton* 1238 Lib, *Eseningeton* c.1250 SHC 1950-1 15, *Esenitigeton, Eseningent* 1255 (1798) Shaw I xviii, *Essington* 1271 For, *Essinton* 1279 Ass. 'The *tun* of the family or followers of a man called Esne'. Esne was a common OE personal name, and although it had various meanings, including 'servant' and 'young man', was often borne by men of high rank: an ealdorman of that name is a witness to 16 charters between 764-802 (Featherstone 2001: 32). Essington is a rare Staffordshire example of an *-ingatun* name: see Gelling 1988: 178. The place lies on high ground of over 600'. It may be noted that the DB entry for this place is repeated similarly (though not exactly, as suggested in VCH Wa I 332 and Darby & Terrett 1971: 164) in folio 243 of the Warwickshire section: VCH IV 55. Cf. Isombridge, Shropshire (PN Sa I 164).

**ESSINGTON HOUSE FARM** on the north side of Alrewas (SK 1715). Perhaps to be associated with *Esstinton* 1260 SHC X NS I 274, *Estintun* 1260 ibid. 280, *estington (field)* 1617 Alrewas ParReg, *Essington meadow, Essington field* c.1750 TSSAHS XX 1978-9 loose map. Possibly from OE *east* 'east, with OE *hean-tun* '*tun* situated on high ground', perhaps associated with *Hayntun* recorded in 1259: SHC X NS I 267.

**ESTMORE** (unlocated, perhaps near St Thomas's priory, Stafford) *Estmore* 1346 SRO D938/10. 'The moor to the east'.

**ETCHING HILL** a 454' hill topped by a sandstone outcrop 1 mile north-west of Rugeley (SK 0318). *Echulhul* 1408 SRO DW1781/6/1/7, *Eychilhill* 1504 VCH IV 54, *Echynge hill* 1554 SRO DW1734/2.3.43, *le Echin, Ichinhill* 1584 Comm, *Eaching Hill*

1678 SHC 1927 13, *Itching Hill* 1698 Fiennes, *Hitching Hill* 1798 Yates, 1834 O.S. Probably from OE *\*ecels* 'an addition, land added to an estate', a common element in the North Midlands: cf. Etchells, Cheshire and Derbyshire; Hichells, Yorkshire.

**ETHNESDICH** (unlocated, possibly in the Blithfield area) *Ethnesdich* c.1250 SRO D986/41. An interesting name, possibly from OE *hæðen*, *dic* 'heathen's (i.e. Dane's) ditch', or from the ON female name Eithne (see Cheney 1970: 132).

**ETINTON** (unlocated, near Lichfield) *Etinton* 1679 SHC 1914 125.

**ETOCETO, LETOCETUM** – see **WALL**; **LICHFIELD**.

**ETRURIA** in Burslem, 1 mile north-west of Stoke-on-Trent (SJ 8647). *Etruria* 173 BLARS L30/14/315/2, 1775 Yates, 1836 O.S. The village was created in 1769 by Josiah Wedgwood (d.1795) for workers from his ceramic factory and so-named in allusion to Etrurian (Etruscan) style pottery produced here from the second-half of the 18th century. Wedgwood's own house, built in the 1760s, was called Etruria Hall. The place was evidently called *Ridghouse* when acquired by Wedgwood: Erdeswick 1844: 20.

**ETTINGSHALL** 2 miles south-east of Wolverhampton (SO 9396). *Ettingeshale* 996 (for 994) (17th century, S.1380), *Etinghale* 1086 DB, *etingeshal'* c.1155 SHC 1941 61, *Ettingehal* 1175, *Etinghale* 1196 P, *Ettingeshale* 1261 FF, *Ettingestal* 1295 TSAS 3rd Series IX 1909 33, *Eltyngeshal Park* 1322 SHC 1911 351, *Odynsall* 1532 SHC 4th Series 8 115. From OE *\*etting* 'a pasturage, a grazing-place', with OE *halh*, or '\*Etting's *halh*', from an unrecorded personal name. The *-ing* formation is compounded in the genitive singular: see PN Ch V (ii) 281.

**EWDNESS** 3 miles north-west of Worfield (SO 7398). *Hendinas* 13th century Misc, *Ewdness* (p) 1315 Jury List, *Eudenas* 1327 SR, *Eudenas* 1360 AD iv, *Hewdeness* 1532 SHC 4th Series 8 117, *Yewdnes* 1597 Worfield ParReg, *Eudnes* c.1640 SA 5460/7/3/7, *Ewdness* 1752 Rocque. According to Ekwall 1936: 202 from Welsh *hyn* (feminine *hen*) 'old', and *dinas* 'city, fortification', where the correct Welsh form would be *Henddinas*. The spelling *Ew-* for *En-* is explained by the misreading of *–n-* as *–u-*. See also Coates & Breeze 2000: 326, where this derivation is categorised as 'confident'. However, that explanation does not seem entirely satisfactory on philological or archaeological grounds, and the name may well be English (see Gelling 1976: 204; 1979: 114), perhaps OE *eow wudu* 'yew wood', or OE *eow dun* 'yew hill' (cf. Eudon, Shropshire), with the addition of OE *ness* 'a nose, a promontory': the place lies on a pronounced headland. In Shropshire since the 12th century.

EYESWALL (obsolete) a former moated house on the west side of Eccleshall (SJ 8229). *Uireswell* 1265 SHC NS III 83, *Uselwalle* c.1270 SHC 1914 52, *Uselwalk* 1276 SHC NS III 84, *Ulshale* 1281 SHC 1911 36, *Ulsale* 1306 ibid.' 66, 1320 SHC IX (i) 79, *Isewell* 1314 SHC NS III 91, *Iselewall* 1323 ibid.' 93, *Isewold* 1568 SRO D260/M/T/5/138, *Illsalls* 1618 SHC VI NS (i) 59. By tradition from 'a little perrenial spring traditionally said to be good for sore eyes' (a belief probably associated with a folk etymology attributing the name to the word 'eyes') which fed the moat to the old house (SHC III NS 82 82), but the spellings suggest a derivation from OE *usel-* or *ysel-*, for which no explanation can be offered, with Mercian OE *wælle* 'a spring', and (sometimes) 'a stream'. See also WMANS 25 1982 65-6. The name is remembered as Usulwall Close in Eccleshall.

**FAIRBOROUGHS** 1 mile south-east of Heaton (SJ 9560). *Ferreborowes* 1291 (1798) Shaw I xxviii, *Feyreybrowes* c.1291 Tax, *Fajreborowes(weye)* 1318 SHC 1911 433, *the*

*Fairbreders, the Feyerbyroughs* 1532 StarCh, *Feirebarous* 1532 SHC 1912 57, *Fitzboredz* 1539 MA, *feirebarons temp*. Henry VIII StarCh, *Feburroughes* 1641 Leek ParReg, *Faerborough* 1663 StV, *Feaborrow* 1695 Leek ParReg, *Fairboroughs* 1842 O.S. From OE *fæger* 'pleasant' with OE *beorg* 'hill', so 'the pleasant hills'.

**FAIRFIELD** ½ mile south-east of Bradnop (SK 0154). *Fayerfeildes* 1596 Okeover T697, *Fairfields* 1699 ibid. T757. From OE *fæger* 'fair, beautiful', so 'the beautiful fields'.

**FAIRFIELDHEAD** – see **FAWFIELD HEAD**.

**FAIROAK** 2½ miles south of Ashley (SJ 7632), *the fayre oke* 1553 SHC 1926 30, *Feare Ocke* 1600 Eccleshall ParReg, *Fayr Ocke* 1609 *et freq*. ibid, *Fairoak* 1686 Plot; FAIR OAK (obsolete) 2 miles north-east of Huntington (SJ 9815), *Fair-Oake* 1686 Plot, *Fair Oak* 1834 O.S. 'The fair or fine oak-tree'.

**FALLINGS PARK** 2 miles north of Wolverhampton (SJ 9200). A 'garden city' housing development created in the later 1920s in the area long known as Old Fallings was given the name Fallings Park. *Olde Falinge* 12th century Duig, *Oldefalling* 1271 SHC 4th Series XVIII 93, *Oldefallyngge* 1286 SHC V (i) 173, *Holdefallinge* 1332 SHC X 126, *Oldefallyngh* 1342 SHC 1913 93, *Oldefallynges* 1350 SHC 1928 78. Probably from Mercian OE *fælling* 'a felling of trees, a clearing', or from Mercian OE *fælging* 'a piece of ploughed or newly cultivated land', sometimes used in the region as a measure of land: see Ekwall 1923: 23. The place was within Cannock Forest, close to its western boundary. Fallings Heath is 1 mile north of Wednesbury, but the history of the name has not been traced.

**FAR BROOK** a tributary of the river Dane.

**FAREWELL** Ancient Parish 2½ miles north of Lichfield (SK 0811; SMR 02074 places a deserted post-Conquest settlement at SK 08301170). *Fagrovella* 1129x1147 (17th century) EEA 14 19, *Faurwelle* 1149x1159 (17th century) ibid. 54, *Fagerwell* 1200 Ch, *Faierwelle* 1200 SHC II 91, *Faurewell* 1251 Cl, *Faverwell* 1261 SHC IV 148, *Fayrwell, Fagerswell, Fagerwelle, Farewell* 13th century Duig, *Feirwalle* 1322 BCA MS3415/152, *Faurwell, Fagrovella* 1375 Dugd, *Ferwall* 1532 SHC 4th Series 8 111, *Farwal* 1561 HLS, *Farroll otherwise Farwall* 1586 SHC XV 168. 'Clear spring', from OE *fæger* 'clear, beautiful, pleasant', and Mercian OE *wælle* 'a spring', and (sometimes) 'a stream': a strong spring lies on the west side of the church here: see StEnc 211. There is a Farewall (*Far Wall Wells* 1836 O.S.) 1 mile north-east of Waterhouses (SK 1051) which probably has the same derivation, but early spellings have not been traced. A priory was founded at Farewell near Lichfield in 1129-48: VCH III 223.

**FARGELOW** 1 mile east of Alton (SK 0842). *Fargelow* 1836 O.S., 1877 SRO D240//K/14a & b, 1891 O.S.

FARLAWE, FERLAWE (unlocated, possibly in the Tamhorn/Hints area) *Ferlowe* 1199 SHC III (i) 54, *Farlawe* 1202 ibid. 75, *Fernlawe* 1204 ibid. 95, *Ferlawe* 1208 ibid.' 100. From OE *fearn-hlaw* 'the ferny mound or tumulus'

FARLEHAM (unlocated, possibly near Darlaston near Stone) *Farleham* ? 13th century SHC VI (i) 8, 10, 21.

**FARLEY** in Alton parish, 4 miles north-east of Cheadle (SK 0644), *Fernelege* 1086 DB, *Farley* 1274 SHC 1911 160, 1646 SHC 4th Series I 259, 1686 Plot, *Farlegh* 1297 SHC VII 48, 1311 SHC 1911 303, *Farleg* late 13th century SRO 3764/93, *Farylee* 1309 ibid. 301, *Farleye* 1327 SHC 1913 9, *Fernelay* 1339 ibid. 77, *Fernelay* 1339 SHC 1913

77, *Fayrley* 1572 SHC XIII 291, *Pharley* 1666 SHC 1925 201, *Farley* 1706 SRO D240/A/2/13/1-3; **FARLEY FARM** 1 mile north of Great Haywood (SK 0024), *Farley* 1775 Yates, 1836 O.S. From OE *fearn-leah* 'the ferny *leah*'.

**FARMCOTE** 1½ miles south-west of Claverley (SO 7891). *Farnecote* 1209 Eyton 1854-60: III 96, *Farnecote* (p) 1255, *Farncote* (p) 1326 (both from Bowcock 1923), *Farncote* 1507 (1801) Shaw II 208, *Fornecott* 1532 SHC 4th Series 8 125, *Farmcote* 1836 O.S. 'Fern cottage', from OE *fearn cot*. In Shropshire since the 12th century.

FARNAM (unlocated, possibly not in Staffordshire) *Farnham* 1240 SHC 1911 8, 1266 SHC IV 161, 1332 SHC X (i) 114, *Farnam* 1274 1911 162.

**FATHOLME** on the river Trent 1 mile south-east of Barton-under-Needwood (SK 2017). ? *Flatteholme* c.1535 SHC VI (ii) 166, *Fat Holme* 1834 O.S., *Fatholme* 1832 Teesdale. If the earliest spelling (indexed in SHC VI (ii) as *Flatterholme*) relates to this place, which is likely, the first element is probably ME *flat* (from ON *flöt*) 'flat', or possibly OE *fleot* 'a stream, a creek': otherwise perhaps OE *fætt* 'fat' or OE *fæt* 'vat, vessel, jar, cup', used in some topographical sense, with ON *holmr* 'an isle, a small island, a water-meadow'. The place lies on the west side of the river Trent. Flat Holme in Somerset is held to be 'the island of the fleet' (from ON *floti*, OE *flota*), commemorating the use of the island as a base by Viking fleets (Ekwall 1960: 181). Such a derivation can be ruled out for Fatholme, although the place lies less than 10 miles from Repton, also on the river Trent, where the Danes wintered in 873-4.

**FAULD** 2 miles west of Tutbury (SK 1828). *Felede* 1086 DB, *Falede* 1236, 1242 Fees, *Falete* 1252 Rolls, *Fauld, Fald, Feld, Felde* 13th century Duig, *Falde* 1326 SHC 1911 107, *Fawlde* 1619 SHC NS VI (i) 51, *Fald* 1686 Plot. From OE *falod* 'a fold, enclosure for animals'. *Falde Holmes*, evidently near Fauld, is recorded in 1619 (SHC NS VI (i) 51). *Holmes* is from ON *holmr* 'a small island, a piece of land by a stream': Fauld lies near the river Dove. Cf. Fould.

**FAWFIELDHEAD** 1½ miles south-west of Longnor (SK 0763). *Fanfeld* (? for *Faufeld*) 1308 SHC XI NS 257, *Faufilde hill, Fawfeild Hill* ?c.1555 SHC 1910 72-3, *Fawfeild hill Heydde, Fawfeild Hedd, Fawfeild hill, Fawfeld Head* 1571 SHC 1931 130, *Fawfieldhead Hill* 1603 Alstonefield ParReg, *Foefield Heade* 1623 SHC NS X (i) 56, *Fawfieldgreen* c.1632 VCH VII 28, *Fawfield Way* 1651 Rental, *Fairfield Head* 1695 Morden. Since 1679 often recorded as Fairfieldhead, but which has now reverted to its older form, which probably derives from OE *fah* 'variegated', with variously OE *hyll* 'hill' and OE *heafod* 'head', usually in place-names 'the top or end', both meaning 'the elevated place with the multi-coloured open land'. It may be noted, however, that Fawfieldhead lies on the south side of the river Manifold, with Fawside (q.v.) to the north, and Fawfieldhead may, notwithstanding the differing early forms, share the same root as Fawside.

**FAWLEY** 1 mile north-east of Hamstall Ridware (SK 1220). *Falday (park) temp.* John (1798) Shaw I 155, *Faldhay* 1798 Yates, *Fawley* 1798 Shaw I 155. Curiously, Shaw seems uncertain whether the earliest spelling relates to this place, although *Faldhay* appears on Yates' map in Shaw 1798: I. Perhaps from OE *falod* 'fold', with Mercian OE *(ge)heg*, so 'the enclosure with the animal fold'.

**FAWSIDE** 1 mile south-east of Hollinsclough (SK 0765). *Foosyde* 1419-21 DRO D2375M/1/1, *Foe Syde* 1599, 1601 Alstonefield ParReg, *fforside* 1626 *Rental, Foo syd* 1662 Alstonefield ParReg, *Fawside* 1775 Yates. This curious name is unexplained: a derivation from OE *fah* 'variegated' (see Fawfieldhead, for a *Foe-* spelling; cf.

Vowchurch, Herefordshire for *Fowe-* spellings; cf. Fawside, Durham), is not supported by the forms (but cf. Foden, Cheshire: PN Ch I 198). A derivation from a cognate of OBreton *fau, fou* 'beeches' (from Latin *fagus* – Faou occurs as the name of a stream in Brittany: Ekwall 1928: 164; 1960: 185; cf. River Fowey, Cornwall), can probably be ruled out, since beeches are not found this far north until well into the second millennia, so possibly from OE *foh* 'measurable, moderate', here in the latter sense. However, the expression *foo cragge* is found in the 14th-century poem *Sir Gawain and the Green Knight*, which has been translated as 'forbidding crag' (Elliott 1984: 146), though the authority is unclear. The second element is OE *side* 'slope of a hill, especially one extending for a considerable distance'. Fawside lies under the point of a sharp projecting nose with steep sides above the river Manifold, with Fawside Edge 1 mile to the west. It is not impossible that Fawfieldhead (q.v.), which lies opposite Fawside on the south side of the river Manifold, shares the same root as this name (which might conceivably be a pre-English name of the river Manifold), notwithstanding the disparate forms. See also Fole.

**FAZELEY** 1 mile south of Tamworth (SK 2001). *Faresleia* c.1120 (1801) Shaw II 2, *Faresleia, Farisleia* c.1142 Dugd, *Faresleye* 1269 SHC IV 175, *Faresleye* 1335 Ch, *Faresleye* 1357 SHC XII 162, *Fareslee* 14th century Duig, *Faresley* 1532 SHC 4th Series 8 23, *Faseley* c.1540 Leland, *Phaseley* 1590 SHC 1930 76, *Faseley* 1686 Plot 399. The first element is OE *fearr* 'bull, ox', with OE *leah*, so 'the *leah* with the bulls'.

**FEATHERSTONE** 3 miles north of Wolverhampton (SJ 9405). *Feoþer(e)stan* 996 (for 994) (17th century, S.1380), *Ferdestan* 1086 DB, *Federestan* 1187 SHC I 131, *Fetherston* 1271 Duig, *Fetherstan* 1280-90 Wodehouse, *Ferestan* 1292 SHC VI (i) 271, *Feverstan* 1292 ibid. 260, *Fayrston* 1292 ibid. 283, *Fethurstone* 1395 SHC XV 68, *Foderstone* 1414 SHC XVII 9, *Federstonne* 1506 SHC 1928 115, *Federston* 1532 SHC 4th Series 8 73. Seemingly from OE *feoðer-* 'four', used only in compounds, in this case with OE *stan* 'stone', generally held to mean 'the four stones or tetralith', and assumed to refer to a prehistoric cromlech of three uprights and a capstone (see, for example, Ekwall 1960: xxxi), although there is no evidence of any such monuments here or at Featherstone in Yorkshire (West Riding), and two stones are (or were) at *Fether stones* in Featherstonehaugh in Cumberland: Gough 1806: II 445. If this derivation is correct it is noteworthy that the forms give *stan* in the singular (unless a complete monument was called a *four-stone*, on which cf. Fourstones, Northumberland, *Fourstanys* 1236, *Fourestanes* 1256, apparently referring to a tetralith, with the plural *stanes*). The local stone in this area is sandstone, which is easily worked but which does not weather well. Notwithstanding the earliest spelling (which is in any event from a corrupt copy of the original), it seems possible that the name may be from OE *feðer* 'feather', meaning 'the feather-shaped standing-stone'. It may be noted that all the places named Featherstone appear to lie on or near Roman roads or Roman sites. *The Four Stones* is ½ mile north-west of Clent (cf. Fourstones, Northumberland), from OE *feower* 'four'. An area on the edge of Cannock Chase was evidently called *the Fethersty* in 1292: see SHC VI (i) 296. Cf. *the Fetstone in the Mosse* (in Quarnford) 1564: SHC 1938 99, possibly to be associated with *Four Stones* recorded in 1720; *4 Stones* recorded in 1730 (Alstonefield ParReg). Cf. Feasby, Yorkshire; Featherstone Castle, Northumberland.

**FEENIE LEE** ½ mile north of Swinscoe (SK 4813). *the Finayeis temp.* Edward I Okeover T48, *Finhas temp.* Edward II ibid. E191, *Ffyneshas* 1324 ibid. T51, *Feenie Lee* 1885 O.S.

**FEGG HAYES** 1 mile north-east of Tunstall (SJ 8753). *Faghays* 1836 O.S. Halliwell 1850: 351 gives *feg* to mean 'rough dead grass', Jackson 1879 as 'long, rank grass', and Wilson 1974 as 'grass', from ME *fogge* 'rank grass'. *Hay* is from Mercian OE *(ge)heg* 'a clearing or enclosure'. See also Fenneshay.

**FEIASHILL** 1 mile south of Trysull (SO 8492). *Fiershill Farm* 1722 VCH XX 187, 1775 Yates, 1834 O.S., *Feersale* 1774 SRO 466/M/21, *Fearshall* 1895 O.S., *Feiashill* 1908 O.S. Possibly from OE *fyrs* 'furze' (Gorse Lane runs off Feiashill Road), or perhaps OE *fearr* 'bull', with OE *hyll* 'hill', or OE *halh*.

**FELTHOUSE** 1½ miles south-west of Grindon (SK 0753), *Felthouses* 1327 SHC VII (i) 221, 1576 FF, *Felhouse* (p) 1333 SR, *Felhouse* 1333 SHC X 114, *Felthouse* 1660 SRO D924/7/3, *Felt House* 1838 O.S.; **FELTHOUSE (FARM, LANE & WOOD)** 1½ miles south-east of Cheddleton (SJ 9750), *felthous* 1327 SHC VII (i) 217, *Felt House* 1700 ParReg, 1837 O.S.; FELTHOUSE WOOD (obsolete) on the north side of Bagot's Park (SK 0828), *Felthouses Wood* 1836 O.S. OE *felt* occurs as the first element of some plant-names, such as *feltwurma* 'wild marjoram', and *feltwyrt* 'wild mullein', and there may well have been a plant-name derived from *felt*, for example *felte*. There is a slight possibility that such word may be found here, but the word in combination with *-house(s)* indicates that the places are likely to have been the site of felt-making, so from OE *felt hus*, which appears to have been a standard compound meaning 'a house or building where felt was made'. Felt was created from compacted wool fibres, often by water-powered hammer-mill, and was used mainly for hats. The trade seems to be unrecorded before the 12th century, and was not widespread before the 16th century: Blair & Ramsay 1991: 343. Cf. Feltham, Middlesex; Feltwell, Norfolk. See also Fields Farm.

**FELTYSITCH** on Morridge (SK 0359). For the first element see Felthouse. The second element is OE *sic* 'stream, watercourse': the place lies on the river Hamps close to its head.

FEMLEY PITS (obsolete) 1 mile south of Lichfield (SK 1107). *Fedmelya* 1176x1182 EEA 16 62, *Fedmeleya* 1252 SHC VI (ii) 117, *Filumleye* 1314 SHC IX (ii) 44, *Odmoleges* 1473 (1798) Shaw I 306, *Fedemeleys* c.1535 ibid. 166, *Femley Pitts* 1736 SRO D15/11/14/34, *Femley Pits* 1834 O.S. A curious name for which no derivation can be suggested, other than OE *faðm*, ME *fedme* 'fathom': for *fodome* used in this sense temp. Henry VIII see SHC VIII (ii) 123, though the precise meaning is uncertain. See TSAS 3rd Series 1901 147-50 and 282 for an implausible theory associating this area (Footherley) with the battle between Ceawlin and Cutha and the Britons in 584 AD at *Fethan leag* (on which see Stenton 1971: 29-30).

**FENNEL PIT FARM** in Essington (SJ 9503). *Fennallspitts* 1514 *Vernon*, *Fenalls pytte* 1590 ibid. From OE *fenn, wælle* 'the spring at the fen or marsh', with *pits*. The name is associated with *Fenwalle* 1312 *Vernon*, *Fen(n)wallemor*, *Fenwallehal*, *Fen(n)wallehul'* 1296 ibid, *Fenwallefurlong* 1321 ibid, *ffenell flatte* 1635 *Survey*, *Fennel(l)moores* 1639 *Vernon*. Possibly from OE *fenn* 'fen, marsh' (a rare element in Staffordshire), with Mercian OE *wælle*, usually in the West Midlands 'a spring', sometimes 'a stream', and (rarely) 'a well'.

FENNY LODGE (unlocated, perhaps near Sinai Park: *Fenny* may be a particularly corrupt spelling of Sinai due to the misreading of a manuscript form) *Fenny lodge* 1607 Kip, *Fenny lodg* 1610 Speed. It is not uncommon for early maps to perpetuate errors.

FENSADE (unlocated, adjoining Knutton) *Fensade* 1227 SHC XI NS 240.

FENSAY (unlocated, near Abbots Bromley) *Feneshay* 1218-23 SHC 1937 35, 1332 SHC X (i) 93, *Fenneshai, Fenneshay* c.1225 SHC XI NS 88, *Fenneshay* 1227 SHC IV 64, 1327 SHC VII (i) 231, *Feneshay* 1332 SHC X (i), SHC V (i) 46, *Fenneshay at the Queche* 1376 SRO DW1721/3/23/9, *le Fensay* 1402 SHC XI NS 208. The first element of the name may be an unidentified personal name; *hay* is 'a clearing or enclosure', from Mercian OE *(ge)heg*. A document dating from the early to mid 13th century records '...Fenneshay on sides of Middlehay bro[c], in Ashbroc...': SRO D986/42. Ash Brook flows from the north-east to the south-east of Abbots Bromley into the river Blithe, and Ashbrook Lane runs east from Abbots Bromley. The suggestion that Fenneshay lay south-west of Bagot's Park in PMA 31 1997 31 is likely to be incorrect, as is the identification of this place with Fegg Hayes (q.v.) in SHC IV 64. See also JNSFC XLIII 1909 145. The 1376 form shows that the place lay near Squitch House (q.v.), but other evidence suggests that it may have been near Dunstal (SRO D(W)1721/3/15/3). *Caremon fennshay* is recorded in the Abbots Bromley area in 1436 (SRO DW1721/3/32A/8), and *Caremonsfenshay* in 1471 (SHC XI NS 194), perhaps incorporating the OE personal name Carman. *Le Fennes*, which seems to have been between Abbots Bromley and Uttoxeter, is recorded in 1291 (SRO DW1733/A/2/38). A place called *Fenneshay* may have existed near Beam Hill on the west side of Burton upon Trent: SHC 1937 35. See also Burne 1915: 11.

**FENTON** Stoke-on-Trent (SJ 8944). *Fentone* 1086 DB, *Fentona* 1250 Fees, *Fenton* 1271 SHC 1914 8, 1274 SHC 1911 161, *Fenton Kylvert* 1276 SHC VI (i) 79, *Culverdesfenton, Fenton, Fenton-Vivian, Fenton-Culvart* 13th century Duig, *Fenton-Kylward* 1478 SHC NS (i) 119, *Fenton-Kilward* 1588 SHC XV 185, *Fenton Culwarte* 1589 SHC 1934 7, *Fenton Vivian otherwise Litell Fenton* 1593 SHC 1930 359. From OE *fenn* 'fen, marsh' (a rare element in Staffordshire), and OE *tun*. Culverd is a local family name: see Culverdislow. The 1588 form incorporates a corrupt version of Culverd. In 1241 possession of Fenton was given by the king to Vivian of Standen: SHC 1914 8; VCH VIII 211. See also SHC 1911 394; SHC 1914 8-9.

**FERNHILL** 1 mile north-east of Forton (SJ 7522). *Fernyhale* 1229 to 1232, 1245 to 1265 Deeds (Tw), 1334 Deed (VCH IV 107). 'The fern covered *halh*'.

**FERNYFORD** 2 miles north-west of Warslow (SK 0661). *Ferniford* (p) 1327 SR, *Fernyforde* 1414 SHC XVII 17, *Ferneford* 1441 SHC II NS 156, *the Fernie Forde* 1589 Alstonefield ParReg, *Fernyford* 1592 SHC 1930 (ii) 342, *Ferny Ford* 1840 O.S. Self-explanatory. The place lies on Blake Brook.

**FERNY HILL** 1½ miles east of Cheddleton (SJ 9952). *Ferneyhill* 1600 PCC, *Fernehill* 1612 FF, *Ferny Hill* 1695 Morden, 1836 O.S. From OE *fearnig, hylle* 'the ferny hill'.

FERNY HOUGH (obsolete) 1 mile west of Endon (SJ 9053). *Fernihalu* c.1278 SHC 1911 430, *Fernihale* c.1282 ibid. 442, *Fernyhalg, Fernihalg* 1307 SHC XI NS 256, 1369 80, *Fernyhalgh* 1325 SHC 1911 366, *? Fernihaleugh* 1327 SHC VII (i) 217, *Fernyhalgh* 1370 SHC VIII NS 249, *Fernehalgh* 1441 (1883) Sleigh 182, *Fernehall* 1535 SHC 1912 77, *Fernehall* 1546 SHC XI 288, *the Fernyhalgh* 1572 AD, *Fearnenaugh* 1589 SHC 1930 100, *? Fernihough* 1619 StSt 10 1998 12, *ffernihow* 1683 Norton-in-the-Moors ParReg, *Ferny Hough* 1836 O.S. The second element is from OE *halh*, so 'the ferny *halh*'. See also Ferny Hill.

**FIELD** on the river Blithe, 4½ miles west of Uttoxeter (SK 0233). *Felda* 1114-8 Burton Survey B, *Felda* 1130 P, *Feldam* c.1177 SHC III (i) 227, *Field* 1686 Plot. From OE *feld* 'open land'. Cf. Church Leigh.

**FIELDHOUSE** (unlocated, perhaps near Worfield or Claverley) *feldhous* 1327 SHC VII (i) 252; TSAS 3rd Series V 1905 244, *Feldushous* 1332 SHC X 129, *Feldehous* 1332 ibid. 130, *Feeldhouse* 1405 SA 5735/2/7/4. This may be the same place as *Fieldhouse of Whittimere*, recorded in 1323 (VCH XX 71), and perhaps to be associated with *la Feelde*, recorded in 1306 (SA 52/4). 'The house at the field'. *le Feeldhowse in Over Bradnop* [Upper Bradnop] is recorded c.1345: SRO D3272/5/13/5.

**FIELD HOUSE (FARM)** 2 miles north-west of Penkridge, on the south side of Levedale (SJ 8916), *Fieldoes* (p) 1578 Penkridge ParReg, *Fieldhouse* (p) 1598 ibid, *Fieldhowse* (p) 1607 ibid.; **FIELDHOUSE FARM** 1 mile east of Eccleshall (SJ 8428), *ye Fieldhouse* 1682 Chebsey ParReg, *The Fieldhouse* 1836 O.S. 'The house at the field'.

**FIELDS FARM** 1½ miles south-west of Horton (SJ 9355). *Felthouse* 1658 PCC, *The Fields* 1718 to 1794 Reade, 1837 O.S. The earliest spelling suggests that the name is from 'the house where felt was made', with the name reducing via *feltus* to *fields*. See also Felthouse.

**FIGHTING COCKS** on the east side of Wolverhampton (SO 9196). *Adjoining a close called Cockshutts...a little or small pleck of land used for a garden and commonly called the Fighting Cocks* 1737 WALS DX-240/7, *Fighting Cocks* 1834 O.S. Presumably where gaming birds were raised or fought.

**FILANCE (BRIDGE)** over the Staffordshire & Worcestershire Canal in Penkridge (SJ 9214). *le f(f)ylond* 1538 and 1541 Ct, *Fyland* 1598 VCH V 126, *Filand* 1653 ibid, 18th century SRO D260/M/T/5/73, *Fylands* 1841 TA. EDD gives for *filands* 'tracts of unenclosed arable land', i.e. field-lands. The name was attached to one of the Penkridge open fields: SRO D260/M/T/5/73.

**FILDESDALE** (unlocated) *Fildesdale* undated (1801) Shaw II 288.

**FILILODE** (unlocated; said to be in Claverley parish: TSAS 4th Series VIII 1920-1 Misc. viii; probably near Astley: Eyton 1854-60: III 153) *Fililodes-welle* 1305 Eyton 1854-60: III 153, *Fylilode* 1327 TSAS 3rd Series VI 1906 127, *Fyllilod* (p) 1341 Eyton 1854-60: III 153, *Fillode* 1386 TSAS 4th Series VIII 1920-1 Misc. viii. Possibly from OE *fylle* 'thyme' or some such plant, with OE *lad* 'watercourse, crossing'.

**FILLEY BROOK** – see **PHILLEY BROOK**.

**FINCHFIELD** 2½ miles west of Wolverhampton (SO 8898). *Fynchenesfeld, Fynchingefeld* 13th century Duig, *Fynchyngefeld* 1323 SHC VI (ii) 184, *Fynchenefeld* 1327 SHC VII 252, *Fynchenfeld* 1336 SHC 1928 34, *Finchfeilds* 1648 Wolverhampton ParReg, *Finchfields* 1662 BCA MS3307/Acc1927-020/337303. 'The *feld* or open land of the finches', from OE *finc*, ME *finch*, This may be the *feld* mentioned in a charter of 985 AD (12th century, S.860): Hooke 1983: 63-5. OE *feld* had the meaning 'open ground'.

**FINNERS HILL** ½ mile north-east of Colton (SK 0620). *Finners Hill* 1953 O.S.

**FINNEY GREEN** 2 miles north-east of Madeley (SJ 7946). *Finney Green* 1699 HOK 56, 1798 Yates, 1833 O.S. Perhaps from OE *fynig* 'moist, marshy', or OE *finig* 'a heap', often 'a wood-pile', or from a personal name. The Green element suggests a squatter settlement. See also Finney Lane.

**FINNEY LANE** 2 miles south-east of Leek (SJ 9953). *Harvey's Riddinge alias Finney Lane* 1596 (1883) Sleigh 172, *Fynney Lane* 1602 SHC 1935 426, *Fynney Layne* 1608 SHC 1948-9 67, *Fynney-lane, als: Harvey Riddinges* 1654 (1883) Sleigh 172, *Finney*

*Lane* 1836 O.S. From the manor of Fynney (obsolete, in Cheddleton, adjoining Ashenhurst: *Fyneye* 1320 SHC 1911 93), said to have been a gift from William the Conqueror to his kinsman Fenis, or de Fiennes, from whom the estate presumably took its name: see Erdeswick 1844: 497-8. *Philip del fyneye* is recorded here in 1320 (Tooth 2000b: 98), *Ricardus Fenay* in 1451 (AD), *Wm Feyney de Feyney land temp.* Henry VIII (Survey), *Wm Feynney de olde Basseford* 1523 (Rental), *Wm Fynny of Fynney Lane* in 1556 (AD 5). See also SRO D538/A/2/4. It is unclear whether *Finees*, recorded in 1290 (SHC XI NS 177), is to be associated with this place: the context points to a place near Cheadle Moor.

**FISHERWICK** 3½ miles north-east of Lichfield (SK 1709; SMR 02082 places a deserted post-Conquest settlement at SK 17550854). *Fiscerwic* 1167 SHC 1923 295, *Fischerewich* 1176 SHC I 78, *Fisherwic* c.1224 (1798) Shaw I 212, *Fisareswik* 1242 Fees, *Fishereswyk* 1279 SHC V (i) 140, *Fysherwyk* 1281 ibid.' 121, *Fyshereswyke* 1293 ibid. 244, *Fisscherwyk* 1309 WL 103, *Fissherwyke* 1356 SHC 1913 169, *Fysshereyyk* 1421 SHC XVII 73, *Fisscherwyke* 1482 SHC VI NS (i) 148, *Fecherwyk* 1532 SHC 4th Series 8 72, *Fisherwicke* 1588 SHC 1927 176, *Fishrike* 1607 Kip, *Fisherwick* 1686 Plot. From OE *fiscere* 'a fisherman', with OE *wic* 'dwelling, village, a farm, a dairy farm, a building or collection of buildings for special purposes', hence 'fishermen's *wic*', in this case perhaps in the sense 'building associated with a trade', here presumably processing of fish, possibly by salting: excavations here in the 1970s produced evidence of large-scale salt production from saline springs in the Iron-Age, which may have been taking place in the area into the Middle Ages – see WMA 22 1979 96; BAR 61 52-7. Fisheries are recorded here in 1419 (SHC XVII 73) and almost certainly existed much earlier. The name may therefore be 'the place where fish were processed and salted'. See also Wigford.

**FISHLEY** 3 miles north of Walsall (SK 0104). *Thistley, Fishley, Fistley, otherwise Thistly Ridding, Lower Fistley otherwise Thistling Ridding, Thistley field* 17th century Duig. The area was part of Essington Wood, within Cannock Forest, and is perhaps to be associated with *Thystlymor*, recorded in 1286: SHC 4th Series XVIII 129. The name is from OE *thistel* 'a thistle', here used adjectivally. *Fissle* and *Fistle* are dialect forms of thistle; *ridding* means a clearing or a cultivated area of wild land.

**FIVESTONES HEATH** (obsolete) on Yarnfield Heath. *Fivestones Heath* 1664 SRO D641/5/T/2/39-46. A name with possible archaeological associations – see Featherstone.

**FLASH** 4 miles north-west of Longnor (SK 0267), and said to be the highest village in England (its school stands at 1526': VCH VII 49). ... *a close or pasture called The Flasshe heade in Wharneforde* ... 1568 DRO D2375M/190/7, *Flasche* 1598 SHC XVI 170, *the Flashe* 1598 Alstonefield ParReg, *the Flash* 1599 DRO D2375M/25/18, *The Flashe* 1601 Alstonefield ParReg, *le Flasshe* 1605 SHC 1946 236, *the Flase* 1664 Alstonefield ParReg, *Flash* 1683 ibid, *the Flass* 1732 ibid, 1686 Plot. The distribution of *flash* in place-names, mainly found in the northern counties, shows that the word is probably of Scandinavian origin, from ODan *flask* 'a swamp, swampy grassland, shallow water, a pool', leading to ME *flassche, flosshe*, with the substitution of ME *sh* for ON *sk* (EPNE i 175): cf. Bell Flask, Water Flash, Flash Dales, Flashley Carr, Yorkshire; Flash, Lincolnshire and Northumberland; Flass, Durham (*Flaskes* 1313, *Flassh* 1382 Ekwall 1960: 181); The Flash, Cumberland. *Flash* and *plash* are onomatopoeic synonyms (cf. Plaish, a parish 5 miles north-east of Church Stretton, Shropshire – *Plesc, Plæsc* 963 (12th century, S.723): see PN Sa I 236-7). OED gives a 1440 reference to 'Plasche or flasche, where reyne water stondythe', while Halliwell gives for *flash* 'a common term for a pool'; for *flosche* 'a pit or pool'; for *plash* 'a pool of water, a large puddle'; and defines *ploshett* as 'a swampy meadow (Devon)'. The words seem to be applied to grassy land where water lies after rain and gradually disappears: in Shropshire *Flash* is a common field-name with the meaning 'shallow pool formed by floodwater': Foxall 1980: 19. The modern word *splash* is the older *plash* with an added *s*. *Plashes Farm* (*the Plash(e)* 1675 Dep) is to the west of Bednall. *le Plashe* in Hednesford is recorded in 1362 (SHC I 338: the place lies on the south side of Splash Lane, between Hednesford and Wimblebury – VCH V 56-7), and a field there is *Middle plashes* in 1841: Oakden 1984: 63. The fact that Flash itself formerly lay at the meeting point of the boundaries of Staffordshire, Cheshire and Derbyshire; Flash House 1 mile east of Barthomley in Cheshire (SJ 7753), 2 miles north-west of Audley, is close to the Staffordshire-Cheshire border; Flash Farm 1 mile north-west of Woore (SJ 7243) lies on the Staffordshire-Cheshire boundary; *The Flash* (1834 1" O.S.) lay north of Oldbury on the Staffordshire-Worcestershire boundary; and *Flashcroft Coppy* (*Flashcroft* in the earlier 16th century (VCH VII 186), *Flascrofte* 1543 (1883) Sleigh 71, *Flash Crost* (*sic*) Yates 1775) is shown on the county boundary 1 mile north-east of Rushton Spencer (SJ 9463) on the first edition 1" O.S. map of 1842 (and very many other places called Flash elsewhere in the country lie on or very close to boundaries) is unexplained and may be no more than coincidence, but the name may in some cases be connected in some way with boundaries: cf. flash, the term commonly applied to the waste material around the edge of a casting, unrecorded by OED. In Cheshire there appears to be a particular association with salt-workings: PN Ch V 1, 180. Flash has been identified with *flosche* mentioned in the 14th century poem *Sir Gawain and the Green Knight* (Elliott 1984: 145-7; see also Flash; Ludchurch; The Roaches). It may be added that there is no basis for the popular folk-belief that this place took its name from the counterfeiting of coins, but see JNSFC LXVI 1931-2 184; VCH VII 51.

**FLASHBROOK** in Adbaston parish, 4 miles north of Newport (SJ 7425). *Fletesbroc* 1086 DB, *Floscebrok* 1240 SHC 1924 191, *Flocebroc, Flocesbroc* 1243 SHC 1911 396, 401, *Flotesbroc* 1253 SHC 1911 122, *Flotesbrok* 1271 SHC IV 185, *Floshbroc* 1278 SHC 1911 170, *Flotusbrok* 1284 SHC 1910 298, *Floxbrok* 1293 SHC VI (i) 282, *Flotesbroc* 13th century Ronton, *Flosbroc* 1303 SHC 1937 76, *Flosbrok* 1309 SHC 1911 301, *Flossebrok* 1315 SHC 1911 88, *Flosse Broke* 1327 SHC XII (ii) 14, *Flosbroke* 1421 ibid.' 73, *Flashbrooke* 1666 SHC 1921 125, *Flassenbroke* 1679 SHC 1914 72, *Flosebroke, Flossebroke, Flosbrook, Flossbrook, Flossebrook* c.1565-70 SHC IX NS, *Flashbrooke otherwise Flotesbrooke* 1589 SHC XVIII 3. The derivation of this name is uncertain, but the spellings (some of which seem to incorporate the not uncommon misreading of manuscript *t* as *c*) show that the first element is not from Flash (q.v.), and is unlikely to be from OE *fleot* 'stream', a word which has a restricted use inland and is rarely found in Staffordshire (but see Fleetgreen; The Flete). OE *flot-weg* 'floating way' is recorded, and has been proposed as the first element in Flotterton, Northumberland (Ekwall 1960 182). Sometimes the middle element of a name was dropped before DB, and it is possible (though unlikely) that the original name here was *Flotwegesbroc*, 'the brook at the floating way', perhaps connected in some way with the Roman road from Water Eaton to Chester (Margary number 19) which runs through the area, forming the county boundary: the 1833 O.S. map shows *Flashbrook* to the north of the road (in Staffordshire) and *Flashbrook Heath* to the south (in Shropshire). Gelling & Cole 2000: 9 suggests a derivation from *flotes*, from OE *flot*, perhaps some sort of causeway. An alternative possibility is an unrecorded OE form which developed into ME *flotise* 'froth, scum', giving 'the brook (i.e. Lonco Brook) with the scummy water'. Halliwell 1850: 365 gives 'Flotis: The foam or froth of anything boiling, etc.', and suggests the word is OE in origin (cf. OE *flotsmeru, flotsmeoru* 'floating fat; scum'), although it is not recorded in

Bosworth-Toller. OED has *flotesse, flotyce, flotyse, flotes, flattesse, flats* (obsolete), 'scum or grease floating on the surface of a liquid', possibly an unrecorded French derivative of *floter* 'to float', first recorded c.1440, or from the plural of OE *\*flot* or ON *flot* 'scum'. The second element is OE *broc* 'brook'. It has been noted that '... with so much unenclosed land at Flashbrooke Heath the county boundary may well have been a little uncertain. At the present day [1758] there is a conflict of opinion between the Staffordshire and Shropshire officials as to the exact course of the boundary at Whitley Ford Bridge ...': JNSFC LXII 1926-7 40. See also SHC 1914 72-6. Cf. Forsbrook.

**FLASH HEAD** on the east side of Flash (SK 0267). *Flashhead* 1714 VCH VII 53. From OE *heafod* 'head', used topographically of 'a headland, summit, upper end, source of a stream'. In this case probably meaning 'headland or upper end at Flash (q.v.)'.

FLAXDALE (unlcated, near Blithfield) *Flaxdale* late 13th century SRO D938/57.

FLAXLEY HILL (unlocated, in or near Bagot's Bromley). *Flaxlegh* mid-13th century SRO D986/48, *Flaxley Hill* late 13th century SRO DW01721/3/4/5. 'The *leah* where flax was grown'.

FLEDISFORD' (unlocated, possibly in or near Calwich) *Fledisford'* c.1235 SHC 4th Series IV 200.

**FLEETGREEN, UPPER & LOWER** 2 miles east of Upper Hulme (SK 0561). *Fleetgreen* 1514 VCH VII 27, *Fleet Greve* 1566 Deed, *Fleetgreen* 1592 SHC 1930 (ii) 328, *Fleete greene* 1598 Alstonefield ParReg, *Fleite Greene* 1606 ibid, *Fled Green* 1703 Leek ParReg, *Fleet Green* 1775 Yates. From OE *fleot* (or ON *fljót*) *grene* 'the grassy open land by the rivulet or stream'. There is a stream to the west.

FLETBRIDGE (unlocated, in Burton upon Trent) *Fletbrugge* 1306 SHC VII 155. Perhaps where Fleet Street crossed an arm of the river Trent known as *the Flete* (q.v).

FLETCHEAM (unlocated, perhaps near Handsworth) *Fletcheam* 1526 SHC XIII 183, 1527 SHC XI 266, SHC XII 183.

FLETCHER'S FARM (obsolete, in Leacroft, Cannock) *Fletchers Farm* 1635 SRO D603/N/16/3, *Fletcher's Farm* 1651, 1653 CC, *Fletcher's alias Rumer Hill Farm* 1622x1777 SRO D260/M/T/6/130. See also Rumer Hill.

FLETE, THE (obsolete) a short arm of the river Trent running west in Burton upon Trent: see VCH IX 23, 37. *(Thomas Juxta-) la-Flet* 1188x1197 SRO D603/A/Add/36b, *Le Flet temp.* Edward I SHC 1937 108, *Flet* 12th/13th century SHC V (i) 47, *Floet* 1257 SHC 1937 59, *the water of the Flete* c.1330 VCH III 212. From OE *fleot* (or ON *fljót*), here meaning 'an arm or earlier channel of a river': cf. *Fleet* (1576), an arm of the river Trent north of Newark (Ekwall 1928: 159). The name is remembered in Fleet Street in Burton upon Trent. See also Fletbridge.

FLITLEY (obsolete), on the river Blythe, 4½ miles west of Uttoxeter (SK 0233). 'Flitlegh seems to be another name for Field, but it is now extinct': SHC XII 281. *Flitleg'* 1222 SHC IV 19, *Flitleye* c.1272 SHC XII 281, *Flyteleye* 1294 SHC VII (i) 9, *Flutteleye* 1327 ibid.' 222, 1377 SHC XIII 128, 1379 SHC XIV 128, *Fretelegh* 1332 SHC X 97, *Flyttley* 1574 SHC XVII 220. From OE *(ge)flit* 'strife, dispute', with OE *leah*, so 'the disputed *leah*'.

**FLORENCE** on the south-west side of Longton (SJ 9142). So-called from Florence (d.1881), eldest daughter of the 3rd Duke of Sutherland of Trentham Hall who developed the area in the 1860s: TNSFC 1970 85.

**FLOTHERIDGE** (obsolete) 1 mile south of Swindon (SO 8688). *Flederich* 1330 SHC 1913 24, *Flotheridge (pool)* 1834 White ccix, *Flotheridge (Basin)* 1834 O.S. Perhaps from northern dialect *flother* 'a boggy place, a swamp', also recorded with the meaning 'foam, froth': (EDD). The second element appears to be OE *hrycg* 'a ridge, a long narrow hill', in which case giving 'the ridge at the boggy place'. The name (recorded from 1296: SHC 4th Series XVIII 188) was applied to a royal fishpond on or near Smestow Brook: VCH II 347 fn.96.

**FOKER FARM** 1 mile north-west of Leek (SJ 9757), *Foker* 1333 VCH VII 194, *le Fowker* 1539 MA, *Foocare* 1576 Loxdale, *Fooker* late 16th century WSL Sleigh Scrpbk ii f.107v, *Leek Fowker* 1615 SHC IV NS 90, *Fooker in Leekefrith* 1642-60 ChancP, *Foker, Fowker* c.1645 Leek ParReg, *Lower Folker, Upper Foker* 1842 O.S., *Lower Foker Farm* 1850 SRO SA 20/7/1850 8; **FOKER GRANGE** 1 mile north-west of Leek (SJ 9757), *ffoker-graunge* 1543 (1883) Sleigh 71, *Fowkers or Fowker Grange* 1694 VCH VII 237; **FOKER MOOR** 1 mile north-west of Leek (SJ 9758), *Fokermor* 1394 *Deed* (1883) Sleigh, *le Fokermore* c.1539 LRMB, *Fowkermore Side* 1589 SHC 1934 7, *Fokar More* 1591 SHC 1930 167, *Foker Moor(e)* 1811 *EnclA, New House or Foker Moor* 1842 O.S. A derivation for these associated names from OE *ful* 'foul, dirty, filthy', and ME *ker* 'a bog, a marsh, especially one overgrown with brushwood' (from ON *kjarr* 'brushwood'), hence 'filthy marsh', describing an area of common waste in the south-west corner of the township of Leekfrith, as suggested in VCH VII 194, seems improbable in the absence of any intermediate *-l-* and the persistent *-o-* in the various spellings. There is some uncertainty about the existence of granges of Dieulacres abbey at both Foker, probably on the north side of the abbey, and at Fowlchurch (VCH VII 194, 237), which makes it very difficult to establish which spellings relate to which place – see Fowlchurch, with which Foker almost certainly shares the same derivation. A farmhouse on the southern edge of the waste at Foker was known as *Lower Foker* in 1770 and *Foker Grange* by the end of the 19th century (VCH VII 194): the house now known as Foker Grange (SJ 966575) would seem to have no connection with any ancient grange in the area. For the reasons given above, some of the cited spellings may relate to what is now Fowlchurch (q.v.), and vice versa..

**FOLD, THE** ½ mile south-west of Sheen (SK 1160). *Fould* 1842 O.S. From OE *fal(o)d* 'a fold, an enclosure for animals'.

**FOLE** on the river Tean, 4 miles north-west of Uttoxeter, in Leigh parish (SK 0437). *Fowall* c.1260 SHC VII NS 146, *Fowale, Foowale* c.1272 Dieul, *Fowell* 1290 SHC 1911 196, *Fowall* 1332 SHC X 112, *Fole* 1538 Ipm, *Foale otherwise Fole* c.1619 SHC VII NS 208, *Fole alias Foley* 1675 SRO 1057/A/1/9/1-7, *Fole* 1680 SHC 1919 272. A difficult name. Possibly from OE *foh* 'measurable, moderate', here in the latter sense, with Mercian OE *wælle* 'a spring', and (sometimes) 'a stream', perhaps in this case the stream running from the hills to the north into the river Tean, so 'the moderate stream'. Or perhaps from OE *feo, wælle* 'the cattle spring'. *Overfoale* and *Netherfoale* (Upper Fole and Lower Fole) are recorded c.1622: SHC X I 20. See also Fawside.

**FOOTHERLEY** 1 mile south-west of Shenstone (SK 0903). *Fulwardlee* 12th century Duig, *? Fulwerhel* 1209 SHC 1923 277, *Folverle* c.1276 SHC XVII 261, *Fulverleye* 1311 SHC 1911 77, *Fullerleye, Folverleye, Fulverleye, Fulfordleigh* 14th century, *Fotherley otherwise Fulderley* 16th century Duig, *Foterley* 1686 Plot, *Footherley Hall*

1834 O.S. The various forms are inconsistent, but 'Folcweard's *leah*', suggested in Duignan 1902: 62, seems the most likely explanation on the available evidence.

**FORD** 1½ miles south-east of Onecote (SK 0653). *Forde* 1240 *Deed* (1883) Sleigh, c.1265 SHC 1911 438, 1281 ibid.' 178, 1477 SHC VI NS (i) 107, 1484 ibid.' 155, *la fford* 13th century Dieul, *Fourde* 1558 BM, *Ford* 1646 SHC 4th Series I 308. From the ford across the river Hamps here before the present bridge was built early in the 20th century.

**FORD BROOK** a tributary of the river Tame; **FORD BROOK** a tributary of the river Trent, *ffordebrooke* 1571 Ct. 'The brook with the ford'.

**FORD FARM** 2 miles south-west of Cheddleton (SK 9551). *Colepytford* 1609 QSR, *Colpitfurde* 1624 PCC, *Colepitford* 1696 ParReg, *Colepitsford* 1664 StV. 'The ford near the coal pits', with reference to former coal shafts, possibly including *Colepit* recorded in 1663 (SHC I (ii) 63).

**FORD GREEN** ½ mile south-west of Norton-in-the-Moors (SJ 8950). *Fordegreene* 1590 SHC 1930 99, *Fourde Greene* 1609-9 SHC 1948-9 5, *Fourdgreene* c.1630 SHC II (ii) 17. Probably from the Ford family who were notable in this area: StEnc 220. The Green element suggests a squatter settlement.

**FORDHOUSES** 3 miles north of Wolverhampton (SJ 9103). *(atte)forde* 1327 SHC VII (i) 251, *Forde of Bysshebury* 1349 SHC 1910 319, *la Fordhouses* 1480 SHC VI NS (i) 142, *Furdehousyn* 1538 SHC VII NS 63, *Fordhouses* 1591 SHC 1930 181, *Fordehowses* 1597 SHC XVI 168, *Fourde Howses* 1600 SHC 1935 338, *Fourd Howss* 1608 SHC 1948-8 72, *Fourdehowses* 1608 ibid.' 73, *Forde otherwise Fordhouses* 1617 SHC NS VI (i) 14, *Fordhouses* 1686 Plot. As explained by John Huntbach c.1695: '[the place] hath its name from the houses erected upon the roadway leading from Wolverhampton towards Stafford near the ford of Wybaston [Wobaston] brook': SHC 1982 72. There are records in 1529 and 1555 of *Fourdhouse* in Wombourn: Shaw 1801: II 215, 216; in 1554 of *the Fordhowse in Womburne*. Wodehouse.

**FORDS RICE** 1 mile north-west of Upper Mayfield (SK 1447). *Fords Rice* 1795 SRO D3437/10/1-10, 1836 O.S. Earlier spellings have not been traced, but probably from OE *hris* 'brushwood', with a personal name. Yates' map of 1775 shows two places named *Rice* in the area, one at or near what is now Harlow Farm, the other on the north-west side of Upper Mayfield. *Mayfeild Ryse* is recorded in 1538 (Survey), *Yields or Healds Rice* is recorded in Mayfield in 1777 (Okeover F425), and *Bank Rices* in 1810 (SRO D3437/11/1-6). Those places probably record brushwood in the area. *Fords Rice Wood* lies to the south-west of Fords Rice on the 1836 1" O.S. map

**FORD WETLEY** 1½ miles south-east of Onecote (SK 0553). *Ford Wetley* 1639 Leek ParReg, 1836 O.S. Perhaps to be associated with *Wurtleg*, recorded in 1222 (Pleas).

**FOREBRIDGE** a former village immediately south of the bridge crossing the Sow outside the South Gate at Stafford (SJ 9222), and another name for Stafford manor, otherwise Castle manor: VCH V 83, 86. *For(e)brigge, For(e)brugge* 1221 FF, *Forburgg by Stafford* 1289 SHC VIII 43, *Forbrugge* 1295 SHC VII 25, 1401 SHC XV 97, *Forbrygg* 1392 SHC XV 41, *Forde Bridg alias Stafford Grene* c.1540 Leland v 21. 'The bridge in front of the town', which later developed into an alternative 'the bridge at the ford': see SHC VIII (ii) 42-5, 50. There was also a Forebridge in Patshull: VCH XX 162.

**FOREGATE** on the east side of Stafford (SJ 9123). *Foregate* pre-1290 SRO D938/193, *Stafford Foriat* 1511 SRO D938/221, *Stafford Forziat* 1526 SRO D938/222. 'The gate at the front of the town'.

FOREIGN – see **WALSALL**.

FORELHEYE (unlocated, perhaps in the Bramshall area) *Forelheye* 1261 SHC IV 154.

**FORGE HOUSE** 1 mile south-east of Brewood (SJ 9007). *Forge houses* 1686 Plot, *Brewood Forge* 1686 Plot 311, 1834 O.S. From 17th century iron-works which lay on the river Penk to the west of Coven: Horovitz 1992: 289-95.

**FORKHILL** 1 mile north-west of Warslow (SK 0759). *? Folkishull temp.* Henry III SHC 4th IV 196, *? Folkeshull'* late Henry III ibid. 109, *? Forkehull'* c.1295 ibid. 243. It is not certain that these forms relate to this place, but the spellings suggest 'the hill of the people', perhaps denoting a meeting-place, or '*Folc's hill', *Folc being the short form of OE names beginning Folc-.

**FORSBROOK** in Dilhorne, 3 miles south-west of Cheadle (SJ 9641). *Fotesbroc* 1086 DB, *Fotesbroc* c.1187 SHC II 261, *Fottesbroc* 1199 FF, *Focebroc* (p) 1200 P, *Fotebroc* (p) 1201 P, *Foddesbroc* 1227 Ass, *Fotesbroc* c.1230 SHC VI (i) 11, *Fotisbroc* 1258 ibid.129, *Fotesbrok* 1276 ibid.' 31, *Fossebrok* c.1290 ibid. 436, *Forsebroke* c.1355 NA DD/FJ/4/26/7, *Fotisbrok* (p) 1458 Ipm, *Forsebrocke* 1586 QSR, *Frostbrocke* 1596 SHC 1932 246, *Fossbrook* 1686 Plot, *Fossbrook* 1775 Yates, *Forsbrook* 1837 O.S. From the ON personal name Fot, so 'Fot's brook': *Fosse-* is a natural development from *Fotes-*. The *r* only began to appear in the name from the 16th century, and an alternative local pronunciation is Fossbrook.

FORSUONEBUTTS, FORSWORENBUTTES (unlocated, perhaps near Halfhyde) *Forsuonebutts, Forsworenbuttes* 1255 SHC IV 123. Perhaps from *foresworn* 'perjured; falsely sworn', with ME *butt* 'thicker end; a mound for archery practice; a ridge dividing ploughed land; a small piece of land', possibly here with the last meaning, indicating an area of land acquired by falsehood.

**FORTON** Ancient Parish 2 miles north of Newport (SJ 7521). *Forton* 1198 Fees, 1274, 1307 Ass, 1291 Ipm, *et freq.*, *Fortone* 1292 Ipm, *Forton* 1563 SHC IX NS 147, *Fauton* 1610 Speed, *Forton* 1686 Plot. Probably 'the *tun* by the ford', the *-d-* having disappeared as often the case with this common name. The place is on the river Meece.

**FOSSEWAY** a road running from south of Lichfield towards Pipehill. The road is recorded as *Falseway* in the late 15th century (VCH XIV 8), and that name, which may not be much older than the 15th century (in the 13th century the road was described as 'the way to Aldershawe': ibid.), the meaning of which (if not itself a corruption) is uncertain, was evidently infuenced by the well-known Roman roads bearing the name Fosse Way. There is no evidence that the road is of Roman origin. VCH XIV 8 identifies Fosseway as the later Birmingham road, but the modern Birmingham Road runs south from Lichfield, and the present Fosseway is an east-west road.

**FOSTON** on the north side of the river Dove, 2 miles north-west of Tutbury (SK 1831). *Farulstun* 1086 DB (entered in Derbyshire), *Farleston'* 1331 AddCh, *Fostun, Fostona(e)* (p) c.1138 Okeover, 1162 (p) Tutbury, *Foteston'* early 14th century SaltCh. Evidently 'Farulf's *tun*', from the OE personal name Farulf, and 'Fot's or Fótr's *tun*' (Fot is an OE personal name, Fótr an ON personal name). PN Db 560 suggests that these were two close settlements, and Foston became the dominant. Cf. Foston (PN Y (North Riding) 39; PN Y (East Riding) 91). The administrative history of the place is complex. Historically in both Staffordshire and Derbyshire, part of the Derbyshire area was transferred to Staffordshire for civil purposes in 1844, and the Staffordshire part was taken into

Tutbury, Tatenhill and Yoxall in 1890, leaving the remainder in Derbyshire: see Youngs 1991: 411.

FOTHEACRES (unlocated) *Fotheacres* 1166 SHC IV 280.

FOUCHERS POOL – see **FOWLCHURCH**.

**FOULD** 2 miles north-west of Leek (SJ 9758). *Fold, Fould* 1634 Leek ParReg, *? the Foult* 1694 Leek ParReg., *? the Fouls* 1691 ibid, *? Fould* 1695 Morden, *Fould* 1842 O.S. From OE *fal(o)d* 'a fold, enclosure for animals': cf. Fauld. The place is said to have been formerly known as *Austen's Tenements*, from a 17th century occupier or owner, and is also recorded as *Sheephouse* (q.v.): VCH VII 194.

FOULESHURST (unlocated) *Fouleshurst* 1367 SHC XIII 58.

**FOUR ASHES** 1½ miles east of Brewood (SJ 9108). *Four Ashes* 1683 PCC; **FOUR ASHES** 1½ miles north-west of Enville (SO 8087). *iiii Ashes* 1590 SHC 1930 54, *Quator Asshes* 1586 SHC 1929 150, *Quatuor Aches* 1590 SHC 1930 80, *Foureashes* 1590 ibid.' 81, *4 Ashes* 1686 Plot, *Four Ashes* 1775 Yates. By local tradition the first place was so-named from an inn which had four ash trees. Four ash trees are recorded at the second place in 1496, and there were still four ash trees there in 1817: VCH XX 94. Cf. Six Ashes 1 mile north-west.

**FOUR CROSSES** an inn on Watling Street, 2 miles south-west of Cannock (SJ 9509). *Fowre Crosses* 1611 SRO D260/M/T/4/48, *the 4 Crosses* 1674 Wolverhampton ParReg, *Fower Crosses* 1682 Browne, *Foure Crosses* 1693 SHC 1938 220, *4 Crosses* 1686 Plot, *The Four Crosses* 1700 PCC, 1775 Yates. From a public house here, possibly so-named because it stands at the point where two roads intersect, although four crosses form the arms of the see of Lichfield. The 1834 1" O.S. map shows *Four Crosses* at Walsall Wood (SK0403). Jackson 1879: 160 cites public houses of this name at Bicton and Baschurch in Shropshire, with the crossroads derivation.

FOUR STONES – see **FEATHERSTONE**.

FOWLCHURCH (obsolete, on the north side of Leek (SJ 9857)) *Focher* 1240 *Deed* (1883) Sleigh, *? Fonake* 1291 (1798) Shaw I xxii, *Foucher* (p) 1293 SHC 1911 215, *Fowthers* 1538 (1883) Sleigh 62, *le Litell Fowchers, Graunge de Fowchers* c.1539 *LRMB, Fowcher's-grange* 1543 ibid.' 72, *Fowchers grainge* 1561 SHC 1938 174, *Fowchers grange* 1552 VCH VII 237, *Fowchars* 1560 Pat, *Fouchers* 1597 (1883) Sleigh 26, *ffowcher* 1604 (1883) Sleigh 58, *Fowchurch* 1637 Leek ParReg, *Foo-Church* 1705 ibid, *Fowlchurch* 1775 Yates, 1842 O.S., *Fowcher, Fowchurch* 1883 Sleigh 46 & index. The place-name history of Fowlchurch and Foker (q.v.) is confusing. VCH VII 237 records a grange at Fowlchurch in 1246 called in 1542 Foker Grange. However, whilst Fisher 1969: 19, 24 mentions a grange at Foker, 'the largest of the Staffordshire granges of Dieulacres', he does not include Fowlchurch as a grange on the map on p.13. It seems likely that both Fowlchurch and Foker were named after Clemencia de Fougeres or Feugeres (cf. *Roberto de Feugeres* recorded in 1194: SHC III (i) 26-7; see also SHC 1937 10, 13, 17, 84), second wife of the founder of Dieulacres, Ranulph III de Blundeville, Earl of Chester (who is mentioned in 'The Vision of Piers Plowman'). Clemencia, the daughter of Ralph de Fougeres/Feugeres (SHX IX NS 293), died in 1253 and was buried at Dieulacres. The only miracle recorded at Dieulacres involved a blind monk whose sight was restored at her tomb where he prayed daily (Fisher 1969: 17). The place-names Fowlchurch and Foker are almost certainly both from Fougeres in Ille-et-Villaine. The alternatives *Focher* (Fowlchurch) and *Fo(w)ker* (Foker) may have been rival pronunciations with -*ch*- and -*k*-, perhaps because -*ch*- is written for -*k*- in early Anglo-Norman records, for example DB. For the reasons given above it is uncertain whether all the cited spellings relate to Fowlchurch: it is possible is that those with -*ch*- relate to this place, with those with -*k*- are properly to be associated with Foker (but note that *ffowchers, alias ffowkars* is recorded in 1604 (1883) Sleigh 58). The later association with *church* in the place-name is fanciful. *Fouchers Pool*, 1 mile west of Greensforge, is recorded in 1834 (O.S.), but the history of the name is unknown.

**FOWLEA BROOK**  a tributary of the river Trent. *aqua de Foulehee* 1414 Rental, *ffowley broke* 1533-9 LRMB, *Fowley brooke* 1538 Survey, *the fowle lea* 1635 Stoke on Trent CA, *ffouley brooke* 1648 ibid. From OE *ful-ea* 'the dark-looking (literally dirty) river', with the second element later becoming confused with OE *leah*. A fishery called *Fowleye*, probably near Tutbury, is recorded in 1327 (1798): Shaw I 40.

**FOXEARTH** 1 mile south of Cellarhead (SJ 9546). *Foxholes* 1836 O.S. Early spellings are not available, but there was an early compound *fox-e(a)rþ*, generally held to contain *earth* in the sense 'lair, burrow', although the word *earth* meaning burrow is not recorded until the 16th century. It is possible that this and other similar names (cf. Foxholes; Foxt) were used in some places for cave-dwellings or rock-houses: the Kinver parish register of 1671 mentions *Maragaret of the fox earth*, and Plot 1686: 172 *Mag a Fox-hole* with reference to cave dwellings (known as *Nanny's Rock* by the 1880s: VCH XX 122) cut into the sandstone there: Will'o Fox, recorded in Kinver in 1332 (SHC X (i) 86) may have been a dweller in these caves. It may be that the modern name Foxearth was considered more refined than Foxholes.

FOXHALE  (unlocated, in Claverley) *Foxhale* 1309 SA 2089/2/2/3, 1388 ibid. 2089/2/2/8, 1435 ibid. 5735/2/7/6/4, *Foxhales* SA 573/2/7/6/3.

**FOXHILLS, THE**  1 mile north of Himley (SO 9288). Perhaps to be associated with *Foxeleislade* and *Foxeleie*, recorded in Baggeridge Wood in 1320: SHC 1928 31. See also SHC 1934 88.

**FOXHOLES** in Hanbury (SK 1627), *? Foxriddyng* 1323 SHC 1911 99, *Foxholes* 1836 O.S.; **FOXHOLES** in Talke (SJ 8153), *Foxholes* 1733 SHC 1944 70, *Fox Holes* 1775 Yates; FOXHOLES (unlocated) in Audley, *Foxholes* 1642 Audley ParReg, 1800 SRO D997/1/1; FOXHOLES (obsolete) in Fawfieldhead, *Foxholes* 1428 DRO D2375M/1/1, *Foxholes* 1591 DRO D2375M/100/27, *Foxehooles* 1602 Alstonefield ParReg, *Foxhole* 1605 SHC 1940 220, *Fox holes* 1662 Alstonefield ParReg; FOXHOLES (unlocated) in Eccleshall, *Foxholes* 1351 SHC 1913 144; FOXHOLES (obsolete) on the north-east side of Bull Bridge, Penkridge (SJ 9214), *Foxholes* 1890 O.S. OE *fox-hol* had the meaning 'a fox-hole, a fox's earth'. See also Foxearth; Foxt.

**FOXLEY, LOWER FOXLEY**  2 miles north of Audley (SJ 7953). *... a wood in the township of Awdeley, called Foxley* ...1553 SHC 1912 203, *Foxley* 1629 Audley ParReg, *Foxey* 1795 SRO D3272/1/22/7/1; **FOXLEY**  1 mile south-west of Standon (SJ 8133), *Foxley* 1833 O.S. 'Fox-*leah*'.

FOXLOW (obsolete) on the south-west side of Crowborough (SJ 9056). *Foxlow* 1891 O.S. Without earlier spellings it is not possible to say whether the second element is OE *hlaw* 'burial mound' or OE *leah*: the former is suggested by the 19th century spelling, but the latter is more likely to be found with the first element.

**FOXLYDIATE**  1½ miles west of Redditch (SO 0167). *Foxhuntleyates* c.1300 Pat, *Foxhunte Ledegate* 1386 SHC I 352, *Foxenlydeyate* 1464 Pat, *Foxlydiate* 1591 Will, *Fox*

*Liddet* 1675 Ogilby. From OE *hlidgeat* 'swing-gate', so 'the swing-gate used by the fox-hunters'. The bounds of Feckenham Forest in 1300 (Pat) include a reference to *Foxhuntwey voc. le Ruggeway*, i.e. 'the way used by the fox-hunt called the Ridgeway'. In Tardebigge parish, forming part of Staffordshire from c.1100 until 1266, in Warwickshire until 1844, and since that date in Worcestershire.

**FOXT** 3 miles north-east of Cheadle (SK 0348). *Foxwiss* 1176 FF, *Foxiate* 1253, *Foxwist, Foxwyst* 1293 Ipm, *Foxiate* 1293 CoramR, *Foxwyss* 1327 SHC VII 217, *Foxxwyist* 1311 SRO D1229/1/4/39, *Foxhurst* 1532 SHC 4th Series 8 37, *Foxt* 1578 Ipstones ParReg, *Foxeweist otherwise Foxewist* 1609 SHC III NS 27, *Fox* 1682 Browne, *Foxton* 1775 Yates. The second element is from OE *wist* 'lair, dwelling', hence 'foxes' burrow'. Curiously, the 1253 form appears to have as the second element OE *geat* 'hole, opening, gap', so 'foxes' hole', with the same meaning as 'foxes' burrow'. The 1775 form explains the modern form which is probably a reduction from *Foxewist*, wrongly expanded to *Foxton*. See also Foxearth, Foxwyst. Cf. Foxwist, Cheshire.

FOXWYST (unlocated, probably in or near Casterne) *Foxwyst* 1562 SRO D1229/1/4/40. From OE *wist* 'dwelling', hence 'foxes' burrow'. Cf. Foxwist, Cheshire. See also Foxt.

**FRADLEY** 4½ miles north-east of Lichfield (SK 1513). *Fodresleye, Foderesleye* 1262 SHC V (i) 136, *Frodele* 1269 SHC 1910 127, *Frodeleye* 1286 Duig, *Fredeleg* 1295 SHC 1911 236, *Frodley* 1532 SHC 4th Series 8 169. The forms are inconsistent, but perhaps from the OE personal name Frod, so 'Frod's *leah* ', or (as proposed by CDEPN) from OE *fodor, leah* 'the clearing where fodder was obtained'.

**FRADSWELL** 7 miles south-east of Stone (SJ 9931). *Frodeswelle* 1086 DB, *Frodesvella* 1155 BM, *Frodeswell* 1177 SHC XII NS 278, *Fredeswelle* 1206 SHC III 171, *Frotheswell* 1221 SHC IV 221, *Fradeswell* 1249 LLRRO 26D53/1292, *Frodesuel* c.1275 SHC 1911 165, *Frodeswall* frequently 13th century Duig, *Frodyswall* 1532 SHC 4th Series 8 59. 'Frod's spring or (more likely) stream', from Mercian OE *wælle*. Fradswell Brook and Dutton's Brook run through the village and unite before joining Gayton Brook.

**FRANKLINS** 2 miles north of Leek (SJ 9859). *Franklinhayes* 1641 Leek ParReg, *Frankelinges Hey* 1658 ibid, *Franklyns* 1663 ibid, *Francklins Heages* 1703 ibid.' From ME *frankelein* 'franklin, freeholder', with Mercian OE *(ge)heg*, so 'the hay or enclosure of the freeholder'.

FRANKWELL (obsolete) a vill near Ellenhall (SMR 00817 places a deserted post-Conquest settlement at SJ 84702544) which John de Frankville gave to the prior of Ranton (otherwise Ronton): SHC 1914 83; SHC IV 276. *Frankeuill'* c.1206 SHC 1928 280, *Franchewyle* c.1270 SHC 1937 74, *Frankvyle, Frankvylle* 1283 SHC IV 276, *Frankeville* 1288 SHC VI (i) 217, *Frankevile* 1292 SHC IV 276, *Frankvile* 1294 SHC 1911 228, *Fraunkeville* 1306 SHC VII (i) 26, 1318 SHC X 25, 1339 SHC XI 82, *Frankvyle* 1348 SHC 1913 127, *Frankeville* 1359 IPM, *Frankwell* 1564 SRO D3089/1/4, *Frankwell (Moor)* 1586x1636 SRO D798/1/6, *Frankville* 1679 SHC 1914 92, *Frankwell (Farm)* 19th century JNSFC LXIII 1928-9 165. The name is French and means 'the free vill or the vill of the Frenchmen' (OFr *franc, ville*), with the substitution of *well* for *ville* occurring from c.1600. For a more detailed discussion of this name see Introduction, pages 54-56.

FREARSALE (unlocated) *(the parish of) Frearsale* c.1540 SHC X NS (i) 127. No parish of this name is recorded in Staffordshire, and the spelling is evidently corrupt.

**FRECHEWOODS** (unlocated, possibly near Madeley in Checkley) *Frechewoods* 1276 SHC 1911 31.

**FREDA'S GRAVE** on Cannock Chase, ½ mile south-east of Brocton (SJ 9718). The burial place of Freda, a dalmatian bitch adopted as a mascot by the Fifth Reserve of the New Zealand Rifle Brigade stationed at Brocton Camp from 1917 to 1920. Freda died in 1918. See also StEnc 224.

**FREEFORD** 1½ miles south-east of Lichfield (SK 1307; SMR 01904 places a deserted post-Conquest settlement at SK 13260799). *Fraiforde* 1086 DB, *Freiford* 12th century Duig, *Freford* 1242 Fees, 1271 Ass, 1311 SHC 1915 31, 1329 SHC XI 10, 1342 SHC 1913 88, 1343 SHC XVII 277, 1565 ibid.' 215, *Ferford* 1273 SHC 1910 123, *Frefort* 1297 SHC 1911 241, *Feyrforde* 1332 SHC X 108, *Freyforde* 1470 OSS 1936 45, *Freyfforde* 1532 SHC 4th Series 8 184, *Friesfords, Freysforde* 1567 SHC XVII 216, *Freyford otherwise Frefords otherwise Freaforde* 1582 SHC XVII 231, *Friesforde* 16th century Duig, *Freysford otherwise Frayford* 1602 SHC XVIII 21, *Freaford* 1670 SHC 1923 225. The place lies on an ancient road between Lichfield and Tamworth where it crosses Darnford Brook on the edge of the former Whittington Heath. The forms suggest 'ford where no toll was payable'. A more detailed discussion of the name will be found in the Introduction, pages 30-31. Freeford may in early times have been of considerable extent, incorporating perhaps part of Whittington, Packington and Tamhorn, and reaching perhaps the river Tame: Shaw 1801: II app. 14.

**FREEHAY** 2 miles south-east of Cheadle (SK 0141). *Free Hay* 1775 Yates, *Huntley Bushes als. Freehay* 1791 SRO D1203/C/9/1-10, *Free Hay* 1836 O.S. A new parish formed out of the old parish of Cheadle in 1847: SHC 1926 153; Youngs 1991: 411. From Mercian OE *(ge)heg* 'a fence, an enclosure', so meaning here perhaps 'the enclosure which could be used free from service or rent'.

FREEZELAND (obsolete) on the north side of Millfields, east of Ettingshall; FRIEZELAND (obsolete) in Walsall Wood; FRIEZELAND (obsolete) in Tipton. A relatively common name, indicating the existence of ancient heathland, derived from OE *fyrs* 'furze', which by metathesis (the shifting of the *r*) became *frise-land*.

FRENCHMAN'S STREET (unlocated, between Shatterford and Upper Arley) *Frenchmanstreet* 1686 Plot, *Frenchman S* 1695 Morden, *Frenchman's Street* 1749 Bowen, 1801 Shaw II 253, *Frenchmans Street* 1775 Yates. The origin of the name, which is recorded in a deed of the 16th century (VCH Wo III 5), is uncertain, but since the forms relate to Frenchman in the singular, presumably from someone of French origin who once lived here.

FRETHINGDENE (unlocated, perhaps near Marchington) *Frethingdene* 1306 SHC VII 154.

**FRIAR PARK** 1½ miles north-east of Wednesbury (SO 9996). *fryer parke (Smythie)* c.1553 Dilworth 1976: 69, *Frier Parke* 1590 SHC 1930 69, 1609 SHC 1948-9 169, *Fryars Park* 1606 Duig. Ede 1962: 109 could trace no connection with either monks or friars, and suggested the name might be from a surname, but VCH XVII 6 states that the place belonged to Halesowen Abbey in the Middle Ages:

FRISEN HILL (unlocated, in Dilhorne) *Frisen Hill* 1567 SRO D1229/1/4/3.

FRITH (obsolete) in the valley to the south of Blithfield rectory: SHC VIII 143-4, Corrigenda before title page SHC X, SHC 1914 125, SHC 1919 7; SRO 50/7894,

52/7894, 53/7894, *le Frithe* 1253 SHC VIII (i) 144, *Le Frit* 1315 SRO D938/55. See The Frith.

FRITH, THE  2 miles north-west of Hollinsclough (SK0569), *Le Frith* 1534 ValEccl (or may refer to Leekfrith (q.v.)). A name found in north Staffordshire, e.g. Leekfrith (q.v.), Frith Bottom (q.v.), and elsewhere in the country, but not in the south of the county, which is generally applied to woods or uncultivated land forming part of ancient hunting forests. Found in OE as *fyrhth*, *fyrhthe* or *gefyrhth*, usually meaning 'poor woodland': see Gelling 1984: 191-2. The word was adopted into Welsh as *ffridd*, which in the 14th century meant 'barren land', and in moorland areas such as north Staffordshire may have meant 'mountain pasture'. Elliott 1984: 122 suggests that the term 'could be anything from a hedgerow to brushwood, and from a stretch of woodland to a deep forest'. See also Bassetfrythe.

**FRITH BOTTOM**  ½ mile north-east of Meerbrook (SJ 9961). See The Frith.

**FROBISHER HALL**  1 mile west of Swynnerton (SJ 8335). One of a number of Government establishments built in the area in the early 1940s and named after naval heroes. See also Beatty Hall, Drake Hall, Duncan Hall, Howard Hall, Nelson Hall, Raleigh Hall, Rodney Hall.

**FROGHALL**  2 miles south of Ipstones, in a deep valley on the river Churnet (SK 0247). *Frogholle* 1434 Banco, 1481 SHC VI (i) NS 133, *Froghole* 1435 SHC XVII 152, *Frogholn* 1593 SHC 1932 37, *Froghole* 1608-9 SHC 1948-9 5, *Froghall* 1612 SRO D1057/A/1, *Froggall* 1639 Leek ParReg, *Frog hall* 1686 Plot; **FROG HOLE** 1½ miles south-east of Longnor (SK 0962), *Frog Hole* 1775 Yates, 1840 O.S. Perhaps from a relatively common field-name, from OE *frogga* 'frog', and OE *hol* 'small valley, hollow, depression, hole', sometimes upgraded into *Frog Hall*. It has been suggested that the name is found applied to marshy or muddy fields, or perhaps to 'land haunted by frogs': PNEF 86. There is a record of smithymen and colliers at Froghall in 1435 (SHC XVII 152), and it seems possible that the name was a colloquial expression attached to man-made excavations. Plot (1686: 399, 448) shows *Froghall* or *Frog(g)-Homer* near Watling Street, between Knaves Castle and Norton ('near the junction of Old Chester Road with Watling Street at Brownhills': MidA I 1883 123), an area which had then apparently been quarried for gravel. This is presumably the same place as *Far Frog Hall* in Little Wyrley recorded in 1815: SRO D3697/3/20. *Froghall* in Baswich is recorded in 1740 (SRO D240/B/1/50) and in 1891 (Census), and *Frog Hall Leasow* in Walton near Baswich c.1800 (SRO D615/D/33). Crop-marks of high-status timber halls dating probably from the Dark Ages have been traced in a field named Frog Hall at Atcham, Shropshire: WMANS 19 53-4. That name may be from a later quarry pit which has destroyed part of the site.

FROGMOOR  (unlocated, in Dilhorne) *Frogmore* 1214x1243 SHC 1914 91, *Froggemor* c.1271 SRO D938/533, *Frogmoor* 1672 D615/D/49. *Frogmore field* in Gatacre is recorded in 1657: SA 2028/1/2/135.

FULBRIDGE  (unlocated, in Wednesfield) *Fulebrig'* 1262 SHC 4th Series XVIII 36, *Folebrigge* 1271 ibid. 92.

FULE (RIVER)  (obsolete) A tributary of the river Churnet with its source at Meerbrook. *Fulee* 1219-32 Dieul, CEC 385, *Ful(e)*, *Fulhee* 13th Century ibid, *le Fuylhe* 1275 SHC XI 334, *Fulhe* 1330 Ch, 1345 Coram R, 1346 Pat. From OE *ful-ea* 'dark-looking (literally foul) river'. Ekwall 1928: 163-4 notes that the form *Fuylhe* is remarkable, and may indicate that the name was really OE *Fȳle*, a derivative of *ful* with an *ion*-suffix. SHC 4th Series 19 fn.6 identifies *Fulhe* as Dingle Brook. See also Fowlea Brook, *supra*.

**FULFEN** 3 miles east of Lichfield (SK 0510). *Folfen* 1152x1155 EEA 14 64, *Fulfon* 1327 SHC VII (i) 227, *Fulfen* 14th century Duig, *Fulfen* 1470 SRO DW1734/2/1/598 n53, 1576 ibid. 612; FULFEN (unlocated, in Blurton), *Fulfen* 1338 SRO D593/B/1/23/3/2/11. From the OE adjective *ful* 'foul, dirty, filthy, muddy', and OE *fenn* 'fen, marsh, mud, mire', hence 'foul or muddy fen'. For Fulfen in Blurton see SHC XI 312.

**FULFORD** 4 miles north-east of Stone (SJ 9538). *Fvleford* 1086 DB, 1166 SHC 1923 297, *Fuleford* 1272 SHC IV (i) 199, *Fulleford* 1276 SHC VI (i) 80, *Fouleforde* 1280 SRO D593/B/1/23/4/1/10, *Folford*, *Fuleford* 13th century Duig, *Fulford* 1583 SHC II (ii) 39, 1686 Plot, *Foalford* 1614 SHC II (ii) 39. 'Dirty or muddy ford', from the OE adjective *ful* 'foul, dirty, filthy, muddy', normally found linked to elements denoting water, and contrasting with Fairford.

**FULL BROOK** a tributary of the river Trent. Probably from OE *ful* 'foul, filthy, dirty', rathe than from OE *full* 'full, deep'. It has been suggested that the name means more particularly 'a stream with muddy banks': see JEPNS 23 1990-1 48. Other streams of the same name are recorded at Barton under Needwood (*Fulbroke* 1414 Rental), and associated with the river Pipe (*Foulbrouk* 1286 For.). See also Fullbrook.

**FULLBROOK** 1 mile south of Walsall (SO 0196). *fulan sitere* 957 (12th century, S.574), *ffulbrooke water* 1617 Willmore 1887: 439, *Fullbrook* 1834 O.S., *Full Brook* 1895 ibid.' This place takes its name from Full Brook, a tributary of the river Tame, the earliest form of which is from the OE adjective *ful*, oblique *fulan*, 'foul, dirty, filthy', and OE *\*scitere* 'a sewer, a channel or stream used as an open sewer'. *Fullmore*, recorded here c.1572 (SHC 1938 19), and *Fulmores*, *Fulmares More* 1588 (SRO D260/M/F/1/2/f8d) are probably associated with this name. See also Full Brook.

**FULLMOOR** 2½ miles south-east of Penkridge (SJ 9411). *Fulmore* 1505 Hatherton, *Fulmore Green* 1641 Penkridge ParReg, *Fullmore leasowe* 1657 *Survey*, *Foomoor (Road and Lane)* 1775 *EnclA*. From Fullmoor Brook – *þone fule broc* 996 (for 994) (17th century, S.1380), *Foulmire* 1686 Plot, *Full Moor Brook* 1754. From the OE adjective *ful* 'foul, dirty, filthy, muddy', perhaps here meaning 'dark looking', with OE *mor* 'moorland', but see also Fullbrook. See also Gailey.

FUNDLESS (unlocated, perhaps in Sedgley or Upper Penn) *Fundemesleye* 1286 SHC VI (i) 162, *Fundomesle* c.1290 SHC 1928 23, *Fundemesle* 1291 SHC 1911 201, *Fundumesley* 1292 ibid. 209, *Fundesley* 1398 SHC 1928 277, 1425 (p) SHC XVII 118, 1428 SHC XVII 118, *Findesley* 1425 (1798) Shaw II 221, *Fundsleyes, Fundsley, Fudsley (leasow)* 1585 SHC X (ii) 56, *Funleys* 1635 Sedgley ParReg, *Founless* 1639 ibid, *Fudles* 1660 ibid, *the ffundlegs* 1664 ibid, *Fundlesse* 1664 ibid, *Funsley* 1670 Will, *Fundless* 1674 Sedgley ParReg. A curious name, perhaps from an unidentified personal name with OE *leah*. Hackwood 1898: 14, 22 notes that 'Fundemesle is evidently the same as 'Fundesley' (SHC X ii 55) which is variously spelt Funles, Fundles or Funsley. It was probably an estate in Sedgley. In 1350 mention is made of one Hamon de Fundousley; there was also a Jo. Findesley de Seggesley...Fundesles is also variously spelt Fundsleyes, Fudsley, Fudles, Fundlesse, Fundless; but the exact whereabouts of the place is now unknown and its very name forgotten'. It has been suggested (SHC X (ii) 55; Underhill 1941: 128) that this place may have been in Himley, but the evidence is not persuasive.

FURLONG, FURLONG LANE (obsolete, in Tunstall (SJ 8652)) *Forlonge* 1607 JNSFC LXIII 76, *Furlong* 1636 Wolstanton ParReg, *the Furelong* 1664 SHC 1934 (ii) 35, *the Furlong lane* 1743 Burslem ParReg. Self-explanatory. 'Furlong, generally called Smithfield from 1790, or Greenfield from 1800': printed Burslem ParReg III iv (index). The name is preserved in Furlong Road.

**FURNACE GRANGE** 1 mile north of Trysull (SO 8496). *ye furnace* 1641 Penn ParReg, *the furnnice* 1652 Trysull ParReg, *Grange Furnace* 1670 TWHS NS 13 7, *Grange, Furnace* 1798 Yates, *Grange Farm, Furnace Hill* 1834 O.S., *Furnace Grange* 1895 O.S. From a furnace established here in the 17th century. The Grange element is evidently from the Grainger family, who are recorded in the area from at least 1539 (SHC VI (i) NS 70), and occupied Grangemill in 1627: Trysull ParReg; SHC 1923 70. By 1636 Richard Foley had an iron-smelting furnace here: StEnc 227. See also SHC 1923 102.

FURNACE POOL (obsolete) 1 mile north-east of Hednesford (SK 0013). *Furnace Pool* 1649 SRO D1336/1, 1743 SRO D1057/F/2/2, 1834 O.S. From a furnace that stood here: see Shaw 1801: II 315.

FYNSPADE or FYNSPATHE (obsolete) in West Bromwich, probably what is now Wednesbury Bridge over the river Tame: see Dilworth 1976: 103-4; TSSAHS VII 1965-6 23. *Fynspade* 1415 SHC XVII 56, *Fynspathe* 1432 ibid. 141, *Fynspath* 1515 SHC 1928 90, *Fynspade* 1532 SHC 4th Series 8 163, *Fynchpath* c.1575 SHC IX NS 207, *Finspothill* 1686 Plot, 1749 Bowen. Evidently associated with the bridge over the river Tame, recorded as *Wynchespathebrigge, Wystibrigge, Pontem de Pynchespath (for Fynchespathebrigge)* 1287 SHC VI (i) 170, *vill of Vynspade* 1457 SRO 3764/118[27572]. Ede 1969 [108] states that the hamlet of Finchpath was at Hill Top, and Finchpath Hill was the ascent to it from Wednesbury Bridge, but see Dilworth 1976: 103-4 who locates the place on each side of the Wednesbury to Birmingham road from the foot of Holloway Bank (SO 989941 to SO 995923) extending some 700 yards to the west and 200 yards east of the road. Probably from 'the path frequented by finches' (cf. Finchfield), although Finch was a local surname from at least the early 17th century: Roper 1980. See also VCH XVII 6, SHC VI (i) 50, 166; 1928 277.

GADS HILL (obsolete) 2 miles south-east of Biddulph (SJ 9055). *Gadshill* 1775 Yates, 1880 Kelly, *Gadsill* 1815 *EnclA, Gads Hill* 1837 O.S., *Gadshill* 1890 O.S. Possibly from the personal name Gadd or God (DES 181), but Stenton 1970: 286 fn.2 considers that names of this type are likely to mark a re-naming of hills formerly devoted to heathen worship: see Introduction, page 31. The place lies on a long west-facing hill-side. See also Godstone.

**GAGS HILL** 1 mile south-east of Quatford (SO 7590). *Gagshill* 1833 O.S., *Gags Bank* 1839 map, *(Near) Gags Hill (Field)* c.1840 T.A. The spellings are too late to suggest a derivation, but perhaps from the surname Gagg(e) (see DES 181), or possibly from OE *geaces-hyll* 'cuckoo hill'. In Shropshire since the 12th century.

**GAIA** on the north-west side of Stowe Pool in Lichfield (SK 1110). *Gaya* 1160x1182 EEA 16 60, *La Gaia* 1200 SHC III (i) 65, *Geya, Gaya* c.1300 SRO D948/3/15-16, *Gaye Majoris* 1313 WL 94, *Gaia Majori* c.1358 SHC I 243, *Gaia Minor* 1370 SHC VIII 236, *Gey(field)* 1386 *Deed, the Gayfield* 1483 SRO 3764/80. Gaia Major and Gaia Minor were two hamlets in St. Chad's parish, Lichfield (Shaw 1798: I 292) which existed before 1279 and formed the cathedral prebend of Gaia, created probably by 1150: VCH XIV 68. Duignan's suggestion that the name is from the Latin (and OFr) word for 'a jay' cannot be supported, but no other derivation can be suggested, though there is a reference to *Gaya, outside the [Lichfield cathedral] close* in 1323 (SRO D948/3/37-8), which suggests that the word, which appears to be a field-name, might have been a term used in the West Midlands for an area adjoining land pertaining to a religious foundation: a parallel would seem to be Gay (Meadows) between the abbey and the river Severn in Shrewsbury: see PN Sa IV 76-7. The origin of the name Gay applied to two roads in Much Wenlock remains unexplained. It is unclear whether a field called *Le Geye* in Lower Penn, recorded in 1332 (SRO D593/B/1/17/1/1/2) has the same derivation.

**GAILEY** in Penkridge parish, 4 miles west of Cannock (SJ 9110). *Gageleage* 1002x10044 (11th century, S.906; 11th century, S.1536), *Gragelie* 1086 DB, *Galeweye* 1270 SHC IV 180, *Galewey, Gaule, Gaueleye* 13th century Duig. From OE *gagel* 'bog-myrtle' (Myrica Gale), with OE *leah*. The bog-myrtle, a shrub of the family Myricaceae, is known by various names, such as gale, gaul-bushes, gaul, sweet willow, bog-myrtle, moor-myrtle, etc., and usually grows from two to four feet high, with numerous twiggy branches, narrow short-stalked fragrant leaves bearing catkins and a dry berry. The plant was used for animal feed, brewing in place of hops, dying wool yellow and tanning leather. The boiled catkins produced a waxy scum used for candles. GM 1786, Pt i 408, records: 'At a place called Foulmire [i.e. Fullmoor q.v.], about a mile from the Four Crosses, an aromatic shrub of the myrtle kind grows spontaneously. It is called gale or sweet gale, and gives its name to a hamlet near it. Where it flourishes is a black morassy ground between two copses, greatly sheltered from the bleak winds, which no doubt contributes greatly to its safety. It thrives not anywhere else, and seems confined to this small spot of a few acres'. Gailey Hay was one of the Hays of Cannock Forest (q.v.). See also Gauledge.

**GAINSBOROUGH HILL FARM** 1½ miles south-west of Shenstone (SK 0702). *Greensberry Hill* c.1774 TSSAHS IX 1967-8 4, *Greensbury Hill* 1794 Sanders 1794 336, *Greensbury hill* 1801 Shaw II 41, *Greensborough Farm* 1824 ibid. 7, *Greensbury Hill* 1834 O.S. Perhaps from OE *byrig* (dative singular of OE *burh*) 'fortified place', so 'the mound or fortification of the man named Green', or from OE *beorg* 'a hill, a mound, a tumulus'. In 1824 a rock-cut grave was found here, and nearby a hoard of 21 Bronze-Age objects, including swords, spearheads and palstaves: TSSAHS IX 1967-8 1-16.

GAITS HILL (obsolete) on Cowall Moor (SJ 9056). *Gaits Hill* 1836 O.S. Early spellings are not available, but possibly from OE *gat* 'goat', with Scandinavianised pronunciation.

GALLESTONES (unlocated, possibly near Leek) *Gallestones* 1286 SHC 1911 432.

**GALLOWS BRIDGE** 1 mile south-west of Branston (SK 2119). *Galoghbrugge* 1395 SRO DW1734/2/1/103[vi]m47, *Galowe brigge* 1503 Ct, *Galowbridges* 1578 SHC 1927 180, *Gallow Bridges* 1608 SHC 1948-9 81, *Gallowbridge* 1836 O.S. Presumably from gallows which stood on the bridge over the stream from Tattenhill to the river Trent. '... the highway called Gallowbridges which leads from the market town of Burton upon Trent to the city of Lichfield ...' is recorded in 1586: SHC 1927 180.

**GALLOWS GREEN** ½ mile south of Alton (SK 0741). *Gallows Hill* 1834 White 724. Where felons were hanged. It is unclear whether *Galowestewe*, recorded in 1415 (Hardy 1908: II 95) is to be associated with this place. See also Olive Green. Shaw mentions *Gallows Green* in Hamstall Ridware: 1798: I 152 fn. The Green element suggests a grassy area.

**GALLOWS KNOLL** (obsolete) on Wredon, 1 mile north of Ramshorn (SK 0846). *Gallows Knoll* 1686 Plot 404. From OE *cnoll* 'a hill top, the summit of a large hill', later 'a knoll, a hillock', so 'The hill top with the gallows'.

**GALLY MORE** (unlocated, in Shenstone) *Busseys Gallymore* 1629 SRO D4363/C/5/1, *Gally More* 1636 SRO DW1784/2. *Busseys* is Bosses (q.v.). See also Allimore Green.

**GAMBALLS GREEN** 1 mile north-east of Flash, on the Staffordshire-Derbyshire border (SK 0367). *Gamew Green* 1514 DRO D2375M/1/3, *Gambus Greve* 1566 *Deed*, *Gamon grenehed* 1564 SHC 1938 99, *Gambushe Green* 1599 Alstonefield ParReg, *Gambush greene* 1612 DRO D2375M/108/27, *Gambush Greave* 1612 DRO D2375M/52/1, *Gambush(e) Greane* 1651 *Rental*, *Gamble's Green* 1720 VCH VII 39, *Gamballs Green* 1775 Yates, *Gamboles* 1851 White. A puzzling name, with confusingly inconsistent spellings. Perhaps from the common OE personal name Gamel (ON Gamall), more usually found within the Danelaw (see Eyton 1881: 75-6), or from ME *\*gamen-busch* 'sport or game bush', i.e. a bush marking the place where games were held (cf. Upper Gambolds and Lower Gambolds, Worcestershire), with OE *grene* 'a grassy spot'. See also *Gammthorn*, which may be an alternative version of Gambush, whatever its meaning.

**GAMESLEY BRIDGE** 1 mile south-west of Great Bridgeford (SJ 8726). From Gamesley Brook (q.v.).

**GAMESLEY BROOK** a tributary of the river Sow running through Seighford. *Gamelesei* ?13th century SHC IV 269. From the common ON personal name Gamel, with OE *ea* 'river, stream', or OE *eg* 'island, land by a stream or between streams', so giving 'Gamel's stream', or 'Gamels land near the stream', which then gave its name to the stream. Perhaps associated with *Gamyssey Lane ende* in Great Bridgeford recorded in 1538: SHC 1910 44. The name Gamesley is also found in Charlesworth, Derbyshire (PN Db 69), and in Lancashire (PN La 48).

**GAMMTHORN** (unlocated, probably near Moreton/Wolseley) *Gammthorn* 1284 FA. See Gamballs Green.

**GANDER WELL** on the south-west side of Ramshorn (SK 0845). Early spellings have not been traced, but presumably associated with *Gandergrange*, recorded in 1681 (Alton ParReg), perhaps from OE *gand(d)ra* 'a male goose'. The *grange* element is unexplained.

**GANNILDES MER** (unlocated, in Apeton) *Gamuldesmere* 1302 SHC VIII 141, *Gannildes Mer* 1356 SRO 137[7900]. Perhaps from the Scandinavian personal name Gunnhildr (feminine), with OE *mere* 'a mere, a pool'.

GARARDESTHORP (unlocated, possibly on the south-west side of Abbots Bromley, perhaps underlying Blithfield Reservoir: see SRO D(W)1721/3/32/4, which implies that the place may have lain near Mickledale) *Gerardsthorpe* 1342 SRO D(W)1721/3/30B/10, 1367 SRO (D(W)1721/3/32/14, *Garardesthorp* 1483 SHC XI 242. A name of particular interest. It cannot be Gerrard's Bromley, for that place was only acquired by the Gerard family during the time of Elizabeth I. The name is clearly 'Gerard's or Garard's thorp'. The name Gar(r)ard or Ger(r)ard is an OFr personal name of OG origin (cf. Gerard's Bromley). Roger Gerard is recorded in the Loxley area *temp.* Edward I (SRO D(W)1733/A/2/6, Richard, son of Adam Gerard, is recorded in Abbot's Bromley in 1342 (SRO D(W)1721/3/30B/4), and John Gerarde is recorded in Rugeley in 1381 (SHC XVII 186). The second element *thorp*, which may be from OE *þorp* or *þrop*, 'farm, hamlet, dependant farm', but is very often from Danish *thorp*, is usually found only in the Danelaw areas, and rarely in the north-western counties, and seems to have been applied to insignificant places, where the very location is now lost, which is true of this place. The general meaning is probably 'farm', perhaps a dependant or outlying farm belonging to a village or manor. In many cases an original place-name Thorp has been given a distinguishing first element, often an English or Norman personal name, in this case perhaps to distinguish it from Neuthorp (q.v.) rather than Thorp Constantine (q.v.).

GARMELOW (obsolete) ½ mile south of Cop Mere (SJ 801284; SMR 02620 places a deserted post-Conquest settlement at SJ 79542771). *Garmilowe* 1665 SRO D5566/9/1-4, *Garmilow* 1775 Yates, *Garmelow* 1778 SA 1045/742, *Garmeylow* 1810 Gnosall tombstone, *Garmalow* 1834 White 1834: 635, *Garmelow* 1872 P.O. This place is not marked on modern maps, but was a hamlet at the junction of the lane from High Offley to Eccleshall and the lane south from Cop Mere. The forms suggest the name is from an unidentified OE personal name beginning Gærm or Garm, with OE *hlaw* 'mound or tumulus'.

GARRUSLEY (unlocated, possibly near Berryhill, north-east of Stafford) *Garrusley* 19th century SRO E(A)2/17.

**GARSHALL GREEN** 4 miles east of Stone (SJ 9634). *Geringeshalew, Geringeshalow* 1327 SHC VII (i) 202-3, *Gerynshale (lately called Grendonshale)* 1376 SHC 1909 159, *Geringeshalgh, Geryngeshalgh, Geryngeshawe* 14th century Duig, *Geryngsale* 1542 SHC XI 283, *Garringshall otherwise Garshall* 1601 SHC 1935 399, *Garsall* 1608 SHC 1948-9 12, *Milwich Garingshall* 1679 SHC XII NS 156. The first element may be from an OE personal name *Gæring. The terminal is OE *halh*. The *Milwich* addition to the 1679 spelling is doubtless to help locate the place, rather than distinguish it from another place of the same name. Ekwall 1960: 193 and CDEPN cite *Garnonshale*, recorded in 1310 (Ipm) for this place, from the surname Garnon, but the later spellings do not support such derivation.

**GARSTONES** 1½ mile east of Bradnop (SK 0254). *Gastones* c.1288, 1302 Loxdale, *Gatestones Green* c.1288 SHC 1911 443, *The Gastones* c.1302 ibid, *Garstons* 1656 Okeover T669, *Gaston* 1686 Plot 154, 1747 Bowen, *Garston Rocks* 1836 O.S. The c.1288 spelling suggests a derivation from OE *gat* 'goat' or *geat* 'gate'. Although the *stones* element perhaps points towards the latter, *geat* might be expected to evolve into *Yat-*. The remaining spellings would be consistent with a derivation from OE *gærstun* 'meadow, grassy enclosure', sometimes with the meaning 'grazing farm': cf. Garston, Hampshire.

**GATACRE** 2½ miles south of Claverley (SO 7990), since the 12th century in Shropshire. *Gatacra* 1160 Eyton 1854-60: III 86, *Gathacre* 1176 Bowcock, *Gattacra* 1195 Cur, *Gatacre* 1208 FF, 1337 SHC VIII 59, 1298 TSAS LXXI 1996 27, *Gatacr'* 1313 SA 2089/2/2/5, *Gatakere* 1380 SHC XIII 154. From OE *gat-æcer* 'goat field': OE *geat* 'gate' suggested in Ekwall 1959: 193 might be expected to have evolved into *Yat-*. *Æcer*, modern 'acre', probably had the very specific meaning 'a small piece of cultivated land on the margin of a settlement' (EPNE i 3 suggests that it probably meant here 'acre of land'), and examples tend to fall into three categories according to their relationship to heath, marsh, or high moorland. Gatacre is an example of a heathland site. There is evidence that in Staffordshire the 12th century acre was the same size as the modern acre: SHC 1911 418.

**GATEHAM** 1 mile north-west of Alstonefield (SK 1156). *? Gaham* c.1200 SHC 1921 5, *Gateham* 1253 VCH VII 17, *Gatham* 1626 DRO D2375M/52/1, *Gaytom* 1671

Alstonefield ParReg, *Gatom* 1678 ibid, *Gatham* 1703 DRO D3155/6495, *Gateham* 1775 Yates, 1840 O.S. From OE *gat* 'goat', and OE *ham*, giving 'goat village'. It is possible that the first element is OScand *geit* 'goat', though less likely to be found mixed with an OE element. The earliest form may relate to Gotham, Nottinghamshire. It seems unlikely that *Gottham*, recorded in 1259 (SHC 1911 130) relates to this place.

**GATEHAM GRANGE** 1 mile north-west of Alstonefield (SK 1156). *Graunge de Gateham* 1655 PCC, *Gateham Grange juxta Austenfield* c.1680 SP, *Gateham Grange 1839 EnclA*. See Gateham. This was a grange of Combermere Abbey, Cheshire: VCH VII 13; 17.

**GATHERWYND FARM** ½ mile north-west of Blymhill (SJ 8012). *Gatherwynd* 18th century Oakden 1984: 130, *Ryecorn Hill or Gatherwind* 1833 O.S., *Gatherwind* 1836 Blymhill ParReg. 'The hill exposed to the winds'.

**GAULEDGE** on west side of Longnor (SK 0865). *Gorlage* 1415 VCH VII 42, *Gorlege* 1564 Pat, *Gorlidge, Gorledge* 1600 to 1610 ParReg, *Gozledge* 1626, 1651 Rental, *Gorlige* 1669 Alstonefield ParReg, *Gawlid* 1848 TA, *Gauledge* 1840 O.S. Possibly from OE *gagel* 'bog-myrtle', with OE *ecg* 'edge', used in its dialect sense 'a ridge, a steep hill or hillside'. The place lies at the foot of a pronounced ridge. Aderivation incorporating dialect *gorl, goal* 'violent wind, howling wind' (EDD) cannot be discounted completely, but is unlikely.

GAYFORD (unlocated, possibly near Pelsall) *Gayford* 1310 SHC 1911 307.

**GAYTON** Ancient Parish 6 miles north-east of Stafford (SJ 9828). *Gaitone* 1086 DB, *Gaiton* 1203 SHC III 104, *Geiton* 1204 ibid.' 126, *Gaidon* 1227 Duig, *Gayton* 1285 SHC 1910 299, 1306 SHC VII 167, 1324 SHC X 45, *Gaytton* 1532 SHC 4th Series 8 58. Probably '*Gæga's *tun*', from an OE personal name related to OE *gēgan* 'to turn aside', or (less likely) from the same name adopted as a stream name for Gayton Brook here. There is no trace of a ME Gatton that would be expected from OE *gata-tun* 'goat *tun*', but a derivation from ON *geit* 'goat' is not impossible. The name is found in Cheshire and several Midland and eastern counties: see Ekwall 1960: 194.

GAYWODE HALL (unlocated, in Stretton, near Penkridge) *Geiwode* 1203 SHC III 122, *Gaywode temp.* Edward I SRO DW1733/A/2/72, *Gaywode Hall* c.1317 VCH IV 167, *? Gaywode* 1332 SHC X (i) 112, *Geywode Hallestede* c.1337 VCH IV 167, *Geywode Hallestede juxta Annwalle* c.1338 ibid. Possibly 'Gæga's wood'. *Hallestede* is from OE *stede* 'place, site of a building', so 'site on which the hall stood'. *Annwalle* is unlocated, perhaps from OE *an* 'one', *ana* 'lonely', and Mercian OE *wælle* 'a spring, and (sometimes) a stream', so 'the isolated spring'. See also SHC XIV (ii) 10.

GENDALL'S COPPICE (obsolete) ½ mile west of Blount's Green (SJ 0732). *Gendall's Wood* 1865: 267, *Gendall's Coppice* 1887 O.S.

**GENTLESHAW** ½ miles north-west of Lichfield (SK 0512). *Gentylshawe* 1505 SHC XI 268, 1528 SHC 1939 77, *Gentleshave* 1589 SHC 1931 250, *Gentleshore* 1788 SHC 4th VI 167. The name was originally attached to a grove of ancient oaks on a high part of Cannock Chase at Longdon. The terminal is OE *scaga* 'a grove, a copse'. In 1338 John Gentil was steward to the Bishop of Lichfield: SHC III NS 97. See also SHC XI 120; SHC 1912 250; SHC 1939 82.

**GERRARD'S BROMLEY** – see **BROMLEY, GERRARD'S**.

**GIBBET WOOD** 1 mile north-east of Whittington (SO 8683). *Gibbet Wood* 1895 O.S. By tradition from a gibbet, a tree or post from which the bodies of executed criminals were displayed, used in 1813 for displaying the body of William Howe, alias John Wood, for the murder of Benjamin Robins of Dunsley Hall (VCH XX 125). He was not, as sometimes claimed, the last man to be gibbeted in England. The name does not appear on the 1834 O.S. map. See also StEnc 224.

**GIBRALTAR** in Kinver, to the south of Dunsley (SO 8583), *Gibraltar* c.1780 VCH XX 124, *Gibraltar Rock* 1834 O.S.; **GIBRALTAR** south of Knightley Grange (SJ 8023), *Gibraltar* 1833 O.S.; **GIBRALTAR** on the east side of Cheadle (SK 0113), *Gibraltar* 1891 O.S.; **GIBRALTAR FARM** 1 mile south of Rudyard (SJ 9556). A transferred name, which is not uncommonly found applied to places which were remote or had rocky features. The first place, accessible only by foot and canal, lies at the foot of a sandstone cliff. The second place lies close to the Gnosall parish boundary. Some of the names may have been adopted during the siege of Gibraltar 1779-83: see Field 1993: 151.

**GIB RIDING (WOOD)** 2 miles north-east of Cheadle (SK 0344). *Gibbe Ruydinges* 1291 Chester 1979: 4, *Gibbe Ruyding* 1309 CroxdenChr, *(bosca de) Gibbe rydinges* 1345 ibid., *(wood called) Gibberydynge* 1539 MA, *Gib-riding* 1836 O.S From ME *gibbe* 'a hump', with with OE *\*ryding* 'a clearing, an assart, land taken into an estate from waste'.

**GIB TORR** 2½ miles north of Upper Hulme (SK 0264). *Gibtor, Gybtorr* 1481 DRO D2375M/1/1, *le Gybtorre* 1515 DRO D2375M/1/6, *Gybtor* 1559 DRO D2375M/1/3, *Gybter* 1564 SHC 1938 99, *Gib Torr* 1564 VCH 7 33, *Gyb(be)torre* 1566 Deed, *Gybbe Torre* 1584 Alstonefield ParReg, *Gybtoore* 1605 ibid, *Gibb Tarr* 1769 Alstonefield ParReg, *Gybtor* 1775 Yates, *Gibtar* 1832 Teesdale. From ME *gibbe* 'a hump', with OE *torr* 'hill', usually in moorland areas meaning 'a rocky outcrop', hence 'the rounded hill with a rocky outcrop'. Gib Torr Rocks (*Gib Torr Rocks* 1842 O.S.) lie a quarter of a mile west of Gib Torr.

**GIFFARD'S CROSS** 1 mile south of Brewood (SJ 8707). *Gyffarde's Crosse* 1569 Ct, *Jiffards Cross* 1816 O.S. By tradition this ancient wooden cross commemorates an exploit in the early 16th century when a member of the Giffard family (the 1816 spelling reflecting the soft *G-*) from nearby Chillington Hall killed with a bow or crossbow a panther which had escaped from a menagerie at the Hall and was about to attack a mother and child. The legend has not been traced back earlier than the 1840s, and is almost certainly apocryphal: see Horovitz 1992: 168-72. The original cross was replaced by a replica in the 1980s, and is now at Chillington Hall..

**GIGGETY** on the south side of Wombourne (SO 8692). *Giggatree* (field name) 1840 TA, 1832 Teesdale, *Giggetty* 1895 O.S. Possibly from gibbet-tree, a tree or post from which the bodies of executed criminals were displayed.

**GIGHALL BRIDGE** over the river Dane. *Gig Hall* 1880 Kelly. From the local dialect word *gigge* 'a hole in the ground for a fire over which flax was dried' (EDD). Cf. Gig Hole, Kingswood, Cheshire. The *hall* element is from OE *halh*: there is no record of any hall here.

GILBERDE (unlocated, in Marchington) *Gilberde* 1520 SRO DW1733/A/3/11.

GILBERT BRIDGE (obsolete) The bridge carryng the Leek-Ashbourne road over the river Hamps at Waterhouses (SK 0850). *Gilbert Bridge* 1764 JNSFC LXXXIII 1948-9 45.

GILBERTS (unlocated, in Farley) *Gilberts* 1753 SRO D240/D/291.

**GILBERT'S CROSS** ½ mile west of Enville, at the junction of Morfe Lane and the Stourbridge-Bridgnorth road (SO 8186). *Gilbert's Cross* 1834 O.S. If the name is ancient, perhaps from Gilbert who held Enville in DB (VCH IV 54; VCH XX 94), or from Gilbert, son of John Fitz Philip II, who held land in Bobbington in the seond half of the 13th century: Eyton 1854-60 : III 171. The cross was perhaps a boundary mark between Enville and Morfe.

**GILL BANK** 1 mile south of Kidsgrove (SJ 8453). *Gilbank* 1625 SRO D1229/1/2/6, 1649, 1672 Wolstanton ParReg, *Gillbank* c.1727 SRO D997/1/1. Possibly from ME *gille*, from ON *gil* 'ravine, narrow valley' (see Elliott 1984: 103-4, where a dialect usage 'wooded valley' is postulated): there is a narrow valley in the hillside here. See also Gillow Heath; Guild of Monks. *Gill Lane* and *Gillfield* are recorded in Audley in 1612: SHC 1944 82.

GILLBRIDGES (unlocated, in Croxton) *the Gillbridge* 1817 SRO D1798/685/114, *Gillibridges* 1775x1863 SRO D1192/24/1-41. Perhaps from gill used in the sense 'brook or rivulet', recorded in OED from 1625.

**GILLEAN'S HALL** 1 mile south-east of Abbots Bromley (SK 1022). It is unclear whether this place is to be associated with *Gyllians Hayes* 1651 SRO D603/E/2/119, *Gillens Hays* 1730 SRO D603/E/2/165-6, *Gillian Hays* 1780 SRO D603/A/3/44-57. *Gillians Bower* in Wolseley is recorded in 1698x1770 (SRO DW1781/9/2/62/1-5), and as *Gillings Bower* in 1849 (SRO DW1781/10/1/56). The surname Gillean and similar is found in various parts of the county, for example Gyllyans 17th century: Dudley CA.

GILLITY GREAVES (obsolete) 2 miles east of Walsall (SO 0197). *Gyllot in greves* 1525 VCH XVII 175, *Gillot in le Greves* 1614 SRO D260/M/F/1/2/f147d-148, *Gillott in the Goraves* 1617 Willmore 1887: 439, *Jennity Greave* 1832 Teesdale, *Ginity Greaves* 1834 O.S. Duignan 1902: 67 states that in the 14th and 15th centuries the place is regularly recorded as *le greve*. The name is from OE *græfe* 'grove, copse, thicket', a place-name element common in the West Midlands, especially Staffordshire. The name is later found as *greves* and *greaves* ('woods'), and later still as *Gillott o' th' Greaves* (Duignan 1902: 67), presumably from a family of that name who lived there, or perhaps from *gyllot* 'hussy': OED.

**GILLOW HEATH** 1 mile north-west of Biddulph (SJ 8758). *? Gilleloh* 1227 SHC XI NS 240, *Gillow(e)* 1279 SHC XII NS 12-3, *Gylloowe Hetht* 1427 Ct, *? Gillow* 1551 SHC 1924 3, *Gylowe* 1576 Biddulph ParReg, *Gille heathe* 1660 ibid, *Gillow, Gilloe* 1663 ibid, *Gilloe heath* 1675 ibid, *Gillow* 1676 SHC XII NS 12, 1755 SRO D593/B/1/20/20/12, *Gillow Heath* 1744 BCA MS917/1258, 1842 O.S. The first element is uncertain. Gillow in Herefordshire is thought to be from Welsh *cil* 'nook, retreat' and *lwch* 'pool' (Coates & Breeze 2000: 307), hence 'the nook by the pool' , but this can be ruled out here. ON *gil* 'ravine, valley' is possible (but unlikely) as the first element (this place lies near two streams in marked valleys), perhaps with OE *hlaw* 'mound, hill, tumulus', or possibly OE *hoh* 'a hill spur, ridge end'. It may also be noted that Plot 1686: 203 mentions the *stock-Gillo-flower*, evidently Gillyflower, which are recorded as 'clove gillie flowers' paid as rent in the early 14th century: see SHC 1912 346. See also Gill Bank; Guild of Monks. For *Gillihay* (unlocated) see Erdeswick 1844: 167.

GILPIN'S MILL (obsolete) 1½ miles south of Cannock (SJ 9609). *Gilpin's Basin* 1792 *EnclA*. From a mill which lay between the canal bridge and Watling Street, said to date from the foundation of William Gilpin's edge tool works here in 1790 (VCH V 49).

**GLASCOTE** 1 mile south-east of Tamworth (SK 2203). *Glascote* 1206 FF, 1262 SHC IV 152, 1276 SHC VI (i) 78, 1292 Ipm, 1307 SHC VII 174, 1432 IpmR, 1568 SHC IX NS 166, 1614 SHC IV NS 64, *Gloscot* 1565 ParReg, *Glascocke* 1667 HT. From OE *glæs* 'glass', and *cot* 'cot, hut, shelter', probably meaning here 'the glass workshop': the element *cot* is often found associated with industrial processes. Another possibility is that the first element may be from a river-name Glas, perhaps the small stream on which the place stands: PN Wa 26. See also Ekwall 1928: 175.

GLASHOUSHAY (unlocated, in Wolseley Wood, possibly near Stafford Brook: see Welch 1997: 30) *Glashouse hey* 1483 Welch 1997: 30, *Glashoushay* 1561 SHC NS IV 212. In this area were a number of early glassworks (see Welch 1997), from which this place took its name. *Hay* is from Mercian OE *(ge)heg* 'enclosure'.

GLASS HOUSE (obsolete) on the north side of Chesterton (SJ 8250). *Glassehouse* 1683 JNSFC LXV 1930-1 46, *Glass House* 1833 O.S. From early glass-works here: see JNSFC LXVIII 1933-4 74-121.

**GLASS HOUSE** on the Staffordshire-Shropshire border, 2½ miles south-west of Ashley (SJ 7432). *the Glassehowse* 1600 JNSFC LXV 1930-1 47, *the glass house* 1675 SRO 1/279/73, *the Glashoush* 1679 Eccleshall ParReg, *the Glasshouses, the Glasshowse* 1704-5 SRO 828/17-19, *Glass House* 1833 O.S. From glass-making carried out here in the 16th century: SHC 1945-6 12; 130; JNSFC LXV 1930-1 45-54; LXVIII 1933-4 74-121. This may be *Brass Hall* recorded in the Hearth Tax Roll of 1674 (SHC 1945-6 156 fn.), but see SHC 1945-6 175. The county boundary here has been modified since 1833.

**GLASSHOUSE FARM** 2 miles south-west of Marchington (SK 1029). *Glasshouse Bank* 1724 *Survey*. From medieval glassworks in Bagot's Park to the south: see PMA 31 1997 1-60.

**GLASS LANE** running south from Bromley Wood, 1 mile south-east of Abbots Bromley (SK 1022). *le Glaslone* 1266 (1798) Shaw I 204, *Glass Lane* 1836 O.S. From glassworking carried out in this area. Curiously, Shaw (1798: I 204) identifies *le Glaslone* as 'the present Hickberry-lane which crosses the two clear streams of the Blythe'. The lane formerly continued south across the river Blithe to join the road between Hamstall Ridware and Blithbury. The 1836 1" O.S. map shows Hickbury House at the southern end of that lane at SK 0919.

GLASTWOOD (unlocated, probably in the east or north-east of the county) *Glastwood* 1414 (1798) Shaw I 43. Probably from OE *\*glæste* 'glade, sunny place', so 'the wood with the glade(s)'.

**GLAZLEYFIELD** 1 mile north-east of Barlaston (SJ 9139). *Glaseleye* 1293 SHC 1911 49, 1327 SHC VII (i) 241, *Glasleye* 1325 SHC VIII (i) 150, *Glaseley Field* 1548 SRO D1810/f.193, *Great Glazie, Long Glazie* (field-names) 1631 SRO D476/1/6/3, *Glazley Field* 1836 O.S. Ekwall 1960: 198 indicates that the etymology for Glazeley in Shropshire (with *Gles-* early spellings: see SHC VI NS 82, 86, 88; Ekwall 1960: 198) is doubtful, but suggests the possibility of a derivation from OWelsh *gleis*, Welsh *glais* 'stream': cf. Glasshampton, Worcestershire; Glaze Brook, Devon and Lancashire; Glaisdale Beck, North Yorkshire; see also Ekwall 1928: 175-6. The word is found in several stream-names in Wales: Ekwall 1928: 175. That derivation seems to be preferred by PN Sa I 136-7. This place lies near the headwaters of two streams which unite to flow north-west, so such a root cannot be ruled out here: the 1631 spellings suggesting a stream-name *glais-ea* 'the stream called Glais' must be discounted, however, in the light of the earlier forms, which show that the second element is OE *leah*. An alternative derivation might be OE *glæs* 'glass', giving 'the wood or clearing where glass was made', but the 1293 spelling would be early for glassmaking in Staffordshire (although *Glascroft* near Lichfield, perhaps from *glæs*, is recorded in 1215: VCH XIV 120). However, OE *glæs* was also an adjective meaning 'clear, bright, shining', and since early glass was often blue-green, there is sometimes a colour sense involved in its usage (see Ekwall 1928: 175), so possibly 'the bright wood' (i.e. with the trees well-spaced), or 'the wood or clearing with the green-blue foliage or vegetation'. For Celtic *glas* 'green growth' see Coates & Breeze 2000: 160-1. The Glaseley or Glazeley family held land at Barlaston from at least 1359 (SHC IV 71-2), and doubtless took their name from this place.

GLEADLEY (unlocated, possibly near Biddulph) *Gleadeleay* 1662 Biddulph ParReg, *Gleadley* 1665 ibid, *Gleadleay* 1670 ibid.' Probably from OE *gleoda* 'kite', with OE *leah* 'a clearing', so 'the clearing with the kite'.

GLEDENHURST (unlocated, probably at Woodhead (q.v.) in Cheadle: SOT D1798/179b) *Gledenhurst* 1331 SHC 1913 36, *Glodonhurst in Hundchedull* 1539 MA, *Gledemhurst* (p) Cheadle ParReg, *Gledenhurst* (p) 1586 ibid, *Gledenhurst* 1675 SOT D1798/179b. Probably from OE *gleoda* 'kite', with OE *hyrst* 'wooded hill', so 'the wooded hill with the kites'.

**GNOSALL** Ancient Parish (pronounced know-sul [nəʊsəl]) 5 miles south of Eccleshall (SJ 8220). *Geneshale* 1086 DB, *Gnowesala* 1140 MRA, *Gonw(e)shal(e)* c.1149x1206 MRA (14th century), *Gnouuesh* 1160x1182 EEA 16 100, *Gnousale* c.1165 Fr, *Gnou(e)shale* c.1181x1184 StCart (14th century), 1189 MRA, *Gnoweshalia temp* Henry II Bk, *Gnodweshal, Gnodeshall* 1199 Ass, *Gnoushal'* 1221 Ass, *Gnoushale* 1221 Ass, 1395 SHC 1927 101, *Gnossal* 1222 Ass, *Cnoushale* 1223 Bracton, *Gnoshale* 1227 Ass, 1307 GDR, *Gnoushala* 1230 Ch, *Gnos(se)hal(le)* 1242 Pat to 1521 LP, *Gnousal* c.1255 RH, *Gnoweshale* 1286 SHC V (i) 155, *Knoshale* 1292 SHC VI (i) 272, *Knoushalla* 1321 Inq, *Gnowsale* 1348-1363 Pap, *Knousale* 1365 SHC VIII NS 25, *Knoshale* 1414 Ch, *Gnowsall* 1462 FF to 1526 StarCh, *Gnostall* c.1502 Bod. 28, *Knossall* 1532 SHC 4th Series 8 94, *Knashall* 1607 Kip, *Nosall* 1643 StComm. Ekwall 1959: 199 tentatively suggests a derivation from an OE personal name formed from OE *gneap* 'niggardly', but concedes that the absence of early spellings where -*a*- might be expected poses difficulties. The various spellings, supported by the modern pronunciation, point towards a derivation from OE *\*Gnoweshalh*. The root of the name is OWelsh *genou*, from a British plural form *\*genoues* 'mouth, opening of a valley'. The identification of *genow*/*geneow* as a Welsh loanword, one of a number in the West Midlands dialect, is widely accepted: see N&Q 238 [NS 40] I 13-4. The word is found with its Welsh spelling *genou* in the Book of Llandaff c.1135-40, and in the 13th century Black Book of Chirk the spelling *geneu* is standard. Borrowed Welsh words are normally anglicised by stress on the first syllable, but in cases where this does not occur the intervening vowel may disappear: cf. Cannock. Since *gn*- is found in Welsh, the first vowel may have disappeared in this case when the area was still Welsh-speaking. It seems therefore that the place may have been known originally as Geneu by the native British, anglicised as Gnow and with the second element added later by Anglo-Saxon settlers. The meaning in place-names may be 'the mouth (in some topographical sense); the constricted valley; the narrow passage', here perhaps with reference to the topography around the junction of Hollies Brook and Doley Brook on the south-west side of the village, or 'the narrowing of the valley with the low-lying land near the stream', applicable perhaps to the most striking feature of this place, a broad flat stream valley on the west which suddenly becomes a narrow stream valley between Gnosall and Gnosall Heath: see also Doley Common; Doley Gate. The second element of Gnosall is clearly OE *halh*, perhaps here in the sense 'a piece of low-lying land by a river'. It is possible that *geneu* was applied to a constricted passage which suddenly opens into a wide valley (or vice versa): see Coates & Breeze 2000: 184-92. The same root is to be found in Gannow Farm & Gannow Wood (Inkberrow), and Gannow Farm & Gannow Green (Bromsgrove), Worcestershire; Gannah, Herefordshire; Gannow, Lancashire; Gannaway, Warwickshire; and perhaps field-names such as *Gannoweslonde* and *Gannowestockyng* (see PN Wo 342). Staffordshire dialect *gennel, jornal* mean 'narrow passage' (see Wilson 1974: 36, 40), may have the same root. Cf. Ganarew, Herefordshire. *Knowsales House* is recorded in the Handsacre/Armitage area in 1552: SHC XII (i) 209.

**GNYPE** – see **TURNER'S KNIPE**.

**GOAL BUTTS** on the south-west side of Eccleshall (SJ 8228). *Gillbutts* 1672 Eccleshall ParReg.

**GODLEY BROOK** a tributary of the river Tean. *Godl(e)y Brook* 1837 O.S. In the absence of earlier forms a derivation from 'Goda's *leah*' must remain uncertain, but cf. Godley, Cheshire (PN Ch I 306).

**GODSTONE** 1 mile south-west of Church Leigh (SK 0134). *the Godstones* c.1680 SRO D1203/B/3-6, *The Godstone* 1789 SRO D543/C/7/10, *Godstone* 1832 Teesdale, 1836 O.S. A curious name, perhaps associated with a legendary origin for a rock outcrop here, but see also Gads Hill.

**GOGESMORRE** (unlocated, possibly near Beffcote, perhaps Goosemoor (q.v)) *Gogesmorre* 1707 BCA MS3145/63/1a&b.

**GOLDEN** (unlocated) *Golden* 1281 SHC 1911 37.

**GOLDEN BANK** (obsolete) on the west side of Pattingham (SO 8199). *Golden Bank* 1942 Brighton 1942: 19. In 1780 a small gold ingot was found here, possibly associated with a gold torc found nearby in 1700: Erdeswick 1844: 364. The name is evidently 'the bank or hillside where gold was found'.

**GOLDENHILL** 2 miles north of Tunstall (SJ 8553), *Goldenhill* 1670 VCH VIII 83, *Golden hill* 1686 Plot; **GOLDENHILL** 2 miles south-west of Ashley (SJ 7432), *Gold Hill* 1833 O.S. Perhaps from OE *golde* 'marigold', meaning 'hill where marigold grew', or 'hill golden from buttercups': an alternative name for the buttercup was the gold-cup. Or possibly 'the hill where gold was found'. For Golden Hill 2 miles west of Butterton (SK 0556), see Lousey Bank. *Goldenhill* is recorded in Sedgley ParReg in 1657.

**GOLDIE BROOK BRIDGE** 1 mile north-west of Shareshill, on Saredon Brook (SJ 9308). *Godyenebrugg* 1307 SHC 4th Series XVIII 190, *Goldy Bridge* 1749 Bowen, *Goodybridge* early 18th century *Terrier*, *Goldybridge* 1804 SRO D3186/8/1/30/8. From OE *golde* 'marigold' and OE *eg* 'island, raised land in wet ground', giving 'raised ground in wet land where marsh marigolds grow'. The brook gave its name to field-names *le gooldylond, le gooldelyhadelond* recorded in 1441 (*Vernon*).

**GOLDS GREEN** 2 miles north-west of West Bromwich (SO 9893). *Golds Green* 1834 O.S. From the Golds family who are recorded here in 1332, and were probably living here by the later 13th century: SHC VIII 86; VCH XVII 6; Ede 1962: 27. The Green element suggests a squatter settlement.

**GOLDSITCH** 2 miles north of Upper Hulme (SK 0064). *Goodsich Fall* 1564 Ch, *Goldsich, Gouldsich* 1643 Leek ParReg. From OE *sic*, 'a small stream, especially one in flat marshland', and often applied to the marshy pasture land bordering such a watercourse. If the earlier spelling is reliable, it would seem that *Gold-* may be a late development. Otherwise, it may refer to yellowish water or to marigolds (OE *golde*): the place has an abundance of marsh marigolds (called locally golds or goulds) in spring: see Goldenhill, above.

**GOLD'S WOOD** 1 mile west of Middle Mayfield (SK 1344). Of unknown age and derivation (earlier spellings have not been traced), but possibly associated with nearby Hordle Sprink (q.v.).

**GOLDTHORN (HILL)** 2 miles south-east of Wolverhampton (SO 9196). *Goldhord* 1291 SRO D593/B/1/17/1/4/4, *le Goldhord* 1302 SRO D593/B/1/17/1/4/4, *Golthord* 1318 SRO D593/B/1/17/1/4/13, *Golterne* 1589 SHC 1929 339, *Gouldthorne* 1612 SHC III NS 68, *Gouldthorne hill* 1634 Wolverhampton ParReg, *gouthorne hill* 1636 ibid, *Gouterne hill* 1651 ibid, *Goldthorn, Goldthorne* 1686 Plot, *Goldthorn Hill* 1737 WALS DX-240/7, 1834 O.S. From OE *gold-hord* 'gold hoard, gold treasure', applied to places where treasure has been found, possibly to be associated in this case with *Burnildlowe* in Penn recorded in the 13th century (SRO D593/B/1/17/1/3/10; See also WA II 41-2), a name meaning 'Brunhild's tumulus' or 'the hill with the burial-mound': cf. Goltherdesbeuch; Burnhill Green. Although the expression *gold-hord* was also used for a privy (EPNE i 205; Parker 1996: 257-8), that meaning is improbable in early place-names. *Goldthorn Acre* in Pattingham is recorded in 1683: SRO DW1778/V/1330; gold has been found in Pattingham (see Golden Bank, also Shaw I 1798 32-3, II 279; TNSFC 1964 31), but in the absence of early spellings the origins of that name are uncertain. *Goldhorde style* in the Shipley/Rudge Heath area is recorded in 1619 (SA 330/25), and *Golthordeshull*, recorded c.1300 (SHC 4th Series XVIII 185), appears to have been in Kinver Forest, which included Lower Penn (VCH II 343): it is not clear whether it is to be associated with this Goldthorn Hill – the original *Goldhord* may have been some distance from the area now known as Goldthorn Hill, which may be 'the hill on which the hoard of gold was found', or 'the hill at (or near) the place called Goldhord'. For *le Goldhord* in Uppington, Shropshire, see TSAHS L 1939-40 36. The compound is also found in Goldsworth in Woking, Surrey; Goldhard in Godstone, Surrey; *Goldhorde Field* in Chiddingfold, Surrey; Gollard in Amport, Hampshire; Goldsworth in Stoke, Cheshire; and Gaulter in Steeple, Dorset. Many such place-names are near ancient roads or tracks. See also Hulhord.

GOLLING GATE (obsolete, 1 mile north-west of Hollinsclough (SK 0467)) *Golldeayate, Geldleayate* 1634 Leek ParReg, *Goldhaygate* 1635 ibid, *Goldhay-yate* 1637 ibid, *Golling Gate* 1775 Yates, *Golling Gate* 1842 O.S., *Gollingate* 1851 White. From OE *golde, (ge)heg* 'marigold enclosure', with OE *geat* 'gate, enclosure; gap'.

GOLTHERDESBEUCH, GOLTHORDESBEUCH (unlocated, probably in the area to the south of Wolverhampton) *Goltherdesbeuch, Golthordesbeuch* 1296 SHC 4th Series XVIII 185. Perhaps to be associated with Goldthorn (q.v.), or possibly near Trescott. The name seems to be 'the *bece* or pronounced stream valley at the place known as Gold Hoard', i.e. where treasure had been found.

GOMES MILL (obsolete) on the southern side of Longton (SJ 9042). *Gom's Mill* 1632 StEnc 701, *Goms mill* 1746 Stoke on Trent ParReg, *Gomes Mill* 1798 Yates. Said to be the site of a quadrangular moat, with a corn mill erected on Furnace Brook by 1632: StEnc 701. Probably from an unidentified personal name. Cf. Gomshall, Surrey.

GOODCAR (unlocated) *Goodcar* 1577 SRO D(iv)1490/15.

GOOD COW (obsolete) 2 miles east of Biddulph (SJ 9157). *Goodcow* 1815 *EnclA*, 1842 O.S., *Good Cow* 1815 *EnclA.*, 1891 O.S. Self-explanatory.

**GOOD'S GREEN** in Upper Arley. From the personal name *le Gode*, found in the SR of 1327 and 1332, and in a Will of 1584: PN Wo 30. The Green element suggests a squatter settlement. In Worcestershire since 1895.

GOOLDBURYNES (unlocated, in Brewood) *Gooldburynes* 1453 Oakden 1984: 47. An interesting field-name meaning 'burial-mound where gold was found', from OE *byrgen* 'tumulus, burial-mound'.

**GOOSEMOOR** 1½ miles west of Church Eaton (SJ 8217). *Gosemere* 1331 Banco, *? Gosenere(pole)* 1349 SHC XIV (ii) 35, *Gosemer sych* 1349 Deed, *Gosmore* 1674 Gnosall ParReg, *Gausemore* 1679 SHC 1919 221, *Gosmore* 1763 SRO DW1909/A/9, *Gosmoor Heath* 1775 Yates, *Goosmoor Common* 1777 SRO 590/58/1-54, *Goosemoor Sitch* 1838 TA. From OE *gos mere* 'the pool of the geese', with OE *mere* 'pool, wet ground' (with OE *pole* 'pool' added to the 1349 form), later becoming *mor* 'marshland'.

**GOOSEMOOR GREEN** 4 miles south-east of Rugeley (SK 0611). *Gorseforthe Greene* 1584 SHC 1939 112, 1608 SHC III NS 21, *Gorsemoor Green* 1775 Yates, *Goosemoor Green* 1834 O.S. From OE *gorst, gors* 'gorse, furze', with an uncertain second element which has become 'moor, marshland', with ME *grene*, probably here denoting a squatter settlement. See also Gogesmorre.

**GORNAL, UPPER & LOWER; GORNALWOOD** 2 miles north-west of Dudley (SO 9292, 9191). *Gornhal temp.* Henry III, *Goronhale* 1375 BM, *Gornehale* c.1400 DR37/2/Box 122/36, *Gwarnell, Guarnell* 15th century Duig, *Gwornall* 1532 SHC 4th Series 8 115, *Gwornolde* 1565 SHC 1926 143, *Over Gwarnall* 1590 SHC 1930 57, *Nether Gwarnall* 1590 ibid.' 58, *Gornal(wood)* 1659 HRO E12/V1/NB/14, *Guarnall* 1664 SHC II (ii) 51, *Gournal* 1686 Plot. Gelling & Cole 2000: 130 suggests a derivation from OE *cweorn-halh* 'mill *halh*' (usually in early names with reference to a water mill: windmills were not introduced into England until about the 13th century), but Gornal is on high ground, and *cweorn* was often used of places where mill-stones were obtained (cf. Quernmore, Lancashire; Quorndon or Quorn, Leicestershire; Quarrington Hill, Durham; Quarley, Hampshire), and here '*halh* where mill-stones were produced' (cf. modern 'quern') is more likely: Shaw 1801: II 222 states that '... at Cotwall-end, they dig excellent grinding stones ...', and there is a record that a quantity of querns were found here in the 16th century: VCH I 192. Hackwood 1898: 105 states that 'the ancient name of Upper or Over Gornal is Sheep Cotwall as appears by the Court Rolls', but no evidence has been traced to support that statement. See also Cotwall End. *Over* and *Nether* are 'Upper' and 'Lower'.

**GORSEBROOK** 1 mile north of Wolverhampton (SJ 9100). *gos broc* 985 (17th century, S.860), *Gosbroke, Gosebroke* 14th and 15th centuries Duig, *Goosbrook (Mill)* 1708 Bushbury ParReg, *Gosbrook* 1834 O.S. From OE *gos broc* 'goose brook'.

**GORSE, THE** ½ miles east of Yarnfield (SJ 8532). *The Gorst* 1737 Swynnerton ParReg, *Gorse* 1737 ibid. From OE *gorst* 'gorse'.

GORSEGATE (unlocated) A district of Walsall. *Gorsgate* 1798 Shaw I 79.

**GORSE HALL** 1 mile north-east of Barton under Needwood (SK 2019). *Gorse Hall* 1836 O.S.

**GORSEHOLM** (unlocated, in Timmor) *Goseholm* 1241 (1798) Shaw I 375, *Gorseholme* 1550 SHC 1912 191. 'Goose *holm*'.

**GORSEY HILL FARM** on south-west side of Barton under Needwood (SK 1818). *? Gorsthull* 1284 SHC VI (i) 143, *Gorsthull* 1415 Hardy 1908: 72, *Gorstill* 1525 ibid.' 203. 'The hill with the gorse'.

**GORSTHULL(E)** (unlocated, possibly near Burntwood) *? Gorsthulle* 1234 SHC XII 35, *? Gorsthull* 1279 SHC VI (i) 143, 1311 SHC IX 29, 1344 SHC XII 35. From OE *gorst* 'gorse', with OE *hyll* 'hill'.

**GORSTY CROFT (FARM)** 1 mile north-east of Foxt (SK 0449), *Gorstycroft* 1616 Kingsley ParReg, *Gorsty Crofte* 1619 ibid, *Gorsty Croft* 1836 O.S.; GORSTY CROFT (unlocated) in West Bromwich, *White's Croft alias Gorstie Crofte* 1615 Willett 1882: 164, *Gosty Crofts* 1651 SRO D260/M/E/425/1. From the OE adjective *\*gorstig* 'overgrown with gorse', with OE *croft* 'a small enclosure of arable or pasture land, an enclosure near a house', so 'the small enclosure overgrown with gorse'.

**GORSTYE HAYE** (unlocated, possibly near Burntwood) *Gorstye Haye* 1498 OSS 1936 49.

**GORSTY HILL** 2 miles south-west of Marchington (SK 1029), *Gorsty Knoll* c.1250 SRO DW1721/3/4/3, *Le Gorstiknol* c.1290 SHC 1937 85, *Gorsty Hill* 1734 D786/26/5, *Gorsty Hill* 1836 O.S.; **GORSTYHILL** 1 mile north-west of Balterley, on the border with Cheshire (SJ 7450), *Gorstihill, Gorstiehill* 1581 Betley ParReg, *Ghorsty Hill* 1833 O.S. From the OE adjective *\*gorstig* 'overgrown with gorse', with OE *hyll* 'hill', so 'the hill overgrown with gorse'. *Gorsty Hill* in Cheadle is recorded in 1605: SRO D786/26/6.

**GORTON GREEN** 1 mile south-west of Longdon (SK 0712). *Gamton Green* 1798 Yates, *Gorton's Green* 1815 SRO DW1885/4/1/26, *Gortons Green* 1834 O.S. Probably from the family name Gordon or Gorton, recorded in the area in 1461 and 1491 (OSS 1936 42, 47). The Green element suggests a squatter settlement.

**GOSCOTE** 3 miles north of Walsall (SK 0102). *Gersicote* 1284 SHC VI (i) 131, *Gorstycote* 1286 SHC V (i) 174, *Gusecote* 1293 SHC 1911 232, *Gorsticote* 1300 SHC V (i) 179, *Gorsticotte otherwise Goscote* 1589 SHC XV 193, *Gorsticott* 1610 SHC III NS 54. From the OE adjective *\*gorstig* 'overgrown with gorse', with OE *cot* 'cot, cottage, hut, shelter', giving 'cot amongst the gorse' or 'cot on the heath'. The place was formerly within Cannock Forest.

**GOSPEL END** 1 mile west of Sedgley (SO 8993). *Gosepole* c.1400 DR37/2/Box 122/36, *Gospelynd* 1532 SHC XII 185, *Gospell Ende* 1532 SHC 4th Series 8 114, 1574 SHC XIV 182, *Gospelend* 1587 Sedgley ParReg, *Gospellend* 1602 SHC XVIII 20, *Gospel End* 1775 Yates. If the earliest spelling can be relied upon, the name is from OE *gos, pol* 'goose pool', which has developed into gospel, normally denoting places where a reading from the Bible was made during perambulations of boundaries on Rogation days. The word *end* normally meant simply place, rather than a terminal point, and was often applied to squatter settlements built on commons or heathland. See also Gospel Oak; Penwood (Farm).

**GOSPEL OAK** 1½ miles west of Wednesbury (SO 9694). *ye Gospel Oake* 1695 Tipton ParReg, *Gospell oaks* 1704 ibid, *Gospel Oak* 1834 O.S. At the junction of the parishes of Wednesbury, Tipton and Sedgley, the former location of an oak where, as a 'Gospel place', a reading from the Bible was made during perambulations of boundaries on Rogation days. Names of this type are very common: *Gospel Oak* is recorded between

Hoar Cross and Yoxall (Redfern 1886: 47), *Gospel Ash* (1834 O.S.) on a parish boundary 2 miles east of Bobbington (SO 8390), and *Gospel Place* (1834 O.S.) ½ mile north-east of Broadhurst Green on Cannock Chase (SJ 9815). See also Gospel End.

**GOTHERSLEY** 2 miles east of Enville (SO 8586). *Godrichesleye* 1329 SHC 1913 24, *Goderichleye* 1342 ibid.' 91, *Gothersley* 1690 HRO E12/V1/KY/7, *Cothersley* 1825 SA 2161/137, *Gothersley* 1834 O.S. The later spellings indicate a derivation from the OE personal name Godric, hence ' Godric's *leah*'. Gothersley formerly stood on the west bank of Smestow Brook, but the name is now attached to Gothersley Farm to the south-west. The place may be associated with *Gutheresburn* (q.v.) recorded in 1248: VCH XX 123, VCH III 137.

**GRADBACH** on the Staffordshire-Cheshire border, 7 miles north of Leek (SJ 9965). *Gratebache* 1374 SHC XIV (i) 136, *Gratebach* 1414 SHC XVII 13, *Gratbache* 1564 Ch. From OE *great* 'bulky, massive', with OE *bece* 'pronounced stream-valley'. The place stands at the confluence of the Black Brook and river Dane, both of which lie in *bece*-type valleys.

**GRAISELEY** 1½ miles south-west of Wolverhampton (SO 9097). *? Glaseleye* 1259 SHC 1911 132, *Graseley* 1282 Duig, 1327 SHC VII (i) 249, *Greseley* 1332 SHC X (i) 127, *Graseley* 1577 Wolverhampton ParReg, *Gresley Farm* 1820 Greenwood, *Grazeley Hall* 1834 O.S. Perhaps from OE *\*grǣg* 'wolf' (EPNE i 207 and Ekwall 1960: 203 *sub nom* Grazeley gives 'badger', but see NM 96 1995 361-65), so 'the *leah* with the wolves'. See also SHC 1919 167.

**GRANGE, THE** in Croxall (SK 2013). *(atte) Grange* (p) 1309 DbCh. From ME *grange, graunge* 'a grange', originally 'a granary, a barn', later 'a farm', also 'an outlying farm belonging to a religious house or a feudal lord where crops were stored'. Transferred from Derbyshire in 1894.

**GRANGE FARM** on the west side of Winnington (SJ 7238). Early spellings have not been traced, but the place may have been a grange of Combermere Abbey: StEnc 666.

**GRANGE FARM** 1 mile east of Coppenhall (SJ 9219). *Picklestich* 1833 O.S. No suggestion can be offered for the 1833 spelling.

GRASSEHAYE (unlocated, near Blithfield) *(Parke of) Grassehaye* 1349 SHC 1919 13.

**GRATTON** 1 mile south-west of Horton (SJ 9356). *Gretton* 1199 FF, 1252 Ch, *Grytton* 1273 SHC VI (i) 59, *Gretone* 1306 Banco, *Gratton* 1343 ibid, *Gretton in the Moors* 1373 SHC VIII 74, *Gretton super Mores* 1375 IpmR, *Grotton* 1393 ibid, *Grottun* 1486 to 1515 ECP, *Gerton otherwise Gratton* 1572 SHC XIII 287, *Gretton* 1608 SHC III NS 27. Almost certainly from OE *greot, tun* 'gravelly *tun*'; see Gratwich; Great Bridge.

**GRATWICH** Ancient Parish 4 miles south-west of Uttoxeter (SK 0231). *Crotewiche* 1086 DB, *Grotewic* 1176 P, *Grotewis* 1242 Fees, *Gretewyz* 1236, 1242 Fees, *Gretewiz* 1276 SHC 1911 177, *Gretewyk* 1286 SHC 4th Series XVIII 126, *Gretewyc, Gretewych* 13th century Duig, *Grotewyf* c.1300 DW1733/A/2/7, *Gretewiz* 13th-14th century SRO D1798/H.M.Cheywynd/1, *Gratwytthe* 1532 SHC 4th Series 8 60, *Gratwiche, Gratwyche* 1562 SHC 1938 111. The first element is derived from OE *greot* 'gravel', with the second element *wic*, hence 'the *wic* which lies on gravel'. CDEPN gives 'dairy farm by the gravelly stream': the place lies on a minor tributary of the river Blithe. See also JEPNS 31 23. For field-names in Gratwich in 1562 see SHC 1938 111-2.

GREASLEY SIDE (obsolete) 1 mile north of Bucknall (SJ 9047). *? Greyley* 1707 Stoke on Trent ParReg, *Greasley Side* 1836 O.S. Perhaps from OE *greosn-leah* 'gravelly *leah*'. Cf. Greasley, Nottinghamshire. OE *side* meant 'side, slope of a hill, especially one extending for a considerable distance'.

**GREAT BARR** – see **BARR, GREAT**.

**GREAT BRIDGE** 2 miles south-west of Wednesbury (SO 9792). *Grete* 1292 SHC VI (i) 212, 1327 SHC VII (i) 229, *Grete* 14th century Duig, *Gretbridge* c.1564 SHC 1931 155, *Greete* 16th century, *Greet Bridge* 17th century Duig, *Grit Bridge* 1686 Plot. The place stands on a stream formerly called *Greet* which divides the parishes of West Bromwich and Tipton. Greta and Greet are common river names in the north of England: see Ekwall 1928: 185. The name is from OE *greot* 'gravel, grit', meaning 'stream with a gravelly bed'. The name *Great Bridge* ('the bridge over the stream called Grete') was being used by the end of the 17th century: VCH XVII 8.

**GREAT BRIDGEFORD** – see **BRIDGEFORD, GREAT**.

**GREATGATE** ½ mile north-west of Croxden (SK 0540). *Greth* c.1176 StEnc 241, *Gretyatt* 1532 SHC 4th Series 8 111, *Greate Yate, Greteyate* 1539 MA, *Greeteyate* 1596 SRO DD/FJ/1/182/2, *Grityat* 1608 SHC 1948-9 58, *Greeteyate* 1666 SHC 1925 232, *Grityate* 1686 StEnc 241, *Great Yate* 1775 Yates, *Great Yate or Great Gate* 1836 O.S. It is said that the settlement of Croxden had moved to Great Gate by the time Croxden abbey was founded c.1179, perhaps indeed to make way for the abbey (TSSAHS XXXVI 1994-5 47), and since Great Gate lay on the abbey demesne, there may have been a 'great gate' here. However, Croxden Brook runs through a narrow pass in the hills here, and a derivation from OE *geat* 'a gate, a gap between hills', may be topographically appropriate, with the first element from OE *greot* 'grit, gravel, so 'the gravelly gap between hills'. If the earliest spelling can be relied upon, the original name may have been '(place at) the gravels'.

**GREAT HAYWOOD** – see **HAYWOOD, GREAT**.

**GREAT MOOR** 1 mile south-east of Pattingham (SO 8398). *More* 1332 SHC X 131, *le More* 1327 SHC VII (i) 249, *Great Moor* 1514 VCH XX 173, *Great Moore* 1590 SHC 1930 68. Self-explanatory. *Le Petyte More* is recorded in this area in 1338 (Brighton 1942: 160), *little more* in 1627 (Pattingham ParReg).

**GREAT OAK (FARM)** 1 mile north-east of Audley (SJ 8051). *Great Oake* 1668 Audley ParReg, 1733 SHC 1944 48. Self-explanatory.

**GREATWOOD (FARM)** ½ mile south-west of Croxton (SJ 7731), *Gratewood* 1558 SRO DW1837/1, *Gratwoodd* 1563 SHC 1938 66, *Grateswood Heathwarren* 1644 SHC 4th Series I 150, *Gratewood (Heath)* 1655 SRO DW1082/L/2/1-20, *Gratwood (Heath)* 1719 SHC 1931 90, *Great Wood (Lodge)* 1833 O.S.; GREATWOOD (obsolete, near Little Wyrley), *Greatwood* c.1250 SHC VI (ii) 191, 1363 ibid. 192, 1403 ibid.' 193. Self-explanatory. Greatwood Heath, an extensive tract of common, covered some four square miles from the Broughton Road in the north to Offley Hay, and from Croxtonbank to Fairoak: StEnc 242. It was enclosed under the first of 113 Staffordshire Enclosure Awards in 1719: SHC 1941 16. It is likely that *Gracewood*, recorded in 1298 (Spufford 2000: 295), refers to that place, with the not untypical misreading of *c* for *t*.

**GREENDALE** 1 mile south-west of Oakamoor (SK 0443). *Greendale* 1573 Ass. From OE *grene dæl* 'the grassy dale or valley', but see also Grindley.

GREENFORD (unlocated) According to Shaw 1801: II 16, small islands in the river Tame, north-west of Drayton Bassett. *Greenford* 1801 Shaw II 16.

GREENHILL in Lichfield (SK 1209). *Grenhull* 1299 SHC VII 66, *Green-hill* 1322 (1798) Shaw I 305, *The Grenehyll* 1532 SHC 4th Series 8 182, *Grenehyll* c.1535 SHC VI (ii) 166, *Green Hill* c.1567 SHC IX NS 158, *The Grene Hylle* 1659 (1798) Shaw I 313. Self-explanatory. The name is first recorded c.1190: VCH XIV 4, 135.

GREENHILLS 1½ miles north-east of Ipstones (SK 0351). *Grenehullus temp.* Edward I SRO D1229/1/4/21, *Grene Hills* 1542 SRO D1229/1/4/38, *Green Hills* (twice) 1836 O.S.

GREENLOW HEAD on the west side of Butterton (SK 0756). *Greenelowe* 1636 *Deed*, *Greenlow Head* 1689 Butterton ParReg. 'The head or summit with the green *hlaw* or tumulus'.

GREENSFORGE 3½ miles south of Wombourne, on the river Stour (SO 8588). *Greensforge* 1600 VCH XX 208, *Greenes forge* 1674 WHS 13 NS 35. Dud Dudley, the early ironmaster, mentions a *Greens-lodge* here in 1656 (Shaw 1801: II 13) and *Green's forge* is recorded in 1665 (SHC X (ii) 32), evidently from the surname Green: Thomas Green is mentioned here in 1600 (VCH XX 208). Cf. Wall Heath.

GREENWAY HALL 1 mile west of Bagnall (SJ 9150). *Greneway* mid-13th century SRO 3764/33[27574], *Grenewey* 1308 SHC XI NS 261, 1364 SHC IV 72, 1512 Horton Ct, *Greneways* 1404 SOT SD4842/17/2, *Greneway Hall* c.1569 SHC IX NS 95, *Greynwaie Hall* 1577 SHC 1926 50, *Greeneway hall, Greenway hall* 1594 Norton-in-the-Moors ParReg. Self-explanatory. It is uncertain whether *Greneveye*, recorded in 1279 (SHC 1911 178), refers to this place. *Greneway* in or near Stone is recorded in 1391 (SHC XVI 28).

GREENWICH POOL 2 miles north-west of Enville (SO 7988). *Greenage* 1770 VCH XX 94. Possibly 'the green oak', though that derivation might be difficult to explain.

GREETS GREEN in south-west of Wednesbury (SO 9791). *Grit green* 1686 Plot – see Great Bridge.

GREGORY 1 mile north-west of Norbury (SJ 7624). *Gregory* 1908 VCH I 378. A curious name of unknown date and derivation, but possibly associated with a large oval flat-topped mound here: see VCH I 378; VCH IV 155.

GRESBROK HALL (obsolete) the name of the manor house in Shenstone (SK 1104), acquired in 1204 by Bartholomew de Gresebroke: TSSAHS XII 1970-1 25. *Gresebrok* 1269 SHC IV (ii) 107, 1295 SHC VII 42, *Gresbroc* 1275 SHC 1923 275.

GRESLEY HALL (obsolete) ½ mile south-east of Hints (SK 1601). *Gresley Hall* 1798 Yates.

GRESLEY HILL (unlocated) *Gresley Hill* 1601 SHC 1935 347.

GRESLOWE MOR (unlocated, in Stafford Foregate) *Greslowe mor* 1295 SRO D938/188.

GRETWOOD (unlocated, probably in Silkmoor) *Gretwood* 1230 SHC VVV (ii) 46.

GREVELEY (unlocated, near Blythe Bridge) See Erdeswick 1844: 269. Possibly Grindley (q.v.).

GREY FRIARS on the north-west side of Stafford (SJ 9123). *The Gray Freres* c.1540 Leland, *The Graye Freers* 1581 SRO D593/A/2/27/17. From a Franciscan friary at the North Gate or Goal Gate in Stafford established by 1274: VCH III 270-1.

**GRIFF WOOD** 1 mile west of Mayfield (SK 1245). Probably to be associated with *Greof,* recorded *temp.* Edward I (Okeover T309). Possibly from ON *gryfja* 'a small deep valley' (found as North Country dialect *griff)*, or ON *gróf* 'a stream, the hollow which a stream makes, a pit': the place lies on Marsh Brook.

**GRIMBLEBROOK FARM** on the north side of Milwich (SJ 9732). *Grimble Brooke House* 1733 SRO D1798/685/207, *Grimble Brook House* 1782 SRO D1798/685/212. A curious name: the place lies on a stream called Wheatlow Brook .

GRIMDITCH (obsolete) ½ mile north-east of Ipstones (SK 0350). *Grimditch Gate* 1780 *EnclA*, *Grimditch* 1836 O.S., *Grimditch otherwise Grimheath* 1897 Eccl.Comm. Early spellings have not been traced (unless the surname *Grymesdych*, recorded in 1535 (1798) Shaw I 412, or *Grymesdiche*, mentioned in 1605 (1883) Sleigh 20 refer to this place), but the name was commonly applied by the Anglo-Saxons to prehistoric ditches or earthworks, probably from the name Grim used as a giant's name in Northern Europe (see Nomina 8 1985 72; PN Ch II 66), rather than from OE Grim, meaning 'the masked one', a nickname for Woden, who by tradition went about in disguise (see PN W 15-6; Gelling 1988: 148-9). The name Crowgutter (q.v.) may refer to this feature (a road from Crowgutter to Grimditch Gate was described as a turnpike road in 1777 *EnclA*), but on the 1836 O.S. map appears to be linked with the stream which flows to the east of Ipstones. See also The Camp, near Adbaston.

GRIMESCROFT (unlocated, near Knighton (SHC 1914 91) or Bishop's Offley (SHC 1914 82)) *Grimescroft* 1220 SHC IV 288, c.1220 SHC 1914 82, *Grymescroft* 1272 SHC 1924 243, 1541 SHC V NS 118. From OE *croft* 'a small enclosed field'. For the first element see Grimditch, or possibly from the surname Grym, Grymm or Grim recorded in this area from an early date: see SHC XII (i) 156; SHC 1914 92, 97, 98, 134, 138-9; SHC 1928 280; note also Francis Grimes, recorded in 1686 Eccleshall ParReg. The 'well known family of Grim' is said to have had its seat at Little Haywood between 1200 and 1350: SHC 1914 134; see also SHC 1928 280.

GRINDER'S MILL – see BUSTLEHOLME.

GRINDLESTONE EDGE (obsolete) 1 miles north of Horton (SJ 9458). *Grindlestone Edge* 1686 Plot, *Grindlestone (House)* 1815 *EnclA*, *Grindle Stone Edge* 1842 O.S. From ME *grindelstone* 'a grindstone', so 'the steep ridge where grindstones were quarried'.

GRINDLEY (obsolete) on west side of Blythe Bridge (SJ 9441). *Grandalesiche, Grandalesichet* c.1250 SHC XI 314, *Grudalesiche* ' 1337 ibid. 306, *Grindley (Hill)* 1798 Yates, *Grindley* 1836 O.S. A curious name, possibly from OE *grene dæl* 'green valley', or perhaps from Grendel, the name of the monster in the OE epic Beowulf: in both cases the change of the first *e* to *a* or *u* would be unusual, but Grendal brook in Devon has become the Grindle or Greendale Brook: Chambers 1959: 309. *-siche* is from OE *sic* 'a small stream, especially one in flat marshland' (the 1836 O.S. map shows a small tributary of the river Blithe here), suggesting another possible derivation from OE *\*grendel* 'gravelly place or stream', though the early spellings make that less likely. The name is now found in Grindley Lane. Possibly the same place as Greveley (q.v). See also *Grinslowe Grange*. For *grendelsmere* in the bounds of Oldswinford (15th/16th century, S.579) in 951x955 see Stenton 1970: 285; Hooke 1990: 164-5.

**GRINDLEY** 5 miles south-west of Uttoxeter (SK 0329). *Grenleg* 1251 Ch, *Grenleg, Grentleg* pre-1290 SRO 594[7937], *Grinley* ? 13th century SHC VIII 169, *Greneleye* frequently 13th century Duig, *Grenlee* 1338 SHC 1913 71, *Grenleygh, Greneleygh* 1341 SHC 1921 18-9, *Grenley* 1374 SRO D(W)1721/3/32/20, 1473 SHC NS IV 183, *Grynley* 1619 SHC VII NS 195, *Grimley* 1679 SHC 1914 124, *Grindley* 1836 O.S. From OE *grene* 'green', hence 'the green *leah*'.

**GRINDON** Ancient Parish 5 miles north-west of Ilam (SK 0854). *Grendone* 1086 DB, *Grendon* 1188x1197 SRO D603/A/Add/36b, 1327 SHC VII (i) 221, *Grendon on le Morys* 1444 SHC 1939 85, *Gren'* 1532 SHC 4th Series 8 6, *Grynne* 1590 SHC 1930 (ii) 203, *Gryndon* 1592 SHC 1930 287, *Grin* 1599 Smith, *Gryn* 1607 SHC III NS 10, *Grin* 1647 Ellastone ParReg, *Grinne* 1656 Leek ParReg. From OE *grene dun* 'green hill'.

**GRINSLOWE GRANGE** (unlocated, perhaps to be associated with Grindley (q.v.)) *Grinslowe Grange* 1592 NA 157DD/2P/19/1.

**GROSVENOR (HIGH), GROSVENOR'S CROSS** (obsolete) 1 mile south-east of Worfield (SO 7693). *Gravenovere* 1293 SHC VI (i) 239, *Gravenor* 1462 SHC IV NS 122, *High Gravenor* 1638 Claverley ParReg, *high-gravener* 1663 SA 2038/1/3/21, *High Gravenor, Gravenors Cross* 1752 Rocque, *High Grosvenor* 1833 O.S. From Fr *gros veneur* 'great or chief huntsman': Robert le Grant-Venur is recorded in 1293 (SHC VI (i) 246); Henry Graven(or) is listed in the SR of 1525. Grosvenor's Cross is now known as The Cross. It is unclear whether *Grandenaue*, recorded in 1260 (SHC 4th Series 13 8), is to be associated with this place.

**GROUNDSLOW FIELDS** 2 miles north-east of Swynnerton (SJ 8637; SMR 00591 places a deserted post-medieval settlement at SJ 86543755). *? Grauntsele* 1281 SHC VI (i) 120, *Grimeslow Fields (House)* c.1698 SRO D593/B/1/22/25, *Groundsley Fields* 1749 JNSFC XLV 1910-11 210, 1798 Yates, *Groundslow Fields* 1836 O.S. Perhaps from OE *grund* 'foundation, ground, bottom', also 'a stretch of land', and later 'an outlying farm, outlying fields'. The more widely found distribution of the particular application to 'field' rather than 'foundation' may well point to a derivation from ON *grund* 'earth, a plain', as in OIcelandic *grund* 'a flat grass-grown plot of land': see EPNE i 211. The second element is almost certainly OE *hlaw* 'low, tumulus', so giving 'the flat grassy land with the tumulus': at least two burial mounds have been recorded here, at SJ 867375 and SJ 867373: StEnc 247.

**GROUNDWYNS** (unlocated) *Groundwyns* 1391: VCH XX 94.

**GRUBBERS HILL** 2 miles north-east of Keele (SJ 8147). *Grobershill* 1576 Audley ParReg, *Grobershill otherwise Grobersasche* 1592 SHC 1930 263, *Grubbers* 1630 Wolstanton ParReg, *Grober's Ash* 1634 ibid, *Grubbers Ash* 1686 Plot 121, *Grabbers Ash* 1707 Keele ParReg, *Grubbers Ash* 1833 O.S. Possibly from ME *grubbere* 'digger' (an occupation frequently recorded in mining areas: see for example Will'o le Grobber 1332 SHC X (i) 95), or the surname Grobbere (see SHC 1913 329, 333; Shaw 1798: I 72) with the same derivation, with OE *hyll* 'hill' and OE *æsc* 'ash-tree'.

**GRUB STREET** ½ mile south of High Offley (SJ 7825). *Grub Street* 1833 O.S. This place, remembered in the name of a lane running south from High Offley, may have taken its name from the lane. The age and derivation of the name are uncertain (it may date from the construction of the Shropshire Union Canal, which ran into considerable difficulties in this area, where the cutting collapsed, and near Shelmore, where the embankment repeatedly failed), but from the 17th century Grub Street has been a term of disparagement applied allusively to the authors of literary works of little merit: Grub

Street was the former name of Milton Street in Moorfields, London, inhabited by writers of 'small histories, dictionaries and temporary poems': OED.

GRUETS WOOD (unlocated, at Wooliscroft) *Gruets Wood* 1136 SHC XII NS 154. See also SHC VI (i) 22 for *Gruet.*

GRUMSDALE (unlocated, near Blithfield) *Grumsdale* late 13th century SRO D938/57.

GRYMESWORTH (unlocated, in Little Wyrley) *Grymesworth* 1395 SHC VI (ii) 193. For *Grymes*-see Grimditch, Grimescroft. *Worth* is OE *worþ* 'an enclosure'.

GRYMSELL (unlocated, possibly near Coton Clanford) *Grymsyll, Grymsull* 1541 SHC V NS 118, *Grymsell* 1541 SRO D590/662. 'Grim's hill'. For the name see Grimditch, Grimescroft. The place evidently belonged to Ranton priory: SRO D590/662. Grinshill in Shropshire is recorded as *Grymsell* in 1587, but belonged to Haughmond abbey: SA 1574/400. There was a Grimshill in Coleshill, Warwickshire.

GRYNGLEY (unlocated) *Gryngeleye* 1327 SHC VII (i) 228, *Gryngley* 1592 SHC 1930 (ii) 295. Possibly Grindley (q.v.).

GUENDELAWE (unlocated, near Tixall) *Guendelawe* 1220 SHC VIII (i) 193. The first element is likely to be an unidentified personal name (or possibly a corruption of OE *cwene* – see Queen's Low and Quennedale), with OE *hlaw* 'mound, tumulus'. The tumulus may be that known as Queen's Low, or one that is said to have been destroyed by ploughing at the end of the 18th century: see Clifford 1817: 86-7. See also King's Low.

**GUILD OF MONKS** 2 miles west of Gnosall (SJ 7820). It has been suggested that the place, which lay at Old Guild, 300 yards west of the present modern farmhouse called Guild of Monks (VCH IV 107), on the north side of the Roman road from Pennocrucium to Chester (Margary number 19), is Shrewsbury Abbey's *manor de loc Sancti Johannis Baptiste* in the wood of *Suthon* (i.e. Sutton, in Forton), a hermit's habitation recorded in 1256 (SHC 1921 189, also SHC 1913 317; 1923 305; VCH III 136). Other references are *land called the Gyle de la Monks* 1487 Rental, *Monks gyle* 1487 *Rental, le Gyle als Gilham Monkes* 1533 Rental, *Gyll a monks* 1545 Ct, *Gyll* 1573 *Rental, (mansion house at) the Gill a monks* 1605 Ipm, *Gillamonkes, Gallamonkes* 1604 SHC 1946 200-1, *Gild of Monkes* 1693 *Terrier, Upper Gill, Lower Gill* 1775 Yates, *Gill* 1808 Baugh, *Guild of Monks* 1832 Teesdale (with *Old Guild* to the west and *New Guild* to the north-west). The present Guild of Monks Farm is shown as *Lower Gill* on Smith's map of 1747, and later as *Guild Farm*. The derivation remains unknown – the complete absence of *-d* in early forms indicates that there is little likelihood that it is connected in any way with OE *gild* 'guild, society' (a detailed analysis of OE *gild* in place-names can be found in PN Wo 124-5). For Gilcrux, Cumberland (*Gillecruz* 1230, *Gillecruce* 1272, *Gillecruice* 1230) Ekwall 1960: 195 suggests the possibility of a Welsh *cil* 'back, corner, retreat' for the first element, with *G-* due to British lenition, citing Culcheth, Lancashire; Kilquite and Colquite, Cornwall; and Cilcoit, Monmouthshire. That seems a possibility here, but Mawer 1929: 16, 78-9 discusses 'the strange word gill, so frequent in the Wealden area that it is recognised in the dialects of both Surrey and Kent' in the sense 'narrow wooded valley', and concludes that it is probably from a Germanic stem *\*gulja*. There is a possibility that the name is from ON *gil* 'ravine, narrow valley': see also Gill Bank and Gillow Heath. For completeness, it may be noted that *gyle* was a variant ME spelling for *guile* 'a deceipt, a stategem, a trick', and the possibility that the name records some dubious property acquisition cannot be dismissed entirely: SRO 3764/99 states that the hermitage of Ranton abbey was said to be at Sutton. Cf. The field-name *Gylecroft,* recorded in Kinvaston in 1507 (Oakden 1984: 126). Guild of Monks lies on the side of a small valley through which runs a stream flowing into Aqualate Mere. New Guild lies ½ mile north-west. See also Gylle. Cf. Kilpeck, Herefordshire.

**GUN** a 1223' hill on high moorland 4 miles north-west of Leek, known locally as The Gun (SJ 9761). *Gonedone* 1229x1232 CEC 385, *Gunedun, Gonedun, Gonedan* c.1230 SHC IX NS 316, *the waste of Gondon* 1318 SHC 1911 433, *? Gernedon'* 1327 SHC VII 216, *? Gunne* 1332 SHC X 115, *Gundon* c.1539 LRMB, *Dunne* 16th century Erdeswicke 1844: 494, *the Gun* 1673 Blome, *a hill called the Gun* 1686 Plot 115, *Dun Mountain* 1747 Bowen, *Gun (Stone Pits)* 1842 O.S. The place-name *Dun*, perhaps in this area, is recorded in 1278 (SHC VI (i) 86). The earliest spellings suggest a derivation from the ON personal pet-name *Gunna, from Gunnhildr, borrowed into OE as Gunna (the latter found in BCS 1130, not listed in Sawyer 1979), giving Gunne in ME, with OE *dun* 'hill, mountain', so 'Gunna's or Gunne's hill'. Will'o Gunne is recorded in Bradnop in 1332 (SHC X 115), perhaps the same person as William Gunne of Bradnop recorded in 1356 (SRO D3272/5/13/12). However, the element *gun* is found attached to a number of hills and high ground in North Staffordshire and the adjoining counties (cf. Gun Farm, Gun Hill, Gun Hills, all in Derbyshire), and it is the not inconceivable (though highly improbable) that the name is of Celtic origin, from OWelsh *guoun* (Welsh *grawn*, Breton *gueun, geun*, Cornish *gun, gon*, found especially in Cornwall as *goon*: Coates & Breeze 2000: 356), meaning 'moor, downland, plain, unenclosed land', e.g. Cornish *gunran* means 'moorland, part of a parish or property' (cf. Gunend; Gunside). In that respect it may be noted that another element, *tor*, which may be of Celtic origin (EPNE ii 184), is found especially in the South West and in the North Midlands: EPNE i xxviii. Against this possibility is the fact that initial *g* only appears in words beginning with earlier *w* from the later 8th century onwards, probably too late to be found in Staffordshire. The name is also found in Gun Common (1731 *Letter* (1883) Sleigh); Gun End (q.v.); Gun Farm (1775 Yates); Gun Gate (1684 Leek ParReg); Gun Mires (1831 Survey); Gun Moor (1775 Yates); Gun Road (1811 *EnclA*); Gun Rock (1820 *EnclA*); Gun Side (q.v.); Gun Hall (1344 Ipm); Gun Heath (1731 *Letter* (1883) Sleigh).

**GUNEND** ½ mile north-east of Heaton (SJ 9662), *Gun End* 1842 O.S.; GUN END (obsolete, on the west side of Huntley (SK 9941)), *Gun End* 1836 O.S.; **GUNSIDE** 1 mile south-west of Meerbrook (SJ 9860), *Gun Side* 1565 Deed, *Gunn-Side* 1698 Leek ParReg, *Gunside* 1798 Yates, *Gun Side* 1842 O.S. See Gun. Side is from OE *side* 'side', later 'slope of a hill, especially one extending for a considerable distance'. The word *end* normally meant simply place, rather than a terminal point, and was often applied to squatter settlements built on commons or heathland.

**GUNSTONE** 2 miles south of Brewood (SJ 8704). *Gonestona* 1176x1184 SHC V NS 214, *Gunnistona* c.1199 SHC III (i) 30, *Gunniston' temp.* Richard 1 Cur, *Gunneston* c.1260 Giffard, *Guneston(e)* 1240 FF, 1250 Banco, *Gounstoun* 1317 Giffard, *Gon(e)ston(e)* c.1176x1184 St Cart, *Gunneston* 1334 SHC XIV 38, *Gunstone* 1341 to 1482 Banco, *Gonston* 1532 SHC 4th Series 8 85. 'The *tun* of Gunni' (a short-form of Gunnhildr, an ON personal name). The place-name is of a type known to philologists as a 'Grimston-hybrid' (or a 'Toton Hybrid': see Cameron 1996: 74-5), i.e. a name in which OE *tun* is combined with an ON personal name. It is possible that the name may date from the taking over of established English settlements by the victorious Danes of the great army of 865 A.D. (see Gelling 1988: 232-4; also Parsons 2001: 308-9), but it is quite possible that it is more recent, and even post-Conquest. Cf. Gunton, Norfolk and Suffolk. It may be significant that the place lies on a lost Roman road running south from Pennocrucium (Water Eaton): Horovitz 1992: 34-5. See also Rolleston. The field-name

*Gunnyngesleke* in Leekfrith is recorded in 1394 (Deed (1883) Sleigh), and is probably from the personal name Gunni, perhaps with ON *slakki* 'a small shallow valley, a hollow in the ground', or (more likely) with OE *\*lece* 'a stream flowing through boggy land, a bog'.

GUTHERESBURN (unlocated) A hermitage in Kinver Forest. *Gutheresburn* 1248 VCH III 137; VCH XX 123. It is unclear whether this place lay in Staffordshire or Worcestershire, but possibly associated with Gothersley (q.v), in which case the name may have applied to what is now Spittle Brook: Gothersley lies in the south-west angle of Spittle Brook and Smestow Brook, and the name is now attached to Gothersley Farm to the south-west. The single spelling, however, if trustworthy, points towards a derivation from the OE personal name Guþhere, with OE *burna* 'stream'.

GYBSTONE (unlocated, possibly near Stone) *Gybstone* 1480 SHC V NS 132, SHC VI NS (i) 126.

GYLLE (unlocated, possibly in the Myton/Levedale area, or perhaps associated with Guild of Monks (q.v.)) *Gylle* 1565 SRO D1798/579/25.

HACONDALE (unlocated, possible in the Blithfield area) *Hacondale* late 13th century SRO D986/41. Possibly from the ON personal name Hákon, with OE *dæl* or ON *dalr* 'valley'.

HADDESORE (unlocated, possibly near Dilhorne) *Haddesore* 1331 SHC 1913 27.

**HADDON** ½ mile north of Rushton Spencer (SJ 9463), *Haddon* 1842 O.S.; HADDON (obsolete) on the north-east side of Maer (SJ 7839; SMR 02636 places a deserted post-Conquest settlement at SJ 80003840), *Haddon* 1833 O.S. From OE *hæþ-dun* 'heathy-hill'. The place near Maer (said in TSSAHS XII 1970-1 35 to have been at SJ 801385) is remembered in Haddon Lane.

**HADEMORE** 1 mile east of Whittington (SK 1708). *Horton Hademore* 1635 VCH XIV 248, *Hademore( fields)* 1640s ibid, *Hademore* 1760 VCH XIV 240, 1834 White 105. Probably from OE *hæð mor* 'heather-covered moor or fen'. See also Horton. It is unclear whether *Heydemere*, recorded in 1308 (SHC 1911 300) is to be associated with this place.

**HADEN HILL, HADEN CROSS, HIGH HADEN** on the Staffordshire-Worcestershire border, 1 mile north of Halesowen (SO 9685). *? Hadenhull, Hodenhull* 1199 SHC III (i) 44,*? Hodenhull* 1273 SHC 1911 152, *Haudene* 1299 Wilson-Jones 79, *Haueden, Haueden, Hadene* 1388 StEnc 251, *Heydon Cross* 1686 Plot, *Haden Cross* 1721 Rowley Regis ParReg, *High Haden, Haden Hill, Haden Cross* 1834 O.S. It has been suggested that *Handen, Handden* 1227 (SHC IV 66) may refer to this place (ibid.). From OE *hean dun* 'high hill', to which the tautologous OE *hyll* 'hill' has been added.

**HADLEY END** 2½ miles north-east of King's Bromley (SK 1320), *Haddeleye* 1301 SHC VII 89, *Hadleye* 1318 SHC IX (i) 73, *Hedle End* c.1599 SHC 1935 199, *Hadley End* 1772 SRO DW3222/245/1-23, 1836 O.S., *Hadley Plain* 1830 Act, *Hadley End* 1830 O.S.; **HADLEY FARM** 1 mile south-west of Draycott in the Clay (SK 1427), *Hadley Plain Cottage* 1836 O.S., *Hedleyplain Barn* 1888 O.S. Probably from a shortening of OE *hæð* 'a heath, heather', with OE *leah*, giving 'the heathy clearing'. The word *end* normally meant simply place, rather than a terminal point, and was often applied to squatter settlements built on commons or heathland. See also SHC II 257; SHC XVI 280 fn.2. The *Plain* element is unusual and unexplained. Shaw 1798: I 224 mentions a place anciently called *Hadley*, then *Tymmorshey*, in King's Bromley. *Wulphateshadleg'*, perhaps near Colton, is recorded in the late 13th century (SRO D938/108), and *Hadley Moor* near Burntwood in 1724 (SRO D217/M/5).

**HADMORE** 1 mile north-east of Bobbington (SO 8192). *Hadmore* 1892 O.S.

**HAGLEY** on south side of Rugeley (SK 0417). *Hageleia* 1130 P, *Hagelega* 1169 ibid, *Haggleges* 1166 RBE, *Hagg(e)ley(e)* 1300 For, *Haggele* 1242 Fees, *Hagley* 1500 Ipm, *Hagley* 1513 VCH V 156, 1571 SHC 1938 162, 1801 Shaw II 325, *Hageley* 1570 *Survey*, *Hagley otherwise Haggeley* 1606 SHC XVIII 67. A manor believed to have been created in the time of Henry II by the keepers of Rugeley Hay in Cannock Forest: VCH V 155. From OE *\*hacga,* a form of *haga* 'haw, fruit of the hawthorne', with OE *leah*, giving '*leah* with the haws'.

**HAILSTONE, THE** (obsolete, a colossal 60' pillar-like outcrop of crystalline hornblende rock which lay on the west side of Turner's Hill, ½ mile north-west of Rowley Regis (SO 9688)) *Hailstone* 1798 Shaw I 122 fn\*, *Rowley hail stone* 1817 Pitt, *Hailstone Hill* 1832 Teesdale, 1834 O.S., *Rowley Hailstone* 1845 SRO D716/5/17-18. Plot 1686: 175 records: 'At Rowley Regis ... I met with the same [very hard black shining stone] again, and scattered here and there all over the Towne: whereof yet there is one more remarkable than the rest, about half a mile N.W. of the Church; as big, and as high, on one side, as many Church Steeples : at the bottom of which on the highest side, if one stamp with ones foot, it returns a hollow sound as if there were a Vault, which made me suspect that some great person of ancient times might be buryed here, under this natural Monument (for I scarce think so great a thing could be put here by art, it much exceeding those of Stonehenge or Aubrey [Avebury] in Wilts) but digging down by it as near as I could (where the sound directed) I could find no such matter.' Scott 1832: 437 described Hailstone as '...a vast cubical pillar...Surrounding it on all sides and scattered in great profusion through the coppice which spreads over the slope, and strewed in multiform fragments at its bases are innumerable blocks...'. The age of the name is not known, but if ancient (which is unlikely) perhaps from OE *halig-stan* 'holy stone': cf. Hailstone, Wiltshire. Or perhaps from the globular or hail-like formations found in the outcrop. The pillar was destroyed by explosives in 1879. See also Wilson-Jones 48; StEnc 252-3.

**HALDESALESMOR** (unlocated, in the Lichfield area) *Haldesalesmor* 1272 SHC 1910 103.

**HALES** Ancient Parish 3 miles east of Market Drayton (SJ 7133). *Halas* 1086 DB, *Hal in Lima* c.1217 Rees 1997: 118, *Hales in Lima* c.1250 ibid. 119, *Hales* 1291 Tax, *Hales under Lyme* 1293 SHC VI (i) 228, *Hales* 13th and 14th centuries Duig, *Draytton Haylles* c.1570 SHC 1931 131, *Blorehales* 1598 SHC XVI 184, *Hales in Tirley* 1705 DRO D3155/C227. The nominative plural of OE *halh*, perhaps here meaning 'dry ground in a marsh', but the use in this way of the place-name element remains puzzling. For *Lima*, *Lyme* see Lyme. *Blorehales* is from nearby Blore (q.v.). The area around Hales was formerly known as *Drayton Hales* (1833 O.S.), from nearby Market Drayton, to distinguish it from other Hales. *Hales* is also recorded in 1562, possibly near Alstonefield: SHC XV 140. See also Aston near Madeley.

**HALEWEHULL** (unlocated) *Halewehull* 1272 SHC IV 195. Perhaps associated with Halugh, Le (q.v.).

**HALFCOT** 2 miles west of Amblecote (SO 8685; SMR 02621 places a deserted post-Conquest settlement at SO 86708570), and presumably the place which appears as *Halford* in Duignan 1902: 71. *Haffecote* 1332 SHC X 86, *Oldeforde* 1343 Duig (presumably *Le Oldeforde* near Stapenhill, recorded in 1342 SHC 1913 90), *Halcote* 1434 SRO D1197/8/1, *Hafecote* 1446 Ch, 1609 SHC 1948-9 118, *Haffcot* 1532 SHC 4th Series 8 16, *Haftcoate* 1656 Sedgley ParReg, *Hafcott* 1669 WHS NS 9 (i) 69, *Hafcot* 1686 Plot. The place is near the river Stour. It is unclear why Duignan 1902: 71 associates *Oldeforde* with this place, but if he is correct it would seem that this place had two names from an early date. The word *old* had two meanings, 'ancient' (implying another, newer, ford nearby), and 'disused'. Both meanings would explain why the name became obsolete. The present name is evidently from OE *halh-cot*, from OE *halh*, here probably meaning 'a piece of low-lying land near a river', and OE *cot* 'a cottage, hut, shelter'. See also Hastecote.

**HALFHEAD**  1 mile north-east of Chebsey (SJ 8729). *Halfhyde Helie* 1164 SHC 1914 63, *Halvahida Helye* 1167 SHC I 49, *Halvehyda Helyæ* 1172 ibid. 65, *Halvehyda Helye* 1179 ibid.' 88, *Halfhide* 1227 SHC IV 53, *Halfehyde, Halvehyde* 1227 SHC XI NS 240, *Halfhida* 1228 ibid.' 219, *Halfhyde* 1288 SHC 1911 194, *Halfhide* 1328 SHC XIV 4, *Halfhyde* 1332 SHC X 91, 1549 SHC IV NS 114, *Halfehedde* 1580 ibid.' 213, *Halferd* 1621 SRO DE615/EX/1. 'So called for that it was anciently rated at half a hide of land': Chetwynd 1679: SHC 1914 63. *Helye* is from a former possessor, Helyas or Helias: SHC I 51, 158; SHC 1914 63. See also Pershall.

HALFMORE (unlocated, in Clent) *Halfmore* 1590 SHC 1930 (ii) 68.

**HALFPENNY  GREEN**    (pronounced Haypnee [heɪpnɪ]) 1 mile north-east of Bobbington, on the Staffordshire-Shropshire border (SO 8291). *Halfpenny Green* 1448, 1532, 1536 VCH XX 65. By local tradition the name is from the payment made for water drawn from a well on the Green, but it is more likely that it alludes to the rent payable for land here, possibly by drovers for overnight feed for their animals (the place lies on a medieval route from Chester to south-west England, in use until c.1800: VCH XX 65) or, as in the case of many field-names containing the word halfpenny (e.g. Halfpenny Butts, Baswich, *Halfpeny buttes* 1570 *Survey*), was a derogatory reference to the poor quality of the land. However, Rocque's 1752 map of Shropshire shows *Halfpeney Lake* (which is recorded as early as 1448: VCH XX 65, and as *Halferpenny Lake* in 1660: Claverley ParReg) in the approximate position of this place, with Halfpenny Green not named, suggesting that the place may have been named originally after a pond or pool shaped like an early (silver) halfpenny, i.e. half a circle, in an area of former waste and marsh: VCH XX 65. A record of land at nearby Whittimere, together with other property, rented for one halfpenny, is to be found in Shaw 1801: II 212. *Halfpenny House* appears on the first edition 1" O.S. map of 1836 1 mile south-west of Newborough (SK 1225), but the history of the name is not known.

**HALFWAY HOUSE**  1 mile west of Wolverhampton (SO 9099). *the half-way house* 1749 WA II 42-3. From a cottage of this name, so-called because it lay mid-way between Wolverhampton and the western town boundary at Newbridge. The cottage was also known as *Rose Cottage.* WA II 42-3.

**HALING**  on east side of Penkridge (SJ 9214). *Haling Grove* 1890 O.S. A relatively modern name from *haling-path* (hauling-path), with reference to the towpath of the Staffordshire & Worcestershire canal here, along which narrow-boats were drawn by horses.

**HALL DALE** – see **STANSLOW**.

**HALL GREEN** on the east side of Coseley (SO 9494). *Hall Greene* 1655 Assessment, *Hallgreen-in-Briarly* 1661 Sedgley ParReg, *Hall Green* 1887 O.S. Perhaps (with Hall Fields) named after Bradley Hall: see StEnc 255.

**HALLHILL** 1 mile south-west of Longnor (SK 0763). *Hallhill* 1626 Rental, 1775 Yates, *Hall Hill* 1645 SRO QSR f.10v. Probably from OE *halh, hyll* 'the hill by or with the *halh*'.

**HALLON** on the west side of Worfield (SO 7595). *Halene* 1238x1250 Eyton 1854-60: III 112, *Alen* 1256 ibid. 112, *hallon* 1499 Worfield CA, *halon* 1515 ibid, *Halyn* 1525 SR, *Hallon* 1532 SHC 4th Series 8 117, *Halling* 1638 D593/3/2/2/3. Hallonsford (*Hallonsford* 1833 O.S.) lies to the north-east on the river Worfe. In Shropshire since the 12th century. *Hallons Heath* in Alton parish is recorded in 1722: SRO D240/D/302. Possibly from the dative plural of OE *halh*.

**HALL O' TH' WOOD** in Balterley (SJ 7650). *(a house called) the hall of woodd* 1611 BCA MS3810/125, *Hall of the Wood* 1629 SRO D641/5/T/1/22, *Hall a Wood* 1686 Plot, 1799 Faden, *Hall of Wood* 1733 SHC 1944 42. A heavily-timbered house built in 1557 by George Wood, a Chester judge and a member of the Wood family recorded in the area from at least 1344: SHC V 325-6; SHC XII NS 235fn.

HALLOWES – see **COMPTON HALLOWES**.

HALL WATER (unlocated; in Endon: SHC 13 1973 29) *Hall Water* 1706 SCH 13 1973 29.

**HALMER END** 1½ miles south of Audley (SJ 7949). *? Harmershale* 1328 SRO DW1082/A/4/2, *Halmore* 1493 SHC 1912 256, *Halmoreende* 1514 JNSFC LX 1925-6 41, *Halmend* 1547 SRO D3155/WH46, *Halmore Ende* 1577 SHC XIV 194, 1617 SHC VI NS (i) 42, *Halmer End* 1579 SHC XIV 188, *Honmerend* 1600 SHC XVI 200, *Homerende* 1602 ibid. 212, *Halmerend* 1733 SHC 1944 1. The conflicting forms preclude any certain derivation, but the earliest spellings suggest 'the moor with the *halh*'. The word *end* meant not a terminal point, but simply 'a place', and was often applied to squatter dwellings on the outskirts of a settlement.

HALSEY (unlocated, probably near Longdon: see BCA 3415/164; *Halsey Lane* is recorded in 1722 on the south-east side of Beaudesert Park (SRO D603/E/2/149), presumably *Halseylone* recorded in 1318 (BCA MS3415/150)) *Halesey* 1286 BCA MS 3415/140, *Halesseye* 13th century BCA MS3415/135, *Halsey(e)* 1306 SHC 1911 287, *Halfseye* 1327 SHC VII (i) 231, *Halsey, Halseye, Alsey* 1332 SHC X (i) 111, *Halsey* 1367 VCH III 223, 1377 SRO D1734. From OE *hals*, ON *hals* 'rock', used in a transferred topographical sense. ON *hals* meant 'projecting part of something, a narrow piece of land'. ME *hals* had the meaning 'a narrow neck of land or channel of water'. The second element may be OE *eg* 'island', land by a stream, land between streams'. Halsey becomes or is replaced by *Hawkersuchende* in the Paget papers (SRO D1734): see Hawkewallsych.

HALSTEADS, THE (obsolete) an 'entrenchment of a square figure' on the south side of Okeover (SK 1547): see Plot 1686: 449; VCH I 192. *the Hallsteds* 1686 Plot 449, *The Halsteads* 1908 VCH I 192. Perhaps from OE *heall* 'a hall', with OE *stede* 'place, site of a building', giving 'the site of the hall'. Lewis 1849: II 280 mentions the considerable remains of a large moated residence, approached by an ancient bridge, and Erdeswick 1844: 488 fn. mentions 'barrows of Hallsteds'.

**HALUGH, LE** (unlocated, possibly in Baswich or Walton) *le Halgh* post-1290 SRO D47[7935], *Le Halewe* c.1300 SRO D81[7991], *le Halugh* 1300 SHC 1911 57. From OE *halh*. See also Halewehull, Halweton.

**HALWETON** (unlocated) *Halweton* 1284 FA. Perhaps asociated with Halugh, Le (q.v.).

**HAMILL** – see **SNEYD GREEN**.

**HAMLEYHEATH** 1 mile north of Colton (SK 0421). *Hamley Heath* 1775 Yates, *Hambley Heath* 1836 O.S., *Hanby Heath* c.1850 StEnc 256. See Hamley (House).

**HAMLEY (HOUSE)** ½ mile north of Colton (SK 0521). *Hom[m]ineley* ?late 13th century SRO D938/108, *Homeley under Wyrdeshay* early 14th century SRO 3764/1[27574], *Homeleyemor* 1314 SRO 110[7936], *Homeley* 1343 SRO 111[7936], SHC 1919 104, 1402 SHC XI NS 201, 1437 SHC 1919 107, *Homeley at Wirdshay* 1635 WCRO CR1908/30/1-2, *Hemley house* 1749 Bowen, *Hamley Stone Farm* 1822 DW1721/2/48, *Hamley House* 1839 DW1721/2/49, *Hambley House* 1836 O.S. Perhaps from OE *hamm, homm* 'an enclosure, a meadow, a water meadow' (the place lies near a stream), rather than OE *ham* 'a dwelling place, a manor, a village, an estate', which is rarely found in the West Midlands, with OE *leah*. *Wyrdeshay* may be Wilderly (Barn) (q.v.). See also Marchington.

**HAMLEY PARK** (obsolete) 1 mile south-west of Pattingham on the Staffordshire-Shropshire border (SO 8198). A medieval park known as *Armeley Park* by 1452 and later as *Armeley Park* (VCH XX 178), *Emley Park* 1662 SA 2028/1/5/17, *Amley Park* 1832 Teesdale. Ekwall 1960: 13 suggests for Armley, West Yorkshire, a derivation from OE *earm* 'wretched', perhaps in the sense 'outlaw', in which case 'the *leah* of the outlaw(s)', an appropriate name for this place on the county border where outlaws could flee from the jurisdiction of one county to the other.

**HAMMDEN, HAMUNDON** (unlocated, perhaps in the Broom Hall/Horsebrook/Coven area, near Brewood, but not traced in Oakden 1984, and possibly outside Staffordshire) *Hammden* 1149x1165 (17th century) EEA 14 62, *Hamundon*', *Hammondona* 1152x1155 (17th century) ibid. 59, *Hamunden* 1434 SHC 1910 202.

**HAMMERSLEY** (unlocated, possibly on the north side of Cheadle: *Hammersleyhays* is recorded 1 mile north of Cheadle (SJ 0144) in 1890: O.S., remembered in Hammersley Hayes Road) *Hemerusley* 1323 SHC IX 94, *Homereslegh* 1324 SHC X 53, *Homersley* 1333 ibid. 116, *Homeresleye* 1344 SHC XII 36, 1370 SHC XIII 75, *Homeresle* 1370 ibid.' 76, *Homresley* 1386 SHC ibid. 194, *Homersley* (p) 1591 SHC XVI 113, *Homersley* 1664 SHC V (ii) 177. It is suggested in SHC 1917-8 99 that 'the name comes from Ombersley, Worcestershire; but it has been naturalised as Hamersley in north Staffordshire and the Potteries since the time of [William Hamersley of Bottom, born c.1320] this first successful lawyer M.P.', and that the family first lived at Botham Hall, Cheddleton, before moving to Basford: see Johnstone 1946: 25; PN Wo 268. That would explain *Hom'sley House* at Botham Hall recorded in 1630 (SHC II (ii) 17), *Hammersley House al. the Wood* recorded in Cheddleton in 1640 (SRO D538/A/5/16), the *ancient capital messuage called Hamersley House of the Wood* in the parish of Cheddleton recorded in 1698 (SRO D538/A/5/21), and why Wetley Moor was also known as Homersley Moor. The name may possibly be a doubtlet of Hombersley in Worcestershire (i.e. *Ombersley* above), which may be 'Ambr's *leah*' as suggested by Ekwall 1960: 349, but could be from OE *amer* 'a bunting', found as *omer, amore, emer, emær* (see VEPN I 13-4; also Parson & Styles 1995-6: 5-13), and OE *leah*, so 'the *leah* with the buntings'. *Hammersley farm*, possibly associated with *Homersle*, mentioned in 1349 (SHC XII 77), is recorded in Fenton Vivian in 1751 (SRO D1788/67/33), but has not been located. It is possible that some of the above spellings may relate to the place near Cheddleton or Fenton Vivian.

**HAMMERWICH** 3 miles south-west of Lichfield (SK 0607). *Humeruuich* 1086 DB, *Hamwich Frankalingorum* 1166 SHC 1923 295, *Hamerwich* 1191 P, *Homerwiz* 1203 SHC III 110, *Hamerwic* 1220 SHC IV 10, *Hamerwick* 1248 SHC IV 241, *Homerwyk* 1281 SHC VI (i) 114, *Homerwys* 1301 SHC VII 74, *Homerwich* 13th century Duig, *Homerwych* 1532 SHC 4th Series 8 184, *? Homerridge* 1601 SHC 1935 388, *Homeriche* 1602 ibid.' 475, *Hammerwich* 1686 Plot. The second element is OE *wic* 'a dwelling, a building or collection of buildings for special purposes, a farm, a dairy farm, a saltworks', later alternating with *ridge* and *bridge*, with OE *hamor* 'hill', from ON *hamarr* 'hammer-shaped crag, a steep rock, a cliff' (which would not seem to fit the topography here, although the place is on high ground), or from OE *hamor* 'a hammer', perhaps in the sense of 'place with a forge or smithy', or (perhaps the most likely derivation) 'the place with the hand-tool workshops': see for parallel 'tool' names Cotwic and the lost *Lootwic*, both in Worcestershire (Parsons 2002: 182). *Frankalingorum* in the 1166 spelling is from ME *frankelein* 'franklin, freeholder'. In 1262 a Forest jury determined that 'a certain heath was burnt by the vill of Hammerwich, to the injury of the King's game': VCH XIV 198. The inference (from a reference in DB to *Duae Humeruuich*) that there were in 1086 two places of that name here, an Upper and a Lower Hammerwich, is rejected as particularly improbable by Thorn 1997: 371, where it is there suggested that the expression *duae* indicates separate holdings or manors that were administratively part of the same vill. Erdeswick 1844: 298 states: 'Anciently the village was divided into Nether and Over; and its division seems to have been at the hill ...'. For *Hammerwich* or *Hambridge* or *Homebridge Brook* (perhaps to be associated with *Hombridge otherwise Hamerwiche* 1610 SHC III NS 37; *Homebridge* 1679 SHC 1919 268), on the north side of Walsall, see Duignan 1880 61; VCH XVII 143 fn.14.

**HAMPS, RIVER** *Hanespe* c.1200 (?14th century) Burton, *Honsp* 1223, 1227 Harl, *Hans* 1577 Saxton, *Hans, Hansley* 1577 Harrison, *the Honsleie water* 1584 Harrison, *Hunsye* 1610 Speed. From Welsh *haf* 'summer' and *hysp* (feminine *hesp*) 'dry, barren', so literally 'summer-dry', applied to streams that dry-up in summer: cf. (Nant) Hafhesp, a tributary of the river Dee; (Aber) Hafesp, a tributary of the river Severn in Montgomeryshire (Ekwall 1928: 190). *Hafhesp* represents OWelsh *\*Hamhesp*, appearing as *Hanespe* in the earliest spelling, which might become *\*Hamspe*, from which derives the present name. The name may have its origin in British *\*Samosispa* (see Jackson 1953: 218-20; 486), adopted by the English in the later part of the 6th century or the beginning of the 7th century (ibid.). The name is particularly apposite, for the river flows partly underground, but in winter or flood also flows along the surface. This surface watercourse is often dry in the summer.

HAMPTON (unlocated) in Newton, Draycott in the Moors, *Hampton* 1251 SHC 1914 167, *Hampton Wood, Hampton Hayes, Hampton Dale* 1677 Survey ibid; HAMPTON (obsolete, at Stansley south of Newton near Blithfield: SRO D603/A/Add/117-8), *Hantona* c.1129 SHC 1916 223, *Hamton* 1199 SHC III (i) 41, SHC XI NS 17, *Hampton (Meadow)* c.1250 SRO D986/41, *Haunton* 1252 SHC 1937, *Hamtone* 1280 SHC 1911 177, *Hampton* 1271 SRO D603/A/Add/202, 1284 SHC VI (i) 132, 1322 SHC 1937 121, 1332 SHC X 89, *Hampton (dale)* 1347 SRO D986/39, *Hampton Dale* c.1634 D(W)1721/3/170. The first is perhaps from OE *ham-tun* 'home farm', and the second may be from OE *(æt þæm) hean tun* 'high *tun*'.

**HAMPTON LOADE** 1 mile south of Quatt, on the river Severn (SO 7486), opposite Hampton on the west side of the river. *Hempton* 1391 Ipm, *Hamptons Lood* 1594 Gelling & Cole 2000: 82, *Hampton Boat (sic)* 1752 Rocque. From OE *(æt þæm) hean-tun* 'high *tun*', with OE *lad* 'road, path, watercourse'. *Loade* (here probably meaning 'passage', in the sense 'river crossing': cf. Cricklade, Wiltshire; Winslade, Berkshire) was added to distinguish the place from other Hamptons. A ferry has long existed here. The earliest evidence for Modern English *lode* meaning 'ferry' is a record of 1480 of *the loode* at Apley in Stockton, Shropshire, where in 1494 the ferry is 'the fery whyles called the loode of Apley with the ware [weir] to the same fery or lode belongyng': Gelling & Cole 2000: 82. Hampton Loade has been in Shropshire since the 12th century. Cf. Aqualate. See also Lye Hall.

**HAMSTALL RIDWARE** – see **RIDWARE, HAMSTALL.**

**HAMSTEAD** 3 miles north-west of Birmingham (SP 0493). *Hamestede* 1213 SHC II 162, *Hamsted* 1227, 1293 Ass, *Hamstede* 1276 SHC VI (i) 80, *Hamstud* 1284 FA, *Hampstude* 13th century SHC XVII 261, *Hampstede, Hamstid* 14th century Duig, *Hampstead* 1564 SHC 1938 185, *Hamstead* 1657 SHC 1910 294, *Hampsted* 1686 Plot 105. From OE *ham-stede* 'homestead, the site of a dwelling': EPNE i 232. The name seems originally to have been applied to single dwellings or farmsteads, for many place-names containing this element have fallen into disuse.

**HAMSTUDE** (unlocated, possibly in Drointon: SRO D938/495) *Hamstude* 1203 SRO D938/495. From OE *ham-stede* 'homestead, the site of a dwelling'.

**HANBURY** Ancient Parish 5 miles north-west of Burton-upon-Trent (SK 1727). *Hamb[ury]* c.1185 Fr, *Hamburi* 1190x1247 SHC 4th Series IV 80, *Hambur* 1251 Ch, *Hanberyate* 1284 SHC VI (i) 131, *Hamburi, Hambyri, Hamberi, Hambery, Hambury* 13th century Duig, *Hampbury* 1352 SRO D4038/A/5/3, *Hanbury, Hambury* 14th century, *Hambury* 1430 Duig, *Hanbere* 1532 SHC 4th Series 8 81. From OE *heah, burh* (dative *hean byrig*) 'high *burh* or fortification': the place stands on a 474' hill overlooking the river Dove. The 1284 spelling is 'Hanbury gate'. *Heanbirig*, recorded in 664 (S.68), has not been identified, but may be this place. The chronicler John ('Florence') of Worcester (d.1118) records that St. Werberga was buried at *Heanbirig*, which has been identified with this place: SHC 1916 134; see also Bradshaw 1887. It may be noted that *Werburghwic* 'Werberga's *wic*', associated with royal visits in 823 and 840, has not been identified (see also Gelling 1992: 155-6). Stenton 1970: 320-21 holds that one specialised meaning of OE *burh* was 'monastery', citing as an example Malmsbury (and cf. also Glastonbury, Fladbury and Tilbury, all of which possessed both a monastery and a name incorporating *burh* by the earlier 8th century), perhaps from the earthwork enclosure (*vallum monasterii*) which surrounded monastic buildings, and that meaning may apply here, since no fortification has been identified. Hanbury in Staffordshire may form part of the group of early administrative and ecclesiastical centres in the north-west Midlands incorporating the element *burg*, including (in Cheshire) Astbury, Bromborough, Bunbury, Prestbury and Wybunbury, and (in Shropshire) Alberbury, Chirbury, Maesbury and Lydbury: see VCH Ch I 246. Since *hean* also had the meaning 'chief, important', there is a possibility that the name here is 'chief fortification', or 'chief monastery'. ASC 'E' records *Heanbyrig* in 675 (11th century), which has been identified as Hanbury, Worcestershire: see Ekwall 1960: 216; PN Wo 321-2; also S.68 and S.72.

**HANBURY WOODEND** 1 mile south-west of Hanbury (SK 1626). *Hanbury Woodend* 1658 (1798) Shaw I 73. The place lies on the road to Needwood Forest, and is to be distinguished from Woodend (q.v.).

**HANCH, HANCH HALL**   3 miles north-west of Lichfield (SK 0913). *Haunchall* c.1522 (1798) Shaw I 226, *Hanch Hall* 1747 Bowen, *Haunch* 1747 Poll, *Haunch Hall* 1834 O.S. From ME *hanche*, derived from OFr *hanche*, often found in field-names, with the meaning 'land having the shape of a haunch'. The second element is probably from OE *heall* 'hall', a rare element in Staffordshire, rather than OE *halh*, although the spellings are too late for any certainty. Shaw (1798: I 226) suggests that the place was known since the time of Edward I as Aston Hall or similar from the Astons of Haywood.

**HANCHURCH** in Trentham parish, 3 miles south of Newcastle under Lyme (SJ 8441). *Hancese* 1086 DB, *Henchurche* 1203 SHC III 94, *Hanchurche* 1212 Fees, *Hannecherche* 1272 SHC IV 204, *Hanchirch* 1275 SHC VI (i) 51, 1327 SHC VII 202, 1477 SHC VI NS (i) 111, *Hamcherch* 1377 SHC 4th Series VI 13, *Hanchurch* 1686 Plot. A puzzling name. The spellings suggest that the name is 'high church', from OE *hean*, the weak oblique form of OE *heah* 'high', with the 1086 terminal *cese* representing OE *cirice* 'church' (the DB spelling with the not unusual representation of the Anglo-Norman sound represented by *s* for OE *c*), or the first element may be OE *hane*, from *hanum*, dative plural of *hane* 'stone, rock': cf. Hanford. The place lies on the side of a pronounced hill, 'on the summit of which is a square plot of ground, surrounded by venerable yew trees, and supposed to be the site of some ancient church or religious house': White 1834: 692 (see also SHC XII NS 73), from which the place is said to have been named. The square plot is to be identified with the 'enclosure about 200' square lined by ancient yew trees' on which stands a house called Hanchurch Yews: StEnc 260. However, apart from the name itself there appears to be no evidence, whether historical, archaeological or documentary, of any early church here (but see also TNSFC LXVI 1932-3 126-7; LXIX 1934-5 28-30, LXXII 1937-8 116). The element *church* is sometimes found (particularly in the name Churchhill, or Church with a hill-word, for example Churchdown, Gloucestershire) to be from British *\*cruc*, Welsh *crug*, OE *cryc* can appear in OE itself as *cyrc, cyric* 'hill, mound, tumulus' (PN Wo 108), and since the present church is not ancient, but traces of earthworks have been recorded, it seems possible that the original name of the hill here, or a tumulus on the summit, was known as Cyrc or Cyric. It may be noted that Hanford, 2 miles to the north-west, has the same first element. *Hanestowe*, the second element of which meant 'place, enclosed place, place of assembly', or sometimes 'holy place' (see Stowe), is recorded in the area in 1334 (SHC 1913 230) and whilst not located appears to have lain between Whitmore and Madeley (or near Knutton), and may (as *(æt þære) hean stowe* 'high (holy) place') be associated in some way with this place. Hanchurch has given its name to Hanchurch Hills to the south. *Hanchurche Hey* is recorded in Cannock in 1540: SHC 1910 53. See also Hanford.

**HANDSACRE** in Armitage parish, 4 miles north-west of Lichfield (SK 0916). *Hadesacre* 1086 DB, *Handesacra* 1166 SHC 1923 295, *Hendesacra* 1195 SHC II 45, *Handesacr'* 1242 Fees, *Hondesakre* 1271 SHC V (i) 148, 1360 SHC VIII NS 7, *Hondesacre* 1395 SHC XV 66, *Honnesacre, Hunnesacre* 1386 SHC XVI 25, *hondesacre* 1420 Signet Letter C81/1365/23, *Honsakur* 1483 SHC NS VI (i) 133, *Honneshacre, Hanneshacrye, Honshacre, Hansacre* 1484 ibid. 151, *Handysaker* 1562 SHC XIII 219, *Hansacre* 1686 Plot. The first element may be from an OE personal name *\*Hand*, seemingly a nickname from OE *hand*, and which formed its genitive – unlike the word for the body part – with *-es*. *\*Hand* appears in the West Midlands as *\*Hond*, and even *\*Hund*. The second element is OE *æcer* 'field or newly cultivated ground'. *Æcer*, modern *acre*, has been found to have the very specific meaning 'small piece of cultivated land on the margin of a settlement', and examples of names containing the word can be grouped into three catagories according to their proximity to heath, marsh, or high moorland. This place is on a low promontary in the marshes of the Trent valley. *Pypehansaker* (presumably for Pipe-Handsacre) is recorded in 1572 (SHC XIII 286), but the location is uncertain. Land called *Handsacre* is recorded in Claverley in the 13th century: SA 5735/2/7/1/1.

**HANDS WOOD** 1 mile south-west of Newcastle under Lyme (SJ 8344). *Hands Wood* 1891 O.S. See Handsacre.

**HANDSWORTH** Ancient Parish 3 miles north-west of Birmingham (SP 0490). *Honesworde* 1086 DB, *Huneswordne* 1209 SHC III 175, *Huneswurth* 1212 SHC XI NS 17, 1222 Ass, *Hunneswrht'* 1236 Fees, *Unesworth* 1251 SHC XVII 123, *Honesworthe* 1242 Fees, c.1270 SHC 1924 352, *Hounesworth* 1276 SHC VI (i) 74, *Honesworth* 1333 *et freq.* SHC 1913 41, *Hunddisworth* 1532 SHC 4th Series 8 73, *Honsworth* 1610 Speed, *Hansworth* 1686 Plot. 'Hun's *worþ*. The late form *Hands-* is probably the result of influence from Handsacre. OE *worþ* meant 'fence or enclosure', and developed into 'enclosure around a homestead', and eventually 'homestead'. The place (the 1659 bounds are given in Shaw 1798: I 108) is now in Birmingham. There is another place of this name in Yorkshire.

**HANESHIRME, HAVERSHINE** (unlocated, in Stubbylane) *Haneshirme* 1587 SRO D786/21/3, *Havershine* 1587 SRO D786/21/5.

**HANESTOWE** – see **HANCHURCH.**

**HANFORD** in Trentham parish, 3 miles south-west of Stoke-on-Trent (SJ 8742). *Heneford* 1086 DB, *Honeford* 1212 SHC 1933 (ii) 11, 1234-4 TestNev, *Hanneford* 1250 SHC XI 319, *Honfort* 1299 SHC XII NS 71, *Heneford* 1307 SHC VII 175, *Handford* 1327 SR, *Honford* 1327 SHC XII NS 71, *Haneford* 1357 SHC XI 306, *Honford* 1399 SHC XV 88, *Henford* 1474 SHC NS IV 191, *Handford* c.1565 SHC 1938 176, 1775 Yates, *Handforde* 1589 Trentham ParReg, *Hanford* 1610 Speed, *Hondford* 1664 SHC II (ii) 61. The place stands on a hill (with natural springs on the summit), near the foot of which the Trent is crossed by what is said to be an ancient road. It is possible that the crossing gave its name to the place. The name is possibly from 'Hana's ford' (Hana being a name which in the West Midlands appears as Hone: see Handbridge, Cheshire), or perhaps 'cock's ford', from OE *hana* 'cock, wild bird': there are three Hanafords in Devon, all believed to be from OE *hana* – see Ekwall 1936: 216. A derivation from OE *hean ford* 'high ford' (in place-names OE *hean* generally becomes *han*), is improbable, since the two elements do not easily fit together, although *hean* could also mean 'chief, important', so here possibly 'the main ford'. However, it may be noted that Hanchurch, 2 miles to the south-east, has the same first element, which may be common to both places, in which case it may be *hane*, from OE *hanum*, dative plural of *han* 'stone, rock': cf. Hanford, Dorset. The supposed existence here of an ancient road (see for example StEnc 262 & 597, and the name *Old Road*, frequently mentioned in Barlaston ParReg from at least 1724, 1 mile north-east of Barlaston at SJ 8839) also makes a derivation from Welsh *hen-ffordd* 'old road, replaced road' (see TSAHS LXXV 2000 109-110) not impossible. See also PN Ch V (II) 346-8.

**HANGING BRIDGE** on the river Dove, marking the Staffordshire-Derbyshire border 1½ miles south-west of Ashbourne (SK 1545). *Le Hongindebrugge* 1296 RadCh, *Hongyndebrugge* 1330 Ass, *Hongyngbryge* 1417 MinA, *Hangynge Bridge* 1568 Lanc, *Hankinbridge* 1607 MinA. OE *hangra* 'a slope' might be considered appropriate for the very steep bank on the Staffordshire side of the river here, but the forms rule out such derivation, and it seems likely that the first element is from OE *hangende* 'hanging, over-hanging', describing some type of hanging or suspension bridge (OE *brycg* 'bridge'). The association of this name with the execution by hanging of felons and others is the result of folk mythology, and has no historical or etymological basis. For *Hongingebruig* near Whitmore recorded in the early 14th century, see SHC 1913 243.

**HANGING HILL** ½ mile north of Wigginton (SK 2007). Perhaps to be associated with *Hongyhille* 1624 (1798 Shaw I 314).

**HANGING STONE** an overhanging rock outcrop 2 miles north-east of Heaton (SJ 9765). *Hanginde stone* 1227 Harl, *the hanging stone* 18th century (1801) Shaw II Adv. ix, *the Hanging Stone* c.1708 ibid.'1, 1874 Brocklehurst. From OE *hangende, stan* 'the (over-)hanging stone'.

**HANGMAN'S OAK** on the north side of the Rugeley to Lichfield road, 3 miles west of Rugeley (SK 0018). *? Hangeman(strete)* 1570 Survey, *Hangmans Oak in ye Road* 1720 Bowen, *Hangman's Oak* 1834 O.S. Doubtless the site of a gallows, but local folklore associates the name with a certain Humphrey Aycocke, a suspected sheep thief who hid in the tree and died when he slipped and was strangled by his own scarf: StEnc 263.

**HANLEY** 1½ miles north of Stoke on Trent (SJ 8847). *Henle* 1212 Fees, 1234x1240 TestNev, *Hanleg* c.1217 Eyton 1881 92, *Hanlih* 1227 Ch, *? Anlegh'* 1239 CurReg, *Hanle* 1250 SHC XI 319, *Hanley* 1592 SHC XVI 117. '(Place) at the high *leah*', from OE *heah* 'high', in this case in its dative form *hean*: the place lies on a 500' headland The alternative name *Hanley Green* appears for this place by at least 1647 (SRO D3272/1/17/4/32), and was still in use in the mid 19th century: VCH VIII 142.

HANLEY HILL FARM (unlocated, near Cannock) *Hanley Hill Farm* 1623 SRO D603/A/3/63-73. See Hanley.

HANLEY PARK (obsolete) a medieval park on the east side of Endon, between Park Lane and Endon Brook (SJ 9253), recorded in 1341, disparked by c.1550: VCH VII 182. See Hanley.

HANNELL (obsolete, 1 mile north-west of Heaton (SJ 9663)) *Hanewelle* (p) 1259 *TallAss, Hennele* 1309 Banco, *Hannell (poole)* 1564 (1883) Sleigh 65, *the Hannell* 1617 SRO Swythamley MSS, *Hannel* 1842 O.S., *Annel* 1880 Kelly. The first element is from OE *hana* 'cock, wild bird', or the personal name Hana, with Mercian OE *wælle* 'a spring', and (sometimes) 'a stream'. Cf. Hanwell, Middlesex.

**HANYARDS** 1 mile north-west of Tixall (SJ 9624). *Haenegate* c.1220 SHC VIII 161, *Hagenegath* 1244x1261 SRO D938/398, *Hagonegate, Hageneyate* 1227 Duig, *Hagengate* 1228 SHC VIII 139, *Hanegate* 1295 SHC VIII 191, *Hanberyate, Hanyate* 13th century Duig, *Havenyate* 1305 SHC 1911 65, *Upper Hanyard, Lower Hanyard* 1836 O.S. The first element is OE *\*hægen* 'enclosure', with OE *geat* 'gate, opening, gap'. The word is often applied to breaks in earthworks, gates through town walls, entrances to parks, etc. The place is an entrance to Tixall Park.

**HARBORNE** Ancient Parish 3 miles south-west of Birmingham (SP 0284). *Horeborne* 1086 DB, *Horeburn* 1221 SHC IV 17, *Holeburn* 1229 ibid. 76, *Horebourn* 1278 SHC 1911 33, *Ourbor* 1284 SHC 1910 298, *Horburn* 1301 SHC VII 84, *Horbourne* 1342 SHC XI 152, *Horborne* 1600 SHC 1935 234, *Harborne* 1600 ibid.' 243. From OE *horu-burna* 'dirty stream'. On the element *burna* see Bourne Vale. The place is now in Birmingham. See also The Hurstage.

**HARDEN** 2 miles north of Walsall (SK 0101). *Haworthyn* 1300 SHC V (i) 178, *Hawerthyn* 1327 SHC VII 224, *Hawerdyn* 1338 SHC XV 40, *Hawardyn* 1381 SHC XVII

163, *Haworthyn, Hawardyn,* 14th century Duig, *Hawardyne* 15th century Duig, *Herden* 1532 SHC 4th Series 8 145, *Harden* 1588 SHC 1927 177, *Hawrden otherwise Horden* 1589 SHC XV 193, *Hawrden, Hawredene* 16th century, *Hawrden* 1611 SHC III NS 54, *Haverdon* 1616 SHC VI NS (i) 15, *harding* 1632 Walsall ParReg, *Harden* 1648 Duig, *Harding otherwise Hawarden* 1686 Plot 188. Possibly *heah-worþign* 'high farm or estate'. The place is on high ground which formerly lay in Cannock Forest. Hawarden in Flintshire has the same root, and is pronounced Harden: PNEF 60.

**HARDINGS BOOTH** 1½ miles west of Longnor (SK 0664). *Hardingesbothe* 1327 SHC VII (i) 218, *Hardyngbothe* 1397 SHC XV 78, *Hardynggesbothe* 1440 B.L. Woll. ch iii 9, *Hardingsbooth* 1608 Leek ParReg, *Hardens booth* 1658 Alstonefield ParReg. From ODan *boð* 'bothy, temporary shelter', with the personal name Harding, a common surname in this area.

**HARDINGS WOOD** 1 mile west of Kidsgrove, on the Staffordshire-Cheshire border (SJ 8254). *Hardingeswood* 1597 SHC 1935 14, *Hardingswood (House)* 1657 Wolstanton ParReg. 'Harding's wood'.

**HARDIWICK (FARM, HEATH & GROVE)** 1 mile north of Sandon (SJ 9332), *? Herdewyke* 1275 SHC VI (i) 55, *Hardewik'* 1237 Cur, *Herdewyke* 1288 SHC 1911 43, *Herdewik, Herdewyk* c.1290 SHC 1921 25, 33, *Herwick* 1549 SRO D641/1/2/288, *Hardeck* 1564 SHC 1938 88, *Hardywicke* 1601-2 SHC 1934 4, *Hardwick* 1686 Plot 288; **HARDIWICK** 1 mile north-west of Dilhorne (SJ 9544), *Herdewyke(gorstes)* late 13th century SRO 3764/2[27574], *Herdewicke* 1333 SHC XII NS 128, *? Herdewyke* 1362-3 JNSFC 1924-5 60, *Hardwick(e)* 1639 Caverswall ParReg, *Hardy-wick* 1836 O.S.; **HARDWICK** 2 miles south-west of Aldridge (SO 0698), *Hardwyke* 1570 SHC XVII 217, *Hardwick* 1834 O.S.; HARDWICK (obsolete, on south side of the Great Pool, Patshull (SO 8099; SMR 01135 places a deserted post-Conquest settlement at SO 80369933), see VCH XX 173), *herduic* c.1155 SHC 1939 182, *Herdewyk', Herdewycke* 1301 ibid.' 1939 187, *Herdewych* 1311 Brighton 1942 159, *Herdwyke* 1412 SHC 1939 191, *Herdewyke* 1448 ibid.' 193, *Hardwick* 1798 Yates. From OE *heord(e)-wic* 'livestock farm'. Gould 1957: 138 suggests that Hardwick near Aldridge may be from the family name of the Countess of Shrewsbury, who held land at nearby Drayton, but the place is said to be recorded in the 15th century (Anon. 1984: 17), which makes a derivation from OE *heord(e)-wic* very likely. *Le Herdewikemor* in Stowe or Chartley is recorded in 1318: SRO 493[7910]. See also Pipehill.

**HARECASTLE** 1 mile south-east of Talke (SJ 8352). *Harecastle* 1585 Speake 1972: 25, 1644 SHC 4th Series I 204, 1656 Wolstanton ParReg, *Hare Castle* 1733 SHC 1944 59, 1775 Yates, 1833 O.S. The first element may be from the OE adjective *har* 'grey, hoary', or from OE *hara* 'a hare'. The second element (if the name is ancient) is OE *castel* 'a castle, a camp', perhaps referring to some ancient earthwork or structure on the hill here, or possibly OE *ce(a)stel* 'a heap of stones'. There are two canal tunnels here, one 2,897 yards long built in 1770-7, the other 2,882 yards long, opened by Thomas Telford in 1827: see SHC 1934 (i) 110.

**HARE HILLS** in Beaudesert Old Park ½ mile west of Upper Longdon (SK 0514). *Harehull* 1340 SHC 1921 24. 'Hare hill'.

**HAREGATE** 1 mile north-east of Leek (SJ 9957). *Hareyate* 1544 (1883) Sleigh 72, *Hare yate* 1634 Leek ParReg, *Hare Yate* 1747 Poll, *Hare Gate* 1842 O.S. The first element is uncertain: possibly OE *hara* 'hare', but OE *har* 'hoary, grey' or OE *\*hær* 'a rock, a heap of stones' (cf. Swedish *har* 'stony ground': Ekwall 1960: 218, but on the existence of *hær see Harewood). *Yate* is from OE *geat*, 'gate', possibly here in the sense 'pass': the place lies on the south side of the steep valley of the Churnet.

**HARETHORN** (unlocated, in Bradley near Stafford; see also Harrethorn) *Harethorn* c.1200 SRO D986/27.

**HAREWOOD** 1 mile north-west of Cheadle (SK 0044), *Harwode* 1483 SHC VI NS (i) 148, *the longe harwood, the grete Harwood or the Cote Harwood* 1610 SHC 1934 (ii) 40, *Harewood* 1616 SHC VI NS (i) 9, *Horwoods* 1762 SRO D1203/B/12/1-3, *Harewood Hall* 1836 O.S.; HAREWOOD (obsolete, in Perton), *Harewude* 1258 SHC IV 138. From OE *hara wudu* 'hare wood', or (more likely, since hares are animals usually found in open areas, and no other recorded place-name with *wudu* has the name of a wild or game mammal as its first element: Coates 1997: 454) 'grey (hoar) wood'. In principle the first element of these names could be *\*hara*, genitive plural of a supposed *\*hær* 'stone', but doubt remains about such word: see Coates 1997: 454. A semi-technical British term for surviving areas of wildwood may have been Letocetum (see Wall), the translation of which survives in the name Harwood (from OE *har* 'grey') found in several counties: see Coates 1997: 453-8; PN Ch IV 227-8; VCH XX 29. See also Lordsley.

**HARLASTON** 4½ miles north-east of Tamworth (SK 2110). *Héorlfestun, Heorelfestun* 1004 (11th century, S.906; 11th century, S.1536), *Horuluestone* 1086 DB, *Heorlauestun* 11th or 12th century Sawyer 1979a: xxxvii, *Herlaueston* 1165 SHC I 39, *Herlaveston'* 1242 Fees, *Herlaweston* 1288 SHC VI (i) 175, *Herliston* 1324 SHC 1911 363, *Herlaston* 1393 SHC XV 50, *Hauston* 1510 SHC 1912 6, *Horlaston* 1532 SHC 4th Series 8 71. 'Heoruwulf's or Heorulaf's *tun*'.

HARLEY (unlocated, near Lea or Drointon or Bagot's Bromley) *Harele, Harleia* early 13th century SRO D986/43, *Harlee, Harley, Harley broc* 13th century SHC VIII 164, *Harleye* 1280 SRO D938/588, *Hareleye, Harlee', Harleg'* late 13th century ibid. 581-2[7912], *Harleye* 1324 ibid.' D603/A/Add/439, 1349 ibid. 567[7922], *Herley* 1401 SRO D4038/E/1/1, *Harley Farm* 1661 SRO D(W)1721/3/170. Perhaps from OE *har* 'hoary, grey; stone, stony ground', or OE *hara* 'hare', with OE *leah*. See also SHC XI NS 170. *Great Harley* in Croxton us recorded in 1797: SRO D5721/1/24/1-14. See Harley Thorns.

**HARLEY THORN (FARM), HARLEY THORNS** 2 miles north-west of Swynnerton (SJ 8439). *? Harleghe* 1301 SHC VII 97, *Harleye* 1334 SHC XI 50, *(Shelton) Harnage* 1368 SRO D641/5/T/1/1, *(Shelton under) Harley* 1381 SHC XIII 160, *Horeley* 1617 SHC 1934 (ii) 52, *Hairleythorn* 1777 SRO D641/5/E(V)/10, *Harleythornwood* 1796 SRO D641/5/E(c)/36, *Harley Thorns* 1833 O.S. Perhaps from OE *hara, leah* 'the wood (or clearing) with the hare', but the 1617 spelling suggests the possibility of a derivation from OE *horh, horu* 'filth, dirt', in place-names often 'mud'. *Harnage* is evidently an error. *Thorns* suggests an abundance of thorn bushes. *Harelyhead* and *Harely head* are recorded in the Harley area in 1732-3: Swynnerton ParReg.

**HARLOW (WOOD)** ½ mile west of Mayfield (SK 1446). *Harlow* 1775 Yates, *Harlow Wood* 1836 O.S., *Harlow (Farm)* 1847 SRO D1134/7/1. Possibly from OE *har* 'hoary, grey; stone, stony ground', with OE *hlaw* 'mound, tumulus', so 'the grey mound or tumulus', or 'the tumulus on stony ground': Harlow Greave (*Harlow-greave* 1686 Plot 404, 1798 Shaw I 33) was a large mound here, probably artificial, now destroyed: StEnc 270; JNSFC 3 1965 44. However, the proximity of the place to Mayfield (q.v.) suggests that the first element may be OE *here* 'army, host, multitude', but also used for 'the whole people', so possibly 'the mound of the people', denoting a meeting-place. Cf. Harlow,

Essex; Harlow Hill, Northumberland. *Greave* is from OE *grafe* 'a grove, a copse', often found in Staffordshire as *greave*. The surname Harlow is recorded in association with this area 1664x1699 (SRO D514/M/15). See also Fords Rice.

**HARPER CLOUGH** (obsolete) 1½ miles south-west of Heathylee (SK 0362). *Harper Clough* 1842 O.S. See Clough. Harper may be from a family or occupation name: see Harpersend.

**HARPERSEND** 1 mile north of Upper Hulme (SK 0162). *Harpersend* c.1870 *Rental*. From the former Harpur-Crewe estates here: VCH VII 4-5.

**HARPER'S GATE** at the south end of Rudyard Lake (SJ 9557). *Harpers Yate* 1568 SHC 1938 154, *Harpurs Gate* 1816 SRO DW1909, 1842 O.S. Evidently from a personal or occupational name; e.g. Robert le Harpere is recorded in 1340 (SHC 1913 79). Harpers Farm (*Harpers Farm* 1842 O.S.) lies 1 mile north-west of the north end of the Lake. After Rudyard Lake became a popular tourist resort in the middle of the 19th century, Harper's Gate expanded and became known as Rudyard. See also Rudyard.

**HARPING** (unlocated - according to Shaw 1801: II 73 a hamlet in Walsall) *Harping* 1801 Shaw II 73.

**HARPLOW** 1 mile south-west of Cheadle (SJ 9941). *Harplow* 1668 SRO D593/8/2, 1836 O.S. It is unclear whether *Orpley otherwise Arpley*, recorded in 1599 (SHC XVI 41), refers to this place.

**HARRACLES HALL** 2 miles north-west of Leek (SJ 9557). *Harecheles* 1279 SHC VI (i) 100, *Harachils* 1313 Banco, *Harecels* 1470 (1883) *Deed* Sleigh, *Harracles* 1559 SRO DW1702/2/6, 1605 Sleigh 1883 20, *Harrackles* 1568 SHC 1938 153, *Harekells* 1583 SHC III (ii) 7, *Haracles* 1635 Leek ParReg, *Herracles* 17th century Duig, *Harracles* 1798 Yates. The first element would seem to be from OE *har* 'grey, hoar', which may have come to mean 'boundary' (the place is near a parish boundary), with OE *\*ecels* 'an addition, land added to an estate', so giving 'land added to an estate near the (parish) boundary'.

**HARRETHORN** (unlocated, possibly near Hednesford; see also Harethorn) *Harrethorn* 1339 SHC 1931 241.

**HARRISEAHEAD** 1 mile south-west of Biddulph (SJ 8656). *Harrishey head* 1662 Wolstanton ParReg, *Harrishey-head* 1671 ibid, *Harrowsey Head* 1798 Yates, *Harrisea*, *Harriseahead* 1811 SRO D997/viii/6, *Harrisea Head* 1836 O.S. Seemingly 'the head or top of the hay or enclosure of Harry': the place lies on a spur of high ground extending from Mow Cop.

**HARSTON ROCK, HARSTON WOOD** ½ mile north-west of Whiston (SK 0347). *Harston Rock & Wood* 1814 *plan*, 1837 O.S., 1840 TA. From OE *har, stan* 'the grey stone or the boundary stone', from a prominent upright stone which lies on the boundary between the townships of Whiston and Foxt. See also Warstones.

**HARTESMERE** (unlocated, in Hamstall Ridware) *Hartesmere* 1596 SBT DR18/1/1875, *Hart(e)smere* 1596 SBT DR18/1/1874-1875, *Hartsmere (farm)* 1618 SBT DR18/1/1899, *Hartysmeere (Farm)* SBT DR18/1/193. 'The stag's pool or wet ground', from OE *heort, mere*.

**HARTLEBURY** 1½ miles north-west of Worfield (SO 7497). *Hartlebury* 1833 O.S., 1839 T.A., *Artlebury* 1841 PRO HO107/908. Early spellings have not been traced, and the name may be of no great age, perhaps transferred from Hartlebury, Worcestershire, 18 miles to the south-east. If ancient, perhaps 'Heortla's fortification or manor', from OE *burg*. The name Heortla is only evidenced in Hartlebury, Worcestershire: see PN Wo 242-3. In Shropshire since the 12th century.

**HARTLEY GREEN** 1 mile north-west of Gayton (SJ 9729). *Hartley Grene* 1565 SHC 1926 79, *Harteley Greene* 1608 SHC 1948-9 112, *Hartley green* 1686 Plot 105, *Hartley-Greene* 1706 SHC 1938 231. Possibly from OE *heorot-leah* 'stag wood or clearing'. ). The Green element suggests a squatter settlement.

**HART'S FARM** 1 mile south-east of Abbots Bromley (SK 0923); **HART'S FARM**, **HART'S COPPICE** 2 miles north of Abbots Bromley (SK 1008). Probably to be associated with the Hart family, recorded in this area since at least the 17th century: 'lands in Bromley Great Park occupied by Thos. Hart': SRO D742/A/2/10.

**HARTSHILL** 1 mile south-east of Newcastle under Lyme (SJ 8645). *? Herthull* 1272 SHC XVII 201, *Hardeshull* 1373 SHC VIII NS 65, *H(er)tishille, hertishil* 1420 Signet Letter C81/1365/26, *Hartshill* 1550 SRO D593/B/1/14/4/7, *Harteshill* 1584 SRO D593/B/1/14/4/13, *Harts Hill* 1732 Stoke on Trent ParReg. From OE *heor(o)t* 'hart, male red deer', giving 'the hill of the hart'.

**HARTWELL** 4 miles north of Stone (SJ 9139). *? Hortwell* 1154-94 SHC XI 332, *Hurtwall* 1293 SHC VI (i) 266, *Hertwell* 1337 SHC XI NS 26, *Hertwall, Hertevalle* 1347 ibid.' 189, *Hertewall* 1366 SHC VIII NS 33, *Hertwalle* 1396 SHC XV 80, *Hartwall* c.1562 SHC IX NS 31, *Hartwell* 1583 SHC 1924 53, 1592 SHC 1930 287. From OE *heorot* 'hart, male red deer', with Mercian OE *wælle* 'a spring', and (sometimes) 'a stream'. *Hertewallfeld* is recorded in 1304, possibly near Admaston (SRO D938/58), evidently to be associated with *Hertwalle*, recorded in Adbaston in 1437 (SHC 1919 104).

**HARVILLS HAWTHORN** 2 miles north-west of West Bromwich (SO 9893). *Humvill* 1255 Fees, *Heranvyl* 1294 SHC 1911 219, *Herunwyll* 1338 SHC 1913 66, *Hervyle* 1419 SHC XVII 67. From the Heronville or Harvill family who held land here from at least 1271: see for example SHC 1928 279; Ede 1962: 25-6. A tree called *Harvyl's Oke* is recorded in 1531. The district was known as *Harvills Oak* by the mid 18th century, and *Harvills Hawthorne* by 1816, although the northern stretch of Dial Lane was still called *Harvills Oak*: VCH XVII 7.

**HARVINGTON BIRCH** 1½ miles west of Brewood (SJ 8508). *? Harrington Birch* 1825 Brewood ParReg, *Harvington Birch* 1834 O.S. The Harvington element is unexplained (unless transferred from Harvington Hall in Worcestershire, a Catholic house with hiding places: the Giffards of Chillington, on whose estate Harvington Birch stands, were staunch Catholics, associated with the escape of Charles II after the battle of Worcester – see Boscobel), and the name may not be ancient: the farmhouse here appears to date from the 18th or 19th century. *Harrington Parke*, recorded in 1601 (Codsall ParReg), is probably to be associated with Harrington Hall, 1 mile west of Beckbury in Shropshire (recorded from 1251: see Eyton 1854-60: II 131; see also TSAS 3rd Series IX 59; 73-4; the Harrington family are first recorded in the local parish registers in 1574: ibid. 93). The Birch element is probably from OE *bryce* 'breaking', used of 'newly cultivated ground', normally found in or close to ancient forests and wastes: Harvington Birch lay deep in Brewood Forrest. By metathesis, or shifting of the *r*, the word becomes *burche* and later *birch*. Cf. Long Birch, near Brewood; Breach Mill near Hagley; the Breach in Halesowen; the Breach near Belbroughton; the Bratch near Enville; and the Bratches in Norton Canes.

302

HARWOOD (seemingly an alternative name for Lordsley (q.v.) near Mucklestone), *Harwodde* 1529 SHC 1910 19, HARWOOD (unlocated, in Cheadle), *the harwood* 1614 SHC 1934 30. See Harewood.

HASALLHURST (unlocated) *Hasallhurst* 1590 SHC 1930 (ii) 51. From OE *hæsel* 'hazel', and OE *hyrst* 'wooded eminence, copse, wood'.

HASELBACHE (unlocated, in Waterfall) *Haselbache* 1185 (1798) Shaw I 3. From OE *hæsel* 'hazel', and OE *bece* 'a well-defined stream valley', so 'the steep sided valley with the hazels'.

HASELEY (obsolete) south-east of Radford Bridge, east of Stafford: VCH V 6 (SJ 9421). *Haseleye* after 1290 SRO D938/47, *Campo de Halseyley* c.1297 VCH V 3, *Haseley* 1474 ibid.' 6, c.1532 ibid, 1613 SHC NS IV 16, *the manor of Haseley beside Berkswyche* c.1480 SHC VII NS 269, *Haseley Farm* 1677 SRO D1979/H.M.Drakeford/3a. *Haseley Manor or Haseley Farm* 1732 SRO D260/M/E/353a. Perhaps the same derivation as Halsey (q.v.), with OE *leah*.

**HASELOUR** 4 miles north of Tamworth (SK 2010). *Hazeloure* 13th century SHC XVIII 61, *? Halsemor* 1309 WL 103, *Haselwor* 1369 SHC VIII NS 232, *Hawlore* 1373 BCA MS3878/28, *Haselovere* 14th century Duig, *Haslore* 1417 SHC XVII 61, *Haslhow* 1532 SHC 4th Series 8 71, *Hasulhowre* 1539 SHC NS IV 217, *Haislor* 1577 BCA MS3878/120, *Hassleore* 1644 SHC 4th Series I 213, *Haselover or Haselor* 1796 Duig. From OE *hæsel* 'hazel', and OE *ofer* 'a flat-topped ridge with a convex shoulder', hence 'the flat-topped ridge with the hazels': the place lies at the southern end of such a feature, which is clearly marked by hachuring on the first edition 1" O.S. map of 1834. Cf. Haselor House and Harvington, Worcestershire (PN Wo 106, 238); Haselor, Warwickshire (PN Wa 211).

HASHALL (unlocated, in or near Audley, possibly Ash Hall (q.v.)) *Hashall manor* 1492 SHC 1912 257.

HASTECOTE (unlocated, in or near Kinver) *Hastcote* 1294 SHC VII 9, *Hascote* 1474 SHC NS IV 196, *Hascote* 1578 SA 2089/2/3/1, *Hastecote, Hascott* 1602 SHC 1935 445, 462, *Hascott* 1602 SHC 1935 462, *Hastecote* 1617 SHC VI (i) NS 42. It seems possible that this is the same place as Halfcot (q.v.).

HASWIC (unlocated) in Seisdon Hundred. *Haswic* 1086 DB. Almost certainly Ashwood (q.v.).

**HATCHLEY** 2 miles north of Dilhorne (SJ 9845). *Hatchley* 1574 SHC XIII 297, 1609 BCA MS3810/196, 1610 SHC III NS 39, *New Hatchley* 1798 Yates, *Hatchley* 1836 O.S. Possibly from OE *hæcc* 'a hatch', generally meaning 'a gate, especially in a forest', but sometimes with the meaning 'floodgate, sluice'. This place is near a stream. The second element is OE *leah*.

**HATELEY HEATH** 1½ miles north of West Bromwich (SO 0093). *Hateley Heath* 1577 Willett 1882: 216, 1654 BCA MS3145/96/1, 1834 O.S. The place is said formerly to have been called *Longmore* (Willet 1882: 209), or *Longnolre*, and also known as *Hackle Heath* (StEnc 272). The name is from OE *hǣðiht* 'heathy', with OE *leah*. The Hateley or Hayteley family is said to have been recorded in the area from the early 15th century: StEnc 272.

**HATHERTON** 2 miles north-west of Cannock (SJ 9510). *Hageþorndune* 996 (for 994) (17th century, S.1380), *Hargedone* 1086 DB, *Hatherdon* 1203 Ass, *Hatherdene*,

*Hetherdon* 1292 SHC VI (i) 233, *Hatherdone* 13th century Duig, *Hathurdon* 1365 Banco, *Haderton* 1532 SHC 4th Series 8 88. 'The hill where hawthorn grows', from OE *haguþorn* 'hawthorn, whitethorn', with OE *dun* 'hill'.

**HATTONS, THE HATTONS** 2 miles south-east of Brewood (SJ 8804; SMR 01900 places a deserted post-Conquest settlement at SJ 88800470). *Hadton* 1227 Ass, 1424 SHC XVII 94, *Hattone* 1292 SHC VI (i) 237, *Hatton* 1302 SHC VII 96, *Hatton Hall* 1712 SA 5380/1/13-15; **HATTON** in Swynnerton, 5 miles north-west of Stone (SJ 8337). *Hetone* 1086 DB, *Hatton'* 1206 Pleas, *Hadton* 1227 SHC IV 43, *Atton* 1263 SHC IV 157, *Aiton* 1689 StSt 13 2001 51. From OE *hēð- tun* 'The *tun* on the heath'. There are *Upper* and *Lower Hattons* at both places.

HAUEKESLYH (unlocated, near Norton-in-the-Moors) *Hauekeslegh* 1227 SHC XI NS 240. From Mercian OE *heafoces-lege* 'the fallow land of the hawk'.

**HAUGHTON** Ancient Parish 4 miles south-west of Stafford (SJ 8620), *Haltone*, *Halstone* 1086 DB, *Halgetona* 1161x1182 SRO (1/7972), *Haluchton* 1189 SHC I 173, *Haleton* 1201 Ass, *Haldeton* 1227 SHC IV 53, *Halixton* 1236 Fees, *Haleweton* 1284 SHC VI (i) 154, *Halweton* 1284 FA, *Halington*, *Halechtone*, *Halctone*, *Halegtone*. *Haluchtone* 13th century Duig, *Haleughton'* 1327 SHC VII 214, *Halughton* 1336 SRO D(W)1721/3/3/18, *Haleughton* 14th century Duig, *Haghtton* 1532 SHC 4th Series 8 67, *Hawghton* 1603 SHC 1940 38, *Haughton* 1686 Plot; HAUGHTON (obsolete) 2½ miles north-west of Ramshorn (SK 064483), *Haughton* 1798 Yates. 'The *tun* by the nook of land or water meadow', from OE *halh*, of which *halch* and *halech* are ME forms. The place near Ramshorn now appears on maps as Windy Harbour.

HAUKESCLYF (unlocated, possibly near Talke) *Haukesclyf* 1282-3 SHC XII NS 251, 1298 SHC XI NS 251. From Mercian OE *heafoc* 'a hawk', with OE *clif* 'a cliff, a bank', so 'the hawk's cliff'.

HAUKESHILL (unlocated) In Bagot's Bromley. *Haukeshill* 1306 SHC XI NS 24. From Mercian OE *heafoc* 'a hawk', with OE *hyll* 'hill'.

HAUKESMOR (unlocated, in Blithfield), *Haukesmor* 1402 SHC XI NS 203; HAUKESMOR (unlocated, in Rickerscote), *Haukesmor* 1346 SRO 85[7904]. From Mercian OE *heafoc* 'a hawk', with OE *mor* 'Hawks' moor'.

**HAUNTON** 5 miles north-east of Tamworth (SK 2310). *Hagnatun* 941 (14th century, S.479), *Honegeton* 1231 SHC VI (i) 34, *Hagheneton* 1249 FF, *Auneton* 1259 SHC IV 139, *? Hainton* c.1260 SHC 1937 62, *Hauneton* 1271 Ass, *Hanneton*, *Anneton*, *Hagan*, *Hagana*, *Haguna*, *Hagene* 13th century Duig, *Hamton* 1532 SHC 4th Series 8 70, *Hawnton* 1565 (1798) Shaw I 412, 1599 SHC 1935 188, *Haunton* 1695 Morden. Possibly 'Hagona's or Hagene's *tun*', but a more likely derivation is from OE *hagena*, genitive plural of OE *haga* 'hedge, enclosure', with OE *tun*, so 'the *tun* with or at the hedges or enclosures'. *Haunton*, recorded in 1252 (SHC 1937 47), is Hampton near Stansley (q.v.).

HAWFORDE (unlocated, possibly in Eccleshall) *Hawforde* c.1565 SHC 1938 73.

HAWKBACH (obsolete) in Upper Arley (SO 7682). *Auchebech* pre-1172 SHC 1924 152, *Haukebache*, *Hawkebach* 1360-98 PN Wo 31, 1547 Pat, *Howkebaiche* 1551 BM, *Haukebach* 1577 Saxton, *Hawkebache* 1603 SHC XVIII 34, *Haukbach* 1686 Plot, *Hauke bach* 1695 Morden, *Hawkbach* 1756 (1801) Shaw II 254, 1834 O.S. From Mercian OE *heafoc* 'a hawk', and OE *bece* 'well defined valley with a stream', the second element particularly common in the Shropshire hill country but less common in Staffordshire.

Hawkbach was in the part of Staffordshire which lay on the west of the Severn, and may have been the site of an early river crossing: StEnc 274. The name survives in Hawkbatch Valleys (SO 7677). In Worcestershire since 1895.

HAWKESWALL CLOUGH (unlocated, possibly in the Audley/Balterley area) *Hawkeswall Clough* 1599 SHC XVI 194. From Mercian OE *heafoc* 'a hawk', with Mercian OE *wælle* 'a spring', and (sometimes) 'a stream', with OE *cloh* 'a small valley with steep sides'.

**HAWKESYARD (PRIORY)** in Armitage (SK 0616). *le Haukeserd* 1337 SHC 1913 319, *Haukesherd* 1367 SHC VIII NS 37, *Haukesort* 1395 SHC XV 66, *Haukesyerd* 1414 SHC 1921 10, *Haukeserthe* 1418 SHC XVII 66, *Hawkes yarde upon Trent* c.1540 Leland, *Haukesley* 1566 SHC 1925 128. A Gothick house built c.1760, known originally as Armitage Park, renamed c.1839 after the medieval house which once stood here. The house was left in 1893 to the Dominican Order, who built Hawkesyard Priory above the house, on completion of which the house was renamed Spode House. The original name of the place (which stands on rising ground above the river Trent) is from OE *heafocscerde*, probably from OE *heafoc* 'hawk', with OE *\*scerde* 'a gap, a cleft, a pass', giving 'the gap with the hawk': there is a pronounced gap adjoining the west side of this place. However, Field 1993: 76-7 prefers to interpret OE *heafocscerde* as 'a woodland clearing in which hawks were flown for falconry'. A derivation from the OE personal name Hafoc cannot be ruled out, but is improbable. Cf. The Hawkshutts, Hawkbach. *Haukesyerd* near Danebridge is recorded in 1347 (Eyre): see also Danebridge.

HAWKEWALLSYCH a manor called '...Hawkeswelsich is the West and North-west part of [Longdon] parish...': Shaw 1798: I 223; SHC 1925 6. *Hakewallsiche* 1311 SRO 3764/90[27574], *Hackewell* 1321 (1801) Shaw II 321, *Hawkewallsych* 1442 (1798) Shaw I 224, *Harkwellsuchende* 1469 OSS 1936 44, *Hakewall suche* 1487 ibid. 46, *Hackwell* 1549 SRO DW1734/2/3/112b(18), *Hackewell* 1550 (1801) Shaw II unpublished plates 321, *Hawkeswelsich* c.1598 Erdeswick 1844: 243, *Hawkwell Sych End* 1613 SRO D603/E/2/110-132, *Haukeswel* 1798 Shaw I \*211, \*212, 223fn. From Mercian OE *heafoc* 'a hawk', and Mercian OE *wælle* 'a spring', and (sometimes) 'a stream', with OE *sic* 'watercourse'. See also Halsey.

HAWKLEYS, THE (unlocated, between Codsall and Albrighton) *Upper Hanckesleye* 1310 SRO D593/A/2/5/1, *Hankeley* 1393 SRO D593/A/2/5/11, *Hauckleys* 1678 Codsall ParReg, *the Haughtleyes* 1687 ibid, *the Hawkleys* 1755 ibid, *Hawkleys Farm* 1788 SRO D802/33. The 1310 and 1393 forms may be mistranscriptions for *Hau-*, since *-n-* and *-u-* are similar, if not identical, in early documents, in which case from Mercian OE *heafoc leah* 'the wood or clearing with the hawk'.

**HAWKSHILL (FARM)** in Scounslow Green (SK 0929). *Hawkeshul* 1307 SRO D986/51, *Hawkshill Farm* 1887 O.S. From Mercian OE *heafoc, hyll* 'the hill of the hawk'.

**HAWKSHUTTS, THE** 2 miles north-west of Brewood (SJ 8509). *Hawkeserde, Haukeserde* 1362, 1383, *Hodgehead* (p) 1601 Brewood PerReg, *Hawkshed* 1640 SRO 590/464, *Hawkesyard Farm* 1674 SRO D590/154, *the Hawkesyard* 1683 Oakden 1984: 47, *Haukshead howse* 1679 SHC 1919 241. From OE *heafocscerde* 'a woodland clearing in which hawks were flown for falconry' (see Field 1993: 76-7): cf. Hawkesyard Priory. The place was in Brewood Forest. *Hawkeheath* in Brewood, recorded in 1608 (SHC 1948-9 16), may be associated with this place.

**HAWKSLEY FARM** ½ mile east of Heaton (SJ 9662). *Haukeslyh* 1227 Ch, *Hawkeleye* (undated) (1883) *Deed* Sleigh, *Hawks Ley* 1842 O.S., *Hawksley* 1891 O.S. From OE *heafoc leah* 'the wood or clearing with the hawk'.

**HAWKSWELL (ROUGH)** (obsolete) on the north side of Himley Park (SO 8892). *Hawkswell Rough* 1834 O.S. From OE *heafoc* 'a hawk', with Mercian OE *wælle* 'a spring', and (sometimes) 'a stream'. *Rough* is a term applied to uncultivated pieces of land, especially slopes going down to a stream: Foxall 1980: 10.

**HAWK'S YARD** 3 miles north-west of Warslow (SK 0663), *Hauekeserd* 1281 SHC 1911 178, *Hauckheserd* 1284 SHC 1911 186, *Haukeserd* 1322 ibid. 98, *Hawkesyerd* 1407 DRO D2375M/1/1, *Hawkesyearde, Hawkesyerth* 1568 DRO D2375M/55/2, *Hawkesyerd otherwise Hawkesearthe* 1616 SHC VI NS (i) 6, *Haukesyard* 1681 Alstonefield ParReg, *Hakesyard* 1687 ibid; HAWKESYARD (unlocated, in Dilhorne), *Hawkesyard* 1393-1503 SRO D260/M/T/7/5/134 – see Hawkesyard Priory.

**HAYES WOOD** 2 miles east of Cheadle (SK 0444). *Hasewall above Parke* 1597 SHC 1934 11. Perhaps from OE *hæsel-wælle* 'hazel spring': cf. Haswell, Durham and Somerset.

**HAYHILL** 1 mile west of Biddulph (SJ 8757). *Heay Hill* 1665 Biddulph ParReg, *The Hay Hill* 1836 O.S. Perhaps from Mercian OE *(ge)heg* 'fence, enclosure', with OE *hyll* 'a hill', so 'the enclosure at the hill', but the 1665 spelling leaves such derivation uncertain: see also Hey House.

**HAY HOUSE** 2 miles north-west of Penkridge (SJ 9017), *Le Heyhouse* 1547 VCH V 124, *the Hay House* 1585 ibid, *The Hayhowse* 1609 Penkridge ParReg, *Hayhouse* 1618 ibid, *The Hayhouse* 1654 Bradley ParReg, *Hayhouse Farm* 1676 SRO D948/2/2/1; **HAY FARM** 1 mile south of Bobbington (SO 8088), *(Atte)hay* 1327 SHC VII (i) 252, *Hay* later 14th century VCH XX 69, *The Hays* 1833 O.S. From Mercian OE *(ge)heg* 'fence, enclosure', but see also Hey House.

**HAYS, HAYES HEAD** 2 miles north of Butterton (SK 0860)). *Hayheade* 1608 SHC 1948-9 100, *Hays, Hays Head* 1842 O.S. A common name, from Mercian OE *(ge)heg* 'fence, enclosure'. Curiously, the 1842 O.S. map shows two *Hays* a short distance apart here. See also Haysgate.

**HAYSEECH** 1½ miles south-west of Rowley Regis (SO 9584). *Haysitch* 1812 Plan, *Hayseech* 1851 White. The name is preserved in Hayseech Road. Perhaps from Mercian OE *(ge)heg* 'fence, enclosure', and OE *sic* 'watercourse'.

**HAYSGATE** 1 mile north-east of Warslow (SK 0959). *Heysgate* 1593 ParReg, *Heyes Yate* 1592 Alstonefield ParReg, *Hayes Yat* 1608 SHC 1948-9 61, *Hees Gate* 1769 ParReg, *Heys Gate* 1798 Yates, *Hays Gate* 1840 O.S. Perhaps from Mercian OE *(ge)heg* 'fence, enclosure', with OE *geat* 'gate', possibly here in the sense 'pass', so 'the enclosure at the pass'.

**HAYWOOD, GREAT** 5 miles north-west of Rugeley (SJ 9922). *Haiwode* 1086 DB, *Hegwde* 1198x1200 EEA 17 75, *Heiwode* 1198x1208 ibid. 74, *Haywode, Heywood, Heiwode* 12th and 13th centuries Duig, *Heywde* 1253 SHC 4th Series IV 117, *Magna Heywode* 1311 SHC 1911 79, *Heywode* 1311 SHC 1939 74, *Heywode Magna* 1428 SHC 1939 84, *Heywod'* 1461 HAME 468, *Heywod* 1538 SHC 1939 84. 'The wood with the enclosure' or 'enclosed wood', from Mercian OE *(ge)heg* 'fence, enclosure', often found as the latinised *haia*, meaning 'a part of the forest fenced-off for hunting', with OE *wudu* 'wood'. The place was a park in Cannock Forest, enclosed by the bishops of Lichfield. *Great* is a later addition, doubtless dating from when Little Haywood was created.

**HAYWOOD, LITTLE** 4½ miles north-west of Rugeley (SK 0021). *Little Haywode* 1432 Duignan, *Little Haywood* 1836 O.S. See Haywood, Great.

**HAZEL BARROW** 1½ miles north of Upper Hulme (SK 0163). *Haselbarrow* 1626 Rental, *Haslebarrow* 1651 ibid, *Haselburrough* 1656 ParReg, *Haselburrow* 1712 Alstonefield ParReg, *Hazelbarrow* 1851 White. Probably 'the grove with the hazel trees', from OE *hesel, bearu*, dative *bearwe* 'grove, wood', rather than OE *beorg* 'hill, mound'. A burial-mound or barrow is generally found as *low* (or sometimes *ley*) in Staffordshire place-names, from OE *hlaw*.

**HAZELHURST BROOK** a tributary of the river Churnet. From OE *hæsel-hyrst* 'hazel wood'.

**HAZEL MILL** 1 mile north-east of Penkridge (SJ 9414). *Le Haselnemulne* 1342 VCH V 128, *Hasyll' Mill* 1598 ibid, *Haye Mill greene* 1627 Penkridge ParReg, *Hazel Mill Greene* 1627 ibid. The mill is recorded c.1280, but had disappeared by 1754: VCH V 128. From OE *hæsel, myl(e)n* 'hazel mill'.

**HAZELS** 1 mile north-west of Kingsley (SK 0048). *Haseles* 1302 SHC 1911 59, *Hazeles, Hazelees* 1356 SHC 1913 165, *? Haseles* 1369 SHC VIII NS 227, *Hazles* 1836 O.S. From OE *hæsel* 'hazel', so 'the hazel trees'.

**HAZEL SLADE** 3 miles south-west of Rugeley (SK 0212). *Hazell slade* 1682 *Dep*, 1834 O.S. From OE *hæsel, slæd* 'the valley with the hazel trees'.

**HAZEL STRINE** 1½ miles south-east of Stafford (SJ 9420). *Hazel Strine* 1775 Yates, 1836 O.S. From OE *hæsel* 'hazel', with ME *strind* 'a stream', often found as dialect *strine*, so 'the stream with the hazel trees'.

**HAZELTON CLUMP** A distinctive conical hill 1 mile north-west of Blore (SK 1249). *Hazel Clump* 1798 Yates, *Hazelton Plantation* 1836 O.S. The name is almost certainly from OE *hæsel, dun* 'the clump of trees at the hill associated with hazel trees'. Initial (and terminal) *d* and *t* frequently interchange in Staffordshire names.

**HAZELWOOD (HOUSE)** ½ mile north-west of Turner's Pool (SJ 9863). *Haselwo(o)d(e)* 1240 (1883) Deed Sleigh, *Haselwode* 1343 SHC XIV 60, *le hasilwod* c.1539 *LRMB*, *Hassell-woode* 1587 (1883) Sleigh 108, *Haselwood* 1600 SHC XVI 204, *Hasselwood* 1611 SHC III NS 60, *Hasslewood* 1621 SHC 1934 (ii) 24, *Hazlewood* 1698 Leek ParReg, *Hazelwood House* 1842 O.S. From OE *hæsel-wudu* 'hazel wood'.

**HEADLESS CROSS** 1 mile south of Redditch (SO 0365). *Hedley* 1275 (p) SR, 1294 (p) Ipm, *Smethehedley* 1300 Pat, *Hedley Cross* 1464 Pat, *Hedles Crosse* 1549 Pat, *Headleys Cross* 1789 *canal map*, *Headless Cross* 1832 O.S. From OE *hæð leah* 'the *leah* of the heath', which developed into Hedley Cross, of which the present name is a corruption. In Tardebigge parish, forming part of Staffordshire from c.1100 until 1266, in Warwickshire until 1844, and since that date in Worcestershire. *Le hedeles Cross* between Lichfield and Freeford is recorded *temp.* Edward III (SHC VI (ii) 186), with the meaning 'cross with the head or crosspiece broken off'.

HEAD POOL (obsolete) ½ mile south of Wolseley Hall (SK 0109). *Cattail Pool* 1887 O.S. Seemingly an artificial pool, now dried up, which appears on the Colwich tithe map of 1839: see PMA 31 1997. The 1887 name is of interest, since it suggests a derivation from '*halh* frequented by wild cats' (cf. Cattal, West Yorkshire), but much earlier forms would be needed for certainty. Wild cats were found in Staffordshire well into the medieval period: in 1267 a licence was granted to hunt foxes, badgers and wild cats in the King's Forests of Shropshire and Staffordshire (Eyton 1854-60: II 243).

HEAKER (unlocated) *Heaker* 1590 SHC 1930 (ii) 99. Possibly High Carr (q.v.).

**HEAKLEY (HALL FARM)** 2 miles north-west of Bagnall (SJ 9051). *Heekleigh* 1240 (1798) Shaw I xxvi, *Heycley* c.1560 SHC 1938 24, *Heycle, Heycley* 1567 ibid. 1938 47, 33, *Nether Henkley* 1567 SHC XIII 261, *Heyckley* 1572 ibid. 287, *Heckeley* 1572 BCA MS917/1515, *Heckley* 1586 SHC 1927 160, *Haickley* 1592 Norton-in-the-Moors ParReg, *Haickeley* 1595 ibid, *Overheackley, Netherheackley* 1601 SHC XVI 208, *Over Hecle, Nether Hecle* 1613-4 SHC 1934 31, *Over Heckley* 1657 BCA MS917/1664. A curious name (see also Heakley Heath and Heakley Mill, which appear to have similar forms) for which no convincing derivation can be offered. It is not certain that all of the spellings refer to this place.

HEAKLEY HEATH (obsolete) between Wednesbury and West Bromwich (? SO 9992). *Heakley Heath* 1775 Yates. The absence of earlier spellings leaves the name uncertain; see also Heakley (Hall Farm).

HEACKLEY MILL (unlocated, in Trysull) *Heykeleye Mulne* 1357 SHC XII 150, *Heyclif mill* 1412 VCH XX 192, *Heackley Mill* 1648 SRO D740/8/13. No derivation can be offered for this name; see also Heakley (Hall Farm). From a mill probably on Smestow Brook.

**HEAMIES** 2 miles north-east of Eccleshall (SJ 8531). *Haymees* 1334 SRO D59[7941], *Heymys* 1414 SHC XVII 53, *? Haymese* 1428 SHC XI 229, *Heymes* 1462 SHC VII NS 253, *? Heymes* 1569 SHC XVII 217, *Hemies* 1590 Eccleshall ParReg, *Heymies* 1597 SHC 1932 344, *Haymis* 1608 SHC 1948-9 99, *Heymes* 1646 SHC 4th Series I 267, *Heymeece* 1673 SRO D590/261, *Hamys* 1679 SHC 1919 228, *Heamis* 1679 SHC 1914 58, *Upper Heamis, Lower Heamis* 1775 Yates. The place is on a 386' hill, so perhaps from OE *heah* 'a high place, a height', with the second element OE *meos* 'a moss, a marsh, a bog' or from the river Meece which runs at the foot of the hill. The place is 2½ miles south-east of Millmeece, and 1½ miles south of Coldmeece, and it likely that it formed 'High Meece' as part of a group of places with this element.

HEATHCOTE GRANGE unlocated, but 'probably situated in the neighbourhood of the modern Grange Farm in the north-west of Seighford parish': VCH III 254. *Heythehouse Grange* 1535 VCH III 254, *Heathcote Grange* 1538 SHC 1914 94, 1608 SHC III NS 15, *The Grange upon the Heyth* 1539 SHC V NS 322, *Hethcote grange* 1539 MA. 'The grange at the house or cottage on the heath'. The place was a grange of Ranton priory.

**HEATH HAYES** 1½ miles north of Norton Canes (SK 0110). *Hethhey* 1570 Mills 1998: 173, *Heathy Hays* 1834 O.S. From OE *hǣð (ge)heg* 'the enclosure at the heath'. The place lay in an area called Wildmore – see Hollies near Heath Hayes.

**HEATH HILL** 1 mile north of Sheriffhales (SJ 7614). *Heathull* 1250 Eyton 1854-60: XI 146, *Hethe Hyll* 1532 SHC 4th Series 8 99, *Hethehyll* 1547 SHC XI 292, *Heath Hill* 1655 SRO D4092/C/1/13, *Heathill* 1657 Sheriffhales ParReg, 1666 SHC 1927 70, *Heathill* 1713 SHC 1938 238. Self-explanatory. Transferred to Shropshire in 1895. *Heath hill cross*, recorded in 1782 (Sheriffhales ParReg), is an ancient sandstone column which lies to the south of Heath Hill.

**HEATH HOUSE** 1 mile north of Horton (SJ 9458). *Hethes House* 1562 SRO DW1702/3/8/9, *Heath House* 1842 O.S. In this case perhaps from the family named Heath.

**HEATH HOUSE (GRANGE)** 1 mile south of Cheddleton (SJ 9651). *le Hethe Graunge* 1328, 1362 Su, *Hethhowse grange* 1524 Rees 1997, *Heathhouse Grange* 1539 MinA, 1558 BM, *le Hethehouse* 1549 Ct, *Hethe Howse graunge* 1553 Pat, *Heathouse Graunge* 1679 SHC 1919 226, *Heath House* 1728 ParReg, 1775 Yates. 'The house on the heath', with ME *grange* 'a grange', originally 'a granary, a barn', later 'a farm', also 'an outlying farm belonging to a religious house or a feudal lord where crops were stored'.

**HEATH TOWN** 1 mile north-west of Wolverhampton (SO 9399). *The Heath* 1532 SHC 4th Series 8 159, *Heathe* 1557 SHC 1928 132, *Heythe* 1555 SHC XII 217, *Heth* 1561 SHC 1938 131, *Heath houses* 1747 Bowen, 1798 Yates. Self explanatory.

**HEATHTON** 2 miles south-east of Claverley (SO 8192). *la Hethe* 1255 Eyton 1854-60: III 97, *Hethton* 1256 ibid.' 97, *Etthon, Hethton* 1262, *Hetton* 1274 Eyton 1854-60: III 76, *Hetthon'* 1309 SA 2089/2/2/3, *Hethton* 1400 ibid. 52/9. From OE *hǣð tun* 'heath *tun*'. In Shropshire since the 12th century.

**HEATHYLEE** 3 miles north-east of Upper Hulme (SK 0464). *? Hethileg* 1242 SHC XI 315, *Hethelogh* 1399 DRO D2375M/1/6, *Hethelegh* 1444 DRO D2375M/1/1, *Over & Nether Heathelie* 1599 ParReg, *Heathie Lee als Lea* 1601 QSR, *Heathelee* 1602 Leek ParReg, *Heathie Lee* 1605 SHC 1940 196. From OE *hǣð leah*, probably here meaning 'the clearing in the heath'. The place lay in the former Malbanc Forest. The 1242 spelling appears in a charter of Hulton Abbey printed in Ward 1843: app. ii which is almost certainly a much later forgery: Tomkinson 1994: 73-102.

**HEATLEY** 1 mile north of Bagot's Bromley (SK 0627). *Haytelega* c.1200 SHC NS XI 148, *Haitele* c.1225 SHC XI NS 150, *Hetleg'* 1236 Fees, *? Hotteley* 1286 SHC 4th Series XVIII 132, *Hayteleye* 1293 SRO D986/50, *Haddeleye* 1345 SHC X (i) 62, *Hayteley* 1392 SHC XV 49, 1408 ibid. 121, c.1435 SHC XI NS 42, 1473 SHC IV NS 187, *Haytle* 1403 SHC XV 104, *Hateley, Hayteley* 1474 SHC NS IV 191, *Hayteley* 1478 SHC VI (i) NS 120. Probably from the OE adjective *hǣðiht*, 'heathy', with OE *leah*, so 'the heathy clearing': see PN Ch V (I:1) xxiv.

**HEATON** 4 miles north-west of Leek (SJ 9562). *Heton* (p) 1230-2 StCart, c.1240 SRO DW1761/A/4/9-10, 1266 FineR, *Hethon* 13th century Dieul, *Heyton* 1534 ValEccl, 1560 Pat, 1616 SHC VI NS (i) 19. 'The high farmstead', from OE *heah, tun*.

**HEATONLOW** ½ mile north of Heaton (SJ 9563). *The Low* 1842 O.S. From OE *hlaw* 'mound, tumulus': see Lowe.

HECKLEY – see **BROMLEY, ABBOT'S; HEAKLEY (HALL FARM)**.

HEDESDALE (unlocated) *Hedesdale* 1423 SHC XVII 91.

**HEDNESFORD** (pronounced Hensford [hɛnzfəd]) 2 miles north-east of Cannock (SK 0012). *Hedenedford* c.1153 Dugd v 447, *? Ernesford* 1307 SHC VII 185, *Ed(e)nesford* 1323 CoramR, *Hedenusford* 1339 SHC 1931 241, *Hedenesford* 1343 SHC XI 154, *Hednesford(e)* 1362 Fine, *Hendusford* 1381 SHC XVII 185, *Adnesford* 1545 SHC XI 288, *Haddenford* 1599 Smith, *Hedgford or Hedsford* 1653 PCC, *Hedg(e)ford* 1666 SHT, *Hedgeford* 1695 Morden, *Hednesford* 1834 O.S. Possibly '*Heddin's ford', *Heddin representing a diminutive of the OE personal name Headda, or from the personal name (possibly ON) Heoden: cf. Hensall, West Yorkshire.

**HEIBRIDGE** (obsolete) in Lower Tean (SJ 0138). *Heibridge* c.1569 SRO Chetwynd bundle 9.

**HEIGHLAY GREENE** (unlocated, near Ranton/Ellenhall) *Heighlay Greene* 1531 SHC 1912 46.

**HEIGHLEY, HEIGHLEY CASTLE** 4½ miles west of Newcastle-under-Lyme (SJ 7747 & SJ 7746; SMR 02478 places a deserted Anglo-Saxon settlement at SJ 77554655). *Heolla* 1086 DB, *Helyh (Castle)* 1227 Ch, *Heleye* 1273 Ipm, *Heleye* 1274 SHC 1911 160, *Heley* c.1540 Leland, *Helay Castle* c.1565 SHC 1938 113, *Heyley* 1587 SHC XV 182, *Heyley Castle* 1686 Plot. Perhaps from OE *heah-leah*, 'high clearing or wood', notwithstanding the DB spelling, which is clearly aberrant. The place lies at a pronounced hill. A hybrid name incorporating Welsh *heol* 'a road, a way' is unlikely, but not impossible.

**HELLECUMBE** (unlocated, near Swinscoe) *Hellecumbe* c.1260 Okeover 231M. The single spelling precludes any firm derivation for the first element, but the second is OE *cumb* 'a coomb, a spoon-shaped valley'.

**HELL HOLE** 1 mile south-west of Colwich (SK 0019). *Hell Hole* 1887 O.S.

**HEM HEATH** 1 mile east of Trentham (SJ 8841). *Heath Hem* 1576 Trentham ParReg, *Hemme Heath* 1668 ibid, *Hem Heath* 1671 ibid, 1799 Faden, *Hemheath* 1836 O.S. From OE *hemm* 'hem, border', with OE *hǣð*, giving 'the place at the edge of the heath', or 'the heath at the boundary'. The place lies to the north of a parish boundary. *le Hem* in Little Barr is recorded in the late 13th century (SRO 3764/2[27572]), presumably associated with *Hemend, Hem Cross* and *Hemend Moore* recorded in 1654 (BCA MS3145/96/1), and *the Hemme* in Colwich is recorded c.1535: SHC NS X (i) 165.

**HEMLOCK'S FARM, HEMLOCK'S BRIDGE** 1 mile north-east of Cannock (SJ 9910). *Hemelok'* 1354 Ct, *Humlokyate* 1370 St, *Astons hemlockes, Hie hem lockes, Thorney hemlockes* 1520 Survey, *Hemlock* 1568 StSt 12 2000 70, *le hem lock* 1580 Anglesey Ch, *Hemlocks Farm* 1834 O.S. Probably from ME *hemeloc* (from OE *hymlice, hymblice*) 'hemlock': the plant has been identified at Fisherwick in a pre-Roman context: WMA 22 1979 96.

**HEMP HOLME** (obsolete) 1 mile north-west of Mavesyn Ridware (SK 0717). *Hempholm* 1325 (1798) Shaw I 176, *Hemp Holme* 1834 O.S. '*Holm* where hemp (OE *henep*) grew': cf. Hempholme, Yorkshire East. For *holm* see Hulme.

**HEMPSTALLS** on the north side of Newcastle-under-Lyme (SJ 8446). *Hempstalls* 1836 O.S. Perhaps 'the stalls or stables at the place where hemp (OE *henep*) was grown'. Remembered in the names Hempstalls Lane, Grove and Court.

**HENBACHES** (unlocated, in Upper Tean) *Henbaches* 1647 SRO D1203/A/4-6. Perhaps from OE *henn* 'wild birds' with OE *bece* 'pronounced stream-valley', so 'the stream valleys with the wild birds'.

**HEN CLOUD** a gritstone outcrop on high moorland ½ mile north-west of Upper Hulme (SK 0161). *Hen-Cloud* 1686 Plot 171. Probably from OE *heah* (oblique *hean*), *clud* 'high cloud', with *cloud* meaning here 'mass of rock, outcrop, hill': cf. Cloud. It is unclear whether *Clowde*, recorded in the parish of Leek in 1451 (SRO DW1761/A/4/29), refers to this outcrop. *Clude* is recorded in the 14th-century poem Sir Gawain and the Green Knight, parts of which may have been set in this area: see Elliott 1984: 64; 95.

HEN HOLE (obsolete) 2 miles east of Biddulph (SJ 9157). *Hen Hole* 1815 *EnclA*, 1842 O.S., *Han Hole* 1880 Kelly. From OE *henn, hol* 'a hollow where hens were kept'.

**HENHURST** ½ mile south of Anslow (SK 2124). *Hennehurst* 1327 SRO DW1734/2/1/101B, *Henhurste* 1601 SRO D603/E/1/61, *Henhurst* 1709 SRO D603/L71. From OE *hyrst* 'a hillock, knoll, copse, wooded eminence', and OE *henn* 'hen, wild bird', so 'the wooded hill with the hens or wild birds'.

**HENRIDDING FARM** on north-west side of Endon (SJ 9153). *Hen Riddings* 1704 Leek ParReg, *Hen Ridding* 1816 1816 SRO DW1909/E/9/1, *Henridding* 1836 O.S., *Penrhydding* 1932 JNSFC LXVI 190. Perhaps from ME *hen ryding* 'high clearing', or 'the clearing with the hens'. The 1932 Welsh-influenced version of the name is unexplained. *Henridding*, recorded in 1687 (SRO D239/M/T/731), is near Thorpe Cloud, Derbyshire.

**HENWOOD** in Tettenhall Wood, 2 miles west of Wolverhampton (SO 8899). *Henwood (field)* 1517 VCH XX 29, *Hernwood Leasow* 1672 SRO D4092/C/1/49. The name appears to have been attached to a field adjoining Lower Green, Tettenhall: VCH XX 7.

**HERBAGE** ½ mile north of Upper Elkstone (SK 0559). *Harebache* 1439 VCH VII 27, *Herbach* 1660 Alstonefield ParReg, *Hie Herbacth (sic)* 1666 ibid, *Harbitch* 1686 ibid, *Hairbage* 1749 Bowen, *Herbage* 1775 Yates, 1840 O.S. Notwithstanding the later forms, a derivation from ME *erbage* 'vegetation, especially grass, used as pasture', seems unlikely, so perhaps from OE *har* 'grey, stony', or OE *hara* 'hare', with OE *bece* 'well defined stream-valley', so giving 'the stony stream valley' or (less likely) 'the stream valley with the hares'. *Hie Herbacth* is High or Upper Herbage, implying the existence of a Lower Herbage.

HERBERDESMULNE (unlocated, in Gnosall) *Herberdesmulne* 1321 SHC 1911 347. Perhaps from the OFr personal name Herbert, introduced by the Normans (DES 228), so 'Herbert's mill'.

**HERMITAGE** on the north-west side of Froghall (SK 0247). *Armitage* 1656 Ipstones ParReg, *Hermitage* 1836 O.S. *The Hermitage* is also recorded in 1531 in the Ranton/Ellenhall area (SHC 1912 46), *Hermitage* at Manwoods in Handsworth in 1649 (BCA MS3145/62/2), and *the Hermitage* in Mayfield c.1875 (SRO D514/M/1). Self-explanatory.

**HERMITAGE, THE** sandstone caves ½ mile north-east of Bridgnorth (SO 7498). *Hermitage of Athewildston* 1328 Eyton 1854-60: III 352, *Hermitage of Adlaston* 1333 ibid, *Hermitage of Athelardeston* 1335 ibid, *the Heremitage* c.1540 Leland ii 86, *The Hermitage* 1833 O.S. Eyton 1854-60: III 352 mentions the tradition (see e.g. Gough 1806: III 19; also TSAS I 1878 159-72) that the brother of King Athelstan (c.895-939), king of Wessex 924-39, ended his days here in retirement from the world. The legend is recorded by Leland c.1540 ii 86: 'In this forest or wood (as some constantly affirme) kynge Ethelstane's brother ledde in a rokke for a tyme an heremite's lyfe. The place is yet sene and is caullyd the Heremitage'. The age of the tradition is unknown, but Athelstan had four brothers, Ælfweard, Edmund, Edred and Edwin. The first died young, the second was banished and drowned, and the other two both took the throne. Three of Athelstan's sisters were nuns. The place has been in Shropshire since the 12th century. For other hermitages in Staffordshire see VCH III 136-7.

HERONVILLE – see **HARVILLS HAWTHORNE**.

HERTESHORN (unlocated, possibly near Crakemarsh/Creighton) *Herteshorn* 1337 SHC 1913 59. Perhaps from OE *heor(o)t* 'stag', possibly with OE *horn* 'a horn-like projection; a spit of land', so giving 'the projecting piece of land in the shape of a stag's horn', or 'the projecting piece of land associated with the stag'.

HERTINDONE (unlocated, in Croxden) *Hertindone* n.d. Shaw 1798: I 155.

HETELSDALE (unlocated, possibly Huddale) *Hetelsdale* 1199 SHC III 169. Perhaps from an unidentified personal name with OE *dæl* 'valley'.

HEUSE (unlocated) *Heuse* 1509 SHC 1923 319.

HEWELL GRANGE (unlocated, possibly in Weston Coyney or Dilhorne, perhaps to be associated with Highhaugh (see Delph House)) *Huwanhale* 1264 SHC 1924 137, *Hunethalen* 1291 Tax, *Hunehalgh* 1313 SHC XII NS 278, *Hivall* 1539 MA, *Hyvall* 1676 SHC 1914 92. Shaw 1801: II viii locates the place on Wetley Moor, as does Erdeswick 1844: 495: 'On Wetley Moor is Hewell, or Hyvall, Grange, called, in the Lichfield tax-book Hunehalgh'. VCH III 253 says the place is in Dilhorne. The forms are inconsistent, and no suggestion can be offered for the derivation. The place was a grange of Ranton priory. It is unclear whether *Hunkall*, recorded c.1646 (1801 Shaw II 5), refers to this place.

**HEWELL GRANGE** 3 miles south-east of Bromsgrove (SO 0068). *Hewell(e)* 1275 *Ass*, 1275 PN Wo 363, 1291 Tax. Perhaps from OE *heah, wælle* 'high spring '. In Tardebigge parish, forming part of Staffordshire from c.1100 until 1266, in Warwickshire until 1844, and since that date in Worcestershire.

**HEXTALL** 1½ miles west of Seighford (SJ 8525). *Hegstal* 1176 P, *Hehstall* 1227 Ass, *Hegestall* 1272 Ass, *Hekstule* 1273 SHC VI (i) 58, *Heckstal* 1295 SHC VII 28, *Heghsale* 1347 SHC VIII 90, *Extolls* 1851 White. Possibly from OE *hege-steall* 'place with a hedge or enclosure' (EPNE i 241, where 'Hextells, St[affordshire]', presumably this place, is cited), the two elements being found together in a charter of 844x848 (11th century, S.205) and in another undated charter no later than the 10th century (11th century, S.1591), both relating to Crowle in Worcestershire. Since the place lies on the north side of Clanford Brook, there is the possibility that the second element has the alternative meaning 'pool in a river', so 'enclosure at the pool in the river'.

**HEXTONS FARM** 1½ miles north of Upper Arley (SO 7582). *Hekstane* 1227 LyttCh, *Heyston* 1293 SHC VII (i) 172, *Heckston* 1306 SHC VI (i) 217, *Hexston* 1312 SHC IX 43, *Hecstal* 1327 SHC VII (i) 247, *Hekston* 1520 FF, *Heck-stones* 1686 Plot 168, *Heckstones* 1801 Shaw II 254. A difficult name, but possibly from OE *hæc* 'hatch', or from OE *heah* 'high', with with OE *stan* 'stone'. *Hæcce* is found as a landmark in an Anglo-Saxon charter of Rolleston of 1008 AD (14th century, S.920), and may be from *hæcc*, which often had the specific meaning of a structure (often of wattle) across a river to trap debris above a ford, or serve as a floodgate or sluice, or more often to serve as a fish trap – and this place lies near the river Severn. A possible derivation from the well-recorded OE personal name Heahstan, with a second element such as *tun* or *stan*, which had disappeared at an early date (a phenomenon paralleled elsewhere, cf. Hexton, Hertfordshire), is also put forward in PN Wo 31. However, Plot 1686: 168-9 records that the place provided particularly good quality stones for sharpening scythes, knives, etc., so perhaps from *hack*, a word for tools used for cutting and digging, although OED has no evidence of early use. In Worcestershire since 1895.

**HEY HOUSE** 1 mile south of Madeley (SJ 7743). *Heyhouse* 1513 JNSFC LX 1925-6 46, *Heyhous* 1514 ibid.' 38, *Hay House* 1833 O.S. Perhaps from Mercian OE *(ge)heg* 'fence, enclosure', or from OE *heg* 'hay', with OE *hus* 'house', giving 'the house at the enclosure', or 'the hay-house', i.e. the building where hay was stored, but a derivation from OE *heah* 'high' is also possible. See also Hay House.

HEYLAYTHUL (unlocated, probably near Longdon) *Haylaythul* 1313 SHC 1921 7. Perhaps associated with Cleat Hill (q.v.).

HEYLEY, DEEP (unlocated, in Sedgley parish) *Deep Heeley* 1715 SRO DW3222/295/1-2, *Deep Hayley* 1814 Himley ParReg, *Deep Heyley* 1816 ibid.

HEY RIGGE (unlocated, near Hillswood north of Leek) *Hey Rigge* 1613 SRO DW1702/1/20. Possibly 'the high ridge'.

**HEY SPRINK** 1½ miles south-east of Madeley (SJ 7842). From Mercian OE *(ge)heg* 'fence, enclosure', often meaning 'a part of the forest fenced-off for hunting', with ME *spring, spryng*, 'a copse, a young plantation', from OE *springan* 'to burst forth', sometimes found in the region as *sprink*.

**HEYWOOD GRANGE** 2 miles north-west of Dilhorne (SJ 9645). *Hyghwalgrange* 1464 SHC IV NS 135, *Hynealgraunge* 1469 SHC IV NS 167, *Hye Hall* 1518-29 ECP, *Heywall Grange* 1655 PCC, *? Highwall* 1691 SRO D1326/12, *Haywood Grange* 1778 Cheadle ParReg, 1837 O.S. The modern name is evidently a corruption of 'the grange surrounded by the high wall', or possibly 'the grange at the high spring', from Mercian OE *wælle* (see also High-Hall-Hill). The place was a grange of Ranton priory: VCH III 253.

HIDE FIELDS (obsolete, 1 mile south-west of Stafford (SJ 9121)) *Hidefield* 1540 SRO D1810/f226, *Hide Fields* or *High Fields* c.1548 SHC VIII (ii) 143-4, *Hydefeldes* 1559 SHC IX NS 7, *the Hyde feildes* 1564 ibid.' 11, *Hydefieldes* 1608 SHC 1948-9 26. Evidently sometimes confused with Highfields (q.v.), this place was 'the field(s) near Hyde (Lea)'.

HIGGE POOL (unlocated, in the far north of the county) *Higge poole* 1662 Alstonefield ParReg, *Higpoole* 1681 ibid.

**HIGH ASH** 1 mile south-west of Hollinsclough (SK 0465). *Hie Ashe* 1671 Alstonefield ParReg, *Hye Ashe* 1678 ibid, *Hie Ash* 1668 ibid, 1851 White; **HIGHASH** on the south-west side of Abbots Bromley (SK 0724), *Hieasthe* (? for *Hieasche*) SRO D(W)1721/3/33/5, *High Ash* 1836 O.S., *Highash* 1887 O.S. Presumably from conspicuous ash-trees.

**HIGH CARR** 3 miles north-west of Newcastle-under-Lyme (SJ 8350). *High Carr Ridges* 1601 SRO D3272/5/25/1-30, *Hey Carr* 1626 Wolstanton ParReg, *Heycar* 1622 Tooth 80, *Heacar* 1623 JNSFC LXIV 1930-1 131, *Hey Carr* 1626 ParReg, *Heycarr* 1666 SHC 1921 162, *The Carr* 1733 SHC 1944 4, *High Car* 1833 O.S. The first element may be from OE *heg* 'hay', or OE *heah* 'high' (the place lies on high ground), or Mercian OE *(ge)heg* 'a fence, an enclosure; a part of a forest fenced off for hunting', with ON *kjarr* 'marsh, wet moor, boggy copse', so giving 'the wet moorland land on which hay was cut', or 'the high wet moorland', or 'the wet moorland enclosure'. See also Heaker.

**HIGHDOWN (COTTAGES)** on the Staffordshire/Worcestershire boundary, on the south-west side of Iverley (SO 8780). *High Down* 1775 Yates, *Down* 1832 O.S. From OE *dun* 'a hill', so 'the high hill'.

**HIGH ELMS** – see SEVEN ASHES.

**HIGHFIELDS**   1 mile south-west of Stafford (SJ 9121). *Westons Highfield* 1644 StComm, *High Fields* 1775 Yates, 1836 O.S.  Notwithstanding a reference c.1548 to *Hide Fields* or *High Fields* (SHC VIII (ii) 143-4), it would appear that High Fields and Hide Fields (q.v.) were two distinct areas (see SHC VIII 144), so 'the high fields'.

**HIGH FOREST**  2½ miles north-east of Heaton (SJ 9865). *High Forest* 1535 Dieulacres Inventory, 1539 SHC IX NS 301, *High-fforest* 1542 (1883) Sleigh 71, *Highe Forest* 1592 SHC 1930 220, *le High Forrest* 1595 SHC 1932 127-8, *High forrest* 1640 Leek ParReg, *High Forest* 1703 ibid, 1842 O.S. Either 'the forest on the higher ground', or 'the northern-most forest': see Middle Forest. Part of this forest was known as *Abbotts Forest* (q.v.).

**HIGHGATE**   2 miles north of Enville (SO 8390). *Highgate (Warren)* 1762 SA 5586/4/2-3, *Highgate, Highgate Forest, Highgate Heath* 1833 O.S.  Self-explanatory. Highgate Farm lies at a height of 324'. Highgate Heath is a large area of heathland to the east.

**HIGHGROVE FARM**  1 mile south-east of Kinver (SO 8582). *Heygrave* c.1200 VCH XX 136, *Heygreve* 1262 ibid, *Haygreave* 1387 ibid, *Highgreaves* 1683 ibid, *The High Groves* 1683 ibid, *High Grove farm* 1796 ibid.' 'The high grove'.

**HIGH-HALL-HILL**  1 mile east of Yoxall, south of Woodhouses (SK 1518). *? Haywall* 1337 Hardy 1908: 24, *? Haiwalles* 1340 ibid. 26, *Gyhewalhyll* 1543 SHC 1910 51, *Hyghwall Hill* 1563 SHC XIII 226, *High-hall hill* 1679 SHC 1919 267. Perhaps from Mercian OE *wælle* 'spring', and (sometimes) 'a stream', meaning 'the spring on the hillside'. Wall House lies 1 mile north-west. The suggestion that 'Highwall Hill was another name for Yoxall' (SHC XII 212) is without foundation.

HIGH HAUGH (obsolete) 1 mile west of Cheadle (SJ 9942). *Huythehalg* c.1311 SHC 1911 436, *Huuehalgh* 1313 Ch, *The Heyghe Halghe* 1380 SHC 1910 265, *the Heyge Halghe* 1538 ibid.' 266, *Hyghehawghe* 1556 SRO D1229/1/4/1, *Highhaugh al. Dephe House* 1615 SRO D1229/1/4/2. Probably 'the high *halh*', perhaps to be associated with Hewell Grange (q.v.), though the two earliest spellings pose difficulties for such derivation. See also Delph House.

HIGHLAND (unlocated, near Cheslyn Hay) *Highland under Cheslyn Hay* 1271 SHC 4th Series XVIII 82.

**HIGHLANDS PARK**   ½ mile west of Tatenhill (SK 1921). *Heylynds Lodge* 1262 Hardy 1907: 132, *Le Haylindes* 1297 SHC 1911 251, *Heghlindes, Heylindes* 1314 ibid. 7, *Hayelyndes, Parcarius del Heyelindes* 1336 ibid. 22, *Heighlyndes* 1422 ibid. 110, *Highelynnes* 1524 ibid. 201, *Highlinges Parke* 1628 ibid. 219, *Heighlyns Park* c.1659 DCL 393, *High Lins Park* 1704 (1798) Shaw I 129, *Highlins Park* 1836 O.S. From ME *linde* 'a lime-tree', with ME *heah* 'high', so 'the lime-trees on the height': the place lies on high ground.

**HIGHLOWS, THE** on the north side of Yarnfield (SJ 8633). *Highlows* 1727 SRO D641/5/T/17, *The Highlows* 1836 O.S. Presumably 'the high lows or burial mounds' (from OE *hlaw*): at least one tumulus is recorded here: StEnc 284. See also Queen's Low.

**HIGH OFFLEY** – see **OFFLEY, HIGH**.

**HIGH ONN** – see **ONN, HIGH**.

**HIGHRIDGE** ½ mile east of Swythamley Hall (SJ 9764). *Hay Rudge* 1621 SHC 1934 24. Seemingly 'the ridge where hay was made'.

**HIGHRIDGES** 1 mile east of Checkley (SK 0338). *Heyridges* 1579 SRO D543/B/1/1/2-3, *Hayridge* 1618 SHC NS VI (i) 31, *Heybridges* 1675 SRO D1057/A/1/9/1-7, *Highridges* 1836 O.S. Possibly 'the high ridges', rather than 'the ridges where hay was made'.

**HIGHSHUTT** 1½ miles east of Cheadle (SK 0343). *High Shutt* c.1600 Chester 1979: 39, 1833 O.S. Probably from dialect *shute* 'a shoot, a steep hill', later 'a steep channel of water': EPNE ii 116.

**HIGH TOWN** on the north side of Cannock (SJ 9912). *Highton* 1644 SHC 4th Series I 100. Self-explanatory.

**HIGHWAY FARM** 1 mile south-west of Keele (SJ 7944). 'The high way or passage'. The place is recorded in 1331 (HOK 16).

**HIGHWOOD** 1 mile south of Uttoxeter (SK 0931). *Uttoxeter Wood or Hight Wood* 1699 SRO DW786/15/1, *the Highwood* 1711 D786/15/2. 'The high wood'.

**HILCOTE** 1 mile north-west of Chebsey (SJ 8429). *Helcote* 1086 DB, *Ulecote* 1227 SHC IV 59, *Hulcote* 1326 SRO 3764/28[27576], *Hylcote* 1419 SHC XVII 67, *Hillcourte* c.1540 Leland ii 172. From OE *hyll-cot* 'cottage or shelter at the hill'. See also Newbold. It is recorded in 1679 that 'Newbold and Hilcote – these were anciently 2 distinct places … but by reason of their lying together, and continuing for so many ages in ye possession of ye same family (ye house at Newbold being also decayed and gone) they have in these latter times generally passed under ye name of Hilcote only': SHC 1914 66-8.

HILDERSHOLME (unlocated, near Pipe Ridware) '… Hildersholme, which was formerly surrounded by two arms of the Trent; but, the South arm being diverted, this holme now lies on the South side of the river …': Shaw 1798: I 170. The name may well be from the OE personal name Hildebald: a meadow in Ridware is recorded as *Hildebaldesholme* (ibid.'), of which Hildersholme may well be a contraction. Holme is from OE *holm*; ON *holmr, holmi* 'a small island, a piece of dry land in a marsh; a piece of land partly surrounded by streams or by a stream'.

**HILDERSTONE** 3 miles east of Stone (SJ 9434). *Heldvlvestone, Hildvlvestune* 1086 DB, *Hyldillveston* 1136 SHC VI (i) 22, *Hildulueston* 1227 Ass, *Idolveston* 1227 SHC IV 53, *Hildeleston, Hildeliston* c.1250 SHC 1911 426, *Hyndolveston* 1277 ibid. 167, *Hyldelweston* ?13th century SHC VI (i) 21, *Hildelveston, Hyldeleston, Hyldreston* 13th century Duig, *Hilderston* 1577 Saxton. From the OE personal name Hildewulf, hence 'Hildewulf's *tun*'. Cf. Hindolveston, Norfolk, which has the same derivation.

HILKLOW (unlocated, possibly near Chesterton in Worfield) *Hilklow* 1602 SA 2028/1/5/8.

HILL, HILL HALL – see **OFFLEY GROVE**.

**HILL CHORLTON** – see **CHORLTON, HILL**.

**HILLEND** at the northern end of the ridge of Abbot's Castle Hill (SO 8195). *Hill End* 1709 Claverley ParReg. Self-explanatory.

**HILL FARM, THE HILL** 1 mile north-west of Butterton (SK 0657). *The Hill* 1749 Butterton ParReg.

**HILLFIELDS HOUSE** in Upper Arley (SO 7881). *la Hulle* (p) 1327, 1332 SR. 'The hill'. In Worcestershire since 1895.

**HILL HALL (FARM)** 1 mile east of Wall (SK 1206). *? Hul* 1317 SRO 3764/53, *? le Hull* 1368 SRO 3764/65, *Hill Hall (Farm)* 1664 SRO DW3222/82-87, *Hill Hall* 1649 (1801) Shaw II 30*, 1798 Yates, *Hill Farm* 1801 Shaw II 53, *Hill Hall Farm* 1834 O.S. From a 367' hill here.

**HILL HOUSE** on south side of Upper Elkstone (SK 0558). *Hill House* 1675 Alstonefield ParReg. 'The house on the hill': the place lies on the side of a 1394' ridge.

**HILLIARDS CROSS** 2 miles north-east of Lichfield (SK 1511). *Hillards Cross* 1810 SRO D615/D/185, *Hilliards Cross Farm* 1896 SRO D615/ES/1/11. Perhaps from the personal name Hilliard, Hillyard, from OGerman Hildigard, Hillyard (DES 321). SHC 1916 139 implies without citing sources that the early name was *Elards*.

**HILL RIDWARE** – see **RIDWARE, HILL**.

**HILLSDALE** 1 mile south-east of Butterton (SK 0855). *Hildesdale* 1203 SHC III 108, *Hildelesdale* 1327 SHC VII (i) 221, *Hydleysdale* 1415 SHC XVII 59, *Hyddysdale* 1417 Banco, *Hillesdale* 1457 SHC XI 237, *Hilsdale* 1583 SHC III 15. 'Hyd(d)el's dale or valley'.

**HILLS FARM** 2 miles south-east of Cheddleton (SJ 9950). *the Hills* 1586 AD, *Hills Farm* 1699 ParReg, 1810 *EnclA*, 1836 O.S. Self-explanatory.

**HILLSWOOD, NORTH & SOUTH** 1 and 2 miles north of Leek (SJ 9858). *Hellis wood* 1340 VCH VII 197, *Helleswode* 1345 SHC IX NS 297, 1346 SHC XIV 66, *Hilliswode* 1534 (1883) Sleigh 122, *Hilleswod(d)e, Hilliswod(d)e* c.1539 LRMB, *Hilliswood, Hyllyswode* 1542 (1883) Sleigh 71, *Helleswood* c.1596 SRO 3764/47[27574], *Helswood* 1610 Speed, *Helleswood* 1613 SRO DW1702/1/20, *Helleswood* 1619 (1883) Sleigh 93, 1709 ibid. 93, *Helswood* 1634 Leek ParReg, *Hilswood* 1645 ibid, *Helswood* 1673 Blome, *Jollyfes Hillswood* c.1728 SRO DW1702/1/20, *Helswood End, Elsewood End* 1775 Yates, *Hills Wood, North Hills Wood, Hills Wood End* 1842 O.S. The name originally applied to an extensive area to the north of Leek, and a number of derivations can be suggested. Perhaps from the OE personal name *Hille, a pet-name from Hilger or Hillary, or from Hille, a pet-form of Helen or Ellis or similar (see DES 231), or possibly from OScand *hiallr, hjalli* in the sense of 'a ledge on the side of a hill', as found in Norwegian place-names (cf. Helsby, Cheshire): the places lie on the side of a pronounced hill to the west of the river Churnet. Or possibly from ON *hellir* 'a cave', found in place-names with the meaning 'a cave-like ravine or hollow', perhaps referring here to the steep-sided valley of the river Churnet to the east (and note the part natural and part man-made cave below Hillswood at Dieulacres Abbey: see Fisher 1969: 14), though it is perhaps unlikely that a Scandinavian element would be compounded with 'wood': ON *lund* might be expected. *Jollyfe* is from the Jolliffe family: see Erdeswick 1844: 249; Sleigh 1883: 33, 35; see also Joliffe's Banks. VCH VII 197 records that Hills Wood was called Abbey Wood by the late 19th century. The 1836 1" O.S. map shows a *Hills Wood* 2 miles south-east of Cheddleton (SJ 9950), on the south-west side of what is now Hills Farm. The history of the name has not been traced. Cf. Helsby, Cheshire.

**HILLYLEES** on south side of Swythamley (SJ 9764). *Hillyleis* 1542 (1883) Sleigh 71, *Hilley leyes* 1662 Leek ParReg, *Hilly Lees* 1698 ibid, 1842 O.S. 'The hilly meadow', from OE *lǣs*.

**HILTON** near Featherstone, 5 miles north-east of Wolverhampton (SJ 9505; SMR 02481 places a deserted Anglo-Saxon settlement at SJ 95200542), *Hylton* 996 (for 994) (17th century, S.1380), *Haltone* 1086 DB, *Hulton* 1262 For; **HILTON** in Shenstone parish, 1 mile south of Muckley Corner (SK 0805), *Hiltun* 996 (for 994) (17th century, S.1380), *Iltone* 1086 DB, *Hulton* 1332 Duig, *Hilton*, *Hilltown* 1794 Sanders 254; **HILTON FARM** 1 mile north of Sheriffhales (SJ 7613), *Hulton* 1327 SHC VII (i) 245, *Hylton* 1532 SHC 4th Series 8 99, 1546 SHC XI 292, *Hilton* 1651 Sheriffhales ParReg, *Hilkton* (*sic*) 1666 SHC 1927 70, *Hilton Farm* 1833 O.S.; **HILTON** 1 mile south-east of Worfield (SO 7795), *Hulton* 1256 Eyton 1854-60: III 112, *Hulton'* 1298 TSAS LXXI 1996 27, *Hillton* 1752 Rocque, *Hilton* 1833 O.S. A very common name, from OE *hyll-tun* '*tun* on or by a hill'. Early maps (e.g. Smith 1599, Kip 1607, Speed 1610, Blaeu 1648) show *Hilton* near Kiddemore Green to the north-west of Chillington (SJ 8508), presumably *Hiltons Kerrimore* recorded in 1650 (SRO 590/466; cf. *Hilton* (p) 1629 Brewood ParReg)), which is almost certainly to be associated with the hamlet now known as Coldham, evidently with a change of name to avoid confusion with Hilton near Sheriffhales and Hilton near Featherstone. Hilton near Worfield has been in Shropshire since the 12th century.

**HIMLEY** Ancient Parish 5 miles south of Wolverhampton (SO 8891). *Himelei* 1086 DB, *Humelilega* 1185 P, *Humilega* 1187 SHC I 131, *Humelele* 1242 Fees, *Hulmelegh* 1271 SHC V (i) 154, *Himelegh* 1286 SHC VI (i) 163, *Hemele* 1306 SHC VII 149, *Humeleye*, *Hemeleye* 1323 SHC 1911 355, *Humeley* 1361 SHC VIII NS 15. From OE *hymele*, *leah* '*leah* where *hymele* grows'. *Hymele* may have been the hop or a similar plant (EPNE i 276). Cf. Himbleton, Worcestershire, *Hymeltun* 884 (S.219), *Hymeltune* c.977 (1373), *Himeltun* 1086 DB.

HINE HYLLE recorded in the boundary clause of a charter of Marchington of 951 AD (14th century, S.557), has been identified by Hooke 1983: 103 as near Marchington Cliff on the boundary of Hanbury (?SK 1329), but the similarity of the name to Houndhill (q.v.) is noteworthy. Hine Hylle would appear to be from OE *hiwan*, *higan*, Mercian OE *hine* (plural) 'a household, the members of a family; a religious community (monks or nuns)'. When the element occurs in place-named of pre-Conquest origin, it is likely to be a reference to ecclesiastical ownership. The place was probably on the border with nearby Hanbury (see Shaw 1798: I 85-6) where a nunnery associated with St Werburgh is said to have been founded c.680 (Hibbert 1908: 10; VCH III 135), and Hine Hylle may have been granted to the nuns there. Cf. Hinton and Hine Heath, Shropshire. Hine Hill, Beckbury (on the Staffordshire-Shropshire border) is recorded as *Hinell* c.1735: VCH Sa XX 240-1; PN Sa III 105.

**HINKSFORD** 2 miles north of Kingswinford (SO 8689). *Henkeston* 1271 SHC V (i 179, *? Hymoksford* 1296 SHC 4th Series XVIII 186, *Hinkesford* 1300 SHC V (i) 179, *Hincksford* 1749 Bowen, 1834 O.S. From OE *hengest* 'a horse, a stallion, a gelding', or from the OE personal name Hengest or Hynca, with the second element *tun*, later *ford*. The place is on the river Smestow and a tributary stream.

**HINTS** Ancient Parish 6 miles south-east of Lichfield (SK 1503). *Hintes* 1086 DB, 1199 FF, 1220 Ass, *Hintas* 1139 (15th century) EEA 14 92, *Hyntys*, *Hyntis* 1469 SHC IV NS 159, *Hynce* 1532 SHC 4th Series 8 187, 1577 Saxton, *Hints als Hence* 1539 (1801) Shaw II 15, *Hynse* 1601 SHC XVI 208, *Hynts* 1686 Plot. The name has been held from at least the end of the 18th century (see Shaw 1801: II 14) to be the English plural of Welsh *hynt* 'road', from British *\*sento-*, an element rare in Wales and Cornwall, but found in Breton coastal names (Padel 1985: 132), in which case indicating that the name must have been used by Welsh speakers through the period when *S-* changed to *H-*, usually considered to be the middle or second half of the 6th century (Jackson 1953: 521), and evidence of the relatively late survival of Welsh speech in the Lichfield area, but it should be noted that Jackson 1953: 519 considers such derivation 'quite uncertain', and the name possibly pre-Celtic. The terminal *-es* and *-se* in the early spellings is not necessarily an indication that the name was plural: the Normans often added *-s* to English place-names, particularly shorter names (see for example early spellings for Stone), but the consistent endings make a plural word probable. The name might possibly be the Welsh equivalent of English Stretton. The place lies on a very pronounced hill on Watling Street with far-reaching views both east and west along the Roman road: Shaw (1798: I 14) claims to have heard the name *Hendon* (i.e. OE *hean dun* 'high hill') used by antiquaries for the place. Duignan (MidA 1882 135) suggests that 'At this place there is an apparently ancient road crossing the Roman way'. It is perhaps surprising that this place was not named after the prominent natural mound (named *Golds Clump* in SHC 1916 208; see also JNSFC 5 1965 39) which stands on the summit of the hill on the south side of the road here. There is another Hints, almost certainly with the same derivation, 3 miles west of Neen in Shropshire.

**HITCHETT HILL**  on the north side of Draycott in the Clay (SK 1529). *Hatchell Hill* 1775 Yates, *Hitchett Hill* 1836 O.S.

**HIXON**  in Colwich and Stowe parish, 5½ miles north-east of Stafford (SK 0025). *Hustedone* 1086 DB, *Huchtesdona* 1130 SHC I 2, *Heisteduna* 1178 ibid. 93, *Huntesdun* 1228 SHC IV 50, *Huhtesdon* 1239 Ass, *Huccesdon* c.1276 SRO 597[7911], *Huncisdone* 1276 SHC 1911 177, *Huntesdon* 1284 SHC VI (i) 141, *Huncesdon* 1287 SHC VI (i) 173, *Huytesdon* 1289 Ass, *? Hutteedon* 1297 SRO DW1734/J2268, *Hughcesdon, Hoghcesdon* 1303 SHC VII 112, *Hughtesdon* 1306 SHC 1911 61, 1310 ibid. 73, 1349 SRO D938/91, *Huccisdone* 1318 SRO D938/493, *Hughcesdon* 1327 SHC VII 71, 1345 SHC XII 43, *Hughtelsdon* 1327 SHC 1913 119, *Hughleston* 1363 SRO DW1781/1/48, *Huxton* 1428 SHC XVII 121, *Hykstone* 1477 SHC VI NS (i) 106, *Hycston* 1532 SHC 4th Series 8 75, *Hyxton* 1586 SHC 1927 153, *Hickston otherwise Hixeton* 16th century Duig, *Hickson* 1686 Plot, *Hixton* 1778 Yates. Probably (though the forms are inconsistent) from the rare OE personal name Hyht, found in a Worcestershire charter of 963 AD (S.1303; Hooke 1990: 256-61), with OE *dun* 'hill'. The personal name Hyht may also be found in the placename Heighington, Lincolnshire: CDEPN. The place lies on the north-west side of a large rounded hill of over 400'.

**HOAR CROSS**  4 miles east of Abbots Bromley (SK 1323). *Horcros* 1230 P, *Harecros* 1236 SHC 1911 403, *Horecross* 1255 ibid. 125, *Horecreys* 1255 (1798) Shaw I xvi, *La Croiz* 1262 SHC IV 152, *Horecros* 1247 SHC 4th Series IV 80, 1251 Ch, *la Croz* 1263 ibid.' 153, *Orcross* 1267, *Harecres* 1248, *Horecros* 1268, *Whorecrose* 1513 Duig, *The Horcrosse* 1532 SHC 4th Series 8 174, *Horecross* 1577 Saxton, *Hore-cross* 1686 Plot. 'Grey or boundary cross', from OE *har* 'hoary, grey', or 'boundary'. Needwood Forest in Elizabethan times was divided into four wards or bailiwicks, Tutbury, Marchington, Yoxall, and Barton. According to the perambulations, all the wards met at Hoar Cross.

**HOARSTONES**  1 mile west of Warslow (SK 0758). *? (the dryestones, neire unto) the marestone* 1571 Alstonefield ParReg, *Meerstone* 1840 O.S., *Hoarstones* 1891 O.S. 'Boundary stones', from OE *har* 'hoary, grey; boundary'. The ealier name is from OE *(ge)mǣre stan* 'boundary stone'.

**HOBBERGATE**  2 miles north of Stone (SJ 9137). *Hazeburzeate, Hacheburggate, Hobbergate* 1192 SHC 1911 417, *Habberyate* 1288 SHC VI (i) 174, *Hacheburgata* c.

13th century SHC ibid. 17, *Hobler Gate* 1832 Teesdale, *Hobber Gate* 1836 O.S. The spellings indicate a derivation from OE *hæcc burh-geat* 'the hatch by the gate or entrance to the manor or fortification' (see VEPN I 85-6), though no archaeological or historical evidence has been traced of any fortification here.

**HOBBLE END** 1½ miles south-east of Great Wyrley (SK 0005). *Obbeleye* c.1300 SRO D1790/A/10/2, *the ob(b)leye(s), the ob(b)lies* 1302 Vernon, 1548 to 1617 Survey, *the Obleyes* 1563 SHC 1931 226, 1617 Willmore 1887: 440, *Hobble End* 1834 O.S. From OE *\*hobb(e)* 'a tussock, a hummock', with OE *leah* 'wood, clearing', and *ende*, giving 'the wood or clearing with tussocks at the end or outskirts of the place'. The word *end* was often applied to squatter dwellings on heathland or waste.

**HOBCROFT (FARM)** in Warslow (SK 0858). *Hobcroft* 1737 Alstonefield ParReg. From ME *hob* 'a sprite, elf, hobgoblin', with OE *croft* 'a small enclosure of arable or pasture land, an enclosure near a house'.

**HOBHILL** 1½ miles west of Bramshall (SK 0433); HOB HILL (obsolete), 1 mile south-west of Abbey Hulton (SJ 9148), *Hob Hill* 1836 O.S.; **HOB HILL** near Knightley (SJ 8024), *Near Hob Hill* 1839 (TA), *Hob Hill* 1891 O.S.; **HOBS HILL** 1 mile south-east of Rugeley (SK 1606). A common name, from ME *hob* 'a sprite, elf, hobgoblin', so 'goblin hill'. Hob Hill near Knightley (517') may have an earthwork on its summit (StEnc 292): hobs, sprites and goblins are often associated with pits, holes, hills and ancient earthworks.

HOB HOUSE (obsolete) 1 mile north-east of Upper Hulme (SK 0361). *Hobhouse* 1634 Leek ParReg, *Hob House* 1733 Alstonefield ParReg, 1842 O.S. From ME *hob* 'a sprite, elf, hobgoblin'.

**HOBMEADOWS** 1 mile south-west of Onecote (SK 0354). *Hobmeadow* 1695 Leek ParReg., *Hobsmeadow* 1841 O.S. Perhaps 'the meadow haunted by the hobgoblin', from ME *hob* 'a sprite, elf, hobgoblin'.

HOBRIDING (obsolete), 1 mile south-west of Anslow (SK 2023)) *Hobberudding* 1546 SRO DW1734/2/3/9f25, *Hobberobins* 1550 SRO DW1734/2/3/112b, *Hobriddin* 1737 Burton upon Trent ParReg, *Hobridding* 1821 WSL 73/22/43, *Hobriding* 1834 O.S. From ME *hob* 'a sprite, elf, hobgoblin' with OE *\*ryding* 'a clearing, an assart, land taken into an estate from waste', here possibly 'the clearing frequented by hobgoblins'.

**HOBS HOLE** 1 mile north-east of Aldridge (SK 0601), *Hob's Hole* 1834 O.S.; **HOBS HOLE** 2 miles north of Wednesbury (SO 9896), *Hobs Hale* 1775 Yates, *Hob's Hole* 1834 O.S.; HOBSHOLE (unlocated, in Freeford), *Hobshole* 1729 SRO D661/4/5. ME *hob* is 'a sprite, elf, hobgoblin', and ME *hole* 'a hollow, dingle, or small valley'. Cf. *Goblins pit Wood* and *Goblins pit Farm*, 1834 O.S., in Walsall Wood; *Hobbe Hey brooke* in Bloxwich 1597, Dent & Hill 1896: 104. Hobs, sprites and goblins are often associated with pits, holes and ancient earthworks.

**HOBSTONE HILL** 1 mile south of Farewell (SK 0710). *Hobbestone* 1392 VCH XIV 202, *Hobbestone (Lane)* 1571 ibid, *Hobstone Hill* 1834 O.S. Probably from the surname Hobb, with OE *stan* 'stone'. The 1571 spelling probably refers to what is now Hobstone Hill Lane.

**HOCCUM** 1½ miles south-west of Worfield (SO 7493). *Hoccumb* 1272 Eyton 1854-60: III 112, *Hoccumbe* 1292 ibid. 113, *Ocumbe* 1292 ibid. 216, *Hoccom* 1292 Jury List, *hocoumbe* 1301 Rees 1975: 249, *Hoccombe* 1327 SR, 1525 SR, *hokkum* 1502 TSAS 3rd Series III 120, *hawcum* 1555 TSAS 3rd Series IX 1909 124, *Hoccom, Hoccum* 1562

Worfield ParReg, *Hoecom* 1602 SA 2028/1/5/8, *Hoccome, Hoccam* 1661 SA 5586/1/464, *Hocham* 1752 Rocque. Perhaps from OE *hocc-cumb* 'the coomb or short spoon-shaped valley where hocks or mallows grew': cf. Oakham. The place lies on a ridge with a valley on both sides, a stream flowing through the one on the east. The *cumb* may be the valley on the west (though it is admittedly not noticeably spoon-shaped), or perhaps the subsidiary valley that runs off it to the west. It is unclear whether *Hockmull*, recorded in 1283 (Eyton 1854-60: III 113), is to be associated with this place. Cf. Hockham, Norfolk.

**HOCKER** (obsolete) 2 miles west of Longnor (SK 0564). *Hocker Head* 1683 Alstonefield ParReg, *Hocker* 1840 O.S. See Hockerhill.

**HOCKERHILL** ½ mile west of Brewood (SJ 8708). *The Hockerill, The Hockerhill* 1799 SRO QS B, *Hockerill Farm* 1834 OS. A not uncommon name of uncertain origin. *Hocker* may represent an old word for 'a hill' or 'a hump', from OE *\*hocer*, cognate with German *hocker* 'a knob, a hump' (Gelling & Cole 2000: 193), giving a meaning 'hill with a hump' (cf. Ocker Hill (q.v.)). There is no hill here, though 18th century maps (e.g. Yates 1775) show a post windmill, which may have stood on an earlier mound, now vanished. It is also possible that OE *\*hocer* was a nickname applied to poor land: cf. Hockerill, Hertfordshire. *Hochull* is recorded c.1270 (SHC 1928 15), and may refer to this place, but is more likely to be in the Wombourne area. See also Hocker.

**HOCKLEY** on the south side of Uttoxeter (SK 0933). *Hockley* 1834 White. Perhaps from OE *hocc-leah* 'the *leah* where hocks or mallows grew'. Redfern 1865: 375 notes that 'The Hockley, or Muckle Brook' ... has the name [to the north of Uttoxeter] of Stony-ford Brook ...'.

**HODDESDONE** (unlocated, in Blurton) *Hoddesdon* c.1200 SRO D593/B/1/23/3/2/2, *Hoddesdone* ?13th century SHC XI 310. From the OE personal name *\*Hod, so *\*Hod's dun* or hill': cf. Hoddesdon, Hertfordshire.

**HOFTON'S CROSS** at Cauldon Lowe (SK 0748). *Hoftons Bank* 1748 SRO D240/D/98, *Haughton* 1775 Yates, *Hoftons Cross* 1800 SRO D240/D/139, 1836 O.S., *Offtman's or Hofton's Cross* 1886 Redfern 1886: 46. Perhaps from the personal name Houghton or similar. The cross, described by Redfern 1886: 46 as a large unhewn stone in a meadow at Caldon Low, may be the rough stone cross on a green facing The Crosses Inn: StEnc 204.

**HOGS HILL** a 283' hill 1 mile south of Harlaston (SK 2209). *Hogshill* 1684 (1798) Shaw I 402, *Hoggshall* 1686 Plot, *Hogs Hill* 1798 Yates, *Hogs hill* 1834 O.S. The first element is probably OE *hogg* 'hog', or (perhaps less likely) *\*Hogg, an unrecorded personal name: cf. Hoggeston, Berkshire. The second element is *hill*, notwithstanding the 1686 spelling.

**HOLBEACHE** 1½ miles south-east of Upper Arley (SO 7878). *Holbeache Farm* 1889 O.S. Perhaps a relatively modern name: it does not appear on the 1834 O.S. map. If ancient, the derivation is likely to be as Holbeche (q.v.): the place lies at the head of a stream valley.

**HOLBECHE, HOLBEACH** 5 miles south-west of Wolverhampton (SO 8890). *Holebache* 1300 and 1327 SHC 1913 8, SHC VII (i) 247, *Holebache* 1300 SHC V (i) 179, 1323 SHC 1911 358, *Holebacke* 1327 SHC 1913 6, *Holbach* 1333 SHC X 87, *Holbeach* 1686 Plot, *Bolbatch* [*sic*] 1822 Himley ParReg. From OE *hol* 'a hole, a hollow', and usually found in place-names as an adjective 'lying in a hollow, sunken', with OE *bece* 'a steep-sided valley with a stream' (cf. Hawkbach). This place would appear to be the *ebles bece* mentioned in a charter of 996 (for 994) (17th century, S.1380). Holbeche Mill was formerly identified as Hubbals Mill, but the latter, later known as Harpsford Mill, lies on Mor Brook at Morville: see WMA 41 1998 63.

HOLDEN, THE (obsolete) 1 mile east of Burslem (SJ 8849). *Houldon* 1656 Norton-in-the-Moors ParReg, *The Holden* 1836 O.S. Perhaps from OE *hol* as an adjective 'lying in a hollow, sunken', with OE *denu* 'hollow (i.e. deep) valley': there is a pronounced stream valley here.

**HOLDITCH**    south-west of Chesterton, near Newcastle-under-Lyme (SJ 8348). *Holdedich* 1307 SHC XI NS 266, *Holdych* 1485 SHC VI NS (i) 158, *Holdich* 1522 SHC XI NS 8, *Holdyche* 1582 SHC XVII 228, *Holditch* c.1685 SHC 1941 124. Probably from OE *hol* 'a hole, a hollow', usually found in place-names as an adjective 'lying in a hollow, sunken', with OE *dic* 'ditch, dyke', so 'deep ditch' (cf. Holditch, Dorset), in which case the 1307 spelling is aberrant. The name may be connected with traces of a Roman settlement discovered here (see StEnc 294), or a medieval ditch traced over the Roman site, or from the mining of ironstone, which began here as early as the second century A.D: see VCH II 108.

**HOLE**    1½ miles north-west of Butterton (SK 0657). *Hole* 1695 Leek ParReg, 1840 O.S. From OE *hol* 'a hole, a hollow'.

**HOLE BROOK**    a tributary of the river Churnet, *? Holdebrook* 1282-3 SHC XI NS 259, *Holbrooke* 1636 *Deed*; **HOLE BROOK**    a tributary of the river Tame, *The brook of Holebro(o)k* 1286 For, 1505 Peramb, *Holbrook* 1617 Willmore 1887: 439. From OE *hol* 'a hole, a hollow', usually found in place-names as an adjective 'lying in a hollow, sunken', hence 'stream in a hollow'. 'Another Holbrook ran into the Tame at Perry Barr, and marked the bounds of Sutton Chase from Bolestile': Dent & Hill 1896: 137.

**HOLE CARR** 1 mile south-west of Hollinsclough (SK 0565). *Holehouse* 1414 VCH VII 33, *Hole Carr* 1568 ibid. 33, *the Hole* 1657 Alstonefield ParReg, *Carr Hole* 1840 O.S. From OE *hol* 'a hole, a hollow', usually found in place-names as an adjective 'lying in a hollow, sunken', with ON *kjarr* 'brushwood', ME *ker* 'a bog, a marsh, especially one overgrown with brushwood', replacing OE *hus* (or ON *hús*) 'house', hence 'the boggy place with brushwood in the hollow'.

HOLEDALE (unlocated, near Hyde Lea, Stafford, possibly near Moss Pit: see SHC VIII (ii) 128) *Holedale* 1166 SHC I 181, 1194 SHC II 266, 1203 SHC III 118, c.1210 SHC XI NS 125, c.1225 SHC II 275. Hyde Lea (near Stafford) and Holedale seem to have been parts of the same estate: see SHC II 266; SHC VIII 128. Holedale may also have been called *Holeden* (ibid.), or *Holedene* (SHC VI (i) 24). The name is evidently from OE *hol* 'a hole, a hollow', usually found in place-names as an adjective 'lying in a hollow, sunken', with OE *dæl* and *denu*, both meaning 'a valley', so 'the deep valley'. See also Holindale.

HOLEGODE (unlocated) *(Honour of) Holegode* 1426 SHC XVII 112. An honour is defined as 'a seigniory of several manors held under one baron or lord paramount' (ME): OED.

**HOLE HOUSE** 1 mile north of Endon (SJ 9254). *the Wholle Howse* 1568 SHC 1931 219, *Hole House* 1697 Leek ParReg, *Holehouse* 1744 Stoke on Trent ParReg, *hole House* 1803 SHC 1933 149, *Hole House* 1836 O.S. Possibly from OE *hol* 'a hole, a hollow', usually found in place-names as an adjective 'lying in a hollow, sunken', so 'the house in the hollow'.

**HOLINDALE** (unlocated, possibly near Penkridge, perhaps Holedale (q.v.)) *Holindale* 1203 SHC III 119. From OE *holegn dæl* 'holly valley'.

**HOLLIES COMMON** 1 mile north-west of Gnosall (SJ 8121), *le holyes* 1327 SHC VII (i) 238, 1332 SR, *Holyes* 1342 SHC 1913 86, *Parva Holneze* 1350 SHC XIV (ii) 35, *the Holies* 1381 SHC XVII 202, *Holys* 1451 Ct, *le(z) hollies* 1585 Ct, 1595 QSR, 1621 and 1658 PCC et freq, *(The) Hollies* 1679 SHC 1919 222; THE HOLLIES (unlocated, 1 mile north-east of Heath Hayes (SK 0310), *? Hollies* 1586 SHC 1927 132, *Wildmoore hollies* 1686 Plot, *Wildmore Hollies* 1749 Bowen, *The Hollies* 1834 O.S.; HOLLIES (unlocated) in Weston Jones parish, *the Holyes* 1308 WSL Deed. '(Place at) the hollies'. *Parva* is 'little', implying another nearby place of the same name which was larger (*Magna*). *Wildmore* was evidently from the extensive heath on which lay Heath Hayes.

**HOLLINGTON** 4 miles south-east of Cheadle (SK 0538). *Holyngton* 13th century Duig, 1408 SHC XVI 65, *Hollington* 1580 SHC XV 129. From OE *holegn* 'holly tree', with OE *tun*.

**HOLLINHALL** 1 mile north of Heaton (SJ 6395). *Holynhall* c.1539 LRMB, *Hollin Hall* 1842 O.S. 'Holly hall'. It is unclear whether *Hollinknolle* and *Hollin Knowle*, recorded in Heaton in 1646 and 1649 (SRO 322/M/10, 322/M/13a-b) relate to this place.

**HOLLIN HAY (WOOD)** 2 miles south-west of Leek (SJ 9653). *Holynehay* c.1220 StSt 5 9, *Hollin Hay Wood* 1836 O.S. From OE *holegn-(ge)heg* 'enclosure with the hollies'.

**HOLLIN HOUSE** 1 mile north-west of Endon (SJ 9154). *Holin House* c.1562 SHC IX 42, *Hollin House* 1836 O.S. From OE *holegn* 'the holly tree'.

**HOLLINS** on north side of Talke (SJ 8353), *Hollen Wood* 1733 SHC 1944 65, *Hollins Wood* 1799 Faden, 1833 O.S.; THE HOLLINS (obsolete), 1 mile south-east of Market Drayton (SJ 6932), now in Shropshire, *Hillins* 1694 SRO D681/E/5/21, *Le Hollins* 1707 SRO D681/E/5/21; **HOLLINS (FARM)** 2 miles south of Audley (SJ 7949), *the Hollens* 1733 SHC 1944 22, *Hollins* 1799 Faden, *The Hollins* 1833 O.S.; **HOLLINS** 1½ miles south-east of Consall (SJ 9947), *Holyns* 1320 SHC 1911 92, *Hollins* 1599 SHC 1935 98, *Hollyns* 1602 ibid. 445, *the Hollins* 1704 Kingsley ParReg; HOLLINS (obsolete) 2 miles east of Biddulph (SJ 9157), *Holyenis* (p) c.1225 StCart, *(ye) Hollins* 1666 ParReg, 1815 EA, 1842 O.S. From OE *holegn* 'holly', so 'wood with the hollies' and '(place at) the hollies'.

**HOLLINSCLOUGH** 2 miles north-west of Longnor (SK 0666). *Howelsclough* c.1395 VCH VII 37, *Howesclogh* 1472 Banco, *Howelles Cloughe* 1565 FF, *Howellas Cloughe* 1565 SHC XIII 240, *Howelscloughe* 1570 FF, *Howelsclough* 1574 SHC XIV 169, *Hollesclough(e)* 1596 ParReg, *Holesclough, Holes Clough* 1600 ibid, *Hollins Clough* 1775 Yates, *Hollinsclough* 1831 CensusRet. The modern spelling, which occurs only since 1775, implies a connection with OE *holegn* 'holly' (perhaps influenced by nearby Hollins Hilll and Hollins Farm on the opposite side of the river Dove in Derbyshire), and Oakden held the root to be OE *hol* 'a hollow' (TSSAHS IX 1967-8 34), but the early forms indicate a derivation from a personal name. PN Ch V (II) 288-9 records *Howeliscloutht* (*sic*) 1287, an unlocated place in Macclesfield Hundred, derived from the OWelsh personal-name Houel, and that seems the most likely explanation here. Cf. Howsen, Worcestershire (PN Wo 117). The *clough* ('narrow valley with steep sides') element is from the short ravine here in which a stream flows north to the river Dove. See also Hores Clough.

**HOLLOWAY FARM** 2 miles south-west of Madeley (SJ 7640). *? Holywall* 1606 SHC XVIII 61, *hollywall (lane)* 1615 SHC 1934 (ii) 29, *Holloway Lane Farm* 1823 SRO D798/1/11/11, *holloway (lane)* 1833 O.S. Other forms would be needed for certainty, but perhaps 'the spring or well at the holly', from Mercian OE *wælle* 'a spring', and (sometimes) 'a stream'.

**HOLLOW MILL FARM** 1 mile west of Wall Heath (SO 8689). *Hollow Mill* 1678 VCH XX 213, 1834 O.S. Evidently associated with *Hollow Moor*, recorded in this area in 1690 (HRO E12/V1/KY/7), probably from OE *hol(h)* 'a hole, a deep place in water', also 'a hollow, a depression in the ground'. Hollow Mill Farm lies on the west bank of Smestow Brook, which may suggest that the name is from a deep part of the river here.

HOLLY BANK FARM (obsolete) ½ mile south of Uttoxeter (SK 0932). *Hollingbury Hall* 1585 SRO D786/20/10iii, *Hollinberie Hall* 1587 SRO D786/20/8iv, *Hollingbury Hall* 1602 SRO D786/2/19, *Hollenbery Hall* 1611 SRO D786/2/26, *Hallyn Hall* 1616 SRO D786/8/1, *Hollingbury-Hall* 1686 Plot 274, *Hollin or Hollinbury Hall* 1721 SRO D786/8, *Hollingbury Hall* 1836 O.S. From OE *holegn*, with OE *burh* 'fortified place, manor', or possibly OE *beorg* 'a hill, a mound', 'the manor or fortification or hill with the holly tree or trees': the place lies on a hill. See also Redfern 1865: 20-21, 250.

**HOLLY BUSH** 1 mile north-west of Newborough (SK 1326). *Holly-bush, Hollybush* 1798 Shaw I 68, 94, *Holly Bush (Hall)* 1836 O.S. Self-explanatory. The place was a hunting lodge in Needwood Forest: Shaw 1798: I 68, 94.

HOLLYCOTE BATCH (unlocated, probably in or near Trentham) *Holycotes Bache* 1585 SA 2922/11/1/23, *Holicotes Bache* 1615 SA 2922/11/1/56, *Holycotte bache, Hollycote batch* 1624 SA 2922/11/1/72. Probably ' the cottages by the hollies at the stream valley', from OE *holegn, cot, bece*.

HOLLYFORD (obsolete, on the river Sow near Shugborough) The ford was replaced by a bridge (now vanished) in the 18th century: SHC 1970 90. It is unclear whether this place is to be associated with the remains of an ancient bridge discovered between Great Haywood and Tixall in 1938: StEnc 591.

**HOLLY WALL FARM** 2 miles south-east of Kidsgrove (SJ 8552). *the Halywalle* 1366 JNSFC LIX 1924-5 64, *Haliwalle, Halywalle* 1366 Ct, *Halliwell* 1586 SHC XVII 235, *? Hallywall(feilde)* 1597 SRO D1463/1, *Holywell* 1623 Wolstanton ParReg, *Halliwall* 1657 ibid, *Holly Wall* 1836 O.S. From OE *halig-wælle* 'Holy spring': cf. Halliwell, Lancashire; Haliwell, Middlesex, and Holywell in various counties. The 1366 record mentions a chapel at this place, and VCH VIII 93 suggests that there was also a hermitage. See however VCH III 136 which concludes that the hermitage was probably at Tunstall in Wolstanton.

**HOLLYWOOD** 1½ miles south-east of Stone (SJ 9333). *Hollywood Gate* 1775 Yates, *Holly Wood* 1798 Yates. Self-explanatory.

**HOLM (FARM, COTTAGE)** ½ mile north-west of Alton, on the south side of the river Churnet (SK 0642), *The Holme* 1608 Chester 1979: 45, *Hulmes* 1770 SRO D240/D/236, *Holme* 1836 O.S.; **HOLME FARM** ½ mile south-west of Mayfield (SK 1445); HOLM, LE (unlocated, in Levedale), *le Holm* 1294 SRO D260/M/7/5/137; HOLM, LE (unlocated, in Hamstall Ridware), *le Holm* 1297 SHC XVI 296. See Hulme.

HOLNEY (unlocated, near Lower Rule: see VCH IV 84) *Holneypol* 1312 SHC IX 33, *Holney* c.1341 SHC XII (i) 291, *Parva Holneze* 1349 SHC XIV (ii) 35. Perhaps originally 'holly pool', from OE *holegn pol*, with *pol* later replaced by *eg* 'island, land on a stream or between streams'. Possibly to be associated with Reulemill Pools, south-west of Lower Reule Farm (SJ 842190). *Parva* is Latin for 'small', implying another, *Magna* ('great'), Holney.

**HOLT HILL** ½ mile north-west of Newborough (SK 1226). *Holt-hall* 1311 (1798) Shaw I 94. See also Kingsley Holt. If the early spelling is correct, from OE *holt* 'a wood', so 'the hall at the wood'.

**HOLY AUSTIN ROCK** at the north end of Kinver Edge (SO 8383). *Holy Austin Rock* 1801 VCH XX 122. Presumably from a former hermitage in the sandstone caves here: ME Austin is the vernacular form of Augustine. *le Ostyn redyng* (perhaps 'the Augustinians' cleared land', from OE *\*ryding*) recorded in 1444 (VCH XX 122) is probably to be associated with this place.

**HOLYOAKE'S FARM** 2 miles north-west of Redditch (SO 0168). *le Haliok* 1255 Ass, *Holiok* 1275 SR. From OE *halig-ac* 'holy-oak', or 'holy cross': see Hooke 1990: 405-6. Cf. Holyoakes, Leicestershire. In Tardebigge parish, forming part of Staffordshire from c.1100 until 1266, in Warwickshire until 1844, and since that date in Worcestershire.

HOLYWELL PARK (unlocated, in Castle Church) *Halwelpark* 1439 Oakden 1984: 80, *Halywell parke* 1460 ibid, *Holywell parke* 1462 ibid.' 'The park of the holy spring or well', from OE *halig* 'holy', and Mercian OE *wælle* 'a spring', and (sometimes) 'a stream'. The spring is also mentioned in the Castle Church field-name *le holiwallefeld* 1364: Oakden 1984: 80.

HOMBRIDGE (obsolete) The lower part of Ford Brook in Walsall was called Hammerwich or Hambridge Forge Brook (VCH XVII 143), or Wombrook (SRO D260/M/F/1/2/f43d). *Wombrugg* 1282-3 SHC XI NS 263, *Wombridgeford* 1590 SHC 1930 (ii) 116, *Wombrokford* 1597 SRO D260/M/F/1/2/f43d, *Wombridge Ford* 1591 VCH XVII 174, *Wombrooke fforde* 1617 Willmore 1887: 440, *Homridge Pelsall* 1640 SHC XV 213, *Hombridge* 1665 Wolverhampton ParReg. Perhaps from OE *wamb* 'womb, belly', perhaps with reference to former pool here, or possibly a bulging topographical feature, with OE *brycg* 'bridge'. Cf. Wombridge, Shropshire; Wombwell, Yorkshire. See also Wombourne.

HOMBRIDGE or HOMEBRIDGE – see **HAMMERWICH**.

**HONEYWALL FARM** 1 mile west of Keele (SJ 7945). *Honeywall* c.1708 SRO D1798/579/3/1, *Honey Wall* 1805 Stoke on Trent ParReg, 1833 O.S. The late spellings make a derivation from 'the spring with agreeable water', from Mercian OE *wælle* 'spring, stream', or 'the bee-hive wall' from OE *weall* equally possible: see TSSAHS XIII 1971-2 43-5. *Honey Wall Meadow* in Penkhull is recorded c.1811: SRO D3272/7/2/2/67-8.

HONGENDEHUL (unlocated, perhaps near Drointon) *Le Hongendehul* pre-1284 SRO 543[7902]

HONGGERSHILL (unlocated, perhaps at Hilderstone) *Honggershill* 1593 SHC 1930 (ii) 340. See also Hungry Hill.

**HOO or HOE** A common place-name, of which there are several examples in Staffordshire (e.g. The Hoo, 1 mile north-west of Enville), generally of hamlets or homesteads. The name comes from OE *hoh* 'a heel, a hill-spur'. A settlement called *Hoo* recorded (as *Ho*) in 1271 (SHC V (i) 141), 1293 (SHC VI (i) 283), and 1336 (SHC 1913 53), and *La Hoe* in 1371 (SA 2089/2/2/24) is probably Blundies (*Blunders* 1827 O.S.) north-east of Enville, named after the Blundel family: VCH XX 93. Hoo Farm (*? Howe* 1562 SHC 1931 184) lies to the north on a hill-spur. *Hoo*, recorded in 1457 (SHC IV NS 100), has not been identified. *The Hoo* in Woodhouses, Stone parish, is recorded 1638x1738: SRO 49/7/44.

HOO (unlocated, at Bradley near Wednesbury) *la Hoo* 1290 Ipm, 1308 SHC X 4, *the Hooes* 1659 SRO D260/M/T/92. The place gave its name to an ancient house called The Hoo, which is said to have been at or near the site of Bradley Hall near Wednesbury (Shaw 1801: II 105; WA II 90). The name appears to have derived from the Hoo family, said to have been of Norman descent: Ede 1962: 89. See also Melleshohe; Maleshou.

**HOO BROOK**   a tributary of the river Manifold. *Harbrocke* 1434 (17th century) Survey, *Howbrook or Holebrook* 1586 Harrison, *le Holbroke* 1593 QSR, *Howbrook, How-brook* 1686 Plot 105. The conflicting early forms make any derivation uncertain, but perhaps from OE *hoh* 'a heel, a spur of land', or ON *haugr* 'a natural height, a hill, a heap, an artificial mound, a burial mound' (perhaps with reference to Ossoms Hill, around which the stream flows on the north), or from OE *hol* 'a hole, a hollow', commonly found in stream-names as an adjective 'deep, lying in a hollow': the brook runs in a deep valley.

**HOOKGATE**   1 mile south-west of Ashley (SJ 7435). *Hook Gate* 1731 Salt 1888: 132, 1833 O.S. CDEPN suggests 'the huckster's road', from late ME *hukker*, 'a petty dealer', but that is unlikely. Probably in this case 'the gate with the hook', although there are other parallels with this name, including Hookagate in Shrewsbury, recorded as *Hucke hey Gate* in 1598, on which see PN Sa IV 150-1.

HOO MILL   on the river Trent, 1 mile south-east of Ingestre (SJ 9923).   *Hore-mulne* 1302 SHC VIII (i) 197, *Howemulne-grene* 1331 SHC XII 291, *Hoo Mill* 1425 SRO D240/B/1/3, 1775 Yates, 1836 O.S., *Howe Mill* 1887 SHC VIII (i) 197 fn.1. This may be the half-mill for which Ingestre is credited in Domesday Book: see Derby & Terret 1971: 206. Perhaps from OE *hoh* 'a heel, a spur of land': there is a long and broad ridge of land here with a heel-shaped profile. The earliest spelling may well be a mistranscription, but if correct, the name is from OE *horu* 'filth', although it might be surprising to find that element associated with OE *mylen*.

**HOPE**   ½ mile south-west of Alstonefield (SK 1255). *Hope* 1371 SHC VIII NS 264, *Hooper* 1540 (1798) Shaw I *156, *Hope* 1512 NA DD/P/CD/140, *Hoope* 1551 SHC 1912 183, 1585 SHC XVII 233, *Hope* 1596 SHC XVI 154, 1695 Morden. From OE *hop*, in the West Midlands probably meaning 'enclosure in a marsh or enclosure in heathland', and in some cases, particularly in the west of the region, with the particular meaning 'remote valley', as here: the valley of Hopedale (q.v.) is notably hidden and secluded. See also Bradnop.

**HOPEDALE**   1 mile south-west of Alstonefield (SK 1255). *Hopedale* 1512 NA DD/P/CD/140, *Hopedale in Aystenfield* 1657 Okeover Deeds, *Hopedale* 1775 Yates. 'The valley of Hope village': see Hope. The secluded valley may well have been the site of the earliest settlement of Hope. There is a Hopedale in Derbyshire; the 1512 spelling may refer to that place.

**HOPESTONE FARM**   1 mile east of Ipstones (SK 0349). *Hope Stone* 1775 Yates, *Hopestone Farm* 1880 Kelly. From OE *hop*, in the West Midlands probably meaning 'enclosure in a marsh or enclosure in heathland', an example of the former being Hopwas, and of the latter Hopton, and in some cases, particularly in the west of the region, with the particular meaning 'remote valley', as in Bradnop and Hope. Here the meaning is probably 'the plot of enclosed land in a marsh', with OE *stan* 'stone': there is a large rock outcrop in marshy land here. See also Hopstone; Ipstones.

HOPPYNGS (unlocated, near Highlands Park, Tatenhill) *Hoppyngs* 1330 (1798) Shaw I 113.

**HOPSTONE** ½ mile north-west of Claverley (SO 7894). *Hopestan* 1209 Eyton 1854-60: III 97, *Hopstan* 1370 SA 2089/2/2/23, *Hopston* 1532 SHC 4th Series 8 123, *Hopstone* 1566 SA 4597/3. From OE *hop*, in the West Midlands probably meaning 'enclosure in a marsh or enclosure in heathland', and in some cases, particularly in the west of the region, with the particular meaning 'remote valley'. It is difficult to know which meaning applies here: there is a small secluded valley on the south side of the village. The second element is OE *stan* 'stone', which here may simply indicate that the underlying sandstone is exposed at the surface. In Shropshire since the 12th century. See also Hopestone Farm.

**HOPTON** 2 miles north-east of Stafford (SJ 9425). *Hotone* 1086 DB, *Hoptuna* 1166 SHC 1923 297, *Hoppeton* 1203 SHC III 133, *Hopton* 1203 ibid. 77, 1242 Fees, *Opton* 1253 SHC IV 126, *? Okton* 1295 SHC VII 25, *Chepton* 1377 SHC 4th Series VI 14, *Hopton* 1686 Plot. OE *hop-tun* '*tun* or settlement in a valley', from OE *hop*, meaning in the Midlands 'a small secondary or blind valley', but here perhaps 'an enclosure in heathland', although the place lies in the bend of a narrow side-valley of the river Sow. The battle of Hopton Heath (an area shown on old maps as *St. Amon's Heath*, and recorded as *La Bruera, Bruera iuxta Hopton* late 13th century (SRO 413-4[7912]), *Haya de Hopton* 1291 (SRO 416[7912]) was fought nearby to the east on 19 March 1643. *Bruera* is from Latin *brueria* 'heath, heathland'. Waste ground or common known as *Hopton Outwood* is recorded in 1548 (SHC 1912 169), 1550 (SHC 1910 77), *Hopton Outwood alias Hopton Short Wood* in 1552 (SRO D1798/H. M. Chetwynd/38), and *Hopton Owtwood* c.1560 (SHC 1910 77).

**HOPWAS** 2 miles west of Tamworth (SK 1704). *Opewas* 1086 DB, *Hopewæes* 11th or 12th century Sawyer 1979a: xxxvii, *Upwas* 1203 SHC III 113, *Hopewas* 1271 SHC V (i) 153, *Hopwas* 1286 ibid.'175, *Hopper* c.1540 Leland, *Hopwaies* 1577 Saxton, *Hopwais* 1607 Kip, *Hoppas* 1686 Plot. From OE *hop*, here probably with the meaning 'an enclosure in fenland' (see Gelling 1981: 10), with OE *wæsse* 'land liable to sudden flooding and drying-out' (cf. Alrewas; Pur Brook; *Wassebroc*; Buildwas; Sugwas). Hopwas lies in a small side-valley of the river Trent, but is an atypical *hop*. Hopwas Hay was one of the Seven Hays of Cannock Forest (q.v.), and extra-parochial until 1857.

HORDEN (unlocated) *Horden* 1240 SHC IV 237. Probably from OE *horh-denu* or *horu-denu* 'the muddy valley'.

HORDLE SPRING (obsolete) 1 mile north-west of Middle Mayfield (SK 1345). *Hordle Sprink* 1836 O.S. Possibly from OE *hord-hyll* 'treasure mound' (see also Gold's Wood), with *sprink* representing a form of *spring*, 'newly-planted trees'. Cf. Hordle, Hampshire.

HORES CLOUGH (unlocated, probably on the north-west side of Hollinsclough (?SK 0567)) *Hooscloughe* 1575 SHC XVII 222, *Hoscloughe* 1580 SHC XIV 212, *Hoarse Clough* 1583 ParReg, *Hoos(e)clough* 1586 FF, *Hores Clough* 1731 *Letter*. The first element may be 'horse', but ME *hors*, which became Modern *hoarse*, was applied to the cry of a raven (OED). Clough is from OE *cloh*, ME *clough* 'stream valley', so perhaps here 'the stream valley with the horse', or 'the stream valley where ravens are heard'. It is not impossible that this place is properly to be identified with Hollinsclough.

**HOREWOOD** – see **COMPTON HALLOWES**.

**HORNINGLOW** 1 mile north of Burton-upon-Trent (SK 2325). *Horninlowe* c.1225 BL Stowe Ch 82, *Horninglow, Horninglawe* 13th century frequently Duig, *Hornyglowe* 1316 SHC 1911 89, *Horninglow temp.* Henry I Burton, *Horninglowe* 1327 SR, *Hurnynglowe* 1332 SHC X (i) 106, *Hornyngslowe* 1450 HLS, *Hornyng Lown* 1532 SHC 4th Series 8 154. The first element is probably from OE *horning* 'a bend, a corner, a spit of land, a headland' (the place lies at the east end of a narrow curving hill-spur overlooking the river Trent – Shaw 1798: I 24 describes '... an angular hill projecting into that expanse of meadows near where the two great rivers of this county, the Trent and Dove, wind into union'), or, perhaps more likely, from a hill-name *horning*, derived from *horn* and meaning 'horn-like hill or peak', with OE *hlaw* 'hill, burial mound.

**HORNINGLOW CROSS** (obsolete) on the south side of Rolleston (SK 2326). Hart 1985: 217 suggests that Horninglow Cross was formerly *dottes hlawe, dotdes hlaw* 'Dot's or *Dottr's' *hlaw*, from an ON personal name which occurs in DB, mentioned in a charter of Rolleston of 1008 AD (14th century, S.920). The name recurs as *Dodeslawe* in the 12th century Burton Abbey survey of Wetmore (SHC 1916 221), *Doddeslowe* in 1290 (NSJFS 12 1972 56), *Dodduslo* in the late 13th century (SRO DW1734/J1615), and is found as the field-name *Dodslow* (near Horninglow Cross) in the Rolleston tithe map of 1837: Hooke 1983: 95.

**HORSE BRIDGE** over the river Churnet 1 mile south-east of Longsdon (SJ 9653). *Horseyate Bridge* 1604 QSR, *Horse Bridge* 1815 *EnclA.* 'The bridge for horses', formerly with a gate.

**HORSEBROOK** 1 mile north of Brewood (SJ 8810). *Horsebrok* 1149x1155 (17th century) EEA 14 62, *Horsebroc* 1262-72 Brewood Ch, *Hossebroke* 1478 Ipm, *Horshbrooke* 1608 SHC 1948-9 128. 'The brook frequented by horses'. See also Bell Field Farm.

**HORSECROFT FARM** (obsolete, ½ mile north of Leek (SJ 9857)) *the Horsecroft* c.1540 *AOMB,* 1619 Deed, *Horsecroft yate* 1639 ParReg, *Horse Croft Gate* 1842 O.S. From OE *hors* 'horse', with OE *croft* 'a small enclosure of arable or pasture land, an enclosure near a house', so 'the small enclosure with the horse'.

**HORSELEY** 1½ miles south-west of Eccleshall (SJ 8128; SMR 02622 places a deserted post-Conquest settlement at SJ 81602790). *Horseley* c.1299 SHC XI 325, *Horseleg* c.1270 SHC 1921 36, *Horselegh* 1289 SHC VI (i) 185, *Horseleye* 1302 SHC VII (i) 97, *Horselega* 1303 ibid. 109, *Horsle* 1348 SHC 1913 126, *Horseley* c.1540 Leland. From OE *hors* 'horse', with OE *leah.*

**HORSELEY FIELDS** 1 mile east of Wolverhampton (SO 9398). *? Horselawe* 1204 SHC III (i) 143, *Horseley Fyld* c.1538 SHC 1912 114, *Horseley Fylde* 1560 BCA MS3145/118/1, *Horseley Field* 1615 SRO D593/B/1/26/11/8, *Orsley Field* 1770 Sketchley. Shaw (1801: II 150) gives, in undated chronological order, *Horslow, Horselowe-field, Horsehull-field, Horseley-field.* From OE *hors* 'horse', with OE *hlaw* 'tumulus, burial mound', one of several tumuli which stood in this area: see Shaw 1801: II 150. Perhaps associated with Horseley More (q.v.).

**HORSELEY HEATH** 2 miles north-west of West Bromwich (SO 9692). *Horseley-Heath* 1686 Plot 122. From OE *hors, leah* 'the *leah* with the horse'. It is unclear whether *Horseleye*, recorded in 1327 (SHC VII (i) 234) is to be associated with this place.

HORSELEY MORE (obsolete) in Prestwood (SJ 9401). *Horsley More, Horseley More* 1661 TSSAHS XXI 1979-80 18. Perhaps associated with Horseley Fields (q.v.).

HORTSELEWELLE (unlocated, possibly in the West Bromwich area) *Hortselewelle* c.1227 SHC II (i) 275.

**HORTON** 3 miles west of Leek (SJ 9457), *Horton* '1239 CurReg, *Herton* c.1240 SRO DW1761/A/4/11, 1252 Ch, *Hyrton* 1273 SHC VI (i) 59, *Hortoneshay* 1307 SHC XI NS 255; HORTONE (unlocated, in Offlow Hundred, possibly on the higher ground of Hademore: SMR 0263 places a deserted Anglo-Saxon settlement at SK 18190840), *Hortone* 1086 DB, *Horton* c.1235 VCH XIV 247, 1513 OSS 1936 55, 1539 BCA MS3878/70, *Henton, Herton* 1439 LLRRO 44'28/384-5. From OE *horh-tun* or *horu-tun* 'the *tun* on muddy land'. For Horton in Offlow (which still existed in 1377) see Shaw 1798: I 379; SHC 4th Series VI 10; VCH XIV 240; TSSAHS XX 1978-9 loose map.

HORTON HAY (obsolete) on the north side of Biddulph Moor (SJ 9160). *the haye of Horton* 1282 Ipm *et freq., Hortones hay* 1307 ibid, *Horton Hey* 1507 Ipm, 1528 StarCh, 1658 ParReg, 1686 Plot, 1775 Yates, *Horton heyes, Horton hayes temp.* Elizabeth I Chanc, 1616 FF, *Horton Hay (House)* 1842 O.S. From Mercian OE *(ge)heg* 'enclosure', so 'the enclosure belonging to Horton (q.v.)'. The place is now within Biddulph parish.

HORWOOD (unlocated, possibly near Draycott in the Clay) *Horwood* 1660 DRO 157/DD/P/37/1. From OE *horh-wudu* 'the dirty (i.e. muddy) wood'.

**HOSE WOOD** 1 mile south-east of Fulford (SJ 9737). *Hose Wood* 1798 Yates, 1836 O.S. Seemingly from OE *hohas*, plural of *hoh* 'a spur of a hill'. Cf. Hose, Leicestershire.

HOSINGTON (unlocated, possibly near Newcastle) *Hosington* c.1565 SHC 1931 206.

HOSYLEYE (unlocated, possibly near Alton) *Hosyleye* 1284 FA.

**HOUGH, THE** 1 mile north-west of Eccleshall (SJ 8129), *the Hough* 1655 Eccleshall ParReg; HOUGH, THE a former hamlet south of Forebridge, Stafford (SJ 9222; see VCH VI 194, SHC VIII (ii) 44-5), *le Halgh* after 1290 SRO D938/47, *Halgh* ?13th century SHC VIII (i) 134, *le Halgh* 1310 SHC 1911 75, *the Haugh* c.1358 ibid. 187, *the Hough* 1405 SRO D641/1/2/46, 1709 SRO D260/M/T/5/122, *The Hough Ho* 1836 O.S. From OE *hoh* 'heel, a spur of a hill'. See also VCH VI 194.

**HOUGHER WALL** on the south side of Audley (SJ 7950). *Houghwall* 1668 Audley ParReg, *Hough Wal, Haughawall* 1733 SHC 1944 6, 15, *Ougherwall* 1890 O.S. From a spring or well of this name, mentioned in 1733 (SHC 1944 15), almost certainly from OE *hoh* 'heel, a spur of a hill': the place lies at a pronounced headland. The name is remembered by Hougher Wall Road.

**HOUGHWOOD** 1 mile north-west of Bagnall (SJ 9250). *Hooghe Wood* c.1562 SHC 1938 24, *Hough Wood* 1836 O.S. From OE *hoh* 'spur of a hill': the place lies on the end of a pronounced hill.

**HOUNDEL** – see **OUNSDALE**.

**HOUNDHILL** a 320' conical hill at Marchington, 3½ miles south-east of Uttoxeter (SK 1330). *Hugenhill* 1204 SHC III (i) 93, *Howenhull* 'c.1260 SHC 4th Series IV 89, *Hounil'*,

*Honnul temp.* Henry III ibid.' 127-8, *Howenhull'* c.1260 SHC 4th Series IV 89, *Hoenil,*
*Hoenul* c.1260 (1798) Shaw I 85, *Hounhull* 1262 ibid, *Unenhull* 1290 SHC VI (i) 204,
*Honhull* 1292 (1798) Shaw I 85, *Howenille* 1294 SHC VII 26, *Hounhul* 1300 (1798)
Shaw I 85, *Hunhyle, Hunhyl, Hogenhull* 13th century Duig, *Hornhull* 1301 SHC 1911
270, *Hunhull* 1306 SHC VII 163, *Howenhull* 1309 SHC 1911 296, *Houghtenhull* 1324
ibid. 361, *Howaull* 1329 (1798) Shaw I 86, *Hounhull* 1357 (1798) ibid, *Houenhull* 1386
(1798) ibid, *Howenhull, Hounhull, Hounhul* 14th century Duig, *Houndhyll, Houndhill*
1460 (1798) Shaw I 86, *Hownehyll* 1539 SHC XI 279, *Howndhill* c.1569 SHC IX NS
101, *Hugenhull* 1608 SHC III 93, *Hound Hill* 1836 O.S. The terminal is OE *hyll*, 'hill'.
The first element seems to be from ON *haugum*, dative plural of *haugr*, 'a natural height,
a heap, an artificial mound, a hillock, a hill, a barrow', frequently used of hills or hilltops
resembling artificial mounds which were, as in Scandinavia and Iceland, adopted as
observation points or meeting places (EPNE i 235-6). The name doubtless attached to the
hill itself, in which case the plural form is problematic, but paralleled in Hoon (*Hougen*
1086 DB, *Howen(e)* 1275, *Houn(e)* 1330 PN Db 573), in Derbyshire, 2 miles north of
Tutbury and 6 miles east of Houndhill, which has the same derivation, but only a single
tumulus is recorded, Hoon Mount, a mound on a ridge defined by a square ditch which
may be a platformed bowl-barrow: Gelling 1988: 138; Gelling & Cole 2000: 174; JEPNS
35 2002-3 45-8. A tumulus is recorded in a field at Moreton near Houndhill: Redfern
1886: 38. Hoon lies a mile or so north of the river Dove, Houndhill a mile or so south of
the Dove. Possibly this name was applied to a larger area which included both places,
hence the plural form, although the form may have been used simply to avoid confusion
with Hoon. Cf. Howe, Norfolk; see also Bignall Hill, Bignall End. Holinshed records
from an unknown sourse that on 13th November 1002, during the reign of King Ethelred,
a massacre of the Danes began at Wellowyn in Hertfordshire, or 'at a place in
Staffordshire called Hown Hill' (Redfern 1865: 338 gives 1012). This event is the St
Brice's Day massacre, where Athelred ordered the killing of every Dane who lived in
England except the Anglo-Danes of the Danelaw, which certainly led to great slaughter
in the south of England. PN Ch V (I:1) 54 proposes a derivation from ME *hane* (from OE
*han*) 'a rock, a boundary stone' for Houndbridge, Cheshire, which has early forms with
*Hone-, Hond-, Hune-, Howne-, Hun-*, suggesting also that the name may be from *haugr*.
See also Hine Hylle.

**HOWARD HALL** in Yarnfield (SJ 8633). One of a number of Government
establishments built in the area in the early 1940s and named after naval heroes. See also
Beatty Hall, Drake Hall, Duncan Hall, Frobisher Hall, Nelson Hall, Raleigh Hall, Rodney
Hall.

HOWSESTYDDES (unlocated, possibly near Ravenscliffe) *Howsestyddes* 1579 SRO
D1229/1/3/62. From OE *hus-stede* 'the site of a house'.

**HUDDALE (FARM)** 1 mile east of Cauldon (SK 0949). *Huddedale* c.1220 SHC V (i)
51, *Huddesdale* 1227 SHC IV 61, *Hud(e)lesdale* 1227 CoramR, *Hudeldale* 1229 Ass,
*Hyd(d)(e)lesdale* c.1240 Okeover, *Hutlesdale* 1254 ibid, *Hudlesdale* 1299 SHC VII 63,
*Hudlesdal* c.1310 SRO D1229/1/4/50, *Hudlisdale temp.* Edward II D1229/1/4/48,
*Hudlesdale* 1324 SRO D1229/1/4/50, 1333 SHC X 114, *Huddesdale* 1458 SHC XI 237,
*Hudhill* 1832 Teesdale, 1844 TA. The forms are not consistent (and it may be noted that
SHC XIII 39 identifies *Hudlesdale*, recorded in 1345, as Hillsdale (q.v.)), but probably
from the OE personal name *Hud(d)el, a diminutive of Hudd (DES 242), with OE *dæl*
'valley'. John Hudde is recorded in this area in 1449: SHC III NS 185. See also
Hetelsdale, Huddlesford.

**HUDDLESFORD** 1 mile north-west of Whittington, near Lichfield (SK 1509). *Huddlesford* 1634 SRO D15/11/26/9, *Hudlesford* 1686 Plot, 1749 Bowen, *Huddlesford* 1834 O.S. Evidently '*Hud(d)el's ford': the personal name was a diminutive of Hudd: DES 242.

**HUG BRIDGE** over the river Dane 2 miles north-west of Heaton, and the name of an associated manor on the Staffordshire-Cheshire border (SJ 9363). *Huggebridge* 1230 SHC X 115, *Hugebruge* 1275 SHC V (i) 120, *Huggebrugge* 1332 SHC X 115, *Hokebrugge* 1431 PN Ch I 55. By tradition from the Christian name of Hugh le Dispenser, an early landowner who may have been responsible for building or rebuilding the old bridge, a medieval *passagium* or toll-road. A compilation of other place-names beginning Ug(g)-, Ig-, Hig-, Uck-, has been used as evidence for an OE element *\*ucga*, *\*(h)ycg(a)*, usually identified as a personal name \*Ucga, \*Hycga, and a ME element *(h)ugge-, (h)ug(g)-*, supposedly the personal name Hugge (supposedly a pet-form of the OE personal name Uhtræd: DES 241-2 *sub nom* Huck, Hug), which developed into Hugh: see PN Ch III 45; V (I:1) xv), or for some otherwise unrecorded OE noun *\*hucg*, *\*hycg* 'mound, hill', cognate with OE *hygel*, ON *haugr* 'hill, mound', originally 'a heap': see PN Ch V (I:1) xv. Cf. Hollands Mill, Worcestershire, *Huggesbrig mylne*, PN Wo 44.

HUGGEFORD (unlocated) in Hilderstone. *Huggeford* 1278 SHC 1911 33, c.1396 SHC 1910 306, *Hugford* 1319 SHC X 30, *Huggeford* 1316 ibid. 64. It seems likely that the place took its name from the Huggeford family, which had acquired Hilderstone by 1272: SHC XII NS 155; SHC 1911 399.

HULHORD (unlocated, in Wombourne) *Hulhord* 1336 SHC 1928 34. The word-order precludes a derivation from OE *hord, hyll* 'hill where treasure was found', and no alternative can be suggested.

**HULL** a 420' hill 1½ miles south-west of Uttoxeter (SK 0632). *Hull* 1414 DW1733/A/2/113. From OE *hyll* 'a hill'.

**HULLOCK'S POOL** 1 mile north-east of Audley (SJ 8051). *Hullokespole* 1298 SHC XI NS 250, *Hulkocuspel, Ullokuspel* 1307 SHC XI NS 264, *Olokkespole* 1493 SHC 1912 256, *Hullocke Poole* 1596 Audley ParReg, *Whillocks Pooll* 1733 SHC 1944 50, *Hullocks Pool* 1833 O.S. The first element would appear to be from the ME surname Hulcok or similar, a diminutive of Hulle, a pet-form of Hugh. There are a number of small pools in this area.

**HULME** 4 miles east of Stoke on Trent (SJ 9345), *Hulme* 1203 FF, 1227 ibid, *Holm'* 1208 Cur, 1218 FineR, *(H)ulmo* 1225 Cur, *Hulm'* 1225 Bracton, *Ulmo* 1226 SHC IV 39, *Hulm, Holm under Kevermund* 1293 SHC VI (i) 239, *Hulm Weston* 1293 ibid. 239, *(bridge of) Holm* late 13th century SRO 3764/21, *Hulm(e) juxta Weston* 1309 FF, *Hulm-by-Weston* 1309 SHC 1911 75, *Holm* 1331 SHC 1913 27, *Hulme-next-Weston* 1428 SHC XI 229, *Holme* 1601 SHC 1934 (ii) 6, *Hoome* 1616 SHC NS IV 88, *Home otherwise Howme* 1619 SHC VII NS 192, *North Hulme, Old Holm* 1749 Bowen, *North, Middle and Old Holm* 1775 Yates, *Hulme* 1836 O.S.; **HULME, UPPER** 3 miles north-east of Leek (SK 0160), *Hulm* c.1214 Dieul, *Holme* 1218 Pat R 1216-25 168, *Hulm* c.1245 SHC 1911 439, *Huln* 1247-8 SHC NS IX 318, *Ovre Hulme* 1284 SHC 1911 187, *Holm under la Roche* 1358 SHC XII (i) 162, *Hulme* 1395 SHC XV 72, *Ouhulme* 1648 Leek Par. Reg, *Upper or Over Hulme* 1775 Yates; **HULME, MIDDLE** 2½ miles north-east of Leek (SJ 9960), *Middehulm* 13th century Dieul, *medulhulme* 1548 PRO SC2/202/65, *Middleholme in Leek Fryth* 1574 SRO D3272/5/13/27; HULME, NETHER (obsolete) 2½ miles north-east of Leek (SJ 9960), *Netherhulm* 13th century Dieul, *Nether Hulm(e)* 1240 Deed

(1883) Sleigh, 1284 Inq aqd, *Nether(e)holm(e)* 1284 Ipm, 1284 Ipm; **HULME END** 1½ miles south-west of Sheen (SK 1059), *Hulme* 1227 Mills 1998: 190, *Hulme End* 1840 O.S.; HULME (unlocated, in Billington near Stafford), *Holm* 1208 SHC III (i) 142, *Hulme* 1209 ibid.' 175, SHC 1914 86, *Holm* 1307 SHC VII 126. The place-name Hulme is generally held to be from ODan *hulm* 'a small island, a piece of land on a stream, dry ground in a marsh', but recent research suggests that *hulm* may be an English dialectical form of ON *holmr*, with the same meaning: see Fellows-Jensen 1997: 79-81. For *Kevermund*, see Carmounthead. *Weston* is Weston Coyney. Upper Hulme is on the river Churnet (*la Roche* is The Roaches (q.v.)). Nether is from OE *neoðera* (or ON *neðri*) 'lower': Nether Hulme was also known as New Grange (q.v.), and was submerged when Tittesworth reservoir was extended c.1960: SHC 4th Series 19 fn.9. *Lower Hulme* in Caverswall is recorded in 1681: SRO D660/8/11. Bates Farm, on the east side of Maer (SJ 7838) appears as *The Holme* on the 1833 1" O.S. map. See also Holm.

**HULMEDALE FARM** 1 mile south of Werrington (SJ 9446). *Hulme Dale* 1836 O.S. See Hulme.

**HULTON** – see **ABBEY HULTON**.

HULWARE, LE (unlocated, in Milwich) *le Hulware* ? 13th century SHC XII NS 173. An intriguing name, probably from OE *hyll* 'hill', with OE *ware* 'dwellers', hence 'people who lived at the hill', but the derivation must remain speculative in the absence of other spellings.

HULY, MAGNUM (unlocated, in Horton) *Magnum Huly* 1239 CurReg.

**HUMESFORD BROOK** a stream which enters the eastern end of Aqualate Mere, also known as *Guild Brook* or *Gill Brook*: see Robinson 1988: 46.

HUMPAGE GREEN (obsolete, 1 mile south-east of Eccleshall, north-east of Pybirch Manor on the Stafford-Eccleshall road (SJ 846283)) *Heuntenbach* c.1220 SHC 1914 68, *Huntenbach* c.1266 SHC III 214, 1270 SHC V NS 217, *Huntinbach* 1272 SHC IV 194, *Huntebache* 1282 SHC VI (i) 152, *Huntenbach* 1293 ibid.' 263, *Huntenebache* 1306 SHC 1911 67, *Hontenbach* 1375 SHC XIII 121, *Huntebache* 1413 SHC XVII 44, *Huntbatche* 1601 Eccleshall ParReg, *Huntbache* 1607 ibid, *Humpidge Green* 1775 Yates, *Humpage Green* 1836 O.S., 1891 O.S. Possibly from OE *huntena*, the genitive plural of *hunta* 'huntsman', with OE *bece* 'stream in a steep-sided valley', so giving 'the huntsmen's stream in the well-defined valley'. Humpage Green (the Green element suggests a grassy area or a squatter settlement) lay on the north side of a slight ridge of higher land running roughly parallel to and to the south of the Stafford-Eccleshall road, which would not be topographically appropriate for a *bece* name, and it must be assumed that Humpage itself lay nearby: at some date after c.1795 Huntbatch was identified as Walton Villa (SJ 849284: 1891 O.S.; SRO D5800/1/4), which lay on the north side of what is now Walton Hall school.

**HUNDRED ACRES** on the west side of Peasley Bank (SJ 8929). *Hundreacre* (*sic*), *Hundredesacre* 1251 SHC 1913 183. Nearby Pirehill is one of the Staffordshire Hundred meeting places, and the early spellings show that the name of this place (the only example of the name Hundred Acres traced in Staffordshire) originated from OE *æcer* 'field, ploughed land', so perhaps 'the ploughed land associated with the Hundred meeting place'. Field 1993: 260-1 cites a collection of Hundred Acres names from throughout England, almost all relatively recent (two early examples may not incorporate the word 'hundred'), and suggests that they are ironic names for small fields. In the case of this name, however, there can be no doubt (given the early date and the genitive -*es*) that the name is to be associated with the Hundred meeting place (which may have been nearby, rather than on what is now called Pire Hill (q.v.)), and forms a notable exception to the general rule. Hundred Acres is inexplicably marked as Whitgreave on the first edition 1" O.S. map of 1836, with Whitgreave to the south shown correctly as Whitgreave. See also Pirehill.

**HUNGRY HILL** 1½ miles west of Brewood (SJ 8507), *Hunger Hill* 1660 Blount, 1808 SRO D590/17/23-26, *Hungry Hill* 1775 Yates, *Hungary Hill* 1834 O.S.; **HUNGER HILL** Hampstall Ridware (SK 0918), *Hunger Hill* 1834 O.S.; **HUNGERSHEATH (FARM)** 2 miles west of Maer (SJ 7638), *the Hungerheath* 1438 SRO D1798/H.M.Aston/2/7, *Hungeryheathe* 1583 Betley ParReg, *Hungersheath* 1664 SHC V (ii) 163, *Hunger Heath* 1833 O.S.; HUNGER HILL GATE (obsolete) ½ mile north of Bucknall (SJ 4848), *Hunger Hill Gate* 1836 O.S.; HUNGER HILL (obsolete) on the north-west side of Kidsgrove (SJ 8455), *Hunger Hill* 1840 O.S. *Hungry* is a common derogatory adjective, found in many parts of Staffordshire, including Teddesley, Wolstanton, Burston, and Horsebrook, for a hill or heath with poor or 'hungry' soil, but in some cases possibly from OE *hangra* 'a wooded slope'. An Anglo-Saxon charter of 1012 (13th century, S.930) mentions *hunger hylle* in Wetmoor; *Hungrehul, Hongerhulll* (undated) is recorded in Whitmore (SHC 1913 241, 244; see also Honggershill); and *Hungarhill* in Broad Street, Leek, is recorded in 1621: SRO D4645/A/1/1-25.

HUNTEBRIGE (unlocated) *Huntebrige* 1227 SHC IV 44.

HUNTECROFT (unlocated) *Huntecroft* 1338 SRO D593/B/1/23/3/2/13.

**HUNTINGTON** 2 miles north of Cannock (SJ 9713). *Estendone* 1086 DB, *Huntendon'* 1167 P, 1198, 1236 Fees, *Huntingdon* 1262, 1271 and 1300 Duig, *Hontyndon* 1333 SHC 1939 74, *Huntenton* 1532 SHC 4th Series 8 187, *Huntington otherwise Ramshorne* 1616 SHC VI NS (i) 10, *Huntington alias Ramshorne* 1616 SHC 1928 143. From OE *hunta* 'a hunter, a huntsman', with OE *dun* 'a hill', so 'the huntsmen's hill', or from the OE personal names Hunta, with *-ing* connective. The place was in the heart of Cannock Forest. The DB form suggests (if correct) an older form based on OE *eastan dun* 'the hill to the east'. For the identification of the DB form with this place see SHC 1923 24-8. The 1616 alternative is curious and unexplained, but see Ramshorn, Ramsor.

**HUNTLEY** 2 miles south of Cheadle (SK 0041). *Huntley* 1332 SRO D1229/1/4/12, 1472 SHC IV NS 181, 1584 SHC XV 151, 1594 SHC 1932 109, 1600 SHC 1934 6, 1656 Leek ParReg, *Hunttley* 1426 SHC III NS 168. Probably 'the *leah* of the huntsman', or from the personal name Hunta. See also Castle Croft.

**HUNT'S FARM** 2 miles north-west of Lichfield (SK 1012). *Hunts Farm* 1834 O.S. Probably from the Hunt family, recorded in the area in 1760: VCH XIV 234.

**HURDEN (HALL)** 1 mile east of Barlaston (SJ 9039). *Hurden Hall Farm* 1913 SRO D997/XI/2A&B, *Hurden Hall* 1930 O.S. Early spellings have not been traced.

**HURDLOW** ½ mile east of Upper Hulme (SK 0260). *Hordelowe* 1539 *AOMB*, *Hurdelowe* 1542 (1883) Sleigh 71, *Hordelowe, Hordlowe* 1546 SHC 1912 350-1, *Hurdelow(e)* 1599 Dep, *Hurdlow(e)* 1607 QSR, 1639 Leek ParReg, *Hurdlow* 1842 O.S. From OE *hord, hlaw* 'tumulus with the treasure-hord'. Cf. Hurdlow, Derbyshire: PN Db 366.

**HURSTAGE, THE** ½ mile north of Hilderstone (SJ 9435). *The Ostriches* 1920 O.S. The name is said to be found locally as *Horesych* and pronounced locally as The

Ostriches: TNSFC 1908 132; 1922 169. Possibly therefore from OE *horu-sic* 'filthy brook': cf. Harborne.

**HURST HILL** 1 mile east of Sedgley (SO 9394). *Hurstemore* 1273 SHC 1911 156, *Hurst' Hill* 1537 Inq, *Hurst Hill* 1581 Sedgley ParReg, 1582 SRO D260/M/T/1/114, 1834 O.S. From OE *hyrst* 'hillock, copse, wooded eminence', originally with OE *mor* 'a moor', here probably 'high waste-land'.

**HURST WOOD** 1 mile east of Colton (SK 0620). From OE *hyrst* 'hillock, copse, wooded eminence'. Perhaps associated with *Holihurst*, recorded in 1327 (SHC VII 217).

**HURT'S WOOD** 1½ miles north of Ilam (SK 1353). *Hurt's Wood* 1838 O.S. From the surname Hurte: Nicholas Hurte of Castern is recorded in 1618 (FF).

**HUSPHINS** 1 mile west of Codsall (SJ 8404). *Ursphants* 1652x1725 SRO D802/32, *Ursfins* 1730 Codsall ParReg, *Ursfins* 1738 ibid, *Urspins* 1744 ibid, *the Ursfins* 1778 *et freq.* ibid, *Ursins* 1804 ibid, *Husphins Farm* 1788 SRO D802/33, *Husphins* 1834 O.S. This curious name has not been traced earlier than the seventeenth century, and no suggestions can be offered for its derivation.

HUSSEY HALL (obsolete) a former moated site 1 mile south-east of Penkridge (SJ 935135). *Husseis Hall* 1558 SHC 1931 263. There were two manors of Penkridge at DB, one held by the King, the other by the Church. The King's manor was granted to the family of Hose or Hussey in 1155 (SHC 1931 254) or 1207 (SHC 1950-1 9-10).

HUSTANS (unlocated, possibly in the Alton area) *Hustans* 1599 SHC 1935 195.

**HUTTS FARM** 1½ miles north-east of Ellastone (SK 1244). *The Hutts* 1779 Ellastone ParReg, 1836 O.S. Possibly from ME *hutte* 'a heap', or ModE *hut*: see PN Db 3 528; PN La 110.

**HYDE, THE** 1 mile south-west of Brewood (SJ 8707), *la Hide*, *Hyde* 1199 Ass, *la Yde* 1211 SHC III 193, *la Huyde* 1317 SHC 1924 32; **HYDE, THE** north-east of Kinver (SO 8484), *Hyde* 1293 VCH XX 123, 1749 Bowen. From OE *hid* 'hide of land': see Introduction, page 61. The name is not uncommon: there seems to have been a Hyde in Freeford (Shaw 1801: II app. 14); *Lehide* within Penkridge parish recorded in 1553 (SHC XII (i) 214); a hamlet of *Hyde* near Butterhill (q.v.) in Bradley, near Stafford, probably cut out of the larger manor of Coppenhall in the mid-12th century (*Hidecopenhall* 1516 SHC XIII 182; see also VCH V 138-40, 142; Darlington 2001: 19); land called *Hyde* at Weston under Lizard ( SHC II NS 134); and *Hida* in Sheriffhales is recorded in c.1175 (Eyton 1854-60: IX 163; Rees 1985: 273) and as *the Hide* in 1631 (SRO D593/B/1/19/6/3), evidently on Burlaughton Common: *The Hide or Burlawton Common* is recorded in 1622 (SRO D593/E/5/3.

**HYDE LEA** 2 miles south of Stafford (SJ 9120). *Hida* c.1187 SHC II 261, 1224 StoneCh, *le Hyde*, *la Hyde*, *Hide* c.1225 RH, 1267, 1327 Pat, *Le Hydelea otherwise Hydeley* 1601 SHC 1935 348. From OE *hid* 'hide of land'. The addition of *leah* occurs comparatively late, and is from the manor of *la Leye* 1261 (SHC 1950-1 52), *Lega* 1321 (ibid.), which appears to have adjoined and become joined with Hyde. Cf. The Hyde.

HYDENHALL (unlocated) *Hydenhall* c.1255 SHC 1913 266.

HYDE PARK (obsolete) a park near Hyde near Butterhill first mentioned in 1372: Darlington 2001: 16.

HYGHT (unlocated) *Hyght* 1563 Church Eaton ParReg.

HYNDEBADESHALL (unlocated, in Bentley near Walsall). *Hyndebadeshall temp.* Henry III (1801) Shaw II 93, *Hyndebadeshull* n/d SHC 1910 198. Perhaps from the OE personal name Hildebeald or similar. If Shaw's spelling is correct, the final element is OE *hall* 'a hall, a large residence', a rare element in early Staffordshire names, so the other spelling, with OE *hyll* 'a hill' is probably more accurate..

ICFORD, YCFORD (unlocated, possibly outside Staffordshire – Icford in Oxfordshire is recorded in 1711 (ORO CJ/IV/11), or the spellings may refer to Ditchford, Warwickshire: see SHC I 163). *Icford* 1199 SHC III (i) 52, *Ycford* 1200 ibid. 66.

**IDLEROCKS** ½ mile north-east of Moddershall (SJ 9337). *Idlerocks* 1915 SA 4629/1/1915/105, 1946 O.S. Early spellings have not been traced, but if the modern spelling is correct, 'idle-headed' is recorded from 1598 for someone who was crazy (OED), so perhaps here 'the crazed or jumbled rocks', or from the dialect *idle* 'soft stone used for whitening stone floors, etc.': EDD. It has also been suggested that the name arises from a mistranscription of the first two letters of Sale Rocks (possibly recorded in *la Sale*: SHC VII 208) , which take their name from nearby Sale Brook: see StEnc 307.

IDMORE HEATH – see EADMORE HEATH.

**ILAM** Ancient Parish 4 miles north-west of Ashbourne (SK 1350). *Hilum* 1004 11th century (11th century, S.906; 11th century, S.1536); 1227 Ass, *Hylum* 1150x1159 SRO D603/A/Add/19, 1183-4 SHC 1937 19, *Ylum* 1176x1182 (13th century) EEA 16 13, 1197x1213 SRO D603/A/Add/45, *Ylam, Ylum* ?12th century SHC V (i) 21, 27, *Inliem, Hilun, Ylun* 1203 SHC III (i) 103, 109, *Hilim* 1203 ibid. 108, *Ylum temp.* Henry I Burton, 1208 SHC III (i) 172, 1286 SHC V (i) 117, *Ilum, Illum* 1227 CoramR, *Hylum* 1256 SRO D603/A/Add/131, *Illum, Hylum* c.1260 SHC VII NS 146, *Ylam* 1280 AngleseyCh, *Ilam* 1293 AngleseyCh, 1547 Pat, *Ylume* 1293 SHC 1911 215, *Ile* 1294 SHC VII (i) 20, *Ylom* 1312 WL 70, *Ilom* 1327 SHC VII (i) 220, *Ilum* 1331 SHC XI 30, *Ylom* 1333 SHC X 114, *Ylom(e)* 1339 Pat, *Ilom* 1460 SHC XI 237, *Ylum* 1523 Rental, *Ilom* 1532 SHC 4th Series 8 12, *Ilam* 1567 SHC XIII 267. The spelling is generally *Ilam* from the 16th century. Possibly from a British name of the river Manifold, *Hile*, related to Irish *silim*, 'drop, distil, sow, spit', Welsh *hil* 'seed, spawn', which has been held to mean 'trickling stream', with the plural form analogous to OE *Liminum* 'Lympne', from *Limen* 'river' (see Ekwall 1928: 207; Ekwall 1960: 262; and Coates & Breeze 2000: 335, where some doubt is expressed about this derivation), or perhaps 'at the pools, at the deep places in the river', from ON *hylum*, the dative plural of ON *hylr* (see Wrander 1983; cf. Healam, North Yorkshire). The particular phenomenon at Ilam is the re-emergence in 'boiling holes' at the foot of the limestone rock on which stands Ilam Hall of the rivers Hamps and Manifold, the Hamps having run underground from near Waterfall (q.v.), and the Manifold having taken a subterranean course from just south of Wetton Mill near Grindon, and it might be expected that any name referred specifically to such a striking feature, called 'the spring at Ilam' by Plot (1686 89). If from *Hile*, therefore (and for British *S* appearing as English *H* see Hamps, ME *Hanespe*, from British *\*samosispa*), possibly with the particular reference to the emerging rivers, and meaning 'the discharging or erupting waters'. Ekwall 1928: 216 also records that 'a river-name stem *Il-* is found in various names ... no certainty can be attained as regards the etymology, because there are several possible derivations ... [including] the root of W[elsh] *ilio* 'to ferment', which is held to belong to a root *il* 'to swell ... all that seems certain is that [the name of the River Isle in Somerset] is pre-English'. A name suggesting fermentation or swelling would seem particularly apposite for emerging underground rivers. Other possible derivations include Welsh *llwym* 'bare, bleak, poor' (see Padel 1985: 151-2), or

OE *lum(m) 'a pool', with an unidentified first element, perhaps OE ig 'piece of land in a marsh' (cf. Iden, Sussex); or OE ig 'a yew tree'. Finally, the suggestion in Duignan 1902: 83-4 of a derivation from OE hillum, the dative plural of OE hyll 'hill', giving '(at) the hills', can be rejected since (as noted in TSAS 4th Series I 1911 XXXIV 17 fn.59) the forms do not support such a root. A derivation from the OE dative plural *hyglum, from OE *hygel 'a small hill', so giving '(the place at) the hills' (CDEPN 329), is perhaps the simplest solution, and would fit the topography here.

ILAM MOOR (obsolete) moorland to the north of Ilam (SK 1351). Ilome More 1443 Okeover E414, (Grendon on) le Morys 1444 SHC 1939 85, Ilam More 1543 SHC XI 285, 1586 SHC XV 169. The name is remembered in Moor Plantation.

**INA'S ROCK** 1 mile east of Alton Towers, on the north side of the Churnet Valley (SK 088428). Early spellings have not been traced, but possibly from the legendary local connection of this area with King Ine: see Slain Hollow.

**INGESTRE** Ancient Parish 4 miles north-east of Stafford (SJ 9824). Gestreon 1086 DB, Iggestroud 1161x1182 SRO 1/7972, Ingestrent pre-1184 SHC XII (i) 271, Ingkestrent 1184x1228 ibid.' 271, Higgestrend c.1200 ibid. 273, Ingestreon 1236 SHC 1911 397, Ingestre 1236 Fees, Ingerstrent 1242 Cl, Ingestreon 1243 SHC 1911 394, Ingestret, Higestront 1242 Fees, Ingestraund 1250 Ass, Ingestrond c.1250 SHC VIII 135, Ingustre, Inggustre 1305 SHC XII (i) 285, Ingestre 1371 SHC X NS (ii) 47, Inglystre 1373 SHC XIII 101, Yngstre 1529 SHC 1910 16, Engulstre 1532 SHC 4th Series 8 57, Inglestre, Ingestre c.1540 Leland, Ingestrye 1605 SHC 1940 320, Ingtastrey 1616 BCA 3369/Acc1961-9/80, Inkestre 1577 Saxton. The DB spelling suggests that the first two letters were treated as a preposition and omitted, a feature which can be traced elsewhere, e.g. Ilkerton, Devon, is Crintone in DB, but Incrintona in the Liber Exoniensis, from which the DB form was extraced. The scribe entering the Staffordshire name may additionally have been influenced by the word in 'in' which precedes all the place-names in this section of DB, and believing the first two letters to be mistakenly repeated, ignored them. The derivation of the name has long posed difficulties. The place lies on gently rising ground to the north-west of the confluence of the rivers Trent and Sow. Ekwall 1960: 264 proposed a derivation from the puzzling OE element *ing, perhaps meaning 'a hill, a peak' (though the hill here is hardly pronounced), with OE (ge)streon (and its variant (ge)streond) 'treasure, property, wealth, riches, gain, profit'. A more likely derivation, however is from a Brittonicised *engyst (cf. Welsh ing, yng) from Latin angustiæ (in vulgar Latin angustie) 'the narrows', found in Ingst in Gloucestershire (Coates & Breeze 2000: 48, 54-7), which lies a mile or so from a narrowing of the river Severn, and in continental place-names such as Angoisse (Dordogne). Such a name would have been taken into OE as *Engest, and the substitution of i for e is not an irregular philological phenomenon. The flood plain of the river Trent narrows opposite Ingestre, and evidence of episodic flooding has been revealed in archaeological evaluations: see ArchIng. The second element is clearly from the name of the river Trent, but in some spellings appears to have been influenced by OE -straund, -rand 'river bank', OE (ge)streon(d) 'treasure', and OE treon, the plural of OE treo 'tree'. For a full discussion of the name see Horovitz et al 2003. Town Field (The Town Field 1836 O.S.) on the south side of Ingestre (SJ 977244) may mark the site of the original village. For a brine spring at Ingestre see Marsh, The.

INGHAMTHORPES MOOR (unlocated, in Rugeley). Inghamthorpes Moor 1712x1768 SRO D615/D/81. Possibly from a family name taken from Inghamthorpe in Kirk Deighton, Yorkshire East Riding.

**INGLE HILL** between Freeford Manor and Swinfen Hall (SK 1036). *Ingle Hill* 1637 SRO DW1738/C/6/4. Possibly from OE *\*ing* 'hill', with OE *hyll* 'hill' (though the existence of OE *\*ing* is not certain), in which case the interesting 'hill-hill-hill' (cf. Ingleton, West Yorkshire), but the Inge family are recorded in this part of the county (see Shaw 1798: I 406, 409), and the name may be 'Inge hill'.

**IPSTONES** 4 miles north-west of Cheadle (SK 0249). *Ypestans* (p) c.1175 Okeover, *Yppestan* 1175 SHC I 73, *Ipestanis* 1201 Cur, *Ipestains* (p) 1201 Cur, *Hipestan*, *Hippestan*, *Hipstanes* 1204 SHC III (i) 105-6, *Ipestane* 1205 Ass, *Ipestanes* 1206 Cur, *Ipstone* 1220 Ass, *Ibestane* 1227 ibid, *Yppestanes* c.1240 SHC 4th Series IV 117, *Ippestanes* 1244 FF, *Iweston*, *Ipestane*, *Ipestanes* 1244 SHC IV 102, *Hypestanis*, *Uppestan* 1261 SHC 1911 426-7, *Ippiston* 1284 SHC I 206, *Hippestans* 1291 Blymhill, *Ipstones* 1310 Banco *et freq*, *Ypstonus* 1395 SHC 1939 14, *Ippistones* 1347 SHC XII 54, *Ypstons* 1532 SHC 1912 87, *Ibston* 1532 SHC 4th Series 8 36, *Ibstone* 1655 Leek ParReg. A derivation from an OE personal name *\*Ippa* with OE *stan* 'stone', so '\*Ippa's stone', is quite possible here, but the forms and topography point towards a derivation from OE *yppe*, *yppa* a derivative of *upp* 'up', meaning 'a raised place, a platform', perhaps also in place-names 'a look-out platform; a hunting dais' (often associated with royal Forests), or 'an upper place, a hill' (although Ekwall 1960: 266 noted the absence of spellings with *u*, and felt a derivation from the personal name was more likely), with OE *stan* 'stone', perhaps here with the meaning 'look-out place at the stone or outcrop': there are a number of rock outcrops hereabouts, including Hopestone 1 mile to the east, and standing stones said to be known as the Sun Stone (q.v.) on Ipstones Edge (StEnc 311), one mile to the north. Cf. Ibstone, 9 miles west of High Wycombe, *Ypestan* in 1086 (DB). See also Turner's Knipe; Sharpcliffe.

**ISEWALL** – see **EYESWALL**.

**ISLE FARM** 1 mile east of Heaton (SJ 9662). *Isle* 1842 O.S.

**IVERLEY** 3 miles south-east of Kinver (SO 8781), on the Staffordshire-Worcestershire border. *Iverley* 1293 VCX XX 126, *Iverley* 1603 SHC 1940 65, *Iverley (House Farm)* 1895 O.S. Probably from OE *yfer* 'ledge, steep slope', with OE *leah*. One of the Hays of Kinver Forest. Cf. Iver, Buckinghamshire.

**IVETSEY (BANK)** on Watling Street, 4 miles east of Brewood (SJ 8310). *Uvetshay*, *Oveyhotes haye*, *Ovetts hay*, *Uvetshay* 13th century Duig, *Quyotesha* 1326 SHC 1911 370, *Overzateshaye* 1412 SHC 1910 311, *Ovyhetteshay*, *Ovioteshay* 14th century Duig, *Evettes Hayes* 1563 SHC XIII 225, *Ivittsay*, *Ivettshay* 17th century Duig, *Ivetsea Bank* 1767 (1801) Shaw II xvii. The terminal *hay* is from Mercian OE *(ge)heg* 'a fence, an enclosure', probably in this case a forest enclosure: the place was formerly in the north-west corner of Brewood Forest (VCH II 336). The first element is the OE personal name Ufegeat, a late development of the OE name Wulfgeat, which had by DB become Ulu(u)iet, Ulviet and by 1204 Oviet (see Oakden 1984: 169; SHC III (i) 110). It is possible that the place is to be associated with Wulfgeat, a thegn recorded in 963 who held an estate at Upper Arley that descended to his kinswoman Wulfrun who used it to endow the foundation of Wolverhampton (see Whitelock 1930: 54-7, 163-7; Hart 1975: 366), or perhaps with Wulfgeat whose will dated perhaps c.1006 mentions Wolverhampton, Donington, Kilsall, Penkridge, and Tong, all places not far distant from Ivetsey (see Whitelock 1930: 54-7, 163-7), but the name Wulfgeat is by no means uncommon: ibid. 164. The Bank is a high point on Watling Street, which was lowered during road improvements c.1985. The Bradford Arms Hotel marks the height of the former road.

IVINDON (unlocated, possibly south-west of Wootton, near Eccleshall) *Ivindon* 1298 Spufford 2000: 295.

**JACKFIELD** near Smallthorne (SJ 8750). *Jacparok* 1408 VCH VIII 119 note 54, *Jacks field*, *Jacksfields* 1696 Burslem ParReg. 'Jack's field'. *Paroc* is from ME *pearroc* 'paddock'.

**JACK HAYES** 1 mile south-west of Bagnall (SJ 9249). *Jack Hay* 1816 SRO DW1909/E/9/1, *Jack Hays* 1836 O.S.

**JAMAGE** (obsolete, 1 mile south of Talke (SJ 8251)) *? Gamaches* 1194 Pipe, *? Gamages* c.1198 SHC II 74, 1203 SHC III 95, 1212 SHC 1911 385, *Gamages temp.* Henry III ibid. 148, *Gemetts* 1479 SHC 1944 57, *Gammots* 1720 Bowen, *Jamitch* 1733 SHC 1944 57, *Gamitch* 1733 ibid.' 72, *Jammage* 1777 SRO D4842/15/2/10, *Gem Edge* 1799 Faden, *Jamage* 1833 O.S. A curious name, possibly transferred from Gamaches-en-Vexin in Eure, Normandy, recorded as an Anglo-Norman family-name: see for example Matthew de Gamages 1198-1214 who held land in Staffordshire: SHC II 74; Eyton 1854-60: IX 38. It may be noted that this place lies some three miles from Domvilles (q.v.). Cf. Mansell Gamage, Herefordshire.

**JAMES BRIDGE** 1½ miles south-west of Walsall (SO 9897). *James Bridge* 1541 SRO D1810/f.49, 1576 Homeshaw 1955: 50, 1617 Willmore 1887: 439, *James Bridge (Green)* 1669 SRO DW1921/3, *James bridge* 1775 Yates, 1834 O.S. A bridge on the Walsall-Darlaston road over the Sneyd (or Bentley) Brook recorded in the 1330s (VCH XVII 167-8), although the age of the name is uncertain, as is the identity of the man after whom it was apparently named.

JARCUMVILE (unlocated) *Jarcumvile* 1276 SHC VI (i) 77, 1289 ibid.' 185. Possibly to be associated with *Scarpenhull*, recorded in the Whitgreave area in 1252 (SHC VIII 200), and *Jarpunnull*, recorded in 1307 (SRO DW1733/A/3/25).

**JEFFREYMEADOW** 1 mile south of Alton (SK 0740). *Jeffry Meadow* 1748 D240/D/96, *Jeffery Meadow* 1775 Yates, *Jeffery Meadow* 1836 O.S. Evidently from a personal name. Hugh Jafres is recorded in this part of the county in 1342: SHC 1913 92.

**JEFFRYNS HAYS** 1 mile east of Balterley (SJ 7749). *Jeffernes Feild* 1602 AD, *Jeffreus heys* 1663 Audley ParReg, *Jeffrons heyes* 1664 ibid, *Jeffronsheyes* 1667 ibid, *Jeffrons Hays*, *Gefrons Hase* 1733 SHC 1944 7, 37, *Jeffrons Hays Farm* c.1736 SRO D1788/A5/i, *Jefferen Hayes* 1833 O.S. From a descriptive surname such as Jevon, Jeavons, Jevons (from OFr *jovene*, Latin *juvenis* 'young'), with Mercian OE *(ge)heg* 'enclosure'.

JOANNE BRIDGE (obsolete) where Charlemont Road crosses the river Tame in West Bromwich (SO 0293). *Jone Bridge* 1526 Dilworth 1967: 53, *Joanne Bridge* 1577 ibid. 53, 1804 EnclA., *Joane Bridge* 1684 SRO D564/3/1/23. The name is associated with Jone or Joanne or Joan Mill which lay in Wigmore Lane, West Bromwich (SO 0293), *Jone Miln* 1526 Dilworth 1967: 53, *Jane Mill* 1602 ibid. 54. The derivation of the name is unexplained, though a personal name Joan or Joanes (DES 255) is likely.

**JOHNSON HALL** ½ mile south-west of Eccleshall (SJ 8228). *Johanneston* c.1233 Rees 1997: 82, *Jonestan* 1228 SHC 1924 167, *Joneston* 1314 SRO D938/39, *Jonestone* 1327 SHC VII (i) 212, 1359 SHC VIII 102, *Joneston* 1347 SHC 1913 122, *Jonson Hall* 1601 Eccleshall ParReg, *Johnson Hall* 1834 O.S. Evidently 'Johannes (John's) *tun* ', with the later addition of *hall*. See also SHC II (ii) 92 fn.

**JOHNSON'S WOOD (FARM)** 1½ miles south-west of Almington (SJ 7132). From Thomas Johnson who died in 1635 in a house he had built in Johnson's Wood: SHC 1945-6 183. The place was also called *Hawkhurst*. SHC 1945-6 24.

**JOLIFFE'S BANK** (unlocated, near Bridge End on the west side of Leek (? SJ 9757)) *Joliffe's Banks* c.1800 SRO D3272/1/4/3/17-20. *Jollyfe* is from the Jolliffe family: see Erdeswick 1844: 249; Sleigh 1883: 33, 35; Jolliffe 1892; see also Hillswood.

**JOL POOL, JOLPOOL BROOK** ½ mile north of Sandon (SJ 9431). *Jolpool (Bank)* 1844 TA. A curious name. Early spellings have not been traced, but if ancient, perhaps from the personal name Jol, Johel, Joel, Juel, of French (possibly Breton) origin: DES 256. The surname Joly is recorded in 1465 (SHC NS IV 135), and Joll, Jolle is recorded in Lichfield 1552 (SHC 1912 194-6), perhaps to be asociated with Yolls Lane (1575), Joles Lane (1599), Joyles Lane (1610), the earlier name of George Street: VCH XIV 42. Or possibly from ME *iuel, iuwele, iuell*, from OFr *joel* 'treasure, jewel, gem', perhaps 'the jewel-like stream', or possibly in a literal sense: eight Roman coins were discovered in the bank of the stream in 1979: SMR 01810. The line of a supposed Roman road from Blythe Bridge to Stafford runs ½ mile to the west. It has been suggested that the name of the pool may have been Jelpel, or was at least pronounced as such: JNSFC XLII 1907-8 143.

**JONES'S WOOD** 1 mile north-east of Swinscoe (SK 1448). *Jones's Wood* 1885 O.S. Perhaps to be associated with *Janneyes Barn*, recorded in 1702: Okeover E5089.

**KEELE** 3 miles west of Newcastle-under-Lyme (SJ 8045). *Keel* 1156 SHC XII NS 53, *Kiel* 1169 ff, 1185 SHC I 119, 1203 SHC III 114, *Kyel* 1173 SHC I 67, 1230 P, *Kell* 1199 SHC III 42, *Kele* 1199 ibid. 57, 1511 HLS, *Kel* 1211 FF, 1277 SHC VI (i) 91, 1282 ibid.' 121, *Kell* 1250 SHC XI 318, *Kyll* 1286 SHC VI (i) 167, *Kekle* (*sic*) 1532 SHC 4th Series 8 52, *Keel* 1686 Plot, 1833 O.S. Probably from ME *kye*, the Northern plural of OE *cu* 'cow' (Southern ME plural *kyn(e)*), with OE *hyll* 'hill', hence 'cows' hill', as proposed by Ekwall 1960: 269, though other examples of names incorporating *kye* are much further north (Kiddal, West Yorkshire; Kyroe, Northumberland; Kyo, Durham, and cf. Coole in Cheshire with *C-* spellings), and a derivation from ON *kjolr* 'a keel', in the sense of 'a ridge with the appearance of an upturned boat, or which separates the waters' (the English word 'keel' derives from the Scandinavian word) cannot be ruled out completely, since there is a ridge on the hill here (clearly indicated by hachuring on the first edition 1" O.S. map of 1833; see also Gelling 1981: 11, where the hill is described as having 'a very narrow summit') which forms the Severn/Trent watershed. Supporting evidence may be detected in the nearby names Kettlemoor (q.v.), which might incorporate an ON personal name, and Nabbs (q.v.). It has been suggested (see for example StSt 8 1996 2) that the Anglo-Saxons would not have realised that Keele lay on a watershed, but sections of the pre-Conquest Staffordshire-Shropshire boundary on the south-west of the county follows the watershed, probably quite deliberately: see TSAS 4th Series VI 1916-7 123-6. For completeness, it may be noted that Welsh *cel* means 'hiding, cover, concealment' (perhaps in place-names meaning 'shelter': Padel 1985: 46), but also 'keel', and Gildas records in the late 5th or early 6th century that the leaders of the Saxones had come across the sea in three *cylis* (keels), a word said to be of Germanic origin: Yorke 2001: 14. Cf. Keelby, Lincolnshire.

**KEELING FORD** (obsolete, a ford across the river Dove, ½ mile north-west of Marchington (SK 1331)) *Kylford* 1323 SHC 1911 355, 1327 SHC VII (i) 223, *Keeling Ford* 1836 O.S. The derivation is uncertain, but the surname Keeling is found, e.g. John

Kelyng recorded in 1301 (SHC 4th series XVIII 195); John Keeling, recorded in 1600 (SHC 1935 293).

**KELMESTOWE** (obsolete, in Clent) *Kelmstowe* 1327 SHC VII (i) 253, *Kenelmestowe* 1372 BCA MS3279/351247, *Kelmystowe* 1462 SHC IV NS 124. Shaw 1801: II 242 states: '... there was anciently a certain district adjoining to, if not surrounding St Kenelm's chapel, called Kelmestowe ...'. Kenelm (properly Cynehelm) was the son of Coenwulf of Mercia, who is said to have become king while still a child on the death of his father in 821, and by tradition was murdered in Clent Forest at the instigation of his jealous sister Cwoenthryth or Cwenthryth. Although the body at first lay hidden, its whereabouts were miraculously revealed to the pope as he celebrated mass in Rome. Once exhumed, Kenelm's remains were transferred to a shrine at Winchcombe, Gloucestershire, where they became a focus for miracles, and by the later 10th century or earlier he was venerated as a saint. The story is first recorded c.1150, but the origins of the cult are unclear. A real Mercian ætheling named Cynehelm is known to have existed, but little is known of him, and he is not recorded after 812 AD: Williams *et al* 1991: 98; see also Love 1966: lxxxix-cxxxix, 50-89; Ashley 1998: 260, 801; Preest 2002: 199, 201. The name incorporates OE *stow* 'a (holy) place, site of periodic assembly, typically associated with a saint'. See also Spelstowe.

**KELSON** (obsolete) 1 mile south-east of Dilhorne (SJ 975424). *Kelson* 1775 Yates, 1836 O.S.

**KEMLOW, THE** (unlocated, in Blurton) *the Kemlow* 1787 SRO D4842/11/1/140-141. The second element would appear to be OE *hlaw* 'mound, tumulus', usually with the latter meaning in Staffordshire.

**KEMPSAGE (FARM & LANE)** 1½ miles north-east of High Offley (SJ 7927). Early spellings have not been traced, but perhaps from the surname Kemp(e): John Kempe of Levedale is recorded in 1377 (SHC 4th Series VI 15). Or possibly associated with Kemsey (q.v.). The second element may be OE *ac* 'an oak-tree'.

**KEMSEY (MANOR)** 2 miles south-west of High Offley (SJ 7624). *Kemeseye* 1278 SHC IV 283, 1316 SHC XII (ii) 57, 1324 SHC X 55, *Kemesey* 1314 SHC VI NS (ii) 79 fn.1, *Kemeseie, Chemesey* 1326 SHC X 62, *Kemesheye* 1351 SHC 1913 143, *Kemmesey* 1408 SHC XVI 64, *Kemsey* 1569 SRO 279/38, 1614 SHC 1940 30, 1616 SHC VI NS (i) 10, 1679 SHC 1919 235, 1833 O.S. The first element is uncertain, but may be associated with the root of Kempsage (Farm and Lane) (q.v.). The second element is perhaps OE *eg* 'raised land in marshland, land by a stream': the place lies on a stream.

**KENDAR WOOD** to the north of Okeover (SK 1548). *Kendar* 1799 Okeover T31. Possibly associated with (pasture called) *Kendall* recorded in 1538 (SHC VII NS 63) and 1547 (Okeover F18). An interesting name. The place lies on a small tributary of the river Dove. Although early spellings for this name have not been traced, Kendall in Westmorland is held to be from 'the *dæl* or valley of the river Kent', a river-name identical with Kennet, probably of British origin, but it may also be noted that the name of the county of Kent may derive from Celtic *canto-*, Welsh *cant* 'rim, border; border land' (though there is no agreement between philologists on this point), and this place lies against a stream running into the river Dove, which forms here the Staffordshire-Derbyshire boundary.

**KENT HILL (FARM)** ½ mile west of Audley (SJ 7950). *the pasture called the Kenthill* 1612 SHC 1944 83.

KENTSHAY (unlocated, in Keele) *Kentshay* 1410 HOK 21.

KERNSLEY (unlocated, possibly near Kingsley) *Kernsley* 1711 DRO D3155/WH472.

**KERRY HILL** an 810' hill 1 mile south of Bagnall (SJ 9249). *? Kery* 1194 Pipe, *? Kyryaule* 1310 SHC IX (i) 21, *? Kyry* 1331 SHC 1913 27, 1343 ibid.' 103, *Kerealhull* 1434 SHC 1933 (ii) 36, *? Kyvyell-hill* 1454 SHC XII NS 73, SHC VII 251, *Kerealhyll* 1537 Ct, *Kyrrehill* 1599 SHC XVI 196, *Kerry Hill* 1803 SHC 1933 149, 1836 O.S. There is a possibility that the name has as its root a British river-name Kyre (see Kiddemore Green), which may have had a Celtic base *\*kour-* or similar, connected with Welsh *ceu-* 'hollow' (cf. Welsh *ceunant*, etc.) – the place lies on the north side of Wetley Moor, from which drain a number of streams – but a more likely drivation is the surname Kerrial, found in Croxton Kerrial, Leicestershire, granted to Betram de Cryoil in 1242, and Nicholas de Kyriel in 1247 (Ekwall 1960: 134). The surname is from Criel in Seine-Inferiere. See also Kyrywilyhaile.

KETELBERNESTONA (unlocated) *Ketelbernestona* 1208 SHC II 148. A significant name of the type known to philologists as Grimston-hybrid, or Toton-hybrid, where an ON personal name is combined with OE *tun*. The meaning is probably 'the *tun* of Ketilbiôrn', or 'the *tun* of the Kettle Brook or Burn' (Ketelbearne, glossed as Chetelbert, is recorded in 1072 as a witness to a grant of Wrottesley to Evesham abbey, and was brother of Turkil, who may have held land at Syrescote in 1086 (VCH IV 52), and son of Alwin, Sheriff of Warwickshire who died before 1086, who both also witnessed the grant: SHC II 179.) It is very possible that the place lay outside Staffordshire, and may indeed be *Ketelburstone*, recorded in Suffolk (SHC XVI 229).

KETELSHUL (unlocated, in or near Chell) *Ketelshul* 1272-90 SHC 1911 446. Probably from the ON personal name Ketill, so 'Ketill's hill'. See also Kettle Hill.

**KETTLEBROOK** on the south-east side of Tamworth (SK 2103). *Ketelbroke* 1436 Ct, *Kettle Brook* 1770 EnclA., *Kettlebrook* (*Colliery*) 1834 O.S., *Kettel* 1845 Map. The place takes its name from Kettle Brook, a tributary of the river Tame, perhaps derived from ON *ketill* (or the Scandinavianised form of OE *cetel*) 'a kettle', but when found in place-names connected with water (here OE *broc* 'brook', which runs though a wide curving valley to join the river Tame) perhaps meaning 'bubbling', so 'the bubbling brook'. The place was on the Staffordshire-Warwickshire border until transferred into Staffordshire in 1965.

KETTLE HILL (unlocated, in Knutton) *Kettle Hill* 1862 SRO D3272/3/6/21. Perhaps associated with Kettlemoor (q.v.). See also Ketelshul.

KETTLEMOOR (obsolete) 1 mile north-west of Keele (SJ 8046). *Kettlemoor* 1833 O.S.; *Kettlesmoor* 1836 TNSFC 1963 57. Early spellings have not been traced (but cf. *Big, Far & Near Kettle Hill (fields)*, recorded in 1862 in Knutton: SRO D3272/3/6/21), but if ancient perhaps from ON *ketill* (or a Scandinavianised form of OE *cetel*) 'a kettle', found in place-names with the meaning 'a deep valley surrounded by hills' (which would be appropriate for this place). However, the 1836 spelling points towards a personal name, and Adam Ketel is recorded in Keele in 1327 and 1332 (SHC VII (i) 199; SHC X 83), and John Ketel in 1338 (HOK 21 fn.47). Perhaps therefore from the ON personal name Ketill, in which case perhaps supporting evidence for a Norse derivation for the name Keele (q.v.); see also Nabbs; Ketelshul; Kettle Hill. There is a Kettlemoor Lane in Sheriffhales, but the history of the name has not been traced.

KEVERMUNT, KEVERMONT – see **CARMOUNTHEAD**.

**KEYWELL** (obsolete) 2 miles north-west of Butterton (SK 0458). *Kewall Green* 1677 Leek ParReg, *Keywell Green* 1842 O.S. The first element is uncertain: a derivation from OE *cȳ* 'cows' is possible, but see discussion of this element under the entry for Keele. The second element is Mercian OE *wælle* 'a spring', and (sometimes) 'a stream'. Green indicates a grassy area in woodland, heathland or moorland, or sometimes a squatter settlement. The place is now called Lower Green Farm.

**KIBBLESTONE** 2 miles north of stone (SJ 9136; SMR 02624 places a deserted post-Conquest settlement at SJ 91403630). *Cublesdon* ?12th century SHC VI (i) 22, 1271 ibid.' 49, 1293 ibid. 276, 1380 SHC XIII 154, *Cubblesd'* c.1224 Rees 1997: 65, *Cublestone* 1288 SHC XIII 174, *Cubblesdon, Cubeleston* 1329 SHC XI 4, *Cubbeleston'* 1333 SHC X 92, *Kybleston* 1383 SHC XIII 203, *Kebiston* 1476 SHC VI NS (i) 104, *Kebylstone* 1478 ibid.' 115, *Kybbulston* 1567 SHC XIII 264, *Kebullston* 1586 SHC 1927 166, *Kibleston* 1643 SHC 4th Series I 27, *Kibbleston* 1666 SHC 1921 80, *Kibblestone (Hall)* 1922 O.S. Erdeswick 1844: 30 states that 'Cubleston is a goodly large manor, containing these hamlets following: viz. Mayford, Oldinton (vulg. Olton), Berryhill, Cotwaldeston, Mathershall, the Spot-Grange, Snelhall, and Woodhouses'. The same source (p.31) also suggests that the place was 'called also Culmsdon'. *Cobintone* 1086 (DB) has been variously identified with this place and Cubbington in Warwickshire (VCH IV 46), but the latter seems more likely: see SHC 1916 169; see also Cobintone. The name Kibblestone (which does not seem to appear on any early map) is evidently from the OE personal name *Cybbel, so '*Cybbel's *dun*' or *Cybbel's *tun*': the second element varies in the earliest spellings. Cf. Kibblesworth, Durham. See also Cuttesdon.

**KIDDEMORE GREEN** 1½ miles west of Brewood (SJ 8508). *Kudimor, Kudymor* 1308 Ipm, *Kyrremore* 1383 Ct, *Kyr(r)ymore* 1387 ibid, *Kerrymore* 1723 ibid, *Kerrimore-greene* 1657 WCRO CR1291/190, *Kadimore* 1659 Deed (NCBrewood), *The Kerrimores* 1661 Lease (NCBrewood), *Kerrimore (lane)* 1681 Will, *Kiddimoore green* 1686 Plot. No definite derivation can be given for this name, for which there appear to have been alternative first elements, but it is unlikely that the name is connected with ON *kjarr* (ME *ker*) 'a bog, a marsh, particularly one overgrown with brushwood', as suggested in Oakden 1984: 38. The root may be a British river-name Kyre, related to Cury, Curry, Cory (see Kerry Hill). There is a small stream in a shallow valley here. Cf. Kyre, Kyrewood, Kyre Brook, Little Kyre, Kyre Magna and Kyre Wyard, all in Worcestershire (PN Wo 55-6); Kyrebach, Herefordshire, and see Ekwall 1928 97-8, 233; also the river Piddle in Dorset, which may have had the British name Car(e)y: Watts 1979: 131 fn.12. The present spelling of Kiddemore (which, curiously, reflect the earliest spellings, which would otherwise appear to be aberrant) has evidently been influenced by Kidderminster, some 20 miles to the south. The second element is OE *mor*, meaning in the Midlands 'marshland', hence 'marshland of the stream called Kyre'. The *Green* element suggests a squatter settlement: the place lay deep in Brewood Forest, a name of part-British origin. *Kyry*, recorded in 1344 (SHC 1913 103; also Richard Kyry 1331 ibid. 27, and the surname Kyrre, associated in 1467 with Pillaton Hall: SHC 1928 155) may be connected with this place. Cf. Kerry Lane, on the west side of Eccleshall (SJ 8228). See also Kerry Hill; Willenhall.

**KIDDLESTICH** (obsolete) 1 mile north-west of Uttoxeter (SK 0733). *Kidlesick* 1658 Redfern 1865: map, *Kiddlestich* 1658 Redfern 1886: 89, *Kiddlestich* c.1752 SRO D4156/1-35, 1834 O.S., *Kiln Sich* 1775 Yates, *Kidlestick* 1886 Redfern 1886: 353. A curious name of uncertain age. If ancient, the first element might be from *kiddle*, 'a wicker fish trap; a weir or barrier in a river with an opening for catching fish; a person in charge of a fishing-weir', with OE *stycc* 'a piece of land', or OE *sic* 'a watercourse'. Since the place lay on high ground but within half a mile of the river Tean, perhaps 'the piece of land occupied by the keeper of the fish weir'. The curious 1775 spelling offers an alternative derivation 'the *sic* or stream with the kiln', but is almost certainly an aberration. EDD records kiddle as meaning 'to dribble or slaver' in the counties around the Welsh border, and this may simply be 'the dribbling *sic* or watercourse'. A gibbet is said to have stood at Kiddlestich, and it is not impossible that Redfern's spelling might reflect a colloquial expression for such a structure.

**KIDSGROVE** 6 miles north-west of Stoke on Trent (SJ 8354). *Kydcrowe* c.1596 SHC 1932 324, *Kidcrow, Kidcrowe* 1656 Wolstanton ParReg, *Kiderew* 1680 SHC 1919 264, *Kidcrow* 1686 Plot, *Kidcrew* 1695 Morden, c.1733 SHC 1944 42, 62, *Kidcrew* 1747 Bowen, *Killegrew* 1763 SHC 1934 (i) 67, *Kid Crew* 1775 Yates, *Kidsgrove* 1807 SHC 4th III 12, *Kidcrew* 1832 Teesdale, *Kidsgrove* 1836 O.S. The various versions of this name – the place has developed rapidly in recent centuries, particularly after it became the base for workers on the Harecastle Tunnel built in 1766-77 – show that the root is an obsolete dialect word *crew(e)*, also found as *creuh, crow, crough, crue*, apparently of British origin (cf. PrWelsh *\*crou*, Modern Welsh *crau*, Cornish *crow*, Breton *kraou*, Irish *cro*), all with the meaning 'pen, sty, hut, hovel'. In this case the name is from *kidcrow, kidcrew*, used in the Cheshire area for a calf-crib (Halliwell). The meaning is 'the place of the stall or fold of the calves'. See also EDD, and Jackson 1879, where *crew* is explained as 'a pen for ducks and geese', also citing Bailey 1782: 104: 'Swine-crue – a swine-sty or hog-sty. An old word'. Evidently during the later eighteenth or early nineteenth century *grove* was felt to be a more refined element than the earlier *crew*. See also Wakelin 1969: 273-81.

KIDSHOUSE (unlocated, possibly near Wolstanton) *Kydhowses* 1565 SHC XIII 240, *Kidshouse* 1647 Wolstanton ParReg. Possibly the same meaning as Kidsgrove (q.v.).

KILBY (obsolete, at Blithbury Bank, ½ mile north of Blithbury (SK 0820)) *Kylby* 1379 SHC XVI 173, *Kilbynshall* 1307 (1798) Shaw I 201, *Kilby-hall* 1332, 1407 ibid, *Kileby hall* 1395 ibid, *Kilbelondes* 1411 ibid, *Kylbye* 1459 Parker 1897: 366, *Kibbihall* c.1598 Erdeswick 1844: 246. '... though the name of Kilby-hall is lost, several of the neighbouring enclosures still (in 1797) retain the name of Kilby or, corruptedly, Gilby fields': Shaw 1798: I 201. From Sir William de Kileby (of Kilby, Leicestershire), who married the daughter of Sir William de Malveysin (of Mavesyn Ridware) at some date between 1100 and 1135: ibid. See also JNSFC 4 1964 366; StEnc 319.

KIMET (unlocated, possibly near Trysull) *Kimet* late 14th/early 15th century (1801) Shaw II 208.

KINCHALE (unlocated, possibly near Pipe Grange Farm, 1½ miles south-west of Lichfield (SK 0908)) *Kynchehal* 1325 SHC 1939 93, BCA MS3415/158, *Kinchale, Kinchalen* 1367 SHC 1939 99, *Kynchale* 1401 ibid. 83. The Kynchall family held property here in 1299: VCH XIV 201. It has been suggested that Broad Lane near Pipehill may be *Kynchall Lane*, recorded in 1412: VCH XIV 202. *Kynchall moor* in the Edial area is recorded in the early 16th century (VCH XIV 212), presumably *Kynchalmor* recorded in 1410 (BCA MS3415/213).

KINGESMERE (unlocated, between Broughton, Bobbington and Gatacre: Eyton 1854-60: III 101) *Kingesmere* 1292 Eyton 1854-60: III 101. 'The king's mere or pool'.

**KING'S BROMLEY** – see **BROMLEY, KING'S.**

**KING'S CLUMP** a smoothly rounded tumulus some 80 yards in diameter on the south-west side of Swynnerton (SJ 846352); see WMANS 1976 11. The age of the name has not been ascertained.

**KING'S BRIDGE** (unlocated, on the river Trent) *King's bridge* 1510 Dent & Hill 1896: 115.

**KINGS BRIDGE** (unlocated, in Moseley near Wolverhampton) *Kings Bridge* 1693/4 SRO D118/10/4. Perhaps associated with the escape of Charles II after the battle of Worcester in September 1651: the king took refuge in Moseley Old Hall during his escape. The bridge over the river Penk on the west side of Coven, known as *Jackson's Bridge*, was formerly *King's Bridge* (Horovitz 1992: 260), perhaps for the same reason. The identity of Jackson is unknown, but it may be noted that Charles II adopted the alias William Jackson during his escape: ibid. 131.

**KINGS HILL** on the north side of Wednesbury (SO 9895), *Kyngeshulleslone, Kyngeshull[forlong]* 1315 SHC 1911 322, Ede 1962: 31, *Kingshill (Field)* 1684 BCA MS3145/91/2, *Kingshill, Kingshill field* 17th century (1801) Shaw II 88; KINGSHILL (obsolete) ½ mile south-west of Sedgley (SO 9195). *Kingshill* 1834 O.S. 'The king's hill'. Of the place near Wednesbury, Ede 1962: 108 fn.43 observes: 'No certain explanation of this name can be given; a derivation from a Saxon king or battle must be rejected. Possibly it was manorial demesne land when the King held the manor or it may indicate a part of Wednesbury, a royal manor, as opposed to Darlaston'.

**KINGSLEY** (obsolete) a manor in Tettenhall parish, on the summit of Tettenhall Wood (Shaw 1801: II 200), 1 mile south-west of Tettenhall (SO 8798). *Kyngeleye* 1286 SHC V (i) 168, *Kyngesleye* 1300 ibid.' 180, 1346 SRO D593/A/2/16/4, *Kingsley otherwise Kinfare* 17th century VCH XX 21-2. 'The King's *leah* or clearing'. The place was at one time held by the king: SHC V (i) 168; see also Tettenhall (Regis). Kingsley Wood (which became Tettenhall Wood) was a detached portion of Kinver Forest: VCH XX 29, hence *Kinfare*: see SHC VI (i) 258; Jones 1894: 268.

**KINGSLEY** Ancient Parish 2 miles north of Cheadle (SK 0146). *Chingeslei, Chingesleia* 1086 DB, *Kingeslegh* 1227 Ass, *Kingeslee* 1232 Eyton 1854-60: X 71, *Kingesle* 1248 SHC IV 110, *Kynchesley* 1385 SHC XIII 191, *Kingsley* 1686 Plot. 'The King's *leah* or clearing'. Kingsley Moor, 1½ miles west of Kingsley (SJ 9946) is recorded as *Kingley Moor* in 1628: Okeover T752. See also Kingstone.

**KINGSLEY HOLT** 1 mile south-east of Kingsley (SK 0246). *? Hout* 1247 Cl, *? Holt* 1359 SHC XII 164, *the Hoult* 1594 Kingsley ParReg, *Holt* 1775 Yates. From OE *holt* 'wood'. The identification of the earlier spelling with this place is uncertain – Holt Hill, 1 mile north-west of Newborough (SK 1226) is another possibility

**KINGSLOW** 2 miles west of Pattingham, on a pronounced hill (SO 7998). *Kyngeslow* 1283 Eyton 1854-60: III 113, *Kynggeslowe* 1327 SR, *Kyngeslowe* 1376 SHC XIV 141, *Kyngelowe* 1542 SRO D593/A/2/11/9, *Kingslowe* 1562 Worfield ParReg, *Kingslow* 1602 SA 2028/1/5/8, *Kingslow* 1833 O.S. From OE *hlaw* 'hill, burial mound', presumably from a tumulus on the hill, so 'the king's tumulus', or 'the tumulus associated with a family called King': Rog'o le Kyng or Kynge is recorded in Pattingham in 1327 (SHC VII (i) 249) and 1332 (SHC X (i) 131). Nearby to the north is Stanlow (q.v.). In Shropshire since the 12th century.

**KING'S LOW** a tumulus near Tixall (SJ 9523); see WMA 1987 30 38-9. *King's Low* 1844 Erdeswick 1844: 70. From OE *hlaw* 'hill, burial mound'. A stone cross from South

Wales is said to have been erected here in about 1803: Erdeswick 1844: 70. See also Guendelawe; Queen's Low.

**KINGSNORDLEY** 2½ miles south-west of Bobbington (SO 7787). *Nordlege* 1086 DB, *Norley* c.1086 Eyton 1854-60: I 109, *Nordley* c.1295 SA 2028/1/2/2, *Northleye* 1305 Eyton 1854-60: 153, *Nordeley Regs* 1525 SRS 3 109, *Nordley Regis* 1695 Morden. 'The north *leah*' (perhaps in relatioship to Alveley) with the later addition of Regis to show that it was at one time held by the king. Since the 12th century in Shropshire.

KINGS POOL (obsolete) on north-east of Stafford (SJ 9323). *Kyngespole* 1292 SHC VI (i) 250, *Kyngespol* 1350 SHC 1913 135, *Le Kyngespoole* 1419 SRO D938/376, *Kyngs pole* 1495 SRO (216/7902), *Kings poole (Medes)* 1610 Speed, *Kingston Pool* 1775 Yates. From the royal fish pool which existed here from at least 1157 until c.1600: VCH VI 210-11. See also SHC VIII (ii) 105. *Medes* is from OE *med* 'a meadow', very common in OE, ME and later field-names.

**KINGSTANDING** a mound, 3 miles south-east of Aldridge (SO 0895), possibly a tumulus, 15' in diameter and 3' high, said to have been destroyed in 1814 and later reconstructed, possibly nearby (see TSSAHS XXXII 1990-1 90; StEnc 321; Hodder 2004: 25), and marked by a circular clump of trees in 1818 (see Scott 1832: 312), *Kingstanding (Warren)* 1811 BCA 3145/58/1, *Kings Standing* 1834 O.S.; **KING'S STANDING** a ploughed-out mound (see StEnc 321) in Needwood Forest (SK 1624), *the King's standing* c.1580 (1798) Shaw I 64, *King's Standing* 1658 ParSurv, *Kingstanding* 1681 Edwards 1949: 159, *King's-Standing* 1798 Shaw I 66, *Kingstanding* 1836 O.S.; **KING'S SRANDING** 3 miles north of Lichfield (SK 1213), *King's Standing* 1887 O.S. By tradition Kingstanding near Aldridge was so-called after the mound was used as a platform by Charles I when reviewing troops on 16 October 1642, and although Needwood Forest was visited by James I in 1619, 1621 and 1624, and by Charles I in 1634 and 1636 (VCH II 356), and King's Standing is said to have been a favoutite resting place of James II (VCH II 356), 'the name of King's-Standing [in Needwood Forest] certainly existed early in Elizabeth's reign at least, and most probably had its origin from [a visit by] Henry VII': Shaw 1798: I 66. Both names derive from ME *stand* 'a hunter's station or stand from which game was shot'. OED quotes Digby MS 182 xxxv (c.1400): 'Thenne shulde the maister of the game…meete the kynge and brynge hym to his stondynge…'. The word developed to mean a kind of roofed grandstand, two or more stories high, with open sides through which the progress of the hunt could be observed. A standing lodge differed from a hunting lodge in that women and older men who were following the hunt would rest at a standing lodge, sometimes for a whole day, and the hunters would use it as a base between chases, while a hunting lodge would be used as overnight accomodation. One of the most famous standings is the misnamed Queen Elizabeth's Hunting Lodge, built by her father, Henry VIII, as a standing at Chingford, Essex. See also AJ lxxviii 33-3. The place near Aldridge (which was associated with Sutton Chase, a royal forest extending westwards almost to Walsall, until granted by Henry I to Roger, Earl of Warwick) has given its name to a district to the south. At King's Standing near Lichfield is a mound in an open area within Ravenshaw Wood, but the name does not appear on the 1834 O.S. map, and its history is untraced. Cf. King's Stand Farm, Nottinghamshire; King's Standing, Sussex.

**KINGSTONE** Ancient Parish 3 miles south-west of Uttoxeter (SK 0629). *? Cunegeston temp.* Henry III SHC I 223, *Kingeston* 1166 P, 1227 SHC IV 61, 1242-3 Fees, 1275 SHC V (i) 118, *Kingestan* 1199 SHC III (i) 48, *Keneston* 1403 SHC XV 108, *Kynston* 1532 SHC 4th Series 8 68, *Kyngstonne* 1602 SHC 1935 475, *Kingston* 1663 SHC II (ii) 41,

*Kingstone* 1953 O.S. From OE *cyninges-tun* 'the king's farm or manor'. The name Kingston(e) is found in at least 15 counties, and this place is the most northerly. The first spelling (which may not refer to this place) indicates Scandinavian influence from ON *kunung*. This place is not recorded in DB, but it has been suggested that the name implies an early royal interest: in the Anglo-Saxon period the king owned in every shire a *cyninges tun*, a key element in the organisation of justice and finance managed for the king by a reeve, where food rents and the income from cases heard in the Hundred courts were collected: see JEPNS 20 13-37. These legal structures may have been in place by the early 7th century: Hough 1997: 55-57. However, other research suggests that places of this name were probably subordinate berewicks, rather than royal vills: Faith 1997: 150-1. Youngs 1990: 415 gives Kingston as the usual civil parish parish spelling, with Kingstone as the ecclesiastical parish form, but the O.S. retains Kingstone. See also Kingsley.

**KINGSTONE HILL** 1 mile east of Stafford (SJ 9523), on the south side of the Stafford-Weston road. *Kinesdun* ?c.1209 SRO D938/228, *Kinesdonehul* c.1298 ibid 938/277, *Kinesdonehull* 1300 ibid. D938/279, *Kynesdone hul* 1317 ibid. 938/294. 'Cyne's *dun* or hill', to which the tautological 'hill' was added. The place is to be associated with Kingstone Brook (q.v.), which runs to the east of the hill southwards into the river Sow near its junction with the river Penk, and Kingstone Pool (*Cynespol* 1320 SRO D938/297; see also WMA 2001 44 147-8), a licence for the creation of which, with a mill, was given 1161x1182 (SRO D938/1). The pool is shown on the 1836 first edition 1" O.S. map, but appears to have become woodland before the end of the 19th century. Names associated with this area are *Kenesdonewey* 1292 SRO D938/271 and *Kinesdonefeld* c.1298 ibid. D938/277.

**KINGSTONE BROOK** 1 mile east of Stafford. *Kenesbroc, Kynesbroc* 1261 SHC 1914 120-1, *Kenesbroc* c.1350 SRO D938/10. (*Kinesbroc*, supposedly recorded c.1178 SHC VIII 172, 1161x1182 SRO D938/1, is from a 14th century forgery: EEA 16 90.) Evidently 'Cyne's brook': see also Kingstone Hill.

**KING STREET** – see CUTTLESTONE.

**KINGSWINFORD** Ancient Parish 4 miles west of Dudley (SO 8888). *Swinford* 951x955 (15th/16th century, S.579), *Suinesford* 1086 DB, *Sueneforda* 1130 SHC I 2, *New Swyneford Regis* c.1322 1911 353, *New Swyneford* 1325 Inq, *Kyngeswynford* 1394 SHC XV 62, *Kyngessynford* 1433x1450 Plmnt Pet C1/2/18(21), *Kynges Swenford* 1532 SHC 4th Series 8, *Swyndford Regis* 1563 SHC IX NS 34. 'The ford of the swine'. The ford occurs as s*wynford*, a boundary mark in the charter of 951x955 (15th/16th century, S.579), on the river Stour, 2½ miles to the south, and the name may have been applied to the north-south road which crossed the ford, which would explain the origin of Kinswinford and Oldswinford (q.v.). A bridge, mentioned in 1255, replaced the ford, and gave its name to Stourbridge. *Kings* is a medieval addition to distinguish Kingswinford from Oldswinford and other Swinfords: it was a royal manor from at least 1066.

**KINGSWOOD** 1 mile west of Oaken, in Codsall parish (SJ 8402), *Kingswood* 1403 VCH XX 84, 1686 Plot, *Kyngeswode* 1468 SHC IV NS 155, *Kyngeswoodde* 1599 Codsall ParReg; KING'S WOOD (obsolete) 3 miles south-east of Cannock (SJ 9908). *Kyngeswode* 1355 Ct, *Kynges Wood* c.1543 SHC NS X (i) 133, *the Kyngeswood* 1543 StarCh, *Kynges-, Kingeswodde* 1570 Rental, *Kings Wood* 1834 O.S.; **KING'S WOOD** in Trentham Park (SJ 8639), *Kings Wood* 1329 SRO D593/B/1/23/7/1/2, *Kings Wood* 1663 Trentham ParReg, *Kingswood* 1666 ibid, *Kingswood Bank* 1836 O.S.; KINGS WOOD (obsolete) in Hatherton, 2 miles south-east of Cannock (SJ 9908),

*Kyngeswode(yate)* 1354 Ct, *Kyngeswode(heth')* 1369 Ct, *King's wood* 1518 SHC I 343, *Kynges Wood* c.1542 SHC X NS (i) 133, *Kingeswoode(gate crofte)*1570 Survey, *the Kingswood* 1643 VCH V 56, *Kingswood (Heath)* 1690 SRO D260/M/T/5/99, *Kings Wood* 1775 Yates, 1834 O.S.; **KING'S WOOD, KINGSWOOD BANK** 1 mile south-west of Trentham (SJ 8639), *Kingswood Bank* 1836 O.S. Self-explanatory. Kingswood near Oaken is an area of heathland, formerly an area of waste in Brewood Forest, disafforested in 1204: VCH V 18 fn.2.

**KINVASTON** 1 mile south-west of Penkridge (SJ 9012; SMR 01917 places a deserted Anglo-Saxon settlement at SJ 90791230). *Kinwaldestun, Kineuoldeston* 996 (for 994) (17th century, S.1380), *Chenwardestone* 1086 DB, *Kinaldeston, Kineldeston* 1203 SHC III 118, 126, *Kyneswaldestan* 1227 Ass, *Keneston* 1403 SHC XV 108, *Kenaston* 1532 SHC 4th Series 8 90, *Kinnerston* 1686 Plot, *Kinvaston* 1798 Yates. 'Cynewald's *tun*'. A reference to *Stonewall alias Kinwaston* in 1553 (Mander & Tildesley 1960: 54) is unexplained, but a possible, if enigmatic, clue is found in Penkridge ParReg, which records a baptism in 1596 of '/        ] Wall of Kinvaston'.

**KINVER** Ancient Parish 4 miles west of Stourbridge (SO 8483). (the wood called) *Cynibre* 736 (17th century, S.89), *Cynefare(s-stane)* 964 (17th century, S.726), *Chenevare* 1086 DB, *Chenefara* 1130 SHC I 14, *Kinefara* 1177 SHC I 85, *Kenefara* 1183 ibid.'107, *Kynefare* 1262 SHC V (i) 154, 1300 ibid.' 179, *Kenefare* 1271 ibid.' 138, 1331 SHC XI 21, *Kyngfare* 1596 SHC XVI 158, *Kinfare* 1834 O.S. An ancient name that has not been satisfactorily explained. Ekwall 1936: 266 felt that the earliest spelling is an adaptation of a Welsh *Cynfre*, with the first element influenced by OE *cyne-* 'royal'. Jackson 1953: 647 suggests a possible derivation from British *\*Cunobriga*, adopted in the second half of the sixth century, developing into OE *Cynibre, Cynefare*. The element Cuno- is not uncommon in early place-names, but the precise meaning is unresolved (see Rivet & Smith 1979: 328-9), although Ekwall 1959: 37 suggests that the name may mean 'dog hill' (from Celtic *\*cuno* 'hound', well attested in Celtic names: it has been noted that the Celts admired their hunting dogs greatly, and *Cuno-* in their personal name carried similar inferences to the word *lion* today: the element was born by kings (such as *Maelgwn* 'princely hound') and saints (*Kentigern* 'hound-like lord'): see Coates & Breeze 2000: 126-8, and TSAHS LXXVI 2001 76-7, where Cound in Shropshire is held to drive from a British river-name *Cun-*. The meaning 'high', often cited, has been shown to be erroneous: see EPNE i 120. There may be a connection with the name Kinver and the probably Celtic name Kinder, the highest hill in the Peak District, held to be from *\*cönderc* 'place with wide views, look-out point': see Coates & Breeze 2000: 165-6. Whatever the first element, it seems to have been rationalised into the OE adjective *cyne-* 'royal', or OE *cyne, cene* 'bold, keen, fierce, warlike'. The second element is normally assumed (since at the end of the Roman period *b* was pronounced rather like *v*) to be the mutated form *fre* of the Welsh (and Cornish) *bre* 'hill', from British *\*briga*, Late British *\*brega* (Kinver lies at the foot of the 543' hill of Kinver Edge), which appears to have changed at an early date via *-ver* (for parallels cf. Dinnever, Cornwall, and Mellor, Lancashire and Derbyshire: see Padel 1985: 30, 163), to a form more readily intelligible to the Anglo-Saxons, OE *fare, fær* 'road, (difficult) passage', hence 'royal (? or public) highway' (cf. OE *cyne-stræt* 'public road'), although *fær*, which seems to have gone out of use as a place-forming term at a very early date, is found in only a handful of names in the east of England. Such development of the second element might have been influenced by the medieval route, known in 1300 as Chester Way, between the south-west of England and Chester, which ran along Kinver High Street and remained in use until c.1800 (VCH XX 94, 126; presumably *Chester roade* in Kinver recorded in 1679: SA 1045/145): early medieval kings are known to have visited Kinver. It is of interest, however, that Shaw (1801: II 262), amongst other possible derivations, considers 'Cyne, Chine, or Chene, royal, great, etc., and Fare, a road, as in thorough-fare ... a great or royal road ... so that the name might not have regard to the hill, but to the Roman road which lay across this forest and passed near the town ...'. (Two miles south-west of Kinver is Kingsford, *cenunga ford* in a charter of 964 AD (17th century, S.726), which is recorded as *Keniggeford* in 1262 (SHC V (i) 139), *Kynyngford* and *Kyngesford* in 1300 (SHC V (i) 180), said to be from 'Ford of the people of C(o)en or C(o)ena': PN Wo 259; see also Hooke 1990: 172. The first edition 1" O.S. map of 1833 shows two Kingsfords some half a mile apart to the west and south-west of Kinver Edge, one in Staffordshire, the other in Worcestershire. See also Kingsley.) The general use of the spelling *Kinver* is recent. The form *Kinfare* evolved from medieval usage, and remained the usual spelling until the 19th century. Saxton used the form on his maps (1577), and later cartographers until Browne (Plot 1686) used the same spelling. In the 18th century the alternative spelling *Kinver* appeared, and had become established by the 1840s, although the use of *Kinfare* lingered on for many years. The first edition 1" O.S. map uses Kinver and Kinfare on adjoining sheets in 1833 and 1834. The place gave its name to the vast Kinver Forest. It may also be noted that the name *Kinefolka*, 'royal folk', is recorded in a Worcestershire Survey c.1150: see PN Wo 18; Ekwall 1960: xiv. Above Kinver village lies an iron-age hillfort, recorded as *Bury* in 1293 (VCH XX 122), and *le Bury* in 1342 (SHC 1913 91) and 1456 (VCH XX 141), with which the place-name may in some way be connected. It has been suggested that *Cynefares-stane* ('The Stone of Kinver', possibly a boundary stone: see WMANS 18 1975 9; Hooke 1990: 172; StEnc 721), perhaps associated with John atte Stone, recorded in 1324 (SHC 1911 362), was at Start's Green on the south-west boundary of Kinver, or that it was Vale's Rock or a vanished hoarstone: VCH XX 119 (but see also Hooke 1990: 172, where a location at SO 824822 is suggested). See also Boltstone. *Novo burgo* ('new town') is recorded in Kinver in 1227: SHC IV 70.

**KINVER EDGE** the high ground to the west of Kinver (SO 8383). *? Egge* 1300 SHC SHC V (i) 180, *Kinver Edge* 1686 Plot 398, 1833 O.S.

**KINVER FOREST** so-named by 1168: VCH XX 118; VCH II 343. An area designated as Royal hunting ground that covered the south-west corner of the county and extended into Worcestershire. At its greatest extent it included Seisdon, Trysull, and part of Tettenhall, Lower Penn, Wombourne, parts of Himley, Kingswinford, Amblecote, Wollaston and Oldswinford, Pedmore, Hagley, Broom, Chaddesley Corbett, Churchill, Hurcott, Kidderminster, Wolverley, Upper Arley, Feckenham, Tardebigge, Enville, Morfe and Bobbington. After c.1327 it shrank to an area roughly between Smestow on the north and the county boundary on the south, with Tettenhall Wood a detached portion on the north. The three Hays were Ashwood, Chasepool and Iverley. See also Kingsley.

**KIPLASS (LANE & FARM)** 1½ miles south-east of Stone (SJ 9431). A curious name for which early spellings have not been traced and for which no suggested derivation can be offered.

KIPPAX (unlocated, possibly near Burton upon Trent) *Kippax* 1349 SHC XII 77. A place of this name is in West Yorkshire, and is perhaps from the OE personal name *Cyppa, with OE *æsc* 'ash-tree', partly Scandinavianised to -*ask*, and so producing -*ax*: Ekwall 1960: 279; CDEPN 349.

**KIRKSTEADS** 1½ miles north of Grindon (SK 0856). *Kirksteds* 1700 SRO D1132/1/17, *Great and Little Kirksteads* 1727 D1132/1/14, *Kirksteads* 1917 JNSFC 1916-7 81. Perhaps from a partly Scandinavianised form of OE *ciric-stede* 'site of a church', with OE *cirice* replaced by the cognate ON *kirkja*: cf. Kirkstead, Lincolnshire. *Kirk Flat*, recorded as a Grindon field-name in 1839 (TA) may be associated with this place. The field-name *Kirkmedwe* is recorded in Stapenhill in 1404 and 1441: SHC 1937 154, 168.

**KITCHEN BROOK** a tributary of the river Dove in Alstonefield. *Kitchinbrook* 1593 DRO D2375M/106/27, *Kytchen brukke*, *Kitchenbrocke* 1611 DRO D2375M/57/1. Evidently from OE *cycene* 'a kitchen', but the sense in which it is used here in unclear.

KNAVES CASTLE (obsolete) an earthwork, now destroyed, on the south side of Watling Street at Ogley Hay, 1 mile north of Brownhills (SK 0406). In the 17th century it was recorded that: '[i]t is circular and hath some three ditches about it. I believe the diameter of it is not above twenty yards at most. The midst of it is not above two or three yards square, and hath a breastwork about it in the nature of a keep. One gate, or entrance south': Aubrey 1980: 388, see also VCH I 345-6; JNSFC 5 1965 45. *Cnaven castle* c.1308 (1801) Shaw II 58, *Knaves Castle* 1686 Plot, *Knave's-castle* 1752 (1798) Shaw I 11, *Knaves Castle* 1882 MA 1882 171. From OE *cnafa*, ME *cnave*, *knave* 'boy, servant'. Earthworks were commonly called castles, and perhaps when modest in size the ironical 'earthwork of the boy or servant'; cf. the recurring Maiden's Castle, a name not uncommonly applied to prehistoric earthworks. Perhaps to be associated with *Knaves Hay* recorded in 1576: SRO D603/E/2/1-16. The place is remembered in Knaves Castle Avenue, Castle Close and Old Castle Grove on the north side of Watling Street. Knaves Castle is also recorded as a field-name in Shawbury Shropshire: Foxall 1980: 53.

**KNENHALL** 3 miles north-east of Stone (SJ 9237). *Knenoll* 1532 SHC 4th Series 8 135, *Kuenall* 1567 SHC XIII 264, *Kneuall* 1577 SHC 1926 49, *Knevall* 1577 SHC XVII 224, *Kneuall* 1577 SHC 1924 49, *Kenanelle* 1604 SHC XVIII 40, *Kuerhall* 1620 SHC VII NS 211, *? Nemwall* 1644 SHC 4th Series I 38, *Knenhall* 1644 ibid.' 178, *Knenhall* 1686 Plot, *Knewhall* 17th century SHC XII NS 86, 1836 O.S. A puzzling name of uncertain derivation: the early spellings show typical transcription confusion of *n* and *u*. The first element may be a short-form of a personal name, but if the second element is OE *cnoll* 'a hill top, a knoll, a hillock', the first element might be OE *cneo(w)* 'knee', perhaps used topographically for 'a knoll or hillock with a knee-like bend'. Or the second element may be OE *halh*, or possibly OE *hyll* 'hill': the place lies on the west side of an irregular hill. Erdeswick 1844: 30, 32 records *Snelhall* c.1600 in Kibblestone, but does not mention Knenhall, and it is clear that *Snenhall* is a transcription error for Knenhall.

**KNIGHTLEY** in Gnosall parish, 3 miles south-west of Eccleshall (SJ 8124), *Chenistelei* 1086 DB, *Cnitteley*, *Cnittelegh* 1203 Ass, 1238 Lib, *Cnettle* 1212 SHC III 158, *Knyttele* c.1260 SHC 1924 105, *Knicteleg* 1275 SHC XI NS 243, *Knycteleye* 1294 SHC 1911 219, *? Knythesleye* 1324 SHC X 50, *Kneyteley* 1402 SHC XI NS 213, *Knightley*, *Knightley Park* 1682 Browne; **KNIGHTLEY PARK** 1 mile north-west of Tatenhill (SK 1923), *Knytheley* 1325 Hardy 1908: 19, *Knightley Park* 1836 O.S. From OE *cnihta leah* 'the woodland glade of the retainers, young men or knights'. Places containing this name in Gnosall are spread over a wide area and include Knightley Gorse, Knightley Grange, Knightley Dale and Lower Knightley. Cf. Knighton.

**KNIGHTON** 1 mile south-west of Adbaston (SJ 7427). *Chnitestone* 1086 DB, *Knichton* 1222 Ass, *Knython* 1273 SHC VI (i) 64, *Knyhton* 1305 SHC VII 141, 1314 D938/39, *Knyghton next Adbaston* 1582 SHC XVII 227, *Knighton (Grange)* 1641 SRO D3067, *Knighton* 1682 Browne, 1833 O.S.; **KNIGHTON** on the Staffordshire-Shropshire border, 1 mile west of Willowbridge (SJ 7240), *Chenistetone* 1086 DB, *Knihtetun* c.1205 Rees 1997: 84, *Knythton* 1306 SHC VII 153, *Kneghton in le Hales* 1341 SHC XII 4,

*Knighton* 1682 Browne. From OE *cnihta-tun* 'the *tun* of the retainers or knights'. In place-names *cniht*, normally only found in the South Midlands and South of England, probably means 'a household servant of a lord, a knight'. Cf. Knightley. A grange of Ranton Abbey lay at Knighton near Adbaston: VCH III 253-4. The county boundary which follows the river Tern deviates dramatically to encompass Knighton near Willoughbridge.

**KNIGHTSFIELDS** – see **KNIGHTSLANDS.**

**KNIGHTSLAND** 2 miles south-west of Uttoxeter (SK 0830). *Knyghteslond* 1306 SHC VII 149, 1306 SHC 1911 65, *Knights Lant* 1724 D3259/Add/1, *Night Lands* 1775 Yates, *Knightsland* 1832 Teesdale, *Knights Land* 1836 O.S. From OE *cniht* 'retainer, young man, knight', with OE *land, lond* 'land'. Knightsfields (*Knights Fields* 1832 Teesdale, 1836 O.S.) lies 1 mile to the north.

**KNIPE WOOD** ½ mile south-east of Belmont Hall (SK 0049). *Knypewood* 1707 Okeover T760, *Knipe Wood* 1890 O.S. Probably from ON *gnipa* 'a steep rock or peak, an overhanging rock in a valley': see also Turner's Knipe. Perhaps to be associated with the wood in Ipstones called *Sherwin Knipe*, recorded in 1750: ERYARS DDCC/126/70.

**KNIVEDEN** 1 mile north-west of Bradnop (SK 0056). *Kniveden* c.1275 StCart, *Kynueton* c.1241 SHC 1924 218, *Knyveton'* 1262 TutCart, *? Kneveton* 1284 SHC 1910 299, *? Naveton* 1422 SHC XVII 40, *Knyveton* 1450 ibid, *Knyveden* 1535 SHC 1912 78, *Knyffeden* c.1539 LRMB, *Knivedon, Gnivedon* 1635 ParReg, *Knifden* 1656 PCC, *Knivdan* 1798 Yates, *Kniveden* 1840 O.S. Perhaps from the OE feminine personal name Cengifu (with the stress shifted at an early date), with an uncertain second element, probably OE *tun*, although the earliest spelling suggests OE *denu* 'valley': the place stands on a hill of 869', with a stream valley on the east. Knayton, Yorkshire, is *Cneveton* in 1223, and Kneeton, Nottinghamshire, is *Knivetun* in 1236: Ekwall 1960: 281; PN Nt 226-7. The stream in the valley to the east is recorded as *Knyvenbroc* in 1223 and 1227: Harl.

**KNOTBURY** 1 mile north-west of Flash (SK 0168). *Knotbury* 1727 Alstonefield ParReg, 1765 StSt 8 1996 80, 1842 O.S, *Notbury* 1775 Yates. Probably from OE *cnotta*, ME *knot* 'a hillock, a rocky hill, a cairn', more often found in the North West, with OE *burh* 'a fortified place', here perhaps referring to some ancient earthwork or other archaeological feature, or a natural feature having such an appearance. The place has been tentatively identified with *a knot*, mentioned in the 14th-century poem *Sir Gawain and the Green Knight* (see Elliott 1984: 61, 92-3; also Castle Cliff Rocks; Hen Cloud; Ludchurch; The Roaches). *Knotbury* is marked on the 1890 O.S. map 1 mile north-east of Ipstones (SK 0351), but the history of the name is untraced. See also Nutborough.

**KNOWLES** ½ mile north-east of Upper Hulme (SK 0161), *Knolles* 1308 SHC NS XI 258, *lez Knolles* 1432 DRO D2375M/126/2/11, *le Knollys* 1476 DRO D2375M/53/8, *Knowles* 1548 DRO D2375M/190/3, *? Knowlles* 1565 SHC 1938 76, *the Knowells* 1575 DRO D2375M/53/8; **KNOWLES FARM** 1 mile north of Endon (SJ 9254), *The Knowls* 1836 O.S.; **KNOWLE STYLE** in Biddulph (SJ 8856), *Knole* 1663 Biddulph ParReg; **KNOWL END** 1½ miles east of Audley (SJ 7751), *Le Knoile* 1298 SHC XI NS 248, *Knol* 1377 SHC 4th Series VI 12, *Knole Ende* 1599 SHC XVI 194, *Knoll* 1575 Audley ParReg, *Knollend* 1682 Browne, *Knowl End* 1833 O.S.; **KNOWL BANK** 1 mile east of Betley (SJ 7749), *le Knoile* 1282 SHC XII NS 213, *Knowle* 1592 Betley ParReg, *Knowle end* 1695 ibid.'; **KNOWLE FARM** 1 mile south of Lichfield (SK 1207), *Le Knoll temp.* Edward II SHC 1939 94, *Knoll* 1571 SHC XVII 218; **KNOWLE FARM** 1 mile south-east of Stowe (SK 0125), *Knol* 1292 SHC 1937 95, *the Cnolle* 1347 SHC XII 66, *Knowle*

*Close* 1741 SRO D1798/HM47/17, *Knowl* 1775 Yates; **KNOWLESWOOD** 1 mile south-west of Hales (SJ 7333), *Knollwood* 1562 SHC 1945-6 117, *Knowll Woodde* 1570 ibid. 146; **KNOWL, LONG** 3 miles north-east of Wolverhampton (SJ 9401), *le Knolle* 13th century Duig, *Long Knowle* 1834 O.S., *Knowles Wood, Knowle Wood* 1687 SRO D681/E/5/21, *Knowlwood* 1772 SRO 828/37, *Knowle Wood* 1833 O.S.; **KNOWL WALL** 2 miles north of Swynnerton (SJ 8539), *the No Wall* 1732 Trentham ParReg, *Nowall* 1739 Tooth 2000b: 39, *Knowl-wall, Stone* 1753 Barlaston ParReg, *Know wall* 1768 Trentham ParReg, *New Wall, New Hall* 1812 Tooth 2000b 39, *Knoll Walls* 1836 O.S. From OE *cnoll* 'hill top, summit, hillock', the last place probably with Mercian OE *wælle* 'a spring', and (sometimes) 'a stream'. There is a conspicuous knoll rising to nearly 600 feet at Knowleswood, and Gelling & Cole 2000: 157 notes that Knowle Farm near Lichfield lies on a spacious hill which has a small, flat-shaped summit rising from it immediately behind the farm.

KNUTTSHALL (unlocated; possibly intended to be Gnosall) *Knuttshall manor* 1648 (1798) Shaw I 282.

**KNUTTON** in Wolstanton parish, 1 mile north-west of Newcastle-under-Lyme (SJ 8346). *Clotone* 1086 DB, *Cnoton* 1212 SHC 1933 (ii) 11, *Cnutton* 1227 Ch, *Knocton* 1255 SHC 1911 124, *Knotton* 1256 Ch, *Knottoun* 1333 SHC 1913 229, *Knootton* 1594 Eccleshall ParReg, *Knutton* 1682 Browne. Possibly 'Cnut's *tun*', although the possessive *s* might have been expected in at least some of the forms, so perhaps more likely to be a Scandinavianised form of OE *cnotta* 'a knot, a hillock', so giving 'the *tun* at the hillock': see Fellows-Jensen 1990: 13-21.

**KNYPERSLEY** (pronounced neepuz-lee) on the south of Biddulph (SJ 8856), *Kniperslee* c.1247 SHC 1911 419, *Knypersleye, Kniprislega* 1272-3 SHC XI NS 245, *Kniprislega* 1278 ibid. 245, *Knybereleye* 1298 ibid. 247, *Knypersleye* 13th century Duig, *Knyperesley* 1315 SHC 1911 85, *Knypresley* 1362 SRO 49/8043; **KNYPERSLEY (HALL)** 2 miles south-west of Marchington (SK 1029). *Kympresleye* 1323 SHC IX (i) 95, *Knypersley* 1414 SHC XVII 16, *? Knypersley* c.1477 SHC VII 276, *(a pasture called) Knypersley* c.1484 SHC XI NS 60, *Knipersley* 1775 Yates, 1798 Shaw I 92, *Knypersley* 1836 O.S. The first element of these names is uncertain. A connection with ON *gnipa* 'steep, overhanging rock', or Norwegian *knip* 'narrow place' is improbable, even though there are gritsone features at the first place, including the so-called *Gawton stone* (see Plot 1686: 106), a huge rock fallen from the cliff in Knypersley Park (which may have formed an overhanging rock before it fell), since the medial -*r*-in the spellings would need to be accounted for. The terminal is OE *leah. Knepresleye*, recorded in 1327 (SHC VII (i) 217) may refer to either place. Cf. Turner's Knipe.

KNYPYNGESTYLE (unlocated, in Tipton) *Knypyngestyle* 1444 (1801) Shaw II 229. Perhaps from a constricted stile that 'nipped'.

KOCKYLEYE (unlocated, probably in the far north of the county) *Kockyleye* 1333 SHC X 115.

KORMODESTUN (unlocated) *Kormodestun temp.* Henry III SHC 1939 85. Possibly associated with Carmounthead (q.v.).

KYNESLEY (unlocated, in or near Loxley, south-west of Uttoxeter). *Kynardesey, Kynardesle* 1324 SHC X 47, 53, *Kynardesleye* 1324 SHC 1911 363, *Kenardeseye* 1349 SRO DW1733/A/20/20, *Kynardesley* 1381 Hardy 1908: 62, *Kynarseye* 1414 SHC XVII 20, *Kynnardesley* 1465 SHC IV NS 138, *Kynnersley* 1474 SRTO DW1733/A/1/1, *Kynnyshley temp.* Henry VIII SHC X NS I 146, *Kynesley* 1564 SHC XVII 212, *Kynsley*

1566 SHC 1938 168, *Kynnesley* 1567 ibid. 33. If ancient perhaps 'the *leah* of Cyneheard' (cf. Kinnersley, Shropshire), but it would seem that the Kynnersley family of Staffordshire descended from John de Kinnardesleye (c.1200-1275) of Kinnardesleye Castle in Herefordshire. In 1327 Loxley manor passed by marriage to the Kinnersleys of Kinnardesleye Castle when John Kynnersley married Joanna de Ferrers, sister and heiress of Thomas de Ferrers, Lord of Loxley. The family name became attached to the place, which is not mentioned by Erdeswick, Plot or Shaw. See also Erdeswick 1844: 512-3. It is unclear whether *The Old Town*, which appears on the first edition 1" O.S. map of 1836 (at SK 058304; *Old Town* 1887 O.S.), is connected with this place. Other places called Kinnersley are recorded in Shropshire, Surrey and Worcestershire: DES. For notes on the early history of the Kynnersley family see SRO DW1733/F/6.

KYNGESLEYE HETH (unlocated, at Teddesley) *Kyngesleye Heth temp.* Elizabeth I SHC 1939 123. See Kyngesoke Heth. 'The heath at the *leah* of the king'.

KYNGESOKE HETH (obsolete) ½ mile north of Huntington (SJ 9713). *Rough Hills and Kyngesoke Heth adjoining Teddesley Hay temp.* Elizabeth I SHC 1939 123. 'The heath with the king's oak'. Probably to be associated with Kyngesleye Heth (q.v.).

KYRKELOND (obsolete) a field-name in Ilam. *Kyrkelond* 1538 *Survey*. From ON *kirkja* 'church', with OE *land* 'land, estate, tract of land', so 'the church land'.

KYRKESLEYE (obsolete) a field-name in Consall. *Kyrkesleye* 1327 SHC 1913 16. From ON *kirkja* 'a church' (which may have replaced OE *cirice*), with OE *leah*, so 'the *leah* belonging to (Cheddleton) church'.

KYRYWILYHAILE (unlocated) *Kyrwelehelye* (p) 1286 SHC 4th Series XVIII 176, *Kyrywilyhaile* 1306 SHC VII 166, perhaps to be associated with *Kyryaule* 1310 (1798) Shaw I 21 (cf. Robert Kyryaul, recorded in 1298: SHC VII (i) 48). SHC VII 166 identifies this place as Willenhall, and the earliest spelling is associated with the Rushall/Aldridge area. *Kyryellesmedo* is recorded in Yoxall in 1421: SRO D170/M/2.

**LACHES, THE** in Brewood parish, 1 mile east of Coven (SJ 9207). *Lece broc* 996 (for 994) (17th century, S.1380), *the Laches* 1686 Plot 403, *The Laches* 1834 O.S. From OE **lece* 'a stream flowing through boggy land, a bog': the field-name *Lachebrok* is recorded in this area c.1290 (SHC 1928 109), with adjoining land called *Lachewalleburne* (ibid; see also Forsberg 1950: 72). Cf. Lechlade, Gloucestershire.

**LADDEREDGE** 2 miles south-west of Leek (SJ 9654). *Latherich* 1338 *Su*, *Ladderedge* early 16th century VCH VII 202, *Laderhedge* 1538 Ct, *Ladderedge (Court)* 1556 SRO D593/B/1/24/2, *Latherich (Court)* 1610 SRO D593/B/1/24/2, *Ladderidge* 1632 SRO D593/E/6/15, *Ladderich, Ladderitch* 1648 Leek ParReg, *Lathridge* 1653 ibid., *Ladderedge* 1698 ibid, *Latheredge* 1813 Bourne, *Ladderedge* 1832 Teesdale, 1836 O.S. Perhaps from ON *hlaða* (found as dialect *lathe*) 'barn', with OE *hrycg* 'ridge'.

**LADFORDFIELD** 1 mile east of Ellenhall (SJ 8626). *Lotford* 1209 SHC III (i) 175, 1611 SHC III NS 67, *Ladford (Pool)* 1836 O.S. Possibly from ME *lote* 'to lurk, to lie concealed', so 'the hidden (overgrown ?) ford': the place lies on Gamesley Brook.

**LADY BRIDGE** a bridge across the river Tame in Tamworth, close to its junction with the river Anker (SK 2003). *Lady Brugge* 1355 SRO D187/1/4, *Lady Bridge (Bank)* 1702 SRO D260/M/T/2/62. Dedicated in honour of Our Lady, the Virgin Mary.

**LADY DALE WELL** a spring with a 19th century stone structure on the south side of Leek (SJ 9855). *Lady Wall Dale* 1587 VCH VII 88, *Ladaway-dale* c.1750 (1883) Sleigh

146, *Lady Way Dale* 1873 SRO D4855/3/1/1-12, *Ladderrmedale* 1883 Sleigh 146. A spring named in honour of Our Lady, the Virgin Mary: see VCH VII 88.

**LADY EDGE**  3 miles south-west of Longnor (SK 0562), *the Ladie Edge* 1645 SRO QSR f.10v, *Lady Edge* 1839 *EnclA*, 1840 O.S.; **LADYEDGE**  1 mile north of Ipstones (SK 0251), *? Lady Ridge* 1695 Leek ParReg, *Lady Edge* 1836 O.S. 'The sharply ridged land dedicated to Our Lady'. VCH VII 29 suggests that the first place is recorded in the later 14th century, though the footnote gives the date 1571.

LADYHURST  (unlocated, in Knutton) *Lawedihurst temp*. Henry III SHC 1913 230, *Ladyhurste* 1334 ibid. 230. 'The hurst or wooded eminence dedicated to Our Lady, the Virgin Mary'.

**LADYMEADOWS**  1 mile south-east of Bradnop (SK 0253), *Lady Meadowes* 1656 Ipstones ParReg, *Meadows-place* 1665 Leek ParReg, *Lady Meadow* 1695 ibid; LADY MEADOW (obsolete, in Yeatsall), *Lady Meadow* 1349 SRO D(W)1721/3/2/24, 1353 SRO D(W)1721/3/1/20, 1647 SRO D(W)1721/3/254. 'The meadows dedicated to Our Lady, the Virgin Mary'.

**LADY MOOR**  2 miles south-east of Biddulph (SJ 9055). *Ladie More* 1547 SHC 1912 161, *Ladie Moore* 1568 SHC 1931 219, *Ladymare yate* 1625 JNSFC LX 1925-6 72, *Ladymoor Gate* 1659 Biddulph ParReg, *Ladymoor, Ladymoor Gate* 1799 Faden, *Lady-Gate* 1836 O.S.; LADYMOOR  (obsolete) on the south-east side of Ettingshall (SO 9395), *Ladymoor* 1810 SRO D695/1/9/33-41, *Ladymoor (Colliery)* 1834 O.S. 'The moorland dedicated to Our Lady, the Virgin Mary'. See also Lordsmore.

**LADYSMITH FARM**  2 miles south-east of Abbots Bromley (SK 1122). *Gullet* 1775 Yates, *Gulletts* 1832 Teesdale, *The Gullets Farm* 1836 O.S. Originally from ME *golet* (from a diminutive of OFr *goule* 'throat'), here in the sense 'a water-channel, a gully': there is a short stream passing through the place falling steeply to nearby Pur Brook. Re-named to commemorate the relief of Ladysmith on 28 February 1900 by General Buller during the Boer War.

**LADYWELL (WOOD)**  A spring with supposedly curative properties cut into a sandstone outcrop on Orton Hill on the south-west of Wolverhampton (SO 8794). *Ladywell (Hill)* 1840 TA, *Ladywell* 1840 Census. Perhaps the same place as *Wodewell* ('the spring at the wood') recorded in the 13th century: VCH XX 200. The name suggests a later association with the Virgin Mary. See also SHC IX (ii) 107.

**LAMBER LOW**  ½ mile north-east of Waterhouses (SK 0850). From OE *lambra*, genitive plural of *lamb* 'lamb', with OE *hlaw* 'mound, tumulus', so 'the lambs' tumulus'

**LAMMASCOTE**  on the east side of Stafford (SJ 9323). *Le Lombercote* 1273 SHC 1911 151, 1317 SRO D938/292, *Lanpucotes, La Lamputtes* after 1352 (16th century) SRO 96[7904], *Lambercote* 1433-4 SRO D641/1/2/53 rot. 2, *Lambircote* 1439 MinA, *Lomburcote* c.1445 Ipm, *Lambcotts* 1537 SRO DW1721/1/1 f.140, *Lambercotes* 1548 Survey, *Lambercoats* 1550 VCH VI 207, *Lambercotte* c.1610 map, *Lammascote* 1775 Yates, *Lamberscot* 1788 BL Eg. MS 2862 ff.52v-53, *Lamberscott* 1846 VCH VI 207. From OE *lamb*, genitive plural *lambra*', hence 'the cottages where lambs were reared'.

LAMSLOUGH, LAMSLOW  (unlocated, possibly near Lapley) *Lamslow* 1327 SHC VII (i) 243, *Lamslough* 1332 SHC X 123.

**LANDER'S WOOD**  1 mile south-west of Milwich (SJ 9630). *Launder's Wood* 1836 O.S. From the Landor family which owned property here: SHC XII NS 126.

**LANDYWOOD** 1 mile south-east of Cheslyn Hay (SJ 9906). *Londewood* 1657 Wolverhampton ParReg, 1670 Ct, *Landywood* 1695 Morden. *Launde* is a ME word derived from OFr *lande, launde*, which is the root of modern *lawn*. It originally meant 'an open ride or glade in a wooded area'. The place lay within Cannock Forest. It is unclear whether *Londehall*, recorded in 1234x1240 (TestNev), refers to this place

**LANE END** (obsolete, in Longton (SJ 9043)) *Meare Lane ende* 1564 SHC XIII 231, *le Meare land end* 1597 SHC 1935 IV 13, *Mearlane end* 1585 SHC XV 160, *Meir Lane End* 1679 SHC XII 59, *Mairlane end* 1679 SHC 1919 259, *Lane End* 1836 O.S. Shaw 1798: I 34 mentions '...a place called Lane-end, the road there being closed up.' *Mear Lane* is the section of Roman road (Margery numbers 70a and 181) between Blythe Bridge and Longton on the first edition 1" O.S. map of 1840 (*Meer Lane* 1775 Yates), evidently from OE *(ge)mǣre* 'boundary' (see Meir): a parish and Hundred boundary run along this length of road. See also SHC XII 59.

**LANE GREEN** on the south side of Bilbrook (SJ 8703). *Loan Green* 1741 Codsall ParReg, *Lone Green* 1747 ibid, *Lane Green* 1834 O.S. Possibly from the dialect word *loan* 'arable land', or *lone* 'lane', but the forms are too late to be certain. The green here is recorded c.1640: VCH XX 12. The first edition 1" O.S. map of 1833 shows a *Lane Green* 2 miles north-east of Alveley (SO 7885).

**LANESFIELD** 1½ miles south-east of Wolverhampton (SO 9395). *Lanesfield* 1834 O.S. Said to derive from the Lane family who lived at Rookery Hall: StEnc 448.

**LANEY GREEN** in Shareshill, 1 mile west of Cheslyn Hay (SJ 9606). *Loany Green* 1704 Penkridge ParReg, *Lanes Green* 1775 Yates, *Lowney Green* 1834 O.S., *Lowney Green* 1838 SRO D351/M/179. Possibly from the dialect word *loan* 'arable land', or *lone* 'lane', but the forms are too late to be certain, and since Green suggests a squatter settlement, it may be from a nickname.

**LANGLEY** 1 mile south-east of Wolverhampton (SO 8696). *Langley* 1798 Yates. From OE *lang-leah* 'the long wood or glade'. The name is preserved in Langley House and Langley Road.

**LANGLEY LAWN** 3 miles south-west of Brewood, near the Staffordshire-Shropshire border (SJ 8406). *Longeley(medewe)* 1349 SRO D593/A/2/5/7, *Longeley Cawende (? Lawende)* 1507 SHC V NS 231, *longeley land* 1569 Ct, *Launlyland* 1668 Codsall ParReg. From OE *lang-leah* 'the long wood or (as here) glade', with ME *launde* 'an open space in woodland, a forest glade, woodland pasture', usually meaning in this area 'an open passage through woodland'. Cf. Blymhill Lawn, Coven Lawn, Oaken Lawn. Another Langley is recorded in Eccleshall (perhaps near Horseley) c.1523: SHC 1938 5.

**LANGOT** 1½ miles north-west of Bishop's Offley (SJ 7631), *Langett* 1610 SHC III NS 27, *? Langall* 1683 SRO 26/6/2/22, *Langot Lane* 1930 O.S.; LANGOTT (unlocated, in Dunwood), *Langett* 1621 SRO D4908/2/4/1-14, *Langott* 1712 ibid; LANGUET (unlocated, in Cheadle), *Languet* 1293 SHC VI (i) 245, *Langett* 1610 SHC III NS 27; LANGOT (unlocated, in Tutbury), *the Langot* 1601 (1798) Shaw I 56; LANGET (unlocated, in Anslow), *Langett* 1563 HLS 555, *the Langet* 1563 ibid. 238. From OE *\*langet* 'a long strip of land': cf. the Langet, Herefordshire.

LANGWAY (unlocated) *Langway* 1599 SHC 1935 99.

**LAPLEY** Ancient Parish 4 miles south-west of Penkridge (SJ 8712). *Lepelie* 1086 DB (listed in Northamptonshire), *Lapeleia* 1130, *Lappeleia* 1200 P, *Lappale* 1286 SHC VI

(i) 170, *Lapley* 1682 Browne. Possibly 'Læppa's *leah*', derived from the OE personal name *Hlæppa or Hlappa (PN Wo 298 ) or *Lappa (DES), but it is more likely that the first element is from OE *lappa* meaning 'a lap, the skirt of a garment', and in a topographical sense 'district', and perhaps 'land at the edge of an estate or parish'. The place is on the eastern border of Wheaton Aston parish. Cf. Lapworth in Warwickshire. See also Forsberg 1950: 62.

**LASK EDGE** 2 miles east of Biddulph (SJ 9157). *Laxege* 1239 VCH VII 72, *Laskedge* 1673 SRO DW1761/A/4/197, *Lasco Edge* 1807 SHC 1960 11, *Lask Edge* 1810 SRO DW1909/N/2, *Lax Edge* 1811 Bourne, *Lask Edge* 1815 *EnclA*, 1842 O.S. The earliest form points towards a derivation from the adjective *lax* 'loose', meaning when applied to stones, etc., 'loose in texture', so here 'the escarpment with the loose stones', *lax* having developed into *lask* by metathesis. It may be noted, however, that the surname Lax is recorded, e.g. in Derbyshire in the 18th century: DRO D258/2/34/6. *Laxege Liulfneslode* recorded in 1237 (Cur) would appear to refer to this place, which lies at the southernmost part of a pronounced ridge; the second word is unexplained.

**LATHBURY'S HILL** on the south-west side of Draycott in the Clay (SK 1528). *Lathbury* 1290 DRO D5236/3/8, 1367 SRO D(W)1721/3/20/20, 1403 DRO D5236/15/2, *Lathebury* 1368 SHC XI 176. From OE *lætt burg* 'the fortification or manor house made of laths': cf. Lathbury, Berkshire. Richard Lathbury is recorded in Fauld in 1557: SA 513/2/18/7/2; see also SRO D786/5/6.

**LATHERFORD** 4 miles south-west of Cannock (SJ 9307). *Loddersford* 1300 OpenFds, *Lod(d)er(e)ford* 1343 to 1511 *Vernon*, *Lodresford* 1358 PlR, *Lotherford* 1442 *Vernon*, *Ladderford* 1534 Ct, *Latherford* 1834 O.S. From OE *loddere* 'beggar', hence 'the ford frequented by beggars'.

LAUGHTON HOUSE (FARM) (unlocated, possibly near Mayfield or Rocester) *Laghenhous* (p) 1423 SRO D3272/13/16, *Laughtonhous(e)* (p) 1494 SRO D786/2/2, *Loughtonhouse (farm)* 1538 SHC 1912 125, *Laughtonhouse* (p) 1494 SRO D786/2/2, *Laghtonhowse* (p) 1539 SHC NS VI (i) 80, *Loughtenhouse* (p) 1539 SHC V NS 305, *Laughton House Farm* 1745 SRO D1134/21/2.

**LAUND (FARM)** 2 miles north-east of Ipstones (SK 0451), *Laund* 1836 O.S.; LAUND FARM (obsolete), ½ mile east of Consall (SK 9948), *the Launde* 1655 WYAS MD244/189, *Laund Farm* 1890 O.S. From ME *launde* 'an open space in woodland, a forest glade, woodland pasture'. See also Lawn.

LAWN (obsolete) 1 mile south of Bucknall (SJ 9145), *Lawnds* 1691 Okeover T762, *Lownd* 1705 Stoke on Trent ParReg, *ye Lawn* 1752 ibid, *Lawn* 1836 O.S.; **LAWN FARM, LITTLE** 1 mile south-east of Bagnall (SJ 9350), *Laund Farm* 1836 O.S.; **LAWN FARM** ½ mile east of Consall (SJ 9948), *the Laund* 1696 ParReg, *The Laund* 1836 O.S.; **LAWNS FARM** ½ mile east of Tatenhill (SK 2122), *The Lawnes* c.1760 SRO D615/M/7/4. From ME *launde* 'an open space in woodland, a forest glade, woodland pasture'. See also Laund (Farm).

**LAWNHEAD** 1 mile south-west of Ellenhall (SJ 8324). *Lawn Head* 1829 SRO D615/ES/4/9/2, 1833 O.S. Evidently associated with *the Launde* recorded in 1585 (Ellenhall ParReg), and perhaps with *Launde* recorded in 1537 (VCH III 254), from ME *launde* 'an open space in woodland, a forest glade, woodland pasture'.

LAWTON (unlocated, in Alton) *Lawton* 1754 SRO D240/D/106. See also Lawton Park.

**LAWTON GRANGE** ½ mile south-east of Wall (SK 1005). Early spellinmgs for this name have not been traced.

**LAWTON PARK** (unlocated) *Lawton Park* 1686 Plot 166. From the reference in Plot it is evident that the place lay in a lead-producing area.

**LAYTON** (unlocated, perhaps near Bagot's Bromley, but see also Leighton) *Layton in Pyrell Hundred* 1539 SHC NS V 315.

**LEA, RIVER** a tributary of the Checkley Brook leading into the river Weaver. *The Lea Brook, the water Lee* 1612, *the Lea* 1656 PN Ch I 18. Lea Brook was the earlier name for Checkley Brook. The Lea Brook, the river Lee and The Lea are topographically connected with Betley Water, so-called from the place-name Betley. The name is a back-formation from Lea Hall, the name of a place on Checkley Brook. The river-name *Lee flu* (Saxton 1577) may have been applied to this stream, though Ekwall 1928: 241 connects it to a brook that flows to the Weaver a little north of Checkley Brook and runs not far from Lea Hall.

**LEA, THE** 1 mile north of High Offley (SJ 7728). *La Lee* 1282 SHC 1914 77, *Lee* 1339 SRO D(W)1082/L/10/3, *Gibbetsley* 1569 SHC 1945-6 113, *Gybbettesley, Gybbattes Ley* 1572 SHC XIII 292, *Gybbottes Ley* 1590 SHC XVI 100, *Ley* 1608 SHC 1948-9 99, *Gebertsley otherwise Jeberdsley* 1610 SHC III NS 39, *Gilberts Lea or Joubberts Lea* c.1680 SHC 1919 248, *Gibbatts Gibbett Ley* 1709 SRO D1788/42/7, *Gilbert's Lea* 1833 O.S. The first element is uncertain, but may be from OE *leg* 'fire', here in the sense 'beacon fire' (the place lies on the north-east slopes of a 420' hill, marked by prominent hachuring on the first edition 1" O.S. map of 1833), or OE *lea*, dative of OE *leah* 'wood or clearing', with the later addition of what may be ME *gibet* 'gibbet' (here in the plural), giving 'the beacon or clearings where the bodies of felons were displayed'. But the 1572 spellings come from records which sometimes give fanciful spellings, and it is possible that the 19th century name is more accurate, so perhaps to be associated with Gilbert who held Loynton, 3 miles to the south-west, as well as Chipnall and Cheswardine, at DB: VCH IV 31. Cf. *Gilberdescroft* (undated): SHC IV 277.

**LEA, THE** (obsolete) an ancient estate in Graiseley on the south-west side of Wolverhampton (SO 9097): see WA I 389. *la Leye* 1261 SHC 1950-1 52, *Lega* 1321 ibid. 52, *Lee* 1441 SHC NS VI (ii) 208, *The lea* 1577 Saxton, *Ye Lee* 1599 Smith, *The Leu (sic)* 1610 Speed, 1690 sellar, *Lea-hall Farm* 1790 Sale Partics. An old house known as The Lea existed until the 1840s: StEnc 332. The name is preserved in Lea Road. See also SHC 1919 167; TSSAHS XXIV 1982-3 57.

**LEACROFT** 1½ miles south-east of Cannock (SJ 9909). *Lecroft* (p) 1327 Mis, 1346 (p) Banco, *Leecroft, Lee Croft* 1327 SHC VII (i) 245, *Leycroft(e)* 1493 Ipm, *Lecroft* 1432 SHC XVII 141, *Leycroft* 1532 SHC 4th Series 8 187, 1567 SHC 1938 90, *Leacroft* 1599 Smith. From the OE adjective *læge* 'fallow, unploughed, untilled', found in OE only in compounds, with OE *croft* 'a small enclosure of arable or pasture land, an enclosure near a house', hence 'the fallow croft'. Leacroft Hall (*Lea Croft* 1836 O.S.) lies 1 mile north-east of Fulford (SJ 9639), but the history of the place is unknown. It is possible that some of the spellings above relate to that place.

**LEADENDALE** 2 miles east of Barlaston (SJ 9239). *Leaden Dale* 1798 Yates, 1836 O.S. The age of this name is uncertain, but if ancient there are several possible derivations, including OE *hleo-denn dæl* 'pasture with a *hleo* or shelter in the valley' (cf. Lydden, Kent), 'Leoda's dale or valley' (cf. Leadenham, Lincolnshire), or a derivative of OE *leod* 'reed meadow grass'. OE *leaden dæl* 'valley of lead' is improbable, although 'iron pyrites and sulphuret of lead (galena) are common in the coal measures': see Langford 1872: I 23. There is a pronounced stream valley here. It is possible that the place is to be associated with the unlocated Leighton (q.v.), in which respect it may be noted that *Leyden House* lies at the end of Leadendale Lane, a cul-de-sac.

**LEA FARM, LEA HEATH)** 1 mile south-east of Drointon (SK 0225), *Lee* 1248 SHC VIII 163, c.1250 SHC 1919 42, 1284 SRO D938/15, 1309 SHC 1911 75, *La Le* c.1256 SRO (576/7922), *La Lee* 1293 SHC 1924 189, *Lea* 1836 O.S.; LEA FARM (unlocated, near Grosvenor (SO 7693)), *Lega* 1221 Eyton 1854-60: III 101, *La Lee* 1272 ibid, *the Lea* 1333 ibid, *Le Lee* 1697 SA 5586/1/157. From OE *leah* 'wood, woodland clearing'.

**LEAFIELDS (FARM)** 1 mile north-west of Brewood (SJ 8709), *leefylde(s), leefeildes* 1535 Survey, *Leyfields* 1775 Yates; **LEAFIELDS (FARM)** on the west side of Abbots Bromley (SK 0724), *Lee Field* 1329 SRO D(W)1721/3/3/9. From the OE adjective *lǣge* 'fallow, unploughed, untilled', found in OE only in compounds, with OE *feld*, which meant originally cleared open land, and later enclosed land, hence 'the open fallow land'.

**LEA HALL** ½ miles north-west of Armitage (SK 0616). *Lee Hall* 1583 (1798) Shaw I *210 fn.3, 1699 SRO D1161/1/1/3, *Lee* 1653 (1798) Shaw I 210*, *Lea Hall* 1712 SRO D1929/7/1. The late spellings do not allow a derivation to be suggested.

**LEA HEAD MANOR** on the Staffordshire-Cheshire border, 2 miles south-west of Madeley (SJ 7542). *Leehedd* 1485 SHC VI NS (i) 158, *Lea Head* 1798 Yates, 1833 O.S. The place lies at the head of the river Lea.

**LEA LAUGHTON** ½ mile west of Horton (SJ 9357). *Lee early* 19th century VCH VII 65, *Lea Lathton* 1841 VCH VII 65, *Leigh Ho.* 1842 O.S., *Leigh* 1891 O.S.

**LEAMONSLEY** 1 mile west of Lichfield (SK 1009). *Leomondsley (Moggs)* 1514 SRO DW1851/8/50, *Leomondsley* 1676 SRO DW1851/8/51, *Leomanslay* 1714 SRO D15/10/1/36, *Lemonsley* 1780 SHC 4th Series VI 152, *Leomansley* 1781 SRP D15/11/14/49, *Lemonsley (Mill)* 1834 O.S., *Leamonsley* 1887 O.S. A difficult name, perhaps from a British river-name associated with Welsh *llif,* Cornish *lif* 'flood, stream' (probably from *lim*): cf. River Leam (Warwickshire and Northamptonshire, *(on)Leomenan, (of) Leomanan* 1033 (c.1225) S. 967: see Hooke 1979: 119), Lyme (Devon and Dorset), or from an ancestor of a Welsh stream-name Llyman or Llymon 'keen, sharp' (Thomas 1938: 73-4), or from an Anglo-Celtic *lemo* 'an elm' as part of its first element, with OE *leah* (cf. Lemon Brook; see VCH XIV 37; Gelling 1979: 112; VCH XIV 37), or from a personal name such as Leoman (Searle 1897: 336; a reference to John Lemon of Lichfield in 1311 (SRO D948) may also be noted). However, none of the foregoing explain the *-d-* in the earliest spellings, which may point towards a derivation from an unrecorded personal name *Leofmund or similar. The element Moggs in the 1514 spelling is unexplained. See also Lemon Brook.

**LEAMORE** 1 mile south of Bloxwich (SJ 9900). *leymore* 1420 Sims 1882: 14, *leymore* 1525 SRO D260/M/T/7/2, *Leamer* 1834 O.S. Perhaps from OE *lǣge* 'fallow', with OE *mor* 'moor, waste', so 'the uncultivated waste'.

**LEATON** ½ mile south-east of Bobbington (SO 8190). *? Laetonia* 1086 VCH Sa II 124, Eyton 1854-60: I 109, *Leton* 13th century SA 5735/2/22/5/1, *Leeton* 1540 SHC XI 283, *Leton* 1554 SRO 3764/92[31759], *Leiton* 1679 SA 4572/6/2/31, *Leaton or Lawton* 1749 Bowen. From OE *leah*, with OE *tun*. Early spellings may be *the Lee*, recorded in the later 13th century (VCH XX 70), *Lee, la Lee* recorded in 1327 (SHC VII (i) 252), and *la Lee, La Leye*, recorded in 1332 (SHC X (i) 129).

**LEAWOODS, LEAWOOD FARM** 1 mile north-west of Norbury (SJ 7724). *La Lee Wode* 1325 SRO D1564/7, *Ley Woods* 1724 D1717/A/1/41, *Leawoods* 1732 1732 D1717/A/1/48. From OE *leah*, with OE *wudu* 'a wood'.

**LEE HALL** (unlocated, in Barlaston: SHC XII NS 82) *la Lee* 1418 SHC XII NS 82, *Lee Hall* 1460 ibid, 1608 ibid.

**LEEK** Ancient Parish 10 miles north-east of Stoke on Trent (SJ 9856). *Lec* 1086 DB, *Lech* c.1100 Chester, 1188 SHC I 136, 1199 SHC II 83, *Lecu* 1165 ibid. 252, *Lech* 1189-90 ibid. 83, *Lec* 1199 ibid. 36, *Leech* 1220x1223 CEC 381, *Leych* c.1240 SRO DW1761/A/4/9-10, *Leik* 1244 SHC I 295, *Leke* 1247 Ass, 1297 SHC I 212, *Leike* 1298 SHC XI NS 251, *Leck* 1318 (1883) Sleigh 51, *Lake* 1425 SHC XVII 100, *Lyk* 1426 ibid. 113, *Lyek* 1474 SHC IV NS 189, *Leke* 1532 SHC 4th Series 8 29, *Leike* 1577 Saxton. An interesting name. Sleigh 1883: 1, 86 put forward a derivation from Welsh *llech* 'rock, crag; a flat stone' (which he seems to associate with 'shrine'), which was taken up by Duignan 1902: 90-1. Ekwall 1928: 246-7 felt that the name 'must have been the name of the upper Churnett or an arm of it, on which are Leek or Leekfrith', and subsequently suggested that the name derives from a conjectural OE *\*lece*, possibly meaning 'a brook', derived from OE *\*lecan* 'to drip, leak', corresponding to ON *leka* (1960: 292). Other authorities (see for example EPNE ii 26; JNSFC XXI 5-6; Gelling 1984: 25; Gelling & Cole 2000: 21; VCH VII 84) have put forward a derivation from ON *l? kr* 'brook, stream', perhaps referring to Spot (i.e. Spout) Water, which gave its name to Spout Lane (*Spoutyate, Spoutgate* 1643 Leek ParReg, now Brook Street), from ME *spouten* 'to discharge liquid', or its tributary fed from a spring, now dry, on the edge of St Edward's churchyard, which ran along the west side of St Edward Street, formerly Spout Street. Another spring near the church may have fed *le Kirkebrok* or *Kyrkebroke* (possibly an early name for Ball Haye brook), a name with Scandinavian influence, recorded in 1281 (SRO A/4/168[10/3]; SRO DW1761/A/4/68; VCH VII 84). It seems likely that the root of this name is indeed OE *\*lece* 'to drip, to leak, to dribble' (cf. Modern *leak*, and *leach* 'to cause liquid to percolate through some material'), with reference to the spring itself, which was strikingly situated at one of the highest points in the area, in a place of early religious activity, and perhaps to be associated with fragments of pre-Conquest crosses with fragmentary runic inscription and other early stonework in St Edward's churchyard (VCH VII 137; see also Lekebourne). It has been argued (see Gelling 1984: 25; Gelling & Cole 2000: 21) that there is little evidence for the existence of OE *\*lece*, but its rarity might be explained if the word had a narrow meaning such as 'leak', and the adjective *hlec* 'leaky' is found in OE (Bosworth-Toller Supplement 550). Although it has been suggested that 'Leek would probably not have been anybody's first choice for a settlement' (Gelling 1992: 137), the place lies on a well-watered hill in a loop of the river Churnet, and it is now clear that in the Anglo-Saxon period a large estate was centred on Leek and perhaps Rudyard: SHC 4th Series 19 1995 5-12. The township of Leek is more properly Leek and Lowe (e.g. *Leek and Lowe* 1583 SHC XV 145), but the origin of Lowe is uncertain: it may be from Lowe Hill (q.v.), or Catteslowe (q.v.), or from a mound in which a cremation burial was found in Birchall Meadows to the west of the Cheddleton road in 1859 (VCH VII 85), but the lowest part of Mill Street was at one time known as *Lowe Hammill* (e.g. *Loe Hammel* 1704 Leek ParReg), from *hamel*, a dialect word for hamlet – *Low Hamlet* is recorded in 1666: SHC 1925 165, from Lowe Hill: JNSFC XXXIX 1905-5 159. Cf. Leake (Lincolnshire, Nottinghamshire and Yorkshire), and Leck, Lancashire. Leek Wootton, Warwickshire, probably has the vegetable name.

**LEEK FOREST** (obsolete) recorded c.1170 (VCH VII 197). The extent of the Forest is not known, but it appears to have been detached from Macclesfield Forest in the 13th century following the grant of the manor of Leek to Dieulacres abbey (VCH Ch II 178), and included Leekfrith and Rushton Spencer (VCH VII 80), and perhaps Gun and Wetwood (VCH VII 197). The Forest was more properly a chase, from OFr *chace* 'chasing, hunting, a hunting ground, wild park-land', which was a term applied to an area subject to Forest law but not held by the king: see VCH II 335. In this case the Forest rights were held by the Earls of Chester: VCH VII 80; also VCH Ch II 178.

**LEEKFRITH** an area 3 miles north of Leek (SJ 9861), formerly of much greater extent. *The Fyrthe* 1532 SHC 4th Series 8 29, *the hamlet of the Fryth called Leekefryth* 1539 SHC IX NS 300, *Leeke Frith* 1598 StSt 13 2001 53, *Leekfrithe* c.1620 SHC VII NS 220, See Frith.

**LEES** – see **(WHISTON) LEES**.

**LEESE FARM** on the east side of Billington (SJ 8920). *the Lees, Leys, Lea Leighes, the Leyse* (undated) SHC VIII (i) 122, *Lees Farm* 1562 VCH V 92, *Lees House* 1695 ibid, *Leys Farm* 1798 Yates, *Lees Farm* 1836 O.S. Perhaps to be associated with *Legh*, recorded in 1290 (SHC VI (i) 199.

LEES GRANGE (obsolete) St Thomas' priory (q.v.) was known as Lees Grange at the Dissolution: VCH III 265.

**LEES HILL** 1 mile north-west of Kingstone (SK 0530). *? Lega* c.1286 NA DD/FJ/1/180/1, *Leyes* 1293 SHC VI (i) 239, *Kyngeston' Leges* late 13th century SRO (146/7931), *Kyngestoneleyes* 1340 SRO D938/147, *Lyghes* 1359 SRO DW1721/3/1/26, *Leyshill* 1493 SRO DW1733/A/2/33, *The Lees* 1532 SHC 4th Series 8 76, *Lees Hill* c.1550 Erdeswick 1844: 261, 1660 SHC 1935 475, *Lees hill* 1680 SHC 1919 272, *Lees Hill Farm* 1706 SRO D3259/1/1. Sometime known as *Loxley Leyes* and *Kingston Leyes*: Erdeswick 1844: 261; SHC 1935 475. From OE *lǣs* 'pasture, meadow-land', or the plural of OE *leah*, or possibly from OE *leg* 'fire', used here in the sense 'beacon': cf. Leysdown, Kent. The place lies at the north end of a short rounded ridge.

**LEES HOUSE FARM** on the north-west side of Okeover (SK 1448). *(pasture called) Leez* 1538 SHC VII NS 63, *the Lees* 1547 Okeover F18, *Lees House* 1836 O.S. From OE *lǣs*, the plural of OE *lǣge* 'pasture, meadow-land'.

**LEESE HOUSE FARM** 1 mile south-west of Draycott-in-the-Moors (SJ 9738). *Leyam* c.1160 SHC III 224, *Leyesheuese, Leyeheuese* 1331 SHC 1913 32, *Draycotelyes* 1334 SHC XVI 6, *Leys* ?14th century SRO DW1733/A/2/71, 1522 SHC 1925 121, *Lees Howses* 1611 SHC IV NS 13. From OE *leagum*, dative plural of OE *leah* 'a wood, a clearing in a wood', so 'the house at the clearings'.

**LEESIDE** on north of Rudyard Lake (SJ 9260). *? Leyland* 1327 SHC VII 233. 'Pasture or meadow-land', from OE *lǣge* 'fallow', with (if the spelling relates to this place) OE *land* 'land, estate, a tract of land'.

LEETECH (obsolete) on the north side of Coven (SJ 9007). *Leet Each* 1739 Brewood ParReg, *Leetech* 1834 O.S., *Leer-each* 1832 Teesdale. Early forms have not been traced, but perhaps from OE *(ge)lǣt* 'water-course' (perhaps with reference to former ironworks: VCH V 20-21; Horovitz 1992: 289-95), with an unidentified second element, possibly ME *ache* 'a umbelliferous plant; properly smallage; parsley' (OED). There is no evidence to show that *Leight Each*, recorded in 1707 (SA 112/1/2792), refers to this place, but the similarity of the names is noteworthy.

**LEIGH, CHURCH** Ancient Parish 5 miles north-west of Uttoxeter (SK 0235). *Lege* 1002x1004 (11th century, S.906; 11th century, S.1536), *Lege* 1086 DB, *Leyam* c.1160 SHC III (i) 224, *Leia* c.1177 ibid. 227, *Leye* 1256 SHC 1911 19, *Leyes* 1294 SHC 1925 89, *Leyam* 13th century ibid. 73, *Leigh* 1724 D3259/Add/1. From OE *leah*, in the dative form *lēge*. By the 12th century the place-name *Field* (q.v.) 1½ miles to the south is found attached to the name: VCH IV 44 fn. The place was also known as *Malbanc's Leigh*, from the name of an early owner: SHC III 224. See also Bokeley(e); Dods Leigh.

**LEIGHTON** said to be a forest in the Meirheath area, enclosed in the 18th century: Shaw 1798: I 44. From OE *leac-tun*, '*tun* where vegetables were grown'. The word *leac* meant literally 'leek', but in place-names is generally taken to mean 'vegetables'. It is possible that *Layton in Pyrell Hundred*, recorded in 1539 (SHC NS V 315) is to be identified with this place, although the context in which the name appears suggests a location in the Bagot's Bromley area. See also Leadendale; Leyton.

**LEIGHTON HAY** (unlocated, *in Overs Longsdon* (Upper Longsdon): SRO DW1761/A/4/257) *Leghton Heye* 1494 SRO DW1761/A/4/261, *Leightonhey* 1609 SRO DW1761/A/4/256, *Leighton Hey* 1618 SRO DW1761/A/4/257.

**LEKEBOURNE** (unlocated, a tributary of the river Churnet) *Lekebourne* (p) 1321 Coram. An interesting stream name, which appears to be 'Leek stream', or perhaps 'the stream fed by the spring on the hill at Leek' (see Leek). The OE element *burna* (which tends to be associated with larger streams) is rare in Staffordshire, and all other examples come from the southern part of the county, so here perhaps from metathesised ON *brunnr* 'a well, a spring'. See also Ludburn.

**LELEHEVED** (unlocated) *Leleheved* 1271 SHC IV 183. From OE *heafod* 'a head or end (of anything)', in this case probably a topographical term in the sense of a headland, perhaps with OE *lēl* 'twig, withe', so 'the headland with the twigs'.

**LEMON BROOK** (obsolete) a former tributary of the river Churnet, north of Ipstones, marked on Yates' map of 1775 and Stockdale's map of 1795, probably what is now known as Coombes Brook: Yates' map shows *Lemon House* near what is now Bottomlane Farm (SK 0251). No early forms are available, but the derivation may be identical with Lemon Brook, Devon; Lem Brook, Worcestershire; the river name Leam, found in Northamptonshire and Warwickshire; and the river Lymm in Lincolnshire. Probably from a British word *Lemana* 'elm', found in OIrish *lem*, Irish *leamh*, and (with an irregular long vowel) Welsh *llwyf* (see Ekwall 1928: 244-5), or perhaps from a Welsh stream-name Llyman or Llymon 'keen, sharp' (see Thomas 1938: 73-4). A derivation from British *\*Lim-* identical with Welsh *llif* 'stream, flood', or from *\*lim-* 'marsh' is also possible: see Rivet & Smith 1979: 385-6. Less likely is a derivation from ME *leomen* 'give light, shine', so 'the shimmering brook', or from OE *leofmon*, ME *leman* 'beloved'. *Leman Sych*, mentioned by Shaw (1798: I 232), *Lemansyche* recorded in the late 13th century (VCH XIV 57), evidently a stream near Lichfield cathedral, may be associated with Leamonsley (q.v.). *Sych(e)* is from OE *sic* 'a watercourse'.

**LEOMINCHISTRETE, LEOMINCHESTRETE** (obsolete) Seemingly an ancient road from Shifford's Grange to Oakley Park (SJ 6935): see TSAS XI 1927-8 132; SHC 1945-6 26, 30; Palliser 1976: 40. *Leominche Street* 12th century JNSFC LXXII 1934-5 117-8; JNSFC LXIII 1948-9112, *Leominchistrete* 1298 SHC 1945-6 30, *Leominchestrete* 1447 ibid. 26. A curious name, possibly containing a personal name such as Leofman, perhaps associated with Little Manchester (q.v.). The terminal appears to be OE *strete*, *strēt* 'a paved road, a Roman road', though no Roman road is recorded in this area.

**LEPER HOUSE, LEPER WELL**  2½ south-west of Brewood (SJ 8704). *the lepre house* 1597 Codsall ParReg, 1652 Ct, *Leopard House* 1827 O.S., *Leper House* 1834 O.S. From a sulphur well said to have been frequented by lepers for its curative powers. By tradition a house or hospital for lepers existed nearby, possibly on or near the site of the present Leper House Farm: VCH V 20; VCH III 136. *The Lazar House* recorded in Lapley in 1838, evidently to be associated with Lazarus field between Ivetsey Road and Bellhurst Farm (SJ 8411), has the same derivation: Weate 1972; Oakden 1984: 172. Lazar Lane runs south from Milford to Cresset Pool and Wood, but the history of the name is not known. See also Freeford.

LETOCETUM, ETOCETO – see **HAREWOOD; LICHFIELD; WALL**.

**LEVEDALE**  2 miles north-west of Penkridge (SJ 8916). *Levehale* 1086 DB, *Levedehal* 1198 SHC III (i) 51, 1208 FF, 1242 Fees, *Levedhale* 1199 SHC III (i) 170, *Levedenhal'*, *Levedeshale* 1242 Fees, *Levedhal, Levedhale* 12th century Duig, *? Lovedale* 1397 SHC XV 76, *Leydall* 1532 SHC 4th Series 8 89, 1562 SHC XVII 211, *Leavedall* 1603 Penkridge ParReg. Perhaps 'Leofede's *halh*' (here perhaps in the sense 'a projecting corner of land', with reference to its location in Penkridge parish), possibly influenced by ME *levedi* 'lady', i.e. the Virgin Mary, from OE *hlæfdige*, though CDEPN prefers a derivation from the OE feminine personal name Leofgyth, and it may be noted that DB records Levild holding Shushions 4½ miles south-west of Levedale (VCH IV 58). No record has been traced of land here dedicated to the Virgin, but 2 acres of land here were held by St Thomas' priory at Stafford (SHC VIII 173), and a medieval chapel is recorded here in 1552 with a meadow in Bradley belonging to it called *St. Laurence meadow.* SHC 1915 207. *Chapel Yard*, recorded in a perambulation of 1772 (Penkridge ParReg) may mark the site. See also Tividale.

LEVENODESHAY (unlocated, probably near Hayend, ½ mile north-west of Hamstall Ridware (SK 0919)) *Levenodeshay temp.* John Shaw 1798: I 155. Possibly from the OE personal name Leofnoþ, with Mercian OE *(ge)heg* 'enclosure', so 'Leofnoþ's enclosure'. *Lawrence's Wood*, recorded in 1836 (O.S.), now Hayend Wood, may incorporate traces of the name.

LEWKENORE, LEUKENORE (unlocated) a manor in Ronton. *Lewkenore* 1454 SHC 1914 95; see also SHC III NS 213. Named after the Lewkenor or Lewknor family (whose name is from Lewknor in Oxfordshire, *(æt) Leofecanoran* c.994, 'Leofsa's *ham*'), who held the place from the late 14th century: SHC 1914 96; see also SHC III NS 213. The place was formerly known as Doyle Manor: SHC 1914 95; SHC III NS 213. See also Doyle.

**LEYCETT**  (pronounced lee-set [liːset]) 2 miles north-east of Madeley (SJ 7946). *Leveringsete* 1275 SHC VI (i) 66, *Loversete* 1278 Ct, *Loveresete* 1307 SHC XI NS 256, *Leveresheved* 1327 SR, *Leversete* 1334 SHC 1913 44, 1335 SHC XI 57, *Lesate* c.1376 Pape 1928: 150, *Levereshed* 1397 SHC XV 79, *Levershede* 1474 NSJFS 3 1963 49, *? Lysot* 1475 SHC VI NS (i) 92, *? Lysatt* 1528 SHC 1928 179, *? lysott* 1528 ibid. 260, *? lyssatt, lissatt* 1528 ibid. 265, *Leycytt* 1548 SHC 1912 169, *Lycett* c.1560 SHC 1938 176, 1562 SHC XIII 219, 1600 SHC 1935 208, *Lycette* 1573 SHC 1931 178, *Lysett* 1582 SHC XV 143, 1602 SHC XVI 219, 1616 SHC VI NS (i) 23, *Lycet* 1605 SHC 1940 249, *Licett* 1611 SHC III NS 55, *Leycett* 1679 SHC XII NS 205, 1747 Bowen, 1833 O.S., *Leasitt, Leasit* 1733 SHC 1944 5, 26. The earliest spellings, which without doubt relate to this place (see TNSFC 1963 48-9), *contra* the suggestion (Tooth 2000b: 120) that they relate to Ladderedge (q.v.), appear to be from the OE personal name Leofhere, with the OE suffix *-ing*, giving the place-name Levering, and OE *heafod* 'headland, summit, upper end, source of a stream', varying with OE *(ge)set* 'dwelling, place of residence; place where animals are kept, fold', hence 'Levering headland', or 'the dwelling-place or animal-fold at Levering'. The shorter forms are clearly a contraction of the earliest spellings.

LEYES GRANGE (unlocated) in Crakemarsh. *the Grange of leyes* c.1251 VCH III 226, *Lee Grange* 1538-9 ibid.' 228. Perhaps from OE *lǣs* 'meadow, pasture'. *Lee Grange or Stichbrooke Grange* is recorded in the 18th century (SRO D260/M/F/3/3): see Stychbrook.

LEYS 2 miles east of Kingsley (SK 0347). *Leys* c.1291 Tax, *Whiston Lyeseuse* 1328 Ipm, *Leyesheuese* 1331 SHC 1913 32, *Lyeshenese* 1331 ibid, *Lees* 1616 FF *et freq*, *Leghes* 1335 (p) Banco, *Whiston Lees* 1608 FF. From OE *lǣswe* (the dative singular of *lǣs*) 'at the pasture'.

LEYTON (unlocated) *Leyton* 1592 SHC 4th Series IX 57. See Layton; Leadendale; Leighton.

**LICHFIELD** Ancient Parish 15 miles north of Birmingham (SK 1109). *Letoceto* 4th century IA, *(On)licitfelda, (An)liccitfelda, Lyccitfelda* c.715 (11th century) Life St. Wilfred, *Lyccidfelth, Liccidfeld* c.737 Bede, *Liccedfeld, Liccetfeld* c.890 OE Bede, *Lecefelle, Licefelle* 1086 DB, *Lichesfeld* 1130 SHC I 3, 1164 ibid. 38, *Lichefeld* 1140s VCH XIV 38, *Licefeld* c.1148 VCH XIV 38, *Lychefild* c.1540 Leland, *Lichfeild* 1610 Speed. The name is not English but derives from British *\*Letocaiton* meaning 'grey or brown wood' (cf. Welsh *llwyd* 'grey, brown', from British *\*leto-*, and Welsh *coed*, from Celtic *\*caito-*, 'wood': the usual rendering of the first element is 'grey', but the word 'crosses the English perceptual boundary between grey brown' (JEPNS 12 73; Briggam 1998)), which became PrWelsh *\*Letged*, giving OWelsh *Luitcoyt*, developing at some time in the 7th century into OE *\*Lycced*, (Jackson 1953: 327, 332-4) to which was added OE *feld*, generally interpreted as 'open land', but at the time of the English incursions perhaps with the special meaning 'common pasture', and applied when the English began to cultivate that pasture: see Gelling 1981: 14. The spellings c.715 with the prefix *on-* (or *an-*) suggest (with parallels elsewhere) an area of some size called 'in Lichfield', and that 'Lichfield' came to mean the principal place in that area, in which case *Lyccidfelth* is likely to have been a late 7th century development of the existing name of a large area, gradually applied more particularly to the cathedral and its immediate area: VCH XIV 38; see also Introduction, pages 18-21. A full account of the history of the name Lichfield can be found in VCH XIV 37-9 and TSSAHS XXVIII 1986-7 1-13; see also Coates 1997: 453-8. On Caer Lwytgoed and Letocetum see TSSAHS XXXIII 1981-2 7-10. See also Christiansfield; Harewood. For Lichfield street-names see VCH XIV 40-42.

LICKSHEAD (obsolete) 5 mile west of Ramshorn (SK 0844). *Likshead* 1705 Alton ParReg, *Lickshead* 1753 SRO D240/D/229, *Lixhead* 1834 White, *Lickshead* 1834 O.S. A curious name of uncertain origin. See also Licks Wood.

**LICKS WOOD** ½ mile south-west of Ramshorn (SK 0844). *Lick's Wood* 1834 O.S. See Lickshead.

LIDGETT, THE (obsolete, ½ mile north of Beacon Hill, 2 miles sout-west of Penkridge (SJ 8913). *the new lyddeatt* 1610 Ct, *The Lidgett* 1834 O.S. From OE *hlid-geat* 'swing gate'. See also Lydiate.

LIGHT GATE (obsolete) a lost farm between Waterfall church and hall (SK 0851): NSJFS 12 1972 123. *Light's Yate* 1606 Ellastone ParReg, *Ligh Gate* (*sic*) 1743 NSJFS 12 1972 123, *Light Gate* 1846 ibid. Seemingly 'the gate of the man named Light'.

**LIGHT OAKS** 1 mile south-west of Bagnall (SJ 9250). *Light Oaks barn* 1816 SRO DW1909/E/9/1, *Oaks Barn* 1836 O.S. Evidently to be associated with *Lichokesegge* recorded in 1597: SRO 3764/47.

**LIGHTS, THE** on the north side of Watling Street, in Lapley parish, 1 mile south of Wheaton Aston (SJ 8511). *Near, Middle & Far Lights Meadow* 1838 TA. In 1548 it was recorded that unidentified donors had given lands worth 8d. net per annum to maintain a candle or light before the rood in Lapley church: SHC 1915 144. This place was evidently that land.

**LIGHTWOOD** 2 miles south-east of Longton (SJ 9241), *Lyhtwude* c.1230 SHC 1921 18, *? Litlewode* 1277 SHC XI 308, *Lichwode* c.1280 SRO D593/B/1/23/7/1/1/, *Lyghtwode* 1306 SHC 1911 65, *Lyghtwode* 1325 SRO D714/1, *Lightwoode* c.1374 Pape 1928: 147, *? Littilwode* 1391 SHC XI 198, ... *Lightewood heathe alias Meare heathe* ... 1544 SHC 1910 74, *Lightwood heath* c.1576 SHC 1931 186, *Lightwood Heath alias Trepwood alias Meare Heath* 1577 SRO D593/B/1/23/23, *Lightwood Forest otherwise Cocknage Bank* 1736 SRO Q/RDm/14b; **LIGHTWOOD** on north-east side of Cheadle (SK 0143), *Lytwode* c.1300 CroxdenChr, *Litwood* 1303 Chester 1979: 7, *Lyttewode* 1338 SRO D1275/3, *Lyghtwod* 1539 SHC XI 280, *Lightwoode (Croft)* c.1560 SHC 1938 172; LIGHTWOODDE HEATHE (unlocated, at Keele), *? Lichwodehet* ? 13th century SHC XI 321, *Lightwoodde Heathe* 1592 SHC 1930 (ii) 285; **LIGHTWOODFIELDS** 2½ miles north-west of Uttoxeter (SK 0534), *Lyghtwode* 1306 SHC 1911 67, *Lightwood field* 1636 SRO D5684/8, *Lightwood Field* 1836 O.S.; LIGHTWOOD on the western edge of Penn Common (SO 9080), *Lightwood* 1717 StEnc 359. From OE *leoht, wudu* 'the wood with the well-spaced trees, i.e. allowing light to penetrate', (perhaps used with reference to silver birch: VEPN I 103), alternating (in the case of Lightwood near Longton, if the forms relate to that place) with 'little wood'. For Lightwood near Longton, see also Shooters Hill. Lightwood Forest (*Lightwood Forest* c.1714 SRO D593/H/3/30), which included land in Meirheath, Normacot, Blurton Common and Cocknage Banks, was enclosed under an Act of 1734: see SHC 1931 90. For *Trepwood* see Threapwood Head.

LILLEBORN (unlocated, possibly near Thickbroom, or perhaps the name of the stream running north-east between The Bodnetts and Dunstall Farm (SK 1703)) *Lilleburne* (p) pre-1147 SHC 1910 312, *Lilleborn* 1234x1140 (1798) Shaw I xxvi, *Lilleburn* 1237 SHC 1910 295, 1240 SHC IV 236, 1281 SHC 4th Series XVIII 159, *Lilleborn'*, *Lilleborne* 1242-3 Fees, *Lileborn* 1243 SHC 1911 403, *Lillebourne* 1301 SHC VII 82, c.1325 Shaw 1798: I xxvi, *? Lilleburne* 1347 SHC XVIII 269. Perhaps from OE *lytel, burna* '(the place by) the little burn or stream', although the absence of -*t*- is surprisingly consistent. If the place was near Thickbroom (which the context of the 1240 and 1242-3 spellings suggests may be the case), the second element might be connected with Bourne Brook (called Black Brook west of Hints, and presumably *Burne* mentioned in 1235 (SHC 1910 295), *la borne* 1300 (SHC 1939 91)), perhaps 'the little bourne or burn', meaning a tributary, possibly what is now called Littlehay Brook. This place may be associated with the unlocated DB *Litelbech* (q.v.). See also Little Hay.

LIME CROFT (obsolete, in Alton Park (SK 0843)) *Lincrofteys* 1274 SHC 1911 160, *Lime Croft* 1836 O.S. Probably from OE *lin* 'flax', with OE *croft* 'a small enclosure of arable or pasture land, an enclosure near a house', giving 'the small piece of arable land used for growing flax'.

**LIMEPITS** 1 mile north-west of Whitmore (SJ 7941). *Lymputtes* c.1300 SHC 1913 239, 1327 SHC VII (i) 199, *Lin Pitts* 1668 Trentham ParReg, *Lime Pitts* 1742 SRO DW1082/C/6/1-4, *Lim Pits* 1833 O.S. Self-explanatory.

**LIMES, THE** on high ground 2 miles south-west of Newcastle (SJ 8243). *(Botterton-juxta-) Lyme* 1208 SHC V (i) 234, *ye Lymes* 1687 Trentham ParReg, *Limes* 1759 ibid, 1799 Faden, *Lymes* 1833 O.S. See Lyme. *Botterton* is Butterton.

**LINBROOK** a tributary of the river Swarbourn running into the river Trent. *Limbreuk* 1286 For, 1300 SHC V (i) 176, *Lynbroke* 1540 Ct, *Lynbrooke* 1611 Survey, *Lintbro(o)ke* 1650, 1658 Parl Survey, *? Limbrooke* 1798 Shaw I 60. The first element is probably OE *hlimme* 'stream, torrent', or OE *hlynn* 'noise, din', hence 'the noisy brook', particularly appropriate for this swift-flowing stream.

**LINCHFORD** (unlocated) *Linchford* 1288 SHC 1911 194.

**LINDORE FARM, LINDORE WOOD** 3 miles south-west of Gnosall (SJ 7919). *Lyndover(e)* 1323 CoramR, 1575 FF, *Lyndover* 1582 SHC XVII 227, *Lindooer* 1597 SHC XVIII 11, *Lindar* 1677 Gnosall ParReg, *Lindor* 1680 ibid. From OE *lind* 'lime-tree', and OE *ofer* 'hill-slope', meaning 'the slope with lime-trees'.

**LINTHURST (FARM)** 2 miles west of Tatenhill (SK 1722). *Linthurst Banks* 1658 DCL 380. Probably from OE *lind* 'lime-tree' and OE *hyrst* 'a copse, a woode eminence', so 'the copse of lime-trees', with ME *banke* 'bank, ridge'.

**LION'S DEN** a lane ½ mile south-east of Hammerwich (SK 0706). *Lion's Den* 1881 VCH XIV 259. From the late middle ages called *Elder Lane*, the name is seemingly from Thomas Lyon, who lived here in the 1840s:VCH XIV 259. See also Mottley Pits.

LITELBECH (unlocated, in Offlow Hundred, but see also Lilleborn) *Litelbech* 1086 DB. If the spelling is accurate (and many Domesday names are very corrupt) the first element of the name is evidently 'little', with the second seemingly from OE *bece* 'beech tree', or OE *bece* 'steep-sided stream valley'. However, a derivation from OE *bece* can probably be discounted because beech, for climatic reasons, were found only in the south of England in the OE period. The place cannot now be identified. It is recorded in DB (which provides the only known reference to the name) in association with *Burouestone* (q.v.) and *Weforde* (Weeford), probably because all three places were held by the same person as members of the Bishop of Lichfield's manor of Lichfield. The place may have lain near Weeford, but might also have been some distance away, possibly associated with another holding of the bishop. *la Bech* (unlocated) in Offlow Hundred is recorded in 1179 (SHC I 93) and 1199 (SHC III 43), and may be the place of the same name south-east of Stowe in Lichfield recorded in the 12th or 13th century (VCH XIV 7; SMR 02625 locates the deserted Anglo-Saxon settlement of Litelbech at SK 1260950), which was held by one of the members of Lichfield (VCH XIV 7; SHC I 93), and is presumably to be associated with *Beech Lone*, recorded in 1361 (*Deed*), *Bechefyld temp.* Edward II (SHC VI (i) 185), *Bechefield temp.* Edward III and Richard II (SHC VI (ii) 185), *Bechefeld* 1336 and 1374 (SHC 1939 94, 100), which is close to Borrowcop Hill (q.v.), and may be associated with *Burouestone*. If the second element in *Litelbech* is to be identified as *bece*, the watercourse may be the stream joining Trunkfield Brook from the higher ground to the east, though the word is normally found attached to well-marked stream valleys. Shaw 1798: I app. xiii identifies *Litelbech* as Littlebench, which has not been traced, but Pipe Wolferson suggested that 'Littlebech may prove to be Little-heth, i.e. the same hamlet, just beyond Thickbrome, in Weeford, now spoken Littlehay': Erdeswick 1844: 300; 545. See also Little Hay near Shenstone.

LITLEBROK (unlocated, in the Stone area). *Litlebrok* late 13th century SRO 3764/87.

**LITLEY** on the south-west side of Cheadle (SJ 9942). *Luthlehaie* 1203 SHC III 92, *Lutlehay* 1276 SHC 1911 168, *Lytlehay, Litlehay, Littlehaye* 1297 SHC VII 42-3, *? Littemay* (*sic*) 1327 SHC VII (i) 215, *Liutlehay* 1337 SHC 1913 65, *Littilhay* 1377 SHC XIII 141, *Lutley* 1601 SRO D538/A/5/53. *Lutley* 1601 SRO D538/A/5/53, *Littley* 1609 SHC III NS 52, 1836 O.S., *Litley (Hay Farm)* 1770 SRO D1229/1/4/19. 'The little hay or clearing'. *Litteleg* in Blithfield is recorded in 1252 (SHC 1937), and as *Luttelye* in 1325 (ibid. 124), but has not been located.

### LITTLE ASTON – see **ASTON, LITTLE**.

**LITTLE HAY** 1½ miles south-east of Shenstone (SK 1202), *? Littleshai* 1203 SHC III 90, *Luttelhay* frequently 13th century Duig, *Lutlehay* 1269 SHC IV 170, *Littlehay* 1327 SHC VII 198, *? Luttylhay* 1379 SHC XVI 174; LITTLE HAY obsolete, on the south-east side of Colton (SK 0520), *Littlehay* 1322 SHC 1914 159, *Luttellhay* 1325 SHC IX (i) 109, *Letylhay* 1542 SHC 1916 331; LITTLEHAY (unlocated, possibly near Shobnall or Branston), *Littlehay* c.1250 (1798) Shaw I 22, 23; LITTLEHAY (unlocated, possibly in Anslow), *Littlehay, Lithlehaya* c.13th century SHC V (i) 48-9, *Luttelhay* 1341 HLS, *Littlehay* n.d. Shaw 1798: I 35. From OE *lytel* 'little', and Mercian OE *(ge)heg* 'a fence, enclosure', here meaning 'the small enclosure'. Until the early 18th century the place near Shenstone was part of a vast heath. Erdeswick 1844: 300 states that it was known c.1800 as *Littleheth*. For Little Hay near Colton see see Parker 1897: 162-4; SHC XI NS 27 fn.1, 45; SHC 1914 153. *Luttelhaysiche* (with OE *sic* 'a watercourse') is recorded in Heatley in 1304 (SRO D(W)1721/3/19/16), evidently to be associated with *Littlehay* recorded in 1670 (SRO D4038/E/11/7). See also Litelbech; Litley; Morehay.

**LITTLEHAY BROOK** a tributary of the Crane Brook running into Black Brook, which flows into the river Bourne. *Littlehay Brook* 1784 Survey – see Littlehay.

**LITTLE HEATH** 2 miles north-west of Penkridge (SJ 9017), *The lyttle heath by Leavedall* 1609 Penkridge ParReg; **LITTLE HEATH GREEN** ½ mile west of Almington (SJ 6934), *Little Heath* 1684 SRO D861/E/5/21. Self-explanatory.

### LITTLE ONN – see **ONN, LITTLE**.

**LITTLEPARK** on the Staffordshire-Derbyshire border, ½ mile north of Okeover (SK 1548). *the Little Park* 1640 SRO D3155/6881. Self-explanatory.

### LITTLE SANDON – see **SANDON**.

### LITTLE STOKE – see **STOKE-BY-STONE**.

LITTLETON – see **BESCOT**.

LITTLEWELL (unlocated,in Baltereley) *Littlewell* 1705 BCA MS3558/72.

**LITTLEWOOD** 1 mile north-east of Cheslyn Hay (SJ 9807), *Luttelwode* 1380 Banco, *Little Wood* 1834 O.S.; LITTLEWOOD (unlocated) in Okeover, *Luttulwde* c.1225 SHC 4th Series IV 105; LITTLEWOOD (unlocated, in Barlaston), *Litlewode* 1277 SHC XI 308. Self-explanatory.

**LITTLEWORTH** on the west side of Stafford (SJ 2223), *Little Worth* 1775 Yates; **LITTLEWORTH** on the south-west side of Woodseaves (SJ 7925), *Littleworth* 1833 O.S.; **LITTLEWORTH** on the south-west side of Hednesford (SK 0111), *Littleworth* 1834 O.S.; **LITTLEWORTH** 2 miles west of Rocester (SK 0738), *Littleworth* 1836

O.S.; **LITTLEWORTH** on the east side of Stafford (SJ 9323), *Littleworth* 1794 SRO D240/E/F/8/20. A common name, a self-explanatory derogatory term for poor land.

**LITTYWOOD** an ancient moated manor house in Bradley parish, 3 miles north of Penkridge (SJ 8818). *Lvtiude* 1086 DB, *Lutiwude* 1203, 1204 P, *Litewude* 1206 Cur, *Litlewude* 1230 SHC IV 229, *Luttewd* c.1251 SRO D59[7909], *Luttywode* 1289 SHC 1911 45, *Lottewode* late 13th century SRO D938/76, *Luteywode* 1301 SHC 1911 270, *Luttewode* 1315 ibid. 1911 85, *Lutywode* 1334 SRO D938/78, *Lutelwode* 1390 FF, *Luttelwode* 1406 SHC XVI 49, *Lytlewod* 1592-3 Eliz ChancP, *Lyttywood* 1601 SHC 1934 (ii) 4, *Littlewood, Littywood* 1624 FF. The principal feature of this place is a great circular double moat, 650' in diameter, which may be pre-Conquest, perhaps developed from a prehistoric earthwork, and possibly the original caput of the de Stafford family: VCH IV 74-5, 79-80. A puzzling name with an uncertain first element. The derivation proposed in Ekwall 1936: 126, OE *litel, lytel* 'little' with the second element lost, does not seem entirely satisfactory, despite the spellings with a medial *l*, so possibly connected in some way with OE *lutian* 'to hide, to lurk, to ambush' (northern dialect *lute* 'to lurk, to lie hid': EDD), used in some topographical sense such as 'sanctuary, refuge', meaning 'the wood where refuge was sought', or from the associated OE *lytig* 'crafty, cunning', perhaps here 'the place or earthwork into which animals were driven and trapped'. A further possibility is that the first element is from OE *hlyð* (plural *hliðu, hleoðu*) 'a slope, hillside, declivity', giving 'the wood on the hillside': the place (which is on a watershed) lies on the slopes of Butter Hill (503'): cf. Lythwood, Shropshire. That would not, however, explain the medial *-i-* or *-e-*. Whatever the root, the first element clearly evolved into ME *lutel, luttil* 'little'. The second element is OE *wudu* 'wood'. It may be noted that the name *Littimore*, which appears to be associated with this place, is recorded before 1261 (SHC IV (i) 220-1), and in 1299: SHC VII (i) 61. *Lightiwode*, possibly in Marchington, is recorded in 1306 (SHC VII 149), but may be associated with Lightwoods (q.v.). The unlocated *Lutheburgh* (q.v.) may have the same root as Littywood, and might refer to an ancient earthwork.

**LIZARD** in Shropshire, 7 miles west of Brewood (SJ 7809), included here because of its association with Weston-under-Lizard (q.v.). *Lusgerde* 664 (12th century, S.68), *Lusgeard* 680 (12th century, S.72), *Lusegarde* 1199 (1285) CH, *Lusyard, Lusard* 1199 Rees 1997: 11-2, *Lusard* 1247 ibid. 1997: 94, *Lusghart, Lusegard* 1265 ibid. 136-7, *Lousyerd* 1282 SHC VI (i) 154, *Lusgarde, Lisgarde* 1291 Tax, *Loseard* c.1298 Rees 1997: 95, *Lousyord* 1307 SHC VII 180, *Lousyerd* 1324 SHC 1911 111, *Luseord* 1404 SHC XV 111. Formerly thought to be from Welsh *llys-garth* 'hall by a hill' (Johnston 1914: 348, 502; Ekwall 1960: 301), the OE spellings *Lus-* show that *llys* cannot be the first element. The 7th century forms make a derivation from OE or OWelsh equally possible, and plausible derivations can be offered in both languages. Possibly therefore from Welsh *llus* 'bilberries', with Welsh *garth* 'mountain ridge, promontory, hill; wooded slope, woodland, brushwood, thicket, uncultivated land', so giving 'bilberry hill' (see Coates & Breeze 2000: 195; but note observation of Dr Oliver Padel, 'unconvinced by ... *llus* 'bilberries' ... mainly because ... [little] evidence for the use of this word in place-names, particularly as the first element in compound names ... [but] it would at least fit phonologically': personal communication 2 December 1995). The name was adopted (or coined) by the Anglo-Saxons when they reached this area in about the mid 7th century, and it is possible that they took the name from its phonological similarity to be (or coined the name anew) *lus geard*, from OE *lus* 'louse' and OE *geard* 'a fence, a hedge, an enclosure' (a rare element in place-names, and especially so in early names): Lizard Hill is a great low rounded hill (with a number of Roman forts on the northern side, to the south of Watling Street, and a lost Roman road running north to south on the east side of the hill), with an unusual fold or cleft to the south of the summit and a smaller hill to the north, with the profile of the hill, particularly from Watling Street, indeed suggestive of a louse, with the 'fold' (or the smaller hill) separating the body from the head. That the name is derived from the hill can be deduced from the name Weston-under-Lizard, for that place is not known to have been a manorial part of Lizard, so the reference to Lizard must be presumed to be to a topographical feature. Near the hill are a number of smaller rounded (? louse-like) knolls. For completeness it may be added that there is evidence that the word *lus* is often found with words denoting 'barrow, hill', where the meaning 'louse infested' would be inappropriate, and it has been suggested that it may have been used to describe something small and insignificant: Luston, Herefordshire, is believed to incorporate the element in that sense. But it seems possible that the association is related to the shape (and possibly size) of the hill or mound, i.e. 'louse-shaped', and *lus* may have been a nickname for such hills or tumuli, cf. *lusdune* 'louse-hill' mentioned in a charter relating to Oldswinford, Worcestershire, of 946x955 (15th/16th century, S.579; see Hooke 1990: 165). Cf. Loosebarrow, Dorset; Lousehill, Somerset; Luscott, Devon (see EPNE i 198). See also Luzlow. *Drayton subtus Lusyerd* is recorded in 1365 (SHC VIII NS 24), and is almost certainly a lost settlement now remembered in Drayton Lodge, 2 miles north-east of Shifnal in Shropshire (SJ 7509).

**LLEPERISDALE** (unlocated, in Barlaston) *Lleperisdale* 13th century SHC XI 324. Seemingly 'the valley of the leper(s)'.

**LLOYD, THE** 1 mile south-east of Almington (SJ 7133). *le loyds-yardes* c.1570 SHC 1945-6 148, *a pasture called the Loyd* 1585 SHC 1945-6 192, *the Loyd* 1623 ibid.' 191, *the Lloyde* 1750 SRO 828/28, *The Lloyde* 1833 O.S. The place lies on a stream, so almost certainly from OE *hlȳde*, a derivative of *hlud* 'loud', with the meaning 'a torrent or swift noisy stream' (cf. Ludlow and Ludford, Shropshire; Ludwell, Derbyshire). See also Lloyd (Brook, Hill & House).

**LLOYD (BROOK, HILL & HOUSE)** 3 miles south-west of Wolverhampton (SO 8894). *hlȳde broc* 985 (12th century, S.860), *Ludebroc* 1206 SHC III (i) 219, *Lude* c.1220, 1242 SHC 1928 12-14, *Lydebrok* 1294-5 ibid.' 23, *Luithulle* 1317-8 ibid. 30, *the Lude* 1353 (1801) Shaw II 222, *la Luyde* 1406, 1410, 1416-17 SHC 1928 63-5, *Lodbroke* 1424 Wodehouse, *Lidbroke* 1442 SHC 1928 47. An ancient and well-recorded name which is linked to both a brook (now Lyde Brook) and a hill, the latter recorded as *Monte de la Lude* in 1292 (SHC 1911 209) and *the Hill of la Lude* 1317 SHC 1928 28. A derivation from OE *hlið* 'a slope, a hillside' (EPNE i 252) might be supported by the earliest spelling and would fit the topography here, with its particularly prominent hillside. If that is correct, the stream will have taken its name from the feature. However, the later forms point strongly towards a derivation from OE *hlȳde*, a derivative of *hlud* 'loud', with the meaning 'a torrent or swift noisy stream' (cf. Ludlow and Ludford, Shropshire; Ludwell, Derbyshire). The stream here appears to have been called *Smalbroke* in 1416 (SHC 1928 65). A derivation from OE *hleda*, *hlȳda* 'a seat, a bench', perhaps in the sense 'ledge', cannot be ruled out completely, and would suit the local topography. There is little likelihood of a derivation from the OE personal name *Hlud. See also Lloyd, The.

**LOACHES BANK** (obsolete) the name of a former earthwork, now levelled, near Bourne Pool, Aldridge, about 100 yards to the west of the Old Chester Road: Willmore 1887: 12-13 (SO 0799). *Loaches Banks* 1831 Lewis. Willmore 1887: 12-13 describes earthworks which covered 2 acres, and there is a plan of the camp (which shows two superimposed earthworks) in Shaw 1798: I plate A, 10-11, but the site is not named. The site may have originated as an Iron Age enclosure, or could be medieval: see TSSAHS XXIY 1982-3 34; TSSAHS XXXII 1990-1 90-1; Hodder 2004: 45. Ric'o de Erdbury is recorded in nearby Aldridge in 1327 (SHC VII (i) 230), and *Arbora Meadow* and *Arbory Meadow* (unlocated) are recorded in Aldridge in the 17th century (SOT D260/M/T/12), presumably associated with *Rough Arbora Wood* recorded in 1676 (SRO D260/M/T/2/12), and *Harborough Meadow* recorded in 1864 (SRO D1317/1/15/2/2), perhaps from Old English *eorþburg* 'earth fortification', an early reference to this earthwork. The name Loaches is perhaps from the surname of an occupier: the surname Loach is believed to be from OE *loche*, 'loach, an edible freshwater fish ' (DES 282), and it may be noted that this place adjoins Bourne Pool. Or perhaps from OE *\*lece* 'a stream flowing through boggy land, a bog': c.f. The Laches. There seems little likelihood that this name is to be associated with Luitcoyt, from which the name Lichfield (q.v.) derives. Cf. Land called *Loshes* (1536), *Laysshes, Laisses, Loshes* (1598), later *Loches, Lowches, Lowchers, Lawshes, Losshes*, in Earls Colne, Lincolnshire.

**LOCKWOOD (HALL)** 1 mile south-east of Kingsley (SK 0245). *Locwode* 1274 SHC 1911 159, 1311 ibid. 436, 1327 Okeover T19, *Lokwode* 1331 SHC 1931 31, *Locwood* 1399 SHC 1921 14, *Lockwood* 1599 SHC 1935 98. From OE *loc-wudu* 'enclosed woodland'. Cf. Lockwood, Yorkshire.

**LODE (HOUSE)** 1 mile east of Alstonefield (SK 1455). *The Load* 1658, 1675, 1679 Alstonefield ParReg, *the Loade* 1678 ibid, *Load* 1834 White. From OE *lad* 'watercourse', or perhaps in some cases 'crossing': the place lies in a valley running down to Load Mill on the river Dove. See also Hampton Loade.

LODYNGTON (unlocated: the context in which the name appears suggests as association with Dudley, but the place is perhaps Loddington, Leicestershire, possibly held by Roger de Somery from his second wife Annabel who brought lands in Leicestershire including Great Dalby, which in 1086 was jointly held with Loddington by Robert de Bucy). *Lodynton* 1273 SHC 1911 153, SHC IX (ii) 25, *Lodyngton* 1305 SHC VII 135. 'The *tun* associated with Luda'.

LOGES (unlocated, at Rodbaston: see SHC 1911 260; SHC 4th Series XVIII 79) *Loges* 1209 SHC III (i) 209, c.1238 (1798) Shaw I xxvii, 1298 SHC 4th Series XVIII 21, 1300 SHC 1911 262. From ME *log(g)e* 'a lodge, a hut, a cottage', perhaps here meaning 'a house in a forest for temporary use', perhaps here the residence of the Forester of Cannock Forest. *Les Logges*, recorded in 1399 and associated with the keeper of Kinver Forest, appears to have been constructed c.1373 at *Coppidhull* (unlocated) within Kinver Forest: VCH II 347.

**LOGGERHEADS** a hamlet based on a crossroads 1 mile south-east of Mucklestone (SJ 7335). *Loghead* 1657 Newcastle ParReg, *Logerheads* 1775 Yates, *Loggerheads* 1798 Yates, 1808 Baugh, *The Logger Heads* 1833 O.S. The age of this place-name is unknown, but although, as noted in Duignan 1902: 96, Loggerhead is a dialect word for the knapweed, *Centaurea nigra*, found on wet ground (EDD), it is almost certain that the name is from a public house predating the present Loggerheads public house (*Logger Heads* 1872 P.O.), which until recent times was the only building here: it appears from Baugh's 1808 map of Shropshire to have stood on a wedge-shaped island at the junction of the four roads which meet here. A loggerhead was a term for 'a blockhead, a dull stupid person' (EDD), probably from dialect *logger* 'heavy block of wood', and the words 'We three loggerheads be' was the inscription on a common public-house sign, in which two wooden heads were shown, the unsuspecting spectator being the third (OED), a jest probably alluded to by Shakespeare in 1588 when he speaks of 'the picture of we three' in 'Love's Labour's Lost', and in Twelfth Night, Act II, scene 1. The expression 'to go to loggerheads' later came to mean 'to fight or squabble' (Halliwell), but is not recorded in that sense until 1831 (Chambers), and is not relevant here. It is likely that the public house was The Loggerheads, with a sign showing two clowns or fools (see Yonge 1923: 167-8), and was replaced by a triangular inn-sign in the gable of the public-house which pictured three jovial bumpkins with the legend 'We three loggerheads be', with a Staffordshire knot (JNSFC LXVI 1931-2 186), which destroyed the nature of the joke. Gough 1968: 177 mentions in 1706 an alehouse at Shrewsbury in 1642 called 'the Loggerheads', which was known as 'ye Loggerheads in Baylie' in 1521 (Lloyd 1942: 37; PN Sa IV 39): the three leopards' heads in the arms of Shrewsbury are said to have been known as loggerheads. Loggerheads is also recorded near Mold, Flintshire, and in Wiltshire: see Larwood & Hotton 1866: 39; 458-9. Cf. *Loggerhead Farm*, Great Wyrley, 1792 Ct.

**LONCO BROOK**   a tributary of the river Mees. The stream forms the Staffordshire-Shropshire border north-west of Forton. This river-name is found in the field-names *Longhale* 1332 to 1364, *Lonkehall medowe* 1487, *Lyncoll feilde* 1493, *Lyncoll medowe* 1527, *Lonco platt* 1599, *Loncolne hill, moor, pits* 1618, *Loncoll field buttes* 1618, *Loncale flatts* 1693, *Loncall* 1720 (forms from Oakden 1984: 150; 152), *Loncall field* c.1700 (SHC 1932 71). Oakden 1984: 150 suggests that the derivation of the field-names is from OE *lang halh* 'long halh', with the unvoicing of *-g* becoming *-k*. However, Welsh *llwnc*, Cornish *\*lonk*, Breton *lonk* 'gulp, gullet' (see Padel 1985: 153), may be the root of this name. The form *Wlonkeslowe* is linked with this name in SHC 1916 142, without date, source, or explanation, but is a 13th century spelling for Longslow, 1 mile north-west of Market Drayton: see PN Sa I 183. *Platt* is from ME *plat* 'a flat place, a footbridge', here probably in the latter sense. See also Carncoe (Rough), which may incorporate the same element.

**LONDONDERRY**   to the west of Smethwick (SO 0187). *Londonderry* 1834 O.S. Transferred from Londonderry in Ireland, a name that dates from the early 17th century: in 1610 the London Companies were involved in the plantation of the old county of Coleraine and parts of adjoining counties, and the exercise was commemorated by adding the name London to the older name of Derry. The reason for its adoption near Smethwick may be due to the influx of Irish immigrants.

**LONDON, LITTLE**   on north side of Willenhall (SO 9698). *Little London* 1658 Wolverhampton ParReg, 1721 WA II 36, 1749 Bowen, 1834 O.S. Perhaps from land here which is said to have been owned by City of London Companies, including the Merchant Taylors' Company (StEnc 363), or from an area set aside as a resting place for drovers and cattle on the drove-road from mid- and north-Wales to London, as also places of the same name in Alveley, Munslow and Oswestry (Shropshire): see Foxall 1980: 24; PN Sa III 176, which describes the place in Alveley as a squatter settlement. Other places called *Little London* are recorded from the 19th century in Walsall and near Harlaston: StEnc 362. Watts 2004: 379 suggests that the name (recorded in Chichester in 1454) is ironic, and that there are many other examples from Buckinghamshire to West Yorkshire.

LONE (unlocated, on the south side of Wolverhampton) *la Lone* 1293 SHC VI (i) 283, 1321 BCA MS3145/288608, 1329 SHC I 325, 1349 SHC XII 89, 1368 BCA MS3279/351231, *le Lone* 1319 SHC X (ii) 29, 1327 SHC VII (i) 249, *la Lona* 1350 (1801) Shaw II 204. A common name: 'the lane'. There are many references to the name, which appears to have given its name to the Lane family of Bentley (see SHC 1910 141), in the 14th and 15th centuries. *le Lone* is recorded in Caverswall in 1327: SHC VII 215.

LONEDALE (unlocated, in Tillington) *Lonedale* (undated) SHC VIII (i) 122. Perhaps to be associated with *Londehall*, recorded c.1238 (1798) Shaw I xxvii.

**LONG BIRCH** 2 miles south of Brewood (SJ 8705). *le longebruch* 1425 Ct, *Longbryche* 1540 Deed, *Longburch* 1682 Browne, *Long burch* 1686 Plot. From OE *bryce* 'breaking', used of 'newly cultivated ground'. By metathesis, or shifting of the *r*, the word becomes *burche* and later *birch*. *Long Birch* means simply 'long piece of newly-cleared land'. Strangleford Birch and Harvington Birch are 2 miles west of Brewood. The country hereabouts was formerly part of Brewood Forest (q.v.). The name is not unique in Staffordshire: see e.g. *Long Birche* in Eccleshall, recorded in 1570 (SHC 1926 78).

**LONG COMPTON** – see **COMPTON**.

LONGCHURCH (unlocated, near Shifford Grange) *(vallem de) Longchurch* 1447 SHC 1945-6 28, 30. A curious name which remains unexplained.

**LONGCROFT (FARM)** 1 mile north-west of Yoxall (SK 1420). *Longcroft temp.* Henry III (1798) Shaw I 102, 1817 Pitt. From OE *lang* 'long', with OE *croft* 'a small enclosure of arable or pasture land, an enclosure near a house', so 'the long enclosure of arable land'.

LONGDOLES (obsolete) 1 mile south-west of Weston Coyney (SJ 9144). *Longdoles* 1836 O.S. Perhaps to be associated with *Longedale*, recorded in 1199 (SHC IV 282), in which case 'the long valley'. Otherwise 'the long pieces of land forming shares in the common field', from ME *dole*.

**LONGDON** Ancient Parish 4 miles north-west of Lichfield (SK 0814). *(æt) Langandune* 1002x1004 (11th century, S.906; 11th century, S.1536), *Langedun* 1158 P, *Longedon* 1166 SHC 1923 295, *Langedon* 1195 P, 1242-3 Fees, *Langedun* 1198x1208 (13th century) EEA 17 74, SRO D938/103, *Longdon* 1268 SRO D603/A/Add/189, *Langgedn*, *Langgedon*, *Langedon* 13th century BCA MS3415/132-3, *Longodon* 1346 BCA MS 3415/174, *Lo'ggedon* 1350 BCA MS3415/184, *Lankedon* 14th century BCA MS3415/144. 'Long hill', from OE *langan-dun*, presumably the high ridge between Longdon and Rugeley. Plot 1686: 406 mentions an ancient square earthwork, later known as Longdon Camp, at the east end of Longdon church (see also VCH I 346; StEnc 366); cf. *Berridun* in this area, possibly from *byrig-dun* 'hill with the fortification', recorded in 1350 (BCA MS3415/184), and *le Berywey*, recorded in 1382 (BCA MS3415/197)), perhaps to be associated with Borough Lane, a road running south-west from Longdon shown as *Burrough Lane* (from OE *burh* 'fortification, manor') on Yates' map of 1775, presumably associated with *Burrow Lane*, marked as a place on the O.S. map of 1834 to the south of Upper Longdon, and *Burwaye Lane* recorded in 1506 (OSS 1936 51), although the *burh* may be Castle Ring (q.v.): see also Burwey. Shaw 1798: I 227 could detect no trace of any earthwork at Longdon, but note *Castle Croft* in Longdon, near Russell's Bank, recorded c.1737 (SRO D260/M/T/5/125). TSAS 4th Series I 1911 xxxiv 18 fn.82 observes, without explanation, that *Langandune* is not this Longdon, but offers no other identification. Upper Longdon, 1 mile west of Longdon (SK 0614), appears as *Longdon Upper End* in 1834 (O.S.). See also Broughton.

LONGEBRUGG (unlocated, near Fisherwick). *Longebrugg* 1309 WL103. 'The long bridge'.

**LONGFORD** a local name for the section of Watling Street between Churchbridge and Four Crosses, south of Cannock (SJ 9609). *? Langeford* 1310 SHC X 7, *? Longeforde* c.1418 (1801) Shaw II 94, *Longford House* 1895 O.S. *Long strete* is given in a charter of 996 (for 994) relating to Hatherton (17th century, S.1380: see Hooke 1983: 78, 82), and is believed to relate to Watling Street here: see also *Longgofordeshet* (Longfordheath) undated, SHC 1928 143. (All the other forms in Oakden 1984: 60 relate to Longford in Shropshire: see PN Sa I 181-2). The second element is OE *ford*, often found where the word is best translated as 'causeway', and that meaning may apply here. The name Longford is not uncommonly applied to Roman roads, for example, south-west of Market Drayton for a length of the road (Margary number 19) between Pennocrucium (Water Eaton) and Chester, and it has been suggested that such names incorporate Welsh *fford* (borrowed into Welsh from OE *ford*), with the meaning 'road', indicating a Roman road (see particularly Jermy 1992: 228-9; Jermy & Breeze 2000: 109-10), but in such cases the names probably incorporate OE *ford*, applied to sections of Roman road which happen to have been constructed on a causeay. An early record relating to the name Longford in Shropshire is found in 1319, when the Sheriff of Shropshire acknowledged that the Royal Road called *Longeford*, between Bletchley and Newport, was dilapidated and impassible from water overflowing from adjacent marshes, and a levy of pontage was granted for necessary repairs: TSAS 2nd Series I 155. *Longefordeweye* in Alrewas is recorded in 1328 (Alrewas Ct), and may refer to Riknild Street, or to a *weg* ('way') leading to that road.

**LONGHAY** (unlocated, in Anslow) *Longhay* c.1240 (1798) Shaw I 35, *Longhaye* 1292 ibid. From OE *lang, (ge)heg* 'fence, enclosure', so 'the long enclosure'.

**LONGINDON** (unlocated, in Amington: SHC 4th Series XVIII 97) *Longindon* 1271 SHC 4th Series XVIII 97.

**LONG LOW** 1 mile south-east of Wetton (SK 122539). From OE *lang-hlaw* 'long burial-mound', with reference to two neolithic round barrows linked by a bank, an arrangement which may be unique in England: see NSJFS 1965 56.

**LONGNOLRE** (unlocated, but possibly what is now Hateley Heath (q.v.), 1½ miles north of West Bromwich (SO 0093)) *Longnolre by Nortune* n.d. Shaw 1798: I 173, *Longmore* n.d. Willet 1882: 209. From OE *lang* (weak dative singular *langan*) 'long', with OE *alor* 'alder', hence 'the tall alder tree', or, more probably, 'the long alder-copse'. *Nortune* is unexplained, but may be evidence that the place was actually near Norton Canes.

**LONGNOR** 10 miles north-east of Leek (SK 0864). *Longenovre* 1277 SHC 1911 169, *Longnour* c.1549 SHC 1910 73, *Longnor* 1682 Browne. From OE *lang* (weak dative singular *langan*) 'long', with OE *ofer* 'flat-topped ridge with a convex shoulder'. Longnor lies at the lower end of a long flat-topped gradually rising ridge. The west the south-east summit ends in a steep concave slope, which would not be expected of an *ofer*. The slope may have been affected by landslips, of which plentiful evidence can be detected on the sides of the bank, a possibility perhaps reinforced by a slight convex curve below the convave summit slope, which may mark a landslip deposit. However, the north-west end of the hill terminates in a conspicuously rounded profile above Nab End (q.v.) . Cf. Longnor, Shropshire.

**LONGNOR** in Bradley parish, 3 miles west of Penkridge (SJ 8614; SMR 01905 places a deserted Anglo-Saxon settlement at SJ 86601400). *Longenalre* 1086 DB, *Langenalre* 1242 Fees, *Lungenalre* 1285 FA(p), *Longenolre* 1327 Duig. From OE *lang* (weak dative singular *langan*) 'long', with OE *alor* 'alder', hence 'the tall alder tree', or, more probably, 'the long alder-copse'.

**LONGPORT** on the west side of Burslem (SJ 8649). *the long bridge* 1544 SHC 1910 247, *Longe Bridge* 1569 JNSFC LX 1925-6 63, *Long Bridge* 1680 SRO D4842/14/1/2-3, *Longbridge* 1766 Simms 1894: 383, *Longport* 1783 BCA MS917/1391, 1836 O.S. It has been suggested that at least some places with this name held markets along their main street, hence 'long port or market' (see Ekwall 1936: 182-3), but Ward 1843: 155-6 saya that 'Formerly, the few cottages standing here had the name of Longbridge, from a foot-bridge of planks, which extended about one hundred yards along the side of a wash or brook-course, through which the old high-way, from Burslem to Newcastle, passed, before the making of the Turnpike Road; but on completion of the canal, and the erection of several houses and manufactures on its banks, the place acquired its present name.' See also SHC 1910 247; SHC 1934 (i) 32; VCH VIII 109.

**LONGRIDGE** 1 mile north-west of Penkridge (SJ 9115). *Langerig(g)e* 1199 Ass, 1274 (p) FF, *Langerig*, *Lungridge* 1253 SHC IV 125, *Longrigge* 1276 SHC VI (i) 74, *Longrigge*, *Langerugge* 13th century Duignan, *Longerygg* 1307 SHC VII 174, *Langrigge* 1399 Pat, *Longerugge* 14th century Duig, *Long Riche* 1532 SHC 4th Series 8 89. From OE *lang*, and OE *hrycg* 'long ridge'.

**LONGSDON** 2 miles south-west of Leek (SJ 9654). *Longusdon* (p) c.1223 Chell, *Longeston* 1240 Harl to 1290 Ipm, *Longesdon* 1242 Fees, 1252 Ch, 1331 SRO D1337/1, *Langesdun'* c.1246-61 StCart, *Longrisdon* 1274 Ipm, *Langesdon'* 1275 Cl, *Long(e)don* 1327 SR to 1560 Pat, *Longisdon Syde* 1532 SHC 4th Series 8 32, *Longysdon* 1547 Ct, *Longsdon als Longston* 1612 FF. The possessive *s* suggests a derivation from the OE personal name Lang, with OE *dun* 'hill', hence 'Lang's hill', but the place is on a long ridge, which may have been called *Long*, to which was added an explanatory *dun*. Cf. Longstone, Derbyshire. Syde is from ME *side* 'a slope of a hill, especially one extending for a considerable distance'. *Nether Longsdon*, recorded in 1278, may have been in what is now Dunwood (q.v.): VCH VII 203; 205.

**LONGSHAW** ½ mile south-east of Bradnop (SK 0154), *Longshawe* 1241 SHC 1911 438, *Longeshaghe* 1337 SHC XI 143, *Longschawe* 1511 Okeover T696; **LONGSHAW** 1 mile north-east of Oakamoor (SK 0745), *Longshaw* 1746 SRO D240/B/3/35. From OE *lang-scaga* 'the long copse'.

LONGSTONE (unlocated, possibly near Leek) *Longstone* 1634 Leek ParReg.

**LONGTON** one of the towns of Stoke on Trent (SJ 9043). *Langetun'* 1242 Fees, *Langeston* c.1249 SHC 1911 146, *Longeton'* 1251 Fees, *Langeton* 1304 SHC VII 124, *Longelton* 1316 SHC IX (i) 52. 'Long *tun*'. This common name often denotes a settlement strung out along a pre-existing ancient road. In this case the place lies on the Roman road (Margary number 181) which runs from Rocester to Stoke on Trent.

LONGWEY (unlocated) *Longwey* 1598 SHC 1935 80.

LOQUIKEHACH (unlocated, perhaps in the Saverley area) *Loquikehach* 1284 SRO D1790/A/12/23.

**LORDSHIRE** on the north-east side of Werrington (SJ 9447). *Lordshare* 1836 O.S., *Lordshire* 1889 O.S. Probably 'the lord of the manor's share or piece of land'.

**LORDSLEY** 1 mile east of Mucklestone (SJ 7437). *Lordes ley*, *Lordisley otherwise Harwodde* 1529 SHC 1910 19, *Lordys Ley* 1530 SHC 1912 33, *Lordsleys Spring* 1790

D240/E/F/8/37, *Lordsley* 1830 Moule, *Lord* 1834 O.S. 'The *leah* of the lord'. *Spring* is a newly-planted wood or a coppice, or a spring of water.

LORDSMORE  (unlocated, near Uttoxeter), *Lordsmore* c.1737 SRO D260/M/T/5/125; LORDSMORE  (unlocated, in Ettingshall), *Lordsmore* 1673 SRO 1237/62. 'The moorland belonging to the lord'. See also Ladymoor.

**LORDSPIECE**  1 mile north-west of Upper Mayfield (SK 6546). *Lords Piece* 1836 O.S. Possibly from the surname Lord(e): Henry Lorde of Caldon is recorded in 1452: SHC VII NS 56.

**LORD'S WELL**  a chalybeate well at Sinai Park (SK 2223). Shaw 1798: I 24 mentions an inscription in stone recording that the well was rebuilt by William, Lord Paget, in 1701.

LOSKESFORD  (unlocated, possibly near Almington) *Loskesford* 1327 SHC VII (i) 203.

LOUDON    (unlocated, possibly near Hamstall Ridware) *Loudon* 1686 SBT DR18/22/7/6.

LOUECOKESHULL  (unlocated, by the river Blithe near Caverswall) *Louecokeshull* ?c.1270 SHC VIII (i) 150.

**LOUNT FARM**  1 mile north-west of Colton (SK 0321), *Lund sub Colewele* 1198x1208 EEA 17 74, *Lund* c.1200 SRO 103[7934], c.1225 SHC VIII (ii) 155, *le Lounde* c.1230 ibid.' 156, *Le Lound* pre-1260 SRO 107[7925]; **LOUNT FARM**  1 mile south-west of Rolleston (SK 2126), *The Lount* 1304 Shaw 1798 I 31, *Lant* 1798 Yates, *The Lount* 1832 Teesdale, *Launt* 1836 O.S.; **LOUNTS, THE**  (obsolete) 2½ miles south of Madeley (SJ 7641), *The Lounts* 1833 O.S.; **LUNT, THE**  1 mile south of Willenhall (SO 9696), *? Hampton Lunt* 1601 SRO D4407/48[SF83],.*Bilston Lunt* 1754 SRO D562/8, *The Lunt* 1863 Lawley 1893: 220-21.  From ON *lundr* 'a grove, a copse' (in some cases 'sacred grove'): cf. Lound, Lincolnshire; Lund and Lunt, Lancashire. For *Colewele* in the first spelling see Coley. *le lunde* is recorded in Field near Leigh in the 14th century: Tooth 2000b: 176.

LOUSEY BANK  (obsolete) 1 mile west of Butterton (SK 0556). *Lousey Bank* 1840 O.S., *Lousy Bank* 1891 O.S. Perhaps from OE *hlose* 'a shed, a shelter', later 'pig-sty', or ME *lowsy* 'lousy, infested with lice; mean, contemptible'. The name was presumably considered unattractive, which may explain why the name Golden Hill appears on modern maps.

**LOW, THE**  1 mile south-east of Elford (SK 1909), *low hill* 1760 WSL 114/31, *The Low* 1834 O.S.; **LOW, THE**  2 miles north-west of Sheen (SK 0862), *le Low* 1399 VCH VII 27, *Lowe* 1413x1415 i*bid*, *the Lowe* DRO D2375M/189/14; **LOW, THE**  on the south side of Worfield (SO 7695), *the lowe* 1522 Worfield CA, *The Lowe* 1567 Worfield ParReg, *The Low* 1833 O.S.; A common name, from OE *hlaw* 'hill, mound, and (the usual meaning in Staffordshire) burial mound'. The Low near Worfield has been in Shropshire since the 12th century. See also Heatonlow.

**LOW HILL**  2 miles north of Wolverhampton, on Bushbury Hill (SJ 9201), *Lawe* c.1240 WA II 95, *Lawia* c.1240 ibid.' 96, *La Lowe* 1287 SHC 1911 193, *Lowe* 13th and 14th centuries, *le Lowe Hyll* 1545 SHC XI 289, *Le Lohill* 1612 SHC IV NS 38, *Low* 1686 Plot; **LOWE HILL**  a 770' hill 1 mile south-east of Leek (SJ 9955), *Lowe* 1240 (1883) Deed Sleigh, 1332 SHC X 115, 1538 (1883) Sleigh 17, 1583 SHC XV 145, 1608 SHC 1948-9 106, *Leeke Low* 1614 FF, *Lowe at Leek* c.1619 SHC VII NS 205, *Leeke Hyll*

1622 FF, *Low Hill* 1798 Yates. From OE *hlaw* 'hill, mound, and (the usual meaning in Staffordshire) burial mound', with OE *hyll* 'hill'. A large tumulus (now destroyed) is recorded on the hill at Bushbury: Plot 1686: 403; NSJFS 1965 59.

LOWTON HALL (obsolete, on the south side of Little Onn (SJ 8415)) *Lowton Hall* 1808 Baugh.

**LOXLEY (UPPER & LOWER)** 2½ miles south-west of Uttoxeter (SK 0630 & SK 0532). *Locheslei* 1086 DB, *Lochesl'* 1177 SHC XII NS 278, *Lockesley* 1227 SHC XI NS 18, 1236 Fees, *Lockesleye* 1292 SHC VI (i) 239, *Churchlockesleye* c.1300 SRO DW1733/A/2/11, *Lokesle* 13th century Duig, *Locksley* 1353 SRO D(W)1721/3/1/20, *Lokkusley* 1375 SHC VIII NS 293, *Loxley* 1473 HLS, 1605 SHC 1940 301. Perhaps from the OE personal name *Loc(c), or *Loxa, with OE *leah* 'clearing', but OE *lox* meant a lynx, and the name may possibly be linked to a similar animal. Cf. Loxley, Warwickshire (PN Wa 235); Loxley, Surrey (PN Sr 236). The *Church* element is unexplained. *Loxley Cottage* appears on the first edition 1" O.S. map of 1836 2 miles north-east of Yoxall at what is now Scotch Hill (SK 1622), but the history of the name is unknown.

**LOYNTON** 1 mile north-west of Norbury (SJ 7724). *Levuntona* 1080 SHC I 183, *Levintone* 1086 DB, *Levuntona*, *Livintuna* 1162 SHC I 183, *Laenton'* 1191 Pipe, *Livinton* 1199 SHC III 169, *Lavenden'* 1242-3 Fees, *Levynton* 1281 Ass, *Levynton* 1309 SHC X (i) 4, *Lemynton* 1317 SHC IX 62, *Levynton* 1325 SRO D1564/7, *Leynton* 1380 SHC XVII 193, *Loyton* 1576 SHC 1926 39, *Levington* 1598 SHC XVI 184, *Laynton* 1686 Plot 209, *Loynton* 1719 WSL 78/43. 'Leofa's *tun*'. See also SHC VI NS (ii) 79 fn.1.

**LUCEPOOL** ½ mile north-east of Yoxall (SK 1519). 'Pike pool', from ME *luce* 'a pike'.

**LUD BROOK** Oakden 1984: 13 suggests that this is a lost tributary of the river Churnet, but it is probably that stretch of the river Churnet that flows around Leek: StSt 5 1993 9. *Luddebroc* 1330 Ch, 1345 Coram R, *Luddebrok* 1346 Pat, *Lodebroc* 13th century Dieul. From OE *hlud-broc* 'the loud brook': perhaps to be identified with Luddebeche. See also Luddebeche; Ludchurch.

**LUDBURN** 1½ miles north-west of Sheen (SK 0962), on the west bank of upper reaches of the river Manifold. *Ludbourne* 1566 Deed, 1582 SHC XV 140, *Ludburne* 1599 DRO D2375M/106/27, *Luddburne* 1600 Alstonefield ParReg, *Ludburn* 1671 ibid. Perhaps from OE *hlud-burna* 'the loud spring'. The element *burna* is rarely found in the north of the county, and would not apply to a sizeable river such as the Manifold. Since there is no stream here, the second element may be a metathesised form of ON *brunnr* 'a well, a spring' (see VEPN II 50-1). See also Lekebourne.

LUDDEBECHE (an unlocated stream in the Leek area) *aquam de Luddebeche* 1217x1272 Barraclough 1988: 378. From OE *hlud*, often applied to streams, meaning 'loud, noisy', with OE *bece* 'a pronounced stream-valley', so 'the well-defined valley with the stream called Lud', or 'the stream that flowed through the well-defined valley called Luddebeche'. The stream evidently lay in the area given by the earl of Chester for the founding of Dieulacres abbey (see Elliott 1984: 45; Barraclough 1988: 378), and may indeed be the same watercourse as Lud Brook (q.v.). See also Lud Brook; Ludchurch.

**LUDCHURCH or LUD'S CHURCH** 1 mile south-east of Swythamley, near the Cheshire border, in Rushton Spencer parish (SJ 9865). *Ludchurch* 1686 Plot, *Lud Church* 1747 Bowen. A deep natural chasm in the millstone grit which, from of its moss-covered rocks, and descriptions of holes at either end, together with other topographical evidence from the poem to support a North Staffordshire setting, has been persuasively identified as the Green Chapel in *Sir Gawain and the Green Knight*, written by an unknown author c.1400: see Elliott 1984: 45ff; Elliott 1997: 105-30; NSJFS 17 20-49. The description of the Green Chapel in Gawain reads: 'Hit hade a hole on þe ende and on ayþer syde, and ouergrowen with gresse in glodes aywhere, And al watz hol? inwith, nobot an olde cave, or a creuisse of an olde cragge…Þhis oritore is vgly, with erbez overgrowen…': Tolkien & Gordon 1967: 60. It can be no more than coincidence that in mythology one of the chief gods of the Britons was Lludd or Nudd, a legendary king of the British mentioned by Geoffrey of Monmouth, supposedly the brother of the historically real Caswallon, which would place Lud's existence at about 60BC. Lludd is recognisable in later times as the mythical King Lud, perhaps to be associated with Lud in 'Lludd and Llefelys' in the Welsh Mabinogion (Ashley 1998: 69), or King Lot or Loth, alluded to frequently in Arthurian legend and romance as king of Lothian and Orkney and husband of Arthur's sister Margawse or Morgause, and the father of Gawain himself. The *church* element in the name is difficult to explain, but if the identification of Gawain's Green Chapel is correct (and the chapel is also decribed in the poem as a *kirk*, the Northern, from Scandinavian, word for 'church'), the poem may have adopted an existing local name, or perhaps the local people recognised the association with the poem and thereafter named the chasm *church*, and the reason for the name became forgotten, although a descriptive *Green* or *Gawain's* might have been expected, and legends of this kind tend to survive or even expand. By popular local tradition, totally unsupported by evidence, Ludchurch was used as a secret place of worship by the Lollards, a religious sect who were the followers of John Wycliffe (1330-84). The legend records that soldiers attracted to the place by singing at a religious service killed a young girl called Alice, who was buried at the entrance. Alice was said to be the grand-daughter of Walter, called de Lud-Auk, a prominent Lollard, and the 'church' was supposedly named after Lud-Auk or the Lollards, abbreviated to Luds. The origin of the tradition (and any evidence of the existence and identity of Lud-Auk) is untraced, but the legend is said to have been recorded by Sir William de Lacey in 1546 (Rathbone 1974: 9), the date given in Hackwood 1906: 18. It is of passing interest, but unrelated to the derivation of the name of this feature, that in an ancient poem *Moliant Cadwallawn* ('Eulogy for Cadwallon', identified as Cadwallon ap Cadfan, king of Gwynedd, d.634), Cadwallon is styled *luydawc Prydain*, '?battle-hosted one of Britain/ruler of the armies of Britain': Kirby 1977: 34, and Maxen Wledic (? Lud-Auk), identified as Magnus Maximus, a Roman who proclaimed himself Emperor during his command of troops in Britain and after defeating Gratian in Gaul in 383 was recognised as Theodosius, Emperor of the East, but was defeated and executed by Theodosius for ordering his troops to invade Italy, was a great folk hero in Celtic folklore (cf. Middle Welsh *gwledic* 'leader, ruler, lord, emperor'), and in some sources is named as the father of the legendary King Arthur. Ludchurch ravine is also said to have been known as *Trafford's Leap*, supposedly from a former occupier of Swythamley Hall whose horse leapt the cleft during a hunt. The absence of early spellings for Ludchurch – the age of the name is unknown, though it was evidently well-established by the later 17th century (Oakden 1984: 13 inexplicably suggests that it is Victorian) – precludes any certain derivation, but the element Lud is found in the names Luddebroc (see Lud Brook) and Ludebeche (q.v.), both of which are recorded in the original charter by which Ranulph de Blundeville, earl of Chester, gave land for the building of Dieulacres abbey near Leek, and the area in which Ludchurch lies was part of the abbey's endowment: Elliott 1984: 45. Lud in place-names is normally associated with OE *hlud* 'loud', frequently linked to an element denoting water (cf. Ludebeche; Lud Brook; Ludburn; also Ludbrook, Devon; Ludlow, Shropshire; Ludwell, Derbyshire), but might here be appropriate with reference to echoes, or from the personal name Luda (cf. Luddenham, Kent; Luddington, Lincolnshire; Ludham, Norfolk). Plot 1686: 173 gives the following account of the place: '... the stupendous cleft in the rock between Swithamley and Wharford commonly call'd Lud-Church, which I found by measure 208 yards long, and at different places 30, 40 or 50. foot deep; the sides steeped and so hanging over, that it sometimes preserves Snow all the Summer, whereof they had signal proof at the Town of Leek on the 17 of July their Fair Day, at which time of year a Wharnford Man brought a sack of Snow thence, and poured it down at the Mercat Cross, telling people that if any body wanted of that commodity, he could quickly help them to a 100 load on't'. For a list of publications on legends associated with Ludchurch, see Elliott 1984: 51 fn.39-40. Litchurch in Derbyshire is recorded as *Ludecerce* in 1086 (DB), and *Ludchurch* in 1234, perhaps meaning 'Luda's church' (PN Db 452; the derivation 'small church', from OE *litel*, *lytel* 'little, small', given in Ekwall 1960: 300 is most improbable, notwithstanding a handful of spellings such as *Litlecherche*, recorded in 1197). The parish of Yr Eglwys Lwyd, in Pembrokeshire, is in English 'Ludchurch': the earliest reference to the place is 1324, but there is uncertainty whether the name is of English or Welsh origin: see Charles 1992: I xxv; II 509.

**LUDINTON** (unlocated, possibly Luddington, Wawickshire) *Ludinton* 1206 SHC III (i) 139.

**LUDSTONE** 1 mile north-east of Claverley (SO 8095). *Ludeston* 1163 SHC II (i) 190, *? Ludesdon* 1190-3 SHC III 217, *Luddesdon* 1250 Cl, *Loddesdona, Loddesdon* 1271 SHC 4th Series XVIII 71, *Luddesdon'* 1333 SR, *Ludston* 1530 FF 8. From the personal name *Hlud, with the second element either OE *dun* 'hill' or OE *tun*, with the earliest form favouring the latter, but the majority favouring the former: the place lies on the west flank of a rounded ridge rising to 357'. Since the 12th century in Shropshire.

**LUD-WALL** (unlocated, between Longton and Normacot) *Lud-Wall* 1679 SHC XII 59. Perhaps from OE *hlude* 'loud, noisy', with Mercian OE *wælle*, usually in the West Midlands 'a spring', so 'the noisy or bubbling spring'.

**LUFFULLEWODE** (unlocated, possibly near Pipe: see VCH XIV 214) *Luffullewode* 1537 SHC XI 276, *Luffulewode* 1598 SHC XVI 175, *Loughfulwood* 1624 SHC X NS (i) 64. Oakden 1984: 65 considers that the name relates to fields in Cannock, and suggests a derivation from OE *ful wudu*, 'the foul or dirty wood', prefixed by OE *luh* 'pool', so the pool at the dirty wood', but *luh* is a very rare element, and such derivation most improbable. The place is probably to be associated with *Leefhull* between Cannock and Lichfield recorded in 1307 (WL 100), and *Lefful*, recorded in 1309 (WL 103).

**LUM EDGE, LUM POOL** 2 miles north-west of Warslow (SK 0859); *Lumpoole* 1626 Rental, *The Lum, Lum Pool* 1842 O.S.; **LUM, THE** 1 mile north-west of Madeley (SJ 7645). Perhaps from the dialect word *lum* 'woody valley, deep pool in a river' and (in Derbyshire) 'a small wood or grove': see PN Db 33. However, *lum* is a dialect word found in the northern counties for (i) 'a chimney, a vent' (EDD), (ii) 'a small wood or grove' (EDD), and a term used in north Staffordshire metal mining areas for 'a lode, vein or fissure' (Ekwall 1960: 307 gives 'a well for the collection of water in a mine'), and the 1833 O.S. map shows *Furnace Mill* close to The Lum: Lum Pool was perhaps associated with lead mining. Cf. Lumb, Lancashire; Great Lumley, Durham.

**LUNT** – see **LOUNT**.

**LUPIN** 2 miles north-west of Alrewas (SK 1416). *Loppe* (p) 1259-60 SHC X NS I 272, *Loopin Chapel* 1660 Erdeswick 1844: 316, *Looping (brook)* 1752 SHC X NS I 272, *Loopin (House)* 1775 Yates, *Lupin (Gate)* c.1800 SRO D615/M/1/8, *Lupin Bank* 1834 O.S. Perhaps from ON *hlaup* 'a leap' (cf. Loups, Yorkshire North), or more likely the dialect word *loup* 'a place where a river becomes so contracted that it can be easily leapt' (EDD): the place is where the road from Alrewas to King's Bromley crosses the Bourne Brook.

**LUTH BURN** a tributary of the river Trent. Probably identical in derivation to Ludburn (q.v.).

**LUTHEBURGH** (unlocated) *Lutheburgh* 1269 SHC IV 170. Possibly Lytlebiri (q.v.).

**LUTHELEYE** (unlocated, posibly near Heatley) *Lutheleye* 1336 SRO D(W)1721/3/3/12.

**LUTLEY** 1½ miles north-west of Enville (SO 8188; SMR 02626 places a deserted post-Conquest settlement at SO 81598850). *Luctelega* 1166 SHC 1923 298, *Luteleg* 1199 P, 1221 SHC IV 221, *Liutteleio* 1208 SHC III 143, *Luttelega, Liutelega* 1262 For, *Lotteleye* 1271 SHC V (i) 140, *Lutley, Lutteleye* 13th centuries Duig, *Lutteley* 1332 SHC X (i) 129, 1472 SRO C140/520 27. Perhaps from an OE personal name such as Luda or Luta. Lutley in Worcestershire (*Ludele(ya)* 1169, *Ledeleye, Lod(e)leye* 1275, *Lotteleye, Lutteleye* 1291, *Lutleye* 1327 (all PN Wo 298-9), *Upper Lutley, Lower Lutley* and *Lutley Mill* 1834 O.S.) lies 1 mile west of Halesowen (SO 9483), and some of the above spellings may relate to that place. See also Litley; Little Hay.

**LUZLOW** (obsolete, 1 mile south-east of Bagnall (SJ 9349)) *Lusse-Lees (otherwise Lusselows)* 1803 SHC 1933 149, *Luzlow* 1837 O.S. Perhaps OE *lus, hlaw* 'louse hill or tumulus', perhaps with the meaning 'the louse-shaped burial-mound' (see also Lizard), or possibly from OE *hlose* 'pigsty', probably originally 'a shed, a shelter', found in dialect *lewze, looze* (EPNE i 253), so 'the burial-mound at the shelter'.

**LYCHEHALE** (unlocated, perhaps near Rocester) *Lychehale* 1359 SHC XII (i) 160.

**LYCHEWODE** (unlocated) a wood in Almington. *Lychewode* 1247 SHC 1945-6 23, *? Lychewode* 1298 SHC XI NS 251, *Lychwod* c.1300 SHC 1945-6 31. See SHC 1945-6 23. Perhaps from OE *lic wudu* 'the wood associated with the corpse'.

**LYDE BROOK** a tributary of the river Smestow. *Hlyðe broc* 985 (12th century, S.860), *Ludebroc* 12th century Wodehouse, *Lydbroc, Lydbrok* 1294-5, 1315-7 Wodehouse, *Lodbroke* 1424-5 ibid. The stream is associated with Lloyd House and Lloyd Hill (q.v.), and is probably from OE *\*hlýde* 'the noisy stream'. Cf. Lydbrook, Gloucestershire. See also Lloyd Hill; Lloyd House.

**LYDIATE** (unlocated, in Wednesbury), *Lydeyate* 1280 SHC VI (i) 147, *Lydyate* 1415 SHC XVII 58, 1575 SHC XIV 178; **LYDIATES HILL** on the south-east side of Baggeridge Wood (SO 8992), *Lidget Hill* 1834 O.S. From OE *hlid-geat* 'a swing-gate'. The Lydyatt family, recorded in 1548 (SHC 1928 132), may be associated with the second place. See also Lidgett.

**LYE HALL** on the north side of Hampton Loade (SO 7486). *Legh* 1232 Eyton 1854-60 III 191, *La Leghe* 1256 ibid. 192, *Legh Shireford* 1259 ibid. 193. From OE *leage*, dative singular of OE *leah*, here probably with the meaning 'a piece of open land, a meadow'. *Shireford* is from the family of that name, associated with the place from at least 1232: Eyton 1854-60 III 191. It is unclear whether the surname is connected with Hampton Loade, for which Shireford may have been an earlier name.

**LYME** A name not yet fully explained by philologists, which came to be applied to a district, a forest and a river. In Staffordshire the element is (or was) found in, or associated with, Burslem, Butterton, Chesterton, Hales, Madeley, Newcastle under

Lyme, Whitmore under Lyme, and the unlocated *Shortelyme*, *Lymecrofte* and *Lymehalwe*. The element is perhaps of similar origin to a river Lyme which gave its name to Lyme Regis in Dorset (*Lim*, *Lym* in charters of 774 (S.263) and 938 (S.442)). Jackson 1953: 555 believed the root incorporated an English *m*, substituted from a Welsh sound perhaps in the seventh or late sixth century. There are frequent references in the 13th and 14th century to *boscus de Lyme*. The name had been thought (for example by Lucius the Monk, quoted in Gough 1806: 530) to refer to a forest, but more recent theories are that the forest connection is a secondary development from a term which originally applied to the southern part of the Pennines, i.e. land over 400' stretching from east Cheshire to north-eastern Derbyshire, and from south Lancashire to north Shropshire. The earliest recorded reference to the name Lyme is c.1125 (although Lyme Hanley, Audlem and Burslem are all recorded in DB, proving that the element is pre-Conquest in date), and an analysis of places incorporating the name suggests that it was applied to a long narrow strip of land, perhaps the original name of the escarpment running along the north-west border of Staffordshire. A more detailed survey and analysis of the name is given in Gelling 1992: 63-5, and PN Ch I 2-6; see also Pape 1928: 4-5. Dodgson has observed that 'the significance of the form of the place-name, regardless of the meaning, is that it shows English m substituted for Pr. Welsh lenited m, as contrasted with the substituted English v in R. Dane...': PN Ch V (II) 294. A derivation from a root *\*lim-* connected with modern Welsh *llif* 'flood' has been proposed in preference to a British stem *\*lemo-*, meaning 'elm' (Ekwall 1928: 274 suggests that a forest-name Lyme may well have been derived from a river-name), but the origin is far from certain: a derivation from *\*lemo-* is improbable given the scarcity of that that species in the region, and the possibility that the name is from an as-yet unidentified root, perhaps meaning 'bare or exposed district' or similar (cf. Mod Welsh *llwm* 'bare, exposed, destitute, poor', ModIrish *lomm* 'bare': see PN Ch I 5-6) can be rejected since nominal derivatives of *llwm* are unknown in Welsh and the word is seemingly absent from Welsh toponomy: see Coates 2001. One possibility is that the name is an unexplained Celtic, or even pre-Celtic, name 'denoting part – usually the western edge – of the southern Pennines, the Peak, and the rim of the Cheshire plain': see Coates & Breeze 2000: 1 335. However, the strongest philological explanation is that the name was borrowed into Brittonic direct from Latin *limes*, *limitum* 'limit', *limen* 'threshhold, lintel', or *llmes* field-balk, limit, boundary' (see Coates 2001; JEPNS 36 2003-4 39-50), perhaps used metaphorically to mark an upland area considered particularly inhospitable or unpopulated, or to mark a boundary zone: Higham 1993 notes that The Lyme lay between the Anglian-period territories of the Pecsæte and of groups which may have constituted the northern part of the territory held by the Wreocensæte, although not claiming any connection with that fact and the name. See also Shortelyme.

**LYME BROOK** a tributary of the river Trent. *Lyme Brook* 1686 Plot. Probably a back-formation from Newcastle under Lyme, through which the stream runs (cf. Burslem). Since early examples have not been found, it seems possible that the name is due to a misunderstanding of the addition *under Lyme*, found in Newcastle under Lyme, etc. – see Lyme. If the name Lyme derived from a river-name, it is unlikely that it was that of this stream, which is insignificant (see Ekwall 1928: 274). An earlier name for the watercourse may have been *Are*: see Newcastle under Lyme. There is a reference to *Limbreuk* 'between Hopewas wood and Tunstall wood' in 1300 (SHC V (i) 176), but the name is probably from OE *hlimme* 'a stream, a torrent': see Linbrook.

**LYMECROFTE** (unlocated, in Acton near Swynnerton) *Lymecrofte* 1589 SRO D1798/663/138.

LYMEHALWE (unlocated, in Leek) *Lymehalwe* c.1287 SHC 1911 429.

LYME HEATH (unlocated, in Tunstall) *the Lyme heath* 1603 JNSFC LXIII 1928-9 52, *The Lymeheath* 1613 D1229/2/4/1, *The Lyme Heath* 1614 SHC 1931 70. 'The heath at Lyme (q.v.)'.

**LYME HOUSE** 1 mile south-east of Horton (SJ 9556). *Lyme* 1414 VCH VII 203, *Limehouse* 1639 Leek ParReg, *Lyme House* 1659 Horton ParReg, *Lymehouse juxta Horton* 1663 SHC II (ii) 63, *? Lommaz* 1696 Leek ParReg, *? Lymmat* 1696 ibid, *Lime House* 1775 Yates, 1842 O.S., *Lym House* 1815 Horton *EnclA*. 'A building used for burning lime; a lime-kiln'.

LYMESYE (unlocated, possibly near Lichfield) *Lymesia* 1223 SHC IV 27, *Lymeseye* 1281 SHC VI (i) 118, 1286 ibid. 160, *Lymescy* c.1312 SHC 1941 173, *Lymesye* 1323 SHC IX 128. Perhaps transferred from Limesy in Seine-Maritime or Seine-Inferieure, France.

LYMEWELL (unlocated, between Hopwas and Dunstall: Shaw 1798: I 433) *Lymewell* 1798 Shaw I 433.

LYME WEY (unlocated, in Little Wyrley) *Lyme Wey* 1395 SHC VI (ii) 193.

LYMFORD (obsolete) 1 mile north-east of The Cloud (SJ 9164). *? Lymme* 1298 SHC XI NS 256, *Lymford* 1332 SHC X (i) 115, 1539 SHC VI NS (i) 84, *Lymforde* c.1565 SHC 1938 76, *Lymeford* 1619 SRO DW1761/A/4/157[9], *Limb Ford* 1775 Yates, *Lymford* 1842 O.S. Possibly from OE *hlimme* 'a stream, a torrent' (the place lay near the river Dane), or from *The Lyme* (q.v.), as suggested in PN Ch I 56. See also Lymm Bridge. *Lymeforde*, recorded c.1565, appears to have been in Alstonefield: SHC 1938 76.

LYMINGFORD (unlocated, in Kinver Forest) *Lymingford* 1262 SHC V (i) 138.

LYMM BRIDGE (obsolete) on the river Dale 1½ miles north-east of The Cloud (SJ 9165). *Lymm Bridge* 1775 Yates. Possibly from OE *hlimme* 'a stream, a torrent', but see also Lymford.

LYNACRE (unlocated, in the north-west of Pipe Ridware: see Shaw I 163, 166*) *Linacre* late 12th century (1798) Shaw I 166*, *Lynacre* late 13th century SRO 3764/8[31724]. From OE *lin* 'flax', with OE *æcer* 'field, ploughed land', so 'the field where flax was grown'.

**LYNCROFT** 1 mile north-west of Lichfield (SJ 1010). *Lyncroft* 1356 Duig, *Lincroft* 1812 *EnclA*, 1834 O.S. From OE *lin* 'flax', with OE *croft* 'a piece of enclosed land used for tillage or pasture, a small piece of land adjacent to a house'. Possibly to be associated with *Lindenescroft*, recorded in 1203 (SHC III 116).

**LYNDON** 1½ miles north-west of Kinver (SO 8285), *Lyndon* c.1290 SRO D1485/6, 1440 SRO Tp 1273 r.h. safe box, 1456 SRO D1485/6, 1545/6 SRO Tp 1273/12/1 No 9, *La Lynden* 1371 SA 2089/2/2/24, *Lynne Hall* 1583 Erdeswick 1844: 430; LYNDON (obsolete) in West Bromwich (SO 0092), *Line* c.1692 StSt 11 1999 63, *Lyndon* 1834 O.S. From OE *lin* 'flax', with OE *dun*.

**LYNE HILL** 1 mile south of Penkridge (SJ 9212). *Linhull'*, *Lynhull'* 1237 Cl, 1251 FF, 1308 Ass, *Loynhill* 1271 For, *Lynhul* 1327 SHC VII 242, *Lynhull'* 1332 SHC X (i) 119, *Lynehill (marche)* 1596 SHC 1932 203, *? Lynell* 1661 Wolverhampton ParReg, *Linehill* 1686 Plot. 'The hill where the flax was grown', from OE *lin* 'flax'.

**LYNN** 4 miles south-west of Lichfield (SJ 0704). *la Lynd* 1262 SHC V (i) 139, 1271 For, *Lynn, Lynda* 1274-5 SHC 1923 275, *la Lynde* 1286 SHC V (i) 173, 1315 SHC XVII 289, 1375 SHC XIII 126, *Lynde* 1311 (1798) Shaw II 55, 1348 SHC XVII 289, *Lynde otherwise Lynne* 1592 SHC 1930 217; *Lynne or Lyndon* 1801 Shaw II 55. From OE *lind* 'linden or lime-tree'. In medieval poetry the word *linde* was frequently used for trees in general, and the plural *lindes* for a grove.

**LYNTERSWOOD** (unlocated, perhaps near King's Bromley) *Lynterswood* 1559 SHC 1931 156. See Lyntus, with which this name may be associated.

**LYNTUS** 2 miles north of Lichfield (SK 1312). *Lynton'*, *Lenton* 1327 SHC VII (i) 234, *Lynton'* 1332 SHC X 84, *Lenton* 1332 ibid.' 105, *Lyntus* 1834 O.S. Perhaps from OE *lind-tun* 'lime-tree *tun*', later *lind-hus* 'the house at the lime-tree'. In 1834 O.S. the name is attached to a copse, which is now Big Lyntus. Another smaller copse nearby to the east is Little Lyntus.

**LYONS, THE** 1 mile west of Enville (SO 8186). *The Lyons* 1834 O.S. Perhaps from the name Leon: Henry III granted Leon de Romelegh a licence to assart in Horwood within Kinver Forest in 1268 (SHC V (i) 82, 158; SHC 1911 140), and his son, also Leon, who was Sheriff of Staffordshire in 1286-8, was granted a similar right by Edward I (ibid., also Eyton 1854-60: III 202). See also SHC VI (i) 82. For John Lyoynne (1301) see SHC VII 81. It is likely that *Liones meduwe* recorded in 1300 (SHC V (i) 180, *Liones medwive* in Jones 1894: 29), is to be associated with this place: VCH XX 94.

**LYONS LODGE** (unlocated, possibly near Biddulph) *Lyon Lodge* 1661 Biddulph ParReg, *Lyons Lodge* 1666 ibid.

**LYSWAYS** 4 miles north-west of Lichfield (SJ 0913). *Lisuis* 1167 Duig, *Lisewis* 1199 SHC III (i) 41, 1243 SHC 1911 402, *Liswis* 1242-3 Fees, *Lysewys* c.1250 SHC 1924 76, *Lesewys, Liswys* 13th century Duig, *? Lychewyz* 1307 SHC XI NS 265, *Lisewis* 14th century SHC 1921 34, *Lyswis* 1686 Plot 157 Evidently transferred (with anglicised pronunciation) from Lisieux in Northern France, so explaining why the c.1250 spelling is indexed *Lisieux* in SHC 1924 388. Perhaps to be associated with Hugh de Nonant, bishop of Coventry and Lichfield, who came from a prominent Norman ecclesiastical family and owed his advancement to his uncle, Arnulf, bishop of Lisieux, was himself archdeacon of Liseux from 1169 or earlier until 1181 or 1182: EEA 17 xxvi. Lisieux in France is from the northern Gaulish tribe, the Lexovii: see JEPNS 23 1990-1 11. See also Clark 1995: 272. Shaw 1798: I 223 mentions '...*Leswes (afterwards called Arblaster) hall...*'.

**LYTLEBIRI** (unlocated, possibly in the Sandon area) *Lutilbeire* 1206 SHC III (i) 36, *Lyttlebury* 1279 SHC VI (i) 142, *Lyttlebyri, Litelbyri* 1284 ibid. 133-4, *Lytlebiri* 1286 ibid.' 161. The name would appear to mean 'the small fortification', from OE *lytel byrig*.

**LYVERSEGGE** (unlocated, in Walsall) *Leverich* 1327 SHC VII 224, SHC 1928 179, *Levereshed* 1355 (1801) Shaw II 304, *Lyversegge* 1425 SHC XVII 101, *Lyveriche (Field)* 1554 SOT D260/M/T/1/1/22, *Lyverich* 1616 SOT D260/M/T/1/49, *Leverich (Field)* 1633 BCA D260/M/T/1/33, *Liverich (Fields)* 1696 BCA D260/M/T/1/119a. The spellings suggest a derivation from an OE personal name such as Leofric or Leofhere, with the second element variously OE *hrycg* 'a ridge, a long narrow hill', OE *heafod* 'the end of a ridge, the upper end or top', and OE *ecg* 'edge, the crest of a sharply pointed ridge, a steep hill', all being elements which could apply to the same type of feature.

**MADELEY** Ancient Parish 4½ miles west of Newcastle-under-Lyme (SJ 7744), *Madanlieg* 975 (11th century, S.801), *Madelie* 1086 DB, *Maddell'* 1177 SHC XII NS

278, *Maddeleye-under-Lyme, Madelegh, Madeleye* 13th century Duig, *Madeleye-subtus-Lynam* 1333 SHC 1913 228, *L Madeley, G Madeley, Madeley Manor* 1682 Browne; **MADELEY (FARM)** in Checkley parish, 3 miles north-west of Uttoxeter (SK 0537; SMR 02016 places a deserted Anglo-Saxon settlement at SK 05383690), *Madelie* 1086 DB, *Madeleye* 1176 FF, *Maddeleg'* 1242 Fees, *Maddeley*, later *Madeley Ulfac* or *Madeley Alfac* VCH IV 51 fn., *Maddeley Alfac* 1293 SHC 1911 47, *Maddeley Alfogh* 1332 SHC X 112, *Madeleg' Alfachk', Madeley Alfogh* 1377 SHC 4th Series VI 7, *Madeleyhome* 1415 SHC XVII 29, *Madleyholme otherwise Madeley Socke* 1559 SHC 1924 139, *Medley Holme* c.1564 SHC 1938 148, *Madeley-holme* 1644 (1798) Shaw I 68, *Madeley House, Madeley Wood* 1836 O.S.; MADELEY WOOD BARN (obsolete) 1½ miles north-east of Colton (SK 0622), *Madeley Wood Barn* 1836 O.S. From a OE personal name *Mad(d)a, so '*Mad(d)a's *leah* or clearing'. *Ulfac* was the name of the tenant in DB, and the name was added to the place-name to distinguish it from other Madeleys, but the meaning of *Socke* is unclear, unless from OE *soc* 'suck, sucking', perhaps used in the sense 'drain, drainage' (see EPNE ii 133), or from late ME sock, used in the north of England for 'a ploughshare', and sometimes meaning 'ploughing' (OED), or perhaps a corruption of Madeley's Oak. *Holme* is from ON *holmr* 'an isle, a small island, a water-meadow': Madeley Farm lies on the river Tean. Any doubt about the identification of *Madanlieg* with Madeley in Staffordshire is removed by the reference in the same charter to *wriman forda*, which is associated with nearby Wrinehill. *Maideleye* in Cannock Forest is recorded in 1346: SHC 1939 75.

**MAER** 6½ miles south-west of Newcastle-under-Lyme (SJ 7938). *Mere* 1086 DB, 1242 Fees, *Mare* 1198 SHC III 29, *Meer* 1291 Tax, *Mere* 13th century Duig, *Mayer* 1471 SHC IV NS 178, *Meire* 1586 SHC 1927 128. From OE *mere* 'lake, mere': there is a large pool here (presumably the '60 acres under water' recorded in 1562: SHC XII 236), feeding the river Tern which flows into Shropshire.

**MAERWAY LANE (FARM)** 2 miles south of Madeley (SJ 7640). *The Meyre Lones* 1532 SHC 4th Series 8 44, *Merewelane* 1610 SHC III NS 34. 'The lane which is the way to Maer'.

MAGHILLS (unlocated, perhaps near Keele) *Maghelles field* 1565 SHC 1938 38, *Meghills* 1678 StSt 11 1999 67, *Maghills* c.1692 ibid. Possibly from the surname McGill or similar, recorded as Macgeil in 1231 (DES 292), or perhaps from OE *mægðe hyll* 'mayweed hill'. The hard *g* would not normally be expected from *mægðe*, but may have been regarded as correct in later times in the light of earlier spellings. Cf. Maghull, Lancashire.

**MAIDENSBRIDGE** at Wall Heath, ½ mile south of Himley (SO 8790). *Maidensbridge* 1727 SRO D1132/1/14, *Maiden Bridge* 1834 O.S. The bridge lies on Maiden's Brook, which formed the boundary between Himley and Kingswinford, and may have taken its name from OE *(ge)mære* 'boundary', perhaps with OE *dun* 'hill', so 'the stream of the hill at the boundary', but OE *mægden* 'maiden' is a common element in place-names, usually applied to lanes, bridges, fords and similar places supposedly frequented by young unmarried women: cf. *of mægidna brycge* 11th century, S.1591. *Meidenesford* is recorded in the bounds of Bilston and Wednesfield in 985 (12th century; S.860): see Hooke 1983: 72, 74, 113. See also Maiden's Well.

**MAIDEN'S WELL** on the south side of Uttoxeter (SK 0932). *Meadenswall (Close)* 1623 SRO D786/10/1, *Maydenswall (Close)* 1646 SRO D786/10/3, *Maidenswall (Close)* 1665 SRO D786/10/5. 'The spring frequented by maidens': see Maidensbridge. Redfern 1886: 54-5, 347 suggests that the well was anciently *Marian's Well* or *Maiden's Wall*

*Well, wall* probably from Mercian OE *wælle* 'spring', with the later addition of 'well'. The name is evidently to be associated with Maiden Field, one of the open fields of Uttoxeter: ibid.

MALBANK or MAUBAN FRITH – see ALSTONEFIELD FOREST.

MALESHOU – see MELLESHOHE.

MALTON (unlocated) *Malton* 1547 SHC 1950-1 41.

MANCHESTER, LITTLE (obsolete, ¼ mile south-east of The Arbour in Mucklestone (SJ 711374)) A curious name, said to have existed from at least c.1854 (JNSFC XLII 1907-8 110; see also SHC 1945-6 26; TSAS XLIX 1937-8 88), perhaps derived from Leominchistrete (q.v.) as a result of 19th century antiquarianism. The element *chester* in ancient names is usually from OE *cester* 'a city, a (Roman) town, an old fortification'. A large low mound enclosed by a wall of huge rough stones forming a rectangle has been recorded here: JNSFC LXXII 1937-8 117-8; JNSFC LXXIII 1938-9 113. However, the name has not been traced in early records (*Manchester*, recorded in 1351 (SHC 1913 146), probably relates to the city), and appears to have been relatively recent and short lived.

**MANIFOLD, RIVER** a tributary of the river Dove. *Water of Manifould* 1434 Survey, *Manifold* c.1540 Leland, *aqua de Manifo(u)ld* 1573, 1618 Ct, *Manyfold(e)* 1577 Saxton, 1586 Harrison, 1686 Plot, c.1598 *Manifold(e)* Erdeswick 1844: 480. From OE *manigfeald*, literally 'many folds or turns', perhaps with reference to the disappearance of the river underground at Wetton Mill near Grindon, to emerge at Ilam, i.e. vertical as opposed to horizontal folds: see Ekwall 1928: 278. Cf. Mangfall, a tributary of the river Inn in Germany. The pre-English name of the river is unknown, but see Fawside.

MANLEY HALL – see **THICKBROOM**.

MANNESMORE (unlocated, in Rushton Grange: VCH III 235) *Mannesmore* 1223 SHC XII NS 30, *Manesmore* 1223 VCH III 235, *Maunesmor* 1256 Ch, *Monsmore* 1539 VCH VIII 116, *Mansmore* 1838 ibid. 116. Possibly from the OE personal name Mann, or the ON personal name Man, so 'Man's or Mann's moor'.

**MANSTY HILL** 2 miles south-east of Penkridge (c.SJ 9512), *Manston Hills* 1537 SRO D260/M/T/5/102, *le Manstonshill* 1547 SHC 1950-1 40, *(le) Manstone(s)hill* 1548 Survey; MANSTY HEAD (obsolete, in Hatherton), *Mansty(e) Head* 1682 Dep; **MANSTY WOOD** 2 miles south-east of Penkridge (SJ 9512), *Mansty Wood* 1834 O.S.; **MANSTY FARM** 2 miles south-east of Penkridge (SJ 9512), *Manstie* 1589 SHC 1928 164; MANSTY POOL (obsolete, in Hatherton), *Mansty Pool* 1682 Hatherton. It would seem that Mansty Head and Mansty Farm are from OE *(ge)mǣne* 'common', and *stig* 'a path, a narrow road', normally applied to an ascending path or road, meaning 'the communal ascending path': cf. *mǣne weig* 'common way' recorded as a boundary mark in a charter relating to Bedintun (believed to have been near Pillaton) dated 996 (S.879). Mansty Hill, which perhaps originated as 'the hill with the Manstone', whatever that might have been, has evidently been influenced by the other names.

MANWAY FIELD (obsolete, on the north side of Wednesbury (SO 9795)) *Moneway* 1325 SHC 1911 323, *Monway Field* 1684 BCA MS3145/91/2, *Manway Feild* 1686 BCA MS3145/114b, *Manway Field* 1834 O.S. Perhaps from OE *manig, monig* 'many', with OE *weg* 'a way, a path', so 'the place with the several paths', or from OE *(ge)mǣne* 'common', so 'the communal path'.

**MANWOODS** on north side of West Bromwich (SO 0292). *Manwoodes* 1649 BCA MS3145/62/2, *Manhoods* (*sic*) 1733 SHC 1944 53, *Manyards* 1775 Yates, *Manwoods* 1798 Yates, *Manwood* 1834 O.S. Early spellings have not been traced, but perhaps 'common wood', from OE *(ge)mǣne* 'common, communal'. Cf. Manhood, Sussex; Monwood Lea, Warwickshire (PN Wa 76).

**MAPLE BROOK** forming the southern boundary of Longdon – see Chesthall.

**MAPLE HAYS** 1 mile west of Lichfield (SK 0909). *Mabbley hays* 1498 VCH XIV 211, *Mabberley Hay* c.1530 SHC VI (ii) 166, *Mapel Hey* 1674 ibid, *Maple Hayes* 1704 ibid, *Pipe or Maple Hays Farm* 1728 ibid, *Maple Hayes* 1834 O.S. The name may have been originally 'the enclosure (from Mercian OE *(ge)heg*) at Malbert's *leah*' (Malbert is an OFr personal name of German origin: cf. Mablethorpe, Lincolnshire: Ekwall 1960: 310) which developed into 'the enclosure with the Maple tree'. *Maple bridge* is recorded here in 1597: VCH XIV 195. *Pipe* may be from the conduit which ran from here and supplied Lichfield with water: see Pipehill; Pipe Grange.

**MARCHINGTON** in Hanbury parish, 3½ miles south-east of Uttoxeter (SK 1330; SMR 03752 places a deserted post-medieval settlement at SK 13763076). *æt Mærcham* 951 (14th century, S.557), *Mærchamtun* 1002x1004 (11th century, S.906; 11th century, S.1536), *Merchametone* 1086 DB, *Merkintona* 1179 SHC I 93, *Mercinton temp.* Henry II Derby, *Mercington* 1230 Ass, c.1235 Rees 1997: 67, *Merchinton'* 1242-3 Fees, *Mersinton* 1276 SHC VI (i) 80, *Marchaunton* 1605 SHC 1940 297. Since the place lies on the south side of the river Dove, which forms the boundary of Staffordshire and Derbyshire, the derivation has been taken to be 'the *tun* of the Mærcham people', who took their name from OE *mearc* 'boundary', and OE *ham* 'home, village': see Ekwall 1960: 314; CDEPN. However, it is very likely that the name is derived from OE *merece* 'smallage, wild celery (Apium graveolens)', with OE *ham(m), hom(m)* 'an enclosure; a meadow, especially a flat low-lying meadow on a stream; flat land on a river or in a river bend', so giving 'the flat land by the river of the people dwelling by the wild celery': see Cole *et al* 2000: 141-48. The plant grows in salt water, and a saline spring is recorded near Draycott Mill, 1 mile north-north-east of Draycott in the Clay (Redfern 1865: 15), and Salt Brook runs from Needwood Forest through Draycott in the Clay and into the Dove. In the same area are Saltbrook Cottage (at SK 168298) and Saltbrook Lane. See also Nomina 23 2000: 141-7. This name is a rare example of an English name to which the *tun* element was added later (but see Ashburton in Devon, which has OE *tun* added to an earlier Æscburna), seemingly between 951 and 1004 AD. The place is also the only probable example of a *ham(m)* name in Staffordshire, but see also Trentham and Hamley. Cf. Marcham, Berkshire; Marchwood, Hampshire; Marchamley, Shopshire, and see particularly PN Sa I 194-6. See also Marsh Barn Farm; Salters Croft.

**MARCHINGTON WOODLANDS** 1 mile south-west of Marchington (SK 1128). *Marchington Wodelands* 1472 (1798) Shaw I 86, *Woodland, Marchington Woodland* 1586 SHC 1927 131. Self-explanatory – see Marchington. The township included the north-western corner of Needwood forest.

**MARE BROOK** a tributary of the river Dove, *Mare brook* 1804 Map; **MARE BROOK** a tributary of the river Swarbourn, *mæran broc* 1008 AD (13th century, S.920), *Marbrouk* 1286 For, *Merebroke* 1338 Ipm, 1379 Banco, *Marebroke* 1434 Rental; **MARE BROOK** a tributary of the river Tame, ? *Marebrocke* 1610 Alrewas ParReg, *Mare Brook* 1847 TA. All these names derive from OE *gemǣre* 'boundary', with *broc*, hence 'stream which forms a boundary'. *Marbroke(house)* is recorded near Leek in 1535 (Dieulacres

Inventory), presumably to be associated with *Marbroucke* in Leek recorded in 1597 (SHC 1935 IV 99).

MAREFORD (unlocated, perhaps near Perton) *Mareford* 1327 SHC VII (i) 253. Perhaps from OE *gemǣre* 'boundary', so 'the ford at the boundary'.

MARING (unlocated, possibly near Mixon) *Maring* 1256 Ch.

**MARLEY HOLLOWS (FARM)** 1 mile south-east of Fulford (SJ 9637). It is unclear whether the place is to be associated with *Marledhey*, recorded in 1537 (MA), from Mercian OE *(ge) heg*, so 'the hay or enclosure where the land was treated with marl'. It seems more likely that the name is a corruption of Morrillow (q.v.), now known as Mount Pleasant, which adjoined it to the north.

**MARNSHAW HEAD** 2 miles west of Longnor (SK 0564). *Marinshawe* 1566 *Deed*, *Merneshawes* 1626, 1651 *Rental*, *Mount Shaw*, *Mount Shaw Head* 1840 O.S. Perhaps from the personal name Marrin (DES 299), with OE *scaga* 'a small wood, a copse'.

MARSH, THE (unlocated, near Castlechurch) *Le Mersh* 1231 SHC XII 5, *Marisco* 1235 SHC VIII (ii) 129, *Merse* 1279 SHC VI (i) 110, *Marisco* 1281 ibid. 106, 1290 SHC 1911 198, *Marreys* 1301 ibid. 271, *Le Mershe* 1316 SHC VIII (ii) 45, *(Castrum cum) Marisco* 1327 SHC VII 244, *Mersh* 1331 SHC 1913 27, *le Marshe* 1349 ibid. 46, *the Mershe* 1403 ibid. 46, *le Marshe juxta Stafford* 1462 Ct. From OE *mer(i)sc* 'marsh'. VCH V 88 describes the place as a vill which paid rent or custom to the manor of Stafford. A place called Marsh (*Mersche* 1332 SHC X 90) where brine comes out of the ground is recorded by Plot 1686: 97 at Ingestre, presumably the place shown as *Brin* to the north of Ingestre on Bowen's map of 1749, and another Marsh is recorded south-west of Pattingham, perhaps to be identified with Rushy Marsh. *Mershh* on the east side of Penkridge (where the famed Penkridge horse fairs were held by 1754: VCH V 129) is recorded in 1344 (SHC 4th Series XVIII 211), and as *the Marshe* in 1614 (Penkridge ParReg).

**MARSH BARN (FARM)** on north side of Handsacre (SK 0916). *Marchbane* 1585 SRO D(W)1721/3/170, *Marchbarne, Merchbarne, M'chbarne* 1586 SHC 1927 130-1, 135, *Marchebarne* 1590 SHC 1930 (ii) 197, *Marchbarne* 1594 SHC 1932 39, *Muchbarne* 1597 ibid. 301, *Marshbarne* 1608 SHC 1948-9 41, *March Barnes* c.1644 SRO 793/7, *Marche Barne* 1676 WCRO CR1291/169, *Marsh Barn* 1814 Armitage ParReg, *March Barn* 1834 O.S. Although the place lies a few hundred yards south of the river Trent, which forms a parish boundary here, a derivation from OE *mearc* 'boundary' can almost certainly be ruled out on philological grounds, and the root is almost certainly OE *merece* 'smallage, wild celery' (see Marchington), giving 'the barn where the wild celery grew'. The name was applied in the 19th century to a building, possibly the barn itself, now vanished, to the east of the present farm.

**MARSTON** 6 miles west of Penkridge (SJ 8314), *Mersetone* 1086 DB (listed in Northamptonshire), *Merston* 1203 SHC III 87, *Mershton* 1316 FA; **MARSTON** 2½ miles north of Stafford (SJ 9227; SMR 02504 places a deserted Anglo-Saxon settlement at SJ 92102750), *Merestonam* 1081 SHC 1914 104, *Mertone* 1086 DB, *Mershton* 1316 FA, 1440 SHC 1914 155. From OE *mersc-tun* '*tun* by a marsh'. *Mersetone* recorded in 1086 (DB) has been identified as Amerton (q.v.), but may well be another entry for Marston near Stafford.

**MARTEN HILL, MARTIN HILL** in Swinscoe (SK 1447). *Mortons Hill* 1743 Okeover E17, *Martin's Hill* 1781 Okeover T82-3, *Martin Hill* 1797 ibid.' T87. Seemingly from a personal name. The O.S. uses the name Martin Hill for the farm, and Marten Hill for the hill.

**MARTIN'S LOW** 1 mile north-west of Waterfall (SK 0752). *Martins lowe* 1631 Senior, *the Common or Moore called Martins low* 1631 ibid, *Martinslow(e)* 1764 ParReg, *Martings Low* 1775 Yates, *Martin's Low, Martinslow Farm* 1836 O.S. Perhaps from the OE personal name Mærtin or Martin, so 'Martin's or Martin's *hlaw* or burial mound', or from ME *marten* 'a weasel, a marten', or possibly from OE *(ge)mærtun* 'boundary *tun*': the place lies on a parish boundary.

MAUBAN or MALBANK FRITH – see ALSTONEFIELD FOREST.

**MAVESYN RIDWARE** – see **RIDWARE, MAVESYN**.

MAWPLECROSSE (unlocated, in Wood End, Yoxall) *Mawplecrosse* 1640 SRO D4533/1/4/1.

MAYDENES (unlocated) *Maydenes* 1345 SHC XII 44.

MAYDINLEGH (unlocated, in or near Needwood Forest: Shaw 1798: I 156) *Maydinlegh* c.1215 (1798) Shaw I 156, *Maydenlegh (spring)* n/d (1798) Shaw I 109. Seemingly 'the maiden's *leah* or clearing'.

**MAYFIELD** Ancient Parish 2 miles south-west of Ashbourne (SK 1545). *Medevelde* 1086 DB, *Machelfeld, Malefeld* c.1150 SHC VII NS 128, *Matlefelt* c.1175 ibid. 134, *Matherfeld* c.1203 SHC 4th Series IV 42, 1252 Ch, 1269 Ass, *Methelfeld* 1183-4 SHC 1937 19, *Mathelfeld* 1227 SHC IV 58, 1309 SHC 1911 73, *Mathelefell, Mathelefelt* 1275 SHC V (i) 118, *Mathelfeld* 1292 VI (i) 265, *Over Matherfeud* 1293 SHC 1911 47, *Maafelt, Maynfelt* ?13th century SHC VII NS 139-40, *Madderfeld* 1304 WL 37, *Malefield* 1307 SHC VII 181, *Mathelfeld* 1309 SHC 1911 73, *ovemastemathelfeld* 1324 ibid.' 103, *Overmast Matherfeld* 1327 SHC 1912 253, *Matherfeld* 1327 SHC VII 201, *Quemas Mathefeld* 1329 SHC I 298, *Kirkematherfeld* 1386 SHC 4th IV 220, *Mathefeld, Mathfeield, Mayfield* 14th century Duig, *Malefeld* 1448 SHC NS III 179, *Malfeld* 1532 SHC 4th Series 8 130, *Over Mathefylde, Mydle Mathefylde, Nether Mathefylde* 1583 SHC XV 150, *Maifield* 1656 Leek ParReg, *Mathfield* 1686 Plot 404, *V[pper] Mathfield, N[ether] Mathfield, Mathfield* 1695 Morden, *Upper Mayfield, Mayfield Cottage, Mayfield Hill, Middle Mayfield* 1836 O.S. It has been suggested that the name incorporates OE *mæddre* 'madder' (see e.g. Ekwall 1936: 110, 1960 318; Gelling & Cole 2000: 275; CDEPN), but botanists say that the plant is unlikely here, and the original *-l-* (and the proximity to Harlow Wood (q.v.) and Motcarn (q.v.) may be noted) makes a derivation from OE *mæthel* 'meeting, council' (Clark 1995: 224 suggests that the element is one of the rarest), with OE *feld* 'field, open land' (as proposed by Duignan 1902: 100-1) certain. The DB spelling suggests Norman pronunciation difficulties, and the *–ch–* spelling of c.1150 is doubtless due to the common misreading of *c* for *t*. *Quemas* in the 1329 spelling is unexplained. The 1324 and 1327 forms appear to be Overmost Mathelfield (i.e. Uppermost Mathelfield, from ME *overmast*), and the 1386 form with a Scandinavianised form of OE *cirice* 'church' (which it probably replaced), is evidently Church Mayfield: DB records a priest at Mayfield, which implies that a church existed at that date. It may be noted however that in 1620 *Nether Mathfield* is described as *otherwise Churchfield*: SHC VII NS 199. The meeting place was perhaps the *hlaw* of Harlow, but see also Marten Hill and Motcarn. Early maps (e.g. Bowen 1749, Yates 1798, O.S. 1836) show Church Mayfield in larger and bolder print, whereas Mayfield is shown on modern maps centred on Hanging Bridge. *Maufeld Smeethe* (from OE *mæþ* 'mowing', i.e. Mayfield's Meadow: see also Okeover) is recorded in 1420: Okeover T22: see also Okeover, and *Mafeild Ryse* is recorded in 1538 (Survey), evidently incorporating OE *hris* 'brushwood': see Fords Rice. Cf. Matlock, Derbyshire; Matlask, Suffolk; Malton, West Yorkshire. Another Mayfield lies ½ mile south-west of Tutbury (SK 2027), but the history of the name is untraced.

**MEADLEYS, THE**   in Patshull Park, 1 mile north-west of Pattingham (SJ 8100). *Meduleye* 1327 SHC VII 251, 1332 SHC X 130, *Medoley* 1401 (1801) Shaw II 282, *Medley* 1494 SHC XI 253, *Medleys* 1539 ibid. 280, *Medlies* 1740 Pattingham ParReg. Probably from OE *mǣd* 'meadow', with OE *leah. See* also Shaw 1801: II 282.

**MEAFORD**   (pronounced Meff-ford [mɛfəd]) 1½ miles north-west of Stone (SJ 8835). *Meford, Metford* 1086 DB, *Medford* 1175 P, *Medford* 1250 SHC IV 123, *Mefford* c.1280 SHC VIII (ii) 139, *Mayford* 1792 Andrews 1936. Probably 'ford at the junction of streams', from OE *(ge)mȳþe* 'stream junction' – the Trent is joined by a tributary at this place – or from OE *mǣd* 'meadow', giving 'meadow-ford'. There are said to have been two manors of this name here: SHC VIII (i) 139. *Meaford Moor alias Hooks Green* is recorded in 1707: SRO D593/B/1/19/2.

**MEAL ARK CLOUGH**   1 mile north-east of Heaton (SJ 9563) A clough is a ravine or narrow valley with steep sides, usually forming the bed of a stream: here the name is applied to a steep-sided valley through which flows a tributary of the river Dane. A meal ark was a meal-chest or flour-bin (EDD), and the clough was presumably so-called from some topographical feature here which resembled such a piece. Early spellings of the name have not been traced, and it does not appear on the 1890 6" O.S. map.

**MEASE, RIVER**   a tributary of the river Trent. *Meys* 1247, 1272 Ass, 1347 Pat, *Water of Mays* 1279 SHC VI (i) 99, *Mese* 1578 BCA MS3878/122, 1586 Harrison, *Messe* c.1600 Erdeswick. From OE *meos* 'moss, bog, marsh'. See also Aqualate Mere.

MEDENALE, MEDALL GREEN   (unlocated, in Compton near Kinver) *Medenale, Medall Green* 1453 and 1562 VCH XX 139. Possibly from OE *mægden halh* 'maiden's halh'. Perhaps associated with Meddins Lane on the west side of Kinver, recorded in 1854: VCH XX 140.

**MEDLEYWOOD BARN**   1 mile south-east of Admaston (SK 0622). *Middlehay* 1278 SRO D986/36, 1347 SHC 1919 12, *Madeley Wood Barn* 1836 O.S. 'Middle hay or enclosure', from OE *(ge)heg*.

**MEECE BROOK**   a tributary of the river Sow. *Mesebrock* 1272 FF. Probably from OE *meos* 'moss, bog, marsh': it is more likely that Millmeece, Coldmeece and Heamies (q.v.) took their names from the river, than the river was named after those places.

**MEERBROOK**   3 miles north of Leek (SJ 9860). *Merebroc* 1229x1232 CEC 385, 1294 Banco, 1330 Ch, *Merebroc(e)* 1327 (p) SR, *Merebroke* 1338 Misc, *Marbrok(e)* 1577 Saxton, *Marbrouke* 1599 SHC 1935 99, *Marbrook(e)* 1604 ParSurv, *Merchbrook* 1755 Bowen, *Meerbrook* 1775 Yates. 'Boundary brook', from OE *(ge)mǣre* 'boundary'. The place is near a stream (*Meer Brook* c.1220: VCH VII 193) of the same name, which served as an ancient boundary. See also Elliott 1984: 57-8.+

**MEER OAK**   1 mile north of Pattingham, at the junction of Deer Leap and Hollies Lane (SJ 8300). *Le Tyndede Meer Oke* 1298 SHC VI NS (ii) 55; Jones 1894 158, *Meer Oak* 1686 Tettenhall ParReg, 1750 Codsall ParReg, 1834 O.S., 1891 O.S. An oak-tree mentioned in 1298 as the boundary mark between the manors of Pattingham, Wrottesley and Perton (VCH XX 10), but long lost by 1894: Jones 1894: 158. From OE *(ge)māre* 'boundary'. *Tyndede* is unexplained, but possibly from the past participle of OE *tind* 'to set fire to, to ignite', so perhaps 'the burned oak at the boundary'.

**MEESE, RIVER**   a tributary of the river Tern. *Mees* 1266 Ch, 13th century Dugd vi 390. From OE *meos* 'moss, bog, marsh'.

**MEGACRE**   1 mile south-east of Audley (SJ 8150). *Magacre* 1733 SHC 1944 69. The most likely derivation is 'the ploughed land with the magpies', from OE *aecer* "field, ploughed land'.

**MEG A FOX HOLE**   (obsolete) a cave on Kinver Edge (SO 8483). *Mag-a-Fox Hole*, *Meg a fox-hole* 1686 Plot 172, 414, *Meg o' Fox Hole, Meg a Fox Hole* c.1750 Wilkes; VCH XX 122 fn.71. Perhaps from Margaret-of-the-fox-earth whose death is recorded on 8 June 1617: StEnc 441.

**MEGCROFTS**   ½ mile north-east of Ipstones (SK 0250). *Meg Croft* 1837 O.S., *Meg crofts* 1872 P.O., *Meggcrofts Farm* 1880 Kelly. 'Margaret's croft or enclosure', from OE *croft* 'a small enclosure of arable or pasture land, an enclosure near a house'. It is unclear whether *Meggscrofte*, recorded in 1494, and *Meggercrofts*, recorded in 1514 (OSS 1936 48, 56) refer to this place.

**MEIR**   2 miles south-east of Longton (SJ 9342). *Mere* 1242 SHC XI 315, 1261 MRA *et freq.* to 1564 Pat, *La Mere* 1250 SHC XI 319, *le Meere* 1447 SHC III NS 177, *Mere juxta Caryswall* 1535 VE, *Meyre* 1564 FF, *Meir* 1656 ParReg, *Meare* 1695 Morden, *Mear* 1836 O.S. Almost certainly from OE *(ge)māre* 'a boundary, a border': the place lay on the boundary dividing Normacot and Stone from Blurton and Trentham (SHC 1910 74-5; see also SHC XII 59), and Meer Lane appears on Yates' map of 1775 on the Roman road between Stoke and Rocester (Margary number 181, 70a) where the road forms the boundary between Pirehill and Tormonslow Hundreds. However, a derivation from OE *mere*, usually meaning 'pool' is etymologically possible. The place lies in a valley, and although there is no record of any sizeable pool here (see however Ward 1843: 559), there is slight evidence of a causeway, suggesting the existence of wet ground (see SHC 1912 219 fn.1), to which the term *mere* was sometimes applied: PN Sa III 244. It may also be noted that Gilbert de la Mere is recorded in this area c.1250: SRO D593/B/1/23/4/1/2.

**MEIRHEATH**   2½ miles south-east of Longton (SJ 9240). *Lightewood heathe alias Meare heathe* c.1545 SHC 1910 74, *Mereheath* 1585 SHC XVII 228, 1592 NA 157DD/2P/19/1, *mereheathe* 1596 SHC 1932 236, *Meer Heath* 1677 Caverswall ParReg, *Mere Heath* 1732 Stoke on Trent ParReg, *Mare Heath* 1756 Swynnerton ParReg. See Meir. It has been suggested that the original settlement of Normacot (q.v.) lay at Meir Heath: StASt 10 1997 16.

**MELLESHOHE, MALESHOU**   a lost place perhaps in the Great Wyrley-Essington area: see StSt 10 1998 97. *Melesho* c.1137 VCH III 223, *Melleshohe* c.1135 EEA 14 38, *Maleshou'* 1140x1147 (13th-14th century) ibid. 40, *Maileshou* 1304 Ch, see also Shaw 1801: II 313. The name is said to incorporate the uncommon OE personal name Mula (or ON Muli), found in the field-names *le mowelesbruche* 1296, *le Moulesgrene* 1347 (*Vernon*) in Essington (see Oakden 1984: 55; StSt 10 1998 97), but the evidence is doubtful. The second element is OE *hoh* 'a heel, a spur of land'. This is likely to be the same place as *Midlestehoo*, recorded in 1313 (1801 Shaw II 58).

**MEREBROK** (unlocated, possibly near Amerton or Chartley) *Merebrok* 1338 SHC 1913 71. Probably 'boundary brook', from OE *(ge)mǣre, broc*, or from OE *mere* 'a pool, a lake'.

**MERE FARM, MERE HALL** 1½ miles north of Enville (SO 8289). *the Meyre* 1534 SHC 1912 73. From OE *(ge)mǣre* 'a boundary, a border': the places lie near the boundary between Enville and Bobbington.

**MERE HILL** ½ mile south-west of Throwley Hall (SK 1052). The history of the name is not known, but possibly from OE *(ge)mǣre* 'a boundary, a border', or from the mere or pool near Throwley Hall (see Throwley).

**MERETOWN** in Forton parish, 1 mile north-west of Newport (SJ 7520). *Mera* 1086 DB, *Mere* 1189 P, *More* 1284-5 SHC 1910 299, *Meerton, Mearton* 1686 Plot, *Meer Town* 1798 Yates. From OE *mere* 'a pool, a lake'. The place adjoins Aqualate Mere (q.v.). See also JEPNS 24 30-41.

**MERLOW, MERELOW** (unlocated, in Ackleton) *Merlow, Merelow* 1882/3 The Reliquary.

**MERLVALE** (unlocated, possibly in the Hilderstone/Stallington area, but see also Murdeford) *Merlvale* 1326 SHC 1911 371. If the spelling is reliable, perhaps from OE *meargealla, mergelle* 'gentian' (cf. Malborough, Devon; Marlborough,Wiltshire), with Fr *vale* 'valley' (an element rare in Staffordshire), so 'the valley with the abundance of gentian'.

**MERRIDALE** 1 mile south-west of Wolverhampton (SO 9098). *Muriden* medieval Duig, *Merridale* 1420 SRO D593/B/1/26/38/2, *Merrydale* 1516 SRO D593/B/1/26/6/29/11, *Merridale* 1557 WA I 223, Wolverhampton ParReg, *Merydall* 1597 SHC 1935 8, *Merrydale* 1614 SHC IV NS 77, 1634 SRO D260/M/T/7/5, *Merredell* 1670 SHC 1923 56. Probably from OE *myrge* 'pleasant, sweet, delightful, agreeable', with OE *denu* 'valley', which has developed into *dale*. Cf. Meriden in Warwickshire. See also SHC VII 112; SHC 1916 167; SHC 1938 58-9.

**MERRIL GROVE** 3 miles west of Longnor (SK 0464). *Merilgre(a)ve* 1556 *Deed*, 1651 *Rental, Merril Grove* c.1870 *Rental*. From OE *(ge)mǣre, hyll* 'boundary hill', with OE *grǣfe* 'copse', hence 'the copse on the boundary hill'. The place lies at a height of 1290'. Cf. Merril Farm, Derbyshire.

**MERRIL'S HALL or MEROLS** (obsolete) on the north-east side of Wolverhampton (SO 948999: TSSAHS XXIV 1982-3 55). The place is said to be recorded in TA 1838, from the Merrill family recorded in Wednesfield: TSSAHS XXIV 1982-3 55.

**MERRY HILL** in Brierley Hill (SO 9286), *Murihul* 1327 SHC VII (i) 246, *Muryhull* 1332 SHC X 86, *Murihull* 1340 Mander & Tildesley 1960: 8, *Myrry hill temp.* Henry VIII SHC X NS (i) 115, *Merry-hill* 1686 Plot; **MERRY HILL** 2 miles south-east of Wolverhampton (SO 8897), *Murihull* 1340-1 SHC 1928 35, *Muryhulls* 1397 (1801) Shaw II *221, *Merryhill* 1647 Survey, *Merry Hill* 1895 O.S., MERRY HILL (obsolete) 1 mile east of Blymhill (SJ 7812), *Merry Hill* 1833 O.S. From OE *myrge* 'pleasant, sweet, delightful, agreeable'. *Murihurst*, recorded in 1327 (SHC VII (i) 248), and *Muryhurst*, recorded in 1332 (SHC X 87), may be associated with Merry Hill in Brierley Hill.

**MERRYTON LOW** on Morridge (SK 0460). *Meriloneslowe* 1223 SHC 4th Series 19 5 fn.3, *Merryton Low* 1842 O.S. 'The boundary-lane tumulus', from OE *(ge)mǣre, lone,* *hlaw*: there is a Bronze-Age tumulus here which formed a Leek parish boundary-mark (SHC 4th Series 19 5; VCH VII 211).

**MESTY CROFT** 1 mile east of Wednesbury (SO 9995), probably midway between the crossings of the river Tame at Hydes Road and Crankhall Lane: Dilworth 1976: 103-4. *Misty Croft* 1684 BCA MS3145/91/2. Perhaps 'the field or little farm with the mistletoe', from OE *mistel*, with OE *croft* 'a small enclosure of arable or pasture land, an enclosure near a house'.

**MICKLEDALE** 1 mile south-west of Abbots Bromley (SK 0723). *Muckledale* c.1271 SRO D(W)1721/3/30A/6, *Mucledale* 1286 SHC 4th Series XX 29, *Muccledale*, *Micheldale* c.1292 SHC 1937 96, *Mukdale* 1367 SRO D(W)1721/3/32/14. From OE *micel*, or (more likely) ON *mikill* 'much, great, large', and ON *dalr* 'valley': the valley or dale here is now part of Blithfield reservoir.

**MICKLE HILLS** 1½ miles south-west of Lichfield (SK 0908). *Mykyll Hylle, otherwise Pype Hylle* 1659 (1798) Shaw I 312, *Mickle Hill* 1834 O.S. From OE *micel, hyll* 'big hill'.

**MICKLEWOOD** 2 miles south-east of Penkridge (SJ 9411). *Mykelwode* 1467 SHC 1928 145, *Myckulwodde* 1525 ibid.' 146, *Micklewood* 1657 Survey. From OE *micel-wudu* 'the big wood'.

MICKLINGS (obsolete) 1 mile south-east of Alton (SK 0841). *Micklings* 1798 Yates, 1836 O.S. Perhaps from OE *micel* 'big', with OE *ings*, perhaps here with the meaning 'hills': see Ekwall 1960: 265, *sub. nom.* Ingon. The place lies on a junction of parish boundaries, on the side of a 541' hill.

**MICKLOW** 1 mile south-west of Stone (SJ 8832*). Mickelowes* 1539 MA, *the Micklows alias Michael House...alias Mickcloses* 1677 SRO D593/B/1/19/2/21/1, *Micklows (Heath)* 1739 SRO D628/19, *Micklow* 1798 Yates, *The Micklow House* 1836 O.S. Unless *Mukelichale* 1273 SHC 1911 151 refers to this place (which seems unlikely), the name is from OE *micel* 'big', with OE *hlaw* 'hill, mound, tumulus', from some lost burial-mounds which stood here. There are a number of burial mounds recorded in this area.

**MIDDLE CLIFF** ½ mile west of Bradnop (SK 0054). *Midlesclif* 1223 Ward 1843: app. ii, *Midlesleclift* 1227 Harl, c.1266 StCart, *Middelisteclif* 1256 Ch, *Middle Cliff* 1596 Okeover T697, *Middle Cliff(e)* 1676 Leek ParReg. From OE *midlest* 'middlemost', with OE *clif* 'cliff, rock, steep descent, promotory', so 'the cliff in the middle of the area'.

MIDDLE FOREST (unlocated, north of Leek) *? La Foreste* 1308 SHC XI NS 257, *Middle Forest* 1539 SHC IX NS 301, *le Midle Forrest* 1621 SHC 1934 24. See also High Forest.

**MIDDLEHILLS (FARM)** in Caldon (SK 0849). Field-names *Long Middlehill Close, Further Middlehill Close, Mean Middlehill* are recorded in 1664: Okeover T754.

MIDDLESTON (obsolete, in Biddulph) *Middleston* 1453 SHC XII 16, 1679 SHC XII 7. Chetwynd states in 1679 that Middleston was an alternative name for Middle Biddulph: SHC XII 7.

MIDDLETON (obsolete, in Hammerwich (SK 0607)) *? Middelton* 1269 SHC IV 175, *Midleton* 1370 SHC VIII NS 245, *Middleton* 1381 VCH XIV 259, 1393 SHC XV 50. 'The middle *tun*': evidently the place lay between Overton and Netherton (q.v.). The name was still in use in 1871: VCH XIV 259.

**MIDDLETON (GREEN)** 3 miles south-west of Checkley (SJ 9935). *Middleton* 1272 SHC IV 187, *Midelton* 1375 SHC XIII 121, *Middleton Greene* 1607-8 SHC 1948-9 54. 'The middle *tun*'.

**MILES GREEN** 1 mile south-east of Audley (SJ 8049). *Miles Grene* 1539 SHC NS V 270, *Myles Greene* 1558 Audley ParReg, *Miles Green* 1693 Betley ParReg, 1833 O.S. It is unclear whether *Mees Green*, recorded in 1733 (SHC 1944 7, which does not mention the name Miles Green) refers to this place. If it does, it would explain why StEnc 393 gives a drivation from the Mee family. The Green element suggests a squatter settlement.

**MILFORD** 3½ miles south-east of Stafford (SJ 9621). *Myfforde (fflate)*, *Milforde wichale* 1570 Survey, *Milford* 1759 SRO D1798/587, 1770 SRO D1368/3, *Millford* 1798 Yates, *Milton or Milford* 1801 Shaw II 325, *Milford* 1836 O.S. 'The ford with the mill'. *Wichale* in the 1570 spelling is unexplained, but may be from OE *wic halh*: one specialist meaning of *wic* is 'salt-works', and this place lies in an area of salt-springs. According to VCH V 3, Milford near Stafford first occurs in the late 18th century. *Milfordbrooke*, probably in Great Haywood, is recorded in 1611: SRO D603/E/4/5. *Muleford*, recorded in 1271 (SHC V (i) 154) may refer to a place south of Wolverhampton.

**MILK HILL** a 938' hill 1 mile north-east of Cauldon (SK 0949). *Milkhills* 1664 Okeover T754, *Miskill (Gate)* 1717 Ellastone ParReg, *Milkhill* 1713 Okeover T761, *Milkhill Gate* 1747 Poll, *Milk Hill Gate* 1775 Yates, *Milk Hill (Gate)* 1836 O.S. Perhaps from OE *meoluc* 'milk', with OE *hyll* 'hill', presumably from the watercourses coloured by limestone quarrying here, with *geat* 'gate, pass', so 'the opening or pass (or gate) of Milk Hill'. The 1717 spelling is evidently a transcription error.

**MILL BROOK** a tributary of the river Dane. From OE *myl(e)n* 'mill'.

**MILLDALE** 1 mile south-east of Alstonefield, on the river Dove (SK 1354). *le Mylne dale* 1594 DRO D2375M/57/1, *Milndale* 1604 SHC 1940 287, *? Milsdale* 1633 DRO D258/18/18/2-10, *Milldale* 1695 Alstonefield ParReg, *Mil Dale* 1749 Bowen. From OE *myl(e)n* 'mill' and OE *dæl* 'valley'.

**MILL DALE FARM** ½ miles north-east of Balterley (SJ 7650). *Knights Mill House* 1733 SHC 1944 38, *Knights Mill* 1799 Faden. The place lies on Dean Brook.

**MILL FARM** on the river Trent south-west of Sandon (SJ 9429). *Milnehouses* 1305 SHC 1913 201.

**MILL FLEAM** a tributary of the river Dove which flows to the north of Tutbury and forms part of the Staffordshire-Derbyshire boundary. *milne Fleame* 1798 Shaw I 56, *Fleam or Little Dove* 1908 VCH I 357. From OE *myl(e)n* 'mill', with OE *\*fleama* 'river, stream'. The fleam or mill-stream serving Tutbury Mill.

MILL HOLMES (unlocated, near the river Trent in Yoxall Parish) *Mulneholm* 1176 SHC 1914 137, *Milne Holmes* 1567 SHC IX NS 26. From OE *myl(e)n* 'mill', with ON *holmr* 'small island, piece of land surrounded by a stream'.

MILLHOUSE (unlocated, in what is now Patshull Park: TSSAHS VIII 1966-7 49; TSSAHS XI 1970-1 35 suggests at SJ 809010). *Mulnehouse* 1294 SHC VII 10, *Mulnehous* 1327 SHC VII (i) 251, *Mulehouse* 1401 (1801) Shaw II 282, *Milhous* 1479 SHC VI NS (i) 120. The settlement of this name appears to have been cleared during the creation of the Park in the 1740s: VCH XX 162.

**MILL HOUSE FARM** on the south-east side of Cheadle (SK 0142). *Mulnehouse* 1435 SHC XVII 151, *Milnehowses* 1609 SHC NS III 14, *Milnhowse* 1617 SHC NS VI (i) 14, *Mill House* 1836 O.S. From OE *myl(e)n hus* 'mill-house'.

**MILLIAN BROOK** a tributary of the river Sow on the north-west of Stafford. Early spellings have not been traced, but perhaps from OE *myl(e)n* 'mill', influenced by the word *million*.

**MILLMEECE** 3 miles north of Eccleshall (SJ 8333). *Mess* 1086 DB, *Mes* 1208 Cur, *Meis* 1218 Cl (these forms could be for Cold Meece (q.v.)), *Mulnemes* 1289 Cl, *Mulnems* (*sic*) 1390 SHC XV 19, *Mylmes* 1532 SHC 4th Series 8 104, *Mill Meece* 1833 O.S. From OE *meos* 'a moss, a marsh, a bog' or from the nearby river Meece or Meese Brook (q.v.), with OE *myl(e)n* 'mill' (recorded here in 1163: CDEPN). See also Heamies, above.

MILST (unlocated, in or near Leek) *Milst* 1662 *et freq* Leek ParReg. This may be an abbreviated form of Mill Street, which is described as a hamlet in 1548: Leek CtRolls PRO SC2/202/65.

**MILTON** 4 miles north-east of Stoke on Trent (SJ 9050). *Mulneton* 1227 (1843) Ward 1843: app. iv, c.1287 ibid.' 296, *Milton* 1539 MA, *Mylton* 1613-4 SHC 1934 31, *Milton Milne* 1625 JNSFC LX 1925-6 73, *Millton* 1749 Bowen. From OE *myl(e)n-tun* 'mill *tun*'.

**MILWICH** (pronounced Mill-ich [mɪlɪtʃ]) Ancient Parish 5 miles south-east of Stone (SJ 9732). *Melewich, Mvlewiche* 1086 DB, *Mulewich* 1159 (13th century) EEA 14 71, 1166 SHC 1923 296, 1177 SHC XII NS 278, *Mulewyz* c.1192 SHC VI (i) 12, *Millewyz* 1236 Fees, *Mulewis* 1242 Fees, *Moleswyke* 1286 SHC VI (i) 161, *Melewys* 1288 ibid. 174, *Mulewyt* 13th century SRO D798/1/1/6, *Meolewych* 1337 SHC 1913 63, *Mylwyche* 1532 SHC 4th Series 8 59. From OE *myl(e)n* 'mill', with OE *wic*. The place is in an area of salt springs (see e.g. Ingestre, Salt, Shirleywich), and lies on salt strata, but the specialised meaning of *wic* 'salt-working place' is improbable with *myl(e)n*.

**MINNBANK** 3 miles south-west of Madeley (SJ 7540). *Mynn Bank* 19th century SRO D3211/99, *Minnbank* 1920 O.S. Early spellings have not been traced – the place does not appear on the 1833 1" O.S. map – but unless the name is recent, it would seem to be from Welsh *mynydd* 'a hill', which was adopted as OE *myned* (see Coates & Breeze 2000: 335): the place lies on a pronounced hill of 541'. See also Minnie Farm. Cf. Bosley Minn and Wincle Minn, on a high ridge (and parish boundary) between Bosley and Wincle in Derbyshire, near the Staffordshire border (SJ 9466): PN Sa III 229.

**MINNIE FARM** 1 mile south-west of Audley (SJ 7948). Early spellings have not been traced, but if the name is ancient (it does not appear on the 1833 1" O.S. map, but the printed Audley ParReg. suggests that it is found in the registers as *Minno*), perhaps from Welsh *mynydd* 'a hill', which was adopted as OE *myned*: the place lies on the northern flanks of a pronounced ridge. See also Minnbank.

**MIRES BROOK** a tributary of the river Blithe. Perhaps from OE *myrr* 'bog, swampy ground'.

MITHAM unlocated, on the river Trent near Hanford, possibly at the junction with the river Lyme. *the Mitham where the two river meet* 1689 Ward 1843: app. lxii, *the Mytham* 1739 SRO D593/B/1/14/10. From OE *(ge)mȳþe* (a derivative of *muþa*) 'the confluence of streams'. See also Mythaholme.

**MITTON** 2½ miles west of Penkridge (SJ 8815). *Mutone* 1086 DB, *Muton'* (p) 1194 Cur, *Mutton* 1203 SHC III 105, *Muiton'* (p) 1221 Ass, *Mitton* 1236 Fees, *Mutton'* 1242

Fees. From OE *(ge)mȳþe-tun* 'the *tun* near the confluence of streams'. The place lies near the confluence of Church Eaton Brook and Whiston Brook.

**MIXON** 5 miles north-east of Leek (SK 0457). *Myxle* 1167 Eyton 1854-60: X 36, *Mixa* 1199 SHC III (i) 53, *Mixenn, Mixene* 1203 ibid.' 122, *Mixne* 1219 FF, 1227 Ch, *Mixsne* 1256 Ch, *Mixen* 1274 SHC VI (i) 64, *Mixene* 1333 SHC X 116. From OE *mixen* 'the dunghill'.

**MOAT BANK (HOUSE)** between Wall and Muckley Corner (SK 0806). *The Moat House* 1633 SRO D15/10/1/14, *Moat Bank* 1834 O.S. From ME *mote* 'a hillock, mound entrenchment; a moat; a protective ditch filled with water'. According to VCH XIV 292 the name is from a rabbit warren in the form of an embankment with protective ditch which existed near here in 1450, but see also Pipehill; Pipe Grange.

**MOAT BROOK** a tributary of the river Penk forming part of the boundary between Brewood and Codsall. *mot-brooke* 1638 Codsall ParReg, *the Motbrook* 1798 ibid. Probably from a former moated site at Moor Hall (q.v.) or Wood Hall (q.v.), Codsall: VCH XX 79, 82. The stream seems to have been called *Longmeadow brook* in 1411: VCH XX 34. Since Bilbrook (q.v.) is believed to have taken its name from this stream, it is perhaps surprising that the stream is not named *Billerbrook* or similar.

**MOAT FARM** 1 mile north-west of Audley (SJ 7851), *Mott (Meadow)* 1689 SRO DW1826/26, *Mott* 1733 SHC 1944 1, 1799 Faden, *Moathouse* 1890 O.S.; **MOAT HALL, MOAT HILL, MOAT LANE** ½ mile south of Newborough (SK 135246), *le Motte lone* 1499 SRO DW1733/A/3/15, *Moat Hall, Moat Lane Gate* 1836 O.S. From ME *mote* 'a hillock, mound entrenchment; a moat; a protective ditch filled with water'.

**MOBBERLEY** 1 mile south of Cheadle (SK 0041). *Maberley* (p) 1565 SHC XIII 245, *Mobberley* (p) 1575 Cheadle ParReg, 1588, 1597 SHC XV 186, SHC XVI 170, *Mobberlei* 1588 SHC 1934 (ii) 19, *Mobberley (brook)* 1668 SRO D1275/7/15, *Moberly* 1798 Yates, *Moberley* 1836 O.S. If ancient, and not transferred from Mobberley in Cheshire, 7 miles south-west of Cheadle (Cheshire), as proposed by Oakden 1984 15 (the place is marked on Yates' map of 1798, but not on Browne's map of 1682), possibly from OE *gemot-beorg, leah* 'the wood or woodland clearing associated with the assembly mound', rather than OE *gemot-burh leah* 'clearing at the fortification where moots or assemblies were held' (cf. Mobberley, Cheshire; Modbury, Devon; Bodbury Hundred, Dorset), perhaps to be associated with the flat-topped circular mound (at SK 006413) which excavation has revealed to be a natural sand knoll (SMR 00039). The mound is known as Castle Croft (q.v.). *Modborleg*, recorded in the late 13th century (SRO 3764/93), must be Mobberley in Cheshire, since the prior is mentioned: Mobberley Priory is in Cheshire.

**MOBBERLEY BROOK** a tributary of the river Tean. See Mobberley.

MOCKBEGGAR HALL (obsolete) in Essington parish, 4 miles north-west of Walsall (SJ 9704). *Mockbeggar Hall* 1808 Baugh. A curious brick oval-plan three storey building with gothick windows on the first floor (illustration StEnc 398) erected in the 18th century as miners' tenements by the Vernons of Hilton Hall on heathland in Essington Wood. The name Mockbeggar became popular after its use in a poem of 1622 by John Taylor (1580-1653, a seaman, pedestrian and author of doggerel verse who became known as 'the Water-Poet' because he was at one time a Thames waterman), and came to mean 'a house with an inviting external aspect, but within poor and bare, and therefore disappointing to those who come to beg' (EDD), here because travellers on the London road who saw the conspicuous building and made a long detour to beg alms found it occupied by the poor. John Taylor made several journeys around England in the first half of the 17th century, and visited Lichfield and Stone in 1639, but there is no evidence that he was in any way directly connected with this place. Shaw 1801: II (unpublished sheets 320) says of nearby Great Wyrley: 'The collieries here are very flourishing, and give employment to numerous inhabitants of this populous hamlet, yet Wyrley bank is still proverbial for its paupers begging about the distant country'. The building, demolished in 1936, was at what is now known as Springhill, and became known as *Colliers Castle* later in the century: 1895 O.S. Oakden 1984: 50 states that *Colliers Castle* derives from a sarcastic reference to the coaltips here, but it was almost certainly so-called from the miners who lived in the hall. See also Hackwood 1896: 120, who mentions this place and another place of the same name in Harborne. *Mockbeggar's Hall* is also recorded at Rocester: TNSFC 1927 168.

**MODDERSHALL**  2½ miles north-east of Stone (SJ 9236). *Modredeshale* 1086 DB, *Modreshalle* 1305 SHC VIII (i) 162, *Mothersall* 1551 SHC XII 209, 1708 SBT DR10/1401, *Mottershaw, Mothershaw* 1747 Poll, *Moddershall* 1836 O.S. 'Modred's *halh*', with *halh* here perhaps meaning 'small valley'. *Withnall Forest (alias Moddershall Heath)* is recorded in 1696 (SRO DW1742/14-17), and *Mothersall Heath (alias Withnall Forest in Kibblestone)* is recorded in 1808: SRO D593/B/1/20/9-10: see also Withnall.

**MODEN HILL**  at Cotwall End  (SO 9193). *Mouldenhill* 1587 Sedgley ParReg, *Modenhill* 1658 ibid, *Modernhill* 1662 ibid, *Moden Hill* 1847 Hackwood 1898: 37. Perhaps from OE *molda* 'top of the head' (cf. ON *\*moldi*), used in a topographical sense of a hill-top, so here 'the hill with the tops'. *Moldi* is a common name for hills in Norway. Cf. Mouldsworth, Cheshire.

MODIESMOR  (unlocated, possibly near Hanbury) *Modiesmor* 1297 SHC 1911 254.

MOISTY LANE  on the west side of Marchington. *Mistelane* 1617 D4038/A/6/1, *Mistorfield's Lane* 1636 Redfern 1886: 127, *Mistelane (Close)* 1647 SRO D4038/A/6/1/(vi), 1663 SRO D4038/A/6/2, *Mistey Lane* 1679 SRO D4038/A/6/7, *Misteylane* 1685 SRO D4038/A/6/1/(viii). Redfern 1886: 17 mentions 'Moisty or, as old writings call it, Mister Field's Lane ...'. Perhaps therefore 'the lane associated with Mr Field', rather than 'mistletoe lane' or 'the lane with the fields with trees on which mistletoe grows', from OE *mistel*,

**MOLEHOUSE**  2 miles east of Biddulph (SJ 9156). *Moles House* 1842 O.S. Possibly to be associated with *Moll* 1634 Biddulph ParReg.

**MOLINEUX**  on the west side of Wolverhampton (SO 9199). *Mr. Molineuxe's Close* 1751 Taylor. From the family of that name (said to have come to England from Flanders in 1307) who were prominent in the town in the first half of the 18th century. The surname is from *moulineaux* 'the mills', a common place-name in France: Tooth 2002: 158. Molineux House was built by the ironmaster John Rotten in 1744, and was owned by the Molineux ironmaking family from 1754 to 1860: ES 6 August 2003. See also WA II 28-33; StEnc 399. A stadium for Wolverhampton Wanderers Football Club was created here in 1889.

MONETVILE  (unlocated, on the west side of Stafford) Stafford. *Monetvile* 1086 DB. A name found in DB but otherwise unrecorded, unless 14th-century references to a field-name *Munthull(e)* (Darlington 2001: 15) refer to the same place, which seems quite possible. Nigel de *Munevilla* is recorded in Parker 1897: 19, but the place is unidentified. The name Monetvile has long intrigued historians. It has been suggested that the place, which almost certainly formed part of the manor of Bradley (VCH IV 3), was a small estate granted for the upkeep of Stafford Mint (SHC XI NS 227-30; VCH IV 53 fn.), and was from *monetae villa* 'the vill of the mint': the existence of a mint at Stafford in the time of Edward the Confessor is well recorded, and three mint-masters are known from the reign of William I (SHC 1927 210). The Latin word *monet* (OE *mynet*) is associated with minting (Latham 1965: 303), often appearing on early coins in full or abbreviated to mean 'moneyer', and DB has references to mints or moneyers (*de moneta*) in Dorchester, Bridport, Hereford, Leicester, Huntingdon, Lewes, Shrewsbury, Wareham, Worcester, Colchester, Ipswich and Norwich. The DB record for York contains a reference to Nigel de *Monnevile* who held one mansion formerly the property of a certain moneyer (*cujusaam monetarii*). The design of coinage was changed every two or three years, and it was an offence to use obsolete coinage. It seems that moneyers normally operated from their own private workshops which may have been concentrated in a particular area of the town: see Lapidge *et al* 1999: 318. Moneyers were taxed personally in respect of their office until William the Conqueror imposed the tax of the mint (*'geldum monete'*) on the town, or in the case of some towns a tax *de moneta*, which in some Domesday entries is specifically said to be payable by the moneyers. Each town had its fixed stablishment of moneyers, ranging from one-moneyer units up to a seven-moneyer mint at Hereford and Canterbury and an eight-moneyer mint at London. Domesday Book implies that the sheriff or other authorities dealt with the moneyers of a town as a group for revenue purposes. A bishop or abbot might have had a moneyer, as for example in the case of Hereford. See specially Metcalf 1987: 281; 284; 287. *Monetvile* may have been where the Stafford moneyer(s) operated, but it is unclear whether a minting place outside the town has parallels elsewhere. A possible derivation of the name from Fr *mont* 'hill' and *ville* 'town' might be appropriate for a new settlement away from the old low-lying site of Stafford, but no parallels have been traced for either of the nouns *mont* or *ville* used quasi-adjectivally either in French place-names or place-names in England of French origin: Gelling 2001. There is the possibilty that the word *monet* is connected with PrWelsh *monith*, Welsh *mynydd*, OE *munt, myned*, OFr, ME *mont* 'a mount, a hill', meaning 'hill town': the place has tentatively been identified as Castle Church (q.v.), which lies at the foot of the prominent hill on which stands Stafford Castle. It may be noted that the DB reference to 'the vill of Burtone' may be Burton in Castle Church (see VCH IV 49 fn.13), which may have adjoined Monetvile. Castle Hill (*Castelhull'* 1439 MinA.) is the only prominent hill in the area unaccounted for by name by early antiquaries: see SHC VIII (ii) 13-4, 16; also TSHCS 1971-3 11. Cf. Menutton, Shropshire *(Munetune* 1086, *Moneton'* 1272, 1284*)*, and Myndtown, Shropshire *(Munete* 1086, PN Sa I 201, 217; III 228-9); Minton (unlocated), in Worcestershire *(Moneton, Munton* 1275, 1332, PN Wo 49); and *Munentone* in Mold, Flintshire, recorded in DB but unlocated (Davies 1959: 114). However, Gelling 2001: Appendix 1 notes that the ridge on which Stafford castle stands is probably not high enough to deserve the name Mynydd, and if *Munthull(e)* is to be identified with *Monetvile*, the former is described as lying 'next to the road which leads from the castle towards Billington' (Darlington 2001: 15), suggesting that it lay to the west of the castle and some distance from it, and its precise location remains unidentified. For completeness, it may be added that names of this type are said to have been associated with an Anglo-Norman form of low Latin *munita*, for *immunitas*, 'privileged district, one free from seignorial rights': see Johnston 1914: 370. Finally, it must not be forgotten that the name *Monetvile* in this single spelling may be slightly or utterly corrupt, as not uncommonly the case in DB.

**MONKEY GREEN** West Bromwich. The development of this area in the early 19th century was financed by a building society, and it has been suggested that the name may come from a slang expression 'a monkey on the house', meaning a mortgage (VCH XVII

10, see also EDD 150), but *monkey* was also a dialect term for a young hare: see Field 1993: 74; EDD 150. The Green element suggests a squatter settlement.

MONKFORD (obsolete, in Cheddleton) *Monckesford* 1254 SHC 1911 123, *Munecford* c.1254 ibid.' 441, *Munksford* c.1255 ibid. 440, *Mouncford* 1261 ibid.' 427, *Munkford* c.1275 ibid. 442, *Munkeford* 1327 SHC VII (i) 217, *Munkeforde* 1413 SHC XVII 43. 'The ford of the monks', presumably to be associated with Dieulacres Abbey. *Mungesfordeseye* is recorded here in the 13th century (Dieulacres), the final element from OE *eg* 'island'.

MONKSBRIDGE (obsolete) The old bridge where Icknield Street crossed the river Dove north of Burton upon Trent: see Plot 1686: 400. *pontem de Egintona super aquam de Dove* 1255 BurtAbSurv, *pontem Monachorum* 1330 Ass, *le Munkbrugg'* 1383 Cor, *Monkebrigge* 1394 Pat, *Lytulmonkbryge* 1406 Ch, *Monkysbryge* 1465 SHC 1939 128, *Monks-bridg* 1686 Plot 400, *Monks Bridge* 1775 Yates. From OE *munuc, brycg* 'the monk's bridge'. Shaw 1798: I 26 states that the bridge, formerly called the bridge of Egenton (from nearby Egginton in Derbyshire), was so-named because it was erected by John de Stretton, prior of Burton. See also PN Db 459-60.

MONKS WOOD (obsolete) 1 mile south-east of Cheddleton (SJ 9751). *Monkes woode* 1529 StarCh, *Monkes Woode* c.1535 SHC X NS I 144, *Monks Wood* 1836 O.S. Self-explanatory. The place was held by Dieulacres Abbey: SHC X NS I 148.

**MONMORE** 1 mile south-east of Wolverhampton (SO 9397), *Monnemerre* c.1240 WA II 95, *Monnemere* 1291 Tax, *Monnemere* 1327 SHC VII (i) 249, *Mounemere* 1385 SHC 1928 131, *Monmore* 16th century Duig, *Monmore (Greene)* 1707 SHC 1938 229; **MONMORE LANE** ½ mile north-east of Willenhall (SO 9798), *Monnemedewe* 14th century, *Monmerfeld* 1550 Duig, *Mumber Lane* 1834 O.S.; **MONWAY FIELDS** ½ mile east of Wednesbury (SO 9995), *Moummer Field* 1538 SHC 1912 113, *Monway Field, Monway Gate* 1682 SRO DW1813/1/4. Perhaps from the common OE personal name Mann(a) or Monn(e), or OE *mann, monn* 'a man', denoting in the plural 'community', with OE *mere* 'lake, pool, mere' (which frequently becomes *more* in place-names), or *medewe* 'meadow'. All the places are low lying. *Mon-Moor meadow* (unlocated) in Cannock is recorded in Shaw 1801: II unpublished sheets 319.

**MONS HILL** between Tipton and Upper Gornal (SO 9392). *Monshull* 1294 SHC 1911 222, *Mounshull* 1307 SHC VII (i) 173, *Mounsels* 1527 SHC XI 266, *Mounsalls* 1562 SHC XIII 221, *Mouncels* 1563 Erdeswick 1844: 241, *Mons Hill* 1812 mining plan, 1834 O.S. Possibly from the ON personal name Man, so 'Man's hill'.

**MOOR END (FARM)** 1 mile north of Gnosall (SJ 8322). *the Moor* 1381 SHC XVII 203, *Moreend* 1586 SHC 1927 134, *Moore End* 1655 Church Eaton ParReg, *More End* 1677 Gnosall ParReg, *Moor End* 1834 O.S. Self-explanatory.

**MOOR FARM** 1 mile north-east of Tamworth (SK 2205). The place lies on Warwickshire Moor (q.v.), and was in Warwickshire until transferred to Staffordshire in 1965.

**MOORFIELDS** – see RODNEY HALL.

**MOOR HALL** to the west of Pillaton Old Hall (SJ 936131). *Mora* 1199 SHC 1931 261, 1227 Ass, *(la) Mor* 1261 Penkridge Inq, 1285 FA, 1293, 1345 Coram R, 1312 Pat, *Morehall* 1475 SHC VI NS (i) 95, *The Mor Hall* 1532 SHC 4th Series 8 88, *manerium de Morehall* 1548 *Survey, the more Hall* 1578 ibid, *Moorehall* 1598 Ct, *the Morehall* 1616 Penkridge ParReg, *Moor Hall* 1890 O.S. From OE *mor* 'the marshland', with *hall* added later. The moor may have been *ofer ðæne mor* mentioned in the bounds of Bedintun (Pillaton) in a charter of 996 (11th century, S.879): see Hart 1975: 196-8; Hooke 1983: 90-3. Sawyer 1968: 270 gives this charter date as 993. Perhaps the same moor as *Mora*, recorded in 1261: SHC 1950-1 46. The place is remembered in Moor Hall Cottages.

**MOOR HALL** 1 mile east of Bagnall (SJ 9450), *Morehalle* 1415 SHC XVII 52, *Moor Hall* 1775 Yates; **MOOR HALL** 1 mile west of Madeley (SJ 7644), *Moor Hall* 1833 O.S.; MOOR HALL (obsolete) 1 mile north of Gnosall (SJ 8322), a prebendal manor house, *Morehall* 1360 Pat, 1496 SHC 4th Series VII 170; MOOR HALL (obsolete) 2 miles west of Codsall (SJ 8503), *Moor Hall* 1538 VCH XX 82, a formerly moated site, the hall having been demolished by 1796 (ibid.'). From OE *mor*, giving 'the hall on marshy ground', but the first name may be from John le More, prebendary there in 1338: SHC 1927 110. *Moorehall* at Wall, near Lichfield, is recorded c.1562: SHC 1931 159. *Moorhall* or *Moorehall Farm* in Eccleshall parish is recordedc.1800: SRO D1798/596/7.

**MOORLANDS, THE** a district in the north of the county, north-west of Leek. *Mora juxta Lech* 13th century Dieul, *the Moorland* 1329 VCH VII 78, *Morland* 1414 SHC XVII 15, *Moreland(e)* 1414 Coram, SHC XVII 20, c.1540 Leland, *Moreland Hills* 1610 Speed, *Moreland* 1614 SHC 1910 267, *the Moorelands* 1644 SHC 4th Series I 149, *Moreland hills* 1798 Shaw I 88. From OE *mor* 'a high tract of barren uncultivated land', and OE *land* 'a tract of land of large area'.

**MOOR LEYS (FARM)** on the south-east side of Gayton (SJ 9828). *Morlease* 1705 SCH 13 1973 22.

**MOORS** 1 mile north-east of Abbots Bromley (SK 0925). *Moos* late 13th century SRO D938/30, *The Moors* 1665 SRO DW1778/V/1305. Self-explanatory.

**MOORSIDE** ½ mile south-west of Onecote (SK 0454). *Moorside* 1745 SRO D3816/2/5/1. Self-explanatory.

**MOORVILLE (HALL)** 1 mile south-east of Cellarhead (SJ 9546). *Moorville* 1836 O.S. Early spellings have not been traced, and the name is unlikely to be ancient.

MORA (unlocated, in Wolverhampton manor) *Mora* c.1249 SHC 1911 144.

MOREHAY (unlocated, in Colton) *Morehay* 1391 SHC XV 33. See also Littlehay.

MORELESMOR (unlocated) *Morelesmor* 1306 SHC 1911 65.

**MORETON** 1 mile north-east of Colwich (SK 0222; SMR 01650 places a deserted Anglo-Saxon settlement at SK 02402310, and SMR 03867 another at SK 02402300), *Mortone* 1086 DB, *Morton* 1284 FA, *Moreton* 1461 HAME 485, *Malton* 1613 SHC 1931 271, *Moreton* 1798 Yates; **MORETON** 3 miles south-east of Newport (SJ 7817), *Mortone* 1086 DB, *Morton* 1280 SHC VI (i) 148, *Morton, Mortone, Moorton* 1381 SHC XVII 199; **MORETON** 1½ miles south-east of Marchington (SK 1529), *Mortvne* 1086 DB, *Morton* 1244 SHC IV 102, 1610 SHC III NS 51. From OE *mor-tun* '*tun* by a marsh or fen'. The *'manor or reputed manor of Moreton, Rugeley'* recorded in 1705 (SRO 1237/59/60) is presumably Moreton near Colwich. A very common place-name: over 50 are recorded in DB.

**MORETON BROOK** a tributary of the river Trent. *Mortonbrok'* 1395, 1467 Ct. 'The stream by Moreton (q.v.)'.

**MORFE** 1 mile north of Enville (SO 8288; SMR 01911 places a deserted Anglo-Saxon settlement at SO 82798810). *Moerheb, Moreb* 736 (8th century, S.89), *Morve* 1086 DB,

*Morf* 1166 SHC 1923 298, *Morve* 1165 SHC I 42, *Morue* 1166 P, *Morf* 1166 SHC I 49, *Morf* 1235 SHC 1924 387, 1268 ibid. 140, 1271 SHC VI (i) 52, *Overmorffe, Nedermorff* 1470 SHC NS IV 166, *Morfe House, Little Morfe, Morfe Hall, Morfe Heath Farm* 1833 O.S. An interesting name, usually held to be British, perhaps a shortening of PrWelsh *\*mor*, Welsh *mawr* 'big', with *dref* 'village', or from Welsh *morfa* 'a marsh, an upland moor', although the loss of *tr* as early as the 8th century would be surprising (see Coates & Breeze 2000: 334), and the topography makes such a meaning questionable. Professor Richard Coates (personal communication 2 April 2002) tentatively suggests a derivation from British *\*marosamjo* or *\*marosamjon* 'greatsummer-place'; which would give *\*morhev* in Brittonic, and if that survived into OWelsh, possible OE renditions *morheb* and *morhef*, with the *h* dropped as in later Welsh compounds where the stress is on the first syllable, though he questions the plausibility of any such name. The name Morfe is historically attached to the ancient Forest of Morfe (*Morfe forest* 1613 SRO D593/H/14/2/15), which appears to have covered a large area from Enville to Quatford: Eyton 1864-60: III 212 records that it was 'at least 8 miles in length, and perhaps 6 in width. Its northern boundary was the Worfe, and its south-eastern extremity by Morf hamlet'. *The Morf Common* 1732 Rocque and *Morf Farm* 1833 O.S., 1½ miles south-east of Bridgnorth, now Common Farm, indicate the extent of the area. The charter in which the earliest spellings are found is a grant by Æthelbald, king of the Mercians, to Cynebert of land for a monastery at Husmere (see Whitelock 1955: 453-4). The grant mentions *silvam quam nominant cynibre* and *silva moreb*, 'the wood called *Cynibre*' and '*Moreb* wood'. Those places later evolved into the adjoining royal forests of Morfe and Kinver, with Kinver lying in the south-west corner of Staffordshire and part of north-west Worcestershire, and Morfe lying between Kinver and the river Severn. It has been suggested that *Moerheb* is probably the same word as the 12th century *Moref* (Moray, Scotland): see McClure 1910: 260. A lost wood-name *Morezyf, Moreyf* is found in Westbury, Wiltshire, in the 13th century (see Mawer 1929: 2-3; PN Bk xiii), probably from OE *morgen-gifu* 'a morning gift': see Morrey and Morghull. For the suggestion that *Moerheb* is to be identified as Kidderminster Heath see King 1979: 73-91; Dark 1994.

**MORFEVALLEY** 2 miles south-west of Claverley (SO 7791). See Morfe.

MORGHULL (obsolete, to the south-west of Streethay, north-east of Lichfield: see Harwood 1806: 566; SHC VII (i) 227; VCH XIV 275) *Morschale* c.1157 SHC 1924 87, *Morehale* c.1177 ibid.' 83, *Morghale* c.1208 ibid. 89, *Mornhale* post-1268 ibid.' 347, *Mogenhull* 1278 SHC 1911 35, *Morewhale* 1302 SHC 1939 91, *Morwale* 1303 ibid, *Morghale* 1319 ibid. 93, *Morghwhale* 1309 SHC ibid. 120, *Morghale* 1344 SHC XII 35, *Mogghale* 1347 SHC 1939 120, *Morughale* 1348 SHC XII 15, *Morghwhale* 1513 SHC XII 181, *Morfall otherwise Murfall otherwise Morghall* 1571 SHC XVII 218, *Moford* 1571 SHC IV NS 192, *Marfowle* 1588 SHC XVII 235, *Morehaughe otherwise Morghall* 1601 SHC XVI 208, *Morghall otherwise Morehall otherwise Morfall* 1618 SHC VI NS (i) 32, *Morfall* 1678 SRO DW1738/A/1/1, *Morghull* 1798 Shaw I 363. See also SHC VI (ii) 187 fn; Shaw 1798: I 121, 363. Some forms suggest a possible derivation from OE *mor, hrycg, halh* 'moorland ridge *halh*', but the topography alone rules out such derivation. However, the most likely derivation (as suggested in VCH XIV) is from OE *morgen-gifu* 'a morning gift': Anglo-Saxon marriages (which were arranged by parents) were in two stages, the 'wedding' or pledging, at which terms were agreed and the bride-price paid, and the 'gift', or giving-away, accompanied by feasting: see Whitelock 1930: Addenda; Lapidge *et al* 1999: 302-3. After the Conquest a medial *g* with a back vowel invariably gives *w*, hence Modern *morrow*; see also Morrey. The second element is almost certainly OE *halh*. VCH XIV 275 suggests that the place may have been deserted by the 1480s (though Shaw 1798: I 363 describes it as 'a small hamlet', so perhaps repopulated). Its location, possibly near Bexmore Farm, is indicated by Valley Lane (formerly Morughale Lane), on the north-east side of Lichfield. Place-names incorporating *morgen-gifu* are found in various parts of the country: see for example PN Ess 276, PN Wa 224, PN W 186.

MORICEMOR (unlocated, in Dilhorne) *Moricemor* 1679 SHC 1914 91. Said to be from Morice de Stanton, a forester: ibid.

**MORREY** 1 mile west of Yoxall (SK 1218). *Morrey* 1499 (1798) Shaw I 98, 1553 SHC XII 212, 1587 SHC 1929 195, *Morray* 1597 SHC 1935 11, *Morry* 1661 Barton under Needwood ParReg, *Murry* 1686 Plot, *Murrey* 1695 Morden, 1747 Bowen, *Morhay* 1798 Yates, *Morry Hills* 1801 Shaw II 7, *Morrey* 1834 O.S. An interesting name, almost certainly from OE *morgen-gifu* (cf. Modern *morrow*: after the Conquest a medial *g* with a back-vowel invariably becomes *w*, and there are early examples of the shortening of the compound: see Moor Farm, PN Ess 276), literally 'morning gift', meaning 'land given by a man to his bride on the morning after their marriage' (Whitelock 1930: Addenda; Mawer 1929: 19; Lapidge *et al* 1999: 302-3): cf. Morghull, and field-names The Murray and Black Morray in Shropshire (Foxall 1980: 62), The Morrey, Hampshire, and Morrif, Warwickshire (Field 1972: 142). Middle Morrey and Higher Morrey are 5 miles north-west of Market Drayton in Shropshire, but the derivation has not been researched.

**MORRIDGE** a high ridge over 3 miles long, mainly in Onecote, Leek (SK 0254), *Morridge hill* c.1233 BM, *Morrug(ge)* 1227 Harl, 1254 to 1345 Loxdale, 1340 (p) Ipm, 1374 Coram, *Mor(r)age* c.1278 St Cart *et freq.* to 1634 Bradnop Deeds, *Morregge* 1328 SRO D1229/1/4/52, *Moryche temp.* Elizabeth I ChancP, *Mor(r)e(d)g(e)* 1413 ProcJP, 1591 QSR, 1662 Okeover Deeds, *Morgage Syde* 1532 SHC 4th Series 8 31, *Mor(r)idge* 1570 Pat, 1612 FF, 1713 Will, 1659 (1883) Sleigh 3, *? Moreage* 1611 SHC III NS 48, *(Black Mare of) Morridge* 1749 Bowen; **MORRIDGE TOP** 1½ miles south-east of Flash (SK 0365), *Moryche* 1564 SHC 1938 99. 'The moorland ridge', or perhaps in some cases 'the moorland edge' from OE *mor* 'moorland' and OE *hrycg* 'ridge' or OE *ecg* 'edge, the crest of a sharp ridge'. The varied spellings also reflect differing local pronunciations: those with -*u*- reflect the West Midland pronunciation of *hrycg* – see Rugeley. The *Black Mare* of 1749 is Blake Mere (q.v.).

MORRILLOW (obsolete) on the road from Draycott in the Moors to Hilderstone, ½ mile east of Fulford (SJ 9637). *Warilare* 1227 SHC IV 52, *Woralawe* c.1230 SHC VI (i) 11, *Warylowe* 1266 SHC IV 160, *Warrilowe* c.1285 SHC 1911 436, *Warilowe* 1293 SHC VI (i) 239, 1305 SHC VIII (i) 153, 1341 SHC 1921 19, *Werrelow* 1302 SHC 1925 97, *Werrelowe* 1349 SHC XII 81, *Warelowe* 1475 SHC VI NS (i) 94, *Warrlowe* 1522 SHC 1925 121, *Wariloe* 1559 SHC 1926 138, *Warralowe* 1559 ibid.' 139, *Warriloe* 1583 ibid. 53, *Warrylowe* 1611 SHC IV NS 13, *Morryley, Morrelowe* 1645 SRO D1367, *Morrilow* 1742 SRO D1380/1/9, *Morrillow* 1890 O.S. If the early spellings are to be identified with this place (*Worylowefeld* 1332 SRO D1229/1/4/12, *Warrylowefelde* 1413 SRO D1229/1/4/15, *Warrowlo Field* 1655 NA DD/4P/24/2, *Warrilow* 1672 SRO D615/D/149 was in Cheadle: SHC 1934 7), the derivation may be from OE *wearg-hlaw*, where the meaning may be 'felon-mound or tumulus', i.e. where felons were hanged, but *wearg* also meant 'wolf', so possibly 'wolf-mound or tumulus': see JEPNS 27 94-5. The surprisingly late change from *W*- to *M*- is curious, and doubtless due to 17th century mistranscriptions of two easily-confused capitals. The place now appears on maps as Mount Pleasant. See Warrilow Brook.

**MORRILLOW HEATH** 2 miles north-east of Milwich (SJ 9835). *Moralow Heath* 1798 Yates, *Marrillow Heath* 1836 O.S., *Mariley Heath* 1837 TA. 'The heathland associated with Morrillow'.

**MOS** (obsolete, in or near Heatley) *the Mos* 1357 SRO D9W01721/3/1/3. From OE *mos* 'a bog, a swamp'. It is unclear whether this place is to be associated with Moss Farm and The Mosses to the south-west of Kingstone.

**MOSCOTT or MORSCOTT** (unlocated, perhaps in Pelsall or Little Bloxwich) *Moscott, Morscott* 1652 WSL M22/27.

**MOSE** 1 mile south-east of Quatford (SO 7590). *Mose* 1262 Eyton 1854-60: III 194, c.1300 Rees 1997: 154, 1833 O.S. From OE *mos* (associated with OE *meos*) 'bog, swamp, morass'. In Shropshire since the 12th century.

**MOSELEY** in Bushbury parish, 3 miles north of Wolverhampton (SJ 9204), *Moleslei* 1086 DB, *Mollesleg* 1227 Ass, *Mollesle* 1242 Fees, 1286 SHC V (i) 169, *Molesleye* 1255 SHC V (i) 113, *Moleston, Molesle* 1271 SHC V (i) 149, *Mollesley* 1286 ibid. 169, 1332 SHC X 126, *Molesleye, Mollesleye* 13th century Duig; **MOSELEY** 1 mile west of Hollinsclough (SK 0466), *Mollesleye* 1327 SHC VII (i) 250, 1333 (p) ibid.'; **MOSELEY** 1½ miles north of Bilston (SO 9498), *Mollesleye, Mollesley* 12th and 13th century Duig, *Mollesleg* 1227 SHC IV 52, *Molleston* 1273 SHC VI (i) 59, *Mollesleye* 1327 SHC VII (i) 250, *Mollesley* 1333 SHC X 127. From the common OE personal name Moll, hence 'Moll's *leah*' (and in 1273 *tun*).

**MOSS** (obsolete, ½ mile south-west of Shenstone (SK 1003)), *le Mosse* 1305 SHC XVII 269, *Mosse* 1531 (1801) Shaw II 47, 1564 SHC XVII 245, *(Manour of) Mosehouse* 1665 TSAS VII 1918-9 32, *The Moss* 1834 O.S.; **MOSS** 1½ miles west of Talke (SJ 8053), *The Moss House* 1621 SRO DW1082/B/5/1-13, *Mosse House* 1833 O.S.; **MOSS BEDS** on the west side of Uttoxeter (SK 0734), *le Mos* 1306 SHC 1911 67. From OE *mos* 'a bog, a swamp'. The first place, granted to the abbots of Oseney in 1129 (SHC XVII 245), appears to have changed its name c.1900 to Shenstone Court. *Beds* may refer to the flat surface of a wet bog.

**MOSS CARR** 1 mile south-east of Hollinsclough (SK 0765). *Mofcure* 1402-3 DRO D2375M/1/1, *Mofcure (? Moscure)* 1416-7 ibid, *Mosture* 1474-5 ibid, *Moscare* 1566 Deed, *Mosker* 1599 Alstonefield ParReg, *Mescor(brooke)* 1603 ibid, *Moscarr(brooke)* DRO D2375M/106/27, *Moscarr* 1614 DRO D2375M/57/1, *Moss Carr* 1775 Yates. Perhaps from ON *mosi* 'moss, lichen; bog, swamp', with ON *kjarr* 'brushwood', so 'the boggy ground overgrown with brushwood'. The two earliest spellings almost certainly have transcription errors.

**MOSS FARM** ½ mile west of Fairoak (SJ 7532). Probably to be associated with *Pallys Moss* c.1840 TA, so-named from palings around the bishop of Lichfield's park mentioned in 1298: Spufford 2000: 295. Moss is from OE *mos* 'a bog, a swamp'.

**MOSS FIELDS** (obsolete) on the north side of Adderley Green (SJ 9245). *Mossfields* 1679 Caverswall ParReg, *Moss Fields* 1836 O.S. From OE *mos* 'a bog, a swamp', so 'the boggy fields'.

**MOSSHOUSE** 2 miles north of Audley (SJ 8053). *? Mos* 1327 SHC VII (i) 205, *? Mosse* 1332 SHC 1332 SHC X (i) 101, *Mosshouse* 1605 SRO DW3222/4/4, *Moss House* 1833 O.S. From OE *mos* 'a bog, a swamp', so 'the house near the boggy ground'.

**MOSSLEE HALL** 1½ miles north-west of Ipstones (SK 0050). *Moseliye* 1298 SHC XI NS 257, *Moysileg'* 13th century Dieul, *Moselegh* (p) 1311 Ipm, *Moseley* 1583 Visitation,

*Moslee* 1644 SHC 4th Series I 227, *Mosseley* 1609 Antrobus, *Moselie* 1686 Plot, *Mossleigh* 1775 Yates. From OE *mos leah* 'the mossy or boggy *leah*'.

**MOSSLEY** on the west side of Bloxwich (SJ 9802). *Mosley* 1613 SHC IV NS 42, *Mosley Field* 1834 O.S.; **MOSSLEY** on the east side of Rugeley (SK 0417), *two Mosse Leyes* 1662 WCRO CR1908/55/23-25, *the Two Mosseleyes* 1669 WCRO CR1908/6/1/ From OE *mos* 'a bog, a swamp', with OE *leah*. StEnc 405 records the name *Matteslye* c.1300 in association with Mossley in Bloxwich, but no source is given, and any connection with that place seems unlikely.

**MOSS PIT** 1½ miles south of Stafford (SJ 9220). *? le mosse* 1548 SHC 1950-1 41, *Moss Pit Bank* 1725 SRO D856/1, *Mospit bank* 1749 Bowen, *Moss Pit* 1775 Yates, *Mospit* 1836 O.S. From OE *mos-pytt* 'the mossy or boggy pit or hollow'. A large pit or hollow, probably an ancient quarry, lies on the west side of the main Stafford to Wolverhampton road.

**MOSS POOL** Meretown (SJ 7520). *Moss poole* 1686 Plot 46, *Moss Pool* 1833 O.S. From OE *mos* 'a moss, a marsh, a bog'

MOSS WOOD, MOSSWOOD FARM (obsolete) on the south side of Cannock (SJ 9709). *Mosse(furlong)* 1369 Ct, *mosse (flatt)* 1570 Survey, *the mosse* 1570 Survey, *le mosses als le mosse* 1580 Anglesey Ch. From OE *mos* 'bog, swamp'. The names are remembered in Mosswood Street.

**MOTCARN SPRINK** 1 mile east of Middle Mayfield (SK 1345). *Motcarn Sprink* 1836 O.S. Possibly from OE *(ge)mot*, which had two meanings,'a meeting, an assembly', and 'a junction of streams': two streams join here. The second element is uncertain: a drivation from Welsh *carn* 'a heap of stones, a cairn' seems improbable, so possibly from ON *kjarr* 'brushwood'. *Sprink* represents a form of *spring*, 'newly-planted trees or coppiced trees with new shoots', so 'the tree shoots (at the place with brushwood?) where assemblies took place', or 'the tree shoots at the stream junction (with the brushwood?)', but in the absence of early spellings such derivation must remain speculative, and indeed the name may not be ancient. Cf. Mayfield; Harlow Wood.

MOTE, LE (obsolete) a moated site adjoining Lapley Priory (SJ 8712). *le mote* 1338 SHC 1913 70, 1389 Inq. From ME *mote* 'a moat, ditch or trench'. See also TSSAHS XXIV 1982-3 44.

MOTELOWE – see **MOTTLEY PITS**.

MOTES an unlocated manor, possibly near Loxley. *Motes* c.1594 SRO DW1733/A/1/4[7].

MOTTEDELHEUED (unlocated, near Cheadle) *Mottedelheued* c.1282 SHC 1923 39.

**MOTTLEY PITS** at Common Plot, Stonefield, 1 mile north-west of Stone (SJ 8935). *Motley Pits* 1798 Act, *Mudley Pits* 1860 P.O., *Mottley Pits Terraces* 1908 VCH. A name of unknown age (it is unclear whether *Motelowe*, recorded several times in the 14th century, e.g. SHC XII (i) 74, 130; SHC 1913 325, refers to this place), attached to earthworks here which are probably old gravel pits and the remains of ridge and furrow, with lynchets on the boundary bank adjoining an old road: see TNSFC 1881 23; TNSFC 1898 32 133-155; TNSFC 1936 70 91; VCH I 373-4. Possibly from OE *(ge)mot*, here meaning 'meeting, assembly, moot', with OE *leah* (or OE *hlaw*), so 'the *leah* (or mound or tumulus) where assemblies were held': meeting places were often at mounds. There is no evidence to support local tradition (see for example Erdeswick 1844: 38) that the earthworks are associated with the Civil War or the Duke of Cumberland's forces in 1745.

**MOTTY MEADOWS** 1 mile south-west of Marston (SJ 8213). *Mutty Meadow* 1682 SHC I 309, *(Great) Motty Meadow* 1735 SHC II (ii) 145, 1841 TA, *Motty meadow*, *Mutty meadow* 1798 Shaw I 110, 113. An intriguing name with a number of possible derivations. Duignan suggests 'a small lump or mound' (from ME *mote*), and Oakden 1984: 130 gives 'spotted', which may be from the local dialect *motty* 'mottled' (from OFr *mot*), perhaps with reference to the chequered flowerheads of Fritillaries which have long grown in abundance here: see Shaw 1798: I 104, who mentions the 'fritillary, chequered daffodil'. Another possible derivation is the local word *mottow* (evidently from *moiety*, meaning 'a half or small share'): EDD gives the (Staffordshire, obsolete) meaning 'a parcel of ground', and Halliwell 1850: 563 illustrates it with the quotation: 'The rent of a piece of meadow ground, in two parcels or mottows, is to be appropriated to the poor of Bradley in the county of Stafford'; see also SHC 1913 187 fn.1. Bradley is some 5 miles distant, assuming it is not the place of the same name near Bilston. But the vicar of Lapley and Wheaton Aston in the early 18th century was John Mott, and at least some of the meadows were glebe land (Shaw 1798: I 102), which suggests the further possibility that the name may be from a local surname, although the earliest spelling pre-dates that period.

**MOTTY MEADOWS BROOK** a tributary of the river Penk. From Motty Meadows (q.v.).

**MOUNT PLEASANT** 1½ miles south-east of Brewood (SJ 8907), *Mount Pleasant* 1885 O.S.; **MOUNT PLEASANT** in Lapley (SJ 8712), *Mount Pleasant* 1838 T.A; **MOUNT PLEASANT** ½ mile south of Kingswinford (SO 8887); **MOUNT PLEASANT** 1 mile south-west of Fenton (SJ 8844); **MOUNT PLEASANT** ½ mile north-east of Forsbrook (SJ 9741); **MOUNT PLEASANT** in Brierley Hill (SO 9285), *Mount Pleasant* 1834 O.S.; **MOUNT PLEASANT** in Chesterton near Newcastle under Lyme (SJ 8349), *Mount Pleasant* 1833 O.S.; **MOUNT PLEASANT** on the north-east side of Leek (SJ 9956), *Mount Pleasant* 1891 O.S.; MOUNT PLEASANT (obsolete) at Fazeley Bridge (SK 2101), *Mount Pleasant* 1888 O.S.; MOUNT PLEASANT (obsolete) on Lask Edge (SJ 9156), *Mount Pleasant* 1891 O.S.; **MOUNT PLEASANT** 1 mile south-east of Cellarhead (SJ 9646), *Mountpleasant* 1890 O.S.; **MOUNT PLEASANT** ½ mile south-east of Fulford (SJ 9637). It has been observed that places of this name, which is very common and found throughout the country, often lie on or near Roman roads. There is no conclusive evidence to link the name with such roads, but it is of interest that the place near Brewood lies directly on the course of a lost Roman road (Margary number 190) running south from Pennocrucium (Water Eaton); the place in Lapley lies near a Roman road running north-west from Pennocrucium (Margary number 19); the place near Kingswinford is near the Roman sites and roads at Greensforge (Margary number 192); the places near Fenton and Forsbrook are close to the Roman road running south-east from Stoke on Trent (Margary number 181); the place near Newcastle under Lyme is in one corner of the Roman fort at Chesterton; and the place at Fazeley Bridge lies on the north side of Watling Street. The place at Leek may be associated with a possible Roman road between Leek and Buxton. The age of all of the names is unknown, although Mount Pleasant near Fulford appears to have been known as Morrillow (q.v.). until at least the late nineteenth century. There is another Mount Pleasant just inside Cheshire 1 mile south-west of Mow Cop, with a Roman road 2 miles to the south, and a Mount Pleasant just within the Shropshire border 2 miles north of Hinstock, within 1 mile of the Roman road from Pennocrucium to Chester (Margary number 19).

**MOUSEHALL (FARM)** 1 mile east of Amblecote (SO 9185). *Mushal* c.1200 SHC 1928 12, *Moushal* c.1294 SHC 1928 23, *Moushul* (p) c.1378 ibid. 39, *Moushall* 1663 SHC II (ii) 32, *Mousall* 1691 Guttery 1950: 47, *Mousehall Farm* 1834 O.S. From OE *mus-halh* 'mouse-infested *halh*'.

MOUSEHILL (unlocated, in Pelsall: see Shaw 1801: II 94) *Moushul* (p) 1378 SHC 1928 39, *Mousehill* 1694 (1801) Shaw II 94. From OE *mus*, *hyll* 'the mouse infested hill'. Cf. Humphrey Mowshill, recorded in 1601: SHC 1935 400.

**MOW COP** a rocky 977' hill north-east of Kidsgrove on the Staffordshire-Cheshire border, 2 miles north-east of Kidsgrove (SJ 8557). *Mole-Hill (or Mole Copp)* 1192, *Mowl*, *Mowel (rocha de)* c.1270 (all from PN Ch II 308), *Mowul* 1278 SHC XI NS 245, 1286 VCH II 187, *Mowhul* 1286, *Mowell* 1298, *Mouhul* 1313, *Molehelle* 1320, *Molle* 1525 (all from PN Ch II 308), *Mow copp hill* 1577 Saxton, *Mowcopp Hill* 1607 Kip, *Mowle Hill* 1616 SRO D1229/1/3/83, *Mole-cop* 1686 Plot, *Mole Cop* 1798 Yates. Ekwall 1936: 143-4 gives a derivation from OE *muga-hyll* 'a hill shaped like a stack or pile, or the hill with the mound', probably with reference to a boundary cairn on the hill, but possibly from an old name of the hill, Muga, either because the sense 'mound, hill' existed in Old English, or owing to the likeness of the hill to a haystack. However, the early spellings may point towards an earlier derivation from Welsh *moel* 'bald or smooth-topped hill' (in which case a late name because of the implied use of a form dipthongised in Welsh), with the tautological ME West Midland *hull* 'hill', here meaning 'rocky hill', developing into ME *mow* (from OE *muga*) 'a stack, a heap', with OE *copp* 'hill, summit', and perhaps a later change to the jocular 'molecop' (molehill), from late ME *mulle, molle, molehill*, which correspond to MDu *mol, moll(e)*, the origins of which are obscure ('mole' in OE was *\*mold(e)wearp, wond* or *wondeweorpe*), before the reversion in recent times to the earlier form. The possibility of a derivation from ON *múli* 'a headland, a jutting crag' cannot be dismissed. On the summit of the hill is an artificial ruin, representing a ruined castle, built c.1752: AJ 120 1963 247. A pamphlet of 1642 referring to Mow Cop mentions '... a very high hill, called Cop in the English maps but in the old Saxon language Hiperbolian Talke, which is a large hill in English, as Talke on the Hill which signifies a bush on the hill ...': SCH I 1965 21 (cf. Talke). It appears that a tumulus from which Culverdslow (q.v.) took its name was also known as *Mole Cop* (VCH VIII 205), and is probably to be identified with *Moll* 1628 Wolstanton ParReg, *Moule* 1658 ibid, *Mould* 1661 ibid. A place called Mow Cop, 1 mile north of Flashbrook, just within Shropshire (SJ 7224), is recorded as *Mole Copp* in Edgmond ParReg.

MOXHALE (unlocated, possibly near Hoar Cross) *Moxhale* 1307 SHC VII 190.

**MOXLEY** 1½ miles west of Wednesbury (SO 9695). *Mockeslawe* 1259 SHC 1911 131, *Moxlowe* c.1364 SHC VIII NS 196, *Mockeslowe, Moxlowe, Mokkeslowe, Moxelowe* 14th century Duig, *Moxlowe* 1424 SHC XVII 106. The forms suggest a derivation from an OE personal name *\*Mocc*, with OE *hlaw* 'tumulus', changing (as frequently the case) to *leah*. It appears that a place called *Moxlowes* existed at Showell near Bushbury in the 17th century: see SHC 1928 106.

MUCHALL (obsolete) 2 miles south-west of Wolverhampton (SO 9096). *Muhclealis* 1184x1203 SHC III (i) 229, *Mushall* 1190x1206 ibid. 219, 1228 SHC IV 69, *Mushal* c.1200 SHC 1928 12, *? Mukelichale* 1273 SHC 1911 151, *Migehall* 1274 Penn ParReg v, *Moushal* 1294 SHC 1928 23, *Mushalle, Mosehull* 1300 ibid, *Mugehale* c.1300 Mander & Tildesley 1960: 8, *Mucheale* 1332, *Muchale* 1409 SHC XI 110, *Michehale* 1430 SHC XVII 138, *Mucheale* (p) 1484 SHC 1928 51, *Mucheoll* 1486 ibid. 51, *Muchall* 1547 SHC 1950-1 41, *Mochall* 1553 ibid. 42, *Michole* 1587 Penn ParReg, *Mychole* 1619 ibid,

*Mycholl* 1624 ibid, *Muchall* 1686 Plot, *Muckhall* 1775 Yates, *Muchall* 1798 Yates, *Mitchell* 1834 O.S. In ME *much* in its forms of *muche, muchel, michel, mochel, mukil, mikil,* etc., was used as an adjective in the sense of 'great, large' (cf. Much Wenlock), but the spellings show that an earlier name may have derived from OE *mus halh* 'the *halh* infested with mice' (or possibly 'the hall infested with mice': *hall* is a rare element in Staffordshire place-names, but the Chapter Court of Penkridge appears to have been held here in the medieval period: SRO D260/M/F/5/87), which developed into Much- from the medial *-sh-*. The place is remembered in the name Muchall Road.

MUCHEBERGE (unlocated, in Bentley near Walsall) *Mucheberge temp.* Henry III (1801) Shaw II 93. Seemingly from OE *micel, mycel* 'great', with OE *beorg* 'hill, mound, tumulus', probably here 'the great mound or hill': OE *hlaw* would normally be expected in Staffordshire if the name applied to a tumulus. The hill may be Pouke Hill (q.v.).

**MUCKLESTONE** Ancient Parish 4 miles north-east of Market Drayton, the most western parish in Staffordshire (SJ 7237). *Moclestone* 1086 DB, *Mukleston* 1221 FF, *Mokeliston* 1280 SHC VI (i) 104, *Mockeliston* 1306 SHC 1921 17, *Mukleston'* 1332 SHC 112, *Mickleston, Mocleston* 1411 SHC 1936 203, *Mogolston* 1532 SHC 4th Series 8 42, *Muckleston* 1682 Browne, *Muckleton or Muggleston* 1747 Bowen. The forms point towards OE *micel, mycel* 'great, large', and OE *stan* 'stone', perhaps with reference to the two stones a mile to the east known as The Devil's Ring and Finger (q.v.), part of a Neolithic chambered tomb, or (less likely) from the well-recorded OE personal name Mucel, with OE *stan*, so 'Mucel's stone', rather than 'Mucel's *tun*.'. *le Stones*, recorded in 1332 (SHC X 101), may refer to this feature. Mucklestone is shown as *Muxton* on the 1833 O.S. map. The parish of Mucklestone lay in Shropshire until transferred to Staffordshire in 1866. It may be noted that *Mucleston* and *Muleston*, recorded c.1225 have been identified as Milson (? in Shropshire): Rees 1997: 74-5.

**MUCKLEY CORNER** 3 miles south-west of Lichfield (SK 0806). *Mukelay* pre-1250 SHC 1924 297, *Muckeley Heath, Muckley Heath* 1565 SHC 1926 91, *Muckley Corner* 1775 Yates, 1788 SHC 4th VI 166, 1801 Smith. Duignan (MidA II 172) states that in early deeds Muckley is called *Mucklow*, and subsequently (1902: 104) refers to 16th and 17th century forms *Mucklow*, in which case the derivation is probably from OE *micel, mycel* 'great, large', with OE *hlaw* 'tumulus': see Gelling 1988: 134. There is a slight possibility that the derivation may be 'Mucel's *hlaw*'. OE *hlaw* 'tumulus' not unusually becomes *-ley* in Staffordshire. There are various tumuli and earthworks along the course of Watling Street (cf. Knaves Castle; Offlow; Rowley Hill), though no tumulus is recorded here. The place lies at the junction of Watling Street and the road south from Lichfield, hence *Corner*. Shaw 1798: I 356 refers to 'a public house called Muckley Corner'. If the name does not in fact incorporate *hlaw* (and the evidence is far from certain) there is the intriguing possibility that the name contains OE *leah* 'a clearing', but 'the great clearing' is improbable, so perhaps *leah* in its earliest sense 'woodland', denoting an area called 'the great wood', possibly to be associated with Luitcoyt: see Lichfield. It is unclear whether *Muhclealis*, recorded in 1184x1203 (SHC III 229), relates to this place.

MUCLEHOLM (unlocated, near Willowbrook Farm near Alrewas, in the angle formed by Icknield Street and the river Trent (? SK 1815)) *Mucleholm* 1259 SHC X NS I 265, *Mukulholmm* c.1300 TSSAHS XX 1978-9 loose map, *Mickleholme meadow* c.1750 ibid. Seemingly from OE *micel, mycel* 'great, large', with ON *holmr* 'a small island', which produces the odd 'large small island', so perhaps 'the larger islet'.

**MUDGELEYE, MUGELEY** (unlocated, in Alstonefield parish: SHC XI NS 257) *Mudgeleye, Mugeley* 1308 SHC XI NS 257-8.

**MURDEFORD** 1 mile south-west of Bobbington (SO 7988). *Murdivalls* 1617 PN Sa IV 27, *Merdivale* 1654 Claverley ParReg, *Merdivall* 1686 ibid, *Murdeville* 1833 O.S. This is likely to be a rare example of a name of unknown etymology so far recorded only in three places, all in Shropshire: Mardol in Shrewsbury (*Mardevall'* c.1215), and Mardol Road and Mardol Lane in Much Wenlock (*Mardeuole* 1321): see PN Sa III 260; PN Sa IV 26-7. In Shropshire since the 12th century.

**MUSDEN GRANGE** ½ mile north-west of Ilam (SK 1251). *Mvsedene* 1086 DB, *Mosedene* 1178 CroxdenCh, *Mosedena* 1184 CartAnt, *Musden* 1232 SHC IV 89, *Museden* 1234 FF, *Grange de Moseden* 1291 (1798) Shaw I xxiii, *Moseden* 1319 CroxdenCh, *Musden(e)* 1448 Banco, *Musden Gra(u)nge* 1538 LP, *Mousden otherwise Mosden grange* 1584 SHC XVII 231, *Musden* 1598 SHC 1935 147, *Upper Musden Grange, Lower Musden Grange* 1836 O.S. From OE *mus-denu* 'mouse valley', doubtless the long narrow valley in Musden Wood. There was a grange here of Croxden Abbey, and the place was extra-parochial until 1857: TNSFC 1913 62; NSJFS 1961 137. Musden Low (*Musden Low* 1836 O.S.), from OE *hlaw* 'burial mound', lies on the summit of the 1179' hill here.

**MUSEFORD** (unlocated, possibly near Stansley: SRO D603/A/Add/117-8) *Museford* 1252 SRO D603/A/Add/117-8. From OE *mus ford* 'mouse ford'.

**MUSHROOM GREEN** in Brierley Hill (SO 9386). *Mushroom Green* 1820 Greenwood, *Musham* 1834 O.S., *Mushroom Green* 1895 O.S. NSJFS 1915 56 suggests that the name comes from musham, a term used in mining circles for crushed shale, and found in Shaw 1801: II 66 for high quality ironstone, but there is no reason to suppose that the name is not from 'the grassy open space or squatter settlement where mushrooms grew': the element green often denotes a squatter settlement.

**MUXTON** – see **MUCKLESTONE**.

**MYNERS** (unlocated, possibly near Blakenhall near Barton under Needwood: see Shaw 1798: I 60) *Myners* 1306 SHC VII 161, 1323 SHC IX (i) 96, 1326 SHC 1911 105, 1334 SHC XI 56, 1407 SHC XV 121, 1409 SHC XI 219, *Migners* c.1331 SRO DW1733/A/2/19, *Myneres* 1374 SHC XIII 101, *Myners Mote* 1529 ibid. 269, *Myno* c.1594 SRO DW1733/A/1/4[7]. Probably from a surname: the Mynors family are recorded in Uttoxeter and Fisherwick in 1419 (VCH XIV 241), and were of Blakenhall, near Barton under Needwood: SRO Mynors Papers. A pedigree of the family is given in Shaw 1798: I 117.

**MYTHAHOLME** 1 mile south-east of Alrewas (SK 1814). *Mytham* 1601 (1798) Shaw I 138, *the Mytham* 1601 Alrewas ParReg, *Little Mitheholm* c.1699 SRO DW1851/8/8, *Mitham House* 1775 Yates, *Mitha Holme House* 1798 Yates, *Mitham* 1799 SRO D615/M/1/1. From OE dative plural *(ge)mȳðe* 'the mouth of a river where it runs into another, a confluence of rivers', so 'the river junctions'. The place lies close to the junctions of the rivers Trent and Tame, and Trent and Mease. *Mytheholme* in Rocester is recorded in 1664 (SRO D786/25/2), and field-names *Mythams, Mythams alias Muthams,* and *Mytham Meadow* are recorded in Walsall'in the 16th century (SRO D260/M/T/1/14). See also Mitham.

**MYVOD** on the north side of Wednesbury (SO 9995). *Myvod House* 1879 SCHAS M37/1. Seemingly a relatively recent transferred Welsh name.

**NABB BROOK** a tributary of the Alders Brook which flows into the river Dove. From The Nabb (q.v.).

**NABB FARM** 1½ miles south-east of Alton (SK 0840), *? Nabbes* 1608 SHC 1948-9 37, *Nabbs* 1704 Alton ParReg, *The Nab* 1710 SRO D240/D/83, *Nabbs Head* 1717 Okeover T764, *The Nab* 1836 O.S.; NABB (obsolete) between what is now Finney Green and Banktop Farm, Madeley Heath (SJ 7946), *Nabbs* 1833 O.S.; **NAB END** 1 mile east of Hollinsclough on the Staffordshire-Derbyshire border (SK 0766), *Nab* 1600, *Nabbe* 1602 Alstonefield ParReg, *? Nabfoot* 1637 Leek ParReg, *Nab End* 1842 O.S.; NABBS WOOD (obsolete, on north-east side of Kidsgrove (SJ 8454)), *Nabbs Wood* 1836 O.S.; NABBS (obsolete, 1 mile north-west of Keele (SJ 7946), *Nabbs* 1833 O.S. Although early forms are not available, there is little doubt that the names are from ON *nabbi, nabbr* 'a projecting peak, a knoll, a hill', an element occurring chiefly in minor names (cf. Nab, Cumberland; Nab, Nottinghamshire; Nab, Yorkshire), and commonly found in Staffordshire. There are several hills over 500' at Nabb, and Nab End lies near the foot of a rounded headland of over 1000'. Higher Nabbs and Lower Nabbs, 1 mile south-west of Wildboarclough, are just across the border in Derbyshire, and a hill called The Nabs is on the Derbyshire side of the river Dove 2 miles south-west of Alstonefield. A *Nabfoot*, presumably at *Nab Hill* (1842 O.S.), shown as *Nah Hill* (*sic*) on Yates' map of 1775 on the west side of Leek, is recorded as *Nabbe* in 1542 (1883) Sleigh 71, and in 1641 (Leek ParReg), presumably *Nabbe* recorded in 1550 SHC 1928 283; and *Nab's Hill* and *Nab's Leasow* (unlocated) are recorded in Cannock in 1821 (Oakden 1984: 60, 79). *Le parke Nebbe* c.1590 (SHC 1929 354), *Parke Nabbe* 1592 (SHC 1930 276), was in Alstonefield parish, and *Nab feet*, to the north-east of Upper Hulme, appears on Yates' map of 1775. *Crocketts Nabb* in Kingsley (Tettenhall) is recorded c.1716 (SRO D3160/10/5), perhaps the same place as *Nabb or Nabbs Croft* in Tettenhall Wood, recorded in 1823 (SRO D3160/10/15). Tettenhall Wood is on a high escarpment, but it would be surprising to find ON names in this part of the county (though see Gunstone), and the antiquity of the names must be doubtful. See also Cocket Knob.

NAILOR (obsolete) 1 mile north-east of Biddulph (SJ 9057). *Knawloe* 1669 Biddulph ParReg, *Knaloe* 1673 ibid, *Nailor* 1842 O.S., *Knawlow* 1930 O.S. Perhaps from OE *cnearr*, ME *knar* 'a rugged rock', with OE *hlaw* 'mound, tumulus'. See also Naychurch.

NANGERUSSAST (unlocated, in Bagot's Bromley). *Nangarschaft* mid-13th century SRO D(W)1721/3/24/2, *Nangerussast* 1356 SRO D(W)1721/3/1/6. A very curious name for which no suggested derivation can be offered. *Hauwgarsobast* (n.d., SRO D4038/A/1/1/a)/(ii)) is probably this same place.

**NAPLEY HEATH** 1 mile north-west of Mucklestone (SJ 7138). *Mapley Heath* 1686 Norton in Hales ParReg, *Maple Heath* 1692 ibid, *Napeley Heath* 1763 ibid, *Mapele Heath* 1808 Baugh, *Napeley Heath* 1811 EnclA, 1946 O.S. Perhaps 'heath at the *leah* with the maples', but in Staffordshire OE *hlaw*, invariably meaning 'burial mound', often became *-ley*, and the proximity of this place to the prehistoric chambered tomb known as The Devil's Ring and Finger (q.v.) may be noted. See also Mucklestone.

**NARROWDALE** 1½ miles north of Alstonefield (SK 1257). *Narendale* c.1275 SHC V (i) 120, *Norwedale* 1277 SHC 1911 168, *Narewedale* 1293 FF, 1319 Ass, *Nar(u)dale* 1299 Banco, *Narwedal* 1331 DRO D2375M/55/3, *Narrowdale* 1594 DRO D2375M/190/4, 1599 Smith, *Narrow-dale* 1686 Plot. 'The narrow valley'. '[S]o very lofty, that the Inhabitants there for that quarter of the year, wherein the Sun is nearest the Tropic of Capricorn, never see it at all; and at length when it does begin to appear again ... they never see it till about one by the clock, which they call hereabouts, the

Narrowdale noon; using it proverbially when they would express a thing done late at noone': Plot 1686: 110.

**NASH END** 1 mile north of Upper Arley (SO 7681). *le Nasshe Eynde* c.1330, 1412 PN Wo 31, *Nashe Ende* 1602 SHC 1935 456, *Nashend* 1686 Plot, *Nash end* 1747 Bowen. One of many examples of Ash with a transposed *n* from *an*. In Worcestershire since 1895.

**NAYCHURCH** ½ mile north of Upper Hulme (SK 0161). *Knachurche* ?1413 DRO D2375M/1/1, 1432 VCH VII 33, 1580 SHC XIV 212, *Knachurch* 1612 DRO D2375M/106/27, *Knalochurch* 1626 Rental, *Nachurch* 1651 Leek ParReg, *Naychurch* 1682 Alstonefield ParReg, *Kna Church* 1702 Leek ParReg, *Nay Church* 1842 O.S. A puzzling name, possibly from OE *cneo* 'knee or bend', with reference to the slight deviation in the nearby road which is said to be Roman (Margary number 713 and O.S., but the present road appears to have been created c.1765 as a turnpike road, with the earlier route from Leek to Buxton passing between Hen Cloud and The Roaches to Flash: see Bowen's map of 1749; NSJFS 1978-9 NS 4 37-8; VCH VII 195; also Sleigh 1883: 199), or more likely from OE *\*cnearr*, ME *knar* 'a rugged rock' (see also Nailor). The place lies below Ramshaw Rocks. The second element is unexplained. In other places the word 'church' has developed from the British *cruc* 'mound, hill, tumulus' – at an early period *u* became *ȳ*, and became associated with OE *cirice* 'church'. It is not possible to say whether this has happened here. However, the word church is not infrequently added to natural features in this area (cf. Dovedale Church, a rock at Ilam; Ludchurch), and the element may refer to a rocky church-like outcrop by Naychurch Farm.

**NAYSE** (unlocated, appurtenant to Beobridge according to Eyton 1854-60: III 85) *Nesse, Nasse* 1274 Eyton 1854-60: III 85, *Nayse* 1316 ibid. 85, *(atte) Nays* 1327 SHC VII (i) 253, *Nayse* 1327 TSAS 3rd Series V 1905 244, *Nassh* 1333 Eyton 1854-60: III 85, *Naysshe* 1392 SHC XV 48, *Nayss* 1525 Sub. From OE *ness* 'a nose, a projecting ridge'. Cf. The Naze, Essex.

**NEACHELLS** 2 miles east of Wolverhampton (SO 9499). *Echeles* 1293, 1304 SHC 1911 231, 276, 1332 SHC X 127, 1370 SHC XIII 73, *Necheles* 1327 SHC VII (i) 174, *Echels, Escheles* 14th century Duignan, *Necheles* 1462 SHC 1928 187, *Echellis* 1517 BCA MS3145/288615, *The Nichell* 1532 SHC 4th Series 8 159, *Neychelles* 1589 SHC XVIII 4, *Nechels, Echells otherwise Nechells, Nechells* 16th *century* Duig, *Nechilles* 1608 SHC III NS 23, *Neachalls* 1633 Wolverhampton ParReg, *Nechells* 1801 Shaw II 150 From OE *\*ecels* 'an addition, land added to an estate', from OE *eaca* 'addition, increase', a word common in the North Midlands from the 13th century. The *n* is from *atten* 'at the'. The history of Echells between Handsacre and King's Bromley has not been traced.

**NEEDWOOD, NEEDWOOD FOREST** 4 miles west of Burton upon Trent (SK 1724). Not recorded before the Conquest: the earliest reference is in the 1120s, but the forest almost certainly existed before that date: VCH II 349. *Nedwode* 1198x1208 SHC 4th Series IV 43, c.1200 ibid. 77, 1329 SHC XI NS 25, *Neidwode* 1254 SHC 1937 50, *Netewode* 1425 SHC XVII 102, *Chacia nostra de Nedwoode* 1248 VCH II 349, *Nedewode* 1256 MinA, *Neede Wodde* c.1540 Leland. The name perhaps means 'wood in which refuge was sought in time of need' (cf. Littywood), or possibly 'forest from which wood and timber were obtained when needed', or 'forest in which feudal service or duty was undertaken', from OE *nied* 'need, distress, necessity, duty, poverty, hardship'. Strictly the forest was a chase (*the chase of Needwood* 1499 SRO DW1733/A/3/15), since it was held by a subject, originally Henry de Ferrars, rather than reserved for royal hunting. It covered an area between the rivers Trent, Dove and Blithe: NSJFS 8 49-50. It survived, albeit in shrunken form, until 1811 (VCH II 354), and incorporated detached portions of various local parishes within the woodland core: Hooke 1998: 218-9. The Abbots Bromley horn dance may be connected with the villagers' former rights in the forest. The ten parks in Needwood Forest were Stockley, Castle-hay, Shireholt (Sherholt), Hanbury, Highlands, Barton, Agardesley, Castle, Rowley and Rolleston: Shaw 1798: I 60-61. There were 32 gates into the Forest: 20 are named in StEnc 413. See also Castle Hayes Park.`

**NEILD'S FARM** 1 mile north-east of Heaton (SJ 9663). *Neild's Farm* 1892 O.S. From the name Neild (see Nield).

NELSON HALL (obsolete) in Cotes Heath (SJ 8235). One of a number of Government establishments created in the area the early 1940s and named after naval heroes. See also Beatty Hall, Drake Hall, Duncan Hall, Frobisher Hall, Howard Hall, Raleigh Hall, Rodney Hall. See also StEnc 414.

NETHERHOLM (unlocated, perhaps near Bury Bank) *Netherholm*? 13th century SHC VI (i) 8. From OE *neoðera* (or ON *neðri*) 'lower', with ON *holmr* 'island, land by or between streams'.

**NETHERLAND GREEN** 2 miles south-east of Uttoxeter (SK 1030). *Netherland Green* 1775 Yates, 1836 O.S, *Netherland* 1872 P.O. From OE *neoðera*, *lond* 'the nether or lower land'. The *green* element denotes a grassy open space, probably in woodland, or perhaps a squatter settlement.

**NETHERSET HEY** 1 mile south-east of Madeley, immediately to the north of Madeley Great Park (SJ 7843). *Netherstedeplace* 1401-2 NSJFS 3 1963 53, *(a wood called) Netherscydhaye* 1531 SHC 1912 46, *Nethersit Hay* 1833 O.S. Associated with the short-lived Nethersethey Park (c.1395-1401): TNSFC 1963 53-5. The names are perhaps from the OE adjective *neoðera* 'lower', with OE *(ge)set* 'dwelling, place of residence, animal fold' apparently replacing OE *stede* 'place, site of a building', and (in early sources) 'dairy farm', and OE *\*scydd* 'a hovel, a shed'. The Hey element is from Mercian OE *(ge)heg* 'a clearing, an enclosure', found in ME for the latinized *haia*, meaning 'a part of a Forest fenced off for hunting'. See also NSJFS 3 1963 53-5.

NETHERTON (obsolete, in Hammerwich (SK 0607)), *Netherton* 1319 VCH XIV 259; NETHERTON (obsolete, at the foot of Tinacre Hill, south-west of Wightwick (SO 8798)), *Netherton* 1327 VCH XX 10, 1880 O.S.; NETHERTON (obsolete, on the south side of Quatford (SO 7391)), *Netherton* c.1290 SA 2922/2/20. 'Nether or lower *tun*'. The name Netherton in Hammerwich remained in use until at least 1871 (VCH XIV 259), and *Netherton Lane* in Quatford was remembered in the 19th century (Eyton 1854-60: I 114).

**NETHERTOWN** 1 mile north-west of King's Bromley (SK 1017). *Nethertoun, Netherton* 1100x1135 (1798) Shaw I 153, *Nepertoun, Neperetoun*? 13th century SHC XVI 260, *Ridware Netherton* 1323x1377 (1798) Shaw I 151, *Nethertowne* 1414 (1798) Shaw I 35, 1579 SHC XIV 202, 1609 SHC III NS 25, *North Town* 1749 Bowen, *Nethertown* 1834 O.S. 'Nether or lower *tun*'. *Ridvare* (DB) may refer to this place: VCH IV 47. Shaw 1798: I 151 suggests that *Walter's Ridvare* (n.d.) may be Nethertown, the name taken from the holder at the time of DB.

NETLOWS, THE (unlocated, in Mayfield or Swinscoe) *Netlow* 1801 Okeover E391, *The Netlows* 1838 Okeover T123. Possibly associated with *Nettlehale*, recorded *temp.*

Edward I (Okeover T49), *Netlehale temp.* late Henry III (ibid. T41), *Nettilhale* 1490 (ibid. T278).

NETTLES (unlocated, in Mavesyn Ridware) *le Netteles* 1325 (1798) Shaw I: 176, *le Neteles* 1362 (1798) Shaw I 177, *Netelesserd* n.d. (1798) Shaw I: 170, *Nettles (Pit)* c.1654 WaCRO CR1908/16/12. From OE *net(e)le, netel* 'nettle'. It has been noted that names incorporating *net(e)le* are often associated with Roman roads and ancient trackways (JEPNS 35 2002-3 49-58), and the name Ridware may mean 'people who lived by the ford (or bridleway)': see Hamstall Ridware.

NEUTHORP (unlocated, perhaps near Hampstall Ridware, but possibly outside Staffordshire) *Neuthorp* 1413 SHC XVII 47. From OE *þrop*, ON *þorp*, meaning in the Danelaw 'secondary settlement, an outlying farmstead or small hamlet dependant on a larger place', with ME *newe* 'new'. Possibly associated with *Garardesthorp* (q.v.).

NEWBOLD 1 mile north-east of Barton-under-Needwood (SK 2019), *Newanbolde* 942 (14th century, S.484), *Newbold lands* 1682 Browne; NEWBOLT in Chebsey parish, at Hilcote (SJ 8429), *Neubotlea* 1175 SHC I 71, *Newbolt* c.1220 SHC 1914 68, *Neubaude* 1236 Fees, *Neubold* 1288 SHC VI (i) 176, *Newbolt juxta Chebbeseye* 1317 SHC 1914 69, *Newbolde, Newbolt* c.1462 SHC VII NS 254-5; NEWBOLD ENDE (obsolete) in Rugeley (SK 0417), *Newbold Ende* 1570 SRO DW1734/2/3/38. From the OE adjective *niwe* 'new', with OE *bold* 'house, dwelling-place'. The second place later became known as Hilcote (q.v.): SHC VI (i) 176 fn; see also Badenhall, Chebsey. There are two places of this name in Shropshire. PN Sa II 135 notes that compounds with *niwe* greatly outnumber other clases of names containing the related words *botl* and *bold*. Cf. Newbald; Newbuildings Farm; Bold.

NEWBOLDS, THE (obsolete) 2 miles north-east of Wolverhampton (SJ 9300). *Neubold* c.1272 (1801) Shaw II 150, *Newbold* c.1295 Mander & Tildesley 1960: 31, *Neubold* 1355 SHC 1913 158, *Newbolt* 1372 SRO D593/B/1/26/6/20/3, *le Newbold, Newboldesbruche, le Neubolt, Newboldes alias Bayliestyles* 14th century Duig, *Newbolds* 1570 SRO D1790/A/2/144, *Newballs* 1599 BCA MS3145/258/18, 1707 BCA MS3145/63/1a&b, *Newbold Farm* 1775 Yates. From the OE adjective *niwe* 'new', with OE *bold* 'house, dwelling-place'. See also Newbold.

NEWBOROUGH 7 miles south-east of Uttoxeter (SK 1325). *Edgareslege* 1086 DB, later *Agardsley. Neuboreg* 1280 SHC VI (i) 98, *Neuborey* 13th century Duig, *Newburg* 1532 SHC 4th Series 8 76. The older name meant 'Eadgar's *leah*'. It changed to *Newborough* (from the OE adjective *niwe* 'new', and *burh* 'fortified place', or, as here, 'a borough, a market town'), with the creation of the new borough by Robert de Ferrers III in 1263: VCH II 349. See also Agardsley. *Novo Burgo* recorded in 1141 (SHC 4th Series IV 75) and later was seemingly an extension of the borough of Tutbury: VCH II 349; VCH III 331. See also Kinver for *Novo Burgo*.

NEWBRIDGE 2 miles west of Wolverhampton (SO 8999). *Novo Ponte* 1286 SHC V (i) 171, *Novum pontem* 1327, *(Atte)newebruge* 1332 SHC X 127, *Newebrugge* 1379 SHC XIII 150. 'New bridge'. The London to Holyhead road passes over the river Smestow here. The name implies the existence of an earlier bridge.

NEWBUILDINGS FARM ½ mile west of Newton (SK 0326). *Neubyggyng* 1417 SHC XVII 61, *Le Newbygginge* 1464 SHC 1939 128. From ME *bigging* 'a building', later 'an outbuilding, an outhouse', a word not normally found outside the north of England. *Le Newbyggynge* is recorded in Burton upon Trent in 1465 (SHC 1939 128), and as

*Newbyggyng now called le Horninglowstrete*, to be identified as Horninglow Street on the south side of Burton (SK 2423), in 1479 (SRO D4379/5/13).

**NEWCASTLE UNDER LYME**   on west side of Stoke on Trent (SJ 8445). *Nouu Oppidu cu soca sub Lima* ('New Castle with its jurisdiction under the Lyme') 1166 SHC 1923 297, *novum castellum de Staffordshira* c.1142 StEnc 415, *Novum Oppidum sub lima* ('New Castle under the Lyme') 1168 SHC I 55, *Noui Castelli, Nouo Castello* 1190 Ch, *Novi Castro sub Lima* 1235 SHC 1935 296, *Novum Castrum subtus Limam* 1250 Fees, *Novum Castrum Super Are (Nef Chastel Sus Are)* 1305 Chr. & Mem. 98, *(de) Newcastle Super Are* 1316-17 Letter Books of the City of London, *Newcastell* 1489 HLS, *New Castel under Line* c.1540 Leland: 'so cawllid of a brooke renning thereby, or of an hille or wodd thereby, so cawllid'. The name is perhaps from the 'new castle' built in the town possibly at the end of the 11th century (AJ 120 1963 289); the earliest reference is 1149 (VCH VIII 11), or between 1140 and 1146 (StEnc 415), perhaps to distinguish it from the 'castle' or Roman fortification at Chesterton (q.v.), or the old castle at Stafford (VCH IV 26), or one that might have existed at Trentham (JNSFC XLVII 1912-3 144-50; SHC XII 75; VCH VIII 11), or perhaps from a rebuilt castle on the same site (see StEnc 415; 598), although there would appear to be no record of any earlier castle. *Are* would seem to be an old name of Lyme Brook, perhaps identical with Ayr, Scotland, and the river Aire in the West Riding of Yorkshire, perhaps of pre-English origin, possibly OEuropean *\*Ara* 'water': see Ekwall 1928: lx, 1-3; CDEPN 433. The element Lyme is generally believed to refer to Lyme Forest, meaning 'place of the elms' (not limes), but it seems possible that Leland's statement may be correct: the *'wodd thereby, socawllid'* was presumably what is now The Limes (q.v.), or Shortelyme (q.v.). See also Lyme.

**NEW CHAPEL** – see **THURSFIELD.**

**NEWCHURCH**   in Needwood Forest, 6 miles west of Burton upon Trent (SK 1423). *Christ Church on Needwood* 1836 O.S. So-named from Christ Church, built in 1809 for the benefit of disafforested parishioners at the time of the enclosure of Needwood Forest: Erdeswick 1844: 279.

**NEW CROSS**   1 mile north-east of Wolverhampton (SO 9399). *Newcroste* 1670 SRO D4407/74[SF91], *New Cross* 1834 O.S. The nature of the cross is not known, but the name implies the existence of an 'old cross'. The name *Crossbyrches*, recorded c.1272 (Shaw 1801: II 150) may be associated with this name, as 'the land newly broken for cultivation associated with the cross' (from OE *bryce*), but the word 'cross' could also mean 'a field lying athwart another' (Foxall 1980: 9), and that is more likely to be the meaning here when found with *bryce*.

**NEWFIELDS**   1 mile west of Wetley Rocks (SJ 9449). *The Dork* 1836 O.S. The older name is curious, possibly from dialect *dawk*, 'a hollow, a depression, a furrow' (EDD; OED), though the word is only recorded in Yorkshire (PN W Yo III 180), but a more likely derivation may be a British river-name *\*Dorce*, derived from the root *derk-* (in Welsh *drych* 'aspect', OE *torht* 'bright': Ekall 1928: 128-9; 1960: 148), so 'the clean, bright stream': the place lies close to the headwater of a stream running north into Stanley Pool. The same river-name is found in Dorchester, Oxfordshire, and Dorking, Surrey.

**NEW FOREST**   *Novam Forestam, Nova Foresta* c.1199 SHC II 95, 98, 115. A royal forest in the north of the county, extending from Tunstall in the north to Tixall in the south, and from west of the Trent to Painsley in the east, which was disafforested in 1204: VCH II 348-9. The enclosed part of this Forest was called the *Haye of Clive* (SHC

II 98), *haiya de Clive* 1234-40 (TestNev), *haya nostra de Clyf juxta novum castrum subtus Limam* 1271 (SHC V (i) 155). It has been suggested that the Forest was so-named because it was reafforested by one of the Norman or Plantagenet kings after Newcastle had come back into royal hands in the middle of the 12th century: SHC 1923 301-2. See also NSJFS 8 1968 48; StEnc 421.

NEW GRANGE (obsolete) 2½ miles north-east of Leek (SJ 9960). *nova grangia* c.1291 Tax, *New Grange* 1521 Dieulacres Inventory, *(le) New(e)gra(u)nge* c.1539 LRMB, c.1540 *AOMB, New Gra(u)nge* 1560 Pat, 1634 ParReg, 1842 O.S., *Newe Grandge* 1630 SHC II (ii) 14, *Grange* c.1680 SHC 1919 40, *New Grange* 1871 SRO D1798/583/8. Self-explanatory. The place was a grange of Dieulacres Abbey, and appears to have been the same place as Nether Grange or Nether Hulme (q.v.). It was submerged when Tittesworth reservoir was extended c.1960: SHC 4th Series 19 fn.9.

**NEWHALL FARM** 1 mile south-west of Newborough (SK 1224). *? Novahall* c.1155 SHC 1937 13, *? Nova Aula* 1280 ibid. 80, *? Neuhall* 1337 SHC 1913 59. Self-explanatory.

**NEWHAY** 1½ miles north-east of Cheadle (SK 0344), *the Neweheye* 1316 VCH III 227, 1330 CroxdenChr, *Newhey Wood* 1836 O.S.; NEW HAY (unlocated, near Gentleshaw), *(le) New(e)hey, Neuhey, Neuhay(e)* 1348 Anglesey Ch, 1348 SHC 1939 77, 1360 (p) Ipm, 1379 Banco, *Newhay* 1461 SHC 1939 109, *Newhey* 1528 SHC XI 268; NEW HAY (unlocated, in West Bromwich), *Nova Haya* 1223 SHC IV 24; NEW HAY (obsolete) on the north side of Smallthorne (SJ 8851), *Newe Hayes* 1598 Norton-in-the-Moors ParReg; NEWHAY (unlocated, in Kele), *Newhay* 1410 Harrison 1986: 21. From OE *niwe, (ge)heg* 'the new enclosure'.

NEW HOUSE (obsolete) 1 mile south-west of Blythe Bridge (SJ 9338). *Newhouse* 1554 SRO SD4842/18/1, c.1680 SHC 1919 262, *New Ho.* 1798 Yates, *New House* 1836 O.S.

NEW INN (obsolete) in Handsworth (SO 0489), *Newe Inne* 1546 SHC XI 291, 1565 SHC XIII 252, 1615 SHC IV NS 79, *New Inn* 1749 Bowen, *New Inn (Hall)* 1798 Yates, 1834 O.S.; NEW INN (obsolete) two places, two miles apart, one 1 mile north of Claverley (SO 8095), the other 1½ miles north-west of Bobbington (SO 8293): Shaw (1798: I 15) mentions *'...the two new inns...'* (citing Wilkes c.1758; see also *the two New Inns* 1762 SA/5586/14/2-3), and each appears as *New Inn* 1752 Rocque – the reference to *the Newe Inne* in 1604 in Pattingham ParReg, *New Inn* in 1625 in Claverley ParReg, and *New Inn or Bowling Green New Inn* in 1769 (SA 5586/2/1/499) could refer to either; NEW INN (MILL) (obsolete) ½ mile north-east of Trentham (SJ 8741), *New Inne* 1537 MA, *the new inne* 1609 Trentham ParReg, *New Inn (Mill)* 1834 O.S. Self-explanatory.

**NEW INVENTION** 1½ miles south-west of Bloxwich (SJ 9701). *ye new invention neare Snead* 1663 Wolverhampton ParReg, *New Invention* 1747 Bowen, *The New Invention* 1834 O.S. The suggestion in Hackwood 1908: 180-1 that the name derives from a hawthorn bush pushed into a chimney to stop it smoking is almost certainly apocryphal: see Tildesley 1951: 187, where it is suggested that the name may be a corruption of an earlier name. That seems equally unlikely. The word invention in the 17th century had two main meanings, 'a finding or discovery' (e.g. 'the Invention of the Cross', the reputed finding of the Cross by Helena, mother of the Emperor Constantine, in 326 AD), and 'a novelty or original device', as in ModE. There is every possibility that the name does indeed refer to some local finding or discovery, or some novelty or invention, not necessarily made here, perhaps relating to mining or ironworking or the pumping of water, the nature of which remains unknown, or even from a beerhouse of that name, commemorating the site of such a discovery. It may also be noted that OED records as one 17th century meaning of invent 'to bring into use formally or by authority', so a meaning 'the newly authorised activity or business' is not inconceivable, and a public house name for a beerhouse previously unauthorised is a possible explanation. The name is not unique. There are two places called New Invention in Shropshire. One is in a pronounced stream valley 2½ miles south of Clun (on which see N&Q 183 380-1, where it is suggested that the name was formerly *The Vention*, possibly named from a water-driven loom on the Redlake Stream: Foxall 1980: 66 states that it was the first place in the district where spinning was carried out by water power). The other is near High Ercall, and is said to be a semi-punning name for an inn ('inn-vention'): CDEPN 435. Cf. Ragged Invention.

**NEWLANDS** 2 miles south of Abbots Bromley (SK 0721), *Neulond temp.* Edward I SHC 1937 113, *Neweland* 1309 SHC IX (i) 9, 1402 SHC XI NS 210, *Newland* c.1311 ibid. 118, *Newelond* 1337 SHC XI 143, *Newlande* 1425 SHC IV NS 113, 1565 SHC 1931 171, 1597 SHC 1935 23, *Newland* 1836 O.S.; NEWLANDE (unlocated, in Cannock), *Newlande* 1544 SHC 1939 79, *Newelande* 1580 ibid. 79; NEWLAND (unlocated, in Barton under Needwood), *Le Newland* 1330 (1798) Shaw I 113. 'The land newly-cleared for cultivation'. *Newelond* in Ellerton Grange is recorded c.1518: SHC IX NS 95.

NEW LIBERTY (unlocated, perhaps near Tutbury: SHC 1912 222) *New Liberty* n.d. SHC 1912 222.

NEWPLACE (unlocated) *Newplace* 1419 SHC XVII 67.#

NEW POOL on west side of Biddulph (SJ 8756). *Newpoole, New Poole* 1655 Biddulph ParReg, *New Pool* 1749 Bowen. Presumably to distinguish it from *Pool* (unlocated), mentioned in the Buddulph ParReg from 1598. Bowen's map of 1749 shows New Pool as 'The head of the river Trent'.

**NEW SPRINGS** 1 mile south-west of Talke (SJ 8153). *Newspringe* 1661 Audley ParReg, *New Springe* 1733 SHC 1944 71, *New Spring* 1799 Faden, *New Springs* 1833 O.S. The word *spring* meant both a flow of water rising from the ground, and a copse of young trees or coppiced trees with new shoots. The second meaning is more likely here.

NEWSTEAD (obsolete, ½ mile south-west of Blurton (SJ 8941)), *Newstead* 1537 VCH III 259, *Newstede* 1537 MA, 1568 Trentham ParReg, *Newstidd* 1576 SHC 1926 43, *Newsted, Newstidd* 1703 ibid, *Newstead* c.1714 SRO D593/H/3/30, *New Stead* 1798 Yates, 1836 O.S.; NEWSTEAD (obsolete, 1 mile south-west of Cheddleton (SJ 9650)), *Newstead* 1836 O.S. From OE *stede* with various meanings, but often denoting 'a farm or estate'. The element is rare in the West Midlands.

**NEWTON** 1 mile south-east of Draycott in the Moors (SJ 9838), *Niwetone* 1086 DB, *Neutonam, Neutona* c.1160 SHC III (i) 224-5, *Neuton* 1294 SHC 1925 89; **NEWTON** in Blithfield parish, 5 miles north of Rugeley (SK 0325), *Niwetone* 1086 DB, *Neuton* 1252 SHC 1937 47, 1257 SRO D938/62, c.1293 SHC NS XI 165, 1306 SHC VIII i 145, *Neweton* 1433 SHC XVII 145; **NEWTON** 2 miles north-west of Worfield (SO 7397), *Newton* 1525 SR, 1564 Worfield ParReg, *Nowton* 1532 SHC 4th Series 8 118, *Newton* 1752 Rocque. 'The new *tun*'. Newton near Worfield has been in Shropshire since the 12th century.

**NEWTOWN** 2 miles south of Great Wyrley (SJ 9904), *Newtown* 1834 O.S.; **NEWTOWN** 2 miles south-west of Longnor (SK 0663), *Newtown* 1775 Yates, 1840

O.S. Self-explanatory. Newtown near Longnor dates from c.1754: VCH VII 28. It is unclear whether Newton Grange, recorded in 1535 as *Nuton Grange* (StCh) and as *Newton Grange* in 1564 (Alstonefield ParReg) is to be associated with that place.

**NEW YORK** ½ mile west of Upper Elkstone (SK 0459). *New York* 1842 O.S, 1851 White. From a copper and lead mine of that name that closed in 1859: VCH II 267. The name is not uncommonly found applied to remote places.

**NEW ZEALAND** 2 mile south-east of Heaton (SJ 9761). *New Zealand* 1891 O.S. Names of this type were frequently applied to places that were particularly remote, as here.

**NIELD** ½ mile east of Flash (SK 0366). *le Neelde* 1455 DRO D2375M/1/1, *Michael Nyelde's croft* 1566 *Deed*, *Neeldeie* 1582 SHC XV 140, *Neild* 1599 Alstonefield ParReg, *the Neilde Eie* 1599 ibid, *Neilde Eye* 1601 ibid, *Needle Eye* 1603 ibid, *Nield Bank* 1651 *Rental*, *Nield Eye* 1683 ParReg, *Needles Eye* 1744 SRO D1029/2/23, *Nield* 1842 O.S. The forms show that there has been some confusion with 'the eye of a needle', but indicate a derivation from the name Neal(e), Neild, Nield (said to be brought to England by the Normans: see DES 320 *sub nom* Neal, etc.), with OE *eg* 'dry ground in wet land', so 'Neild's dry ground in boggy land'. Rich. Neeld is recorded in 1662 in Leek ParReg. See also Neild's farm.

NIMMINGS PLANTATION (obsolete) in Clent. *Nemmynges* 1429 PN Wo 280. From OE *niming*, a word applied to land taken into cultivation or enclosed. The place was in Staffordshire from the early 13th century until 1844, when it became part of Worcestershire. See also Rumble & Mills 1997: 32.

**NOBUT** 1 mile east of Church Leigh (SK 0435). *? Nabbalt* 1332 SHC X (i) 100, *? Nobyt* 1327 SHC VII 200, *Nobett* 1590 SHC XVI 99, *? Nobert* 1593 SHC 1930 371, *Nobott* 1596 SHC XVI 38, *Nobolt* 1605 SHC 1940 280, *Nobould* 1608 SHC 1948-9 54, *Nobot* 1608 ibid.' 74, *Nobald* c.1619 SHC VII NS 208, *Nobbot* 1680 SHC 1919 269, *Nobut* 1711 Kingsley ParReg. Perhaps OE *niwe* 'new', with OE *bold* 'a dwelling, a house', often found paired together. Cf. The Newbolds.

NODDY FIELD VALLEY (obsolete) on south-west of Cannock Wood (SK 0211). *Noady feild* 1682 Dep, *Noddy Field* 1834 O.S. Oakden 1984: 60 suggests a derivation from the dialect adjective *noddy* 'weak, ailing' (EDD), used in a derogatory sense for poor land, and since early spellings have not been traced, that may be correct, but the possibility of a derivation from a personal name or surname cannot be ruled out: see JEPNS 3 24. It may also be noted that ME *atten ode* (literally 'at the ash-heap or funeral pyre') may become Node, Noad or similar, perhaps marking the site of early beacons. This place lies at the foot of a prominent un-named hill of 787', almost the highest point in central Staffordshire: the only higher land is Castle Ring, which lies one mile to the north-east and is 10' higher, but any beacon at Castle Ring would be obscured across a large southern sector by the hill here. The hill-name may have been Noady Hill or similar, from which Noady Field may have taken its name. No documentary or other evidence for such a beacon has been traced, but a beacon on the hill would have been visible for a considerable distance. It may be noted that a field called *Noddy field* is recorded in Compton (Kinver) c.1712 in association with other fields including *Blazehill Meadow*. SRO D801/2/2. Cf. Noddyshall (Cottages), Surrey JEPNS 3 1970-1 24. See also VEPN I 5-6.

**NODDY PARK** in Aldridge (SK 0601). *Noddy Park* 1768 StEnc 424, *Noddy Park (Lane)* 1777 BCA MS3558/63/8. Noddy Park Farm is said to have existed in the 17th century: StEnc 424-5. For possible derivations see Noddy Field Valley: the place is on high ground. The name is perpetuated in Noddy Park Road. Duignan (MidA II 171) mentions a tumulus or butt in Noddy Park, Aldridge, and a *Noddy Park* near Daw End in Rushall.

**NOONSUN COMMON**  ½ mile west of Ipstones (SK 0149), *Noon Sun Common* 1836 O.S.; **NOON SUN**  2 miles north-east of Upper Hulme (SK 0462), *Noon Sun* 1777 Alstonefield ParReg, 1839 *EnclA*. Perhaps applied to places that remained in shadow until mid-day.

**NOOSE (LANE)**  in Willenhall (SK 9498). *le Nous* c.1272 (1801) Shaw II 150, late 13th century WA II 89, *the Nose* 1514 SRO D593/B/1/19/1, *Noose Lane* 1775 Yates, *Noose (Lane)* c.1800 D593/H/3/402. A derivation from OE *nos(e)* 'a nose, a headland, a promontory' (found in this sense in Beowulf) seems improbable, since the phonology is very problematic, and there is no obvious topographical feature in this area might be so-described. Possibly from ME *atten wase* 'at the muddy or miry place': cf: The Noose, Gloucestershire.

**NORBURY**  Ancient Parish 5 miles north-east of Newport (SJ 7823). *Nortberie* 1086 DB, *Nordbiri* 1198 P, *Northbyri* 13th century Duig, *Norbere* 1532 SHC 4th Series 8 170, *Northbyri* c.1540 Leland. From OE *burh*, dative *byrig*, 'fortified place, manor-house', probably here meaning 'the northern manor-house'. It is unclear what place this is north of: Oulton is to the south, but is not mentioned in DB, so may not have existed in 1086.

**NORE HILL**  1 mile north-west of Pattingham (SO 8199). ? *(Atte)novre* 1295 SHC 1911 237, *? atte Novere* 1317 SHC V NS 223, *Nore* 1323 SHC IX (i) 94, 1327 SHC VII (i) 251, 1410 SHC XVI 71, 1494 SHC XI 253, *Le Nore juxta Pattingham* 1334 (1801) Shaw II 281, *Norehill* c.1585 SRO D1237/49, 1695 Morden, 1752 Rocque, *Noah Hill* 1833 O.S. The earliest spelling, if it relates to this place, indicates a derivation from OE *ofer* 'a bank', more specifically 'a flat-topped ridge with a convex shoulder' (see Gelling & Cole 2000: 203), which sometimes takes an initial N from OE *atten* '(place) at the'. The locations of *Norhulle*, recorded in 1276 (SHC 4th Series XVIII 18, 107), and *Noherhall*, recorded in 1625 (West Bromwich ParReg) are uncertain, but may well have the same derivation. Camden 1674: 125 notes 'Nore, the same with [i.e. as] North'.

**NORMACOT**  4 miles south-east of Stoke-on-Trent (SJ 9242). *Normanescote* 1086 DB, *Normantona, Normacot* 1177 SHC XII NS 279, *Northmannescot* 1227 Ch, *Normancote* 1242 Duig, 1414 SHC XVII 13, *Normaunte, Normauncote* 1251-2 Fees, *Nomannescot* 1256 Ch, *Normecote Graunge* 1472 SHC IV NS 183, *Normicoat* 1733 Stoke on Trent ParReg. Northman was used for 'Norwegian' or an OE personal name, sometimes contracted to Norman, and the derivation is probably 'the Northman's or Norwegian's cot or shelter'. A grange of Hulton Abbey existed here from at least 1242: StEnc 426; StASt 10 1997. It has been suggested that the original settlement of Normacot was originally at Meir Heath: StASt 10 1997 16.

NORMANESWELL  (unlocated, in Lichfield) *Normaneswell* early 13th century SRO D948/3/1, *Noremonnis welle* mid-13th century SRO D948/3/2. 'Norman's (or the Northman's) spring', from Mercian OE *wælle*.

**NORMAN'S WOOD, NORMANSWOOD FARM**  ½ mile east of Stowe (SK 0127). *Normans Wood* 1775 Yates. The name Norman was found in OE, so (if ancient) perhaps 'the wood of the Northman or Norwegian', or 'Norman's wood'. 'No-man's wood', is unlikely, as the place is neither remote nor on a boundary. The place may be associated with Normansle (q.v.). *Normans* is recorded in 1578 (SRO D260/M/T/1/41), and *Normans Meadow* in 1758 (SRO D260/M/T/1/85), but the location is uncertain.

NORMANSLE (unlocated, possibly near Checkley) *Normansle* 1272 SHC IV (i) 187.

NORTHALE (unlocated, possibly the moated site south of Bath Farm, Brewood (SJ 8507): see SHC VI (i) 192; Horovitz 1992: 62-3) *Northale* 1280 SHC VI (i) 150, 1362 SHC SHC XIII 25, 1377 SHC XIV 144, 1426 SHC XVII 113, *Northalle* 1577 SHC XIV 194. Probably 'northern *halh*' rather than 'northern hall'.

NORTHOVERE (unlocated, possibly near Crakemarsh/Creighton) *Northovere* 1337 SHC 1913 58. Evidently from OE *norð ofer* 'the northern flat-topped ridge with a convex shoulder'.

NORTHULL, NORTHERULL (unlocated) *Northull* 1261 SHC IV 149, *Northerhill* 1294 SHC VII 25. See also Northulle.

NORTHULLE (unlocated, perhaps in the Denstone area) *Northulle* 1191x1194 CEC 261. See also Northull, Northerhull.

**NORTHWOOD** 1 mile north-east of Hanchurch (SJ 8542), *Norwerde* 1166 SHC I 153, *Northwude* 1227 SHC VII (i) 6, *Northwud* 1247 SHC IV 239, *Trentham Wode, called Northwode* 1359 SHC XII 171; **NORTHWOOD** ½ mile north-east of Ellastone (SK 1243), *Norwude* 1197 SHC II 68, *Norwerde* 1198 SHC I 136, *Northwode* 1275 SHC V (i) 120, *Norwood* 1666 SHC 1925 191, *Northwode wast* 1671 SRO D4731/1/4, *Northwood* 1749 Bowen; NORTHWOOD (unlocated, in or near Wrottesley), *Northwode* 13th century SHC VI NS (ii) 50. Self-explanatory.

**NORTHYCOTE** 3 miles north of Wolverhampton (SJ 9303). *Northicote* 1199 SHC III 48, *Northicoten'* c.1240 WA II 100, *Northincote* 1255, *Nortkote* 1286 SHC V (i) 169, *Northcote* 1293 SHC 1911 231, *Nordicote* 1327 Duig. '(Place) north of the cot or shelter', the *i* coming from the terminal of the OE adjective *northan*.

**NORTON BRIDGE** 1 mile north-east of Chebsey (SJ 8630). *Norton Bridge* c.1795 SRO DE615/EX/1. Early spellings have not been traced, and the name may not be ancient, but Whitelock 1930: 154 suggests that *Norðtune*, recorded in the will of Wulfric Spot 1002x1004 may be this place, or Norton in Hales, Shropshire, or Norton Cuckney, Nottinghamshire. The place lies on Meece Brook.

**NORTON CANES** Ancient Parish 6 miles north of Walsall (SK 0107). *? æt Norðtune* 951 (14th century, S.554), *Nortone* 1086 DB, *Norton* 1166 SHC 1923 195, 1176x1182 EEA 16 66, *Norton-super-le-Canok* 1289 SHC VI (i) 183, *Norton Juxta Canke* 1532 SHC 4th Series 8 175, *Norton Kains, Norton Kaynes* 1566 SHC 1931 230, *Norton Caynes otherwise Norton Coynes otherwise Norton-on-Canuch* 1579 SHC XIV 205, *Norton-Kaynes* c.1609 SHC III NS 37. 'North town', possibly because it lies on the north side of Watling Street. Birch 1885-93: III 50 No. 891 seems to identify the 951 spelling with Norton Canes, but no evidence has been traced to support the identification: see SHC 1916 91. The derivation of the *Canes* element is unclear, although the southern boundary of the manor is formed by Gain's Brook; *the Gaynes meadow* is recorded in 1693 (SRO D1317/1/10/1/2-3); *Gains Gap* c.1699 (SRO D1317/1/13/1); and a Gains Lane also exists. Canes or Gains perhaps represents the name of a former landowner: Master John de Canes, 'rector of Norzbourgh', is a witness to a deed to which Richard de Bentley of Nortune is a party c.1300: SRO D1790/A/10/2. Arnold de Kanes is recorded elsewhere c.1207 (SHC 1937 27), and John de Kane c.1260 (SHC 1924 153), and the surname Caaines, Cahaignes, Cahaingn', Cahainn' are found in 1221 (Pleas), evidently from Cahaignes in Eure, though no evidence has been traced of any association with this place. For completeness, it may be added that *canes* is Latin for 'dogs', though it is difficult to explain any association here (but note that Cheadle (q.v.) was also called Hound Cheadle or Dog Cheadle, for reasons which remain unknown).

**NORTON, COLD** 2 miles south-west of Stone (SJ 8732; SMR 02100 places a deserted post-Conquest settlement at SJ 88003220). *Coldenorthon, Coldenorthton, Colde Norton* 1227 SHC XI NS 240, *Calde Norton* 1227 Ch, *Coldenorton'* c.1313 SHC 1941 173, *Coldenorton* 1319 SHC V NS 223, *Norton* 1625 JNSFC LX 1925-6 73, *Cold Norton* 1836 O.S. The spellings indicate a derivation from 'cold or exposed north *tun*'. See also Chebsey.

**NORTON FARM** ½ mile south of Cold Norton (SJ 8830). *Norton Farme* 1661 Swinnerton ParReg. The name is from Cold Norton (q.v.).

**NORTON-IN-THE-MOORS** 2 miles north-east of Burslem (SJ 8951), *Nortone* 1086 DB, *Norton subtus Keversmunt* 1218 SHC XII NS 219, *Norton under Kevremunt* 1227 SHC XI NS 240, 1239 SHC IV 90, *? Norton'* 1239 CurReg, *Norton-en-le-Hiles* c.1230 SHC 1911 445, *Northon* 1261 SHC 1924 138, *Northmores* 1279 SHC VI (i) 95, *Norton Super le Mores* 1285 FA, *Norton under Kevermunt* 1288 SHC VI (i) 175, *Norton super moras subtus Kervermund* 1303 Erdeswick 1844: 98, *Nortonothemores* 1315 SHC IX 49, *Norton Woodhouses in le Mores* 1592 SHC 1930 226; **NORTON GREEN** & **LITTLE NORTON** 2 miles north-west of Brownhills (SK 0207); **NORTON GREEN** 1 mile south-west of Brown Edge (SJ 8952). 'The *tun* north of another', in the first case with OE *mor* 'marshland, a high tract of barren uncultivated ground'. There is another Norton Green to the north of Norton-in-the-Moors. For the element *Keuremunt, Kevremunt see* Carmounthead. The Green element suggests a squatter settlement.

**NOTHILL (FARM)** 2½ miles north-west of Uttoxeter (SK 0737). *Knotel* 1764 Croxden ParReg, *the Notel* 1770 ibid, *the knotale* 1772 ibid, *Nothill* 1836 O.S. Perhaps from ON *knottr*, OE *cnotta* 'a hard mass', used topgraphically for 'a hillock, a rocky hill, a cairn', found chiefly in the north-west of England, with OE *hyll*, so meaning here 'the rocky hill'.

NOWELISHEY (unlocated, in Seighford) *Nowelishey* undated SHC 1914 86. From the family of Noel or Noell, who founded Ranton Abbey and held property in the area for several centuries: see e.g. SHC 1914 66-7 (pedigree), 68-9, 83, 85.

NUN FIELDS (obsolete) on the north side of Rudge Heath (SO 7997). *the Nuan feild* 1716 SA 2161/107, *Nun Fields* 1751 SRO D564/3/4/6, *Nunfields* 1765 SA 2161/109, *Nuns Fields* 1840 TA, *Nun Fields* 1857 SA 1190/86/84. Land in this area was held by the nuns of Blackladies priory near Brewood: SHC 1939 182-94. *Nun Brook* flows south from Patshull Great Pool: WJ October 1908 267. Cf. *Nonne Bridge* on Rudge Heath recorded in 1662: SA 2028/1/5/17.

**NURTON** 1 mile east of Pattingham (SO 8399). *Uverton* 1280 SHC VI (i) 106, *Noverton* 1312 VCH XX 173, 1391 SHC XV 36, *Norton* 1532 SHC 4th Series 8 186, c.1560 SHC 1938 115, *Nurton* 1586 SHC 1927 160, 1686 Plot 384, *Nerton* 1695 Morden. Possibly from OE *ofertun* 'ridge-town', with *atten ofertun* 'at the ridge-town', developing via the not unusual mis-division *atte nofertun* to become Nurton.

NUTBOROUGH (unlocated, in Wolstanton near Newcastle) *Noteburgh* c.1391 SRO D641/1/2/35, *Noteburgh* 1427 Cl 1422-9 318, *Nutburh* 1433 SRO D641/1/2/53, *Noteburgh* c.1445 Ipm, *Nutborough* 1453 SRO D641/1/2/72, *Nuttberewe* 1610 PRO

E367/1239. Perhaps from OE *cnotta* 'a hillock, a rocky hill, a cairn', more often found in the North West, with OE *burh* 'a fortified place', here perhaps referring to the lumpy earthworks of the Roman fortification at Chesterton.

**NUT WOOD** 1 mile north-west of Gnosall (SJ 8122). *Nut Wood* 1833 O.S. Perhaps *Nutiwude*, recorded in 1206 (SHC III 136). Self-explanatory.

OAK, THE (obsolete) on the east side of Sandwell (? SO 0289). *ye Oke in ye pish of West Bromwich* 1654 Roper 1980: 97, *The Oak* 1768 Ellis map, 1749 Bowen.

**OAKAMOOR** 3 miles east of Cheadle (SK 0544). *Ocuallemor* 1327 SHC 1913 14, *Okwallemor* 1328 ibid. 17, *Okwallmor* 1331 ibid.' 31, *Okam More* 1573 SHC 1931 179, *Okemore* 1602 SHC 1935 441, *Oakallmore* 1636 Cheadle ParReg, *Oakeall Moor* 1655 WYAS MD244/189, *Oackamore* 1680 ibid, *Oakeymoore* 1686 Plot, *Oakway Moor* 1693 SHC 1947 56, *Oakemoor* 1798 Yates. The early forms show that the original meaning was 'the moor with the spring at the oak', incorporating Mercian OE *wælle* 'a spring', and (sometimes) 'a stream', which has survived only as an intermediate -*a*-. Another Oakamoor (*Okamoor* 1624 SHC 1970 86, *Okemoor* 1624 SRO D615/D/1) seems to have existed to the west of Shugborough: SHC 1970 map 110.

**OAKEDGE** on Cannock Chase, to the south of Colwich (SK 0020). *Oak Edge (Wood)* 1593 SRO DW1781/5/2/1, *Oakedge (Hill)* 1641 SRO DW1781/5/16/1-14, *Oak Edge Estate alias Greenwood* 1771 SRO D615/M/6/42, *Oakedge* 1808 Baugh, *Oakedge (Park)* c.1833 SRO D615/M/6/53. Oakedge Hall was originally called Whitby Wood, built by John Whitby (d.1752). It was eventually acquired by the Ansons of Shugborough who renamed it *Oakedge Hall*, occupied it for a few years during the enlargement of Shugborough Hall, then demolished it: Burne 1961: 23-4. See also StEnc 429.

**OAKEN** 1 mile west of Codsall (SJ 8502). *Ache* 1086 DB, *Aks* 1234x1240 TestNev, *Ak'* 1242 Fees, *Halken* c.1250 SHC VI NS (ii) 49, *Oca* 1253 SHC IV 124, *Hoken* 1292 SHC VI (i) 249, *Ocke, Oke* 1293 ibid. 239, *Oce, Ake, Oken* 13th century Duig, *Oke* 1327 SHC 1913 8, *Okne* 1377 SHC 4th Series VI 8, *Okene* 1378 SHC XIII 149, *Weken* 1462 SHC IV NS 121, *Woken* 1577 Saxton, *Oken alias Oking* 1653 BCA 867/463. Ekwall 1960: 346 gives a derivation from OE *ācen*, a derivative adjective of OE *ac* meaning 'of oaks', hence '(place) of the oaks', but an adjective would not be expected to form a place-name, the A- and O- forms rule out *ǣ*-, and the name would seem to be from OE *acum*, dative plural of *ac*, so '(place of) the oaks'. A name unique in England, which is curious, since oak trees must have been ubiquitous in the Anglo-Saxon period, so perhaps with some specialised meaning, now lost, involving oak trees or oak timber. There was a grange of Croxden Abbey here: VCH III 226. Oaken Lawn (*Oaken lawne* 1691 Codsall ParReg) lies 1 mile to the east of Oaken, on the north side of Kingswood Common (SJ 8403), from ME *launde* 'open space in woodland, a forest glade, woodland pasture'.

**OAKENCLOUGH** 2½ miles south-west of Longnor (SK 0563). *Oconclogh* 1419-21 DRO D2375M/1/1, *Oconcloughe* 1556 Rental, *Okenclough* 1568 DRO D2375M/55/2, *Okenclough(e)* 1599 ParReg, *Okenclough* 1655 DRO D2375M/58/3, *Oakenclough* 1794 Stockdale. From OE *acen, cloh* 'oaks ravine'. A stream here runs through a deep oak-lined ravine. Cf Oaken Clough in Hayfield, Derbyshire. See also Clough.

**OAKENCLOUGH BROOK** a tributary of the river Manifold. *Oakenclough Brook* 1794 Stockdale. From OE *acen, cloh* 'oaken ravine': the stream runs through a deep tree-lined ravine.

**OAKESWELL** in Wednesbury (SO 9894). *Okeswell or Hopkins Newhall Place* 1662 Ede 1962: 78-9; TSSAHS 1988-9 XXX 65, *Oakes-Well-Hall* 1672 (1801) Shaw II 87, *Oakeswell Hall* 1708 Ede 1962: 79. From Mercian OE *wælle* 'a spring', and (sometimes) 'a stream', so 'the spring at the oaks'. A house existed on the site by 1421, and the hall, known as The Rookery in the late 19th century, was demolished in 1961-2: TSSAHS XXX 1988-9 65; see also Ede 1962 77-9.

**OAKHAM** 1 mile north of Rowley Regis (SO 9689). *Hocume* 1674 BCA MS3532/Acc1935-054/444070, *Hocum, Hoocom* 1687 Rowley Regis ParReg, *Hocum* 1698 BCA MS3532/Acc 1935-054, *Holcom* 1723 ibid, *Oakham* 1817 Pitt, 1834 O.S. Seemingly OE *hocc-ham* '*ham* where hocks or mallows grew' (cf. Hoccum), though the forms suggest a long vowel.

**OAK HILL** 2 miles south-west of Newcastle-under-Lyme (SJ 8643). *Ochull, Okhul*? 13th-14th century SHC XI 321-2, *Ochull* 1346 SRO DW 1082/A/4/4, *Oak Hill* 1836 O.S. Self-explanatory. It is unclear whether *Okehill*, recorded in 1308 (SHC 1911 297), refers to this place. There is an Oakhill on the north-west side of Upper Tean (SK 0039), but earlier spellings have not been traced.

**OAKLEY** in Croxall parish, 7 miles north-east of Lichfield (SK 1913; SMR 02529 places a deserted Anglo-Saxon settlement at SK 19201330), *(æt) Acclea* 1002x1004 (11th century, S.906; 11th century, S.1536), *Acle* 1086 DB, c.1187 SHC II 261, 1260 SHC X NS I 275, *Acleia* c.1180 SHC III (i) 205, *Ocle* 1199 ibid. 47, *Ochley* 1272-3 SHC XI NS 243, *Axell* 1294 SHC XII NS 267, *Acleia, Okeley* 13th century Duig, *Okele* 1320 SHC 1911 97, *Ocleye* 1327 SHC VII 215, *Okeley* 1332 SHC X 107, *Oakley* 1798 Shaw I 387, *Oakley Mill* 1834 O.S.; **OAKLEY** in Mucklestone parish, 2½ miles north-east of Market Drayton (SJ 7036), *Aclei* 1086 DB, *Akele* 1236 Fees, *Okle* 1265 SHC IV 159, *Ocleye* 1327 SHC VII (i) 215, 1334 SHC XI 184, *Okeley* 1462 SRO 3764/96, *Oakelyne* 1666 SHC 1921 131. From OE *ac, leah* 'oak wood', or 'glade where oaks grow'. Croxall was transferred from Derbyshire to Staffordshire in 1894, but Oakley ('... an old manor, situated about a mile North-west from the church beyond Elford park, and seems to derive its name from the fine oak-trees still growing in this vicinity ...': Shaw 1798: I 387) has always been in Staffordshire. The fact that both Oakleys lie on the county border is coincidence. Modern maps show Oakley 1 mile west of Brewood (SJ 8608), but the place probably takes its name from the 19th century Oakley Farm (*Oakley House* 1861 SA 4752/19/20): no trace of the name has been found in earlier records.

**OAKS FARM** at Callingwood (SK 1923). *Le Hokes under Rohay temp.* Edward I SHC 1937 113, *the Oaks* 1629 SRO 820/15. 'The oaks under Rough Hay'.

OAT HILL (unlocated, in Trysull) *Othull* 1354 (1801) Shaw II 208, *Watt-hill or Oat-hill* 1603 (1801) ibid. 207. Possibly from ME *ode* 'beacon, bonfire', so 'the hill with the beacon'. Or the name is perhaps associated with William Othull, recorded in the area at an early date (Shaw 1801: II 208).

**OCKER HILL** in Tipton, 1 mile south-west of Wednesbury (SO 9793). *Hocherhill* 1747 Bowen, *Ocker Hill* 1787 Act, *Hockeshill* 1788 Harrison map, *Ochre Hill* 1798 Yates, *Oker Hill* 1808 Baugh, *Ockher Hill* 1809 BCA MS3145/11/1, *Hockerhill* 1834 O.S. Possibly from OE *\*hocer* 'a hump, a rounded hill', or from OWelsh *ochr* 'a side, shelving locality': there is a considerable hill here. However, Plot 1686: 122 mentions 'a reddish sort of earth gotten at Tipton', and outcrops of a reddish orange clay-like earth are remembered at the top of Toll End Road (ES 27 January 2005 23). It is possible that the 1798 spelling accurately reflects the origin of the name, from ochre or ochre-coloured earth found here. Cf. Hockerills Farm, Worcestershire. It is unclear whether *Ochull,* recorded in 1271 (SHC V (i) 143), refers to this place. *Ockerhill in the parish of Meare* is recorded in an undated document: SRO D1229/1/3/33. See also Hockerhill.

**ODDO HALL** on the north-west side of Ipstones (SK 0150). *Odda Hall* 1837 O.S., *Oddo Hall* 1890 O.S. The age of this name is uncertain, but if ancient (which is unlikely) possibly from the personal name Odda, held to be a hypocoristic form of names beginning Ord- (see Insley 1999: 4-5), perhaps betraying a Norse origin (see Björkman 1910: 99-100; and note Oda who held Aston-by-Stone in 1066 (DB): VCH IV 40).

**ODYNSMEDUE** (unlocated) a field-name in Hammerwich. *Odynsmedue* 1360 SHC 1921 38. The likelihood that the name is from the Norse god Oðinn, equivalent to the Anglo-Saxon god Woden (and the proximity of Hammerwich to the other Staffordshire place-names incorporating the names of pagan gods, and to Lichfield, may be noteworthy), is improbable, but the name could be from an Anglo-Scandinavian byname (see Tenvik 1938: 32), or possibly from Welsh *odyn* 'kiln' surviving as a loan word: see Onn. *Odyn's Fee* in Penmark, Dinas Powys, 'would appear to be the personal name Odin, Odyn, from a diminutive form Odinel, of an original Odo, Otho (Otto), OGer. Audo...introduced into England by the Normans. Well-evidenced in DB': Pierce 1968: 189. Or perhaps from the OE personal name Odwine. The second element is ME *medewe* 'meadow'.

**OFFANDYKE** (unlocated, in Trentham, perhaps near Strongford or Kingswood Bank) *Offandyke* c.1909 SHC XII NS 74. 'Offa's dyke', presumably attached to some ancient linear earthwork here, but the date of the spelling and the original source are unrecorded. The name is also said to be recorded as *Offandyne*: StEnc 432.

**OFFLEY, BISHOPS** 3½ miles west of Eccleshall (SJ 7729). *Offeleia* 1086 DB, *Offley Cyprian* 1203 SHC 1914 81, *Offileia* c.1233 Rees 1997: 82, *Offileg' Cyprian* 1242 Fees. Offa was a common OE personal name, hence 'Offa's *leah'*. At the time of DB and afterwards it was held by the bishop of Lichfield. *Cyprian* is from Sir Cyprian de Offley, who held the place in 1203: SHC 1914 81. Cf. High Offley; Offlow.

**OFFLEY, HIGH** Ancient Parish 6 miles north-east of Newport (SJ 7826). *Offelie* 1086 DB, *Hegheoffele* 1293 SHC VI (i) 291, *Hee Offley* 1532 SHC 4th Series 8 175, *High Ofley* 1610 Speed. 'Offa's *leah'*. The place lies on a 434' hill, hence *High*. Cf. Bishops Offley; Offlow.

**OFFLEYGROVE FARM** ½ mile south of Adbaston (SJ 7627). *Hill otherwise Hyllhowse* 1582 SHC XVII 227, *The Hill* 1679 SHC 1919 235, *Hill House* 1749 Bowen, *Hill Hall* 1833 O.S. Formerly 'the house on the hill': the place lies on a 311' hill.

**OFFLOW** 3 miles south of Lichfield, in Swinfen on the northern boundary of Shenstone parish (SK 1205), one of the meeting places which gave their name to the five adminstrative Hundreds into which Staffordshire was divided. *Offelau, Offelaw* 1086 DB, *Offelawehundredum* 1182, 1185 P, *Offelaw'* 1189 Fees, *Offelaue* 1203, *Offelawe* 1227 Ass, *Offloue* 1255 RH, *Offelowe* 1272 Ass, 1327 SR, 1330 Cl, 1402 FA, *Uffelowe* 1307 Ass. 'Offa's mound or tumulus', from OE *hlaw* 'hill, mound', but in Staffordshire almost invariably 'tumulus'. The name Offa was not uncommon, and there is no evidence to connect this place with the king of the same name. A more detailed discussion of this name will be found in the Introduction, page 48. Cf. High Offley; Bishops Offley. See also Offlow in Hamstall Ridware.

**OFFLOW** According to Shaw 1798: I 152 the name of the combined manors of Cowley and Nethertown in Hamstall Ridware. In DB Hamstall Ridware lay in the Hundreds of Pirehill and Offlow, and the part in Offlow may have become the manor of Offlow: Shaw 1798: I *154.

**OGLEY HAY** 5 miles north of Walsall (SK 0506). *Ogintune* 996 (for 994) (17th century, S.1380), *Hocintvne* 1086 DB, *Hogeley* 12th century Duig, *Hoggel'* 1256 SHC 1911 127, *Oggele* 1271 SHC V (i) 154, *Oggeleye* 1300, *Oggeley* 1431 Duig, *Hogley (Lodge)* 1775 Yates. Probably from an OE personal name Hocca, Occa, Ocga or Ogga, with the later terminal *leah* replacing the earlier *tun*. Formerly one of the Hays or Bailiwicks of Cannock Forest.

**OILS HEATH** 1 mile north of Butterton (SK 0758). *Oils Heath* 1840 O.S. Possibly from the family named Oilli, Oyley, Oilly, Oyly recorded in the 12th and 13th century: SHC XVII 240, 252; SHC 1923 259; Sleigh 1883 56. The surname is probably from one of the five Ouillys in Calvados: Ouilly-le-Basset, Ouilly-le-Vicomte, Ouilly-du-Houlley, Ouilly-la-Ribaude and Ouilly-le-Tesson. Or perhaps from the name Hoyle (a south Yorkshire dialectical pronunciation of Hole), with the typical Staffordshire loss of the initial H. See also Doyle; Stokedoily.

**OKEMERE** (unlocated) *the water of Okemere* 1275 Banco. From OE *ac-mere* 'the pool of the oak'.

**OKEOVER** 2 miles north-west of Ashbourne (SK 1647; SMR 00448 places a deserted settlement at SK 15804810). *Acofre* 1002x1004 Whitelock 1930: 48, 157; Sawyer 1979: 51, 55, *Acovre* 1086 DB, *Acoure* 1130 SHC I 3, *Hachoure* 1150x1159 SRO D603/A/Add/21, *Akoura* 1188x97 SHC 1937 20, *Acovera* 1197x1213 ibid.' 28, *Hocovre* 1241 SHC VII NS 16, *Acovre* 1257 SHC XII NS 274, *Acorve* 1269 SHC VII NS 18, *Acove* 1279 SHC 1911 175, *Ockeovre* 1335 SHC XI 140, *Oker* 1507 SHC VII NS 60, 1532 SHC 4th Series 8 12. 'The slope or ridge where oaks grew', from OE *ac* 'oak', with OE *ofer* 'flat-topped ridge with a convex shoulder': the place lies at the mouth of a steep valley to the south of a flat-topped headland with convex shoulder on the west bank of the River Dove. For 16th century field names in Okeover see SHC VII NS 63-6, 73. Waste in Okeover called *Okeover Smethe* (probably from OE *mæþ* 'mowing', so 'Okeover's Meadow': see also Mayfield) is recorded in 1324 (Okeover T16), and *Okovermethende* in 1475 (Okeover T277): see also Mayfield.

OLDALL (unlocated, in Consall) See Wolfdale.

**OLDBURY** (obsolete) 1 mile north-west of Claverley (SO 7893). *Woldebery* (p) 1524 SRS 3 32, *Oldbure* (p) 1524 ibid. 39, *Woldbere* (p) 1524 ibid. 99, *Oldbury* 1697 Claverley ParReg, 1840 TA. From OE *ald-burh* 'old fortification', perhaps referring to some ancient earthworks here: see also Wall Hill. Some or all of the early spellings might relate to Oldbury in Shropshire, 1 mile south-west of Bridgnorth (SO 7191). *Oldbury Moor* in Gornal Wood is recorded in the 18th century (Hackwood 1898: 103); and field-names *Oldbury* are recorded c.1840 south of Hilton near Worfield (SO 7895): TA.

**OLDCOTE** (obsolete) in Kidsgrove (SJ 8553). *Olecote* 1273 SHC VI (i) 59, *Oulecote* 1327 SHC VII (i) 206, 1353 JNSFC LIX 1924-5 49, *Oulekot* c.1360 ibid. 55, *Oldcoote* 1567 SHC 1938 131, *Olcott* c.1575 SHC 1912 204, *Olcot* 1637 Wolstanton ParReg, *Oldcott* 1650 SRO D1229/1/3/26, *Ouldcott* 1679 SHC XII NS 34, *Oldcott* 1679 ibid. 36, *Oldcote* 1836 O.S. Perhaps 'owl cottage', from OE *ule cote*, but that would leave the later *-d-* unexplained, unless from a supposed 'old cot'.

**OLDEFORDE** – see **HALFCOT**.

**OLDENHILL FARM** (unlocated) in Clent. *Hodenhull* 1237 Ipm. Duignan 1905: 120 gives untraced forms including *Aldenhulle* and *Oldenhull*, which suggest 'Ealda's hill'. In Staffordshire from the early 13th century until 1844, when it became part of Worcestershire.

**OLDERSHAWS** 1 mile south-west of High Offley (SJ 7725). *Hildershawes* 1833 O.S., *Oldershire* 1851 White. Early spellings have not been traced, but perhaps from OE *alor scaga* 'alder copse'. As noted by Ekwall 1960: 349, OE *alor* is *owler* in many dialects.

OLD FALLINGS – see **FALLINGS PARK**.

**OLDFIELD** 1 mile north-east of Cheddleton (SJ 9952). *le olde fylde* 1561 AD 6, *Old Fields* 1810 *EnclA*, *Old Field* 1836 O.S. The word *old* is ambiguous, and could mean 'disused' or 'in use for a long time'. The precise meaning here in uncertain.

OLDFORD (unlocated, in Stapenhill) *Le Oldeforde* 1342 SHC 4th Series VI 91, SHC 1913 90. Probably 'old' in the sense 'former, disused'. Transferred from Derbyshire in 1894.

**OLDFURNACE** 2 miles east of Cheadle (SK 0443). *Old Furnace* 1693 SHC 1947 64. There were ironworks here from at least the 17th century: VCH II 116.

**OLD HAG** 1½ miles north-east of Heaton (SJ 9763). *old Hagg* 1663 Leek ParReg, *Ould Hagg* 1697 Leek ParReg, *Old Hag* 1842 O.S. Probably from OE *\*hagga* 'hawthorn', or possibly dialect *hag* 'coppice'.

**OLDHAY TOP** 1 mile north-west of Meerbrook (SJ 9762). *Oldhay* (p) 1332 SR, *Ould Hay* 1698 Leek ParReg, *Old Hay* 1842 O.S. From Mercian OE *(ge)heg* 'enclosure', so 'the top or high part of the old enclosure'.

**OLD HILL** in Rowley Regis (SO 9586); *the Owldhill* 1556 SHC 1936 216, *Old Hill* 1727 Rowley Regis ParReg, 1834 O.S.; OLD HILL (unlocated, in Stone), *Old Hill* 1703 SRO D4913/C/1/2; OLD HILL (WOOD) (obsolete) 1 mile north-east of Audley (SJ8251), *Old Hill (Wood)* 1836 O.S. A name meaning 'old hill' is not readily explicable, so (if ancient) possibly from OE *olde* 'a steep slope' (see Ekwall 1936: 144), or even in some cases from Celtic *\*alt* 'hill', so 'the hill called (by the Angles) Olde, or (by the Britons and afterwards by the Angles) *\*Alt*': see Coates & Breeze 2000: 229-231. The hill from which Old Hill in Rowley Regis takes its name may be the higher ground rising to 431' on the east side of the place, or *Reddall Hill* (1834) on the south side.

**OLDINGTON** 2 miles north-west of Worfield (SO 7397). *Holdington* 1238x1250 Eyton 1854-60: III 112, *Oldinton* 1301 SHC 1939 187, *Oldynton* 1525 SR, *Oldington* 1564 Worfield ParReg, 1651 SRO D4092/C/1/39. In Shropshire since the 12th century. The *-ing-* suggests 'the *tun* of the family or followers of a man called Alda'. See also Olton.

**OLD PEEL FARM** 1 mile south-west of Audley (SJ 7850). *Peele* 1575 Audley ParReg, *Peele apud Betteley* 1587 ibid, *Peele* 1587 Betley ParReg, *the Peelhouse* 1612 SHC 1944 82, *the Peel* 1733 ibid. 4, *Old Peel (& New Peel)* 1799 Faden. A curious name, possibly from ME *pel(e)* 'a pallisaded and moated enclosure', although that use is not recorded in OED until 1596. Peel has also been proposed as a variant of the not uncommon Mockbeggar (q.v.), *pel* being an abbreviated form of *repel*, or from peel in the sense 'rob': see PN Sa III 216.

**OLDRIDGE** 1½ miles south-east of Ipstones (SK 0448). *Olderuge* 1327 SHC VII (i) 217, 1333 (p) SR, *Old(e)rich(e)*, *Old(e)ryche* (p) 1331 Ipm, 1442 AD 5, *Olderych* 1419 SHC XVII 37, *Old(e)rech* (p) 1438 AD 5, *? the Ouldrign* 1693 Kingsley ParReg, *Old Ridge* 1775 Yates. Perhaps from OE *eald* 'old' with OE *hrycg* 'ridge', giving 'the old ridge', though the meaning of such name is unclear, or possibly from Celtic *\*alt* 'hill' so 'the ridge on the hill called Alt': see Coates & Breeze 2000: 229-231.

**OLD ROAD** – see **HANFORD**.

**OLD SPRINGS** ½ mile south-east of Swythamley Hall (SJ 9764). Perhaps to be associated with *le Springe*, recorded in 1621 (SHC 1934 24). Probably 'the long-used spring of water'.

**OLD SPRINGS (FARM & HALL)** 1 mile and 1½ miles south of Almington (SJ 7032). *yolde Spring* c.1570 SHC 1945-6 147, *woods or pastures called Old Springs* 1603 ibid. 168. The word *springs* meant a flow of water or young tree shoots. In this case the word probably has the former meaning: good springs of water are recorded here: SHC 1945-6 129.

**OLDSWINFORD** Ancient Parish 1 mile south of Amblecote (SO 9083). *Swinford* 951x955 (15th/16th century, S.579), *Swineford* 1086 DB, *Swyneford*, *Swineford* 1235 Fees, 1291 Tax, *Old Swynford* 1291 Ipm, *Oldeswyneford*, *Woldiswynford* 1327 SR, 1438 Pat. From OE *swin*, *ford* 'pig or boar ford'. The 'old' distinguishes this place from Kingswinford (q.v.), which sometimes appears as New Swinford. *Wold* is a dialect form of *old*. The ecclesiastical parish is properly Old Swinford. Oldswinford was included in the Staffordshire township of Amblecote, but transferred to Worcestershire in the 19th century: see Youngs 1991: 484-5.

**OLIVE GREEN** 1 mile south-east of Hamstall Ridware (SK 1118). *Gallows Green* 1741/2 SRO D789, *Olive Green* 1775 Yates, *Gallows green* 1798 Shaw I 152, *Olive Green* 1801 Smith, *Gallows Green* 1806 Hamstall Ridware ParReg, 1834 O.S. The place, which appears to have had alternative names, was presumably the site of the manorial gallows. The present name was perhaps preferred by residents as less macabre. The Green element suggests a squatter settlement.

**OLIVER HILL** a 1684' hill, the highest point in Staffordshire, to the north of Flash (SK 0267). *Hill Top* 1798 Yates. Early spellings have not been traced, and the age of the name is unknown, but presumably a family name (for the Oliver/Olliver families in Alstonefield/Fawfield Head in 1666 see SHC 1925 233-6), perhaps from Thomas Oliver, agent of the Harpur Crewe estate in the early 19th century (StSt 8 1996 72-3); Ralph Oliver, farmer and carrier of Flash, mentioned in 1872 (P.O. 696); George Oliver of Heathylee recorded in 1880 (Kelly); or James Oliver, recorded in a Quarnford Rental c.1870, and presumably the donor of a reredos to Flash church c.1901 (VCH VII 55).

**OLIVER'S GREEN** on the west side of Denstone (SK 0940). Presumably to be associated with *Cromwell's Green* recorded by Redfern 1886: 75, a name attached to a double-ditched earthwork, from a supposed but improbable association with Oliver Cromwell. The Green element suggests a squatter settlement.

**OLMETON** (unlocated) *Olmeton* 1234x1240 TestNev.

**OLRENSCHAWE** (unlocated, probably near Uttoxeter or Marchington, or perhaps Ollerenshaw (Hall) in Chapel-en-le-Frith, Derbyshire: see Tooth 2000b: 176) *Olrenschawe* 1306 SHC 1911 67. From the OE adjective *\*alren* 'growing with alders', with OE *sceaga* 'a small wood, a copse'. The place, wherever it lay, may have given rise to the Staffordshire surname Olrenshaw. A meadow called *Holreshawe* is recorded in or near Bagot's Bromley in 1324 (SRO D(W)1721/3/21/32), as *Ollereshawe* in 1338 (SRO D(W)1721/3/7/12), and as *Oldershawe* in 1434 (SRO D(W)1721/3/32A/4.

**OLTON** (obsolete, a settlement which existed 1 mile west of Pattingham until the creation of Patshull Park in the 1740s: VCH XX 162 (SO 7999)) *Oldinton* 1294 SHC VII 10, *Oldynton'* 1333 SHC X (i) 130, *Olton* 1336 VCH XX 167, 1374 ibid. 162 fn.34, *Oldington* 1401 (1801) Shaw II 282, *Oldyngton* 1435 SHC III NS 133, *Woldyngton* 1538 SHC XI 276, *Oldington* 1603 VCH XX 167, *Oulton* 1695 Morden, *Olton* early 19th century VCH XX 162. The *-ing-* in some of the forms suggests 'the *tun* of the family or followers of a man called Alda'. See also Oldington; Oulton.

**OLYNGH, OTYNGH** (unlocated, a park in the manor of Sedgley) *Olyngh, Otyngh* 1291 SHC 1911 202. Perhaps to be associated with Ettingsall Park (q.v.), or with Ettymore (Road) on the west side of Bull Ring.

**ONECOTE** (pronounced Uncut [ʌnkʌt]) 5 miles east of Leek (SK 0455), *Anecote* 1199 FF, *Anecot* 1204 SHC III 104, *Onecot* (p) c.1266 StCart, *Onechot* (p) c.1270 Loxdale, *Hunicote* 1271 SHC VI (i) 51, *Onecote* 1272 Ass, *Honecota* (p) c.1275 Loxdale, *Onecoat* 1298 SHC XI NS 257, *Onyecot* 1306 SHC VII 171, *Ounecote* 1325 SHC 1911 366, *Uncote* 1413 SHC XVII 7, *Oncte* 1553 SHC 1926 15, *Oncott near Whittle* 1599 SHC XVIII 16, *Angcotes* 1631 D476/75-77, *Onkot* 1703 Alstonefield ParReg; **ONCOTE** 1 mile west of Great Bridgeford (SJ 8626), *Onecote* 1381 SHC XIII 166, *Uncote* 1596 SHC XVI 159, *Oncott Hall* c.1727 D615/M/4/15, *Oncote* 1836 O.S., *Oncote (Coppice)* 1995 O.S. From OE *an* 'one, single', with OE *cote* 'cottage', giving 'the lonely cottage', or perhaps from the OE personal name Anna or Onna: cf. Onneley. *Whittle* mentioned in the 1599 spelling is Whitle (Upper and Under) (q.v.). See also Coudry.

**ONECOTE GRANGE** on north-west side of Onecote (SK 0455). *Anecote grange* 1227 Harl, *Onecote Grange* 1539 CtAugm, 1837 O.S., *Oncott Grange* 1657 PCC. 'The grange at Onecote'. The grange belonged to Croxden abbey: VCH VII 213.

**ONINLEG** (unlocated, in Drointon) *Oninleg'* c.1315 SRO D938/555.

**ONNELEY** 1½ miles north-east of Woore (SJ 7543). *Anelege* 1086 DB (entered in Shropshire), *Oneleia* 1185 TpR, *Onilegh* 1211 Cur, *Onileg* 1212 SHC III 153, *Onyleye* 1293 SHC VI (i) 228, 1323 SHC IX (i) 93, *Onelay, Onaylay* 1381 SHC VIII NS 359-61, *Anneley* 1577 Saxton. 'Anna's or Onna's *leah*', or, more likely since the spellings have only one *n*, from OE *ana leah* 'the single or lonely *leah*'. Cf. Onecote; Olney, Northamptonshire.

**ONN, HIGH & LITTLE** in Church Eaton parish, 6 and 7 miles south-east of Newport (SJ 8216 & SJ 8415; SMR 01634 places a deserted Anglo-Saxon settlement at SJ 82611620, and SMR 04622 another at SJ 83981601). *Othnam* 1081 SHC 1914 104, *Otne* (High Onn), *Anne* (Little Onn) 1086 DB, *Onna, Othna* c.1130 Ordericus, *Onne* 1221 FF, *Onna* 1230 Cl, *Oyme* 1233 VCH IV 95, 1293 SHC VI (i) 289, *Olne, Great Onne, Little Onne* 1253 SHC IV 128, *Honne* 1260 ibid. 145, *Parva Honne* 1272 SHC IV 191, *Little onne* 1293 SHC 1911 49, *Oten Edisch* 1343 SHC 1913 106, *Hyonne* 1430 SHC XVII 127, *Great Own* 1532 SHC 4th Series 8 96, *Hyghon* 1577 Plot, *Alta Onne otherwise Highe Onne* 1592 SHC XVI 131, *High On, Little On* 1682 Browne. Although a pre-Celtic word *onna*, *\*onna* 'watercourse, source' has been recorded (Rivet & Smith 1979: 431-2), and both places lie at the head of minor streams which join to the east and flow into Church Eaton Brook, the spellings appear to support a derivation from the plural

MWelsh *odyneu*, ModWelsh *odynau*, probably borrowed from PrWelsh *\*Odenou* after the loss of the British final endings, later with the loss of the *e* through syncope, and weakening of the final *-ou* to OE *-a* or *-e*: see TSAHS 1995-6 XXXVII 139; Coates & Breeze 2000: 197-8. The early Welsh laws mention the rights and obligations relating to kilns, and even distinguish between those with flues and those without. These places – which lie very close to and either side of a lost Roman road from Pennocrucium to Chester (Margary number 19), which may have been in use when the kilns were constructed – evidently borrowed their name from Brittonic in the late 6th century, when the area was first occupied by the English. The plural forms show that there was more than one kiln at these two places, but whether only one at each place, or several which were divided into two estates is not known. Cf. *Odencolc*, recorded in 846 near Thurlstone, Devon, from OCornish *oden* 'lime-kiln'. The second element of the 1343 form is from OE *edisc* 'an enclosure, an enclosed park'. The 1081 spelling is from a charter of William I (from Monasticon II 966), in Erdeswick 1844: 31. See also Odynsmedue and Oulton House Farm for other possible examples of the element in Staffordshire.

ORBETON, HERBETON (unlocated) in Hopton. *Orbreton* 1161x1182 SRO (1/7972), *Orbeton* pre-1182 SHC 1909 145, *Horbrihtona, Horbuhtona* ?c.1180 SRO D938/392, *Erburtun, Orbeton* c.1200 SHC VIII (i) 176-7, *Herbertum* 1203 SHC III 95, *Orbrichton* c.1240 SHC 1913 307, *Herbreton* 1261 SHC IV 148, *Orreberton* c.1350 SRO (10/7962), *Eburton* 1351 SHC 1913 146, *Arborton* 1581 SHC XVII 227, *Herbeton* 1778 D240/E(A)2/222. Described in 1679 as 'Anciently a small village, long since depopulated": SHC 1909 145; 'between St Thomas' Priory and Tixall, not far from Kingston Cover. Now absorbed into Coton': SHC VIII I 177; see also SHC 1913 222-3. The inconsistent forms are confusing, but perhaps from an OE personal name Ordbriht or similar, with OE *tun*.

ORCHARD FARM the northernmost habitation in Staffordshire, 1 mile north of Flash (SK 0168). *Orchard Farm* 1737 VCH VII 52, *Orchards* 1775 Yates, *The Orchard* 1842 O.S. Presumably from fruit trees which grew here, or (since this is high ground where fruit trees might not normally be expected) perhaps from the surname Orchard, recorded in the area in the early 17th century: SHC XVIII 56. The place is associated with coal mining: see StSt 9 1996 71.

ORCHARDS, THE on the south side of Okeover (SK 1547). *the Orchards* 1776 Okeover E26-7, *Orchard* 1798 Yates, *Orchards* 1836 O.S.; ORCHARD, THE (unlocated, in Alstonefield), *the Orchard* 1563 SHC 1931 125. The first place may be associated with *an Orchard in Oker* recorded in 1508: SHC VII NS 64. Presumably from fruit trees, but see also Orchard Farm.

ORDLEY BROOK a tributary of the Tit Brook, which flows into the river Dove.

ORDSEY (obsolete, on Ford Brook in Pelsall (SK 0203)) *Ordeiseie* 996 (for 994) (17th century, S.1380), *Ordesee* c.1175x1208 SRO 3005/1, *Ordescia* 1247 SHC 1911 407. 'Ordheah's island': see Hooke 1983: 76.

ORGREAVE 1½ miles north-west of Alrewas (SK 1416). *Ordgrave* 1195 f. P, *Orgrave* 1203 SHC III 104, *Oregrave* 1260 SHC X NS (i) 276, *Oregraue* 1269 SHC 1910 129, *Ordgrave, Ordegrave* 13th century Duig, *Orgrave* 1682 Browne. The second element is held by Ekwall 1960: 350 to be OE *græfe* 'grove', with an uncertain first element. *Græfe* is difficult to distinguish from OE *graf* 'a trench, a ditch'. The first element may be an unrecorded OE personal name Orda, or be from OE *ord* 'point, sword', which may have had a topographical meaning such as 'shaped like a point'. Gelling & Cole 2000: 228 suggests that the name may incorporate *ord* with reference to the shape of the wood or to an adjacent topographical feature. A derivation from OE *ora græf* 'ore pit' is proposed for Orgrave, Lancashire, and Orgreave, West Yorkshire (see Ekwall 1960: 350), but while there is a history of coal mining here, that etymology would not account for the *-d-*. The place was in Alrewas Hay in Cannock Forest, near the Trent. Cf. Orgreave, Yorkshire West; Orgrave, Lancashire.

**ORSLOW** in Church Eaton parish, 5 miles south-east of Newport (SJ 8015). *Horslage* 1195 P, *Horselawe* 1203 Fine, 1208 Ass, c.1215 Rees 1997: 62, *Hors(e)lowe* 1208 FF, 1285 Ch, 1298 to 1468 Ipm, *Orselawe* 1242 Fees, *Horselegh* 1294 SHC 1911 227, *Horselowe* 1298 ibid.' 236, *Orsolooue* c.1301 Bod. 31, *Orselowe* 14th century Duig, *Horslow* 1532 SHC 4th Series 8 96, *Orslow* 1682 Browne. Perhaps 'Horsa's burial mound', or 'the horses' mound', from OE *horsa hlaw* 'low, tumulus'. CDEPN suggests that the name incorporates the rare OE element *lagu* 'sea, flood, water', but the term is found only in OE poetry, especially of 'the ocean', although ME *lawe* is sometimes found with the meaning 'lake, pool'. There is no lake or pool on modern maps: the place lies on Orslow Brook.

**ORTON** 4½ miles south-west of Wolverhampton (SO 8695). *Overtone* 1086 DB, *Overton* c.1195 SHC III (i) 219, *Orton* frequently 13th century Duig, *Ouerton* 1388 SA 2089/2/2/8, *Monte de Huverton* 1391 SHC XV 36, *Overton otherwise Orton* 1597 SHC XVI 163. Probably from OE *ofer-tun* 'the upper settlement' (Orton Hill reaches a height of 535'), though it is not possible to distinguish OE *uferra tun* 'ridge setlement' and *ofer tun*. CDEPN 466 suggests that Orton was an earlier name for Upper Penn.

**OSCOTT** 4 miles south-west of Aldridge (SO 0794), *Oscote* 1297 SHC VII 43, 1344 Ch, *Oscott* 13th century Duig, 1532 SHC 4th Series 8 74, *Oscote* 1421 BCA MS3145/37/11, c.1566 SHC 1938 57, *Awscote* 1587 SHC 1929 203, *Oscote* 1617 SHC VI NS (i) 24, *Oscott* 1680 SHC 1919 268, *Auscot* 1686 Plot, 1747 Bowen; OSCOTE (unlocated, in Church Eaton), *Auscott, Auskott* 1597 QSR. The terminal is clearly OE *cot* 'cot, cottage, shelter', and the qualifier may be from one of many OE personal names beginning with Os-, such as Osa, Oswald, Osbeorn, Osmod, Oswulf, etc. The first place was originally at what is now known as Kingstanding, and the name has been transferred to the area surrounding Oscott Roman Catholic College 3 miles to the east, known as New Oscott since it was built in 1838.

OSFIELD (unlocated, possibly near Mayfield) *Osfield* 1656 Leek ParReg, *Osfuld* 1657 ibid.'

**OSSOMS HILL** a 1093' hill 1 mile west of Wetton (SK 0955). *Ossomshill* 1836 O.S. Early forms have not been traced, but if ancient possibly from the OE personal name Osmund, which is found in the parishes of Osmaston by Derby and Osmaston by Ashbourne, Derbyshire (see PN Db 595), the latter 10 miles to the south-east of Ossoms Hill, or from the ON personal name Ásmundr. A cave in the hill here is known as Ossoms's Cave, but early spellings have not been traced.

**OTHERTON** 1 mile south of Penkridge (SJ 9212; SMR 01916 places a deserted Anglo-Saxon settlement at SJ 92591220). *Orretone* 1086 DB, *Oðerton* 1166 SHC 1923 295, *Otherton* 1242 Fees, ? *Oderstone* 1421 SHC XVII 77. The forms suggest a derivation from OE *oðer* 'other', giving 'the other, or second, *tun*' (cf. Othery, Somerset).

OTTIWELL (unlocated, said to be between Audley and Heighley Castle: preface to printed Betley ParReg). *Ottiwell* 1619 Betley ParReg. It is said that this name is a corruption of a place called *the Devil's Well*, a sacred well dedicated to St Ottilia, whose French name Odille became corrupted by popular etymology into Old De'il: preface to printed Betley ParReg; see also StEnc 178.

**OULIES, GREAT** (unlocated, probably in Uttoxeter) *the great Owleys* 1651 SRO D786/7, *the Owleys* 1684 SRO D7/3, *Great Oulies* 1735 SRO D786/7/4. Perhaps from OE *ule lǣs* 'the pasture or meadow-land with the owl'.

**OULTON** 1 mile north of Stone (SJ 9135), *Oldington(e)* (p) 1268 Coram, 1370 FF, *Oldeton* 1280 SHC VI (i) 99, *Oldington* 13th century Duig, *Aldeton* c.1346 SRO D938/10, *Oldyngton* 1364 Misc, *Oulton* 1682 Browne; **OULTON, UPPER & LOWER** 1 mile south of Norbury (SJ 7822), *Oldington(e)* (p) 1268 Coram, 1371 FF, *Alta Odynton, Parva Oldynton* 1376 SRO D1717/A/1/1, *Oldyngton* 1364 Misc, *O(u)lton* 1413 Deed (*et freq.*), *Nether Oldyngton, Overoldyngton* 1405 SHC XVI 45, *Wolton* 1532 SHC 4th Series 8 170, *Netherolton* 1608 SHC 1948-9 112. The *-ing* in the various forms suggests 'the *tun* of the family or followers of a man called Alda', rather than *ald tun* 'the old *tun*'. Many of the forms are too imprecise in context to identify which Oulton is meant. *Nether* is from OE *neopera* 'lower'. See also Olton.

**OULTON** 1 mile west of Rushton Spencer (SJ 9262). *Oulton* 1651 SRO DW1761/A/4/150, *Holton* 1798 Yates, *Oulton (Wood)* 1842 O.S., *The Oultons* 1849 SRO D5003/2/3/1. Without earlier spellings it is not possible to put forward a likely derivation.

**OULTON (HOUSE FARM)** ½ mile north-west of Milwich (SJ 9633). *Oten* 1568 SHC 1931 211, *Oulton House* 1836 O.S. Probably 'the old *tun*', but the absence of early spellings makes the derivation uncertain: see Oulton. Whilst the early spelling is probably a local phonetic version of Oulton, it is not impossible that the name derives from Welsh *odyn* 'kiln', from an older PrWelsh *\*otn*. *Odyn* would have been in OWelsh *odin*, with the spelling *\*otin* (cf. OCornish *oden*). The name may have developed into Oulton from knowledge of other places of this more common name. The 1568 spelling may refer to the moated site a short distance north of the present house. See also Odynsmedue; Onn.

**OUNSDALE** on north side of Wombourne (SO 8693). The name would appear to derive from one of the open fields of Wombourn, *Holendenesfelde* c.1314 SHC 1928 22, *Holundenesfeld* 1316 ibid. 26, *Holendenefeld* 1347 ibid. 37, *Hounden field* 1483 VCH XX 209, *Houndel* 1767 SRO D1368/1, *Ounsdale* 1834 O.S., *Houndale (Leasow)* 1840 T.A. The name is from OE *holen denu* 'valley with the hollies', with OE *feld* 'open country', later 'field'. The older name Houndel survives in Houndel Bridge on the Staffordshire & Worcestershire canal at Ounsdale.

**OUSE, THE** (unlocated, in the Uttoxeter/Dove Bridge area) *the Ouse* 1678 SRO DW1733/A//2/56. From OE *wase* 'a muddy place, a marsh, a mire'.

**OUSLEY BROOK** a tributary of the Rangemoor Brook which runs into the river Dove. Possibly from OE *wase* 'a muddy place, a marsh', or OE *osle*, ME *osel* 'blackbird', with OE *leah*. Evidently associated with *Ousley Wood* (1836 O.S.), 1 mile north-east of Ellastone (SK 1244), and Ousley Cross.

**OUSLEY CROSS** 1½ miles north-east of Ellastone (SK 1244). *Oseley Cross* 1629 Ellastone ParReg, *Ouseley Crosse* 1682 ibid, *Ousley Cross* 1836 O.S., *Ouzley Cross* late 19th century SRO D240/E(A)2/187. Possibly from OE *wase* 'a muddy place, a marsh', or OE *osle*, ME *osel* 'blackbird', with OE *leah*, or perhaps a surname: see SRO D644/3/1;

StEnc 440. There is an ancient socketed stone cross base here, possibly from elsewhere: see NSJFS 12 1972 123. See also Ousley Brook.

**OUTCLOUGH** (obsolete) 1½ miles north-west of Norton-in-the-Moors (SJ 8853). *The outclough* 1645/6 SRO QSR A1645/6ff12-3, *The Outclough* c.1715 SRO D5240/1/8-13, *Outclough* 1836 O.S. From OE *ut, ute* 'outside, on the outskirts', used to describe '(a place) lying on the outskirts or futher away from something', but usually as an affix with an adjectival function meaning 'outer, more distant', with OE *\*cloh*, ME *clough* 'asteep sided valley', so 'the further valley or ravine'.

**OUTLANDS** 2 miles north of Adbaston (SJ 7730). *Outlands* 1672 Eccleshall ParReg. 'The land outside the settlement'.

**OUTWOODS** 2½ miles east of Newport (SJ 7818), *Outwoods* 1666 SRO DW1736/iv/1, *the Outwood(s)* 1674 Gnosall ParReg, *Outwoods Common* 1797 *EnclA*, *Outwoods* 1834 O.S.; **OUTWOODS** 1 mile east of Anslow (SK 2225), *Outewoode* 1499 StEnc 440, *owtwoods of Burton* 1560 SRO D603/E/1/7, *the outwood* 1619 HLS 537, *Outwoods* 1709 SRO D603/L71, *Outwoods (Common)* 1771 SRO M/7/21. From OE *ut, ute* 'outside, on the outskirts', used to describe '(a place) lying on the outskirts or futher away from something', but usually as an affix with an adjectival function meaning 'outer, more distant'. The first place is close to the Staffordshire-Shropshire border. *Horninglowe outwoodes* is recorded in 1619 (SRO DW1734/1/4/166), *Hopton Outwood* in 1548 (SHC 1912 169), and *Outwood Gate*, on Biddulph Moor (SJ 9058), in 1842 (O.S.).

**OVEREND** (unlocated, in Handsworth) *Overend* c.1564 SHC 1938 68. Possibly to be associated with *Weuerend*, recorded c.1565: SHC 1938 8.

**OVERLEY** ½ mile north-west of Alrewas (SK 1515). *Overley* 1834 O.S.

**OVERTON** 1 mile east of Checkley (SK 0438), *Overton* 1236 SHC XII 178, c.1559 SHC 1925 139, 1580 SHC XV 129, *Hoverton* 1836 O.S.; **OVERTON** an alternative name for Upper Biddulph (SJ 8961): see Biddulph, *Overton* 1333 SHC X 94, *Ovurton*, *Overton* 1343 SHC XII 25, *Overton* c.1535 SHC 1912 75, 1662 Biddulph ParReg; **OVERTON** on the west side of Hammerwich (SK 0607), *Overton* 13th century VCH XIV 259, 1327 SHC VII (i) 232. From OE *ofer tun* 'ridge setlement' or OE *uferra tun* 'higher settlement' – it is not possible to distinguish the two.

**OXBURY** on south side of Lichfield (SK 1208), *Oxenbury* 1391, 1444 SHC VI (ii) 187. See Borrowcop Hill.

**OXENFORD** (obsolete, said to be the old name of Chapel Ash Farm, Wolverhampton (SO 9098): WA I 128) *Oxenevord'* c.1270 WA I 278, *Oxeford* 1286 ibid.' 279, *Oxneford* 1321 ibid, 1327 SHC VII (i) 249, *Oxeneford* 1337 ibid, *Oxenefordesbruche* 1338 ibid, *Oxenford Burch* 1342 ibid, *Newbridge* [sic] 1701 ibid.' 281. 'The ford of the oxen'. *Burch*, *-bruche*, and *-bridge* are from OE *brece* 'cleared or newly cultivated ground'. By metathesis, or shifting of the *r*, the words become *burche* and later *birch*. See also WA I 278-281; Mander & Tildesley 1960: 31; 42.

**OXFORD** 1½ miles north-east of Norton-in-the-Moors (SJ 8753). *New Oxford* 1836 O.S. Perhaps to be associated with *Oxneford* (unlocated, but possibly near Bucknall), recorded in 1327 (SHC VII (i) 198), and *Oxunford*, recorded in 1332 (Tax), which may have been the 'old' Oxford.

**OX HAY** 1 mile west of Meerbrook (SJ 9760), *Oxehay* 16th century CtAugm, 1725 Okeover, 1842 O.S.; OXHEY (unlocated) *Oxhey* 1461 HAME 485. From Mercian OE *(ge)heg* 'fence, enclosure', so 'the enclosure with the ox'.

**OXLEY** in Bushbury parish, 1½ miles north of Wolverhampton (SJ 9002). *Oxelie* 1086 DB, *Oxelea* 1228 SHC IV 62, *Oxeleg* 1236 Fees, *Oxele* 1242 Fees, 1271 SHC V (i) 154, 1279 SHC 1911 175, *Oxleg* 1262 SHC V (i) 137, *Oxeley* 13th century Duig, *Ox Ley* 1775 Yates. From OE *oxa* 'ox', and OE *leah*.

**PACKINGTON** in Weeford parish, 3 miles north-west of Tamworth (SK 1606; SMR 02536 places a deserted Anglo-Saxon settlement at SK 15990630). *Padintone, Pagintone* 1086 DB, *Pakinton, Pachinton* 1166 SHC 1923 295, *Pakintona* 1177 SHC XII NS 278, *Pakinton'* 1242 Fees, *Packington* 1296 (1801) Shaw II 26, *Packynton* 1335 ibid, *Packington* 1682 Browne. In the absence of a common noun which could account for the first element of this name, it must be assumed to be from an unrecorded OE personal name Pac(c)a, so giving '*Pac(c)a's tun*' or 'the *tun* of *Pac(c)a's* people': see Ekwall 1960: 356. Cf. Packington, Leicestershire.

**PADBURY (LANE)** 1 mile south-west of Chorley (SK 0610). *Padebury temp*. Henry VIII SHC VI (ii) 166, *Padburies* 1561 HLS 582, *the little Padburies and the great Padburies* 1608 SHC 1934 (ii) 43. Perhaps from OE *\*padde* 'toad', with OE *byrig*, dative of *burh* 'fortification, manor house', so possibly 'fortification with the toads'. Perhaps to be associated with Padwall (q.v.).

**PADWALLE** (unlocated, possibly near Longdon) *Padwalle* 1481 OSS 1936 42. Perhaps from OE *\*padde* 'toad', with Mercian OE *wælle* 'spring', and (sometimes) 'a stream', so 'toad spring'. Perhaps to be associated with Padbury (q.v.).

**PADWICK** ½ mile north-east of Sharpcliffe (SK 0152). *Parnwic* c.1245 SHC 1911 439, *Padewick* 1275 SHC V (i) 118, *Padewyk* 1292 SHC 1911 216, 1304 ibid. 433. Possibly 'the *wic* notable for the surfeit of toads', from OE *\*padde* 'toad'.

**PAGET'S BROMLEY** – see **ABBOTS BROMLEY.**

**PAINLEY (HILL)** 3 miles west of Uttoxeter (SK 0333). *Pynlawe* 1272 SHC IV 187, *Paynelowe* 1293 SHC VI (i) 227, *Paynlowe* 1327 SHC VII (i) 222, 1332 SHC X (i) 111, *Paynolow* 1341 SHC 1921 19, *Painley* 1562 SHC IX NS 213, *Painley Hill* 1596 SHC XVI 152, *Paynley-hill* 1686 Plot, *Painley Hill* 1836 O.S. The hill evidently gave its name to Painley, shown to the east of the hill on the 1836 O.S. map. The earliest spelling suggests a derivation from OE *pinn* 'peg, pin', used topographically of narrow ridges, with OE *hlaw* 'tumulus, mound, hill', hence 'the narrow ridge with the tumulus', or possibly 'the ridged tumulus'. An alternative derivation is from the OE personal name Pægna (found in Bede as Pægnalaech): cf. Paignton, Dorset, or from Pain, a French personal name, identical with OFr *paien* 'heathen' (from Latin *paganus*), originally 'villager, rustic', later 'heathen'. If from a personal name, that same name may be found in Paynsley (q.v.).

**PALFREY GREEN** 1 mile south-east of Walsall (SO 0197). *Palfrey Green* 1386 VCH XVII 157, *Palfry-Green-Leasow* 1520 (1798) Shaw I 80, *Palfrey's Field, Palfrey's Green* 1528 SRO D593/A/2/20/35, *Palfrey's Field lying next to Palfrey's Green* 1540 SRO D593/A/2/20/36, *Palfraye Green* 16th century, *Palfrey Green* 17th century Duig. From ME (from OFr) *palfrey* 'riding horse'. The Green element suggests a squatter settlement.

**PALMERS CROSS** 2 miles north-west of Wolverhampton (SJ 8801). *? Cros* 1359 WA I 268, *Palmers Cross Corner* 1613 map of Tettenhall Hay PRO, VCH XX 12 fn, *Palmers Cross* 1788 Codsall ParReg, *Palmer's Cross* 1801 Shaw II 202, 1834 O.S. The OED defines a palmer as 'a pilgrim who has returned from the Holy Land, in token of which he carried a palm-branch or palm leaf; also, an itinerant monk under a perpetual vow of poverty; often simply a pilgrim'. It seems possible that the name is from a pilgrim cross which stood here. The surname le Palmer is not uncommon in Staffordshire and Shropshire: TSAS LXXIII 1998 85. For a lost *Palmerecros* in Derbyshire see PN Db I xl, and for Palmer's Cross in Derby, see PN Db 455.

**PANNIERS POOL** at Three Shires Heads (SK 0068). *Pannyer poole* 1533 Bateman 1861, *Paviner pool 'usually called Panniers'* 1654 ibid. Perhaps because the pool is supposedly shaped like a pannier, or from the panniers of packhorses resting here: a number of packhorse trails met at this point. DES 337 suggests that Panniers may be a surname from OFr *paniere*, ME *panier* 'a basket', which might give rise to 'basket-maker' or 'hawker'.

**PARADISE** 1 mile south-east of Coven (SJ 9206). *le olde paradis* 1338 Ct, *Paradise* 1775 Yates, 1834 O.S. 'The perfect or beautiful place', but perhaps also used (though unlikely here) in a disparaging way of places quite the opposite. The reference to *olde* in 1338 suggests the possibility that the name may refer to some archaeological feature: the place lies on a Roman road (Margary number 190) which ran south-east from Pennocrucium (Water Eaton). The ModE word *parvis* 'the enclosed area in front of a building', sometimes 'the portico or colonnade in front of a church; a church porch', derives from Latin *paradisum*, and might explain some names of this type. It has also been suggested that such names might refer to the growing of Paradise seed imported from Morocco or Tripoli in the 15th century: Foxall 1980: 26; PNEF 33; 152. *Paradise* also occurs 2 miles west of Brewood, near Pearce Hay Farm (SJ 8408), but the age of the name is unknown; 1 mile south of Alton (1798 Yates, 1836 O.S.); at Lane End, Willenhall (1721 WA II 36); at Castle Church (*Paradyse* 1462 Oakden 1984: 81); at Horton (*Paradise* 1652 SRO DW1702/1/25), and in Audley (*the Paradize* 1646 SRO D916, *Parradise* 1659 SRO DW1826/29).

**PARK END** ½ mile north-west of Audley (SJ 7851; SMR 02627 places a deserted post-Conquest settlement at SJ 86055160). Evidently to be associated with the *newly erected messuage called The Parkes*, recorded 1739x1752 SRO D(W)1082/D/4/1-15, *Parkend*, *Parkland* 1733 SHC 1944 3, *Park, Park Lane* 1775 Yates, Park Lane 1833 O.S.

PARKES HALL – see PERSEHOUSE HALL.

**PARKFIELD** 1½ miles south of Wolverhampton (SO 9296). *Parkfeilds* 1659 Sedgley ParReg. 'The fields in the Park (of Ettingshall)'.

**PARKFIELDS (FARM)** on the north-east side of Cheadle (SK 0244). *the Park Field* 1698 SRO D1229/2/2/7.

**PARKHALL** 1 mile west of Cheadle (SJ 9943). *Parkehall* 1340 SHC XII 178, 1583 SHC XV 147, 1499 SHC XII 178, *Parkhall* 1347 SHC XI 162, 1411 SHC XVI 91, 1609 SHC III NS 52, *Parke Hall* 1663 SHC II (ii) 50. 'The hall at the park'. It has been suggested that Parkhall may have been another name for the manor of Cheadle: see SHC XII 178, but also SHC XVI 90-2.

**PARK HALL** on the west side of Weston Coyney (SJ 9244). *Parkhall* 1411 SHC XVI 90, *one new erected messuage or house called The Park Hall* 1589 SRO D5100/42, 1836 O.S. See also SHC XVI 90-1.

**PARK NOOK** ½ mile west of Ranton (SJ 8423). *Park Nook* 1771 SRO D641/6/1/26.

**PARK PALES** ½ mile north-west of Bishop's Wood near Brewood (SJ 8309), *Park Pales* 1836 O.S.; **PARK PALES** ½ mile east of Stockwell Heath (SK 0621). Park pales were 'a special palisade of cleft-oak stakes': Rackham 1990: 153. The first place lies on the boundary of Weston Park, the second to the south of Park Barn Farm.

**PARK SPRINGS (FARM)** 1 mile south-west of Hales (SJ 7333). *Sprinkes* (young tree shoots) are recorded in 1562: SHC 1945-6 117. The place was known as *Goatman's Hill* in 1684 (*Goatmans Hill* 1675 SRO 1/129/73): SHC 1945-6 190.

**PARKS, THE** 1 mile north-west of Uttoxeter (SK 0734). *The Parkes* 1677 SRO D786/13/5, *The Poarkes or Parks* 1686 SRO D786/13/6.

PARLES (unlocated, possibly in Handsworth, but early references to the Parles family are associated with Coleshill: BCA MS3888) *Parles* 1181 SHC I 96, 1194 SHC III (i) 28, 1208 ibid. 148, 1228 SHC 1924 251, c.1238 (1798) Shaw I xxvi, 1279 SHC 1911 173.

PARON'S WOOD (unlocated, in Wolseley Wood) *Paron's Wood* c.1520 SRO DW1781/5/1/1-3.

**PASFORD** 1 mile west of Pattingham (SO 8099). *Paffard (Bridge)* 1542 SRO A/2/11/9, *Pafford (Bridge)* 1543 SRO D593/A/2/11/9, *Pasford (Brooke)* 1716 SA 2161/107, *Basford or Pasford (Mills)* 1717 SRO D564/3/4/9; VCH XX 173. Perhaps from OE *pæþ ford* 'the ford at the path', with *-thf-* developing into *-ff-*: cf. Stafford.

**PATMARSH** 2 miles north-east of Worfield (SO 7498). *Pattmarsh* 1722 SA 5586/2/1/467-468, *the Patmarsh* 1816 SA 5586/5/5/158. In Shropshire since the 12th century. Early spellings have not been traced, but the first element may be associated with the personal name found in Pattingham (q.v.).

**PATSHULL** Ancient Parish 9 miles north-west of Wolverhampton (SJ 8000; SMR 01899 places a deserted Anglo-Saxon settlement at SJ 80500100). *Pecleshella* 1086 DB, *Pecdeshull* 1166 SHC I 166, *Patleshull, Patneshull* 1201 SHC III 70, *Pateshell* 1201 ibid.' 75, *Panteshull* 1227 SHC IV 49, *Patleshul* 1242 Fees, *Patushul, Pattushul* c.1250 SHC VI NS (ii) 49, *Petleshull* 1256 SHC I 167, *Patleshull* 1255 ibid, *Patelshulle* 1276 SHC 1911 31, *Patteleshull* 1306 SHC VII 177, *Patleshull* 1373 SHC XIV 373, *Patteshille* 1427 SHC XVII 118, *Patsyll* 1532 SHC 4th Series 8 186, *Petesey* c.1540 Leland ii 170, *Patsell* c.1565 SHC 1938 165, *Patsyll* 1589 SHC 1929 313, *Pattes hill* 1607 Kip, *Patteshill* 1610 Speed. '*Pættel's hill': the earliest two spellings probably show typical scribal confusion or misreading of the original *-t-* (but note undisclosed reservations about this attribution in Gelling & Cole 2000: 193). The person named Pættel was perhaps related to the individual named P(e)atta from whom nearby Pattingham takes its name, since parents very frequently chose alliterative names for their children, and *Pættel is an */-* derivative of P(e)atta: PN W 158. The name *Pættel is probably found in Petteridge, Kent; Paddlesworth near Dover, and Paddlesworth near Snodland (both Kent): Ekwall 1960: 356.

**PATTINGHAM** Ancient Parish 6 miles west of Wolverhampton (SO 8299). *Patingha* 1086 DB, *Patingeham* 1157 SHC I 25, *Patingeham* 1169 P, *Patingcham* 12th century Duig, *Patingham* 1275 SHC VI (i) 66, *Patyncheham* 1380 SHC XIII 154,

*Patyngeham(home)* 1433 SHC XVII 142, *Patyncham* 1448 SHC 1939 193, *Patengham* 1532 SHC 4th Series 8 185, *Patteingeham* 1574 SRO 2089/2/2/57, *pattingeam* 1583 TSAS 3rd Series X 71, *Pattengem* 1762 Trysull ParReg. 'The *ham* of *P(e)atta's people'. A more detailed discussion of the name will be found in the Introduction, pages 24-26. The 1380 and 1448 spellings are evidence of the soft '-indge-' pronunciation of *-ing-* which was still heard until recent times. The 1433 spelling with *home* incorporates OE *ham* in the sense 'manor', so 'Pattingham manor': see PN Sa II 19. The ancient parish of Pattingham consisted of two townships and manors, Pattingham in Staffordshire and Rudge in Shropshire. Since 1866 each has been a civil parish in its own county. See also Patshull.

**PATYNGEHAMBORNE** (unlocated, in or near Pattingham) *Patyngehamborne* 1424 Brighton 1942: 120. Incorporating OE *burna* 'stream', so 'The Pattingham stream'.

**PAYNSLEY** 1½ miles south of Draycott in the Moors (SJ 9838). *Lvfamesleg* 1086 (DB, formerly identified as Checkley: VCH IV 41), *Payneslegh* 1348 SHC 1925 107, *Paynesley* 1413 SHC XVII 9, *Peynesley* 1529 SHC 1912 31, *Painsle* c.1540 Leland, *Painsley* 1599 Smith, *Paynsley* 1611 SHC IV NS 13, *Painsley*, *Peynsley* 1644 SHC 4th Series I 68-9, *Paynsley* 1686 Plot 107. The DB form is clearly aberrant, or conceivably an earlier name. The derivation would seem to be from a personal name, possibly the same as that proposed as a possible derivation for Painley (q.v.), with OE *leah*.

**PEACOCK'S WOOD** 1 mile south-east of Kingstone (SK 0728). *? Pecokesholme* 1402 SHC XI NS 215, *Peacocus Wood* 1724 *Survey*. Perhaps from Jn. Pekoc (*Jno. of Cannock, called Peacock* in 1360: SRO DW1721/3/19/21), vicar of Abbots Bromley in 1351 (SRO DW1731/3/2/11/12).

**PEAK DISTRICT** the rugged high ground of north-west Derbyshire and northern Staffordshire. *Pecsætna lond* 7th century Tribal Hidage, *? Peacesdele* 1015 (11th century, S.1503), *Peeke* 1326 SHC X 66, *Peek* 1338 SRO D593/B/1/24/1/1, *the Peak* 1686 Plot 135. 'District of the peak dwellers', from OE *\*peac* with OE *sēte* 'settlers, dwellers', genitive plural *sætna*. *Lond* is from OE *lond* 'land'; *-dele* from OE *dǽl* 'a share of land, a district', a rare place-name element. See also Totmonslow.

**PEAKSTONES** 1 mile east of Alton (SK 0542). *Pekston, Pekstone* 1361 SHC XIII 19, *Pekestones* 1414 SHC XI 225, *Peykestonys* 1541 StEnc 447, *Peakstones* 1604 SHC 1940 194, *Pixton otherwise Peakestones* 1612 SHC IV NS 23, *Peackstones* 1686 Plot, *Peak Stones Rock* 1836 O.S. From OE *\*peac* 'hill, peak', the name meaning here 'the pointed stones'. See also Thor's Cave.

**PEARCE HAY** 2 miles west of Brewood (SJ 8408). *? Percehaie* 1207 SHC III 140, *? Percehay* 1370 SHC XIII 62, *Peyrse hey* 1569 Ct, *Pearce Hayside* 1646 Survey, *Peircehay or Priests hay* 1723 ibid, *Pearsey Hay* 1775 Yates. From Mercian OE *(ge)heg* 'a fence, an enclosure', so 'the enclosure of Pearce or Piers'. Pearce and Piers are common personal first names from OFr Piers: DES 351. Although the Bishops of Lichfield had a wood here (see Bishop's Wood), 'priests' hay' is not supported by the other forms, and would in any event be difficult to explain. The place lay within Brewood Forest.

**PEARL BROOK** a tributary of the river Sow. *Sparch Pearl ditch* 1629 VCH VI 185. From OE *pyrle* 'bubbling', a common spring or stream-name, with OE *broc* 'brook'. The meaning of *Sparch* is uncertain: possibly from sparge 'to splash, to sprinkle': OED.

**PEARL WELL** (obsolete) in Lyndon in West Bromwich (SO 0092). *Pearl Well* 1851 White. Described in White 1851: 682 as 'a remarkably fine spring of pure water'. From

OE *pyrle* 'bubbling', with Mercian OE *wælle* 'a spring,' and (sometimes) 'a stream'. Perwallesiche in Bradnop is recorded in 1343, probably from OE *pyrle* 'bubbling', with Mercian OE *wælle* 'a spring,' and OE *sic* 'watercourse'.

**PEASLEY BANK** a 541' hill 2 miles south of Stone (SJ 9029). *Pessehul* c.1210 SHC VIII (i) 200, *? Peasall* 1542 SHC 1939 114, *Peasley Bank* 1836 O.S. Perhaps from OE *pise, piosu, peosu* 'pea', with OE *hyll*, later changing to *leah*. Cf. Peasemore, Berkshire. See also Pershall; Pirehill. *Hundred-acres* which lies on the west side of Peasley Bank is to be associated with the site of Pirehill Hundred meeting place, and may be evidence that the assemblies took place on Peasley Bank.

PEATMOORS, THE (obsolete) a marsh of some 600 acres in Shenstone: Shaw 1801: II 1801 40-1. *The Peatmoors, The Peat-more, The Peatmore* 1801 Shaw II 40-1, 50.

**PEATSWOOD** 1 mile east of Market Drayton (SJ 6933). *Payt's hyll* 1576 SHC 1945-6 86, *Peats Wood* 1699 ibid, *Peats Wood* 1833 O.S. From the Peat, Peyte, or Payt family of Almington who held land here: see SHC 1945-6 136. StEnc 447 suggests that the place was also called *Skelhorne Spring* and *John Preston's Hill*.

PECKSTOWE (unlocated) *Peckstowe* 1327 (1798) Shaw I 40. Possibly from OE *\*peac* 'peak', with OE *stow* 'a place, a place of assembly, a holy place', in which case a name of some interest, but the single form makes any derivation speculative.

PEDELE (unlocated, possibly near Doxey/Seighford) *Pedele* 1304 SHC VII 120.

**PEEL, OLD** 1 mile south-west of Audley (SJ 7850). *Peele* 1586 Betley ParReg, *Peele (Greene)* 1656 ibid, *Old Peel (Carr), New Peel* 1798 Yates, *Old Peel* 1833 O.S. Possibly from OE *pel(e)* 'a stockade, an enclosure formed by a palisade'. *Carr* is from ON *kjarr* 'brushwood', ME *ker* 'marsh overgrown with brushwood'.

**PEGS WOOD (BIG & LITTLE)** ½ mile north of Okeover (SK 1549). *Pegs Posterne* 1640 Ipm, *Peggs Wood* 1752 Okeover E18, *Pegs Wood* 1838 O.S. From the surname Pegg: cf. Joseph Pegg 1775 (Okeover E6). Posterne is from ME *posterne* 'a back door, a secondary gate'.

**PELL WALL** 1 mile south-east of Market Drayton (SJ 6733), in Shropshire. *Pellwall* 1833 O.S. Pell Wall Hall, which was built in the 1820s, is said to have taken its name from one of the fields here: SHC 1945-6 258. The first word may be from the southern dialect word *pell*, meaning 'a deep place, a hole of water' (EDD): a spring of water in a deep cavity is recorded here: SHC 1945-6 258.. The second word is Mercian OE *wælle*, meaning 'a spring', and (sometimes) 'a stream'. *Pill Will*, recorded in the north-west of Onecote in 1845 (VCH VII 215) may have the same derivation, or the first element may be from ON *pill* 'a willow', with the second word meaning the same, indicating that *pill* had become meaningless.

**PELSALL** 3 miles north of Walsall (SK 0103). *Weoleshale* (for *Peoleshale*) 996 (for 994) (17th century, S.1380), *Peleshale* 1086 DB, 1327 SHC VII (i) 249, *Peleshala* 1167 P, *Selsul* (? for *Pelsul*) c.1175x1208 SRO 3005/1, *Poleshale* 1286 SHC V (i) 174, *Pellyshale* 1307 SHC 1911 287, *Pieleshale* 1310 ibid. 307, *Pelleshale* 1311 ibid.' 309, *Peoleshale* 1339 SHC 1913 74, *Pellsall* 1532 SHC 4th Series 8 144. 'Peol's *halh*', the second element possibly meaning here 'land between two streams': see Hooke 1983: 77. The personal name Peol is also found in the name *Peolesford* in the charter of 994 (17th century, S.1380) in which Pelsall is first recorded.

**PENDEFORD** in Tettenhall parish, 3½ miles north of Wolverhampton (SJ 8903; SMR 01903 places a deserted Anglo-Saxon settlement at SJ 89540355). *Pendeford* 1086 DB, *Pendefort* late 12th century Rees 1997: 42, *Penneford* 1222 SHC IV 19, *Pendeford* 1249, 1273, 1278, 1316 SHC 1911, 1282 Ch, *Penford* 1277 SHC 1887 177, *Pondeford* 1284 FA, *Penford* 1577 Smith SRO, *Penford* 1686 Plot, *Penford als Penkford* 1747 Bowen, *Penford* 1775 Yates. The name is generally held to be 'Penda's ford'. The personal name may be connected with the powerful pagan king Penda, who ruled Mercia c.626-655 AD. It is probable that other Mercians bore the same name, but apart from its occurrence in place-names, the only example of the name in early records refers to the king (see PN Wo 223). Place-names believed to incorporate the name – which may possibly be a shortened form of the recorded personal names Pendræd, Pendweald, and Pendwulf – include Penley (Flintshire), Pinbury (Gloucestershire), Peddimore (Warwickshire), and Pinvin and a lost *Pendiford* in King's Norton (Worcesterhire). The names are concentrated in the West Midlands, and may commemorate a particular and early phase of Anglian colonisation (see Brooks 1989: 163-4, which does not include this name, and Jones 1998: 29-62, which does). The place lies on the north side of a pronounced sandstone hill (*Pendeford Rock* 1828: SRO D3186/8/1/30/5) rising from the river Penk. A Roman road (Margary number 190) running south from Pennocrucium (Water Eaton) towards the Roman sites at Greensforge has been traced through Pendeford across the summit of the hill, suggesting the possibility of a derivation from Welsh *ffordd* 'road': see Jermy & Breeze 2000: 109-110, but see also Longford. It may be added that MED suggests that *pende* may have had the meaning 'pound; enclosure; pond', and there is a pool (noted by Plot 1686: 45, 167 as able to foretell rain) in low lying ground close to the river Penk here, making 'ford at the pond' a possibility. TSSAHS XII 1970-1 35 identifies two possible sites of the original settlement of Pendeford, SJ 900050 and SJ 895038. The name Pendeford is now attached to a large modern housing development to the south-west on what was formerly Wolverhampton Airport. A *Penfordelie* (unlocated, but unlikely to be associated with this place) is recorded c.1220: SHC 1921 19. See also Patshull; Pendlestone.

**PENDLESTONE (FORT)** 1 mile north-east of Bridgnorth (SO 7294). *Pendestan* early 13th century Rees 1997: 88, *Pendeston Mill* 1226 Eyton 1854-60: III 116, *Mill of Pendaston* 1227 Eyton 1854-60: I 303, *Pendelstanes, Pendelstanys mulnne* 1298 TSAS LXXI 1996 27, *Pendlestone Rockes, Pendlestone Milles* c.1560 Bellett 1856: 206, *Penstone* 1593 TSAS X 1887 142, *Pendlestone (Mills)* 1739 ibid. 196, *Pendlestone (Rock)* 1833 O.S. The earliest spellings point to 'Penda's stone' (see also Pendeford), later influenced by pendle, 'an overhanging part, natural or artificial', and pendle-stone, 'a local term for various kinds of beds or stone as occurring in quarries': OED; see also Halliwell 1850: 614. The place lies on the banks of the Severn under crumbling sandstone cliffs. The Fort is a sandstone pseudo-Gothic castellated structure built c.1845 on the site of Pendlestone Mill. The place has been in Shropshire since the 12th century. Cf. Pendle (Hill) and Pendlebury, Lancashire.

**PENDRELL HALL** 1 mile north-west of Codsall (SJ 8504). Built c.1870 (and originally known as *Pendryl Hall*: VCH XX 79) and named after the Penderel or Pendrell family of nearby Boscobel, White Ladies and Hubbal Grange, tenants of the Giffard family of Chillington, who assisted Charles II in his escape after the battle of Worcester in September 1651: DNB.

PENECFORD (unlocated, in Timmor) *Penecford* (n.d.) Shaw I 375. A curious name, possibly from OE *pennuc*, ME *penok, pinnok* 'a small pen', found chiefly in field-names in the west of England from the 13th century: EPNE ii 62.

**PEN FARM** on a 1227' hill 1 mile south of Butterton (SK 0755), *The Pen* 1836 O.S., *Pen* 1851 White; **PEN FIELDS** 1 mile south-east of Keele (SJ 8043), *Pennfield* 1378 HOK 17, *Pennefild* 1537 JNSFC LXI 1926-7 36, *Penfields* 1708 Swynnerton ParReg, c.1750 HOK 70, *Pen Fields* 1833 O.S. Possibly from OE *penn* 'a pen, a fold', but the elevated position of both places might point towards the same derivation as Penn (q.v.). *Pen Brook* is recorded in the Thornes/Upper Stonnal area c.1840: TSSAHS V 1963-4 67.

**PENK, RIVER** a tributary of the river Trent. *Penk* 1577 Saxton, *Penk flu.* 1610 Speed. A back formation (perhaps pre-English and due to a misunderstanding of a British expression such as *Dufr Pencruc* 'Pencruc river': see Ekwall 1928: xlvi; liv), from Penkridge (q.v.), i.e. the river (unusually) takes its name from the place. In early times a different back-formation was used (*Penchrich* 996 Mon, *Pencrigh* c.1175 SHC VIII 133, *Pencriz* 1300 For, *Penck* 1567 SHC XIII 270), but at a later later date became the Penk by a wrongly-perceived division of the name. Cf. Painshaw, Durham.

**PENKHOLME** on the north side of the river Trent, south of Pipe Ridware (SK 0917). *Penkholme* 1600 (1798) Shaw I 183, 1648 SRO D260/M/T/5/140. The second element is ON *holmr* 'a small island, a piece of land partly surrounded by a river or rivers'. The first element may be a stream-name transferred from the river Penk (q.v.).

**PENKHULL** 1½ miles south-west of Newcastle under Lyme (SJ 8644). *Pinchetel* 1086 DB, *Pencula* 1177 SHC I 87, *Pencul* 1169 ibid. 56, 1195 SHC II 47, *Penchille* 1200 SHC II 103, *Penkhil* 1230 P, *Penchul* c.1249 SHC 1911 145, *Penkel* 1292 ibid.' 242, *Penkhill* 1300 SHC 1911 57, *Pencle* 1607 Kip, *Penkhull* 1686 Plot. The earliest form is from British *Pencet* 'end of the wood' with OE *hyll* 'hill', giving 'hill by Pencet': see Ekwall 1960: 362; Gelling 1984: 171, 183, 190. The implication is that a British place *Pencet* lay at or near the hill. A field-name *Little Penkshull* appears on the tithe map 1840 adjoining the stream south-east of Claverley and north-east of Bulwardine. See also Penkridge.

**PENKRIDGE** Ancient Parish 5 miles south of Stafford (SJ 9214). *Pencric* 858 (? for 958) (14th century, S.667), c.1000 (11th century, S.1534), *Pancriz* 1086 DB, *Pencrich* 1156, *Peinc(h)riz* 1158 P, *Penkerich(e)* 1360 to 1545 FF, *Penkerage* 1449 Banco, *Pankryd* 1564 Mercator, *Penkrige vulg Pankrage* 1724-47 Defoe, *Penkridg* 1610 Speed, *Penkridge* 1749 Bowen. A British name from *\*penno-* (Welsh *pen*) 'head, end, headland' (and as an adjective 'chief'), and a derivative of OBritish *cruco* (Welsh *crug*, OCornish *cruc*) 'hill, mound, tumulus', so giving British *\*Pennocroucion* 'headland tumulus', 'chief mound', or 'head of the mound'. The name is discussed in more detail in the Introduction, pages 21-23. The name Penkridge is not unique: see for example Penkridge Hall, 3 miles north-east of Church Stretton, Shropshire (built c.1590 by Rowland Whitbrooke, whose wife came from Penkridge in Staffordshire: VCH Shrops X 27), and Penkridge Hall (otherwise Penkridge Lake Farm), recorded in the 13th century, 5 miles east of Runcorn (PN Ch II 155). The reference to *Penchrych-under-Lyme* in 1293 (SHC VI (i) 256) is presumably a transcription error, and doubtless refers to Penkhull (q.v.). Cf. Penchrise, south of Hawick: *Pencriz* 1380 (Watson 1926: 354).

PENKRIDGE WELL (obsolete) the name given to a meadow between Yew Tree Lane and Wrottesley Road at The Wergs (SJ 0601) from which a spring formed the head of the river Penk. *Penkrich Well* 1296x1307 SHC 4th Series XVIII 184; VCH XX 1. The name is commemorated in the road-name Penk Rise.

**PENN HILL** west of Seisdon (SO 8495), *Penn Hill* c.1300 VCH XX 185; **PENN HILL** in Lower Penn, 3 miles south-west of Wolverhampton (SO 8695), *Penn Hill* 1834 O.S.

For the derivation see Penn, Upper & Lower. Penn Hill near Seisdon may have been the meeting-place of Seisdon Hundred: see Seisdon.

**PENNOCRUCIO, PENNOCRUCIUM** a Roman settlement on Watling Street, ½ miles west of Gailey (SJ 9010). *Pennocrucio* 4th century (8th century) Antonine Itinerary. For a discussion of this name see Introduction, 21-23. The spelling *-crucio* derives from the ablative form, giving '(at) Pennocruc'.

**PENN STONES** (obsolete) 1 mile south-west of Wolverhampton (SO 8695). *Penn Stones* 1840 TA. Perhaps from stones marking the boundary of Penn. The name is remembered in Pennstones Lane, formerly Pound Lane: Dunphy 2002: 104, 120.

**PENN, UPPER & LOWER** Ancient Parish 3 miles south-west of Wolverhampton (SO 8995). *Penne* (Lower Penn) 1086 DB, *Penna'*, *Penne* (Upper Penn) 1086 DB, *Penne* 1176 P, *Penne Buffard* (Lower Penn) 1212 SHC 1911 387, 1236 Fees, *Netherpenne* 1271 For, *Church Penne* 1293 SHC 1911 49, 1300 SHC VII 85, *Overpenne* 1318 Ch, *Upper Penn*, *Lower Penn* 1834 O.S. Almost certainly from the British *penn* 'head, end, headland': Upper Penn lies on the highest point of a long ridge, Lower Penn on a distinct headland (*Penn Hill* 1834 O.S.). The names add weight to the evidence that a substantial enclave of Welsh-speakers remained in this area well into the Anglo-Saxon period. *Nether* is from OE *neopera* 'lower'. *Buffard* is from the Buffry or Buffere family who held the place: see SHC I 201-2; SHC II 10; SHC 1941 63. It is possible that the name Penne was applied to a considerable area of high ground to the south of Wolverhampton, and included, for example, Pensnett (q.v.). Evidence for this may be seen in the name *Penwie* (i.e. 'Penn *weg*') found in a charter dated 996 (for 994) (S.1380), and applied to a trackway to the north of Wolverhampton, '... the forerunner of the road to Cannock, which formed the north-western boundary of Wednesfield ...': Hooke 1983: 72-5. CDEPN 466 suggests that Orton (q.v.) was an earlier name for Upper Penn. Penn Fields, ½ mile north-east of Upper Penn, is recorded as *Pennefeld* in 1423: SRO D593/B/1/17/1/6/7.

**PENNYCROFTS** ½ mile east of Stafford (SJ 9323), *Crossapenyes croft* 1546 VCH VI 185, *Pennycrofts* 1670 ibid. 185, *Pennycroft* 1877 ibid. 192; PENNYCROFT WELL (obsolete) on the north-east side of Uttoxeter (SK 0934), *Pennycroft Well* 1836 O.S. From OE *pening* 'penny', a common field-name element indicating the rent payable (Field 1993: 193), with OE *croft* 'a small enclosure of arable or pasture land, an enclosure near a house'. The earliest spelling for the first place indicates that there was a cross (or crossroads) there, so 'the cross(road) at Penny Croft'. Pennycroft Well (from OE *wælle* 'a spring') was a sulphurous spring (Redfern 1865: 15), remembered in the names Pennycroft Lane and Pennycroft Road.

**PENNY MOOR** (obsolete, on the south-west side of Calf Heath (?SJ 9208)) *Penny Moor* 1832 Teesdale.

**PENQUALL LEYS** (unlocated, in Sheen) *Penquall leys* 1673 SRO D633. Perhaps from OE *pennuc* 'a small pen', with Mercian OE *wælle* 'a spring', so 'the leys or meadows at the small pens by the spring'.

**PENSNETT** 2 miles south-west of Dudley (SO 9188). *Pennak* 1272 Guttery 1950: 8, *Pensnet* 1244 Cl, *Peninak* 1247 FF, *Pennak, Pennaly* 1273 SHC 1911 157, *Penniak* 1273 SHC IX (ii) 26, *Pennyoake* 1292 Guttery 1950: 9, *Penynak* 1292 SHC 1911 209, *Peniuak* 1292 ibid. 211, *Pennyoke* 1292 SHC IX 39, *Penyuak, Peniuak* 1292 SHC 1911 210-1, *Pensned* 1322 ibid.' 351, *Pensnede* 1322 ibid.' 353, *Pensenet* 1357 ibid. 458. An intriguing name. The spellings suggest two different names used interchangeably, or possibly for different but close places. In one name the first element is probably from British *penno-, OWelsh penn 'head, end, headland' (possibly the plural pennou), given the elevation of the place, rather than OE penn 'a small enclosure, a fold', with OE snǣd 'detached piece of land, a piece of woodland' (cf. Sneyd), hence perhaps 'wood on Pen hill'. However, the meaning 'a piece of woodland' is derived from a reading of a charter of 843 A.D. (S.293), and is not free from doubt: see Ekwall 1960: 428; EPNE ii 131. Another possibility is that the name may be 'the detached piece of land belonging to the manor of Penn (q.v)': Penn lies some four miles north of Pensnett, but Penn Common extended well to the south-east, as evidenced by the first edition 1" O.S. map of 1834, and Upper Penn, Lower Penn and part of Sedgley appear in consecutive entries in DB, all held by William Fitz Ansculf: VCH IV 54. Ekwall 1960: 362 suggests that the other form with the spelling Peninak may be seen as evidence of Norman influence: Pensnǣd became by sound-substitution Peninet, which came to be read as Peninec, and evolved into Peninak perhaps due to the influence of French names ending in -ac, but the various spellings now available throw considerable doubt on that explanation, with the forms pointing towards a derivation from later OE penig 'penny' (cf. one wood called Pennak, recorded in 1272, identified as a wood which covered the southern slopes of Brierley Hill: Guttery 1950: 8, and certain lands called Penny Oak recorded in Himley in 1587 (SHC X (ii) 56) may be noted), with OE ac 'oak-tree, so 'penny oak-tree', though no explanation can be offered for such name, unless it can be seen as a later rationalisation of an earlier name. See also Penyval.

PENSNETT CHASE (obsolete) 'The chase associated with Pensnett'. A chase was an area subject to Forest law but not held by the king; in this case, the Earls of Dudley. The Chase, which covered a large area, extending to Gornalwood, Sedgley, Dudley Wood, Kingswinford, Himley, Enville and Bobbington, and as far south as the river Stour (Guttery 1950: 9, 11; HRO E12/V1/NB/14), is recorded from at least c.1230 (Guttery 1950: 7) or 1254 (StSt 3 1990-1 25), with the name appearing as Pennyoke Chase 1291 SHC IX (ii) 39, the Chace of Pensned 1322 Guttery 1950: 9, Pencenytt chaice 1536 SHC X NS (i) 113, Pensnet Chase 1610 Speed, Pensnet Chace 1633 Guttery 1950: 20, Pensnet chase 1686 Plot 1686: 98, Pinsnett Chace 1775 ibid. 9, Pensnet Chace 1776 HL HL/PO/PU/1/1776/16G3n80. Duignan 1902: 118 says: 'It may be that Pensnett Chase was once a part of Kinver Forest, which it adjoined, and, becoming the property of a subject, became a 'chase', not subject to the laws of the Forest, and, being detached, acquired the terminal snead'.

PENSNETT SPA (obsolete) at Pensnett (q.v.). In the early part of the 19th century, attempts were made to develop mineral springs here (mentioned by Plot 1686: 98-9) into a spa. '[T]his saline spring is situated near to the margin of a rivulet...Archill-brook, issuing from rising ground of that name, near the Holly-hall...serves as the county boundary...In 1823 a neat row of buildings was erected over the Spring, containing hot and cold baths': Scott 1832: 138-140; see also Guttery 1950: 26. Holly Hall (Holly Hall 1834 O.S.) lay just inside the detached portion of Worcestershire at SO 9089. The spring probably gave rise to the name Spring Mire (SO 9289), which straddles the boundary of Staffordshire and Worcestershire. The name Saltwell is also given to this area.

PENWIE – see **PENN, UPPER & LOWER**.

PENYVAL (unlocated, possibly near Gospel End: Hackwood 1898: 12) Penyval 1273 SHC 1911 156. See Pensnett, with which the name may be associated.

**PENWOOD (FARM)** 1 mile north-west of Sedgley (SO 9094). *Penwood* 1659 Sedgley ParReg, *Penn Wood* 1798 Yates. The area is now part of Penn Common, and in 1659 included Gospel End (q.v.): Sedgley ParReg. See Penn, Upper & Lower.

**[PEPPERHILL** an area of high rounded ground on the Staffordshire-Shropshire border 2 miles north-west of Pattingham (SJ 8100), and the site of a hall probably built by Sir John Talbot [d.1549] c.1519, probably the 'goodly Logge on the High Toppe of Albrighton Parke...on the very egge of Shropshire' as noted c.1540 by Leland v 18. *Pepper Hill* c.1540 Leland v 18, *Pepperhill* 1564 TSAS XI 1899 102, 1577 Saxton, 1599 Smith, 1610 Speed, 1643 (1798) Shaw I 61, 1686 Plot, *Peperhill* 1599 Albrighton ParReg, *Pepper Hill* 1749 Bowen, 1834 O.S. Early spellings have not been traced: WJ January 1905 351 records that 'Colonel Thorneycroft took very considerable trouble to trace its history, but the information he was able to obtain is very meagre, notwithstanding he, at his own expense, had special searches made at the British Museum extending over many weeks'. Peppercombe, Hampshire, and Pepper Ness, Kent, are believed to be from OE *pipere* 'a piper', which seems unlikely here. It is not likely that the name incorporates the OE personal name Pypba or Pyppa, (Pypba was father, or immediate predecessor, of Penda (see Pendeford), and grandfather of Peada (see Patshull, which this place adjoins)), but might if ancient be from a personal name *Pyphere or similar. Various other derivations are possible. Payment of rent in pepper is recorded in Staffordshire in the 14th century (see JNSFC LIX 1924-5 61), and this name may be associated with payment of that type. The topography does not fit a derivation 'hill shaped like a peppercorn', i.e. small and rounded, but would fit a derivation from a Germanic base *pip-*, 'to swell': see PN K 593. However, the most likely derivation for the Staffordshire name Pepperhill is from OE *pipere* 'a stream, watercourse, spring' (cf. Peppering, Sussex): a Wrottesley estate map of 1634 (SRO D3548/1) shows *The Condet head* ('the conduit head') on the north-east corner of the Wrottesley estate above Pepperhill at SJ 832021. Pepperhill is now in Shropshire: the county maps of Saxton (1577), Smith (1599), Browne (1682), and Bowen (1749) all indicate or suggest that the place is in Staffordshire; Speed (1610) and Rocque (1752) include the place in Shropshire. See also Pepper Street.]

**PEPPER STREET** on the north side of Silverdale (SK 8046). TSAS II 1879 321 and 356 suggest that the name Pepper 'occurs upon a great number of fragments of Roman roads in Cheshire and Staffordshire', and that Pepper Street to the north of Keele (*Pepper Street* 1799 Faden) is a Roman road from Chesterton to Bury Walls in Shropshire (*sic*). No evidence has been found to support the suggestion: the surname Pepper appears in the Keele ParReg from the mid-18th century (see also HOK 54, 71), and it seems likely that the lane is named from the family: the lane near Keele appears to have been called previously Haying Lane (HOK 14). There is another Pepper Street in Newcastle-under-Lyme. A detailed discussion of the name Pepper Street (and other names incorporating Pepper, including Pepper Hill), appears in PN Ch V (I:1) 17-18; see also PN Wa 163 for Pepper Lane in Coventry. A supposed Roman road known as Pepper Street near Hawkstone is recorded in TSAS 3rd Series VI 1906 Misc. iii. See also Pepperhill.

**PERRY BARR** – see **BARR, PERRY.**

**PERRY CROFTS** on north side of Tamworth (SK 2004). *Pyrycroft* 1233 SHC XII 23, *Perrycroft* 1548 SRO D641/1/2/288, *Perriecrofts or the Large-perri-crofts, Peare-tree-Croft* 17th century SRO D260/M/T/2/9. From OE *pirige* 'pear tree', with OE *croft* 'a small enclosure of arable or pasture land, an enclosure near a house'.

**PERRY HALL** 3 miles north-east of Wolverhampton (SJ 9600). *Pirie* 1294 SHC 1911 237, *Pyrye* 1332 SHC X (i) 127, *Pyrie, Perye* 14th century Duig, *Perry Hall* 1664 SHC II (ii) 29, 1834 O.S. From an ancient house, dating from at least the 16th century (Mander & Tildesley 1960: 52, 96), which took its name from OE *pirige* 'pear tree'.

PERSBUTT (obsolete, 1 mile north-west of Eccleshall (SJ 8129)) *Pesebutt* 1274 SHC 1921 22, *Persbutt* 1833 O.S. Probably from OE *pese* 'pea', sometimes referring to a wild plant resembling a pea, e.g. marsh trefoil, also called buck-bean, with OE *butt* 'thick end', here perhaps meaning the headland of one of the open fields. The OE word *pese* was singular, and when over time the final *e* was dropped, *peas* was regarded as a plural, and the singular *pea* created by those ignorant of the history of the word: Ekwall 1959: 64. *Peafield Coppice* appears on modern maps 1 mile south-west of Pershall (SJ 8028). See also Pershall.

PERSEHOUSE HALL (unlocated, in Woodseton: WSL 327/276/81) *Parsus Hall* 1582 Sedgley ParReg, *Pershouse hall* 1601 ibid, *Persehall-hall* 1801 Shaw II 222, *Pershouse* 1844 Erdeswick 1844: 369. Roper 1976: 78 states that Perkes Hall was sometimes called Persehouse Hall, and that the family, who had been in Woodsetton from at least 1439, sometimes used the name Parkes. Hackwood 1898: 13 cites a statement: 'The park of Dudley Castle extended as far as Eve Lane, Upper Gornal, including Park Farm, the Wren's Nest and the Old Park between Tipton and Wren's Nest. Most of this would be in Sedgley Parish. Parkes Hall stood between Sedgley and Woodsetton'. See also Shaw 1801: II 222; SHC 1923 94-5. The map in SHC 1944 shows *Park Hall* some 2 miles north-west of Great Barr. A pedigree of the Persehouse family appears in MidA I 133-8, 162-4; Willmore 1887: 295-7.

**PERSHALL** 1 mile north-west of Eccleshall (SJ 8129). *Peleshala Helye, Peleshalla Helyæ* 1167 SHC I 48, 158, *? Pereshulla* 1188 ibid. 137, *? Pessehal* 1203 SHC III 86, *Pesehale* 1205 ibid. 135, *Peshale* c.1233 Rees 1997: 82, *? Pesale, Pesenhale* 1269 SHC IV 173, *Pyshale* 1272 SHC III 18, *Pesehall* 1280 SHC VI (i) 121, *Pessahull* 1284 SHC I 158, *Pessall* 1297 SRO DW1734/J2268, *Peshale* 1342 SHC XI 153, *Peeshale* 1374 SHC XIII 111, *Pershale* 1414 SHC XVII 20, *Peysall* 1532 SHC 4th Series 8 102, 1560 SHC IX 114, *Pershall* 1600 Eccleshall ParReg, *Peshall otherwise Pearshall* 1616 SHC NS VI (i) 3, *Peirshall* 1708 SBT DR10/1401. If the 1167 forms are to be trusted, then as Pelsall (q.v.). But given the other spellings and other nearby names (e.g. Persbutt), probably from OE *pese* 'pea', sometimes referring to a wild plant resembling a pea, e.g. marsh trefoil, also called buck-bean, with OE *hale*, dative singular of OE *halh. Helye* is from a former tenant, Helyas or Helias fitz Geoffrey: SHC I 51; 158; SHC 1914 43. Some of the spellings may relate to Peasley Bank (q.v.). *Peafield Coppice* appears on modern maps 1 mile south-west of Pershall (SJ 8028). *Pesecroft*, possibly near Ranton, is recorded in an early undated deed: SHC IV 276.

**PERTON** in Tettenhall, 3 miles west of Wolverhampton (SO 8598; SMR 02541 places a deserted Anglo-Saxon settlement at SO 85699860). *Pertone* 1086 DB, *Ptona* 1166 SHC 1923 296, *Periton* 1193 P, *Perton* '1198 Fees, *Portun', Porton* '1242 Fees, *Porton* '1250 Fees, *Pirton* 1606 CKS U269/T144, *Purton* 1686 Plot 168. From OE *peru* 'a pear', so *peru tun*, 'pear-tree *tun*', perhaps meaning here 'pear-orchard' (cf. OE *æppel-tun* 'apple-orchard'). 'This parish [Tettenhall] has one singularity in the fruit way; it has produced a peculiar kind of pear, called Tettenhall pear, and known by no other name; many hundreds of the trees grow in this parish, though not, or scarcely to be found at all, at any considerable distance. The tree is large, and a plentiful bearer, the fruit well flavoured, and bakes and boils well, but will not keep long enough, even for carriage, to any considerable distance, unless some time before it is ripe: the average annual produce of this parish is many thousand bushels more than its own consumption. The whole duration of the pear, from its becoming fit for use to its decay, about one month, during which time all the neighbouring markets are so plentifully supplied, that it often scarcely pays for picking and carrying in. This pear makes but an indifferent perry, and is therefore not much applied that way: in plentiful seasons, and it is seldom, otherwise, large quantities are eaten by hogs, which are suffered to pick them up as they fall': Pitt 1791: 89, quoted in Shaw 1801: II 199. A reference to *the castle of Perthon* in 1286 (SHC 4th Series XVIII 142) may refer to Perton Hall, a moated medieval house which stood at Old Perton (SO 860986): TSSAHS XXIV 1982-3 58; VCH XX 10. No other reference to a castle has been traced. The name Perton has been adopted for a modern housing development on the former airfield at Wrottesley Park.

**PESSALL FARM** 1 mile north of Edingale (SK 2113). *Le Peyshylle (Pyttes)* 1541 Derby, *Pessall (pit)* 1581 Brookhill. Seemingly from OE *pese, hyll* 'peas hill'. Transferred from Derbyshire in 1894.

**PETHILLS, PETHILLSHEAD** 2 and 2½ miles respectively north-west of Waterhouses (SK 0652 & 0452), *Puttel(l)s* before 1214 (1883) Sleigh, 1535 VE, *Pethills* 1251 VCH VII 212, *Puthullis* c.1251 VCH III 226, *? Pitlesle* 1272 SHC IV (i) 193, *Puttelles* 1540 VCH III 226, *Pytylls* 1540 AOMB, *Pettell* 1560 Pat, *Pethilles* 1570 SHC XIII 279, *Petchulls, Petchuls, Petehulles alias Petehuls milne holme, Pittefeild* 1571 SHC 1931 126-7, *Pethilles* 1608 SHC 1948-9 26, *Pethills* 1636 Leek ParReg, *Petthills* 1747 Bowen, *Pethills* 1842 O.S.; **PETHILLS** 1 mile south-east of Flash (SK 0365), *Pethills* 1602 ParReg, c.1870 Rental. It has been suggested that Pethills in Sutton Downes, Cheshire, may be from ME *pightel* 'a small enclosure' (PN Ch I 155), but the spellings with *put(t)-* make that derivation for the first two places improbable, even though they are said to have been land owned and enclosed by Croxden Abbey (VCH III 226), so probably from OE *pytt* 'pit, hollow', hence 'the hills with the hollows': *Ironpits* (*Iron Pits* 1836 O.S., perhaps to be identified with *Petchuls mylne hoomes 'which places are as well known by the names of Pittefeild':* 1571 SHC 1931 127) lies to the east of Pethills near Waterhouses, suggesting that the hollows are ancient bell-pits from mining. The late forms for Pethills near Flash make any derivation uncertain: perhaps as Pethills near Waterhouses, or from ME *pightel*. The possibility of a derivation for the place near Flash from ME *pete* 'peat', giving 'the hills were peat was dug', is improbable, since the usual spelling for peat was *peat(e)* or *pete* (cf. the field-names *Peatefeild* 1626, *Peatefeelde* 1651, Hollinsclough; *le Petehiles* in Cheshire, PN Ch V (I:1) xix), and peat is not in any event dug from hills.

**PHEASEY** on the southern side of Barr Beacon (SO 0695). *Veysies Farm* 1610 StEnc, *veysies Farm in Barre* 1648 BCA MS3883/608964, *Pheasey Farm* 1799 StEnc. Said to be from Simon Vesey who held property here in 1559: TSSAHS XX 1978-9 51; StEnc 457.

**PHILLEY BROOK** a tributary of the river Smestow, *Follbrok* 1316 SHC 1911 329; **FILLEY BROOK** a tributary of the river Trent, *Filly Brook* 1774 SRO D3160/2/1-2. The latter appears in the bounds of Darlaston in *þær fulan broc scyt on Trentan* 956 (11th century, S.602: see Hooke 1983: 85-9). The derivation is from OE *ful* 'dirty, foul, filthy, muddy', perhaps meaning in some cases 'dark'.

**PICKARDS FARM** 1 mile north-west of Upper Arley (SO 7681). *Pykaslond* 1357 PN Wo 32, *Pikarslond* 1460 ibid, *le Pykards* 1485 ibid, BCA MS3279/351524. From the surname Picard, recorded from 1276 (Ass), probably originally from Picardy in France. In Worcestershire since 1895.

**PICKMERE, PICMOOR** in Weston-under-Lizard (SJ 7811). *? Pikemere* 1327 SHC VII 209, *Pikemore* 1380 SHC II NS 61, *Pickmore* 1679 SHC 1919 242, *Pikemere* 1752 Rocque, *Pike Mere* 1763 GM 445-6, *Pitmoor Pool* 1775 Yates, *Pitmoor-pool* 1798 Shaw I 106, *Pike More* 1833 O.S. Probably from OE *pic-mere* 'pike-pool' (see Ekwall 1936: 100), or (less likely) 'pool with the pointed end' (the pool shown on Rocque's 1752 map of Shropshire has no pointed end). Another possible derivation is from OE *pic*, with a number of meanings including 'a conical hill': there is a prominent rounded hillock to the west (*The Mount* 1833 O.S.), so perhaps 'the pool by the conical hill'. It may be noted however that a measure of land known as a *pike* is recorded in Staffordshire in 1402 (SHC XI NS 204), though the word does not seem to have been in common use, and is not mentioned in Halliwell. The pool here was an artificial pool which has now been drained. It is possible that the earliest spelling is from OE *mor* 'marshland'. Field-boundaries suggest that at one time Watling Street may have avoided this low-lying area. The name is now applied to the junction of Watling Street and the Newport-Wolverhampton road. Cf. Pickmere, Cheshire.

**PICKNALLS** on the south-west side of Uttoxeter (SK 0833). *the Picknall (later Red Hills)* 1686 SRO D1194/10/4, *the Picknall* 1693 SRO D758/1/1-4, *the Picknoll* 1731 SRO D758/1/6, *the Picknalls* 1741 SRO D758/1/7. Perhaps from OE *pic, cnoll* 'the knoll with the peak or point'.

PICKSTOCK (unlocated, in Levedale) *Pikestoke* c.1238 (1798) Shaw I xxv, *Pickstoke* 1281 SHC XII NS 127, *Picstoke, Picstok* 1284 FA, *Pickestok* 1314 SHC IX 45, *Pykestoke, Pikstoke, Pikestoke* 1315 SHC X 15, *Pixstoke* 1323 SHC IX 89, *Pixstok* 1324 ibid.' 101, *Pikestoke* 1324 ibid.' 106, *Pikestok* 1325 ibid.' 108, *Pykstoke* 1325 ibid.' 110, 1414 SHC XVII 49, *Pixstoke* 1380 SHC XVII 193, *Pycke Stocke hay* 1565 SHC 1938 9, *Pickstock Leyes* 17th century SRO D260/M/T/5/4. Perhaps from OE *pic* 'point, hill' and OE *stoc* 'place', sometimes 'holy place', or 'secondary settlement, outlying cattle farm', here possibly 'the outlying farmstead by the pointed hill': Levedale lies on a 357' hill. *Leyes* is from OE *lǣs* 'meadow, pasture'. *Pigstockheys Covert* (*Pyckstocks heyes* 1598 Ct) lies ½ mile north-west of Levedale (SJ 893175), and may mark the location of this place. Some of the above spellings may relate to Pickstock, 3 miles north-west of Newport (SJ 7223).

PIGGENHOLE (unlocated, in Fawfieldhead) *Pigeng Hole* 1775 Alstonefield ParReg, *Piging hole* 1778 ibid, *Piggin Hole* 1789 ibid, *Piggenhole* 1870 *Rental.* 'The hollow frequented by wood pigeons'.

**PIKELOW** on the east side of Waterfall (SK 0851), *Pike Low* 1775 Yates; PIKELOW (unlocated) on the north side of Upper Cotton, *Pikelow* 1775 Yates. From OE *pic* 'point, hill' with OE *hlaw* 'mound, tumulus', so 'the pointed burial mound'.

PIKERING, PYKERING (unlocated, a fishery in Tillington) *Pikering* 1275 SHC 1925 81, SHC VI (i) 72, *Pykering* 1276 ibid.' 81.

**PILLATON, PILLATON HALL** 1½ miles south-east of Penkridge (SJ 9413; SMR 01793 places a deserted Anglo-Saxon settlement at SJ 94441313). *Bedintona et Pilatehala* 1116x1133 Burton, *Pilethale* 1185 (1798) Shaw I 3, *Pilatonhall* 1271 For, *Pylatenhale, Pilletenhale, Pilatehale, Pilitenhale, Piletehale* 13th century Duig, *Pilotenhale* 1300 For, *Pelitnall* 1532 SHC 4th Series 8 88. A name that perplexed earlier scholars: Duignan 1902: 119 was forced to postulate a derivation from Pontius Pilate. In fact the name means '*halh* where pilled oats grew', presumably a field-name (cf. the field-name Pillar's Croft in Brewood (*Pyllyttes crofte* 1562, *Pyllerscroft* 1576: Oakden 1984: 45; Field 1993: 96)) which by confusion of the second element *hale* and *hall* attached itself to Pillaton Hall, the ancient seat of the Littletons. Pilled oats are those in which the grain is free from husk or glumes. The OE form of the place-name was presumably *Pilatan halh* 'the *halh* with the pilled oats'. It has been suggested that the earlier name of Pillaton was Bedintun (q.v., see TSAS 4th Series I 16), but although Pillaton was held with Bedintun in the 12th century (SHC 1916 31), it was not identical with that place, and is later represented by Pillaton Hall alone (VCH IV 44 fn.). Curiously, Yates' 1775 map of Staffordshire shows *Pile-eaton Hall (sic)* 1½ miles north-east of Penkridge, some distance from *Pile-eaton*. The bounds in a charter of land at Wetmore dated 1012 (13th century, S.930) conclude with a statement that *pylltunes landes* belongs half to Wetmore and half to Rolleston. Although many miles from both of those places, Pillaton belonged to Burton Abbey (Shaw 1798: I 3; SHC V (i) 7, 22, 35, 37, 40, 42; StSt 8 1996 24), and it might be thought that the the name refers to Pillaton, which may then have been a berewick or demesne farm (see Hart 1975: 241), but Sawyer 1979a: xxix shows that *Pylltunes Landes* should read *Willtunes Landes*, referring to Willington in Derbyshire, which was given to Burton abbey by Edward the Confessor between 1042 and 1050 (ibid. xlvi; see also Hooke 1983: 38-9). It may be noted however that *Pylett'holme* near Horninglow is recorded in 1309 (BL Camp. Ch. V 17). The name is not unique: the field-name *Pilatenhalewh'* (1209), *Pilatenhale* (1360), *Pylatenhale* (1360), for example, is recorded in Malpas, Cheshire: PN Ch IV 43.

PILSTONES (obsolete) 1 mile south-east of Swynnerton (SJ 8534). *Pilson* 1722 SRO D641/5/T/8, *Pilstones* 1727 SRO 641/5/T/17, *the Butterstone Pilson* 1746 SRO D641/5/T/8, *Pilstones* 1812 SRO D641/5/T/9, *The Pilsons* 1836 O.S. map, *Pilstones Farm* 1922 O.S. In the absence of early forms no suggestions can be offered for this name, unless (which seems most improbable) there is a connection with Puleston and Puleston Hill, 3 miles north-west of Newport, Shropshire (SJ 7322), found frequently in early records (e.g. *Pivelesdon* 1228 SHC IV 53, *Puywellesdone* 1240 (1798) Shaw I xxv, *Pyvelesdon* 1272 SHC IV 204, *Pyvelesdone* 1275, 1277 ibid.' 284, *Peulesdon* 1275 SHC VI (i) 66, *Pivelesdon* 1275 ibid. 72, *Pynlesdon* 1276 ibid.' 55, *Pyurlesdon* 1277 SHC IV 81, *Pylesdon* 1277 ibid. 85, *Pulesdon* 1300 SHC 1924 129, *Peulesdon* 1303 SHC VII 112, *Pynlesdon* 1307 SHC IV 55). The first element of the Shropshire place-name appears to be a personal name *Peofel, a diminutive of Peof(a), with OE *dun* 'hill', hence '*Peofel's hill': see Gelling & Cole 2000: 172. The absence of any medial -*v*- in this name suggests (if the name is indeed ancient) a derivation from OE *pil* 'stake', although 'stake stones' is not readily explicable. The reference to *Butterstone* in the 1746 form is unexplained.

PINCHLEY (obsolete, 2 miles north of Mayfield (SK 1548)) *Pincheney(e)*, *Pyncheney(e)* c.1240 Okeover, 1294 Banco, *Pynchene Syde* 1443 Okeover E9, *Pyncheney* 1508 SHC VII NS 73, *Pyncheney (meydow)* 1538 Ipm, *the two Pinchlees* 1640 Ipm, *Pinchley* 1838 O.S. From OE *eg* 'island, land partly surrounded by water, well-watered land', with the French surname Pinchon (see DES 352) so 'Pinchon's well-watered land', with *leah* later substituted for *eg*. See SHC VII NS 141.

PINNOCK (obsolete, between Burslem and Tunstall (SJ 8750)) *Pinox* 1775 Yates, *Pinnock* 1799 Faden, *Pinnox (Colliery)* 1836 O.S. Ekwall 1960: 367 suggests that Pinnock in Gloucestershire may be a diminutive *pennuc or *pennoc, formed from British *pen* 'hill', either in British or English, but EPNE ii 62 gives OE *pennuc, ME *penok*, *pinnok*, from OE *penn* 'a small enclosure, a fold', later 'an enclosure for animals', so here perhaps 'a small animal-pen', found chiefly as field-names in the west of England from the 13th century. Cf. Pinnock, Gloucestershire; Pinnocks, Wiltshire.

PIPE (unlocated, 'beside Stafford': SHC 1923 319) *Pipe* 1509 SHC 1923 319.

**PIPE, RIVER** a tributary of the river Trent. *Pipebrouk* 1286, 1300 For. From OE *pipe* 'pipe, water course', with OE *broc* 'brook', meaning here 'the brook with the small channel'.

PIPE HALL (obsolete) 1½ miles south-west of Hamstall Ridware – see **QUINTIN'S ORCHARD**.

**PIPE HALL** 1 mile west of Lichfield (SK 0909). *The Pype Hyll (sic)* 1532 SHC 4th Series 8 183. From Pipehill (q.v.). See also Quintin's Orchard.

**PIPEHAY (FARM)** on the east side of Draycott in the Clay (SK 1528). *Pipe Hay* 1798 Shaw I 83, *Pipehay (Lane)* 1836 O.S. From Mercian OE *(ge)heg* 'a fence, an enclosure', here perhaps in the sense 'an administrative area within a forest area', i.e. Needwood Forest. The Pipe element is from the family who took their name from Pipe near Lichfield and acquired this place c.1295: Shaw 1798: I 83. The place may be associated with Rough Hay Park: see Shaw 1798: I 83; SHC 1912 222.

**PIPEHILL, PIPE GRANGE** 1½ miles south-west of Lichfield (SK 0908). *Pipe* c.1140 VCH XIV 198, *Magna Pipa, Parva Pipa* (Great Pipe and Little Pipe, 1 mile to the north-west) 1166 SHC 1923 295, *Pypa* 1294 SHC 1939 86, *Pipe* 1361 ibid. 115, *Herdewykepipe* 1349 SHC III 280, *Herdewyk, Pypeherdewyk* 1374 SHC XIII 118, *Pype* 1527 ibid.' 82, *Pypehill* 1562 SHC XIII 221, *Pipe-Hardycke otherwise Pipe-Minor* 1589 SHC XV 197, *Pipehill otherwise Pipehardwicke* 1593 SHC XVI 131, *Pipe Hull* c.1598 Erdeswick 1844: 241, *Mykyll Hylle, otherwise Pype Hylle* 1659 (1798) Shaw I 312, *Great Pipe (manor)* 1666 SRO D661/4/1. From OE *pipe* 'conduit, water-pipe, water-course', perhaps referring to a water conduit from Maple Hayes south of Pipe Hall to the cathedral Close which was created between c.1140 and c.1170 when two springs were granted to Lichfield cathedral by Thomas of Bromley (VCH XIV 95; SHC 1950-1 161; AntJ 56 (i) 73-9), although that derivation was doubted by Ekwall 1928: 327, and the conduit at Lichfield may have been copied from that at Christ Church, Canterbury, which was constructed in about the late 1150s (the Bishop of Lichfield was previously prior of Christ Church), which would post-date the earliest recorded reference to the name, so possibly from water found in natural subterranean strata in the area (see Shaw 1798: I app. 9), or from Leamonsley or Pipe Brook which rises near the conduit: VCH XIV 95; 198. Lichfield was for many centuries supplied with water from springs rising in Pipe manor: see Leland ii 100-101. But evidence exists of a mysterious but remarkable oak palisade, and associated ditch some 12' wide and 5' deep, running for over 500 yards at Pipe Place Farm to the north-west of Wall: Shaw I 1798 19-20; TSSAHS XXIII 1980-1 2-3; Wells 1998: 117. The feature, if the description can be relied upon, appears to have followed the natural contour and been aligned with Wall, suggesting that it may perhaps have been associated with a typical Roman aqueduct supplying Letocetum, which was still in existence in the Anglo-Saxon period, and so perhaps accounting for the element 'pipe' which occurs in a number of place-names in a sizeable area hereabouts: see Leah *et al.* 1998: 117; but note also the structure said to explain the name of Moat House (Farm) (q.v.). Pipehill was known as *Hardwick* or *Pipe Hardwick* from the 14th to the 17th century (VCH XIV 286), from OE *heord(e)-wic* 'a livestock farm'. Pipe Grange, held by the prior of St John's Hospital in Lichfield, existed by 1298 and was known as Pipe Grange by 1377: VCH XIV 289. Little Pipe (*Pipa Minor* or *Parva Pipa*) was also known as *Prees* (VCH XIV 69), so-called from 'the impropriated parish of Prees, Shropshire': Shaw 1798: I 292; see also SRO DW3222/290/1-40. See also Handsacre.

**PIPE RIDWARE** see **RIDWARE, PIPE**.

**PIPERS HILL** on the north side of Himley (SO 8891). *Pipers Hill* 1834 O.S. See Pepperhill for possible derivations.

**PIPE WOOD** ½ mile south-west of Blithbury (SK 0819). *Pipe Wood* 1798 Yates, 1836 O.S. Perhaps named from the streams that rise here.

**PIREHILL** 2 miles south of Stone (SJ 8931). *Pirehel, Pireholle, Pereoll(e), Pereholle* 1086 DB, *Pirhill'* 1199 Fees, *Pirehulle* 1201 SP, *Pirhulle* 1203 Ass, *Pirhull'* 1212 Fees, 1230 P, *Pyrhull* 1227 Ass, 1275 Fine, 1283 Ipm, *Pirhul* 1228 Pat, 1252 Fees, *Pirul* 1254 Ipm, *Pyrehull* 1269 Ass, *Pyrhulle* 1281 SHC VI (i) 117, *Pirihull* 1285 Fees, *Pire-hill Hill* 1775 Yates, *Pyre Hill* 1798 Yates. The name of one of the five ancient Staffordshire Hundreds, now attached to a rounded 463' hill on the west side of the river Trent. The name is possibly connected with ME *piren* 'to peer', ModE *peer* 'to look narrowly', Low German *piren* 'to scan', here meaning 'look-out hill' (see Gelling 1988: 211; 1992 144; also Ekwall 1928: 333), perhaps more apposite for a hill with commanding views (as here) than the usual root put forward for such names, OE *pirige* 'pear tree' (see Duignan 1902: 120). In that respect the frequent absence of a vowel after the *r*, and the modern pronunciation, may be significant. The 'look-out' association is also found in Totmonslow, another of the Staffordshire Hundred names. Although Modern *pyre*, from Latin *pyra, pira*, meaning 'bonfire, funeral fire', is not recorded before 1658 (OED), the Latin word is well-recorded from c.950 (Latham 1980: 384), and it is not inconceivable (but improbable) that the name could record the early use of the hill as a beacon or for some other activity involving fire. See also Elmhurst near Stafford.

PIRIE (unlocated, probably in or near Gunstone) *la Pyrye, la Piri* 1284 SHC VI (i) 139, *Pirie* 1422 SHC XVII 77, *Lypyrye* 1425 ibid. 106, *Pirre* 1539 SHC NS VI (i) 75. From OE *pirige* 'pear-tree'.

PIREWASSE a boundary mark on Pur Brook, mentioned in a charter of 951 relating to Marchington: see Hooke 1983: 103-6, who locates the place at SK 122266. *Pirewasse* 951 (14th century, S.557). For the first element see Pur Brook. The second element is -*wæsse*, meaning 'land liable to sudden flooding and draining': cf. Alrewas.

PIRLEWALLSICHE (unlocated, probably near Newcastle-under-Lyme) *Pirlewallsiche* c.1300 SHC NS XIII XIII 239. From OE *pyrle* 'bubbling', a common stream-name, with OE Mercian *wælle* 'a spring', and (sometimes) 'a stream', and OE *sic* 'a small stream', so possibly 'the small stream fed by the bubbling spring'. Cf. Purlwell Lane, Yorkshire (PN West Riding Yorkshire 2 180).

**PITCHINGS FARM** ½ mile north-east of Waterhouses (SK 0950). *Pitchins* 1836 O.S. See also Piging Hole.

PITTENSHALL (unlocated, in Hallon) *Pittenshall* 1775 SA 1045/244-5.

**PLARDIWICK** ½ mile west of Gnosall (SJ 8120). *Plerdewirke* 1199 SHC III (i) 167, *Plerdewike, Plerdewicke, Plerdewyk, Plerdewyck* 1199 FF, 1268 PIR, *Plerdewirke* 1199 SHC III 167, *Plardswicke* 1585 Ct, *Plaidwicke* 1602 SHC 1935 476, *Plordewicke* 1607 FF, *Plordwecke* 1632 SHC V NS 169, *Plardewick* 1833 O.S. The second element is OE *wic*, but the first is uncertain. Coates 2000: 21-2 suggests a derivation from a compound of OE *plega* 'play', and the OE element -*rǣdenn* 'condition; rule, government', producing

*Pleg-rǣdenn*, with a meaning such as 'place where the right to conduct games exists', so '*wic* at or of the place called *Pleg-rǣdenn*', or perhaps from the OE personal name Plegheard. Another possibility is a derivation from OE *plega* with OE *heord(e)wic* 'sheep farm', so giving 'the sheep farm where games were played'. A derivation from OWelsh *paladyr*, plural *pelydr* (with the stress falling on the penultimate syllable: see Watson 1926: 345) 'a spear-shaft', possibly given as a stream-name (and the Doley Brook has a noticeably straight course between Gnosall and Plardiwick) can probably be ruled out, though the element is not unknown in place-names: Geoffrey of Monmouth mentions the 'fortress of Mount Paladur, which is now called Shaftsbury'; see also Spearhill, and a discussion of *paladr* in Coates & Breeze 2000: 74-6, which mentions river-names incorporating Welsh words for 'spear', 'knife' and 'needle'.

**PLAT** 1 mile west of Consall (SJ 9548). Early forms are not available, but presumably from ME *plat* 'a flat place, a footbridge'.

**PLATT BRIDGE** 1 mile north-west of Eccleshall (SJ 8129). *Plat Bridge* 1833 O.S., *Platt Bridge* 1838 SRO D641/3/R/5/5. Although early spellings have not been traced, presumably from ME *plat* 'a flat place, a footbridge', here in the latter sense, the word 'bridge' having perhaps been added when the meaning of plat had become forgotten. Another *Platt Bridge* is recorded in Yoxall in 1684: NA DD/4P/24/109.

**PLECK** 1 mile south-west of Walsall (SO 9997), *The Pleck* 1576, 1617 VCH XVII 157, 1686 Plot 314; **PLECK** 1 mile north of Whitmore (SJ 8142); **PLECKS** (unlocated, in Ettingshall), *Pleckis Meade, the Plexe* 1569 SRO D4407/56-7[SF73-4], *the Plecks* 1667 SRO D440/90[SF107]. From ME *plecche* 'a small enclosure or plot of land'.

**PODMORE** 2 miles south-east of Ashley (SJ 7835). *Podemore* 1086 DB, *Podemor* c.1235 SHC 1911 424, 1288 Ass, *Poddemere* 13th century, *Podemor, Podmore* 14th century Duig, *Pademor* 1300 SHC VII 67, *? Potemour* 1332 SHC X 92, *Podemor* 1332 ibid. 97, *Poddemor* 1362-3 JNSFC LIX 1924-5 61, *Podmore* 1559 SHC 1926 116, *Podmer Gerards* 1695 Mordern. From OE *podde* 'frog, toad', hence 'frog moor (or mere)'. The place adjoins Gerrard's Bromley, which explains the 1695 form. Cf. Podmore, Somerset.

**POINTHORNE** 1 mile south-east of Croxden (SK 0739). *? Pointon temp.* Henry III Shaw 1798: I 172, *Poynthorn* 1694 StEnc 461, *Pinethorn* 1762 Croxden ParReg, *Pinethorn* 1763 ibid, *the Pinthorn* 1779 ibid, *Pointon* 1775 Yates, *Pointhorn* 1836 O.S. *Point* is recorded by EDD as a dialect word for thorn, though recorded only in Norfolk, and for the branch or projection upon a stag's horn in Somerset and Devon. Perhaps here 'the thorn tree with the spiky branches'.

**POLE** (unlocated, possibly near Alton or Denstone) *Pole* 1339 SHC 1913 77. From OE *pol* 'pool'.

**PONES BROOK** on the north-western boundary of Lichfield. *Pones Brook* 1806 Harwood 1806: 357. See Pones Mill.

**PONES MILL** (unlocated, in Nether Stowe (SJ 1110)) *Pone'smylle* 1460 (1801) Shaw II app. 10, *Pownsmyle* 1514 OSS 1936 56, *Ponds Mill* 1678 SRO DW1738/A/1/1, *Pones Mill als Pound Mill* 1744 SRO DW1738/C/5/6. A mill held c.1180 by Gilbert Poun, and in 1261 by Geoffrey Poun: SHC 1914 125. The name is found in *Pownes fields* recorded in 1647, and preserved in Ponesfield Road: see VCH XIV 71. Perhaps to be associated with *Pounes well*, recorded c.1400: SRO DW1721/1/148-9. On the name Poun(e) see also Franklin 1998: xxxix-xl.

**POOL END** 1 mile north-west of Leek (SJ 9658). *Poolend* 1736 Church Faculty, 1811 *EnclA*, 1842 O.S.

**POOL FARM** 2 miles north-west of Meerbrook (SJ 9763). *Le Pole howse* c.1539 LRMB, *Pool House* 1842 O.S. From Turner's Pool here.

**POOL GREEN FARM** in Tatenhill (SK 2022). *Le olde pole grene* 1546 SRO DW1734/2/3/9f24\*v, *the Olde poole grene* 1550 SRO DW1734/2/3/112b, *Old Pool Green Farm* 1744 SRO D603/E/1/696. 'The grassy open area near the old pool'.

**POOL HALL** on the south-east side of Leek (SK 0055). *Pool Hall* 1695 Leek ParReg.

**POOL HOUSE** ½ mile north of Audley (SJ 7951). *Poole* 1584 Audley ParReg., *Poole House* 1833 O.S. Evidently from a former pool here.

**POPINJAY** 1½ miles south-west of Uttoxeter (SK 0732). *Poppingey Fields* 1775 Yates, *Popinjay* 1836 O.S. Redfern 1886: 370 suggests that the old name of this place was Popingham, but no other evidence has been traced to support such statement.

POPLAR STYLE (unlocated, probably near Claverley, since the place is recorded several times in Claverley parish register) *Popler stile* 1641 Claverley ParReg, *poplar* 1651 ibid, *Poppler style* 1658 ibid. 'The stile at the poplar tree'.

**PORTOBELLO** 2 miles east of Wolverhampton (SO 9498). *Porto Bello* 1775 Yates, 1834 O.S., *Portobello* 1850 Trysull ParReg. The age of the name is uncertain, but it occurs in other counties, normally to commemorate the naval victory at Porto Bello in 1739. Cf. Portobello Farm, Warwickshire; Portobello and Portobello Mine, Derbyshire.

PORT VALE (obsolete) in Burslem (SJ 8649). *Port Vale (Tileries)* 1871 SRO DW1885/6/3. A name that became attached to the Burslem football club, which was based at Port Vale House in 1892. The club was known as Burslem Port Vale until it moved to Hanley c.1911. The *port* element is from a wharf on the Trent & Mersey Canal. Cf. *Port Hill* in Wolstanton, recorded in 1820: SRO D1798/536/413.

**PORTWAY** 1 mile north-east of Rowley regis (SO 9788). *Portway* 1641 Rowley Regis ParReg, 1686 Plot, *Mordern* 1695, *Port Way* 1834 O.S. A common name (e.g. *Port Way*, Uttoxeter, 1601 (1798) Shaw I 56; *Porte Waye* (undated), in Haughton/Coton Clanford (PRO C1/1376/14-15; *the portway*, on the west side of Bushbury Hill, recorded in 1588 (Mander & Tildesley 1960: 8 fn.1)), from OE *port-weg* 'a road leading to a town or market'. Nash 1781: I 521 mentions 'a Roman road called Portway' at Oldbury. The *portwey that leads to Bilston*, recorded c.1538 (SHC 1912 113) is almost certainly *portstrete* mentioned in a charter of Bilston and Wednesfield in 985 (S.860): see Hooke 1983: 74, identified by Duignan 1888: 12, n.11 as a continuation of the Portway in Wednesbury. The road from Forton to Newport was called *Portway* in 1618: *Survey*. Shaw 1798: I 29 records *the Portway* in 1257 in the Rolleston/Tutbury area, and Portway Lane runs from Wigginton to Harlaston.

**POSTERN HOUSE FARM** ½ mile north of Tatenhill (SK 2023). *Postern House* 1774 SRO 11/M/02, *Postern House* 1834 O.S. From ME *posterne* 'a back door, a secondary gate', presumably from a park gate here.

**POTHOOKS BROOK** a tributary of the river Penk. Probably from *pot-hook*, an S-shaped iron hook for suspending pots over a fire, to reflect the meanders of the stream.

**POTTAL POOL** on the western edge of Cannock Chase, in Hatherton parish (SJ 9714). *Pottals Slade* 1806 SRO D3186/8/1/30/17, *Pottal Pool* 1814 *EnclA*, *Pottal Pools* 1834

O.S. Perhaps from OE *potte* 'deep pit or hole', with OE *halh* or OE *hol(h)* 'hole, deep place in water, cave, burrow', but OED gives *pottle, pottel* as obsolete words for a measure of capacity equal to two quarts, so possibly an ironic name for a small pit or pits: there are gravel workings in the immediate area. Slade is from OE *slæd* 'low flat marshy land, a valley'.

**POTTERIES, THE**   an expression in use during the latter half of the 18th century (Wedgwood), and according to the OED a name first recorded in 1825 (but earlier examples have been traced, for example *the potteries* on 1794 gravestone of Jonathan Shelley in Barlaston churchyard) for the pottery and porcelain producing area of North Staffordshire, including Stoke-on-Trent, Burslem, Hanley, Longton, Fenton and Tunstall, but not including Newcastle-under-Lyme.

POTTERSLEGA   (unlocated, perhaps near the north side of Bagot's Park) *potteres leage'* 951 (14th century, S.557), *Potterslega* 13th century SHC XI NS 163. 'The *leah* of the potter(s)': *potter* was possibly a term used for not only for a maker of clay pots, but also for a maker of metal vessels: Blair & Ramsay 1991 93-4. Apart from this place-name, the word potter is not recorded until 1284: OED; Ekwall 1960: xxxiii. A *culnpitt* ('kiln-pit') is recorded in *Cundesley* (see Scounslow Green, q.v.) in the late 13th century (SRO D(W)1721/3/4/6), and could be associated with this place.

**POUKE HILL**   2 miles north-west of Walsall (SO 9999), *Poukehill* 1661 SRO D802/25, *Powke-hill, Powk-Hill* 1686 Plot 174, 212, *Pouch Hill* 1775 Yates, *Pouk Hill Quarry* 1834 O.S. From OE *puca* (Welsh *pwca*, ON *puki*), ME *pouke* 'demon, sprite, hobgoblin', a common element in place-names, particularly with reference to hills and pits: cf. Puckeridge, Hertfordshire; Puckington, Somerset; Pucklechurch, Gloucestershire. Pouke Hill was a huge conical hill of hard bluish-black basalt with veins of white quartz, calcite and zeolite, some 300 or 400 yards in circumference, totally destroyed by quarrying by the early twentieth century, in an area which was formerly isolated, wooded and near a large heath: see WJ August 1908 209-11. StEnc 464 mentions pasture here called *Poukeloftons* in 1565, presumably *Powke Laughton in Walsall Foreign* recorded in 1649 (WLHC 276/250). *The Powgh next Powgh lane* is recorded in 1536 (SRO D4407/28[SF65]), and may be associated with this place. Powke Lane in Rowley Regis probably has the same derivation. See also Mucheberge.

PREES – see **PIPEHILL**.

PREOSTESLAND   (unlocated, in or near Wolverhampton) *Preostesland of Wulfrenehamtun* 11th or 12th century Sawyer 1979a: xxxvii. Possibly to be associated with Prestwood or Priestfield(s) (q.v.). This was probably a record of ownership of land rather than a place-name proper, and may be associated with the land, house and rents in Wolverhampton given by Henry I for the support of 6 priests of the church of Wolverhampton: Hooke & Slater 1986: 32.

**PRESFORD BROOK**   a tributary of the river Sow which flows to the north of Coton Clanford. From Presford (q.v.).

**PRESFORD (HOUSE FARM & BRIDGE)**   2 miles west of Stafford (SJ 8722, 8823). *Prestefordde, Prestford* 13th century SHC IV 277, *Presford (Barn)* 1836 O.S. From OE *preost ford* 'priest ford'.

PRESTESMORE, PRESTYSMORE   (unlocated, in Talke), *Le Prestesmore, Prestysmore* c.1492 SHC 1912 256;   PRESTESMORE   (unlocated, in Shareshill), *le Prestesmore*

1441 Oakden 1984: 118; PRESTEMOR (unlocated, possibly near Caverswall: SRO 3764/21), *le Prestemor* late 13th century SRO 3764/21. 'The priests' moor'.

PRESTFORD BROOKE (unlocated, in Brewood) *Presteford broke* 1303 (1801) Shaw II 293, *Prestford broke* 1409 (1801) ibid. 293.

PRESTON 1 mile west of Penkridge (SJ 9014). *Preston* c.1215 HMC (Middleton 57), 1261 PenkridgeInq, c.1340 ECP, 1548 Survey, 1563, 1620 FF. 'The *tun* of the priests', perhaps to be associated with the priests who served the parochia of the minster church at Penkridge.

PRESTWOOD 3 miles north-east of Wolverhampton (SJ 9401), *Prestwode* 1199 SHC III 39, 1286 SHC V (i) 164, 1399 SHC 1910 145; **PRESTWOOD** 3 miles north-west of Stourbridge (SO 8686), *Prestewuda* 1166 SHC 1923 298, *Prestuuode* c.1250 SRO D938/169, *Prestwode* 1276 SHC VI (i) 82, *Prestewode, Prestwode* 13th and 14th century Duig, *Preestwoodes ferme* 1537 SHC 1912 88; **PRESTWOOD** 1½ miles south-west of Ellastone (SK 1042), *Prestewude* 1197 SHC II 68, *Prestwolde* 1307 SHC XI NS 1307, *Prestuolt* 1324 SHC 1911 360, *Prestwood* 1666 SHC 1925 191, 1749 Bowen, 1836 O.S. From OE *preost wudu* 'priest's wood'. The first place lay within Cannock Forest and was held by the monks of Wolverhampton (see also Preostesland, Priestfield(s)), the second in Kinver Forest and held before the Conquest by the bishops of Worcester. The spellings of the place near Ellastone suggest an alternative containing OE *wald* 'a wood', later 'open upland', an element rare in Staffordshire. *Prestwood Rydway* is recorded in 1567 (SHC 1931 166), possibly in the Balterley area, but has not been located.

PRIESTFIELD(S) in Bilston, 2 miles south-east of Wolverhampton (SO 9397). *le Prest Felde* 1458 (1801) Shaw II 170, *Prestfyld* 1571 SRO D4407/18[SF35], *Prest Field* 1708 SRO D260/M/T/5/123. Land which was once divided between the churches of Penkridge, Stretton and Bilston. See also Preostesland; Prestwood.

PRISONS, THE (unlocated, in Bilston) *The Prisons* 1458 Langford c.1872: 334, 1635 D1798/666/22[316], 1801 Shaw II 170. The derivation of this curious name (found as a field-name *Little Prisons* in 1771) is not known.

PUR BROOK a tributary of the river Blithe. *on pire broc, of pirebrok* 951 (14th century, S.557), *in pire broces heafde, æfter pire broc* 996 (11th century, S.878), *Pirebroc* c.1205 HMC Var Coll 11, 290, *Pilesbroc* 1192x1247 SHC 4th Series IV 79, *Pirbroke* 1548 Ct, *per brooke* 1559 Survey, *Perbrooke* 1711 SBT DR18/22/7/21, *Pirebrook* 1844 Erdeswick 1844: 275. Possibly from a word connected with Norwegian *pira* 'to trickle, to peer', although the history of the word (cf. Pirehill) is unclear: see Ekwall 1928: 333. Hart 1975: 207 prefers a derivation from OE *pur* which he gives to mean 'bittern, snipe' (but according to BT 'tern and black-headed gull'), rather than from OE *pirige* 'pear tree', but the spellings make the latter the most likely derivation, with *pire* representing a reduction of *pirige*. *Pire* is found in *Pirewasse* (q.v.) on Pur Brook. See also Purleyhill.

PURLEYHILL 1 mile north of Hamstall Ridware (SK 1020). *Per Hill* 1775 Yates, *Purlieu hills* 1798 Shaw I 152, 155, *Purl Hill* 1832 Teesdale, *Parley Hill* 1836 O.S. The OED defines *purlieu*, first recorded in 1482, as 'a piece or tract of land on the fringe or border of a forest, originally one that, having been (wrongly, as was thought) included within the bounds of the forest, was disafforested by a new perambulation, but still remained in some respects, especially as to the hunting or killing of game, subject to provisions of Forest Law': the place lies to the west of Needwood Forest. However, Purleyhill is on the east side of Pur Brook (q.v.), and the two names may be associated.

**PUTLEY** on the west side of Bearnett, south-west of Wolverhampton (SO 8894). *Puttelye, Puttelyth* 1262 SHC V (i) 139, *Putley* 1271 1271 SHC 4th Series XVIII 72, *Putteleye, Puttesleye* 1286 SHC V (i) 158, *(wood of) Putley* 1286 SHC 4th Series XVIII 114, *Putly(feld)* 1378 SHC 1928 39, *Puttlease* 1666 SRO D260/M/T/4/85, *Putt Lease* 1672 SRO D260/M/T/4/99. '... on the narrow neck of land which connects the Lloyd with the bulk of Lower Penn ...': printed Penn ParReg vi. The name is preserved in various field-names on the south and west side of Bearnett, and Bearnett House was built in 1854 on the the site of Putley Villa: Dunphy 2002: 49. Possibly from OE *pytt-leah*, 'the *leah* with the pit or pits', although some forms suggest a variant with the second element OE *hlip* 'a slope, a hillside, a declivity, a hill', more especially in the West Midlands 'a concave hillside, a hillside with a single large hollow in one side or a concavity at the foot': Gelling & Cole 2000: 182-5. Perhaps the same place as, or associated with, *Putte, Puthous* (q.v.). See also VCH XX 212.

**PUTTE, PUTHOUS** (unlocated, perhaps in Bushbury, or the same place as Putley (q.v.)) *la Putte* 1220 SHC IV 13, 1271 SHC 4th Series XVIII 72, *Putte* 1255 SHC V (i) 112, 1311 SHC 1911 308, 1322 ibid. 352, *(in the) putte* 1327 SHC VII (i) 251, *Puthous* 1417 SHC XVII 60. From OE *pytt* 'a pit, a natural hollow, an excavated hole', with OE *hus* 'house', perhaps meaning 'the house at the pit or excavation'.

**PYAT'S BARN** 1 mile south-west of Rushton Spencer (SJ 9261). *Piaes Barn (sic)* 1842 O.S. From the family named Piatte, recorded in 1580 (SHC XIV (i) 213), Pyott, recorded in 1590 (SHC 1935 97), Pyatt or Pyott, recorded in the late 16th century (VCH VII 219-220), who occupied property in Rushton.

**PYEBIRCH MANOR** 1 mile south-east of Eccleshall (SJ 8428). *Pipebriche* 1298 Survey. Seemingly from OE *pipe* 'conduit, water-pipe, water-course', with OE *bryce* 'breaking', used of 'newly cultivated ground', normally found in or close to ancient forests and wastes. By metathesis, or shifting of the *r*, the word becomes *burche* and later *birch*. So 'the cultivated land with the drainage pipe'.

**PYECLOUGH** 2 miles south-west of Hollinsclough (SK 0464). *Py(e)clough* 1556 *Deed*, 1566, 1561 Rental, *the Pye Cloughe* 1583 Alstonefield ParReg, *Pyeclough(e)* 1591 DRO D2375M/106/27, *Pyeclough* 1635 Leek ParReg. From OE *pie-cloh* 'steep valley or ravine of the magpies', or perhaps 'steep valley or ravine infested by gnats or midges': cf. Pycombe, Sussex (Ekwall 1960: 376). See also Clough.

**PYE GREEN** 2½ miles north of Cannock (SJ 9814). *Pye Green* 1880 Kelly. From ME *pie-grene* 'the grassy place frequented by magpies', although the element green was often attached to squatter settlements, and that may be the case here. An earlier name for this place may have been Deakin's Grave (*Dickens Lodge* 1758 SRO D603/L181, *Dakins Lodge* 1775 Yates, *Dickin's Grave* 1821 map; see StEnc 721). Legends associating this name with a notorious highwayman who disguised himself as a cleric and became known as 'th' Deacon', and who was apprehended by Bow Street Runners (*sic*) near the Wishing Stone near St Chad's Ditch (see e.g. Anon. 1957: 67-8), are doubtless apocryphal. There is another place called Pye Green, also known as Dab Green (*? Teb Green* 1655 Betley ParReg), in Whitmore parish.

**PYKE BURCHE** (unlocated, possibly near Wood Eaton, or perhaps Pyebirch (q.v.)) *Pyke burche* 1586 SHC 1927 179.

**PYKSHILL** (unlocated, near Pipe Hall) *Pykshill* 1688 SRO D650/6/4.

**PYNLUCDON** (unlocated, possibly near Stone or Walton) *Pynlucdon* 1284 FA.

**QUAMENDEHULLE** (unlocated, in Leekfrith) *Quainendehul* 1330 Ch, *Quamendehul* 1346 Pat, *Gaviendhul* 14th century *Deed*, *Quamendehulle* 1467 SHC IX NS 363. Possibly incorporating OE *cwelm* 'spring', with usual West Midland vowel change (see PN Sa III 25), with OE *ende* 'end, the end of an estate, district or quarter of a village or town', and OE *hyll* 'hill', so 'the hill at the end with the spring'.

**QUARNFORD** an area 5 miles south-west of Buxton (SK 0166; SMR 02628 places a deserted post-Conquest settlement at SK 00206630). *Querneford* 1228 SHC IV 64, *Cornford* 1282-3 SHC XI NS 258, *Quernyford* 1307 ibid. 257, ... *a vaccary* [cow pasture] *called Quernyford* ... 1308 Cal IPM, *Cornford* 1308 SHC XI NS 258, *Cornfeu* 1334 SHC 1914 3, *Corneford* 1421 SHC XVII 96, *? Kerneford* 1340 Pat, *Quernford* 1396-7 SHC XV 78, *Corneford* 1447 SHC III NS 174, *Wharnford* 1614 DRO D2375M/106/27. From OE *cweorn* 'mill, millstone', so 'quern ford', i.e. 'ford by a mill', or 'ford by the place where mill-stones were made, or over which mill-stones were carried'. The ford was perhaps the one over the river Dane at Gradbach.

**QUARRY, THE** ½ mile north-west of Audley (SJ 7850). *Quarne* 1744 SHC 1944 4, *Quarrel* 1744 ibid. 15, *The Quarrels* 1744 ibid.' 23, *Quarrell* 1832 Teesdale. The various forms are inconsistent, but point towards a derivation from OE *quarelle* 'quarry', rather than OE *cweorn* 'a mill, a watermill, a place where mill-stones were quarried', although the place does lie on a watercourse.

**QUARRY HEATH** 1 mile east of Penkridge (SJ 9413). *Quarr(e)y Heath* 1598 Ct, 1622 FF, 1634, 1682 ParReg. From ME *quarrere* 'quarry', and OE *hǣð* 'heath'. Perhaps to be associated with *Stondelf* 'the stone diggings' recorded in 1261: SHC 1950-1 47.

**QUATFORD** Ancient Parish on the east bank of the river Severn, 2 miles south-east of Bridgnorth (SO 7391). *Quattford* 1012x1056 Forster 1941: 769, *Quatford* 1086 DB, *Quateford* 1271-2 Ass, c.1540 Leyland, *Quhatford* 1595 PR(H) 3. 'The ford near the place called Cwat(t)': see Quatt. There is a very slight possibility that the name means 'the ford at the cwatt': the mound, usually taken to be a Norman motte (TSAS LVII 1961-4 47-62), on the riverside sandstone cliff known as Camp Hill, was held to be a tumulus in 1829 (Scott 1832: 331, although this report may confuse the excavation of the bailey in 1830-31: see TSAS LVII 1961-4 61-2), and if the motte was a re-modelled tumulus, the mound may have given its name to the ford. In Shropshire since the 12th century. Watford Farm in Hartington Upper Quarter in Derbyshire is recorded as *Quatford* c.1104: SHC 1924 328; PN Db 372, but that name (which is probably 'wheat ford') does not help solve the name of Quatford near Bridgnorth: PN Db 372.

**QUATT** Ancient Parish 4 miles south-east of Bridgnorth (SO 7588). *? Cwatt* 1015 (Searle 1897: 335; Finberg 1972: 225-6), *Quatone* 1086 DB (listed under Warwickshire), *Quatte* 1212 Fees, 1291 Tax, 1363 SHC X NS (ii) 113. A unique and difficult name, for which various derivations have been advanced. Eyton 1854-60: I 104 felt that Cwatt represented Welsh *coed*, earlier *coyt* 'forest', but that derivation presents philological difficulties, since a rounded vowel is not likely to have developed early enough to produce a 9th century OE place-name form *Cwat-*: the sound is likely to have been 8th century in Welsh, and the normal form in English place-names is *Chet-*, *Chit-*, etc. (on which see Gelling 1992: 66-71), and for this derivation the area would have had to be Welsh-speaking in the 8th century, which is unlikely (see Jackson 1953: 327). Furthermore, on philological grounds the root of the name is more likely to be English than Welsh. In considering the personal name Leofwine Cwatt, recorded in 1015 (S.1503 Searle 1897: 335; Whitelock 1930: 60-1, 173), Tengvik 1938: 305 took the view that the byname was probably identical with an early ME and ModE dialect word *quat* 'pimple,

447

boil', formed from the German stem *kwat- with the sense of 'something lumpy, protruding'. Förster 1941: 769 thought a personal name might be involved, and Ekwall put forward a derivation from OE *cwead* 'dirt, mud', but later conceded that such theory was not entirely satisfactory and should be abandoned, observing that it was doubtful, though not altogether impossible, that there could have been a nickname Cwatt, and concluding that the most reasonable explanation seemed to be a derivation from a word *cwatt*, with a meaning liable to give rise to both a nickname and a place-name, and that a word meaning 'a lump, a hump', or similar would be possible, although it appeared to have no analogy in any Germanic language, and suggesting, with typical prescience, further consideration of the local topography (Ekwall 1960: 376-7; TSAS LVII 1961-4 45 fn.14). PN Sa 56-9 concludes that the name is unexplained. The derivation of the name is almost certainly from OE *cwætt* or *cwatt*, ancestor of modern *quat* 'a pimple, a pustule, a small boil, a lump', not recorded before its appearance in Othello (V I) in 1579 (OED), but in common use in several Midland counties in the 19th century (OED; Baker 1842: II 150), found here in a topographical sense: the village of Quatt is centred on a pronounced lump or knoll. The churchyard is a slightly domed roughly square plateau modelled out of the summit of the knoll, with the parish church of St Andrew (incorporating Anglo-Saxon fabric) standing on its highest point: see TSAS LXVI 1989 15. Possibly Leofwine Cwatt was distinguished by some lump or blemish, or his byname was taken from the place-name. Quatt has been in Shropshire since the 12th century. The place called *Cwatbrycg(e)*, otherwise *Brycge*, mentioned in relation to events in 895 and 910 (10th century, ASC 'A'; 978-988 Æthelweard), which evidently incorporates the name, was almost certainly at or close to Quatford, where a bridge is recorded in 1086 (see especially Groom 1992: 19), and there is little evidence, apart from a doubtful reference in John ('Florence') of Worcester), to associate it with Bridgnorth, four miles to the north, as suggested by TSAS LVII 1961-4 38, and PN Sa I 56-9. The existence of any bridge or settlement at Bridgnorth (first known as *Brycge*) before 1101-2, when the borough, castle and bridge at Quatford were transferred there by Robert de Bellême (Eyton 1854-50: I 242), is improbable, and John of Worcester may have conflated the two places. *Quateway* (presumably 'the way or road to Quatt') is recorded in a 16th century deed relating to Upper Arley: VCH Wo III 5. See also PN Sa I 58, 248; TSAS LVII 1961-4 37-46.

**QUEELANE (FARM)** 2 miles south of Uttoxeter (SK 0830). *Quee Lanes* 1775 Yates, 1836 O.S. A curious name. Possibly from the Northern dialect *quee* 'a female calf' (Halliwell).

**QUEEN'S LOW** a tumulus near Tixall (SJ 9623), *Queen's Low* 1686 Plot, 1844 Erdeswick 1844: 70; QUEEN LOW (unlocated) on Ribden, 1 mile south of Cauldon (? SK 0747), *Queen Low* 1686 Plot 404. From OE *hlaw* 'hill, burial mound', with OE *cwene* 'queen', or the OE female name Cwene. For Queen's Low see JNSFC 3 1965 49; WMA 1987 30 38-9. See also Guendelawe; King's Low; Quennedale. Other lows in the area included *Le Heghlow*, recorded in 1346 (SRO D938/350), and *Stanylowe*, recorded in 1347 (ibid. D938/352).

**QUEENSVILLE** on the south-east side of Stafford (SJ 9322). Adopted as the new name for the hamlet of Spittal Brook, so-named from the brook associated with the hospital of St. Leonard (*Spittall* 1487 VCH V 91), to commemorate Queen Victoria's coronation in 1838: VCH VI 194.

QUENNEDALE unlocated, perhaps on Tixall Heath (see VCH III 261 fn.18), but the first form suggests a location in Ingestre. *Quennedale in Iggestroud* 1161x1182 SRO

1/7972, *Quennedale* c.1200 SHC VIII (i) 135, 1261 SHC 1914 121. Possibly from OE *cwene dæl* 'the woman's valley'. OE *cwen* meant 'queen, wife'; OE *cwene* meant 'woman, female serf, prostitute'. Finberg 1975: 200 suggests that *Quyndale* in Lancashire is from ON *\*hvin*, ME *whin*, 'whin, gorse'.

**QUESLETT** 3½ miles south of Aldridge, in Great Barr (SP 0694). *Quieslade, Queeslade* 16th century Duig, *Queislet* 1686 Plot 403, *Queeslet* 1766 ABG, *Queeslet* 1834 O.S. From ME *queest* 'wood-pigeon', with OE *slæd* 'small valley'. The place lies in a hollow.

**QUINTIN'S ORCHARD** 1½ miles south-west of Hamstall Ridware (SK 0818). *Pipehalle* 1319 (1798) Shaw I 166*, *Pypehalle* 1420 ibid, *Pypehal orcharde* 1443 ibid, *Pip-hall* 1560 SHC 1925 27, *? Quintaynes Leasowes or Nevalles Heies* 1654 WaCRO CR1908/16/12, *? Quintons Barn* 1700 SRO D260/M/T/5/134, *Quinton's orchard* ibid, *Quintin's Orchard* 1834 O.S. For Pipe see Ridware, Pipe. According to Shaw (1798: I 166*) the later name is from that of the occupier ('Quinten or Quinton, corruptedly Quintin or Quintyn') in 1606 and later, and (Shaw 1798: I 357): 'Quintyn, or St. Quintyn, is a name and family of note for antiquity and possessions in Wall, Lichfield and Longdon ... derived of French lineage that took their name from St. Quintyn, a town in Piccardy, and most likely settled in this nation in the reign of William I'. In 1286 Ralph Quenten/Quynten/Quyntyn of Fradley was forester of Alrewas and keeper of the prison at Cannock: SHC 4th Series XVIII 110, 128-9, 143. See also Pipe Hall. Shaw 1798: I 166 observes 'there is no reason to believe that the antient manerial residence [of Pipe Hall] was not situate on the Trent, but at or near Linacre...'.

**QUIXHILL** 1 mile north-west of Rocester (SK 1041). *Quikesulle, Quicksall* 1191x1194 CEC 261, *Quhicol, Kuhicel* pre-1236 SHC 1921 37-8, *Witekeshull* 1236 SHC 1911 405, *Wikeshull* 1240 SHC IV 237, *Wytekeshull* 1242 Fees, *Whydekeshull* 1272 SHC IV 211, *Quikeshull* 1272 FF, 1279 Ass, *Quyxhull* 1277 SHC 1911 31, *Wexhull* 1277 SHC V (i) 91, *Wykehull* 1309 SHC X 4, *Quicsul* 1327 SHC VII 217, *Ewykkshull'* 1332 SHC X 113, *Quyxsale* 1387 SRO 3764/98[40001], *Cwykhill* 1529 SHC 1912 35, *Quiksell* 1666 SHC 1925 194, *Quiksill* 1686 Plot, *Quikshill* 1798 Yates. Ekwall 1960: 377 gives 'Cwic's hill', but some spellings point towards an alternative OE form with an OE personal name *\*Wittuc*, which may be found in Uttoxeter (q.v.), less than five miles from Quixhill, suggesting that the same individual may be associated with both places. Whittingslow ('Hwittuc's tumulus'), 3 miles south-west of Church Stretton, and Whixall, both in Shropshire, may incorporate the same personal name. Quixhill is a puzzling name for a hamlet which lies in a valley bottom: the name was evidently taken from that of a nearby hill.

**RAD BROOK** a tributary of the river Churnet. *Rad Brook* 1842 O.S. Almost certainly from OE *read broc* 'the red brook': see Rudyard.

RADBROOK FIELD (obsolete) on south-west side of Rugeley (SK 0316). *Redebroke* 1262 (1798) Shaw I 107, *reddebroke* 1570 *Survey*, *redbrokefield* 1585 *Comm*, *Redbrook Meadow* 1885 *EnclA*. The name is more likely to be from the colour of the water in the brook (perhaps stained from iron) than from OE *hreod* 'reed'. See also Radmore.

**RADDLE FARM** on the east side of Elford (SK 1910), *Raddle Farm* 1834 O.S., 1863 BCA MS3878/966; **RADDLE FARM** 1½ miles north-east of Edingale (SK 2213), *Raddle Farm* 1834 O.S. Perhaps from *raddle* 'to paint or mark with raddle; to colour with red', so 'the red-painted farmhouse', or 'the barn where rams were raddled', presumably associated with the obsolete *Raddle Barn* (1834 O.S.), 1 mile south-east of Harlaston (SK 2209). Cf. Raddlebarn, Worcestershire (PN Wo 349).

**RADDLEPITS** on the Weaver Hills (SK 1045). *Raddlepits* 1836 O.S. *Raddle* is a variant of *ruddle* 'red ochre' (from OE *rudu* 'red or ruddy colour'), used for marking sheep. There is an ancient earthwork here (VCH I 372 mentions a line of trenches 166 yards long; see also SHC 1916 207), and it is unclear whether the name has been attached to the earthworks because they resembled raddle pits, or whether the place was actually used as raddle pits: evidence for the latter may be detected in the name Wredon (q.v.). *Raddlepits* in Ipstones is recorded in 1644 (SHC 4th Series I 213), and Plot 1686: 124 mentions 'Ruddle or Red-Ocher they dig very good at the parish of Ipston'.

RADDLE STICH (obsolete) on the south-east side of Mobberley (SK 0141). *Raddle Sitch* 1775 Yates, *Raddlestick* 1791 SOT SD4842/20/64, *Raddle Stich* 1841 O.S. From *raddle*, a variant of *ruddle* 'red ochre', with OE *sic* 'watercourse', so giving 'the stream coloured by raddle'.

RADECLIFF (unlocated) *Radecliff* 1256 Ch.

**RADFORD** a crossing of the river Penk, 1 mile south-east of Stafford (SJ 9321). *Radfordbregge* 1476 SHC VI NS (i) 98, *Radford bridge* 1576 SHC 1926 33, *Ratforde Bridge* 1600 SHC 1930 270, *Ratford Bridge* 1608 SHC 1948-9 82, *Radford Bridge* 1775 Yates, *Radford* 1836 QS. In most cases the common name Radford probably means 'red ford', often from the colour of the soil or near the ford, but some places of this name may be from OE *radeford* or *rȳdeford* 'horse-riding ford', presumably denoting a ford unsuitable for crossing on foot. The bridge here, recorded *temp.* John (VCH VI 197), was formerly of considerable importance, as shown by its maintenance by the Hundred, and evidently replaced the ford hereabouts (see VCH V 2). The bridge also marked the boundary of the Forest of Cannock: SHC VIII (ii) 133.

**RADLEY MOOR** the name of several fields between Little Aston and Footherley in Shenstone over which runs the ancient Icknield Street (SK 0902). The name might be from OE *rad* 'road', from Icknield Street, but if it is to be identified with *æt reod lege* mentioned in a charter of 957 of land at Aston and Barr (12th century, S.574), then perhaps from OE *read* 'red', denoting the colour of the soil or vegetation, with OE *leah*, hence 'red *leah*' with later *moor*.

**RADMORE LANE (FARM)** 1½ miles west of Gnosall (SJ 7920). *Redamora* 1157 P, *Rademora* 1156, 1158 ibid, *Radmore* 1227 ch, *Radmo(o)re Lane* 1481 Ct, 1668 Survey, 1775 Yates, *Radnor lane* 1695 Morden, *Radmore Lane* 1833 O.S. From OE *read-mor* 'the red marshland', possibly so-coloured from clay or peat-staining.

**RADMORE WOOD** ½ mile north-east of Abbots Bromley (SK 0825), *Redmora* c.1220 SHC 1937 35, *Rodemore* 1275 ibid.' 79, *Radmore (Lane)* 1542 SRO D4038/A/3/2, *Radmore* 1836 O.S.; **RADMORE, RED MOOR** ½ mile south-west of Cannock Wood (SK 0411), *Radenor* 1141-9 BM, *Redamora* 1157, *Rademora* 1156, 1158 P, *Redemore* c.1235 SHC XII 274, *Radmore* 1227 Ch, *Rydemore* 1291 (1798) Shaw I xxi, *Redemoor* 1279 SHC VI (i) 93, *Redde More* 1505 OSS 1936 51. Possibly from OE *hreod-mor* 'marshland where reeds grow', but the spellings also suggest OE *read-mor* 'red moor', from the colour of the soil or vegetation (and Redmoor (*sic*) Stream and Red Gutter stream near Cannock Wood are said to have been stained as they flowed over ochrous outcrops up to 18" thick: MidA III 59; WJ 1904 247). Radmore near Cannock Wood is the site of the Cistercian abbey of Radmore founded in c.1143 which moved to

Stoneleigh in Warwickshire c.1154 (VCH III 225; EEA 14 106-9), the site being converted into a royal hunting lodge (TSSAHS VII 1965-6 31; see also Cannock).

**RADWAY HILL** 2 miles north of Kinver (SO 8586). Presumably from the red colour of the soil here. See also Redford.

**RADWOOD** 2 miles west of Whitmore (SJ 7741; SMR 02629 places a deserted post-Conquest settlement at SJ 77304120). *Radwode* 1272 SHC VIII 151, *Radewode* 1335 SHC XI 50, *Radwood* 1386 SRO DW1082/A/3/3, *(the hall of) Radwood* c.1565 SHC IX NS 180, 198, 1609 SHC III NS 46, *Meare Radwood* 1606 SHC XVIII 61, *Radwood (house)* 1679 SHC 1914 16. Possibly 'the reedy wood': it is very unlikely that the name *hreoditan more* 'the reedy moor', found in the bounds of a charter of Madeley of 975 (11th century, S.801), is to be associated with this place as suggested in Hooke 1983: 106-9; see also Hart 1985: 96. *Meare* is Maer.

RAGGED INVENTION (obsolete) 2 miles west of Codsall (SJ 8504). *Ragged Invention* 1834 O.S. The meaning of this long-disused name, which seems to have been linked to what is now known as Wheatstone Farm, is unknown. Ragged had the meaning 'rough, irregular, jagged, and (of sounds) harsh, discordant', so possibly 'the noisy invention or contraption'. Cf. New Invention.

RAGLIS (obsolete) ½ mile north of Crumpfield (SO 0166). *Raglis* 1832 O.S. Perhaps from the surname *de Wraggeleye*, found in the Subsidy Roll of 1275, from Ragley in Worcestershire: PN Wo 364. In Tardebigge parish, forming part of Staffordshire from c.1100 until 1266, in Warwickshire until 1844, and since that date in Worcestershire.

**RAILS FARM** 2 miles north-east of Biddulph (SJ 9158). *The Rails* 1842 O.S. Said to be from the railed fence that enclosed Horton Hay in the early 15th century: VCH VII 67.

**RAKE END** on north-west of Hill Ridware (SK 0718), *Le Rake* 1334 (1798) Shaw I 200, *Rakeynd* 1523 ibid, *Rake Ende* 1562 SHC XIII 219, *Raikend* 1834 O.S. Perhaps from ME *rake* 'a rough path'. The word *end* meant not a terminal point, but simply 'a place', and was often applied to squatter dwellings on the outskirts of a settlement.

**RAKE GATE** 2 miles north-west of Wolverhampton (SK 9002). *? Bake Gate* [sic] *(alias Wyotts meadow)* c.1635 SHC 1928 106, *Rake Yate* 1649 map TSSAHS XXI 1979-80 16, *... a meadow called Rake gate ...* 17th century SHC 1928 102, *Rake Gate* 1702 Bushbury ParReg, *Rake Yeates* 1712 ibid, *Rake Yate* 1778 WALS DX13/17, *Rake Gate Farm* 1895 O.S., *Rake Gate* 1921 O.S. Perhaps from ME *rake* 'a rough path', but more likely from a gate made from a discarded horse-drawn implement.

**RAKE HILL** on north side of Burntwood (SK 0509). *Rackhill* 1597 VCH XIV 199, *Rakehill Lane* 1670 ibid.' The road now called Rake Hill was known as *Stephen's Hill* in the early 19th century: ibid; 1834 O.S. Perhaps from ME *rake* 'a rough path', or a more recent term for an incline, so perhaps 'the hill with the steep path'.

RAKEMOR, LE (unlocated, near Fisherwick) *Le Rakemor* 1309 WL 103.

**RAKES, THE** 1 mile south-east of Sheen (SK 1159). *Rackes Head* 1657 Alstonefield ParReg, *Raikeshead* 1660 ibid, *the Rakes* 1740 ibid, *Rakes* 1829 SRO D538/A/5/59, 1840 O.S. From OE *hrace* throat', used topographically in the sense of 'a pass', so 'the head or end of the narrow valley': cf. The Rake, Sussex. The place was also known as *Bartine Edge* in 1651: VCH VII 242.

**RAKES DALE** ½ mile west of Alton (SK 0642). *Rakes Dale* 1836 O.S. Perhaps from OE *hrace hrœce* throat', used topographically in the sense of 'a pass', with OE *dœl* 'valley'.

**RAKEWAY** 1 mile south-east of Cheadle (SK 0241). *Rakeway* c.1680 SHC 1919 269, 1798 Yates, 1836 O.S. From OE *hrace* throat', used topographically in the sense of 'a pass', found in 14th century England as *rake* 'a way, a (rough) path', but perhaps here in the more recent meaning 'a slope, an incline', so 'the way or track in the narrow valley', or 'the sloping track': the place lies on a hillside.

**RALEIGH HALL** 1 mile north of Eccleshall (SJ 8330). One of a number of Government establishments built in the area in the early 1940s and named after naval heroes. See also Beatty Hall, Drake Hall, Duncan Hall, Frobisher Hall, Howard Hall, Nelson Hall, Rodney Hall.

RAMENSCLOUGH (unlocated, in Leek) *Ramensclough* 1562 SHC IX NS 113. Perhaps a mistranscription of Raven's Clough (q.v.).

RAMMESHED (unlocated) *Rammeshed* 1305 SHC VII 124. Perhaps the same place as *Rommeseude*, recorded in 1327 (SHC 1913 14).

**RAMSHAW, RAMSHAW ROCKS** 1 mile north-east of Upper Hulme (SK 0162). *Ramshaw* 1842 O.S., *The Rocks* 1842 ibid. The second element is OE *scaga* 'a grove'. For the first element, see Ramshorn or Ramsor. Cf Ramshaw, Derbyshire.

**RAMSHORN or RAMSOR** (pronounced Ramser [ræmzə]) 5 miles east of Cheadle (SK 0845). *Rumesoura* 1196 SHC 1923 41, *Romesovere* 1275 SHC V (i) 120, *Ramnesoure* 1307 SHC XI NS 263, *Roumesovere* 1327 SHC XI NS 186, *Rommeseude* 1327 SHC 1913 14, *Romesovere* 1327 ibid.17, *Romesor* 1309 IPM, *Rommessore* 14th century Duig, *Ramsor* 1538 *et freq.* Ellastone ParReg, *Ramshorn* 1723 ibid, *Romshorn* 1809 ibid. The first element is probably from OE *rammes*, genitive singular of *ramm* 'a ram', or OE *hrœfn* 'raven', or a derived OE personal name. A derivation from OE *hramsa* 'wild garlic' is unlikely, since the plant favours woodland at lower levels. The second element is OE *ofer* (frequently shortened to *ore*) 'slope, hill-side, ridge'. The partly-wooded topography surrounding Ramsor makes it difficult to identify the *ofer*, but it is perhaps the rising ground to the west of Sycamore Farm. See also TNSFC 1972 124; TNSFC 1967 197; TSSAHS 1967 35-6. Another Ramshorn is recorded at Huntington q.v.).

RANASHE (unlocated) *Ranashe* 1414 SHC XVII 23.

**RANGEMOOR** 1 mile north-east of Ellastone (SK 1245). *Rangemoor (Wood)* 1836 O.S. Early spellings have not been traced.

**RANGEMORE** 1 mile north-west of Tatenhill (SK 1822). *Rauenwolmesmor* 1337 Ct, *Rauenesmor* 1337 Hardy 1908: 26, *Rangemoor (House & Wood)* 1836 O.S. The first spelling appears to incoporate OE *hrœfn* 'raven' and Mercian OE *wœlm* 'spring', or the personal name *Hræfn or Hrafn (*Hræfn is an OE personal name, *Hrafn an ON personal name), with OE *mor* 'moor', so 'raven's spring moor', and 'raven's moor' or '*Hræfn's or Hrafn's moor'. The place was also known as *Tatenhill Gate* (1838 O.S.), being one of the gates of Needwood Forest: StEnc 470.

**RANGER, THE** 1 mile south of Oakamoor (SK 0543). StSt 12 2000 70 associates *Ringie* c.1595 (Middleton 5/165/52/3d) with this place, but the spelling suggests that that place may be Ring Hey (q.v.).

**RANSLOW FARM** 1 mile south-west of Salt (SJ 9427). *Ranslowe* c.1785 SRO D240/ER/1/21. The age of this name is not known, and it is unclear whether it is to be associated with *Rollowe*, recorded in 1279 (SHC VI (i) 138). A derivation from the ON personal name Hrani, with OE *hlaw* 'tumulus, burial mound', poses philological difficulties with the first element, but the name evidently incorporates OE *hlaw*. An estate called Ranslow appears to have existed near Lichfield Road, Stafford, c.1790, perhaps near Weeping Cross: see SRO D240/ER/1/21; SRO 5593/9/33a.

**RANTON or RONTON** Ancient Parish 4½ miles west of Stafford (SJ 8524). *Rantone* 1086 DB, *Rantonie* c.1182 SHC II 256, *Rampton, Ramton* 1208 SHC III (i) 142-3, *Ranton* 1209 ibid. 175, *Ronton* 1236 Fees, *Raunton* 13th century Duig, *Routon* 1350 Erdeswick 1844: 136, *Rannton* 1532 SHC 4th Series 8 67, *Ramton* c.1540 Leland, *Runton* 1471 SHC IV NS 175, *Ramton* c.1540 Leland. Probably from OE *rand* 'edge, border, bank', hence '*tun* on the border' (the place is in Pirehill Hundred one mile from the boundary with Cuttlestone Hundred), or (perhaps more likely) '*tun* at the bank': JNSFC XXXVI 1901-2 118 describes nearby Brough Hall as 'a strongly entrenched position on high ground ... connected with Ranton by a remarkable earthwork or vallum about 25' wide, traces of which are also to be seen in the wood to the north of Ranton Abbey on the way to Ellenhall', such earthwork (presumably the 'vallum or raised road in the neightbourhood of Ranton Abbey' mentioned in VCH I 186) perhaps the *Wal* (from OE *weall* 'a wall, a rampart of earth or stone') recorded in 1213 (SHC III 161). The foundation charter of Ranton Abbey, created as a cell of Haughmond Abbey in Shropshire before 1166 (SHC 1914 94), refers to the name as the house of St Mary *des Essarz* (VCH III 251), also recorded as *de Sartis* or *Essars Abbey* (Erdeswick 1844: 136), or *des Essarz, Exsartis* (SHC IV 264, 267), from ME (OFr) *assart, essart* 'a clearing in woodland', often shortened to *sart*, and frequently found in ME and later field-names: 'the monastery had probably been built on assarts from the waste of the manor': SHC IV 264 fn.1. Ranton was extra-parochial until 1857. The Ordnance Survey gives Ranton, but the name is sometimes found as Ronton. The parish registers show that Ranton was generally used for the abbey and liberty, and Ronton for the parish. See also Broad Heath.

RATHERSEATES (unlocated, in Alstonefield) *Ratherseates* 1603 SHC 1946 66, *Radder Seat(e)s* 1607, 1737 Alstonefield ParReg, *Rotherseat* 1661 ibid. Perhaps from OE *hriðer, hrȳðer* 'an ox, cattle', or the ON personal name Hraði, or ON *rauðr* 'red', usually in allusion to the colour of soil (though each of those suggestions poses philological difficulties), with ON *sætr* 'mountain pasture', or ME *sete* 'a house, a permanent residence'. It is unclear whether *Rashets*, recorded in 1586 (SHC 1927 172) is to be associated with this place.

RAUENSHYLL (unlocated, in Bignall) *Rauenshyll in Bignoo* 1492 SHC 1912 256. From OE *hræfn hyll* 'raven hill'.

**RAVEN HILL** ½ mile south-east of Rugeley (SK 0517). *Ryvynghul* 1428 Deed (VCH V 159), *Revynghill* 16th century Survey, *Reavinghill (Leasowe)* 1653 WCRO CR1908/17/1, *Revering Hill* 1659 Erdeswick 1844: 69, *Reeving Hill* 1747 SRO D615/E/6/4, *Ravenhill* 1771 SRO D1161/1/1/7, *Reveing Hill* 1775 Yates, *Ravenghill* 1794 SRO D1161/1/1/9, *Raven Hill* 1834 O.S. Perhaps from OE *hreof* 'rough, rugged', with -*ing* and OE *hyll* 'hill' (cf. Rivington Pike, Lancashire).

**RAVENSCLIFFE** 2 miles north-west of Tunstall (SJ 8352). *Romersclyf* 1307 SHC XI NS 262, *Romesclyf* 1326 JNSFC LIX 1924-5 38, *Romusclyff* 1348 ibid.' 48, *Ramusclyf* c.1360 ibid. 55, *? Rame cliff* 1509 SHC 1923 320, *Raunscliffe* 1567 SHC 1938 131,

*Ranescliffe, Ranescliff* c.1575 SHC 1912 205, *Ravenscliffe* 1579 SRO D1229/1/3/62, *Ramsclyffe* 1589 SHC XV 190, 1590 SHC 1930 197, *Ravensclyffe* 1601 JNSFC LXIII 1928-9 41, *Ravensclyff* 1601 ibid. 43, *Ranescliffe* 1608 SHC 1948-9 76, *Ramscliff(e)* 17th century SHC XII NS 36-7, *Ranscliff* 1836 O.S. Perhaps from OE *hramsa* 'ransom, wild garlic, *Allium ursinum*', with OE *clif* 'cliff', or from an unidentified personal name. Some of the spellings might relate to Ravenscliff west of Kniveton in Derbyshire.

**RAVEN'S CLOUGH** 1½ miles north-west of Rushton Spencer (SJ 9163). *Raven's Clough* 1596 VCH VII 224, *Ravenscloughe* 1604 Eliz ChancP, 1607 SHC III NS 9, *Ravens Clough* 1775 Yates, 1842 O.S. 'The clough or steep-sided valley frequented by ravens'. See also Clough; Ramensclough.

**RAVENSCLOUGH BROOK** a tributary of the river Dane. From Raven's Clough (q.v.).

RAVENSDALE (unlocated, possibly near Waterfall) *Ravensdale* 1292 SHC VI (i) 226, 1327 SHC VII (i) 217, 1344 SHC 1913 108, 1374 SHC XIV 118, 1392 SHC XV 43, *Revensdale* 1292 SHC 1911 215, *Rawennsdale* 1370 SRO D1229/1/4/44, *Ravennsdale* 1376 Okeover T55, *Ravenesdale* 1377 SHC 4th Series VI 7. 'The dale with the ravens'.

**RAVENSHALL** ½ mile south of Betley (SJ 7547). *Ravenshale* 1323 AD, *Raveneshalow* 1327 SHC VII 207, *Ramishall* 1590 Betley ParReg, *Raunshall otherwise Raunsor* 1596 SHC XVI 157, *Ravenhall* 1600 Betley ParReg, *Rannsall* 1611 BCA MS3810/125, *Ransall* 1628 Audley ParReg, *Ranshall* c.1647 SRO DW1082/C/3/1-5, *Ravenshill* 1679 SHC XII NS 212, *Ransell, Rensall* 1690 Betley ParReg, *Ravenshill or Ranshall* 1744 SRO D1461/7/10, *Ravenshall* 1833 O.S. From OE *hræfn halh* 'the *halh* of the ravens', or possibly 'Hrafn's or *Hræfn's halh*': Hrafn is an ON personal name, and *Hræfn an OE personal name.

**RAVENSHAW WOOD** 2 miles north of Lichfield (SK 1213). *Ravenshaw Wood* 1887 O.S. From OE *scaga* 'a copse, a grove, a small wood', so 'the small wood with the ravens'.

**RAVENSHURST** 1 mile north of Harborne (SO 0285). *Ravenhurst* 1656 Dugdale, *Ravens House* 1775 Yates, 1787 Cary, 1804 Smith. From OE *hyrst* 'hill, wooded hill, copse', so 'the wooded hill with the ravens'. The name is remembered in Ravenshurst Road.

RAVENS NEST (obsolete) 1 mile west of Codsall, just within Shropshire (SJ 8304), *Raven Nest* 1741 Codsall ParReg, *Ravens Nest* 1775 Yates; RAVENS NEST (obsolete) 1 mile south of Newborough (SK 1323), *Ravensnest (Gate)* 1788 VCH II 350, *Ravens Nest* 1836 O.S. *Ravensnest* recorded in 1203 (SHC III 95) is unidentified, but may be the place near Codsall.

RAWNESHAWE (unlocated, in King's Bromley) *Rawneshawe* 1599 SHC 1931 156. Possibly associated with *Reyneseshawnes* (undated), recorded in Shaw 1798: I 132.

**RAWNPIKE OAK** from a great oak which, damaged by lightning, stood until 1932 at the foot of Castle Ring on Cannock Chase (SK 0412). *Rawn pike Oak* 1884 MidA III 9 1884 63. *Raunpick, raunpike* and *rampick* are dialect words for a tree beginning to decay at the top from age: EDD. See also Rawnsley. A replacement tree was planted in 2000. See also StEnc 472.

**RAWNSLEY** 1 mile east of Hednesford (SK 0212), *Rawnsley* 1895 O.S.; **RAWNSLEY HILLS** 1 mile north-east of Hednesford (SK 0212), *Rawnsley Hills* 1834 O.S. Oakden

1984: 60 suggests a derivation from OE *hræfn, hrafn* 'raven', with OE *leah*, hence 'the *leah* frequented by ravens', but early forms have not been traced, and the name may not be ancient, in which case it may have the same root as nearby Rawnpike Oak (q.v.).

**RAY HALL or REA HALL**  2 miles south-west of Great Barr (SK 0294). *Rehall* 1215 Duig, *Reahall* 1599 LJRO D187/1/6, *Reahall, Rea Hall* 1609 BCA MS3810/196, *Rea Hall* 1655 Willett 1882: 201, *Ray Hall* 1834 O.S. The place is probably so-named from OE *ea* 'a stream': the river Tame runs here. In OE charters *æt þære ea* 'on the river' is commonly found, and the last word sometimes takes on part of the preceding word to form *rea, ray, rhee*, all of which are found as stream names in various parts of the country. The derivation of this type of name (said to date from about the 12th century) is held by Ekwall 1928: 337 to have been deduced by Duignan 1902: 125-6.

**REA CLIFF FARM**  1 mile north of Horton (SJ 9458). *Rayclyffe, Rayclyff, Raycliffe* 1538 SHC NS X (i) 139, *Pey Cliff (sic)* 1566 SRO DW1761/A/4/79, *Reacliff* 1675 ParReg, *Raecliff Farm* 1842 O.S. From OE *æt þere ea* 'at the river', a common expression which often becomes Rea by misdivision (see also Ray Hall), so here 'the cliff or bank above the river': the place lies on the west side of a valley above what was Dunsmore Brook, dammed c.1793 to create Rudyard reservoir. The 1566 spelling is likely to be a transcription error.

**REAPS MOOR**  2 miles north of Warslow (SK 0861). *Reaps moor* 1595 VCH VII 29, *the Reapes* 1601 Alstonefield ParReg, *Reapsmore, Reaps Moore* 1650 ibid, *Repesmoore End* 1682 ibid, *Repemoor Top* 1775 Yates, *Repemoor* 1787 Cary, *Repsmoor* 1794 Stockdale, *Reaps Moor* 1840 O.S. The first element is uncertain, but the surname Reap is recorded (DES 374), or possibly from ME *repe* 'sheaf of corn' (cf. Reaps, in Charlesworth, Derbyshire: PN Db 71), suggesting that this area was cultivated.

**RED BROOK**  a tributary of the river Swarbourn. *Red(e)broke* 1262 Hardy 1907: 42, *the Redebroke* 1330 (1798) Shaw I 113, 1603 DuLaMb. The first element is either from OE *read* 'ed' or OE *hreod* 'reed'.

**RED BULL**  ½ mile north of Almington (SJ 7035). *Red Bull* 1733 BCA MS3069/Acc1930-022. From a public house of this name.

RED CROSS  (obsolete) on south side of Biddulph (SJ 8856). *Red Cross* 1791 SHC 4th Series 13 137. From a tall square red freestone pillar: ibid. Carvings on the cross are said to date from the 19th century: JNSFC XXIV 1890 26. The pillar is now in Knypersley churchyard: StEnc 473.

**REDDAL HILL**  on the north side of Cradley Heath (SO 9586). *Reddall Hill* 1834 O.S. In the absence of early spellings any derivation must be conjectural, but perhaps if ancient from OE *read* 'red', or OE *hreod* 'reeds', or OE *\*reod* 'clearing', with OE *halh*, with the later addition of hill.

**REDDITCH**  6 miles south-east of Bromsgrove (SO 0467). *de Rubeo Fossato* c.1200, *la Rededich, le Rededych* 1247 PN Wo 364, 1300 Pat, *Reddich, Reddyche* 1394 Pat, *The Rediche, le Redyche* 1446 AD ii, 1464 Pat, *The Redde Dych* 1536 PN Wo 364. From OE *read, dic* 'red ditch', from the colour of the soil here. In Tardebigge parish, forming part of Staffordshire from c.1100 until 1266, in Warwickshire until 1844, and since that date in Worcestershire.

**RED EARTH**  2 miles north-west of Leek (SJ 9759). *Ruudeuorth* 1298 SHC XI NS 248, *Reede-yerth* 1504 (1883) Sleigh 126, *Red Earth alias Overhouse* c.1560 SRO

DW1702/1/6-8, *Redyerth* 1563 SRO DW1702/1/6-8, *the Redde Earth* 1564 ibid et freq., *Readde-erthe* 1569 (1883) Sleigh 120, *Redearth* 1604 SHC 1946 74, *Readhearth* 1609 SHC 1948-9 119. From a remarkable area of red-coloured soil here (see TNSFC 1885 54), which may also be the origin of the name Rudyard (q.v.). The place evidently had the alternative name Overhouse in the 17th century.

**REDFORD** (unlocated, on Smestow Brook near Gothersley) *Redford* 1690 HRO E12/V1/KY/7. See also Radway Hill.

**REDGREET** 3 miles north-west of Eccleshall (SJ 7830). *Ridgreet* 1600 Eccleshall ParReg, *the Reedgreitt* 1603 ibid, *the Readgreite* 1606 ibid. From OE *greot* 'gravel', so 'the place with the red grit or gravel'.

**RED HALL (FARM)** 1 mile south-west of Halmer End (SJ 7847). *Red Hall* 1681 Audley ParReg., 1833 O.S.

**RED HALL** ½ mile south-east of Broom (SO 9078). *Le Rede wall* 1373 PN Wo 378. Perhaps from OE *hreod-wælle* 'the reedy spring': a derivation from OE *read-wælle* 'red spring' seems improbable. Since 1844 in Worcestershire.

**REDHILL** on the south-west side of Tutbury (SK 2028). *Redhill* 1798 Shaw I 58.

**RED HOUSE** 1 mile south of Cauldon (SK 0847). *Red House* 1836 O.S. Presumably from a red house here, but see also Weaver Hills and Raddlepits.

**REDHURST WOOD** 1 mile north-west of Essington (SJ 9604). *le Redehurst* 1351 Vernon, *Redhurst* 1526 ibid.' From OE *read-hyrst* 'the red wooded eminence'.

**RED MOOR** – see **RADMORE**.

**REDMOOR BROOK** a tributary of the river Trent. Probably from OE *read* 'red', from the colour of the soil or vegetation, or OE *hreod* 'reed', with OE *mor* 'marsh, bog'.

**REDSHAW** 1 mile south-east of Heaton (SJ 9661). *Red Shaw* 1820 *EnclA*, 1842 O.S. From OE *scaga* 'a copse, a grove, a small wood', with *red* here probably meaning 'red soil' (see Redearth), so 'the small wood on the red earth'. See also Redshaw Wood.

**REDSHAW WOOD** to the north of Rudyard Hall (SJ 9660). *Red Shaw* 1820 *EnclA*, 1842 O.S.. The name is from OE *scaga* 'wood, copse', so 'the wood near Redshaw (q.v.)'.

**RED STREET** 2 miles south of Talke (SJ 8251). *Red Street* 1585 Speake 1972: 25, 1594 Audley ParReg, *Redstreete* 1608 SHC 1948-9 53, *Red-street* 1641 Wolstanton ParReg, *the Red street* 1644 SHC 4th Series I 205, *the Red Street Lane* 1654 SHC 1934 (i) 36, *Red Street* 1686 Plot 159, 1733 SHC 1944 58, 1747 Bowen. There seems little reason to doubt the statement in Plot 1686: 161 that the name is from a red-coloured iron-ore which was mined here, though red earth is sometimes found at places where burning processes have taken place, and there is some slight evidence that glass-making may have been carried out here: see PMA 31 1997 45. The *street* element, if ancient, may originate from a Roman road associated with the Roman site at Chesterton: with the exception of Roman roads the element is rarely found in Staffordshire names. The only evidence that has been traced to support the suggestion in SHC 1916 140 and the printed Audley ParReg. that the name is from Ridge Street is reference to *Rudge Street* in 1554 (SHC XII 213), *Ridgestrete* in 1572 (HOK 27), and *Ridge Street* in 1585 (Audley ParReg.), which may refer to this place, in which case the derivation self-explanatory, but the change to *Red* noteworthy.

REEVE END on the north-west side of Yoxall: Stuart 1990: 7 (SK 1419). *Reeves End* 1628 BCA MS3558/292, *Reeve End* 1631 NA DD/P/6/3/40, *Reevend* 1665 NA DD/4P/24/109, *Reve End* 1684 NA DD/4P/24/109. Presumably to be associated with land called *a Reevesthinge (sic)* recorded in 1613: NA157DD/5P/9/68: *thinge* is from OE *þing* 'a meeting, assembly, court'. An Elizabethan survey of Tutbury mentions 'twenty eight copyholders, which are called Reeves-places ... every tenant holding by copy of court-roll a tenement ... by the name of Reeves-place, shall be Reve when it cometh to his course, and shall collect the rent of the manor and the profits of the courts, [etc]': Shaw 1798 I 29. See also Rew End.

**REULE, UPPER** 1 mile west of Haughton (SJ 8420), *Over Rewle* 1686 Plot; **REULE, LOWER** 1½ miles south-west of Haughton (SJ 8419), *Roel* 1184x1228 SHC XII (i) 273, *Rohale* 1168 P, *Ruhale* 1199 Ass, *Rughal'* (p) 1222 Cur, *Rowl'* (p) 1221 Ass, *Reul* 1241 SHC 1924 134, *Roewelle* 1285 SHC 1911 189, *Rewyl* 1286 ibid.' 43, *Roule* 1297 SRO DW1734/J2268, *Reuel* 1332 SRO D938/132, *Nether Rewe* 1593 WSL SMS478/19/96, *the rule* 1618 Bradley ParReg, *the Rule* 1650 ibid, *Nether Rewle* 1686 Plot. Roel or Rowell (*Rawelle* DB) in Gloucestershire is held to be from OE *ra-wella* 'roe stream' (Ekwall 1960: 391), but the spellings in that case have *Ra-*, in addition to *Ro-*. This place may have the same derivation, or the first element might be OE *ruh*, with Mercian OE *wælle* 'a spring', and (sometimes) 'a stream', hence 'the rough spring', although such name is not readily explicable. Some spellings suggest the second element may be OE *halh*. It is not inconceivable that the root is OFr *ruelle* (from *rue* 'street' with the diminutive suffix *-elle*) 'a small road, a track, a path' (cf. Rewell, Sussex; see also Mawer 1929: 89), and in that respect it may be noted that Robert de Frankevill' (see Frankwell) witnessed a charter for land here at some date before 1206: SHC 1928 280. It may also be noted that the parish registers invariably use the definite article for the name. The possibility that the name is transferred from Rouelle or Ruelle near Angoulême in Poitou-Charentes, France, cannot be ruled out completely: cf. Frankwell; Desire, Le.

**REVEDGE** 1 mile south of Bradnop (SK 0053). *Ruhegg* 1223 SHC XII NS 30, *Revehegg'* 1227 *Harl*, *Refeggis* c.1255 SHC 1911 427, *Refegg* c.1270 SHC 1911 442, *Refeggmers* c.1278 ibid.' 430, *Refegge* 1304 ibid. 433, *Retheg* c.1302 ibid.' 443, *Revegg'* 1317 ibid. 433, *Ryflugge* 1331 SRO D1337/1, *Reveegge* 1332 SHC X 115, *Rauache* 1414 ProcJP, *Revedge* 1644 Leek ParReg, *Revi(t)ch* 1649 ibid, *Reavidge* 1837 O.S. From OE *hreof, ecg* 'the rough edge': the place stands on a ridge of high ground. Some of the spellings may relate to Revidge (q.v.).

**REVIDGE** a 1312' hill 2 miles west of Hulme End (SK 0759), *Revage Side, Great & Little Revage Hill* 1839 *EnclA*, *Ravage Top* 1840 O.S, *Revidge Lodge, Moor & Side* c.1870 *Rental*. Given the topography, probably with the same derivation as Revedge (q.v.); some of the spellings cited here may relate to that place.

REW END (unlocated, a hamlet in Yoxall: Shaw 1798: I 98) *Rew-end* 1499 (1798) Shaw I 98. Possibly associated with Reeve End (q.v.).

**REWLACH** 2 miles north of Warslow (SK 0961). *Rewlach* c.1422 VCH VII 27, *Rowlach(e)* 1566 *Deed*, 1633 *Rental*, *Rowelash* 1659 ibid, *Rue Lache* 1676 Alstonefield ParReg, *Rowlaytch* 1769 Rental, *Reulach* 1774 ParReg, *Rowlatch* 1840 O.S. A name recorded in the 1420s: VCH VII 27. Possibly from OE *ruh* 'rough or uncultivated ground', with OE *\*lece* 'a stream flowing through boggy ground', here meaning 'the boggy area with the stream in the rough ground'.

REYNOLDS (unlocated) *Reynoldes* 1666 SHC 1921 137.

REYNOLDS HALL (obsolete) 1 mile north-east of Walsall (SO 0200). *Reignalds Hall* 1589 Walsall ParReg, *Reynoulds Hall* 1595 HaRO 44M69/G3/39, *Renols Hawle* 1590x1603 Hackwood 1895: 52, *Reynoulds Hall* 1595 HantsRO 44M69/G3/39, *Reynalds Hall* 1610 SHC 1934 40, *Raynolds Hall* c.1630 SHC II (i) 14, *Reynald's Hall* 1663 SHC II (ii) 50, *Renalds hall* 1679 VCH XVII 146, *Reynolds Hall* 1798 Yates. The hall, named after the Reynalds family (*Raynald* 1302 SRO D1790/134), was an ancient structure, demolished in the late 18th century to mine limestone below (Shaw 1801: II 74-5), which stood near the site of Reynolds Hall Farm, demolished in 1897: VCH XVII 152. See also SHC 1982 19-20.

**RIBDEN** 1½ miles south of Cauldon (SK 0746). *Wrybbedon* 1327 SHC 1913 14, *Wrebedun* 1328 Ipm, *Wrebesdon* 1331 SHC 1913 31, *Wrubden* 1339 ibid. 77, *Ribden* 1608 SHC 1948-9 12, 1686 Plot. Perhaps from an OE personal name *Wrybba, which may have been an umlauted side-form of Wrobba, with OE *dun* 'hill'. Cf. Wribbenhall, Worcestershire (Ekwall 1960: 539).

**RICKERSCOTE** in Castle Church parish, 2 miles south of Stafford (SJ 9220). *Ricardescote* 1086 DB, *Richardescote*, *Richardscote* 1275 SHC VI (i) 72, *Rikardescote* 13th century Duig, *Rycescot* 1564 SHC 1928 156, *Richardscote* 1603 Penkridge ParReg, *Ricarscot* 1686 Plot. 'Ricard's cottage'. Ricard is a name introduced by the Normans.

RICKTHORN (obsolete) on the Staffordshire-Shropshire border, 1 mile south-west of Bobbington (SO 7989). *Pickethorn (sic)* 1316 SHC 1911 329, *Rykethorn* 1316 PRO C143/115/17, *Rikethorn* 1327 SHC VII (i) 252, *Rikethorne*, *Rykethorne* 1332 SHC X (i) 129, *Rickethorn* 1338 SHC 4th Series XVIII 205, *Pykthorn* 1531 SA 5735/2/22/1, *Rycthorn* (p) 1539 SHC VI (i) 68, *Rickthorne* 1608 SHC 1948-9 41, *Richthorn* 1775 Yates, *Rickthorn Farm* 1832 Teesdale, *Rickthorn* 1833 O.S. Possibly from an unrecorded OE *ric* 'stream, ditch', hence 'the thorn-bush at the ditch'.

**RIDDING** – see **STOCKING**.

**RIDDING FARM** ½ mile south-west of Anslow (SK 2024). *Ryding, Ruyding* 1287 SHC VI (i) 168, *New Rydinge* 1297 SHC VII (i) 45, *Rudynges* 1303 SHC 1911 59, *Ruydingis* 1326 HLS 267, *le Ruddyng* 1415 Hardy 1908: 95, *le Ruddynge* 1494 ibid.' 138, *Ryddyng* 1516 ibid.' 177, *Rydding* 1570 SHC XVII 218, *Roding* 1592 SHC XVIII 7, *Anslow Riding* 1836 O.S. From OE *ryding* 'a clearing, an assart, land taken into an estate from waste'. See also SHC 1912 222.

**RIDEWARE BROOK** a tributary of the river Trent. *aqua de Rideware* 1255 Ipm. From the four Ridware villages (q.v.).

RIDEWARE MORHAY (unlocated) *Rydeware Morhay* 1324 (1798) Shaw I *154, *Morhay* (undated) ibid. *153. See also Hamstall Ridware; Hill Ridware; Mavesyn Ridware; Pipe Ridware.

**RIDGACRE** 1 mile south-west of Harborne (SO 0083). *Ruggiacre* 1327 SHC VII (i) 229, *Ridgacre* 1749 Bowen. From OE *hrycg* 'a ridge' and OE *æcer* 'field, ploughed field', so 'the cultivated land at the ridge'.

**RIDGE, THE** 2 miles east of Cheddleton (SJ 9951), *the Ridge* 1572 AD 6, *the ridge in Fernihaulgh* 1586 AD, *The Ridge* 1836 O.S.; **RIDGE** 1 mile south of Longnor (SK 0862), *Ridge* 1687 Alstonefield ParReg. Self-explanatory.

**RIDGEHILL** 1 mile north-east of Madeley (SJ 7845). *Cylethyll otherwise Rydgehyll* 1574 SHC XIV 174, *Ridghill* 1614 SHC 1934 32, *Ridghill* 1644 SHC 4th Series I 55,

*Ridge Hill* 1798 Yates, 1833 O.S. Self-explanatory. The 1574 alternative name is curious and unexplained, unless from OE *clate* 'burdock': cf. Cleat Hill'.

**RIDGEWAY** 1 mile north of Norton-in-the-Moors (SJ 8953), *? Roggeweye* 1298 SHC XI NS 253, *Roggewey* 1308 ibid. 261, *Ruggeweye* 1326 JNSFC 1924-5 38, *Ruggewey* 1333 SHC X 94, 1512 JNSFC LX 1925-6 32, *Rygdwaye* 1459 (1843) Ward 1843: app. v, *Ridgway* 1615 Norton-in-the-Moors ParReg, 1619 BCA MS917/1585, c.1630 SHC II (ii) 20, *Ridgeway, Ridgeway Hill* 1836 O.S.; **RUDGEWAY** the old Chester road between Castle Bromwich and Stonnal, *Rugeway* 1324 (1798) Shaw I *app. 22, *Rudgeway* n.d. Shaw 1798: I 35; app. 22. Ridgeway or Rudgeway are names frequently borne by ancient roads, medieval forms being typically *Ruggeway, la Rugge, Ruggeway*, from OE *hrycg* with OE *weg* 'a way, a path, a road', giving 'the road running along the ridge'. The first place may be associated with *Rugges*, recorded in 1448: SHC III NS 169. An ancient road, possibly Roman, from Chester to Worcester over Rudge Heath (near Rudge, south of Pattingham) is recorded as *la Rugge* and *the Stanwey* ('stone-way'): Duig 126; Shaw 1798: I 34-5.

**RIDWARE, HAMSTALL** Ancient Parish 4 miles east of Rugeley (SK 1019). *Rideware* 1086 DB, **Rideware* 11th or 12th century Sawyer 1979a: xxxvii, **Ridewala* 1155 SHC I 20, **Ridwara* 1169 P, *Hamstede Ridwale* 1236 Fees, *Hamstal, Media Ridewar* 1242 ibid, **Rydeware* 1281 SHC II 257, *Widewarhamstall* 1281 SHC VI (i) 120, *Hamstall Rudwer* 1532 SHC 4th Series 8 5, *Rudware Hampstall* 1586 SHC 1927 181, *Hampstall Ridware alias Hamscall Ridware alias Ridware Hampsall alias Ridware Hamscall* 1629 SBT DR18/1/1910a. The DB form may relate to Cowley (q.v.), near Nethertown, and the forms marked * may refer to any of the other other Ridwares. The first element of this name is from a group of people who took their name from a feature probably derived from British **ritu-*, ancestor of Welsh *rhyd* 'ford', or possibly OE **ride* 'riding-path'. The former is more likely, for the four settlements incorporating the name lie between the rivers Blithe and Trent, where river crossings must have been of particular importance, or the 'ford' may have been a road through wet ground between the two rivers (see Gelling & Cole 2000: 91; Coates & Breeze 2000: 335). The second element is OE *-ware* 'dwellers', hence 'people who lived by the ford (or bridleway)'. Hamstall is OE *ham-stall* 'homestead, residence', meaning in this case 'demesne farm' or similar (*Hamstalmedwe* in Penkridge is recorded in 1357: SHC 1931 256). See also Hill Ridware; Mavesyn Ridware; Pipe Ridware; and Rideware Morhay. See also Nettles.

**RIDWARE, HILL** 4 miles east of Rugeley (SK 0817). For Ridware see Hamstall Ridware. The Hill element is recorded in 1346 as *le Hul* (Shaw 1801: II 199), *Hullo* n.d. (Shaw 1798: I 169), *Hulcausey* 1359 (Erdeswick 1844: 232), and *Alwynehull* (later corrupted into a farmhouse name The Hall in the Hole), from Alwyne who held land on the hill (ibid.). *Causey* is from ME *cause* 'causeway, embankment' (see e.g. Leland c.1540: ii 101, who mentions the *causey* between the two pools at Lichfield), but still used in Northern and Midland dialect with its earlier meaning 'paved path or pavement': OED. See also Hamstall Ridware; Mavesyn Ridware; Pipe Ridware; and Rideware Morhay.

**RIDWARE, MAVESYN** (pronounced Mavis-son [meɪvɪsən]) Ancient Parish 4 miles east of Rugeley (SK 0816). *Ridvare* 1086 DB, *Ridewale Mauvaisin* 1236 Fees, *Mavessen Rudward* 1532 SHC 4th Series 8 62. For Ridware, see Hamstall Ridware. Mavesyn is from the Malveisin or Malvoisin family (one of whom is said to have fought with the Conqueror at Hastings: see Shaw 1798: I 167; 205) who held the manor in the 12th, 13th and 14th century, when it passed to females – Shaw 1798: I 205 records how the name

459

was then '... vulgarly pronounced Mason or Ma'syn Ridware ...'. The DB spelling may refer to Nethertown (q.v.). See also Hamstall Ridware; Hill Ridware; Pipe Ridware; and Rideware Morhay.

**RIDWARE, PIPE**  4 miles east of Rugeley (SK 0917). *Rideuuare* 1086 DB, *Piep Ridewar* 1371 SHC VIII NS 266, *Pipe Ridware* 14th century Duig, *Pyp Rudwer* 1532 SHC 4th Series 8 62. For Ridware, see Hamstall Ridware. Pipe comes from the de Pipe family, from Pipe near Lichfield, who held the manor in the 13th and 14th century. Shaw 1798: I 161 records that Pipe Ridware was 'formerly called Media, or Parva [Little] Rideware'. See also Hamstall Ridware; Mavesyn Ridware; Hill Ridware; Rideware Morhay; Pipe; Nethertown.

**RILEYHILL**  1 mile south of King's Bromley (SK 1115). *Riley Hill* 1798 Yates, 1801 SRO D357/D/10, 1834 O.S. From OE *rȳge* 'rye', with OE *leah*, so 'the hill at the clearing in which rye was grown.'

**RINDLEFORD**  1½ miles north-west of Worfield (SO 7395). *Ryndelford* 1525 SR, *Rendelford* 1532 SHC 4th Series 8 118, *Ryndulford* 1536 SA 5586/1/423, *Ringlefford* 1577 Saxton, *Rindleford* 1752 Rocque, 1833 O.S. Probably from the North Western dialect *rindel* 'a small brook; to trickle slowly' (EDD), so 'the ford across the slow-moving stream'. In Shropshire since the 12th century.

**RING, THE**  ½ mile south-east of Great Haywood (SK 0022). *The Ring* 1836 O.S. From a group of 16 houses, demolished c.1965, built as an octagon to rehouse villagers from Shugborough c.1772: SHC 4th Series VI 88; 109.

**RING HEY**  1½ miles north-east of Cheddleton (SJ 9953). *Ryngheye* 1327 SHC VII 217, *Ryngeye* 1345 SHC XII 37, 69, *Ryngesleye* 1346 ibid. 53, 1409 SHC XVI 66, *Rungehey* 1536 SHC XI 273, *Rungehay* 1565 SHC XIII 266, *Ring Hey* 1836 O.S. The first word would seem to be from OE *hring* 'ring, circle', perhaps connected with some prehistoric stone circle – there may have been a monument called Bride Stones near Cheddleton: Shaw 1801: II 2 quotes from the notes to Holliday's poem 'The British Oak': '... Bride Stones, as they have been called time out of memory, have been found in or near Chedleton ... [t]hese Bride Stones the author has not yet had an opportunity of seeing ...'. No other reference to such supposed monument has been traced. However, from the *-unge* spellings it is not impossible that the element is OE *hrung*, *\*hrynge* 'a rung, pole, stake'. Ekwall 1960: 394 suggests that names with this element are likely to refer to a primitive bridge over marshy ground formed by poles placed close together at right angles to the direction of the road. The second element is from Mercian OE *(ge)heg* 'enclosure'. Cf. East Rounton and West Rounton, Yorkshire; North Runcton and South Runcton, Norfolk. See also The Ranger.

**RING HILL (COVERT)**  1 mile north-east of Codsall (SJ 8804). *Ring Hill* 1842 TA, *Ring Hill (Coppice)* 1888 O.S. It is possible that this name is to be identified with *Rankehull* 1453 Ct, *Rankyll* 1584 Ct, in which case the name is perhaps from OE *ranc*, *hyll* 'the hill with the heavy growth of coarse vegetation'. Otherwise perhaps from OE *\*hringel*, *\*hringels* 'a small ring', possibly with archaeological connotations: the place lies on a south-facing headland with extensive views. It appears that the field-name *Rankellesput* recorded in 1320 (Oakden 1984: 48), which perhaps incorporates the ON personal name Hrafnkell, so giving 'Hrafnkell's pit' (from OE *pytt*), was at or close to Broomhall near Brewood (SHC V NS 225), and is not to be associated with this place.

**RINGSTONE FARM**  ½ mile north-west of Werrington (SJ 9348). *Ringstone Farm* 1995 O.S. The name does not appear on the first edition 1" O.S. map, and early spellings have not been traced, but may well derive from a large stone with a round hole in it mentioned in an account of the perambulation of Bucknall manor boundary in 1803: StEnc 476. OE *hring-stan* had the meaning 'a round perforated stone', and possibly 'a stone circle'.

**RISING BROOK** 1 mile south of Stafford: see SHC VIII (ii) 103 (SJ 9221), *Rysond brooke* 1538 Star Ch, *Risom Brook* c.1571 SHC VIII (ii) 40, *Rysombro(o)ke* 1585 Comm, *Rysembrooke* 1590 SHC 1930 82, *Risingbrook farm* 1658 SHC VIII 125, *Risan brook* 1663 ibid.' 103, *Rising brooke (meadow)* 1669 WRO 705:24/1071, *Risonbrook* c.1680 SHC 1919 212, *Rising brook* 1836 O.S. '... a small estate ... of about 35 acres in the parish of Castle Church ...': SHC VIII (ii) 103; **RISING BROOK** on south-west side of Rugeley, *Rysynge broke(hillis)* 1554 SRO DW1734/2/3/43, *Rysombrook(e)* 1584 Comm, c.1680 GKNB, *Rising Brook or Sneyde Brook* 1742/3 SRO D603/E/204. The first element is possibly OE *hrisen* 'growing with brushwood' (cf. Rising Bridge, Northamptonshire). The same element may be found in *Rysounbrygge* 'brushwood causeway' in the Bobbington area, recorded in 1400: SA 52/9.

**ROACH END** on the north-west side of The Roaches (SJ 9964). *Roach End Farm* 1881 SRO 4974/B/7/50-51. See The Roaches. Several roads and tracks meet here.

**ROACH GRANGE** 2 miles north of Meerbrook (SJ 9963). *Ro(a)ch(e) grange* 1240 (1883) Deed Sleigh, *grange de Rupe* 1246 VCH VII 197, *Roche-graunge* 1406 (1883) Sleigh 51, *Roach grange* 1564 Swythamley MSS, *Rochegraynge* 1624 SHC NS X (i) 65, *Rochgrange* 1634 *et freq.* Leek ParReg. 'The grange (of Dieulacres Abbey) by the Roaches': see The Roaches. *de Rupe* in the 1246 spelling is from Latin *rupes, rupis* 'a rock face; a cliff'.

**ROACHES, THE** a gritstone outcrop on high moorland 4 miles north-east of Leek (SK 0063). *Roches* c.1340 SRO D1333/1, *la Roche* 1358 SHC XII (i) 162, 1361 SHC XIII 18, *Roch* 1637 Leek ParReg, *Leek Roches* 1686 Plot 171, *Rotch* 1697 Leek ParReg. From OFr, ME *roche* 'rock'. (Cf. Roach, Derbyshire; Roche, Cornwall). Perhaps to be identified with *þe rogh rocher* 'the rugged rocky mass' mentioned in the 14th century poem *Sir Gawain and the Green Knight*: see Elliott 1984: 3. Rocks on the west side of The Roaches were called *The Five Clouds* from 1681: VCH VII 194. Sleigh 1866: 173 mentions four jutting rocks, *the Bully Thrumble* (both *bully* and *thrumble* have various meanings in OED and EDD, but it is not possible to put forward a sensible meaning here), *the Marebach Rock* (perhaps from OE *(ge)mǣre, bece* 'boundary stream-valley'), *the Raven's Rock*, and *the Sugar Rock* (possibly from *sugar-loaf* 'a moulded conical mass of hard refined sugar': OED). See also Ludchurch; Roach Grange.

**ROBIN HILL** 1 mile east of Biddulph (SJ 9057). Perhaps to be associated with *Robinstone* 1665 Biddulph ParReg. The bird-name is taken from the personal name Robin, and is not recorded before the mid-16th century: Lockwood 1993: 7.

ROBIN HOOD'S SHOOTING BUTTS (obsolete) a name formerly applied to a group of mounds, believed to be prehistoric tumuli, in the area around Elford Low (SK 1909); *Robin Hood's Shooting Butts* c.1750 (1798) Shaw I 381, *Robin Hoods Butts* 1771 SRO D3720/2/2, *(a remarkable eminence, which is called) Robin Hood's Butt* 1818 Parson & Bradshaw; ROBINS HOOD FORD (obsolete, on the river Trent, adjacent to Tucklesholme (SK 2119)), *Robins hood ford* [*sic*] 1549 (1798) Shaw I 23. Place-names associated with the legendary outlaw are common in many parts of England, but most date from recent centuries: see Holt 1989: 187-90; JEPNS 30 1997-8 43-52. A supposed tumulus at Lowfields, on the south-west side of Combridge (SK 0937) called *Robin*

*Hood's butts* is recorded by Redfern 1865: 25-6, evidently the origin of Lowfields, from OE *hlaw* 'mound, tumulus', but see also JNSFC 3 1965 50.

**ROCESTER** (pronounced Roe-ster [rəustə]) Ancient Parish 4 miles north of Uttoxeter (SK 1139). *Rowcestre* 1086 DB, *Roucestre* 1191x1194 CEC 261, *Roffecestre* 12th century Duig, *Rouecestre* 1208 FF, *Rovecestre* 1225 ibid, *Rocestre* 1246 Ch, *Roffa* 1281 SHC VI (i) 151, *Rowecestre* 13th century Duig, *Raucestr'* 1360 SHC VIII 172, *Rowcetr* 1405 SRO D786/3/4, *Ruggestre* 1414 SHC XVII 16, *Roucett* 1454 Okeover T56, *Racetur* 1532 SHC 4th Series 8 109, *Rowcettre* c.1588 SHC 1927 177, *Rowcester* 1610 Speed. Ekwall 1960: 389 suggests 'Hroþwulf's or Hrof's Roman fort'. However, the spellings do not support those names, Hrof is a very doubtful personal name, and Mills 1998: 289 suggests OE *ruh* 'rough' (as preferred by CDEPN 503), or an unidentified personal name. The weak oblique form of *ruh* is *ruga* which often becomes Row- in place-names. The second element is OE *ceaster* 'fortress, Roman city', pronounced *chester* or sometimes softened to *sester* in the Mercian dialect, is invariably indicative of Roman occupation, extensive traces of which have been found here: see StEnc 479.

ROCHEFORD (unlocated, possibly near Comberford) *Rocheford* 1294 SHC VII 11, 1343 SHC XIV 60, *Rochford, Rocheford* 1342 SHC XVII 274, *Rochford* 1366 SHC XIII 49, *Rochefforde* (p) 1503 OSS 1936 50. From OFr, ME *roche* 'rock', so 'the ford at the rocky outcrop'.

**ROCKHALL** between the Roaches and Hen Cloud (SK 0099). *Rockhall* 1770 VCH VII 194, *Rock Hall* 1842 O.S. From a cave here inhabited from at least the early 17th century, which was incorporated into a Gothic-style shooting lodge c.1900: VCH VII 194.

**RODBASTON** 2 miles south of Penkridge (SJ 9211; SMR 01231 places a deserted Anglo-Saxon settlement at SJ 92161095). *Redbaldestone* 1086 DB, *Rembaldeston* 1195 SHC II 45, *Redbaldeston* 1198 P, *Rembaldeston* 1194-5 SHC X NS I 195, *Rodbaldeston* 1221 Ass, 1236 Fees, *Rothelboldeston* 1282 SHC VI (i) 152, *Rostlaston* 1385 SHC XIII 194. 'Redbeald's *tun*'. At the time of Domesday the place was held by Richard the Forester, and his descendants were chief foresters of Cannock Forest, probably residing at the moated site here (VCH V 120-22), perhaps the land at Rodbaston attached to the office of Chief Forester from before the Conquest to c.1246 (VCH II 339). A *castrum* or castle is recorded at Rodbaston in 1215 (VCH V 120 fn.32), but no castle in the usual sense is known, and the description may have applied to the moated house which stood some 500 yards north-west of Stables Farm at SJ 921124: VCH V 121. See also Loges.

RODBERDES LAND (unlocated, in Offlow Hundred) *Rodberdes land* 11th or 12th century Sawyer 1979a: xxxvii. MidA IV 2 April 1886 113 suggests that the name may refer to Robert de Stafford, but the person is evidently Rober de Ferrers: see SHC 1919 133. Rodbert was a Germanic name introduced via Norman French during the reign of Edward the Confessor and became very popular; see also DES 380. Cf. *Rodbardesfeld* in Shelfield, recorded in 1317: VCH XVII 279.

**RODDIGE** (pronounced Roddidge) 1 mile east of Fradley (SK 1713). *Redich, Redihige* c.1300 TSSAHS XX 1978-9 loose map, *Roddige* 1834 O.S. Possibly 'the red ditch': cf. Redditch, Worcestershire. It is unclear whether *Rodeyre*, recorded in 1310 (SHC 1911 109), relates to this place. See also Rodinge.

RODEFORD (obsolete, 2 miles north-west of Wolverhampton (SJ 8601)) *Rodeford* 1260 SHC 4th Series 13 6, *Rodesford* 1300 SHC V (i) 180, *Redesford* 1327 SHC VII (i)

255, *Reddysford* 1539 SHC VI NS (i) 67. The name was attached to a mill which stood near the present Wergs Hall. Perhaps from OE *rod* 'a rood, a cross', with reference to a cross, known as Bell Cross, which stood at the junction of Wergs Road, Woodhouse Road and Keepers Lane: VCH XX 12.

RODINGE (unlocated; perhaps near Morghall or Elmhurst, or perhaps the same place as Roddige (q.v.)) *Rodinge* 1567 SHC XVII 216, 1598 SHC XVIII 14.

RODMAN (unlocated; perhaps near Tutbury: SHC 1912 222) *Rodman* (undated) SHC 1912 222.

RODNEY HALL (obsolete) at Cotes Heath, 1 mile west of Swynnerton (SJ 8335). A hostel built c.1938 for the Royal Ordnance factory at Swynnerton, but never completed. Named after Admiral Rodney, the 18th century naval hero. The place is now called Moorfields. See also Beatty Hall, Drake Hall, Duncan Hall, Frobisher Hall, Howard Hall, Nelson Hall, Raleigh Hall.

**ROE LANE (FARM)** on south side of Newcastle-under-Lyme (SJ 8443). *? Wrd* 1307 SHC XI NS 262, *le Wro* 1327 SHC VII (i) 199, 1332 SHC X (i) 61, 82, 1360 SHC XII (i) 166, *Rowloune* 1487 SHC XI 329, *Row-lane* 1689 (1843) Ward 1843: app. lxii, *? ye Row* 1731 Swynnerton ParReg, *Roe Lane* 1773 Newcastle under Lyme ParReg, 1836 O.S. From ON *vrá, rá* 'a nook, a corner of land', found in Scandinavian place-names for 'a secluded or outlying place, a patch of cultivated ground projecting from the main part of an estate': see EPNE ii 232. Found quite frequently in ME field-names as *wro, wray, roe* and *ray* 'a nook, a secluded spot, a cattle shelter': see Field 1993: 129. The initial *w* is always retained in ME in the Danelaw. Cf. Roe Farm (Lane), Chaddesdon, Derbyshire (PN Db 545). The 1307 form may be a transcription error for *Wro*. See also Roe Moor; Rowe, The.

ROE MOOR (unlocated, in Norton-in-the-Moors) *Roe Moor Meadow* 1614 SRO D1798/166, *Roe Moor* 1773 SRO MS917/1628. In the absence of earlier spellings the derivation is uncertain, but probably from OE *ra* 'roe deer', with OE *mor* 'moorland, marsh'.

**ROLLESTON ON DOVE** Ancient Parish (pronounced Roll-ston [rəulstən]) 4 miles north of Burton-upon-Trent (SK 2327). *Roðulfeston* 941 (14th century, S.479), *Rólfestun* 1002x1004 (11th century, S.906, 11th century, S.1536), *Roulestune'*, *Rolueston* 1008 (12th century, S.920), *Rolvestvne* 1086 DB, *Rolveston* 1252 Ch, *Rolinstone* 1309 SHC 1911 71, *Rolluston, Rollustone, Rolleston* 1373 HLS, *Rolstone* 1468 SHC IV NS 158, *Rellyston* 1563 HLS. From OE 'Hroðwulf's *tun*', or from the Scandinavian name *\*Hróðulfr*, giving '\*Hróðulfr's *tun*': cf. Rolleston, Wiltshire. If the personal name is Scandinavian, the place-name is of a type known as a 'Grimston-hybrid' (or a 'Toton Hybrid': see Cameron 1996: 74-5), i.e. a name in which OE *tun* is combined with an ON personal name. It is possible that the name may date from the taking over of established English settlements by the victorious Danes of the great army of 865 AD: see Gelling 1988: 232-4. The parish changed its name from Rolleston to Rolleston on Dove in 1983. See also Gunstone.

ROLLOWE (unlocated, possibly near Salt) *Rollowe* 1279 SHC VI (i) 138. Perhaps from a personal name such as \*Hrolla, a short form of Hroþlaf, with OE *hlaw* 'tumulus, burial mound'. It is unclear whether this name is to be asociated with Ranslow Farm, 1 mile south-west of Salt (SJ 9427).

ROMANS GATE (unlocated, in Audley) *Romans-gate* 1612 SHC 1944 82, *land called Romans* 1697 SRO D1788/A7/C, *The Rummans* 1733 SHC 1944 2, *Rummons* 1733 ibid. 4, *Romans-gate* 1733 ibid. 82. Perhaps indicative of archaeological remains or artifacts.

ROMENHALE (unlocated, possibly near Ellastone or Quixhill) *Romenhale* 1293 SHC VI (i) 217. Perhaps from OE *(æt þæm) ruman halh* 'the spacious *halh*'.

**ROMER FARM** 1½ miles south-east of Stone (SJ 9330). *Romere* 1258 SHC 1911 129, *Romer (field)* 1580 SRO 358/1/34, *Romer* 1775 Yates, 1836 O.S. Perhaps from OE *ruh* 'rough', with OE *mor*, usually in the Midlands meaning 'marshland'.

ROMESCUMBE (unlocated, in Arley) *Romescumbe* 1255 SHC V (i) 114. Possibly from the OE personal name *Rum, a short form of names beginning Rum- (see Ekwall 1960: 392 s.n. Romsey), with OE *cumb* 'a short spoon-shaped valley'.

ROMESHELDE (unlocated) *Romeshelde* 1298 SHC XI NS 251, 1307 ibid.' 265. Possibly from the OE personal name *Rum, a short form of names beginning Rum- (see Ekwall 1960: 392 s.n. Romsey), with OE *helde* 'a slope, a declivity'.

**ROMSLEY** 3 miles south of Halesowen (SO 9679). *Rameslege* 1086 DB (listed in Warwickshire), *Rameslea* 1166 SHC 1923 298, *Ramesleg* 1203 Bowcock, *Rammesle* 1207 ibid, *Romesle(ye)* 1270 Ct, 1291 Tax, 1293 Ct, *Rummesleye* 1355 Pat, *Romisley* 1500 Nash, *Ramsley* 1686 Plot. Perhaps from OE *hramsa* 'wild garlic', or OE *ramm* 'ram', or an OE personal name *Hræm or *Ram, with OE *leah*. In Shropshire since the 12th century.

**RONTON** – see **RANTON.**

**ROOST HILL** 2 miles south-east of Leek (SK 0053). *Rusthill* 1589 SHC 1929 145, 1590 SHC 1930 (ii) 54, 1602 SHC 1935 518, 1667 Leek ParReg, *Rowsthill otherwise Rohnstehill* 1619 SHC VII NS 207, *Roost Hill* 1607, 1639 Leek ParReg, 1766 SHC 1931 91, 1837 O.S. 'The rust-coloured hill'. It is uncertain whether *Roost*, recorded in 1327 (SHC VII 198) is to be associated with this place. Roosthill Wood (*Roost Hill Coppice* 1836 O.S.) lies 1 mile south-west of Newborough (SK 1224), but early spellings have not been traced.

**ROPER'S HILL FARM** in Scounslow Green (SK 0929). *Ropers Hill* 1836 O.S., *Roper Hill* 1887 O.S. Possibly to be associated with *Thos. the Roper of Cundesley* [Scounslow Green] recorded in 1302: SRO D(W)1721/3/19/13.

**ROSEBANK** 1 mile south-west of Cheddleton (SJ 9451). *Rosebank* 1627 Deed, 1697 ParReg, 1880 Kelly, *Rose Bank* 1836 O.S. Self-explanatory.

**ROUCH** ½ mile north of Hulme (SJ 9346). *messuage called the Rouch alias The Rowarth* 1661x1780 SRO D4731/6/1-7, *the Rouch* 1689 Ward 1843: lxi, *Rouch* 1705 Stoke on Trent ParReg, *Rouch alias Rowish* 1769x1814 SRO D5378/2/11-15, *Rowch* 1772 SRO D4731/11/1. Possibly from *rouch(e)*, an obsolete form of *rough*: OED. The alternative forms are curious, the first suggesting the addition of OE *eorþe* 'earth', or OE *erþ* 'ploughing, plough-land'.

**ROUGH, THE** ½ mile north of Huntington (SJ 9713). *Rough Hills and Kyngesoke Heth adjoining Teddesley Hay temp.* Elizabeth I SHC 1939 123.

**ROUGHCOTE** 2 miles north-west of Caverswall (SJ 9444). *Rofcote* 1635 Caverswall ParReg, *Roughcoat* 1642 ibid, *Roughcot, Roughcote* 1644 SHC 4th Series I 117, 151, *Roughcote* 1682 Browne, 1836 O.S. 'The rough cottage'. Perhaps to be associated with

*Roug croft*, recorded c.13th century (SHC VIII (i) 150), and *Roughhegga* recorded in the late 13th century (SRO 3764/21) .

**ROUGH HAY** 2 miles west of Burton-upon-Trent (SK 2023). *Ruyhall* 1275 SHC 1937 79, *Rohay* 1284 ibid. 83, 1292 ibid. 86, c.1292 ibid. 97, 1307 SHC VII 176, *Rough hey vulgi Rewy* 1722 Burton upon Trent ParReg, *Ruff hay* 1736 SRO D4219/1/2. From OE *ruh* 'rough, uncultivated', with OE *halh*, replaced by Mercian OE *(ge)heg* 'enclosure'.

ROUGH HAY PARK (unlocated, in Agardsley: SHC 1991 251) *Rif-hay* c.1295 (1798) Shaw I 83, *Haye called Le Refhay* 1296 SHC 1911 251, *Ryfhay* 1306 SHC VII 149, *Ryffhay* 14th century SHC 1912 222, *Riffay house* 1670 SHC 1923 219, *Ruffhey park* 1704 (1798) Shaw I 129, *Ruffay Park* 1774 SRO DL31/231, *Rough Hay Park* 1814 SRO DL31/233. From Mercian OE *(ge)heg* 'enclosure', with an unidentified first element. The place has been associated with Pipehay Farm (q.v.): see Shaw 1798: I 83, 129.

**ROUGH KNIPE** – see **TURNER'S KNIPE.**

**ROUGH PARK** 1 mile east-east of Hampstall Ridware. *Rough Park* 1730 (1798) Shaw I 160, *Roughpark* 1798 ibid. 157. See also Shaw 1798: I 157; app. 24.

ROUGHSTONE HOLE (obsolete) 1½ miles north-west of Ipstones (SK 0153). *Roostonehole* 1717 Okeover E5092, *Roostone Hole* 1743 SRO D694/1-6/13, *Roughstone Hole* 1836 O.S. Perhaps from OE *ruh stan* 'rough stone', with OE *hol(h)* 'a hole, a hollow'.

**ROUGHTON** 1 mile south-west of Worfield (SO 7594). *Riwyton* 1238x1250 Eyton 1854-60: III 112, *Rucheton* 1300 ibid. 114, *Roghtone* 1301 Rees 1975: 249, *Rowhton* 1525 SR, *Roghtton* 1532 SHC 4th Series 8 117. Possibly from Welsh *rhiw* 'hill, ascent', with OE *tun*, so 'the *tun* at the hill called Rhiw'. In Shropshire since the 12th century.

ROUSEND (unlocated, in Longdon) *Rous ende* 1301 BCA MS3415/146, *le Rousende* 1323 BCA MS3415/154, *Le Roussende* 1350 BCA MS3415/184.

**ROUSTER** 2 miles north-east of Heaton (SJ 9764). *Roaster* 1826 SRO 4974/B/2/7, *Rouster* 1842 O.S. An intriguing name, perhaps to be associated with the field-names *Roster Bank* and *Roster Croft* recorded in Leekfrith in 1831 (Survey). *Roaster*, recorded in 1811 (PRO WO97/45/152), probably refers to Rocester.

ROUTHALE (unlocated, in Shenstone parish) *Routhale* 1343 SHC XVII 278, *Rothale* 1345 ibid.' 285.

ROUTHESLEIGH (unlocated, possibly in the Newcastle area) *Routhesleye* 1289 SHC VI (i) 192, *Routhesleigh* 1349 SHC XII 77, *Routheleigh* 1349 ibid. 79.

**ROWDON LANES** in Essington parish (SJ 9303). *Rowdon* 1306, *Roudo(u)n* 1314, 1340 Ct, *le Rowdons* 1549, *Rowden Flatt* 1653 Vernon, *Reudon lanes* 1682 Browne, *Rouden Lanes* 1695 Morden, 1775 Yates, *Rowdon Lanes* 1702 Bushbury ParReg, 1834 O.S. From OE *ruh-dun* 'rough hill'.

**ROWE, THE** 1½ miles south-east of Whitmore (SJ 8238). *the Row* 1729 Swynnerton ParReg, *the Row* 1810 SRO D641/5/E(L)/4, *Row Farm* 1813 SRO D641/5/E(L)/5, *The Rowe* 1920 O.S. Perhaps from ME *row*, applied to things (often trees) arranged in s straight line, or possibly from ON *vrá, rá* 'a nook, a corner of land', found in Scandinavian place-names for 'a secluded or outlying place, a patch of cultivated ground projecting from the main part of an estate': see EPNE ii 232. *The Raw* in Stone parish is recorded in 1708: SRO 49/4/44. See also Roe Lane (Farm).

**ROW HILL** on the north-east side of Coton in the Clay (SK 1729). Probably to be associated with *Rowe (Meadow)* 1698 SA 513/2/31/2/2.

ROWHURST (unlocated) *Rowehurst* 1300 BL AddCh. 46638, *Rowhurst* 1392 BL AddCh. 46643. From OE *ruh hyrst* 'the rough copse'.

**ROWLEY** 2 miles north-west of Yoxall (SK 1221), *Rouueleia* 1086 DB, *Roele temp.* Henry III TutCart, *Roulee* c.1290 SHC 1937 94, *Rouleye* 1296 SHC 1911 250, *Rowlegh* 1324 ibid. 358, *Roweley* 1424 SHC XVII 105, *Rowley* 1686 Plot 205; ROWLEIGH (unlocated, in Gratwich), *Rowleigh* 1562 SHC 1938 111. From OE *ruh leah* 'rough *leah*'. *Rowley* in Lower Penn is recorded in 1624 (WALS DX-240/14), and *Rowley Hill* in Upper Penn in 1753 (WALS DX-240/22).

**ROWLEY** on south-west side of Stafford (SJ 9122), *Roweleye als Roweleg'* 1291 Ipm, *Row(e)lowe* 1300 Banco, *Rowelowe, Rowelawe* 1300 SHC VII 72, *Rughlowe* 1306 SHC 1911 65, *Rovlowe* 1401 StaffAcc, *Rolowe* 1410 SHC XVI 72, *Rowley* 1486 VCH V 90, c.1630 SHC II (i) 21, *Rowlowe* 1539 VCH V 90; **ROWLEY** ½ mile north-east of Worfield (SO 7696), in Shropshire since the 12th century, *Rowlowe* (p) 1466 SA 2089/2/2/35, *Rowelowe* (p) 1494 GRO D2153/441, *Roulowe* (p), *Roeloe* (p) 1524 SRS 3 33, *Rowley* 1602 SA 2028/1/5/8, 1752 Rocque, 1833 O.S. From OE *ruh-hlaw* 'rough mound or tumulus'.

**ROWLEY GATE** 2 miles west of Leek (SJ 9556). *Throwleyate* 1515 VCH VII 203, *Rowley Gate* 1562 HRO B47/S19, 1836 O.S. The first element may be OE *þruh* 'water-pipe, conduit', originally 'a hollowed-out tree trunk', but here perhaps from ON *þrú* 'trough', in the sense 'steep-sided valley': the place lies on the west side of a steep valley in which runs the river Churnet. The second element is OE *leah*, with OE *geat* 'a gate, a pass, a gap between hills'. Yates' map of 1775 shows two adjoining places, the northern *Rowley*, the southern *Gate*. Cf. Throwleigh, Devon; Throwley, Kent.

**ROWLEY HILL** 2½ miles south-west of Penkridge (SJ 9011). *Rolaue* 1203 SHC III 91, *Rowleyfeld* 1284 to 1358 Deeds, *Rowley Hill (Field)* 1606 VCH IV 167, *Rowley Hill* 1798 Shaw I 31, 1834 O.S. From OE *ruh-hlaw* 'rough-tumulus', with OE *hyll* 'hill'. There is a tumulus on a headland here, noted before 1798 (Shaw 1798: I 31), which may be the basis of the name Penkridge (q.v.): see VCH I 376; JNSFC 1965 48; Gelling 1984: 138; Gelling 1988: 41, 243; .

**ROWLEY REGIS** 3 miles south-east of Dudley (SO 9687). *Roelea* 1173, *Reulega* 1174 P, 1182 SHC I 101, *Reuley, Rohele, Reuleg* 12th century Duig, *Ruleg, Rulegh'* 1240-1 Cur, *Ruleye* 1272 Ass, *Rugeleye* 1280 SHC VI (i) 109, *Roweleye* 1291 SHC IX (ii) 31, *Rouleye* 1294 SHC VII 9, *Rughel* 1242 SHC IV 96, *Reuleg, Roule* 13th century Duig, *Rouleye Somery* 1322 SHC 1911 351, *Kinges Rowley* c.1564 SHC 1931 173, *Rowley Regis otherwise Kinges Rowley* 1604 SHC XVIII 20. From OE *ruh* 'rough, uncultivated', hence 'rough *leah*'. Although Erdeswick 1844: 345 states that Rowley Regis 'at the conquest, remained in the king's demesne, and so continued till the 20th of his reign', the place is not recorded in DB, and the *Regis* element does not appear before the 1330s, by which time some part must have belonged to the king, presumably that part granted to Halesowen abbey at the death of John de Somery in 1322. *Somery* is from Sommeri near Rouen in Normandy; the family of that name held the Barony of Dudley towards the end of the 12th century: see SHC 1941; Erdeswick 1844: 345.

ROWLOW (unlocated, between Mayfield and Ellastone.) SHC 4th Series 11 177 fn.29 states that the Staffordshire historian Sir Simon Degge differed from Robert Plot in the interpretation of Rowlow 'between Mayfield and Ellastone'. The name, probably from

OE *ruh-hlaw* 'rough mound or tumulus', may be associated with tumuli called *The Rowleys* recorded in Mayfield in 1908 and 1916: VCH I 376; SHC 1916 208; JNSFC 3 1965 44. A bell barrow has been recorded at SK 13814417.

**ROWNALL** in Cheddleton parish, 6½ miles north-west of Cheadle (SJ 9549). *Rvgehala* 1086 DB, *Ruanhall, Ruhenhal'* 1221 Ass, *Magna et parva Roughenhale* 1272 ibid, *Roughenhale* 1273 SHC VI (i) 54, *Ronnal* 1274 SHC 1911 160, *Rowenhale* 1274x1290 Loxdale, *Rounal(l), Ron(e)hale* 1274 Ipm, *Ronnal* 1274 SHC 1911 160, *Rowenhall(e)* 1279 Ass, *Great Rowenhale, Little Rowenhale* 1284 SHC VI (i) 136, *Rounhale* (p) 1294 Orig, *Rewenhale* 1306 GDR, *Rouwenhale* 1327 Ipm, *Rounhal'* 1328 Ipm, *Rounales* (p) 1380 AD, *Rownall* 1479 AD 1, *Great & Little Rownall* 1558 BM, *Rownall* 1566 SHC IX NS 82, *Rawnall* 1589 SHC 1929 40, *Rawnall in the Moors* 1733 SHC 1944 50, *Rounhall* 1755 Bowen. From OE *ruh* 'rough, uncultivated', hence 'rough *halh*'. See also Erdeswick 1844: 344.

**ROWNEY FARM** 1 mile south-east of Mucklestone (SJ 7335). *? Les Rounales* 1304 SHC XII NS 278, *a wood called Rounhay* 1361 SHC 1945-6 46, *Rowney* 1833 O.S.

**ROYALS FARM** 1 mile north-east of Tamworth (SK 2205). *Rye Hills or Royals* 1834 O.S. Seemingly 'hills where rye was grown'.

**ROYLEDGE** ½ mile north-west of Upper Elkstone (SK 0459). *Royledge* 1648 Leek ParReg, *Rilidge* 1694 Alstonefield ParReg, 1842 O.S, *Ryeledge* 1775 Yates, *Rylage* 1850 TA. Names with *Roy-* can derive from *Ruy-* (cf. Royton, Lancashire), so possibly OE *rȳe*, *hyll* 'rye-hill', with OE *ecg*, giving 'the rye-hill at the sharp edge of the hill or escarpment', but Royle was a family name in north Staffordshire (see VCH VII 200, 202, 202; *Tho. Royle* 1636 Leek ParReg), and so possibly 'Royle's edge'.

RUCK OF STONES (obsolete, in Sandwell (SO 0289)) *the Ruck of Stones* 1617 Willmore 1887: 440, 1732 BCA MS3069/Acc1930-022, 1771 BCA MS3375/445769, 1801 Shaw II 125, 1834 O.S., *Ruck of Stones (Farm)* 1820 BCA MS3375/445769. An interesting example of the Midlands and Northern dialect *ruck* 'a pile, a heap' (EDD). The mound lay on a parish boundary (StEnc 485), and may have originated as a boundary marker. A field-name *Ruck of Stones* or *Rack of Stones* is recorded in Wellington, Shropshire: Foxall 1980: 28.

**RUDGE** 6 miles west of Wolverhampton (SO 8197), *Rigge* 1086 DB (listed in Warwickshire), *Rugge* 1188 SHC I 137, 1303 SHC VII 105, *Ruge* 1194 Bowcock, *Rigge* 1197, *Ruge* 1240 ibid, *Rugge* 1253 ibid, 1324 SHC 1911 104, *Rudge* 1652 Pattingham ParReg; **RUDGE** 1 mile south of Ashley (SJ 7634), *Rigge* 1086 DB, *Rugge* 1166 SHC 1914 14, 1307 SHC XI NS 265, 1307 SHC XII 251, 1512 SHC XII 181, *Rughe(haye)* 1227 SHC IV 73, *Ruges* 1484 SHC VI NS (i) 157, *Rudge* 1585 SHC XV 160, *Rydge* 1617 SHC VI NS (i) 42. From OE *hrycg* 'a ridge, a long narrow hill'. The first place has been in Shropshire since the 12th century. For Rudge near Ashley see also SHC 1914 14-5.

RUDGE BANKE (unlocated, in Ferny Hough) *Rudge Banke* 1600 SHC 1935 214.

RUDGE STRETE (unlocated, possibly in the Burslem area, or perhaps Red Street (q.v.)) *Rudge Strete* 1554 SHC XII 213.

**RUDGEWAY** – see **RIDGEWAY**.

RUDLOW (obsolete, '... at Tatenhill-wood-lane about half a mile West of Callingwood-hall ... supposed to be Rudlow': Shaw 1798: I 37; 110 (SK 1923)) *Roddelowe* 1272 SHC

IV 259, *Rudelowe* 1290 SHC VI (i) 204, *Rodelowe* 1304 SHC 1911 63, *Ruddelowe* 1313 SHC IX 39, *Rudloe* 1618 SHC VI NS (i) 57, *Rudlow* 1686 Plot 414, *Rodelowe* 1798 Shaw I 60. The second element is from OE *hlaw* 'mound, tumulus' (Plot 1686: 414 records a tumulus here), with an uncertain first element – perhaps from OE *rod* 'a clearing in the forest', or from OE *rudig* 'red, ruddy': see Rudyard. Cf. Radlow, Herefordshire.

**RUDYARD** 2 miles north-west of Leek (SJ 9659). *Rudegeard* 1002x1004 (11th century, S.906; 11th century, S.1536), *Rudierd* 1086 DB, *Rudehard* 1199 SHC III 36, *Rodehierd*, *Rodiehierd* c.1255 SHC 1911 426, *Rodehierd* ?1272 SRO NRA list 216, *Rodhord* 1278 SHC XI NS 245, *Rodeyert* 1275 SHC V (i) 117, c.1280 SHC 1911 431, *Rodeyord* 1286 ibid.' 432, *Rodehyerd* c.1290 ibid.' 432, ? *Rudeuorth* 1299 SHC XI NS 248, *Rodezard* 1307 SHC XI NS 255, *Rudeyard* 1330 Ch, *Rydrorte* 1365 (1883) Sleigh 126, *Redeyerd* 1532 SHC 4th Series 8 33, *Rudyerd* 1842 O.S. The second element is OE *geard* 'yard, enclosure'. The first element has caused surprising difficulty. Ekwall 1936: 395 held that the place took its name from the adjoining Rudyard Lake, said to mean 'enclosure where rudd were kept', although the word rudd has not been recorded earlier than the 17th century, and the fish are said to be inedible. That derivation can be discounted since it is unlikely that a word for a fish would be associated with a non-water element, but more importantly because the lake was in fact formed by damming the Dingle Brook to create a reservoir to feed the Caldon canal in about 1799 (Sleigh 1883: 159 gives the date 1793; see also StEnc 485). Bowen's map of 1749 shows 'New Pool being the head of the river Trent' to the north-west of what is now Rudyard, but Yates' map of 1775 shows no pool. It has also been suggested that the name is from OE *rude* 'rue', hence 'enclosure where rue was grown' (Ekwall 1936: 395, EPNE i 198, EPNE ii 88, Oakden 1967-8: 34, Mills 1998: 293, VCH VII 216), or the OE personal name Ruda or Rudda, giving 'Rud(d)a's enclosure' (Duignan 1902: 128, Paffard 1996: 3, VCH VII 216). There can be no doubt that the name derives from OE *rudig, rudu* 'red, ruddy' from the colour of the soil here, perhaps in particular a prominent area of red earth at Red Earth Farm (q.v.): Rudyard Hall adjoins Redearth Wood and Red Earth Farm, hence 'the yard or enclosure with the red soil'. Possibly the name was taken from the coloured area at Red Earth Farm before Red Earth became a separate estate: the earliest spelling that has been traced for Red Earth is *Ruudeuorth* (1298). The geological map shows that Rudyard Hall lies on the junction of the (Triassic) Bunter Sandstone and Conglomerate with the (Carboniferous) Millstone Grit and Culm Measures. The stream which runs into the north end of Rudyard Lake is *Rad Brook* (1842 O.S.), here probably meaning 'the brook with red-coloured water'. See also Redshaw. The hamlet of Rudyard became gradually deserted in the 19th century, particularly after Rudyard Lake (known as *Rudyard Reservoir* until c.1850: StEnc 486-7) became a popular tourist resort in the middle of the century, when the hamlet of Harper's Gate south of the lake expanded and took the name Rudyard. It was at Rudyard that J. L. Kipling and his wife became engaged; their son, born in India, was named Rudyard after the place. It is likely that Rudheath, Cheshire, and possibly Rudford, Gloucestershire, have the same first element as Rudyard.

**RUE BARN (FARM)** 2 miles south-west of Eccleshall (SJ 8027). ? *the Rewes* 1603 Eccleshall ParReg, *the Reu barne* 1627 ibid, *Rue Barn* 1775 Yates, 1833 O.S. Perhaps from late ME *rue* 'the shrub *Ruta graveolens* or similar', formerly much used in medicine, although a derivation from OFr *rue* 'a road, a street' cannot be ruled out completely, since a French presence is recorded in this area from at least the early 13th century (see Frankwell; Reule), and a supposedly ancient road is recorded by early antiquaries at Wootton (q.v.) south of Eccleshall, 1 mile to the east. The place is on a 400' hill.

**RUE HILL** 1 mile south-east of Cauldon (SK 0847). *Rowlow* 1686 Plot, 1775 Yates, *Row Hill* 1798 Yates, *Bue Hill (sic)* 1836 O.S. From OE *ruh-hlaw* 'the rough mound or tumulus', the second element now replaced by *hill*. There is a Rue Hill in Weston Jones, but early spellings have not been traced.

**RUELOW (WOOD)** 1 mile north of Kingsley (SK 0148). *Ruelow* 1836 O.S. From OE *ruh-hlaw* 'the rough mound or tumulus'. See also Rue Hill.

**RUGELEY** Ancient Parish 8 miles south-east of Stafford (SK 0418). *Rvgelie* 1086 DB, *Rug(g)elega* 1156-7 P, 1190 P, *Reggele* 1199 SHC III 60, *Regeley* 1532 SHC 4th Series 8 13, *Puys Baylywicke otherwise Rugeley Baylywicke* 1583 SHC XV 147, *Ridgeley* 1570 SHC 1939 157, *Ridgley* 1586 SHC 1927 129, *Ridgley alias Rudgley* 1636 BCA MS3069/Acc1920-020/288203. '*Leah* on a ridge', from OE *hrycg* 'a ridge, a long narrow hill'. The name is in some respects curious, for though there is high ground on the north-east side of the river Trent, and the high ground of Cannock Chase lies to the south-east, there is no striking obvious topographical feature which might be thought of as a typical *hrycg*, although there is a small hill with a narrow crest to the north-east side of Hagley Farm (SK 0317) which may have given rise to the name. It would appear that an early alternative for this name was *Puteo* or *Puys*. *Puiz* is recorded in 1195 (Fees), 1198 (SHC IV (i) 99) and 1199 (SHC III 40), *Puteo* in 1195 (Pipe), 1225 (SHC IV 37) and 1236 (Fees), *Puysland* in 1359 (SHC VI (ii) 16), and *Puysbaillie* in 1398 (SHC XI 204). However, *Puisland* and *Rugeley* are noted as separate places in 1649: SHC VI (ii) 101. The name is from the bailiwick formerly held by the family of Dupuis or de Puteo, tenants of the bishop of Lichfield at Rugeley and Hereditary Foresters of the bishop (ibid.), or the Peytos or Petos family, originally from Poitou in France (SHC X NS I 212); see also SHC XII (i) 284. See also SHC I 38; SHC VII 92-3; SHC X (ii) 216-7; SHC 1927 13; Oakden 1984: 66.

**RUITON** on west side of Upper Gornal 'occupying a fine lofty limestone eminence, which had formerly a beacon, and commands a most extensive prospect ...': White 1834 (SO 9191). *Ruton'* 1327 SR, *Rowerton* 1684 Sedgley ParReg, *Rewarton* 1685 ibid, *Rewardine* 1702 Roper 1952, *Routon* 1749 Bowen, *Rewarden* 1795 Roper 1952, *Ruiton* 1895 O.S. The inconsistent spellings make any etymology difficult. The first element may be Welsh *rhiw* 'a slope or hill-side', and the second may be OE *worþign* 'enclosure, open place in a village or farmstead', so perhaps 'the village or farmstead at the place called Rhiw'. It is unclear whether *Ruton*, recorded in 1271 (SHC V (i) 142) refers to this place, or to Ryton in Shropshire or Ryton under Dunsmore, Warwickshire.

RUMBELOWS (FARM) (obsolete) in Wednesfield, 1½ miles north-east of Wolverhampton (SJ 9200). *Tromelow* c.1272 (1801) Shaw II 150, *le Thromelowe* 1339 PN Wa 31, *Thromelowe* 1360 ibid, *Romylow* 1420, 1428 ibid, *Tromylow* 1392 SRO D593/B/1/26/6/6/14, *le Rombelose* 1576-7 SHC 1910 164, *le Thromylowes* 16th century Duig, *Trombelowes* 1614 SRO D593/B/1/26/11/12, *Thrombelowe, Tromelow* 1801 Shaw II 150, *Rumballows* 1834 O.S, *Tromelow Farm* 1895 O.S., *Rumbelows Farm* 1907 O.S. The first element is from *þreom*, the dative of OE *þreo* 'three', and the second is OE *hlaw* 'low, tumulus', giving 'the three tumuli'. When the name was prefixed by *æt* 'at', the initial letter of the name became confused with the end of the preposition, which produced *at Romelow*, rather than *at Tromelow*. Alternative forms of the name appear to have been used from the 15th century. Cf. Tremelau Hundred and The Rumbelow, Warwickshire (PN Wa 31, 247). There is a Threelows 2½ miles north-west of Wootton (SK 0746), but early spellings have not been traced. On the surname Rumbelow, see DES 386.

**RUMER HILL**  ½ mile south-east of Cannock (SJ 9809). *Rugemor* 1245-50 MRA, *?Rumere* 1258 SHC 1911 129, *Rouwemer* 1335 Ct, *? le Roomere* 1471 OSS 1936 45, *Romehill* 1570 Survey, *Rowe more hills* 1610 ibid, *Fletcher's alias Rumer Hill Farm* 1662x1777 SRO D260/M/T/6/130, *Rumore Hill Farm* 1630 Okeover, *Romer Hills* 1666 SRO D260/M/T/4/85, *Romers Hill* 1775 Yates, *Reaumorehill (well)* 1817 Pitt, *Rumour Hill* 1834 O.S. From OE *ruh* 'rough', with OE *mor*, usually in the Midlands meaning 'marshland'. *le Ruemor* is recorded in Longton in 1250: SHC XI 318.

**RUMFORD HILL**  2 miles east of Enville (SO 8587). *Rumford Hill & Pool* 1834 O.S. Early forms are not available, but evidently from the OE adjective *rum* 'roomy, spacious', hence 'the wide ford': the place lies above the Smestow Brook.

**RUSCOTE** – see **BROCKHURST**.

**RUSHALL** Ancient Parish 2 miles north-east of Walsall (SK 0201). *Rischale* 1086 DB, *Rushale* 1195 P, 1242 Fees, *Ruishall* 1176x1182 EEA 16 66, *Ruishale, Rushale* 1199 SHC III 49, *Russehall* 1242 SHC IV 96, *Roshale* 1300 SHC VII 76, *Ruysshale* 1335 SHC 1928 79, *Ruyshale* 1438 SHC 1921 29, *Russeshall* 1577 SHC 1939 138. 'The *halh* overgrown with rushes', from OE *rysc* 'rush'. Much of the area here is low lying.

**RUSHLEY**  ½ mile north-west of Ilam (SK 1251). *Rush(e)ley* 1605 QSR, 1777 Blore ParReg, 1838 O.S. From OE *rysc*, *leah* 'the rushy *leah*'. Rushley Bridge (*Rushley Bridge* 1729 Alstonefield ParReg) crosses the river Manifold to the north.

**RUSHTON MARSH**  (obsolete, on the north-east side of Rushton Spencer (SJ 9462)) *The Marsh* 1617 SRO D(W)1761/A/4/101, *Rushton Marsh* 1842 O.S.

**RUSHTON SPENCER, RUSHTON JAMES** 5 miles north-west of Leek (SJ 9362). *Risetone* 1086 DB, *Rixton* 1227 SHC IV 225, *Russton* 1282-3 SHC XI NS 247, *Ruston* 13th century Duig, *Russheton Jamys* 1306 SHC 1911 437, *Rouston'* 1307 SHC XI NS 255, *Russheton Spencer* 1399 SHC XV 88; RUSHTON GRANGE (obsolete, on west side of Cobridge (SJ 8748)), *Risctone* 1086 DB, *Rushton* 1223 SHC XII NS 30, *Rushton grainge* 1653 Burslem ParReg, *Rushton Grange* 1836 O.S. From OE *rysc* 'rush', and OE *tun*, hence 'the rushy *tun*'. The first place, originally the northern part of the manor of Rushton known as Hug Bridge (q.v.), was so-named from marshland in the valley on the east side of the township, and from the early 14th century after the Despencer family, who once held the place (SHC IV 245; SHC IX NS 300; VCH VII 223), the second perhaps after James de Audley (d.1272), an early lord of the manor: VCH VII 219, 221. Rushton Grange, established at Rushton by the Cistercians of Hulton Abbey by 1235 (VCH VIII 107), is preserved in the name Grange Park.

**RUSSELL'S BANK**  in Upper Longdon, 1 mile west of Longdon (SK 0514). *Russell Bank* c.1714 SRO D260/M/T/5/125.

**RUSSELLS HALL**  2 miles north-west of Dudley (SO 9291). *Russelleshalle* 1315 Ipm, *Russel Hall* c.1538 SHC IX (ii) 90, *Russells* 1571 ibid. 92, *Russels* 1577 Saxton, *Russells* 1610 Speed, *Russells Hall* 1834 O.S. 'Russell's Hall', from the Russell family, who held the place from at least 1275 (Ipm). See also SHC IX (ii) 37; StEnc 737. 'Hall' is a rare element in early place-names.

RUTHENDEE (unlocated, in Horton) *Ruthendee* 1239 CurReg.

**RYDER'S HAYES (FARM)** 1 mile north-east of Pelsall (SK 0304). *Rideres Heye* 1298 SHC 1928 161, *Rydders Heys* 1643 Pelsall peramb., *Ryder's Hayes* 1834 O.S., *Rider's Ease* 1841 Census. Perhaps 'the riders' hay or enclosure', from Mercian OE *(ge)heg*.

**RYEBROOK**  a tributary of the river Hamps; **RYEBROOK**  a tributary of the river Trent, *Riebroke* 1504 Ipm. Possibly from OE *ryge* 'rye', hence 'the brook where the rye grows', but for the first place the absence of early forms makes any derivation uncertain. See also SHC 1924 286 for *Rybrok* as the name of Stafford Brook.

**RYECROFT**  1½ miles north-west of Butterton (SK 0558), *Rycroft* 1683 Alstonefield ParReg, *Riecroft* 1686 ibid; **RYE CROFT**  ½ mile west of Rushall (SO 0199), *Rucroft* 1317 WSL A/3/37, *Ruycroft* 1327 SHC 1912 252, 1352 WA II 100, *Rucroft* 1349 SHC 1913 131, *Ryecroft* 1834 O.S., *Rycroft Farm* 1895 O.S. From OE *croft* 'a small enclosure of arable or pasture land, an enclosure near a house', hence 'the small enclosure by the house, where rye was grown'. *Rye halan*, 'rye nook', recorded in a charter of Pelsall in 996 (for 994) (17th century, S.1380), may be Rye Croft: Hooke 1983: 76. *Ruycroft*, recorded in 1326 (SHC IX (i) 112), and *Ruycroft*, recorded in 1327 (SHC VII (i) 174), have not been located.

**RYEHILL**  ½ mile south-west of Little Onn (SJ 8414). *The Ryehill* 1675 Bradley ParReg, *Rye Hill* 1678 ibid. Self-explanatory.

**RYELANDS, THE**  ½ mile north-west of Croxall (SK 1914). *Rye Lands* 1581 Derby. From OE *rȳge, land* 'the estate or piece of land with the rye'. Transferred from Derbyshire in 1894.

**RYKNIELD STREET, RYKNILD STREET or RYKNELD STREET** a Roman road (Margary number 18b, 18c) running through Gloucestershire, Worcestershire, Warwickshire, Staffordshire and Derbyshire. *Hikenildstrete* 1070 SHC 1916 302, *Ad regalum viam quae vocatur Ikenhildestrete; Stratum regiam quae appellatur Ykenild; via regia vel le Ricnelde strete* 12th century, *Rikelinge strete* 13th century, *Rykenyldstrete* 14th century (the above extracts, except the first, relating to parts of the road between Lichfield and Derby), *Stanistrete* c.1203 SHC 1924 67, *le Stantway, Ykenilde stret* 13th century SHC XVII 263, *Hykeneldis strete* c.1257 SHC 1937 56, *Ritling Street* 1684 SRO 115/1/41, *Rigning-way* 1798 Shaw I 19 (all referring to the road in Shenstone parish). The Roman road which runs from Derby and Chesterton near Stoke-on-Trent via Rocester and Draycott in the Moors (Margary number 181), called in part *Mear Lane* between Blythe Bridge and Longton on the 1840 1" O.S. map, is called *Rikenilde Streete* in the supposed foundation charter (almost certainly a much later forgery: see Tomkinson 1994: 73-102) of Abbey Hulton of 1223 (Ward 1843: App ii; SHC 1916 141; VCH I 188; VCH Wa I 241-2; VCH Db I 245-6), *Rikenildestrete* c.1230 (SHC 1921 18), or *Richmilde Street* (Dugdale 1817-30: v 715), the *m* seemingly an error for *n* (see also SHC 1934 38). The *R* is perhaps due to corruption of ME *at there Ikenilde strete* to *at the Rikenilde strete*. Another Ikenild Way, apparently Roman, runs 3 miles south of Burford in Oxfordshire. Thorpe Salvin, 5 miles north-west of Worksop, was anciently *Rikenhildthorp* (Duig), but is not known to have any connection with any ancient road. It seems likely that *Icknield* and variants thereof was a name adopted by the Anglo-Saxons for ancient roads and trackways after the ancient Icknield Way, perhaps Neolithic in origin, which runs from Norfolk to Hertfordhire, the name of which (recorded as *Icenhylt* or *Icenhilde Street* in pre-Conquest sources: VCH Wa I 241) may be connected with the Iceni tribe, which is believed to have inhabited Norfolk. OE *hilde* means 'war, battle', but is only found used in poetry, and no example is known of its use in conjunction with OE *weg* 'way' or *strǣt* 'street, Roman road'. A reference dating probably from 1116x1133 suggests that part of Ryknild Street in Derbyshire may have been known as *Waldwike strete*: SHC 1916 233. See also Shaw 1798: I 16-19; VCH Wa I 241-2; PN Wa 2-3; TBAS 60 1936 42-55; CDEPN 328, 499, 515.

**RYPPELEYELOND, RYPLEY MEADOW** (unlocated, in Gratwich) *Ryppeleyelond* 1348 SHC XII (i) 293, *Rypley* (*meadow*) 1562 SHC 1938 111. Probably from OE *\*ripel*, found only in place-names, cognate with Norwegian *ripel* 'strip', and surviving as dialect *ripple* 'a coppice, a thicket', with OE *leah* and OE *land*, with various meanings including 'estate, district, portion of a village', so perhaps 'the estate of the clearing with the thicket'. There is little likelihood that the name incorporates the OE tribal name Hrype, found in Repton, 20 miles to the east, which was probably founded by members of the northern tribe whose name survives in Ripon: Stenton 1970: 270; Rumble 1977: 169-71. Cf. Roberti de Rippa recorded in 1186-7 (SHC I 131), and Thoma de Rippel recorded in Burton upon Trent in 1327 (SHC VII (i) 226). See also PN Db 23-4; PN Wa 7; PN Wo 2.

**SADDLESALL** on the north-west side of King's Bromley (SK 1117). *Sadlesall* 1564 SHC XIII 235, *Sadleshall* 1591 SRO D(W)1721/1/217/23, *Saddelsall* 1601 SRO D(W)1721/1/223/23, *Saddleshall (Meadow)* 1626 WRO CR1908/48, 1649 SRO D(W)1721/1/232/23, *Saddlesall* 1720 SRO D1101/5/18, 1730 (1798) Shaw I 160, 1834 O.S. The name is attached to a piece of land which is effectively a large island formed where the river Trent divides and rejoins (by a 'new cut' according to StEnc 493; the O.S. marks the southern branch of the river as *New Trent*). The first element appears to be a personal name (Johannes Sedale is recorded in 1402: SHC NS XI 210), o rperhaps OE *sadel* 'saddle', sometimes found in place-names, and applied to a saddle-like ridge or similar feature (cf. Saddleworth, Yorkshire), and might here perhaps refer to the shape of the island, if such is ancient (cf. *Saddle Acre* recorded in Blithfield in 1677: SHC 1919 29), or bend in the river. The second element is probably *hall*, a rare element in Staffordshire place-names, rather than OE OE *halh*, although one meaning of *halh* is 'remote land enclosed by a river', which would be appropriate here.

SALE (BROOK) – see **IDLEROCKS.**

**SALE (SALE FARM, LOWER SALE HOUSE)** 2½ miles south-west of Marchington (SK 1127), *Sale* 1414 (1798) Shaw I 46, *temp.* Elizabeth I (1798) Shaw I 60, *Sale Corner* 1697 SA 513/2/18/11/3, *Over Sale* 1775 Yates, 1801 Smith; **SALE FARM (THE)** 1 mile south-west of Alrewas (SK 1514). *la Sale* 1271 SHC 4th Series XVIII 96, *Sale Farm* 1834 O.S.; SALE CORNER (unlocated, in Fauld), *Sale Corner* 1696 SA 513/2/18/11/3; **SALES FARM** 1 mile north-east of Yoxall (SK 1420), *? the Sale* 1661 DRO D3155/WH13, *Sale* 1558 (1798) Shaw I 46; SALE (unlocated, in Elmhurst: SHC 4th Series XVIII 96) *Sale* 1271 SHC 4th Series XVIII 96. Possibly from OE *salh*, dative *sale*, 'sallow', a small willow-like tree or shrub, so '(the place at) the sallows', though it may be noted that Rackham 1990: 108 states that the word *sale* is also applied to a coppice area within a wood, and such derivation may be more appropriate for some of these places.

**SALISBURY HILL** a pronounced hill on the south side of Market Drayton (SJ 6732). Early spellings have not been traced, but by local tradition (for which no evidence has been found, but see Twemlow 1912: viii; 30) named after Lord Salisbury of the Yorkist forces, who is said to have camped here after the battle of Blore Heath in 1459: SHC 1945-6 93.

**SALLOWE** (unlocated, possibly to be associated with Sallyfield Lane (*Sally Lane* 1836 O.S.) in Stanton (SK 1447): see SHC VIII NS 8), *Sallowe* 1360 SHC VIII NS 8. Probably from OE *salh* 'sallow', mainly used of certain species of low growing or shrubby species of genus Salix (especially Salix caprea), as distinct from 'osiers' or 'willows'. *Sallou*, recorded 1167x1182 (EEA 16 65), and *Sallowe*, recorded in 1191 (EEA 17 39), refer to Sawley in Derbyshire (ibid.).

**SALLY MOOR** (obsolete) 1 mile north of Ramshorn (SK 0846). *Saly Moor* 1798 Yates, *Sally Moor* 1836 O.S. Probably from OE *salh* 'sallow', mainly used of certain species of low growing or shrubby species of genus Salix (especially Salix caprea), as distinct from 'osiers' or 'willows'.

**SALT** on the west side of the river Trent, 3½ miles north-east of Stafford (SJ 9527). *Selte* 1086 DB, *Salt* 1166 SHC 1923 297, *Saute* 1236 Fees, *Saut'* 1242 Fees, *Salte by Trente* 1293 SHC 1911 53, *Sallt* 1532 SHC 4th Series 8 57. From OE *\*selte*, Mercian OE *\*sælte* 'salt-pit' or similar. The name is self-explanatory but puzzling, for whilst there are (or were) salt-springs in the area around Salt (see e.g. Ingestre; Shirleywich), there is no evidence of any kind, archaeological or documentary, of any brine springs or salt processing in or close to Salt. The place-name *Halen*, suggestive of Welsh *halen* 'salt', recorded in 1002x1004 (11th century, S.906; 11th century, S.1536) and previously associated with this place, is now believed to refer to Hawne, near Clent, Worcestershire (see Sawyer 1979a: xxx-xxxi), which derives from OE *healum*, dative plural of *healh. Halenmor*, recorded in 1273 (SHC 1911 152) was also near Clent, preumably adjoining *Halen*.

**SALT BROOK** a tributary of the river Trent that rises in Needwood Forest and runs through Draycott in the Clay, *Saltbroke* 1435 SRO DW1733/A/3/12. Evidently so-called from contamination by salt springs. *Saltbrook Cottages* lie at SK 1629, and the name is also found in *Saltbrook Lane*. See also Marchington.

**SALTERFORD** (obsolete) a ford across the river Churnet to the east of Alton (SJ 0941), presumably on the site of *Saltersford Bridge* recorded in 1608 (SHC 1948-9 82), *Salters Bridge* 1836 (O.S.). *Salterford* 1583 SHC 1929 65. The ridgeway from Alton towards the ford is recorded as *Salterfortherigg*, i.e. Salter's-ford-ridge, in 1339 (SHC 1913 77), and is now known as Saltersford Lane (*Salters Ford Lane* 1836 O.S.). See also NSJFS 12 1972 122. *Salterisford* (unlocated, in or near Branston) is recorded in the late 15th century (SHC 1937 180), and may be the ford recorded in the early 13th century (VCH IX 24), and the same place as *Salterforde* recorded in 1344 (Hardy 1908: 38). See also Saltmoor.

**SALTER'S BRIDGE** over the river Tame, 1 mile south-east of Alrewas (SK 1813). *Salteresbrige* 1293 TSSAHS 1991-2 XXXIII 13, *Saltersbrygge, Saltarbrige* 1341 SHC 4th Series XX 69, 81, *Saltbrugge* 1388 SHC XV 5, *Saltesbrugge* 1389 ibid.' 10, *Salterbrugge* 1390 ibid.'17, *la Sattersbridge, pontis voc' Salters* c.1535 SHC VI (ii) 166, *Saltar's bridge* c.1540 Leland ii 103, *Salters Bridge* 1600 Alrewas ParReg, *Salters bridge* 1601 SHC 1930 413, 1608 SHC 1948-9 82, 1632 SRO Q/SR/205, *Salter's brydges* 1644 (1798) Shaw I 71, *Salters bridg* 1686 Plot 244. From OE *saltere* 'a salter, a salt-worker, a salt-merchant', and OE *brycg* 'bridge'. So-called because it carries the Saltway, an ancient road along which salt was evidently transported. Perhaps to be associated with *Saltersholme*, recorded c.1535 (SHC VI (ii) 166, probably the same place as *Great Salterholme*, a meadow in Alrewas recorded in 1618 (SRO D541), and as *Salterholme* in 1825 (SRO D615/D/148). The bridge was renamed *Chetwynd Bridge*, after the Chairman of the Quarter Sessions at the time of its reconstruction in Coalbrookdale cast-iron in 1824, and appears on modern maps as *Chetwynd or Salter's Bridge*. Salters Bridge is recorded in Moseley, north of Wolverhampton, in 1693/4 (SRO D718/10/4), possibly on the road between Stafford and Wolverhampton. There is a Salters Lane in Walsall Wood, perhaps on the line of a saltway to Birmingham. Shaw Lane (*Shaw Lane* 1547 Ct, 1632 Lease (Tw)) in Forton was earlier known as *Salters Lane* from 1460 to 1516: Oakden 1984: 148. *Salteslone* in or near Stafford is recorded in 1392: SHC 1928 276.

SALTERS CROFT (unlocated, in Hanbury) *Salters Croft* 1699 SA D513/2/18/17/2. See also Marchington.

**SALTERSHALL FARM** on north-east side of Bobbington (SO 8190). *Saltershall Farm* 1840 VCH XX 72, 1892 O.S. The age of this name has not been traced, but the proximity of the place to Salters Park Farm (q.v.) may be significant.

**SALTERS PARK FARM** 1 mile south-east of Bobbington (SO 8189). *Saltershill* 1496 VCH XX 71. 'The hill of the salter or a man named Salter'. See also Saltershall Farm.

SALTERS MEADOW FIELD (obsolete, what is now Chase Terrace (SK 0409)) *Salters Meadow Field* 1834 O.S.

SALTER'S STREET – see **BLAKE STREET**.

SALTER'S WAY (obsolete) in Uttoxeter: Redfern 1886: 58, 83. *Salter's Way otherwise Portway* 1886 Redfern 1886: 83. The existence of such name is doubtful: ibid.

**SALTER'S WELL FARM** on the south side of Bagnall (SJ 9150). *Salters Well* 1836 O.S.

SALTFORD (unlocated, possibly south of Wolverhampton) *Saltford* c.1270 WA I 278, 1287 SHC 1911 193, 1324 ibid. 361, 1327 SHC VII (i) 249, 1386 SHC XVI 27, 1405 ibid. 59, *Saldeford* 1311 ibid. 308, *Saltford* 1347 SRO 3764/122[27573], *Salforde* 1395 SHC XV 67, *Salford* 1539 SHC VI NS (i) 63. There are no recorded salt deposits in the area, but the forms are consistent, so probably 'the ford on the salt-route'. See also Saltmoor.

**SALTHOUSE (FARM)** 1 mile south-west of Werrington (SJ 9446). *Salthouse (Lane)* 1836 O.S.

SALTMOOR (obsolete, probably what is now Dunstall racecourse, north of Wolverhampton (SJ 9001): see TSSAHS XXI 1979-80 16) *sæffan mor* 985 (12th century, S.860), *Saffemor* 1286 SHC 1924 330, *Saffemor* 1401 SRO D593/B/1/26/6/37/2, *Saftmore* 1516 WA II 13, *Salte More* 1569 SHC 1926 105, *Saffemore alias Saltmore* 1618 SRO D590/178, *Saltmore* 1649 TSSAHS XXI 1979-80 16, 1709 SRO D1364/2/22, *Saltmoor* 1801 WALS D/JSR/45/5. Perhaps from an otherwise unrecorded OE personal name *Sæffa (the form evidencing that name in Searle 1897: 406 is from the charter of 985AD (S.860) which relates to this place), the spelling of which became corrupted by the mistranscription of *-ff-* as *-ft-*, which in turn became the more intelligible *-lt-*, with OE *mor* 'marshland, moorland'. *Saffemore* was one of the boundary marks of Cannock Forest in 1286: SHC 1924 330; see also Hooke 1983: 63. See also Saltmoor Meadow; Showell.

SALTMOOR MEADOW (obsolete) in the Wobaston/Oxley/Coven Heath area (?SJ 9104): TSSAHS XXI 1979-80 16, *Saltmore Meadow* 1636 SRO D3377/85, *Saltmoor Meadow* 1801 Coven Heath Indenture WALS; SALTMORE (obsolete) to the west of Wrottesley Lodge, on the Staffordshire-Shropshire border (?SJ 8301), *Saltmore* 1709 SRO D1364/2/22, 1714 Shaw 1801: II 194, 197. Saltmoor appears to have been an ancient name which attached either to two distinct areas to the north and west of Wolverhampton, or to a considerable area roughly in the form of a quadrant extending from the west to the north of Wolvehampton. No salt springs are recorded in the area, and the names are likely to be derived from Saltmoor (q.v.).

**SALTWELL** (obsolete) to the north-east of Rickerscote (SJ 9220). *the Bryne pittes* 1612 SHC 1934 (ii) 36, *Salt Spring* 1836 O.S., *Saltwell* 1946 O.S. Self-explanatory: brine springs rise here.

**SALTWELLS** 3 miles south-west of Dudley (SO 9387). *Saltwell Coppice* 1812 map, *Salt Well* 1834 O.S, *Saltwells House* 1895 O.S. Plot 1686: 98 states: 'In Pensnet chase, s. from Dudley about a mile and a half, there is a weak brine (spring) belonging to the Right Honorable Edward Lord Ward, of which his lordship once attempted to make salt; but the brine proving too weak he thought fit to desist'. This place, which eventually became known as Pensnett Spa, with brine baths which were in use until at least 1919 ('Dudley Chronicle' 19th July 1919), and demolished by 1930, may be *Sallimor*, recorded in 1273 (SHC IX (ii) 26, where the index records the name as *Saltimor*), or *Saltley*, recorded in 1292 (SHC 1911 212). *Saltiswalle* ('salt-spring', or 'spring of the man called Salt') is recorded near Sugnall in about the 14th century: SHC 1921 15. See also Guttery 1950: 26.

**SANDBOROUGH** ½ mile east of Hamstall Ridware (SK 1119). *Sandbarewe* 1300 SHC VII 76, *Sondbarwe* 1327 ibid.' 230, *Sandbrough* 1607 SBT DR18/12/5, *Sandbarrow* 1641 SRO D641/5/T/29, *Sanborough* 1695 Morden, 1724 Hamstall Ridware ParReg, *Sanburrow* 1723 ibid., *Sandborough* 1758 DRO DR18/12/22. The second element is OE *bearu*, dative *bearwe* 'grove, wood', so 'the wood in the sandy place'.

**SANDFORD** (unlocated, possibly near Eccleshall) *? Sanford* 1186 SHC I 126, *? le sondiforde* 1274 SHC 1921 II 22, *? Sontfort* 1333 SHC XIV 33. Self-explanatory. *Sondford* in Brewood is recorded c.1280 (SHC VI (ii) 147), but has not been located.

**SANDFORD BROOK** a tributary of the river Dove. Self-explanatory.

**SANDHILLS (FARM)** on north-west side of Brownhills (SK 0605). Possibly to be associated with *Sandalls*, recorded in 1801: Shaw 1801: II 53.

**SANDFORD (HILL)** 1 mile north-east of Longton (SJ 9244). *Sandford* 1466 SHC IV NS 138. Self-explanatory. The foundation charter of Hulton Abbey of 1223 naming *Sondiford*, printed in Ward 1843: app. Ii, is almost certainly a much later forgery: Tomkinson 1994: 73-102.

**SANDON** Ancient Parish 5 miles south-east of Stone (SJ 9429; SMR 00751 places a deserted post-Conquest settlement at SJ 95532955). *Scandone* 1086 DB, *Sandona* 1166 SHC 1923 297, *Sondown* c.1231 SHC 1911 425, *Sandun* 1236 Fees, *Sandon* 1242 ibid, 1276 SHC VI (i) 79, *Sondona* c.1285 SHC 1921 11, *Sondon, Parva Sondon* 1327 SHC VII (i) 201, *Sanndon* 1532 SHC 4th Series 8 104. From OE *sand-dun* 'sand-hill'. The *c* in the DB form is obviously an error. The place *Parva Sandon* ('Little Sandon') is recorded in 1086 (DB) and as *Lettyl Sanndon* in 1532 (SHC 4th Series 8 106), *L Sandon* in 1749 (Bowen). It no longer survives, but probably lay in the western part of Sandon near Aston and Stoke-by-Stone: VCH IV 49 fn. Sandon Hall may mark the site of Sandon proper. *Sondilowe* 'the sandy tumulus', from OE *hlaw*, is recorded in Little Sandon in 1295: SHC 1921 37. *Sondon* in Uttoxeter is recorded in 1414: Shaw 1798: I 43.

**SANDWELL** in the north-east part of Warley (SP 0289). *Sandewell'* 1255 Fees, *Saundwell, Sandwell* 13th century Duig, *Sandhall* 1600 SHC 1935 260, *Sondall* 1749 Bowen, 1788 Harrison. From OE *sand* 'sand', probably here 'sandstone', with Mercian OE *wælle* 'spring', hence 'the spring that flows from the sandstone'. Shaw 1801: II 128 states that the place takes its name from *Sancta Fons*, or the Holy Well, about a mile south-east of the church; see also Erdeswick 1844: 415, which mentions '*Sandall ... alias*

*Sandywell, or Sandyhill*, presumably Sandwell. The spring survives in the ruins of the priory complex. See also Sarnell.

**SANDYFORD** (obsolete, on the north side of Stafford (SJ 9224)) *Le Sondeford* late 13th century SRO D938/252, 1334 SRO (193/7931), *Sondiford, Le Sondyford* 1419 SRO 376[7930], *Sondeford (Croys)* 1423 SRO 379[7926], *Sandyford* 1890 Cherry 1890: 80. Self-explanatory. Said to be the site, from at least 1546 until 1793, of gallows: VCH VI 228. A broadsheet of the latter date records 'A true and particular account of John Betley, John Biddle and Richard Ellis ... Executed at Sandy-ford, near Stafford ... for the murder of Thomas Ward, Gent.'. It is possible that the Bier or Burial Bridge carrying the road over Sandyford brook (VCH VI 198) took its name from the gallows. *Sondeforde*, recorded in the late 15th century (Oakden 1984: 79) was in Castle Church (cf. Sonde).

**SANDYFORD** north-east of Tunstall (SJ 8552), *Sandyford* 1836 O.S.; **SANDYFORD (FARM)** ½ mile north-east of Swynnerton (SJ 8536), *Sandy Ford* 1798 Yates, *Sandyford* 1836 O.S. Self-explanatory. *Sondiford*, recorded in 1223 (Ward 1843: app. ii) may refer to either of these places. *Sondyforde* in Parva Ridware is recorded *temp.* Edward I: Shaw 1798: I 162. See also Sandford.

**SANDYFORD BROOK** a tributary of the river Sow. *le Sondyford* 1432 St Thomas. From OE *sandig-ford* 'sandy-ford'. The stream evidently took its name from one of the Sandyfords near Stafford (q.v.). In the West Midlands OE *sand* was often pronounced sond and sund: cf. Sonde.

**SANDYFORD DINGLE** 1 mile north-east of Tatenhill (SK 2123). *Sandyford* 1415 Hardy 1908: 77, *Sondyford* 1415 ibid.' 79.

**SAPFORDE** (unlocated, near Kingstone) *Sapforde* 1414 SRO DW1733/A/2/113[1].

**SAREDON BROOK** a tributary of the river Penk. *Searesbroc, Searesbrocesforde* 996 (for 994) (17th century, S.1380), *Sarebrok(e)* 1290 Ch, 1317 Vernon, 1338 Ass, *Sarebro(u)k* 1300 For, *the Sarebrok water* 1338 Ass, *Sarebruck* 1596 QSR. See Saredon and Shareshill.

**SAREDON, GREAT & LITTLE** in Shareshill parish, 3 miles south-west of Cannock (SJ 9508). *Saredone* (Great Saredon) 1086 DB, *Seresdone* (Little Saredon) 1086 DB, *Sardon* 1166 SHC 1923 295, *Sardun* 1236 Fees, *Parva Sar(e)don* 1251 Ass, *Boershardon* 1280 SHC VI (i) 107, *Sardon, Saredune, Beresardon, Beresardun* 13th century Duig, *Magna Sardon* 1316 FA, *Saerdon* 15th century DuigA difficult name. The second element is OE *dun* 'hill': Great Saredon stands on the north-east and Little Saredon on the south-east of Saredon Hill (*Hul* 1286: SHC 4th Series XVIII 147), a conical hill of 505', and both places are by Saredon Brook (q.v.). It has been suggested (see Ekwall 1960: 404; Oakden 1984: 17) that the hill may have had a name incorporating OE *sear* 'sere, withered, dry, barren' (although there is no parallel elsewhere for such a hill-name: see Gelling & Cole 2000: 172-3), and that the brook took its name from the hill, or that the first element may be a OE personal name *Searu (see Searle 1897: 412), hence '*Searu's hill', although there is some doubt whether such a name existed. In fact it seem likely that this name is to be considered with that of nearby Shareshill (q.v.), and in that respect it may be noted that Great Saredon and Shareshill were both held by the same tenant in DB. The *Bere-* and *Boer-* elements are from the le Bere or le Boer family who were, for a time, its lords: see SHC VI (i) 220.

**SARNELL, SARDENHILL** (unlocated, in Sedgley: SPI 62) *Sarnell* 1599 Sedgley ParReg, *Sernall* 1630 ibid, *Sardenhil* 1661 ibid, *Sernill* 1665 ibid, *Sernell* 1665 ibid,

*Sarnell* 1674 SPI, *Sernall* 1676 Sedgley ParReg. Possibly the same place as Soundehill, Saundehill (q.v.). *Sandwall Field* in Sedgley is recorded in 1692 (HRO E12/V1/NC/39).

**SATNALL HILLS** 1 mile south of Tixall (SJ 9720). *Sotnor* 1581 SRO DW1734/2/5/15, *Satner* ?1617 SRO D615/D/6(1-9), *Satnall* 1671 Baswich ParReg, *Satnall Hills* 1686 Plot, 1821 D615/M/6/26, 1836 O.S., *Satnal* 1719 Baswich ParReg, *Sattenhill (Plantation), Sattenhall* 1783 SHC 1970 156-7. A curious name, for which no derivation can be suggested.

**SAUNDEHILL, SOUNDEHILL** (unlocated, on the west side of Walsall) *Soundehill, Saundehill* 1617 Willmore 1887: 439. See also Sarnell, Sardenhill.

**SAVERLEY GREEN** 1½ miles south-west of Draycott in the Moors (SJ 9638). *Seifirleg* 1204 SHC III 117, *Severle, Severled* 1228 SHC IV 55, *Severleg, Severlega* 1228 ibid. 64, *? Suyrleye* 1278 SHC 1911 32, *Severle* 1279 SHV VI (i) 138, *Severleye* 1285 SHC XII NS 94, 1309 SHC 1911 75, *Selverle* 1291 SHC VI (i) 203, *Severleye* 1309 SHC 1911 75, *Severlee, Severle* 1310 SHC IX 15, *Zeveleye* 1327 SHC VII (i) 220, *Severley* 1337 SHC XI 71, 1428 ibid. 229, *? Sconerleye* 1355 SRO D4038/A/4/11, *Seu'ley* (p) 1374 Pape 1928: 148, *Saverley* 1405 SRO D593/A/2/27/11, *Searley* 1571 SHC XIII 282, *Sareley Green* 1664 SHC VI (ii) 342, *Sareley* 1676 SHC XII NS 115, *Sarely alias Seidley* 1683 WaCRO CR1908/6, *Severly Gr, Severley Gren* 1686 Plot, *Severly green* 1749 Bowen, *Staverley Green* 1798 Yates, *Saverley Green* 1836 O.S. The forms point to a derivation from the river-name Severn, from British *Sabrina*, Welsh *Hafren*, OE *Sæfern*, a name of unknown meaning but possibly cognate with Sanskrit *sabar-* (in *sabardhuk*), which may mean 'milk' (see Ekwall 1928: 358ff; Jackson 1953: 271; 519; CDEPN 5370. Sabrina may have been the name of the divinity of the river: CDEPN 537. The loss of *-n* in Saverley is well-evidenced – see for example Thorley from *þorn-leah*, Arley and Earley from *earn-leah*. If not a mistranscription, one of the 1228 forms may incorporate OE *lad* 'watercourse'. Saverley may have been a different place to Saverley Green, but if so its location has not been traced. The place presumably takes its name from a minor tributary of the Blithe, of which there are at least two in this area, including one running to the south of Saverley Green through Fulford. It may be noted that Saverley Green lies some three miles from Cocking Farm (q.v.), a name perhaps also be of British origin. The Green element suggests a squatter settlement.

**SAXONS LOWE** 1 mile north-east of Swynnerton (SJ 8736). *Saxons Low* 1836 O.S. A natural conical hill. The name, of uncertain age but probably relatively recent, means 'the low or tumulus of the Saxon or Saxons', though no archaeological features have been traced here: see JNSFC 4 1964 49. The mound is said to have been known as *Hangman's Hill* in the 19th century (TNSFC 1887 57; TNSFC 1915 115), probably to be seen as evidence of popular folklore rather than its former use.

**SCALDERSITCH** ½ mile north-east of Hulme End (SK 1159). *Sitch* 1775 Yates, *Scaldersitch* 1840 O.S. *Sitch* is from OE *sic* 'watercourse'. A curious name. The first element cannot be identified, but names beginning Sc- suggest Scandinavian origins, possibly here from ON *skjalda* 'to cover (furnish) with a sheild or shields', used in some transferred sense. The second element is OE *sic* 'a watersourse'.

**SCALPCLIFF HILL** a 335' hill on the north side of Stapenhill (SK 2522). *montis Scalecl', montis Scalleclif* c.1150 (13th century), *Skalclyf* 1398, *Scalfclyffhill* 1585, *Skalvecliff hill* 1598, *Scalpecliffe* 1623, *Scalpcliff Hill (Side)* 1717x1796 SRO D877/96/1-7. The name, the early spellings for which are taken from VCH working papers (see VCH IX 207), probably incorporates ON *skjalf* 'shelf', which may have replaced a similar OE word: ibid. The element was used of flat level ground: the hill is noticeably level above the 300' contour. The place was transferred from Derbyshire to Staffordshire in 1894.

**SCARBOROUGH** on the west side of Rugeley (SK 0218). *Scarborough Farm* 1776 SRO DW1781/9/2/76.

SCHARPLOWE (unlocated, in Gayton) *Scharpelowe* 1303 SHC XII NS 173. Evidently from OE *sc(e)arp hlaw* 'the steep or pointed mound or tumulus'.

SCHIRROLDUS (unlocated, in Bagot's Bromley) *Schirroldus* 1448 SRO DW1733/A/3/24. See also Shirrall (Hall).

SCHOLLE, SCHOWLE – see **WHITTIMERE**.

SCHONERLEYE (unlocated, possibly in the Barlaston/Fulford area) *Schonerleye* 1355 SRO D4038/A/4/11.

**SCHOOL CLOUGH** 1 mile south-west of Longnor (SK 0863). *Scoldeclogh* 1331 SHC XI 32, *Scolclose* 1353 SHC XII 122, *Scolecloughe* 1556 Deed, 1775 Yates, *Scholecloughe* 1582 SHC XV 140, *School Clough* 1775 Yates, *School Close* 1840 O.S. Possibly from OWScandinavian *skáli* 'a temporary hut or shed', with OE *\*cloh* 'deep valley, ravine'. A lost place-name *Scoleclogh, Skoleclogh* is recorded c.1340 in Macclesfield Hundred, Cheshire: PN Ch I 54. *Scolhalgh*, recorded in 1348 (SHC XI 162) has not been identified.

SCORTESTONA unlocated, to the west of Bury Bank (? SJ 871361: Hooke 1983: 87-9). *Scortestona* ?13th century SHC VI (i) 10. 'The short stone'. This is *Sceortan stane* recorded as a boundary mark in a charter of 956 of Darlaston (11th century; S. 602; see Hooke 1983: 87-9). See also Cumberstone Wood.

**SCOT HAY** 2 miles north-west of Keele (SJ 7947). *Skotteshay* 1410 HOK 21, *Scott Heyes* 1566 ibid. 39, *Scott hey* 1689 Wolstanton ParReg, *Scot Heyes* 1689 Audley ParReg, *Scothay* 1698 Keele ParReg, *Scothay* 1733 SHC 1944 24, 25, 31. Probably from the a family named Scot recorded in this area from at least 1327 (SHC VII (i) 199; see also SHC 1944 62-4, 72, 74). A derivation from ME *scot* 'a tax, a payment' (see SHC XI NS 235 and 245 for references to *couscout, scuth* and *parvum scout* every third year in this area in 1307-8) is possible but perhaps unlikely to be found with the second element, which is from Mercian OE *(ge)heg* 'fence, enclosure', perhaps to be associated with the *hege* mentioned in a charter relating to Madeley of 975 AD (11th century, S.801; see Hooke 1983: 106-9). *Scott heys* is recorded in the bounds of Walsall manor in 1617 (Willmore 1887: 439), and *Scot Hays farm* in Kings Bromley is recorded in 1794 (SRO D357/A/20).

**SCOTCH HILL** a 366' hill 2 miles west of Tatenhill (SK 1622). *Skoteswallehull* 1337 Hardy 1908: 23. Hardy 1907: 137 also gives the undated forms *Scotchhills, Scotshills.* The name would appear to have originated as 'the hill at the spring of the man named Scot' (from Mercian OE *wælle* 'a spring'): the surname Scot is well-recorded in the area from at least the 14th century: see Hardy 1908: 23, 29, 38.

SCOTESLEI *Scoteslei* 1086 DB. A reference to *Scoteslei* in DB has puzzled historians. Erdeswick 1844: 545 identified the place as 'Scotesley', without the usual *Q[uare]* which appeared against names he was unable to identify, suggesting that he may have known of a place of that name, but no such place has been traced. Shaw 1798: I *12 (index to Domesday Book, etc.) gives no identification. Eyton 1881: 37 believed that *Scoteslei* (which precedes the entry for Moreton in DB) was in or near Colwich, which almost certainly existed in 1086 but is not recorded as such in DB. In 1916 Wedgwood suggested that *Scoteslei* 'can I think be identified with Coley ... The transformation is in accordance with the laws of euphony' (SHC 1916 168), and that identification has been accepted without question since that date: see e.g. VCH IV 42. However, whilst *Scoteslei* may well be Coley, there is little philological reason for the association, and it is possible that *Scoteslei* is to be identified with Colwich itself: DB records that both *Scoteslei* and Moreton were held by the church of Lichfield with the same tenant. All that can be said with any certainty is that *Scoteslei* was held by the bishop somewhere in Pirehill Hundred. In OE *Scotes-* would be pronounced *Shotes-*, and it may be noted that Shugborough (for which early spellings include *Shutborrow* and *Shottboro*) lies in Colwich parish (an ecclesiastical peculiar; i.e. exempt from jurisdiction of the bishop of the diocese), though there is no evidence to associate *Scoteslei* with Shugborough, which does not seem to have become established until the 14th century (see SHC 4th Series VI 89), but note observation in SHC 1913 131 that Shugborough was once a much larger place than Colwich, and is said to have been held by the church before the Conquest: StEnc 516. It may be noted that land called *Shootersoake* is recorded in the Shugborough/Colwich/Bishton area in 1629 and 1701 (SRO DW1781/2/7; DW1781/9/2/64/1-2), and as *Shooters Oaks* in 1697x1771 (SRO DW1781), names (seemingly unparalleled in Staffordshire) which reflect traces of the DB name: DB *Scoteslei* (if not aberrant) might be interpreted as 'shooters' *leah*': cf. Shushions, in DB *Sceotestan*.

**SCOTLANDS, THE** 3 miles north of Wolverhampton (SJ 9301). *Scotland* 1834 O.S., *Scotlands* 1856 WALS DX103/6. A name of uncertain age. Duignan's derivation from OE *sceatlandes* 'corner lands' (1902: 132) is based on his observation that the place was formerly at the corner of a triangular piece of land, bounded on all sides by roads, but the 1834 1" O.S. sheet shows the name attached to what appears to be a copse of irregular shape 1 mile south of Bushbury Hill, and in any event OE *sc-* would normally become *sh-*. If ancient the name may have some connection with ME *scot* 'tribute, payment' (cf. the field-name *Scotland*, recorded in Warwickshire in 1278: PN Wa 329; the field-name *Scotland* recorded in Clent: Tithe Map 1838), or with a field-name denoting remoteness, or with Scot, as a personal name or nationality. A piece of land called *Scotland* is recorded in the parish of Seighford in 1882: SRO D1798/667/29. See also PNEF 132.

**SCOUNSLOW GREEN** 2½ miles south of Uttoxeter (SK 0929). *Cundeslee in Bircheholt* late 12th century SRO D(W)1721/3/11/1, *Coundesley* early 13th century SRO D(W)1721/3/25/3, *Gundesle* 1275 SHC 1911 28, *Gundesleye* 1280 SHC VI (i) 108, *Cundeslegh* 1305 SHC VII 134, *Cundesley, Coundesleye* 1306 SHC 1911 67, *? Condeslegh* 1327 SHC XI NS 187, *Coundesleyh'* 1333 SHC X 88, *Conndesleye* 1342 SRO DW1733/A/2/109, *Cundele* 1359 SHC XI 171, *Scounslow Green* 1775 Yates. Some of the forms may refer to Counslow (q.v.), or (in the case of the earliest spellings), to another place altogether. The place *cundesleage* appears in a charter of 951 (14th century, S.557) of land at Marchington, and since at least 1916 (SHC 1916 90; Hooke 1983: 103-5) has been associated with Scounslow Green (*Sounlow Green* 1798 Yates, *Scounsley Green* 1836 O.S.). It has been suggested that *cund* may have been a river-name, perhaps of the stream passing through Marchington, associated with an extensive area (see Hooke 1983: 103-5: Sawyer 1979: 18 and *Cund* in glossary ibid. 87): *Cundy Field* (*Gundy Fields* 1836 O.S.) is shown on the Marchington tithe map 2 miles east of Scounslow Green at SK 121289. *Cund* would appear to be the same river-name as Kennet, Kennett and Kent, believed to be from a British name of uncertain derivation (see PN Sa I 102; cf. Cound, Shropshire), but which may mean 'hound': TSAHS LXXVI

2001 76-7. The place evidently gave its name to a hay or forest clearing (from Mercian OE *(ge)heg*) at an early date: *haya que vocatur Counderslega sub Potterslega* is recorded in the 13th century: SHC XI NS 163; see also Pottersley. Also Sawyer 1979a: 18; Hooke 1983: 103-5. A field name *Gundesleye* is recorded in Brewood in 1278 (Oakden 1984: 47), perhaps with the same derivation, and *Goundersley* is recorded in Uttoxeter in 1624 (SRO D599/15). The Green element suggests a squatter settlement.

**SCROPTON** Ancient Parish on the north side of the river Dove, 1½ miles north-west of Tutbury (SK 1930). *Scroton* 1086 DB (entered in Derbyshire), *Scropton(e)* late 11th century, c.1141 TutCart, *Cropton* 1260 Ipm, 1298 Cl, *Scroperton* 1380 PN Db 560-1, *Schropton* 1538 ibid, *Scrapton* 1577 Saxton, 1610 Speed. PN Db 560-1 gives the derivation 'Skropi's *tun*' for this place, but the personal name has not been traced elsewhere. If it existed, it was evidently of Scandinavian origin. Historically in both Staffordshire and Derbyshire, there have been many administrative changes in the status of Scropton parish since 1832, and it ceased to have a civil identity after 1866: see Youngs 1991: 421. The place is now in Derbyshire.

**SEABRIDGE** 1 mile south-west of Newcastle-under-Lyme (SJ 8343). *Sheperuge* 1235 SHC VI (i) 5, *seeperug* c.1250 SHC 1911 146, *Sheperingley* 1288 Hibbert 1909: 33, *Sheperingg* 1291 Tax, *Sheprigge* 1292 SHC VI (i) 220, *Scheperug* 1297 SHC 1911 244, *Shepinge* 1297 (1798) Shaw I 40, *Sheprig* 1305 SHC VII 140, *Sheperugge* 1332 SHC X 82, *Shepbrugge* 1381 SHC XIII 165, *Shepruge* 1422 SHC VIII i 219, *Sherbrigge* 1560 SHC XIII 208, *Sheabridge* 1749 Bowen. From OE *scep* 'sheep', and OE *hrycg* 'a ridge, a long narrow hill', or *brycg* 'bridge'. The terminals conflict, but the earlier was evidently 'ridge' (with the common West Midland *-u-* for *-i-*), which in any event seems more satisfactory than 'sheep bridge'. For another example of the name Sheepridge becoming Sheepbridge see PN Bk 189.

SECHEHULLE FOREST – see **CHECKHILL.**

**SECKLEY WOOD** in Upper Arley. *Soegeslea* (dative) 866 (S.212), *Secceslea* 866 (11th century, S.211), *Sechele* 1270 Eyton 1854-60: IV 278, *Seckley Wood* 1801 Shaw II 252. The identification of this place with the early forms is uncertain (see PN Wo 32), but if correct the meaning is probably 'Secg's *leah*' (cf. Sedgley; Seckloe, Berkshire). The place is on the west side of the river Severn, but may originally have included land on the east side: see PN Wo 32. In Worcestershire since 1895.

**SEDGLEY** Ancient Parish 3 miles south of Wolverhampton (SO 9193). *Secges leage (gemære)* 985 (12th century, S.860), *Segleslei* 1086 DB, *Seggeslegh* 1221 BM, *Seggesley* c.1270 SHC 1941 77, *Seggesle* 1275 SHC V (i) 116, *Seggesley* c.1400 DR37/2/Box 122/36, *Seigeslei* 1498 SHC 1928 69, *Sedgeley* 1525 SHC 1941 79. 'Secg's *leah*'. *Soegeslea*, mentioned in a charter of 866 AD (S.212; see SHC 1916 77-9) is now held (see Hooke 1990: 120-1) to be Seckley (Wood) (q.v.), rather than Sedgley as suggested in SHC 1916 77-9. *Sedgley Beacon* is recorded as *Beacon Piece* in 1736 (BCA D3155/WH94), and mentioned in 1801 (Shaw 1801: II 221). *Sedgley Hay*, recorded in 1255, may be Baggeridge Hay or Wood, recorded in the early 13th century: VCH II 344. Sedgley Parke is recorded in 1442 as *Sedgeley Park* (SHC III NS 167), and in 1596 as *Sedgley Parke* (BCA MS3145/258/14); see also Shaw 1801: II 136, 221. Another park called *the Olde parke at Seggeley* is recorded in 1444 (SHC III NS 167), perhaps to be associated with *(Atte)parke* (p) recorded in Sedgley in 1332 (SHC VII (i) 128. For names in the Sedgley district in 1655 see Hackwood 1898: 95-6.

**SEDGLEY HAY** (unlocated, possibly Baggeridge Hay or Wood: VCH II 344 fn.10) *(Hay of) Sedgley* 1255 VCH II 344 fn.10.

**SEEDY MILL** 1 mile north-west of Elmhurst (SK 1013). *Synethimulne, Synethimilne* c.1250 SHC 1924 75, VCH XIV 236, *Cywythi milne, Siwithmilne, Siwethimulne, Sywethmulne, Sewesimuln, Sywwehmlne* 1268x1272 SHC 1910 106, *Sindi Milne or seedy Mill House* 1734 SA 1987/3/3, *seedy Mill* 1775 Yates. Perhaps from OE *sineðe*, an adjective meaning 'very gentle' with an *-ig* ending, which might conceivably be used of a slow-grinding mill.

**SEGGEHALVEFORDE** (unlocated, perhaps near Rocester) *Seggehalveforde* 1359 SHC XII (i) 160.

**SEGGERSLEY (FARM)** on north-east side of Ellenhall (SJ 8426). *? Seggesley* 1445 SHC XI 305, *Seggersley* 1669 Ellenhall ParReg, *Sigersley* 1851 White. The derivation is probably as Sedgley (q.v.).

**SEIGHFORD** Ancient Parish (pronounced Sigh-ford [saɪfəd]) 3 miles north-west of Stafford (SJ 8824). *Cesteforde* 1086 DB, *Sesteford* 1152 EEA 14 96, *Sestesforde* c.1200 Rees 1985, *Seteford* 1208 Cur, *Seteford* 1208 SHC III 142, *Sevettford* 1208 ibid. 143, *Seasteford* 1209 SHC 1914 86, 1209 SHC III (i) 175, *Seaford* 1248 SHC IV 243, *Cestford* ?13th century SHC IV 269-70, *Cesteford* c.1300 SRO D615/D/345/1, *Sefteford* 1314 SRO D1564/4, *Ceysteford, Cesteford* 1327 SHC VII (i) 211-2, *Casteforde* 1330 SHC 1913 23, *Sesteford* 1337 SRO DW1781/4/7, *Cesteford, Sesteford* 13th century Duig, *Sesteford* c.1330 SHC IV 275, *Sefteford* 1468 SRO DW1761/A/4/34, *Sextiford* 1476 SHC VI NS (i) 101, *Seghford* 1532 SHC 4th Series 8 64, *Seyghford* 1541 SHC V NS 118, *Seytford* 1538 SHC 1910 44, *Seythford* c.1565 SHC 1926 140, *Seckford* c.1566 SHC 1938 192, *Cyford* 1644 SHC 4th Series I 168, *Syford* 1663 SHC II (ii) 37, *Seighford or Sightford* 1749 Bowen. The various forms (the spelling *Cesterford*, cited in Ekwall 1960:410, is properly *Cestford*), which are particularly irregular, make any derivation problematic – it is very possible that even by the time of DB, the derivation of the name had been forgotten – but Stevenson observed 'from the later forms it is evident that the D[omesday] *st* here must have its usual value, i.e. it represents an OE *ht*, so that Cesteford might represent OE *Seohtre-ford*, from *seohtre* 'brook, ditch' – the ford of the brook or ditch': Duignan 1902: 132; cf. Spettisbury, Dorset, found as *Prehtesberie, Spesteberie* in DB. Another possibility is that the name incorporates OE *seht* 'agreed; a peace between two parties', though in what sense is uncertain. The forms with *S-* might suggest possible Anglo-Norman influence: in OE *c* before *e* or *i* had the sound represented by Modern English *ch*, but in Norman-French *c* in such cases was pronounced [ts], later becoming *s*, although still represented by *c* in writing. Possibly therefore from a compound of OE *ceast* (*c* being pronounced [tʃ]), meaning 'strife, contention', and OE *ford* 'a ford', as preferred by CDEPN 535-6. However, it is quite likely that the DB form is (as often the case) aberrant. It is tentatively suggested that the first element may be OE *sester, seoxter* 'a vessel, a jar; a measure of capacity', with cognates in other European languages: Fr *sétier* 'two gallons', German *sester, sechter* '16 quarts', Italian *sestere* 'a pint-measure'. The term *sester* is well-recorded in the Anglo-Saxon period (found in the ASC, a Will, and the text *Leechdoms*), and was used for both dry and liquid measures. DB records *sester* in the form of Latin *sextarius*, which varied in size, and in the 14th century the *sester* of London was a measure containing 4 gallons, though it is unclear whether that volume was a recent innovation or dated back to the Anglo-Saxon period.: Harmer 1914: 79-80; see also Earle & Plummer 1892-9 II: 224; Robertson 1956: 273. If the word *sester* is indeed to be found in this place-name, this would appear to be a unique example, but EPNE i xxvii observes that 'Ford-names are ... often combined with words which describe some early recognised feature which marked the site of the ford, but still more often with words describing its capacity or nature ... ... combinations which should remove any hesitation about accepting less usual words ... as likely themes in place-names'. Whatever the root, there is no evidence, either philological, archaeological or historical, of any connection with OE *cester*, denoting a Roman site (*contra* VEPN II 160), though doubtless some forms have been influenced by the element. The ford is over Millian Brook, on the north side of the village. Cf. Cheslyn Hay; Chestall.

**SEISDON** (pronounced seez-dunn [siːzdən]) 6 miles south-west of Wolverhampton (SO 8394). *Saisdon, Saisdone, Seiesdon* 1086 DB, *Saiesdona* 1130 SHC I 2, *Seisdun* 1160x1206 SHC III (i) 215, *Seyston'* 1222 Pleas, *Seisdun* 1227 SHC IV 52, *Seyxdun* 1235 Fees, *Seydon'* 1236 Cl, *Seisdun* 1242 Fees, *Sysdon* 1257 SHC 1911 128, *Seysdone* 1292 Ipm, *Seysdon* 1309 BCA MS3066/Acc1903-003, *Seyseden* 1323 Cl, *Seyason* 1532 SHC 4th Series 8 127, *Seasdon* 1590 SHC XVI 102. The meaning of Seisdon is probably OE *Seax-dun* 'the hill of the Saxons or of Seax'. The change to *sais-* may be due to Norman influence. The name might indicate the presence of Saxons from Hwiccan territory to the south moving into a mainly Anglian community, rather than the presence of English amongst a Celtic community (Gelling 1988: 210). However, it is not impossible that the first element of the name is from Welsh *sais* '(place of the) English', with an OE suffix, since the place is in an area with other names of Celtic derivation (e.g. Penn, Trysull), although it would be surprising to find such advanced Welsh phonology so far east. The village of Seisdon lies on Smestow Brook; the hill from which the place takes its name may be the hill of over 400' to the north-east of the village. Cf. Pensax, Worcestershire, with older phonology, and see Coates & Breeze 2000: 335.

SELLESLEG (unlocated, possibly near Essington) *Sellesleg* 1284 FA.

SELSUL (unlocated, possibly in the Walsall area) *Selsul* c.1175x1208 SRO 3005/1.

SENA PARK – see **SINAI PARK**.

SERLEHOUSE, SERLELANE (unlocated, perhaps near Slymansdale) *Serleshous* 1462-3 NSJFS 3 1963 42, *Serlelane* c.1468 ibid. It is unclear whether these name are associated with the name Slymansdale (q.v.).

SERNILL – see SARNELL, SARDENHILL.

SETESWEY (unlocated, probably near Longdon) *Seyteswey* 1313 SHC 1921 7, *Seteswey* early 14th century ibid. 16. Possibly to be associated with *Seytces*, recorded in the same area c.1270x1298: SHC 1921 8.

SEVEN ASHES (obsolete) between Fulford and Hilderstone (? SJ 9535). *7 asshes* 1577 Saxton, *7 Ashes* 1599 Smith, *the 7 Ashes* 1673 Blome. The place cannot be identified with certainty, but may be what is now High Elms (SJ 9535; *High Alms* 1920 O.S.) on a hill of over 700'. Self-explanatory. Smith's map of 1599 has thumbnail sketch of four (*sic*) trees on a hill.

**SEVEN SPRINGS** ½ mile south-west of Colwich on the south side of the river Trent (SK 0020); **SEVEN SPRINGS** 1½ miles to the south of the first on Wolseley Plain (SK 0018); SEVEN SPRINGS on Cannock Chase at SK 032147 (StEnc 508). The names are self-explanatory, and although their age is uncertain, it is evident that the number seven had particular importance in OE law and custom: now-forgotten folk-lore probably influenced the choice of seven as the number commonly found in association with Mercian OE *wælle* 'a spring', and (sometimes) 'a stream': see EPNE ii 119; PN Wo 35-6. The number is frequently found in place-names connected with various topographical features including springs, although it is improbable that any place had in reality seven springs (cf. Showell; Seawell, Northumberland; Sewell, Bedfordshire; Sowell, Devon). *Seven Oaks* is recorded in 1834 (O.S.) to the north of Beaudesert Park (SK 0314). *Senokestre* 'seven oak trees' (? near Tittensor) is recorded in about the 13th century: SHC VI (i) 21. See also Seven Ashes; Showells. on Cannock Chase.

SEYNESHAULWE (unlocated, possibly near Squitch House) *(land called) Seyneshaulwe* 1402 SHC XI NS 209. Possibly from *Sene, Senyie*, from Fr *sene, senne*, Latin *synodus* 'a meeting of clergy for deliberations, a synod': *seyney-houses* were buildings belonging to monastic houses where breaks ('seyneys') were taken by monks in need of rest and recuperation after the regular bleedings they undertook for health reasons, or after illness: see also Coena's Well; Sinai Park; Wallbridge. The remainer of the name is from OE *halh*.

SEYNT MARS MORE (unlocated, probably near Stafford Foregate) *Seynt Mars More* 1535 SRO D938/224.

SHACKAMORE (obsolete) ½ mile south-east of Shugborough (SJ 9921). *Shackamore* 1624 SHC 1970 86, *Shakemoor* 1624 SRO D615/D/1, *Shackamore Field, Shackamore Pasture* c.1800 ibid. map 110. Possibly from OE *sceacere mor* 'moorland of the robber'. *Shackerley* near Seabridge or Clayton is recorded in 1658: SRO D593/B/1/14/4/25. Cf. Shakerlowe.

**SHAFFALONG** 1 mile west of Cheddleton (SJ 9652). *Shaffurlong* 1775 Yates, *Shafferlong* 1836 O.S. A curious name. The first element may be a corruption of OE *scaga* 'a copse', perhaps influenced by the word *shaft* since coal extraction began here in the 18th century. The field name *Sufferlong*, possibly denoting poor ground, but more likely to be 'south furlong', has also been recorded: see Field 1993: 14. The locations of *Asferlung* recorded in the 13th century (SHC IX NS 317), *Affurlunge* (undated) mentioned in Sleigh 1883: 50, *S(h)ortforlong* 13th century Erdeswick 1844: 15, *Furlong* 1628, 1636, *Forlong* 1656 Wolstanton ParReg have not been identified, but *Ashfurlong* recorded in 1350 (SHC VI (ii) 192) and 1611 (SRO D16/2/20) was in Little Wyrley, and *Ashfurlongs* was near Handsacre (SRO D260/M/T/5/134). Cf. Shuffers Wood south of Betley (SJ 7547); Ashfurlong Hall at Sutton Coldfield, Warwickshire (PN Wa 49).

SHAKELESFORD (unlocated, possibly on Smestow Brook near Gothersley) *Shakelesford* 1342 SHC 1913 91.

SHAKERLOWE (unlocated, perhaps in the Swynnerton/Oulton/Draycott area) *Shakerlowe* 1360 SHC XII (i) 173. From OE *sceacere hlaw* 'robber's mound or tumulus'. Shackeley in Shropshire, 1½ miles south-west of Boscobel (SJ 8106) is recorded as *Shakerlawe* in 1284 (Eyton 1854-60: II 178), *Shakerlow* in 1525 (SA 1781/2/6), *Shackerlow(e)* in 1679 (SA 1781/2/52-86). Cf. Shackamore.

**SHALLOWFORD** 5 miles south-west of Stone (SJ 8729). *Shawford* 1271 PR, SHC 1914 62, *Schaldeford* 1272 SHC IV 190, *Sheldeford* 1299 SHC 1911 251, *Schaldeford* 1278 SHC VI (i) 89, *Saldeford* 1302 SHC VII 123, *Schaldeford* 1322 BCA MS3415/152, *Shawford* 1595 Erdeswick 1844: 139, 1599 Smith, 1599 SHC 1935 81. Self-explanatory, from the OE adjective *\*sc(e)ald* 'shallow', a word not evidenced before the end of the 14th century except in place-names. See also Saltford.

SHALSTONE   (unlocated, possibly an error for Shenstone)  *Shalstone* 1579 BCA MS10998/63.

**SHARESHILL**   Ancient Parish 5.5 miles north-east of Wolverhampton (SJ 9406). *Servesed* 1086 DB, *Sarueshul(l)* 1213 FF, *Sharesweshull* 1225 SHC IV 34, *Saresweshull'* 1225 Cur, *Sarsculf* 1227 Ass, *Saleshul* 1236 Fees, *Sarueshul* 1242 ibid, *Shareweshulf* 1252 Cl, *Saruesculf* 1262 For, *Schelfhulle* 1277 SHC VI (i) 80, *Sarushylf* 1285 FA, *Sharesweshulf* 1293 Ass, *Sarushulf* 1298 Vernon, *Scharesschulf* 1325 SHC 1911 368, *Sarveshulf* 1326 Vernon, *Schareshulfe* 1327 SHC VII (i) 224, *Sarnesculf, Sarneshull, Sharnshull, Shareshulle, Sareshull* 13th century Duig, *Shareshulle* 14th century Duig, *Sharchelf* 1532 SHC 4th Series 8 73, *Shasell* 1608 SHC 1908 1948-9 75, *Shareshill* 1775 Yates. A difficult name: the philogical problems are set out in Oakden 1984: 116-7, where a derivation from an OE personal name \*Scearf (corresponding to ON Skarfr), as proposed by Löfvenberg, with OE *hyll* 'a hill', and a variant with OE *scelf* 'shelf', is confidently asserted. In a discussion on the derivation of Sarebrook, the old name of Saredon Brook (*Searesbroc* 996 (for 994) (17th century, S.1380), Ekwall 1928: 352 considered early spellings for Saredon, and concluded that the first element of Saredon and Sarebrook was the OE personal name Searu (recorded in Searle 1897: 412; doubted as a genuine name in Redin 1919: 23; accepted as genuine by Ekwall 1928: 352, questioned in Gelling & Cole 2000: 172-3), but noted that the genitive form *Searwes* would be expected, and suggested that Saredon and Sarebrook incorporated a hypocoristic (pet-name) form of a name beginning Searu-. Early spellings for Shareshill might be seen to preserve traces of the genitive form not known to Ekwall, who concluded that the -*ea*-, supported by DB *Seres*-, pointed to OE *Sear*, which looked like a Germanic element, probably a personal name, corresponding to or related to Gothic *Sarus*, OHG *Saro*-. The spellings for Shareshill show that the earliest forms are *Sar*-, which continue until the mid fourteenth century, and spellings with *Scar*-, which would be pronounced *Shar*-, do not begin to appear until the early 13th century. Furthermore, Duignan 1902: 133 was undoubtedly correct in his view that the *n* in his cited 13th century spellings resulted from mistranscriptions, and should be read as *v*, those letters (with *u*) being indistinguishable in early handwriting. The various spellings for Shareshill show that the root of the name cannot be \*Scearf, but is almost certainly *Saru*-, *Sarue*- (with the DB form to be read as *Serue*-), and the forms are markedly similar to the first element of early spellings for Sardon Brook (q.v.) and Great and Little Saredon (q.v.). Saredon Brook runs close to the north side of Great Saredon, which lies a mile or so from Shareshill, with Saredon Hill (505') and Little Saredon between the two. It is clear that the names Great and Little Saredon, Saredon Brook and Shareshill are to be considered as a group with a common root (and it may be noted that Great Saredon and Shareshill, together with Stretton, Water Eaton, and Gailey, were held by the same tenant in DB, forming a large estate extending for four miles or so along and straddling Watling Street; until 1866 the ancient parish of Shareshill included the township of Saredon: Youngs 1991: 422; and that Great Saredon, Little Saredon and Shareshill all occupy relatively high ground on a formation of water-bearing Pebble Beds: VCH V 173). It is possible that the root of Saredon and Shareshill is a pre-English estate, wood-, river- or hill-name: the places lie on the south side of Watling Street, but the northern boundary is Saredon Brook, suggesting that the boundary may be of greater age than the Roman road. Many early place-names may have been applied to large estates which often incorporated topographical elements including Celtic topographical names. Ekwall 1928: 352 felt it unlikely that the first element of the place-name was an old name of the brook, but compared such a name with MBret *Sar*, a tributary of the river Blaret. Some clue to the derivation may be hidden in the reference to the *woods of Sarewefeld* are recorded in the area in 1235: SHC 4th Series XVIII 22. It may be noted that Salisbury, *Sorviodunum* to the Romano-Britons, is held to derive from British *\*soruio*, later British *\*serw*, of unknown meaning, and taken into Anglo-Saxon as *\*Seru*, and by 'breaking', *\*Seoru-* and *\*Searu-* (Jackson 1953: 260-1; Rivet & Smith 1979: 461; CDEPN 524). PN W 18-19 raises the possibility that the early name of Salisbury may have become associated later by folk-etymology with the well-recorded OE *searu*, genitive *searwe*, *searwes* 'trick' (also 'device, design, contrivance, art, artifice, wile, ambush'), though the precise meaning there is unclear, and Ekwall 1960: 402 saw some influence from OE *searu* 'armour'. The second element in Shareshill is variously OE *scelf* 'shelf, ledge', and OE *hyll* 'hill'. *Sa-* perhaps developed into *Sh-* in the name Shareshill from the influence of the first letters of *scelf*, an element which could well have been in use long before the earliest recorded spellings. The area behind the church at Shareshill is notably level, and might be considered a shelf. The DB spelling may (as noted in Duignan 1902: 133) point to OE *hēð* 'heath' (*ð* often appears as *d*), but DB forms are frequently corrupt, though in this case the first part of the name would appear to be reasonably accurate. Square earthworks recorded on the north and south sides of Shareshill and once believed to be Roman are now thought to have been medieval moated sites: Shaw 1801: II 308-9; VCH I 192, 346, 348; VCH V 173. Windy Arbour Lane in Great Saredon may be from a colloquial expression for an exposed place adopted relatively recently, but might derive from OE *eorþburg* 'earthern fortification', denoting the existence of earthworks, or OE *here-beorg* 'shelter or protection for a number of men; army quarters' (EPNE ii 244).

**SHARPCLIFFE**  2 miles south of Bradnop (SK 0052). *Scarpcliffe* (p) 1261 StCart, *Charpeclif* 1275 SHC V (i) 118, *Sharpecliff* 1292 SHC 1911 216, *Scharpclif* 1310 ibid. 437, *Sarpeclif* c.1311 ibid. 436, *Sarpeklif* 1311 ibid.' 437, *Sharpclyf* 1409 SHC XVI 72, *Sharpe Clyffe otherwise Sharpclife* 1595 SHC 1934 14, *sharp Cliffe* 1598 SHC 1935 147, *the Sharpe cliefe* 1656 ParReg. From OE *sc(e)arp* 'sharp, pointed, steep', with OE *clif* 'a cliff, a bank, a steep slope': there is a sharp rocky outcrop on high ground here, perhaps the look-out place suggested by the name Ipstones (q.v.).

**SHARPLEY HEATH**  1 mile north-east of Hilderstone (SJ 9635). *Sharpley Heath* 1772 SRO D1462/9/1-30, 1798 Yates, *Sharply Heath* 1836 O.S. from OE *sc(e)arp* 'sharp, pointed, rugged; steep', with OE *leah*.

SHARSEMORE  (unlocated, in Sedgley) *Sharsemore* 1547 TSAS 3rd Series VIII 1908 237.

**SHATTERFORD**  in Upper Arley, 1 mile south-east of Romsley (SO 7981). *Sciteresforda* 996 (for 994) (17th century, S.1380), *Scheteford* 1271 SHC V (i) 146, *Sheteresford* 1286 SHC V (i) 157, *Shutterford* 1577 Saxton, *Shitterford* 1673 Blome. Almost certainly from OE *\*scitere* 'a sewer, a channel or stream used as an open sewer', from OE *scite* 'shit, dung', rather than from OE *scytere*, *sceotere* (*sc* pronounced [ʃ]) 'shooter, archer', hence 'archer's ford': it is difficult to distinguish *scitere* from *scytere*, but the fact that the place is evidently connected with a stream makes the second meaning improbable (cf. Skitterlyn, Northumberland; Skitter Beck, Lincolnshire). In Worcestershire since 1895. Delicacy has led to the inevitable corruption of all names with this root.

**SHAW**  1 mile south-east of Heaton (SJ 9661), *la Schawe* 1325 SHC 1911 366, *Litleton shawe* c.1539 LRMB, *Shaw* 1775 Yates, *White Shaw* 1842 O.S., *Shaw Farm* 1880 Kelly, *Shaw* 1891 O.S.; SHAW (unlocated, in or near Alton), *la Schawe* 1327 SHC VII (i) 216, 1331 SHC 1913 36, *Schagh* 1339 ibid. 77; **SHAW HALL**  2 miles north of Wolverhampton (SJ 9105), *Shaw (furlong)* 1313-4 Parke 1860: 84, *Shaw (Heath)* 1615

*Will, Shaw(croft)* 1657 SHC 1928 123, *Shaw (House)* c.1725 SRO DW1813/1/4, *Shaw Hall* 1834 O.S.; **SHAW HALL** 2 miles north of Cheadle (SK 0145), *la Shaghe* 1281 Ipm, *? the Shawe* 1530 SHC 1910 19, *Shawe* 1598 SHC XVI 185, *Shawe* 1609 SHC III NS 52, *the Shaw* 1644 SHC 1944 23, *Shaw Hall* 1832 Teesdale, *Shaw* 1836 O.S.; **SHAW HALL** ½ mile south of Kingsley (SK 0046), *? Schawe* 1350 SHC XI NS 30, *Shawe* 1659 Kingsley ParReg, *Shaw* 1836 O.S. From OE *scaga* 'wood, copse', probably with modern *Hall* rather than OE *halh*.

**SHAWFIELD** 2½ miles south-west of Longnor (SK 0661). *Shawfield* 1566 *Deed*, 1775 Yates, 1840 O.S. From OE *scaga* 'a small wood, a copse, a strip of undergrowth or wood', with OE *feld* 'open country'.

**SHAW (HOUSE)** 1 mile north-west of Alton (SK 0543). *Schagh* 1339 SHC 1913 77, *The Shaw House* 1750 SRO D240/D/99. From OE *scaga* 'small wood, copse'.

SHAWMOOR FARM (obsolete, 1 mile south-west of Sheen, on the west bank of the river Manifold (SK 0960)) *Shawmoor Farm* 1820 Greenwood. Perhaps to be associated with *Schal moor*, recorded in 1392 (VCH VII 29), in which case the name may be from ON *skál* 'a bowl, a hollow' (used in some topographical sense), or ON *skáli* 'a temporary hut or shed'.

**SHAW WALL** 2 miles south-east of Ipstones (SK 0448). *Shaw Wall* 1781x1785 SRO D240/D/184. From OE *scaga* 'small wood, copse', probably with Mercian OE *wælle* 'a spring', and (sometimes) 'a stream', rather than OE *weall* 'a wall, a rampart': the place lies within the junction of two headwaters of Shirley Brook.

**SHAY LANE** a lane that runs between Shebdon and Forton. *Shee Lane* 1798 Shaw I 34. Possibly a back-formation from *chaise*, 'a carriage', which is sometimes found as *shay* (OED), but OE *scaga* 'small wood, copse' is sometimes found as *shay*, generally in West Yorkshire, but also recorded elsewhere: see PNEF 6.

**SHEBDON** 4 miles north of Newport (SJ 7625). *Schebbedon* 1267 For, 1293 SHC VI (i) 277, *Shebdon Ley* 1572 SHC XIII 292, *Shebben (poole)* 1686 Plot 209, 232. From the OE personal name *Sceobba or *Sceoba (see Shobnall) with OE *dun* 'hill', so 'Sceobba's or Sceoba's hill'. The pool at Shebdon (illustrated in Plot 1686: plate XIX) was drained and enclosed in the early 19th century.

**SHEEN** Ancient Parish 3½ miles south-east of Longnor (SK 1161). *æt Sceon* 1002x1004 (11th century, S.906; 11th century, S.1536), *Sceon* 1086 DB, *Schone* 1226 SHC 1937 37, SHC V (i) 54, *Chone* c.1241 SHC VII NS 142, *Shene* 1265 Ass, *Chene* 1281 Okeover E5104, *Sceone* 1301 SHC VII 90, *Schene* 1344 DRO D2375M/55/3, *Sheine* 1666 SHC 1925 203, *Sheen* 1682 Browne. The basis for the name seems to have been OE *\*sceo*, (nominative plural *sceon*) an unrecorded word probably related to Norwegian *skjaa* 'shed, kiln', or ON *skjál* 'shelter', with the *-n* representing the OE oblique form, possibly here meaning 'shelters, sheds', so '(the place at) the sheds or shelters': see Ekwall 1936: 55-7; CDEPN 541. Cf. Sheen in Surrey, *Sceon* c.950 (S.1526), which has a similar derivation. It may be noted that there is no connection with the word 'shine'. Sheen Hill is a prominent 1247' hill 1 mile to the north, with a sloping summit plateau and the north, west and east sides roughly forming three sides of a truncated four-sided pyramid, a shape sufficiently striking to suggest that it might in some way be associated with the derivation of the name. High Sheen (*Hie Sheen* 1658 Alstonefield ParReg) lies on the southern slopes of Sheen Hill. See also Ekwall 1936: 55-7.

**SHEEPHOUSE FARM** 1½ miles south of Leek (SJ 9854). *Shephouse* 1538 Dieulacres Inventory, 1654 Leek ParReg, *Shiphouse* 1662 ibid, *Shephouse* 1704 ibid, *Sheep House* 1836 O.S. Self-explanatory: the place was Crown property, and also called *ye Kynges folde*: PRO SC6 3353. See also Fould, to which the 1662 spelling may refer: VCH VII 194.

**SHEEPWALKS, THE** an upland area to the south of Enville (SO 8286). VCH IV 54 fn.3 associates this place with *Scipricg* recorded in Wulfrun's grant of 994 (17th century, S.1380) of Upper Arley to the monastery of Wolverhampton (see SHC 1916 107; Hooke 1983: 68, 70), but the evidence is slight.

**SHEEPWASH FARM** 1½ miles north of Caverswall (SJ 9545), *Sheepwash* 1836 O.S.; **SHEEPWASH FARM** 1 mile south-west of Rugeley (SK 0215), *The Sheep Wash* 1834 O.S.; SHEEPWASH (unlocated, on the river Sow near St. Thomas's priory: VCH III 260 fn.4; see also SRO D938/10), *Scepeswach, Scepewas* 12th century SHC VIII (i) 134-5, *Scepewas* 1261 SHC 1914 120, *Scepeswas* c.1350 SRO D938/10, *Shepwasshe* 1414 SHC XVII 50; SHEEPWASH (obsolete) at Hamstall Hall Farm (SK 1019), *Sheepwash* c.1817 Alexander & Binski 1987: 238; **SHEEPWASH** near Great Bridge (SO 9792), *Sheepwash* c.1713 Dilworth 1976: 151. All the places lie close to watercourses, and were presumably places where sheep were dipped.

SHELD (unlocated, in Stafford) *Sheld* 1446 SHC XII 312. Perhaps from OE *sceld* 'a shield, a protector', probably used in place-names of a shelter of some kind.

**SHELFIELD** 3 miles north-east of Walsall (SK 0302). *Scelfeld* 1086 DB, *Schelfhul* 1271 For, *Schelfhulle* 1276 SHC VI (i) 80, *Chelfeld* 1278 ibid. 87, *Sheleftel* 1280 SHC VI (i) 147, *Shelfhulle* 1288 ibid. 174, *Schelfhulle* 13th century Duig, *Shelfhull* 1300 For, *Shelfel* 1320 SHC 1911 95, *Alta Selfeld* 1327 SHC VII 224, *Sheldfyeld* 1590 SHC 1930 86. From OE *scelf* 'ledge, shelf', with OE *hyll*, meaning 'hill with a plateau or shelf'. The place lies on a moderately elevated plateau, sloping on all sides. The *-field* element has, as frequently happens, become interchanged with *-hyll*. *Alta* 'high' implies that there was a 'lower' Shelfield. Cf. Shelfield, Warwickshire.

SHELLESCROFTE (unlocated, in Norton-in-the-Moors) *Shellescrofte* 1607 BCA MS917/1699.

**SHELMORE** 2 miles west of Gnosall (SJ 8021). *Shellmore (Park Wood)* 1830 SOT D615/M/9/11, *Shelmore* 1833 O.S. Early spellings have not been traced, and the name may be relatively modern.

**SHELTON** 1 mile south-west of Hanley (SJ 8746), *Selton* 1263 SHC XII NS 241, *Selton, Shelton* 1686 Plot, *Shelton* 1836 O.S.; **SHELTON UNDER HARLEY** 1 mile south-east of Whitmore (SJ 8139), *Scelftone* 1086 DB, *Schelton* 1189, *Sheltun* 1227 SHC IV 53, *Schelftun* 1253 SHC 1911 121, *Scheston* 1280 SHC VI (i) 103, *Selfton* 13th century, *Shelton Harnage* 1369 SRO D641/5/T/1/1, *Shelton under Harley* 1381 SHC XIII 160, *Schelton* 14th century Duig, *Shelton under Horeley* 1617 SHC 1934 (ii) 52, *Shelton* 1559 SRO D641/5/T/1/10, 1747 Bowen. '*Tun* at the shelf or ledge', from OE *scelf*, presumably with reference in the first place to the long slope to the south of Shelton church, and in the second place perhaps to the flat fields in the valley to the west of Shelton Harley Farm: see Gelling & Cole 2000: 216. Harley is probably from the high ground to the east, part of which is still known as Harley Thorns (q.v.). *Harnage* in the 1368 form is evidently an error for Harley.

**SHELTWOOD FARM** 2½ miles south-east of Bromsgrove (SO 9867). *Sylkwode* 1256 FF, *Siltwood* c.1260 AD ii, *Schiltewode, Shyltewode* 1275 Ass, 1388 Ipm, *Schildwode* 1275 Ass, *Schiltwode, Shiltwode* 1279 RH, 1374 Pat, *Saltwood* 1291 Tax. PN Wo 364 suggests that the first element may be OE *scielet* (dialect *shillet, shilt*), a sort of rock or shale. The place, in Tardebigge parish, was in Staffordshire from c.1100 to 1266, and in Warwickshire until 1844, when it transferred to Worcestershire.

SHEMERBROKE (not located, perhaps near Stonywell or Burntwood) *Shemerbroke* 1561 HLS 582.

**SHENSTONE** Ancient Parish 3 miles south of Lichfield (SK 1104). *Seneste* 1086 DB, *Scenstan* 11th century Duig, *Scenestan* c.1130 Oxf, *Schenestan* 1166 SHC 1923 295, 1193 ibid.' 270, *Shenestan* 12th century Duig, *Senestan* 1270 SHC 1923 266, *Scenestone* 1282 SHC V (i) 136, *Sheneston, Schenestane* 13th century Duig, *Schenestone* 14th century Duig, *Shenstonhame* 1414 SHC XVII 6, *Sheinston* c.1540 Leland. Seemingly from OE *scene* 'bright, shining, beautiful', with OE *stan* 'stone'. The first element is not uncommon in place-names, and is also recorded in combination with *feld* 'open ground', *dun* 'hill', and *leah*, but a meaning 'beautiful stone, shining rock' or similar is not easy to explain. Other than fragments of two cylindrical Roman milestones of local sandstone in the area, one found 1 mile north of Shenstone (at SK 105062: TSSAHS XIX 1977-8 2, 4), and a note in Shaw 1801: II 19 that 'some mile stones have been found by the brook running West of the city [i.e. Wall]', no record or tradition has been traced of any stone in the area, beautiful or otherwise. Although an outcrop of Lower Keuper sandstone is found on the hill here (and many place-names in the West Midlands containing *stan* are connected with rock outcrops), it could hardly be described as bright, shining or beautiful. Roman Watling Street and Icknield Street both run close by, a Romano-British farmstead has been discovered here, and the Roman site of Letocetum is nearby. Perhaps therefore from some lost Roman monument or similar, although given the difficulties with this explanation, a derivation from an OE personal name cannot be ignored: Shenstone, Worcestershire (but with slightly different spellings to this name), is held to be from a personal name (PN Wo 256). For completeness, Plot 1686: 118, 242 records 'At Pipe-hill...I found another shining Sort of Earth...made up in great part, with silver colour'd Laminæ,...guilding the hands if rub'd upon them...'. Pipehill is less than 3 miles to the north-west of Shenstone. Cf. Shengay or Shingay, Cambridgeshire; Shenington, Oxfordshire; Shenfield, Essex. For Shenstone Park see SSAHS 1988-9 XXX 46-8.

SHEPERINGLY (unlocated, near Newcastle under Lyme) *Sheperingly* 1288 Hibbert 1908: 3.

**SHEPHERD'S CROSS** 2 miles north of Biddulph (SJ 8960). *Shepherds's Cross* 1791 SHC 4th Series 13 137. An ancient millstone-grit cross: see NSFC 1908 42 170. The various references to *Crosse* from 1579 in Biddulph ParReg probably refer to the so-called Plague Cross south of the church rather than this place.

**SHEPPY FARM** 1 mile north-west of Haughton (SJ 8521). *Shepey* 1679 SHC 1914 91, *Shippy* 1775 Yates, *Sheepy* 1832 Teesdale, *The Isle of Shippy* 1836 O.S., see also VCH IV 76. The name is probably too early to be transferred from the Isle of Sheppey to denote remoteness, even though the place lies on an isolated hill. An unlocated *Schepedon* is recorded in 1327 (SHC V (i) 240), and might refer to Shebdon or to this place, in which case the name is a corruption of OE *sceap-dun* 'sheep-hill'. Otherwise, perhaps from *sceap* with OE *eg* 'island, land between streams', here in the latter sense, since there are streams to the north and south. Or possibly the name was transferred from Great Sheepy in Leicestershire: see SHC IV 268. *Shepeye* is recorded in 1300 (SHC VII

68), 1323 (SHC IX (i) 90) and 1345 (SHC XII 38), but it is not clear whether it refers to this place.

**SHERBROOK** a tributary of the river Trent which rises in the southern part of Cannock Chase. *Sherbrok'* 1290 MRA, *Shirebroke* 1290 Ch, 1292 SHC VI (i) 296. From OE *scir* 'bright, shining', with *broc* 'brook'.

**SHERBROOK BANK** on Cannock Chase, 1 mile south-east of Brocton (SJ 9818). From Sherbrook (q.v.).

**SHERHOLT** 3 miles north of Alrewas (SK 1620). *Shirholt, Shyrholt* 1256 MinA, *Shirholt* 1425 SHC XVII 102, *Shirall or Sherholt Lodge, Shirall Thorn* 1836 O.S. The first element may be OE *scir* 'clean, bright', but the place lies on a parish boundary, so probably OE *scir* 'shire, jurisdiction, district', with OE *holt* 'wood'. See also Shirrall Hall. A place called *Schyrholt* in Kingstone is recorded in 1317 (SHC XI NS 184), perhaps to be associated with *Shyrholt* recrded in the late 12th century (SRO D(W)1721/3/6/15), and *Shiraldes* recorded in 1425 (SRO DW1733/A/3/27[3]). See also Schirroldus.

**SHERIFFHALES** Ancient Parish 4 miles south of Newport (SJ 7512). *Halam, Halas* 1086 (DB), *Hales* c.1125-38 Rees 1985, *Hales Paunton* 1259 TSAS 2nd Series VI 1894 14, *Schirrenghales, Schirrenchal'* (p) 1271-2 For, *Hales upon Lousyerd* 1282 SHC VI (i) 154, *Shirreuehale(s), Shirreueshales* ibid, *Schyreueshale, Sireuehales* 1291-2 Ass, *Hales Trussell* 1294 SHC 1911 220, *Shiruehales* 1301 Rees 1975, *Shirevehalys, Shirrevehales* 1367 Pat, *Sherreyf Hales* 1398 Pat, *Shereff Halis* 1532 SHC 4th Series 8 99, *Shrevehales* 1539 SHC NS V 248, *Sheryfehales* 1577 Saxton, *Sherifehallse, Sherifehalse* 1598 SHC 1935 168, *Sherrif Hales* 1686 Plot, *Sheriff Hales* 1920 O.S. The name is from OE *halas,* plural of *halh,* here probably meaning 'nooks', one of a cluster of *halh* names around Shifnal and Albrighton: see PN Sa I 262-3. *Lousyerd* is from Lizard Hill (q.v.). The word *Sheriff* comes from Rainold of Bailgiole (Balliol), Sheriff of Shropshire, who held the place in 1086 when it was then in Staffordshire, but it is curious that the name should incorporate a reference to one of the earlier lords rather than a more recent lord. The 1259 spelling commemorates the Pantulfs, later known as the Pantons or Pauntons, who were lords of the place in the 12th and 13th centuries. The 1294 spelling is from Roesia de Trussell: the Trussells held the place in succession to the Pantulfs (TSAS 2nd Series VI 1894 14; SHC 1911 220). Sheriff Hales (*sic*) became part of Shropshire in boundary reorganisations in 1895, and by 1971 the name was officially Sheriffhales: Youngs 1991: 393, 422. There are references to *Little Hales* in the 12th and 13th centuries: SRO D593/B/1/19/2/2/3; note also John Littlehales of Sheriffhales recorded in 1539: SHC V (i) NS 249. The references are doubtless to Little Hales Manor Farm in Shropshire, 1 mile north-east of Lilleshall (SJ 7516), *Little Hales* 1833 (see also Eyton 1854-60: IX 126). Morden's map of Shropshire incorrectly shows two places at Sheriffhales, with *Sheriffe* lying to the north-east of *Hales*.

SHERIFF'S RIDDING (unlocated) *Shyrreveriding* 1252 SHC IV 245, *Sheriffs Ridding* c.1737 SRO 5081/1/78-83. 'The sheriff's cleared land', from OE *scir-(ge)refa* 'a sheriff, the king's chief executive', with OE *\*ryding* 'a clearing, an assart, land taken into an estate from waste'.

**SHERRACOP LANE** which runs north-west from Stockwell Heath (SK 0522). *Semitam Vicecomitis* c.1252 (1897) Parker 1897: 303. Said to be from the Sheriff of Staffordshire who visited the area to determine a dispute c.1250: SHC 1919 8; Parker 1897: 41. Cop is from OE *cop(p)* 'the top of a hill, a summit, a peak', so 'the lane of the sheriff's hill': the road crosses a hill of 342'.

**SHETTESFORD** (unlocated, possibly near Wolverhampton) *Shettesford* 1271 SHC 4th Series XVIII 80. Perhaps from the OE personal name *Sceot, so '*Sceot's ford'.

**SHIFFORD'S GRANGE** on the Staffordshire-Shropshire border, 1 mile north-east of Market Drayton (SJ 6935). *Shipford* 1166 SHC I 227, *Spipford* (*sic*) 1266 SHC 1945-6 317, *Schipford* c.1300 SHC 1945-6 30, *Shifford Grange* 1546 SRO D1553/119, *Shepherd's Grange* 1714 SRO D861/E/5/21, *Shiffords Grange* 1833 O.S. 'The sheep ford': Shifford's Bridge, where the road from Market Drayton to Bloreheath crosses the river Tern, was probably the site of the original ford. Between the 12th century and the Reformation there was a grange here belonging to Combermere Abbey: StEnc 513.

**SHIPLEY** 7 miles west of Wolverhampton, in Claverley parish (SO 8095). *(æt) Sciplea* 1002x1004 (11th century, S.906; 11th century, S.1536), *Sciplei* 1086 DB (listed in Warwickshire), *Shiple* 1242 Fees, *Schipleg'* (p) 1255 RH, *Schypleg'* 1274 RH, *Shiplegh'* 1291 Ass, *Sheppele* 1291-2 Ass, *Schypele(feldes)*, *Schupele(feldes)* 1294 SHC VII 29, *Shippleye* 1334 SR, *Shypley* 1532 SHC 4th Series 8 124, *Shipley* 1577 Saxton. From OE *sceap-leah* 'sheep *leah*'. In Shropshire since the 12th century.

SHIREFORD (unlocated, possibly on the river Dove at or near Ellastone) *Skyreford* 1227 SHC IV 41, *Shireford* 1260 SHC IV (i) 146, 1307 SHC VII 190, *Shyreford* 1272 SHC IV (i) 202, *Schereford* 1274 SHC 1911 161, *Sireford*, *Syrefort* ?13th century SHC VII NS 139-40, *Sireford* ?13th century SHC VI (i) 11, *Shirford* 1303 SHC VII 108, *Schireford* 1312 SHC 1912 81. 'The shire ford': the Dove is the border between Staffordshire and Derbyshire.

**SHIRELAND** in Harborne (SO 0387). *Shireland* 1552 Hackwood 1896: 42-3, *Shireland (Hall)* 1775 Yates. 'The land or estate at the shire boundary': the place lies on the Staffordshire-Warwickshire border, and is commemorated in Shireland Road.

**SHIRE OAK** 1 mile south of Brownhills (SK 0504).... *a place called the Shire ooke near unto Walsall Wode* ... 1534 SHC 1910 35, *The Shire okes* 1577 Saxton, *Shire Oak* 1686 Plot, 1747 Bowen (who shows a single tree here). An ancient tree or group of trees which stood where the boundary between Walsall Wood and Shenstone crossed the Walsall-Lichfield road, ½ mile south-west of the present Shire Oak crossroads. 'A large oak in the valley ... named Shire Oak, from the word Scyre, to divide ...': Shaw 1801: II 53. A drawing of the oak, the last fragments of which disappeared c.1895 (VCH XVII 277) is in Palmer & Crowquill 1846: 295. The name became attached to a nearby hill and a farm on its summit: Shaw 1801: II 53. Cf. Shireoaks, Nottinghamshire, on the border between Nottinghamshire and Yorkshire (PN Nt 108-9).

**SHIRKLEY HALL** 1½ miles north-west of Horton (SJ 9258). *Shearley Hall* 1672 SRO DW1761/A/4/193, *Sherkley Hall* 1798 Yates, *Shirkley Hall* 1816 SRO DW1909/E/91.

**SHIRLEY (FARM, COMMON, HOLLOW, BROOK)** ½ mile south-east of Foxt (SK 0448). *?Shirleye* 1297 SHC 1911 246, *?Scheleye* 1327 SHC VII (i) 204, *?Shirleye* 1339 SHC XI 82, *?Shirley* 1368 SHC VIII NS 217, *Shirley* 1832 Teesdale, *Shurley*, *Shurley Hollow*, *Shurley Cottage* 1836 O.S. Possibly from OE *scir leah* 'bright clearing'.

**SHIRLEYWICH** 1 mile south-east of Weston upon Trent (SJ 9825). *Brine Pits* 1682 Browne, *Brine-pits* 1686 Plot 93, *Shirley wich* 1749 Bowen, *Shirleywich* 1836 O.S. A name that provides an object lesson and warning for place-name scholars. The OE word *wic*, which originated from Latin *vicus*, came to have a variety of meanings, but in its general sense 'building(s) used for specialised purpose' was applied to a salt-works, and when these were common, it became specialised as such: cf. Droitwich, Nantwich,

Middlewich and Northwich. Saline springs existed at Shirleywich, and the earliest recorded name for the place is *Brine Pits* (1682 Browne), after the first successful attempt to manufacture salt on a large-scale commercial basis in Staffordshire by Robert Shirley, Lord Ferrars (1650-1717) in the 17th century, the springs lying in his Weston upon Trent estate, although *Saltwich*, recorded in Stowe-by-Chartley ParReg in 1656 has not been identified, and may relate to this place. By 1690 the 'salt-houses' were a local landmark. Some years after the building of the saltworks (but probably not before the late 1680s, for Plot 1686: 93 does not mention it) the place became known as *Shirleywich* after the Ferrars family name: VCH II 247. The only connection with the OE word *wic* is the likelihood that in the West Midlands *wich* as a place-name element retained or regained its association with salt-working. See WMA 1993 36 71-2. For '... a Saltpan ... built in Wyche [in Burton-upon-Trent] ...' in c.1194 see SHC 1937 20.

**SHIRRALL (HALL)** 5 miles south-west of Tamworth (SO 1699). *? Sheralf* c.1485 TSSAHS XXX 1988-9 45, *Sherral, Sherrolde (parke)* 1801 Shaw II 9, *Shirrall (Gorse)*, *Shirrall (Lodge)* 1834 O.S. The place lies near the boundary between Staffordshire-Warwickshire border, so perhaps from OE *scir holt* or *scir halh* 'shire boundary wood' or 'shire *halh*'. For Shirral Park, recorded in the late 15th century, see TSSAHS XXX 1988-9 45-6. See also Sherholt; Schirroldus.

**SHOAL HILL** a 650' hill 1 mile north-west of Cannock (SJ 9711). *le sholle* 1286, 1300 For, 1587 Ct, *le Shole (copie)* 1610 Ct, *Sholehill Common* 1617 SRO D260/M/T/4/96, *Shore Hill* 1834 O.S. From OE *sceolh* 'twisted, awry', meaning 'the twisted hill' (cf. Shoulton, Worcestershire). The western side of the hill is a long slope to a plain at its foot and lies on the boundary with Hatherton. It has been suggested (see e.g. Nicolaisen *et. al.* 1970: 66; Oakden 1984: 56) that the hill may be the one from which Cannock took its name (but see Cannock). *The Shole* in Morton near Colwich is recorded in the 17th century (SRO DW1781/9/2/34/1-3), and one or more of the above spellings may relate to that place. See also Whittimere.

**SHOBNALL** 2 miles west of Burton-on-Trent (SK 2223). *Sobehal* 1114 (1798) Shaw I 23, 114x1118 SHC 1916 213, *Scopenhal* 1116x1133 SHC 1916 213, *Sobenala* 1114x1150 SRO D603/A/Add/9, *Schobenhale* 1188x1197 SRO D603/A/Add/36b, *Schobenhal, Scobehal* 1247 SRO D603/A/Add/9, *Schobenh'* 1262 SHC 1937 67, *Schobinhale* 1295 ibid, *Shopinhale* c.1345 (1798) Shaw I 24, *Shopenhale* 1406 CalPat, *Shopunhalle* 1441 ibid, *Shobenall* 1532 ibid, *Shepnall (Graunge)* 1584 SHC XI NS 88, *Shopnal* 1645 (1798) Shaw I 24, *Shobnall* 1682 Browne. From the OE personal name Sceob(b)a (see also Shebdon), hence '*Sceoba's or *Sceobba's halh*'. The personal name is also found in Shoppenhangers, in Maldemead, Berkshire: Gelling & Cole 2000: 232.

SHOOTERHILL (obsolete) on Lightwood Common, Meirheath (SJ 9240). *Shooters Hill Estate* 1807 SRO D593/B/1/20/15/1, *Shooterhill* 1922 O.S.

SHOOTERSOAKE – see **COLWICH**.

SHORTELYME (unlocated, in Hanchurch or Acton) *Shortelyme* 1280 SRO D593/B/1/23/7/2/4, *Schertelune* 1284 SHC 1913 256, *Schertelyme, Schort-lyme* 1280 SHC XI 325, *Sherteline* 1297 (1798) Shaw I 40, *Shortelyme* 1326 SHC IX (i) 113, *Schortelyme* 1330 SHC 1913 235, *Shortelyme* 1590 SOT SD4842/17/6, *Shortelyme in Acton* 1615 SHC 1934 (ii) 28. 'A considerable tract of waste land lying partly in the manor of Newcastle and partly in that of Swynnerton', which was the subject of many disputes which were finally resolved in 1279: SHC 1913 226; see also SHC XI 325. The first element appears to be the OE adjective *\*sc(e)ort* 'short'. For -*lyme* see Lyme. The precise meaning of the full name is unclear. The name survives in The Lymes (q.v.) and possibly Shutlanehead (q.v.).

**SHORTFIELDS (FARM)** ½ mile east of Balterley (SJ 7750). *Shorts fields* 1707 Audley ParReg, *Short Fields* 1795 SRO D3272/1/22/7/1. If the 1707 spelling is correct, from the surname Short.

**SHORT HEATH** 2 miles north-east of Wednesfield (SJ 9801). *? Smalhet(grene)* 1360 SRO D593/B//1/26/6/18/5, *Short Heath* 1834 O.S. Self-explanatory. *Sherthethesend* is recorded in 1368 in the Swynnerton area: SRO D641/5/T/1/1.

**SHORTWOOD FARM** 1 mile south of Maer (SJ 7936), *Schertewode* 1338 Salt 1888: 67, *Short Wood* 1742 Standon ParReg, 1833 O.S.; **SHORTWOOD FARM, GREAT** 2 miles north-west of Redditch (SO 0169), *Surthewode* 1249 AD ii, *Sortewode* 13th century AD ii, *Schortewodde* 1535 PN Wo 364. From OE *\*sceort* 'short', with OE *wudu* 'a wood'. The second place was in Tardebigge parish, forming part of Staffordshire from c.1100 until 1266, in Warwickshire until 1844, and since that date in Worcestershire.

**SHOTWOOD HILL** between Rolleston and Tutbury (SK 2228). *Schotwod (Mill of)* 1324 SHC 1911 361, *Schotewode* c.1450 CEC 255, *Shot-at-Hill* 1798 Shaw I 34, *Shotwood Hill* 1836 O.S. Although the place is on a projecting headland, a derivation from OE *\*sceot* 'a steep slope' (see Ekwall 1936: 147-51) cannot be supported on philological grounds, and the name remains unexplained. The wood is evidently not the one mentioned in Shaw 1798: I 34 as '... Newly planted by Sir John Moseley...'.

**SHOWELL, SHOWELLS** 2 miles north of Wolverhampton (SJ 9201), a former moated farm, once a manor, just within the boundary of the ancient Cannock Forest. *seofan wyllan (broc)* 895 (12th century, S.860; see Hooke 1983: 63), *Sewall* 1287 SHC 1911 193, *Sewale* 1286 SHC V (i) 166, *Sewel(felde)* 1293 SHC VI (i) 291, *Sewalle* 13th century Duig, *Seawall, Sewall* 14th century Duig, *Shewells, Seawall, Sewell* 16th century Duig, *Shawell* 1614 SHC II (ii) 44, *Show Hill* 1834 O.S., 1872 P.O. Duignan 1902: 135-6; MidA I 168-9 believed the name to be from a feathered device to scare deer, a derivation that has been much repeated since, but it is an abbreviated form of OE *seofon-wælles* 'the seven springs' (q.v.). The same derivation is found in Seawell, Northamptonshire; Sewall, Derbyshire; Sewell, Bedfordshire; Sewell or Showell, Oxfordshire. It is unlikely that there were actually seven springs here, but the former moat and mill-pond '... appear to have been fed chiefly by springs ...' (SHC 1928 70), and there 'used to be a strong spring running through Seawall moat' (filled in in 1935): Mander & Tildesley 1960: 9 fn. The place may have been the un-named virgate in Bushbury recorded in DB as held by Countess Godiva, widow of Earl Aelfgar, the legendary Lady: see VCH IV 55. The name Sewall (on which cf. Saltmoor, recorded as *sæffan mor* 985 (12th century, S.860), with which this name may conceivably be connected) is remembered in Showell Road and Showell Circus, a large road island at Low Hill, 1 mile to the north-east of the site of Seawall. See also SHC 1928 70-1. *Sewalmedowe, Sewall Medowe*, near Barton under Needwood, is recorded in 1415 and 1494 (Hardy 1908: 136), almost certainly with the same derivation. Showell Grange near Chetwynd lies just within Shropshire (SJ 722249), presumably to be associated with *Showel Mill*, recorded in 1749 (Edgmond ParReg). See also Seven Springs.

**SHOWELL** 3 miles south-west of Wolverhampton (SO 8795). 'The name Showell is also found at Springhill on the road from Penn to Lower Penn': Mander & Tildesley 1960: 9 fn. For the derivation see Showell, Showells. A spring rises at Showell Bank here. It is noteworthy that the two Staffordshire Showells should lie so close together.

**SHRALEYBROOK, SHRALEY HOUSE** 1½ miles east of Balterley (SJ 7849). *Shraley* 1512 JNSFC LX 1925-6 34, *Shraley* 1641 Audley ParReg, *Shraley, Shraley Brooke* 1733 SHC 1944 37, *Shraley Farm* 1836 O.S. A derivation from OE *scir heah-leah* 'the *leah* at the clear height' (see Ekwall 1959: 85) seems improbable.

**SHREDICOTE** 2 miles south-east of Church Eaton (SJ 8716). *Sradekoton* c.1195 SHC XI NS 125, *Scradycote, Shredicot(e), Shradecote* (p) 1221 Cur, *Shradicote* 1290 SHC 1911 198, *Shradycote* 1335 SHC 1913 49, *Schredycote* 1534 SRO D1810/f279. Oakden 1984: 137 suggests a derivation from OE *scread* 'shred', with OE *-ig* and *cot(e)*, perhaps meaning 'the cottage on the cut-off piece of land' (cf. Sneyd), but the name may incorporate a weak noun giving an early *Screadan-cot* or similar (cf. Caldicote), though no OE word *\*screada* has been recorded. Perhaps therefore from an unrecorded personal name *\*Screada or similar.

**SHROPSHIRE BROOK** a tributary of the river Trent. Perhaps derived from the name of a local family.

**SHRUGGS, THE** ½ mile north of Sandon (SJ 9430). *The Shruggs* 1836 O.S. No early forms are available, but the derivation would appear to be from ME *shrogge*, a common word in Nottinghamshire and Yorkshire denoting 'a bush, brushwood', meaning here 'scrub-land': cf. Clipstone Shruggs, Nottinghamshire. A field-name *Shruggs* is recorded in Sheriffhales: Field 1993: 68. See also PN Ch V (II) xviii.

**SHUGBOROUGH** 4 miles east of Stafford (SJ 9922; SMR 01649 places a deserted post-Conquest settlement at SJ 98992170). *Shokeburew* 1285 SHC 1914 32, *Schukeburg* 1328 SRO D3718/5, *Schukkeburgh* 1377 SHC X NS (ii) 55, *Shokkeburgh, Shukburgh, Shutborrow* 14th century, *Shitborogh* 1483 SHC VI (i) 151, *Shutborow* 1465 SRO D615/D/(16), *Shutborrow, Shokesborow, Shukesborow, Shottboro* 1532 SHC 4th Series 8 63, *Showtboro* 1539 SHC V (i) NS 309, *Shocborow, Shuckesbyry, Shuckesborough, Shukborow, Suchborows, Shokesborow* c.1540 Leland ii 169; v 21-2, *Shuchborow, Shulborow* 16th century Duig, *Shutborowe* 1610 SHC NS III 29, *Shutburrowe* 1624 SRO D615/D/1, *Shutborough* 1686 Plot. The first element is from OE *scucca* 'demon, evil spirit, devil', and OE *burh*, in its nominative form *byrig*, 'a fortified place, an ancient earthwork or encampment', implying that the fortification was haunted. The existence of the name Tooters Hill (q.v.) within Shugborough Park, where one of the monuments, the Lantern of Demosthenes, was erected c.1764-71 (SHC 4th Series VI 103) lends weight to the possibility of an early fortification here: see Gelling 1988: 147. Cf. Shuckburgh, Warwickshire (from OE *berg* 'hill': PN Wa 143-4); Shucknall, Herefordshire. *Shuckbarrow Feild* in Hamstead or Perry Barr is recorded in 1670: BCA MS3145/64/2. See also Scoteslei.

**SHUSHIONS** in Church Eaton parish, 5 miles west of Penkridge (SJ 8414; SMR 01059 places a post-Conquest deserted settlement at SJ 84301430). *Sceotestan* 1086 DB, *Shiston* 1283 SHC 1911 39, *Sustone* 1291 ibid. 201, *Scuston* 1293 QW, *Shuston* 1300 SHC VII 72, *Schuston* 1302 ibid. 101, *Chuston als Schuston* 1310 Ipm, *Shuston* 1566 SHC XIII 256, *Shoustones* 1607 Swynnerton ParReg, *Shushton Hall* 1798 Yates. Perhaps from OE *sc(e)ot* 'shooting', or the OE personal name Sc(e)ot with OE *stan* 'stone'. The derivation is thus 'the shooting stone' (whatever that might be), or 'the stone of Sc(e)ot'.

**SHUSTOKE** 1 mile south-west of Great Barr (SO 0395). *Shestock* 1620 (p) Roper 1980: 9, *Shustocke* (p) 1621 ibid. 191, *Shewstocke Farm* 1649 BCA MS3145/62/2, *Shustocke* 1653 BCA MS3145/258/26, *Shustoke, Shustock* 1786 BCA MS3602/275-281, *Shustoke (Lodge)* 1834 O.S. Only two places of this name have been traced, the other

493

some 12 miles to the east, 2 miles north-east of Coleshill in Warwickshire (SP 2290), recorded in DB. Spellings for the place near Great Barr pre-dating the 17th century have not been traced, and it is very likely that the name has been transferred from the place in Warwickshire, which may incorporate OE *sciete*, related to *sceat* 'nook, corner', or an OE personal name such as *Scytta, with OE *stocc* 'stock, stump of a tree': see PN Wa 92-3. Duignan 1902: 138 describes Shustoke near Great Barr as an ancient moated homestead and farm (TSSAHS XXIV 1982-3 50 places the site at SP 036962): if the present name is indeed transferred, the earlier name of the place has not been traced.

**SHUT END** 1 mile south-east of Himley (SO 8989). *Shuttend* 1686 Plot, *Shutt End* 1692 Tipton ParReg, *Shuttend* 1698 BCA MS3532/Acc 1935-054, *Shuttend* 1747 Bowen. From OE *scyte* 'a shute, a steep hill'. The word *end* meant a place, rather than a terminal point, often in heathland or common land.

**SHUT HEATH** ½ mile north-west of Haughton (SJ 8621). *Shut heath* 1836 O.S. From OE *scyte* 'a shute, a steep hill'.

**SHUTLANEHEAD** 1½ miles north-east of Whitmore (SJ 8242). *Shutlane* 1673 SHC VII 145, *Shutland-head* 1686 Plot, 1695 Morden, *Shutlane Head* 1696 SHC VII 145, *Shuttlanehead* 1707 Swynnerton ParReg. Perhaps 'the head or top of the steep lane', from OE *scyte* 'a shute, a steep hill' (the place is on a hill), but possibly a corruption of Shortelyme (q.v.).

**SHUTT GREEN** 1 mile north-west of Brewood (SJ 8709). *le shutegrene* 1320 Giffard, *Shetgrene* 1338 Ct, *Shutt Greene* 1591 SHC 1930 171, *Shuttgreen* 1723 Ct. From OE *scyte* 'a shute, a steep hill', probably with reference to the hill on which stands Broom Hall. The element Green indicates an area of grassland, or perhaps a squatter settlement: the place lay within Brewood Forest (q.v.).

**SHUTTERSHAW** 1½ miles west of Horton (SJ 9257). *Shyttershaw* 1574 SRO D(iv)1490/15, *Shuttershaw* 1675 ParReg, *Shittershaw* 1815 *Enc*, c.1820 SRO DW1909/D/4/3. Probably from OE *\*scitere*, possibly meaning 'a sewer, a channel or stream used as an open sewer', a common element formed from OE *scite* 'dung', with OE *scaga* 'a small wood, a copse, a strip of undergrowth', often with the particular meaning 'a strip of undergrowth surrounding a field'.

[SIBEFORD recorded in Staffordshire in DB (folio 250), is now held to be Sibford Gower, Oxfordshire: VCH IV 55; Darby & Terrett 1971: 164.]

SICHESBROC, SICHELESBROC (unlocated) *Sichesbroc, Sichelesbroc* 1227 SHC IV 52.

**SIDNALLS, THE** 2 miles west of Alvechurch (SO 9972). *Sidenhale* c.1245 *Bodl*, 1265 Wulst, 1275 *Ass*. From OE *(æt þæm) sidan heale* 'broad or spacious corner of land'. In Tardebigge parish, forming part of Staffordshire from c.1100 until 1266, in Warwickshire until 1844, and since that date in Worcestershire. *Sidnall Brook* in Kynesley is recorded in 1569: SRO D4092/C/1/16. The Sidnall (in Shropshire), 2 miles north-west of Cheswardine, is recorded as *Sydhaugh* 1699 (SRO D3212/1/1), *Sidhaw* 1786 (SRO D590/97/1-2).

**SIDWAY** 2½ miles north-west of Maer (SJ 7539), *Sidewei* c.1239 SHC 1911 424, *Sideway* 1327 SHC VII 198, *Sideway* 1433 SRO D641/1/2/53, *Sydweye* 16th century SHC IX NS 198, *Sydwey otherwise Sydwaye* 1589 SHC XVIII 6, *Sidway* 1609 SHC III NS 46, *Sydweye* 16th century SHC IX NS 198; **SIDEWAY** (pronounced Sidderway [sɪdəweɪ]) 2 miles west of Longton (SJ 8743), *? Sydewey* 1327 Tooth 2000b: 61, *Sideway* 1836 O.S. The first element is OE *weg* 'a way, a path, a road'. The second element is perhaps from ME *side* 'side, slope of a hill, especially one extending for a considerable distance', but some names containing the element may refer to land beside a river or wood: see Gelling & Coles 2000: 219. Or possibly from the OE adjective *sid* 'large, spacious, extensive, roomy'. The first place lies on the side of the Maer Hills, the second between hills. Some of the spellings given for the first place could relate to the second.

**SILKMORE**  in Castle Church parish, 1 mile south of Stafford (SJ 9320). *Selchemore* 1086 DB, *Selkemore* c.1198 SHC VI (i) 24, *Selkemor* 1230 Cl, 1303 SHC 1911 37, *Selkemer, Selkmor, Selkmore* 13th and 14th century Duig. Ekwall 1960: 422 proposes a derivation from OE *sioluc*, a derivative of an OE *\*seol, \*siol* meaning 'a drain, a canal' (citing *siolucham(m)* found in a charter of 990 (S.874) of land in Hampshire), so 'a small drain, a rill', with OE *mor* 'moor, marshland', hence 'drain to the moor'. That derivation would be particularly appropriate: the land here is low-lying and very prone to flooding from the river Penk. Cf. Silkstead, Hampshire; Selkley, Wiltshire. See also Gelling & Cole 2000: 60.

**SILVERDALE**  2 miles west of Newcastle under Lyme (SJ 8146). *Silverdale* 1796 VCH II 130, 1833 O.S., *Silverdale Furnace* 1832 Teesdale. Possibly not an ancient name, since no early references have been traced in local parish registers (but note Nicholas Sivedale recorded in 1568 SHC 1938 86), so perhaps from the Silverdale Iron Company started here by the Sneyd family c.1792 (see JHMS xi 4, 10; Simons 1978; HOK 65), meaning 'the valley where riches were found (or to be hoped for)', though it is said that the mine was formerly called Kent's Lane Colliery: StEnc 709. CDEPN suggests that the name may have been transferred from Silverdale in Lancashire.

**SILVER HILL**  ½ mile north-west of Barton under Needwood (SK 1819). *Silverhill Gate* 1812 *EnclA*. Perhaps to be associated with *Silver Flatt House*, recorded 1665x1697: SRO 30/74/42.

**SINAI PARK**  1 mile west of Burton-upon-Trent (SK 2223). *Seyne* 1410 SHC 1937 156, *le Seignes* early 16th century VCH III 203, *le Seygnyes'* 1518 SHC 4th Series VII 27, *Seyny Park* 1545 SHC 1939 116, *Sennye Park* 1549 (1798) Shaw I 24, *Seney Park* 1578 SRO D1734, *seenye Park* 1584 SHC XI NS 88. The park (according to Shaw 1801: II 24 formerly called *Shapenhale park*, i.e. Shobnall Park) is said to have originated c.1320 as 'a place surrounded by a ditch' in Shobnall Park used as a retreat for monks from Burton Abbey undergoing bloodletting: VCH III 203; VCH IX 168-9. The statement in Fuller 1880 that the place was 'at first so named by the abbot of Burton, because [it was] a vast, rough, hilly ground, like the wilderness of Sinai in Arabia ...' is fanciful – the name is from *Sene, Senyie*, from Fr *sene, senne*, Latin *synodus* 'a meeting of clergy for deliberations, a synod': *seyney-houses* were buildings belonging to monastic houses where breaks ('seyneys') were taken by monks in need of rest and recuperation after the regular bleedings they undertook for health reasons, or after illness: see also Coena's Well; *Seyneshaulwe*, Wallbridge. VCH I 189 mentions the remains of a Roman camp on the summit of the hill here, but there is no reference to any earthwork in Shaw. *Sena Park* recorded in Baswich parish in 1735 (SRO D260/M/T/4/106) may have the same derivation as Sinai Park, in which case it is perhaps to be associated with St Thomas's Priory.

**SITTLES**  2 miles south of Alrewas (SK 1712). *Sedhull* c.1300 TSSAHS XX 1979-80 loose map, *? Sydenhall* 1512 OSS 1936 55, *Sithills* 1664 Alrewas ParReg, *Sittels* 1775 Yates, *Sitels Farm* c.1830 SRO DW1851/10/4, *Sittles* 1834 O.S. Perhaps from OE *sidan* *heale* 'broad corner of land'. *Sitchells House* is recorded in Alrewas in 1719 (SRO D165/D/158, and *Sitchells Farm* in 1742 (ibid.), but the location has not been traced: it seems unlikely that the names are to be associated with this place. It is unclear whether *the Sidnalls* 1720 SRO (D201/M/T/10) is to be associated with either place.

SIWARDESMOR    (unlocated, in Burton upon Trent) *Sywardismoor* c.1258 SRO D603/A/Add/155, *Siwardesmore* 1272x1327 SRO D603/A/Add/365, *Siwarmore*, *Siwardmore* 1279 (1798) Shaw I 6. 'Sigeweard's moor'. Shaw 1798 I 6 refers to '*Cattestreet [passing through] the middle of Siwarmore to Hickenelstreet [Ryknild Street]*'. Sigeweard was an OE personal name.

**SKEATH HOUSE FARM**   1 mile south-west of Salt (SJ 9327). *Skeath House Farm* 1610 SRO D240/B/1/25, *Skeath House* 1738 SRO D240/ER/1/21, *Skeath House* 1775 Yates, *Skeath Farm* 1813 SRO D641/5/E(L)/2/3, *The Skeath Barn* 1836 O.S. A curious name. Although Ekwall 1960: 236 observes that 'Horse racing was a favourite sport of the old Scandinavians', a derivation from ON *skeið*, normally interpreted as 'a racecourse', seems unlikely, though the place lies within a mile of the river Trent, which was certainly used by the Danes. Cf. Hesketh, Lancashire; Hesketh Grange, Yorkshire North. However, there is some doubt about the interpretation of the word *skeið* as 'racecourse' (although a racecourse at Stone Flats, on the north side of Stafford, marked on Teesdale's map of 1832 may be noted), and indeed the original meaning was 'boundary', and is more likely to have referred to a stretch of land along a boundary that was used for grazing or left uncultivated: see Ekwall 1928: 361; Mills & Rumble 1997: 88. A study of the element has suggested that it 'should be regarded, not as being sometimes race-course and sometimes boundary, but as normally being both': JEPNS 1977-8 26-39. This place lies very close to a parish boundary. Another possible derivation is from common dialect *skeath, skath, scathe* (and similar) 'injury, damage, hurt, loss, danger; expense', perhaps here with the latter ironic meaning, but it may be noted that the surname Skeath is also recorded (e.g. in 1591 (SRO DW1851/8/118), and 1669 (SRO DW1851/8/18)).

SKULLS HILL    (unlocated, perhaps near Betley) *Skulls hill* 1608 Betley ParReg. A name of particular interest, since names beginning *Sk-* are often of Scandinavian origin. Or possibly 'the hill where skulls were found', with archaeological connotations. It is unclear whether *Scholes*, recorded in 1371 (SHC VIII NS 271), is to be associated with this place.

**SLACK, THE**   1 mile south-west of Longnor (SK 0663). *Slack* 1683 Alstonefield ParReg, *Slacke Farm* 1687 ibid, *Slack* 1840 O.S. Perhaps from ON *slakki* 'a small shallow valley, a hollow in the ground', found as northern dialect *slack* 'a hollow, especially one in a hill-side; a dip in the surface of the ground; a shallow dell; a glade; a pass between hills': EDD. Slack Lane on the west side of Mayfield probably has the same derivation. See also Gunstone.

**SLADE HEATH**   in Brewood parish, ½ mile east of Coven (SJ 9206), *Slade Heath* 1834 O.S.; **SLADE HOUSE**  1 mile west of Ilam (SK 1051), *Slade House* 1660 Ilam ParReg, 1730 ibid. 1801 Alstonefield ParReg, 1842 O.S. From OE *slæd* 'low flat marshy land, a valley'.

**SLAIN HOLLOW**   on the east side of Alton Towers (SK 0713). *Slain Hollow* 1891 O.S. A name of uncertain age, but unlikely to be ancient. By tradition the site of a battle in 716 A.D. between Ceolred, King of Mercia, and Ine, King of the West Saxons: see Plot 1686: 410; Shaw 1798: I 36-7. Although there is no evidence of any kind to support the legend, the O.S. mark the place as a battle site: see VCH I 212-3. See also Bunbury Hill; Ina's Rock; Yornburi.

SLAMFORD – see **CLANFORD**.

**SLATE HOUSE (FARM)** ½ mile north-west of Sheen (SK 1061). *Slate House* 1667 Alstonefield ParReg, 1695 SRO D239/M/T/392-3, 1798 Yates, 1840 ibid.' Possibly from ON *slétta* 'a smooth, level field', which survives as modern *sleat* (see PN Nt 290), but earlier spellings would be required for any certainty. The name *Slate House* 1 mile south-west of Onecote (SK0354) would not appear to be ancient: the place is not shown on the 1840 first edition 1" O.S. map.

SLIDERFORD – see **DANEBRIDGE**.

**SLINDON** 2 miles north of Eccleshall (SJ 8232). *Slindone* 1086 DB, *Sclindon* 1199 SHC III 167, *Slindon* 1242 Fees, *Sclindon* 1253 SHC 1911 122, *Slyndon* 1311 ibid. 77, *Slyndone* 1320 SHC X 31, *Slyndon* 1370 SHC VIII NS 251, *Slinge* 1566 SHC XIII 258, SHC 1914 12, *Biana Slin alias Slindon* c.1811 SRO D802/35. Perhaps from OE *\*slinu* 'slope', with OE *dun* 'hill'. The place lies on a level river terrace which rises on the western side of the village. For *Biana* see Byanna. *Slindon* in Wombourne is recorded in 1507: SA 5735/2/32/1.

**SLITTING MILL** 1 mile south-west of Rugeley (SK 0217), *Slitting Mill* 1775 Yates, 1832 Teesdale; **SLITTING MILL** 1 mile south-east of West Bromwich, *Slitting Mill* 1775 Yates. Places with this not uncommon name developed as hamlets around 17th century iron-slitting mills, which cut iron bar into rods suitable for nail making, a process patented in 1588. There were several other places of the same name in the county at various times, all with the same meaning. The mill near Rugeley was built c.1611: TSAHS 1996-7 XXXVIII 71.

**SLYMANSDALE** 2 miles north-west of Maer (SJ 7840). *Slimesdale* 1851 White. Early spellings are not available and the place is not named on the first edition 1" O.S. map. Perhaps connected in some way with OE *slim* 'slime, mud'. It may be noted that Margeria Slayomegrene is recorded c.1376 in the Newcastle area: Pape 1928: 149. See also Serleshous; Serlelane.

**SMALLBROOK** a tributary of the river Smestow. *Smalbroke* 1416 Wodehouse. From OE *smæl* 'narrow', with *broc*.

**SMALLBROOK (FARM)** to the north of Roughcote (SJ 9445). *Smallbrook* 1679 Caverswall ParReg.

**SMALLRICE, SMALLRISE** 1½ miles south-west of Milwich (SJ 9531). *Smallris, Smalerys* 13th century Duig, *Smalrys* c.1276x1300 SHC 1921 25, 1318 SHC 1911 92, 1364 SHC X NS (ii) 117, *Smalrys* 1423 SHC XVII 11, *Smalrich* 1481 SHC VI NS (i) 135, *Smallrice* 1580 SHC XV 148, *Smaleryse* 1581 SHC XVI 116, *Smalerise* 1591 SHC 1934 18. From OE *smæl* 'narrow, thin', and OE *hris* 'shrubs, brushwood', so 'the narrow piece of brushwood'.

**SMALLTHORN** on the north-east side of Burslem (SJ 8850). *Smallthorne* 1572 SHC XIII 287, 1836 O.S., *Smalthorne* 1595 Norton-in-the-Moors ParReg. From OE *smæl* 'narrow, thin', and OE *þorn* 'a thorn-tree', so 'the narrow belt of thorns'. *Smalethornis*, recorded in the 13th century (SHC XI 324), appears to have been in Barlaston.

**SMALLWOOD** 2 miles south-east of Uttoxeter (SK 1029). *Smalwode* 1382 Rental, *Smallwood Hall* 1652 SRO D1164/3/1, *Smallwood-hall* 1686 Plot, *Smallwood (Manor)* 1836 O.S. From OE *smæl* 'narrow, thin', and OE *wudu* 'wood', so 'the narrow wood'.

**SMEDLEY SYTCH** 2 miles south-west of Longnor (SK 0662). *Snethlesych* 1406 VCH VII 27, *Smetheley sitch* 1566 *Deed, Snyth(e)ley sitche* 1626 Rental, *Snedley siche* 1651 ibid, *Sitch* 1691 Alstonefield ParReg, 1840 O.S. The earliest form suggests that the first element is from OE *snæd* 'something cut off, a detached piece of land', with OE *leah* and OE *sic* 'boggy stream', so 'the detached piece of land with the boggy stream'.

**SMESTOW** 6 miles south-west of Wolverhampton (SO 8591) – see **SMESTOW BROOK**.

**SMESTOW BROOK** a tributary of the river Stour. *Smethestall* 1300, *Smethestalle(smor)* 1301 SHC 4th Series XVIII 188, *Smethestalle(-ford)* 1360 For, *Smestall* 1404 (1801) Shaw II 215, *Smethestall* 1465 SHC 1928 49, *Smestall* 1577 Saxton, *Smestal* 1686 Plot, *Smestow(e) Brook* 1778 *EnclA, The Smestal* 1844 Erdeswick 1844. It is possible that *Smethestall* is the old name of a pool in the river Trysull (q.v.), perhaps near Smestow, where the river Trysull receives a tributary stream, derived from Mercian OE *stæll* 'a place for catching fish' (cf. stall-net, 'a net laid across a river'), or OE *steall* as in *wætersteal* 'stagnant water', with OE *smeðe* 'smooth', or OE *smeðe* 'a smithy' (this area has long been associated with metal-working, as noted by Duignan 1902: 139; see also VCH II 108), and that the name was applied to one branch of the river Trysull (q.v.) and subsequently attached to the main branch: see Ekwall 1928: 272. The derivation *smer* ('small') with *ster* ('Stour') was put forward by McClure 1910: 159 in ignorance of the early spellings, and the first element is in any event unrecorded. *Smethestallesmor*, recorded in 1296 (SHC 4th Series XVIII 188), is evidently associated with this name. See also Cocretone, Cocortone.

SMETHCOTE (unlocated, in Trentham) *Trentham Wood or Smethcote Wood* 1584 SA 2922/11/1/22, *Smethcote* 1624 SA 2992/11/1/71. 'The cottage of the smith'.

SMETHDOWNE WOOD (unlocated, possibly near Norton-in-the-Moors) *Smethdowne Wood* 1598 SHC XVI 157. Perhaps from OE *smeðe* 'smooth, flat, level', with OE *dun* 'hill', so 'the hill with the level summit'.

**SMETHWICK** (pronounced Smethick [smeðık]) in Harborne parish, 3 miles west of Birmingham (SP 0288). *Smedeuuich* 1086 DB, *Smeythewik, Smethewyke* 12th century Duig, *Smethewic* 1229 SHC IV 76, *Smethewik* 1278 SHC 1924 82, *Smethwik* 1327 SHC VII 234, *Smethwick* 1682 Browne. The name is more likely to be from OE *smeðe* 'a smith, metalworker', and OE *wic* 'village', hence 'the metal-workers' village', than from OE *smeðe* 'smooth, flat, level', giving 'the village on the plain', although the place does lie on level ground. The place is now in Birmingham. Cf. Smethwick, Cheshire; Great-, Kirk- and Little Smeaton, Yorkshire; Smethcote and Smethcott, Shropshire.

**SMITH'S WOOD** between Farley and the river Churnet (SK 0644). *Smithe Wood* c.1581 Rental, *Smiths Wood* 1836 O.S. From the iron-working carried out in the area for many centuries.

**SMITHY MOOR (FARM)** ½ mile south-east of Stanton (SK 1345). *Smithy Moor* 1683 Ellastone ParReg, 'The moorland associated with the smithy'.

**SMYTHE'S PLANTATION**  ½ mile south-west of Okeover Hall (SK 1547). *Smyeth* 1539 LRMB, 1571 Ipm. From the OE adjective *smeðe* 'smooth', used of a smooth level piece of land, perhaps later taken to be a personal name.

**SNAILS END**  (obsolete) on the north side of Yoxall (SK 1419). *Snelles-end* 1499 (1798) Shaw I 98, *Snellis End* 1532 SHC 4th Series 8 172, *Snayles End* 1631 NA DD/P/6/3/40, *Snails End* 1724 SRO 5166/1/144-9, *Snail's End* 1836 O.S. Possibly from the OE personal name Snell or an Anglicised form of ON Sniallr

**SNAILS GREEN**  (obsolete, in Great Barr (SO 0495)) *Snails Green* 1686 Plot, 1775 Yates, 1834 O.S., *Sneales Greene Howse* 1693 BCA MS3145/93/1a & b, *Sneal's Green* 1834 White. 'The grassy area with the snails', or 'the squatter's settlement with the smails'.

**SNAPE MARSH**  (unlocated, near Shelton, Hanley) *Snapemarshe* 1597 SHC 1935 10, *Snape Marsh (House)* 1615 SRO D3272/1/17/4/6-8, Ward 1834: App. xlii, *Snape Marsh* 1657 SRO D3272/1/10/4/1. Probably from OE *snæp mersc*: both words had the meaning 'marshland, boggy land', so perhaps 'the marshland called Snape', adopted when the meaning of the first word had been forgotten. Or possibly from the rare ME *snape* 'poor pasture'.

**SNELSDALE**  ½ mile north-west of Mayfield (SK 1546). *Snellesdale* 1227 SHC IV 58, 1240 SHC VII NS 145, *Snelesdale, Swelesdale* 1277 SHC VI (i) 83, *Suellesdale* c.1450 TutCart, *Snellesdale* ? 15th century SHC VII NS 171. 'Snell's dale'. The name Snell (which was not uncommon in Staffordshire) may be from OE or ON, and this place may contain the English Snell or an Anglicised form of ON Sniallr. Snelston in Derbyshire lies 2 miles to the south, and it is quite possible that both places take their name from the same person. Cf. Snelland, Lincolnshire; Snelshall, Berkshire; Snelson, Cheshire. See also SHC 4th IV 229 fn.

**SNEYD**  3 miles north-west of Walsall (SJ 9702). *Sned* 1256 Ch, *Snede* 1410 Duig, *the Snead* 1663 Tildesley 1951: 187. From an OE word *snæd* found only in OE charters, meaning perhaps 'something cut off, a fragment, a piece of land, piece of woodland, clearing' (see Ekwall 1960: 428). The place is a portion of the manor of Essington which projects, wedge-like, into the manors of Walsall and Wednesfield. *Snedhet* ('Sneyd heath') is recorded c.1250x1280 (WSL A/2/35) 'next to the road which leads from *Esintone* to *Walshall*. The word sneyd was a not uncommon place-name in the Midlands (cf. Sneyd Green).

**SNEYD GREEN**  (pronounced Snade [sneɪd]) 1½ miles north-east of Hanley (SJ 8949). (Wood called) *Sneyd* 1223 SHC XII NS 30, *Snede* 1296 SHC 1911 242, *Snedde* 1298 SHC XI NS 253, 1332 SHC X 82, *Sned* 1326 JNSFC LIX 1924-5 37, *Snete* 1414 SHC XVII 19, *Sneade Greyne* c.1572 SHC IX NS 137, *Sneydegreene* 1591 Norton-in-the-Moors ParReg – see Sneyd. VCH VIII 106 suggests that the place was also known as *Hamil* by the 18th century; *Hamile* and *Hamil* are found in the Burslem ParReg in 1639, *Hamill* 1836 O.S. The word is from dialect *hamel* 'hamlet': OED. The Green element suggests a squatter settlement.

**SNIDDLES**  2 miles south-west of Flash (SK 0065). *Snidilles Head* 1842 O.S. From dialect *sniddles* 'sedge, rushes or long grass found in wet ground' (EDD), with OE *heafod* 'a head, the upper end', so 'the boggy headland with the long grass'.

**SNOCKESTONES**  (unlocated) perhaps in the Beech/Swynnerton/Trentham area. *Snarkestone* 1283 SHC 1939 90, *Snockestones* 1293 SHC VI (i) 233, SHC VII 18,

*Snokestones* (p) 1423 SHC XVII 89, *Snokstone* (p) 1444, *Snokystone* (p) 1447 Salt 1888: 94. The first element is possibly a personal name, perhaps *Snaroc, which is found in Snarestone, Leicestershire, but the place-name *Senokestre* near Stone is recorded in the 12th or 13th century (SHC VI (i) 21), and the first syllables may be a shortening of Senokes, probably meaning 'seven oaks'. Snockestones and Senokestre may be associated places. See also Hanchurch; Hanford.

**SNOWDON (UPPER & LOWER).** 1 mile west of Patshull (SJ 7801 & SJ 7800). *Snodden* 1240 SHC IV 237, 1279 SHC XI NS 133, 1293 SHC VI (i) 259, 1401 (1801) Shaw II 282, *Snowdon* 1686 Plot 43, *Snowden* 1798 Yates. The second element is OE *dun* 'hill' (the place lies on the west side of a large, low hill), the first perhaps from the OE personal name Snodd (cf. Snoddington, Hampshire), or possibly from *snod* 'smooth, sleek, even', a word not recorded independently until the 15th century (OED), but from which the personal name Snodd is said to be derived: Ekwall 1960: 429. See also PN Sa III 105.

**SNOW HILL** on the south side of Wolverhampton (SO 9198). *Snow Hill* 1770 Sketchley. The age of this name is not known, but there is some evidence for the existence of OE *snor, derived from a root meaning 'to twist', and used in place-names for a road which deviates from a straight line to negotiate a slight hill: see Gelling 1997: 93-5. However, it is possible that like Snow Hill in Birmingham, the name is of 18th century date in imitation of a London name (ibid. 93), or derives from a surname: John Snow of Wolverhampton is recorded in 1727: WRO 103/1/12/72. Mander & Tildesley 1960: 6 suggests that the hill (513') has been lowered.

SODOM (obsolete) 1 mile east-north-east of Sedgley (SO 9274). *Sodom* 18th century Ct, 1803 WALS DX/241/25, 1834 White. A place of ill-repute, through which ran the notorious Hell Lane (*Hell Lane* 1775 Yates), with a reputation for crime and violence in the 18th century. The name is also recorded in Wombourne in 1867 (VCH XX 201), and found in Sodom Hall in Upper Ettingshall (StEnc 526). From the notorious biblical city of the same name.

**SOHO** 2½ miles north-west of Birmingham (SO 0388). *Soho* 1775 Yates, *Soho Hill* 1834 O.S. 'Soho is the name of a hill ... about two miles from Birmingham; which, a very few years ago, was a barren heath, on the bleak summit of which stood a naked hut, the habitation of a warrener ...': Shaw 1801: II 117. The word is held to be from the cry used by sportsmen to call attention to the hunted animal: OED gives 'An A[nglo-]F[rench] hunting call, probably of exclamatory origin'. Halliwell explains *So-How* as a cry in hunting, when the hare was found, or in hawking as a call to make a hawk return to the lure, and Skeat (N & Q 8 S vi 1894 365) suggests that *so-how* was the English adaptation of the original Anglo-French *saho*, meaning 'come hither'. The place became famous for the Soho factory of Matthew Boulton and John Fothergill which operated here from 1765 (Shaw 1801: II 117), but there is some evidence that the factory was built on the site of a public house with a representation of a hunt on its signboard: MidA III 89 .

SOLDIERS HILL – see BATTLEFIELD.

**SOLES HILL, SOLES HOLLOW, SOLES COPPICE, OLD SOLES WOOD** on Throwley Moor (SK 0952, SK 1053). Early spellings have not been traced, but perhaps from OE *sol* 'muddy place, wallowing-place for animals': cf. Soles in Nonington, Kent. However, it may be noted that ON *sól* is occasionally found in place-names in allusion to sunny hills: see EPNE ii 134.

**SOLOMON'S HOLLOW** where the Leek to Buxton road crosses Tittesworth Brook 2 miles south-west of Upper Hulme (SK 0058). *Edge End Hollow* 1765 VCH VII 235, *Solomon's Hollow* 1890 O.S. The older name was in use until at least 1800: VCH VII 235. The derivation of the present name is uncertain, but perhaps from Solomon Ash who is said to have held land here in the 19th century (StEnc 528), or from a surname derived from OE *sulh-mann* 'plough-man'. Cf. *Selimonescroft* recorded in Castle Church in 1346: Oakden 1984: 81; *Selman's flatt* in Forton in 1635 (ibid. 152); *Selman's Croft* in Weston-under-Lizard in 1782 (ibid. 182).

**SOMERFORD** 1 mile east of Brewood (SJ 9008). *Somerford* c.1086x1117 (17th century) EEA 14 3, c.1123 SHC III 178, 1135x1138 SHC 1916 259, *Sumerford* 1130 SHC I 2, c.1130 SHC III 180, 1181 SHC I 104, 1204 OblR, 1285 FA, *Sonterford* 1333 SR. From OE *sumor* 'summer', in place-names usually in allusion to things which could only be used in summer, hence 'the ford usable only in summer' (possibly, though not certainly, the site of the present Somerford Bridge over the river Penk, formerly known as Stonebridge: see Shaw 1801: II 305), though there is a very slight possibility that the name is associated with ME *somer, summer* 'pack-horse': the 1334 form is suggestive of ME *sumpter* 'pack-horse driver' (cf. Rog' le Somtere, recorded in Rocester in 1327: SHC VII (i) 216). It is of interest that the bounds of nearby Penkridge recorded in 1598 mention *Somer laine*: Oakden 1984: 101. There are 37 *Sumer-* or *Sumrefords* in DB.

SONDE (obsolete) in a loop of the river Worfe, on north side of Worfield (SO 7696). *Sonde* 1510 Worfield CA, *Soond* 1525 SR, *the Sond* 1565 Worfield ParReg, *the Sonde* 1578 ibid, *the Soond* 1611 ibid, *Sonde* 1833 O.S. Probably from OE *sand*, in the West Midlands often pronounced sond or sund.

**SOUTHLOW** ½ mile north-west of Cellarhead (SJ 9548). *Southley* 1705 Stoke on Trent ParReg, *South Low* 1836 O.S., *Southlow* 1890 O.S. Since OE *leah* and *hlaw* sometimes interchange in Staffordshire place-names, and the forms here are inconsistent, it is not possible to be certain whether the derivation is from OE *suð-hlaw* 'southern low or tumulus' or OE *suð- leah* 'the southern *leah*'.

**SOUTH STREET, SOW STREET** a trackway which runs south-west across Cannock Chase from south of Colwich. *Sow Street Lane* 1651x1711 SRO DW1781/9/2/23/1-5, *South or Sow Street* 1998 O.S. The track appears as *South Street (Roman Road)* on the 1887 O.S. map, and SHC 1914 136 records the 'Roman road leading down to a ford in the Trent, called South Street', but there is no evidence to support a Roman origin. See also Sowsbetch; Sutherowe.

**SOW, RIVER** *Sowa* 1118 Flor, *Stouue* c.1130 Symeon, *Sowa* c.1175, c.1200 St Thomas, *Sowe* 1272 Ass, *Souhe* 1274 SHC VI (i) 62, *Sovve* 1401 StaffAcc, *Sow* c.1540 Leland, 1577 Saxton. Ekwall 1928: 375-6 suggests that the name is derived from a British river-name *Souo-* linked to Gaulish *Savus, Sava* and derived from *seu-* 'to flow, liquid', in OE *seaw*, and Welsh *sug* 'juice'. However, Jackson 1953: 372 believes this etymology to be very doubtful. All that can be said is that the name would appear to be pre-English and of unknown origin and meaning.

SOWDELEY (unlocated, in Rushall) *Sowdeley* c.1539 SRO D260/M/T/1/1a/25-30.

**SOWE** *Sowe* 1203 SHC III 95, 1271 SHC V (i) 145, 1592 SHC 4th Series IX 57. SHC 1917-8 349 notes that the manor of St. Thomas was known as Sowe, 'being on that river'. The *manor of Sowe uppon the ryver of Sowe* with *a dwelling howse caled Sowe* are recorded in 1581: SHC 4th Series IX 8. From the 16th century the manor of Baswich (q.v.) was usually referred to as the manors of Sowe and Brocton: VCH V 5. See also SHC XI NS 131.

**SOWFIELD GORSE** (obsolete) 1 mile south of Alrewas (SK 1713). *Sowfield* c.1300 TSSAHS XX 1978-9 loose map, *Sowfield Gorse* 1834 O.S. Possibly 'the south field'.

**SOWSBETCH** (unlocated, probably near Wolseley Bridge) *Sowsbetch* 1593 SRO DW1781/5/5/1. The second element is probably OE *bece* 'a well-defined stream valley'. The first element is uncertain, but perhaps from *sow* 'a female pig', or connected with South Street (q.v.), formerly known as Sow Street.

**SOW STREET** – see **SOUTH STREET**.

**SPADE GREEN** 1½ miles west of Lichfield (SK 0809). *Spade Green* 1538 VCH XIV 198, 1653 SRO D786/12/2, 1737 SHC 1925 53. The Green element suggests a squatter settlement.

**SPARROWLEE** ½ mile east of Waterfall (SK 0951). *Sparrowlee House* c.1777 SRO D514/M/36.

**SPARROWLEE BRIDGE** a footbridge over the river Hamps ½ mile east of Waterfall (SK 0951). The bridge is recorded *temp.* Henry III: SHC 1913 (i) 28.

SPASHBROOKE (unlocated) *Spashbrooke* 1718 SA 465/335.

**SPATH** 1 mile north of Uttoxeter (SK 0835). *Spath Gate* 1763 SHC 1934 70, *Spath* 1775 Yates, 1801 Smith's map, *Sparth* 1779 SA 665/5939, *The Spath, Spath Cottage* 1836 O.S. A curious name also found in Derbyshire and Cheshire (see PN Db 582, and PN Ch I 255), sometimes with the definite article, but for which no early forms have been traced. PN Ch I 255 suggests a derivation from OE *sparð* 'sheep droppings', but that does not seem entirely satisfactory, so perhaps from *spath*, a friable stone like gypsum (OED), first recorded in 1763, a possibility reinforced by the proximity of *Spar Flat Farm* (*spar* given as 'any rock substance with a crystalline appearance': EDD) in Beamhurst 1 mile to the north-west (SK 0636), early spellings for which have not been traced: it is not named on the 1" O.S. map of 1836. If so, perhaps associated with the Tutbury Sulphate seam which runs from near Houndhill to Tutbury, and is up to 4 metres thick, but not uniform, consisting of discontinuous outcropping masses of gypsum and anhydrite separated by silty mudstones with small amounts of alabaster (a very pure form of gypsum) and rock salt. Or the name might perhaps be an abbreviated 'horse path', with the first word reduced to *s*: Yates' map of 1775 shows two places near Uttoxeter named Spath, one at SJ 092356, the other nearby at SJ 081355, which would be explained if the track passing between or through both places bore the name. Curiously, a reference in 1780 mentions *Spath ... on the banks of the Dove near Uttoxeter.* SA 665/5940. The gate in the 1763 form was a toll-gate: SHC 1934 70. It may be significant that both a lane and gate (*Spath Lane* and *Spath Gate*) bear the name in Cheshire (PN Ch I 255). The name Spath is also recorded as a field name in Talke in 1733 (SHC 1944 59, 65), and as a field name in Fawfieldhead in 1870 (Rental).

SPEARHILL (obsolete, 1 mile east of Lichfield, where Cappers Lane joins Ryknild Street (SK 1309)) *Sperehull* 1472 TSSAHS XXVIII 1986-7 9, *Sperehyll* c.1535 SHC VI (ii) 166, *Speare Hill (Field)* 1681 SRO DW1851/8/40, *Spear-hill* 1798 Shaw I 316. Probably from OE *spere* 'a spear, a spear-shaft', found in place-names in allusion to woods where such shafts were obtained (VCH XIV 110), or from the surname Spere (ibid. 38). OE *spere* had the meaning 'sphere', so 'the rounded or dome-shaped hill' cannot be entirely discounted, but is unlikely.

**SPELLOWE FIELD** (obsolete, on the east side of Alrewas (SK 1814)) *Spellowe field* c.1300 TSSAHS XX 1978-9 loose map. An interesting name which appears to incorporate OE *spell* 'speech, discourse', and OE *hlaw* 'mound, tumulus', so giving 'the tumulus where assemblies were held'. Cf. Spellow, Lancashire.

**SPELSTOWE** (unlocated, in Clent and Broom: SHC X 87) *Spelstowe* (p) 1327 SHC VII 253, 1332 SHC X 87. The place has not been identified, but the elements are of interest: as well as *stow* 'a place, a holy place', the name appears to incorporate OE *spell* 'speech, discourse', so giving 'the (?holy) place where assemblies were held', suggesting a possible association with Kelmestowe (q.v.).

**SPICERS STONE** on Cheddleton Heath, 1 mile north of Cheddleton (SJ 9853). *Spicer's Stone* 1771 SHC 1934 136, *Spices Stone* 1836 O.S. Evidently named from someone called Spicer.

**SPITCHILL** (unlocated, in Tean) *Spithill* 1685 SRO D5476/A/2/1, *Spitchill* 1754 SRO D644/3/2A-F. Possibly denoting a hill from which speeches were made, from OE *spec hyll*.

**SPITTLE BROOK** a tributary of the river Smestow. *Spittel broc* 1300 For, *Spitelbrook* 1342 SHC 1913 91, *Spittell brooke* 1609 FF. From OE *spitel* 'hospital, religious house' which evidently stood nearby and is recorded as *Oldspittle* 1296x1307 SHC 4th Series XVIII 187, *Spytel* 1327 SHC VII (i) 252, 1332 SHC X 129, *Hospital* 1332 SHC 1913 39, *La Spytell'* 1371 SA 2089/2/2/24, *Spitele* 1375 Ipm, *Spyttull* 1539 SHC VI NS (i) 69.

**SPITTLEFORD BROOK** a tributary of the Meese Brook. *Spittleford brooke* 1731 Ct. From OE *spitel* 'hospital, religious house', with OE *broc* 'brook'.

**SPON, SPON FARM, SPON LANE** 1 mile south-west of Sandwell (SO 0189). *Sponne* 1244 SHC IV 103, 1343 SHC XII 31, 1344 SHC XVII 9, 1381 ibid. 174, *Spon(howse)* 1560 SHC 1931 167, *Spon Brook* 1585 VCH XVII 9, *Spon Lane* 1694 ibid, *Spon Coppice* 1695 ibid, *Span* 1686 Plot. From OE *spann* 'a hand's breadth, a span', perhaps denoting something narrow or something which joins two things together, such as a footbridge. The word is a common place-name element in the West Midlands. Spon Farm, Spon Drumble and Spon Drumble Farm lie 1 mile east of Milwich (SJ 9932), but the history of the names is not known. Cf. Spond Farm.

**SPOND FARM** ½ mile south-west of Alton (SK 0641). *Sponne* 1271 StEnc 531, 1327 SHC 1913 17, 1331 ibid. 31, *la Sponne* 1327 SHC VII (i) 216, *Spon* 1642 StEnc 531, *Span* 1695 Morden, *Spond-house* 1700 Alton ParReg, *Spond* 1704 Alton ParReg, 1836 O.S. See Spon.

**SPOONLEYGATE** 2 miles south-west of Pattingham (SO 3096). *Spoonley Gate* 1898 SA 2161/188. Possibly to be associated with the field-name *Spoonley piece*, recorded in 1807: SA 2161/92. In Shropshire since the 12th century.

**SPOT, SPOT GRANGE, SPOT ACRE** 1 mile south-east of Fulford (SJ 9436). *Spotte* 1332 SHC X 91, c.1452 SHC 1939 233, 1552 SHC XII (i) 209, *Spotgraunge* 1418 ibid. 309, *Spott* 1532 SHC 4th Series 8 135, 1567 SHC XIII 264, 1577 SHC XVII 224, 1613 Griffiths 1894: 218, 1619 SHC VII NS 205, *Spot Farm, Spot Grange, Spot Acres, Spot Gate* 1836 O.S. Almost certainly from OE *\*spot*, ON *spotti* 'a small piece, a bit', used here as in Norwegian *spott* 'a piece of ground'. This may be the place from which Wulfric Spot, the founder of Burton Abbey, took his name: his Will includes land here – see SHC 1916 6, 35, but see also Sawyer 1979a: xxxi, which concludes that any association with Wulfric is unlikely. It is probable that Wulfric took his appellation (first recorded in the

13th century Burton Annals and a closely related passage in the history of Abingdon abbey, a very late 12th century compilation: TSAS 4th Series I 22; Sawyer 1979: xxxviii) from a skin blemish, perhaps a birthmark. See also StEnc 102. A place called *Wittspot* (1280), *Wytspot* (1299) is recorded near Hanchurch: SHC XI 325, Ekwall 1928 376.

**SPOUT HOUSE**   ½ mile east of Wetley Rocks (SJ 9749), *Spout Farm* 1791 SRO D1123 Add; **SPOUTHOUSE**   in Hamstead (SO 0493), *Spout* 1682 Browne, 1695 Morden, *Spouthouse* 1834 O.S.; SPOUTHOUSE (unlocated, in Kidsgrove), *Spouthouse* c.1800 SRO D997/1/1; SPOUTHOUSE (unlocated, in Betley), *Spouthouse* 1730 SRO D210/M/20a-b. From ME *spoute* 'a spout, a gutter, the mouth of a water-pipe', presumably from a piped spring. The second name is preserved in Spouthouse Lane.

**SPRAGG HOUSE**   on the south side of Norton-in-the-Moors (SJ 8951). *Spragge House* 1597 SRO D1463/1, *the Spragg house* 1613 Norton-in-the-Moors ParReg, *Spraghouse* 1646 ibid, 1747 BCA MS917/1361, *Scrag House* 1836 O.S. The meaning of this name is uncertain: *sprag* had many dialect meanings: see EDD. Perhaps here with the meaning 'bulging' or 'propped', but the surname Spragg is found in Staffordshire: see SRO D786/8/15; BCA MS917/1614. The name survives in Spragg House Lane in Ford Green.

**SPRATS SLADE**   on the south side of Longton (SJ 9042). *Spratesslade* 1607 Trentham ParReg, *Sprates Slade* 1616 ibid, *ye Spratt Slade* 1687 ibid, *Spratts (lane)* 1709 ibid, *Spratslade* 1810 SRO D593/L/1/28, *Pratts Slade* 1836 O.S. Perhaps from the surname Sprat, with OE *slæd* 'low flat marshy land, a valley'. The name survives in Spratslade Drive.

**SPRINKS FARM**   1 mile west of Horton (SJ 9257). Perhaps to be associated with *The Sprink*, recorded in 1566: SRO DW1761/A/4/179. The name is probably from springs which rise here and flow into Horton Brook.

**SPRING HILL**   in Baswich parish, on Milford Common, 1 mile south of Tixall (SJ 9720); **SPRING HILL**   1 mile north-east of Essington (SJ 9704), *Spring Hill* 1834 O.S.; **SPRING HILL**   1½ miles east of Brownhills (SJ 0705), *Springhill* 1895 O.S.; **SPRING HILL**   1 mile south-east of Burntwood (SJ 0508), *Spring Hill* 1895 O.S.; **SPRING HILL** 3 miles south-west of Wolverhampton (SO 8795), *Sprungewall, Spryngwall* 1255 SHC V (i) 111, *? Sprengewell* 1286 SHC 4th Series XVIII 153, *? Springwalle, Spryngwall* 1382 SHC X (ii) 44, *Spring Hill* 1895 O.S.; **SPRING HILL**   1 mile south-east of Walsall (SO 0297), *Spring Hill* 1834 O.S. From OE *spring hylle* 'the hill with or by the young copse' (for 'spring, fountain' the element *wall, well* or similar, from Mercian OE *wælle*, would normally be expected, rather than OE *spryng* or *spring*), but in some names a connection with a water source cannot be ruled out. Spring Hill south-west of Wolverhampton appears to have combined the two elements, giving 'the hill with or by the coppiced trees at the spring of water', but see also Springwall Brook.

**SPRINGSLADE**   2 miles south-east of Bednall, on Cannock Chase (SJ 9716). *Spryngslade* 1271 to 1300 For, *Springslade* 1834 O.S. From OE *spring-slæd* 'the valley with the newly-planted trees or with the coppiced trees with new shoots'

**SPRINGWALL BROOK**   a tributary of the river Penk. *Springewallbrouk* 1286 For. Possibly from OE *spring-wælla* 'the spring in or at the young plantation', with OE *broc* 'a brook': *Springewall* is recorded in 1300 SHC V (i) 177, *Springwalls broke* 1586 Ct, *Springwall* 1841 TA. There are several instances of Springwell in Shropshire (see PN Sa III 248), and it seems likely that there was a particular type of spring so called: that they were all by coppice woods is improbable. See also Spring Hill, Spring Slade.

**SPURLEY BROOK** a tributary of Gamesley Brook. *Spurley Brook (Farm)* 1881 SRO D615/EL/6/126, *Spurleybrook (Cottage)* 1891 O.S.

**SQUITCH HOUSE** in Bagot's Park, 1 mile north of Abbots Bromley (SK 0826). *the Quech* early 13th century SRO D(W)1721/3/16/8, *Quechesterd* 1282-3 SHC XI NS 265, *Le Quech* 1317 SRO D(W)1721/3/12/15, *Quech (Field)* 1333 SRO D(W)1721/3/3/6, *le Quecche* 1401 SHC XI NS 193, *Querche Wodehouses* 1402 ibid. 208, *Queche* 1493 ibid.' 197, *le Queche* 1537 SHC 1916 332, *Squitch* 1798 Yates, *Squitch (Bank)* 1836 O.S. From ME *queche* 'a thicket'. This is a particularly early example of the word: the earliest spelling cited by OED is 1450. The 1282 spelling probably incorporates OE *steort* 'promontory, hill spur': the place lies at the end of a prominent narrow headland. Field-names *le Quetchen* and *the Quatch* are recorded in Brewood in 1598 and 1650: Oakden 1984: 74.

**ST AMON'S HEATH** (unlocated, below Beacon Hill, at Hopton Heath, 1½ miles north-east of Stafford (SJ 9425): see StEnc 493) *St. Amon's Heath* 19th century Erdeswick 1844: 61.

**ST BERTRAM'S WELL** ½ mile north of Ilam (SK 1351). *St. Bertram's well* 1686 Plot 207, 403, *Sir Bertram's Well (sic)* 1798 Shaw I 33. A tree over the well was known as St Bertram's Ash in the 17th century: Plot 1686: 207. Possibly a corruption of Bertelin: the saint supposedly moved to Ilam from Stafford: Plot 1686: 409; VCH III137; VCH VI 186; Oswald 1955. A cave at Beeston Tor is known as St Bertram's Cave.

**ST CHAD'S WELL** a spring on the west side of Chadwell which feeds the pool of Chadwell Mill (SJ 7814). *St Chadds Well* 1833 O.S. 'Ceatta's spring', from OE Mercian *wælle* 'a spring', and (sometimes) 'a stream'. The spring, perhaps the *fonte* (from OE *funta* 'spring') recorded between 1208 and 1236 (SHC 1921 9), was evidently later associated with St Chad, and may be the one which gave its name to Chadwell and Great Chatwell (q.v.).

**ST HELEN'S WELL** in Rushton Spencer (SJ 9462). *St. Hellens well* 1686 Plot 49. Plot records the well as foretelling disaster. *St Hellens Wall* is recorded in 1498 in Newcastle: Pape 1928: 185.

**ST MIGHELLS** (unlocated, possibly near Fisherwick) *St Mighells* 1584 SHC XV 154.

**ST STEPHEN'S HILL** 1 mile north of Colton (SK 0523). *Stenson* 1541 SHC 1919 58, *Stenson's Croft* c.1541 SRO DW1721/3/255, *St Stephens Hill* 1836 O.S. Perhaps to be associated with Styvington (q.v.). The place, otherwise known as Steenwood (q.v.), was long ago merged into Blithfield: SHC 1914 157. It may be noted that Stephen the Forester is mentioned in 1292 in records relating to this place: SHC 1919 100-1.

**ST THOMAS** 1½ miles east of Stafford (SJ 9523). *Sancti Thome* 1174 SHC VIII (i) 132, *abbey de Seynt Tomas* 1414 SHC XVII 51, *St. Thomas* 1605 SHC 1940 320, 1663 SHC II (ii) 40. From the priory of St. Thomas, founded c.1174 on the north side of the river Sow: see SHC 1914 116-29; VCH III 260-7. In 1570 the manor is recorded as *the Manor upon Sowt'*: SHC 1926 103. SHC 1917-8 349 suggests that this place was sometimes called *Sowe* (q.v.), 'being on that river'. See also Sena Park.

**STABLEFORD** 1½ miles south of Whitmore (SJ 8138), *Stapulforda* 956 (11th century, S.602), *Stapelford* 1367 SHC X NS (ii) 199, *Stableford* 1593 SHC XVI 129, *Stapleford(bridge)* 1602 SHC 1935 422, *Stablton [sic]* 1628 SRO D641/5/T/1/10, *Stableford (Bridge)* 1656 Eccleshall ParReg, *Stablefoard (bridge)* 1662 Trentham ParReg, *Stableford Bridge* 1720 Bowen; **STABLEFORD** 2 miles north of Worfield

(SO 7598), *Stapelford* 1272 Eyton 1854-60: III 112, *Stapulford* 1525 SR, 1532 SHC 4th Series 8 118, *Stalfort* 1583 Worfield ParReg, *Stapleford* 1752 Rocque. From OE *stapol* 'stake, pole, pillar', giving 'the ford marked by posts', with the later addition of *bridge* for the first place. Hart 1975: 177 surmises that *Stapulforda* comprised the parishes of Chapel and Hill Chorlton (q.v.), and was the original *ceorla tun* of the *ham* at Darlaston. Stableford near Worfield has been in Shropshire since the 12th century.

**STADMORSLOW** 2 miles north-east of Kidsgrove (SJ 8755). *Stodmarelowe* 1332 SHC X 94, *Stodmorelawe* 1546 SHC 1938 21, *Stodmorlee* 1466 SHC NS IV 138, *Stadmoreslowe* 1586 SHC 1929 147, *Stodmerslowe* 1619 SHC VII NS 204, *Stodmonlow* 1641 Wolstanton ParReg, *Stodmorelow* 1647 ibid, *Stadmorelow* 1649 ibid, *Stadmoor Low* 1775 Yates. The earliest forms suggest ME *stodmere* 'stud-mare', with OE *hlaw*, giving 'the low or burial mound associated with the stud-mares': see PN Nt 200.

**STAFFORD** Ancient Parish and county town of Staffordshire, lying on the river Sow more or less at the centre of Staffordshire (SJ 9223). *Stæf-forda*, *Stæfford, Staffordaburh* 913 ASC, *Stadford, Statford* 1086 DB, *Stephordi* 1102 VCH VI 200, *Stadfort* 1115x1120 CEC 13, *Statford* 1130 SHC I 1, *Stafford* 1162 SHC I 35. A surprisingly difficult name, possibly 'the ford by a *stæþ* or landing-place'. The name and place are considered in more detail in the Introduction, pages 36-41. See also Staffordshire.

**STAFFORDSHIRE** first recorded as *Stæfford(scir)* 1016 ASC (D, E), *Steffordscire*, *Stæffordscire* 1062-6 ASWrits (11th century, S.1140), *Stadfordshire, Statfordshire* 1086 DB, *Stadfordscire* 11th century Sawyer 1979a: xxxv, *Statfordsiræ, Statfordscira* 1130 SHC I 1, 1188 ibid. 140. 'County of Stafford'. The *-shire* element is from OE *scir*, a word with various meanings but in this context 'an administrative district consisting of a group of Hundreds'. See also Stafford.

STAFFORDSHIRE MOOR (obsolete) on the north-west side of Tamworth (SK 2005)) *Staffordshire Moor* 1834 O.S. Warwickshire Moor (q.v.) lay to the north-east of Tamworth, in Warwickshire. See also Tamworth Moores.

STAGDALE (unlocated, in Bradnop) *Stagdale* 1656 Okeover T699. Perhaps from OE *stagga dæl* 'valley of the stag'.

**STALLBROOK (HALL)** 2 miles west of Stafford (SJ 8887). *Stalbrooke* c.1295 DW1721/1/118-120, *Stalbrook* c.1345 SRO DW1721/1/294, c.1366 DW1721/1/42, *Stalbroke* 1385 SHC XIII 189, 1538 SRO D1810/f105d, *Strawbrook (Hall)* 1836 O.S. From OE *stall, steall* 'place; stable, stall; pool in a river', with OE *broc* 'stream, brook', so 'the stream with the pool(s)'.

**STALLINGTON** 1 mile south-west of Blythe Bridge (SK 9439). *Stalinton* c.1154 SHC III 194, 1177 SHC XII NS 279, c.1230 SHC VI (i) 11, *Stalenton* 1251 Ch, *Stalington* 1265 Ass, 1293 QW, *Stalinton* 1293 SHC VI (i) 242, *Stelenton* 1532 SHC 4th Series 8 136, *Stallington* 1590 SHC 1930 116. Ekwall 1935: 436 suggests that that the first element of the name is a folkname, probably Stælingas, possibly derived from OE *stæl* 'place', so giving 'the people or followers at Stæl', but the name is an *–intun* name, not an *–ingatun* formation, and the personal name Stal is unrecorded, although it may be a pet-form of St(e)allere: see Tengvik 1938: 270. Perhaps 'Stæl's *tun*', or possibly from OE *stan-hlinc* 'stony hill', with OE *tun*: the place lies on a bank rising from the south side of a tributary of the river Blithe.

STAMBERLOWE (unlocated, in Pattingham) *Stamberlowe* 1582 SRO DW1807/378, *Stamberloo* 1723 Brighton 1942: 111. From OE *stan, burna* 'stony stream', with OE *hlaw* 'mound, tumulus', so 'the burial mound at the stony stream'. The place is probably to be associated with *Stammerlow* (field), recorded in 1711: SHC 1931 81. Cf. Stambermill, Worcestershire.

STANBURNEFORD (obsolete) in Bromley wood on the border of Teddesley Hay: SHC 4th Series XVIII 131. From OE *stan, burna* 'stony stream', with OE *ford* 'a ford', so 'the ford across the stony stream', or 'the ford over the stream called Stanburne'.

STANCLIF (unlocated, possibly near Stourton or Kinver) *Stanclif* 1342 SHC 1913 91.

**STANDEFORD** 2 miles east of Brewood (SJ 9107). *Stanieford* 996 (for 994) (17th century, S.1380), *Staunford'* (p) 1245x1250 MRA, *Stoniford* 1300 SHC V (i) 177, *Stawntiford* 1506 SHC 1928 115. From OE *stanig ford* 'the stony ford': the place lies where Deepmore Brook is crossed by a lost Roman road (Margary number 190) running south-east from Pennocrucium (Water Eaton). There is a slight possibility that the *ford* element is from PrWelsh *ford* meaning 'Roman road' (see Jermy & Breeze 2000: 109-110). *Standford Moor*, on the east side of Lower Elkstone (SK 0758) appears on the 1840 1" O.S. map, but is likely to be from Stoneyfold (q.v.).

**STANDON** Ancient Parish 4 miles north of Eccleshall (SJ 8134). *Stantone* 1086 DB, *Standon'* 1190 Pipe, *Staundon* 1277 SHC VI (i) 91, 1321 SRO DW1733/A/3/28, 1597 SHC 1935 28, *Stawne* 1655 Wolverhampton ParReg, *Standon (vulg. Stawne)* 1679 SHC 1914 7, *Stawna* 1715 Blymhill ParReg. The DB spelling indicates a derivation from OE *stan-tun* '*tun* on stony ground', or possibly '*tun* at the stone': the exposed sandstone here suggests 'the *tun* at the stoney outcrop'. In view of the consistent later spellings, it seems that the DB form is aberrant, and the second element is OE *dun* 'hill': the place lies on a long rounded ridge, though that could not be considered a *dun*. The variant local pronunciation is preserved in *Staun Wood* on the north-west side of Standon.

**STANDON BOWERS** – see **BOWERS**.

STANFORD (unlocated, near Tutbury), *Stanford* c.1170 SHC 4th Series 4 69; STANFORD (unlocated, on the river Sow or Penk near Water Wending), *Stanford* 1261 (1679) SHC 1914 119. 'The stony ford', or possibly 'the ford crossed by the stony road': see Standeford. *Stanford*, relating to the second place and supposedly recorded c.1178 SHC VIII 172, is from a 14th century forgery: EEA 16 90.

STANFORD BROOK (obsolete) a tributary of the river Penk running from The Wergs to Pendeford . *Stanfordbrok* 1411 SHC XV 124. See also VCH XX 34. 'The brook with the stony ford', from OE *stan, ford.* There is a slight possibility that the *ford* element is from PrWelsh *ford* meaning 'Roman road' (see Jermy & Breeze 2000: 109-110), so 'the brook associated with the stony (? Roman) road'.

STANHALLE (unlocated) probably near Denstone. *Stanhalle, Stanhale* 1235 Fees.

**STANLEY** 2½ miles west of Cheddleton (SJ 9352). *Stanle* 1273 SHC VI (i) 59, *Stanlowe* c.1280 SHC 1911 431, 1332 SHC X 116, *Stanlegh* 1285 SHC VI (i) 157, *Stonilowe* 1327 SHC VII 208, *Stanlowe* 1351 SHC XII 107, *Stanley otherwise Stanlow* 1587 SHC 1929 312. The inconsistent spellings make it unclear whether the derivation is from OE *stan-hlaw* 'stony mound or tumulus', or OE *stan-leah* 'the stony *leah*': in Staffordshire *hlaw* often becomes *-ley*, and vice versa, so a firm derivation is not possible on the available evidence, though *hlaw* is probably more likely. Stanley Pool was constructed as a reservoir for the Caldon canal in 1786, and enlarged in 1840: VCH VII 230.

**STANLEY FIELDS** ½ mile south-west of Bemersley Green (SJ 8853). *Stanlowe* 1360 JNSFC LIX 1924-5 53, *Stanley Fields* 1836 O.S. Probably from OE *stan-hlaw* 'stony mound or tumulus', but in Staffordshire *-ley* often becomes *hlaw*, and vice versa. See also Stanley.

**STANLOW** 2 miles west of Pattingham (SO 7898), now in Shropshire, *Stanlowe* 1272 Eyton 1854-60: III 112, 1327 SR, 1332 SHC X 131, *Stanlowe at Paffard [Pasford] Bridge* 1542 SRO D593/A/2/11/9; **STANLOW (HALL)** 2½ miles south-west of Leek (SJ 9554), *Stanlow* 1210 SHC XI 332, *Stanlewe* 1275 ibid.' 334, *Stonilowe* 1301 SHC VII 90, *Stanlowe* 1325 SHC 1911 366, *Stanlow* 1582 Worfield ParReg, 1836 O.S. From OE *stan-hlaw* 'stony mound or tumulus', Stanlow near Leek almost certainly from the 'huge hexagonal pile of boulders some 200 yards away' from Stanlow Hall: JNSFC 1916-7 142. One or more of the 1210, 1275 and 1301 spellings may refer to Stony Low (q.v.). See also Kingslow.

**STANMORE (HALL)** 1 mile south-east of Bridgnorth (SO 7492). *a new erected mansion house – Stanmore Grove* 1814 SA 1987/19/29, *Stanmore Grove* 1833 O.S. Early spellings have not been traced, and the name may have been coined when the hall was built c.1814. In Shropshire since the 12th century.

**STANSHOPE** in Alstonfield parish, 6½ miles north-west of Ashbourne (SK 1254). *Stanesope* 1086 DB, *Staneshop(e)* 1203 Ass, *Stansope* 1227 SHC IV (i) 43, *Stansop* 1329 SHC 1913 21, *Stanhapp* 1420 Signet Letter C81/1365/26, *Stansoppe* 1598 SHC 1935, *Stanshop* 1603 SHC 1935 477, *Stansop* 1686 Plot. Possibly '*Stan's valley or secluded place', from OE *hop* 'sheltered valley, secluded place', or from OE *stanes* 'of stone', hence 'the valley of the stone', perhaps with reference to the limestone outcrop here. There is no identifiable *hop* at Stanshope, which lies at the head of the steep-sided dale, and it is likely that the original Stanshope lay in Hall Dale, recorded as *Stanshope Dale* in 1623: VCH VIII 8.

**STANSLEY (WOOD)** 2 miles west of Abbots Bromley (SK 0525). *? Leofstanesleg* 1252 SHC 1937 47 (see also SRO D603/A/Add/117-8), *Stanesleye* 1349 SHC 1919 13, *Stainesleye* 1361 SHC XI NS 218, *Stansley Wood* 1639 SRO D4038/B/2/81. If the first spelling relates to this place (which is quite possible, though not certain: the deed from which the spelling is taken refers to *Leofstanesleg between Berleg and Littleleg in the parish of Blithefeld*: SHC 1937 47), the name is Leofstan's *leah* 'the wood by Leofstan's glade', with the later loss of the first syllable. Otherwise 'Stan's *leah*'. See also SHC 1919 100.

**STANSMORE (HALL)** ½ mile north-west of Dilhorne (SJ 9643). *Stanton More alias Stanmore* 1608 SRO D3359/40/1/8, *Stanton More otherwise Stante More* 1609 SHC NS III 22, *Stanton More otherwise Stante Mere* 1610 ibid. 28, *Stansmore (Hall)* 1836 O.S. From OE *stan-tun*, probably '*tun* on stony ground', but possibly '*tun* at the stone', with OE *mor* 'moorland'.

**STANTON** 2 miles north-east of Ellastone (SK 1246), *Stantone* 1086 DB, *Stanton* 1197 SHC II 68, 1242-3 Fees, *Stant'* c.1235 SHC 4th Series IV 200, *Stanton near Whevere* 1315 SHC IX (i) 49, 1328 SHC 1913 16, 1339 ibid. 79, *Stanton* 1798 Yates, 1840 O.S. From OE *stan-tun*, probably '*tun* on stony ground', but possibly '*tun* at the stone'.

STANWEY (unlocated, possibly near Chesterton near Newcastle) *Stanwey* n.d. SHC XI NS 266. 'The stony way'. A name of possible archaeological significance: see Chesterton.

STANYLOND (unlocated) perhaps near Holditch. *Stanylond* 1272 SHC IV 204. 'The stony land or estate'. A name of possible archaeological significance: there are several Roman sites in the Holditch area: see StEnc 294.

**STAPENHILL** 2 miles north-west of Stourbridge (SO 8885), *Stapenhull* 1342 SHC 1913 90, *Stepnall* 1775 Yates, *Stapenhill* 1834 O.S.; **STAPENHILL** Ancient Parish 2 miles south-east of Burton upon Trent (SK 2522), *Stapenh'* 942 (14th century, S.1606), *Stapenhille, Stapenhilla* 1086 DB (listed in Derbyshire), *Stapehille* 1086 Burton, *Stapenhull* c.1180 SHC VII NS 132, *Stapenhul* 1188x1197 SRO D603/A/Add/36b, *Stapunhill* 1316 FA, *Stapenhull(e)* 1330 Ass, *Stapulhull* 1404 Burton, *Stapenhyll* 1449 SHC III NS 185, *Stapynhill* 1452 SHC 1910 321, *Stapenell* 1577 Saxton, 1610 Speed, *Staping hill* 1633 DbA vi. Both names are almost certainly from OE *(æt þære) steapan hylle* 'steep hill', with early shortening of *ea*, for both places lie on pronounced hills, the one near Burton upon Trent somewhat squat but with a steep slope to the river Trent. The suggestion in Duignan 1902: 142 that *stapen* was a corrupt form of OE *stapol* 'stake, pole, pillar', often used to mark the boundary of a manor, estate, etc., and meaning here 'hill of the boundary pillar' (cf. Bassetts Pole) is unlikely (notwithstanding the 1404 form *Stapul-* for Stapenhill near Burton, evidently influenced by *stapol*): it is doubtless no more than coincidence that the first Stapenhill adjoins the border between Staffordshire and Worcestershire, and the second, which was transferred from Derbyshire between 1878 and 1894 (PN Db 662; Youngs 1991: 423) adjoins the border between Staffordshire and Derbyshire. A derivation from the OE personal name Steapa cannot be ruled out completely for these names, but it is improbable that both hills are derived from that name. Stapenhill near Burton upon Trent, which was formerly in Derbyshire, became part of Staffordshire in 1894. The field-name *Hondesacre thyng*, recorded in Stapenhill near Burton upon Trent in 1477 (SHC 1937 178) is likely to incorporate OE *þing* in the sense 'possession', so here meaning 'land belonging to Handsacre'.

**STAR WOOD** on the north side of Oakamoor (SK 0545). *Starwood* 1823 SOT D240/E(A)2/198(1)-(2). Perhaps associated with *the Star*, recorded in 1771: SRO D240/D/119, *The Star, Old Star* 1836 O.S.

STARE BRIDGE (unlocated, in Yoxall) *Stare Bridge* 1732 SRO D4533/2/5/2-3, *Stair Bridge* 1799 (1801) Shaw II 7, *Stair Bridge (*and *Stairfields)* c.1820 WSL M657. Almost certainly from a personal name: John Stare of Joxhale is recorded in 1296: SHC 1911 268.

STARE WOOD (obsolete) 1 mile north-east of Leek (SJ 9957). *Stareholt* c.1539 LRMB, *Starwood* 1823 SRO E/A/2/198, *Stare Wood* 1842 O.S. From OE *stæger, holt* 'wood with the stair-like ascent', with *holt* later replaced with *wood*. It is unclear whether *Starefeld*, recorded in 1542 (1883) Sleigh 1883: 72, is to be associated with this name.

STAREHURST (obsolete) near Knaves Castle (q.v.) Brownhills (? SK 0406). *Starehurst* 1308 (1798) Shaw I 58.

STARESMOR (unlocated) near Rowley village, in Rowley Regis. *Scaresmor* 1327 SHC VII 248, *Staresmore* 1332 SHC X 87, *Staresmor* 1405 SHC XVI 44. Shaw (1801: II 239) states that 'Near Rowley [Regis] is a place called Staresmore'. Possibly the same place as *Staremor*, recorded in 1494 (SHC 1928 225). The surname Starysmore is recorded in 1448: Hackwood 1898: 114. See also Erdeswick 1844: 345.

**STATFOLD** Ancient Parish 3 miles north-east of Tamworth (SK 2307; SMR 01179) places a deserted post-Conquest settlement at SK 23880719). *Statfeud* c.1226 (1798) Shaw I 410, *? Stodford, Stofford* 1242 Fees, *Stotfield* 1291 Tax, *Stotfeld, Stocfeld* 1284 FA, *Stotesfeld* 1293 Ass, *Stotfold* 1326 (1798) Shaw I 410, *Stotfolt* 1327 SR, *Stodefold* 1441 SBT DR37/2/Box89, *Stotfolde* 1514 OSS 1936 56, *Stodefold* 1542 SBT DR37/2/Box91/16. Probably OE *stod-falod* 'stud-fold'. The first element may be from late OE *stott* 'a horse' (or OE *stot*, the exact meaning of which is unclear, possibly 'a horse, an ox'), but which became ME *stott* 'a young castrated ox, a steer'. The place may have continued to be associated with horses over several centuries: Shaw 1798: I 410 cites a marginal note made by Sir Simon Degge in Degge's copy of Plot 1686 8: 'Statfold, or Stotfold, is a place famous for keeping of horses; which circumstance, it is likely, gave name to this place, for Stat or Stade signifies littus, or the shore of a river, which does not answer this situation'. *stod faldes* appears in the boundary clause of a charter relating to Braunston, Northamptonshire of 956 (?11th century, S.623).

**STAWBROKE** (unlocated, possibly near Chapel Chorlton) *Stawbroke* 1565 SHC 1938 73.

**STEEL HOUSE** in Horton (SJ 9457). *Style* late 13th century VCH VII 65, *Steel House* 1561 ibid. 65, *Stele House* 1565 SRO DW1702/8/10, *Stile House* 1775 Yates. Perhaps from OE *stigel* 'a stile, a step over a fence or wall', also used in a topographical sense 'a steep ascent', doubtless the meaning here. Cf. Steel(e) in Northumberland and Shropshire.

**STEENWOOD** ½ mile south-east of Admaston (SK 0522). Evidently associated with *Stivinton* 1199 SHC III (i) 41, *Styphinton* c.1232 SRO D938/50, *Stiventon* 1254 SHC 1911 123, *Stineton* c.13th century SHC VIII 169, 1382 SHC XIII 172, *Styfinton* ? late 13th century SRO D938/107, *Steventon* 1306 ibid. 286, *Styvyngton* 1318 ibid. 85, *Styvynton* 1332 SHC X 89, *Styvington* 1349 SHC 1919 13, *Styventon* 1356 SRO D(W)1721/3/1/23, *Steventon* 1414 SHC XVII 52, *Stenson* 1541 SHC 1919 58, *Stevenson* 1569 SRO D(W)1721/3/255, *Steven Wood* 1575 Parker 1897: 371, *Steen Wood* 1706 ibid. 121. Perhaps from OE *styfic* 'clearing' (EPNE ii 166): cf. Steventon, Hampshire (Coates 1989: 155). See also SHC 1914 157; SHC 1919 12-3. Cf. St. Stephen's Hill. EEA 17 13 records *Steyveston* [Stenson] in Derbyshire 1198x1208..

**STEEP LOW** ½ mile north-west of Alstonefield (SK 1256). *Steeplow* 1840 O.S.

**STEEPLEHOUSE FARM** ½ mile north-west of Ilam (SK 1351). *Steeple House* 1838 O.S., *Stapleshouse* 1851 White. From OE *stepel* 'a house with a tower or on a steep declivity', in this case almost certainly in the latter sense. Perhaps associated with *Steple Dale*, recorded in 1538 (*Survey*). However, the surname Steeple and Steple are recorded (see for example SHC VII NS 72; SHC VIII NS 38), and it is not impossible that this place-name is so derived.

**STERMORE** (unlocated, near Stowe by Chartley) *Sterremor(e)* c.1275 SRO D938/487, *Sterremor* 1284 SRO D938/490, *Staresmore (Yate)* 1605 SHC XI NS 268, *Stermore* c.1679 SHC 1914 124. *Yate* is from OE *geat* 'gate': the place was a gate to Chartley Park: SHC XI NS 268. The first element of the name is uncertain, possibly an unidentified personal name; the second is OE *mor*, here probably meaning 'marshland'.

**STEWPONEY, THE** an area which takes its name from a public house (rebuilt several times and demolished in 2002) 3 miles west of Stourbridge, on the road from Wolverhampton to Kidderminster (SJ 8684). *Stewponey* 1744 VCH XX 124, *Stewponey* 1765 Staffs & Worcs Canal Act, *Stewponey* 1774 Hadfield 1969: 51, *Stew Poney* 1775

Yates, ... *a good inn, called the Stew-poney* ... 1801 Shaw II 267, *Stew Poney* 1808 Baugh, *The Stewponey and Foley Arms* 1868 Burritt 1868: 155. There are many traditions as to the origin of this strange name, which is almost certainly unique. Scott 1832: 173 believed it came from the common public-house name The Pony, with an explanatory reference to a nearby fishpond or stew, and added 'A gentleman who made particular enquiries in the neightbourhood, agrees to this derivation of Stewponey, adding…some particulars of the master of the inn and his pony. The master…was a successful competitor at Stourbridge races, hence, and from the piscatory entertainment of the place [presumably in the nearby river Stour], the house acquired its celebrity'. The Rev. Sabine Baring-Gould, a Victorian antiquarian and extraordinarily prolific writer, set his novel 'Bladys [sic] of the Stewponey' in Kinver, and suggested that 'an old soldier in the wars of Queen Anne [the Peninsular War with France 1708-14], a native of the place, settled there when her wars were over, and, as was customary with old soldiers, set up an inn near the bridge, at the cross roads. He had been quartered at Estepona, in the South of Spain, and thence he had brought a Spanish wife. Partly in honour of her, chiefly in reminiscence of his old military days, he entitled his inn 'The Estepona Tavern'. The Spanish name in English mouths became rapidly transformed into Stewponey': Baring-Gould 1897: 14-5. The story may be literary fiction or based on local folk-etymology: it has not been traced before 1897. Other suggested origins for the name are a derivation from the nearby bridge over the Stour, hence 'Stouri pons'; from a nearby stew or fishpond (cf. *Stanclewe*, possibly 'mound or low at the stank or fishpond', recorded in this area in 1296: SHC 4th Series XVIII 188); or from stepony, a type of ale or a raisin wine: VCH XX 124. (Blount 1656 defines stipone as 'a kind of sweet compound liquor drunk in some places in London in the Summer time'; OED gives stepony, stepponi, steppony, stipone, stiponie, stipony, steponey, stepany, stepney as of obscure origin, in 1770 said to be made from raisins, lemons and sugar, and there are parallels for an inn named after a drink: see Larwood & Hotten 1985: 231). Professor Richard Coates (personal communication) has suggested that since the first syllable of the name carries the stress, the drink derivation may be the more likely derivation, but an Estepona origin need not be ruled out: the word could have been folk-etymologised into Stewponey with stress on the second element, but then treated as an English compound and subject to stress shift. In 1744 the inn was described as 'the house of Benjamin Hallen, being the sign of the Green Man and called the Stewponey' (VCH XX 124), suggesting that the name Stewponey was a local nickname, with the Green Man perhaps providing some clue to the true derivation. Early in the 19th century sellers of simples (plants or herbs used for medicinal purposes) known as 'green men' travelled the country in search of herbs with portable apparatus for distilling essences and extracts, and it has been suggested (Fernie 1897) that inn signs such as 'The Green Man and Still' in London and elsewhere were named after such travellers. The green man was apparently an artistic corruption of the Red Indian supporters of the arms of the Distillers Company ['a fess wavy in chief, the sun in splendour, in base a still, supporters two Indians with bows and arrows']. The Indians were said to have been transformed by painters into wild men or green men. A stew was a vessel for stewing or boiling, and a stewpony may have been the animal carrying the distillation equipment for a 'green man'. Finally, the word stew had various meanings, including 'brothel' (Halliwell), and 'dust; vapour, smoke; an offensive smell; bustle excitement' (EDD), and it may be noted that the spelling of this name is invariably *-poney*, although the word for a small horse has properly been 'pony' from at least the eighteenth century. The expression 'riding the poney' is recorded in the Midlands as an artisan's phrase for receiving money in advance for work not yet completed: Baker 1854: II 128. For completeness, both Stew and Pon(e)y are recorded as surnames in the region: see e.g. for Pon(e)y WRO 1/1/578; 1/1/501/72; 1/1/610; WA II 38; Codsall ParReg 1689;

and for Stew in 1619 WRO 1/1/31/92. *Foley Arms* is from the local Foley family, prominent in this area since at least the 17th century.

STEYNESMOOR (unlocated, in Brewood parish) *Steynesmoor* 1364 Oakden 1984: 49. From the ON personal name Steinn, so 'Steinn's marshland'.

**STILE COP** 1½ miles south-west of Rugeley (SK 0315). *Style Coppe* 16th century SRO DW1734/2/5/68, *Stile-Cop* 1686 Plot, *Stiles Coppice* 1698 Fiennes, *Style Copp* 1776 SHC 4th VI 145, *Stile-coppice* 1801 Shaw II 315, *Stilecop Field* 1834 O.S. Notwithstanding the 1798 form, probably from OE *cop(p)* 'a hilltop, a summit', sometimes 'a mound, a ridge of earth, an embankment', with OE *stigel* 'a stile (and sometimes a steep ascent)': it is uncertain in what sense stile is used here. Plot notes that clay from here was used to make tobacco pipes, and Fiennes (Morris 1959: 167), mentioning the 'fine tufft of trees' here, claims that seven counties – Cheshire, Derbyshire, Gloucestershire, Leicestershire, Shropshire, Staffordshire and Warwickshire – can be seen from the summit, without explaining how each is recognised.

STILLEHAULT (unlocated) *Stillehault* ?early 14th century SRO D798/1/1/23. The context in which the name appears suggests that the place may have been near Coton or Milwich.

STINKENDEMOR (unlocated, perhaps near Longton, possibly associated with Lightwood Forest) *Stinkendemor*?c.1230 SHC 1921 18.

STINKING LAKE (obsolete) where a stream from Wheaton Aston crosses Watling Street (SJ 8410). *Stinking-lake* 1681 Blymhill ParReg, *Stinking Lake* 1834 O.S. From a suphurous spring, of which there were a number in this area: see The Leper Well. Plot 1686: 104 mentions '... that stinking water which crosses Watlingstreet Way, not far from Horsebrook ...'. See also Field 1993: 49.

STOCKENBRIDGE (unlocated, in the Mere/Morfe/Enville area) *Stockenbridge* 1592 SA 5735/2/23/1/20, 1601 SRO D5735/2/23/1/23, 1637 SA 2089/2/2/81, 1673 SA 2089/2/2/90. From OE *stoccen* 'made of logs', with OE *brycg*, so 'the log bridge'. Cf. Stockenbridge, Devon.

STOCKFORD GREEN (unlocated, possibly near Willford (q.v.)) *Stockforth* 1341 SHC 4th Series XX 81, *Stockford greene* 1624 Alrewas ParReg, *Stockford green* 1624 (1798) Shaw I 139, *Stockford (Lane)* 1834 O.S. From OE *stocc* 'a tree-trunk, a stump, a log of wood', so perhaps 'the ford at the tree-trunk', or 'the ford with the footbridge made from a log'. Stockford Lane lies to the west of the river Tame at Sittles (SK 1711).

**STOCKINGS** a name commonly found in the vicinity of areas of former woodland (e.g. The Stockings, 2 miles west of Codsall (SJ 8403), *Stocking Lane* 1308-9 VCH XX 79, *The Stockinges* 1607 Codsall ParReg; *Stockynges* (unlocated, in Kinver) 1569 (SHC 1938 51); Stockings Cote, ½ mile south of Balterley (SJ 7749)) Stocking is from OE *\*stocing* (from OE *stocc* 'tree trunks, stumps, logs'), meaning 'the place grubbed-up or cleared of trees' or 'the place of the tree-stumps', and is synonymous with *ridding* and *birch*.

**STOCKINGS BROOK** a tributary of the river Trent. From OE *\*stocing* 'the cleared land with the tree stumps'.

**STOCKLEY PARK** 2 miles south of Tutbury (SK 2025). *Stochileam* 1170 TutCart, *Stokel'* 1261 ibid, *Stokkeleye* 1296 SHC 1911 251, *Stockelegh* 1324 ibid. 358, *Stackeleye* 1326 HLS 267, *Stockley parke* 1798 Shaw I 60. For the first element see Stockton. The second element is from OE *leah*, which may originally have been *leum*, dative plural of *leah*, but it seems more likely that the spelling of the earliest name has the Latin accusative singular form. The place was one of the hays (bailiwicks or clearings, from Mercian OE *(ge)heg*) of Needwood Forest. There is disagreement whether *stoc legan ford* mentioned in a charter of Rolleston of 1008 AD (14th century, S.920) can be associated with this place: see Hart 1975: 218 and Hooke 1983: 96.

**STOCKTON** on the east side of Weeping Cross, Stafford (SJ 9521). *Stokton* 1284 St Thomas, *Stokken* 1314 SHC IX 48, *Stocton* (p) 1327 SR, 1539 MR to 1586 QSR, *Scottone* 1377 SHC 4th Series VI 20, *Stockton* 1836 O.S. From the common place-name element OE *stoc* 'place', which is unlikely to have become, as once thought, 'monastery, cell, religious place', but may have that particular meaning here, since the place belonged to St Thomas' Priory (VCH V 5), with OE *tun*. The place is perhaps to be identified with Stolben (q.v.). See also SHC 1914 124.

**STOCKWELL END** on the north side of Tettenhall (SJ 8800); **STOCKWELL HEATH** 1 mile north-east of Colton (SK 0521), *Stockewell (Butts)* 1261 (1798) Shaw I *154, *Stockwell Heath* 1775 Yates, *Stockwell Heath* 1836 O.S. Probably from OE *stocc* 'stock, trunk of a tree' (but see also Stockton), with Mercian OE *wælle* 'a spring', and (rarely) 'a stream', so here perhaps 'the spring at the tree stump'. The surnames Stockewall and Stockall are recorded in Tettenhall ParReg in 1611 and 1617, and Stockwell End itself is recorded in the 1640s (VCH XX 7).

**STOKE-BY-STONE, STOKE GRANGE, LITTLE STOKE** 1 mile south-east of Stone (SJ 9133). *Stoca* 1086 DB, *Stocha* 1166 SHC I 48, *Stoke, Stokes* 1200 SHC III (i) 68, c.1230 SHC VI (i) 18, *Stook* 1360 SHC XII (i) 202, *Stook* 1549 SHC XII 202, *Stoke* 1686 Plot, 1836 O.S. From OE *stoc* 'place', and sometimes 'monastery, cell, religious place': religious houses existed at Stone and Burston (StEnc 102, 557), though the age of each is uncertain.

**STOKE ON TRENT** Ancient Parish (SJ 8745). *Stoche* 1086 DB, *Stoches* 1149x1159 EEA 14 50, *Stoch* 1166 SHC 1923 297, *Stokes* 1223 SHC IV 223, *Stoke* 1224 ibid.' 31, *Stokes subtus Limam* 1305 WL 36, *Stok' underlym* 1305 ibid. 37, *Stoke Super Trent* 1686 Plot, *Stoke upon Trent* 1747 Bowen, 1836 O.S. From OE *stoc* 'place', probably here meaning 'dependant settlement', to which the name of the river Trent was later added to distinguish it from the many other Stokes, most of which assumed distinctive additions after the Conquest. Browne's map of Staffordshire, 1682, seems to incorporate the first reference to the appellation *Super Trent* or *Upon Trent* to distinguish this place from other Stokes, including Stoke-by-Stone. The 1305 and 1306 forms incorporate the element Lyme (q.v.). DB has 31 entries for *Stoche* (*ch* = *k*), and 32 for *Stoches*. The six towns of Tunstall, Burslem, Hanley, Stoke, Fenton and Longton were united under the name Stoke on Trent in 1910: VCH VIII 252.

STOKEDOILY (unlocated, possibly near Bagots Bromley) *Stokedoily* 1373 BodCh. From OE *stoc* 'place' with the name d'Oille, d'Oily, recorded in the 12th and 13th century: see Doyle; Oils Heath.

STOLBEN (unlocated) *Stolben* 1284 SHC VIII (i) 136; 1293 SHC VI (i) 243. It is likely that this place, which is incorporated in both sources in a list of places, is a mistranscription of the name Stockton, which is not included in either list (see SHC VIII (i) 136 fn), but the early spellings of Stubbeley (q.v.) should be noted.

**STONE** Ancient Parish 7 miles north of Stafford (SJ 9034). *Stanis* ?1132 RHP 259, *Stan'* 1149x1159 (13th century) EEA 14 70, *Stanes* c.1154 SHC III 194, 1187 P, 1201 Cur, *Stana* 1160x1182 EEA 16 60, *Stanes* 1280 SHC VI (i) 99, 1293 ibid.' 285, *Stonne*

1532 SHC 4th Series 8 131, *Stone* 1610 Speed. From OE *stan* or *stanas* '(place at or by the) the stone or stones'. The place is not listed in DB, though certainly in existence in 1086. The terminal *-s* in the spellings is not an indication that the name was plural: the Normans often added an *-s* to English place-names, particularly to shorter names, for example Staines, Middlesex, and Barnes, Surrey. Various theories have been suggested for the name, which is found in various parts of the country, but its origin remains unknown. There is no evidence to support the tradition (see e.g. Erdeswick 1844: 45) that the name is from a mound of stones created by pilgrims to a church erected by Queen Eormenhild, their mother, to commemorate the place of slaughter of Wulfhad and Ruffin, supposed twin sons of King Wulfhere (659-675), for their secret conversion to Christianity by St Chad (see also Bury Bank, the alternative early name for which may be associated with and derive from that legend). Whilst Walter of Whittesley, who added a note to the Chronicle of Hugh Candidus, a monk at Peteborough who wrote a history of that house during the latter part of the 12th century (see SHC 1916 131-2; Mellows 1949: 140-58), probably based on much earlier evidence, claimed that Wulfhere founded a hermitage at Stone in the 7th century, the roots of the legend involving his two sons appear to lie in a record of similar events involving the execution of two royal youths, brothers of Atwald, king of the Isle of Wight, after betrayal by Werbod, a pagan who hoped to inherit the kingdom by marrying their sister, (St) Werburh, described by Bede (iv c.16) as occurring at a place called *Ad Lapidem*, which formed the basis of a later Mercian version which led to a royal cult centred on Stone in Staffordshire. Bede names neither the martyred brothers nor the guilty king, but from the context the latter may be identified as Cadwalla, or possibly Wulfhere. The first known record of the Mercian version associating the legend with the Staffordshire place, possibly recording local oral tradition, is the *Passio Sanctorum Wlfadi et Ruffini*, a hagiography (described as 'historically valueless' in Thacker 1985: 6) dating perhaps from the 12th century (written after, and perhaps to enhance, the founding of the Augustinian priory of St Wulfhad and St Michael at Stone), known only from a manuscript, probably of 14th century date, now destroyed: Rumble 1997: 314-5. *Ad Lapidem* has been identified as Stone Farm at Fawley in Hampshire, which lies on a peninsula of land between the river Otter and Southampton Water, and at the end of a Roman road, suggesting that the place was a crossing point to the Isle of Wight during the Roman period: see Basset 1989: 90, also fn.56. For a full analysis of the Wulfhad and Ruffin tradition see Rumble 1997: 307-19; also SHC VI (ii) 214; VCH III 240; VCH XIV 240; also SHC VI (i) 1-2; SHC XII NS 100-1, 111 fn., 118; also Eyton 1854-60: II 200, but see also StEnc 553. The Roman name for Staines, Middlesex, was *Pontibus* 'at the bridges', and the name Staines may be associated with the remains of a stone bridge. It has been shown that many names containing the element *stan* in Lincolnshire lie on or close to Roman roads, or can be associated with the site of Roman buildings: see Owen 1997: 365-6. Apart from bridge remains or a Roman connection (which cannot be entirely ruled out here, since the place lies on the river Trent, and Roman sites or artifacts have been recorded at nearby Campfield Wood (q.v.), Swynnerton (see WMA 34 1991 70-1), and Aston by Stone (StEnc 22)), the most likely derivation for most places called Stone is from a prehistoric megalith, Roman milestone, a natural boulder or rock formation, or from 'a place where stone was obtained' (see JEPNS 3 1970-1 13), and a Keuper sandstone outcrop on the north side of Stone, long quarried for building materials, may be the topographical feature from which the place was named. Indeed, many place-names with the element *stan*, including this place, may simply indicate that underlying stone is exposed at the surface It may also be noted that a huge stone or erratic is recorded on Common Plot (JNSFC 1897-8 XXXII 165), and in that respect it is unclear whether *Stone Field* here, one of the open-fields of Stone (*Stone Field* 1665 SRO D3272/5/21/1-9, 1798 Act; see also StEnc 556) is 'the field at Stone' or 'the field with the stone'.

**STONE CROSS** 2 miles north of West Bromwich (SP 0194). *Stone Cross* 1626 Willett 1882: 182. From a wayside cross which was still standing in the 18th century. The base survived as part of a signpost until c.1897.

**STONE CROSS** in Penkridge (SJ 9214). *Stone crosse yate* 1598 Ct. 'Stone cross gate', from a former stone cross, apparently set on circular graded steps, in the Wolverhampton-Stafford road, which is mentioned until at least 1747 and appears on a map of 1754: VCH V facing 104. See also StEnc 556.

**STONE EDGE** (obsolete) 1½ miles south-west of Rushton Spencer (SJ 9262). *Stonyegge* 1304 SHC VII 107, *Stone Edge* 1775 Yates, *Stony Edge* 1842 O.S. From OE *stanig*, with OE *ecg*, 'stony edge'. *Stonihegge* (unlocated, possibly in Madeley) is recorded in 1320 (SHC X (i) 31, and *Stonyegge* (unlocated, possibly in Draycott in the Moors) in 1302 (SHC VII 107) and 1332 (SHC X (i) 93).

STONEGETON (unlocated, possibly near Curborough) *Stonegeton* c.1563 SHC 1938 35.

**STONE HEATH** 1 mile north-east of Hilderstone (SJ 9735). *Stone Heath* 1836 O.S. 'The heathland with the stone'.

**STONEHOUSE (LOWER)** on north side of Brown Edge (SJ 8954). *Stonhouse* 1333 SHC X 94, *Stone House* 1836 O.S. Self-explanatory.

**STONE PARK** 1 mile east of Stone (SJ 9134). *Stone Park* 1658 SRO D4913/B/1/2, *Stone Parke* 1663 SHC II (ii) 35, SHC V (ii) 103, *Stone Park* 1679 SHC 1919 262. 'The park at Stone'.

**STONE TROUGH** 1 mile south-west of Biddulph (SJ 8656). *Stantrough* 1644 SHC 4th Series I 59, *Stone Trow* 1644 Wolstanton ParReg, *Stonetrough* 1658 ibid, *Stone Trough* 1836 O.S. Self-explanatory.

**STONEWALLS FARM** 1 mile north of Dilhorne (SJ 9745). *? Stonwalle* 1319 SHC 1911 344, *Stone Walls* c.1761 SRO D1798/520, 1775 Yates, 1834 O.S. Self-explanatory, although it is possible that *Asternwalle* 'the eastern wall or spring', recorded in the late 13th century (SRO 3764/21[27574] refers to this place. It is unclear whether the name refers to some archaeological feature here (from OE *weall* 'a wall'), or is from 'the spring at the stone', from Mercian OE *wælle*. *Wal*, recorded in 1327 (SHC VII 218), may refer to this place.

**STONEY CLIFFE** ½ miles south-east of Upper Hulme (SK 0160). *? Staniclif* 13th century SHC IX NS 313, *? Staniclyf* 1326 JNSFC LIX 1925-5 42, *Stonycliff* 1438 BodCh, *Stoney Cliffe* 1586 VCH VII 33, *Stonie Cliff* 1645 SHC 4th Series I 266, *Stoniecliffe* 1645 ibid.' 308, *Stonycliffe* 1667 Leek ParReg. Self-explanatory.

**STONEYFOLD** 1 mile north of Butterton (SK 0757). *Stone Fould* 1680 Alstonefield ParReg, *Stonefould* 1691 ibid, *Stone fold Moor* 1723 ibid, *Stonyford* 1840 O.S. From *fal(o)d* 'a pen or enclosure for domestic animals', hence 'the stone-walled stock enclosure'.

**STONEYLOW** 1 mile south-east of Madeley (SJ 7943). *Stonylowe* 1332 SHC X 101, *Stannelowe* 1547 SHC XII 191, *Stonylowe* c.1565 SHC 1938 185, *Stonylow* c.1566 SHC IX NS 88, *Stonnielowe* 1645 SHC 4th Series I 262, *Stannylow* 1679 SHC 1919 255, *Stoney Low* 1833 O.S. 'The stony low or burial mound', from OE *hlaw* 'hill, mound, tumulus'. The place lies on a 497' hill. No record of any tumulus has been traced here (but see StEnc 558), though a large tumulus lies on the opposite side of the valley of the river Lea at Manor Farm, 1 mile to the south-west. See also Kingslow; Queen's Low; Stanlow.

**STONNAL, UPPER & LOWER** 2 miles south-east and 1 mile east of Brownhills (SK 0603 & SK 0803). *Stonehala* 1140 (1801) Shaw II 53, *Stanahala* 1143 Oxf, *Stanhala* 1166 SHC 1923 295, *Stanhale* c.1175x1208 SRO 3005/1, *Parva Stanhale* 1209 SHC 1923 277, *Stonhale* 1271 For, *Stanhale* 1273 SHC VI (i) 61, *Stonhal temp.* Henry III BM, *Stonhale* 1443 SHC XI 255, *Netherstonewall* 1578 SHC XIV 196, *Stonwall* 1590 SHC 1930 87, *Upper Stonnal, Lower Stonnall, Stonnall Chapel, Stonall Gorse* 1834 O.S. 'Stony *halh*', with *halh* here perhaps in the sense 'a hollow, a small valley'. The 1209 form incorporates *parva* 'little', and the 1578 spelling *nether* 'lower'.

**STONY BROOK** a tributary of the Rising Brook which runs into the river Trent; **STONY BROOK** a tributary of the river Blithe which runs between Stowe and Grindley, *Stonybrok* 1332 SR. Self explanatory.

**STONYDELPH** on the south side of Tamworth (SK 2301). *Stanidelf* 1202 FF, 1229x1260 SHC 1937 63, *Staindelf in Tamworth* 1284 Ipm, *Stanydelf* (p) 1327 SR, 1405 SHC XVI 47, *Stonydelph* 1359 SRO 3764/109[31759], *Stonidelfe* c.1360 (1798) Shaw I 15, *Stanydelff* 1542 BSE E18/222/4, *Stonydell* 1569 ParReg, *Stanidelfe* 1656 Dugdale, *Stony Delph* 1666 FF. From OE *stan-(ge)delf* 'a digging, a quarry', so 'the stone quarry'. In Warwickshire until transferred to Staffordshire in 1965. The name *Stondelf* is recorded in the Penkridge area in 1261 (SHC 1950-1 47), and *Stony Delph* in Audley is recorded in 1612 (SHC 1944 82).

STONYFIELDS (obsolete, in Basford (SJ 8546)) *Stony Fields* 1836 O.S. The age of the name is unknown: it became attached to a large house built c.1780: StEnc 558. The place lay on or close to the Roman road from Rocester (Margary number 70a), the stone or gravelling from which may have given rise to the name.

**STONYFORD** 1½ miles north-east of Yoxall (SK 1520), *Stony-ford* 1836 O.S.; **STONYFORD** 2 miles south-west of Hales (SJ 6932), *Stoneyford Yate* 1553 SHC 1945-6 18, *Stonyford* 1635 ibid. 224, *Stoney Ford* 1747 Poll, *Stoney Ford* 1832 O.S.; **STONYFORD (HOUSE, LANE & COVERT)** 1 mile south-west of Blithbury (SK 0719); STONYFORDE (unlocated, possibly in or near Uttoxeter) *Staynford* 1331 SHC XI 132, *Stonyforde* 1596 SHC 1932 151. Self-explanatory. *Yate* means gate: Stoneyford near Hales was one of the gates to Tyrley Park. It is unclear whether *Stanford*, recorded c.1170 (SHC 4th Series IV 69) and 1190x1247 (ibid. 81), which appears to have been in Needwood Forest, relates to Stonyford near Yoxall: see SHC 4th Series IV 7.

**STONYFORD BROOK** a tributary of the river Swarbourn, south of Uttoxeter. Associated with Stonyforde (q.v.).

**STONYSLACK** 1 mile north of Winkhill (SK 0652). *Stony Slacks* 1836 O.S. From ON *slakki* 'a small shallow valley, a hollow in the ground', found as northern dialect *slack* 'a hollow, especially one in a hill-side; a dip in the surface of the ground; a shallow dell; a glade; a pass between hills': EDD.

**STONYWELL** 3 miles north-west of Lichfield (SK 0812). *Stoniwelle* 1272X1307 Erdeswick 1844: 242, *Stoniewell* 1297 SRO D1734, *Stamwell* 1307 SHC 1911 286, *Stoniwalle* 1327 SHC VII (i) 231, *Stonywall* 1332 SHC X 110, *Stoniwalle, Stonywalle* 14th century Duig, *Stoniwel* 1561 HLS, *Stonwell* 1597 SHC 1932 318. From OE

Mercian *wælle* 'a spring', and (sometimes) 'a stream'. Shaw 1798: I 222 attributes the name to 'a stone in the well, situated about a mile south-east of the church, in the road to Farewell ... wherein is at the bottom a large stone, which seems to be no more than a little rock, whence springs the water that supplies that well ... the well above mentioned is a small round piece of water by the road side, and the stone is a very large boulder stone in the middle of it. The common people have been superstitious about its being removed, imagining thereby some injury would befal their cattle'. It has been suggested that New Stoneywell Farm may have been built over the well: StEnc 558.

**STOOP** (unlocated, possibly near Butterton, perhaps Beacon Stoop (q.v.)) *Stoop* 1687 Butterton ParReg. From dialect *stulpe, stolpe*, from ON *stólpi* 'a post, a pillar', sometimes applied to beacon posts (see Beacon Stoop), boundary posts and similar.

**STOUR, RIVER** a tributary of the river Severn. *Stur* 736 (8th century, S.89), *(æt) Sture* 866 (S.212; see SHC 1916 75-6), *Stoura, Stoure, Stowra* 1280 Hales, *Stoure* 1300 For, *Store* 1344 Fine, *Stowre* c.1540 Leland. There are several rivers with this name in other parts of the country. There is no OE *stur*, but Ekwall 1928: 380-2 suggests an Indo-European root meaning 'strong, powerful river', citing the Stura in Italy, but the theory is not free from doubt, and a derivation from an unattested OE relative, *\*stor*, or ON *storr* 'big', is a possibility, particularly since the name appears to attach to larger rivers, but the derivation poses philological difficulties. It may be that the name is from an OE relative of Low German *stur* 'unfriendly', Norwegian *stur* 'gloomy': see EPNE ii 165, 195; PN Wa 15.

**STOURTON** 3 miles west of Stourbridge (SO 8584). *Sturton* 1227 SHC IV 51, 1255 Duig, *Storton* 1271 SHC V (i) 143, 1416 SHC XVII 57, 1532 SHC 4th Series 8 15, *Stawreton* 1539 SHC NS VI (i) 73, *Sturseley, or Stourton Castle* c.1540 Leland v 20. 'The *tun* on the river Stour'. Stourton Castle originated as a royal hunting lodge built in the 12th century, and has been rebuilt several times since: VCH XX 68; 123; 130-2; 145.

**STOW ELM** (unlocated, in Lichfield, probably at or near Stowe (q.v.)) *Stow Elm* 1361 *Deed*.

**STOW HEATH** 1½ miles east of Wolverhampton (SO 9498). *Stowheth* c.1272 (1801) Shaw II 150, *le Stowheth* c.1295 Mander & Tildesley 1960: 31, *Stow Heath* 1303 WSL D593/B/1/26/6/14/1, *Stoweheth* 1327 SHC 1913 8, 1467 SHC NS IV 145, 1537 SHC 1912 93, *Stowheath* c.1565 SHC 1938 187, 1592 SHC 1930 355, *Stowhethe* 1571 SRO D440/18[SF35], *Stoweheathe* 1608 SHC 1948-9 102. The name of a royal manor, one of the two main Wolverhampton manors, which included the east part of Wolverhampton, Bilston and part of Willenhall: see Shaw 1801: II 166; WA II 103; SHC 1911 143-4; TSSAHS XXIV 1982-3 56; Hooke & Slater 1986: 29-44. OE *stow* had several meanings, including 'place, enclosed place, place of periodic assembly', but often with the specialised meaning 'holy place': see Gelling 1982: 187-96. It has been suggested that the name *Stowheath* implies 'the heath of a place called Stow', perhaps incorporating a memory of an earlier minster church than the one founded by Wulfrun at Wolverhampton: see Hooke & Slater 1986: 37. It may also be noted (and may be significant) that Stow Heath lies at or close to one of the sites at which the Mercians and West Saxons are said to have vanquished a Danish army c.910: see Wednesfield and Tettenhall. See also Stowe.

**STOWE** ½ mile north-east of Lichfield (SK 1210), *that sacred spot called Stowe* 13th century SHC 1924 51, *Stowe* 1257 ibid.' 317, 1311 ibid.' 319, 1433 SHC XVII 145; **STOWE BY CHARTLEY** Ancient Parish 6 miles north-east of Stafford (SK 0027),

*Stowea* 1199 SHC III (i) 56, *Stowe* 1242 Fees, 1251 Ch, 1278 SHC 1924 156, 1304 ibid.' 311, *Le Stowe subtus Certeley* 1271 SRO 465/7910, *Stowe subtus Chartele* 1302 SRO D938/471, *Estowe* 1302 ibid. D938/470, *Staw* 1532 SHC 4th Series 8 74, *Stoo* 1567 SHC XIII 263, *Stow* 1686 Plot. From OE *stow* 'place, enclosed place, place of assembly', but often with the specialised meaning 'holy place'. In place-names dating from before the Danish wars the word was used for sites associated with the physical presence of saints, either during their lifetime or as the place of their burial: Gelling 1992: 97. This is certainly the case for the Lichfield Stowe, which is the reputed site of St Chad's 7th century hermitage and where he died (and is recorded as Stowe in the 12th or 13th century: VCH XIV 7), but it is not inconceivable that the place was so-named because it was already a Christian site when the place was chosen for the first Mercian cathedral (cf. Eccleshall and its Celtic Christian history). No evidence has been traced to support such a meaning in the case of Stowe by Chartley. *Stowelond* in Lichfield 'which extended to the king's way leading from Burton to Lichfield', is recorded in 1333: SHC 1939 94. See also Stow Elm; Stowheath; Stowehill.

**STOWEHILL** (unlocated, in Brocton near Stafford: SRO D1798/H.M.Chetwynd/21) *Stowehill* 1677 SRO D1798/H.M.Chetwynd/21. From OE *stow* 'place, enclosed place, place of assembly', but often with the specialised meaning 'holy place', with OE *hyll* 'hill'.

**STRAITS, THE** on the east side of Himley Park (SO 8992). *the Straight* 1672 Sedgley ParReg, *The Streights* 1701-25 Sedgley RentRolls, *the Straight* 1724 SPI, *Strate* 1775 Yates, *The Streights* 1777 SRO D5450/2, *Straight* 1784 SHC 1947 88, *Strait* 1808 Baugh, 1818 Himley ParReg, *Straight* 1832 Teesdale, *The Streights* 1834 O.S. Although early forms are not available, it is certain (notwithstanding the 1784 spelling) that the name has no connection with modern *straight*. It may be from a word of ME origin, derived from OFr *estreit*, spelt *strait* and *streight*, meaning 'narrow passage' (common in the plural), hence the Straits of Gibralter, the Straits of Dover, etc. The OED gives the meaning 'a narrow lane, alley or passage' (1622), and it is likely that this is the meaning here: Duignan 1902: 145 describes the place as 'a steep narrow road between Sedgley and Himley'. Field names containing the word Straight or Strait have been found to mean 'land adjoining a (Roman) road', from OE *strǣt* (see Field 1972: 221 – Streethay, for example, appears in 1563 as *Streighthey*: SHC 1931 229). No Roman road is known here, and the conjectured line of a lost Roman road from Pennocrucium (Water Eaton) to the Roman sites at Greensforge lies at least four miles to the west, but the straight road from Muckley Corner to Walsall Wood is marked on Yates' map of 1799, and may be Roman, perhaps aligned with Greensforge (although a continuation would run south of Greensforge), and a short length of Roman road running north-north-east from the east side of Greensforge is marked on the 1995 O.S. map, although aligned too far north to meet the Muckley road without a change of direction. See also Bassett 2001: 8, 10.

**STRAMSHALL** 1½ miles north-west of Uttoxeter (SK 0835). *Stagrigesholle* 1086 DB, *Stranricheshill* 1199 SHC III (i) 38, *Stranritheshull* 1199 ibid. 62, *Sterangricheshull* 1208 ibid. 173, *Strangricheshall* 1221 FF, *Strangricheshull* 1227 Ass, *Strangersheshull*, *Strongersheshull* 1257 SHC 1911 128, *Strongushull* 1274 ibid. 161, *Strongeshulf* 1269 Ass, *Strongeshull, Strongeshul* 1327 SHC 1913 8, 17, *Strongkeshill* 1339 ibid. 78, *Stronggeshull* 1391 SRO D786/3/1, *Stronsheff* 1415 SHC XVII 56, *Strangsil* 1419 ibid. 68, *Strongeshill* 1425 ibid.' 111, *Strowneshyll* 1566 SRO D786/3/6, 1568 SHC XIII 272, *Strowneshyll* 1566 SRO D786/3/6, *Stroneshill* 1604 SRO D786/3/11, *Stramshall* 1669 SRO D786/3/15. The forms suggest '*Strangric's hill': the place lies on a 300' hill overlooking the river Tean. The reversion to Stran- since the 17th century is noteworthy.

**STRANGLEFORD BIRCH** 1½ miles west of Brewood (SJ 8508). *Strangelford* 1300 SHC 1911 257, 1307 SHC VII 186, 1330 SHC VIII i 216, *Strangilford* 1307 SHC VII 181, 1327 SHC VII 181, *Strangeford* 1308 SHC IX 4, *Straungeford* 1311 ibid, *Strangleford* (p) 1327 SR, 1382 SHC XIII 17, *Strangulford* c.1407 SHC XVI 60, 1420 SHC XVII 73, *Strangleford Byrch* 1748 SRO DW1921/2, *Strangleford Birch* 1749 SRO D547/1/622/. Duignan 1902: 145 notes that *strangle* is a provincial or dialect word for the orobanche and cuscuta plant, also called choke-fitch, chokeweed, strangletare, and other local variants. There is a small stream to the south of this place (see Kiddemore Green) which may at one time have been strangled or overgrown with weed, though OED does not record strangle meaning 'choke' (from OFr *estrangler*) before the 14th century. Another slight possibility is a derivation from the OE personal name Strang(w)ulf. The *Birch* element is evidently a later addition, from OE *bryce* 'breaking', used of 'newly cultivated ground', by metathesis becoming *birch*. The place is in what was Brewood Forest. *Strangle Forde Meadow* is recorded in Linely, Shropshire, in 1639: PN Sa III 169.

STRAW HALL (obsolete) a house on Penn Road near The Royal Wolverhampton School on the south side of Wolverhampton: StEnc 561 (SO 9097). *Straw Hall* 1774 Penn ParReg.

**STREETHAY** 2 miles north-east of Lichfield (SK 1410). *Strethay* pre-1176 SHC 1924 165, 1256 SHC 1911 127, *Strithay* 1216x1272 SRO DW1734/J/1762, *Strethai* 1247 SHC 1911 118, *Stretheye* 1262 For, *Strethay* 1272 SHC 1910 108, *Stretehay* 1470 SHC 1939 122, *Strettey* 1532 SHC 4th Series 8 183, *Streighthey* 1563 SHC 1931 229, *Streethaie* 1601 SHC 1935 346. From OE *strǣt* 'a paved road, a Roman road, a street', and Mercian OE *(ge)heg* 'a fence, an enclosure', so giving 'the hay or enclosure on the Roman road'. Ryknild Street (Margary 18c) passes through the place.

**STREETLY** 4 miles east of Walsall (SO 0898). *strǣt lea* 957 AD (12th century, S.574), *Stretle* 1361 SHC X NS (ii) 72, *Streetly Hill* 1834 O.S. From OE *strǣt*, *leah* 'The *leah* on the Roman road'. The place lies on Ryknild Street (Margary 18c), on the Staffordshire-Warwickshire boundary.

**STREETWAY** Watling Street (q.v.) was known in the 18th and 19th centuries both in formal documents and colloquially as *The Streetway* (*Street-way, called by some Watling-street* ...1798 Shaw I 20; *Streetway* 1704 Penkridge ParReg). The Turnpike Act of 1760 for the Wolverhampton-Stafford road refers to Watling Street as *'the road called the Streetway'*. Farms on the road are frequently named *Streetway Farm*, as in Brewood and Cannock.

STRETFORD, STRATFORD (unlocated, near Chesterton in Worfield) *Stretford near Chesterton* 1583 Worfield ParReg, *Stretford (Field)* 1822 WJ October 1908 267. Almost certainly from the Roman road (Margary 192) that ran from Greensforge to Chesterton (q.v.), to be associated with *stoni-strete* recorded in 1300: VCH Sa I 273. In Shropshire since the 12th century.

STRETFORD (obsolete, where the river Tame is crossed by Watling Street south of Tamworth (SK 2101; SMR 01155 places a deserted post-Conquest settlement at SK 21580160)) *Stratford* 1253 Ch, *Stretford juxta Wilmundecote* 1313 Pat, *Streforth juxta Tamworth* 1375 Ipm, *Stretford* 1554 SRO D5386/2/2, 1656 Dugdale, 1750 K. Dugdale 1817-30: 824 explains that the place 'had its name originally from the situation thereof, upon the great Roman way called Watlingstreet, where it thwarts the River towards Faseley'. From OE *strǣt ford* 'the ford on the Roman road'. *Wilmundecote* is Wilnecote.

**STRETMESLE** (unlocated) *Stretmesle, Stremesle* 1199 SHC III (i) 57, 167. Possibly near Flashbrook, in which case the *Stret-* element is likely to be the Roman road (Margary number 19) from Pennocrucium to Chester. The rest of the name is unexplained, but the resemblance of the spellings to Stretwile (q.v.) suggest that it may be the same place.

**STRETTON** 2 miles north of Brewood (SJ 8811), *Estretone* 1086 DB, *Strattona* 1175 SHC III (i) 226, *Strettona* c.1182 SHC II 256, *Stretton* 1242 Fees, 1286 SHC V (i) 165, *Stretton-juxta-Horsebrook* 1380 SHC NS II 60, *Stretton be Strete* 1433 SHC XVII 146, *Downnys Stretton* 1491 SHC 1931 241; **STRETTON** 2 miles north-east of Burton upon Trent (SK 2526), *Stretton* 941 (14th century, S.479), *Strættun* 1002x1004 (11th century, S.906; 11th century, S.1536), *Stratone* 1086 DB, *Straton, Stratton* 1114 (1798) Shaw I 25-6, *Halfstretton* 1240 (1798) Shaw I xxviii. From OE *strēt-tun* '*tun* on a Roman road'. The former lies not on Watling Street, as often stated (see e.g. Erdeswick 1844: 167), but on the Roman road (Margary number 19) from Pennocrucium (Water Eaton) to Chester, the latter on Ryknild (Icknield) Street. *Downnys* in the 1491 spelling is perhaps from the downs from which Down House Farm (q.v.) took its name (although that place is some distance from Stretton), used to distinguish the Stretton near Brewood from other places of the same name. *Horsebrook* is from the place of this name near Brewood. The first part of the 1240 spelling for Stretton near Burton upon Trent is unexplained. It has been suggested (Gelling & Cole 2000: 65) that places named *strēt-tun* denoted not just that they lay by a Roman road, but that they may have offered facilities to those using the road.

**STRETWILE, STRETWYLE** (unlocated) Probably the point where Watling Street crosses the river Penk (SJ 8910). *Stretwile, Stretwyle* 1300 SHC V (i) 177 (the bridge itself is recorded as *Eton Bridge* – from nearby Water Eaton – in 1273: VCH IV 163, and 1344: SHC XIV (ii) 27). The first element is from OE *strēt* 'a Roman road, a paved road', for the Roman road Watling Street. The second element appears to be OE *\*wil*, literally 'a wile, a trick', but probably used in the sense 'a gin, a trap, a snare', more specifically here in the sense of the cognate ON *vél* 'a device for catching fish', perhaps connected with the artificial channels in the river Penk between Stretton Mill and Watling Street, but ON *vél* also has the meaning 'an engine, a machine', and OE *wil* may have had a similar meaning and been used for some mechanical apparatus associated with a windmill or watermill, perhaps connected with Stretton Mill itself: see Ekwall 1936: 157; PN Sa IV 18-20. The word is found in the north of England, and has been associated with the Danelaw (see EPNE ii 265), but is also found in Berkshire, Hampshire, and in Wyle Cop in Shrewsbury, Shropshire. See also Stretmesle.

**STRINE BROOK** a tributary of the Moreton Brook which runs into Blymhill Brook. From ME *strind* 'a stream'.

**STRINES** 1 mile north-east of Upper Hulme (SK 0361). *Blakemerstrundes* 1270x1286 StCart, *Strines* 1415 VCH VII 33, *Streins* 1566 *Deed*, *Blacmeer Stroynes* 1626 Rental, *Blackmeer Stroynes* 1668 Alstonefield ParReg, *Blackmeere Strynes* 1670 ibid, *Blackmere Strynes* 1670 Rental, *the Strines* 1677 Alstonefield ParReg, *Strines* 1775 Yates. Probably from the plural of ME *strind*, with the meaning 'stream, watercourse' or similar: the place lies at the headwater of a stream that divides here to flow in different directions. Cf. Strines in High Peak, Derbyshire, PN Db 152.

**STRONGFORD** on the river Trent, 1 mile south of Trentham (SJ 8739). *Strongford (Bridge)* 1599 D593/H/3/339, 1658 SRO Q/SR/304, *Stronford (Bridge)* 1836 O.S.

Seemingly from OE *strang* 'firm, compact soil; water with a powerful current', evidently used here in the sense 'the ford with the strong current'.

STUBBELEY (unlocated, possibly in the Milwich/Caverswall area; it is unlikely that there is any connection with Stubby Lea (q.v.)) *Stubbeley* c.1230 SHC 1921 18, 1286 SHC VI (i) 174, *Stubbiley* 1250 SHC XI 319, *Stolbeleye*, *Stobbeleye* 1306 SHC VII 171, 172. From OE *\*stubbig leah* 'the wood or clearing with the stubs or tree-stumps'. It is possible that Stolben (q.v.) is to be indentified with this place.

**STUBBY LANE** between Wednesfield and New Invention (SJ 9600). *Stobby Lane end* 1536 SRO 26/6/8/3. Perhaps 'the lane to the wood or clearing of the stubs or tree-stumps', from OE *stubb* 'a stub, a tree-trunk'.

**STUBBYLANE** 1 mile west of Draycott in the Clay (SK 1428). *Stubbylane* 1479 SHC VI NS (i) 125, 1550 SHC XII 204, *Stubbe Lane* 1532 SHC 4th Series 8 82, *Stubby Lane* 1533 SHC XIII 275, 1601 SHC XVI 220, *Stubbilane* 1587 SRO D786/21/3, *Stubby lane* 1686 Plot. 'The lane to the wood or clearing of the stubs or tree-stumps'. A lane here bears the name Stubby Lane.

**STUBBY LEA (FARM)** 1 mile north-east of Whittington (SK 1809). *(pasture called) Stubby Lee* c.1550 SHC 1912 194. See also Stubbeley.

STUBCROSS (unlocated, in Walsall) *Stubcross* 1539 SRO D593/A/2/20/12. Presumably 'the stump of the broken cross'.

**STUBWOOD** on north side of Rocester (SK 1039). *Stubwood* 1675 Rocester ParReg, 1686 Plot, *Stubwood green* 1728 Rocester ParReg. 'The wood with the tree-stumps', from OE *stubb* 'a stub, a tree-trunk'. *Stubwood* near Ranton, recorded in the 18th century (JNSFC LXIII 1928-9 165), is probably to be associated with Stubbs' Wood (*Stubbs Wood* 1891 O.S.), 1 mile north-west of Ranton (SJ 8425).

**STURBRIDGE** 1 mile north of Eccleshall (SJ 8330). *Sturbridge* 1771x1842 SRO D1192/31, *Stourbridge* 1889 O.S. A puzzling name, for the spelling indicates 'the bridge over the river Stour', yet there is no bridge, river or stream here, and the nearest river, ½ mile to the south, is the Sow. If the name is ancient (*Sturbrugge* 1414 SHC XVII 16, and *Sturbyche*, *Sturbeche* c.1554 SHC 1938 140, although expressly stated to be in Staffordshire, are almost certainly Stourbridge, Worcestershire, which has never formed part of Staffordshire), perhaps associated with a (lost) stream called *Stawbrooke* recorded in Eccleshall in 1565 (SHC 1938 73), or the unlocated *Tunbryge*, recorded in this area in 1565 (SHC XIII 254), or, if fairs were anciently held here, perhaps transferred from the renowned Stourbridge Fair in Cambridgeshire.

**STYCHBROOK** 1½ miles north of Lichfield (SK 1111; SMR 02088 places a deserted post-Conquest settlement at SK 11701160). *Tichebroc* 1086 DB, *Stickelesbroc* c.1200 SHC 1939 87, *Stichelesbrok* 1248 SHC IV 241, *Sticheslesbroch* early 13th century SRO (150/7923), *Stichesbroc* mid 13th century ibid, *Stichesbroke* 1291 SHC 1911 45, *Sichelesbroc*, *Sichesbroc*, *Sticklesbrok* 13th century Duig, *Stikesbrok* 1302 SHC 1911 59, *Schitesbrok* 1325 SHC 1939 93, *Stychebrok* 1363 SHC 1939 98, *Stichebrook* 1394 SHC VI (ii) 188, *Stychebroke* 1410 ibid, *Stichbrooke (Grange)* 1692 SRO D15/8/5/1, *Stich Brook* 1834 O.S. Perhaps from the OE *\*sticel(e)* 'a steep place, a declivity', with the OE terminal *broc* 'brook', presumably what is now Circuit Brook: VCH XIV 229. One meaning of *stickle* given by the OED is 'a place in a river where the bed slopes and the water is shallow and runs swiftly; a rapid', although the earliest recorded use is 1616. Plot 1686: 106 noted a small stream which rose at Stychbrook and left a residue of aluminous sulphate at the spring head, a phenomenon that which might conceivably be reflected in the stream-name. The DB spelling may reflect the Norman difficulty pronouncing St-: see e.g. Tutbury and Nottingham. Cf. Stittenham, Yorkshire. See also Leyes Grange.

STYCHELEYS, OLDE (obsolete, near Canwell: SMR 0485 places a deserted Anglo-Saxon settlement at SK 14880050) *Sticeleia* c.1148 VCH III 214, *Stichesleia* 12th century ibid, *Olde Stycheleys temp.* Henry VIII ibid. Almost certainly from OE *stycce* 'a bit, a piece', with OE *leah*. *Olde* could mean 'disused' or 'long-used'.

**STYCHFIELD (HALL)** on the south side of Stafford (SJ 9222). *Stychfieldes in Castell* 1582 SHC XV 143. Almost certainly from OE *stycce* 'a bit, a piece', with OE *feld*, with an early meaning 'open land' and a later meaning 'enclosed land'.

**STYVINGTON** – see **STEENWOOD.**

SUFFORD (unlocated, near Stonnal: SHC XVII 244) *Sufford* ? *temp.* Edward III SHC XVII 248, *the Suffords* 1635 SRO D15/11/20/18. See also Sanders 1794: 122.

**SUGARLOAF** a hill 1 mile south of Ecton (SK 0956), ? *the Loe* 1600 Alstonefield ParReg, *Lou* 1612 ibid, *The Loe* 1840 O.S., *The Sugar-loaf Hill, or Coplow* 1844 Garner 1844: 73; **SUGARLOAF FARM** near a 480' hill which lies on the boundary between Staffordshire and Worcestershire at Iverley (SO 8881), *Sugarloaf Farm* 1776 VCH XX 139, *Sugar Loaf Farm* 1832 Teesdale; SUGAR LOAF HILL (obsolete) in the Maer Hills, 1 mile north-west of Maer (SJ 7839), *Sugar Loaf Hill* 1833 O.S. From *sugar-loaf* 'a moulded conical mass of hard refined sugar' (OED) in which form sugar was sold well into the 20th century. The first name is from a tumulus (OE *hlaw* 'mound, tumulus') on a rocky mount, the last probably from a tumulus: a number are recorded on the hills around Maer.

**SUGNALL** 3 miles north-west of Eccleshall (SJ 7930). *Sotehelle* 1086 DB, *Sugenhulle* 1222 Ass, *Suggenh'* c.1233 Rees 1997: 82, *Sogenhul, Parva Sogenhul* 1242 Fees, *Sogenhull* 1280 SHC VI (i) 121, *Suggenhale* ?13th century SHC VI (i) 8, *Sugginhille, Sugginhull, Suggenhale* 13th century Duig, *Magna Suggenhull, parva Suggenhull* 1311 SHC 1914 30, *Sugge* ? 14th century SHC 1921 15, *Great Suggenylle* 1472 SHC NS IV 173, *Sogunhill* 1472 BCA MS917/1253, *Suknell Magna* 1532 SHC 4th Series 8 102, *Shoginhill* 1564 SHC 1926 119. The first element may be OE *sucga* 'a bird (possibly a sparrow)', or the OE personal name Sucga, with *hyll*. The DB form is clearly an error. There were evidently a 'great' (*Magna*) and a 'little' (*Parva*) Sugnall. A place-name *Sugge* is recorded in this area in about the 14th century: SHC 1921 15.

SUKARS HALL (obsolete) A prebendal manor house which possibly adjoined St Lawrence's church, Gnosall (SJ 831208): see WMA 36 1993 70. *Seukesworth* 1369 SHC X NS (ii) 127, *Sukerhall* 1395 ChancM, *Seturhall* 1496 SHC 4th Series VII 170, *Sewkeworth alias Sucars Hall* c.1503 SRO DW1449/1, ? *Sukar* 1587 SHC XV 173. From William de Seukesworth (d.1314), a canon of Lichfield and prebendary of Gnosall in 1278 (SHC 1927 110), with OE *worð* 'an enclosure', later replaced with *hall*. The name Francis Sukar or Shuker is recorded in 1595: Ipm. The hall was no longer standing by 1677: VCH IV 115.

SUKERS LODGE (obsolete) 1 mile north-west of Castle Ring on Cannock Chase (SK 0213). *Sugars Lodge* 1798 Yates, 1834 O.S. A former lodge to Beaudesert Hall, named perhaps from Richard Suker, recorded in 1580 (ParReg). It was demolished by 1992: StEnc. 564.

**SUMMERFORD** in Willenhall (SP 9597). The stream here was too modest to be fordable only in summer (see Somerford above), but the name may be explained by a reference in a grant by the Dean of Wolverhampton in 1359 which mentions *Stomfords lone*, which would appear to be the present road running north from Summerford. The 14th century spelling may be a transcription error for *Stoniford*, which has become corrupted into the present name: see WA II 91.

**SUMMERHILL** on the west side of Kingswinford (SO 8788), *Summerhill* 1749 Bowen, *Summer Hill* 1808 Baugh, 1834 O.S.; **SUMMERHILL** ½ mile north-west of Whitgreave (SJ 8828), *Summerhill* 1890 O.S.; **SUMMERHILL** 1 mile south-east of Flash (SK 0367), *Summer Hill* 1775 Yates. A common name, the precise meaning of which is uncertain, but possibly from ME *somer, summer* 'pack-horse'.

SUMMERSTREET LANE (obsolete) a disused ancient lane which ran from Cotwalton towards Spot. *Summerstreet Lane* 1890 O.S. '... a long stretch of broad green lane...sometimes in the middle, sometimes by the side, the line of [a] vallum is clearly traceable for more than a mile': JNSFC XXXVI 1901-2 118. Perhaps to be associated with a Roman road running from Blythe Bridge to Stafford, which appears to have run through Spot.

SUREY (obsolete) on the south side of Abbey Green near Leek (?SJ 9858). *Sury* 1644 Leek ParReg, *Shury* 1705 ibid, *Sturry (Meadow)* 1770 SRO D3272/1/4/3/17-20. The place is recorded in the earlier 17th century (VCH VII 198), and in the Quarter Sessions records of 1724 'the village of Surrey in the Parish of Leek' was held responsible for the upkeep of the road from the end of 'Surrey Pavement to Gun Gate' (StEnc 564-5). Surrey Pavement may have been a paved causeway across the Churnet valley between Dieulacres and Leek (the printed Leek ParReg I 3 states that 'The southernmost houses at Abbey Green are often called now by this name [Sury]. Tradition says that it was because they stood on the Sure-way to relief – the way to the Abbey. But rather it was because the raised road was paved', and the Leek ParReg II 11 suggests that Sary was 'a pavement of cobble stones on the raised road which led to [Dieulacres Abbey gatehouse] from Broad's Bridge'), perhaps to be associated with a deep hollow-way known as The Trusseway (q.v.) which is said to have run from Fauld across Gun: StEnc 564-5, 603. No derivation can be suggested for Surey or Surrey, which may be associated with *The Corsee* in Leek, recorded in 1596 (Okeover T697), from *causee, causey* (see VEPN II 51-2), from ME *cauce* 'a mound, an embankment, a raised way across low wet ground'. Pasture in Eccleshall called *Sarrey* is recorded in 1519: WCRO GR1291/170/1-2.

SUTHEROWE (unlocated, in or near Cannock Chase, perhaps associated with South Street (q.v.)) *Sutherowe* 1546 SHC 1912 348. See also Sowsbetch.

**SUTTON** 2 miles north-east of Newport, in Forton parish (SJ 7622), *Sutton* 1203 to 1209 Ass, 1227 SHC IV 41, *Sauthon* 1254 SHC 1911 123, *Suthon* 1256 SHC 1913 317, 1332, 1346 Ch, *Southetonne* c.1540 Leland; **SUTTON** 1 mile south of Market Drayton (SJ 6631), now in Shropshire, *Sutton* 1583 SHC XVII 228; SUTTON (obsolete) ½ mile north of Claverley (SO 792941), *Sutton* 1255 Eyton 1854-60: III 93, 1385 SA 5735/2/7/8/1, 1743 Shaw 1801: II 272, 1833 O.S. 'The southern *tun*'. Whitelock 1930: 157 identifies *Suðtune*, recorded in the will of Wulfric Spot 1002x1004 as Sutton Maddock, 2½ miles west of Beckbury, Shropshire, but (as suggested by Eyton 1854-60: III 209) it may be Sutton near Claverley: the name *Suðtune* is immediately preceded in the will by *Sciplea* (Shipley q.v.), 1½ miles from Sutton near Claverley; Shipley is 6½ miles from Sutton Maddock. SHC 1916 34 considers both places and prefers to identify the place as Sutton Maddock.

**SWAINSLEY** on the west side of the river Manifold, ½ mile south-west of Warslow (SK 0957). *Swanslow* 1840 O.S. Without early spellings, and with differing recent forms, any derivation must be uncertain, but perhaps from the ON personal name Sveinn (or the same name borrowed into OE as Swegn: see e.g. SHC V (i) 27), with OE *hlaw* 'tumulus', which (not untypically in Staffordshire) has developed into –ley, or possibly 'the burial mound associated with the swans'.

**SWAINSMOOR** 1 mile north-east of Upper Hulme (SK 0261). *Swaynsmor* 1286 Court (p), *Swaynsmore* 1302 SHC 1925 97, *Swaynesmor* 1302 SHC 1911 59, *Sweynesmor* 1348 Banco, *Sweynsmore* 1511 SHC 1935 123, 1522 SHC 1925 121, *Swannes Meyre* 1538 SHC 1939 84, *Swaynesmore* 1650 SRO DW1761/A/4/267[10/105], *Swenesmoore* 1675 Alstonefield ParReg, *Sweanes Moore* 1676 ibid, *Swans Moor* 1798 Yates, *Swainsmoor* 1839 *EnclA*, *Swansmoor* 1842 O.S. From the ON personal name Sveinn (or the same name borrowed into OE as Swegn: see e.g. SHC V (i) 27), with OE *mor* 'a moor, a marsh', here meaning 'high moorland' since the area is not marshy. The name may be evidence of Norse influence in north Staffordshire. See also Swansmoor.

**SWALLOW MOSS** 1½ miles north-west of Warslow (SK 0760). *Swallow Moss* 1775 Yates, 1839 *Enc*, *Swallow Moor* 1840 O.S. The first word may be the bird (OE *swalwe*), or possibly OE *\*swalg* 'a pit, a pool', perhaps in some cases referring to a swallow hole, an opening or cavity through which a stream disappears underground. OE *mor* and *mos* both meant 'marshy land', so 'the marshy land frequented by swallows', or 'the marshy land with the pit or pool'.

**SWAN VILLAGE** 1 mile north-west of West Bromwich (SO 9991). From the Swan public house, recorded in 1655, but perhaps existing at least twenty years earlier: Hackwood 1895: 7, 50; VCH XVII 8. There is another Swan Village 2 miles north of Dudley (SO 9393) which presumably has a similar derivation. The Swan was the badge of the Stafford family, and a common name for public houses.

**SWANCOTE** 2 miles south-west of Worfield (SO 7494). *Swanecot* 1208-9 For, *Swankote* 1512 Worfield CA, *Swancote* 1525 SR, *Swancott* 1571 Worfield ParReg, *Swancot* 1752 Rocque. Either OE *swan* 'herd, swineherd', originally 'young man, servant', with OE *cot* 'cottage', or 'the cottage with the swans'. In Shropshire since the 12th century.

SWANFORD (unlocated) in Dunston, 4 miles south of Stafford. *Suanford, Swanfurd* 13th century SHC VIII (i) 165, *Swanford* 1532 Deed, *Swainford* 1598 Ct. Self-explanatory. *Swanne Medowe otherwise Earles meadow in Dunston* is recorded in 1587: SHC 1927 173. Yates' map of 1798 shows *Swan Lane* at what is now Drayton Manor, 1 mile north of Penkridge: cf. *Swan-Lane-End* 1686 Penkridge ParReg. The 1598 bounds of Penkridge indicate that Swanford lay between Teddesley Hay and the river Penk: Oakden 1984: 102. The name is likely to be from the bird.

SWAN HAY (unlocated, in King's Bromley) *Swan Hay als Coat Leasows* 1613 SRO D357/A/1/1-26, *Swan Hay* 1720 SRO D15/11/4/111.

SWANHAYS (unlocated, in Audley) *Swanhays* 1566 SHC IX NS 84.

**SWANSMOOR** 2 miles north of Colwich (SK 0124), *Swannesmore* 1461 HAME 483, *Swannes Meyre* 1538 SHC 1939 84, *Swaynesmore* 1625 SHC 1914 131, *Swansmore* 1662 SRO DW1871/9/2/25, *Swans Moor* 1798 Yates, *Swansmoor* 1842 O.S. It is difficult to know wehtehr the derivation is 'the mere or pool with the swans', or 'the moorland with the swans'. Modern maps show no pool here, but *Meyre*, in the 1538 spelling, suggests that one may have existed, or there was particularly wet ground. There is little possibility that the name is from a personal name, although there are references c.1235 to Swane the Smith hereabouts (SHC 1919 7), c.1270 to Richard Swein of Blithfield (SHC 1937 76), and in 1299 to Swane le Fevre of Blithfield (SHC VII (i) 54), and in 1066 Suʒin held Milwich (DB). *Swans Moor* in Hatherton is recorded in 1760: SRO D260/M/T/4/65. See also Swainsmoor.

**SWARBOURN, RIVER** a tributary of the river Trent. *Suereburn* early 13th century BM, *Swereburne* 1252 RydewareCh, *Sweb(o)urn* 1337, 1341, 1443 Ct, *Swerborn'* 1192x1247 SHC 4th Series IV 79, *Suereburn'* 13th century BM, *Swarborn(e), -burn(e)* 1414 *Rental*, 1509, 1512, 1516, 1568 1586 Ct, 1686 Plot, *Swerborne brook* 1571 NA DD/4P/24/55, *Swarbourn Brook* 1834 O.S. A derivation from OE *swǣr* 'heavy, oppresive, slow', applicable to a stream with a slow current, is improbable with a minor watercourse considered to be a *burna* 'stream'. Perhaps therefore from OE *sweora* 'a neck, a col' (see Ekwall 1928: 386, who gives the river-name as *Swerbourn*), found in place-names with the meaning 'a neck of land, a hollow on the top of a ridge or hill', and found in the dialect form *swire*. Cf. Sourton, Devon; Swerford, Oxfordshire.

**SWILCAR LAWN, SWILCAR OAK** 1½ miles south of Marchington (SK 1228). *Swilcar lawn oak, Swilcar oak* 1798 Shaw I 66, *Swilker Lawn Oak* 1836 O.S. 'Swilcar oak stands singly upon a beautiful lawn surrounded with extensive woods ...' 1798 Shaw I 69. A curious name. Halliwell (with EDD and OED) gives the meaning 'to splash about' for *swilker*, which may derive from OE *\*swille* 'a sloppy mess', from OE *swillan, swilian* 'to swill, wash', and ON *kjarr* 'marsh, wet moor, boggy copse', with ME *launde* 'an open space in woodland, a forest glade, woodland pasture', so giving 'the boggy open grassland in the wood'. The oak, destroyed by a storm or lightning c.1942, was famed for its size: Shaw 1798: I 66 fn.1, 69 states that it had a girth of 21' at a height of five feet; see also StEnc 566.

**SWINCHURCH** 5 miles north of Eccleshall, in Chapel Chorlton (SJ 8037). *Suesneshed* 1086 DB, *Suinesheved* c.1230 SHC VI (i) 11, *Swyneshevred* (p) 1244x1261 SRO D938/398, *Swyneshd* 1256 SHC XII NS 85, *Swinesheved* 1261 SHC XI 324, *Swinesheuid* c.1266 SHC 1924 360, *Suineshefd, Swyneshefd* c.1270 SRO D935-5, *Suenisheved* 1272-3 SHC XI NS 242, *Swynesheved* 1283 SHC 1911 180, *Sueneshefd* 1287 SHC 1921 16, *Sweneshead* 1373 SHC 1914 37, *Swyneshede* 1425 SHC XVII 111, *Swynshedd* 1569 SA 279/38. The first element is from OE *swin* 'swine, pig', and the second is from OE *heafod* 'a head or end (of anything)'. The second element is often combined with the name of an animal. *Swineshead*, by far the most common of such compounds, is found surprisingly frequently in English place-names, with at least 13 examples known. The *heafod* 'head' element is invariably a topographical term, in the sense of 'a low headland', giving 'a low headland suggestive of a swine's head' (see Gelling & Cole 2000: 175-6), but the distinction between a *heafod* and a *ness* is unclear. The place lies near the foot of a pronounced narrow headland, with another headland nearby to the east, the latter (from hachuring on the first edition 1" O.S. map of 1833) with a rounded and stepped nose (or 'snout'), and is mentioned in the late 16th century as *Swinshead*, a name still used in 1819, but appears as Swinchurch on Greenwood's map of Staffordshire, 1820 (VCH IV 42), presumably because the older name was then considered indelicate. See also SHC 1945-6 107; PN Wo 161-2. The name is found as early as the 7th century: ASC 'E' records Swineshead in Lincolnshire as *Swines hæfed* in 675 (ASC 'E'), *æt Suinesheabde* 786x796 (13th century, S.1412); Ekwall 1960: 457; see also S.68 and S.72. An association with pagan rituals involving animal sacrifice with place-names of this type is now rejected by place-name scholars: see Gelling & Cole 2000: 175-6. See also Boleheved.

**SWINDON**  5 miles west of Dudley (SO 8690). *Swineduna* 1167 SHC 1923 298, *Suindun* 1236 Fees, *Swyndon* 1271 SHC V (i) 141, *Swyneden* 1275 SHC VI (i) 71, *Suyndon* 1300 SHC VII 66, *Swyndon* 1332 SHC X (i) 130, *Sevindon (sic)* 1775 Yates, 1787 Cary. From OE *swin-dun* 'hill of the pig or swine'. The place lay in Kinver Forest, where the pasturage of swine was an important privilege.

SWINESHEAD – see **SWINCHURCH**.

**SWINFEN**  3 miles south-east of Lichfield (SK 1305; SMR 02075 places a deserted post-Conquest settlement at SK 13500600). *Swyneffen* 1232 Ass, *Swynefen* 1252 FF, 1294 SHC VI (i) 294, *Swynfend* 1255 (1801) Shaw II 29, *Swinesfeud, Swynefen, Swynesfen* 13th century Duig, *Wynfyn* 1532 SHC 4th Series 8 184, *Swinefeld* c.1540 Leland. 'The fen of the swine', from OE *swin* 'swine, pig', and OE *fen* 'marsh, fen', and OE *feld* 'open ground'. The area here is low lying.

**SWINNERTON** – see **SWYNNERTON**.

**SWINSCOE**  3½ miles north-west of Ashbourne (SK 1348). *Swinescho* 1203 FF, *Swineskou* 1203 SHC III 109, *Swinestoh* 1203 SHC III 117, *Swineskoc* 1241 Okeover, *Swyneskow* 1248 FF, *Swynsco* 1253 Ipm, *Swinescow* 1254 Okeover, *Swineschoch* (p) 1275 FF, *Swinescohe* 1280 Banco, *Swynescou* 1281 Ass, *Swynescogh* 1295 SHC 1911 55, *Swynescho* 1299 Ass, *Svineskoch, Swyneschouh, Swenesco, Swenescho, Swynescough, Swenestoch* ?13th century SHC VII NS 142-64, *Swinesc(h)o(h)* 1318 Okeover, *Swyneskow(e)* 1414 ibid, *Swynstoo* 1539 MA, *Swynskoo* 1564 Pat, *Swynscoe* 1606 FF, *Swainscow otherwise Swainscoewe, Swynscow* 1605 SHC IV NS 5. The first element is from OE *swin*, or ON *svin* 'pig, swine'. The second element is from ON *skógr* 'wood', the only example traced in Staffordshire, and provides evidence of Scandinavian influence.

SWYNELEYE, SUENYLEYE  (unlocated, 'below Chasepool Hay': SHC 4th Series XVIII 187-8) *Swyneleye, Suenyleye* 1296x1307 SHC 4th Series XVIII 188, *Swynleye* 1342 SHC 1913 91. 'The *leah* of the swine'.

**SWYNNERTON** Ancient Parish 3 miles north-west of Stone (SJ 8535). *Svlvertone* 1086 DB, *Suuinwrtona* 1159 (13th century) EEA 14 71, *Swineduna Helye* 1166-7 SHC I 49, *Swaneforton* c.1195 SHC XI NS 125, *Sinvertona* c.1199 SHC II 95, *Suinerton, Silverton* 1205 SHC III 134, *Swiluerton'* 1206 Pleas, *Soulverton* 1206 SHC 1912 269, *Sumerverton* 1228 SHC IV 74, *Swynnertona* 1230 SHC 1912 270, *Swilverston* 1236 Fees, *Suinnerton* 1242 ibid, *Surlton* 1263 SHC IV 154, *Sonnerton* 1264 SHC 1912 269, *Swynaferton, Swynforton* 1272 Ass, *Silvereston* 1275 SHC I 174, *Swynemerton* 1289 SHC VI (i) 181, *Swonnerton* 1320 SHC 1912 269, *Swinnerton* (frequently), *Swynefarton* 13th century Duig, *Swilveston, Silveston* 13th century SHC I 174, *Swyndverton* 1326 SHC IX (i) 113, *Swineforton* 1355 ibid. 324, *Sonnertone* 1372 SHC 1912 269, *Swynarton* 1404 HLS, *Swenerton* 1532 SHC 4th Series 8 137. Ekwall 1960: 457 gives a derivation from OE *Swinford-tun* 'tun by the pig ford' (supported by Mills 1998: 336), and that is likely to be correct, although the early forms are inconsistent, doubtless indicating that the true derivation had become uncertain at an early date. The *-ver-* element is evidently the commonly-found corruption of *ford*. One explanation for some of the forms is a derivation incorporating an OE adjective *sulig* 'a pig-sty, pig's lair' (of interest because it is found only in the nearby counties of Gloucestershire, Worcestershire, and Warwickshire, suggesting that it was a term used in and around Hwiccan territory that went out of use at an early date: see Ekwall 1936: 56), from OE *sylu* 'wallowing place for animals, miry place', sometimes found as *sol(h)*, later *sil*, perhaps explained by unrounding and fronting (see PN Wa 68, Gelling & Cole 2000: 62-3, and note Sol-, Sul-, Syl-, and Sil- spellings for Solihull). That would give alternative names with similar meaning '*tun* at the ford with the wallowing place', and '*tun* by the pig ford', which may have been in use together. Another possibility is that the name incorporates OE *sulh* (generally found as *sylh*) 'plough', which occurs in several place-names, perhaps in the sense 'furrow' and 'gulley, narrow valley', with ME and modern forms in *Sil-* going back to OE mutated oblique cases (EPNE ii 167): cf. Silverton, Dorset, and see especially Ekwall 1960: 452. Hachuring on the first edition 1" O.S. map shows a pronounced narrow valley with a stream running south from the village. It may be noted that *sulhforda* is recorded in 718x745 (11th century, S.1254) in the first known English lease: Stenton 1971: 485. A ford may seem improbable at a place on high ground, but the original settlement may have lain on lower ground, perhaps to the south of the present village: see StEnc 567. CDEPN identifies from the spellings 'the settlement at Swilford' (from OE *\*swille* 'a sloppy mess'), 'the settlement at Swinford', 'the settlement at Swanford, the herdsmen's or swan's ford' (from OE *swan* or *swan*), and 'the settlement at Somerford, the summer ford'. For the pronunciation of the name as Sonnerton or Sinnerton see SHC 1912 269. *Helye* in the 1166-7 spelling is from the name the Domesday tenant, variously found as Aslen (DB), Eelen, Ehelen, Eelen, Eslenem, Esluem, and Aelem: SHC VII (i) 2. See also SHC I 174.

**SWYTHAMLEY** 6 miles east of Congleton, near the Staffordshire-Cheshire border (SJ 9764). *Swythomlee* 1180, 1283 *Brocklehurst, Swythamley-grange* 1234, *Switholm* c.1291 Tax, *Swythuley* 1406, *Swythunley* 1534, *Swythaley* 1538, *Swythumley-graunge* 1538 (all NSFC LXVII 51-70), *Swytherley* 1540 Pat, *Swithorn le Graunge* 1599 SHC 1935 212, *Swithanly* 1607 Kip, *Swythern-grange, Swythernley* 1614, 1645, 1697 and 1762 NSFC LXVII 51-70. The first element appears to be the dative plural of ON *svíðum* 'at the burnings' or similar (cf. ME *swithin* 'a clearing', related to ON *\*sviðinn* 'land cleared by burning', found in English dialect as *swithin* and *swidden*, from ON *sviða* 'to burn': see Ekwall 1960: 457 *sub nom.* Swithland; JEPNS I 37), with ON *holmr* 'raised ground in marsh-land', with OE *leah*, giving 'the *leah* cleared by burning in the marsh land'. *Parke land*, recorded in 1621 (SHC 1934 24), may be Swythamley Park. For O.S. observations on place-names on the Sythamley estate see SRO 4974/B/4/61. Cf. Swithenthwaite, Cumbria.

**SYDNALL** 2 miles south-east of Market Drayton (SJ 6830), now in Shropshire. *Sydnall* 1583 SHC XVII 228. Probably from OE *(æt þæm) sidan heale* 'broad corner of land'. Cf. The Sidnalls. *Sydenhale* (lost, north of Whittington) is recorded c.1300 (TSSAHS XX 1978-9 loose map), and *Sidenhale* in Loynton in an undated deed (WCRO CR1291/233).

**SYERSCOTE** 2½ miles north-east of Tamworth (SK 2207; SMR 01175 places a deserted Anglo-Saxon settlement at SK 22460768). *Fricescote* 1086 DB, *Siricescotan* 11th or 12th century Sawyer 1979a: xxxvii, *Sireskote* 1236 Fees, *Sirescot* 1242 Fees, *Sidecote* 1292 SHC VI (i) 247, *Shyrescote* 1293 ibid. 288, 1303 WL 36, *Cyrescot'* 1320 WL 172, *Syrescote* 1375 BCA MS3669/Acc1938-049/506569, *Cuescote* 1377 SHC 4th Series VI 11, *Sirescote* 1380 SHC XVII 192, *Sierscotte* c.1566 SHC 1938 170, *Surcote, Surcot* 1566 SHC IX 79, *Sierrot* 1753 BCA MS3878/613. The DB *F-* is crearly an error for *Si-*. The name means 'the cot or cottages of Sigeric'.

**SYTCH HOUSE GREEN** 2 miles south-west of Claverley (SO 7890). From OE *sic* 'a watercourse'. It is unclear whether *Scythe*, recorded *temp.* Edward I (Eyton 1854-60: III

76), is to be associated with this place. The Green element suggests a grassy area or a squatter settlement. In Shropshire since the 12th century.

**TAD BROOK** running south of Kingstone into Blithfield reservoir. *ceabbe broc* 996 (11th century, S.878), *? Capbroc* early-to-mid 13th century SROD986/41, *Tap(pe)broc* 13th century Bagot, *Taldbro(o)ke* 1349 Blithfield, *Tabbrock(medue)* 1402 SHC XI NS 202, 1508 Bagot, 1546 Ct, *Tade Broke* 1543 Ct., *Taddebroke* 1562 SRO DW1734/J/1070. Hart 1975: 207 suggests a derivation from a patronymic *Ceabba, which may have been a pet-form of Ceadda, but it is likely that the earliest spelling confuses *c* and *t*, a common occurrence. The name may derive from OE *tæppa* 'a peg, a spigot' (also found in Tappeley (q.v.)), usually applied to a place where wood for these was obtained, so 'the brook of the wood or clearing where pegs or spigots were got', with OE *broc* 'brook'. See also Hooke 1983: 90.

**TAFT FARM**- see **TOFT**.

TAG DALE (obsolete) 1 mile south-east of Mucklestone (SJ 7336). *Tag Dale* 1833 O.S. The history of the name is not known, but a possible derivation is OE *tacga, tegga* 'a teg or young sheep': see PN Wo 288; Ekwall 1936: 74; Foxall 1980: 43; but note also the surname Tag(g): the Tagge family of Chebsey is recorded in the 16th century (SRO D615/PM/1/4), and the surname appears frequently in Ellenhall ParReg. Perhaps associated with the unlocated *Tag Moor*. SHC 1945-6 24.

**TAGG MOOR** – see DAIRY HOUSE near Market Drayton.

**TALKE, TALK O' TH' HILL** south-west of Kidsgrove (SJ 8253). *Talc* 1086 DB, 1203 SHC III (i) 116, *Talk* 1252 CH, 1280 SHC VI (i) 112, *Talke* 1276 Ipm, 1540 DRO D3155/WH44. Probably from a British hill-name representing the ancestor of Welsh *talcen* 'forehead, brow, gable-end', with loss of *n* in an unstressed syllable. The place lies on a prominent ridge reaching 559' overlooking the Cheshire plain and the valley in which Kidsgrove lies, and the name was probably that of the ridge itself. *Talkhamell* is recorded in Audley c.1571 (SHC 1931 131), from dialect *hamel* 'hamlet' (OED), but the identity of the place is uncertain, unless it is to be read as *Talk-Halmer*. See also Mow Cop.

**TALLASH** 2 miles south-west of Rushton Spencer (SJ 9260). *Tallash* 1842 O.S. Evidently from 'tall ash'. See also Ashmore Heath.

TAMBER (unlocated, possibly near Trentham)  *Tamber* 1584 and 1589 Trentham ParReg.

**TAME, RIVER** a tributary of the river Trent. *Tame* c.1025 Saints, 1228 Ass, *Tama* 1232 Ass, *Tamme* 1285 QW, *Tome* 1379 Ipm, 1414 SRO D593/B/1/26/6/1/6, *Thame* 1282 Banco, 1286 For, 1292 Cl, 1315, 1381 Ipm, 1350 BM, 1509 Rental. A British river name identical with Taff and Taf in Wales, and meaning perhaps 'dark river' (Ekwall 1928: 389-90), which forms part of the name of the Anglo-Saxon tribal territory of the *Tomsætan* 841 (12th century, S.197), *Tonsetorum* 844 for 848 (12th century, S.197, see Stenton 1970: 184 fn.3), *Tomsetna* 849 (11th century, S.199; 11th century, S.1272), 'people of the Tame'. See also Tamhorn; Tamworth.

**TAMHORN** 2 miles north-west of Tamworth (SK 1706; SMR 02081 places a deserted post-Conquest settlement at SK 18030732). *Tamahore* 1086 DB, *Tamehorn* 1167 P, *Tamenhorn* 1179 SHC I 90, *Thamehorin* 1199x1216 D(W)1734/J/1712, *Thamenhor* c.1255 (1798) Shaw I xvi, *Thamenhoren* 1266 SHC 1911 136, *Thomenhorn* 1271 SHC V (i) 145, *Tomenhorn* 1289 SHC 1924 360, *Thamehorne* 13th century Duig, *Toumehorn*,

*Tomnehor* 1371 SHC VIII NS 262, 265. The first element is from the river Tame (q.v.). The second element is OE *horn* 'horn, corner, bend', hence 'the horn-shaped land near the river Tame', or possibly '(the estate at) the bend of the river Tame'. The place lies near a slight bend of the river. The boundary of the tribe which lived by the Tame is recorded as *Tomsetna gemære* 'Tame people boundary' in 849 (11th century S.199).

**TAMWORTH** Ancient Parish 14 miles north-east of Birmingham (SK 2004). *? Tomtun* 675x692 (S.1804; see Gelling 1992 146-7), *Tamouuorthie* 781 (11th century, S.120), *Tamouuorthige* 781 (11th century S.121), *(æt) Tome worðige* 799 (13th century, S.155), *Tomeworðig* 808 (9th century, S.163), *Tomoworðig, Tomoworthin* 814 (11th century, S.195-6), *Tomeuuorðig* 841 (S.198), *Tomweorthin* 855 (11th century, S.207), *Tameworþig* 922 ASC, *Tom[wy]rðin, Tamwurþin* 1002x1004 (11th century, S.1536), *Tamuuorde* 1086 DB, *Tamwurda* 1179 SHC I 90, *Tamewurd, Tamewurde* 1190 Ch, *Thammoth* 1271 SHC V (i) 147, *Thomworye, Thamworye, Thamworyie* c.1280 SHC 1921 4, *Thamworyie* c.1280 SHC 1921 5, *Thamewourthe* 1292 SHC VI (i) 247, *Thamworthe* 1313 SHC 1911 313, *Toneworthe* 1396 SHC XV 76, *Tameworthe, Tamworth* c.1540 Leland. The second element is from OE *worþig* (and its variant *worthign*), an element rarely found outside the south-west, which is usually said to mean 'enclosure, homestead', but which perhaps developed at an early stage in the Midlands a particular meaning synonymous with *burh*, and was applied to places of particular importance: cf. *Northworthig*, the earlier name of Derby. (Hart 1992: 37 fn.37 suggests that *Northworthig* may have been the 'capital' of the people called by Bede the North Mercians, with Tamworth the capital of the South Mercians, comparable with Norwich in Norfolk and Ipswich in Suffolk.) It is almost certain that the name was originally *Tomtun* 'the *tun* on the river Tame' (SHC 1950-1 146fn; Stenton 1970: 184 fn.3; Hooke 1983: 21; Gelling 1992: 146-8), with *tun* perhaps meaning at an early period 'royal vill': see Campbell 1986: 115. Tamworth was the early capital of Mercia, where Offa established a base and fortified the town 757-96, but the place was destroyed by the Danes in 874. The change from the generic *tun* which occurred during the 8th (or late 7th) century perhaps reflected a change in the nature of the place: see Gelling 1992: 147. Offa's fortifications, which probably took the form of an enclosing ditch and bank, may explain and date the *worþig* element: see TSSAHS X 1968-9 32-42. The pre-Conquest spellings, which vary in printed sources (e.g. Hart 1975; Whitelock 1930: 48, 50; Zaluckyj 2001: 218-9), are taken from Gelling 1992 146-7. Cf. Tame; Tamhorn.

TAMWORTH MOORES (unlocated, perhaps to be associated with Staffordshire Moor (q.v.)) *Tamworth Moores* 1647 BCA MS3878/261.

**TANSLEY HILL** 1 mile south-east of Dudley (SO 9589). *Tansley Hill* 1834 O.S. Presumably to be associated with *Tensly*, recorded in 1656: Roper 1980: 109.

TAPMORE (unlocated, in 'Great Loxley': DW1733/A/2/39) *Tapmore* 1337 SRO DW1733/A/2/39. Evidently associated with *Tapforde* recorded in 1430 SRO DW1733/A/2/42. See also Tappeley.

TAPPELEY (lost, in Bagot's Bromley, perhaps near Squitch House: SHC XI NS 15) *Tapelega* 1198 SHC XI NS 146, *Tappelegh* 1199 SHC III (i) 167, *Tappelee* c.1225 SHC XI NS 149, *Tarpele* 1290 ibid. 22, *Tappeley* 1369 SHC XIII 64, 1402 SHC XI NS 208, *temp.* Henry VIII ibid. 15. See SHC XI NS 15. The first element may be OE *tæppa* 'tap' in an earlier sense 'peg', with OE *leah*, so possibly 'the wood from which pegs were obtained', or 'the clearing where pegs were made'. See also Tad Brook; Tapmore.

**TAR HILL**  a 517' hill on south side of Brocton (SJ 9618). *Tar Hill* 1834 O.S. The single late spelling precludes a derivation: OE *torr* 'a rock, a rocky outcrop, a rocky peak' seems unlikely. *Tarr Hill* near Hilton (Worfield) is recorded in 1822: WJ October 1908 267.

**TARDEBIGGE**  3½ miles south-east of Bromsgrove (SO 9969). *Tærdebicgan* c.1000 (11th century, S.1534), *Tyrdebicgan* (11th century, S.1598), *Terdebiggan* 11th century, *Tyrdebicgan* 11th century, *Terdebigan* 11th century (17th century) PN Wo 362, *Terdeberie* 1086 DB, *Terdebig'* 1160x1176 (12th-13th century) EEA 16 78, *Terde(s)bigga* 1173 SHC I 67, 1266 Ch, *Terdebigga* 1138 BM, 1169-92 P, 1275, 1327 SR, *Terdebig* 1230 Cl, *Therdebigge* 1258 FF, *Terdebig* 1270 SHC 1923 266, *Tertebigge* 1275 Ass, *Terdebygge* 1293 SHC VI (i) 261, *Terbygge* 1486, 1499 Pat, 1589 CKS U386/T107, *Tarbick* 1675 Ogilby, *Tarbeck* 1680 FF. A puzzling name for which no satisfactory derivation can presently be offered. The -*cg*- in the suffix might point to an English rather than a Celtic origin, but Mawer 1929: 3 suggests 'probably Celtic', and Coates 1988: 57-64 suggests Celtic *tarp pig* 'spring at the hill': the place lies on the north-east side of a hill of over 525' with streams flowing between small spurs on all sides. However, the name is included with reservations in the corpus of Celtic names which appear in Coates & Breeze 2000: 341, and there are philological difficulties associated with that derivation, including the many early spellings with *Te*-. Another possibility is that the first element may be associated with a personal name Tyrdda, believed to be found in Tredington, Worcestershire, some 24 mile south-east of Tardebigge: Tyrdda is recorded in a charter of 757 (11th century, S.55) as a previous holder of the estate: see PN Wo 172, 362; Gelling 1988: 178. It is possible that the T- is a ghost of the preposition *æt*: early documents generally used the preposition with village names which were originally toponymics, but not with names that were originally habitative or which still retained names of natural features, or old British names: EPNE i 6; cf. Cf. Tawdbridge, Lancashire, from *at Ald-brycg*. A minor unsolved name in Willoughby, Warwickshire, is recorded as *Turdebigge* c.1250, c.1280, 1321 *Magdalen Deeds*, *Turdebice* 1375 ibid, *Tardebigge* 1349 ibid, and has perhaps the same derivation as Tardebigge: see PN Wo xliii. Tardebigge was in Staffordshire from c.1100 to 1266 (see also SHC 1944 88) and in Warwickshire until 1844, when it transferred to Worcestershire.

**TATENHILL**  Ancient Parish 2½ miles south-west of Burton (SK 2022). *Tatenhyll* 941 (14th century, S.479), *Tatenhala* 1093 CEC 4, *Tattenhull* c.1180 SHC VII NS 132, *Tatenhell* 1188x1197 SRO D603/A/Add/36b, *Tatenhala* 1188 SHC I 140, *Tatenhulle* 1227 SHC IV 43, *Tattenhull* 1251 Ch, *Tatenyll* 1421 SHC XVII 94, *Tattnell* 1532 SHC 4th Series 8 141, *Tatenhall* 1589 SHC XV 190, *Tatenhill* 1680 Browne. From the OE personal name Tata, genitive singular Tatan, hence 'Tata's hill'.

TATERYNGE (unlocated) *Taterynge* 1553 SHC XII (i) 211.

TATTENHAM (unlocated, possibly fictitious: see SHC IX NS 150) *Tattenham* c.1565 SHC IX NS 150.

TAYLORS GREEN  (obsolete) ½ mile north-west of Longsdon (SJ 9555). *Taylors Green* 1695 Leek ParReg, 1815 SRO Q/SB. The name is recorded in 1482: VCH VII 203. The Green element suggests a squatter settlement.

TAYLOR'S PARK  (unlocated, near Yoxall) *Taylor's Park* 1723 SBT DR 18/22/7/22. Perhaps from a surname, but Rackham 1990: 107-8 notes that Norman-French *tailz* had the same meaning as *coppice* (also a Norman-French word), found in modern French as *tailiss* 'underwood', and sometimes found in English wood-names as Taylor's.

**TEAN, RIVER** a tributary of the river Dove. *Tene* 1389 Ct, 1686 Plot, *Teine* 1577, 1586 Harrison, *Tayne* 1577 Saxton, *Tayne, Teane, Tene* c.1595 Erdeswick, *Tain(e)* 1613 Drayton, 1617 Keer, 1611 Survey, *Tean(e)* 1610 Speed, 1775 Yates. A British river name, identical with Teign in Devon, held by Ekwall 1928: 397-8 to mean simply 'stream', but now held to be related to the Modern Welsh verb *taenu* 'to spread, to scatter', a sense well-attested in early Welsh: Ekwall 1928: 395. The name here can be translated as 'sweeper, scatterer', that is, a river liable to flood: see Coates & Breeze 2000: 136-7.

**TEAN, UPPER & LOWER** in Checkley parish, 3 miles south of Cheadle (SK 0139 & SK 0238). *Tene* 1086 DB, *Tena* 1240-1 Cur, *Tene* 1242 Fees, *Thene* 1295 SHC 1911 240, *Tene, Teyne* 13th and 14th century Duig, *Tayne* 1592 SHC 1930 304. The place takes its name from the river Tean (q.v.) on which it stands.

**TEANFORD** (pronounced Tenford [tɛnfəd]) 2 miles south of Cheadle (SK 0040). *Teanford* 1698 SHC 1925 145, *Tenford* 1732 SRO D927/4, 1836 O.S., 1870 P.O. 'The ford across the river Tean'.

**TEDDESLEY** 2 miles north-east of Penkridge (SJ 9415). *Teddesl'* 1236 Fees, *Tedesle* 1242 Fees, *Tudeslegh* 1246 Cl, *Teddesleg (hay)* 1252 ibid, *Tedeslegh, Tidesleye* 1275 ibid, *Teddesleye* 1327 SHC VII 228, *Teddesley Head* 1682 Browne. Probably from the OE personal name Tydi, with OE *leah*, so 'Tydi's *leah*', as suggested in Ekwall 1960: 462.

TEGUES WELL (unlocated, in High Onn) *Tegues Well* 1808 Baugh.

TENSEPARK (unlocated, in Needwood Forest: VCH II 349) *Tensetepark* (undated) VCH II 349 fn.13, *Tensepark* (undated) SHC 1912 222. Perhaps to be associated with *Tensetwode*, recorded *temp.* Edward I/Edward II: SRO DW1733/A/2/43.

**TENTERBANKS** on west side of Stafford (SJ 9222). From ME *teyntour* 'a frame for drying and stretching cloth': *le teynter on the walls* is recorded in Stafford in 1468 (VCH VI 189), presumably associated with this name.

**TENTERHILL** 1 mile north-west of Hollinsclough (SK 0467). *Tenter Hill* 1798 Yates, 1794 Stockdale. Almost certainly from ME *teyntour* 'a frame for drying and stretching cloth'.

**TERN, RIVER** a tributary of the river Severn. *Tren* 12th century Taliesin, *Terne* 1232 Ch, 1255 MRA, c.1291 Tax, 1379 Banco, c.1540 Leland, 1577 Saxton, 1686 Plot, *Tirne* 1316, 1360 AD 6, 1319 Pat, *Tirn'* 1228 Ass, *Tyren* 1477 AD 6, *Tyrne* 1322 Pat, *Teryn* 1439 AD 4, *Tyerne temp.* Elizabeth I Chanc P, *Tearne* 1613 QSR, 1617 Keer, *Tirne* c.1200 Gervase, 1256 Ass, *Terne* c.1200 Sa. Deeds. A British river-name derived from Welsh *tren* 'strong, powerful': see Ekwall 1928: 400-1.

**TETTENHALL** Ancient Parish 2 miles north-west of Wolverhampton (SJ 8700). *(æt) Teotaheale* 910 (11th century) ASC (C,E), *(æt) Totanheale* ibid. (D), *Totehala, Totenhale* 1086 DB, *Tettenhala* 1169 P, *Tettenhal* 1173 SHC I 68, *Totenhala* 1186 SHC I 130, 1190 SHC II 12, *Teteneshal'* 1194 Pipe, *Tettenhale* 1195 SHC II 46, 1201 ibid. 108, *Tetenhalle* 1196 ibid. 57, *Totenhall* 1240 (1798) Shaw I xxx, *Totenhale* 1255 SHC V (i) 113, *Tottenhale* 1286 ibid.' 166, *Tetenhaul* c.1540 Leland v 19, *Tetnall* 1577 Saxton, 1610 Speed, *Tettenhall* 1682 Browne. The forms indicate a probable derivation from an OE personal name *Teotta, which is not on record; cf. Teoda, representing pet-forms from names beginning Þeod-. There was evidently another form with *t* for *d*, as here: cf. Teddington, Worcestershire (*Teottingtun* 780), and Tiddingford, Buckinghamshire

(*Teotanheale* ASC): see PN Bk 81. The commanding views eastwards from the abrupt sandstone bluff here encouraged Skeat to suggest that 'if we take the words as they stand (Anglo-Saxon Totanhale, Domesday spelling Totehala), then Anglo-Saxon totan heall means 'tout's corner'; i.e. a corner (or convenient spyplace) whence a spy looks out. Totan should be To'tan with long o, and is the genitive case of To'ta, a spy, or lookout man. Mod. Eng. Tout for custom. It means the Hall or Dwelling on a look out hill. We should call it Spy Hall if we had to make up the word nowadays': Jones 1894: 8. That derivation cannot be entirely dismissed, notwithstanding the *Teot-* spellings in ASC, which are not found in other records: cf. Tutbury. Tettenhall was held by the clergy of Wolverhampton, which explains references to *Tettenhall Clericorum* (see for example SHC XVIII 160), and the king held a manor in Tettenhall Regis (q.v.). Tettenhall is the name of the civil parish, Tettenhall Regis the ecclesiastical parish. *Tetenhalehome* or *Totenhalehome*, recorded in 1337 (SHC VI NS (ii) 94-5), incorporate OE *ham* in the sense 'manor', so 'Tettenhall manor': see PN Sa II 10. ASC and chroniclers including Æthelweard record a great battle at Tettenhall (or Wednesfield) on 5th August 910 AD, at which the Saxons vanquished the Danes: Earle & Plummer 1892-9: i 94-7. The battle is commemorated by *Dane's Court* (*Danescourt* 1922 O.S.) an area north-west of Tettenhall (SJ 8800), a name which is not ancient but comes from the name, inspired by local antiquarianism, of a large house built there in 1864 and demolished in 1958: VCH XX 8. See also Perton. *Tettenburn* is recorded in a charter of 739 AD of land at Crediton, Devon (11th century, S.255): see Whitelock 1955: 256.

**TETTENHALL REGIS** Ancient Parish 2 miles north-west of Wolverhampton (SJ 8700). *Kinges Tetnolde* c.1560 SHC 1938 158. See Tettenhall.

**TETTENHALL WOOD** – see **TETTENHALL** and **KINVER**.

TEYERTON (unlocated) *Teyerton* 1421 SHC XVII 82.

THACHILEYE (unlocated, possibly near Cheadle) *Thachelee* 1295 SHC 1911 55, *Thachileye* 1323 SHC 1911 98.

**THACKER'S CROP** in the north-west part of Bagot's Park (SK 0728). *Thackers Crop* 1724 *Survey*, *Thacker's Crops* 1836 O.S. Perhaps to be associated with *Thacherfild*, recorded in 1402: SHC XI NS 209. The name perhaps refers to the field from which thatchers took their long straw.

**THATCHMOOR** 3 miles north-east of Lichfield (SK 1510). *Thatchmores* 1583 SHC XVII 229, *Thatchmoore* 1649 Barton under Needwood ParReg, *Thachmoor* 1695 Morden, *Thatch Moor* 1775 Yates, 1801 Smith. 'The moor which produced thatching material'. The material may have been heather, often used for thatching. *Thatchmore* in Yoxall is recorded c.1710 (SRO D820/1). *Thackmore* in Rolleston, recorded in 1675 (SRO D1553/107), has not been located.

THERLEYEMOR (unlocated, possibly near Curborough or Hampstead) *Torleymor* 1176x1182 EEA 16 66, *Thelemore, Thorleymore, Thurleymore* 1262 SHC 4th Series XVIII 28, 44, 51, *Therleyemor* 1320 SHC 1911 344, *Thorleymor* 1327 SRO 3764/91, 1332 SHC X 105, *Therlaymor* 1336 SRO 3764/59. Perhaps from OE *þyrre leah mor* 'the moorland at the dry *leah* '.

THEVESBY (unlocated, in Horton) *Thevesby* 1239 CurReg. An important place-name, the only instance traced in Staffordshire of a name which would appear to incorporate OWScandinavian *bý* 'a farmstead, a village', evidently with the plural of OE *þeof* 'thief', so 'the thieves' village'. The combination of an OE and Scandinavian element is not unusual (cf. Denby, West Riding of Yorkshire and Derbyshire), but remains unexplained: see Coates & Breeze 2000: 152 fn.1. It has been suggested that in some cases *bȳ* might have been an element borrowed into English: Fellows-Jensen 1972: 189.

**THICK WITHINS** 1 mile west of Hollinsclough (SK 0466). *Thickwithins, Thickwethins, Thycke Witheshead* 1600 Alstonefield ParReg, 1651 *Rental, Thickwithins Head* 1626, 1651 *Rental, Thick Withins* 1842 O.S. From OE *picce* 'thick, dense', and OE *\*wiðign* 'a willow, a willow copse', found as dialect *withen* 'a willow holt', giving '(the place with) the abundance of willows'.

**THICKBROOM** 1 mile south-west of Weeford (SK 1203; SMR 02083 places a deserted post-Conquest settlement at SK 12990380). *Tichbrom* c.1175x1208 SRO 3005/1, *Tichebrom* 1199 SHC III (i) 35, *Tichebrome, Titebrome* 1201 SHC II 105, *Tykebrom* 1227 SHC IV 41, 1286 SHC V (i) 162, *Thikebrom* 1256 SHC 1911 127, *Thyckeborne* 1271 ibid. 149, *Thickebrom* 1327 SHC VII (i) 233, *Tikkebrome* 1425 SHC XVII 151, *Thyckbrome* 1566 ibid. 215. From the OE adjective *picce* 'thick, dense', and OE *brom* 'broom', hence 'the broom thicket'. The country around was formerly heathland. Thickbroom Manor which appears on the 1834 O.S. map was replaced c.1836 by Manley Hall (SHC 1942-3 216; StEnc 382; SHC 4th Series 14, plate 5), built in tudor style by Thomas Trubshaw for Admiral J. S. Manley. It was demolished in the early 1960s.

**THICKNALL FARM** 1 mile west of Clent (SO 9079). *Thyckennaile* (p) 1304 Ct, *Thickoll* 1592 Wills. Of uncertain origin, but *Thikenolre* (1327 SR), *Thickenalre* 1339 (SHC XI 84), which appear to relate to this place, point to a meaning 'the alder thicket'. In Staffordshire from the 13th century until 1844, when it became part of Worcestershire.

THICKNES Erdeswick 1844: 23 notes 'Thicknesse, a place not observed in maps of Staffordshire', and considerable uncertainty has surrounded the precise location of this place, which existed in the Balterley/Audley/Podmore/Apedale area from at least the 13th century until at least 1565 (Shaw 1798: I *411, 412), or 1630 (SRO D948/1/1/2/2). The introduction (by Mr Ralph Thicknesse) to the printed volume of Betley ParReg states that Balterley Hall (SJ 764499), which was held by the Thicknesse family until 1790 (SHC XII NS 233-5 fn.), was known as the manor of Thicknesse (*manor of Thiknes* 1378: SHC VIII NS 135), and was never part of the manor of Balterley. Although no evidence is given for that statement, the Thickness family certainly held lands in Balterley and Betley *temp.* Henry III (SHC XII NS 236), and *Rauff Ralph Theken(e)s, gent, at Balturley* is recorded in 1485, 1493 and 1495 (DRO D231M/E210; DRO D231M/E213-4). In 1320 William de Thicknes was married in the oratory of his manor house of Thicknes (SHC VIII NS 135), perhaps an earlier building on the site of Balterley Hall. Spellings for the name include *Thwykenesse* 1271 SHC VI (i) 51, *Thycknes near Auddeleg* 1272-3 SHC IV (i) 189, *Thycknes* 1273 SHC XI NS 243, 1295 SHC VII 26, *Thicknesse* 1282-3 SHC XI NS 248, 1296 ibid.' 35, *Thicknesse* 1296 SHC VII 35, *Thicnisse* 1296 SRO D(W)1082/A/1/1, *Tykeneshe* 1299 SHC XII 28, *Thicknesse* 1302 ibid.' 102, *Tychnes* 1307 SHC XI NS 266, *Theckness* 1309 ibid. 267, *Thycnes* 1312 SHC IX (i) 122, *Thycknes* 1313 SHC X 13, *Thickenes* 1327 SHC XI NS 205, *Thicknes* 1327 SHC VII 201, *Thyckenes, Thickenes* 1332 SHC X (i) 101, *Thiknesse* 1363 SRO D(W)1082/A/1/13, *Thyknes* 1367 SHC VIII NS 40, *Thikkenes* 1387 SHC XIII 202, *Thykkenes* 1419 SHC XVII 70, *Thicknes* 1565 SHC XIII 245; SHC 1925 27, *Thickens* 1595 SRO SD4842/42/48, *Thicknes* 1630 SRO D948/1/1/2/2. The etymology of the name is almost certainly from OE *picce* 'thick' and Mercian OE *nes(s)* 'headland', so 'the wide headland': cf. Amounderness and Furness, Lancashire; Holderness, East Yorkshire; Skegness, Lincolnshire. Balterley Hall lies to the north-west of a conspicuous hill marked by hachuring on the 1833 first edition 1" O.S. map, but it is unclear whether the place gave its name to the family, or the family (which might have originated elsewhere, although no other place called Thickness has been traced) gave its name to the place. The introduction to the printed Betley ParReg suggests that the name sometimes appears as Thickwithies in the early Plea Rolls, but any such spellings are unlikely to point towards a different derivation.

**THIEVES DITCH**   a ditch behind North Walls, Stafford (SJ 9223). *Thevesdych* c.1401 VCH V 94, *pratum de thevesdiche* 1401 StaffAcc, 1462 Ct, *Thevesdich'*, *Thevesdych* 1399 Ipm, *le Thevys dych* 1500 Egerton, *Theeves ditch* c.1610 plan. Presumably 'the ditch frequented by the thieves'.

THORALDESWOD (unlocated, in Alton: SHC 1913 79) *Toraldeswade* 1328 SHC 1913 16, *Thoraldeswod* 1339 ibid.' 79. Þoraldr is an ON personal name (cf. Thorlby, Yorkshire; Torrisholme, Lancashire), so 'Þoraldr's wood'. Or the second element may be OE *(ge)wæd* 'a ford', perhaps across the river Churnet.

THORLEY ACRES   (unlocated, in Norton-in-the-Moors)   *Thorley acres* 1615 BCA MS917/1723. From OE *þorn-leah* 'the thorny *leah*'.

**THORNBURY HALL**   2 miles north-east of Cheadle (SK 0245). *Thornbire* 1203 SHC III (i) 92, *Thornbury* 1250 SHC 1911 123, 1298 SHC VII 48, 1325 SHC X 59, 1482 SHC VI NS (i) 144, *Thornebiri, Thorneburg* 1278 SHC VI (i) 98-9, *Thornbury* 1381 SHC XVII 201, *Thornebury (Hall)* 1609 SHC III NS 52. 'The manor-house or fortification at the thorns'.

**THORNCLIFF**   2 miles north-east of Leek (SK 0158). *Thorn(e)cley, Thorn(e)clay(e)* 1230-2 StCart, *Thorntileg'* 1250x1259 StCart, *Thorenteleye* 1279 SHC 1911 35, *? Thorenteleye* 1279 SHC 1911 35, *Thorntileg* ?13th century VCH VII 233, *Thorneteley, Thornteley* 1476 SHC VI NS (i) 102, *Thorneteley* 1479 Banco, *Thornecley* 1548 PRO SC2/202/65, *Thorncliffe* c.1600 ibid, *Thownecliffe* 1650 SRO DW1761/C/29, *Thorn Cliff* 1695 Morden. The original name (mistranscribed as *Yombele, Yomberley, Yombelega* in SHC NS IX 319: VCH VII 233 fn.28) was evidently from OE *þorn* 'thorny'. Although it is possible that the earliest spellings are mistranscribed, with *t* read as *c*, so 'the thorny *leah*', the 1548 spelling suggests that there may have been variant spellings, or two nearby places, one incorporating OE *cla* 'land in the fork of a river' (the place lies at the junction of two streams), or, as suggested by CDEPN from OE *\*thornett* 'a thorn-tree copse', with OE *leah*. By the end of the 16th century the present generic had been adopted, *cliff* meaning here 'a steep bank or slope', from the deep ravine to the north in which runs the Tittesworth brook. See also Coppedlowe Cloughs.

**THORNES**   2 miles north-east of Aldridge (SK 0703). *Aldethornes* 1209 SHC 1923 277, *le Thoynes* 1343 SRO 3005/2, *Thornes* 1348 SHC XVII 289, 1470 (1801) Shaw II 53, 1532 SHC 4th Series 8 169, *Thorns* 1651 Aldridge church monument. A shortened form of an original 'old thorns'.

**THORNEY HILLS**   2 miles south-west of Marchington (SK 1127). *Thornihul* 1190x1247 SHC 4th Series 80, *Thornhill* 1227 SHC IV 44, *Thornihull* 1255 SHC 1911 125, *Thornyhill* c.1300 SRO D(W)1721/3/14/11, *Thornyhull* 1309 SHC IX (i) 23, *Thornyhul* 1340 SRO D4038/A/5/1, *le Thornyhull* 1353 SRO D4038/A/5/3, *Thornyhylles* 1546 SRO D4038/A/5/7-8, *Thorny hill* 1577 Saxton, *Thorney Hills otherwise Thorne Hill Lanes* 1605 SHC XVIII 53, *Thornie Hills* 1656 SRO D4030/C/6-7, *Thorney Hill* 1663 SHC II (ii) 48. 'The thorny hill'. See also Thorney Lanes.

**THORNEY LANES** a lane that runs from Gorsty Hill (SK 1029) to Hoar Cross (SK 1223), the lower part known as Thorney Lane. *Thornylanes, Thurnylanes* 1602 SHC 1935 459, *Thorney Hills otherwise Thorne Hill Lanes* 1605 SHC XVIII 53, *Thornilanes* c.1666 SRO D1504/1/1, *Thorney Lane* 1671 SRO D786/22/1, *Thorny-lanes* 1686 Plot. An abbreviated form of Thorney Hills Lane: see Thorney Hills. *Thorny Lanes Mill* (1836 O.S.) lay south-west of Newborough (SK 1224).

**THORNHILL, HIGHER & LOWER** 1 mile north-west of Madeley (SJ 7645). *Thornall* 1601 SHC XVI 207, 1608 SHC 1948-9 69, *Thorneall* 1608 ibid.' 89, *Thornhall, Thornall, the Thornalls* 1644 SHC 4th Series I 249, 277. 'The *halh* at the thornbush' is perhaps more likely than 'the hall at the thornbush'. It is uncertain whether *Thornhill, Thornhull, Thornbiri* (1227 SHC IV 59, 227) relate to these places.

**THORNYLEIGH** 1½ miles east of Heaton (SJ 9762). *Thorniliegh, Thornylee* 1538 Dieulacres Inventory, *(le) Thornylegh* c.1539 LRMB, 1539 MinA, *Thorneley* 1535 StarCh, *Thornelie, Thornilee* 1613 QSR, *Thorneley* 1682 Browne, *Thounyleigh* 1692 Leek ParReg, *Thornyleigh* 1842 O.S. From OE *þornig* 'thorny, growing with thorns', with OE *leah*, so 'the thorny *leah*'.

**THORPE CONSTANTINE** Ancient Parish 5 miles north-east of Tamworth (SK 2608; SMR 01181 places a deserted post-Conquest settlement at SK 25860879). *Torp* 1086 DB, *Thorp Costentin* c.1245 Cl, *Thorpe Constantyn* 1318 SHC IX (i) 73, *Thorp* 1395 SHC XV 65. From OE *þrop*, ON *þorp*, meaning in the Danelaw 'secondary settlement, an outlying farmstead or small hamlet dependant on a larger place' hence 'the outlying farm of the Costetin family'. The derivation here is probably from the ON word. Galfrid de Costetin (from Constantine in Normandy) held land in Thorp in 1212 (Fees). See also Gararardesthorp.

**THOR'S CAVE** ½ mile south-west of Wetton (SK 0954). ... *Thurse-house or Thursehole, sometimes call'd Hob-hurst Cave* .. 1686 Plot 172, *Thorshouse (Tor)* 1775 Yates, *Thyrsis's Cavern (Thor's House Cavern)* 1817 Pitt 198, *Thyrsis or Thor's House* 1831 Lewis, *Thor's Cave* 1836 O.S. From OE *thyrs* 'giant, demon': Thor's House is a common name for caves – see Dickins 1947: 9-23. Cf. Thirst House in Chelmoston, Derbyshire (PN Db 75). Hob is a word used for hobgoblins, sprites and elves. Hurst is from OE *hyrst* 'a wood, a copse, a wooded eminence'. Plot 1686: 172 mentions a hollow in the rock called *the Thurse-house* near Peakstones, perhaps to be identified with Rock Farm: see StEnc 447, 584.

**THORSWOOD (HOUSE)** on a 1103' hill 3 miles south-west of Ilam (SK 1147). *Thorswood* 1639 Ellastone ParReg, 1733 SRO D240/D/308, *Thorns Wood* 1836 O.S. See Thor's Cave.

**THREAPWOOD HEAD** 2 miles east of Cheadle (SK 0442). *? le Trepwode* 1266 SHC 1913 317, *Threapwood* 1760 SRO D240/D/111. From OE *þreap* 'dispute, quarrel', hence 'the disputed wood': cf. Threapwood, Cheshire; Threepwood, Cumberland and Yorkshire. Another Threapwood is recorded adjoining Wall Grange, Leek (*Threpwode* 1242 SHC XI 314, 1313 SHC XI 334, *le Therepwode' de la Wal* 1275 VCH VII 208, see also SHC XII NS 75), but the 1242 spelling is from a charter of Hulton Abbey printed in Ward 1843: app. ii which is almost certainly a much later forgery: Tomkinson 1994: 73-102. An unidentified *Threepwood* (n.d.) is recorded in or near Croxton: SRO D240/E(A)2/73. See also Lightwood near Longton.

THREE FARMES (unlocated, in the Mill Meece/Chorlton area) *Three Farmes* 1666 SHC 1921 119.

THREELOWS – see RUMBELOWS (FARM).

THREE MEER STONES (obsolete) ½ mile north-east of Flash (SK 0367). *the Three Sheres* 1533 Bateman, *the Three merestones* 1564 SHC 1938 99, *the Three sheres* 1564 ibid.' 99, *The 3 Stone Mere* 1577 Saxton, *Three Shires' Mear* c.1595 Erdeswick 1844: 476, *The 3 Shyre Mere* 1599 Smith, *ye 3 Shire stone* 1673 Blome, *3. Shire-heads* 1682 Browne, *3 Shire Head or Stones* 1749 Bowen, *Moor Stones* 1775 Yates, *Three Meer Stones* 1842 O.S. The history of the meeting point of the counties of Staffordshire, Derbyshire and Cheshire is confused, partly because of the innacuracy of early maps, and partly by difficulties caused by the word meer, which can mean 'a pool' (OE *mere*, in this case probably Panniers Pool) or 'a boundary' (OE *(ge)mǣre*). From at least the 16th century, and perhaps much earlier, the junction of the three counties seems to have been marked by three stones, presumably one in each county, but the junction (and the stones) appears not to have been static, and many of the names are now difficult to unravel. Much confusion is due to the identification of the boundary meeting point as (variously) the head of the river Dove, or the head of the river Dane, or the head of the river Manifold. Erdeswick 1844: 476 states that the spring forming the head of the river Dove (presumably Dove Head Farm, *Dove Head Spring* 1842 O.S.) formed the meeting point of the boundaries, and the 1599 form appears on Smith's MS map of Staffordshire alongside a thumbnail drawing of three squared stone pillars on a rounded hill, and the same elements appear on Blome's 1673 map of Staffordshire labelled *ye 3 Shire stone*, and on Morden's 1696 map of Shropshire, labelled *The three Shire Stones*, but the same place on Morden's map of Staffordshire (without the drawing) is labelled *3 shire heads*. According to VCH VII 49 Panniers Pool or Three Shire Heads was held in 1533 to be the meeting point, and *Three Shires' Mear* mentioned by Erdeswick, the location of the spring forming the head of the river Dove, was presumably the same place, but (according to VCH VII 49) by the early 17th century three stones (still existing in the early 19th century) on the top of Cheeks Hill were thought to mark the spot, although Pococke 1888-9: 42 refers to *the three shire stones where the Dove rises* in 1750. However, Bowen's map of 1749 appears to place *3 Shire Head or Stones* on Cheeks Hill and *Three Meer Stones* which appears on the first edition 1" O.S. map of 1842 is ½ mile north-east of Flash, at the head of the river Manifold. It would appear that until the 19th century this high and remote moorland was an area of dispute between the parishes of Alstonefield (in Staffordshire) and Hartington (in Derbyshire). In 1804, when the common waste of Hartington (Derbyshire) was enclosed, Panniers Pool at Three Shire Heads (q.v.) was confirmed as the meeting point of the three counties: VCH VII 49. In 1599 a plan was drawn of this locality: PRO MPC 214. See also SHC 1938 99; StEnc 585.

**THREE MILE OAK** West Bromwich (SJ 0289). *Three-mile Oak* 1851 White. An ancient oak which stood near the boundary of West Bromwich and Smethwick, 1 mile south of Sandwell Hall. By the 1830s it had disappeared: VCH XVII 10. It is unclear where the tree was three miles from.

**THREE SHIRE HEADS** 1½ miles north-west of Flash (SK 0068). *the Three Sheres* 1533 Bateman, *the Three sheres at the Dane hed* 1564 SHC 1938 99, *three shire heads* 1686 Plot 110. Where the boundaries of Staffordshire, Derbyshire and Cheshire meet – see Three Meer Stones. Three Shires Bridge here was known as *Galleyford bridge* in 1599: map PRO MPC 214. *The Three sheres* recorded c.1565 (SHC 1938 99) now lies in Derbyshire to the north of Cheeks Hill.

**THREE SHIRES OAK** An ancient tree, cut down in 1904 (VCH XVII 96), which stood in what is now Bearwood, Smethwick, at the junction of Three Shires Oak Road, Thatchers Hill (now Abbey Road), and Love Lane (now Wigorn Road), where Staffordshire adjoined the detached portions of Shropshire and Worcestershire (SO 0286): StEnc 585. *The Oak* 1747 Bowen, *3 Shires Oak* 1834 O.S., *Three Shires Oak* 1895 O.S. For *Three Shires Fields* in Mucklestone, at the junction of Staffordshire, Cheshire and Shropshire, see Foxall 1980: 22.

**THRIFT, THE** a wood 3 miles south-east of Bromsgrove (SO 9866). *atte Frithe* 1275 SR. From OE *(ge)fyrhð*, usually meaning 'the poor woodland': see also Frith. In Tardebigge parish, forming part of Staffordshire from c.1100 until 1266, in Warwickshire until 1844, and since that date in Worcestershire.

**THROSTLE NEST** ½ mile south-east of Bradnop (SJ 0154). *Throske Nest (Wood)* (*sic*) 1725 Okeover T705, *Throstle Nest* 1770 ibid. E5017, *Throstle Nest Farm* 1774 SRO D231M/E5023. The age of the name is unknown, but throstle is the dialect word for a song thrush, from OE *\*þryscele, þrostle*: for a full etymology see Lockwood 1993: 154-5. The 1725 spelling is probably a transcription error. It may be more than coincidence that the place adjoins Birdsgrove Farm (*Birds Grove* 1836 O.S.).

**THROWLEY** 2 miles north-west of Ilam (SK 1152; SMR 02630 places a deserted post-Conquest settlement at SK 11005250). *Treulega* 1185 Burton, *Treule* 1201 P, 1208 FF, *Truleg'* 1227 Ass, *Throelega* c.1240 Okeover, *Trowilegh* 1278 *Antrobus, T(h)rouleg'* 1306 SaltMSS, *Throuleg* 1306 SHC 1921 17, *Throw(e)ley(e)* 1306 GDR, 1367 to 1383 Banco, *Trouleye* 1324 SHC X 53, *Throuley(e)* 1332 SR, 1336 to 1438 Banco, *Throughley* 1343 Erdeswick 1844: 482, *Throghley* 1343 SHC 1921 27, *Throweley* 1400 SHC XV 92, *Throley* 1414 SHC XVII 19, *Throwesley* 1473 AddCh, *Throughley* c.1540 Leland, *Thorley als Throwley* 1565 FF, *Throwley (Park)* 1571 SHC 1931 194, *Trowley* 1614 Stowe, *Throwley* 1682 Browne. The terminal appears to be *leah*. The specific may be from OE *þruh* 'tomb, coffin, grave, conduit, water-pipe'. There are many tumuli in the area (see StEnc 18; 585), and the name may be from a stone cist, or may refer to a box-shaped valley: cf. Throwleigh, Devon, and Througham, Hampshire and Gloucestershire. The conduit meaning in place-names often refers to a deep valley: Throwley lies above the deep valley of the river Manifold. Or possibly associated with the pool to the south of Throwley Hall, an unsusual feature in limestone country, explained by the underlying bedrock here, (Dinantian) Hopedale limestone with an adjacent outcrop of argillaceous Widmerpool formation: SAHS XXXIX 2001 28.

**THURSFIELD** 2 miles south-west of Biddulph (SJ 8654). *Tvrvoldesfeld* 1086 DB, *Thurfredsfeld* 1212 SHC XII NS 36, *Thurinodesfelde* 1217x1227 CEC 393, *Turnesfeld* 1227 SHC XI NS 243, *Thuredesfeld* 1227 ibid.' 240, *Thurmedesfeld* 1227 Ch, *Turnedesfeld* 1236 SHC 1911 391, *Thurfedesfeld* 1240 (1798) Shaw I xxx, *Thuresfield* 1253 Ward 1843: app. iv, *Torefeld* 1273 SHC VI (i) 59, *Thurnesfeld* 1272-3 ibid, *Thurnesfeld* 1279 SHC VI (i) 141, *Thuresfeld* 1293 ibid.' 276, *Tornorsfeld* 1306 SHC 1911 67, *Thurfeld* 1306 SHC VII 165, *? Thoresfeld* 1377-8 JNSFC LIX 1924-5, *Thursfeilde* 1608 SHC 1948-9 113, *Newe Chappell* 1611 Norton-in-the-Moors ParReg, *New Chapel* 1646 Wolstanton ParReg, *Thrusfeild* (*sic*) 1666 SHC 1921 161, *Thursfield als New Chap.* 1747 Bowen. Probably from the ON personal name Þorfreðr: the DB form could be aberrate, but might incorporate the ON personal name Þorvaldr, of which traces may be detected in other spellings. The terminal is OE *feld* 'open space'. Thursfield gave its name to a chapelry, later called Newchapel: VCH IV 56 fn. The name is preserved in Thursfield Lodge.

TILBURY CAMP – see **CASTLECROFT** near Wolverhampton.

**TILED HOUSE** 1 mile east of Kingswinford (SO 9088). *the Tile-house* 1672 Sedgley ParReg, *the Tile-house at Bromley* 1686 Plot 374, *the Tyle House* 1700 WALS DX-240/20, *Tiled House* 1763 SRO 4664/A/1/1/1-33, 1808 Baugh, 1834 O.S. Of uncertain derivation. Plot (1686 374) records that in the later 17th century the place was used for turning Spanish or Swedish iron into steel, suggesting that tiles may have formed some type of fireproofing. The name is preserved in Tiled House Lane. See also Wiggen de Tilehouse. It appears that *the Tild house* (*sic*) recorded in 1671 (Sedgley ParReg), and *tild hous*, recorded in 1697 (Underhill 1941: 101) may have been in Lower Gornal.

**TILLINGTON** 2 miles north-west of Stafford (SJ 9125; SMR 02580 places a deserted Anglo-Saxon settlement at SJ 91302570). *Tillintone* 1086 DB, *Tillinton* 1242, *Titlingeston'* 1236 Fees, *Tillinton'* 1242 Fees, *the Mount of Tilinton* 1277 SHC 1911 170, *Tylintone* 1304 SHC ibid. 61, *Tilynton* 1334 SRO D938/193, *Telenton* 1532 SHC 4th Series 8 80. From the OE personal name Tilli, so 'the *tun* associated with Tilli'. Cf. Tillington, Sussex.

TIMMOR (obsolete, north-east of Fisherwick, not located with certainty, but possibly at SK 178082: in about 1550 Stubby Lea (q.v.) is recorded in Timmor (SHC 1912 194; SHC X NS (i) 125-6; VCH XIV 239, 245-6; see also SRO D661).) Shaw 1798: I 375 records that 'Tymmore ... was on the right side of the road between Whittington and Elford, opposite to Fisherwick Park'. *Timmor* 1086 DB, *Tymor* 1167x1183 Rees 1997: 83, *Timor* 1206 SHC III 137, *Tymor* 1227 SHC 1939 123, *Tynmor* 1256 SHC 1911 127, *Temor* 1263 SHC IV 156, *Tymmor* 1271 SHC 1924 53, 1284 FA, 1289 SHC 1924 360, *Tympmore* 1306 SHC VII 161, *Tymmore* 1367 SHC VIII NS 35, *Tymor* 1373 BCA MS3878/28, *Tinmore* 1374 (1801) Shaw II 204, *Tymore iuxta Ellesford* 1387 DRO D5236/9/12, *Tymover* c.1532 SHC NS X (i) 120, *Tymoner* 1539 SHC 1912 133, *Tymhorne* 1539 SHC IV NS 230, *Tympehorn, Tymhorne* 16th century VCH XIV 239, *Tymore* 1609 BCA MS3810/196. The 16th century spellings show confusion with Tamhorn. See also SHC 1924 53; VCH XIV 239. Although the place probably lay within the mile of the river Tame, the spellings do not suggest a derivation from the river name. The intervocalic consonants may show that the preceding vowel was short, pointing to a possible derivation from OE *teo* 'boundary, boundary line', derived from OE *teon* 'to draw' (cf. Teffont Ewyas & Teffont Magna, Wiltshire; Tyburn, Middlesex), with OE *mor* 'moor', so perhaps 'the moor at the boundary', though what boundary might explain such derivation is unclear, unless the Tame itself. Or possibly from OE *tige* 'goat' (cf. Tyneham, Dorset). Or there may have been a first element with a final consonant, as the 1256 and 1306 forms seem to suggest, but no suggestions can be offered if that is the case. Cf. Tymburhale, which lay in Timmor, and may share the same root in its name. The settlement of Timmor was seemingly in decay by the end of the 16th century, and by the later 18th century the name was used only for a notional manor: VCH XIV 239. For *Tymmorshey* in King's Bromley see Hadley End.

**TINKER'S GREEN** ½ mile south of Wilnecote (SK 2200). *Tinker's Green* 1834 O.S. The surname Tinker has not been traced in the parish, so the place-name may be from encampments made by itinerant tinkers, but the Green element suggests a squatter settlement.

TINKERBOROUGH (obsolete) ½ mile south of Salt (SJ 9526). *Tinker Borough* 1836 O.S, *Tinkerburrow* 1908 Cherry 31, *Tinkerborough* 1922 O.S. The site of a former row of cottages built against and into a sandstoone outcrop, perhaps dating from the 18th century and created for workers in the nearby quarries, and abandoned in the early 20th century. Some of the occupants were by tradition tinkers; the *borough* element was perhaps an ironical addition to reflect the small size of the place, or had the appropriate meaning *burrow*. See also StEnc 586.

**TINKER'S CASTLE** on the Staffordshire-Shropshire border on the crest of the escarpment of Abbot's Castle Hill (SO 8294). *Tinker's Castle* 1886 O.S., 1905 Hackwood 1905 150. WJ August 1908 127 cites a local resident recalling that the house from which the name appears to derive was 'built in the early [eighteen-] forties by a Seisdon gentleman, and I think it was occupied by one of his farm men before John the Tinker'. The house was described in 1903 as a single-storey cottage (WJ 1903 218), and in 1905 as 'a rude [rough] cottage' (Hackwood 1905: 150), later castellated, with a cellar incorporating a rock dwelling (SMR 2690), with the castle element (and castellations) doubtless influenced by the name Abbot's Castle, or merely an ironic name for a humble dwelling.

**TINSELL BROOK** a tributary of the river Dove. OED has a number of meanings for *tinsel*, including 'brushwood for fencing and hedging' and 'sparkling, glittering', so perhaps 'the brook flowing through the brushwood' or 'the sparkling brook', but the stream flows through Tinsell Brook, 1 mile south-west of Stanton (SK 1145), and it seems likely that the stream took its name from the wood, 'the wood where brushwood for fencing and hedging was obtained'. For *Tinsell* (pasture) in Bemersley in 1635, *Tinsill Park* 1775 Yates, see SHC 1910 251.

**TIPPETY GREEN** on the north side of Rowley Regis (SO 9687). *Iberty* 1682 Browne, 1695 Morden, *Ibetty* 1788 Harrison, *Tivity Green* 1775 Yates, *Tipety Green* 1834 White, *Tippity Green* 1834 O.S., *Tipety Green* 1851 White. The derivation of the name is uncertain, but the Green element suggests a squatter settlement.

**TIPTON** Ancient Parish 3 miles west of West Bromwich (SO 9592). *Tibintone* 1086 DB, *Tibynton* 1139 (15th century) EEA 14 92, *Tibinton Tibintuna* 1152 ibid. 96, 1242 Fees, *Dippidon* 1259 SHC 1911 131, *Tybeton* 13th century Duig, *Tybrython* 1355 SHC 1913 161, *Tybynton* 1393 SHC XV 61, *Tybington Schalenwebb* 1444 (1801) Shaw II 229, *Tybton* 1456 SHC NS IV 96, *Typtone* 1461 OSS 1936 41, *Typtoune* 1461 ibid. 42, *Tepton* 1532 SHC 4th Series 8 17, *Typton* 1546 TSAHS 3rd Series VIII (ii) 237, *Tipton* 1587 Sedgley ParReg, *Tynton alias Tibington* c.1692 StSt 11 1999 63. From the OE personal name Tibba, so 'Tibba's *tun*'. The curious form recorded in 1444 is unexplained, unless a corruption of ONFr *calenge* 'challenge, dispute', with OE *wudu* 'a wood': cf. Callingwood. VCH IV 43 fn. suggests that the old name Tibbington is recorded as late as 1872, but the name still appears on modern maps as an area to the north-west of Tipton.

**TIT BROOK** a tributary of the river Dove near Ellastone. *Tipp(Bridge)* 1655 NSJFS 12 1972 124, *Tipp(bridge)* 1671 Ellastone ParReg. The spellings (attached to the bridge crossing the stream on the north-east side of Ellastone) suggest that the name might possibly be from OE *yppe* 'a raised place, a platform', with the *T*- taken from the OE preposition *æt*: the stream runs between high hills. Or perhaps from some particular type of bridge. Cf. Tipalt Burn, Northumberland.

**TITTENSOR** 3½ miles north-west of Stone (SJ 8738). *Titesovre* 1086 DB, *Tichesoura* 1167 SHC I 174, *Tidesovre* 1200 SHC III 68, *Titneshovere* c.1200 SHC VI (i) 8, *Tineshovere, Tinneshore* 1203 SHC III 86, *Tiddesore, Tiddesor'* 1203 Pleas, *Titnihouir* c.1248 SHC 1911 420, *Titneshovere* 1236, *Titnesovere* 1242 Fees, *Tythenesovere* 1293 SHC VI (i) 279, *Tytteneshouere* 1294 SHC 1911 219, *Titnesoure* 1296 SHC 1911 239, *Tyntnesoure* 1351 SHC 1913 146, *Titensouere* 1366 SHC VIII 33, *Tetenshows* 1532 SHC

4th Series 8 134, *Tutensar* 1607 Kip, *Tentenhall otherwise Tentenshale otherwise Tytenshall otherwise Tittensor otherwise Titensore* 1617 SHC 1934 50, *Tittensor* 1682 Browne. '*Titten's slope', from OE *ofer* 'hill, slope, ridge': the place occupies a shelf at the lower end of a long ridge of high ground.

**TITTESWORTH** 2 miles north-east of Leek (SK 9959; SMR 02631 places a deserted post-Conquest settlement at SK 00305850). *Tetesword'* 1203 P, *Tatteswarhle* 1203 SHC III (i) 93, *Thetesworthe* c.1246 Dieul, *Thetiswurthe* 1250 SHC 1911 428, *Thethisurt* c.1250 ibid. 426, *Tetiswurthe* 1250x1259 StCart, *Tetisworth, Tetesworthe* 1274x1279 SHC 1911 430, *Tetesworth(e)* 1302 (p) Ass *et freq.* to 1614 StV, 1635 Leek ParReg, 1686 Plot, 1755 Bowen, *Teseworth* 1477 SHC VI (i) NS 102, *Teysorthe* 1538 StarCh, *Tetysworthe* c.1539 LRMB, *Tettysworthe* 1539 CtAugm, *Teesworth* 1540 *AOMB*, *Tosworth* 1565 FF, *Tedsworth* 1560 Pat, *Tetesworth* 1686 Plot, *Titseworth* 1798 Yates. Possibly from the OE personal name *Tetti, with OE *worþ* or *worþig* 'farm, homestead, enclosure', though doubt must remain about the first element. Upper Tittesworth is found as *Upper Thetiswurthe* 1250-9 StCart, and Lower Tittesworth as *nether Thetesworth* 1240 (1883) Deed Sleigh, *Nether Tetesworth* 1292 (SHC VI (i) 205). Tittesworth reservoir was created in 1858, and greatly enlarged in 1962: VCH VII 235.

**TIVIDALE** 1½ miles north-east of Dudley (SO9790). *Tividale* 1641 Rowley Regis ParReg, *Tivy dale* 1695 ibid, *Tividale* 1717 ibid, 1798 Yates, 1834 O.S. The age of this name is not known, and there is no obvious derivation. Possibly from OE *teafor* 'red pigment', from the colour of the earth in this area: Plot 1686: 121-2, for example, mentions reddish earth at Tipton (see also Shaw 1801: II 85), and Halliwell gives *tiver* to mean 'red ochre', although Hackwood 1915: 24 observes without further explanation that Tividale is said to be an importation – and a corruption – of Teviotdale, recorded c.1420 as Tevidale: TSAS LXXVI 2001 21; Lang 1879: 89. Or possibly from the obscure OE *tifet or *tyfet of unknown meaning, probably to be found in Tivetshall St Margaret, Norfolk. A derivation from the stream-name Teifi, a Welsh name (found for example in Aberteifi, the Welsh name for Cardigan) recorded from the second century, but of obscure origin (Nicolaison *et al* 1970: 68) seems improbable, although the place lies in a stream-valley. *Dale* (from OE *dæl* or ON *dalr* 'a valley') is a relatively rare element in Staffordshire, and may be comparatively recent here. *Dividale* appears on the 1834 1" O.S. map 1 mile south of Wordsley, on the Staffordshire-Worcestershire boundary, but other spellings for the place have not been traced. There are various references to a place called Tivedale or Tevedale in the parish registers of places to the south of Stafford, e.g. *Tivedale* 1686 Coppenhall ParReg, *Tevedale* 1673 Bradley ParReg, 1730, 1776-7 ibid. They may be misreadings of Levedale, but the forms appear over a long period, and a place of this name may have existed in or near Bradley.

**TIXALL** Ancient Parish 4 miles north-east of Stafford (SJ 9722). *Ticheshale* 1086 DB, *Tikeshala* 1166 SHC 1923 297, *Tykesal* early 13th century SRO D938/447, *Tykessal* 1237 SRO D938/599, *Thikeshalle* c.1240 SHC VIII (i) 193, *Tikeshale* 1242 Fees, *Tycsall* 1284 SHC VI (i) 136, *Tykeshale* 1286 SHC V (i) 165, *Tyschale* 1351 SHC 1913 146, *Tyxsall* 1532 SHC 4th Series 8 56, *Tixhaul* c.1540 Leland. The first element is OE *ticce*, a shorter form of OE *ticcen* 'kid, a young goat' (EPNE ii 178), with OE *halh*, here probably in the sense 'small valley', so 'the small valley with the kid'. Cf. Ticknall, Derbyshire.

**TOFT, THE** in Dunston, 4 miles south of Stafford (SJ 9018), *le Tofte* 1460 StaffAcc, *le Toft* 1595 QSR, *Tofte* 1596 SHC 1932 170, *the Toafe* 1598 Ct, *Toft* 1775 Yates, *The Toft* 1834 O.S.; **TOFT FARM** 1 mile south-west of Trentham (SJ 8539), *Toftes* ? 14th century SHC XI 326, *Toftus* 1526 ibid. 314, *Tofte* 1575 Trentham ParReg, *Toftes* 1579 ibid, *Toft* 1663 ibid, *Toft Green* 1757 SHC 1931 91, *The Toft* 1836 O.S.; **TOFTGREEN (FARM)** near the Cheshire Border, 3½ miles north-east of Biddulph (SJ 9064), *Toft Green* 1798 Yates, 1801 Smith; **TOFTHALL** ½ mile east of Heaton (SJ 9562), *? Toft* 1548 PRO SC2/202/65, *Toft Hall* 1775 Yates, *Toft Ho.* 1842 O.S.; **TAFT FARM** on the east side of Bishton (SK 0220), *the Toft* 1295 SRO DW1781/1/19, 1316 (p) WL 106, 13th or 14th century SRO DW1781/1/23, *the Tofthouse* 1675 SRO DW1781/9/33/1-7, *Taft Farm* 1887 O.S. From ON *toft*, originally 'site of a house', later 'land adjoining a house, deserted site'. The change from -*o*- to -*a*- in the last place is noteworthy. *Toftes* 'in Eccleshall manor' is recorded in 1275 (SHC 1913 229, 266), *Toft House* in Leek parish in 1768 (BCA MS917/1263). ,

**TOLL END** 1½ mile north-east of Tipton (SO 9693). *Tole ende* 1596 SHC 1932 228, *Tole end* 1598 SHC 1935 161, 1602 SHC 1935 161, *Tole-end* 1686 Plot 261. The name predates the Turnpike Acts, so possibly from the dialect *toll, tolt* 'a clump of trees' (see EDD; PN Ch V (II) xxii), rather than from ME *toll* 'a tax, a toll', often in the sense 'the market boundary', or 'point beyond which market tolls were not levied for the sale of goods'. More likely is a derivation from a personal name: the Tolle/Tole family of Dippidon (i.e. Tipton) are recorded in 1259 (SHC 1911 131), and frequently in the late 16th century: SHC 1935 14, 112; SHC 1930. The word *end* meant not a terminal point, but simply a place, often a squatter settlement in heathland. See also StEnc 594. *Toles Inn* in Tipton is recorded in 1668: BCA MS3810/98.

**TOLLDISH** 1 milesouth-east of Hixon (SK 0023). *Tolldish* 1833 SRO D240/E(A)2/98, 1836 O.S., 1845 SRO D679/1. Early spellings have not been traced, but presumably from *toll-dish*, 'a dish or bowl of stated dimensions for measuring the toll of grain at a mill' (OED), perhaps used in some topographical sense, or because grain was measured here. Tolldish Hall in Warwickshire and Toddishall in Essex are both said to have a similar topographical derivation: PN Wa 111; PN Ess 446.

**TOMHAY (WOOD)** 1 mile north of Lichfield (SK 1113). *Tom Hay* 1796 SRO D5510/A/1/17/i, 1801 SRO D357/D/10.

**TOMHILL COPPICE** (obsolete) 1 mile west of Wombourne (SO 8392). *Tomhill Coppice* 1890 O.S. The name is to be associated with Tom Lane, which runs to the south of Upper Whittimere Farm.

**TONGUE LANE FARM** 2 miles north-west of Endon (SJ 8954). *Junglane* (*sic*) 1614 Norton-in-the-Moors ParReg, *Tonglane* 1672 ibid, *Tong Lane* 1832 Teesdale, *Tongue Lane* 1836 O.S. If, as seems likely, the 1614 spelling is a mistranscription, perhaps from OE *\*tong, tang* 'forceps', possibly with reference to the parallel Judgfield Lane to the north, which curves away to the north-west, whilst Tongue Lane turns south-west, so forming a forceps-like or tong-like feature, and containing a tongue-like area. Cf. Tong, Shropshire (PN Sa I 293).

**TOOT HILL** in Alton, 4 miles east of Cheadle (SK 0742), *Toothill* 1704 Alton ParReg, 1737 DRO D240/D/86, *Toothill, Toothills* 1754 SRO D240/D/98/101-103; **TOOT HILL** ½ mile south of Hollington (SK 0537), *Toothill* 1890 O.S.; **TOOT HILL** 1 mile south-east of Uttoxeter (SK 1031), *Toot-hill* 1798 Shaw I 34, *Toot Hill* 1836 O.S. From OE *totian* 'to look out, spy', OE *tote* 'look-out place', often applied to isolated and conspicuous hills, hence 'look-out hill' (cf. Tuters Hill; Tutbury). There is a tumulus on Toot Hill, Hollington, recorded in 1750 (SRO D1109/2); Shaw 1798: I 34 mentions '…a remarkable eminence, called Toot-hill, supposed to be a tumulus' at Toot Hill near

Uttoxeter (see JNSFC 3 1965 49). Another Toot Hill at The Old Field (1836 O.S.) on the north side of Uttoxeter (SK 0834) is recorded by Redfern 1865: 352; *Toothill* in Heaton parish is recorded in 1864 SRO D3566/1/28; the field-name *Tothills* is recorded in 1858 (TA) in Onecote, *The Toot* in Lapley in 1838 (TA), and *Totehill* in The Wergs, Tettenhall in 1518 (SRO D593/A/2/16/6; a *Tooters Hill* is recorded at Shugborough (possibly associated with a *burh* or fortification there) where one of the monuments, the Lantern of Demosthenes, was erected c.1764-71: SHC 4th Series VI 103. For references to *Totmoore* (1654) and *Tot more* (1655) in Ettingshall (which lies on high ground on the east side of Cinder Hill: 1834 O.S.) see Hackwood 1898: 92, 95; Underhill 1941: 101; SOT D695/1/9/33-41. See also Gelling 1988: 146-7. Cf. Toot Hill, Nottinghamshire; Toothill, Yorkshire; Tothill, Lincolnshire and Middlesex; Tuttle Hill, Warwickshire.

**TOR** used in north-east Staffordshire and adjoining parts of Derbyshire (and other moorland areas) for 'a pile of rocks, a rocky heap', from OE *torr* 'hill'. The place-name *le Thorres* is recorded in 1291 (SHC 4th Series IV 224) in Wetton, but has not been identified. It may refer to the hill 1 mile south of Ecton known as Sugarloaf which appears as *The Tor* on the first edition 1" O.S. map of 1842. *High Tor* is shown on the west side of Brown Edge on the 1836 1" O.S. map.

**TOTMONSLOW** in Draycott in the Moors, 2 miles south-west of Cheadle (SJ 9939), on the Roman road (Margary number 181) from Stoke on Trent to Rocester. *Tatemaneslav*, *Tamenaslau*, *Tateslau* 1086 DB, *Tatesmannislawa* 1175 P, *Tatemanneslawehundredum* 1185, *Tatemanneslawa* 1187 P, *Tatemaneslawe* 1199 Fees, 1227 Ass, *Thatemanneslowe* 1204 SHC III (i) 92, *Tatemanelawe* 1253 Misc, *Tatemanneslowe* 1262 Pat, *Tatemonnelowe* 1272 Ass, *Tatemoneslowe* 1293 Ass, 1316 FA, *Tatemandeslaw* 1320 SHC VII (ii) 21, *Tatemonlowe* 1327 SR, *Tatmanneslawe* 1327 Pat, *Tammeslowe* 1338 Cl, *Tattemanneslowe* 1356 Fine, *Tatmondeslowe* 1402 Fees, *Tatemondeslowe* 1510 SRO D231M/E215. 'Tatmann's *hlaw* or tumulus', from OE *hlaw* 'burial mound'. A more detailed discussion of the place-name will be found in the Introduction, pages 47-48.

TOTMORE (unlocated, in Ettingshall) *Totmore* 1654 Underhill 1942: 174, *Tot More* 1655 ibid. 78. OE Perhaps from OE *tote* 'look-out place', with OE *mor* 'moorland', so 'the moorland with the look-out place'.

TOUK, TOK (unlocated, possibly near Colton) *Touk* 1241 SHC VII NS 16, *Took* 1283 SHC VI NS (ii) 182, *Touk* 1296 ibid. 247, *Touk, Tok* 1324 ibid. 357, 361. The name may be associated with the family of Tok, recorded in Anslow before 1300: Hardy 1908: 139.

TOWALL (unlocated, possibly near Tean, perhaps a reference to Fole (q.v.)) *Towall* 1269 SHC IV 170.

**TOWER HILL** 2 miles south-east of Great Barr (SO 0592), *Tower hill* 1590 BCA MS/3887/260.SM50, *Towerhill* 1602 SHC 1935 427, *Towrehill* 1604 SHC 1940 179, *Tower hill* 1682 Browne, 1747 Bowen, *Tower Hill* 1775 Yates, 1834 O.S.; TOWER HILL (obsolete) 1 mile west of Biddulph , *Tower Hill Farm* 1729 SRO D997/VIII/1/1, *Tower Hill* 1842 O.S. 'The hill with the tower', or possibly (in the case of Tower Hill near Biddulph), 'tor hill', from OE *torr* 'high rock, rocky peak', and *hyll* 'hill'. Tower Hill near Great Barr may have taken its name from a folly erected as an eye catcher on the slope leading to the former Perry Wood by the Wyrleys of Hamstead Hall (StEnc 595), but *Tourhill* is recorded in 1331 (SHC 1913 25), and may possibly be associated with that place.

**TOWNHOUSE** ½ mile north of Audley (SJ 7951). *Townhouse* 1704 Audley ParReg, *Town House* 1733 SHC 1944 2.

**TOYS, THE** 1 mile north-west of Enville (SO 8087). *Toys* 1833 O.S. The place was in existence by 1496 according to VCH XX 94. Probably from a surname: Ph'o Toye is recorded in this area in 1327 (SHC VII (i) 252) and 1332 (SHC X 129), but the derivation of the surname (which 'occurs frequently in the early registers of Hagley, and the adjoining parishes': MidA II 5 Sept 1883) is not known: DES 452.

**TRENT, RIVER** *Trisantona* 115-17 Tacitus, *Treanta, Treenta* c.730 Bede (II 16; III 24; IV 21), *Trahonnini fluminus* c.800 HB, *Treontan stream* (obl.) c.890 OEBede, 942 ASC, *(on) Trentan* 956 (11th century, S.602), *Trente* 1086 DB, *Taranhon* 12 Taliesin, *Trent, Treant* c.1540 Leland. A British river name *Trisantona*, PrWelsh *Trisantona*, of doubtful meaning but perhaps formed from *tri* 'through, across', and *santon*, a word related to Welsh *hynt* 'road, way', Latin *semita* 'footpath', and Fr se*ntier* 'path'. The earliest reference to the name is in the Annals of Tacitus in the second century (XII 31 as emended by Bradley). The Roman name *Trisantone* had the prefix *tri*, an intensive prefix. This is the same Roman name as for the river Arun, which was originally the Tarrant. The modern spelling of Trent is an abbreviated form of the original. The name perhaps means 'trespasser, intruder', in the sense of a river prone to flooding. See Ekwall 1928: 415-8; Jackson 1953: 502-3, 524-5; Rivet & Smith 1979: 478.

**TRENT VALE** ½ mile north-west of Hanford, (SJ 8643). *Trent Vale* 1836 O.S. Self-explanatory. White 1834: 535 suggests that the place was previously known as *Black Lion*, but both places are mentioned in 1813 (SRO D593/T/1/34), *Black Lion* presumably after an inn of that name.

**TRENTHAM** Ancient Parish 3 miles south of Stoke on Trent (SJ 8641). *Trenham* 1086 DB, *Trentham* c.1145 SHC XI 322, *Tengham, Trentham* 1153 CEC 132-3, *Trentham* 1140x1147 (13th-14th century) EEA 14 40, *Trenteham* 1166x1176 EEA 16 104, *Trentham* 1250 SHC XI 319, 1330 ibid. 326, 1380 ibid. 328, 1487 ibid. 329, 1526 ibid. Probably *'ham* or village on the river Trent', but possibly from OE *ham(m), hom(m)* 'meadow, especially a flat low-lying meadow on a stream, a water meadow; an enclosed plot, a close', since the place lies on low ground on the river Trent. The elements *ham* and *ham(m)* are difficult to distinguish unless early spellings with -*mm*- or -*o*- are available, but it may be noted that all the names incorporating *ham(m)* listed in Ekwall 1936: 214 are in the south of England. John ('Florence') of Worcester (d.1118) records that St. Werberga died at her monastery at *Triccingeham* or *Tricengham*, which was formerly held to be Trentham (Erdeswick 1844: 26; SHC XI 295; SHC 1916 74; 134), but is now identified as Threckingham, Lincolnshire: VCH III 255. No evidence has been traced to support the suggestion (SHC 1909 74) that an earlier name of Trentham was *Trytenham*.

**TRESCOTT** 4 miles west of Wolverhampton (SO 8497). *(æt) Treselcotum* 985 (12th century, S.860), *? Cote* 1086 DB, *Trescote* c.1195 SHC III (i) 221, 1271 SHC V (i) 143, *Trescota* 1200 SHC III (i) 68, *Tressecot* 1259 SHC 1911 132, *Tressecote, Tresshecote* 1332 SHC X (i) 126. The second element is OE *cot* 'a cottage, a hut, a shelter', so '*Tresel* cottage'. For *Tresel* see Trysull and Trysull river. The DB entry for Cote is generally held to refer to this place (SHC 1916 104; VCH IV 45), although Coton (q.v.), south of Wolverhampton, is closer to Bushbury and Tettenhall, the entries for which precede and follow it in DB, and spellings for Trescott incorporate the river name *Tresel* before and after 1086. No other example has been traced where the place is recorded as *Cote*. Trescott was granted in 985 to the monastery of Wolverhampton, but not included within the OE boundary clause of the charter to the church, merely mentioned in the Latin introduction (Hooke 1983: 63-5). Coton lies within the area granted by the 985 charter: ibid. It is quite possible that the DB name is properly to be identified with Coton.

**TRESCOTT GRANGE** 4 miles west of Wolverhampton (S0 8596). *the abbey grange at Trescott* 1271 SHC 4th series XVIII 70, *Trescott Grange* 1563 SRO D260/M/T/7/5, *Trescottgrannge* 1573 WRO 705:349/1296/476894, *Trescote Grange* 1644 SHC 4th Series I 320, *The Grange* 1691 Penn ParReg. William Buffery, lord of Lower Penn, granted land here for a grange of Combe Abbey, Warwickshire, c.1195 (Erdeswick 1844: 364; SHC III 221; SHC V (i) 143; SHC VI NS (ii) 336; StEnc 602), and a convent existed here in 1414 (SRO D3835/4).

**TRIANGLE, THE** 1 mile south of Ellenhall (SJ 8324). It is unclear whether *Trygle*, recorded in 1365 (SHC 1921 31), is to be associated with this place. Evidently from the shape of the land, bounded by a road, a track and a footpath.

**TRICKLEY COPPICE (FARM)** 2 miles west of Drayton Basset (SK 1699). The Coppice straddles the Staffordshire-Warwickshire border; the farm lies within Warwickshire. Duignan 1912 suggests that the place was also known as *Crickley Coppice*.

TRILLEMILL (obsolete, in Orton near Penn: SHC VI NS (ii) 176)) *Trillemulne* 1337 SHC VI NS (ii) 110, *Trille* 1463 ibid. 208, *Trylmyll* 1501 ibid. 251. Perhaps from ME *trill* 'to flow in a slender stream' (Trill and Trull are found as stream-names in south-west England: Ekwall 1928: 409), so here 'the mill on the narrow stream, or the stream called Trill'. However, the river-name is outside the area where the river-name is found, and the same word also had the meaning 'to roll, trundle, to revolve, to rotate, to spin', which may have referred to the water-wheel here. Other Trill Mills have been recorded, including one in Shrewsbury: see PN Sa IV 82; PN O lii, 10.

TROMELOWE – see RUMBELOWS (FARM).

**TROUGHSTONE FARM** 1½ miles north-east of Biddulph (SJ 9059). The place lies on a hill of 1017' which has long been used as a stone quarry, so perhaps 'the place where the stone for troughs was obtained'. It is unclear whether *Thruff Stones*, recorded in 1809 (Kennedy 1980: 51) refers to this place.

**TRUBSHAW** 1 mile north-east of Kidsgrove (SJ 8555). *Trubbeshawe* 1231 SHC XII 5, *Trumpeshawe* 1298 SHC XI NS 252, 1308 ibid. 261, *Trubbeshagh* 1340 SHC XI 107, *Trubbeshawe* 1343 SHC XII (i) 5, *Trobeshawe* 1353 JNSFC LIX 1924-5 49, *Trobeschawe* 1372 ibid. 73, *Trubschawe* (p) 1451 HLS 386, *Trubshawe* 1465 SHC IV NS 138, *Trubshawe* (p) 1489 HLS 558, *Trubsha* 1532 SHC XI 272, *Trubshaw* 1658 Wolstanton ParReg, *Trabshaw* 1775 Yates, *Trubshaw* 1836 O.S. The second element is from OE *scaga* 'thicket, grove, small wood', but the first remains uncertain. Curiously, DES has no entry for this common surname which may derive from this place.

**TRUBSHAW CROSS** in Longport (SJ 8549). *Trubshaw's Cross* 1763 SHC 1934 (ii) 69. From an ancient cross here, the plinth of which was rediscovered in 1949, and on which a new cross based on the Anglo-Saxon Ruthwell Cross in Dumfrieshire was erected in 1977: StEnc 602-3.

TRUMWYN (obsolete) *the bailiwick of Hegghe Cank which is called Trumwyn's baillie* 1375 SHC NS II 72, *Tromwyns bayley* 1600 Oakden 1984: 66. One of the two bailiwicks of Cannock Chase, also known as High Cannock (*Hyghe Cank* c.1357 SHC X NS I 215), from the Trumwyn family, who held Forest office from at least the reign of William the

Conqueror: Erdeswick 1844: 200 fn; VCH V 55; SHC X (ii) 215. The other bailiwick was Puys (i.e. Rugeley): VCH V 55, 59.

TRUSSEWAY (unlocated, a trackway which is said to have run from Clulow Cross in Cheshire to Fould, and via Leek to Leekbrook, Cheddleton, and Basford Green, its course marked by a series of crosses: StEnc 603) *Trussewey* 1229x1232 CEC 385. Perhaps from ME *trusse* (from OFr *trouse*) 'truss, pack, bundle', with OE *weg*, so 'way or track along which packs were carried (?by pack-horse), packway'. CEC 385 suggests the trackway lay in the Heaton/Gun End area; see also SHC IX NS 316. See also Surey.

TRYMPLE (unlocated) *Trymple* 1271 SHC V (i) 148.

**TRYSULL** (pronounced Treesul [triːsəl]) 5 miles south-west of Wolverhampton (SO 8594). *Treslei* 1086 DB, *Tresel* 1176 P, *Treshill* 1204 SHC III (i) 90, *Trisel* 1236 Fees, *Tresell* 1251 Ch, *Tresale* 1283 SHC 1911 185, *Tresseleye* 1293 SHC VI (i) 239, *Tresel* 1295 Misc, *Tresil* 1577 Saxton, *Tryste* 1613 BCA MS3307/Acc1927-020, *Treesle* c.1646 SRO D3449/1, *Treasle* 1686 Plot, *Treosle* 1775 Yates. The name is a back-formation, the place taking its name from the river Trysull (q.v.). See also Smestow; Trescott.

**TRYSULL (RIVER)** since Smestow Brook is the only stream that flows through Trysull, the river name must have been an earlier name for the brook. *(on) Tresel* 985 (12th century, S.860), *Tresel* 996 (for 994) (17th century, S.1380), *Tresel* 13th century Wodehouse, *Tresel water* 1300 For, *Tressul* 1307 Pat, *Trysull brook* 1617 SA 5735/2/27/6, *Treesle* 1690 HRO E12/V1/KY/7. Ekwall 1928: 420 suggests that this unique name is a British river name of similar origins to the river Test, Welsh *tres* 'toil, labour', with a derivative suffix *-el*, as in Welsh *tawel* 'silent', giving a meaning such as 'officious, busy', but that derivation (described by Jackson as 'speculative': Oakden 1984: 22) has been refined and explained as a Celtic river-name meaning 'contentious one, tumultuous one, noisy one', from Early Welsh *tres* 'uproar, commotion', with a British feminine form *\*-ella*, found not only in such Welsh river-names as *Crafnell* ('The Scratcher') in Brecknock, *Crychell* ('The Rusher') in north Powys, and *Llynfell* ('The Smooth One') in Brecknock and Glamorgan, but also in Gaulish river-names such as *Mosella*, the Moselle, running from France into Germany: see StSt 10 1998 77-8; Coates & Breeze 2000: 213-4. If that is correct (and considerable doubt must remain, since as Ekwall 1928: 420 observed, 'the map indicates that [Trysull] cannot be a swift stream…A meaning 'the powerful, strong river would hardly be suitable"), the name may provide evidence for Celtic survival in the area after English occupation in the later 6th century, with the final *a* of the name being lost after borrowing into English. The brook was known as *Smestow* (q.v.) by 1576, but in the late 18th century *Trysull brook* was still being used as an alternative name: VCH XX 185. See also Ashwood; Cocortone; Little Burbrook; Smestow; Trescott.

**TUCK HILL** on the Staffordshire-Shropshire border, 2 miles south-west of Bobbington (SO 7887). *? Tuckenhale* 1292 SHC 1911 204, *Tuckhill* 1541 VCH XX 70, 1670 Claverley ParReg, 1678 VCH XX 70, 1833 O.S., *Tuckehill* 1623 Alveley ParReg, *Tackhill* 1686 Plot, *Tuck Hill* 1775 Yates. Probably from the family named Tuk or Tukke recorded in the area from the later 14th century: VCH XX 70. Cf. Thomas Tukke recorded in 1365 (SHC 4th Series XVIII 215), John Tukke recorded in 1369 (SA 2089/2/2/23); see also SA 5735/2/22/1/21; SRO 3764/45. The surname may be from the Anglo-Scandinavian *\*Tukka*, a pet-form of Þorketill (DES 456). If the 1292 spelling relates to this place, the original *hale* (from OE *halh*), has developed into *hill*. Tuckash lies 1 mile north of Upper Arley, on the Shropshire-Worcestershire border, but the history of the name is not known.

**TUCKLESHOLME (FARM)** 1½ miles south-west of Branston (SK 2119). *Tokkesholme* 1415 Hardy 1908: 80, *Tukkulsholme* 1523 ibid. 199, *Tuckulholme Close* 1531 ibid. 75, *Tukulsholme* 1537 ibid. 210, *Tokilshulme* 1588 SRO DW1734/J1133, *le Tuckelshomeleys* 1591 SHC 1930 174, *Tucklesholm* c.1760 SRO D615/M/7/4, *Trucklesome* 1836 O.S. From the ON personal name Þorkell, with ON *holmr* 'small island, a piece of dry land in a fen, a piece of land partly surrounded by streams', here probably in the latter sense: the place lies on the west side of the river Trent in an angle formed by streams.

**TUENEBROK** (unlocated, possibly near Walsall) *Tuenebrok* 1283 SHC 1911 187, *Twynebrock* 1286 SHC 4th Series XVIII 143. The first element would appear to be the OE preposition *(be)tweon* 'between, amongst', frequently found combined with elements for streams or rivers (VEPN II 93-4), here with OE *broc* 'brook', so here perhaps '(the place) between the streams'.

**TUNSTAL SYTCH** on the east side of Caverswall (SJ 9543). *Tunstalshyche* late 13th century SRO 3764/21[27574]. For the first word see Tunstall. The second is from OE *sic* 'water-course'.

**TUNSTALL** 4 miles north-west of Stoke on Trent (SJ 8651), *Dunstall'* 1162 VCH III 136, *Tunstal* 1212 Fees, *Tonestale* c.1225 SHC XI NS 150, *Tunstall* 1227 Ch, *Tonstal* 1242 Fees, *Dunstall* c.1250 SHC XI 303, *Tonestall* 1278 SHC XI NS 262, *Tonstall* 1280 SHC 1911 172, *Tunstall* 1282 SHC 1914 77, *Tonsthalle* 1356 SHC XII 147, *Dunstall* 1402 SHC XI NS 207; **TUNSTALL** near Adbaston, 1 mile north-west of High Offley (SJ 7727), *Tunestal* 1086 DB, *Tonstal* 1243 SHC 1935 401, *Tunstall* 1267 Ch, *Tunstalle* 1267 For, *Tounskall* 1284-5 SHC 1910 298, *Tonstal* 1293 SHC VI (i) 262, *Tounstall* 1532 SHC 4th Series 8 49, *Tonshell* 1569 SA 279/38, *Tunstall* 1695 Morden. From OE *tun-stall* 'site of the farmstead', *t* and *d* commonly interchanging (cf. Dunstall). The name is a common field-name, often found on the edges of ancient wastes, and perhaps used in OE with the meaning 'abandoned farmhouse': see Duignan 1902: 53. A hermitage of the well at Dunstall near Trentham is recorded in 1162 (VCH III 136), found as *Over, Middle* and *Nether Tunstall* in 1537: ibid.

**TUNSTEAD** ½ mile north-west of Longnor (SK 0765). *? Tunstedes* 1340s SRO D 3272/5/13/5, *Tunstidd(e)*, *Tunstyde* 1600 Alstonefield ParReg, 1609 FF, *Timsteede* 1605 Alstonefield ParReg, *Tunstidd* 1609 SHC III NS 29, *Tunste(a)d* 1775 Yates. From OE *tun-stede* 'the farmstead'.

**TUPPENHURST** ½ mile east of Handsacre (SK 0915). *Tubney* 1682 Browne, 1695 Morden, 1749 Bowen, *Tuppenhurst* 1691 SRO 513/2/19/2, *Tapenhurst* c.1705 SA 513/2/18/18/2, *Tuppenhurst* 1732 SA 513/2/18/21/3, 1771 SRO D3924/1/13, *Tupping Hurst* 1834 O.S., *Tappenhurst* 1834 White. If the earliest spellings can be trusted, possibly from an OE personal name *Tubba, which might be a short form of Tunbeorht, with OE *eg* 'island, piece of dry land in marshland'. Hurst is from OE *hyrst* 'a copse, a wooded hill'.

TURKEYSHALL (obsolete, said to be the name of a mill in Rugeley: StEnc 724) *Turkysall flatt, Turkyshall meade* 1570 Survey, *Turkysall lane* 1671 Ct, *Turkeysill* 1840 TA. The name appears to have originated as one of the eight common meadows of Rugeley (VCH V 159), and from the possessive *-ys* is likely to be from a personal name Turk or similar (cf. Henry Turc recorded in Penkridge in the early 13th century: SRO D260/M/T/5/139), or from someone of the Moslem faith, for which Turk was the general term. Or perhaps from the word *turken* or *turkis*, meaning 'to twist or turn about' from

Old Fr *torquir*, a by-form of *torquer* 'to twist' (see N & Q 1882 6 S v 165), which would be appropriate for a stream-name, or perhaps for a mill. Land called *Tyrkeslake* 'Tyrkes stream', from OE *lœcc* 'a stream', is recorded in Essington in 1521 (Oakden 1984: 55).

**TURKILLE** (unlocated) *Turkille* c.1255 SHC VIII 155.

**TURLS HILL** in Sedgley (SO 9293). *? Terhull* 1273 SHC IX (ii) 28, 1290 Ipm, 1295 SHC 1911 239, *? Turhull* 1333 SHC X 87, *Turleshill* 1580 Sedgley ParReg, *Turles hill* 1585 ibid, *Turls Hill* 16th century ManRolls, *? Turle hills (Moore)* 1654 Roper 1980: 97, *Turles Hill* 1668 SRO DW1871, 1834 O.S. Possibly from the ME preposition *atter*, from OE *at þer* 'at the', so giving '(the place) at the hill'. However, a family named Turle or similar (later Tyrley) is recorded in the area from at least the 15th century (e.g. *Tyrull* c.1472 SHC 1928 180, *Turle* 1498 ibid. 179), and the place may take its name from such name: StEnc 605. See also Turner's Hill near Sedgley.

**TURMANNYSHILL** (unlocated, in Walsall) *Turnamshill* 1513 SRO D593/A/2/20/27, *Turmannyshill* 1531 SRO D593/B/1/26/6/8/8-9. The surname Turnham is recorded: DES 458.

**TURNDITCH (FARM)** ½ mile south-west of Alton (SK 0641). *Turn-Ditch* 1706 Alton ParReg. The absence of early spellings makes any derivation uncertain, but perhaps from OE *þyrne, þorn* 'thorn-bush', so 'the ditch with the thorn-bushes'.

**TURNER'S HILL** an 876' hill, the highest ground in South Staffordshire, between Rowley Regis and Oldbury (SO 9788). *Turner's Hill* 1798 Shaw I 122; 1801 Shaw II 240, 1834 O.S. Presumably from a personal name.

**TURNER'S HILL** 1 mile south-west of Sedgley (SO 9092). *? Tours Hill* 1636 Sedgley ParReg, *? Turshill* 1661 ibid, *Tyreshill* 1690 SRO 327/267/81. Some of these spellings may relate to Turls Hill (q.v.).

**TURNER'S KNIPE** in the valley of the river Churnet, 1½ miles north-east of Consall (SJ 9949). *? Gnytwode* 1271 (1883) Sleigh 80, *? Le Gwypp* 1292 SHC 1911 216, *? Le Gnypp* 1292 SHC I 296, *Gnype* 1394 SHC 1928 276, *Knipe Wood* 1842 O.S., *Turner's Knipe* 1995 O.S. The 13th century spellings are recorded in association with nearby Sharpcliffe, Whitehough and Padwick, and it may be assumed that both relate to the same place or feature: the first may be a transcription error. *Gnype* is recorded in 1394 as an estate in Ipstones (SHC 1928 276). The derivation is from ON *gnípa* 'a steep rock or peak, an overhanging rock in a valley'. *Turner's* is from the family of that name which was associated with the area for many centuries. *Rough Knipe* appears on the 2½" O.S. map about ½ mile south-east of this place, perhaps with a similar derivation, or from Norwegian *knip* 'a narrow place'. Cf. Knipe Wood; Knypersley; Knipe Close and Knipe Scar in West Yorkshire.

**TURNER'S POOL** 1½ miles north-east of Heaton (SJ 9763). *Thurnehurst-pole* c.1540 TNSFC 1932 58-9, *Poole otherwise Turnehurst Poole in Fryth otherwise Leekefryth* 1621 SHC 1934 (ii) 24, *Turner's Pool(e)* 1670 Leek ParReg, *Turners Pool* 1798 Yates. From OE *þyrne* 'thornbush', and OE *hyrst* 'hill, wood, wooded hill', giving 'the pool at the copse of thornbushes'.

**TURNHURST (HALL FARM)** 2 miles east of Kidsgrove (SJ 8654). *Turnhurste* 1539 StEnc 605, *Turnehurst* 1604 JNSFC LXIII 56, *Turnhurst* 1608 SHC 1948-9 100, 1626 Wolstanton ParReg, 1836 O.S, *Tarne Hurst* 1609 Norton-in-the-Moors ParReg, *Turn Hurst* 1798 Yates. Possibly from a personal name: Stephen *Turne* is recorded in 1298 (SHC XI NS 256). *Hurst* is from OE *hyrst* 'hillock, copse, wooded eminence'.

TURTON'S HALL (obsolete) in Wolverhampton (SO 9198). *Turton's Hall* 1718 Wolverhampton CA. The later name of the medieval moated site known as Great Hall or Old Hall, from the Turton family who lived here in the 17th century: see Shaw 1801: II 163. The site is marked by the name Old Hall Street. See also StEnc 241. Great Sugnall Hall at Sugnall may at one time have been known as Turton Hall: StEnc 564.

**TUTBURY** Ancient Parish 4 miles north-west of Burton upon Trent (SK 2128). *Toteberie* 1086 DB, *Tuttesbir', Tutteburie, Tuttesbur'* 1087x1100 SHC 4th Series IV 63, *Stutesberia* 1139x60 FrD, *Totesbery* 1140x50 FrD, *Totesberie, Stutesburiæ* 1141 ibid., *Tuttesburie* c.1150 SHC VII NS 128, *Stutesb', Tutesbir', Tutesbyr', Tuttesbir'* 1152x1158 (13th-15th century) EEA 14 46, *Stuteberia* 1176 SHC I 78, *Stutesb'ia* late 12th century BL Cott. Ch. V, *Tottebury* 1308 WL 64, *Tutesbery* c.1180 SHC I 13, *Tuttebury* 1200 FF, 1293 SHC VI (i) 241, *Tutbury* 1255 SRO D258/27/1/26, *Tuttebyr* 1255 (1798) Shaw I 39, *Tuteburi* 1287 SHC VI (i) 168, *Tuttebur'* 1305 DRO D5236/8/2, *Tutt'* 1318 DRO DD/FJ/8/4/2, *Tutburg* 1384 HLS, *Teutbery* 1564 Mercator, *Tutbery* 1610 Speed, *Tutbury* 1682 Browne. The terminal element is OE *byrig*, dative singular of OE *burg*, with various meanings including 'fortified place, manor house', and (rarely) 'monastery', but the inconsistent spellings make the first element uncertain. The evidence is ambiguous, and points towards two OE personal names, Tutta and *Stut, so 'Tutta's *burg*' or '*Stut's *burg*'. It is unclear from this whether there may have been alternative names for the same place, or the names of two different places in close proximity became confused when the places merged. Possibly one name was applied to a prehistoric earthwork now obscured by the castle on the summit of the hill, and the other to an Anglo-Saxon *burh*, possibly evidenced by a bank and ditch protecting the town centre (Palliser 1976: 146). It seems from the forms beginning with *St-* that the name may have been originally *Stoteberie* or similar, with the loss of the initial *s* perhaps due to Norman influence, as *Snotengaham* became Nottingham (CDEPN 445), but the absence of early spellings with *St-* is curious. The postulated personal name *Stut, perhaps deriving in some way from OE *stut* 'gnat', is not recorded in Searle 1897, but is evidenced in place-names (cf. Stuchbury or Stutsbury, a deserted village in Northamptonshire (*Stoteberie* DB, *Stutesbiria* c.1157, *Stuttesbyri* 1228: Ekwall 1960: 451), and Stuston in Suffolk (*Stutestuna* DB: Ekwall 1960: 452; CDEPN 588)). However, OE *stut* is also found in place-names with the meaning 'a hill, a stumpy hillock', related to ON *stútr* in the sense 'stumpy thing, the butt end of a horn' (EPNE ii 165; cf. Stutton in Suffolk, which lies on a prominent hump or hill overlooking the river Stour: CDEPN 588; also Stowting, Kent). Bosworth-Toller 930 notes the use of OE *stutere* in a reference *On stuteres hylle*, perhaps 'on the hill with the stumpy projection'. Tutbury lies on a very prominent headland with a stumpy summit in a strategic position above the river Dove, and the meaning 'the fortification at the stumpy headland' would fit the Tutbury topography well. It is unlikely that the name incorporates the ON word *stútr*, replacing an OE personal name or element. The far-reaching views from the hill suggest another possible derivation from OE *tot* 'a look-out'. In early times advance warning of the approach of hostile or uninvited forces will have been a matter of considerable importance, particularly at the time of Danish threat, and it seems reasonable to suppose the existence of a network of look-out points. Several names in Staffordshire beginning Tat-, Tet-, To(o)t-, Tut- lie on high ground, in some cases with particularly commanding views, and it is possible that if such names do not indicate a look-out point *per se*, they derive from the occupational name of an individual who undertook look-out duties, so here 'the fortification of the look-out man called Tutta', or simply 'the fortification of the look-out man'. See also Newborough.

**TUTERS HILL** ½ mile west of Pattingham (SO 8199). *Tootershill* 1683 Pattingham ParReg, *Tattershill* 1686 Plot, 1695 Morden, *Tutor's-Hill* 1731 SA 330/14, *Tutershill* 1731 Pattingham ParReg, *tutarshill* 1731 ibid, *Tatters Hill* 1747 Bowen, *Tutores Hill (Meadow)* 1762 SRO D32221, *Tutasall* 1773 Boningale ParReg, *Totterhill* 1775 Yates, *Tuters Hill Bank* 1834 O.S., *Tutor's Hill Bank* 1832 Teesdale. Perhaps from OE *\*tot-ærn hyll* 'the look-out-house hill' (*ærn* often becomes -*er*- in modern forms: see Foxall 1980: 31, 54; Gelling 1988: 147), or OE *\*totere* 'look-out', related to OE *totian* 'to peer, peep, look out', used in place-names for places with far-reaching views, as here, and often indicative of an ancient hill-fort: see VCH XX 173; Gelling 1988: 147. There are no traces or record of any hill-fort here (although the topography is an appropriate location), so the possibility of a derivation from the tutoring or dressing of hemp cannot be ruled out: Hemp Yard is recorded 'at the back of [Pattingham] church', and Hemp Field is also recorded: Brighton 1942: 16-7. The hill is named as *Buchstone Bank* on the 1827 first edition 1" O.S. map of Shropshire. A field-name *Tutors Hill* is recorded on the south-east side of Furnace Grange (SO 8496) in 1839 (TA); *Tutors Hill* is recorded in 1840 south-west of Claverley (TA); and *Totters Bank* at Chesterton (q.v.) near Worfield. See also Toot Hill; Tutnall.

**TUTNALL** 2 miles east of Bromsgrove (SO 9970). *Tothehal* 1086 DB, *Tottenhull* 1262 For, *Totenhull* 1275 Ass, *Totynhyll* 1542 LP, *Toutnell* 1675 Ogilby. Probably 'Tot(t)a's hill', but a derivation from OE *\*tot-ærn hyll* 'the look-out-house hill' (see Tuter's Hill) cannot be ruled out completely: the place is on a hill. In Tardebigge parish, forming part of Staffordshire from c.1100 until 1266, in Warwickshire until 1844, and since that date in Worcestershire.

**TWAMLOW** (unlocated, possibly near Hamstall Ridware) *? Tuaml(ow)* 1214-7 SHC 1913 315, *Twamlow* 1666 SHC 1921 161, *Twamlowe, Twanlowe* (p) 1666 SHC 1923 117. From OE *twǣm hlaw* 'the two mounds or tumuli'. See also Twillow; Twirlow. The first spelling may relate to Twemlow, Cheshire.

**TWELL** – see TYWALL GREEN.

**TWICHILLS** 1 mile south-east of Yoxall (SK 1518). *Twychele* 1341 SHC 1913 84, *Twichels Rough* 1834 O.S. From OE *twicen(e)* 'the fork of a road, cross-roads', later usually (as dialect *twitchel* or similar) 'a narrow passage, a narrow footpath between hedges, a blind alley, a short cut' (EDD): the place lies on a track between Upper Blakenhall and Wychnor Park. *Twichills* meadow is recorded in Pipe in 1608 (SHC 1934 (ii) 43); an unlocated *Twifhel* is recorded in 1307 (SHC XI NS 259); *Twychele* 1341 (SHC 1913 84), evidently *Twichell Brook* (1805 *Survey*) on the boundary of Essington and Bushbury; *(a lane called) Twychull* is recorded in Stretton near Burton upon Trent in 1477 (WSL DD110/104(part); and *Tuechele* is mentioned in 1333 (SHC X (i) 126), possibly associated with *Twychenewey* (and *Quichenewey*) recorded near Baggeridge c.1250 (SHC 1928 13-4). Cf. Twitchill Farm, Hope, Derbyshire (PN Db 120), Twitchell, Nottinghamshire (PN Nt 21).

**TWILLOW** 1 mile north of Bradnop (SK 0155). *Twillow Bottom* 1591, 1613 *Deeds*, *Twillow* 1645 Leek ParReg, *Twillow Heath, Twilloe heath* 1662 ibid, *Treillow* 1694 Okeover T704, *Twillow Heath* 1766 SHC 1931 91, 1837 O.S. Perhaps from OE *twi* 'double, two', with OE *hlaw* 'mound, tumulus', so giving '(the heath with) the two lows or tumuli'. See also Twirlow (Farm); Twamlow.

**TWIRLOW (FARM)** 1 mile west of Milwich (SJ 9532). *? Thirle, Tunlawe ?* c.13th century SHC VI (i) 17, *Turnlyes* 1480 SHC VI (i) NS 129, *(Asshedales Lytell) Turlowe*

1591 SHC 1934 (ii) 18, *? Tuerley (Wood)* 1675 SHC V NS 190, *Turley* 1775 Yates, *Twirlow* 1836 O.S., *Twerlow* 1920 O.S. The spellings are inconsistent, but the first element may be OE *\*turn* 'circular', with OE *hlaw* 'mound, tumulus', so 'the circular burial mound'. See also Twillow; Twamlow.

**TWIST, THE; TWISTGREEN** ½ mile south-west of Butterton (SK 0656). *Twys* (p) 1394 Tooth 134, *Twysse* 1424 ibid, *Twisgreene* 1434 (17th century) Survey, *Twysse* 1506 Ipm, *Twist* 1775 Yates, 1838 O.S., *Twiss* 1851 White. From OE *twist* 'something twisted', denoting in ME 'something (e.g. a hinge) working in two parts; a twig, a branch', and perhaps used topographically for 'fork', perhaps in this case with reference to the Hog Brook which divides into three branches here. Cf. Twisgates, Devon; Twishy, Sussex; Twist, Devon; Twist, Sussex.

**TWO GATES** 1½ miles south-east of Tamworth (SK 2101). *Two Gates* 1770 EnclA. Self-explanatory.

**TWYFORDS** (unlocated, in Yoxall), *? Twyford* c.1331 SRO DW1733/A/2/19, *Twyfords* c.1615 SRO D877/91/1-17; **TYFORD** (unlocated, in Wolverhampton), *Tyford* 1359 SHC XII (i) 156. 'The two fords', from OE *twi* 'two, double'.

**TYKYNGTON** (unlocated) *Tykyngton* 1455 SHC 1912 234.

**TYMBURHALE** (unlocated, in Timmor) *Tymburhale* 1305 (1798) Shaw I 375. Seemingly from OE *timber* 'building, timbered house', with OE *halh*, so 'the *halh* with the building', although the name may be a supposed rationalisation of the name Timmor (q.v.). See also Timmor.

**TYNSALL** (unlocated, in Tardebigge parish) *Tuneslega* 1086 DB, *? Campis de Tunesnosham* 1199 SHC III 39, *Tinneshal(l)* 1230 Ch, *Tuneshale* 1230 Cl, 1244 FF, *T(h)eneshale* 1266 Ch, *Tunsale* 1327 SR (p), *Tynsall Filde* 1535 PN Wo 363. The second element is evidently OE *halh*. The first element is an OE personal name *\*Tynni*: PN Wo 363-4. The place was demesne land of Bordesley Abbey, and was listed under Bromsgrove in DB (PN Wo 363). In Staffordshire from c.1100 until 1266, in Warwickshire until 1844, and since that date in Worcestershire.

**TYRLEY** on the east side of Market Drayton (SJ 6833). *Tirelire* 1086 DB (listed under Shropshire), *Tyrle (Wood)* 1247 Ass, *Thyrlegh* 1248 SHC IV 243, *Tyrlegh* 1256 Eyton 1854-60 IX: 192, *Tireleye* 1267 ibid. 193, *Tyrelegh* 1283 Cl, *Trileg* 1284 Eyton 1854-60 IX 193, *Tyrle, Tyrlegh, Tireleye* Duig. The place lies on the river Tern, and is generally held to have conserved the river name without the *n* (see for example Ekwall 1960: 484; see also Saverley Green), so 'the *leah* on the river Tern'. However, spellings for the name have not been found with *-n-*, which is generally present in spellings for the river-name (Ekwall 1928: 400-1; Oakden 1984: 21), and the derivation may be from OE *þyrre leah* 'the dry *leah*'. The place was formerly in Shropshire. The place has been in Staffordshire since about the ime of Henry I: VCH IV 59 fn.

**TYWALL GREEN** (unlocated, near Hardwick, Pattingham: VCH XX 178) *Tywall Green* 1405 VCH XX 178. Possibly 'the two springs or streams', from Mercian OE *wælle* 'a spring', and (sometimes) 'a stream', with OE *twi* 'two, double'. It seems likely that *Twell*, recorded in 1409 (Brighton 1942: 160) is to be associated with this place. The Green element suggests a squatter settlement.

**UBBERLEY** 2 miles north-east of Fenton (SJ 9146). *Ubberley* c.1300 SHC 1911 441, *Abbiley temp.* Edward III SHC XII NS 33, *Ubbeley* 1586 SHC 1927 128, *Ubley* 1586 ibid. 172, *Vbbeley* 1601 SHC 1934 (ii) 5, *Ubbyley* 1661 SRO D3272/1/14/1/9, *?Bud-*

*Heleigh* 1679 SHC XII NS 212, *Ubberley* 1836 O.S. A curious name. The second element is evidently OE *leah.* The first element is uncertain, but perhaps from an unrecorded personal name. See also SHC XII NS 33.

ULMSMORE (unlocated, in Leekfrith: Sleigh 1883: 71) *? Ulnesmor* 1332 SHC X 107, *Ulmsmore* 1542 (1883) Sleigh 71. Possibly from the OE personal names Ulmar or Ulfmær, with OE *mor* 'moorland, marsh'.

**UNDERHILL** 3 miles north of Wolverhampton (SJ 9302). *Vnderhulle* 1308 SHC 1928 75, Thomas-*under-the-hull,Underhull* (p) 1327 SHC VII (i) 250, *Hunderhulle* 1342 SHC XII 9. Self-explanatory. The place lies to the east and at the foot of Bushbury Hill.

**UPFIELDS (FARM)** on the north-west side of Rugeley (SK 0318). *Upfeld* 1353 Ct, 1548 Survey, *the Upp felde* 1570 ibid, *la Upfeild* 1584 Comm. 'The higher fields'. One of the old open fields of Rugeley.

**UPPER HULME – see HULME, UPPER.**

**UTTOXETER** Ancient Parish 13 miles north-east of Stafford (SK 0933), said to be pronounced 'U-toxeter, Ukseter by middle class, Utchiter by working class': StEnc 616. *Wotocheshede* 1086 DB, *Huttokeshal* 1135x1139 SHC 4th Series IV 71, *Huttoshal* c.1155 ibid. 65, *Uttokishedere* 1175 P, *Utochashadra* c.1180 SHC 4th Series IV 13, *Wittokeshather* 1242 Fees, *Huttokeshagh* 1242 SHC IV 96, *Uittokesather, Huttokesather* 1251 Ch, *Huctekeshall, Huttokeshal* 1261 SRO D938/11-12, *Huctekeshall, Huttokeshal* 1261 SRO D938/11-12, *Hutokeshall* 1261 SHC VIII (i) 178, *Uttoczachere* c.1275 SHC 1921 6, *Ottokhathere* 1275 SHC VI (i) 56, *Hottokesacre* 1292 ibid.' 250, *Uttoxhather* 1306 WL 39, *Hockeshatre* 1306 SHC VII 160, *Ottokeshare, Huttokeshare* 1310 WL 90, *Uttokeshather* 1346 NA DD/FJ/1/186/1, *Ottoxhatre* 1413 SHC XVII 6, *Huttokkeshatre* 1420 ibid. 73, *Uttoxhatr* 1423 BCA MS3878/37, *Uttoxatre, Uttoxhather, Uttoxeshather, Uttockcester, Utcheter* 15th century Duig, *Utcester, Utseter, Uttecester* 16th century Duig, *Uttok Cestre, Vttoxcester* c.1540 Leland v 19, *Utcyter* 1561 SHC XVII 211, *Toksetor* 1562 SHC 1938 111, *Vttoxeter* 1577 Saxton, *Utcetter* 1603 Longleat DU/Box iv/87, *Uttoxater* 1686 Browne. The first element is from an OE personal name *Wittuc, a derivative of Witta, or (more likely) a side-form Wuttuc (Ekwall 1960: 488; see also Redin 1919: 152), possibly found in Quixhill (q.v.), less than five miles from Uttoxeter, suggesting that the same individual may be associated with both places. The place-name *Wattiches æces*, which may incorporate a similar name, is recorded in a charter of Rolleston of 1008 (14th century, S.920; see Hooke 1983: 93-7). The second element of the modern name might suggest Mercian OE *cester*, referring to a Roman camp or fort, but notwithstandin various reports to the contrary, there appears to be no evidence that there was ever a Roman settlement at Uttoxeter, though the modern spelling has doubtless been influenced by genuine well-known *cester* (*-chester*) names. The early forms show that the second element is from OE *hǣddre* 'heather', a word not recorded before the 14th century but found in place-names from at least Domesday, in some cases perhaps from a derivative of *hǣð* such as *hǣðra* or *hǣðor*, literally 'a heath, a tract of uncultivated ground' (cf. *mora de Huttokeshal* recorded c.1137: SHC 4th Series IV 71): it has been observed that the presence of the element in place-names will refer primarily to the quality of the ground, rather than the presence of the plant: see Gelling & Cole 2000: 279. Ekwall 1960: 488 suggests that the second element may have been OE *hǣþærn* 'house on the heath', citing Seasalter (Kent) for the loss of the final *n*. Another possibility is a derivation from ON *heiðr* 'heath', though perhaps less likely in combination with an OE personal name. The *-c-* of the terminal *acre* in the 1292 form is probably to be read as

*-t-*, *c* and *t* being often indistinguishable. Cf. Whiteoxmead, Somerset (*Witochesmede* DB).

VALEMERSHE (unlocated, possibly in Trescott) *Valemershe* 1464 SRO D3835/4.

**VALENTIA WOOD** in Upper Arley. From Viscount Valentia, who held Arley in the 18th century: VCH Wo III 6. In Worcestershire since 1895.

**VALLEY FARM** 1 mile north-west of Dunston (SJ 9118). *Valleys* 1332 SHC X 122. From ME *valeie* 'a valley', an element surprisingly rare in Staffordshire place-names.

VERDON (unlocated, in Ellastone) *Verdon maner in Ellaston* 1391 SHC SHC XI 198; VCH III 228. 'Ellaston was a double manor, one part being held by the Verdons, of Alton, and the other part by the Longfords, of Longford, co, Derby': SHC XI 198; see also SRO 5269. It is unclear whether *Verdium* 1130 (SHC I 4), *Verdon* 1199 (SHC III 38), 1392 (VCH III 228), *Verdun* 1203 (SHC III 106), 1275 (SHC V (i) 117), *Werdoun* 1288 (SHC VI (i) 174), *Verduyn* 1300 (SHC XI 307), *Werdoun* 1288 (SHC VI (i) 174) refer to this place. For the de Verdon family see SHC 1933 (ii) 126-130, 133.

VERDOUN'S MANOR a lost manor in Handacre, so-named by Shaw 1798: I *208, who notes that it is un-named in a grant he mentions. White 1851 states that the manor was granted in 1318 to Sir Robert de Verdon, son of Sir Henry Mavesyn.

**VIATOR'S BRIDGE** 1 mile south-east of Alstonefield (SK 1354). A 16th century narrow stone packhorse bridge which takes its name from one of the characters in a section contributed by Charles Cotton to Isaac Walton's 'Compleat Angler': VCH VII 10. It is likely to be *Alstonefield bridge* recorded in the late 1420s: ibid.

**VICAR'S COPPICE** 3 miles north of Lichfield (SK 1113). *Vicar's Coppice* 1679 LRO D30/10/5/3. The place was held by the Dean and Chapter of Vicar's Choral of Lichfield Cathedral. It is unclear whether *Vikers*, recorded in 1372 (SRO D1734), is to be asociated with this place. *Vikers*, possibly in Loxley, is recorded in 1327 (SHC VII 221), and 1378 (SHC XIII 154), *Vykeres* in 1332 (SHC X 113), and *the Vykeres* in 1414 (SHC XVII 52), but their location is uncertain. The Vicars family is recorded in Longdon in 1549: OSS 1936 41; see also SHC XV 115.

VIGO (obsolete) on the south-east side of Walsall Wood (SK 0502). *Vigo* 1805 VCH XVII 278. The derivation of the name is unknown: there were brickworks here known as Vigo Brickworks, but it is unclear whether they took their name from this place, or vice versa. The battle of Vigo was fought in 1702 when a combined British and Dutch fleet destroyed a Franco-Spanish fleet, and may be commemorated in this place-name.

VOLATIE (unlocated, possibly in Swynnerton) *Volatie* 1281 SRO D4842/17/1.

VOLVANECHAE (unlocated, in Chartley) *Volvanechae* 1276 SHC 1911 165.

WADDELEY (unlocated, possibly near Hollington) *Waddeley* 1509 SHC XII 188. See also Waddune.

**WADDEN (FARM & LANE)** ½ mile south of Gayton (SJ 9827). *Wadden Farm* 1887 O.S. An interesting name, for which early spellings have not been traced (although there is an intriguing reference c.1188x1204 to *Widden*, which seems to have been in the Hopton/Salt area, and evidently to be associated with *Widden Salt* recorded 1913x1956: SRO D938/395; SRO D240/D/46), since if ancient – the farmhouse has the appearance of 18th/19th century date, and may be the building marked on Yates' map of 1775 (though there is no building on Yates' map of 1798) – possibly associated with Waddune (q.v.).

**WADDENSBROOK** a tributary of the river Tame, on the east and south-east of Wednesfield. Early spellings have not been traced (unless to be associated with *Wademorebroke*, recorded in 1524: SRO D593/B/1/26/6/38/14; cf. *Wadmore* 1380 SRO D593/B/1/26/6/36/3), but see Waddune. StEnc 622 suggests that the name is also found as *Waddamsbrook*: cf. the surname Wadams, recorded in 1574 (Sedgley ParReg), 1655 (Roper 1980: 101.

**WADDUNE** (unlocated) *(æt) Waddune* 1002x1004 (11th century, S.906; 11th century, S.1536). The place is mentioned in the Will of Wulfric Spot and from its context may have lain somewhere between Sheen and Eccleshall (Sawyer 1979a: xxxiv), but has also been tentatively identified as Whaddon, Gloucestershire (ibid.; Whitelock 1930: 159; Whitelock 1955: 543 fn.11; see also TSAS 4th Series I XXXIV 1911 16 fn.42). The name *Waddune* is listed in the boundary clause of a grant of 1015 by Æthelred of land at Chilton, Berkshire, to the bishop of Sherborne, confiscated from Wulfgeat, a king's thegn (S. 934), possibly the Wulfgeat who was a kinsman of Wulfrun, who bequeathed 10 hides for his soul 'on account of his offences' (Whitelock 1930: 164-5). Wulfrun was the mother of Wulfric Spot, but it is unlikely that a boundary mark could be identified as the estate referred to in Wulfric's Will. *Wædedun*, also mentioned in Wulfric's will, is not believed to be the same place as *(æt) Waddune*: Sawyer xxxiv. See also Waddeley; Wadden.

**WADE LANE (HOUSE)** in Hill Ridware (SK 0717). *Wadelone* 1393 (1798) Shaw I 200, *Wade-lane-house* 1798 Shaw I 200. Wade Lane runs south from Hill Ridware towards Armitage, and would have crossed the river Trent, suggesting that the lane is named from OE *(ge)wæd* 'ford'. Shaw 1798 I 200, having mentioned 'Wade-lane-house, an ancient stone building', observes that Wade Lane is so-named 'probably from its being occasionally so watery as hardly to be passed dry-shod, the whole village being annoyed by the well-springs near the surface'. That derivation is improbable. Shaw (ibid.) also notes that Henry Wade is recorded in Ridware *temp.* Edward II, but the surname is likely to be from a ford across the Trent.

**WAGGERSLEY** 1 mile north-east of Swynnerton (SJ 8637). *Waggersley* c.1646 SRO D593/B/1/20/22/24, 1836 O.S., 1922 O.S. A derivation from dialect *wagger* 'quaking-grass' (EDD) is unlikely, since it appears to be recorded only in Yorkshire, so possibly from the surname Wagg(e) or similar (see DES 471), with OE *leah*.

**WAGGS BROOK** a brook running south from Knighton Reservoir which forms the boundary between Staffordshire and Shropshire. Early spellings have not been traced, but perhaps from a surname. See also Waggersley; Weags Bridge.

**WAL** (unlocated, possibly near Dilhorne) *Wal* 1327 SHC VII 218, *Wall* 1331 SHC 1913 27, *? Wall* 1586 SHC 1927 135. John atte Walle of Cheadle is recorded in 1430 (SHC XVII 129), and may have been associated with this place. Perhaps from some archaeological feature (see e.g. Chesterhurst), or possibly to be associated with Stone Walls Farm (q.v.), or Wallmires Farm (q.v.).

**WALDRESLOWE, WARDRESLOWE, WALRESLOWE** (unlocated, probably near Normacot, near the river Blithe) *Waldreslowe, Wardreslowe* 1242 SHC XI 314, SRO D593/B/1/23/2/7, *Waldreslowe' in Blitheforde* (see Blythe Bridge) c.1250 ibid. 306, *Walreslowe* 1255 ibid. 315. Perhaps from the OE personal name Waldhere, so 'Waldhere's *hlaw* or tumulus'. It should be noted that the earliest spellings appear in a charter of Hulton Abbey printed in Ward 1843: app. ii which is almost certainly a much later forgery: see Tomkinson 1994: 73-102.

**WALES END** a farmhouse c.1450 (see TSSAHS XXXVIII 1996-7 46) in Wales Lane, Barton-under-Needwood (SK 1818). *Walessend* 1509 StEnc 622. The age of this name is uncertain, but if ancient it would appear to be from OE *walas* 'the Welsh', so giving 'the end of the place where the Welsh lived': cf. Waleswood and Wales, West Yorkshire. It is unclear whether *Walesh*, recorded in 1255 in association with places near Hoarcross (SHC 1911 125), and *Waleshe*, recorded c.1275 (SHC 1928 73), are to be linked to this place. Walesend in the parish of Bedworth, Warwickshire, is recorded from the late 17th century: CA PA 87/34.

**WALFORD** ½ mile south-east of Standon (SJ 8133). *Waleford* 1199 SHC III 166, c.1200 SHC VI (i) 8, 1288 ibid. 183, 1292 SHC 1911 212, 1307 SHC VII 173, 1314 SHC IX (i) 44, 1326 ibid.' 111, 1332 SHC X (i) 95, *Walford* 1327 SHC VII 197, *Walleford* 1361 Salt 1888: 77, *Waford* 1566 SHC XIII 258, *Walford* 1579 SHC XIV 208, 1614 SHC IV NS 65, *Wallfort* 1801 Shaw II 5. The earliest spellings from the Stone Cartulary (SHC VI (i)) consistently have a medial -*e*-, indicating that the meaning here is probably 'the ford of the Britons or of the (British) serfs', from OE *walh* 'Welshman', to be preferred to later spellings which suggest that the first element may be from Mercian OE *wælle* 'a spring', and (sometimes) 'a stream', which would in any event give an improbable 'ford at the spring or stream'. OE *wal* is a form of *weall* 'wall, rampart', but would not be found with the medial -*e*- of these spellings. The place (which does not appear in the corpus of OE place names with *walh* in JEPNS 12 or JEPNS 14) is on the Chatcull Brook. A reference in 1342 (c.1600, D(W)1721/1/1/15) to *Great Walford* and *Little Walford* suggests that there were two places of the name at that date. See also Walton. Cf. Walford, Herefordshire.

**WALK MILL** a common lane, e.g. Walk Mill Bridge 1½ miles south of Cannock, south-west of Bridgtown (SJ 9708), *Walk Mill* 1775 Yates; *le Walkmulne* recorded near Styvington in 1349 (SHC 1919 13); *walk mill* recorded in Rugeley in 1564 (SHC 1938 146); *the Walk Myll* at Kingslow near Pattingham recorded in 1565 (Worfield ParReg); Walk Mill 1 mile east of Bishop's Offley (SJ 7929), *Walkemill* recorded in 1655 (Eccleshall ParReg); *Walkmill* at or near Almington recorded in 1693 (SRO D861/E/5/21); *Walke Mylne* in Leek, recorded in 1548 (PRO SC2/202/65); Walk Mill 2 miles north-west of Stone (SJ 8836); *Walk Mill* recorded 1736 (Swynnerton ParReg, 1836 O.S.). The name is particularly widely recorded on the north of England, from the ME verb *walke* 'to full', with ME *mille* 'mill', so giving 'fulling mill'. All Walk Mills were at some period cloth or fulling mills. Cf. Walkley Bank; Walkern, Hertfordshire; VCH II 216-8.

**WALK, THE** on the Weaver Hills (SK 0947, but see StEnc 622). *The Walk* 1836 O.S. Early spellings have not been traced, but possibly from *walk* 'a tree-lined avenue', or if ancient perhaps from OE *(ge)weorc* 'a work, a building, a structure', but in place-names sometimes with the meaning 'a fortification', perhaps referring to some ancient earthwork (see also Raddlepits), although *(ge)weorc*, probably pronounced something like 'walk', would not be expected to develop into the spelling Walk. OE *wealcere* 'fuller' may also be noted, a name that might be associated with the name Raddlepits. Cf. Walkwood, Worcestershire.

WALKLEY BANK (unlocated) in Forton parish. *Walkley Bank* 1689 Ct. Possibly from from OE *walc-leah* '(place) where fulling or dressing of cloth took place' (cf. Walk Mill), or perhaps from OE *(ge)weorc* 'a work, a building, a structure', but in place-names often with the meaning 'a fortification', so perhaps referring to some ancient earthwork

(*Buryhill*, recorded in 1487 (Oakden 1984: 151) provides evidence for a fortification at Forton), but it would be unsafe to rely on a single spelling as late as 1689.

**WALL**   2 miles south-west of Lichfield (SK 0906). *Walla* 1167 SHC I 47, SHC 1923 295, *Wal* 1201 SHC II 105, *Wal, Walle* 1228 SHC IV 70, *La Wal* 1242 Fees, *le-Wal-extra-Lichefeld* 1272 SHC IV 188, *Wall* 1273 SHC VI (i) 61, *la Val* 1243 SHC 1913 402, *le Wal, le Walle* 13th century Duig, *Walle* 1307 WL 101, *The Wall* 1532 SHC 4th Series 8 184. From OE *weall* 'wall, rampart', frequently found associated with Roman forts and walls. The place is on the site of the Romano-British settlement *Letocetum* (see under Lichfield), where from the 4th century a 9' thick wall backed by a turf rampart and fronted by three ditches enclosed an area of c.5 acres astride Watling Street: Gould 1993a: 1. Part of the wall was still standing in the 18th century, and early antiquaries (e.g. Camden, Plot, Shaw) record substantial masonry remains: Horsley 1732: 40 noted that '[Wall] had its name from certain walls which encompass about two acres of ground called Castle Croft', and Stukeley 1799: II 21 noted parallel walls 12' apart, 3' thick and 12' high forming square rooms. Some of the masonry was still standing in 1817, but had disappeared by 1872, when excavations revealed remains of a wall 150' long and 11' thick: TSSAHS V 1963-4 1. Bede III ch. 21 records the baptism of Peada, the first Christian king of the Mercians, *ad Murum* ('at the wall') c.654, but the context suggests a Northumbria connection, and there is no evidence that *ad Murum* refers to this place, although *Letocetum* may have been the site of the original episcopal seat which transferred to Lichfield in 669, a dozen or so years after its foundation. In that respect it may be noted that bishop Rabel Durdent held land at Wall in 1164: SHC I 50. *le Walles*, recorded in 1342 (SHC 1913 91) perhaps refers to the Roman site at Greensforge (see Greensforge; Wall Heath). A personal name *del Walle of Aston* (near Stone) recorded in 1345 (SHC XII 42) may refer to a Roman marching camp recorded at Aston (SJ 915311; SMR 04606).

**WALL ACRE**   1 mile south-east of Butterton (SJ 0856). *Wall* 1798 Yates. Without earlier spelling it is unclear whether this is from OE *weall* 'wall, rampart, or Mercian OE *wælle* 'a spring', and (sometimes) 'a stream'.

**WALLBRIDGE**   1 mile south-west of Leek (SJ 9755). *Le Wall* 1244 Dieul, 1311 SHC 1911 310, *le Wal* 1252 SHC XI 304, *the great bridge at Wall* 1257 Dieul, *Wal juxta Leek* 1293 StSt I 1996 18, *Walbrugge* 1298 SHC XI NS 250, *Wal juxta Lek* 1293 QW, *Walle* 1415 SHC XVII 56, *Walebridge* 1603 SHC 1940 45. Almost certainly from Mercian OE *wælle* 'a spring', and (sometimes) 'a stream': there is a spring (or springs) here recorded in the 1870s as *Coena's well* (q.v.): VCH VII 203. One spring runs into a pool south-west of Wall Grange Farm. The springs were evidently prolific, in 1908 described as 'The Springs at Wall Grange pouring out over 2 million gallons daily, supplied to the Potteries': VCH I 22.   It should be noted however that a Roman road from Leek is believed to run south from here to Blythe Bridge (although it does not appear on the 1994 O.S. map of Roman Britain), and the name is possibly from OE *weall* 'a rampart of earth or stone', with reference to some wall-like archaeological feature. In 1244 an agreement resolved a dispute between the convents of Dieulacres and Trentham and allowed the former to build a bridge here over the river Churnet (SHC XI 333; VCH VII 99, correcting the date in SHC; VCH III 232). The bridge was reportedly only a wooden horse bridge until the early 18th century: SHC 1913 28. It is unlikely that any of the spellings above relate to *Wall Farm* (1836 O.S.), 2 miles east of Leek (SK 0056), which does not appear on modern maps, and which may have a relatively recent history: the area was the subject of the local 1811 EnclA, and the name was almost certainly adopted after that date. See also Coena's Well.

**WALL FARM** – see **WALLBRIDGE**.

**WALL GRANGE** 1 mile south-west of Leek (SJ 9755; SMR 02632 places a deserted post-Conquest settlement at SJ 97805500). *Valgrange* 1539 MA, *Woolgraung* 1604 SHC 1940 284, *Wall graunge* 1686 Browne. The place was a Grange belonging to Trentham Priory: VCH VII 203; 205-6. The name is recorded in 1510, and may have been in use by 1439: SHC VII 205. See Wallbridge.

**WALL HEATH** 1 mile south of Himley (SO 8889). *Kingswallhuth* 1330 SHC 1913 24, *Wall Heath* 1834 O.S. From OE *weall* 'wall, rampart', generally held to refer to the Roman earthworks (possibly *le Walles* recorded in 1342: SHC 1913 91, and in 1362: SHC 4th Series XVIII 214) at Greensforge on Ashwood Heath, although they lie 1½ miles to the south-west (SHC 1927 185-206; see also Shaw 1801: II 233): Wall Heath, which lies within the former Kinver Forest, was extensive, and covered this area. The first part of the 1330 spelling indicates that together with adjoining Kingswinford it was held by the king. Cf. Ashwood; Greensforge; Knowl Wall.

**WALL HILL** 1 mile north-west of Claverley (SO 7893). *Wallhill* 1628 Claverley ParReg, *Wall* 1695 Morden, 1752 Rocque, *Wall Hill* 1833 O.S. Seemingly from OE *weall* 'wall, rampart': field-names *Oldbury* are recorded here in 1840 (TA), probably from OE *ald-burh* 'old fortification', and perhaps associated with early earthworks or Roman remains: a Roman road from Greensforge to central Wales is said to have run to the south of Wall Hill: see TSAHS LVI 1957-60 237.

**WALLHILL** a steep hill 1 mile north-west of Rushton Spencer (SJ 9363). *Wal(l)hill* 1597 QSR, *Walhill* 1616 SHC VI NS (i) 19, *Wallhill* 1641 Leek ParReg, *Wall hill* 1644 SHC 4th Series I 80, *Warhill* 1842 O.S. Probably from OE *weall* 'wall, rampart', giving 'the hill with the fortification'.

WALLINGE (unlocated) *Wallinge* 1583 SHC III (ii) 9.

**WALLINGTON HEATH** on the west side of Pelsall (SJ 0002). *Wallenton Heathe* 1576 Homeshaw 1955: 50, *Wallington Heath* 1763 ibid., 1791 Bloxwich ParReg, 1818 ibid., 1834 O.S. Wallington Heath is said to be recorded in the 13th century, with an early form *Wale(n)ton* (StEnc 624), in which case the derivation is perhaps from a personal name *Waling, derived from OE *w(e)alh* 'Welshman', with OE *tun*, giving 'the *tun* to be associated with Waling', or from the personal name W(e)alh (with the same derivation) with *-ing*, an OE connective particle linking an appellative or personal name to a final element, so 'the *tun* to be associated with W(e)alh', perhaps the person who is likely to be remembered in the name Walsall (q.v.), 2½ miles to the south-east. The precise location of Wallington, after which the heath was named, has not been traced.

**WALLMIRES FARM** 1 mile south of Werrington (SJ 9446). *Wallmires* 1836 O.S. See also Wal. Perhaps from Mercian OE *wælle* 'a spring', and (sometimes) 'a stream', with ME *mire* 'a piece of wet boggy ground; muddy ground', so 'the areas of boggy ground at the spring'.

**WALLS, THE** an area south of Enville (SO 8286) which is said to have taken its name from a house surrounded by a high brick wall forming a square with sides of 400 yards created by the Earl of Stamford in 1728 on part of Enville Common: VCH XX 93. *Le Walles*, recorded in 1342 (SHC 1913 91), appears to have been in the Kinver area, possibly the Roman camp at Greensforge.

**WALLS, THE** a 22-acre Iron Age quadrangular earthwork on the south side of Chesterton, 7 miles west of Wolverhampton (SO 7896): see VCH Sa I 277; 377-8. *Walls*

1695 Morden, *The Walls* 1719 Gale 1780: 123, *the Walls of Chesterton* 1798 Shaw I 30. From OE *weall* 'wall, rampart' (cf. Chesterton; Wall). *The Walstone* or *The Walston*, recorded several times in Worfield CA (eg. *the walston* 1513) has not been located, but may be 'the stone (or possibly the *tun*) at The Walls'. In Shropshire since the 12th century.

**WALLS, THE** ½ mile west of Audley (SJ 7950). *Well* 1598 Audley ParReg, *Wall* 1695 Morden, *The Wall Estate* 1733 SHC 1944 5, *The Walls* 1833 O.S. From Mercian OE *wælle* 'a spring', and (sometimes) 'a stream', possibly here from a spring forming the headwater of Dean Brook.

**WALSALL** Ancient Parish (pronounced Wawl-sull [wɔːlsəl]) 8 miles north-west of Wolverhampton (SP 0198). *Waleshale* 11th or 12th century Sawyer 1979a: xxxvii, 1163, 1276 SHC VI (i) 80, 1291 SHC 1911 47, *Walveshull* 1125 (1887) Willmore 1887: 32, *Walesala* 1159 (1887) ibid. 32, *Waleshala* 1169 P, *Waleshal'*, *Waleshala* 1190 Ch, *? Wullsile* 1199 SHC III (i) 36, *Waleshal* 1201 Cur, 1212 SHC 1911 385, *Wellyall*, *Wellchall* 1231-2 (1887) Willmore 1887: 32, *Walessall*, *Walessal* 1226x1247 SSAHS XVII 1975-6 67, *Walesale* 1283 SHC 1911 187, *Walsall*, *Wallsall*, *Waleshale* 1309 SSAHS XVII 1975-6 71, *Walshale* 1313 ibid. 317, *Walsale* 1324 ibid. 101, *Walsall Foreyne* 1583 SHC III (ii) 20, *Wallsall fforren* 1637 SHC 1931 118, *Walshall Burrough*, *Walshall Forreine* 1664 SHC II (ii) 64. The word *walh* is sometimes found (in the genitive plural) in place-names, with the meaning 'Welshmen', but here the element occurs in the genitive singular, suggesting a derivation from 'W(e)alh's *halh*', the personal name probably denoting a Briton, or at least a Celtic strain in the population at the date at which the name was coined: see JEPNS 12 1979-80 46 (cf. Walton). The same personal name may be found in the name of Wallington Heath (q.v.), 2½ miles north-west of Walsall. *Foreign* is an expression (probably from Latin *forinseca* 'lying outside the bounds') denoting 'that part of a town which lies outside the borough or parish proper': OED, which cites Walsall as one of the places to illustrate the meaning, referring to Plot 1686: 314, who mentions '... the Town or Burg of Walsall; and in all the Villages and Hamlets belonging thereunto; ... which they call the forraigne ...'. Shaw (1801: II 73) mentions '... the town part, which is called the Borough, and the country part, called the Foreign ....'. Both parts were incorporated as the borough and foreign in 1627: VCH XVII 143. The word *foreign* is also found occasionally with reference to other places in Staffordshire, including Brewood, Burton upon Trent, Dudley, Eccleshall, Kinver and outlying parts of Wolverhampton (see SHC 1931 117; SRO DW1823/6; DR37/2/Box 122/36), and the Bailiff of the foreign of Newcastle ( *'Ballivus forinsecus'* ) is recorded in 1306 (SHC VII 169; see also SHC 1912 223). Although Walsall is not mentioned by name in DB (see VCH XVII 169), it is possible that it was an appendage to Wednesbury, and that the original settlement was in the Townend area to the north-west of Walsall church (TSSAHS XXIV 1982-3 1-7), technically within Cannock Forest (q.v.). That would better explain the element *halh* (perhaps here in the sense 'small valley, hollow') in the name, which would be unlikely to attach to the hilltop settlement, for which there is no evidence pre-dating 1159: ibid. 5. *(æt) Waleshó*, mentioned in Wulfric Spot's Will in the early 11th century (11th century, S.906; 11th century, S.1536), is now believed not to refer to Walsall, but to Wales (*Wales*, *Walis*, *Walise* 1086 DB), a parish and village in South Yorkshire: see Whitelock 1930: 155; Sawyer 1979a: xxvi. For other early spellings of Walsall see Willmore 1887: 32.

WALSALL PARK (obsolete) in Walsall. *Waulleshal Parke* c.1540 Leland. The park is recorded in 1528 (Willmore 1887: 89, which gives bounds), and appears to have been disparked by 1557 (ibid. 97).

**WALSALL WOOD** 2 miles north-west of Aldridge (SK 0403). *bosci de Waleshale* 1199-1200 SHC II (i) 94, *(wood called) Waleshale* 1271 SHC V (i) 153, *Walsall Wode* 1535 SHC 1910 35. The name was in use by 1200 when the wood of Walsall was a distinct part of Cannock Forest (VCH XVII 277-8), with Walsall itself just outside the Forest boundary (JNSFC 8 1968 45). In the 17th century the place seems to have had an alternative name: *Walsall Wood also known as Ediall Spring* is recorded in 1612 (SRO DW1851/8/51), and *Edidle Springe* is recorded in 1665 (SRO D260/M/T/7/6) – see Ediall.

WALSTON(E) – see **THE WALLS**.

WALTER'S RIDVARE – see **NETHERTOWN** near King's Bromley.

**WALTON** 2 miles south-east of Eccleshall (SJ 8627), *Waletone* 1086 DB, *Walton* 1242 Fees, 1285 FA, *Waletone* 1302 SHC VII 101; **WALTON** 1 mile south of Stone (SJ 8933), *Waletone* 1086 DB, *Waletona* c.1130 BM, *Waleton* 1279 SHC VI (i) 91, 1282-3 SHC XI NS 246, 1276 SHC VI (i) 84, *Walton juxta Stanes* 1285 F, *? Wheleton* 1406 SHC XVI 49, *Walshale (sic)* 1534 SRO D1810/f.279; **WALTON-ON-THE-HILL** 3 miles south-east of Stafford (SJ 9520), *Waletone* 1086 DB, *Waletona* c.1166 StCart, *Waleton* 1199 Ass, *Walton super Canoke* 1326 StThomas, *Walton on the Hill* 1812 RegDiss. Probably 'the *tun* of the Britons or of the (British) serfs', from OE *walh* 'Welshman'. The OE personal name Walh and the adjective *welisc* are found in witness lists in Anglo-Saxon charters. It is possible that such names were given by Welsh parents to reflect their origins, or were nicknames of those with Welsh connections (cf. modern 'Taffy'). It has been observed that names incorporating *walh* are often compounded with OE *tun*, *cot* and *worþ*, suggesting that most arose after 700 AD (Gelling 1992: 55), and that they are often near Roman or important Anglo-Saxon settlements, supporting the theory that the English occupied the better land and allowed Britons to occupy areas on the margins of that land: see JEPNS 12 1-53. *Waltunmenemedwe* in Wednesbury is recorded in 1315 (SHC 1911 323), and *Walton* ford in Barton under Needwood in 1279 (ibid. 248, presumably to be associated with *Walton Leas* n.d. SRO D877/2/1-18). It may also be noted that *to wala crofte* appears in 951x955 (15th/16th century, S.579) in the bounds of Oldswinford: JEPNS 12 1-53 41; Hooke 1990 162-7. See also Watford Gap.

**WALTON GRANGE** 2½ miles south-west of Gnosall (SJ 8017). *Waltone* 1086 DB, *Walton* 1291 Tax, 1292 Ch, *Walton Graunge* 1485 SHC VI NS (i) 161, *the Grange* 1666 SHC 1927 70. Since OE *wal* was a form of *weall* 'wall, rampart' (and in Mercian OE *wælle* 'a spring', and sometimes 'a stream'), different Waltons may have different derivations. It is likely that the spellings *Waltone* without a medial *e* (as here) are not derived from *walh*, but denote 'the *tun* with ramparts or by the wall': the significance of the meaning in this case is unclear, but the place lies ½ mile south of the Roman road from Pennocrucium (Water Eaton) to Chester (Margary number 19), and the name may refer to some archaeological feature. The place was given to Buildwas Abbey in Shropshire *temp.* Henry II, and became a grange of the abbey: VCH IV 122. See also Walton Hill, Walton House & Pool, and Walton-on-the-Hill.

**WALTON HILL** 1 mile east of Clent (SO 9479); **WALTON HOUSE & POOL** ½ mile south-east of Clent (SO 9378), *Walton* (p) 1275 SR, 1545 Wills. PN Wo 280 suggests a derivation from *Walh-* (on which see Walton-on the-Hill), but in the absence of early spellings with *Wale-* the names must be considered to have the same derivation as Walton Grange (q.v.). Both Walton Hill and Walton House were in Staffordshire from the early 13th century until 1844, when they became part of Worcestershire.

**WALTONFIELDS** ½ miles south of Walton Grange (SJ 8116). *Waltonfield* 1632 SHC NS V 163. 'The field at Walton (Grange) (q.v.)'.

**WANDON** on Cannock Chase, 1 mile south of Rugeley (SK 0314). *Wondon* 1796 SHC 1925 23, *Wandon* 1834 O.S. Perhaps to be associated with *Wandlengrene*, recorded in the 16th century (SRO D(W)1734/2/5/68. Wandon Lodge is one of the lodges to Beaudesert Hall.

**WANFIELD** 1 mile south-west of Kingstone (SK 0429). *Wanfield* 1667 SRO D7861/14/1, *Windy Hall* 1798 Yates, *Winfield Hall* 1836 O.S. Perhaps from OE *wægen* 'wagon', so 'the field with the wagon'. See also Redfern 1886: 432.

**WARD HILL FARM** 1 mile north-west of Dilhorne (SJ 9544). *Ward Hill* 1836 O.S., *Wardhill* 1889 O.S. Early spellings have not been traced, but perhaps from OE *weard*, *hyll* 'watch or look-out hill', or from OE *\*wearde-hyll* or *\*wearda-hyll* 'beacon hill': the place lies on the north side of an 810' hill.

**WARDLE BARN FARM** on the north-east side of Leek, on the east side of the Churnet valley (SJ 9957). Probably from the Wardle family, noted Methodists in the area in the late 19th century. A derivation from OE *weard*, *hyll* 'watch or look-out hill', or from OE *\*wearde-hyll* or *\*wearda-hyll* 'beacon hill' seems unlikely. The place is shown but un-named on the 1890 6" O.S. map.

**WARDLOW** on the north side of a 1211' hill 1½ miles south-east of Cauldon (SK 0947). *Wardlow* 1775 Yates, 1836 O.S. Early spellings have not been traced, but if the name is ancient, perhaps 'look-out hill or tumulus', from OE *weard* 'watch, look-out', often found in association with words for 'hill', or OE *\*wearde*, *\*wearda* 'beacon', similarly associated with hills, with OE *hlaw* 'mound, tumulus', presumably the tumulus on the summit of the hill.

WARE (obsolete) on the river Churnet, 1½ miles south-west of Ellastone (SK 0942). *Ware* 1836 O.S. From OE *wer*, *wær* 'weir'. *Le Ware* recorded in 1295 (Ipm) appears to have been on the river Mease near Croxall Mill.

**WAR HILL** ½ mile north-west of Maer (SJ 7839). *War Hill* 1833 O.S. Early spellings have not been traced, but perhaps from OE *weard*, *hyll* 'watch or look-out hill', or 'beacon hill' from OE *\*wearde*: the place lies on a pronounced hill.

**WARLEY** a name adopted in 1966 for a new Worcestershire county borough which included Oldbury, Smethwick (except for a part of the Albion area on either side of Halford's Lane) and Rowley Regis, and which in 1974 became part of the metropolitan borough of Sandwell: VCH XVII 10, 120. Warley Wigorn was formed from the manors of Cradley, Warley Wigorn and Witley which remained in Worcestershire when the remainder of the manor of Halesowen was transferred to Shropshire c.1109, and consisted of about 16 isolated areas, many under 5 acres, most of them detached islands of land within Shropshire before 1832: PN Wo 302-3. Early spellings for Warley Wigorn include *Werwelie* 1086 (DB), *Werueslea* 1185, *Weruesley* 1212, *Worveleg*, *Waveleye* 1235-6, *Whernelege*, *Weruele(y)e* 1255, *Worneleigh*, *Wernelegh* 1291, *Worley* 1500, *Wareley* 1763 (all PN Wo 302). PN Wo 302 suggests that the meaning of the name is 'the *leah* associated with a stream name Worf or similar', but Dodgson 1987: 129 prefers a derivation from OE *weofeslege* or *weorfalege* 'cattle clearing', suggesting that the DB form may be explained if the -*f*- in the original form, written in miniscules, was mistaken for the OE letter 'wynn'. *Wigorn* is Latin for Worcester.

WARREN HILL (obsolete). Ogilby 1675 shows *Warren hill* on the road between Sandyford and Stableford Bridge (SJ 8236), evidently from a rabbit warren here.

**WARRILOW BROOK** a tributary of the river Dove. From OE *wearg-hlaw* 'felon-mound or tumulus' or 'wolf-mound or tumulus', with *broc*, hence 'stream of the felon's tumulus' (presumably where they were hanged), or 'stream of the mound of the wolf': cf. Warrilowhead, Cheshire (PN Ch I 129). Fields named *Near Warrilow* and *Far Warrilow* (*Warelowe* 1574 Survey) are recorded in Acton Trussell and Bednall (Oakden 1984: 29), *Lytle Warrelowe* is recorded in Combridge in 1528 (Croxden Chronicle) and as *Litle Warylowe* in 1539 (MA), and the field-name *Warrelow* is recorded in Penkridge: SRO D260/M/T/5/3. Redfern 1865: 351 mentions tumuli in the vicinity of Checkley called *Werlows*. See also Morrilow.

**WARSLOW** 7 miles east of Leek (SK 0858). *Wereslei* 1086 DB, *Werselaw'* 1198 Fees, *Werselone, Werselow(e)* c.1220 SHC 4th Series IV 164, *Verselowe* 1275 SHC V (i) 117, *Werslegh* 1290 SHC VI (i) 197, *Werselowe* 1300 SHC VII 23, *Worselowe* 1302 SHC 1925 97, *Weselowe* 1332 SHC X 116, *Warcelowe* 1477 SHC VI NS (i) 107, *Wars(e)lowe* 1566 Deed, *Warslow(e)* 1592 Survey, *Worseloe* 1604 Leek ParReg, *Warlow* 1658 PCC, *Waslowe* 1666 SHT, *Warnslow* 1839 EnclA, *Warslow or Warnslow* 1851 White. Ekwall 1960: 499 suggests a derivation from OE *weardsetl-hlaw* 'hill or tumulus with a watch-tower' (perhaps connected with a beacon: see PN Wo 253). The place has far-reaching views down the Manifold valley, and the phonological contraction of *weardsetl* and *hlaw* with the double *l* coming together is quite natural, though taking place at an early period. Or from OE *hlaw* with the OE personal name Wǣr or Ware. Cf. Warshill Top Farm, Wassell Wood, Wassel Grove, and West Hills, all in Worcestershire. See also Oakden 1967-8: 32.

**WARSLOW BROOK** a tributary of the river Manifold. From the place of the same name.

**WARSTONE** 3 miles south of Cannock (SJ 9605), *Harstan* 994 (17th century, S.1380), *Horeston* 1300 For, *Wereston* 1428 FA, *Horestones* 1775 Yates; **WAR STONE** in Bobbington parish, at the southern end of Abbot's Castle Hill (q.v.) (SO 8392), *horston'* 1327 SHC VII 252, *Horstone* 1332 SHC X (i) 129, *the Whorestone* 1695 Gibson, *The Whore Stone* 1695 Morden, *the Hoar Stone* 1801 Shaw II 278; **WARSTONES** 3 miles south-west of Wolverhampton (SO 8895), *Whorestone fielde* 1598 SHC 1934 (ii) 11. A common name, from OE *har-stan* 'the grey or boundary stone'. The word *har* meant 'grey, hoary, old, lichen covered', and is believed to have come to mean 'boundary marker': place-names Warstones, Hoarstones and Whorestones often attach to places on ancient boundaries. One of several huge boulders at Warstone is on the boundary between Hilton and Essington, and the War Stone ('...a triangular great stone, standing erect...': Shaw 1801: II 210; see also VCH I 192; StEnc 635) lies on the boundary between Trysull, Bobbington and Swindon, and the former boundary of Staffordshire and Shropshire. It is not mentioned in a perambulation of 1298 (TSAS LXXI 1996 27), suggesting it post-dates the 13th century, but Bertram *Atteharstone* is recorded in 1327 (SHC VII 252). Warstones lies on the Wolverhampton boundary, and is remembered in the names Warstones Road, Drive, Crescent and Gardens. For a boundary-stone north-east of Pattingham giving its name to Whorestone field and Warstone Hill Road, see VCH XX 172; SHC 1934 (ii) 11. *Horestonescnol* is recorded in Sandon c.1300 (SHC

1921 39), *Horeston* is recorded in Little Wyrley in 1395 (SHC VI (ii) 193), and *Hore Stone* is marked north-east of Ipstones on Yates' map of Staffordshire 1798.

**WARTON (GRANGE)** 3 miles north-east of Newport (SJ 7623). *Waverton* 1242 Ch, 1273 SHC VI (i) 54, *Wavertune* 1272 FF, *Warton* 1405 SHC XVI 45, *Wartton* 1675 Weston-under-Lizard ParReg, *Warton* 1686 Plot 212, *Worton* 1749 Bowen, *Warton* 1833 O.S. The root of this name is said to be a conjectural OE word *\*wæfer*, *wæfre* 'restless, wandering, flickering', in the sense of a waving or swaying tree – Duignan 1902: 160 suggests the aspen poplar (*populus tremula*) – hence '*tun* at the swaying tree', but the explanation does not seem entirely convincing. Ekwall 1960: 502 suggests that OE *waver*, well-evidenced in Continental names and seemingly associated with woodland, probably had the meaning 'brushwood' or similar (see also PN Ch V (I:1) xl), and it is possible that meaning may apply to this place. Warton, near Polesworth in Warwickshire (close to the Staffordshire boundary), and Woore in Shropshire (close to the Staffordshire boundary), are both *Wavre* in DB, suggesting the possibility that the names imply ancient debates or disputes as to the jurisdiction in which they lay: Woore, for example, lies on the watershed (PN Sa I 325) which elsewhere defines the Staffordshire boundary: see TSAS 4th Series VI 1916-7 123-6.

WARWICKSFORD, WARWICK FORD (obsolete, where the road between Moseley and Brinsford crosses a minor tributary of the river Penk at SJ 923050: see Mander & Tildesley 1960: 8) *Warewyke Fordesmedue* 1305 SHC 1928 110, *Warrewickford* 1506 ibid. 115, *Warwicke Forde* 1608 ibid. 120. The ford lay on what was the road from Stafford to Warwick: see SHC 1928 96, 108, 115, 120-1.

WARWICKSHIRE MOOR (obsolete, 1 mile north-east of Tamworth (SK 2305)) *Port More or Warwickshire Moor* 1584 Dep, *Portmoor alias Warwickshire Moor* 1780 SRO 5269/6/1. *Port* (from OE *port* 'town, harbour') refers to the town of Tamworth. The place lay in Warwickshire until transferred to Staffordshire in 1965. Staffordshire Moor (q.v.) lay to the north-west of Tamworth: 1834 O.S. See also StEnc 635.

WASH BROOK (unlocated) A tributary of the river Churnet. *Washeye Brook* 1676 Leek ParReg, *Washey Brooke* 1693 ibid. From OE *wæsce* 'a place for washing', with OE *ea* 'river', the latter element having been lost.

**WASH DALE** 1 mile north of Stone (SJ 9036). *Wash Dale* 1922 O.S.

**WASHERWALL** 1 mile west of Cellarhead (SJ 9347). *Washywall* 1705 Stoke on Trent ParReg, *Washer-Wall* 1843 Ward 1843: 529, *Washerwall* 1836 O.S. Perhaps from OE *\*wæsce wælle* 'the spring used for washing', or 'Waessa's spring or stream'. Ward 1843: 529 mentions a very copious spring here sufficient to supply the requirements of Hanley and Shelton.

**WASSAGE COVERT** 1 mile south-east of Seighford (SJ 8823). *Washage Covert* 1890 O.S.

WASSEBROC (unlocated, near Hanchurch) *Wassebroc*? 14th century SHC XI 326. Evidently from OE *wæsse* 'land which floods and drains rapidly': see Alrewas.

**WASTE FARM** 1 mile west of Ellastone (SK 0943), *Wast* 1610 SHC III NS 28, *The Waste* 1643 Ellastone ParReg, 1836 O.S., *Waste* 1798 Yates; WASTE, THE obsolete, on the south side of Barlaston (SJ 9038). *The Waste* 1711 Barlaston ParReg, *Waste* 1798 Yates. A name applied to rough uncultivated ground, but if ancient (and both places lie on high ground, though not the highest in the area) perhaps from OE *weardsetl* 'the hill with the guard-house or watch-house': see Gelling 1988: 146. Cf. Waste Farm near Alsager Bank; Wast Hills, Worcestershire.

**WASTE FARM** on the south side of Alsager Bank (SJ 8048). *? Waste* 1628 Audley ParReg. See Waste Farm near Ellastone.

**WASTEGATE** 1 mile south-west of Draycott in the Moors (SJ 9738). *Waste Gate* 1836 O.S. Early spellings have not been traced, but if ancient perhaps from OE *weardsetl-geat* 'the gap or pass in the hill with the guard-house or watch-house' (see Gelling 1988 146): the place lies in a notch or short valley on the north-west side of a 544' hill and has far-eaching views. Cf. Wast Hills, Worcestershire.

**WASTE WOOD** 1 mile south-west of Kingsley (SK 0046). Early spellings have not been traced, but if ancient possibly from OE *weardsetl* 'guard-house or watch-house': see Gelling 1988: 146. Perhaps to be associated with Robert de *la Warde*, recorded in Kingsley in 1274 (SHC 1911 160). Cf. Wast Hills, Worcestershire.

WATELEG (unlocated, in Penkhull) *Wateleg* 1297 SHC 1911 242. Shaw 1801: II 93 mentions *Wateley* in Bentley near Walsall; see also SHC 1910 198.

**WATER EATON** – see **EATON, WATER.`**

**WATERFALL** Ancient Parish 8 miles south-east of Leek (SK 0851). *Waterfala* c.1116x1127 SHC 1916 225, 248, *Waterfalle, Waterfall* 1191x1194 CEC 261, *Wateraval* c.1200 DRO D258/27/1/6, *Waterfale* 1201 SHC II 104, *Waterfal, Waterfale* 1201 SHC III 70, *Waterfathe* 1228 SHC IV 55, *Faterfal* 1259 SHC VI NS (ii) 44, *Waterfall* 1272 Ass, *Waterfale* 13th and 14th century Duig, *Waterfalle* 1374 SHC XIII 118. From OE *wæter-(ge)feall* 'waterfall'. There is no waterfall here (though a stream falls east to the river Hamps 1 mile to the east), and the name is almost certainly from the place where the river Hamps flows underground to reappear at Ilam, so perhaps with OE *fealle* in the sense 'trap' (cf. OE *\*wulf-fealle* 'wolf-pit, wolf-trap'; *mus-fealle* 'mouse-trap'), giving here 'the water trap'. The c.1200 spelling suggests the second element may have alternated with Norman Fr *aval* 'below', which would be appropriate for this distinctive phenomenon, although an English/French combination would not normally be expected. The name *Wadenesfale*, recorded in 1337 (SHC VI NS (ii) 158), is likely to be mistranscription of the name of this place, but poses the intriguing possibility that the place may also have been known as 'Woden's pit'.

WATER GAPS (obsolete) between Warslow and Upper Elkstone (SK 0719). *Water Gap* 1774 Alstonefield ParReg, *Water Gaps* 1840 O.S.

**WATERHEAD BROOK** a tributary of the river Penk. 'The source or head of a small stream', from OE *wæter-heafod* 'stream head', with *broc*.

**WATERHOUSES** 5 miles south-east of Leek (SK 0850). *Over water house, Overwaterhowse* c.1571 SHC 1931 127, *le Upper or Over Waterhowses* 1580 DRO D2375M/3/1, *Waterhouses Hay* 1586 DRO D2375M/3/5, *Over Waterhouses* 1591 DRO D2375M/54/3/13, *the Over Water Howses* 1612 DRO D2375M/57/1, *Over and Nether Waterhous(es)* 1621 ElizChancP, *the water Houses* 1686 Plot 88, *Water Houses* 1798 Yates. The place is on the river Hamps, but a meaning 'house near the water' seems improbably simplistic, for countless places lie on stream and rivers, yet the name is rare. Since the river Hamps famously disappears underground here, it would be surprising if the place-name did not reflect that phenomenon, so perhaps *house* here meaning 'to receive, to accomodate, to give shelter to; to harbour or lodge' (e.g. cow-house, dove-house, green-house, hen-house, etc.). OE *hus*, from which modern *house* is derived, is associated with the verbal root *hud-*, from *hydan* 'to hide', so perhaps 'the place where the water is housed or hidden, i.e. disappears': it may be noted that the earliest spellings refer to house in the singular.

**WATERHOUSE** 1 mile north-west of Onecote (SK 0355). *Waterhouse* 1775 Yates, 1840 O.S.; **WATERHOUSE** on the east side of Meerbrook (SK 9960) *Waterhouse* 1958 O.S. Perhaps 'the house by the water', but countless houses lie near streams and rivers, and the name, though not rare, is uncommon. Perhaps therefore denoting a house built at least in part over a watercourse: a Clothworkers' Company plan of 1612 shows a building labelled 'a water house' which incorporates a bridge over Fleet Ditch in London: Quiney 2003: 252. The first place lies between two stream junctions, but the second, according to maps, has a watercourse which runs to the property. It is unclear whether *Waterhouse*, recorded in 1638 (Leek ParReg) refers to one of these places.

**WATERINGS** 1 mile south-west of Ilam (SK 1249), *Waterings meadow & pasture* 1631 *Senior, Waterings* 1838 O.S.; WATERINGS (obsolete) to the west of Harlow (SK 1445), *Waterings* 1836 O.S. From *watering* 'a ditch for draining a marsh; a piece of land drained by such a ditch': OED.

WATERLOO (obsolete) on south-east side of Normacot (SJ 9242). *Waterloo, Waterloo Plantation* 1836 O.S. Most places of this name commemorate the battle of 1815, but in this case the name may be associated with *Wardreslowe, Waldreslowe, Walreslowe* (q.v.).

**WATERSLACKS** ½ mile south-east of Butterton (SK 0856). *Water Slack* 1775 Yates, *Waterslacks* 1836 O.S. From ON *slakki* 'a small shallow valley, a hollow in the ground', found as northern dialect *slack* 'a hollow, especially one in a hill-side; a dip in the surface of the ground; a shallow dell; a glade; a pass between hills': EDD. *Water* is probably the nearby river Manifold.

WATERSTOKE (unlocated, probably in or near Tillington) *Walkeres Stoke* late 13th century SRO 3764/114[36347], *Waterstoke* 1655x1676 SRO D1798/662/5A. Perhaps from a personal name rather than OE *wealcere* 'fuller', with OE *stoc* 'a place, a religious place, a secondary settlement'. Presumably associated with *Walkeriston*, recorded in Tillington *temp*. Edward I: Okeover T746.

WATER WINDING (obsolete, at the junction of the rivers Penk and Sow to the north of Baswich (SJ 9422): SHC VIII 134) *Water Wending* 1310, 1355 VCH IV 54 fn.40, 1335 SRO D1734, *Water Winding* 1836 O.S. From ME *wending* 'a turning, a bend in the road', here referring to the river Sow. *Watur Wending*, recorded in a deed supposedly 1176x1182 SHC VIII 90, 134, is from a 14th century forgery: EEA 16 90.

**WATFORD GAP** 1½ miles south of Shenstone (SK 1101). *Walford Gap* 1775 Yates, *Watford Gap* 1826 BCA MS20/315, 1834 O.S. The name is difficult to explain: it predates the construction of the nearby railway, and so unlikely to be a transferred name from the more famous place in Northamptonshire. Early spellings are not available (unless *Watford* 1268 SHC VIII (i) 118, 1341 SHC XI 114, 1390 SHC XV 18, refer to this place), and the 1775 form is likely to be an error, so perhaps (as Watford, Hertfordshire) from OE *wað* 'hunting', giving 'the ford used when hunting'. The word *gap* is ME for 'breach, opening', for example in a wall, fence or road, perhaps in this case with reference to a gap in a perimeter boundary of that part of Sutton Chase known as Colefield or Coldfield, in which case increasing the likelihood that the first element is *wað*. The place, which is on the Staffordshire-Warwickshire border, lies on an ancient road called Blake Street which crosses the Birmingham to Lichfield road at right angles, and may be associated with the *forþ, forda* mentioned in a charter of 957 (12th century, S.574: see Hooke 1983 102-3): OE *(ge)wæd* means 'a ford' (see e.g. Wade Lane; Wade Street, Lichfield, VCH XIV 42), and the name may be a tautological 'ford-ford', with the second element added when the meaning of the first had been forgotten. However, evidence of a water crossing here is lacking, although there a stream-crossing over Blake Street ½ mile south-west, so the *ford* element may perhaps here be translated as 'causeway', and if the 1775 spelling is correct (and the place is shown to the south of the county boundary), a derivation from OE *w(e)alh ford* 'Welsh ford, ford of the Welsh' (the word *Wealh* being used here in the uninflected form) cannot be ruled out completely. Cf. Watford; Walton.

WATHERFELD (unlocated) *Watherfeld* 1227 SHC IV 54.

WAT HILL – see OAT HILL.

WATLANDS (obsolete) on north-east side of Dimsdale (SJ 8448). *? Watteslond* c.1535 SRO D641/1/2/108, *? Wuttonsfonds* 1532 SHC VIII 179, *Watlands* 1615 Ward 1843: xlvii, 1836 O.S., 1877 SOT SD4842/16/1/6.

**WATLING STREET** *Wætlingestræt* 880 Ekwall 1960: 501, *Watling strete* 1070 SHC 1916 302, *Weatlinga-Streate* c.1015 (14th century, S.912), *Watlinge Streete* 1149x1155 (17th century) EEA 14 62, *Watheling-strete* 12th century Duig, *Waltlynggestrete* 1260x1270 SHC 1939 29, *Wattelingestrete* 1300 SHC V (i) 176, *Watlinstrete* 1315 SHC IV (ii) 106, *Watteling strete* 1621 Penkridge ParReg. The Roman name for the road (Margary number 1h, 19) is not known, but a late ninth century text (S.912) refers to *Wætlingstræt*, which is likely to have been the section of the Roman road from London to St. Albans, at that date known as *Wætlingceaster*, meaning 'the Roman fort of the *Wæclingas, a group name (mentioned by Bede as the Væclingas) formed from a compound of an OE personal name *Wacol or Wæcel with the collective suffix *-ingas*. The group or tribe gave their name to what the Romans knew as Verulamium (*Uæclinga cæstir* in Bede), and now St. Albans. Nothing is known of the *Wæclingas or of the reasons which caused their name to be given to the road. In a similar way to Ermine Street and Icknield Way, the name Watling Street is also applied to other ancient roads, for example the Roman road between York and Corbridge, and parts of Roman roads between Manchester and Ribchester and between Ribchester and Poulton le Fylde in Lancashire. Watling Street in Staffordshire was known in the 18th and 19th century (and probably earlier) both colloquially and formally as *The Streetway* (q.v.). See also PN Wa 7-8; CDEPN 656.

**WEAGS BRIDGE** 1 mile south-west of Wetton (SK 1054). *waigh Brig (pasture)* 1631 SRO D593/3/92, *Weags Bridge, Weags Barn* 1836 O.S., *Weegs Bridge and Wood* 1901 SRO D3359/58/7/12. A curious name, for which early spellings are not available, and for which no derivation can be suggested, given the breaking of the vowels. See also Waggs Brook.

**WEATHERWORTH** 1 mile east of Bradnop (SK 0254). *Witherwode* 1223, 1227 *Harl, the wode of Wytherward* 1256 Ch, *Witherswood* c.1270 Loxdale, 1297 SHC 1911 442, *Witherwork, Witherworth* 1696 Leek ParReg. A number of possibilities may explain this name, including OE *wiþer 'a castrated ram, a wether', with *wudu* 'wood' or *weard* 'watch, protection', so giving 'the wood where wether-sheep were kept', or 'the place where a watch was kept on the wether-sheep', or from ON *vithr* 'a wood', with the later addition of the tautalogical 'wood', so 'the wood called *Vithr*'. OE *wiþer* was also an adjective meaning 'against, opposite', and the name may be '(the place) against the wood'.

Walter Wyther is recorded in Ilam in 1319 (SHC X 12), Walt'o Whyther is recorded in Okeover in 1332 (ibid. 114), and Lord William Wyther in 1300 (Okeover T276), which suggests that a family name may have been attached to the wood. It may also be noted that Wither derives from the ON personal name Viðarr, and a derivation from such name is possible here.

**WEAVER HILLS** a collective name for a range of hills south of Cauldon, including The Walk (1,217'), Wredon and Cauldon Hills (q.v.) (SK 0946) mentioned in Plot 1686: 404. *Suth Wevere* 1315 SHC IX (ii) 52, *Whevere* 1315 SHC IX (i) 49, *Wevere in Stanton* 1315 Banco, *? Wavre* 1316 SHC 1911 88. Ekwall 1928: 443-4; 1960: 503 suggested that the river Weaver in Cheshire derived its name from OE *wefer(e)* 'winding or weaving stream' (or from a British river-name, probably identical with Wipper in Germany, derived from the same root as Latin *vibrare*, a derivation first proposed by E. C. Quiggin in TPS 1911-14 99ff). More recent research (see Coates & Breeze 2000: 81-2) suggests that the name may be Celtic, from *gwefr* 'yellowish-brown or amber-coloured', although that meaning, put forward in Johnston 1914: 498, was considered unsatisfactory in Ekwall 1928: 443. There are no major rivers at or near the Weaver Hills (though it may be noted that Shaw 1798: I xii mentions *Vale of Weaver*), but the name may perhaps be from amber-coloured minor streams, or from standing amber-coloured water here: the 1836 1" O.S. map shows Weaver Pools at approximately SK 099471, although no pools appear on the modern map. The denotion of colour may be linked with Raddlepits (q.v.), and Red House (q.v.). *Wevre*, recorded in 1279 (SHC 1913 226) and as *Weaver* in 1640 (Trentham ParReg), was evidently waste between Swynnerton and Newcastle: SHC 1913 226.

**WEAVER'S HILL** a circular eminence of fine sand 1 mile east of Aqualate Mere (SJ 7920). *Big Wiver's Hill* 1855 Robinson 1988: 47.

**WEAVERSLAKE** ½ mile north-west of Yoxall (SK 1319). *Weaver Lake* 1828 SRO Q/SB 1829 Easter, *Weavers Lake* 1836 O.S.

**WEBB STONE** in Bradley near Stafford (SJ 8717). One of three glacial boulders to which various legends have been attached (see VCH IV 76; StEnc 637), but probably named after the Webb family, local farmers whose land it once marked: StEnc 637.

**WEDGES MILL** 1½ miles south of Cannock (SJ 9609). *Wedges Mill* 1711 SHC 1934 (ii) 81, 1754 VCH V 61, *Wedges Mills* 1775 Yates. From the river Wedge.

**WEDGWOOD** 1 mile north of Tunstall (SJ 8753). *Wegewode* 1278 SHC XI NS 253, *Wegeswod* 1307 ibid. 255, *Weggewode* 1307 ibid.' 262, 1370 Ward 1843: 198, *Wegwode* 1327 SHC VII 206, *Weggevode* 1456 SHC IV NS 96, 1515 SHC XVI 49, *Wedgewood* 1509 SHC 1923 320, *Wygewode* c.1537 SHC X NS I 139, *Wedgewoode* 1595 Norton-in-the-Moors ParReg, *Wedgwood* 1836 O.S. Seemingly from OE *wecg-wudu* 'the wedge-shaped wood', and the place from which the name of the famous pottery manufacturer originated. *Weggeslowe*, recorded in 1332 (SHC X (i) 95) in this area, suggests a personal name (Wege is recorded in DB), with OE *hlaw* 'mound, tumulus', but there is no genitive *-s* in the spellings for Wedgwood.

**WEDNESBURY** Ancient Parish 5½ miles south-east of Wolverhampton (SO 9894). *Wadnesberie* 1086 DB, *Wodnesbyri* 11th or 12th century Sawyer 1979a: xxxvii, *Wodnesberi* 1164x1169 SHC II 39, 46, 55, *Wadesburi* 1166 SHC I 48, *Wadesb'i* 1166 SHC 1923 295, *Wodnesbi* 1167 SHC II 52, *Wodnesberia* 1169x1209 SHC I, *Wodnesbia* 1177 SHC I 88, *Wodnesb'ia* 1182 SHC II (i) 106, *Wodnesburia* 1184x1186 ibid. 119, *Wotnesbiri* 1226 SHC IV 38, *Wednesbiri* 1227 ibid. 62, *Wonnesbury* 1255 (1798) Shaw I xxvii, *Wodnesbyri* 1271 SHC V (i) 154, *Wodnesbery* 1280 SHC VI (i) 102, *Wodnesburi* 1280 ibid. 116, *Wodenesburi* 1286 SHC V (ii) 165, *Wadnesbury* 1288 ibid.' 177, *Wodnesburi* c.1325 SHC I 168, *Wednesbury* 1327 SHC VII 229, *Wynnesbury* 1397 SHC XVI 31, *Wenesbury* 1405 ibid.' 47, *Wodesbury* 1406 BCA MS3279/351312, *Weddysbere* 1532 SHC 4th Series 8 83, *Weddisburie* 1564 ESRO GLY/1403, *Weddisboroughe* 1566 SHC 1931 200, *Weddsborrow* 1695 Morden. From OE Woden, a heathen German god, corresponding to Oðinn and nicknamed Grimm, with OE *burh* 'fortification', so 'Woden's fortification', implying that the fortification was associated with or protected by the pagan god or dedicated to him. The early spellings are not unequivocal, but the later spellings make it most unlikely that the first element is OE *beorg*, a word which has been shown to have the particularly precise definition 'rounded hill, tumulus': see Gelling & Cole 2000: 145. Cf. Wednesfield; Tettenhall; Weeford. A more detailed discussion of the name and place will be found in the Introduction, pages 26-36.

**WEDNESBURY OLD PARK** (obsolete) A park is recorded in 1484, but not after the 17th century: Ede 1962: 109.

**WEDNESFIELD** 2 miles north-east of Wolverhampton (SJ 9400). *Vuodnesfelda* c.975 (11th century) Campbell 1962: 53, *Wodnesfeld, Wodnesfelde* 985 (12th century, S.860), *Wodnesfeld* 996 (for 994) (17th century, S.1380), 1227 Ass, *Wodnesfelde* 1086 DB, *Wodnesfeld* 1227 SHC IV 69, *Wedingfeld* 1248 ibid. 111, *Wudesfeud* 1250 ibid. 120, *Wednesfeld* 1251 Ch, *Wonnesfeud* 1255 (1798) Shaw I xxvii, *Wodenesfeud* 1300 SHC V (i) 178, *Wedffeld* 1532 SHC 4th Series 8 159, *Weddesfeld* 1538 SHC 1912 107. 'Woden's *feld* or open land'. A more detailed discussion of the name and place will be found in the Introduction, pages 26-36.

**WEEFORD** on Watling Street, 4 miles south-east of Lichfield (SK 1403). *Weforde* 1086 DB, *Weford* 1200 P, 1242 Fees, 1293 SHC VI (i) 290, *Wyford* 1227 SHC IV 66, 1354 SHC XII (i) 128, *Weoford* 1291 Tax, *Weyford* c.1360 SHC 1913 321, *Wifford* c.1540 Leland ii 103, *Wefurde* 1578 BCA MS3375/430071. The first element is generally held to be from OE *wig, weoh* 'an idol', and perhaps 'holy place, shrine'. The name (also probably found in Wyfordby, Leicestershire) is likely to be 'ford by the heathen temple'. For a more detailed discussion of this name see the Introduction, page 29. For Weeford Park see TSSAHS XXX 1988-9 49-50. See also Freeford.

**WEEPING CROSS** 2 miles south-east of Stafford (SJ 9421). *Weeping Cross* 1668 ParReg, 1686 Plot, 1695 Morden, 1719 Baswich ParReg, 1775 Yates. A wooden cross (perhaps a boundary marker) stood here in the reign of Edward VI where penitents offered their devotions: VCH V 2. The place may have been connected with the leper house of Radford: see SHC VIII (ii) 32, 114; VCH III 289-90. Pennant 1782: 78 says: 'After leaving the town [Stafford] I crossed the Wolverhampton Navigation at Radford Bridge ... A little further is Weeping Cross; so stiled from its vicinity to the antient place of execution'. Pennant's derivation is doubtless apocryphal, but folk-memory tells of criminals bound to hurdles and dragged to Weeping Cross before execution near the top of Radford Bank (see OSS 1932 44). No evidence has been found to support the tradition, but *Great Silkmore or Gallows Field*, recorded in 1617 (SRO Drakeford papers 63) and the 19th century field-names *Gallows Flat* and *Gallows Leasow* north of the Lichfield Road at Queensville (VCH VI 228) indicates the site of gallows. The earliest example of this name, traced on the outskirts of five other towns (Banbury, Bury St. Edmunds, Ludlow, Salisbury and Shrewsbury), may be *Wepincros*, recorded in Much Wenlock in 1284: PN Sa III 262-3. A proverbial phrase 'To come home by Weeping Cross' (first recorded in 1579) meant to suffer grievous disappointment or failure (OED).

Similar sayings were 'to make our prayers at whining crosse' (1602), and 'to come home by broken cross' (1662): OED. See also Hackwood 1924: 87; JNSFC LXXII 1937-8 48; PN Sa III xiv-xv, 262-3; CDEPN 659.

WELDE (unlocated, perhaps near Tutbury: SHC 1912 222) *Welde* (undated) SHC 1912 222.

WELDFORDBRIGGE (unlocated, in or near Penkhull) *Wolfordbridge* 1387 Pape 1928: 119, *Welfordbrigg* 1422 SHC 1912 219, *Weldfordbrigge* 1428 Pape 1928: 191. Perhaps from OE *wulf* 'wolf', or the OE personal name Ulf or Wulf, with OE *ford* 'a ford', to which was later added *brycg* 'bridge'.

**WELLINGTON** in Hanley (SJ 8847). An ecclesiastical parish created in 1845, perhaps named after the Duke of Wellington, and now known as Hanley St. Luke: Youngs 1991: 428.

**WELLINGTON FARM** 2 miles east of Leek (SK 0256). *Wellington* 1834 White. Perhaps named after the Duke of Wellington.

WELSH HARP (obsolete, a former coaching inn on the Old Chester Road at Stonall (SK 0603)) *Welch-harp* 1732 Penkridge ParReg.

**WELSH HOUSE** ½ mile west of Harborne (SO 0184). Said to have been named after an owner called Welch: VCH Wa VII 23.

**WEMBLETON BROOK** one name of the the tributary of Bloredale Brook on which the battle of Blore Heath is said to have been fought. *Wambrim(e)brok* 13th century SHC 1945-6 7; 26; 30, *Wemberton Brook* late 15th century ibid, *Wembleton Brook* 1713 ibid. The stream appears to have been known by various names, including *Hemphill Brook, Stow Brook, Sow Brook,* and *Tern Brook*: StEnc 280, 646. *Wembleton* seems to incorporate OE *wamb* 'womb', used in a topographical sense 'womb, belly', perhaps referring to a hollow or pool, possibly Daisy Lake, or a bulge-like topographical feature.

WENFORDEHEUED (unlocated, near Cheadle: SHC 1923 39) *Wenfordeheued* c.1282 SHC 1923 39.

WENWES, WENWE (unlocated, in Keele). *Wewes* ? 1281 SOT SD4842/17/1, *Wenwes, Wenwe* 1307 SHC VII (i) 181. Cf. *Wayes field, Waysfeild* recorded in Keele in 1565: SHC 1938 38-9.

**WENTLOW** on the north side of Upper Tean (SK 0040). *Wentlow* 1685 SRO D5476/A/2/1. The name is remembered in Wentlows Road and Wentlows Avenue. Halliwell 1850: 923 gives various meanings for *went*, including 'a passage; a furlong of land; the teasel'. The second element is OE *hlaw* 'a mound, a tumulus'.

**WERBURGH'S WOOD** 2 miles east of Madeley (SJ 8044). *Warbow Wood* 1833 O.S. Although earlier forms have not been traced, the place is possibly associated with *wilburge wege* 'Wilburh's way' mentioned in a charter of the bounds of Madeley of 975 AD (17th century, S.801; SHC XII NS 202). It may be noted that in 1232 Henry III granted rent to the church of St Werburgh in Chester from the town of Newcastle-under-Lyme: SHC 1939 114. The name Werburge is also recorded in 1352 in association with Overton and Gillow: SHC XII NS 16, and in 1514 a woman called Werburga Whitall of Chesterton is recorded (Ct., JNSFC LX 1925-6 41), suggesting continuity of tradition in this area from an early date. Werburgh was the daughter of king Wulfhere: see Bury Bank. This place lies on the south-west side of a hill rising to 585'. If the 1833 spelling can be relied upon, the first element may be from OE *weard* 'watch, guard', or OE *\*wearde, \*wearda* 'beacon'. See also Wenwes; Wenwe.

**WERETON** on the south-west side of Audley (SJ 7950). *Werrington* 1890 O.S. A name of uncertain age: it is not mentioned in Parrott's account of Audley of 1733 (SHC 1944), and does not appear on the first edition 1" O.S. map.

**WERGS, THE** 3 miles north-west of Wolverhampton (SJ 8600). *Withegas* 1202 FF, *Withoges* 1260 SHC 4th Series 13 6, *Wygges* 1306 SHC VII 157, *Wytheges* 1306 Ass, *Withegis* 1327 SR, *Wyrges* 14th century Duig, *Wythegys* 1403 SHC XV 102, *Wythegus* 1418 SHC XVII 64, *Wygheges* 1420 ibid. 76, *Wrgys* 1472 SHC IV NS 186, *Wythegyffe* 1484 SHC VI NS (i) 152, *Wrigges* 1586 SHC 1929 112, *le Wirgges* 1608 SHC 1948-9 128. Usually from *werg*, a dialect form from OE *wiþigas* 'willows, withies'. Part of the area remains low-lying and wet. The element invariably carries the definite article: cf. *The Wirge* (1628), East Hendred, Berkshire; *The Wergs* (1652), Longford, Shropshire; *The Wirg* 17th century, The Wergs, Stratfieldsaye, Hampshire. *Wythergs* is recorded in Ipstones c.1311 (SHC 1911 436), and as *Wythegas* in 1332 (SHC X (i) 117). *Wytheges* in Blurton is recorded in the 14th century (SHC XI 312). See also Field 1993: 66.

**WERRINGTON** 4 miles east of Hanley (SJ 9447). *Werinton* 1259 SHC 1911 133, *Woningtone* 1267 SHC XII (ii) 101, *Werynton* 1272 SHC IV (i) 196, *Werington* 1297 FF, *Wenintone* 1307 SHC XII (ii) 102, *Wonytone* 1309 ibid.' 102, *Weryngton* 1330 Pat, *Wonyton* 1321 SHC XII (ii) 103, *Wonyngton* 1363 SHC XIII 33, 1375 ibid.' 125 *et freq.*, *Weryngton* 1438 SHC III NS 138, *Werrington* 1775 Yates, *Wherrington* 1841 Census. The forms are inconsistent, but suggest '*tun* associated with a man called \*Wer' or some similar personal name (cf. Essington). See also PN Wo 246. See also Wereton.

**WEST BROMWICH** – see **BROMWICH, WEST**.

**WEST BROOK** a tributary of the river Hamps. The western arm of the river.

**WESTBEECH** ½ mile north of Pattingham (SJ 8200). *Westbache* 1312 Brighton 1942: 60, *West Batch (Common)* 1634 map SRO D3548/1, *West-bach* 1686 Plot 394, *Westbitch* 1762 SRO D3221, *Westbitch Common* 1780 Ct, *Westbeach* 1833 O.S. From OE *bece* 'a stream in the well-defined valley': the place lies at the head of a valley with a stream running west into Patshull Pool.

**WESTCROFT** 1 mile north-east of Bushbury (SJ 9302). *Warlow Westcroft* c.1280 Homeshaw 1955: 25, *Werlascroft* 1286 SHC V (i) 171, 1312 SHC 1911 307, *Warlawestcroft* 1302 SHC VII 95, *Werlascroft* 1315 SHC 1928 112, *Worlascroft* 1426 ibid. 83, 1460 ibid. 88, *Worlescroft* 1462 ibid. 92, *Westcroft* 1594 SHC 1932 4, 1834 O.S.; *Westcroft Hall* c.1629 SRO D1790/A/2/170. It is not certain that the earliest spellings refer to this place (they appear to relate to a place in or near Essington), but if they do the derivation may be from OE *wǣrloga* (in later forms *warlagh* or *warlaw*: see Ekwall 1959: 95) 'traitor', used as a nickname: cf. Warlaby, North Yorkshire. The second element is OE *croft* 'a small enclosure of arable or pasture land, an enclosure near a house'.

WESTLOWE (unlocated) *Westlowe* 1523 SHC 1925 122. From OE *west hlaw* 'the western mound or tumulus'.

**WESTON** 1 mile south-east of Maer (SJ 8036). *Westone* 1086 DB, *? Weston* 1212 SHC III 158, *Merweston* 1252 SHC IV 239, *? Weston* 1586 SHC 1927 172, *Weston-juxta-Standon* 1679 SHC 1914 12. From OE *west-tun* '*tun* west of another'. The first element in the 1252 spelling is from nearby Maer, to distinguish this Weston from other places of the same name. The place has been associated with a lost medieval village of Weston Hawes (see TSSAHS XII 1970-1 36), but that place is likely to be Weston Hues or Hughes: see Weston under Lizard. StEnc gives the name of this place as Weston-in-the-Hedge, but this form has not been traced elsewhere, and the source is unclear. See also Weston Meres (Farm).

**WESTON COYNEY** in Caverswall parish, 2 miles north-west of Caverswall (SJ 9343). *Westone* 1086 DB, *Westona* 1166 SHC 1923 296, *Weston under Kevermont* 1236 SHC 1911 403, *Weston Subt' Kavernoc* 1240 (1798) Shaw I xxvi, *Weston sub Keveremont* 1242 Fees, *Western Houme* 1285 SHC 1911 403, *Weston under Couremunt* 1287 SHC VI (i) 167, *Weston Cun(e)y(e)* 1448 Banco, *Weston subtus Caversmounte* 1595 Erdeswick 1844: 246, *Weston Coyney* 1775 Yates. From OE *west-tun* 'western *tun*, *tun* west of another' (perhaps Caverswall). The place was held by Thomas Coyney (*Tome Cuinniea*) in 1164-5: SHC 1911 417; 1923 296. Coyney is from a French family name Coignet: StSt I 1987 232. For the element *Keveremont*, etc., see Carmounthead. *Houme* is from Hulme, 1 mile north of Weston Coyney.

WESTON HARALD, HARALDESWESTON – see **WESTON UNDER LIZARD**.

**WESTON JONES** in Standon parish, 3 miles north-east of Newport (SJ 7624). *Weston* 1242 Fees, *Weston' Johannis* 1236 Fees, *Weston* 1325 SRO D1564/7, *Weston Jhones* 1327 SR, *Westona* 1380 SHC 1914 183. From OE *west-tun* 'western *tun*, *tun* west of another place', presumably Norbury. The *Jones* element is from Johannis (John), the name of an early owner. Weston Jones may be an un-named 2 hides held by Robert de Stafford in DB: VCH IV 52 fn.51. See also SHC 1914 182-3.

**WESTON MERES (FARM)** 1 mile south-west of Chapel Chorlton (SJ 7937; SMR 02595 places a deserted Anglo-Saxon settlement at SJ 80443642). *Weston* 1457 SRO D938/153, *Weston Maer* 1748 SRO D3272/5/15/5-31, *Weston Meir* 1792 SRO D3272/5/15/32. 'Weston near Maer'.

WESTON MOOR (unlocated, in or near Weston Coyney: SHC 1912 38) *Weston Moor* 1529 SHC 1912 37. The moor evidently lay to the south of Wetley Moor: SHC 1912 37.

**WESTON ON TRENT** Ancient Parish 6 miles north-east of Stafford (SJ 9727). *Westone* 1086 DB, *Weston-upon-Trent* 1293 SHC 1911 49, *Weston-on-Trentham* (*sic*) 1544 SHC XI 288, *Weston* 1682 Browne. The DB entry is held to relate to this place, although entered under Pirehill Hundred. From OE *west-tun* 'western *tun*, *tun* west of another'. The place (west of Amerton) is on the river Trent. Sometimes called *Weston by Stafford*.

**WESTON UNDER LIZARD** Ancient Parish 7 miles south-east of Newport (SJ 8010). *Guestona* 1081 Ord, *Westone* 1086 DB, *Weston* c.1150 St Thomas, c.1247 Rees 1997: 94, *Westona* 1166 SHC 1923 296, *Weston subtus Brewod(e)* c.1255 RH, *Weston-under-Brewode, Weston Hewes* 1327 SHC II NS 37, *Weston under Lusezerd* 1340 SHC II NS 40, *Weston subtus Lus(e)yord* 1349 FF, *Weston subtus Luz(e)yerd, Weston subtus Luzyert, Weston subtus Luzyard, Weston subtus Luzhord* 1352 Ass, *Weston juxta Blumenhull(e)* 1359 Banco, *Weston Howes* 1377 SHC 4th Series VI 16, *Weston Hues* 1379 PollTax, 1381 SHC II NS 48, *Weston huos* 1512 SHC I 364, *Weston Hughwes* 1534 Deed, *Weston Huys* 1547 TSAS 3rd Series VIII 1908 138, *Westonhewes, Weston Hewes* 1586 SHC 1927 169-70, *Weston subtus Lyzyard* 1672 ParReg, *Weston under Lizard* 1833 O.S. From OE *west-tun* 'western *tun*, *tun* west of another', perhaps with reference to Brockhurst (which lies between this place and Wheaton Aston), or with reference to Brewood, of which parish it was formerly a part: at least as early as 1254, nearly a century before it was first recorded as *Weston-under-Lizeard*, it was known as *Weston-under-Brewood* (SHC 1916 196). The additions to the name served to distinguish it from other places called Weston. The place is north-east of Lizard (q.v.), and since there is no evidence that it formed part of the Lizard estate in the 14th century – or at any other time – the reference to Lizard must be to Lizard Hill. *Blumenhull(e)* is Blymhill (q.v.). *Hewes* is from Sir Hugh de Weston who held the place in or about 1240-2 and died in 1305 (see also Weston near Maer): SHC I 214; SHC NS II 37 fn.2. Oakden 1984: 180 identifies *Haraldeswestone* 1410 *Cur*, *Haraldesweston* 1414 SHC XVII, and *Weston Harald* 1424 SHC XVII 106 as this place, but the evidence is unclear. The identity of Harald has not been established.

**WESTON (WOOD)**  ½ mile north-east of Norbury (SJ 7924; SMR 02633 places a deserted post-Conquest settlement at SJ 79402430). *Westonwood* 1891 O.S.

**WESTSIDE (MILL)**  on the east bank of the river Manifold, 1 mile east of Warslow (SK 1058). *Wessyd* 1656 *et freq*. Alstonefield ParReg, *Wessyd Milne* 1668 ibid, *Westsyde* 1675 ibid, *Westside* 1689 ibid, *Weside, Wesside Mill* 1694 ibid. Perhaps so-named because it lies on the west side of Archford Moor. There was a mill here by 1584: see VCH VII 11.

**WESTWOOD**  1½ miles west of Leek (SJ 9656). *Westwode* 1291 (1798) Shaw I xxii, 1292 SHC VI (i) 220, 1298 SHC VII 44, *Graunge de Westwood(e)* 1539 MinA, *Westewood* 1539 MA, *Westwod* 1543 (1883) Sleigh 73. 'The wood to the west' (of Dieulacres Abbey, which had a grange here: VCH III 223; VCH VII 85; 101). *Westwode* in or near Loxley is recorded in 1439: SRO DW1733/A/2/30.

WETHAL, WETHALES  (unlocated, in Mytton) *Wethal* 1209 SHC XI 311, *Wethale* 1284 SHC VI (i) 140, *Whethales* 1289 ibid. 191, *Whetales* 1290 SHC XI NS 23, *Wethale* 1297 SRO DW1734/J2268, *Wethales* 1302 SHC XI 311, 1320 SHC 1911 346, *Whethales* 1349 SHC XII 80, *Whetenhall* 1537 SRO D590/662. The inconsistent spellings make it uncertain whether the derivation is from the plural of OE *halh*, with OE *wet* 'wet', or 'the *halh* where wheat is grown'. It is unclear whether *Wetenale*, recorded c.1299 (SHC XI 325), and *Whetenhall*, recorded in 1537 (SRO D590/133), are to be associated with this place.

**WETLEY MOOR**  2 miles south of Bagnall (SJ 9248). *Wetley Moor* 1529 SHC VIII 8, *Wetley More* c.1529 SHC NS X (i) 183, *Watleymore* c.1529 ibid. 149, *Watley More* c.1531 ibid.' 149, *Whatley More* c.1540 ibid.' 175, *Whitle moore* c.1540 Leland, *Wettelye moor* 1586 AD 6, *Wotley More otherwise Homersley More at Chedleton* 1599 SHC 1935 131, *Wetley Moore* 1604 SHC 1940 196, *Whitely Moore* 1735 Stoke on Trent ParReg, *Wetley Moor* 1810 *EnclA*. Probably from OE *wet-leah* 'wet *leah*'. The name seems originally to have been applied to an extensive area to the south and south-west of Leek, and in 1529 covered about 1000 acres: JNSFC XCII 1957-8 68. *Wythemor*, recorded in 1226 (JNSFC XCII 1957-8 68), is said to refer to Wetley Moor, but if accurate would give 'the moor with the withies', and is unlikely to be an early spelling for Wetley Moor. A survey of the boundary of the Moor was made in 1605: SRO D5590/1/10/3; see also Browne's map of 1682 and Yates' map of 1775, also StEnc 654. It is unclear whether *Wetelea*, recorded in 1182 (SHC I 103) refers to this place. See also Ford Wetley; Hammersley; Weston Moor.

**WETLEY ROCKS** 5 miles south-east of Leek (SJ 9649). *ye Rocks* 1734 ParReg, *Wetley Rocks* 1784 SHC 1947 79, 1792 Andrews 1936: 124, 1836 O.S. From OE *wet-leah* 'wet *leah*': the place lay in Wetley Moor (q.v.). The rocks are a gritstone outcrop.

**WETMOOR** (unlocated) in Stretton parish, 3 miles south-west of Penkridge, *Wetemore* 1453 Banco, *Wetmore* 1455 SHC III NS 217; **WETMOOR FARM** ½ mile north of Gayton (SJ 9829), *Wetmoor* 1836 O.S. From OE *wet-mor* 'the wet marshland'. Cf. Hungry Hill.

**WETMORE** 2 miles north-east of Burton-upon-Trent (SK 2524). *Withmere* 1012 (14th century, S.930), *Witmere* 1086 DB, *Witmere* 11th century Sawyer 1979a: xxxv, *Wismera* 1113 Burton, *Withmere* 1114 ibid, *Wictm* 1197x1213 SRO D603/A/Add/45, *Wichtmere* c.1235 BM, *Withmere* 13th and 14th centuries (frequently) Duig, *Wyghtmer'* 1394 SRO DW1734/2/1/103(vi)m.46, *Weghtmer* 1532 SHC 4th Series 8 156, *Wightmere* 1538 SRO DW1734/1/4/24, *Whitmere* 1606 SHC 1939 125, *Weightmore* 1663-4 SHC 1910 35, *Weetmoore* 1686 Plot, *Withmere, Wightmere, Whitmore* 1798 Shaw I 19. The first element is from OE *wiht* 'bend, curve', a word found only in place-names, with OE *mere* 'a pool', hence 'pool by the river bend'. The place is on an island formed by two branches of the river Trent. Confusingly, Wetmoor *(sic)* Hall Farm lies to the north (in Derbyshire). VCH IV 43 incorrectly identifies the DB form as Wetmoor near Gayton, unlike the map in the same volume.

**WETTON** Ancient Parish 7 miles north-west of Ashbourne (SK 1055). *Wetindona* 1188-94 SHC 4th Series IV 31, *Wettindun* 1252 Ch, *Wettindon'* 1253 SHC 4th Series IV 36, *Wetindon'* 1255 ibid. 37, *Wetton* 1327 SR. From OE *wet* 'wet, damp', hence 'wet *dun* or hill', probably explained by springs which rise on the hill. There is a tumulus on Wetton Low (OE *hlaw* 'mound, tumulus') ½ mile to the south. A Grange of Tutbury Abbey existed at Wetton: SHC VII 180.

**WETWOOD** 4 miles north-west of Eccleshall (SJ 7733), *Wetewode* 1291 Tax, *Wetwode* 1312 SHC IX (i) 32, *Wettwod* 1532 SHC 4th Series 8 103, *Wetwood* 1563 SHC 1938 66, 1833 O.S., *Wetwood (Cross)* 1727 LJRO ExD&CB/A/21/1/68; **WETWOOD** 1 mile north-west of Merbrook (SJ 9861), *Wethwode* 1229x1232 CEC 385, *Wethwood* c.1231 SHC 1911 425, *Wetwath* 1248 (1883) Sleigh 48, *Wetewode* c.1291 Tax, *Wethwod* 13th century Dieul, *Wetwo(o)d* 1537 Deed, 1775 Yates, *Whetwoode, Wettwood* 1607 QSR, *Wettwood* 1634 Leek ParReg, *Wetwood, Wet Wood Farm* 1842 O.S. From OE *wet-wudu* 'wet or boggy wood'. Cf. Weetwood, Northumberland.

**WEYMOUTH** to the west of Willoughbridge (SJ 7440). *Womworthyn* 1293 JNSFC 4 1964 62, *Wommerthin (wood of)* 1363 SHC XIV 110, ibid.' XII NS 255, *Weymouth Cottage* 1833 O.S. Perhaps from OE *wamb* 'womb, belly', used in a topographical sense 'a hollow, a bulge', with OE *worþign* 'an enclosure', so giving 'the enclosure at the hollow or bulge', which has evidently been influenced by the name of Weymouth in Dorset.

**WHEATLOW BROOKS** 1 mile north of Milwich (SJ 9734). *Wheatley Brooks* 1787 Deed, *Wetley Brooks* 1813 SRO D1798/616/40/ii, 1836 O.S., 1838 T.A., *Wheatlow Brooks* 1891 O.S. The forms are inconsistent, but perhaps '*leah* where wheat was grown'. The place lies between two stream-junctions.

**WHEATON ASTON** – see **ASTON, WHEATON**.

**WHETSTONE GREEN** (obsolete, on the east side of Bushbury (SJ 9202)) *Whetstone Green* 1834 O.S. Perhaps from OE *hwet-stan*, generally found in later place-names to indicate places where stone suitable for whetstones was obtained. The name is remembered in Whetstone Road and Whetstone Grove. The Green element suggests a squatter settlement.

**WHISTAMERE**   on the north side of Farmcote Hall (SO 7892). *Wystanmere* 1255x1265 Eyton 1854-60: III 94, *Wystanesmere* 1298 Peramb, *Wystannesmer'* 1327 SR, *Whistamere* 1981 O.S. 'Wigstan's or Winestan's mere or pool'. In Shropshire since the 12th century. See also Eyton 1854-60: III 100, who records fields called *Whistimore* near Farmcote.

**WHISTON**   2 miles west of Penkridge (SJ 8914), *Witestun* 1002x1004 (11th century, S.906; 11th century, S.1536), *Witestone* 1086 DB, *Witestona* 1116-33 Burton, *Wistona* c.1176x1184 SHC III (i) 205, *Wistun'* c.1255 RH, *W(h)ystone* 1291 Ipm, *Wyston'* 1240 (p) FF, *Whyston* 1251 Ass, *Whiston* 1333 Banco; **WHISTON** 2½ miles north of Cheadle (SK 0347), *Witestone* 1086 DB, *Whystan* 1277 SHC VII NS 20, *Wyston* early 13th century Okeover T319, *Whytston* 1306 GDR, *Wytston* 1328 SHC 1913 17, *Whishton* 1357 Pat. In the first case 'Hwit's *tun* or *Witi's tun*', but OE *hwit-stan* 'white stone' is the derivation for Whiston near Cheadle: the place is noted for the exposed rocks which produced white sand used in the manufacture of cosmetics.

**WHISTON EAVES**   2 miles south-east of Kingsley (SK 0346). *Whyston Eves, Whiston Eves* 1456 Banco, *? Ebys* 1565 SHC XIII 246, *Eves* 1585 SHC 1927 128, *Whiston Eves* 1613 SRO Q/SR/127, *Eaves* 1686 Plot. For the first word see Whiston. The second word is from OE *efes* 'eaves; an edge or border, especially of a wood', and in place-names 'the brow of a hill, the edge of a precipice or bank', or as here, 'the place on the edge of the township'. It is possible that some or all of the later spellings relate to The Eaves (q.v.), 1 mile south of Cheadle.

WHISTON'S MOORE   (unlocated, in Bishop's Wood near Brewood) *Whiston's Moore* 1801 Shaw II 303.

WHISTONWICK (unlocated) *Wystaneswyk* 1344 SHC 1913 211, *Whistonwick* 1736 SRO D641/5/T(S)/T/18a&b. The context in which this name appears suggests that it lay near Standon or Chebsey, and there is a reference in 1674 to *'Aston, Walton, Burston, Stoke, Fulford, Chebsey and Wicke'* (SRO D909): the last name (which is otherwise unidentified) may refer to this place, but see also Wynstansley. However, Wistanswick in Shropshire lies 3½ miles south of Market Drayton, and it is possible that the spellings refer to that place.

**WHITACRE FARM**   1 mile east of Brownhills (SK 0705). *Whitacres* 1300 SHC V (i) 178. From OE *hwit* 'white', with OE *æcer* 'field, ploughed land'.

WHITBY WOOD – see **OAKEDGE HALL**.

**WHITE CHIMNIES**   3 miles east of Biddulph (SJ 9156). *Whit Chimnies* 1666 SRO D.4069/1/1, *White Chimney* 1775 Yates. Self-explanatory.

**WHITECROSS**   ½ mile north-east of Bobbington (SO 8191). *le Whitcrosse* 1400 SA 52/9, *White Cross* 1892 O.S. 'The white cross'.

**WHITE LEE FARM**   2 miles north-west of Onecote (SK 0256). *Whytley Leke* 1519 Dieulacres Inventory, *Whitelee* c.1538 VCH VII 213, *White Lee* 1615 SHC NS IV 69, 1836 O.S. Evidently 'the white *leah*'. *Leke* is Leek. See also The Whitelowe.

WHITELOWE, THE (unlocated) *the Whitelowe* 1646 SHC 4th Series I 308. From OE *hwit hlaw* 'the white mound or tumulus'.

**WHITEHOUGH** 1 mile north-west of Ipstones (SK 0151). *Whythalk* c.1253 SHC 1911 428, *Whytehalg* 1281 ibid.' 178, *Le Whytehalgh* 1292 ibid. 216, *Le Whitehalg* 1292 SHC I 296, *Whythalk* 1293 ibid. 428, *Whitehalgh* 1348 SRO DW1761/A/4/164, 1380 StEnc 657. 'The white *halh*', with later confusion with the ending *-hough* from OE *hoh* 'a heel, a steep ridge, a spur of land'.

**WHITEHURST** 1½ miles north of Dilhorne (SJ 9745). *Whytehurst* 1281 SHC 1911 178, *Wythehurst* 1294 ibid. 225, *Whytehurste* 1295 ibid. 54, *Whitehurst* 1329 SHC XI 7, *Whitchurst (sic)* 1695 Morden, *Whitehurst* 1726 BCA MS917/1286, *White Hurst* 1837 O.S. From OE *hwit* 'white', with OE *hyrst* 'hillock, knoll, copse, wood, wooded eminence', so 'the white copse or wooded hill'.

**WHITEMOOR, THE** 1½ miles west of Brewood (SJ 8508), *Wytemore* 1276 SHC 1936 200, *Wytemore* 1286 SHC V (i) 163, *Alba Mora* 1292 SHC VI (i) 221, *Hwytemore* 1295 SHC 1911 237, *Whitemere* 1327 (p) SR, 1373 Ct, *Wyt(e)mor* 1327 (p) SR, *le Whitemor* 1334 SHC XVI 6, *Whitemor* 1348 SHC XII 83, *long whitemore* 1390 Ct, *Whytemore* 1538 ParReg; WHITE MOOR (obsolete) near the south-east corner of Bagot's Park (SK 1026), *Whitemoor* late 13th century SRO D(W)1721/3/11/6, *Wytemor* c.1345 SRO D986/81, *White Moor* 1724 *Survey*; **WHITEMOOR** 1 mile north of Biddulph (SJ 8860), *White Moor (Wood)* 1798 Yates. From OE *hwit-mor* 'white moor'.

**WHITEMOOR HAYE** ½ mile south-west of Alrewas (SK 1713). *Whytemore hey* 1568 Alrewas ParReg. From OE *hwit-mor* 'white moor', with Mercian OE *(ge)heg* 'a fence, an enclosed piece of land', often in Staffordshire and Shropshire with the meaning 'an administrative bailiwick within a Forest area', arising from an amalgamation of Latin *haia*.

WHITES BRIDGE – see **BRIDGE END**.

**WHITESICH BROOK** a tributary of the river Penk. Probably from OE *hwit-sic* 'light-coloured stream'.

**WHITE WOOD FARM** 1 mile north-east of Yoxall (SK 1520). *le Whyte Wode* 1337 Hardy 1908: 23, *Whitewood* 1812 *EnclA*. Self-explanatory, although the precise meaning is unclear.

**WHITFIELD** 2 miles north-east of Tunstall (SJ 8852). *? Quitefeld* 1212 SHC III 158, *Whytefeld* 1273 SHC VI (i) 59, *Whitfield* 1307 SHC XI NS 262, *Wytfeld* 1327 SHC VII 207, *Wytefeld* 1332 SHC X 94, *Whitefeyld* 1614 SHC 1934 32. From OE *hwit feld* 'white open ground'.

**WHITGREAVE** 4 miles north-west of Stafford (SJ 8928). *Witegrave* 1193 P, *Witegrave* 1203 Cur, *Whytegrave* 1227 Ass, *Witegreve* 1251 Misc, *Wytegrave* 1251 Fees, *Witegrefe* c.1271 SHC VIII (i) 193, *Wytegreve* 1292 SHC VI (i) 239, *Qwytgreve* 1447 SHC III NS 178, *Wythegyffe* 1483 SHC VI (i) NS 152, *Wytgreff* 1532 SHC 4th Series 8 66. From OE *hwit* 'white', with OE *graf* 'grove, thicket', a common element in the West Midlands.

**WHITLE, UPPER & UNDER** 1 mile south-east of Longnor (SK 1064 & SK 1063). *Whittle* 1599 SHC XVIII 16, *Under Whitle* 1711 VCH VII 241, *Whittle* 1840 O.S. 'White hill', from the rounded limestone headland here. The place is mentioned in the early 15th century: VCH VII 241 fn.29.

**WHITLEYGREAVES** 2 miles south-west of Eccleshall (SJ 8126). *? the Cashes* 1723 SRO D1798/643/7, *Cash* 1775 Yates, 1833 O.S., *Cash* 1851 White, *Whitleygreaves* 1963 O.S. See Cashheath Farm.

**WHITLEY HEATH** 2 miles south-west of Eccleshall (SJ 8126). Evidently to be associated with *Whitwell* 1531 SHC 1912 46, *Whitle Well* 1530 ibid. 43, *Whitley Welle* 1526 ibid. 25. From OE *hwit wælle* 'the white spring', which was probably a reduction from *hwit leah* 'the white clearing', with OE *wælle*. See also Cash Farm.

**WHITLEYFORD** 1½ miles south of Knighton, on the Staffordshire-Shropshire border (SJ 7423). *Whytley ford, Whitley ford, Whitley forth* 1487 to 1523 *Rental, Whitley Ford* 1833 O.S. 'The ford at the white *leah*'. *(Fontem de) Witewell*, recorded in an early undated deed (SHC IV 274), may be associated with this place. *Fontem* is from Latin 'spring'.

**WHITLEYGREAVES** – see **CASH FARM**.

**WHITMORE** 4½ miles south-west of Newcastle-under-Lyme (SJ 8140). *Witemore* 1086 DB, *Witemere* 1176 EEA 16 104, *Whytemore* 1227 Ass, *Wytemore-under-Lyme* 1242 Duig, *Wytemore* 1299 SHC XI 311, *Whitemor* 1333 SHC 1913 228, *Wetemere, Wetemore* 1425 HLS, *Whitemore* 1450 ibid, *Whyttemore* 1511 ibid.' Although the DB form suggests the possibility of a derivation from the OE personal name Wita, the later spellings and the records of limeworking here (see Limepits) indicate that the name is from OE *hwit-mor* 'white moor', though CDEPN suggests the name might denote 'a place where snow lies long'.

**WHITMORE REANS** 1 mile north-west of Wolverhampton (SO 9099). *Whitmore-ends* 1801 Shaw II 165, *Whitmore End House* 1842 TA, c.1850 Brigden map, *Whitmore Reans* 1895 O.S. Whitmore is from OE *hwit-mor* 'white moor'. The second part may be from the dialect word *rean*, from ON *reinn*, an element commonly found in Shropshire with the meaning 'drainage channel': Shaw 1801: II 65 says of the land here: 'As a striking instance of the effects of improved cultivation, the fine and highly-productive tract of meadows, now called ... Whitmore-ends, was, in the 16th century, nearly a morass, and, on account of its poverty [poor quality], distinguished by the name of the Hungry Leas'. In the 19th century the area was also known as *New Hampton* to reflect the rehousing of residents from the slum areas of Wolverhampton, a name preserved in Newhampton Road.

**WHITTIMERE** 2 miles north-east of Bobbington (SO 8292; SMR 02684 places a deserted post-Conquest settlement at SO 82999273). *Wytemere* 1296 SHC 1911 266, *Wythemere* 1298 TSAS LXXI 1996 27, *Whitemere* 1375 Ipm, *Whittymer* 1643 Claverley ParReg, *Whitimore* 1834 O.S., *Whittimere* 1895 O.S. 'The white moor'. The names *le Scholle* (1286 SHC 4th Series XVIII 119; 1443 VCH XX 71), *Schowle* (1544 VCH XX 71), and *Scolle or Whittimere Hall* (late 16th century VCH XX 71) are associated with this place, from the OE adjective *sceolh* 'twisted, awry', seemingly used for 'the twisted hill': see Shoal Hill.

**WHITTINGTON** Ancient Parish 3 miles east of Lichfield (SK 1608), *Witinton* 1182 P, *Wytinton* 1242 SHC 1924 70, *Withinton'* 1242-3 Fees, *Whitinton* 1309 WL 103, *Whytynton* 14th century Duig, *Whetyngton* 1482 SHC VI NS (i) 141, *Whittington* 1686 Plot, 1798 Yates; **WHITTINGTON** 4 miles south-west of Stourbridge (SO 8582), *Quitenton* 1203 Selden Soc. lxxxiii 66, *Whitinton* c.1255 SHC V (i) 110, *Wytyndon* ibid.' 159, *Wytinton* 1286 ibid.' 157, *Whytynton, Whitenton* 13th century Duig, *Withynden* 1414 SHC 1921 27, *Wyttenton* 1532 SHC 4th Series 8 15, *Whittington* 1686 Plot, 1798 Yates;

**WHITTINGTON** 4 miles north-west of Eccleshall (SJ 7933), *Wytindon* 1339 SRO D(W)1082/L/10/3, *Wytinton* 1348 SRO DW1082/A/2/3, *Whittyngton* 1524 SHC XI 264, *Whittenton, Whittington* 1676 SHC 1914 29, *Whittington* 1833 O.S. 'The *tun* associated with a man called of Hwita', or possibly 'Hwita's *tun*' or 'white *tun*'. *(æt) Hwitantune*, recorded in 925 (14th century, S.395), is probably to be identified with Whittington near Chesterfield, in Derbyshire: see Sawyer 1979: 5.

**WHITTON (COMMON)** (unlocated, in Cotton, 'near the road to Cheadle': D240/D/294) *Whitton* 1766 D240/D/294.

**WHORROCKS BANK** on the south-west side of Rudyard Lake (SJ 9458). *Horhoc* later 13th century VCH VII 67, *Horchok* 1308 SHC XI NS 255, *Whorrocks* 1607 VCH VII 67, *Horrocks Bank* 1664 ParReg, *whorrocks bank* 1815 *EnclA*. The spellings point towards a derivation from OE *horh* 'dirt', with OE *ac*, so 'the oak-tree in the dirty or muddy place', rather than from OE *har-ac* 'grey or hoary oak'. The place is now known as Horton Bank, but the name is preserved in Whorrocks Bank Road: StEnc 661.

**WHYTACRE** (unlocated, possibly on the Staffordshire-Warwickshire border) *Whitacre, Whytacre* 1262 SHC IV 150, 152. From OE *hwit æcer* 'the white area of cultivated land'.

**WHYTLEY** (unlocated, possibly near Onecote in Leek) *Whytley* 1539 VCH III 228, *Whyteley* 1539 MA. Perhaps to be associated with *Whitlegh*, recorded in 1339 (SHC 1913 79). OE *hwit leah* 'the white clearing'.

**WIBBILDE MOOR** or **WIBBELLE MOOR** (obsolete, near Knowle on the southern boundary of Lichfield: VCH XIV 111) *Wibbildemor* 1198x1208 EEA 17 93, *Wibbilde Moor, Wibbelle Moor* c.1200 VCH XIV 111. Perhaps from the OE personal name Wibald or similar.

**WICHINHAM** (unlocated) *Wichinham* 1190 Pipe.

**WICKEN LOW** 1 mile south-east of Flash (SK 0366). *Wickenlow* 1842 O.S. Perhaps from OE *\*cwicen hlaw* 'burial mound with or at the rowan or mountain ash'.

**WICKEN WALLS** ½ mile west of Flash (SK 0167). *Wicken Wall* 1842 O.S, *Wicken Walls* 1870 *Rental*, 1880 Kelly. *Wicken* may be from OE *\*cwicen* 'rowan or mountain ash'. Plot 1686: 223 records the belief that the *Quicken-tree* warded-off evil spirits, with some countryfolk keeping boughs by their beds or carrying sticks made from the wood. The variants *Quicken* and *Wiggin* are sometimes found, e.g. *Quitens greene* 1615 Alstonefield ParReg. *Walls* is from Mercian OE *wælle* 'a spring', and (sometimes) 'a stream', hence 'the spring(s) by the mountain ash'. See also Wickenstone Farm.

**WICKENSTONE FARM** on the east side of Biddulph (SJ 8956). *Wickenstone* 1836 O.S. From the Wicken Stones (*Wicking Rocks* 1888 TNSFC 1888 68), a long narrow gritsone ridge here. The age of the name is uncertain. *Wicken* is from OE *\*cwicen* 'rowan or mountain ash', so 'the stone at or near the mountain ash'.

**WICKEYTREE** ½ mile west of Loggerheads (SJ 7235). *the Wickey tree* 1681 SHC 1945-6 180, *Whicky tree* 1689 Ward 1843: lxi. From OE *\*cwicen* 'mountain ash', so 'rowan or mountain ash tree'.

**WIDNESS** on the south side of the river Churnet, 1 mile north-west of Alton (SK 0543). *Withness* 1702 Alton ParReg, 1706 SRO D240/D/79, *Widness* 1768 SRO D707/2/1, *Widneys* 1836 O.S. From Mercian OE *wid, nes(s)* 'wide headland or projecting ridge'.

The ridge is evidently that between the Churnet and the stream that flows through Dimmings Dale.

**WIGENHALL** (unlocated, possibly south-east of Copmere) *Wigenhall* 1298 Spufford 2000: 295.

**WIGFORD** on the east side of the river Tame, on the north-west side of Dosthill (SK 2000). *Wycford* 1326 (1798) Shaw I 411, *Wyc-ford* 1526 ibid. 412, *Wigford* 1327 SHC VII 232. From the ford across the river here, which is marked on Yates' map of 1798 and the first edition 1" O.S. map. Probably from OE *wic*, so 'ford by the hamlet or dairy-farm'. The latter element sometimes had the specialised meaning 'associated with salt-working' or 'dependant place with a specialised commercial function' (see Nomina 22 1999 88), and it is possible that the name here has the meaning 'ford associated with the saltings' or similar: see Fisherwick. It may also be noted that the element *wic* is associated with Roman sites (see Gelling 1988: 247-8; Nomina 22 1999 110-11), and this place lies ½ mile south of Watling Street. The possibility that the first element is from a people known as the *Hwicce, whose name is believed to be associated with Wychnor (q.v.), cannot be ruled out completely, since the hard g may be a relatively recent development.

**WIGGEN DE TILEHOUSE** (unlocated, probably near Haughton) *Wiggen de Tilehouse* 1616 SHC VI NS 21. No suggestions can be offered for this odd name, which is recorded as a place-name but appears to be a personal name. *Villehouse* (p) is recorded in Penkridge ParReg in 1597, and may refer to this place. See also Tiled House.

**WIGGENSTALL** 1½ miles south-west of Sheen (SK 0960). *Wigginstall* 1396 VCH VII 28, 1695 Morden, *Wyginstalle* 1566 *Deed*, *Wigganstaff* 1775 Yates, *Wiggenstall* 1840 O.S. The second element is evidently from OE *stall* 'a place', particularly 'a stall for cattle' and 'a place for catching fish', but in this case probably 'Wicga's cattle stall'.

**WIGGINTON** 2½ miles north of Tamworth (SK 2006). *Wigetone* 1086 DB, *Wicgintun* 11th century Duig, *Wichenton* 1173 SHC I 71, *Wikenton* 1174 ibid. 73, *Wigenton* 1188 ibid. 136, *Wigenton'* 1190 Pipe, *Wigginton* 1242 SHC 1911 11, *Wyginton* 1260 SHC X NS I 284, *Wygynton* 1472 SBT DR3/575 at 570, *Wegenton* 1532 SHC 4th Series 8 24, *Wiginton* 1682 Browne. Probably from the common OE personal name Wicga, hence 'Wicga's *tun*'.

**WIGHTWICK** (pronounced Wit-tick [wɪtɪk]) in Tettenhall parish, 3 miles west of Wolverhampton (SO 8698). *Wisteuuic* 1086 DB, *Wyttewik* 1290 SHC 4th Series 13 8, *Wystewyk, Wytewyk* 13th century Duig, *Whistwyke, Whistewykeford* 1300 SHC V (i) 180, *Whitewyke* 1307 SHC VII 178, *Wightwyk* 1539 SHC VI NS (i) 64, *Wyghtwyke temp.* Elizabeth I SHC IX NS 31. Possibly from the OE personal name *Wihta, with OE *wic* 'village', hence '*Wihta's village'. There is a small stream here, known as Wightwick Brook, but no marked bend that would justify a meaning from OE *wiht* 'bend, curve', although it is possible that one existed formerly. The *s* in early spellings is a typical Norman attempt to reproduce ME *-gh-*, and when the sound became lost before consonants the letter disappeared.

**WIGMORE** 2 miles south-east of Wednesbury (SO 0193). *Wigmore (Field)* 1608 SRO D564/3/1/3, *Wigmoor* 1834 O.S., *Wigmore* 1887 Willmore 1887: 29, *Wigmore (Schools)* 1895 O.S. Willmore 1887: 29 states that this is 'the ancient name of the valley lying to the east of Wednesbury'. The first element is OE *wicga*, recorded as a term for an insect, found in modern earwig, perhaps meaning 'something which wiggles, a quaking object'.

The second is OE *mor* 'marsh'. It has been suggested (Gelling 1984: 56; Gelling & Cole 2000: 59) that OE *\*wicga-mor* was a term for an unstable marsh in which wet mounds erupt and disappear: cf. Wigmore, Herefordshire; Wigmore, Shropshire. See also PN Sa I 314-5.

**WIKENESLOWE** (unlocated, possibly near Almington) *Wikeneslowe* 1332 SHC X (i) 100. Perhaps 'the low or tumulus with the Rowan tree', from OE *\*cwicen hlaw*.

**WILBERSTONES** (unlocated, in Burslem: SRO D4842/14/1/35) *Wilboures Stoones* 1607 JNSFC LXIII 1928-9 78, *Wilberstones* 1707 SRO D4842/14/1/35. Cf *wilburge weye* 'Wilburg's Way' recorded in 975AD near Madeley (11th century, S.801).

**WILBRIGHTON** in Gnosall (SJ 7918; SMR 02093 places a deserted post-Conquest settlement at SJ 79501840). *Wilbrestone* 1086 DB, *Wilbritone* 1166 RBE, *Wibertona* 1198 SHC III (i) 29, *Wilewich* 1200 SHC III 69, *Wylbriton* c.1225 SHC 1921 9, *Wylbrhiton* before 1236 SHC 1921 38, *Wilbrixton* 1236 Fees, *Wilbricton* 1242 Fees, *Wilbruyghton* c.1290x1313 Bod. 33, *Wyllbrytton* 1341 SHC 1913 85, *Willbrighton* 1686 Plot 395, *Wilbrington (Heath)* 1672 Blymhill ParReg. 'Wilbriht's *tun*'.

**WILDBOARSEGREAVE** (unlocated, possibly north-west of Leek: CEC 385 suggests in the Heaton/Gun End area) *Wildboarsegreave* 1229x1232 CEC 385. From OE *græfe* 'grove, thicket, copse', and in some case 'trench, pit', so 'the thicket (or pit) of the wild boar'.

**WILDECOTE** on the north side of Patshull Park, just inside Shropshire (SJ 8001). *Weldecote* 1311 TSAS XI 1899 100, *Wylderdecot'* 1327 SR, *Wildicote, Wyldicote* 1598 Albrighton ParReg, *Wildecote* c.1875 SRO D1517. Possibly from OE *wilde* 'wild, waste, uncultivated', so 'the cottage on the waste land'.

**WILDERLY (BARN)** 1½ miles north-west of Colton (SK 0422). *Wylderdeleg* 1277 SHC 1911 169, *? Wyldesdele* 1283 ibid. 182, *Wilderdelaye temp.* Edward I SHC 1919 43, *Wilderley(hull)* c.1348 SHC 1919 43, *Wylderley* 1359 SRO D9W01721/3/1/26, *Wilderly Barn* 1836 O.S. Perhaps from OE *wildeor* 'a wild animal, a deer', often found in OE in the contracted form *wildres*, but the medial *d* in the early forms would be difficult to explain (unless from Mercian OE *del* 'a pit, a hollow'), so possibly therefore from an unidentified personal name (see PN Sa I 316; III 198). The terminal element appears to be OE *leah*.

**WILDGOOSE (FARM)** ½ mile north-west of Bradnop (SK 0055). *Wildegos* 1270x1286 SRO DW1761/A/4/161, c.1275 SHC 1911 429, 1327 (p) SR, *Wylgose* 1480 SHC VI NS (i) 128, *Wylgouse-house* 1540 *AOMB*, *Wylgoose House* 1546 SHC 1912 350, *Wildgoose House* 1635 Leek ParReg. From the surname Wildgoose.

**WILDHAY** 1 mile south-east of Stanton (SK 1145). *Will Hay* 1742 Ellastone ParReg., 1840 O.S. The spellings suggest a derivation from OE *(ge)heg* 'the enclosure', with an uncertain element, perhaps a local term for 'willows'.

**WILDHAY BROOK** a tributary of the river Dove: see Wildhay, which lies on this stream.

WILDMORE – see **HOLLIES** near Heath Hayes.

**WILKINSPLECK** ½ mile north of Whitmore (SJ 8242). *Wilkins Pleck* 1737 Swynnerton ParReg., *Wilkings Pleck* 1766 SRO D593/B/1/21/1/2, *Wilkinspleck* 1920

O.S. Evidently from ME *plecke* 'a small enclosure or plot of land', with the personal name Wilkin(s).

**WILLENHALL**   3 miles west of Walsall (SO 9698). *Willanhalch* c.732 (15th century, S.86 and S.87), *Willenhale* 996 (for 994) (17th century, S.1380), *Winenhale, Winehala* 1086 DB, *Wilinhale* 11th or 12th century Sawyer 1979a: xxxvii, *Willenhal* 1166 SHC 1923 295, *Willenhale* 12th century Duig, *Wulenhale* c.1227 SHC II 275, *Wylenhale, Walenhull* 1286 SHC V (i) 169-70, *Walwenhalle* 1293 SHC VI (i) 235, *Wylenhale* 1304 SHC 1911 277, *Wyllenhale* c.1310 SHC 1928 129, *Wilnoll* c.1564 SHC 1931 155, *Willnall alias Willenhall* 1596 WALS DX-240/34. 'Willa's *halh*'. *Hale* is the dative form of *halh*, perhaps here in the sense 'small valley, a hollow'. The earliest spellings are found in two grants by King Æthelbald of Mercia to Mildred, abbess of Thanet and her family, of tolls of two ships at London, the grants having been made at *Willanhalch*. Shaw 1801: II 8 mentions *Willenhall Spaw* in 1801, so-named from several springs here. It may be noted that Willenhall, Warwickshire, has a single *l* in early spellings, and is probably from OE *\*wilegn* 'willow'.

**WILLIFORD**   2 miles north of Whittington (SK 1610). *Wilnifort, Wilifort* 1159-81 VCH III 340, *Wyliford* 1288 SHC VI (i) 178, *Weliford* c.1290 SHC VI (ii) 151, *? Withyford* 1379 SHC XIII 161, *Wyllyford (Hey)* 1456 SRO D1798/685/177, *Willeford (meadow)* 1508 BCA MS3878/55, *Willoford, Willoford (Hays in Fisherwick Wood)* 1514 SRO D1798/685/180, *Wyllyfourde* 1600 SHC 1935 230, *Williford* 1614 SHC IV NS 64, *Willeford* 1686 Plot 241, 1695 Morden; WILLIFORDE (unlocated, between Bitham and Moor Hall in Penkridge), *Wylyford* 1317 SHC 1931 246, *Williforde* 1598 Ct. From OE *wilig* 'willow', hence 'willow ford': the first place lies near the river Tame.

**WILLINGSWORTH**   1 mile south-west of Wednesbury (SO 9794). *Willingsworth* 1555 SHC IX (ii) 105, 1663 SHC II (ii) 50, SHC IX 120, 1666 SHC 1923 96, 1669 Erdeswick 1844: lviii, 1670 SHC 1923 96, 1686 Plot, 1749 Bowen, 1775 Yates, 1834 O.S.; *Wellingworth* 1686 Plot, 1801 Shaw II 85. From the OE personal name \*Willing, with OE *worð* 'an enclosure', so 'the enclosure of Willing'.

**WILLOUGHBRIDGE**   8 miles south-east of Newcastle under Lyme, on the Staffordshire-Shropshire border (SJ 7440). *Willibrydge Parke* 1547 SHC 1912 180, *Wyllynbrydge* 1564 SHC XVII 213, *Willobridge* 1570 SHC 1931 133, *Wylotbridge otherwise Wyllottesbridge otherwise Wylloughbrydge* 1585 SHC XV 166, *Willowghbridge (Park)* 1590 SHC 1930 95, *Willughbridge* 1611 SHC III NS 60, *Willowbridg* 1686 Plot, *Willowbridge wells* 1747 Bowen. From OE *wilig* 'willows', with OE *brycg* 'bridge', meaning 'the bridge at the willows': the place lies on the river Tern. In the late 17th century unsuccessful attempts were made to develop mineral springs here into a spa: Plot 1686: 103. This accounts for the *wells* element in the 1749 spelling.

**WILLSLOCK**   2½ miles south-west of Uttoxeter (SK 0730). *Willeslocke* 1590 SHC 1930 69, *Willeslock* 1611 SHC III NS 69, *Wills Lock* 1798 Yates, 1834 White 764, *Willstock* 1836 O.S. The spellings suggest a derivation from the ME personal name Wille or similar, a short-form of William (DES 493), with OE *loc(a)* 'enclosure', so 'Will(e)'s enclosure'. The name may be recorded in 1356: see SRTO D543/B/1/2/1-15.

**WILNECOTE**   (pronounced Wincut [wɪnkʌt]) on the south-east side of Tamworth (SK 2201). *Wilmundecote* 1086 DB, *Wilmundecota* 1166 P, *Wilmundicote* 1221 Ass, *Wilmondecote* 1272 Ipm, *Wilmenkot al. Wilmecote* 1326 Ipm, *Wymencote* 1326 Pat, *Wylyncote* 1336 ibid, *Willemondcot* 1274 Ipm, *Wilmindecote* 1290 FF, *Wilmcota, Wilmecote* 1217 Bracton, *Wilmencote* 1298 Ipm, *Wilmendecote* 1315 Ipm, *Wilnecote*

578

1316 FA, *Wilmecote al. Wilnecote* 1607 FF, *Wincote* 1656 Dugdale, *Wilmundecote al. Wilmecote al. Wilnecote* 1663 FF. From the OE personal name Wilmund, with OE *cot* 'cottage, shelter, hovel', so 'Wilmund's cottage'. The absence of the genitival *s*, and the appearance of the *-i-* in the 1221 and 1313 forms show that the original name may have been *Wilmundingcot(e)*, with a medial *-ing-* representing an alternative to the genitive inflection. Formerly in Warwickshire, the place became part of Staffordshire in 1965.

**WILSHAW** 1 mile west of Hollinsclough (SK 0566). *? Wylchar* 1313 SHC 1911 314, *Wilshewe 1566 Deed*, *Wylshawe* 1583 DRO D2375M/190/4, 1602 Alstonefield ParReg, *Wil(l)shaw(e)* 1626, 1651 *Rental*, *Wiltshaw Bottom* 1840 O.S. From OE *wilig, scaga* 'willow copse'.

WILSTANSWUDE (unlocated, probably near Northwood, 2 miles south of Newcastle-under-Lyme) *Wilstan(e)swude* 1227 SHC IV 48, ibid. 239, SHC VII (ii) 6, *Wulstaneswude* 1247 SHC IV 239. 'Wulfstan's wood'. See also Wolstanton.

**WIMBLEBURY** 1 mile south-east of Hednesford (SK 0111). *Wimblebury* 1834 O.S. Early spellings have not been traced (StEnc 665 suggests that this may be the same place as *Wildmore Hollies*, recorded on Browne's map of 1682), so possibly a relatively recent name from dialect *whimberry*, a version of *winberry*, another name for the bilberry or whortleberry: the Cannock Chase Berry (*Vaccinum x intermedium Ruthe*) is a hybrid of common bilberry (*V. mytillus L.*) and cowberry (*V. vitis-idaea L*).

WIMERSLEY (obsolete) 1 mile north-east of Butterton (SK 0857). *Wimersley* 1840 O.S. Seemingly (if ancient) from an unidentified personal name, with OE *leah*.

WIMUNDESLIE (unlocated, possibly in the Sandon area) *Wimundeslie* ?c.1220 SHC 1921 19. Perhaps from the OE personal name Wigmund or similar, with OE *leah*.

**WINCHESTER FARM** ½ mile south-east of Claverley (SO 7992). *Winchester* 1840 TA. Without early spellings the first element is uncertain, but if the name is ancient (and since early spellings have not been traced it is more likely to be a recent transfer) the second is from OE *ceaster* 'Roman fortification', presumably from some ancient earthworks here: see Webster 1981: 79; Webster 1991: 63-4. The place lies on what is said to be a lost Roman road (Margary number 193) from Greensforge to Central Wales, a section of which has been excavated here: TSAS LVI 1957-60 237; 241. In Shropshire since the 12th century.

**WINDGATES** between The Roaches and Hen Cloud (SK 0061). *le Wyndgates* c.1539 *LRMB*, *Wyndyate* 1542 (1883) Sleigh 71, *Wynyattes* 1604 SHC 1946 265, *Wynyates* 1608 SHC 1948-9 89, *Wynigates, Winnigates* 1635 Leek ParReg, *Windegates* 1655 PCC, *? The Wyneyards* 1659 Leek ParReg, *Winneatte* 1678 Alstonefield ParReg, *Winyates* 1686 Plot, *Win Yard* 1775 Yates. From OE *wind-geat*, literally 'gate for the wind', meaning 'a windswept pass or gap'. Cf. Wingate, Durham; Wingates, Northumberland; Compton Wynyates, Warwickshire. Sleigh 1883: 118 states 'Up to about 100 years ago Windgates was called Bourke-grange', quoting 'Late John Millward', and records (p.210) *Windyhay-cross* on Ipstones Edge.

**WINDLESDALE (HOLLOW)** between Wetton and Alstonefield (SK 1155). *Windle dale* 1776 Alstonefield ParReg, 1840 O.S. Perhaps from OE *windel* 'a basket', so 'the basket-shaped dale or valley' (which would explain *Hollow*), or from the OE personal name Windel. Cf. Windsor, Berkshire.

WINDMILL, THE (obsolete) a district 1 mile south of Walsall (SO 0197), perhaps named from the windmill recorded in Walsall in 1304 (Willmore 1887: 66). *Windmill*

*Field* 1554 SOT D260/M/T/1/1/22, *Wynmylne fylde* 1559 Willett 1882: 32, *Windemylle Field* 1610 SOT D260/M/T/7/8, *1 windmill in Walsall* 1628 SRO D260/M/T/1/117a, *W Mill* 1798 Yates, *the Windmill* 1798 Shaw I 73. Self-explanatory. Yates' map of 1775 does not show the name, but gives a representation of a windmill.

WINDSWELL POOL  (obsolete) 1 mile north of Forton, on the east side of the road to Shebdon (SJ 7522). *Wyne(e)Wall, Wyns(e)well pool* 1242 Ch, 1487, 1573 Rental, 1493, 1520 Survey, *Windswall pool* 1618 Survey, *Windswell Pool* 1833 O.S. Perhaps from the OE personal name Wine, meaning 'Wine's spring', from Mercian OE *wælle* 'spring', and (sometimes) 'a stream'.

**WINDY ARBOUR** ½ mile north-west of Madeley (SJ 7645), *Windey harbor* 1733 SHC 1944 59, *Windy Harbour* 1775 Yates; **WINDY HARBOUR** in Cheadle (SK 0143), *Windyarbor* 1890 O.S., **WINDYHARBOUR** ½ mile west of Denstone (SK 0841), *Windyharbour* 1891 O.S.; **WINDY HARBOUR** 1 mile north-west of Cauldon Lowe (SK 0648); WINDY ARBOUR (obsolete) on the north-west side of Cocknage (SJ 9041), *Windy Arbour* 1836 O.S.; WINDY ARBOUR  (obsolete) between Talke Pits and Chatterley (?SJ 8351), *Windy Harbor* 1703 Audley ParReg, *Arbor* 1798 Yates, *Windy Arbour* 1819 SOT DS4842/42/31. A common name for high or exposed places: *arbour* means 'shelter or retreat'. Such names are usually of late origin, but sometimes derive from OE *eorþburg* 'earthern fortification', denoting the existence of earthworks, or OE *here-beorg* 'shelter or protection for a number of men; army quarters' (EPNE ii 244). The field-name *Windy Harbour* is recorded in Claverley, Shropshire (Foxall 1980: 55), and there is a Windy Arbour Lane in Great Saredon. See also Haughton.

**WINKHILL** 1 mile north-west of Waterhouses (SK 0651). *Wycleshull* 1278 SHC VI (i) 86, *Wykynghull* 1307 SHC VII 174, *? Wynkeshull* 1329 SHC XI 14, *Wynkyll* c.1538 SHC 1912 121, *Wyncle, Wynkill, Wyckill, Wynckill* c.1585 SHC 1929, *Winkle-hill* 1686 Plot. Perhaps from OE *\*wincles*, genitive singular of OE *wincel* 'nook, corner', well-established as a topographical term, often with the meaning of 'a sharp bend in a river': this place lies in a bend of the river Hamps: cf. PN Buckinghamshire 203-4. Some of the spellings may refer to Wincle in Cheshire, just across the Staffordshire border 3 miles north of Heaton. See also PN Ch I 164-5.

**WINNINGTON**  in Mucklestone parish, 4 miles north-east of Market Drayton (SJ 7238). *Wennintone* 1086 DB, *Woninton* 1273 Ipm, 1306 SHC VII 153, *Wynynton* 1293 SHC 1911 47, *Wenenton* 1532 SHC 4th Series 8 42, *Winington* 1682 Browne. Ekwall 1960: 524 suggests a derivation from the OE personal name Wynna, so 'the *tun* of Wynna's people', but the early spellings for this place have inconsistent vowels, and there must be some uncertainty about the personal name.

**WINNOTHDALE** 2½ miles south-east of Cheadle (SK 0340)). *Vinath Dale* 1769 Croxden ParReg, *Windworth Dale* 1775 Yates, *Vinadale* 1791 Croxden ParReg, *Winworth Dale* 1832 Teesdale, 1836 O.S., *Winnott Dale* 1883 SRO D637/13/1. An interesting name, but the inconsistency in the available spellings and the absence of early forms precludes any suggested derivation.

**WINSCOTE** 2 miles north-west of Worfield (SO 7396). *Wynescote* 1564 Worfield ParReg, *Winscott* 1602 SA 2028/1/5/8, 1674 SA 5586/1/472, 1731 SA 5586/1/509, 1752 Rocque, *Winscote Hopes* 1833 O.S., *Winscote* 1891 O.S. Perhaps 'the cottage of Wine': see Winshill. *Hopes* in the 1833 spelling is unexplained. Winscote Hills, the name of the escarpment on the east bank of the river Severn to the north-west of Winscote appears as *Wynscothylles* in 1536 (SA 5586/1/423), and *Wynscote hylse* in 1541 (SA 5586/1/428).

**WINSHILL** 2 miles east of Burton upon Trent (SK 2623). *(on) Wineshylle* 1002x1004 (11th century, S.906; 11th century, S.1536), *Wynesh'* early 11th century (13th century) *Peniarth*, *Wineshalle* 1086 DB (listed in Derbyshire), *Wineshulla* 1113 Burton, *Winishil* (p) c.1150 Okeover, *Wineshill* (p) 1150x1159 Burton, *Wineshella* 1159x1175 Burton, *Wineshulle*, *Wisnehulle* 1188x1197 SRO D603/A/Add/36b, *Wyncehulle* 1316 Pat, *Wynsul* 1322 SHC 1924 306, *Wynsell* 1521 Burton, 1532 SHC 4th Series 8 154. 'Wine's hill'. 'The name Wine, earlier Wini, somewhat rare before the tenth century, was common from thence to the conquest': PN Bk 75. The name, a short form of an uncompounded name like Winefriþ (see Stenton 1970: 88) is widely distributed: cf. Winslow (Buckinghamshire); Winsley (Shropshire); Winston (Suffolk). The place was transferred from Derbyshire to Staffordshire in 1894.

**WINTER SIDE** (obsolete) on north-west side of Hollinsclough (SK 0567). *? Wytursyde, in Bassetfryth* 1401 SHC XVI 82, *Wintersyd, Winterside* 1566 Deed, *Winter Side, near Hoarse Clough* 1683 Alstonefield ParReg, *Winter Side* 1840 O.S. Perhaps with a similar meaning to OE *winterdun* 'a tract of upland on which sheep could be pastured in winter to keep them free from soggy land lower down', or 'the tract of upland for sowing winter corn': see Rumble & Mills 1997: 301-6. Ekwall 1959: 83 notes that *winter* may be an unrecorded OE loanword from Latin *vinitrium* 'vineyard', observing that a corresponding word is found in place-names, but it is hardly likely that this place could be associated with the growing of grapes. Cf. Averill Side.

**WIRDSHAY** (unlocated). *Wyrdeshay* early 14th century SRO 3764/1[27574], *Wirdshay* 1635 WCRO CR1908/30/1-2.

**WISTAN** (unlocated, in Huntington) *Wistan* 1262 SHC 4th series XVIII 37.

**WISTY** unlocated, in West Bromwich or Wednesbury. *Wisti, Wistibrigge* 1286 SHC V (ii) 165, *Wysti(brigge)* 1287 SHC VI (i) 170. Ede 1969 [108] states that ground near the stream in Hydes Road is frequently recorded as *Wisty* until recent times, with *Wystibrigge* the bridge carrying Hydes Road over the Tame, an identification accepted by Dilworth 1976: 103-4.

**WITHINGTON** 4 miles north-west of Uttoxeter (SK 0335). *Wythinton* 1272 SHC IV 187, *Withyngeton* 1590 SHC XVI 99, *Withington* c.1602 SHC 1935 436, 1607-8 SHC 1948-9 54, 1836 O.S., *Whittington* (*sic*) 1789 SRO D543/C/7/10. From OE *wiþign* 'willows, wet land where willows grow', and OE *tun*.

**WITHNALL FOREST** (obsolete) in Moddershall (SJ 9236). *Whitnall Forest* 1636 SRO DW1742/28, *Withnall Forest (alias Moddershall Heath)* 1696 SRO DW1742/14-17, *Wicknall Forest* 1732 Okeover T770, *Mothersall Heath (alias Withnall Forest in Kibblestone)* 1808 SRO D593/B/1/20/9-10. The name would appear to be from OE *hwitan-halh* '(at the) white *halh*' or 'Hwita's *halh*'. It is unclear whether *Whitnall mylle*, recorded in 1531 (SHC 1910 21), and as *Whytnall Mill* in 1602 (SHC 1935 457) is to be asociated with this place. The name is remembered in *Whitnall Cottage* (1890 O.S.), on the south-west side of Spot Acres (SJ 9337). See also Moddershall.

**WITHYMOOR MILL** (obsolete) in Rowley Regis (SO 9587). *Wythiemore Mill* 1627 Roper 1980: 25, *Withermore* 1674 ibid. 132, *Withymere Mill* 1834 O.S. From OE *wiþig mor* 'the moorland with the withies or willows', but since the place lay on the boundary Staffordshire/Worcestershire boundary, possibly from OE *(ge)mǣre* 'boundary'.

**WITHYSITCH** 1 mile north-east of Milwich (SJ 9833). *Withysitch* c.1687 SRO DW1826/X2, *Withy Sitch* 1775 Yates, 1832 Teesdale. From OE *sic* 'watercourse', so 'the stream with the withies or willows'.

**WITHYSTAKES** ½ mile east of Cellarhead (SJ 9547). *Withy Stakes* 1836 O.S. 'The land with the upright withies'.

**WITNELLS END** in Upper Arley, 1 mile south-east of Romsley (SJ 7981). *Whytenhull, Wytenhull, Whytehull* (p) 1295 PN Wo 33, 1325 (p) Ipm, 1332 (p) SR, *? Whitenhulle* 1403 SHC XV 112. From OE *hwitan-hylle* '(at the) white-hill'. In Worcestershire since 1895.

**WITTON** in Handsworth, an ecclesiastical parish created in 1926: Youngs 1991: 429.

**WIVERSALL (HOUSE)** on the west side of Abbots Bromley (SK 0724). *? Wilbardeshalough* 1367 SRO D(W)1721/3/32/14, *Wilversall* 1686 SRO D832, c.1795 SRO D832/10/1, *Wilversallfield* 1851 White. If the earliest spelling is indeed to be associated with this place, the derivation would appear to be 'Wilbard's (or similar) haugh' (the personal name Wilbard is not otherwise recorded), from OE *halh*, perhaps here in the sense 'a piece of flat alluvial land by the side of a river, forming part of the floor of the river valley' (OED): the place lies on the east side of the river Blithe and the valley which now holds Blithfield Reservoir. See also Wyversale. Possibly associated with *Wyvelesle* (undated), recorded in Shaw 1798: I 155, 156, 171.

WLFHUL (unlcated, near Cheadle: SHC 1923 39) *Wlfhul* c.1282 SHC 1939 39.

**WOBASTON** 3 miles north of Wolverhampton (SJ 9003). *Wibaldestun* 1227 SHC IV 51, *Wybaston* 1275 SHC VI (i) 56, 1327 SHC VII (i) 251, *Wybaldeston* 1276 ibid. 91, *Wylaston* 1286 SHC V (i) 172, *Wobaston* 1377 SHC 4th Series VI 16, 1608 SHC 1948-9 41. 'Wigbald's *tun*'.

WOBURNSHAWE (unlocated, in Chartley) *Woburnshawe* early 14th century SRO 3718/3. From OE *woh* 'crooked', with OE *burna* 'stream' and OE *scaga* 'shaw, copse', so 'the copse at the winding stream', or 'the copse on Woburn stream'. Cf. Woburn, Bedfordshire.

**WODEHOUSE** (pronounced Woodhouse [wʊdhaʊs]) an ancient house ½ mile north-east of Wombourne (SO 8893), in existence from at least 1242: VCH XX 205. *Wodehous* 1332 SHC X 130, 1347 SHC XII 67, *The Wodhows* 1532 SHC 4th Series 8 185. From the 17th century the spelling was *Woodhouse* or *Woodhouses*, which reverted to the earlier form *Wodehouse* c.1875 after the remodelling of the house: VCH XX 205. The name meant not 'wooden-house', since almost every house was so-constructed, but 'house near the wood' (cf. Woodhouse), and may refer to buildings which housed those who performed functions associated with the management of woodland: see Gelling & Cole 2000: 258.

WODEHURES (unlocated, possibly near Checkley) *Wodehures* 1272 SHC IV (i) 187; SHC IV (ii) 107. Possibly from ME *hewer* 'one who cuts, fells or brings down', so '(the place of) the woodcutters'. VCH Wo II 5 mentions *Wudres* in a 15th century deed relating to Upper Arley.

WODEWARDINGTON (unlocated, probably Wolverton in Warwickshire (see PN Wa 228), but if in Staffordshire possibly near Eccleshall or Chapel Chorlton) *Wolverdinton* 1268 SHC 1914 36, c.1270 SHC 1921 36, *Wulvurdiston* 1272 SHC 1914 35, *Wolverderton* 1291 SHC 1913 245, *Wodewardington* 1293 SHC VI (i) 296. The place was also known as *Wulfatton* (SHC 1914 37). If in Staffordshire probably from the OE personal name Wulfheard, or possibly from the OE personal name Wulfhere: see Bury Bank. See also *Welvedale, Woledale* under the entry for Wolfdale. See also SHC I 164.

WODINGES (unlocated, in or near Longdon) *Wodingis* 1199x1215 D(W)1734/J/1712, *Wodingges* 1216 SHC 1921 31, *Wodengis* c.1305 (1798) Shaw I 223, *Wodynges* 1327 SHC VII (i) 231, 1334 BCA MS3415/163, *Wodynge, Wodynges* (undated) 1801 Shaw I 223. The names *Wodingfeld, Wodynye* (Shaw 1798: I 223) are probably to be associated with this place.

**WOLF LOW** 1 mile south of The Cloud (SJ 9161). *Wlvelagh* 1241 SHC 1911 438, *? Wolvel'* c.1275 ibid.' 429,*? Wolfelowe* early 14th century VCH VII 219, *Wolveleye* 1377 SHC 4th Series VI 14, *Woolfe Low* 1649 Leek ParReg, *Woof-Lowe* 1766 Rowlands 1766: 319. From OE *wulf hlaw* 'wolf tumulus'.

**WOLFDALE** 1 mile south-west of Heaton (SJ 9461). *Parvum Wulvedale* 13th century Dieulacres, *? Wolvedale* 1291 (1798) Shaw I xxii, *Wlwedale* 1322 SRO D1229/1/4/57, *Wulvedale* 1322-3 SHC 1911 437, *Wulfdale* (p) 1327 SR, *Wolfdale* 1327 SHC VII 215, 1331 SHC 1913 27, 1383 SRO D1229/1/4/59, 1451 SRO DW1761/A/4/29, *Wolfdale(hey)* 1485 *Antrobus*, 1489 SRO DW1761?A/4/39, *Wolf Dale* 1798 Yates. 'Valley of the wolves', though a personal name cannot be ruled out completely: Nicholas Wolfe is recorded in 1489: SRO DW1761/A/4/39. *Parvum* 'little' implies the existence of a larger and a smaller place of this name. It would appear that that Oldall Grange in Consall was known as *Wolvedale* in 1313 (SHC XII NS 278; see also Sleigh 1883: 137). Erdeswick 1844: 496 states 'Oldall Grange, or Wolvedale, is, in the Lichfield tax-book, called Wlvedale'. That place was a grange of Ranton priory (VCH III 253), and is recorded as *Oldall* in 1539 (MA). Chetwynd mentions *Welvedale, Woledale* in Weston, near Chapel Chorlton, in 1679: SHC 1914 92; see also Wodewardington. *Wolvedalebruche* in Barlaston is recorded in the 13th century: SHC XI 324.

WOLFEDALEHEY (unlocated, in Rushton James: SRO DW1761/A/4/39) *Wolfedalehey* 1489 SRO DW1761/A/4/39. Seemingly 'the hay or clearing at Wolfedale', from Mercian OE *(ge)heg*. Wolfedale may in this case be from a surname: Nicholas Wolfe is recorded here in 1489: SRO DW1761/A/4/39.

WOLFELEGA (unlocated) *Wolfelega* c.1200 (SHC VI (i) 8). The place appears to have been in the Stone area, perhaps near Bury Bank (q.v.), and may (from the medial -*e*-) incorporate the personal name Wulfhere, but is more likely to be from OE *wulf-leah* 'the clearing with the wolf', or possibly from the OE personal name Ulf or Wulf, with OE *leah*. See also Wolferley.

WOLFERLEY (unlocated, between Lane End and Chesterton: Shaw 1798: I 72) *Wolferley* c.1758 (1798) Shaw 1798: I 34. A name perhaps incorporating the personal name Wulfhere (see Bury Bank), or from OE *wulf-leah* 'the clearing with the wolf', or possibly from the OE personal name Ulf or Wulf, with OE *leah*. See also Wolfelega; Wolfesbrigg.

WOLFESBRIGG (unlocated, in Uttoxeter, possibly near Knightsland) *Wolvesbrugge* 1306 SHC 1911 65, *Wulfvesbrugge* 1317 SHC XI 68, *Wolvesbrugge* 1319 ibid.' 45, *Wulfresbrugge* 1336 ibid. 68, *Wolfesbrigg* 1426 SHC XVII 111. 'The bridge of the wolves', or possibly (from the 1336 spelling) 'Wulfhere's bridge': see Bury Bank; Wolferley.

WOLFHAY (unlocated, in Leekfrith) *le Ulfe haye* c.1539 LRMB, *Ufehey* 1681 ParReg, *le Wolffe haye* c.1540 *AOMB*, *Wolfhay* 1695 Leek ParReg, *? Woolf's Hay* 1811 EnclA. From ON *ulfr* 'a wolf', with OE *(ge)heg*, so 'the forest enclosure with the wolves'.

WOLFOTEBRIDGE (unlocated, possibly in Penkhull) *Wolfotebrugge* 1332 SHC II (ii) 103, *Wylfotebrugge* 1332 SHC 1913 38, *Wilfotebrige* 1336 SHC XII (ii) 25, *Wolfotesbrigge* 1365 (1843) Ward 1843: app. lxvi. Possibly from Wolfote, a ME form of the personal name Wulfhad (one of twin sons of King Wulfhere, by tradition martyred by their father for their Christian faith, who became a local saint associated with Stone (q.v.); see Rumble & Mills 1997: 312), so 'Wolfote's bridge', though considerable doubt must remain. See SHC 1924 64 for the personal name Wlfet, after 1254.

**WOLGARSTON** 1 mile south-east of Penkridge (SJ 9313; SMR 02604 places a deserted Anglo-Saxon settlement at SJ 93501410). *Tvrgarestone* 1086 DB, *W(u)lgareston'* (p) 1167 SHC 1923 296, 1261 Penkridge Inq, *Wolgar(e)ston* 1215 MRA, *Wolgareston* 1442 SHC 1928 147, *Owgaston* 1608 SHC 1948-9 132, *Woolgaston alias Woogaston* 1609 SHC 1928 165, *Wollgaston* 1686 Plot, *Woolgarstone (Farm & Mill)* 1834 O.S. 'Wulfgar's *tun*'. The DB spelling with *T-* is clearly an error. *Wulfgares more* 'Wulfgar's marsh', near Pillaton, is recorded in a charter of 994 (S.860; Hart 1975: 197; Hooke 1983: 92) and *Woolgars Hill* is said to be recorded as a field-name in Whiston in 1775 (Oakden 1984: 104; though *Huggas or Wulgas Hill*, formerly *Woolgas Hill*, recorded in the 18th century (SRO D260/M/T/5/74) was probably in or near Pillaton): it is possible that these names refer to the same person. Wulfgar was a common Anglo-Saxon name (held by at least one bishop of Lichfield: see Hart 1975: 365), and it would be unwise to attempt to identify the particular individual who gave his name to this place. However, it is of interest that a Wulfgar was a Mercian thegn who served Edgar during the latter's brief rule as king of Mercia 957-959 and king of the re-united English from 959 to 975. Edgar visited Penkridge as king of Mercia in 958 where a charter (14th century, S.667) was attested recording *in loco famoso qui dicitur Pencric* 'that famous place which is called Pencric'. Wulfgar attested many charters between 958 and 969, though not the one executed at Penkridge: see Hart 1975: 366.

**WOLLASTON'S COPPICE** on the north side of Heatley (SK 0627). *? Wolaveston* 1199 SHC XI NS 16, *Wolaston* 1341 SRO D(W)1721/3/7/7, *Wolaxton* ?14th century ibid. 175, *Wolastons Coppice* 1724 *Survey*. The spellings suggest a derivation from the OE personal name Wulflac, so 'Wulflac's *tun*'. Perhaps associated with Thomas de Wolaston who held land in Dunstall in 1402: SHC XI NS 207.

WOLLESBRUGGE (unlocated, possibly near Hanbury) *Wollesbrugge (manor of)* 1487 D1798/H. M. Aston/10/27.

WOLLFORDES MARSHE (unlocated, in Pershall: SHC 1934 52, possibly Elford Heath (q.v.)) *Wollfordes Marshe* 1603 SHC 1934 52.

**WOLMORE (FARM)** on the Staffordshire-Shropshire border, 1 mile west of Seisdon (SO 8194). *Wolemere* 1292 SHC XII 64, *? Waldemor* 1314 SHC 1911 319, *Wollemere* 1343 SHC VI NS (ii) 159, *Wollemere* 1401 SHC XV 115, *Willmoor* 1775 Yates, *Wildmoore* 1801 Trysull ParReg, *Wildmoor* 1827 O.S. The forms are inconsistent, but perhaps from OE *w(e)ald* 'high forest land; open upland ground' (cf. *la Wolde* 1314 (1801) Shaw II 212), which often develops into *wilde*, with OE *mor* 'moor', so 'the moor on the high upland ground', although *wald* is a very rare element in Staffordshire: see also Wymundeswolde. The place lies on the Staffordshire-Shropshire boundary on the edge of the high escarpment of Abbot's Castle Hill.

**WOLSELEY**  in Colwich parish, 2 miles north-west of Rugeley (SK 0220). *Vlselei* 1086 DB, *Wasselega* 1166 SHC 1923 295, *Wulfsieslega* 1175 ff.P, *Wulffieslege* 1178 SHC I 92, *Wolsileye, Wolseleye* 1195 SHC 1914 138, *Wulsislea, Wolseslea* 1199 SHC III (i) 32, *Wolselee* 1199 ibid.' 57, *Wullsile* 1199 ibid. 36, *Wulfsiesley, Wulfsiesleg* (frequently) 12th century Duig, *Wulseleg* 1200 SHC III 66, *Wulsileya* 1227 SHC IV 58, *Wulseleye* 13th century Duig, *Wolselegh* 1301 SHC VII 91, *Wukseleye* 1305 SHC VII 136. 'Wulfsige's *leah'*.

**WOLSELEY BRIDGE**  2 miles north-west of Rugeley, on the river Trent (SK 0220). *Worseley Bridge* c.1540 Leland, *Wolseley Bridge* 1593 SRO DW1781/5/5/1, *Ousley Bridge* 1675 Ogilby. See Wolseley. A bridge has existed here from at least 1281, when *Bridggend* is recorded (SHC 1914 144), *(Atte)briggende* 1279 (SHC VI (i) 106). The bridge, which replaced a ford mentioned in the 12th century as *Vadum de Wolseley*, from Latin *vadere*, OE *wadan* (SHC 1914 144), 'a wade, a ford', evidently had a chapel at the northern end: SHC VI (i) 106.

**WOLSELEY PLAIN**  1 mile south-west of Wolseley Park (SK 0018). *Wolseley Plain* 1651x1711 SRO DW1781/9/2/23/1-5.

**WOLSTANTON**  Ancient Parish 2 miles north of Newcastle under Lyme (SJ 8548). *Wlstanetone* 1086 DB, *Wulstanestona* 1199 SHC II 79, *Wlstaneston* 1200 Ward 1843: app. i, *Wolstanneston* 1233 SHC XII 25, *Wlstonton* c.1249 SHC 1911 145, *Wurstynton* 1456 SHC IV NS 96, *Ulsynton* 1532 SHC 4th Series 8 42, *hulstenton* 1535 SHC 1910 246, *Wolsington* 1586 SHC 1927 131, *Wolstanton* 1598-9 SHC 1930 6, *Woolstington* 1609 SHC III NS 34. 'Wulfstan's *tun'*. See also Wilstanswude.

**WOLSTANTON MARSH**  in Wolstanton (SJ 8547). *Gosegreen* 1297 SHC 1911 243, *Wolstanton Marsh* 1836 O.S. Evidently Goose Green was the alternative or earlier name for this wet ground.

**WOLVERHAMPTON**  Ancient Parish 12 miles north-west of Birmingham (SO 9198). *æt Heantune* 985 (12th century, S.860), *Hamtun, Hantone* 996 (for 994) (17th century, S.1380), *Hampton'* 10th century (14th century, S.1155; see Harmer 1989: 403-7), *Heantune* c.1000 (11th century, S.1534), *Hamptun* 1053x1062 SHC 1916 125, *Hantone, Handone* 1086 DB, *Wolvrenehamptonia* 1070x1085 VCH III 322, *Wlrunehamton* c.1078 Mander & Tildesley 1960: 12, *Wlfrunehamtona* 1096x1117 EEA 14 5, *Wulfrenehamtun* 11th or 12th century Sawyer 1979a: xxxvii, *Wolveroveshampton* c.1139 Reg 170, *Wlfrunton* 1145x1153 Letter to pope Eugenius, *Wulfrunehanton* 1169 P, *Wulverne-Hampton* c.1175 SHC VIII 133, *Wulfrenhamptune* c.1175 SHC 1941 73, *Wulfronhamton* 1181 SHC I 96, *Wulverune Hampton* 1161x1182 SRO D938/1, *Wulfrunehant'* 1190 Ch, *Wolfrehampton* 1199 SHC III (i) 36, *Wulfrunehamtun, Wolverenhampton, Wolvernhampton* 12th century Duig, *Wulvernhanton* 1203 SHC III (i) 95, *Wulvunhanton* 1204 ibid. 99, *Wlfrunehamtune* c.1275 Seal, *Wulfrenhampton* 1262 SHC V (i) 139, *Wolvrenhampton* 1288 SHC VI (i) 181, *Wollerhampton* 1424 SHC XVII 95, *Wolvorhampton* c.1540 Leland, *Wolverhampton otherwise Wemerhampton* 1619 SHC VII NS 209, *Wolverhampton* 1610 Speed. From OE *hean-tun* '(at the) high *tun'*: *hean* frequently becomes *ham-* or *han-* in the West Midlands. The place, which stands on elevated tableland which forms part of a long undulating range of hills running from Rowley and Dudley in the south to Bushbury in the north (with the porch of St Peter's church at 529'), was given in 985 to Wulfrun who later granted it to the monastery at Wolverhampton. Wulfrun was a wealthy noblewoman of Mercia, whose estates seem to have lain chiefly in Staffordshire. She is perhaps to be identified with Wulfren, the only hostage who is known to have been taken when Olafr Gothfrithson captured Tamworth c.943 (ASC D). The name of her husband is not known, but the fact that her son, Wulfric Spot, is recorded as Wulfric son of Wulfrun suggests that she was of higher rank than her spouse. Her name was not added to the place-name until it became necessary to distinguish this Hampton from many others. It may be noted however that whilst Wolverhampton is on high ground, other nearby places are higher, with Bushbury Hill 590', Goldthorn Hill 610', and Colton Hills 608'. *Hean* also had the meaning 'chief, important', and there is some slight evidence that Wulfrun may have held land to the east of Wolverhampton, i.e. an early undated reference to the *fossatum Wulfrini* 'the entrenched place of Wulfrini' (Mander & Tildesley 1960: 28), who is almost certainly to be identified as Wulfrun, the name Wulfrini being otherwise unrecorded, which may perhaps show that she held the same site, possibly associated with Stow Heath (q.v.), making a meaning 'chief *tun*' not inconceivable. Indeed, Stenton: 1970: 317 and Harmer 1989: 404 translate *heah tun* (dative *into Heantune*) as 'chief manor'; see also Whitelock 1930: 152; 164. For details of Wulfrun (sometimes confused with her daughter-in-law of the same name) see Searle 1897: 418-9; WA I 289-91; SHC 1916 55-7; Hart 1975: 373-4; Sawyer 1979a: xl; Williams, Smyth & Kirby 1991: 241; Swanton 1996 [295].

**WOM BROOK**  a tributary of Smestow Brook. The older name would appear to have been *Wombourne* (q.v.), for which no early forms are recorded. *Wom Brook* is evidently a back-formation from Wombourn(e). see also Wembleton Brook.

WOMBEWELL (unlocated, perhaps near Abbots Bromley) *Wombewell* 1385 SHC 1937 146. Perhaps from OE *wamb* 'womb, belly', probably with reference to a pool or a bulge-like topographical feature, with Mercian OE *wælle* 'a spring', and (sometimes) 'a stream', so 'the spring at the pool'.

**WOMBOURNE** Ancient Parish 5 miles south-west of Wolverhampton (SO 8793). *Wambvrne* 1086 DB, 1271 SHC V (i) 154, *Wamburna* 1166 SHC 1923 298, 1167 SHC I 48, *Wamburn* 1175 P, 1224 SHC IV 223, *Womburne* 1236 Fees, 1260 SRO D938/168, *Womborne* 1242 ibid, *Wombeburne*, *Wamburn* 13th century Duig, *Wonburne* 1319 SHC 1924 192, *Womburne* 1445 SHC XI 305, *Womberone* 1457 SHC 1928 49, *Wamborn* 1577 Saxton. Ekwall 1960: 531 considers that the name is from OE *(æt) won-burnan* 'the winding stream', from the OE adjective *woh* 'twisted, crooked', found mainly in place-names, with OE *burna* 'stream' (for which element see Bourne Vale), and identical with the origin of Woburn, Bedfordshire. But the topography, the absence of forms with *-n-*, and the fact that the stream here, with the (tautologous) name *Wombornebroc* in 1322 (SHC 1928 33), is not noticeably winding, show that the suggestion in Duignan 1902: 175 that the first element is OE *wamb* 'womb, belly', probably with reference to a hollow or former pool (cf. Wombridge, Shropshire; Wombwell, Yorkshire) or a bulge-like topographical feature, is more likely: see also Wom Brook. The element *wamb* is relatively uncommon, and it is unclear whether it would have been applied to a concave or convex topographical feature: the word means usually 'belly': see PN Sa III 44. Wombourne is found with and without a final *e*. The O.S. adopted an inconsistent policy on its maps until c.1980, when it adopted Wombourne as the correct form. Cf. Oborne, Dorset, *(æt) Womburnam* c.974 (12th century, S.813), *Wonburna* 998 (12th century, S.895). See also Wombridgeford.

**WOMERE** an upland bog on Cannock Chase, 1 mile south-east of Brocton (SJ 9817). *Womeer* 1834 O.S. The age of the name is unknown, but it may have the same root as Wombourne (q.v.), in which case OE *wamb-mere* 'the mere or pool in the womb-like hollow'.

**WOOD EATON** – see **EATON, WOOD**.

**WOODCROFT** on south-west side of Leek (SJ 9755). *Wo(o)dcroft(e)* 1539 MinA, *Wodcrofte* 1539 MA, *Woodcrofte Grange* 1552 Pat, *Woodcroft* 1560 SHC XIII 207, c.1569 SHC IX NS 73, *Wood Croft* 1836 O.S. From OE *wudu* 'wood', with OE *croft* 'a small enclosure of arable or pasture land, an enclosure near a house', so 'the small enclosed field at the wood'.

**WOODCROSS** on the north-east side of Sedgley (SO 9294). *Woodcrosse* 1614 Inq, *Woodcross* 1895 O.S. Presumably 'the cross at the wood' or 'the wooden cross'.

**WOODEND** 1 mile south of Hanbury (SK 1726), *The Wodend* 1532 SHC 4th Series 8 81, *Fauld Woodend* 1620 SA 513/2/18/16/1, *Wood End* 1776 SRO D240/ER/1; **WOOD END** 2 miles north-east of Wolverhampton (SJ 9401), *le Wodehende* 1348 SHC XVI 8, *le Wodende* 1428 SHC XVII 121, *Wodend* 1470 SHC IV NS 171. 'The remote part of the village near the wood'. The first place is to be distinguished from Hanbury Woodend (q.v.).

**WOODFIELD** on the north side of Claverley (SO 7993). *Woodfield* 1833 O.S. In Shropshire since the 12th century.

**WOODFORD** 2 miles south-east of Uttoxeter (SK 1131), *Wodford* 1440 (1798) Shaw I 86, *Woodford* 1453 ibid. 86, *Woodeforde* 1560 SHC 1938 159, *Woodford* 1603 SHC XVIII 32, 1798 Shaw I 86; **WOODFORD GRANGE** 1 mile north-west of Wombourne (SO 8593; SMR 02634 places a deserted post-Conquest settlement at SO 85609350), *Wudeford* c.1160 SHC I 200, c.1180 ibid. 198, *Wdeford, Wdeforda* 1160x1206 SHC III (i) 215, *Woddeford* 1271 SHC V (i) 142, *Wodeford* 1286 ibid.' 158, *Woudefeud* 1306 SHC VII 157, *Woodford Graunge* 1559 SHC 1938 152. From OE *wudu-ford* 'The ford at the wood'. *Wodefeud, Woudefeud* is recorded in 1306 (SHC VII 157), and may relate to either of the above places. Woodford Grange, extra-parochial until 1900, was a grange of Dudley Priory: VCH XX 225.

**WOODGATE** 1 mile south-east of Uttoxeter (SK 1032). *the Wood gate* 1626 SRO D786/20/15. 'The gate to the wood'.

**WOOD GREEN** on the north-east side of Wednesbury (SO 9995). *Woodgreen* 1724 SA 2089/4/2/13-14. 'The grassy area at the wood'.

**WOOD HALL** 1½ miles west of Codsall (SJ 8404). *the Hall in the Wood* 1601 VCH XX 82, *The Wood Hall* 1625 Codsall ParReg. 'The hall at the wood'. The site is moated (cf. Bilbrook; Moat Brook), and may have been the location of the Forest Court of Brewood Forest (q.v.): Forest Courts elsewhere are known to have borne this name: see Ekwall 1960: 212, 531. The hall was demolished by 1835: VCH XX 82.

WOODHAM GREEN (unlocated, posibly in the Kingstone area) *Woodham Green* c.1810 SRO D240/E/C/1/15/1-74.

**WOODHEAD** 1 mile north-east of Cheadle (SK 0144), *Woodheade* 1586 SHC 1927 135, *Woodhead* 1598 SHC XVI 185, 1600 SHC 1935 203, 1609 SHC III NS 52, *Woodhead (Close)* 1779 SRO D240/A/2/14, *Wood Head (Colliery)* 1836 O.S.; **WOODHEAD** 1 mile north-east of Waterfall (SK 0951), *Woodhead* 1712 Ilam ParReg, *Wood Head* 1798 Yates. 'The head or end of the wood'. See also Gledenhurst.

**WOODHOUSE** 1 mile east of Tamworth (SK 2304; in Warwickshire until transferred to Staffordshire in 1965), *Wodehouses* 1540 BM, *Wood House* 1834 O.S.; **WOOD HOUSE** 1½ miles north-east of Stone (SJ 9235), *Woodhouses* c.1680 SHC 1919 262; **WOODHOUSE** on the north side of Biddulph (SJ 8958), *Wood House* 1798 Yates,

*Woodhouse* 1842 O.S.; **WOODHOUSE** 1 mile north-west of Whitmore (SJ 7942), *Woodhowson* 1590 WCRO CR1291/176; WOODHOUSE (unlocated) in Castern, *Wodhows* c.1450 SHC 4th Series IV 253; WOODHOUSE (unlocated) in Penkridge, *Woodhowse* 1646 Penkridge ParReg, *Wudhowse yate* 1650 ibid; WOODHOUSE (unlocated) in Paynsley, *Woodhouse* 1522 SHC 1925 121; WOODHOUSE (unlocated) near Okeover, *la Wodehuse* 1306 SHC VII 182, *Wodhouses* 1510 SRO D231M/E215; WOODHOUSE (unlocated, at Morfe), *Wodehous* 1326 SRO 1485/7/3/2, *Wodehous otherwise Rumpney landes* 1442 SRO 1485/7/4/1; **WOODHOUSE FARM** 1 mile north-west of Haughton (SJ 8521), *Wodford* 1546 SHC XI 291, *Wodehous* (p) 1327, 1332 SR, *Wood Ho* 1775 Yates, *The Wood House* 1836 O.S.; **WOODHOUSE FARM** 1 mile south-west of Tutbury (SK 2028), *Tuteburi Wodehouses* 1287 SHC VI (i) 168; *The Woodhouse, Tutbury Woodhouses* 1601 (1798) Shaw I 56, 58, *Woodhouse* 1836 O.S.; WOODHOUSES (unlocated) near Mayfield, *le Wudehuses* ?13th century SHC VII NS 153, *Woodhouses by Mathelfeld* 1309 SHC 1911 73, *Woodhowses* 1600 SHC 1935 263; **WOODHOUSE FARM** 1 mile east of Croxden (SK 0839), *Wodehuses* 1176 VCH III 226; **WOODHOUSE FARM** 2 miles south-east of Upper Arley (SO 7678; in Worcestershire since 1895), *Woddus, le Wodehouse* 1387, 1460 PN Wo 33, *Woodseaves* 1686 Plot, *Woodsease* 1695 Morden, *Woods Ease* 1752 Rocque, *Woods Eaves* 1775 Yates; **WOODHOUSE GREEN** 1½ miles west of Rushton Spencer (SJ 9162), *Wodehous(feld)* 1359 SRO DW1761/A/4/15, *Woodhouse Gre(e)ne* 1448 Antrobus, *Wodehouse* 1539 MinA, *Wo(o)dhowse* 1559 Pat, *Woodhouse Green* 1616 SHC VI NS (i) 19, 1725 SRO DW1761/B/3/128, *Woodhouses* 1793 Cary; WOODHOUSES (unlocated, in Marchington), *Woodhowsen* 1586 SHC 1927 158; **WOODHOUSES** 2 miles west of Lichfield (SK 0809), *Woodhousleye* 1374 VCH XIV 202, *Woodhousegreen* 1433 ibid, *Wood Houses* 1834 O.S.; **WOODHOUSES** ½ mile north-east of Swinscoe (SK 1348), *Wodehowsis temp.* Edward I Okeover T274, *Blore Wodhowse* 1439 SHC VII NS 50, *Wodhuse* 1507 ibid. 60, *Bloore Wodhouses* 1600 SHC 1935 296; **WOODHOUSES** ½ mile north-west of Pattingham (SO 8199), *Wodehouse* 1315 Brighton 1942: 159, *Woodhouses* 1920 O.S.; WOODHOUSES (unlocated, near Harley), *Wodehouses* 1324 SRO D603/A/Add/439, *Wodehuses* 1327 SRO D603/A/Add/451; **WOODHOUSES** 1 mile east of Yoxall (SK 1519), *Woodhouse* 1499 (1798) Shaw I 98, *The Wodhowse* 1532 SHC 4th Series 8 173. From OE *wudu-hus* 'house by or in the wood', a very common name, sometimes added to the name of a village, for minor places created by the assarting of woodland at some distance from the village. The name may in some cases denote buildings which housed those who performed functions associated with the management of woodland: see Gelling & Cole 2002: 58. Cf. The Wodehouse. *Norton Woodhouses* in Norton-in-the-Moors is recorded as *Norton Woodhouses in le Mores* in 1592 (SHC 1930 226), and in 1625 (BCA MS917/1670).

**WOODLAND** 1 mile south-east of Uttoxeter (SK 1031). *? Woodland* 1586 SHC 1927 131, *Woodland* 1666 SHC 1923 214, *Woodlands (Hall)* 1836 O.S., *Woodlands* 1872 P.O. An extensive area of former common land apportioned under the Enclosure Acts. From OE *wudu-land* 'the newly cultivated land near a wood', rather than 'an area of trees' in the modern sense.

**WOODLANDS** in Weston-under-Lizard parish (SJ 7910). *Wodelands* 1380 Blymhill, *the Wadlandes* 1666 SHC II NS 344, *Wadland (Meadow)* 1782 Weston, *Wadland* 1840 TA. From OE *wad-land* 'ground where woad was grown'. Before 1600 woad, with green leaves which produce a yellow dye which turns blue on oxidisation in air, was the only source of blue dye in Europe, and in the 15th century was the second most important import. The name *Wadeleye* (perhaps 'forest clearing where woad could be found', or 'clearing with the ford' from OE *(ge)wæd*) is recorded in Coppenhall in 1217-37: Oakden 1984: 84. See also PN Wo 221; PN Sa III 214-5.

**WOODROFFE'S**  a close-studded lobby entrance house built c.1622 3 miles south-east of Uttoxeter (SK 1129). *Woodruffe* (p) 1558 SRO D786, *Woodroofe* (p) 1666 SHC 1923 215, *Woodroffes (Cliff)* 1836 O.S. From *wudu-rofe*, 'woodruff', the herb *Hasta regia, hastula, legiscus, asperula odorata*. The name is recorded as a personal name from at least the 12th century: see for example SHC 1929 76; DES 500-501. At one time ladies carried sweet-smelling woodruff with their prayer-books when attending church. The plant name may have been given as a nickname to those who used perfumes – or perhaps in an ironical sense to those who did not. It is unclear whether this place took its name from the plant or the personal name.

**WOODSEAVES**  2 miles north-west of Norbury (SJ 7925), *Woodseaves* 1572 WCRO CR1291/175, *Woodease, Wooddease* 1594 SHC 1934 14-5, *Woodeseves* 1612 ibid.' 38, *Woodes Eves* 1613 ibid. 35, *The Woods Ease* 1679 SHC 1919 235, *Woodseaves* 1747 Bowen; WOODS EAVES  (obsolete) on the west side of the river Severn, 1 mile south-west of Upper Arley (SO 7578), *Woodseaves* 1686 Plot, *Woods Eaves* 1798 Yates. From OE *efes* 'eaves; an edge or border, especially of a wood', and in place-names 'the brow of a hill, the edge of a precipice or bank', with OE *wudu* 'a wood'. *Wodeseves* in Wolverhampton is recorded in 1460: SHC 1928 88.

**WOODSETTON**  1 mile south-east of Sedgley (SO 9393). *Wodeston* c.1400 SBT DR37/2/Box 122/36, *Wodsetton* 1532 SHC 4th Series 8 115, *Woodsetton* 1537 Inq, 1581 Sedgley ParReg, *Woodsetten* 1620 SHC VII NS 224, *Woodsatton* 1686 Yates, *Woodsutton* 1695 Morden. Possibly from OE *wudu, seten* 'a plantation', so 'the woodland plantation', but perhaps more likely to be the dative of Woodsetts/Woodseats (which occurs three times in the West Riding of Yorkshire and once in Derbyshire), which is held to be 'the animal fold in woodland'.

WOODSHUTS  (obsolete) on the north-east side of Talke (SJ 8254). *Woodseats* 1733 SHC 1944 66, 1799 Yates, *Woodshuts* 1833 O.S. Possibly from OE *wudu*, with OE *(ge)set* 'dwelling, place of residence; place where animals were kept, fold', so giving 'the house or fold in the wood'.

**WOODWALL GREEN**  3½ miles north-west of Eccleshall (SJ 7831). *Woodwall Green* 1651 SRO D1798/685/89, *Wood Wall Green* 1691 Eccleshall ParReg, 1833 O.S. Evidently to be associated with a field called *Wodewallefilt*, recorded in an undated (? 14th century) deed: SHC 1921 15. The name is from OE *wudu-wælle* 'spring at the wood'. The Green element suggests a squatter settlement.

**WOOLISCROFT**  2 miles east of Stone (SJ 9334). *Willanes-croft* 1136 SHC XII NS 154, *Willanes croftum* 1136 SHC VI (i) 22, *Willianescroft, Willanescroft* 1311 SHC IX (i) 24, *Wyllanescroft* 1310 SHC X 7, *Willanescroft* 1310 SHC 1911 75, 1321 SHC XII NS 113, *Williamescroft* 1314 SHC IX (i) 45, *Willianscroft* 1314 SHC 1911 80-1, *Wyllarddyscrofte* 1377 SHC 4th Series VI 13, *Wyllardcrofte* 1377 ibid. 12, *Wylascroft* 1442 SHC XI 233, *Wollascroft* 1488 SHC 1921 3, *Weylescrofte* 1549 (1801) Shaw II app. 12, *Weylescrofte* 1557 SHC XII 202, *Willescrosse* 1564 SHC XIII 231, *Wollascroft otherwise Willowescroftes otherwise Wyllerscroftes* 1605 SHC IV NS 6, *Wollowescrofts alias Willocrofte* 1621 SA 11/68, *Wallescroft* 1836 O.S. It seems likely that the earliest spellings have transcription errors which (not unusually) confuse -*n*- and -*v*-, and that the name is from Willavescroft 'Wiglaf's croft', from OE *croft* 'a small enclosure of arable or pasture land, an enclosure near a house'. The medial -*es*- and -*ys*- represent the possessive 'his'. See also Gruets Wood. One of the open fields on the south side of Keele was called *Wolanuscroft* in 1385 and 1398: HOK 16, 21.

**WOOLLASTON** in Bradley parish, 7 miles south-west of Stafford (SJ 8615; SMR 02000 places a deserted post-Conquest settlement at SJ 86001630). *Vllauestone* 1086 DB, *Wolaveston* 1199 SHC XI NS 16, *Wullaveston* 1200 SHC II 95, *Wollaveston* 1203 SHC III 86, *Welaston* 1280 SHC 1911 37, *Wolaston* 1368 SHC 1921 28, 1380 SHC XIII 153, *Wollaston* 1616 SHC VI NS (i) 21. 'Wulflaf's farmstead'. Cf. Wollaston and Wollashill, Worcestershire (PN Wo 196, 311).

**WOOLLEY** ½ mile south-west of Brewood (SJ 8707). *Wolveley* 1199 Ass, *Wlvelega* c.1200x1210 SHC 1939 9, *Wolvenelegh* 1280 SHC VI (i) 105, *Wulveley* 1289 SHC ibid.' 186, *Wolveleye* (p) 1313 Giffard, *Wulveley iuxta Hyde* 1273 FF, *Wolfley* c.1680 SP. From OE *wulf*, 'wolf', genitive plural *wulfa*, with OE *leah*. The Hyde adjoins this place, which lay within Brewood Forest, and references to *the Leye* in association with The Hyde in 1387 (SHC 1910 201) and 1501 (ibid. 200) may refer to Woolley. A field-name *the Wholley* in Penkridge (*Wolley* (p) 1582 (Penkridge ParReg)), may have the same derivation as Woolley: Oakden 1984: 124-5. *Wolleye*, recorded in 1291 (SHC 1911 203), is Weoley, Worcestershire: SHC IX (ii) 26. See also Wolf Low.

**WOOTTON (UNDER WEAVER)** 5 miles west of Ashbourne, in Ellastone parish (SK 1045), *Wodetone* 1086 DB, *Wotton, Watton* 1191x1194 CEC 261, *Vutton* 1275 SHC V (i) 119, *Wotton* 1274 SHC 1911 160, *Wutton* 1316 ibid. 334, *Wotton under Wever* 1424 SHC XVII 96; **WOOTTON** 1½ miles south of Eccleshall (SJ 8227), *Wodestone* 1086 DB, *Wotton* 1253 Ch, *Woderton* 1305 SHC VII 165, *Wodeton* 1341 HRO 44M69/C/93, *Wottone* 1380 SHC XVII 203, *Wotton Palment* 1599 *et freq*. Eccleshall ParReg, *Wotton Palmente* 1609 ibid, *Wootton Pavement* 1623 ibid, *Wotton, Wotton Pavement* c.1680 SHC 1919 229, *Wotton* 1686 Plot; **WOOTTONS** 1 mile south-east of Croxden (SK 0738), *the Woottons* 1656x1778 SRO D3272/5/18/1-19, *Whottons* 1836 O.S.; **WOOT(T)ON** 1 mile north-east of Quatt (SO 7688), *Wodeton in Foresta de Morf* 1255 Rotali Hundredorum (RC), 1812-8, *Le Wodeton* 1296 SHC 1911 267, *Wodeton'* 1298 TSAS LXXI 1996 27, *Wootton Green* 1833 O.S. From OE *wudu-tun* '*tun* in or by a wood'. The element *tun* is characteritic of non-forested areas, which suggests that a *wudu-tun* was near a wood, rather than in it. PN Sa I 325-5 and Gelling & Cole 2000: 258 tentatively propose 'settlement which performs some function in relation to a wood' for places of this name. The first place lies under the Weaver Hills. For the second place, the early parish registers suggest that two distinct places existed, Wootton and Wootton Palment. *Palment* is presumably to be read as Pavement, almost certainly with reference to 'the high paved way ... a part of one of these [supposed Roman roads], which seems to have been made by reason of any wet or dirty way, it being raised between two other deep ways, which lye dry enough ...' recorded in Wootton by Plot (1686 402; see also VCH I 192), evidently on the line of Wincote Lane, on the north side of which lay *'the Pavement Crofts'*: Burne 1913: XVII; see also Rue Barn Farm. Considerable doubt remains about the antiquity of this feature: the 'paved way' is said to have run east-west, but Shaw 1798: I 34 suggests that it had the appearance of continuing to Forton and Darlaston, which would put it on the line of the main road south from Eccleshall), but supposedly Roman material has been recorded from Wootton Lodge here: JNSFC LIX 1924-5; see also StEnc 622. If *Wotenhull, Wootenhulle* recorded in 1232 (SHC XII 15), *Wotenhull* in 1360 (SHC VIII 8), *Wootenhull* in 1342 (SHC XII 15) is a Staffordshire place-name, it may be associated with the 496' hill to the west of Wootton near Eccleshall, or the 382' hill to the east of Woottons. VCH III 247 fn.4 gives an early reference to *Wootton*, either the place near Ellastone, or Woottons near Croxden. Wootton in Quatt has been in Shropshire since the 12th century.

**WORDSLEY** 2 miles north of Stourbridge (SO 8886). *Wuluardeslea* 12th century, *Wolwardele* 13th century Duig, *Wolleye* 1299 SHC 1911 203, *Wordeslei* 1509 SA 2089/2/2/39-40, *Wordsley* 1578 SA 2089/2/3/1. 'Wulfweard's *leah*'.

**WORFE, RIVER** *Wrhe* c.1211 Ekwall 1928: 470, *Wurgh* 1227 ibid, *Wrgh* 1247 ibid, *Wornh* 1248 ibid, *Worth* 13th century ibid, *Worgh* 1298 TSAS LXXI 1996 27. Early spellings indicate a derivative of OE *wyrgan* 'to strangle', modern 'worry' (see PN Sa I 327): the river takes a particularly convoluted course. Finberg 1972: 148 suggests that an earlier name of the river may have been Kenn.

**WORFIELD** Ancient Parish 3 miles north-east of Bridgnorth (SO 7595). *Wrfeld*, *Guruelde* 1086 DB, 1265 Pat, *Woresfeld* 1167 P, *Wurefeld'* 1177 P, *Wurrefeld'* 1230 P, *Wurefeud* 1242 SHC 1911 11, *Wrofeld* 1271 For, *Worfield alias Worvel Holme* 1703 Shifnal ParReg, *Worfield Holm* 1747 Poll. 'Open land on the river Worfe (q.v.)': there is an expanse of relatively level land surrounding the village: PN Sa I 327. A royal charter of 1477 records *Wolueresford alias dicto Worfield* 'Wolueresford otherwise called Worfield' (SAS 2nd Series XI 1899 1-4), but no other reference to Wolueresford has been traced. In Shropshire since the 12th century. The reference to *Holm(e)* is from OE *ham* in the sense 'manor', so 'Worfield manor': see PN Sa II 10.

**WORMHILL** 1 mile north-west of Heaton (SJ 9363). *Wormhill* 1333 SHC XIV 32, *Wormhulle* 1413 SHC XVII 42, *Wormehulle* 1403 SHC XV 108, *Wormhale* 1655 Leek ParReg, *Worm Hill als Wormhough* c.1702 SRO D1260/1/1-4, *Worm Hill* 1775 Yates. From OE *wyrm* 'a reptile, a snake, a dragon' (a derivation from OE *wurma*, a purple dye and the plant from which it was extracted, is improbable, but see PN Ch I 54), with OE *hyll*, so 'the dragon's hill'. See also Woundon; Wormslow; Wormhough.

WORMHOUGH (obsolete, on the west side of Wormhill (SJ 9363)) *Wurmildehalch* 1248 (1883) Sleigh 50, *Wurnulde halh* c.1248 SHC NS IX 318, *Wormehalgh* c.1539 LRMB, *Wormehalgh* 1605 SCA MD 5649, *Wormehaughe* 1615 QSREnr, *Wormhough* 1891 O.S. The two earliest forms suggest a derivation from an OE pseronal name *Wurmild or similar, but no such name is recorded. The later spellings point towards OE *wyrm*, *halh* 'the dragon's *halh*'. See also Wormhill; Wormslow.

**WORMLOW FARM** 1½ miles north-east of Bradnop (SK 0256). *Wormlow* 1768 VCH VII 211, *Warmlow* 1842 O.S. Early spellings have not been traced, but almost certainly from OE *wyrm-hlaw* 'dragon's tumulus'. See Wormhill; Wormhough.

**WORSTEAD GREEN** 2 miles south of Walsall (SO 0196). *Walsterwode* 1271 SHC V (i) 150, *Walfteswod* 1271 SHC 4th Series XVIII 89, *Walstwude* 1300 SHC V (i) 178, *Wastewede* 1403 SHC XV 106, *Walstwode* 1419 SHC XVII 73, *Walstode* 1463 SHC IV NS 125, *Walstode, Walstead, Walstede* 15th century Duig, *Walstede Delves* 1542 SHC XI 285, *Wallestede Delvishe, Walsteddeluyshe* 1546 TSAS 3rd Series VIII (ii) 238, *Walsted* c.1560 SHC 1926 108, *Worstead (Hall)* Duig 176. Duignan 1902: 49 suggests that the first part of this name is from a family called Walstead, originally the OE personal name Wealhstod (ibid. 176), but the spellings point towards a derivation from an OE personal name such as Wælgist, with OE *wudu* 'a wood', with later corruption. *Delves* and *-deluyshe* are from OE *(ge)delf* 'a digging, a trench, a pit, a quarry': the place lies in an area of early mineral and coal mining. The place appears to have been called *Delves Green* in the 19th century (1834 O.S.), and Walstead Hall lay on the north side of the common there. The Green element suggests a squatter settlement. See also Delves.

**WORSTON**  5 miles north-west of Stafford (SJ 8727; SMR 02635 places a deserted post-Conquest settlement at SJ 87802780). *Wiveredeston* c.1193 SHC 1924 80, 1205 SHC III 136, *Wyfrideston* ?12th century SHC VI (i) 22, *Wodiston* 1271 SHC V (i) 149, SHC XII (ii) 11, *? Wymereston* 1279 SHC VI (i) 115, *Wyverston* 1286 SHC V (i) 173, *Wiverstone* 1292 SHC VI (i) 221, *Worflestone, Wythtrestone* ?13th century ibid. 24, *Wiveleston, Wyverstone, Wyfridestone, Worflestone* 13th and 14th century Duig, *Worston otherwise Wevereston* 1582 SHC XVII 229, *Worston otherwise Weaverstone* c.1737 SRO D1499. The early spellings are inconsistent, but perhaps from the OE personal name Wilfriþ or similar, with OE *tun*.

WOTTONS-LAND (unlocated) *Wottons-land* 1531 SHC XI NS 9.

**WOUNDALE**  1½ miles west of Claverley (SO 7793). *Wundenwall, Wundewell* 1221 Eyton 1854-60: III 97, *Wundenewell* 1235 ibid. 98, *Wondewalld* 1525 SR, *Wondwall* 1625 Claverley ParReg, *Woundwall* 1627 ibid, 1808 Baugh, *Woundwell* 1820 Greenwood, *Wondell* 1833 O.S. From the OE past participle *wundan* 'twisted; winding', in this case meaning perhaps 'windlass', with OE Mercian *wælle* 'a spring', and (sometimes) 'a stream', but (since there is no stream here) perhaps with its less-usual meaning 'well', giving 'windlass well'; or 'the twisted spring', though what the latter might mean is unclear. In Shropshire since the 12th century. Cf. Woundale, Shropshire.

WOUNDON, WERMDON, OUNDON (obsolete, an early name of Dunstall Hill, north of Wolverhampton (SJ 9100)) *Woundon* 1258 SHC IV 136, 1327 SHC VII (i) 249, *Wermdon* 1283 SHC 1911 186, *? Wondon* 1286 SHC V (i) 171, *Wormdon* 1300 ibid. 178, *Ounehil* 1353 (1801) Shaw II 175, *? Wyndon* 1545 SHC XII 189, *Wounden Hill* 1707 BCA MS3145/63/1a&b, *Wernden* 1745 WA II 79, *Ouen* 1802 ibid.' See also SHC 1919 167, SHC 1911 167. Some of the earliest spellings suggest OE *wyrm* 'a reptile, a snake, a dragon'. In combination with OE *dun* 'a hill', it is likely that the name was 'the hill of the dragon': see Wormhill. See also SHC 1919 167. WA II 79 suggests that c.1938 the place was remembered as *Woon-hills*.

**WREDON**  a hill 1 mile north of Ramshorn (SK 0846), which forms part of the Weaver Hills. *Reedon* 1686 Plot, *Raydon Hill* 1838 O. S. Perhaps from OE *read-dun* 'the red hill': see Raddlepits. The W is a recent affectation, perhaps influenced by names such as The Wrekin and Wrockwardine. It is unclear whether *Reydon*, recorded in 1281 (NA DD/FJ/1/298/4), is to be associated with this name.

**WRENS NEST**  a prominent heavily-quarried hill of alternate layers of lime-rich Wenlock shale and Upper and Lower Silurian Limestone 1 mile north-west of Dudley (SO 9391). *Wrosne* 1248 SHC IV 243, 1278 FF, *Wrosene* 1291 Tax, *ate Wrosne* 1293 Ipm, *atte Wrosome* 1395 ibid, *Wrennesnest* 1554 StEnc 157, *Wren's Nest (House)* 1642 SHC IX (ii) 120, *Wrens nest* 1674 WHS NS 9 (ii) 35, *Wrens Nest* 1798 Yates. PN Wo 290 and Stenton 1970: 296 fn.1 endorse Skeat's suggestion of a derivation from OE *wras(e)n* 'a band, a tie, a chain, a fetter', found as a gloss for Latin *nodus*, and probably used in some topographical sense as 'bent or twisted', with reference to what might be described as a 'hill-knot'. Places containing the element are often associated with either ancient roads or earthworks, and the meaning here is probably 'an irregular hill with the appearance of a pile of chain', possibly with reference to irregular natural weathering or the scars left by ancient quarrying. The name seems to have become modified to a more intelligible 'Wren's nest' at some time before or during the 16th century. In Worcestershire since 1844. Cf. Grimsworth Hundred, Herefordshire. See also Bowland Knotts, Yorkshire; Blawith Knott, Lancashire.

**WRESTLERS FARM, WRESTLERS WOOD** 1 mile north-east of Blymhill (SJ 8213). *the Restlars* 1798 Shaw I 110, *The Rostlers* 1832 Blymhill ParReg, *Wrestlers Barn* 1833 O.S., *Ristler's Meadow* 1841 TA. Probably to be associated with Thomas Wrestler, associated with property in Wheaton Aston in 1469: WLHC Walsall Town Chest 276. It is of interest that a mysterious 'great wrestling' is recorded at Burlaughton near Blymhill in 1289 (SHC 1921 185), and that a crowd of 20,000 gathered at Boscobel for a prize fight in 1828 (WC 29 April 1828). This area close to the Staffordshire-Shropshire boundary may by long tradition have been favoured for such gatherings.

**WRINEHILL** 5 miles west of Newcastle-under-Lyme (SJ 7547). *Wrinehull* 1225 Cl, *Wryneford* 1273 SHC VI (i) 58, *Wrime* 1278 SHC XI NS 260, *Wryme* 1298 ibid.' 249, 1307 ibid. 265, 1332 AD, *Le Wryme* 1430 AD, *le Wrimehull* 1486 AD, *Wrymhyll* c.1540 SHC X NS (i) 174, *Wrynehill otherwise Wryneford* 1593 SHC XVI 133. The first element is found in *wriman forda* 975 (11th century, S.801), *Wryneford* 1273 SHC VI (i) 58, *le Wrineford* 1322 Ipm, *Wrymford* 1377x1390 SRO Chetwynd bundle 9, *Wryngeford* 1396 SHC XV 73; see also Hooke 1983: 106. *Wryme* was the early name of the area around Wrinehill on the Staffordshire-Cheshire border. Duignan 1902: 176 suggests that the first element of *Wriman ford* was a personal name Wrim(a). Ekwall 1960 539 considers that *Wryme* might be the old name of Checkley Brook. Hart 1975: 96 postulates a Celtic derivation. Dodgson (PN Ch III 56-8), in a detailed analysis of the name, concludes that Wrinehill is named from its position on a modest ridge between Cracow Moss and Checkley Brook. Wrinehill Bridge, which is probably on the site of the ford, crosses Checkley Brook on the county boundary. *Wryme* seems to have been the name applied to a tract of land around the junction of Checkley Brook and the river Lea. *le Wryme Syche* ('the Wryme stream', from OE *sic*) is identified as a watercourse draining into Cracow Moss. The high ground at Randilow deflects these streams from their westwards course, and it is likely that this deflection explains the name, from a stem *wrig-*, from the OE verb *wrigian* 'to tend, to go forward, to bend' (cf. Modern *wry*), with the Primitive Germanic noun-forming suffix *–ma(n)*, which could produce an OE *\*wrima* 'a bend', giving *\*Wrima*, (*\*æt) Wriman* '(at) The Bend', with the genitive singular represented in *Wriman ford*, which would develop into ME Wryme, Wrime, with the spellings Wrine and Wryne perhaps due to scribal confusion or otherwise explicable philologically. It should ne noted that footnote 50 in PN Ch V (ii) 275 is from an article first published in 1967, and is superseded by the above derivation.

**WROTTESLEY** (pronounced Rotters-lee [rɒtəzliː]) in Tettenhall parish, 4 miles north-west of Wolverhampton (SJ 8501; SMR 01901 places a deserted Anglo-Saxon settlement at SJ 85000160). *(æt) Wrotteslea* c.1000 (11th century, S.1534), *Wroteslea* 1080 SHC I 182, *Wrotolei* 1086 DB, *Wrotteslega* c.1162 SHC I 183, 1167 SHC 1923 296, *Wrotteslee* 1199 SHC III 36, *Wrotele* 1221 SHC I 183, *Wurtlega* 1222 SHC IV 20, *Wrokesley* 1256 SHC VI NS (ii) 41, *Wortteslewe* c.1250 ibid.' 47, *Wrotesley* 1271 SHC V (i) 148, *Wroctesley, Wrottele* 1284 ibid.' 134, *Wrotkesley* 1285 SHC I 183, *Wrotesmere* 1286 SHC V (i) 157, *Wrottesleye* 1310 SHC 1911 74, *Roddesley* 1414 SHC XVII 25, *Rocheley* 1567 SHC IX NS 226, SHC 1925 90, *Wrottesley* 1686 Plot. Probably '\*Wrott's *leah*'. To the west of this place, at Wrottesley Lodge Farm on the Staffordshire-Shropshire border, is the supposed 'ancient city' some 3 or 4 miles in circumference recorded by Plot 1686: 394, 415 and many subsequent historians, and marked as 'Site of Supposed British Town' on the 1886 6" O.S. map. Subsequent investigations have produced no evidence of any such remains, or that they ever existed: VCH I 331.

**WROTTESLEY LODGE (FARM)** 1 mile north-east of Pattingham (SJ 8301). *Le Logge Park* 1382 Jones 1894: 195, *Logge* c.1540 Leland ii 170, *The Lodg* 1634 SRO

D3548/1. From ME *log(g)e* 'a house in a forest for temporary use; a house at the entrance to a park'. The lodge (formerly moated) to one of the three parks at Wrottesley. The 16th century spelling may refer to Patshull. See also Wrottesley.

WULFCESTRE, WULFECESTRE, WELFERCESTER, WULFERCESTER – see **BURY BANK**.

WULFHAMPTON (unlocated, in Whittimere, 2 miles north-east of Bobbington (SO 8292): VCH XX 65, 71) *Wulfhamton* late 13th century VCH XX 65, 71, *Wulfhampton* n.d. Eyton 1854-60: III 166, Shaw 1801: II 208. 'The village of the wolf', from OE *wulf* 'wolf', with *hamtun*, a relatively rare element meaning 'the village proper', to distinguish it from the outlying parts, or even 'the chief manor of a large estate'.

WULFURSYDE (unlocated, in or near Audley or Bignall) *Wulfursyde, Southwolfursyde, Northwolfursyde* 1492 SHC 1912 256-7. Perhaps incorporating the name of the Mercian King Wulfhere (658-75; see Bury Bank), with OE *side* 'side', ME *side* 'slope of a hill, especially one extending for a considerable distance' (Ekwall 1960).

WULNELEG (unlocated, in Milwich: LLRRO 26D53/1292) *Wulneleg* 1249 LLRRO 26D53/1292.

WULREDESTON (unlocated) *Wulredeston* 1214 SHC III 163. Perhaps from the OE personal name Wulfræd with OE *tun*, so ' Wulfræd's *tun*', but possibly to be associated with *Wulverdistone* (q.v.).

WULSIESHOLM (unlocated) *Wulsiesholm* 1227 Ch. Perhaps from the OE personal name Wulfsige. *Holm* is generally held to be from ON *hulm* 'a small island, a piece of land on a stream, dry ground in a marsh'.

WULVERDISTONE (unlocated, possibly near Newcastle under Lyme) *Wulverdistone* 1272 SHC VIII Ii) 151. Perhaps from the common OE personal name Wulfheard with OE *tun*, so ' Wulfheard's *tun*', but possibly to be associated with *Wulredeston* (q.v.).

**WYCHDON LODGE** 1 mile north-east of Ingestre (SJ 9825). *Wichdon Lodge* 1836 O.S. Built in 1818 for William Moore, owner of Shirleywich salt works. The name is evidently coined from the name Shirleywich (q.v.).

**WYCHNOR** 1 mile north-east of Alrewas (SK 1716; SMR 00128 places a deserted medieval settlement at SK 17781625). *Wicenore* 1086 DB, *Hwiccenofre* 11th or 12th century Sawyer 1979a: xxxvii, *Wychenofere* 1216x1272 (1798) Shaw I 125, *Wichenovere* 1236 Fees, Wiccenor 1251 SHC 1934 (i) 25, *Wychnore* 1261 SHC X NS I 293, *Wycchenovre* 1280 SHC 1911 172, *Wiethenouere* 1282 BCA MS3878/16, *Whichenovre* 1291 (1798) Shaw I 119, *Wichenovere, Wycchenore* 13th century Duig, *Wycchenovere* 1300 SHC V (i) 177, *Wochenoure* 1325 SRO DW1733/A/3/26, *Whychonore* 1329 BCA 3669/Acc1938-049/506539, *Whytchenore, Whitchnore* 1366 SHC VIII 26, *Phwychenor temp.* Edward I (1798) Shaw I 152, *Whichenore* 1476 SHC VI NS (i) 103, *Wychenowr* 1532 SHC 4th Series 8 107. From OE *ofer* 'flat-topped ridge with a convex shoulder' (see Gelling & Cole 2000: 200): the place lies on the south side of a particularly prominent flat-topped headland with a convex shoulder. The first element is almost certainly to be associated with a people known as the *Hwicce, based in Gloucestershire, Worcestershire and Warwickshire, recorded by Bede as *prouincia Huicciorum* c.730, the name Wychnor possibly suggesting a detached sub-group of the main people, or from a weak personal name Hwicca, formed from that folk-name: see PN Wo xv; Stenton 1970: 270. Cf. Whichford, Warwickshire; Whiston, Northampton. But OE *hwicce* also meant 'box, chest, coffer', and the term may have been applied to pagan

Anglian inhumation burials found here (see StEnc 691; Losco-Bradley & Kinsley 2002), where the word may be interpreted as 'cist', or it may have been applied in a topographical sense to the headland here, perhaps in the sense 'coffer'. Cf. Wychwood, Oxfordshire; Wicklewood, Norfolk; Wichenford and Whichbold, Worcestershire; Whichford, Warwickshire; and Witchford, Cambridgeshire. See also Wigford.

**WYDENHALL** (unlocated, possibly in or near Utoxeter) *Wydenhall* 1324 SHC 1911 357.

**WYKE** (unlocated, in Worfield parish) *Wyke* 13th century SA 5735/2/1/1/2, *Ewyke* 1327 SR, *Wyke* 1525 Sub. From the ME plural form of OE *wic* 'dairy farm'. In Shropshire since the 12th century. It is clear that Wyke and Wyken (q.v.) were two separate places: see SRS 3 101-2.

**WYKEN** ½ mile south-east of Worfield (SO 7694). *Wykyn* 1512 Worfield CA, 1525 SR, *Wyken* 1752 Rocque. From OE *wicum*, dative plural of *wic* 'dairy farm'. In Shropshire since the 12th century. See also Wyke.

**WYMERSTRETE, WRIMESTRETE** (unlocated) *Wymerstrete, Wrimestrete* 1281 SHC VI (i) 113. William and Henry Wymer de Stafford are recorded: the family name may be found in this place: SHC VI (i) 23 fn., 114.

**WYMONSALLE** (unlocated, perhaps in or near Madeley) *Wymonneshale* 1346 SRO DW1082/A/4/4, *Wymodesshale* 1359 SRO DW1082/A/4/6-7, *Wymonsalle* 1365 SRO DW1082/A/4/7. 'Wigmund's or Widmund's *halh*'.

**WYMUNDEWOLDE** (unlocated, possibly in Bramshall, or associated with *Wimundsway*, said to be the road from Anslow through Rough Hays (StEnc 665), although that road appears on modern maps as Hopley Road. Or perhaps Wymeswold, Leicestershire). *Wymundewolde* 1280 SHC VI (i) 110. 'Wigmund's *wald* or wood'. If from OE *wald* ('high forest land; open upland ground'), perhaps the only place in Staffordshire incorporating the element, but see also Wolmore.

**WYNBROOK, WYNBANK FARM** 1 mile south of Audley (SJ 7949). *Wynbrook (House)* 1628 SHC 1910 223, *Wane Brook, Waine Brook* 1733 SHC 1944 15, *Win Brook* 1833 O.S. Possibly 'the brook of the wagons', from OE *wægen* (wagons were driven through water so that the wooden wheels would swell to retain their iron rims), and 'the bank or steep slope at wagon brook'. Or perhaps from OE *wagen* 'quagmire' (see EPNE i 151), giving 'the boggy brook'. Wynbrook lies on Dean Brook, which would seem to be a more recent name.

**WYNDFORD MILL** in Blymhill parish (SJ 8014). *Molendinum de Waynford* c.1290 Giffard, *Waynford* 1272 FF, *Windford Mill* 1833 O.S. Perhaps from OE *wægen* 'cart, waggon', with OE *ford* and (later) 'mill', meaning 'the mill at the ford used by waggons or carts': wooden-wheeled vehicles were driven into water in dry weather to ensure that the wheels did not shrink, allowing the iron rims to loosen. Or possibly from OE *wagen* 'quagmire' (see EPNE i 151), giving 'the boggy ford'. This place (and Wyndford Pool) lie on *Wyndford Brook* (1816 VCH IV 70).

**WYNFORD BROOK** a tributary of Dean Brook. See Wynbrook. *Windeford Brook* is recorded in 1817 (BM), probably from OE *wægen-ford* 'wagon ford', notwithstanding the spelling with -*d*-.

**WYNRESTON** (unlocated) *Wynreston* 1251 Ipm.

**WYNSTANSLEY** (unlocated, in Eccleshall) *Wynstansley* 1329 SRO D(W)1082/A/2/1. Perhaps 'Wyn(n)stan's *leah*'. The place may be associated with Wistanswick (q.v.).

**WYPPERSLEY** (unlocated) *Wyppersley* 1269 SHC 1910 111.

**WYRE HALL** (obsolete, on the east side of Penkridge (SJ 9214): VCH V 107) *Wyr(r)all end croft* 1598 Ct. From OE *wir*, *wȳr* 'bog myrtle (*Myrica gale*)', with OE *halh*. Cf. Wyrley; Gailey.

**WYRLEY BANK** – see **CHESLYN HAY**.

**WYRLEY, GREAT & LITTLE** in Norton Canes parish, 6 and 5 miles north of Walsall (SJ 9907 and SK 0105). *Wireleia* (Little Wyrley) 1086 DB, *Wirlega* 1170, 1176 P, *Werlaye*, *Werley* 1279 SHC VI (i) 93, *Great Wyrleye* 1300 For, *Little Wyrle* 1293 Ass. From OE *wir*, *wȳr* 'bog-myrtle (Myrica Gale)', with OE *leah*. Another OE word for the same plant was *gagel* (cf. Gailey).

**WYSO** (unlocated, possibly in Bramshall) *Wyso* 1280 SHC VI (i) 110.

**WYSTANSLEY** (unlocated, in Eccleshall parish). *Wystansley* 1329 SRO D(W)1082/A/2/1.

**WYTENACRE** (unlocated, in the Sandon/Smallrise area) *Wytenacre* 1285 SHC VI (i) 146. The place has been identified as Wheatenacre (SHC VI (i) 146), which has not been traced.

**WYVERESHALE** (unlocated, in Leek) *Wyvereshale* 1324 SHC X 53. See also Wiversall.

**YARLET** 4½ miles north of Stafford (SJ 9128; SMR 01760 places a deserted Anglo-Saxon settlement at SJ 91452915). *Erlid* 1086 DB, *Erlida* 1166 SHC 1923 297, *Erlide* 12th century, *Erlode* 1280 SHC VI (i) 111, *Erlide*, *Erlyde*, *Herlide* 13th century, *Erlede*, *Erlide* 14th century, *Erlid* 15th century Duig, *Yarlett otherwise Yerlydclaye* 1566 SHC XIII 259, *Yerlett otherwise Yerletclay* 1590 SHC XVI 103, *Yarlett* 1591 SHC 1930 164, *Erlid*, *Yerlett* 16th century Duig. The first element is probably OE *ear* 'gravel', or *earn* 'eagle'. The second element is more difficult. Ekwall 1960: 542 suggests OE *hlid* 'slope', a side-form of OE *hlid* 'hill-slope', an element only rarely found in place-names, which would be most appropriate for this place, which lies on a particularly long slope, but PN Sa I 189 suggests that such a meaning is unlikely, and points out that the only recorded OE word *hlid* is the ancestor of modern *lid*, and sometimes had the further meaning 'door, window-shutter', but a topographical meaning is unclear. *Clay* appears to be from OE *clæg* 'clayey'.

**YARNFIELD** 2 miles west of Norton Bridge (SJ 8632). *Ernefeld* 1266 Duig, *Hernef* 1272-3 SHC XI NS 242, *Ernefen* 1300 SHC VII 75, *Yernefyn* 1558 SHC XII 233, *Yernfyn*, *Yarnefylde* 16th century Duig, *Ernefeild otherwise Ernfen* 1610 SHC NS III 24, *Yarnfield* 1646 SHC 4th Series I 281, *Earnefield(e)* 1663 *et freq.* Swynnerton ParReg. Probably from OE *earn* 'an eagle', with OE *feld* 'open land', alternating with OE *fen* 'fen, marsh', giving 'open land with the eagles' and 'marshy land with the eagles'. The eagle to which the name refers is likely to have been the white-tailed eagle (*Halitus abicilla*): see Gelling 1987: 173-181. It has been noted that OE *gearn* 'yarn' may have had some transferred meaning 'guts', perhaps in place-names 'something extended in length', or 'something long and narrow', and that element might be incorporated in Yarnfield, Wiltshire: Kristensson 2000: 4-5.

**YATTON** (unlocated, possibly near Rushton Spencer) *Yatton* 1479 SHC XI 241. Probably 'the *tun* at the gate or pass', from OE *geat tun*.

**YATE** (unlocated, in Cheddleton) *Yat* c.1540 SHC X NS I 174, *Yeate* 1666 SHC 1925 222. Yat is from OE *geat*, 'gate, pass': it is not possible to say in what sense the word was used here.

**YEATSALL** 1 mile west of Abbots Bromley (SK 0624). *Achesale* 1307 SHC NS XI 183, *Aythesal* c.1313 SRO D986/52, *Hathesale* 1336 ibid.' 187, *Attesale* 1350 SHC XI NS 30, *Yateshall* 1374 SRO D(W)1721/3/32/20, *Attesole* 1381 ibid.' 32, *Adesale* 1402 ibid. 44, *Atesale* 1402 ibid.206, *Atteshale* c.1435 ibid. 42, *Atsale* 1493 ibid. 197, *Adesall* 1616 SHC VI NS (i) 35, *Adsall* 1747 Bowen, *Yeatsall or Adsall* 1836 O.S. Possibly from 'Ætti's *halh*'. The farmhouse known as Adsall is said to have burnt down before 1831, and when rebuilt was named Yeatsall: StEnc 6, 695.

**YELD HOUSE** (obsolete) on the west side of Tunstall (SJ 8451). *la Helde* 1327 SHC VII (i) 206, *Yeld Hill* 1628 VCH VIII 92, *Yeld House* 1836 O.S. From OE *helde* 'a slope'.

**YELLS FARM** 1 mile south-east of Shareshill (SJ 9505), *le held* 1562 Ct, *The Yell* 1834 O.S.; **YELL BANK** 2½ miles north-west of Gnosall (SJ 2380), *le helde* 1583 Ct, *Yell Bank* 1833 O.S. A common name in Staffordshire, from OE *helde* 'a slope'. See also SHC 1921 30 fn.

**YELPERSLEY TOR** on the west bank of the Manifold Valley at Wetton (SK 0955) '...where River Manifold enters it...': Plot 1686: 174; 'In and about the second Inlet of Manifold, under Yelpersley Tor...': Plot 1686: 175. *Yelpersley Torr* 1686 Plot 172, 174, *Yelpursley Ptar* 1750 Pococke 1888-9. A curious name for the place where the river Manifold starts its subterranean journey to Ilam. Halliwell gives *yelper* to mean 'a young dog, a whelp', and EDD 'avocet' and 'redshank', but the derivation remains uncertain. The 1750 spelling of Tor is evidently a Pococke eccentricity: he mentions '... Ptar or cliff ...': Pococke 1888-9: 214.

**YELSWAY or YARLSWAY LANE** On the north-east side of Caldon (SK 0849). Early spellings have not been traced, but there can be little doubt that the name is from OE *eorl* (ON *jarl*), with OE *weg* 'way', giving 'the earl's way', the *via comitis* of the Earls of Chester (or from the pre-Conquest Earls of Mercia: SHC 4th Series 19 11), also found in Earlsway House, Rushton Spencer (q.v.), and Yerley Farm and Yerley Hill, Okeover (q.v.). Yelsway or Yarlsway is the name in local use; the formal name is Earlsway. See also VCH II 279.

**YENBROOK FARM** ½ mile south-west of Abbots Bromley (SK 0723). *Byondebrok* 1324 SHC 1937 123, *Byendeyebrok* c.1335 ibid.' 121, *Byendebrok* 1330 ibid.' 127, *Byndbrok* 1333 ibid.' 129. From OE *begeondan* '(place) beyond, on the other side of', with OE *broc* 'brook'. The place lies on the west side of Mires Brook: Abbots Bromley lies to the east of the stream.

**YEOLBRIDGE FORD** (unlocated, in or near Walsall) *Yeolbridge Ford* 1599 DRO D260/M/T/1/32.

**YERLEY FARM** ½ miles north-west of Okeover (SK 1548). *Urlesweye temp.* Edward II SHC VII NS 7, *Yell(e)y* 1538 Ipm, 1547 Okeover, *Early* 1775 Yates, *Yerley* 1799 Okeover T31. From OE *eorl* (ON *jarl*), with OE *weg* 'way', giving 'the earl's way', the *via comitis* of the Earls of Chester, also found in *Earlsway House*, Rushton Spencer; *Yelsway* or *Yarlsway Lane* (q.v.) in Caldon; and *Urlesweye temp.* Edward II (Bodleian) in Okeover. See also VCH II 279.

**YEWTREE FARM**  2½ miles south-east of Madeley (SJ 8043). *Hewtree* 1601 JNSFC LXIII 1928-9 41, *Ewe Tree* c.1630 SHC II (ii) 14, *Ewtree* 1644 SHC 4th Series I 204, *Yew Tree* 1718 Keele ParReg. Self-explanatory.

**YIELDFIELDS HALL**  1 mile north of Bloxwich (SJ 9903). *Yeld feldes* 1549 VCH XVII 179, *Yieldfields* 1596 ibid. 179. From OE *helde* 'a slope'.

YOMBERLEY  (unlocated, perhaps near Tittesworth) *Yombele, Yombelega, Yomberley* (medieval, undated) SHC IX NS 319.

YORNBURI  (unlocated, in or near Alton) *Yornburi* 1275 SHC V (i) 119. A curious name. If not a mistranscription of Bunbury (q.v.), the first element is uncertain, but may be from OE *eorne* 'a duel, combat', or from OE *earn* 'an eagle', with OE *burh* 'a fortification', often applied to earthworks, here perhaps the name of the iron-age hillfort largely obliterated by the construction of Alton Towers: see VCH I 334. See also Ina's Rock; Slain Hollow.

**YOXALL**  Ancient Parish 7 miles north-east of Lichfield (SK 1419). *Iocheshale* 1086 DB, *Yoxhal* 1222 Ass, *Jokeshale* 1236 SHC 1911 404, *Iokeshale* 1242 Fees, *Joxhale* 1252 Rolls, *Yoxhale* 1284 SHC 1911 40, *Jokesal* 13th century Duig, *Yokeshale, Joxhale* 14th century ibid, *Oxall* c.1570 SHC 1931 216, *Yoxsall* 1589-90 SHC 1930 5, *Yoxal* 1682 Browne. A puzzling name. The first element is perhaps from OE *geoc*, with several meanings including 'yoke, yoke of oxen', 'a measure of land' (i.e. the area a pair of oxen could plough in a day, notionally a quarter of an acre or sulung, or, according to Ekwall 1960: 261 *sub nom.* Ickham, 50 to 60 acres, or 'a small estate or manor', or (more likely given the medial -*es*-) from the ON personal name Jókell, with an early reduction to Jóke (cf. Yokefleet, Yorkshire). The generic is OE *halh*, perhaps here in the sense of 'a small valley', or 'a piece of low-lying land by a river'. The place, which lies in the valley of the river Swarbourn forming a narrow side-valley opening off the Trent, lay within Needwood Forest. Yoxley lies 1 mile south-east of Hixon (SK 0124), but early spellings have not been traced. Cf. Yockleton, Shropshire.

**YOXALL LODGE**  2 miles north-east of Yoxall (SK 1522). *Yoxall Lodge* 1658 DCL 380, 1771 DRO D3155/C5227, 1786 Barton under Needwood ParReg. 'Reputed an extra-parochial place in the Forest of Needwood': ibid.' *See* also Erdeswick 1844: 279.

**YOXLEY**  1 mile south-east of Hixon (SK 0124). See Yoxall.

## Elements found in Staffordshire place-names

The Anglo-Saxons, whose language we now call Old English, had a very sophisticated vocabulary which included a vast range of precise topographical terms used as elements in the formation of place-names. Although many appear to be synonymous, it is likely that an Anglo-Saxon peasant living much closer to the land would have readily distinguished them. Many developed to describe specific physical features of the landscape for which we have no one-word equivalent today. Modern research and fieldwork is slowly unlocking those precise meanings. The following are some of the principal Old English (and other) place-name elements found in Staffordshire place-names, their meanings, and place-names which are likely to incorporate the elements:

*æcer*, modern *acre*, has been found to have the very specific meaning 'small piece of cultivated land on the margin of a settlement', and examples of names containing the word can be grouped into three categories according to their proximity to heath, marsh, or high moorland. Found in Acre Head, Batchacre, Fotheacres, Handsacre, Hundred Acres, Lynacre, Megacre, Spot Acre, Whitacre Farm, *Whyacre*, *Wytenacre*, Upper Acre, and Wall Acre.

*bearu*, dative singular *bearwe* 'a grove or small wood', found in modern spellings as -barrow, -bear, beare, -ber, -borough.[488] Found in Ackbury, *Barewehull*, Barrow Hill, Barrow Moor, Bearwood, Crowborough, Fairboroughs, Hazel Barrow, and Sandborough.

*bece* 'a stream or steep-sided stream valley', and traced in the elements -batch, -badge, -bage, -beach, -bech.[489] Found in *Batchley*, Beech, *Comberbach*, *Goltherdesbeuch*, Gradbach, *Haselbache*, Hawkbach, Henbaches, Herbage, Holbeche/Holbeach, Humpage Green, *Sowsbetch*, and Westbeech.

*beorg* 'a small rounded hill', often found as -barrow, -beare, -ber, -berry, -borough, -burgh, -bury.[490] Found in Barrow Hill, Barrow Moor, Broughton, Fairboroughs, Gainsborough Hill, Hazel Barrow, Hollingbury Hall, *Mucheberge*, and Windy Arbour. Although in Staffordshire Old English *hlaw* was normally applied to tumuli, or hills or mounds with the appearance of tumuli, in some cases Old English *beorg* may have been used: cf. Barrow Hill, on the Staffordshire-Worcestershire boundary.

*bold, botl, boðl* 'a dwelling, a house'.[491] Found in Bold, Booden Farm, Bull Bridge, Newbold, Newbolds, and Nobut.

*bre* (PrWelsh) 'hill',[492] found in Brewood.

*brec*, ME *breche* 'land broken up for cultivation' (literally 'breach'). Names incorporating this element are usually of ME origin,[493] and seem to indicate heathland rather than

---

[488] Gelling & Cole 2000: 221-2.
[489] Gellin & Cole 2000: 3-4, where Holbeche is cited as perhaps the only Staffordshire example.
[490] Gelling & Cole 2000: 145-50. Campbell 1986: 113-4 suggests that in some cases the meaning may be 'a superior hall, a castle, a mansion', and a possible derivation from Offa's law code, since the element is found in place-name formations in the North and Midlands but not in Wessex.
[491] EPNE i 43-5.
[492] Gelling 1984: 128-9.
[493] Gelling & Cole 2000: 266-7.

woodland reclamation.[494] Found in The Bratch, Breach Mill, The Breach, The Breatch, Breech Coppice, and Pyebirch Manor.

*broc* a very common place-name element in Staffordshire meaning 'a minor watercourse characterised by a muddy bed', often with grass-like plants, where the water is likely to come from rainfall run-off.[495] Apart from numerous stream-names, the element is found in Brockley Moor, Brockmoor, Brockton, Brockton Grange, Broughton, and Stallbrook.

*bryce* 'breaking', used of 'newly cultivated ground', found in Long Birch, Harvington Birch and Pyebirch Manor.

*brycg* 'a bridge', sometimes 'a causeway', giving -bridge, -brig.[496] Found in Beobridge, Bridgeford, *Bridgewood*, Bull Bridge, Combridge, Dove Bridge, Hanging Bridge, Maidensbridge, *Monksbridge* (Egginton Bridge), Salter's Bridge, *Stare Bridge*, *Stockenbridge*, *Welfordbridge*, Willowbridge, and *Wombridge(ford)*.

*burh, burg*, dative singular *byrig, byrg* 'a fortified place, an ancient earthwork or encampment, a Roman station or camp, an Anglo-Saxon fortification, a castle of post-Conquest date, a fortified house or manor, a monastery'[497] and, later, 'a manor', 'a fortified town' and, later still, 'a town, a market town, a borough'. The meaning 'a fortified place' is the most common in older place-names, with 'manor' the most likely meaning in post-Conquest place-names.[498] Found in *Aldeburge, Archberry*, Berry Hill, Berry Ring Farm, The Berth, Blithbury, Bloomsbury, Borrowcop Hill, *Borway Ford*, Broadfields, Brough Hall, Broughton, Bunbury Hill, *Bur Walls*, Little Burbrook, Burcot, Burf Castle, *Burford*, *Burhall*, Burleyfields, *The Burleys*, Burston, Burton, *Burtone*, Burton upon Trent, *Burwey*, Bury, Bury Bank, Bury Farm, Bury Hill, *Bury Hill*, Bushbury, Curborough, Gainsborough Hill Farm, Hanbury, Hobbergate, Knotbury, *Newborough, Oldbury*, Padbury (Lane), Shugborough, Tutbury and Wednesbury.

*burna* in Old English had the meaning 'a stream or brook', generally denoting minor watercourses with clear water, often derived from springs, and with gravelly beds and grass-like plants. The same word is found in modern Scottish 'burn'. The element is found in Boney Hay, Bourne Brook, Bourn Brook, *Burnaham/Bornam, Gutheresburn*, Harborne, *Lekebourne, Lilleborn, Ludburn, Patyngehamborne, Stamberlowe, Stanburneford*, Swarbourn, *Woburnshawe*, and Wombourne.

*camp* in its earliest use perhaps 'uncultivated land on the outskirts of a Roman villa or town', but later to mean 'an enclosed piece of land, a field',[499] and later applied to earthworks, especially those considered to be Roman, found in Camp Farm, Camp Hill, Campfield (Wood), and *Tilbury Camp*.

*car, carr* a name found only in the north of the county, from ON *kjarr* 'marsh, wet moor, boggy copse'. The name is so often found linked to that of the alder that the word may

---

[494] PN Sa IV xvii-xviii.
[495] Gelling & Cole 2000: 67.
[496] Gelling & Cole 2000: 67-70.
[497] Stenton 1970: 320-1 notes that one meaning of *burh* was 'monastery', perhaps from the enclosure which surrounded the monastic buildings; see also VEPN II 77.
[498] EPNE i 58-62.
[499] Gelling 1988: 75-8.

mean 'wet place with alders'.[500] Found in Alder Carr, Carry Coppice, *Car House*, Car House, Carr Wood, *Carr*, Carry Coppice, High Carr, Hole Carr, and Moss Carr.

*cester* 'a walled town, a city, a (Roman) town, an old fortification',[501] found in Chesterfield, *Chestrehurst*, Chesterton, and Rocester.

*clif* 'a cliff, a bank, a rock', and giving the suffixes -cliff, -cliffe, -ley. The element is generally applied to slopes exceeding 45 degrees, with *helde* used for less steep slopes,[502] and is found in *Castle Cliff, Cliff*, Cliff Vale, Cliff Wood, Clifton Campville, The Clive, *The Clyves, Haukesclyf*, Middle Cliff, Ravenscliffe, Rea Cliff Farm, Sharpecliffe, Stoney Cliffe, and Thorncliff.

*clough* a common name in the North Staffordshire moorlands and with variants in the northern counties, but not found south of Stone. From Old English *cloh*, not independently recorded, which Duignan[503] believed to mean 'a ravine or narrow valley with steep sides', usually forming the bed of a stream, but which seems to have been applied to less pronounced or secondary features.[504] The old pronunciation was as in 'bough', but is now 'cluff'. Examples include Bullclough, Clough Head, Clough House, Harper Cloughh, Hawkeswall Clough, *Hoarse Clough*, Hollinsclough, Oakenclough, *Outclough*, Pyeclough, Raven's Clough, and School Clough.

*cot*, dative singular *cote*, dative plural *cotum* 'a cottage, a shelter, a hut',[505] found in Allscott, Amblecote, Beffcote, Bescott, Bitterscote, Brancote, Broncott, Calcot Hill, *Cherlecot*, Cotes, Coton (6 places), Cotwalton, Dalicott, Draycott (5 places), Dun Cow's Grove, Glascote, Goscote, Halfcot, Normacot, Northycote, *Oscote*, Swancote, Syrescote, and Trescott.

*croft* 'a piece of enclosed land, a small piece of arable'.[506] The element is found in Badger's Croft, Bancroft, *Calverecroft, Calver Croft*, Castle Croft, Castlecroft, Crow Crofts, Gorsty Croft, Grimescroft, Hobcroft (Farm), *Horsecroft Farm*, Huntecroft, Leacroft, *Lime Croft*, Longcroft (Farm), Lyncroft, Mesty Croft, Penny Crofts, Ryecroft, Rye Croft, *Salter's Croft*, Westcroft, Woodcroft, and Wooliscroft.

*cruc, crȳc* (cf. PrWelsh *crug*) 'a hill, a tumulus', found as crich, crook.[507] The element is found in Penkridge.

*cumb* 'a short, broad valley, usually with three fairly steep sides', also 'a cup, vessel', perhaps found in place-names in a transferred topographical sense. The word is generally held to derive from the ancestor of Welsh *cwm*, but may be associated with Old English *cumb* 'cup, vessel', used in a topographical sense influenced by the Welsh word.[508] The names Combes Brook, Combridge, Compton (6 places), Congreve, The Coombes, Coombesdale, Cumbwell Brook, Hoccum and Romescumbe probably incoporate the element.

---

[500] A meaning favoured by Rackham 1990: 108. Camden 1674: 121 notes 'Car, a low watery place where alders grow, or a pool'.
[501] Gelling 1988: 151-3.
[502] Gelling & Cole 2000: 153-6.
[503] 1902: 42.
[504] Gelling 1984: 88; Gelling & Cole 2000: 101-2.
[505] EPNE i 108-9.
[506] EPNE i 113.
[507] Gelling & Cole 2000: 159-63.
[508] Gelling & Cole 2000: 103-7.

*cutte* (ME) 'a cut, a water-channel',[509] perhaps found in Cotwall End.

*dæl* (ME *dale*) 'a dale, valley'. Names incorporating Dal- and -dale tend to be found in areas of Scandinavian influence, and will usually contain Old Norse *dalr*, Old Danish and Old Swedish *dal* 'valley'. It has been noted that late Old English *dæl* may also have meant 'pit, hollow'.[510] Found in *Bugsdale*, Dagdale, Dale Brook, Dale House, *Dale Torr*, The Dale, Dalesgap, Dearndales, Dearnsdale, Dimming's Dale, Dimmingsdale, Dimmins Dale, Greendale, *Grindley*, *Hetelsdale*, *Holedale*, *Holindale*, Huddale (Farm), Leadendale, Mickledale, Milldale, Mill Dale Farm, *Quennedale*, Rakes Dale, and *Stagdale*.

*denu* 'a main valley', found as -dean, -den, as in Croxden and Musden Grange.[511] Found in Ballington, Croxden, Dean Brook, Dydon, *The Holden*, *Horden*, Kniveden, Merridale, Musden Grange, and Ounsdale.

*dun* 'a hill'. The various names used by the Anglo-Saxons to mean 'hill' were not synonyms, but each described a particular type or shape of landscape feature. A *dun* is generally (but not invariably) of moderate height with a relatively flat area on or near the summit offering a preferred settlement site. The element would indicate to an Anglo-Saxon both the topography and the probability that the settlement would be of a high status.[512] Found in Acton Hill, Ankerton, Brereton, Butterton, Calton, Cauldon, Hill Chorlton, *Cuttesdon*, Down House Farm, Dunwood, Elkstone, *Elmdon*, Endon, Grindon, Haddon, Haden Hill, Hatherton, Highdown (Cottage), Hixon, *Hoddesdone*, Huntington, Knivedon, Longdon, Longsdon, Ludstone, Lyndon, Ribden, Rowdon Lanes, Sandon, Saredon, Seisdon, Shebdon, Slindon, Snowdon, Standon, Swindon, Wetton, Woundon, and Wredon.

*ea* 'a major river', as opposed to a minor watercourse,[513] found in Water Eaton (but not Church Eaton, where the first element is *eg*, denoting a settlement on a raised area in wet land).

*ecels* 'an addition, land aded to an estate', a common element in the North Midlands,[514] and found in Echells and Neachells.

*eg* 'island' in the conventional sense, but also raised ground which provided safe habitation sites in marshland. The element used in this sense almost certainly dates from the earliest Anglo-Saxon period – of all the elements found in place-names recorded before c.730, it is the most common – and the gradual disuse of the term after that period probably accounts for the use of various other words (such as *dun*, *hamm*, *halh*, *hop*, and ON *holmr*) to describe this type of site.[515] Examples of these early names are Andressey, Anglesea, Bradney, Bradney Wood, *Broad Eye*, Chebsey, Doxey, Church Eaton, Gamesley, Goldie (Brook), Kemsey Manor, *Mungesfordeseye*, Nield, Pinchley, Tuppenhurst and Wood Eaton. All the places are close to rivers or streams, or form higher points of dry land in wet ground.

---

[509] EPNE i 120.
[510] Gelling & Cole 2000: 110-2.
[511] Gelling & Cole 2000: 113-5.
[512] Gelling & Cole 2000: 78, 164-7.
[513] Gelling & Cole 2000 14-6.
[514] EPNE i xxviii.
[515] Gelling & Cole 2000: 37-40.

*feld* 'open land' (cf. Modern *veldt*), probably used by the Anglo-Saxons from the earliest period for uncultivated areas used for common pasture, and incorporated into settlement names when arable encroachments forming part of new settlements were made.[516] Found in Alstonefield, Ashfield Brook, *Bernefeld, Betsfeilde*, Black Field, Blithfield, *Bowelles Felde*, Broadfields Farm, Brownsfields, Chatfield, Chesterfield, Christiansfield, *Cokefeld, Cumberfield*, Enville, Fauld, Fawfieldhead, Field, Field House (Farm), *Fieldhouse*, Fieldhouse Farm, Finchfield, Highfields, Leafields (Farm), Lichfield, Mayfield, Ounsdale, Priestfield(s), Shawfield, Shelfield, Statfold, Stychfield, Thursfield, Upfields, *Watherfeld*, Wednesfield, Whitfield, Worfield, Yarnfield, Yieldfields Hall.

*fen* perhaps a special term for a linear marsh,[517] found in Fennel Pit Farm, Fenton, Fulfen, Swinfen, and Yarnfield.

*ford* an element in use from an early date which has not changed its meaning, although sometimes it may have been used in the sense 'causeway', and (as one of the most common elements) found in numerous Staffordshire place-names. The element *(ge)wæd* was also used for fords, but the only trace of the word found in Staffordshire is Wade Lane.[518] Names incorporating the element *ford* include Apesford, *Asshforde, Ayleslade Ford*, Blithford, *Boreway Ford*, Great Bridgeford, Brindley Ford, *Broad Ford, Chelsford*, Clanford, Comberford, Crateford, *Crawford*, Darnford, Denford, *Drakeford*, Elford, Elford Heath, Ford Brook, Ford Farm, Ford Green, Fordhouses, Ford Wetley, Hanford, Hollyford, Keeling Ford, Kingswinford, Latherford, *Lymford, Mareford, Robins Hood Ford, Rocheford, Salterford, Saltford*, Sandyford, Seighford, Somerford, Stableford, Stafford, *Stanburneford*, Standeford, Stanford Brook, *Stockford Green*, Stonyford, *Stonyforde*, Strongford, Swynnerton, Teanford, Walford, *Warwicksford*, Watford Gap, *Weldfordbridge*, Whitleyford, Wigford, *Wombridge(ford)*, Woodford, Woodford Grange, Wyndford Mill, Wynford Brook, and *Yeolbridge Ford*.

*grǣfe* meaning 'coppice', probably a prominent managed wood of modest size which could be described by its colour (e.g. Whitgreave), shape (? Orgreave), or relationship with a topographical feature (e.g. Congreve). The element, particularly common in the West Riding of Yorkshire and West Midlands,[519] is also found in Staffordshire in Boarsgrove, The Chuckery, Dun Cow's Grove, Gillity Greaves, Merril Grove, and *Wildboarsegreave*. The word *grǣfe* (also found as *graf, grafe*, and *grafa*), is probably associated in some way with Old English *grafan* 'to dig', *grafa, grǣf* 'pit, trench'. It has been suggested that 'grove' is the more likely unless the first element suggests 'pit' or similar, or there is such a feature in the vicinity.[520]

*(ge)heg* 'a fence, an enclosed piece of land', a very common element in Staffordshire and Shropshire, often with the meaning 'an administrative bailiwick within a Forest area', arising from an amalgamation of Latin *haia*.[521] Found in *Bennetshay*, Boney Hay, The Bosses, Bowhill Farm, Bryan's Hay, Calf Heath, *Childerhay, Cockshut Hay*, Cowhay, Fawley, Fegg Hayes, *Fensay*, Freehay, *Glashoushay*, Golling Gate, Hay House, Hayhill, Hays, Hayes Head, Haysgate, Great Haywood, Little Haywood, Heath Hayes, Hey House, Hey Sprink, High Carr, Hollin Hay, Horton Hay, Ivetsey, Jeffryns Hays,

---

[516] Gelling & Cole 2000: 310-11.
[517] Gelling & Cole 2000: 44-6.
[518] Gelling & Cole 2000: 71-2.
[519] Gelling & Cole 2000: 226-8.
[520] Gelling & Cole 2000: 226-30.
[521] EPNE i 214-5; PN Sa III 66.

*Levenodeshay*, Little Hay, *Longhay*, Maple Hays, Medleywood Barn, Newhay, New Hay, Oldhay Top, Ox Hay, Oxhey, Pearce Hay, Pipehay, Ring Hey, Rough Hay, Rowney Farm, Ryder's Hayes, Scot Hay, Stockley Park, Streethay, Whitemoor Haye, Wildhay Brook, *Wolfedalehey*, and Wolfhay.

*hǣð* 'heath, heather',[522] found in Haddon, Hadley End, Hamleyheath, Hatton, The Hattons, Heathylee, Headless Cross, Heath Hayes, Heath Hill, Heath House (Grange), Heath Town, Heathcote Grange, Heathton, and Heathylee.

*halh*, dative singular *hale* related to *holh*, modern 'hollow', and usually meaning in the Midlands 'a remote narrow valley or sunken place', commonly described as 'a nook or corner', but also with other quite different meanings. In an administrative sense the word can mean 'a piece of land projecting from, or detached from, its main administrative unit', although there are no obvious examples in Staffordshire. Another meaning is 'a nook, taking the form of land between rivers or almost enclosed by a river bend, a piece of low-lying land by a river, a haugh', or perhaps 'raised ground in marshland',[523] for example Edingale, north-east of Lichfield. It is worth repeating the observations of Mawer,[524] who felt that 'it is at times difficult to feel that it necessarily means anything so precise ... and one is inclined to be as general as Simeon of Durham was when he speaks of *Hearrahhalch, quod interpretari potest locus Dominorum*, and makes it mean simply place'. *Halh* is a very common element, and is found in Abnalls, *Aluredeshale/Haluredeshale, Anewardeshale*, Baden Hall, Black Hough, Blackhalves, Blakenhall, Blundies, Broom Hall, Broomhall Grange, Brough Hall, Bucknall, Chestall, Codsall, Coppenhall, Cracow Moss, Cronk Hill, Croxall, Eccleshall, Edingale, Ellenhall, Ettingshall, Ferny Hill, Gighall Bridge, Gnosall, Gornal, Haughton, Lonco Brook, Moddershall, Morghull, Muchall, Pelsall, Pillaton, Pottal Pool, Ravenshall, *Romenhale*, Rownall, Rushall, Sheriffhales, Stonnal, Tettenhall, Thornhill, *Tymburhale*, Walsall, Wethal(es), Willenhall, Wormhough and Yoxall.

*ham* 'a village, manor, homestead'.[525] Place-names likely to include the element are Audnam, *Burnaham/Bornam*, Gateham, Hamley (House), *Hampton*, Hamstead, Marchington, Oakham, Pattingham, Hamstall (Ridware), Trentham, and *Wulfhampton*. Sometimes in south Staffordshire (e.g. Claverley, Pattingham, Tettenhall, Worfield) the element was added to a name with specific reference to the manor: see PN Sa II 19.

*hamm, homm* 'an enclosure; a meadow, especially a flat low-lying meadow on a stream',[526] perhaps more specifically 'a place hemmed in by some feature of the topography, often by water or marsh',[527] often found as -ham, and easily confused with *ham*. Found in Barthomley, Hamley (House), and perhaps Berkhamsych (Upper and Lower).

*hangra* 'a slope, a wood on a slope', found in Clayhanger.[528]

---

[522] EPNE i 219-20.
[523] See Gelling & Cole 2000: 123-8.
[524] 1929: 43.
[525] Gelling & Cole 2000: 47-9.
[526] Gelling 1984: 41-9.
[527] Gelling 1988: 114-15.
[528] Gelling & Cole 2000: 230-1, which notes that the element is vey rare in Warwickshire, Worcestershire and Herefordshire.

*heafod* 'a head or end (of anything)', found as a topographical term in the sense of a headland. The element is found in Fawfieldhead, *Leleheved*, and Shareshill.

*helde* 'a slope', found in *Romeshelde*, Yeld House, Yells Farm and Yieldfields.

*hlaw* 'hill, burial mound'.[529] In the Midlands the term usually (and in Staffordshire almost invariably) means 'tumulus' (or perhaps 'a mound with the appearance of a tumulus'), a characteristic landscape feature of the Bronze Age and late 6th and early 7th centuries AD, and indicating either an Anglo-Saxon burial mound, or a prehistoric tumulus with Anglo-Saxon secondary inhumations. If a tumulus cannot be linked to a *hlaw* name, it is likely that the feature has been detroyed, although in some cases the word may possibly refer to a natural hill-spur or abrupt hill. Found very frequently in Staffordshire, including Ablow, Beelow Hill, Blakelow, Boothlow, Botteslow, *Buglawe*, *Castlow Cross*, Catshill, Catteslowe, Cauldon Lowe, Clulow, Copley, *Coppedlowe Cloughs*, Counslow, Cowlow, Wakelowe, *Derneslowe*, *Dorueslau/Doresley*, Drakelow, Elford Low farm, *Farlawe/Ferlawe*, Garmelow, Gillow Heath, Greenlow Head, Groundslow Fields, *Guendelawe*, Harlow Wood, Heatonlow, Horninglow, *Horninglow Cross*, Horseley Fields, Hurdlow, King's Low, Lamber Low, Long Low, Low Hill, Lowe Hill, The Low, *Luzlow*, Martin's Low, Merryton Low, Morrilow, Moxley, Muckley Corner, Nailor, Offlow, Orslow, Painley Hill, Pikelow, Queen's Low, Ranslow Farm, *Rollowe*, Rowley, Rowley Hill, *Rudlow*, Rue Hill, Ruelow Wood, *Rumbelows (Farm)*, *Shakerlowe*, Southlow, *Spellowe Field*, *Stamberlowe*, Stanley, Stanley Fields, Stanlow, Stanlow Hall, Stoneylow, Totmonslow, *Twamlow*, Twillow, *Waldreslowe*, Wardlow, Warrilow Brook, Warslow, Wentlow, *Westlowe*, *The Whitelowe*, Wicken Low, Wold Low, and Wormlow Farm. Cases also occur where a *-leah* name has developed into *-low* (e.g. Anslow). It should be noted that in Scotland and the northern counties as far south as Derbyshire, Nottinghamshire and Northamptonshire the dialect word *low(e)* (from ON *loge*, Danish *lue*) was used for 'a flame, a blaze, a light',[530] and may have been attached to hills used as beacons. It is possible that some *-low* places in north Staffordshire have been so named, but it would be very difficult to differentiate the meanings if the hill had a known tumulus: tumuli and beacons are both associated with hilltops. Although in Staffordshire Old English *hlaw* was normally applied to tumuli, or hills or mounds with the appearance of tumuli, in some cases Old English *beorg* may have been used: cf. Barrow Hill, on the Staffordshire-Worcestershire boundary, which may indicate Hwiccan influence.

*hoh* 'a hill-spur' (literally 'a heel', cf. the hock of a horse).[531] The element, common in other parts of the country but less common in Staffordshire, is found in Bignall Head, Birchall, Black Hough, Ferny Hough, Hoo, Hoe, Hoo Brook, Hoo Mill, Hose Wood, The Hough, Houghwood, *Melleshohe/Maleshou*, and Whitehough.

*hol* 'a hole, a hollow', found in *Brockholes*, *Foxholes*, Froghall, *Hen Hole*, Holbeche/Holbeach, *The Holden*, Holditch, Hole Brook, Hole Carr, Hole House, and *Holedale*.

OE *holm*, ON *holmr*, *holmi* 'a small island, a piece of dry land in a marsh; a piece of land partly surrounded by streams or by a stream', found as modern Holme and Hulme. Spellings with -u- are unlikely to be Danish rather than Norwegian, but dialectical

---

[529] Gelling 1988: 134-7; Gelling & Cole 2000: 46-9. Details of 57 Staffordshire tumuli can be found in VCH I 374-8; JNSFC 3 1965 20-63.
[530] EDD.
[531] Gelling 1988: 167-9; Gelling & Cole 2000: 186-8.

variants.[532] In the Danelaw the term is used sometimes of 'an island', but mostly for 'higher ground in marshland', or perhaps 'cultivated land on the edge of moors', and more generally as 'a piece of flat ground'.[533] The term is not necessarily to be taken as direct evidence of Scandinavian settlement, and place-names with *–hulm* are especially problematic: it is thought that the element represents a late OE or early ME confusion of late OE *holm* (from ON *holmr*) with *humm*, a north-west Midlands variant of OE *homm* or *hamm* (q.v.),[534] but from the geographical distribution in Staffordshire it may not be unreasonable to see the element as evidence of Scandinavian influence, with English settlers using a borrowed term.[535] The element is very common in field-names along the river Trent in the east of the county. Many names, including places which cannot now be located, have been traced, including *The Holme* (now Bates Farm), Bearda, Bustleholme, Cat Holme, *Chaddesholm*, *Chartley Holme*, *Cokkesholm*, Fatholme, *Gorseholm*, *Hemp Holme*, Hildersholme, *Holm-Bidolf* (Biddulph), Home Farm and Cottage, Holme Farm, *Le Holm*, *Hulme*, Upper Hulme, Hulme End, Hulmedale Farm, Madeley Holme, *Mill Holmes*, *Mucleholme*, Mythaholme, *Netherholm*, *Nether Hulme*, Penkholme, *Worfield Holm*, and *Wulsiesholm*.

*holt* 'a wood', usually when found in place-names meaning a wood of one species,[536] found in Bagnall, Birchall, Crowholt, Holt Hill, Kingsley Holt, Sherholt, Shirrall (Hall), and *Stare Wood*.

*hop* in the West Midlands probably meaning 'enclosure in a marsh or enclosure in heathland', an example of the former being Hopwas, and of the latter Hopton, and in some cases, particularly in the west of the region, from the ME period with the particular meaning 'remote valley', or 'hidden place', as in Bradnop.[537] Other examples of name incorporating the element include Hope, Hopedale, and Hopstone.

*hyll* 'a hill', generally applied to hills with pointed summits or with broken edges, perhaps belonging to the later stages of Anglo-Saxon name formation, and usually less attractive as habitation sites than places with the element *dun* in their names.[538] *Hyll* is found in Blymhill, Keele, Patshull, Penkhull, Shelfield, Stapenhill, Stramshal, Sugnall, Tatenhall and Winshill.

*hrycg* 'a ridge',[539] found in Baggeridge, *Barridge Moor*, Cobridge, Coldridge Wood, Combridge, *Flotheridge*, Ladderedge, Longridge, *Lyversegge*, Morridge, Oldridge, Ridgacre, Ridgeway, Rudgeway, Rudge, *Rudge bank*, *Rudge Strete*, Rugeley, and Seabridge.

*hyrst* 'a wood, a copse, a wooded eminence, a hillock'.[540] The element is found in *Ashenhurst*, Barnhurst, *Baxstonehurst*, Beamhurst, Brakenhurst (Farm), Brieryhurst, Brockhurst (Farm), Burndhurst Mill, *Chesterhurst*, *Cobshurst*, Copshurst, *Edritheshurst*,

---

[532] Gelling & Cole 2000: 55.
[533] Gelling 1984:51.
[534] VCH Ch I 258-9.
[535] It should be emphasised, however, that the great majority of names cited here are extracted from Shaw 1798 and 1801, which (together with the 'unpublished sheets' included in the 1976 facsimile reprint by E. P. Publishing Limited) cover only the eastern, southern, and some of the central parts of the county, so creating an imbalance which will inevitably distort the analysis.
[536] Gelling & Cole 2000: 233.
[537] Gelling & Cole 2000: 133-9.
[538] Gelling & Cole 2000 192.
[539] Gelling & Cole 2000: 190.
[540] Gelling & Cole 2000: 234-5.

Elmhurst, *Gledenhurst*, *Hasallhurst*, Hazalhurst Brook, Henhurst, Hurst Hall, Hurst Wood, Ravenshurst, Redhurst Wood, Tuppenhurst, Turnhurst, and Whitehurst.

*-ing* an ending in place-names meaning 'place or river',[541] as perhaps in Easing Farm.

*-inga* an Old English suffix added to a personal name which developed to mean 'the dependents of or people of', strictly a group-name rather than a place-name proper, but a formation widely found as place-names.[542]

*-ingatun* 'the tun of x's people', a rare element in Staffordshire. Examples are Essington and Werrington.

*-ingtun* '*tun* associated with...'. Examples are Ackleton, Bobbington, and possibly Almington.

*launde* (ME, from French) 'open space in woodland, a forest glade, woodland pasture',[543] found with the meaning 'grassy ride in woodland' most notably in a cluster of names in south-west Staffordshire: Blymhill Lawn, Coven Lawn, Landywood, Langley Lawn, and Oaken Lawn, and also in Laund (Farm), *Lawn*, and Lawnhead.

*leah* 'a wood, woodland, a rough open space or clearing in a wood, a glade; woodland clearing, especially one used for pasture or arable', and later 'a piece of open land, a meadow'.[544] A very common element, giving the endings -ley, -leigh, -le. It should be noted that in Staffordshire the common element *hlaw* very often becomes -ley, and if a tumulus or tumuli are known to exist in a place with a modern *ley* ending or similar, there is a strong possibility that the derivation is from *hlaw*. *Leah* is found in Abbots Bromley, Adderley, Adderley Green, Agardsley, *Alburley*, Alder Lee, Aldersley, *Aldredeslega*, Alveley, Andersley, Anslow, Areley Kings, Ashley, Aspley, Astley, Audley, Baddeley, Bagot's Bromley, Balterley, Barnsley, Barthomley, *Batchley*, Bemersley Green, Bentilee, Bentley, Betley, *Birchinlee*, Birdsley Farm, Bishop's Offley, Black Lees, Blakeley, Blakeley Green, Blakeley Lane, Boley Park, Boosley Grange, Bordesley, Bradley, Bradley Green, Bradeley, Bradeley Farm, Bradley in the Moors, Brandy-Lea, *Bridley Moor*, Brierley Hill, Brindley Heath, *Brockley Moor*, Bromley, Bromley Farm, Bromley Green, Bromley Wood, *Burgardeslee*, Burleyfields, The Bursleys, Chartley, Chatterley, Cheadle, Checkley, Chestall, Chorley, Claverley, Cockley, Coley, Coley Mill, Copley, Coseley, Cowley, Crackley Bank, Cradley Heath, Crawley, *Cromsley*, Darley Oaks, *Dawley*, Dods Leigh, Doley Common, Doley Gate, Dudley, Dunlea Farm, Dunsley, Eardleyend, Farley, Fazeley, *Flitley*, Footherley, Farlea Brook, Foxley, Fradley, *Fundesley*, Gerrard's Bromley, *Gleadley*, Godley Brook, *Gothersley*, Graiseley, *Greasley Side*, Great Wyrley, Grindley, Hadley End, Hagley, Hamley (House), Hamleyheath, *Hamley Park*, *Hammersley*, Hanley, *Harley*, Harley Thorn (Farm), Hartley Green, *Haseley*, Hatchley, *The Hawkleys*, Hawksley Farm, Headless Cross, Heathylee, Heatley, Heighley, High Offley, Himley, Hobble End, Hockley, Horseley, Horseley Heath, Huntley, Hyde Lea, Iverley, King's Bromley, Kingsley, *Kingsley*, Kingsnordley, Knightly, Knypersley, *Kynesley*, *Kyngesleye Heth*, *Kyrkesleye*, Langley, Langley Lawn, Lapley, The Lea, Lea Farm, Lea Heath, river Lea, Leacroft, Leamonsley, leaton, Lees Hill, Leese Jouse Farm, Church Leigh, Litley, Loxley, Madeley, Maple Hays, *Maydinlegh*, The Meadleys, Mobberley, Moseley, Mosslee Hall, Mossley, Mottley Pits,

---

[541] EPNE i 282-98.
[542] Cameron 1996: 66-72.
[543] EPNE ii 17. Camden 1674: 124 notes: 'Laund, a plain among trees'.
[544] Gelling & Cole 2000: 237-9.

Moxley, Napley Heath, Oakley, Ogley Hay, Onneley, Ousley Brook, Ouseley Cross, Oxley, Paynsley, Peasley Bank, Pinchley, *Pottersley*, Putley, Radley Moor, Rawnsley, *Rileyhill*, Romsley, Rowley, Rowley Gate, Rowley Regis, Rugeley, Rushley, *Ryppeleyelond/Rypley Meadow*, Saverley Green, Scounslow Green, Seckley Wood, Sedgley, Sharpley Heath, Shipley, Shirley, Smedley Sytch, Stanley, Stansley (Wood), Stockley Park, Streetly, *Stubbeley*, Swythamley, *Tappeley*, Teddesley, Thornyleigh, Throwley, Tyrley, Ubberley, Upper Arley, Waggersley, *Walkley Bank*, Warley, Wetley Moor, Wetley Rocks, White Lee Farm, Whitley Heath, Whitleyford, Wilderley (Barn), *Wimersley, Wimundeslie, Wolfelega,* Wolseley, Woolley, Wordsley, Wrottesley, and *Wystansley.*

*(ge)mǣre* 'boundary',[545] found in Mare Brook, Meer Oak, Meerbrook, Meir, Mere Farm, Mere Hall, and Mere Hill.

*mere* 'a mere, pool, lake' (in the south of England used for pools of almost any size, and in the Midlands and Lake District to larger pools and lakes), and sometimes 'wetland'.[546] The element often becomes *more* or *moor* in Staffordshire, and is found in Aqualate Mere, Ashmore Brook, Blakemore (House), Bradmore, Brockmoor, Cop Mere, Cromer Hill, Goosemoor, *Kingesmere,* Maer, Mere Hill, Meretown, Monmore, Pickmere/Picmoor, Swansmoor, Wetmore, and Whistamere.

*mersc* 'a marsh',[547] found in Crakemarsh, The Marsh, and Marston.

*mor* applied not only to low-lying marsh, but also to bleak upland areas, both wet and dry.[548] Most names incorporating the element in Staffordshire refer to low, wet sites, such as the three Moretons and Silkmore on the south-east outskirts of Stafford, but Whitmore, south-west of Newcastle, may use the word in its other sense. Other places incorporating the element are Ashmore Brook, Ashmore Heath, Ashmore Park, Audmore, Barrow Moor, *Bexmore Farm,* Boscomoor, Bullmoor Lane, Caldmore, *Cippemore, Cobb Moor,* Cranmere Farm, Cranmoor, Fullmoor, Hademore, Hurst Hull, Kiddemore Green, Leamore, Moor Hall, The Moorlands, Moreton Brook, Morridge, Norton-in-the-Moors, Radmore Lane, Radmore Wood, Redmoor Brook, Romer Farm, Rumer Hill, *Saltmoor, Shackamore,* Swainsmoor, *Wetmoor,* Wetmore, The Whitemoor, Whitmore, Whitmore Reans, Wigmore, and Wolmore Farm.

*myln, mylen* 'a mill',[549] found in Dam Mill, Hazel Mill, Mill Brook, Mill Fleam, *Mill Holmes,* Mill House Farm, Milldale, Millian Brook, Millmeece, Milton, and Milwich.

*ofer, ufer* 'a bank, a flat-topped ridge with a convex shoulder', or (for spellings without -*u*-), perhaps 'river-bank',[550] found in Ashmore Park, *Birchover,* Haselour, Lindore Farm, Longnor, Nore Hill, *Northovere,* Nurton, Okeover, Orton, Overton, Ramshorn/Ramsor, Tittensor, and Wychnor

*penn* (PrWelsh), *pen* (Welsh) 'head, headland, end' and (as an adjective) 'chief'. The word has long been interpreted by place-name scholars to mean also 'hill', but there is continuing debate as to whether the word was ever used in that sense for English place-

---

[545] EPNE ii 33-4.
[546] Gelling & Cole 2000: 21-6.
[547] Gelling & Cole 2000: 36-7.
[548] Gelling & Cole 2000: 58-9.
[549] EPNE ii 46.
[550] Gelling & Cole 2000: 199-202.

names.[551] In Staffordshire the element is probably found in Pen Farm, Pen Fields, Penn, and Penn Hill.

*pyrle* 'bubbling', often applied to springs and streams,[552] found in Pearl Brook and *Pirlewallsiche*.

ME *queche* 'a thicket', found in Squitch House near Abbots Bromley.

*\*ryding* 'a clearing', found as -ridding, -riding.[553] The element is found in Coldriding Farm, Henridding Farm, and Ridding Farm.

*scaga* 'a copse, a grove, a small wood'.[554] Found in *Ametesawe*, *Ashnough*, Blackshaw Moor, Bradshaw, Coldshaw, Colshaw, Gentleshaw, Longshaw, Marnshaw Head, Oldershaws, Ramshaw, Redshaw, Shaw, Shaw Hall, *Shawmoor Farm*, Shaw House, Shay Lane, Shuttershaw, Trubshaw, Wilshaw, and *Woburnshawe*.

*scelf* 'a rock, a ledge, shelving land'.[555] The element is found in Bramshall, Shareshill, and Shelton.

*\*scerde* 'a gap, a cleft, a pass',[556] found in Hawkesyard (Priory).

*seten* 'a plantation', perhaps found in Woodsetton.

*slæd* 'a flat-bottomed, especially a wet-bottomed valley',[557] found in *Ayleslade*, Hazel Slade, Queslett, Slade Heath, Slade House, and Spring Slade.

ON *slakki* 'a small shallow valley, a hollow in the ground', found as northern dialect *slack* 'a hollow, especially one in a hill-side; a dip in the surface of the ground; a shallow dell; a glade; a pass between hills', found in The Slack, Stonyslack and Waterslacks.

*spring, spryng* (ME), 'a copse, a young plantation, coppiced trees with new shoots', from Old English *springan* 'to burst forth', sometimes found in the region as *sprink*. Old English *spring* also meant 'place where water issued from the ground',[558] but in Staffordshire Mercian Old English *wælle*, ME *walle*, was generally used for a spring of water. The element is found in Hey Sprink, Spring Hill, and Springslade.

*stan* 'stone', found in *Bacstonesley*, *Baxstonehurst*, Beeston Torr, Featherstone, The Hailstone, Hanging Stone, Harston Rock and Wood, Hexton's Farm, Hopestone Farm, Ipstones, Mucklestone, *Roughstone Hole*, Shenstone, Shushions, Stallington, *Stamberlowe*, *Stanburneford*, Standeford, Standon, *Stanford Brook*, Stanley, Stanley Fields, Stanlow, Stanlow Hall, Stanshope, Stansley (Wood), Stanton (Wood), *Stone Edge*, Stonydelph, Stony Brook, *Stonyfields*, Stonyford, Stonylow, Stonyslack, Stonywell, Warstone, War Stone, and Warstones. In many cases the element may simply indicate that underlying stone is exposed at the surface.

*steort* 'promontory, hill spur', found associated with Squitch House in Bagot's Park.

---

[551] Gelling & Cole 2000: 211.
[552] PN Sa III 130.
[553] Gelling & Cole 2000: 244.
[554] Gelling & Cole 2000: 245-6.
[555] Gelling & Cole 2000: 216-8.
[556] EPNE ii 108.
[557] Gelling & Cole 2000: 141-2.
[558] Gelling & Cole 2000: 278.

*stoc* 'a place, a religious settlement, a secondary settlement',[559] found in Stockton, Stoke by Stone, Stoke Grange, Little Stoke, Stoke on Trent, *Stokedoilly*, and *Waterstoke*.

*stow* 'a place, a holy place, a place of assembly',[560] found in *Kelmestowe*, *Peckstowe*, Stow Heath, Stowe, and Stowe by Chartley.

*strēt, stret* The Anglo-Saxons seem to have applied the term *strēt, stret* to 'a paved (Roman) road, a street',[561] but whether those who used the term realised that the roads to which the name applied were invariably Roman in origin is unclear. *Strēt* is from late Latin *strata*, for *via strata* 'paved road'. An Anglo-Saxon *strēt* was a paved road, which in reality will have limited the term to Roman roads, for very few others (if any) will have been paved. The element *strēt, stret* is found in *Stratford*, Streethay, Streetly, Stretton, and *Stretwile/Stretwyle*.

ME *strind* 'stream, watercourse, ditch, channel' or similar. The element is found in the 14th century poem *Sir Gawain and the Green Knight*, and appears to be found mainly in northern and western England.[562] It is found in Hazel Strine, Strine Brook, and Strines.

OE *þorp*, ON *þorp*. An element usually found only in the Danelaw areas, and rarely in the north-western counties, which seems to have been applied to insignificant places, where frequently the very location is now lost. The general meaning is probably 'farm', perhaps a dependant or outlying farm belonging to a village or manor. In many cases an original place-name Thorp has been given a distinguishing first element, often an English or Norman personal name,[563] of which *Garardesthorpe* is a Staffordshire example. The element is also found in Thorpe Constantine

*torr* 'a rock, a rocky peak',[564] a word generally found only in the South West and the West Midlands,[565] and considered to have derived from Brittonic *torr*,[566] found in Apes Tor, Beeston Tor, *Dale Torr*, Gib Tor, and Yelpersley Tor.

*tun* the precise meaning of this element varied during a long evolution from the original Germanic concept of a fence or hedge, to the modern English 'town'. The word is by far the most common element in English place-names, and can be translated variously as 'a dwelling, an enclosure, a farmstead, a manor, an estate, a vill, a hamlet, a village, and (rarely, and later) a town'. The word was used in the formation of place-names for a very long period, as early as the 7th century meaning 'a village', and even perhaps in some cases 'royal vill',[567] and increasingly until after the Conquest, in those examples normally meaning 'manor', but it is probably safer to interpret the element as 'estate'.[568] Few *tun* names are likely to predate the 7th century. The use of the word *tun* in connection with that of a topographical feature or a personal name is particularly common in the West

---

[559] EPNE ii 153-6.
[560] EPNE ii 158-61. A perambulation of Walsall manor in 1617 refers several times to *a perle* in the sense 'a stream': Willmore 1887: 440.
[561] EPNE ii 161-2.
[562] Elliott 1984: 142-3.
[563] Ekwall 1960: xvi; 468-9.
[564] EPNE ii 184.
[565] EPNE i xxviii.
[566] See Padel 1985: 221, who notes the rarity of the element in Cornish place-names, and the absence of the Cornish cognate with the meaning 'tor', which is *carn* in Cornish.
[567] See Campbell 1986: 115, who bases his conclusion on an analysis of Bede's vocabulary and cites Tamworth as an example..
[568] Cameron 1996: 143.

Midlands, but it is generally impossible to determine the exact meaning in a particular place-name. The common use of this word has been dated to the period c.750 to c.950, and when the first element is a personal name, such as Darlaston, it is likely that the person named was a king's thegn who was given the overlordship of the manor at a relatively late date in its history, probably replacing an earlier topographical name. *Tun* names were not used for settlements in a forest area, which indicates that the areas in which they lay were relatively clear of woodland,[569] but are widely found in the West Midlands in a compound formed by the possessive case of a personal name, such as Darlaston and Admaston. The element is found in Abbey Hulton, Acton, Adbaston, Admaston, Almington, Alstone, Alton, Amerton, Amington, Ankerton, Apeton, *Armiston*, Aston, Astonsitch, *The Austrells*, Barlaston, Barton, Barton under Needwood, *Bedintun*, Beeston Tor, Beighterton, *Berdingeston*, Billington, Bilson, Bilston, Bishton, Blurton, Bobbington, The Brampton, Branston, Brineton, Brockton Grange, Brockton, Broughton, Burston, Burton, *Burtone*, Burton upon Trent, Butterton, Calton, Catton, Chedleton, Chesterton, Chillington, Chapel Chorlton, Hill Chorlton, Clayton, Clifton Campville, *Cobintone*, Cold Norton, Colton, Compton, Cotwalton, Creighton, Croxton, Darlaston, Denstone, Derrington, Drayton, Drayton Bassett, Drointon, Dudmaston, Dunstall, Dunston, Church Eaton, Water Eaton, Wood Eaton, Ecton, Ellastone, Ellerton Grange, Elston, Engleton, Enson, Essington, Essington House Farm, Fenton, Forton, Foston, Garstones, Gayton, Gratton, Gunstone, *Hampton*, Hampton Loade, Harlaston, Hattons, Haughton, Haunton, Heathton, Heaton, Hextons Farm, Hilderstone, Hilton, Hinksford, Hollington, Hopton, Horton, *Hortone*, Johnson Hall, *Ketelbernestona*, Kibblestone, Kingston, Kinvaston, Knighton, Kniveden, Knutton, Leaton, *Leighton*, *Leighton Hay*, *Lodyngton*, Loynton, Ludstone, Lyntus, Marchington, Marston, *Middleton*, Middleton Green, Milton, Mitton, Moreton, Mucklestone, *Netherton*, Nethertown, Newton, Norton-in-the-Moors, Norton Green, Nurton, Ogley Hay, *Olton*, *Orbeton/Herbeton*, Orton, Otherton, Oulton, Oulton House Farm, Overton, Packington, Perton, Preston, Ranton/Ronton, Rodbaston, Rolleston, Rushton James, *Rushton Grange*, Rushton Spencer, Scropton, Shelton, Stallington, Standon, Stanton (Wood), Stockton, Stourton, Stretton, Sutton, Swynnerton, Tamworth, Tillington, Tipton, Tunstall, Tunstead, Walton, Walton Grange, Walton Hill, Werrington, Weston, Weston on Trent, Weston Coyney, Weston Jones, *Weston Moor*, Weston under Lizard, Whiston, Whiston Eaves, (Whiston) Lees, Whittington, Wigginton, Wilbrighton, Winnington, Withington, Wobaston, Wolgarston, Wollaston's Coppice, Wolstanton, Wolverhampton, Woodsetton, Wootton, Wootons, Wootton under Weaver, *Wulfhampton*, *Wulredeston*, and *Yatton*.

*\*wæsse* related to, but not identical with, modern 'wash'. The word is only found in place-names, and the precise meaning is a low-lying plain, adjacent to a river or stream, but usually a significant river, prone to sudden flooding and draining. Examples in Staffordshire include Hopwas on the Tame, Alrewas on the Trent, and the unlocated *Pirewasse* on Pur Brook, a tributary of the Blithe, and *Wassebroc*, near Hanchurch.

*wald, weald* a district name, probably of early date, meaning 'woodland, forest, high forest-land', later 'open upland'.[570] No place-names certainly incorporating the element have been traced in Staffordshire, but it might be found in Cotwalton, Prestwood near Ellastone, Wolmore (Farm), and *Wymundewalde*.

---

[569] Gelling 1988: 128.
[570] Gelling & Cole 2000: 253-6.

Mercian Old English *wælle*, usually in the West Midlands 'a spring', sometimes 'a stream', and (rarely) 'a well',[571] probably found in Ablewell, *Anc's Hill(s)*, Blackwell, Bradwell, Canwell, *Caudy Fields*, Caverswall, Chadwell, Great Chatwell, Coldwall, Cotwall End, Cotwalton, Creswell, *Cronkhall*, Cumbwell Brook, Eastwall, *Eyeswall*, Farewell, Fennel Pit Farm, Fole, Fradswell, *Hannell*, Hartwell, *Hawkeswall Clough*, *Hawkewallsych, Hawkswell (Rough)*, Hayes Wood, Hewell Grange, Heywood Grange, High-Hall-Hill, Holloway Farm, Holly Wall Farm, *Holywell Park*, Honeywall Farm, *Keywell Green*, Knowl Wall, *Lud-Wall*, *Normaneswell*, Oakamoor, Oakeswell, *Padwalle*, *Pearl Well*, Pell Wall, *Pirlewallsiche*, Red Hall, Sandwell, Scotch Hill, Showell/Showells, Spring Hill, St Chad's Well, Stockwell Heath, Stockwell End, Stonewalls Farm, Stonywell, Strangleford Birch, *Tywall Green*, Walford, Wall Acre, Wallbridge, Walton Grange, Washerwall, Wicken Walls, *Windswell Pool*, *Wombewell*, Woodwall Green, and Woundale.

*wic*, dative plural *wicum* 'a dwelling, a building or group of buildings used for special purposes, a trading place, a farm, a dairy farm', and, in the plural, 'a hamlet, a village'. Names incorporating *wic* can be divided into two main categories, those with palatalisation (for example Aldridge, Bromwich, where the element is pronounced [ɪtʃ]), and those without (for example Smethwick, with the pronunciation [ɪk]). The palatalised and assibilated forms probably contain *wic* in the singular, and the unpalatalised in the plural, though this is not always the case. It is usually not possible to determine the precise meaning of *wic* in a place-name, but it has long been observed that the names of many salt-producing places have the ending *-wic*, for example Droitwich, Nantwich, Northwich, Middlewich. This is a specialised use of the 'trading place' meaning, and was well-established by the Middle Saxon period. The element is found in Aldridge, Baswich/Berkswich, Bloxwich, West Bromwich, Calwich, *Chetewik*, Colwich, Fisherwick, Gratwich, Hammerwich, Hardiwick, Millwich, Padwick, Plardiwick, Shirleywich, Smethwick, Wigford, *Wyke*, and Wyken. No trace has so far been found in Staffordshire of the place-name compound *wic-ham*, to which place-name scholars have given particular attention, since the term has been found to be associated with Roman roads and Romano-British settlement sites.

*wincel* 'nook, corner', perhaps found in Winkhill.

*wamb* 'womb, belly', found in Hombridge, Wombewell and Wombourne, and possibly Wemberton, Weymouth and Womere.

*worþig* 'farm, homestead, enclosure',[572] found in Tamworth and Tittesworth, though the latter may incorporate the Old English synonym *worþ*.

*worþign* 'farm, homestead, enclosure',[573] probably found in Bulwardine, Cheswardine, Harden, Ruiton, and Weymouth.

*wudu* 'a wood, a forest, timber', generally used for larger stretches of woodland, but rarely used before c.700.[574] The element is found in Ash Wood/Ashwood, Bearwood, Brewood, *Bridgewood*, Brockwood Hill, Dunwood, Wood Eaton, *Harewood*, Great Haywood, Hazelwood (House), *Horwood*, Lightwood, Lightwoodfields, *Lightwoodde Heathe*, Littywood, Lockwood (Hall), Micklewood, Prestwood, Shortwood Farm,

---

[571] Gelling & Cole 2000: 31.
[572] EPNE ii 275-6.
[573] EPNE ii 277.
[574] JEPNS 8 1975-6 43; see also Gelling & Cole 2000: 257-8.

Smallwood, Weatherworth, Wedgwood, Wetwood, Woodcroft, Woodend/Wood End, Woodford, Woodford Grange, Woodgate, Wood Green, Wood Hall, Woodham Green, Woodhead, Woodhouse, Woodhouse Farm, Woodhouses, Woodland, Woodseaves, Woodsetton, *Woodshutts*, Woodwall Green, Wootton, Woottons, Wootton (under Weaver).

**Distribution maps of elements found in Staffordshire place-names**

# Staffordshire place-name elements - key to symbols.

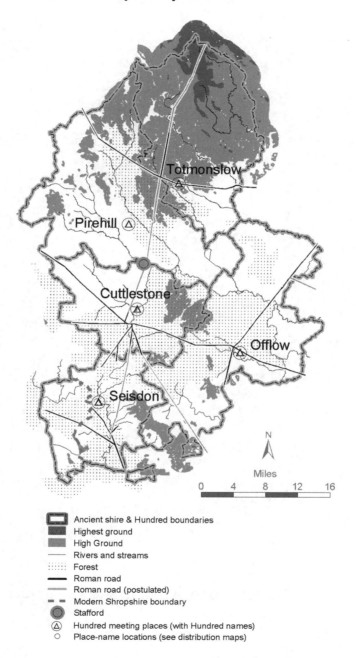

Ancient shire & Hundred boundaries
Highest ground
High Ground
Rivers and streams
Forest
Roman road
Roman road (postulated)
Modern Shropshire boundary
Stafford
Hundred meeting places (with Hundred names)
Place-name locations (see distribution maps)

# Staffordshire place-names
## incorporating the element *burh/burg*.

BERRY HILL
BERRYHILL FARM
BERTH, THE
BLITHBURY
BLOOMSBURY
BORROWCOP HILL
BROUGH HALL
BROADFIELD
BROUGHTON
BUNBURY HILL
BURBROOK, LITTLE
BURCOTE
BURF CASTLE
BURLEYFIELDS
BURSTON
BURTON
BURTON UPON TRENT
BURY
BURY BANK
BURY FARM
BURY HILL
BUSHBURY
CURBOROUGH
GAINSBOROUGH HILL FARM
HANBURY
HOBBERGATE
KNOTBURY
NEWBOROUGH
OLDBURY
PADBURY
SHUGBOROUGH
TUTBURY
WEDNESBURY

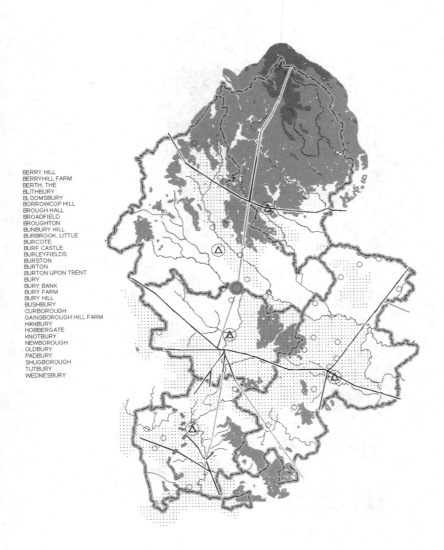

## for scale and symbols see key map

# Staffordshire place-names
incorporating the element *cester*.

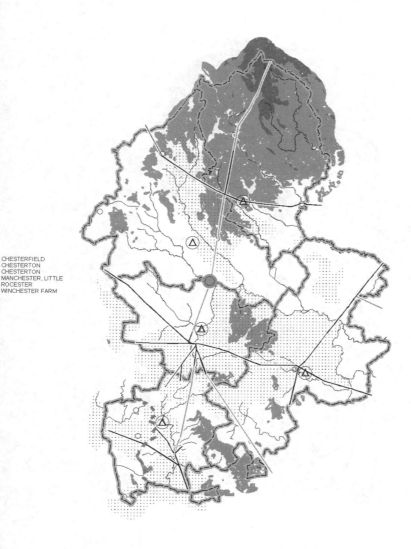

CHESTERFIELD
CHESTERTON
CHESTERTON
MANCHESTER, LITTLE
ROCESTER
WINCHESTER FARM

## for scale and symbols see key map

# Staffordshire place-names
## incorporating the element *cumb*.

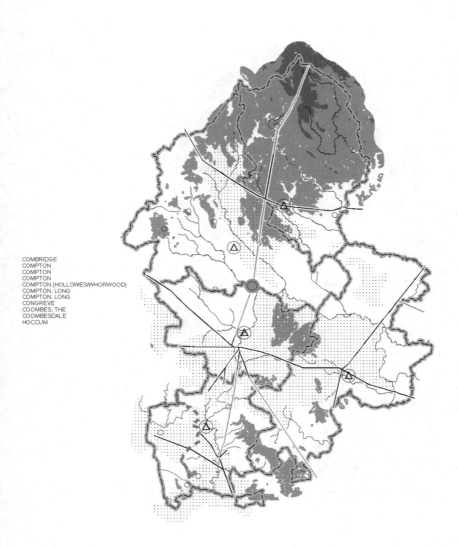

COMBRIDGE
COMPTON
COMPTON
COMPTON
COMPTON (HOLLOWES/WHORWOOD)
COMPTON, LONG
COMPTON, LONG
CONGREVE
COOMBES, THE
COOMBESDALE
HOCCUM

## for scale and symbols see key map

# Staffordshire place-names
## incorporating the element *dūn*.

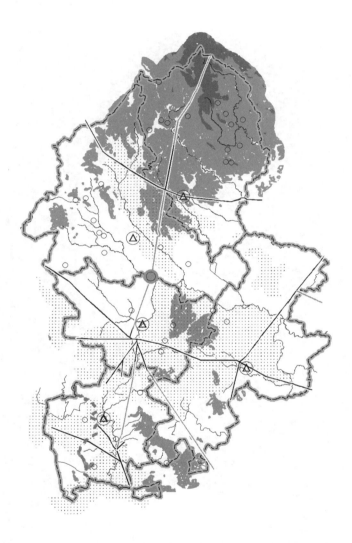

ACKTON HILL
ANKERTON
BRERETON
BUTTERTON
BUTTERTON
CALTON
CAULDON
CHORLTON, HILL
DOWN HOUSE FARM
DUNWOOD
ELKSTON
INDON
-RINDON
HADDON
HADDON
HADEN HILL
HATHERTON
HIGHDOWN
HIXON
HUNTINGTON
KNIVEDEN
LONGDON
LONGSDON
LUDSTONE
LYNDON
MBDEN
OWDON LANES
RANDON
RAREDON
REISDON
SHEBDON
SLINDON
SNOWDON
STANDON
SWINDON
UETTON
YOUNDON
YREDON

## for scale and symbols see key map

# Staffordshire place-names incorporating the element *feld*.

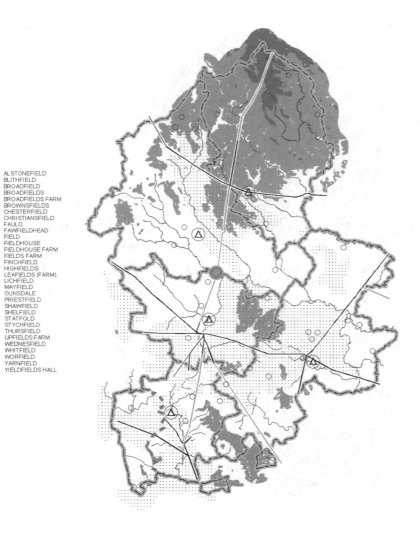

ALSTONEFIELD
BLITHFIELD
BROADFIELD
BROADFIELDS
BROADFIELDS FARM
BROWNSFIELDS
CHESTERFIELD
CHRISTIANSFIELD
FAULD
FAWFIELDHEAD
FIELD
FIELDHOUSE
FIELDHOUSE FARM
FIELDS FARM
FINCHFIELD
HIGHFIELDS
LEAFIELDS (FARM)
LICHFIELD
MAYFIELD
OUNSDALE
PRIESTFIELD
SHAWFIELD
SHELFIELD
STATFOLD
STYCHFIELD
THURSFIELD
UPFIELDS FARM
WEDNESFIELD
WHITFIELD
WORFIELD
YARNFIELD
YIELDFIELDS HALL

for scale and symbols see key map

# Staffordshire place-names incorporating the element *halh*.

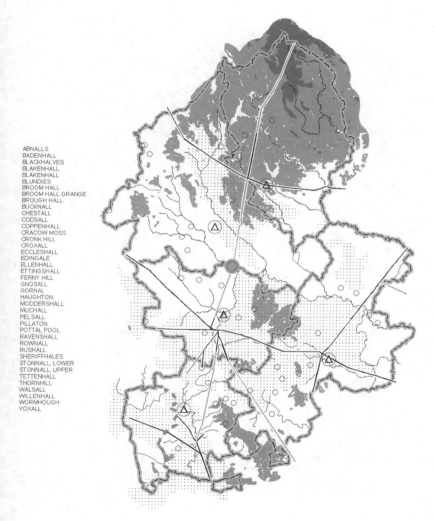

ABNALLS
BADENHALL
BLACKHALVES
BLAKENHALL
BLAKENHALL
BLUNDIES
BROOM HALL
BROOM HALL GRANGE
BROUGH HALL
BUCKNALL
CHESTALL
CODSALL
COPPENHALL
CRACOW MOSS
CRONK HILL
CROXALL
ECCLESHALL
EDINGALE
ELLENHALL
ETTINGSHALL
FERNY HILL
GNOSALL
GORNAL
HAUGHTON
MODDERSHALL
MUCHALL
PELSALL
PILLATON
POTTAL POOL
RAVENSHALL
ROWNALL
RUSHALL
SHERIFFHALES
STONNALL, LOWER
STONNALL, UPPER
TETTENHALL
THORNHILL
WALSALL
WILLENHALL
WORMHOUGH
YOXALL

## for scale and symbols see key map

# Staffordshire place-names
incorporating the element *hām*.

AUDNAM
BERKHAMSYTCH
GATEHAM
HAMPTON LOADE
HAMSTEAD
MARCHINGTON
OAKHAM
PATTINGHAM
RIDWARE, HAMSTALL
TRENTHAM
WULFHAMPTON

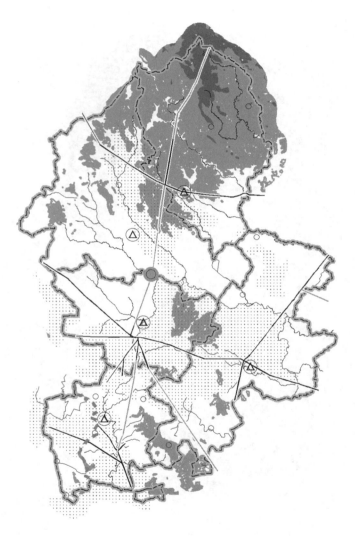

## for scale and symbols see key map

# Staffordshire place-names incorporating the element *hlāw*.

BLOW
EELOW HILL
LAKELOW
LAKELOW
LAKELOW
LAKELOW
LAKELOW
OOTHLOW
OTTESLOW
ATSHILL
AULDON LOWE
OPLEY
OPPEDLOWE CLOUGHS
OUNSLOW
OWLOW
RAKELOW
FORD LOW FARM
ARMELOW
LLOW HEATH
REENLOW HEAD
ROUNDSLOW FIELDS
ARLOW WOOD
EATONLOW
ORNINGLOW
ORSELEY FIELDS
IRDLOW
NG'S LOW
MBER LOW
NG LOW
W HILL
W, THE
WE HILL
ZLOW
RTIN'S LOW
RRYTON LOW
ORRILOW
OXLEY
ICKLEY CORNER
FLOW
SLOW
NLEY (HILL)
KELOW
EEN'S LOW
NSLOW FARM
WLEY
WLEY
WLEY
WLEY HILL
OLOW
E HILL
ELOW WOOD
MBELOWS FARM
UTHLOW
ELLOWE FIELD
ANLEY
ANLEY FIELDS
ANLOW
ANLOW HALL
ONEYLOW
TMONSLOW
LLOW
RDLOW
RSLOW
NTLOW
KEN LOW
RMLOW FARM

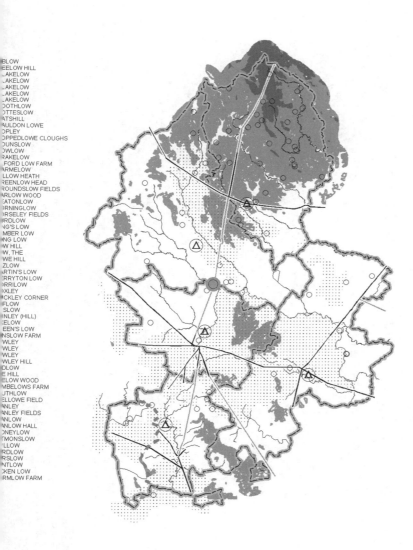

for scale and symbols see key map

# Staffordshire place-names incorporating the element *lēah*.

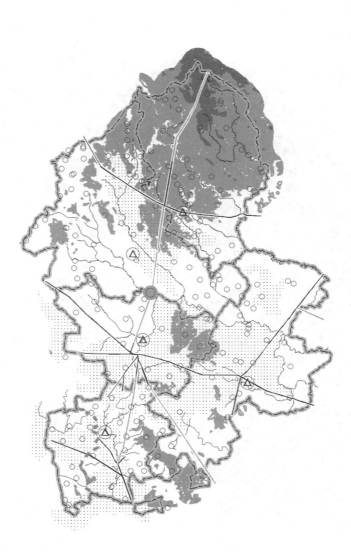

ADDERLEY
ADDERLEY GREEN
ADDERLEY GREEN
AGARDSLEY
ALDER LEE
ALDERSLEY
ALVELY
ANSLOW
ARELEY KINGS
ARLEY, UPPER
ASHLEY
ASPLEY
ASPLEY
ASTLEY
AUDLEY
BADDERLEY
BALTERLEY
BARNSLEY
BARTHOMLEY
BEMERSLEY GREEN
BENTILEE
BENTILEE PARK
BENTLEY
BETLEY
BIRDSLEY FARM
BLACK LEES
BLAKELEY
BLAKELEY GREEN
BLAKELEY LANE
BOLEY PARK
BOOSLEY GRANGE
BORDESLEY
BRADELEY
BRADES VILLAGE
BRADLEY
BRADLEY
BRADLEY GREEN
BRADLEY IN THE MOORS
BRANDY-LEA
BRIERLEY HILL
BRINDLEY HEATH
BROMLEY
BROMLEY
BROMLEY FARM
BROMLEY FARM
BROMLEY GREEN
BROMLEY WOOD
BROMLEY, ABBOTS
BROMLEY, BAGOT'S
BROMLEY, GERRARD'S
BROMLEY, KING'S
BURLEYFIELDS
CHARTLEY
CHATTERLEY
CHEADLE
CHECKLEY
CHESTALL
CHORLEY
CLAVERLEY
COCKLEY
COLEY
COLEY MILL
COPLEY
COSELEY
COWLEY
COWLEY
CRACKLEY BANK
CRADLEY HEATH
CRAWLEY
CROMSLEY
DARLEY OAKS
DODS LEIGH
DOLEY COMMON
DUDLEY
DUNLEA FARM
DUNSLEY
EARDLEYEND
FAIZELEY
FARLEY
FOOTHERLEY
FOXLEY
FRADLEY
GRAISELEY
GREASLEY SIDE
GRINDLEY
GRINDLEY
HADLEY END
HAGLEY
HANLEY
HARLEY THORNS
HARTLEY GREEN
HATCHLEY

HAWKSLEY FARM
HEADLESS CROSS
HEATHYLEE
HEATLEY
HEIGHLEY
HIMLEY
HOBBLE END
HOCKLEY
HORSELEY
HORSELEY HEATH
HUNTLEY
HYDE LEA
IVERLEY
KINGSLEY
KINGSLEY
KINGSNORDLEY
KNIGHTLEY
KNYPERSLEY
LANGLEY
LANGLEY LAWN
LAPLEY
LEA (FARM)
LEA, THE
LEA, THE
LEACROFT
LEAMONSLEY
LEATON
LEES HILL
LEES HOUSE FARM
LITLEY
LOXLEY
MADELEY
MAPLE HAYS
MEADLEYS, THE
MOBBERLEY
MOSELEY
MOSELEY
MOSELEY
MOSSLEE HALL
MOSSLEY
MOSSLEY
MOTTLEY PITS
MOXLEY
NAPLEY HEATH
OAKLEY
OAKLEY
OGLEY HAY
ONNELEY
OUSLEY CROSS
OXLEY
PAYNSLEY
PEASLEY BANK
PINCHLEY
PUTLEY
RADLEY MOOR
RAWNSLEY
RILEYHILL
ROMSLEY
ROWLEY
ROWLEY
ROWLEY
ROWLEY GATE
ROWLEY REGIS
RUGELEY
RUSHLEY
SAVERLEY GREEN
SCOUNSLOW GREEN
SECKLEY WOOD
SEDGLEY
SHARPLEY HEATH
SHIPLEY
SMEDLEY SYTCH
STANLEY
STOCKLEY PARK
STREETLY
SWYTHAMLEY
TEDDESLEY
THORNYLEIGH
THROWLEY
TYRLEY
UBBERLEY
WAGGERSLEY
WETLEY MOOR
WETLEY ROCKS
WHITE LEE FARM
WHITLEY HEATH
WHITLEYFORD
WILDERLY BARN
WMERSLEY
WOLSELEY
WOOLLY
WORDSLEY
WROTTESLEY
WYRLEY, GREAT

for scale and symbols see key map

# Staffordshire place-names incorporating the element *penn*.

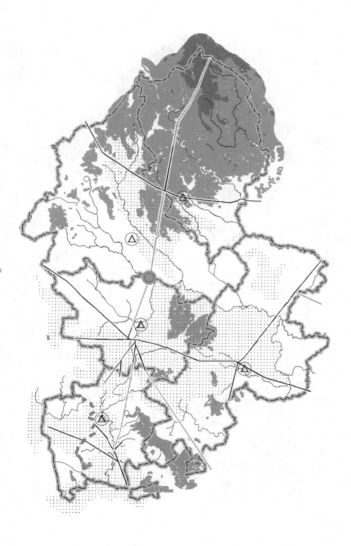

PEN FARM
PENFIELDS
PENN HILL
PENN HILL
PENN, UPPER & LOWER

for scale and symbols see key map

# Staffordshire place-names incorporating the element *salt*.

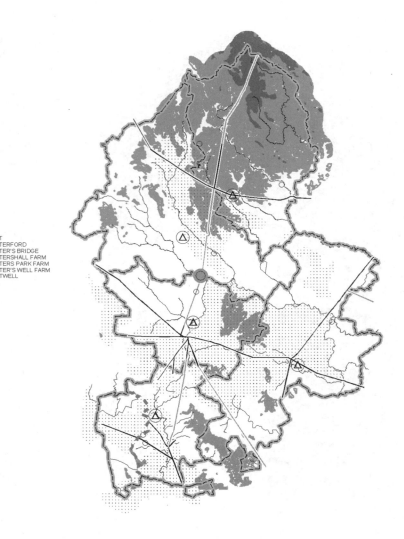

SALT
SALTERFORD
SALTER'S BRIDGE
SALTERSHALL FARM
SALTERS PARK FARM
SALTER'S WELL FARM
SALTWELL

## for scale and symbols see key map

# Staffordshire place-names
## incorporating the element *strǣt*.

STRATFORD
STREETHAY
STREETLY
STRETTON
STRETTON

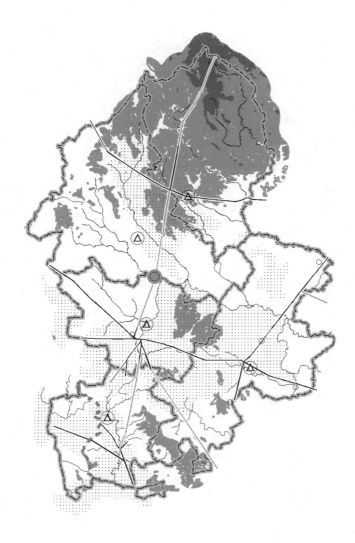

## for scale and symbols see key map

# Staffordshire place-names incorporating the element *tūn*.

ABBEY HULTON
ACTON TRUSSELL
ADBASTON
ADMASTON
ALMINGTON
ALSTONE
ALTON
AMERTON
AMINGTON
ANKERTON
APETON
ASTON
ASTON
ASTON
ASTON
ASTONSITCH
BARLASTON
BARTON
BARTON UNDER NEEDWOOD
BEDINTUN
BEESTON TOR
BEIGHTERTON
BILLINGTON
BILSON
BILSTON
BISHTON
BLURTON
BOBBINGTON
BRAMPTON, THE
BRANSTON
BRINETON
BROCKTON
BROCKTON GRANGE
BROUGHTON
BROUGHTON
BROUGHTON
BURSTON
BURTON
BURTON UPON TRENT
BUTTERTON
BUTTERTON
CALTON
CATTON
CHEDDLETON
CHESTERTON
CHESTERTON
CHILLINGTON
CHORLTON, CHAPEL
CHORLTON, HILL
CLAYTON (GRIFFITH)
CLIFTON CAMPVILLE
COLTON
COMPTON
COMPTON
COMPTON
COMPTON (HOLLOWES/WHORWOOD)
COTWALTON
CREIGHTON
CROXTON
DENSTONE
DERRINGTON
DRAYTON
DRAYTON BASSETT
DROINTON
DUDMASTON
DUNSTALL
DUNSTALL
DUNSTALL
DUNSTALL
DUNSTON
EATON, CHURCH
EATON, WATER
EATON, WOOD
ECTON
ELLASTONE
ELLERTON GRANGE
ELSTON
ENGLETON
ENSON
ESSINGTON
ESSINGTON HOUSE FARM
FENTON
FORTON
FOSTON
GARSTONES
GAYTON
GRATTON
GUNSTONE
HAMPTON LOADE
HARLASTON
HATTONS
HAUGHTON
HAUNTON
HEATHTON
HEATON
HEXTONS FARM
HILDERSTONE
HILTON
HILTON
HILTON
HINKSFORD
HOLLINGTON
HOPTON
HORTON
JOHNSON HALL
KIBBLESTONE

KINVASTON
KNIGHTON
KNIGHTON
KNIVEDEN
KNUTTON
LEATON
LOYNTON
LUDSTONE
LYNTUS
MARCHINGTON
MARSTON
MARSTON
MIDDLETON
MIDDLETON GREEN
MILTON
MITTON
MORETON
MORETON
MORETON
MUCKLESTONE
NETHERTON
NETHERTON
NETHERTOWN
NEWTON
NEWTON
NEWTON
NORTON IN THE MOORS
NORTON, COLD
NURTON
OGLEY HAY
OLTON
ORTON
OTHERTON
OULTON
OULTON
OULTON HOUSE FARM
OULTON, UPPER & LOWER
OVERTON
OVERTON
OVERTON
PACKINGTON
PERTON
PRESTON
RANTON/RONTON
RODBASTON
ROLLESTON
RUSHTON GRANGE
RUSHTON JAMES
RUSHTON SPENCER
SCROPTON
SHELTON
STALLINGTON
STANDON
STANTON (WOOD)
STOCKTON
STOURTON
STRETTON
STRETTON
SUTTON
SUTTON
SUTTON
SWYNNERTON
TAMWORTH
TILLINGTON
TIPTON
TUNSTALL
TUNSTALL
TUNSTEAD
WALTON
WALTON
WALTON GRANGE
WALTON HILL
WERRINGTON
WESTON
WESTON COYNEY
WESTON JONES
WESTON ON TRENT
WESTON UNDER LIZARD
WHISTON
WHISTON
WHITTINGTON
WHITTINGTON
WHITTINGTON
WIGGINTON
WILBRIGHTON
WINNINGTON
WITHINGTON
WOBASTON
WOLGARSTON
WOLLASTON'S COPPICE
WOLSTANTON
WOLVERHAMPTON
WOODSETTON
WOOTONS
WOOTTON
WOOTTON UNDER WEAVER
WULFHAMPTON

## for scale and symbols see key map

# Staffordshire place-names incorporating the element *wīc*.

ALDRIDGE
IASWICH/BERKSWICH
ILOXWICH
IROMWICH, WEST
CALWICH
COLWICH
IISHERWICK
IRATWICH
IAMMERWICH
IARDIWICK
CARDIWICK
IILWICH
IADWICK
ILARDIWICK
HIRLEYWICH
IMETHWICK
IIGFORD
IYKEN

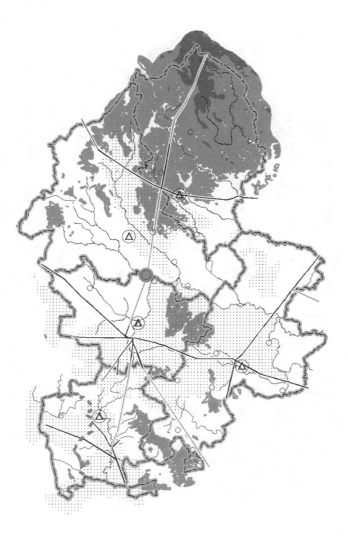

for scale and symbols see key map

# Staffordshire place-names incorporating Celtic elements.

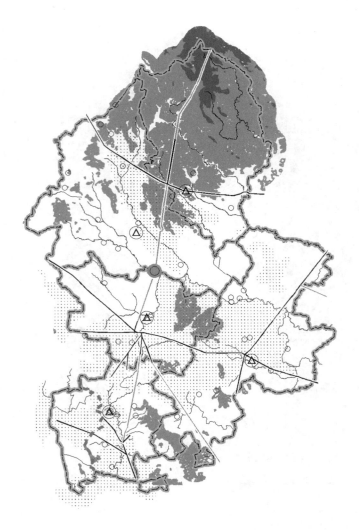

BAR HILL
BARR, GREAT
BREWOOD
CHATTERLEY
CHEADLE
COMBERFORD
CREIGHTON
ECCLESHALL
GNOSALL
HINTS
INGESTRE
KIDDEMORE GREEN
KINVER
LEAMONSLEY
LICHFIELD
LIZARD
MINNBANK
MORFE
MOW COP
OCKER HILL
PENKHULL
PENKRIDGE
PENN HILL
PENN WOOD
PENN, UPPER & LOWER
RIDWARE, HAMSTALL
RIDWARE, HILL
RIDWARE, MAVESYN
RIDWARE, PIPE
SAVERLEY GREEN
SEISDON
TALKE
TRYSULL
WALFORD
WALSALL
WALTON
WALTON

## for scale and symbols see key map

# Staffordshire place-names incorporating Scandinavian elements.

BASFORD
BEASLEY
BEESTON TOR
CAR HOUSE
CAR HOUSE
CATTON
CROXALL
CROXDEN
CROXTON
DAGDALE
DENSY LODGE
DROINTON
FORSBROOK
FOSTON
GAYTON
GILL BANK
GUNSTONE
HILLSWOOD
HOUNDHILL
HULME
HULME END
HULME, UPPER
HULMEDALE FARM
KEELE
KETTLEMOOR
KINGSTONE
KNIPE WOOD
KNYPERSLEY
LADDEREDGE
MICKLEDALE
NAB END
NABB
NABB FARM
NABBS WOOD
OSSOMS HILL
RANSLOW FARM
ROCESTER
ROE LANE FARM
ROLLESTON
ROWE, THE
SNAILS END
SNELSDALE
SWAINSMOOR
SWINSCOE
THORPE CONSTANTINE
THURSFIELD
TUCKLESHOLM FARM
TURNER'S KNIPE
WINNOTHDALE
YOXALL

for scale and symbols see key map

# Staffordshire place-names
# incorporating pagan elements.

FREEFORD
WEDNESBURY
WEDNESFIELD
WEEFORD

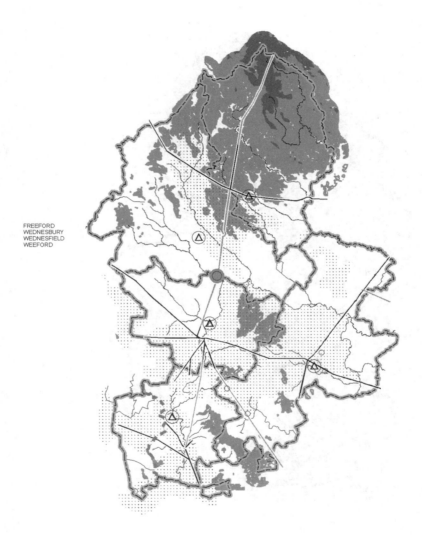

for scale and symbols see key map

## Personal names found in Staffordshire place-names

Although masculine forms are generally given in the place-name entries, it is in many cases impossible to distinguish masculine and feminine personal names in place-names. The following list gives masculine names, with certain feminine names marked (*f*).

ContG = Continental Germanic
Fr = French
ME = Middle English
OE = Old English
OFr = Old French
OG = Old German
ON = Old Norse
< = derived from

| | | |
|---|---|---|
| Ab(b)a | OE | Ablow, Abnalls, Apeton |
| *Ægel | OE | Ayleslade Ford |
| Ælf | OE | Elston |
| *Ælfa | OE | Alton |
| Ælffled (*f*) | OE | Alfledeway, Alfledeford |
| Ælfgȳþ (*f*) | OE | Alveley |
| Ælfheah | OE | Allscott |
| Ælfred | OE | Aluredeshale, Haluredeshale, Alstone |
| Ælfstan | OE | Alstanesax, Alstonefield |
| Ælmer | OE | Elmhurst |
| *Æmela | OE | Amblecote |
| Æþelheard | OE | Ellerton Grange |
| Ætti | OE | Yeatsall |
| Alda | OE | Olton, Oulton |
| Aldgȳþ (*f*) | OE | Audley |
| Aldred | OE | Adderley (Green), Aldredeslega |
| Aluburg or Alhburg (*f*) | OE | Alburley |
| *Amma or Earnmund | OE | Amington |
| Andrew | ME | Andersley, Andressey |
| Annot | ME | Annot's Dale |
| Aquart | OFr | Abbot's Castle Hill |
| Arblaster | ME | Arblaster |
| Archer | ME | Archford Moor |
| Austyn | ME | Austin's Wood |
| Bada | OE | Baden Hall |
| *Baldþrȳþ (*f*) | OE | Balterley |
| *Bad(d)ing | OE | Batchacre |
| Badeca or Bacga | OE | Bagnall |
| Batman | ME | Batman's Hill |
| B(e)adda | OE | Baddeley (Green) |
| Beda | OE | Bednall |
| Beffa | OE | Beffcote |
| Beloc | ME | Belockes Bridge |
| *Benning | OE | Bennystoils, Benynscrosse |

633

| | | |
|---|---|---|
| Beorcol | OE | Baswich/Berkswich |
| *Beord | OE | Beardmoors |
| *Beording | OE | Beardmoors, Berdingeston |
| Beorhthere | OE | Beighterton |
| Beorhtmund | OE | Bescot |
| Beorn | OE | Barnsley |
| Beornwulf | OE | Barlaston |
| Bernart | OFr(<OG) | Barnswood, Barnardscroft, Bernardsmore |
| Bertran(t) | OFr(<OG) | Bertramscote |
| Bette | OE | Betsfeilde |
| ? Bic(c)a | OE | Bickford |
| ? Bill or *Billa | OE | Billington |
| ? Biscop | OE | Bushbury |
| ? *Blocc | OE | Bloxwich |
| Bochard | OFr(<OG) | Badger's Croft |
| Börkr or Barkr | ON | Basford |
| Bøsi | ON | Beasley, Beeston Tor |
| Bourchier or Bourcher | OFr | Bourchiers Wood |
| *Brant | OE | Branston |
| Brian | ME | Bryan's Hay |
| *Bridd | OE | Briddeshus, Briddeshall |
| Brihthere or Beorhthere | OE | Bitterscote |
| Brun | OE | Brinsford |
| ? Brun(a) or Brȳni | OE | Brineton |
| Bubba | OE | Bobbington |
| Bucca | OE | Bucknall, Bukkeleg, Bukkelyh |
| *Bula | OE | Bullclough |
| *Bunt | OE | Bunster |
| Burgweard | OE | Burslem |
| ? *Can(n)a | OE | Canwell |
| *Catt | OE | Castern, Cat Hayes, Catshill, Catstree, Catteslowe |
| *Ceacca | OE | Checkley |
| *Ceatta | OE | Chatwell, Chatcull |
| *Cebbi | OE | Chebsey |
| ? *Cilla | OE | Chillington |
| *Cocc | OE | Cokesalle or Coxall, Cokkesholm |
| Croc | OE | Croxall, Croxden, Croxton |
| *Cruda | OE | Crowdicote |
| Cuþwulf | OE | Cudel(s)ford, Cuttlestone |
| *Cybbel | OE | Kibblestone |
| Cyne | OE | Kinesbroc |
| Cyneheard | OE | Kynesley |
| Cynehelm | OE | Kelmstowe |
| Cynewald | OE | Kinvaston |
| ? *Dæda | OE | Deadman's Green |
| Dealla | OE | Dawley |
| *Dene | ON | Densy Lodge, Denstone |
| Deorlaf | OE | Darlaston |
| *Docc | OE | Doxey |

| | | |
|---|---|---|
| Dod(d)a | OE | Derrington |
| Dot (?Dottr) | ON | Dodslow (Horninglow Cross) |
| Dudda | OE | Dudley |
| Dudeman | OE | Dudmaston |
| Dunn | OE | Dun Cow's Grove, Dunsley |
| Dunnic | OE | Dunkford |
| Eadbald | OE | Adbaston |
| Eadgar | OE | Agardsley |
| Eadlac or *Æþelac | OE | Ellastone |
| Eadmund | OE | Admaston |
| Eadþryþ (f) | OE | Edritheshurst |
| Ealac | OE | Elkstone |
| *Ean | OE | Enson |
| Eanbriht | OE | Amerton |
| Eanswiþ | OE | Anslow |
| *Earp | OE | Apesford |
| Ecca | OE | Ecton (Hill) |
| Eden or Eadwine | OE | Edingale |
| Ella | OE | Elford |
| Esne | OE | Essington |
| Farulf | ContG | Foston |
| ?*Folc | OE | Forkhill |
| Fot or Fótr | OE/ON | Forsbrook, Foston |
| Frod | OE | Fradley, Fradswell |
| *Gæga | OE | Gayton |
| ?*Gæga | OE | Gaywode Hall |
| *Gæring | OE | Garshall Green |
| Godric | OE | Gothersley |
| ? Gunne | ON | Gun Hill |
| Gunni | ON | Gunstone |
| Guþhere | OE | Gutheresburn |
| Hákon | ON | Hacondale |
| ? Hana | OE | Hanford |
| Hagona or Hagene | OE | Haunton |
| *Hand | OE | Handsacre |
| *Heddin | OE | Hednesford |
| Hengest or Hynca | OE | Hinksford |
| ? Heortla | OE | Hartlebury |
| Herbert | OFr(<OG) | Herberdesmulne |
| ? Hiddi | OE | Hedesdale |
| Hildebald | OE | Hildersholme |
| Hildewulf | OE | Hilderstone, Oldershaws |
| ?*Hille | OE | Hillswood |
| *Hlud | OE | Ludstone |
| *Hod | OE | Hoddesdone |
| ? Horsa | OE | Orslow |
| Houel | OW | Hollinsclough |
| ? *Hræfn or Hrafn | OE/ON | Rangemore |
| Hrani | ON | Ranslow |
| ?Hraði | ON | Ratherseates |

635

| | | |
|---|---|---|
| Hroþwulf or *Hróðulfr | OE/ON | Rolleston, Rocester |
| Hud(d)a | OE | Huddale (Farm) |
| *Hud(d)el | OE | Huddelsford |
| Hulcok | ME | Hullocks Pool |
| Hun | OE | Handsworth |
| ?Hunta | OE | Huntley, Huntingdon |
| Hwita | OE | Whittington |
| ?Hwita | OE | Withnall |
| ?Hwit or *Witi | OE | Whiston |
| Hwittuc | OE | Quixhill |
| Hyd(d)el | OE | Hillsdale |
| Hyht | OE | Hixon |
| ?*Ippa | OE | Ipstones |
| Jevon | ME | Jeffryns Hays |
| Jókell | ON | Yoxall |
| Johannes | ME | Johnson Hall |
| Káti or *Catta | ON/OE | Catton |
| Lang | OE | Longsdon |
| Leofa | OE | Loynton |
| Leofede | OE | Levedale |
| Leofhere | OE | Leycett |
| Leofnoþ | OE | Levenodeshay |
| Leofric or Leofhere | OE | Lyversegge |
| Leofstan | OE | Stansley Wood |
| *Mad(d)a | OE | Madeley |
| Malbert | OFr | Maple Hays |
| *Mocc | OE | Moxley |
| Modred | OE | Moddershall |
| Moll | OE | Moseley |
| Mucel | OE | Mucklestone |
| *Ord | OE | Ordsey |
| ?Orda | OE | Orgreave |
| ? Ordbriht | OE | Orbeton |
| ? Osmund or Ásmundr` | OE/ON | Ossoms Hill |
| *Pættel | OE | Patshull |
| *P(e)atta | OE | Pattingham |
| Penda | OE | Pendeford |
| *Peofel | OE | Pilstones |
| Peol | OE | Pelsall |
| Pinchon | OFr | Pinchley |
| Redbald | OE | Rodbaston |
| Rodbert | OFr(<OG) | Rodberdes Land |
| *Sceob(b)a | OE | Shebdon, Shobnall |
| *Sceot | OE | Shettesford |
| ?*Searu | OE | Saredon, Shareshill |
| Secg | OE | Seckley Wood, Sedgley |
| Sigeric | OE | Syerscote |
| Sigeweard | OE | Siwardesmor |
| Skropi | ON | Scropton |
| Snell or Sniallr | OE/ON | Snail's End, Snelsdale |
| *Stan | OE | Stanshope |

| | | |
|---|---|---|
| Steinn | ON | Steynesmor |
| *Strangric | OE | Stramshall |
| Sveinn | ON | Swainsmoor |
| Tata | OE | Tatenhill |
| Tatmann | OE | Totmonslow |
| *Teotta | OE | Tettenhall |
| *Tetti | OE | Tittesworth |
| Þoraldr | ON | Thoraldeswod |
| Þorkell | ON | Tucklesholme |
| Þorfreðr, Þorvaldr | ON | Thursfield |
| Tibba | OE | Tipton |
| Tilli | OE | Tillington |
| *Titten | OE | Tittensor |
| Þorfreðr | ON | Thursfield |
| Tot(t)a | OE | Tutnall |
| ? Tubba | OE | Tuppenhurst |
| Tutta or *Stut | OE | Tutbury |
| *Tynni | OE | Tynsall |
| ? *Ucga or *Hycga | OE | Hug Bridge |
| Ufegeat | OE | Ivetsey |
| ?Wælgist | OE | Worstead Green |
| Waldhere | OE | Waldreslowe |
| *Wer | OE | Werrington |
| Wicga | OE | Wiggenstall |
| Wigbald | OE | Wobaston |
| Wiglaf | OE | Wooliscroft |
| Wigmund | OE | Wymundewolde |
| Wigstan or Winestan | OE | Whistamere |
| *Wihta | OE | Wightwick |
| Wilbriht | OE | Wilbrighton |
| ? Wilfriþ | OE | Worston |
| Wille | ME | Willslock |
| Willa | OE | Willenhall |
| *Willing | OE | Willingsworth |
| Wilmund | OE | Wilnecote |
| Wilþrȳþ | OE | Wilderley (Barn) |
| ? Wine | OE | Winscote |
| Wine | OE | Windswell Pool, Winshill |
| Woden | OE | Wednesbury, Wednesfield |
| Wolfhad | OE | Wolfotebridge |
| *Wrott | OE | Wrottesley |
| Wulfgar | OE | Wolgarstone |
| Wulfhere | OE | Wulfherecester |
| Wulfrun | OE | Wolverhampton |
| Wulfsige | OE | Wolseley |
| Wulfstan | OE | Wilstanswude, Wolstanton |
| Wulfweard | OE | Wordsley |
| *Wuttuc or Hwittuc | OE | Uttoxeter |
| ?Wynna | OE | Winnington |

# Bibliography

Abrams, L. (2001) 'Edward the Elder's Danelaw' in Higham & Hill 2001 128-43.

Alecto (1988) *County Edition of Domesday Book for Staffordshire*, 3 vols and maps, London: Alecto Historical Editions.

Alexander, J. & Binski, P. (1987) *Age of Chivalry: Art in Plantagenet England 1200-1400*, London: Royal Academy of Arts/Weidenfeld & Nicolson.

Anderson, O. S. (1934) *The English Hundred Names*, Vol. 1, Lund : Lunds Universitets.

Andrews, C. B. (ed.) (1936) *The Torrington Diaries*, London and New York: Methuen and Barnes & Noble.

Anon. (1957) *A Portrait of Cannock Chase*, Bloxwich: The Friends of Cannock Chase.

—— (1984) *Aldridge and District Yesterdays*, Walsall: Walsall Archive Service.

—— (1985) *The Past in the Pipeline: Archaeology and the Midland Pipeline*, London: Esso Petroleum.

Ashley, M. (1998) *British Monarchs*, London: Robinson Publishing.

Aubrey, J. (1980) *Monumenta Britannica*, M.S. facsimile (2 vols), Sherborne: Dorset Publishing.

Bailey, N. (1782) *Universal Etymological English Dictionary*, 24th edition, Edward Harwood (ed.), London.

Baker, A. E. (1854) *Glossary of Nottinghamshire Words & Phrases*, 2 vols., London: John Russell Smith; Northampton: Abel & Sons, and Mark Dorman.

Ball, H. (1991) *Place-Names in the Moorlands*, Leek: Harry Ball.

Baring-Gould, S. (1897) *Bladys of the Stewponey*, London: Methuen & Co.

Barnes, T. (1907-8) 'Celtic Survivals in the Place-Names of North Staffordshire', *Journal of the North Staffordshire Field Club & Archaeological Society*, Vol. 42 126-48.

—— (1910-11) 'Some Further Notes on Celtic Place-Names', *Journal of the North Staffordshire Field Club & Archaeological Society*, Vol. 45 159-68.

Bartlett, R. (ed. and trans.) (2002) *Geoffrey of Burton 'Life and Miracles of St Modwenna'*, Oxford: Oxford University Press.

Barraclough, G. (1988) *The Charters of the Anglo-Norman Earls of Chester c.1071-1237*, The Record Society of Lancashire and Cheshire, Vol. CXXXVI, Gloucester: Alan Sutton Publishing.

Bassett, S. (ed.) (1989) *The Origins of the Anglo-Saxon Kingdoms*, London and New York: Leicester University Press.

—— (1992) 'Church and Diocese in the West Midlands: the Transition from British ot Anglo-Saxon Control', in Blair & Sharpe 1992: 13-40.

—— (1996) 'The Administrative Landscape of the Diocese of Worcester in the tenth century', in Brooks & Cubitt 1996: 147-73.

—— (2000) 'How the west was won: the Anglo-Saxon takeover in the West-Midlands', in *Anglo-Saxon Studies in Archaeology and History* 11 107.

—— (2001) 'Birmingham Before the Bull Ring', in *Midland History*, XXXVI 1-33.

Bateman, T. (1861) *Ten Years' Diggings in Celtic and Saxon Grave Hills in the Counties of Derby, Stafford, and York, from 1848; with Notices of Some Former Discoveries, Hitherto Unpublished, and Remarks on the Crania and Pottery from the Mounds*, Derby: W. Bemrose & Sons; London: J. R. Smith.

Beaver, S. H. & Turton, B. J. (1979) *The Potteries: A Description of the O.S. 1:50,000 sheet 118*, London: Geographical Association 13.

Beaver, S. H. (1945) 'The Black Country', in J. Myers, *The Land of Britain: the Report of the Land Utilisation Survey, Part 61: Staffordshire*, London: Geographical Publications.

Bellett, G. (1865) *Antiquities of Bridgnorth*, London: Longmans, Brown, Green & Longmans.

Beresford, W. (1883) *The Diocescan History of Lichfield*, London & Brighton: SPCK.

Bigmore, P. (1979) *The Bedfordshire and Huntingdonshire Landscape*, London: Hodder & Stoughton.

Birch, W. de G. (1885-93) *Cartularium Saxonicum*, 4 vols., London: Whiting & Co.

Björkman, E. (1910) *Nordische Personennamen in England*, Halle.

Blair, J. & Ramsay, N. (1991) *English Medieval Industries: Craftsmen, Techniques, Products*, London & Rio Grande: The Hambledon Press.

Blair, J. & Sharpe, R. (eds) (1992) *Pastoral Care Before the Parish*, London and New York: Leicester University Press.

Blount, T. (1656) *Glossographia, or a Dictionary Interpreting all such Hard Words, Whether Hebrew, Greek, Latin, Italian, Spanich, French, Tentonick, Belgick, British or Saxon, as are now Used in our refined English Tongue*, London: Tho. Newcomb.

—— (1660) *Boscobel, or The History of His Sacred Majesties Most Miraculous Preservation after the Battle of Worcester, 3 Sept 1651*, London: Henry Seile.

Blunt, C. E. (1973) 'The Origins of Stafford Mint', *Otium et Negotium*, Stockholm, 13-22.

Blunt, C. E., Stewart, B. H. I. H., & Lyon, C. S. S. (1989) *Coinage in Tenth-Century England., from Edward the Elder to Edgar's Reform*, Oxford & New York: Christopher Evelyn.

Bocock, E. W. (1923) *Shropshire Place Names*, Shrewsbury: Wilding.

Bonney, D. (1979) 'Early Boundaries and Estates in Southern England', in Sawyer 1979: 41-51.

Bonser, W. (1932) 'Survivals of Paganism in Anglo-Saxon England', *Transactions of Birmingham Archaeological Society* Vol. 56 37-70.

Bradley, H. (1886) 'Etocetum or Letocetum', *Academy*, 756, 30th October.

—— (1889) 'Etymology of Lichfield', *Academy*, 914, 9th November.

—— (1928) *The Collected Papers of Henry Bradley*, Oxford: The Clarendon Press.

Bradshaw, H. (1887) *Life of St Werberge*, London: Early English Texts Society.

Branston, B. (1974) *The Lost Gods of England*, London: Thames and Hudson.

Brewer, D. & Gibson, J. (eds) (1997) *A Companion to the Sir Gawain Poet*, Woodbridge: Brewer.

Breeze, A. (2001) 'The Name of Cound, near Wroxeter', in *Transactions of the Shropshire Archaeological and Historical Society* Vol. LXXVI 2001 76-7.

Briggam, C. P. (1998) *Grey in Old English: An Interdisciplinary Semantic Study*, London: Runetree Press.

Brighton, F. (1937) *Tales of Ipstones*, Dudley: Blocksidge.

—— (1942) *Pattingham*, Dudley: Blocksidge.

Bromwich, R. (ed.) (1972) *The Beginnings of Welsh Poetry: Studies for Sir Ifor Williams*, Cardiff: University of Wales Press.

Bronnenkant, L. J. (1985) 'Place-Names and Anglo-Saxon Paganism', in Nomina 8 72.

Brooks, N. (1989) 'The Formation of the Mercian Kingdom', in Bassett 1989: 159-70.

—— (2000a) *Communities and Warfare 700-1400*, London: Hambledon Press.

—— (2000b) *Anglo-Saxon Myths – State and Church 400-1066*, London: Hambledon Press.

—— (2000c) (with M. Gelling and D. Johnson), 'A New Charter of King Edward', in Brooks 2000b: 217-38.

Brooks, N. & Cubitt, C. (eds) (1996) *St Oswald of Worcester: Life and Influence*, London and New York: Leicester University Press.

Brown, M. P. & Farr, C. A. (eds) (2001) *Mercia: An Anglo-Saxon kingdom in Europe*, London & New York: Leicester University Press.

Burne, S. A. H. (1913) *Excursions into Staffordshire History*, Tunstall: Edwin H. Eardley.

—— (1915) *Archaeological Papers with Special Reference to Staffordshire*, Tunstall: Edwin H. Eardley.

—— (1957) 'Historical Notes on Cannock Chase', in Anon. 1957 9-13.

—— (1961) *Occasional Papers derived from Manuscript Sources in The William Salt Library* (copy in WSL).

—— (n.d.) *More Occasional Papers derived from Manuscript Sources in The William Salt Library*, privately printed (copy in WSL, bound with Burne 1961).

Burritt, E. (1868) *Walks in the Black Country and Its Green Borderland*, London: Sampson Low.

Cameron, K. (1976) 'The Significance of English Place-Names', *Sir Israel Gollancz Memorial Lecture, Proceedings of the British Academy* Vol. 62 Oxford: Oxford University Press.

—— (1979-80) 'The Meaning and Significance of Old English *walh* in English Place-Names', Journal of the English Place-Name Society 12 1-46.

—— (1987) (studies collected by), *Place-Name Evidence for the Anglo-Saxon Invasion and Scandinavian Settlements*, Nottingham: English Place-Name Society.

—— (1996) *English Place-Names*, 2nd edition, London: Batsford.

—— (1998) *A Dictionary of Lincolnshire Place-Names*, with contributions by J. Insley, Nottingham: English Place-Name Society.

Camden, W. (1674) *Remains Concerning Britain*, J. Philipot & W. D. Gent (eds), 7th impression 1870 London: John R. Smith.

Campbell, A. (ed.) (1962) *The Cronicle of Æthelweard*, London: Thomas Nelson & Sons Limited.

Campbell, J. (1986) *Essays in Anglo-Saxon History*, London & Ronceverte: The Hambledon Press.

—— (1995) 'The late Anglo-Saxon state: a maximum view', *Proceedings of the British Academy*, 87 39-65.

Carter, W. F. (1882-7) *The Midland Antiquary*, 4 vols, Birmingham: Houghton & Co., London: Simpkin, Marshall & Co.

Cartwright, J. J. (ed.) *The Travels through England of Dr Richard Pococke*, 2 vols., London: Camden Society.

Cavill, P., Harding, S. E. & Jesch, J. (2000) *Wirral and its Viking Heritage*, Nottingham: English Place-Name Society.

Chambers, R. W. (1959), *Beowulf: An introduction to the study of the poem with a discussion of the stories of Offa and Finn*, (with a Supplement by C. L. Wrenn), 3rd edition, Cambridge: Cambridge University Press.

Chaney, W. A. (1970) *The Cult of Kingship in Anglo-Saxon England: the Transition from Paganism to Christianity*, Manchester: Manchester University Press.

Charles, B. G. (1992) *The Place-Names of Pembrokeshire*, Aberystwyth: National Library of Wales.

Cherry, J. L. (1890) *Stafford in Olden Times*, Stafford: J. C. Mort.

Cherry, J. L. & Cherry, K. (1908) *Historical Studies Relating Chiefly to Staffordshire*, Stafford: J. C. Mort.

Chester, H. O. (1979) *The Iron Valley; Eight Centuries of Iron Making and Ore Mining in the Churnet Valley*, Cheadle: Churnett Valley Press.

Clapham, A. R., Tutin, T. G. & Warburg, E. F. (1962) *Flora of the British Isles*, Cambridge: Cambridge University Press.

Clark, J. (1919) 'Survivals of the Dialect and Customs of the Moorlands', *Transactions of the North Staffordshire Field Club* Vol. 54 919-20.

Clark, C. (1993) 'Domesday Book – A Great Red Herring: Thoughts on some Late-Eleventh-Century Orthographies', *England in the Eleventh Century*, Stamford: Paul Watkins.

Clifford, R. (1817) *A Topographical & Historical Description of the Parish of Tixall in the County of Stafford*, Paris: privately published.

Coates, R. (1988) *Toponymic Topics: Essays on the Early Toponomy of the British Isles*, Brighton: Younsmere Press.

—— (1989) *The Place-Names of Hampshire*, London: B. T. Batsford.

—— (1997) 'The Scriptorium of the Mercian Rushworth Gloss: a Bilingual Perspective', *Notes & Queries* 242 (NS 44) 1997 4 453-8.

—— (2000) 'Plardiwick', in *Journal of the English Place-Name Society* 32 1999-2000 21-2.

—— (2001) *The Lyme* [an unpublished paper delivered at the Annual general Meeting of The English Place-Name Society on 11th July 2001 in memory of Professor Kenneth Cameron].

Coates, R. & Breeze, A., with a contribution by D. Horovitz (2000) *Celtic Voices, English Places: Studies of the Celtic Impact on Place-Names in England*, Stamford: Shaun Tyas.

Cockin, T. (2000) *The Staffordshire Encyclopaedia: A secondary source index on the history of the old county of Stafford, celebrating its curiosities, peculiarities and legends*, Barlaston: Malthouse Press.

Cole, A. (1987-8) 'The Distribution and Usage of the Place-Name Elements *botm, bytme*, and *botn*', *Journal of the English Place-Name Society* 20 38-46.

Cole, A., Cumber, J., & Gelling, M. (2000) 'Old English *merece* 'Wild Celery, Smallage' in Place-Names', in *Nomina* 23 2000 140-148.

Colgrave, B. (ed.) (1927) *The Life of Bishop Wilfrid by Eddius Stephanus*, Cambridge: Cambridge University Press.

Copley, G. (1986) *Archaeology and Place-Names in the Fifth and Sixth Centuries*, British Archaeological Reports, British Series 147.

—— (1988) *Early Place-Names of the Anglian Regions of England*, British Archaeological Reports, British Series 185.

Coplestone-Crow, B. (1989) *Herefordshire Place-Names*, British Archaeological Reports, British Series 214.

Cox, T. (1720) *Magna Britannica*, 6 vols., Part IV, London: Elizabeth Nutt.

Cox, B. (1971-2) 'Leicestershire Moot Sites: the Place-Name Evidence', *Leicestershire Archaeological and Historical Society*, Vol. 47.

—— (1973) 'The Significance of the Distribution of English Place-Names in –ham in the Midlands and East Anglia', *Journal of the English Place-Name Society* 5 15-73.

—— (1975-6) 'The Place-Names of the Earliest English Records', *Journal of the English Place-Name Society* 8 12-66.

Cramp, R. (1977) 'Schools of Mercian Sculpture', in Dornier 1977: 191-231.

Cronne, H. A. & Davis, R. H. C. (eds) (1968) *Regesta Regum Anglo-Normannorum 1066-1154*, Oxford: Clarendon Press.

Darby, H. C. (1977) *Domesday England*, Cambridge: Cambridge University Press.

Darby, H. C. & Terrett, I. B. (1971) *The Domesday Geography of Midland England*, 2nd edition, Cambridge: Cambridge University Press.

Dark, K. R. (1994) *Civitas to Kingdom: British Political Continuity 300-800*, London and New York: Leicester University Press.

Darlington, J. (1994) *Stafford Past, A Guide to the Archaeological and Historical Sites of the Stafford Area*, Stafford: Stafford Borough Council.

—— (ed.) (2001) *Stafford Castle: Survey, Excavation & Research 1978-1998, Vol. I – The Surveys*, Stafford: Stafford Borough Council.

Davies, E. (1959) *Flintshire Place-Names*, Aberystwyth: University of Wales Press.

Davies, W. (1977) 'Annals and the Origin of Mercia', in Dornier 1977: 17-29.

Davies, R. H. C. (1982) 'Alfred & Guthrum's Frontier', *English Historical Review*.

Dent, R. K. & Hill, J. (1896) *Historical Staffordshire*, Birmingham, Leicester and Leamington: Midland Educational.

Desborough, D. (1991) 'An Introduction to the Staffordshire Domesday', in *County Edition of Domesday Book for Staffordshire*, 3 vols and maps, London: Alecto Historical Editions.

Dickins, B. (1947) 'Yorkshire Hobs', *Transactions of the Yorkshire Dialect Society*, part xlviii, vol. vii.

Dilworth, D. (1976) *The Tame Mills of Staffordshire*, London & Chichester: Phillimore.

Dodgson, J. McN. (1987) 'Domesday Book: Place-Names and Personal Names', in Holt 1987: 121-38.

Dornier, A. (ed.) (1977) *Mercian Studies*, Leicester: Leicester University Press.

Dudley, D. (1665) *Metallum Martis; or, Iron made with Pit-coale, Sea-cole, etc.*

Dugdale, W. (1730) *The Antiquities of Warwickshire*, 2nd edition, 2 vols, London: J. Osborn and T. Longman.

Dugdale, W. (1817-30) *Monasticum Anglicanum'*, 6 vols, London.

Duignan, W. H. (1880) 'On Local Etymology' (parts I, II and III), *The Midland Magazine*.

—— (1882-3) 'On some Midland Etymologies', *The Midland Antiquary*, Vol. I (c.1882) 131-5; Vol. II (1883) 121-3.

—— (1888) *The Charter of Wulfrun to the Monastery at "Hamtun" Wolverhampton*, Wolverhampton.

—— (1894) 'On some Shropshire Place-Names', *Transactions of the Shropshire Archaeological Society*, 2nd Series 6 1-18.

—— (1897) 'On some Shropshire Place-Names', *Transactions of the Shropshire Archaeological Society*, 2nd Series 9 385-394.

—— (1897-8) 'Etymology of Local Place-Names', in Hackwood 1898: 3-7.

—— (1902) *Notes on Staffordshire Place-Names*, London & New York: Henry Frowde.

Dumville, D. N. (1989) 'Essex, East Anglia and the Expansion of Mercia in the South-East Midlands', in Bassett 1989: 123-40.

—— (1992) *Wessex and England from Alfred to Edgar: Six Essays on Political, Cultural and Ecclesiastical Revival*, Woodbridge: Boydell.

Dunphy, A. (2002) *Tales from Penn Forge* [private printing].

Dyer, C. (2002) 'The Urbanizing of Staffordshire: The First Phases', in *Staffordshire Studies* 14 2002 1-31.

—— (2004) 'The Historical Evidence', in *Medieval Tanning and Retting at Brewood, Staffordshire: Archaeological Excavations 1999-2000*, Birmingham: Birmingham Archaeology.

Earle, J. (1888) *A Handbook to the Land-Charters, and other Saxonic Documents*, Oxford: Oxford University Press.

Earle, J. & Plummer, C. (1892-9) *Two of the Saxon Chronicles Parallel*, 2 vols, Oxford: Oxford University Press.

Ede, J. F. (1962) *History of Wednesbury*, Wednesbury: Wednesbury Corporation.

—— (1969) *Corrigenda to History of Wednesbury 1962*, Wednesbury: Wednesbury Corporation.

Edwards, N. (1949) *Medieval Tutbury*, Lichfield: A. C. Lomax's Successors.

Ekwall, E. (a) *Namn och Bygd*, xli.

—— (1923) *English Place-names in -ing*, Lund.

—— (1928) *English River-Names*, Oxford: Oxford University Press.

—— (1936) *Studies in English Place-Names*, Stockholm: Wahlström & Widstrand.

—— (1959) *Etymological Notes on English Place-Names*, Lund Studies in English 27.

—— (1960) *The Concise Oxford Dictionary of English Place-Names*, 4th edition, Oxford: Oxford University Press.

—— (1963) S*elected Papers*, Lund Studies in English 33.

Elliott, R. W. V. (1964) 'A Runic Fragment at Leek', *Medieval Archaeology* 8 213-4.

—— (1984) *The Gawain Country: Essays on the Topography of Middle English Poetry*, Leeds Texts & Monographs, New Series No. 8, Leeds: University of Leeds.

Elliott, R. (1997) 'Landscape & Geography', in Brewer & Gibson 1997.

Erdeswick, S. (1844) *A Survey of Staffordshire*, ed. by T. Harwood, London: J. B. Nichols.

Eyton, R. W. (1854-60) *The Antiquities of Shropshire*, London: John Russell Smith; Shifnal: B. L. Beddow.

—— (1881) *Domesday Studies: An Analysis and Digest of the Staffordshire Survey*, London: Trubner.

Faith, R. (1997) 'Hyde Farms and Hyde Place-Names: Summary Report of Work', *Medieval Settlement Research Group Annual Report* 10 20-3.

Faull, M. G. (1984) *Studies in Late Anglo-Saxon Settlement*, Oxford: Oxford University Department for External Studies.

Featherstone, P. (2001) 'The Tribal Hidage and the Ealdormen of Mercia', in Brown & Farr 2001: 23-34.

Feilitzen, O. von F. (1937) *The Pre-Conquest Personal Names of Domesday Book*, Nomina Germanica, Uppsala: Almqvist & Wiksell.

Fell, C. (1984) *Women in Anglo-Saxon England, and the impact of 1066*, London: British Museum.

Fellows-Jensen, G. (1972) *Scandinavian settlement-names in Yorkshire*, Copenhagen: Institut for Navneforskning.

—— (1990) 'Place-Names as a Reflection of Cultural Interaction', *Anglo-Saxon England* 19 13-21.

—— (1997) 'Scandinavians in Cheshire: a Reassessment of the Onomastic Evidence', in Rumble & Mills 1997: 79-81.

Fernie, W. T. (1897) *Herbal Simples, Approved for Modern Uses of Cure*, 2nd edition, Bristol: Wright & Co.

Field, J. (1972) *English Field Names, a Dictionary*, Newton Abbot: David & Charles.

—— (1993) *A History of English Field Names*, London & New York: Longman.

Finberg, H. P. R. (1957) *Gloucestershire Studies*, Leicester: Leicester University Press.

—— (ed.) (1964) *Lucerna*, 159.

—— (1972) *The Early Charters of the West Midlands*, 2nd edition, Leicester: Leicester University Press.

—— (ed.) (1975) *Scandinavian England: Collected Papers by F. T.Wainwright*, Chichester: Phillimore.

Fisher, M. (1969) *Dieulacres Abbey, Leek, Staffordshire*, Leek [ no publisher].

Fitzhugh, W. W. & Ward, E. I. (eds) (2000) *Vikings: The North Atlantic Saga*, Washington & London: Smithsonian Institution Press.

Fletcher, W. D. G. & Auden, H. M. (1907) *The Shropshire Lay Subsidy Roll of 1327*, Shrewsbury: Salop County Council.

Ford, W. J. (1979) 'Settlement Patterns in Warwickshire', in Sawyer 1979: 143-163.

Forsberg, R. (1950) *A Contribution to a Dictionary of Old English Place-Names*, Nomina Germanica, Uppsala: Almqvist & Wiksell.

—— (1970) 'On Old English ad in English Place-Names', *Namn och Bygd* 58 20-82.

—— (1987) 'The Boundaries of BCS 987 Once Again', *Namn och Bygd* 75 82-9.

Forster, K. (1981) *A Pronouncing Dictionary of English Place-Names including standard local and archaic variants*, London: Routledge.

Förster, M. (1941) *Der Flussname Themse und seine Sippe*, Munich.

Foxall, H. D. G. (1980) *Shropshire Field Names*, Shrewsbury: Shopshire Archaeological Society.

Franklin, M. J. (ed.) (1997) *English Episcopal Acta 14: Coventry and Lichfield 1072-1159*, Oxford: Oxford University Press for The British Academy.

—— (ed.) (1998a) *English Episcopal Acta 16: Coventry and Lichfield 1160-1182*, Oxford: Oxford University Press for The British Academy.

—— (ed.) (1998b) *English Episcopal Acta 17: Coventry and Lichfield 1183-1208*, Oxford: Oxford University Press for The British Academy.

Fuller, T. (1840) *The History of the Worthies of England*, 3 vols, London: Thomas Tegg.

Gale, R. (1780) 'Reliquiæ Galeanæ; or a miscellaneous pieces by R. & S. G., etc', in J. Nicolls *et al*, *Bibliographica Topographica Britannia*, Vol 3.

Garner, R. (1844) *The Natural History of the County of Stafford*, London: John Van Voorst (Supplement 1860).

Gelling, M. (1975) 'Further thoughts on pagan place-names', in Cameron 1987          .

—— (1976) 'The Evidence of Place-Names' in Sawyer 1976: 201-11.

—— (1977) 'The Early History of Western Mercia', in Dornier 1977: 184-201.

—— (1979) 'The Evidence of Place-Names I', in Sawyer 1979: 110-121.

—— (1981) 'Some thoughts on Staffordshire Place-Names', *North Staffordshire Journal of Field Studies*, 21 1-20.

—— (1982) 'Some Meanings of Stow', in Pearce 1982: 187-96.

—— (1984) *Place-Names in the Landscape*, London & Melbourne: Dent.

—— (1987) 'Anglo-Saxon Eagles', *Leeds Studies in English*, New Series 18.

—— (1988) *Signposts to the Past; Place-Names and the History of England*, 2nd edition, Chichester: Phillimore.

—— (1989) 'The Early History of Western Mercia', in Bassett 1989: 184-201.

—— (1990) *The Place-Names of Shropshire, Part I*, English Place-Name Society 62/63

—— (1992) *The West Midlands in the Early Middle Ages*, London and New York: Leicester University Press.

—— (1997) 'The Hunting of the Snor', in Rumble & Mills 1997: 93-5.

—— (2001) 'The Name *Monetvile* in Darlington 2001: Appendix 1.

Gelling, M. & Cole, A. (2000) *The Landscape of Place-Names*, Stamford: Shaun Tyas.

Gibson, E. (1695) *Camden's Britannia newly translated into English: with Large Additions and Improvements*, Oxford: Edmund Gibson.

Gough, R. (trans. and enlarged by) (1806). *Britannia, or, a Chorographical Description of the Flourishing Kingdoms of England, Scotland, and Ireland, and the Islands Adjacent, from the Earliest Antiquity,* by William Camden, 2nd edition, 4 vols, London: John Stockdale.

Gough, R. (1968) *History of Myddle*, Fontwell: Centaur.

Gould, J. (1957) *Men of Aldridge, a Local History of the area now included in the Irban District of Aldridge*, Bloxwich: Geof J. Clark (reprinted with additions 1983, Stroud: Alan Sutton).

—— (1967-8) 'First Report of the Excavations at Tamworth, Staffs., 1967 – The Saxon Defences', *Transactions of the Lichfield and South Staffordshire Archaeological and Historical Society* 9 18-29.

—— (1987) 'Old English *ad* and the Bounds of Barr', *Namn och Bygd* 82-9.

—— (1991-2) 'Lwytgoed: its significance in Early Medieval Documents', *Transactions of South Staffordshire Archaeological & Historical Society* 33 7-8.

—— (1993) 'Lichfield – Ecclesiastical Origins', in Martin Carver (ed.), *In Search of Cult: Archaeological Investigations in Honour of Philip Rahtz*, Woodbridge & Rochester: Boydell.

—— (1993a) 'Lichfield before St Chad', in Maddison 1993: 1-10.

Graham-Campbell, J., R. Hall, J. Jesch & D. Parsons (eds) (2001) *Vikings and the Danelaw: Select Papers from the Proceedings of the Thirteenth Viking Congress, Nottingham and York, 21-30 August 1997*, Oxford: Oxbow Books.

Greenslade, M. W. & Stuart, D. G. (1984) *A History of Staffordshire*, Chichester: Phillimore.

Grierson, P. (1987) 'The Monetary System under William I', in Alecto 1988: 76-7.

Griffiths, G. (1894) *History of Tong & Boscobel*, 2nd edition, Newport, Market Drayton & Stone: Horne & Bennion; London: Simpkin, Marshall, Hamilton, Kent & Co.

Griffiths, D. (2001) 'The North-West Frontier', in Higham & Hill 2001: 167-87.

Groom, J. N. (1992) 'The Topographical Analysis of Medieval Town Plans: the Examples of Much Wenlock and Bridgnorth', *Midland History*, 17 16-38.

Gunstone, A. J. H. (1971) *Sylloge of Coins of the British Isles (17): Ancient British, Anglo-Saxon and Norman Coins in Midland Museums*, London: British Academy.

Guttery, D. R. (1950) *The Story of Pensnett Chase*, Dudley: Dudley County Borough.

Hackwood, F. W. (1895) *A History of West Bromwich*, Birmingham: The 'Birmingham News' and Printing Co. Ltd.

—— (1896) *Some Records of Smethwick*, Smethwick: Telephone Printing Co.

—— (1898) *Sedgley Researches*, Dudley: reprinted from *Dudley Herald* 6 March 1897 to 19 February 1898.

—— (1902) *Wednesbury Ancient and Modern*, Wednesbury: Ryder and Son, reprinted from *Wednesbury Herald* 29 June 1901 to 5 July 1902

—— (1905a) 'Some Staffordshire Place-Names', *Staffordshire Chronicle* 154-64.

—— (1905b) *Staffordshire Curiosities & Antiquities*, Stafford & Birmingham: Staffordshire Chronicle.

—— (1906) *Staffordshire Stories Historical and Legendary: A Miscellany of County Lore and Anecdotes*, Stafford & Birmingham.

—— (1908) *Annals of Willenhall*, Wolverhampton: Whitehead.

—— (1915) *Oldbury and Round About*, Wolverhampton: Whitehead Bros; Birmingham: Cornish Bros Ltd.

—— (1924) *Staffordshire Customs, Superstitions and Folklore*, Birmingham: Mercury Press.

Haden-Jones, J. (n.d.) *The History of the Black Country*, Halesowen: Janus Books.

Hadfield, C. (1969) *The Canals of the West Midlands*, 2nd edition, Newton Abbott: David & Charles.

Hald, K. (1963) 'The Cult of Odin in Danish Place-Names', in A. Brown & P. Foote (eds), *Early English and Norse Studies: Presented to Hugh Smith in Hnour of his Sixtieth Birthday*, London: Methuen & Co Limited.

Halliwell, J. O. (1850) *Dictionary of Archaic and Provincial Words*, London: John Russell Smith.

Hamilton, N. E. S. A. (1870) *De Gestis Pontificum Anglorum*, Rolls Series.

Hardwicke, W. (1822) 'Parochial Histories of Shropshire', (MS) Vol. IV, WSL

Hardy, R. H. (1907) *A History of the Parish of Tatenhill in the County of Stafford*, London: Harrison.

—— (1908) *Court Rolls of the Parish of Tatenhill in the County of Stafford*, London: Harrison.

Harmer, F. E. (1989) *Anglo-Saxon Writs*, 2nd edition, Stamford: Paul Watkins.

Harrison, C. (ed.) (1986) *The History of Keele*, Keele: University of Keele.

Harrison, C. J. (2004) *A Bibliography of the History of Staffordshire*, Keele: University of Keele.

Harrison, W. (1877-1909) *William Harrison, A Description of England in Shakspere's Youth, being the second and third books of his Description of Britaine and England, from the first two editions of Holinshed's Chronicle AD 1577, 1587*, 3 vols., London: Bungay.

Hart, C. R. (1975) *The Early Charters of Northern England and the North Midlands*, Leicester: Leicester University Press.

—— (1977) 'The Kingdom of Mercia', in Dornier 1977: 43-61.

—— (1992) *The Danelaw*, London & Rio Grande: The Hambledon Press.

Harthorne, C. H. (1841) *Salopia Antiqua*, London: John W. Parker.

Harvey, S. P. J. (1979) 'Evidence for Settlement Study: Domesday Book', in Sawyer 1979: 105-9.

Harwood, T. (1806) *The History and Antiquities of the Church and City of Lichfield*, Gloucester.

Haslam, J. (1993) 'Market and Fortress in England in the reign of Offa', in *World Archaeology* 19 76-93.

Hearne, T. (ed.) (1723) *Hemingi Chartularium Ecclesiæ Wigorniensis*, 2 vols., Oxford: Theatro Sheldoniano.

Hibbert, F. A. (1909) *Monasticism in Staffordshire*, Stafford: J. C. Mort.

Higham, M. C. (2002) 'The Problems of the Bee-Keepers', *Journal of the English Place-Name Society* 34 2001-2 23-8.

Higham, N. J. & Hill, D (eds) (2001) *Edward the Elder 899-921*, London & New York: Routledge.

Higham, N. J. (1992) 'King Cearl, the Battle of Chester and the Origins of the Mercian "Overkingship"', *Midland History* 17 1992 1-15.

—— (1993) *The Origins of Cheshire*, Manchester: Manchester University Press.

—— (2002) *King Athur: Myth-Making and History*, London & New York: Routledge.

Hill, D. (1984) *An Atlas of Anglo-Saxon England*, Oxford: Basil Blackwell.

—— (2001) 'The Shiring of Mercia – Again', in Higham & Hill 2001: 144-59.

Hjelmslev, L. H. (1968) *Prolegomenes a une theorie du langage*, Paris: Paris Minuit.

Hodder, M. (2004) *Birmingham: The Hidden History*, Stroud: Tempus Publishing Limited.

Hodgkin, M. T. (1939) 'Domesday Water Mills', *Antiquity* 13 26-79.

Hodgson, R. H. (1939) *A History of the Anglo-Saxons*, 2nd edition, Oxford: Oxford University Press.

Holman, K. (2001) 'Defining the Danelaw', in Graham-Campbell *et al.* 2001 1-11.

Holt, J. C. (ed.) (1987) *Domesday Studies*, Woodbridge & Rochester: Boydell.

—— (1989) *Robin Hood*, London: Thames & Hudson.

Homeshaw, J. & Sambrook, R. (1951) *Great Wyrley*, Bloxwich.

Hooke, D. (1983) *The Landscape of Anglo-Saxon Staffordshire: The Charter Evidence*, Keele: Keele University.

—— (1990) *Worcestershire Anglo-Saxon Charter-Bounds*, Woodbridge: Boydell.

—— (1998) *The Landscape of Anglo-Saxon England*, London and New York: Leicester University Press.

—— (1999) *Warwickshire Anglo-Saxon Charter Bounds*, Woodbridge: Boydell.

Hooke, D. & Slater, T. R. (1986) *Anglo-Saxon Wolverhampton: the Town and its Monastery*, Wolverhampton: Wolverhampton Borough Council.

Hope, R. C. (1893) *Holy Wells, Legends & Traditions*, London: Elliot Stock.

Hopkinson, B. (1994) *Salt and the Domesday Salinae at Droitwich A.D. 674 to 1690: A Quantitive Analysis*, Stroud: Alan Sutton Publishing.

Horovitz, D. (1992) *Some Notes on the History of Brewood*, 2nd edition, Brewood: David Horovitz.

Horovitz, D., Coates, R. & Potter, S. (2003) 'Ingestre, Stafford', *Nomina* 26 65-82.

Horsley, J. (1732) *Britannia Romana*, London: J. Osborn & T. Longman.

Hough, C. (1997) 'The place-name Kingston and the laws of Æthelberht', *Studia Neophilogica* 69 55-7.

—— (1998) 'Place-Name Evidence for Old English Bird-Names', in Journal of the English Place-Name Society 30 1997-8 60-76.

—— (2001) 'Place-name evidence for an Anglo-Saxon animal name: OE *Pohha/*Pocca 'fallow deer'', in *Anglo-Saxon England* 30 1-14.

Hough, C. & Lowe, K. A. (eds) (2002) *'Lastwords Betst': Essays in Memory of Christine E. Fell with her unpublished writings*, Donington: Shaun Tyas.

Hutchinson, S. W. (1893) *The Archdeaconry of Stoke on Trent*, London: Bemrose.

Insley, J. (1999) 'Old English Odda', *Notes & Queries* 244 (NS 46), No. 1, 4-5.

Jackson, G. (1879) *Shropshire Word Book: a Glossary of Archaic and Provincial Words, etc., used in the County*, London: Trubner.

Jackson, K. H. (1938) 'Nennius and the Twenty-Eight Cities of Britain', *Antiquity* 12 44-55.

—— (1953) *Language and History in Early Britain*, Edinburgh: Edinburgh University Press.

Jackson, P. (ed.) (1995) *Words, Names & History: Selected Writings of Cecily Clark*, Cambridge: D. S. Brewer.

James, S. B. (1878) *Worfield on the Worfe*, London & Derby: Bembrose.

Jeayes, I. H. (1906) *Descriptive Catalogue of Derbyshire Charters in Public and Private Libraries and Muniment Rooms*, London and Derby: Bemrose & Sons Limited.

Jermy, K. E. (1992) 'Longford and Langford as Significant Names in Establishing Lines of Roman Roads', in *Britannia* 23 228-9.

Jermy, K. E. & Breeze, A. (2000) 'Welsh *ffordd* "road" in English Place-Names, particularly as an indicator of a Roman road', *Transactions of the Shropshire Archaeological & Historical Society* 75 109-10.

Jewell, H. M. (1972) *English Local Administration in the Middle Ages*, Newton Abbot: David & Charles.

Johansson, C. (1975) *Old English Place-Names Containing Leah*, Stockholm: Almqvist and Wiksell International.

John, E. (1964) *Land Tenure in Early England*, Leicester: Leicester University Press.

—— (1992) 'The Point of Woden', in *Anglo-Saxon Studies in Archaeology and History* 5 127-34.

Johnson, C. & Cronne, H. A. (eds) (1956) *Regesta Henrici Primi 1100-1135*, Oxford: Clarendon Press.

Johnston, J. B. (1914) *The Place-Names of England & Wales*, London: John Murray.

Johnstone, J. D. (1946) *Werrington; Some Notes on its History*, Cheadle.

Jolliffe, H. G. H. (1892) *The Jolliffes of Staffordshire and their Descendants down to the year 1865, compiled from Family Papers and Other Sources*, London and Aylesbury: Private Printing.

Jones, J. P. (1883) 'Tettenhall Place and Field Names', *Midland Antiquary*, 2, No. 6, Dec 1883 83-6.

—— (1894) *A History of the Parish of Tettenhall*, Wolverhampton: John Steen.

Jones, G. (1998) 'Penda's Footprint? Place-Names containing Personal Names associated with those of early Mercian Kings', *Nomina* 21 29-62.

Kemble, J. M. (ed.) (1876) *The Saxons in England*, London: Bernard Quaritch.

Kennedy, J. (ed.) (1980) *Biddulph ('by the Diggings'): a Local History*, Keele: University of Keele.

Kenyon, D. (1986) 'The Antiquity of *ham* Place-Names in Lancashire and Cheshire', in *Nomina* 10 11-27.

Keynes, S. (2001) 'Edward, King of the Anglo-Saxons', in Higham & Hill 2001: 40-66.

Keynes, S. (2001a) 'Mercia and West-Sussex in the Ninth Century', in Brown & Farr 2001: 310-28.

Kimball, E. G. (1959) *Shropshire Peace Roll 1400-1414*, Shrewsbury: Salop County Council.

King, P. W. (1996) 'The Minster *Aet Sture* in Husmere and the Northern Boundary of the Hwicce', *Transactions of the Worcestershire Archaeological Society* 3rd Series 15 73-91.

Kirby, D. P. (1977) 'Welsh Bards and the Border', in Dornier 1977: 31-42.

Kitson, P. R. (1995) 'The nature of Old English dialect distributions, mainly as exhibited in charter boundaries', in J. Fisiak (ed.), *Medieval Dialectology*, Trends in Linguistics Studies and Monographs 79 (Berlin and New York), 43-135.

Kristensson, G. (1970) *Studies on Middle English Topographical Terms*, Lund: Berlingska.

—— (1987) *A Survey of Middle English Dialects 1290-1350: the West Midland Counties*, Lund: Lund University.

—— (2000) 'The Place-Name Yarnfield in Wiltshire', in *Notes & Queries* 245 (NS 47), No. 1, 4-5.

Lang, D. (ed.) (1879) *Androw Wyntoun's The Orygynale Cronykil of Scotland* III.

Langford, A. (1872) *Staffordshire & Warwickshire Past & Present*, 4 vols, London & Birmingham: William Mackenzie.

Lapidge, M., Blair, J., Keynes, S., & Scragg, D. (eds) (1999) *The Blackwell Encyclopaedia of Anglo-Saxon England*, Oxford: Blackwell.

Larwood, J. & Hotten, J. C. (1985) *English Inn Signs: Being a Revised and Modernized Version of History of Signboards*, Exeter: Blaketon Hall.

Latham, R. E. (1980) *Revised Medieval Latin Word-List*, (with Supplement), Oxford: Oxford University Press.

Lawley, G. T. (1893) *History of Bilston*, Bilston: J. Price.

Leah, M. D., Wells, C. E., Stamper, P., Huckersby, E., & Welch, C. (1998) *The Wetlands of Shropshire and Staffordshire*, Lancaster: Lancaster University.

Levitt, J. H. (1968) *North Staffordshire Speech*, Keele: Keele University.

—— (1987) 'Charles Henry Poole and the Study of the Staffordshire Dialect', in P. Morgan (ed.), *Staffordshire Studies: Essays Presented to Denis Stuart*, Keele: Keele University.

Lewis, S. (1849) *Topographical Dictionary of England*, 7th edition, 4 vols, London: Lewis.

Lias, A. (1991) *Place Names of the Welsh Borderlands*, Ludlow: Palmers Press.

Liebermann, F. (1898-1916) *Die Gesetze der Angelsachsen*, 3 vols, Halle.

Lloyd, L. C. (1942) *Inns of Shrewsbury*, Shrewsbury: Shrewsbury Circular Printing & Publishing.

Lockwood, W. B. (1993) *The Oxford Dictionary of English Bird Names*, Oxford & New York: Oxford University Press.

Losco-Bradley, S. & Wheeler, H. M. (1984) 'Anglo-Saxon Settlement in the Trent Valley: Some Aspects', in Faull 1984: 101-14.

Losco-Bradley, S. & Kinsley, G. (2002) *Catholme: An Anglo-Saxon Settlement on the Trent Gravels of Staffordshire*, Nottingham: Trent & Peak Archaeological Unit.

Love, R. C. (ed. and translated) (1996) *Three Eleventh-Century Anglo-Latin Saints' Lives*, OMT.

Lyon, S. (2001) 'The Coinage of Edward the Elder', in Higham & Hill 2001: 67-78.

Madge, S. J. (1938) *The Domesday of Crown Lands: A Study of Legislation, Surveys and Sales of Royal Estates under the Commonwealth*, London: George Routledge & Sons Limited.

Maddison, J. (ed.) (1987) *Medieval Archaeology and Architecture at Lichfield*, British Archaeological Association Conference Transactions XIII, Leeds: Maney Publishing.

Mander, G. P. & Tildesley, N. W. (1960) *A History of Wolverhampton to the Nineteenth Century*, Wolverhampton: Wolverhampton Borough Council.

Mander, G. P. (1933-45) *Wolverhampton Antiquary: Being Collections for a History of the Town*, 2 vols, Wolverhampton: Whitehead Bros.

—— (1944) 'The Stafford Mint', *Transactions of the Old Stafford Society* 5 13-8.

Margary, I. D. (1973) *Roman Roads in Britain*, 3rd edition, London: John Baker.

Mason, J. F. A. (1961-4) 'The Norman Castle at Quatford', *Transactions of the Shropshire Archaeological Society* Vol. 57 37-62.

Mawer, A. & Stenton, F. M. (1927) *The Place-Names of Worcestershire*, Cambridge: English Place-Name Society.

Mawer, A. (1929) *Problems of Place-Name Study*, Cambridge: Cambridge University Press.

McClure, E. (1910) *Place-Names in their Historical Setting*, London, Brighton & New York: SPCK.

Meaney, A. L. (1995) 'Pagan English Sanctuaries, Place-Names and Hundred Meeting Places', *Anglo-Saxon Studies in Archaeology and History* 8 1995 29-42.

—— (1997) 'Hundred Meeting Places in the Cambridgeshire Region', in Rumble & Mills 1997: 195-240.

Mellows, W. T. (ed.) (1949) *The Peterborough Chronicle of Hugh Candidus*, London: Oxford University Press.

Metcalf, D. M. (1987) 'The Taxation of Moneyers under Edward the Confessor and in 1086', in Holt 1987: 279-94.

Miller, M. H. (1891-1900) 'Dialect of the District', *Olde Leeke*, 2 vols, Leek: Leek Times.

Mills, A. D. (1998) *A Dictionary of English Place-Names*, 2nd edition, Oxford & New York: Oxford University Press.

Morris, C. (ed.) (1959) *The Journeys of Celia Fiennes*, London: Cresset Press.

Morris, J. (ed.) (1976) *Domesday Book: Staffordshire*, Chichester: Phillimore.

Nash, T. R. (1781) *Worcestershire*, London: T. Payne

Nicholls, R. (1934) *Dialect Words and Phrases used in the Staffordshire Potteries*, 2nd edition, Hanley: James Heap.

Nicolaisen, W. F. H., Gelling, M., & Richards, M. (1970) *The Names of Towns and Cities in Britain*, London: Batsford.

Niles, J. D. (1991). 'Pagan survivals and popular belief', in M. Godden & M. Lappidge (eds), *The Cambridge Companion to Old English Literature*, Cambridge: Cambridge University Press, 126-41.

North, J. J. (1994) *English Hammered Coinage*, Vol. I, Spink & Son: London.

North, R. (1997) *Heathen Gods in Old English Literature*, Cambridge: Cambridge University Press.

Northall, G. F. (1894) *Folk Phrases of Gloucestershire, Staffordshire, Warwickshire and Worcestershire*, English Dialect Society 30, Part 11.

Oakden, J. P. (1967-8) 'W. H. Duignan's Notes on Staffordshire Place-Names, 1902: a Reassessment', *Transactions of the Lichfield & South Staffordshire Archaeological & Historical Society* 9 31-6.

——— (1984) *The Place-Names of Staffordshire, Part I, Cuttlestone Hundred*, Nottingham: English Place-Name Society.

Ogilby, J. (1675) *Britannia ... or an Illustration of the Kingdom of England and Dominion of Wales: By a Geographical and Historical Description of the Principal Roads thereof*, London.

Orton, H. (1969) *Survey of English Dialects, The West Midland Counties*, Leeds: English Dialect Society.

Oswald, A. (ed.) (1955) *The Church of St Bertelin at Stafford and its Cross*, Birmingham: Birmingham Museum.

Owen, H. W. (1994) *The Place-Names of East Flintshire*, Cardiff: University of Wales Press.

Owen, A. E. B. (1997) 'Roads and Romans in South-East Lindsey: the Place-Name Evidence', in Rumble & Mills 1997: 254-268.

Ozanne, A. (1962-3) 'The Peak Dwellers', *Medieval Archaeology* 6-7 41-7.

Padel, O. J. (1985) *Cornish Place-Name Elements*, Nottingham: English Place-Name Society 56/57.

Paffard, M. (1996) 'Staffordshire Place-Names', *Staffordshire Studies* 8 1-23.

Palliser, D. M. (1976) *The Staffordshire Landscape*, London, Sydney, Auckland & Toronto: Hodder & Stoughton.

Palmer, F. P. & Crowquill, A. (1846) *Wanderings of a Pen and Pencil*, London: Jeremiah How.

Palmer, C. F. R. (1845) *The History of the Town and Castle of Tamworth in the Counties of Stafford and Warwick*, Tamworth: J. Thompson.

Pantos, A. (2003) '"On the edge of things": the boundary location of Anglo-Saxon assembly sites', in D. Griffiths, A. Reynolds & S. Semple (eds), *Boundaries in Early Medieval Britain: Anglo Saxon Studies in Archaeology and History* 12 Oxford: Oxford University School of Archaeology.

Pape, T. (1928) *Medieval Newcastle-under-Lyme*, Manchester: Manchester University Press.

Parke, W. (1860) *Notes & Collections relating to Brewood*, Wolverhampton: W. Parke.

Parker, R. F. (1897) *Some account of Colton and of the De Wasteneys Family*, 2nd edition, Birmingham: Hudson.

Parker, D. W. (1996) 'Old English Goldhordus: A Privy or just a Treasurehouse ?', *Notes & Queries*, Sept. 1996 257-8.

Parson, W. & Bradshaw, T. (1818) *Staffordshire General & Commercial Directory*, J. Leigh.

Parsons, D (2001) 'How long did the Scandinavian language survive in England. Again', in Graham-Campbell *et al.* 2001299-312.

—— (2002) 'Old English *\*lot*, Dialect *loot*, a Salt-Maker's 'Ladle", in Hough & Lowe 2002.

Parsons, D & Styles, T. (1995-6) *'Birds in amber: the nature of English place-name elements'*, Journal of English Place-Name Studies, Vol. 28 1995-6 5-13.

Parsons, D. & Styles, T. (eds) (1997 and continuing) *The Vocabulary of English Place-Names*, Nottingham: Centre for English Name Studies.

Pearce, S. M. (1982) 'The Early Church in Western Britain and Ireland', *British Archaeological Reports* 102.

Pennant, T. (1782) *The Journey from Chester to London*, London.

Phillips, A. D. M. (ed.) (1993) *The Potteries: Certainty and Change in a Staffordshire Conurbation*, Stroud: Allan Sutton.

Pierce, G.O. (1968) *The Place-Names of Dinas Powys Hundred*, Cardiff: University of Wales Press.

Pitt, W. (1817) *A Topographical History of Staffordshire*, Newcastle under Lyme.

Plot, R. (1686) *The Natural History of Staffordshire*, Oxford: The Theatre (and Browne's map of 1682 in that volume).

Poole, C. H. (1880) *An Attempt Towards a Glossary of the Archaic and Provincial Words of the County of Stafford*, Stratford upon Avon: Saint Gregory's Press.

Preest, D. (2002) *William of Walmsbury: The Deeds of the Bishops of England (Gesta Pontificum Anglorum)*, Woodbridge: The Boydell Press.

Quiney, A. (2003) *Town Houses of Medieval Britain*, New Haven & London: Yale University Press.

Rackham, O. (1980) *Ancient Woodland: its History, Vegetation and Uses in England*, London: Edward Arnold.

—— (1986) *The History of the Countryside*, London & Melbourne: Dent.

—— (1990) *Trees & Woodland in the British Landscape*, revised edition, London & New York: Phoenix Press.

Rahtz, P. (1968) 'The Defences of Hereford', *Current Archaeology* 9 (July) 242-6.

Rathbone, C. (1974) *Dane Valley Story*, 2nd edition, Macclesfield: Macclesfield Press Limited.

Reaney, P. H. (1985) *The Origin of English Place-Names*, London: Routledge & Keegan Paul.

Reaney, P. H. & Wilson, R. M. (1997) *A Dictionary of English Surnames*, 3rd edition, Oxford: Oxford University Press.

Redfern, F. (1865) *History of the Town of Uttoxeter with Notices of Places in its Neighbourhood*, London: J. Russell Smith.

—— (1886) *History and Antiquities of the Town and Neighbourhood of Uttoxeter, with Notices of Adjoining Places*, Hanley: Allbut and Daniel; London: Simpkin, Marshall & Co.

Redin, M. (1919) *Studies in Uncompounded Personal Names in Old English*, Uppsala: U B Akademiska Bokhandeln.

Rees, U. (ed.) (1975) *The Cartulary of Shrewsbury Abbey*, Aberystwyth: National Library of Wales.

—— (ed.) (1985) *The Cartulary of Haughmond Abbey*, Cardiff: Shropshire Archaeological Society & University of Wales Press.

—— (ed.) (1997) *The Cartulary of Lilleshall Abbey*, Shrewsbury: Shropshire Archaeological & Historical Society.

Reno, F. D. (2000) *Historic Figures of the Arthurian Era*, Jefferson, North Carolina, and London: McFarland & Company, Inc, Publishers.

Rivet, A. L. F. & Smith, C. (1979) *The Place-Names of Roman Britain*, London: Batsford.

Rivet, A. L. F. (1964) *Town and Country in Roman Britain*, 2nd edition, London: Hutchinson.

Roberts, J. & Kay, C., with Grundy, L. (1995) *A Thesaurus of Old English*, 2 vols, London: King's College.

Robertson, A. J. (ed & trans.) (1956) *Anglo-Saxon Charters*, Cambridge: Cambridge University Press.

Robinson, P. H. (1968-70) 'The Stafford Moneyers 924-1165', *Transactions of the Old Stafford Society*.

Robinson, D. H. (1988) *The Sleepy Meese*, Albrighton: Waine Research Publications.

Rodwell, J. S. (ed.) (1991) *British Plant Communities: Volume I – Woodlands and Scrub*, Cambridge: Cambridge University Press.

Rolleson, A. A. (ed.) (1899) *The Old Non-Parochial Registers of Dudley*, Dudley: Dudley Herald.

Roper, J. S. (1952) *History of Coseley*, Coseley: Coseley Urban District Council.

—— (1966) *Wolverhampton, the Early Town & its History* [no publisher].

—— (1976) *History of Coseley*, Dudley: Dudley Technical Centre.

—— (1980) *The Dudley Churchwardens' Book 1618-1725*, Redditch: Ace Stationery.

Rowland, J. (1990) *Early Welsh Saga Poetry: a Study and Edition of the Englynion*, Woodbridge: Brewer.

Rowlands, H. (1766) *Mona Antiqua Restaurata, An Archaeological Discourse on the Antiquities, Natural and Historical, of the Isle of Anglesey, the Antient Seat of the British Druids*, 2nd edition, London: J. Knox.

Rowney, I. (1980) 'Medieval Chroniclers and the battle of Blore Heath', *North Staffs Journal of Field Studies* 20 1980 9-17.

Rumble, A. R. & Mills, A. D. (eds) (1997) *Names, Places and People: An Onomastic Miscellany in Memory of John McNeal Dodgson*, Stamford: Paul Watkins.

Rumble, A. R. (1977) "Hrepingas' reconsidered', in Dornier 1977: 169-71

——— (1986) review of Hooke 1983, *Nomina* 10 177-9.

——— (1997) *Ad Lapidem in Bede and a Mercian Martyrdom*, in Rumble & Mills 1997: 307-19.

Salt, E. (1888) *The History of Standon: Parish, Manor and Church, with two hundred years of registers*, Birmingham: Cornish Bros.

Salzman, L. F. (1950) *Buildings in England Down to 1540*, Oxford: Oxford University Press.

Sanders, H. (1794) *The History and Antiquities of Shenstone in the County of Stafford*, London: Nichols (but written before 1774).

Sandred, K. I. (1987) 'Ingham in East Anglia: A New Interpretation', *Leeds Studies in English*, New Series 18 231-40.

Sawyer, P. H. (1956) 'The Place-Names of the Domesday manuscripts', *Bulletin of the John Rylands Library*, Vol. 38, No. 2, March 1956.

——— (1957-8) 'The Density of the Danish Settlement in England', *Birmingham University Historical Journal* 6 1-17.

——— (1968) *Anglo-Saxon Charters: An Annotated List and Bibliography*, London: Royal Historical Society.

——— (1971) *The Age of the Vikings*, 2nd edition, London: Edward Arnold.

——— (ed.) (1979) *English Medieval Settlement: Continuity and Change*, London: Edward Arnold.

——— (1979a) *Charters of Burton Abbey*, Oxford: Oxford University Press.

Scott, W. (1832) *Stourbridge and its Vicinity*, Stourbridge: J. Heming.

Searle, W. G. (1897) *Onomasticon Anglo-Saxonicum*, Cambridge: Cambridge University Press.

——— (1899) *Anglo-Saxon Bishops, Kings and Nobles*, Cambridge: Cambridge University Press.

Shaw, S. (1798/1801) *The History & Antiquities of Staffordshire*, 2 vols, London: Nichols, reprinted in facsimile (with unpublished text and plates) by E. P. Publishing, Wakefield, in collaboration with Staffordshire County Library, 1976.

Shoesmith, R. (1982) *Hereford City Excavations II: Excavations On and Close to the Defences*, Council for British Archaeology Rescue Report 36.

Sidebottom, P. C. (1994) 'Schools of Anglo-Saxon Stone Sculpture in the North Midlands', unpublished Ph.D thesis, University of Sheffield.

―――― (1996) 'The North-Western Frontier of Viking Mercia: The Evidence from Stone Monuments', in *West Midland Archaeology* 39 1996 3-15.

Simms, R. (1894) *Bibliotheca Staffordiensis*, Lichfield: A. C. Lomax.

Sims, R. (1882) *Calendar of the Deeds and Documents belonging to the Corporation of Walsall, in the Town Chest, extending from the reign of King John to the end of the reign of James II (1688)*, Walsall: T. Kirby.

Simons, E. P. (1978) 'Silverdale Iron, 1792-1902: A Study of the Effects of Management and Technological Changes on the Fluctuating Fortunes of a North Staffordshire Iron Company', unpublished MA thesis, University of Keele.

Sketchley (1770) *Sketchley's and Adam's Thadesman's True Guide; or an Universal Directory for the Towns of Birmingham, Wolverhampton, Walsall, Dudley, etc.*, 4th edition.

Slade, C. F. (1958) *The Staffordshire Domesday*, Victoria History of the County of Stafford, Vol. IV, London: University of London.

Sleigh, J. (1883) *A History of the Ancient Parish of Leek, in Staffordshire*, 2nd edition, London & Derby: Bemrose.

Smith, W. (1932-3) 'Place-Names of the Staffordshire-Cheshire Border', *Journal of the North Staffordshire Field Club & Archaeological Society*, 67 51-70.

Smith, A. H. (1956) *English Place-Name Elements*, 2 vols, English Place-Name Society 25-6, Cambridge: Cambridge University Press.

―――― (1990) 'Place-Names and the Anglo-Saxon Settlement', Sir Israel Gollancz Memorial Lecture, 29th September 1956 (reprinted in *British Academy Papers on Anglo-Saxon England*, Oxford: Oxford University Press, 1990, 204-26).

Speake, R. (ed.) (1972) *Audley: an out of the way quiet place*, Keele: Keel University Department of Adult Education.

Spufford, M. (2000) '*Eccleshall, Staffordshire: A Bishop's Estate*', in Thirsk 2000: 290-306.

Stafford, P. (1985) *The East Midlands in the Early Middle Ages*, Leicester: Leicester University Press.

Stenton, D. M. (ed.) (1970) *Preparatory to Anglo-Saxon England: Being the Collected Papers of Frank Merry Stenton*, Oxford: Oxford University Press.

Stenton, F. M. (1971) *Anglo-Saxon England*, 3rd edition, Oxford: Oxford University Press.

Stratman, F. H. (1891) *A Middle English Dictionary*, new edition by Henry Bradley, Oxford: Clarendon Press; London: Henry Froude.

Stubbs, W. (1880) *The Constitutional History of England*, 3 vols., Oxford: Clarendon Press.

Studd, R. (1993) 'Early Medieval Settlement in North Staffordshire', in Phillips 1993: 53-68.

Stuart, D. (1980) *Medieval Bromley*, Keele: Keele University.

—— (1990) *A Social History of Yoxall in the Sixteenth and Seventeenth Centuries*, Keele: Keele University.

Stukeley, W. (1776) *Itinerarium Curiosum*, 2nd edition, London: Centuria.

Styles, D. (1936) 'The Early History of the King's Chapels in Staffordshire', *Transactions of the Birmingham Archaeological Society* 60 56-95.

Swanton, M. J. (trans.) (1997) *The Anglo-Saxon Chronicle*, London: Dent.

Sweet, H. (1885) *The Oldest English Texts*, London: Early English Text Society.

Sweet, R. (2004) *Antiquaries: The Discovery of the Past in Eighteenth-Century Britain*, London & New York: Hambledon and London.

Taylor, C. (1979) *Roads & Tracks of Britain*, London: J. M. Dent & Sons Limited.

Taylor, C. S. (1957) 'The Origin of the Mercian Shires', in Finberg 1957: 17-51.

Taylor, S. (ed.) (1988) *The Uses of Place-Names*, Edinburgh: Scottish Cultural Press.

Tengvik, G (1938) *Old English Bynames*, Uppsala: Almqvist & Wiksell.

Thacker, A.T. (1984) 'Chester and Gloucester: Early Ecclesiastical Organisation in Two Mercian Burhs', *Northern History*, 18 199-211.

Thacker, T. (1985) 'Kings, Saints, and Monasteries in Pre-Viking Mercia', *Midland History*, 10 1985 1-25.

Thirsk, J. (ed.) (2000) *English Rural Landscape*, Oxford: Oxford University Press.

Thomas, R. J. (1938) *Enwau afonydd a nentydd Cymru*, Cardiff: Gwasg Prifysgol Cymru.

Thorn F. R. (1991) 'Hundreds and Wapentakes', in Alecto 1988.

—— (1997) 'Another Seaborough, The Other Dinnation: Some Manorial Affixes in Domesday Book', in Rumble & Mills 1997: 345-77.

Thompson, H. V. (1916-17) 'Scandinavian Place-Names in North Staffordshire', *Journal of the North Staffordshire Field Club & Archaeological Society*, 51 77-83.

Tildesley, N. J. (1951) *A History of Willenhall*, Willenhall: Willenhall Urban District Council.

Todd, M. (1991) *The Coritani*, Stroud: Allan Sutton.

Tolkien, J. R. R. & Gordon, E. V. (1967) *Sir Gawain and the Green Kinight*, 2nd edition, revised by N. Davis, Oxford: Oxford University Press.

Toller, T. N. (ed. and enlarged by) (1898) *An Anglo-Saxon Dictionary based on the Manuscript Collections of the late Joseph Bosworth*, Oxford: Oxford University Press, with T. N. Toller, *A Supplement to An Anglo-Saxon Dictionary based on the Manuscript Collections of the late Joseph Toller*, 1921, Oxford: Oxford University Press, with A. Campbell, *Enlarged Addenda and Corrigenda*, 1954, Oxford: Oxford University Press.

Tomkinson, J. L. (1994) 'The Documentation of Hulton Abbey: Two Cases of Forgery', in *Staffordshire Studies*, 6 73-102.

Tooth, E. (2000a) 'The Survival of Scandinavian Personal Names in Staffordshire Surnames', in *Staffordshire Studies*, 12 1-16.

―― (2000b) *The Distinctive Surnames of North Staffordshire*, Vol. I, Leek: Churnet Valley Books.

―― (2002) *The Distinctive Surnames of North Staffordshire*, Vol. II, Leek: Churnet Valley Books.

Toulmin Smith, L. (ed.) (1906-10) *The Itinerary of John Leland*, London 1906-10, London: Centaur Press (reprinted 1964).

Trubshaw, S. (1867) *Old Dick Slee's Cave* (copy in WSL).

Twemlowe, F. R. (1912) *The Battle of Bloreheath*, Stafford: Whitehead Bros.

Underhill, E. A. (1941) *The Story of the Ancient Manor of Sedgley*, Tipton: E. A. Underhill.

Underhill, C. H. (1976) *History of Burton upon Trent*, 2nd edition, Burton upon Trent: Tresises.

Wacher, J. S. (1966) *The Civitas Capitals of Roman Britain*, Leicester: Leicester University Press.

Wager, S. J. (1998) *Woods, Wolds & Groves: The Woodland of Medieval Warwickshire*, British Archaeological Reports British Series 269, Oxford: Oxbow Books.

Wainwright, F. T. (1942) 'North-West Mercia AD 871-924', *Transactions of the Historic Society of Lancashire and Cheshire* 94 3-55, reprinted in Cavill, Harding & Jesch 2000: 19-42.

Wakelin, M. (1969) 'Crew, cree and crow: Celtic Words in English Dialect', *Anglia*, 87 273-81.

Walker, A. J. (1976) 'The Archaeology of Stafford to 1600 A.D.', unpublished M.A. dissertation, Bradford University.

Walker, I. W. ( 2000) *Mercia and the Making of England*, Stroud: Sutton.

Wallenberg, J. K. (1931) *Kentish Place-Names*, Uppsala.

―― (1934) *The Place-Names of Kent*, Uppsala.

Ward, J. (1843) *The Borough of Stoke-upon-Trent*, London: W. Lewis.

Watkins-Pitchford, W. (1932) *Morfe Forest and some of its people*, Shepshed: F. B. Foxall.

Watson, W. J. (1926) *The History of the Celtic Place-Names of Scotland*, Edinburgh & London: Blackwood.

Watts, V. E. (1979) 'Comment on 'The Evidence of Place-Names II" in Sawyer 1979: 122-32.

―― (ed.) (2004) *The Cambridge Dictionary of English Place-Names*, Cambridge: Cambridge University Press.

Weate, M. (1972) *The Parish of Lapley-with-Wheaton Aston*, Lapley: Lapley Parish Council.

Webster, G. (1981) *Rome Against Caractacus*, London: Batsford.

—— (1991) *The Cornovii*, 2nd edition, Stroud: Allan Sutton.

Wedgwood, J. (n.d.) *Staffordshire Pottery and its History*, London.

Welch, C. (1997) 'Glass-making in Wolseley, Staffordshire', in *Post-Medieval Archaeology* 31 1997 160.

—— (2001) 'Elizabethan Ironworking and the Woodlands of Cannock Chase and the Churnet Valley, Staffordshire', in *Staffordshire Studies* 12 17-73.

White, W. (1851) *History and Directory of Staffordshire*, 2nd edition, Sheffield: Robert Leader.

White, W. (1860) *All Round the Wrekin*, London: Chapman & Hall.

Whitelock, D. (ed.) (1930) *Anglo-Saxon Wills*, Cambridge: Cambridge University Press.

—— (1952) *The Beginning of English Society*, Harmondsworth: Penguin Books.

—— (1955) *English Historical Documents, c.500-1042*, London: Eyre & Spottiswoode (2nd edition, 1979 London).

Whybra, J. (1990) *A Lost English County: Winchcombshire in the Tenth and Eleventh-Centuries*, Woodbridge: Boydell & Brewer.

Willett, M. (1882) *A History of West Bromwich*, West Bromwich: The Free Press Co.

Williams, A., Smyth, A. P. & Kirby, D. P. (1991) *A Biographical Dictionary of Dark Age Britain: England, Scotland and Wales c.500 – c.1050*, London: Seaby.

Willmore, F. W. (1887) *A History of Walsall*, Walsall & London: Simpkin, Marshall.

Wilson, D. (1974) *Staffordshire Dialect Words – a Historical Survey*, Hartington & Little Haywood: Moorland.

Wilson, R. A. (1907) *The Register or Act Books of the Bishops of Coventry and Lichfield*, Vol IV, Stafford: William Salt Archaeological Society.

Woodiwiss, S. (ed.) (1992) *Iron Age and Roman salt production in the medieval town of Droitwich: Excavations at the Old Bowling Green and Friar Street*, CBA Research Report No. 81.

Woolgar, C. M. (1993) *Household Accounts from Medieval England, Part 2*, Oxford: Oxford University Press.

Wrander, N. (1983) *English Place-Names in the Dative Plural*, Lund Studies in English 65, Lund.

Wright, J. (1898) *The English Dialect Dictionary*, London & New York: Henry Frowde.

Yonge, W. E. V. (1923) *Bye-Paths of Staffordshire*, 3rd edition, Market Drayton: Bennion.

Yorke, B. (1995) *Wessex in the Early Middle Ages*, London & New York: Leicester University Press.

—— (2001) 'The Origins of Mercia', in Brown & Farr 2001: 13-22.

Youngs, F. A. (1991) *Guide to the Local Administrative Units of England, Vol. II (Northern England)*, London: Royal Historical Society.

Zaluckyj, S. (2001) *Mercia: The Anglo-Saxon Kingdom of Central England*, Logaston: Logaston Press.